Cell Migration

A subject collection from *Cold Spring Harbor Perspectives in Biology*

OTHER SUBJECT COLLECTIONS FROM *COLD SPRING HARBOR PERSPECTIVES IN BIOLOGY*

Telomeres and Telomerase
Evolution and Development of Neural Circuits
Machine Learning for Protein Science and Engineering
Glia, Second Edition
Speciation
The Biology of Lipids: Trafficking, Regulation, and Function, Second Edition
Synthetic Biology and Greenhouse Gases
Wound Healing: From Bench to Bedside
The Endoplasmic Reticulum, Second Edition
Sex Differences in Brain and Behavior
Regeneration
The Nucleus, Second Edition
Auxin Signaling: From Synthesis to Systems Biology, Second Edition
Stem Cells: From Biological Principles to Regenerative Medicine
Heart Development and Disease
Cell Survival and Cell Death, Second Edition
Calcium Signaling, Second Edition
Engineering Plants for Agriculture
Protein Homeostasis, Second Edition

SUBJECT COLLECTIONS FROM *COLD SPRING HARBOR PERSPECTIVES IN MEDICINE*

Parkinson's Disease, Second Edition
Developmental Oncology: Principles and Therapy of Cancers of Children and Young Adults
Type 1 Diabetes: Advances in Understanding and Treatment 100 Years after the Discovery of Insulin, Second Edition
Cancer Metabolism: Historical Landmarks, New Concepts, and Opportunities
Modeling Cancer in Mice
Angiogenesis: Biology and Pathology, Second Edition
Retinal Disorders: Approaches to Diagnosis and Treatment, Second Edition
Breast Cancer: From Fundamental Biology to Therapeutic Strategies
Aging: Geroscience as the New Public Health Frontier, Second Edition
Combining Human Genetics and Causal Inference to Understand Human Disease and Development
Lung Cancer: Disease Biology and Its Potential for Clinical Translation
Influenza: The Cutting Edge
Leukemia and Lymphoma: Molecular and Therapeutic Insights
Addiction, Second Edition
Hepatitis C Virus: The Story of a Scientific and Therapeutic Revolution
The PTEN Family
Metastasis: Mechanism to Therapy
Genetic Counseling: Clinical Practice and Ethical Considerations

Cell Migration

A subject collection from *Cold Spring Harbor Perspectives in Biology*

EDITED BY

Michelle Starz-Gaiano
University of Maryland, Baltimore County

Peter N. Devreotes
Johns Hopkins University School of Medicine

COLD SPRING HARBOR LABORATORY PRESS
Cold Spring Harbor, New York • www.cshlpress.org

Cell Migration

A subject collection from *Cold Spring Harbor Perspectives in Biology*
Articles online at www.cshperspectives.org

Executive Editor	John Inglis
Project Supervisor	Barbara Acosta
Editorial Assistant	Danett Gil
Permissions Administrator	Carol Brown
Production Editor	Diane Schubach
Production Manager/Cover Designer	Denise Weiss
Publisher	David Crotty

Front cover artwork: Top-left corner inset: Polarized HL-60-derived neutrophils migrating along an fMLP gradient. Cells express the F-actin biosensor, LifeAct-miRFP703. Note prominent F-actin-rich lamellipodia at the front (false colors) and elongated, contractile uropod at the rear. (Image provided by Dhiman Sankar Pal, CUNY Hunter College.) *Top-left:* Developed *Dictyostelium* cells chemotaxing toward a cAMP-filled micropipette. Cells express F-actin biosensor, $LimE_{\Delta coil}$-mCherry (yellow), seen at the leading edge of the phase-contrast imaged cells (blue). (Image provided by Tatsat Banerjee, Massachusetts Institute of Technology.) *Upper-right:* Differentiated HL-60 neutrophils overexpressing PIP5K1B-GFP. Note the localization of PIP5K1B at the cell rear. (Image provided by Yu Deng, Johns Hopkins University.) *Lower-left:* MDA-MB-231 cells in spheroids overexpressing PIP5K1C-GFP. Conditions favor dissemination, which is prevented by the PIP5K1C expression. (Image provided by Yu Deng, Johns Hopkins University.) *Lower-right:* Chicken neural crest cells, expressing surface marker, TFAP2A E1::GPI-eGFP (blue), and F-actin biosensor, F-tractin-Tdtomato (magenta), delaminating from an explant. (Image provided by Yiyan Lin and Michael Piacentino, Johns Hopkins University.) *Lower-right corner inset:* Zoomed-in view of one neural crest cell expressing F-tractin-Tdtomato. (Image provided by Yiyan Lin and Michael Piacentino, Johns Hopkins University.)

Library of Congress Cataloging-in-Publication Data

Names: Starz-Gaiano, Michelle editor | Devreotes, Peter N. editor
Title: Cell migration : a subject collection from Cold Spring Harbor perspectives in biology / edited by Michelle Starz-Gaiano, Peter N. Devreotes.
Description: Cold Spring Harbor, New York : Cold Spring Harbor Laboratory Press, [2026] | Series: Cold Spring Harbor perspectives in biology | Includes bibliographical references and index. | Summary: "Cell movement is a critical part of normal physiology, embryonic development and wound healing. This volume explores the molecular mechanisms by which directional motion is generated, the roles of the process in different contexts, and innovative strategies for advancing this research"-- Provided by publisher.
Identifiers: LCCN 2025020270 (print) | LCCN 2025020271 (ebook) | ISBN 9781621825319 cloth | ISBN 9781621825326 epub
Subjects: LCSH: Cell migration
Classification: LCC QH647 .C4415 2025 (print) | LCC QH647 (ebook)
LC record available at https://lccn.loc.gov/2025020270
LC ebook record available at https://lccn.loc.gov/2025020271

All World Wide Web addresses are accurate to the best of our knowledge at the time of printing.

For a complete catalog of all Cold Spring Harbor Laboratory Press publications, visit our website at www.cshlpress.org.

Contents

Preface

Cell migration is a fascinating, essential feature of cellular behavior. For more than a century and a half, we have appreciated the immune cells that patrol organisms to neutralize foreign substances. Over recent decades, it has become clear that nearly all cells can move, often directionally, and our understanding of how and why they do so continues to evolve. Without migration, embryos would not take shape, the nervous system would fail to wire, wounds would not close, and immune defenses would falter. Yet migration also plays a sinister role in pathological inflammation and cancer metastasis, a reminder that the same motile capabilities that sustain life can, in certain contexts, undermine it. Thus, it is critical to figure out how cells migrate. This book demonstrates that our knowledge about this process is rapidly advancing and substantiates the idea that it is a very exciting time to work in this field.

Cells inhabit a spectrum of social states. Many free-living and tissue-resident eukaryotic cells exist as single, autonomous entities that only transiently interact with neighbors. Some alternate between solitary movement and collective behavior, joining or leaving coordinated groups as demanded by development, homeostasis, or repair, while others function almost exclusively as collectives. The basic mechanisms that underlie these behaviors have largely been delineated in single-cell studies, especially in systems such as the social amoebae *Dictyostelium* and human neutrophils, which are amenable to experimental manipulation and amazing to observe directly under the microscope. More recently, attention has been paid to investigating cell clusters and group interactions within tissues. While many molecular motifs are conserved across cell types, individual cells exhibit idiosyncrasies that equip them for specific tasks and environments.

This volume begins with the foundations of single-cell motility, goes on to explore collective migration and tissue-scale dynamics, and then examines new, diverse experimental and computational approaches to study cell migration. The opening section on single-cell migration lays the conceptual, mechanistic, and methodological groundwork for key features of directed movement: cell polarity and memory, excitability and inhibition of signaling, cytoskeletal force generation, adhesion and traction, and the quantitative tools needed to analyze and model these processes. The second section builds on this base to explore how cells coordinate in groups, how tissues remodel and respond, and how these phenomena play out in developmental contexts. The third section highlights how technological advances and experimental tools have advanced our understanding of cell motility, and the fourth section reviews multiple computational and mathematical models that have provided insights, and previews how machine-learning techniques are poised to accelerate these discoveries. Throughout, readers will find recurring themes—the interplay of biophysics and molecular signaling, the influence of the mechanics and signals from the microenvironment, the value of standardized assays and quantitative imaging, and the power of computational modeling to unify disparate observations and generate testable predictions. Our hope is that the book offers both a coherent map of the field and signposts for future discovery.

Section I: Cell Biology of Single-Cell Migration

Single-cell migration is the elemental unit of motility that underlies the survival of single-cell organisms, as well as animal immune surveillance, organ morphogenesis, wound healing, and the earliest steps of metastasis. A classic framework is that cells evaluate chemicals in the environment and move toward or away from these cues, a process called chemotaxis. Section I charts the logic of how cells polarize, sense, decide, and move in complex environments, and how intrinsic dynamics interact with external cues to produce persistent, directed behavior. van Haastert and Kortholt illuminate the persistence of amoeboid motility of *Dictyostelium,* which is driven by dynamic cytoskeletal signaling networks at the front and the back of a cell that result in positional memories and sensitive chemotaxis in shallow gradients. Similarly, Li et al. examine regional actin protrusions and actomyosin contraction with a focus on the governing, intrinsic excitable signaling networks that can sustain motility, even in the absence of external cues. Gong and Meyer advance the conceptual models for directed migration by combining an activator–inhibitor signaling mechanism with one that accounts for structural asymmetries due to cytoskeletal and organelle positioning. Badu-Nkansah and Collins highlight findings from genomic and optogenetic methods that reveal a requirement for negative regulators to hold excitatory receptor signaling in check, which enables leukocytes to achieve robust chemotaxis. Lerche and Waterman address how local cytoskeletal dynamics allow neutrophils to navigate diverse environments by flexibly adopting two different migration strategies: mesenchymal movement, characterized by actomyosin and adhesion regulation, or bleb-based amoeboid motility. Strickland et al. bridge single-cell behavior to collective outcomes by comparing neutrophil swarming with *Dictyostelium* aggregation, arguing that signal relays and self-generated gradients offer efficient, robust strategies for organizing migration in any environment.

Across the first section, a coherent picture emerges: excitable networks and inhibitory circuits shape when and where protrusions form; structural polarity and cytoskeletal architecture provide memory and persistence; local adhesion and actomyosin forces couple signaling to movement; and polarized sensitivity and self-generated gradients enable cues to interpret external cues effectively. Readers will find that insights from one chapter resonate with others. Structural asymmetries discussed by Gong and Meyer complement the polarity "memories" in van Haastert and Kortholt. The inhibitory logic articulated by Badu-Nkansah and Collins refines the excitability described by Li et al., and the mechanistic guidance in Lerche and Waterman sets the stage for the collective phenomena explored by Strickland et al. By articulating shared assumptions, highlighting open questions, and pointing to standardized metrics and methods, the section aims to foster comparability and underscore cumulative progress. The presentation of these foundational principles of single-cell migration equips readers with the conceptual vocabulary, mechanistic intuition, and quantitative understanding needed to interpret the complex biological systems that arise in the next section and to consider strategies to alter cell behaviors.

Section II: Coordinated Cell Migration within Multicellular Organisms

Within a multicellular organism, motile cells rely on signals from other cells to navigate complex, dynamic environments. To understand these coordinated events, researchers have studied a variety of cell types in different species to exploit different experimental strengths, for example, cell-type-specific genetic manipulation and direct visualization of cell migrations within their normal tissue. These strategies have pinpointed common mechanisms that control migratory cells intrinsically, regardless of context, and have begun to uncover the effects of key chemical and mechanical signals provided by substrate cells and the larger environment of a tissue, organ, or embryo. Section II explores how migratory cells coordinate with each other and respond to chemical, mechanical, and other information from surrounding tissues. Weijer analyzes the dramatic movements of large sheets of mesodermal and endodermal cells during vertebrate gastrulation, which downregulate stable junctional components and adopt mesenchymal properties. The chapter elucidates the need for feedback between cell-level and tissue-level dynamics to ensure accuracy and precision in embryogenesis. Schrope et al. explore the inflammatory response and delve into how mammalian leukocytes modulate their motility strategies based on their surroundings, switching between amoeboid-like and coordinated mesenchymal migrations depending on the chemical signals and the mechanical properties of the local environment. Barton and Lin describe the complex migratory route and varied interactions of the primordial germ cells in the fruit fly embryo, which undergo an amoeboid-like migration using cortical actin flows to respond to an array of biomolecules and cell types along their path. Le and Mayor focus on cell-on-cell migration (versus cell-on-matrix), comparing three cell types that move collectively during development: the border cells of the fruit fly ovary, the lateral line sensory primordium in fish, and neural crest migration in frogs. The comparison suggests interesting and likely conserved molecular and biophysical explanations for clustered cell behaviors. George and Starz-Gaiano expand on the migration of border cells, considering the interplay between tissue structure, cell movement, and signaling, and highlighting recently identified genetic regulators. Examining the lateral line primordium, Nechiporuk and Knaut stress the importance of patterning, cell–cell communication, and transmission of forces within a collective, which all provide robustness to changing cues as cells move through a complex environment, and they discuss the means for signaling systems to tune responses, which amplifies information from initially shallow chemoattractant gradients. Gao and colleagues compare the migrations of the neural crest and associated placodes in chick, fish, and mouse, with an emphasis on cranial neural crest and the lateral line placode. They emphasize that physical and molecular interactions are required both for proper spatial distribution of the neural crest and for cell identities.

While this section features diverse cell types, common threads become clear. In all cases, motility is conferred through intrinsic programs that modulate polarity and actomyosin dynamics, mirroring the mechanisms discovered for single cells in Section I. Within the complex, changing environment of an organism, though, migratory cells benefit from additional strategies to navigate effectively. Cells have different genetic programs to respond to signals from different substrate cells, which provide directional information, as described by Weijer, Schrope et al., Barton and Lin, and Gao et al. In some cases, these programs allow cells to alter behaviors depending on their surroundings, as seen for leukocytes and neural crest. Methods that allow deeper interrogation of tissue mechanics reveal how substrate cells may govern cell migration by constraining or promoting movement through changes in stiffness or structure. This idea is brought to light in chapters by Weijer, Schrope et al., Le and Mayor, and George and Starz-Gaiano. Finally, a key concept of this section is that there are advantages to migrating as a group, either in a tightly connected cohort like the lateral line, or as a looser collective like streams of the neural crest, as illustrated very well by Nechiporuk and Knaut and Gao et al. This advantage can be attributed to a longer-range assessment of cues in the environment via communication between cells of the collective, and to improved mechanical properties enabled by coupling cells and distributing forces. It will be essential to learn more about how these communication strategies work to understand how and why cells adopt different migration programs in different contexts. As more experimental approaches become available—some of which are described in the next section—more of these questions can be answered. We anticipate that further work on cell migration in development and immune response will be important in a biomedical context as it will aid in our understanding of congenital birth abnormalities, inflammatory diseases, effective wound healing, and cancer metastasis.

Section III: Novel Approaches to Cell Migration

Studies of cell migration are entering a transformative phase, propelled by tools that let us see, measure, and control motility with unprecedented precision. Genetically encoded fluorescent proteins ignited a revolution, turning the invisible into the observable and revealing local biochemical and mechanical states that previously could only be inferred. Biosensors now report the timing and localization of signaling and enzymatic activity inside moving cells. Synthetic biological actuators—chemogenetic and optogenetic—allow us to trigger specific events at chosen places and times, and to watch how cells respond in real time. At the same moment, the field has expanded to quantify forces generated within cells and imposed by the environment, and to trace reciprocal coupling between mechanics and biochemistry.

Section III focuses on novel and innovative approaches to study cell migration. Together, the chapters show how technology is shifting migration research from descriptive observation to causal dissection and control, and they portend what may be possible as our understanding and ability to manipulate motility continues to advance. Deb Roy et al. present a toolbox of inducible systems—chemogenetic and optogenetic—and design principles for their use in moving cells that enable on-demand control of protein activity with subcellular spatial and real-time temporal resolution. Subramanian et al. focus on how intravital subcellular microscopy has enabled direct visualization of mammalian cells in their native tissue, which allows insight into the biochemical and mechanical cues and cytoskeletal remodeling that powers movement and reveals remarkable plasticity. Zegers et al. conceptualize a framework that links cell biology and soft-matter physics, potentially enabling the use of new models to explain the underlying mechanics. They map collective and single-cell migration in three-dimensional environments across physical regimes and explore transition states from jammed, solid-like epithelial states to unjammed fluid-like motion and gas-like individual movement in mesenchymal and amoeboid modes. Si et al. showcase optogenetics under spatial constraints, revealing ways to provide precise cues to steer cells on demand and to interrogate links between microenvironmental cues and intrinsic protrusion, contraction, and traction signals. Young et al. review how cell and matrix stiffness influences motility, how cells often soften when adopting more migratory states, and how such changes connect to metabolism. They then describe tools to measure stiffness, traction, and forces while using probes and biosensors to monitor metabolic energy usage at single-cell resolution. Zhao et al. explore galvanotaxis: how endogenous and applied direct-current electric fields guide migration. They describe the electric fields in development and wound healing, summarize cell-type-specific responses, and outline the signaling mechanisms that integrate electrical guidance with chemical cues. By highlighting shared assumptions, open questions, and needs for standardized metrics, the section aims to foster comparability and cumulative progress.

Different experimental strategies reinforce recurring themes. First, precision matters: fast decisions at the front and back of a cell require perturbations and sensors that act on matching spatial and temporal scales. Deb Roy et al. make this case explicitly, and Si et al. show how local inputs propagate to reshape whole-cell mechanics under confinement. Second, context matters: Subramanian et al. illustrate how cells reconfigure motility programs in native microenvironments, and Zegers et al. provide a physics-based vocabulary to describe and predict transitions across densities, adhesions, and topologies. Third, coupling matters: Young et al. integrate mechanics and metabolism, reminding us that force generation has an energetic cost and that efficiency can be regulated; Zhao et al. add the bioelectric dimension, demonstrating that electrical cues modulate signaling networks and interact with chemical gradients and geometry. Taken together, the chapters shift the field toward multimodal measurement and control, where light, chemicals, mechanics, and electricity can be combined to elicit and interpret specific behaviors.

The technologies assembled here do more than reveal—they empower. As tools for imaging, perturbation, and modeling converge, and as we learn to precisely manipulate cells in native tissues and complex environments, the prospect of predicting and directing motility—amplifying beneficial migration in repair and tempering harmful migration in disease—comes into view. Exploring the approaches in this section, both sequentially and synthetically, will equip readers with the conceptual vocabulary, mechanistic intuition, and practical toolkits needed to interrogate and influence cell movement across systems and scales. However, these chapters also surface open questions. How can we extend optogenetic and chemogenetic control into different contexts or deep into tissues, overcoming light penetration and delivery barriers without sacrificing spatial precision? Can cells' soft-matter states be dynamically switched on demand using chemogenetic and optogenetic actuators, and what are the energetic constraints? How can we use different inputs—mechanical, metabolic, and electrical—to influence cell migration therapeutically while avoiding aberrant invasive properties? These questions force us to consider what we do not yet understand and cannot yet control, and invite creative new methods to push the field further.

Section IV: Computational Modeling of Cell Migration

While technologies continue to improve and enable increasingly sophisticated answers to biological questions, multiple aspects of cell migration remain elusive. Biology is being flooded with large, detailed, quantitative data sets that can be difficult to interpret. Because of these factors, mathematical and computational modeling has proven invaluable to complement wet-laboratory approaches. Well-constructed models and simulations help researchers interpret results from the bench, make mechanistic predictions, expose limits of understanding, and develop new, testable hypotheses.

Section IV explores several types of cell migration models and reveals how they have advanced our understanding and unlocked new questions. This section is intended to be accessible to readers across disciplines, from biologists to computational researchers, and encourages those new to mathematical modeling to explore these chapters and learn about the advantages of these strategies. Iglesias and Banerjee explain how *Dictyostelium* chemotaxis can be modeled as "excitable systems"—dynamical systems that, upon above-threshold activation, generate a large, predictable response—and how this can be described using partial differential equations (PDEs). The authors step the reader through development of a model that links activator–inhibitor signaling systems and cytoskeletal modules to produce polarized cell activation and chemotaxis. Ueda et al. probe the molecular machinery controlling stochastic cell movement, which underlies the fundamental mechanisms of directional pathfinding in *Dictyostelium*. They describe experimental data in cell biology and genetics that is captured by a two-tier excitable system model and apply the model to cell polarity, protrusion dynamics, movement, and micropinocytosis. Algorta et al. provide a broad overview of mathematical models of cell migration at different scales, building from reaction–diffusion PDE models of cytoskeletal dynamics, to deformable agent-based cellular models, and then to pairs and collectives of motile cells in two and three dimensions. Chitnis gives a primer of different modeling frameworks for those unfamiliar with computational modeling and evaluates multiple models for the collective migration and reorganization of the lateral line primordium in Zebrafish, including illustrative movies of corresponding simulations. Shi and Mogilner assess models to describe how membrane tension can govern aspects of cell shape changes (and vice versa), how electrical fields guide cell movement, and how mesenchymal cells migrate in different three-dimensional platforms with increasing complexity. Baker and colleagues review classic PDE-based models of collective cell migration, then discuss two emerging types of data-driven, machine learning methods that can leverage newly available large-scale, quantitative experimental data. Finally, they discuss challenges that must be overcome in both biology/experimental aspects and on the modeling side to improve understanding of collective migration.

Overall, this section emphasizes the utility of using different frameworks to conceptualize what is going on within a cell as it moves, either singly or as part of a larger collective or tissue. Clearly, diverse and adaptable strategies exist for modeling different cell migrations, and each can provide certain insights (and limitations) depending on model design. Reaction–diffusion models can inform us about signaling events at molecular scales, as described by Iglesias and Banerjee, Ueda et al., Algorta et al., and Chitnis. Agent-based models, such as those highlighted by Algorta et al., Chitnis, and Shi and Mogilner, can reveal key features of how cells move singly or in concert. Simulations from these models do not just mimic biological results but can identify emergent properties of the systems at play, determine key parameters that govern behaviors, and suggest missing components, leading researchers back to the laboratory. While cell migration models historically provided a useful, reductionist framework, increasingly complex levels of understanding and quantitative data compel the field to build new types of models, as argued by Algorta et al., Shi and Mogilner, and Baker et al. These chapters thus introduce how machine learning and more computational power will enable researchers to leverage larger data sets and rapidly push the boundaries of current modeling. From this, it also becomes apparent that more universal frameworks and open-source platforms would benefit the whole field and broadly accelerate progress. It is an exciting time to take advantage of these techniques, and we look forward to what these strategies uncover.

Cell Migration Research Moving Forward

Across the volume, a coherent understanding of cell migration comes into focus. At the single-cell level, excitable signaling, inhibitory logic, and structural polarity form a modular control system that translates environmental information into directed force production, with adhesion and actomyosin coupling providing the mechanical substrate for motion and memory. In multicellular contexts, that same circuitry is redeployed while incorporating larger-scale inputs through communication and mechanics: substrate cells sculpt paths, collectives amplify and buffer cues, and tissues impose constraints that migrating cells adapt to and exploit. Novel measurement and actuation tools now match the spatiotemporal granularity of cellular decisions, enabling causal tests of how biochemical, mechanical, metabolic, and electrical inputs are integrated. Modeling closes the loop, revealing emergent properties, translating rich data sets into potential mechanisms, and guiding experiments toward decisive perturbations.

Together, the chapters here reframe the field's central questions and future directions. How do individual cells integrate multimodal inputs, and how do cell collectives distribute that computation across members and scales? What are the rules for switching between modes—amoeboid, mesenchymal, jammed, or unjammed—and how do energetic budgets and mechanical memory constrain those transitions in vivo? Which standardized metrics best compare polarity, force, and sensitivity across tissues and species? How can intrinsic, chemical, mechanical, and electrical signals be harnessed for tissue repair without enabling adverse movements? By bringing single-cell principles into tissue context, and by connecting precision techniques to intravital imaging, soft-matter frameworks, mechanometabolism, and predictive models, the field is poised to move from description to design: predicting trajectories, steering outcomes, and correcting failures in development, immunity, and disease.

We began with the dual nature of migration, which is indispensable for building and healing organisms, yet capable of driving pathology. The chapters assembled here chart how far our mechanistic understanding and technology have progressed toward balancing that tension. They offer a shared vocabulary and tested toolkits for comparing behaviors, a suite of quantitative and computational approaches for generating and vetting hypotheses, and concrete pathways to intervention. As readers move between single cells and tissues, instruments and models, we hope this volume provides both a map and a means to begin their journeys into the vast field of cell migration.

Acknowledgments

The Editors sincerely thank the authors who contributed their insights and made this a fun and successful project. We also acknowledge Tatsat Banerjee, Yu Deng, Yiyan Lin, Michael Piacentino, and Dhiman Sankar Pal for their striking images of migrating cells shown on the cover. Finally, we thank Richard Sever for his initial encouragement about the *Cell Migration* project, Barbara Acosta and Danett Gill for managing the logistical aspects of working with the authors and publisher, and Brigitte Walsh for administrative assistance.

Michelle Starz-Gaiano
Peter N. Devreotes

Collaboration of Antipodes: Synergy of Branched and Linear F-Actin during Amoeboid Cell Movement and Chemotaxis

Peter J.M. van Haastert and Arjan Kortholt

Department of Cell Biochemistry, University of Groningen, 9747 AG Groningen, The Netherlands

Correspondence: p.j.m.van.haastert@rug.nl

The actin cytoskeleton plays a major role in locomotion of amoeboid cells. The extending pseudopod contains predominantly branched F-actin nucleated by actin-related protein 2/3 (Arp2/3) that is oriented toward the membrane, while the side/back of the cell contains predominantly linear F-actin nucleated by formins that is arranged parallel to the membrane in a contractile network using cross-linkers, membrane anchors, and myosin filaments. During cell movement, branched and linear F-actin have opposite functions: Elongation of branched F-actin filaments leads to pseudopod growth in the front, whereas pseudopod formation is strongly inhibited in areas of the contractile network. On the other hand, branched and linear F-actin also collaborate to optimize locomotion and navigation. Assembly of branched F-actin to induce a new pseudopod in the front also activates linear F-actin in the rest of the cell to inhibit a second pseudopod. Furthermore, linear F-actin at the side/back of the cell and branched F-actin each provide a memory of direction that is highly synergistic to mediate strong persistence of cell movement and sensitive chemotaxis.

Many eukaryotic cells move by making protrusions (Ridley 2011). Upon flow of cytoplasm into the protrusion, the center of mass of the cell displaces, and the cell has effectively moved in the direction of the extending protrusion. These protrusions can be long-lived as in keratocytes, which glide with a single broad anterior protrusion that is continuously extending and filled with cytoplasm (Mogilner et al. 2020). However, in most cells the protrusions are transient with a short phase of extension and filling with cytoplasm, followed by the formation of a new protrusion (Ridley 2011). In amoeboid cells, such as neutrophils and *Dictyostelium*, these protrusions are in the form of pseudopods that extend fast at a high frequency (Skoge et al. 2016).

Pseudopods are extended perpendicular to the cell surface (Van Haastert and Bosgraaf 2009b). Therefore, the path of a moving cell depends on the time and place where a series of pseudopods is formed (Andrew and Insall 2007; Li et al. 2008; Bosgraaf and Van Haastert 2009b; Van Haastert and Bosgraaf 2009a; Fritz-Laylin et al. 2017). Unpolarized cells start new protrusions at nearly random positions, resulting in near Brownian motion (Van Haastert and Bosgraaf 2009a; Makarava et al. 2014). In polar-

Cite this article as *Cold Spring Harb Perspect Biol* doi: 10.1101/cshperspect.a041790

ized cells, new pseudopods are preferentially formed in a stable front close to previous pseudopods, while pseudopods are inhibited at the side and in the rear of the cell. This results in movement with strong persistence, but not in a specific direction (Killich et al. 1994; Li et al. 2008; Van Haastert and Bosgraaf 2009a; Makarava et al. 2014). During chemotaxis, pseudopods are preferentially made in the part of the cell facing the highest concentration of chemoattractant, which allows the cells to move with persistence and in the direction of the gradient (Janetopoulos et al. 2001; Andrew and Insall 2007; Bosgraaf and Van Haastert 2009a; Insall 2010).

Cell movement is mediated by rearrangements of the cytoskeleton, notably of the filamentous actin-based systems and its associated proteins such as myosin filaments (Chugh and Paluch 2018). Actin filament formation can be nucleated by actin-related protein 2/3 (Arp2/3) generating branched F-actin and by formins leading to linear F-actin (Svitkina 2020; Gautreau et al. 2022). Biochemical assays and live cell imaging have shown large rearrangements of F-actin in the extending pseudopod and at the side of moving cells (McRobbie and Newell 1983; Diez et al. 2005). More detailed studies have uncovered the location, structure, and regulation of linear and branched F-actin that together regulate cell movement (Chugh and Paluch 2018; Kadzik et al. 2020; Kelkar et al. 2020; Svitkina 2020). In the past 10 years, we have analyzed the dynamics and mechanism of pseudopod extension in *Dictyostelium*. From the analysis of mutants defective in branched or linear F-actin, we recognized their fundamental different contribution to pseudopod extension, cell movement, and chemotaxis (Van Haastert et al. 2018; Van Haastert 2020, 2021) In this essay, we discuss cell movement from the perspective of the extending pseudopod and the role of branched and linear F-actin.

The components of the cytoskeleton are very diverse in size over several orders of magnitude. Therefore, many schematics in textbooks and reviews are optimized to visualize on paper all components and their interactions. A drawback is often that the components are not shown to scale, which hinders intuitive understanding of sizes, open spaces, diffusion, interactions, and what is small and big. Since science is nowadays read mostly electronically and images can easily be enlarged to see the smallest components, we have decided to present in most figures all components to scale.

BRANCHED F-ACTIN IN THE PSEUDOPOD AND LINEAR F-ACTIN IN THE CORTEX

As discussed above, the two major components of the cytoskeleton of moving cells are branched and linear F-actin filaments. A combination of biochemical, genetic, and localization experiments lead to the following view (Insall and Machesky 2009; Urban et al. 2010; Yang and Svitkina 2011; Ydenberg et al. 2011; Bovellan et al. 2014; Li et al. 2022; Cao et al. 2023). Protrusions contain mainly branched F-actin but also some linear F-actin, whereas the contractile cortex at the side and in the rear contain mainly linear F-actin (Fig. 1A). Branched F-actin in pseudopods is regulated by nucleation of Arp2/3 branch points that generate new filament ends from which polymerization can occur. Cryo-electron tomography of protrusions reveals that branched actin filaments are relatively short with a mean size of 100–200 nm. The Arp2/3 branch points and cross-linkers such as ABP120 lead to a dense network (Fig. 1B). The barbed ends of the filaments are oriented toward the membrane.

Linear F-actin in the contractile cortex at the side of the cell is part of a thin 100–200-nm-thick layer under the plasma membrane (Fig. 1C–F; Bovellan et al. 2014; Jasnin et al. 2019; Muresan et al. 2022; Cao et al. 2023). The ~1200 nm linear actin filaments are ~10 times longer than the branched actin filaments. The linear F-actin filaments in the cortex are oriented approximately parallel to the plasma membrane, in contrast to branched F-actin that is oriented more perpendicular to the membrane. Besides linear F-actin filament, the contractile cortex consists of membrane anchor proteins, small cross-linkers such as fimbrin and 34 kD–binding protein and the longer α-actinin (Fig. 1E,F; Noegel and Schleicher 2000). Polymerization of linear F-actin in the cortex is regulated by formins, ring-like structures that are activated by RacE in *Dictyostelium* (Ramalingam et al. 2015; Litschko et al. 2019).

Conventional myosin (myosin II) is an abundant protein of the actin filament network that plays a role in various cellular processes such as cytokinesis, cell movement, and development. A single myosin II complex is composed of two myosin heavy chains, two essential light chains, and two regulatory light chains. This myosin II heterohexamer can spontaneously assemble into filaments composed of 10–20 myosin hexamers (Clarke and Baron 1987). This reaction is reversed by phosphorylation of myosin heavy chain on its tail. The motor domain of myosin heavy chain can bind to F-actin filaments. Myosin II appears to have two modes of action in *Dictyostelium*: local stabilization of the cytoskeleton by myosin filament binding to the actin meshwork (structural mode) and force generation by ATP-driven movement of the globular head leading to contraction of actin filaments (motor mode). In *Dictyostelium*, cyclic guanosine monophosphate (cGMP) signaling plays important roles in regulation of both the structural and motor roles of myosin II during chemotaxis. cGMP stimulates the translocation of myosin filaments from the cytoplasm to the F-actin of the contractile cortex in the rear of the cell, and it induces the phosphorylation of the regulatory light chain that increases motor activity of the myosin II filament several-fold (Bosgraaf and Van Haastert 2006).

In addition to pseudopods, cells may also extend blebs, which are local bulges of the plasma membrane of the cell. Blebs have a spherical morphology and are formed within 1 sec (Fig. 1G). A bleb is initiated by local weakening or rupture of cortex-membrane attachment and is inflated by intracellular pressure and flow of cytoplasm into the structure (Zatulovskiy et al. 2014; García-Arcos et al. 2024). Blebs might be used as an indicator of membrane-cortex stability. It has been reported that blebs often occur at the base of protrusions in an area of the cell membrane with negative curvature (Collier et al. 2017). The base of a protrusion is often associated with the transition of the organelle-free front and the organelle-rich rear of the cell. Is the cortex-membrane instability due to negative curvature or to properties of the cortex in this transition region? To address this, we carefully analyzed bleb formation in polarized developed *Dictyostelium* cells. In 2 h of movies, we detected 96 blebs with a mean diameter of 2.2 ± 0.4 µm (mean and SD), similar as was shown before (Tyson et al. 2014; Collier et al. 2017). In the frame before the bleb appears, we define the transition point at the cell boundary that marks the transition of the organelle-free front to the organelle-rich rear (Fig. 1G). Next, we measured the distance of this transition point to the front of the cell, to the rear of the cell, and to the center of the bleb; we also measured the membrane curvature at the position where the bleb appears (Fig. 1G). We observed 44 blebs in areas with negative curvature (angle larger than 180°) and 52 blebs in areas with positive curvature (angle smaller than 180°); the mean angle was $175 \pm 25°$, indicating that in these polarized cells blebs are not specifically formed in areas of negative curvature. We then carefully analyzed the formation of the blebs relative to the transition points. The distribution of blebs along the cell perimeter is highly skewed (Fig. 1H). The data are best described by a model with 22 blebs formed randomly along the entire cell perimeter (at a low probability of 0.21 blebs/µm/h) and 74 blebs formed within a distance of ~3 µm from the transition point (at a maximal probability of 6.12 blebs/µm/h). Apparently, the plasma membrane in the transition zone of the organelle-free protrusion and the contractile cortex is not firmly attached to neither the branched nor the linear F-actin. In contrast, blebs are rarely observed in the rear of the cell or near the extending tip of the pseudopod, suggesting that the membrane is firmly attached to the cytoskeleton in these areas. These observations are used below in the discussions on pseudopod growth and memory, respectively.

NUMBERS AND SIZES

A globular actin monomer is ~4 nm in diameter, while an F-actin filament has a diameter of ~7 nm. In the branched F-actin network, the

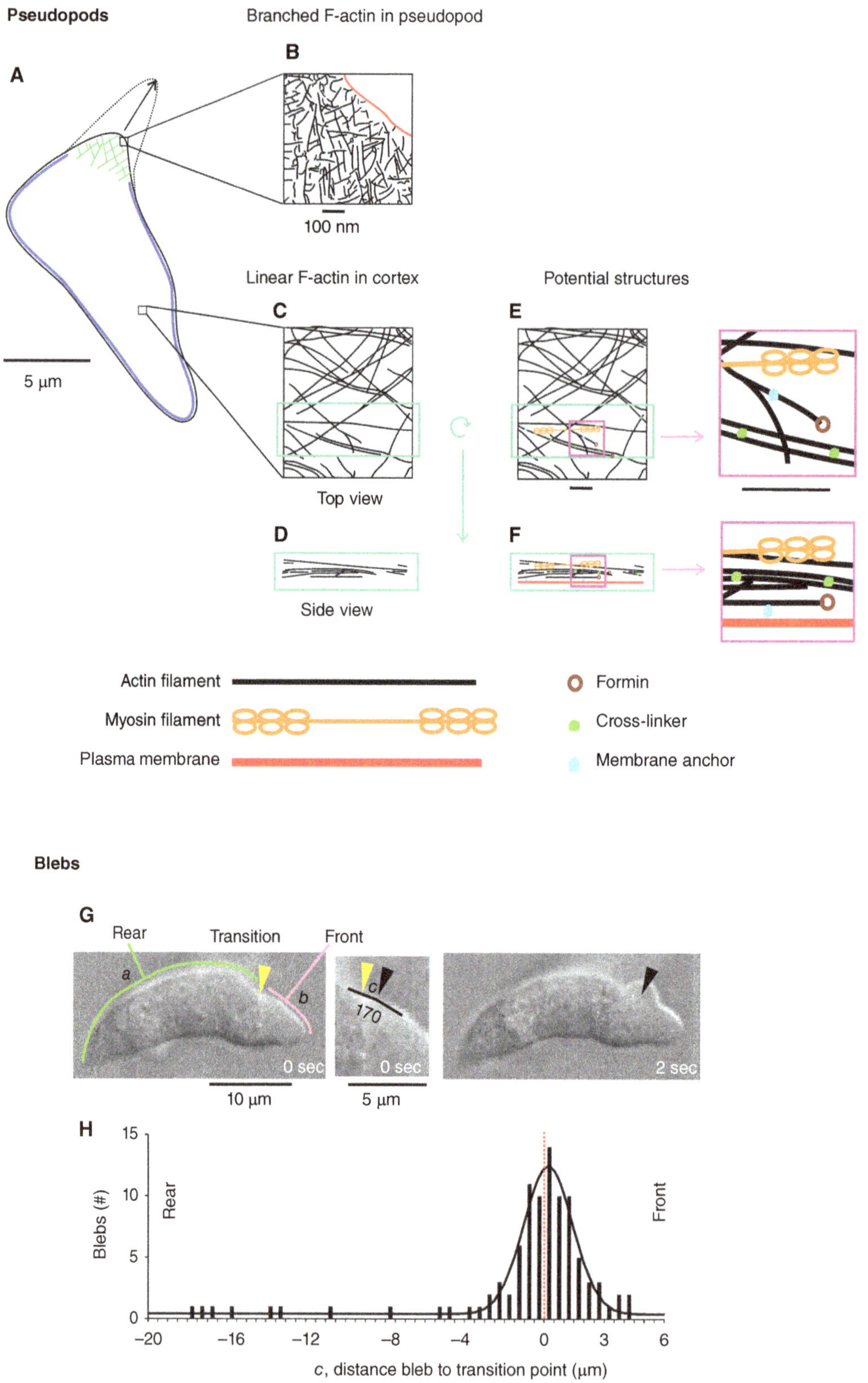

Figure 1. (*See following page for legend.*)

length of the filaments between the branch points is ~120 nm, consisting of ~60 actin molecules. The branched network in extending pseudopods is several micrometers long and is rather dense with an average mesh size of only 45 nm. The density of actin filaments that point to the plasma membrane is ~300 F-actin filaments per μm^2, yielding an average distance between filaments pointing to the membrane of ~60 nm (Fig. 1B; Urban et al. 2010; Yang and Svitkina 2011; Ydenberg et al. 2011). A myosin II filament has a size of 300 × 30 nm, which is much larger than the mesh size of the branched F-actin network in the pseudopod, explaining why myosin II filaments cannot easily penetrate from the cytoplasm into the thick and dense branched F-actin network of the extending pseudopod. As noted earlier, linear F-actin filaments are ~1200 nm, ~10 times longer than branched F-actin. The layer of linear F-actin in the cortex is only 100–150 nm thick. In addition, most filaments run parallel to the plasma membrane, and the density is rather low yielding a network with an average mesh size of ~100–150 nm. The large myosin filaments can thus easily diffuse from the cytoplasm into the thin cortex and land in between linear F-actin filaments.

It is sometimes helpful to convert the very small scales of a cell into the larger scale of our world. So, let us assume that 10 nm is 1 m, and then a cell has a diameter of 1000 m. The plasma membrane is a very flexible and deformable layer with a thickness of ~0.5–1 m. An actin monomer has a size of 0.4 m like a small dog, while a human is similar to a formin monomer or a stack of four actin proteins. A myosin molecule is very large (15 m in length), and a myosin filament is gigantic (30 × 3 m), something like two trucks or a blue whale. The branched actin has actin filaments acting as cables of 12 × 0.7 m and are arranged at an average distance of 4.5 m. The linear actin in the cortex are cables of 120 × 0.7 m, and the average distance between cables is 10 m. The branched F-actin network in a pseudopod extends up to ~500 m into the cell. In contrast, the cortex at the side is only 15 m thick. Assume that a person walks on the membrane at the inside of the cell. In the front, it sees hanging cables that point to the membrane but are not connected to the membrane. The nearest Arp2/3 branch points are ~5–10 m away from the membrane. The cables sweep a little and the flexible membrane goes up and down, both providing the opportunity that an open space emerges at the end of the cable to allow elongation with an actin sphere. On average, the distance between two cables touching the membrane is 6 m, and, therefore, it is very easy to walk around. However, for a myosin filament with the size of a blue whale or a very large truck, it is very hard to maneuver in between the actin

Figure 1. Branched and linear F-actin. (*A–F*) Pseudopod. (*A*) Overview. (*B*) Line drawing of branched actin filaments using a cryo-electron tomography image of the edge of a protrusion of a keratocyte (panel based on data in Urban et al. 2010). (*C*) Line drawing of linear actin filaments using a cryo-electron tomography image of a *top* view of the cortex of *Dictyostelium* cells (panel based on data from Jasnin et al. 2019). (*D*) The green section of the *top* view in *C* is tilted 90° to obtain the *side* view (using their associated movie); this *side* view is ~250 nm deep. (*E,F*) Impressions of potential structures with myosin filaments, formin, cross-linkers, and membrane anchors. The purple sections are enlarged at the *right* side. *A* is schematic; in *B–F*, all structures are drawn to scale. Scale bar, 100 nm. (*G,H*) Blebs. (*G*) Phase contrast images of a cell before and after formation of a bleb. In the *left* image, the yellow triangle indicates the transition point, which is the point at the boundary that marks the transition of an organelle-free front and an organelle-rich rear. Measured were the distance of the transition point to the uropod (*a*, mean 19.64 μm) and to the tip (*b*, mean 5.75 μm). In the *right* image, the center of the bleb is indicated by the black triangle. In the *middle* image, the distance *c* of the center of the future bleb to the transition point is measured, and the angle of the membrane at the center of the bleb. (*H*) Histogram of 96 blebs observed in 2 h. To obtain an equal representation of the cell boundary in the histogram, the distances *a* and *b* were normalized to −20 and 6 μm, respectively. Shown is the normalized distance *c* from the start of the bleb to the transition point. Bin size is 0.5 μm; 52 bins; half a cell. The black line is the optimal distribution with a model composed of 21.54 blebs equally distributed over all bins (at 0.21 blebs/μm/h) and 74.46 blebs with a normal distribution with mean = 0.22 μm and SD = 1.24 μm (at a maximum of 6.12 blebs/μm/h).

cables in the front. When the person walks away from the front to the side of the cell, the layer of branched F-actin becomes thinner and the person will see more membrane anchors with linear F-actin cables attached to them. The cables run approximately parallel to the membrane and are connected to each other by cross-linkers. Arriving at the side of the cell, the layer with cables is only 15 m thick, and the space between the cables is relatively large, 10–20 m. Therefore, a myosin II filament with the size of a blue whale can sink from the cytoplasm toward the membrane and intercalate between two actin cables. Because the cables are firmly attached to the membrane and to each other, myosin binding will further increase the rigidity of the cortex and associated membrane, while motor activity of myosin will generate contraction and tension in the cortex and plasma membrane.

A MODEL FOR PSEUDOPOD EXTENSION BY POLYMERIZATION OF BRANCHED F-ACTIN

Detailed analysis of pseudopod extension has revealed that a pseudopod grows at a constant rate and then suddenly stops (Van Haastert 2020). A conceptual framework of pseudopod extension (Peskin et al. 1993; Mogilner and Oster 1996) was used to explain this observation. Pseudopods are induced by the nucleation and growth of actin filaments in a branched network. The density of branched F-actin is ~300 F-actin filaments per μm^2 pointing to the membrane (Urban et al. 2010; Yang and Svitkina 2011; Ydenberg et al. 2011). Initially the surface area of the emerging pseudopod is ~14 μm^2 containing on the order of 4000 actin filaments pointing toward the membrane (Fig. 2C). To allow further actin polymerization at the end of these actin filaments near the membrane, an "open" space between the end of an actin filament and the membrane should be available for sufficient time (Danuser et al. 2013; Prahl and Odde 2018). This can be achieved by two methods (Fig. 2A): (1) stochastic/thermal bending of the actin filaments thereby creating an open space (Gittes et al. 1993) or (2) stochastic/thermal movement of the membrane, both followed by polymerization in the period of sufficient open space (racket model; Hill 1981; Peskin et al. 1993; Mogilner and Oster 1996). Actin polymerization produces a forward force leading to the growth of the pseudopod. On the other hand, pseudopod growth induces counterforces such as membrane tension and viscous drag, and counteractivities such as depletion of materials (Danuser et al. 2013). The pseudopod keeps extending as long as the forward force is larger than the counterforces/activities. The counterforces/activities are divided over the number of growing actin filaments at the tip ~1 μm of the protrusion (Fig. 2C). As the pseudopod grows, the tension in the membrane at the side and tip increases, the tip gets more narrow with fewer branched F-actin filaments pointing toward the membrane, materials may become depleted, all reducing the probability to create an "open" space (Van Haastert 2020). When polymerization of some actin filaments stalls, the counterforces/activities are divided over fewer filaments, thereby increasing the probability that more filaments fail further extension: the polymerization process collapses and pseudopod extension rapidly stops. This model postulates a sharp transition between constant actin polymerization and its sudden arrest, which explains the observation that pseudopods either move at a constant rate or stop. This model was used to explain why many components influence pseudopod extension, including viscosity of the medium, shape of the pseudopod, defects of mutants, and function of the uropod (Van Haastert 2020).

As described above and shown in Figure 1H, blebs in the extending tip of the pseudopod are very rare. Blebs are formed due to a weak attachment of the cytoskeleton to the membrane. Although the filaments of branched F-actin are not directly anchored to the membrane, the membrane in the extension zone is apparently sufficiently stabilized to prevent bleb formation. Probably some linear actin filaments, cross-linkers and membrane anchors allow sufficient stability of the membrane to prevent bleb formation and at the same time do not prevent sufficient flexibility for forward movement of the membrane by actin polymerization of branched filaments.

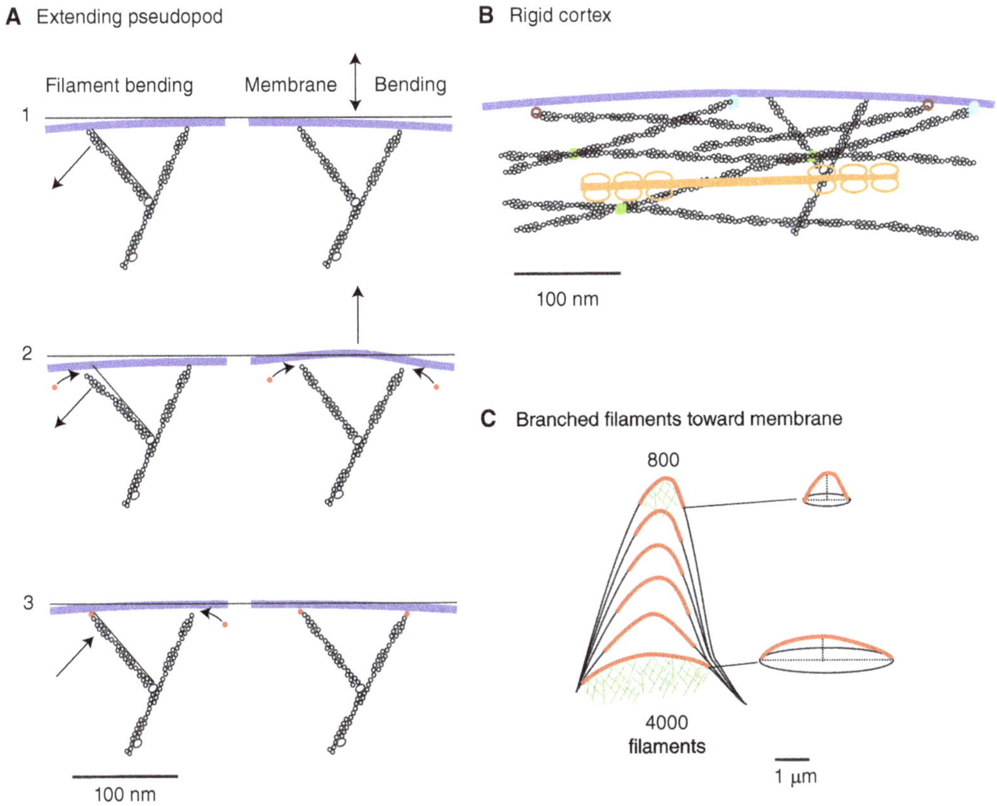

Figure 2. Actin polymerization and pseudopod growth. (*A*,*B*) All structures are drawn to scale, including the bending of the plasma membrane (in blue); thin black lines are helper lines that remain at the same position to see the bending and movement of membrane and filament. (*A*) In a protrusion, filaments from the branch points are not bound to the membrane. In step 1, the filament may bend or the membrane bends, creating an open space. In step 2, actin protein is added to the filament in the open space. In step 3 at the *left*, the filaments bend back pushing the membrane forward; at the *right*, the membrane cannot bend because the filament was extended in step 2. (*B*) The contractile cortex contains few filaments pointing toward the membrane and the membrane is not very flexible, both making it unlikely that a pseudopod can start in the contractile cortex. (See Figure 1 for the identity of the colored structures.) (*C*) Schematic of the shape change of an extending pseudopod predicting the strong decline in branched filaments in the tip; the balance of forward and counterforces occurs in the ~1 µm front area in the tip (indicated by red line).

In the contractile cortex at the side of the cell, the linear F-actin filaments run approximately parallel to the membrane, branched F-actin is scarce, and therefore very few F-actin filaments point toward the membrane at the side of the cell. In addition, F-actin cross-linkers and membrane anchors make the cortex and associated plasma membrane relatively rigid. The low abundance of actin filaments pointing to the membrane in combination with an inflexible membrane explain the strong suppression of pseudopods at the side and the rear of the cell (Fig. 2B).

HOW PSEUDOPODS START IN THE FRONT AND AT THE SIDE OF THE CELL

In *Dictyostelium*, new pseudopods are formed in an coupled excitable medium of F-actin and the small G-protein Ras (Xiong et al. 2010; Kortholt et al. 2013; Van Haastert et al. 2017, 2018; Kuhn et al. 2024). Coupled means that Ras-guanosine triphosphate (GTP) activates the formation of branched F-actin and branched F-actin potentiates Ras activation. Excitable means that small fluctuations of Ras-GTP or branched F-actin are damped, while fluctuations above a threshold are amplified to generate full activation of Ras-GTP and branched F-actin. *Dictyostelium* cells spontaneously form multiple dynamic patches of activated Ras (Fig. 3A,B; Sasaki et al. 2004; Van Haastert et al. 2017). Essentially all pseudopods start in an area of a Ras-GTP patch, usually at the position of the strongest Ras-GTP patch, and then the emerging pseudopod grows by polymerization of branched F-actin (Van Haastert et al. 2017). Experiments with sensors and mutants suggest that pseudopods are induced by a signaling cascade containing activated Ras, Rac, SCAR/WAVE, and Arp2/3, leading to nucleation of branched F-actin (Blagg and Insall 2004; Insall and Machesky 2009; Davidson and Insall 2013; Huang et al. 2013; Van Haastert 2020). Because pseudopods in the front are induced by the coupled excitable Ras-GTP/F-actin, the trigger for a new pseudopod can begin with the activation of any component of the stimulatory signaling cascade containing Ras-GTP → Rac-GTP → Scar → Arp2/3 → branched F-actin → Ras-GTP (Fig. 3C). Using sensitive and fast sensors, we have carefully analyzed the activation of Ras-GTP and F-actin at the location in the cell where later the pseudopod starts (Van Haastert et al. 2017). We observed that first F-actin polymerization starts to increase at −1.9 sec relative to the start of the pseudopod followed by an increase of Ras-GTP at −0.7 sec (Fig. 3D). Thus, a pseudopod in the front starts in a "basal" patch of Ras-GTP with an increase of branched F-actin followed by further activation of Ras-GTP and F-actin.

Although several Ras-GTP patches are present at the side and in the rear of the cell (Fig. 3A) and such patches may induce some branched F-actin, this rarely leads to a pseudopod. The rigid contractile cortex makes it difficult to generate a new protrusion of branched F-actin. We carefully also analyzed how a pseudopod at the side of the cell may start by using sensors that detect specific events before the pseudopod starts (Van Haastert 2021). These experiments suggest that the contractile linear F-actin/myosin must be degraded before branched F-actin can induce a pseudopod (Fig. 3E). The first event at the place of the future pseudopod is the activation of Ras at −6 sec relative to the start of the pseudopod, followed by the activation of another small G-protein Rap1 at −4 sec and the disappearance of myosin at −3 sec. F-actin increases at −1.6 sec relative to the start of the pseudopod (Fig. 3F). This sequence of events is compatible with two Ras-GTP-induced signaling pathways: first, spontaneous excitation of a Ras-GTP patch leads to the activation of Rap1, which induces the disassembly of myosin filaments and locally weakens the contractile cortex. Secondly, Ras-GTP activates Rac, Scar/Wave, and Arp2/3 leading to the formation of branched F-actin that can now start a pseudopod at the side of the cell.

We envision that in the organelle-free front, degradation of linear F-actin is not required to induce a protrusion. First, branched F-actin is far more abundant then linear F-actin. Secondly, myosin filaments are absent in the front, so linear F-actin is not contractile. Third, some linear F-actin in the front is probably important to provide a stable interaction between the branched F-actin network and the membrane, thereby preventing bleb formation in the front (see discussion on blebs above).

EXTENDING PSEUDOPODS INHIBIT NEW PSEUDOPODS, WHICH REQUIRES LINEAR F-ACTIN

We have analyzed the kinetics of how cells that have no extending pseudopods will start a first pseudopod. It appears that the start of the first pseudopod is random in time with a probability of 14%/sec (Van Haastert 2020). This probability is constant until all cells have started a pseudopod. While the cell extends this first pseudopod, the cell may start a second pseudopod—again random—but now with a much lower probability of only 4%/sec (Fig. 4A). In the few cells with two extending pseudopods, a third pseudopod may start, which occurs with again a lower probability of only 1%/sec. Thus, for each extending pseudopod, the start of a new pseudopod is inhibited by a factor of 3.5. Interestingly, a similar three- to fourfold inhibition of the start of multiple pseudopods was observed in neutrophils, mesenchymal stem cells, and the fungus *Batrachochytrium dendrobatidis chytrid* (Van Haastert 2020). Nearly all *Dictyostelium* mutants investigated have this strong inhibition to start multiple pseudopods, with the exception of mutants with a perturbed linear F-actin/myosin cortex, such as myosin II–null cells, the formin *forAHE* triple mutant, and the cortex activating *racE*-null cells (Litschko et al. 2019). In these mutants, all pseudopods start with a high probability, irrespective of the presence of other extending pseudopods, and therefore cells often extend multiple, even up to five, pseudopods. These results reveal that the linear F-actin/formin/myosin II of the contractile cortex is essential to inhibit the start of multiple pseudopods. The molecular mechanism is not exactly known. We reported that immediately (within 1 sec) after the start of the first pseudopod, the start of a second pseudopod is inhibited 3.5-fold (Fig. 4A); careful kinetic analysis reveals that even before the first pseudopod begins extending, the start of the second pseudopod is already inhibited (Van Haastert 2020). As shown above, a pseudopod is induced locally by the sequential activation of Ras, Rac, Scar, Arp2/3, and F-actin, which occurs ~2–6 sec before the start of pseudopod extension (Fig. 3). An attractive hypothesis is that in this time window, the signal that will trigger the formation of branched F-actin to start the first pseudopod also activates the linear F-actin of the contractile cortex to inhibit the start of a second pseudopod (Fig. 4B). A possible candidate in *Dictyostelium* is Rac1-GTP that has been shown to associate with the Scar complex to induce branched F-actin (to start a pseudopod), and to associate with cortexilin/formin/IQGap to activate the contractile cortex (to inhibit a pseudopod elsewhere) (Filić et al. 2014; Litschko et al. 2017; Šoštar et al. 2024).

The property of a high probability to start a first pseudopod and a reduced probability to start a second or third pseudopod results in a strong enrichment of cells extending exactly one pseudopod, and not zero or more than one pseudopod. Obviously, cells move better with one pseudopod at a time than with multiple pseudopods as was shown in the *racE* and *forAEH* mutants that heavily wiggle with many pseudopods and go nowhere (Litschko et al. 2019; Van Haastert 2020).

POLARITY MEMORY WITH LINEAR F-ACTIN/MYOSIN IN THE REAR

Although amoeboid cells are very dynamic and constantly change their shape, these cells do not move in random directions. As early as 1953, it was shown that in the absence of external cues cells exhibit a so-called correlated random walk (Patlak 1953), an observation that has been reproduced for nearly all moving cells (Gail and Boone 1970; Hall 1977; Potel and Mackay 1979; Li et al. 2008; Bosgraaf and Van Haastert 2009b). "Correlated" means that a cell is more likely to move in a direction like its previous direction of movement. This tendency to move in the same direction is called persistence. The function of a correlated random walk versus a random walk has been investigated both experimentally and theoretically. It serves a strategy for optimal foraging and strongly improves chemotaxis (Li et al. 2008; Van Haastert and Bosgraaf 2009a; Van Haastert 2010b). A correlated random walk must imply that moving cells have a memory of direction. The detailed analysis of the direction of series of pseudopods revealed two forms of memory: a relatively long-term memory of a global polarity axis, and a short-term memory of the start position of the previous pseudopod. Branched and linear F-actin are instrumental for these memories as explained below.

Developed *Dictyostelium* cells are polarized with a front and a rear. New pseudopods start predominantly in the front one-third of the cell and are suppressed in the rear two-thirds part of the cell. For one pseudopod event in the large rear, there are six pseudopod events in the smaller front (Fig. 5A). Since pseudopods are extended perpendicular to the cell surface, pseudopods that start in the front lead to the approximate continuation of the direction of movement: polarized cells possess directional memory and ex-

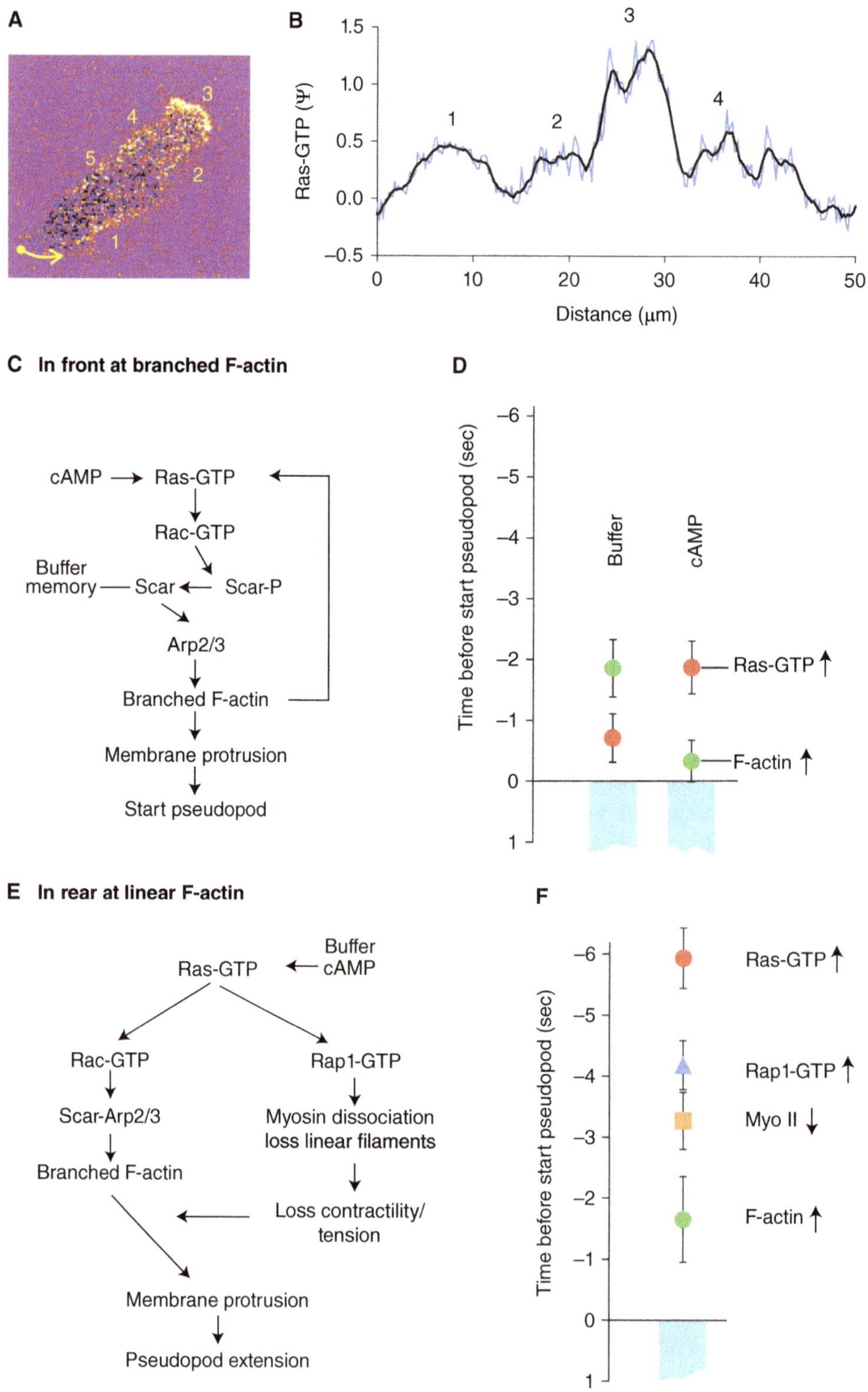

Figure 3. (*See following page for legend.*)

hibit persistence (Van Haastert 2021). Persistence is expressed by the correlation factor for dispersion γ, which is defined as the arithmetic mean of the cosine of angles between two subsequent pseudopods ($\gamma = 0$ for random movement and $\gamma = 1$ if all pseudopods are extended in the same direction; Van Haastert 2010a). The correlation factor for polarized memory is $\gamma = 0.58$. The kinetics of learning a new polarity axis and forgetting the old polarity axis was investigated in cells that start a new pseudopod at the side of the cell, which potentially generates a new polarity axis. It was shown that the strength of the new polarity axis increases with a half-time of ~30 sec (learning), while the strength of the old polarity axis decreases with a half-time of 90 sec (forgetting). Thus, the polarity axis has a relatively long memory of a few minutes; since *Dictyostelium* cells extend about four pseudopods per minute, memory of polarity is the average direction of the last six to 10 pseudopods (Van Haastert 2021).

In *Dictyostelium* and neutrophils, the polarity axis is formed by linear F-actin/myosin II filaments in the rear of the cell that inhibit the formation of branched F-actin-induced pseudopods. *Dictyostelium* myosin II–null cells have strongly reduced polarity and extend many lateral pseudopods. In *Dictyostelium*, myosin filament formation in the cortex and cell polarity is mediated by a cGMP signaling pathway (Bosgraaf et al. 2002). Of interest, in *Dictyostelium* the memory time of cell polarity (90 sec) is similar to the half-time of dissociation of cGMP from the cGMP-binding protein complex (Van Haastert et al. 1982; Van Haastert 2021). *Dic-*

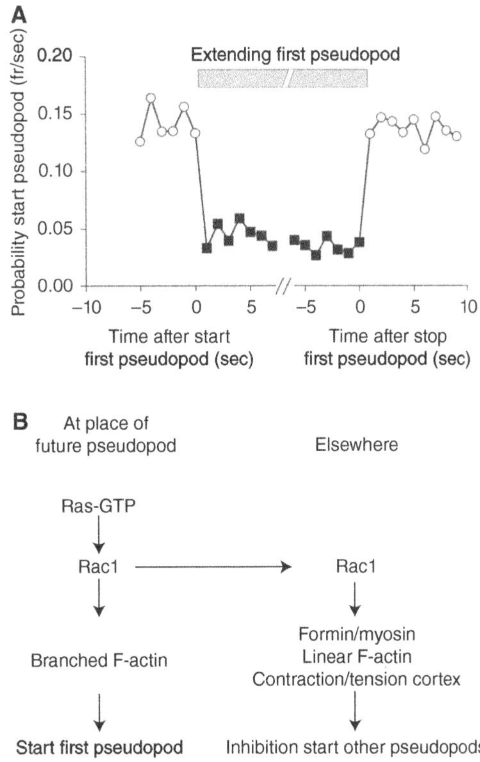

Figure 4. One pseudopod at a time. The induction of new pseudopods inhibits the start of other pseudopods. (*A*) In cells that have no pseudopod, the probability to start a first pseudopod is ~0.14 fraction/sec (open circles) and declines to 0.04 fraction/sec for the start of a second pseudopod; this decline is immediate even before the first pseudopod has moved (reprinted from Van Haastert 2020, under a Creative Commons CC BY 4.0 license). (*B*) Schematic: A component of the activation pathway of a new pseudopod (here Rac1) activates the contractile cortex to inhibit the start of a second pseudopod.

Figure 3. Start of a pseudopod. (*A*) Ras-guanosine triphosphate (GTP) in a polarized cell detected with Raf-Ras-binding domain (RBD)-green fluorescent protein (GFP) and cytosolic red fluorescent protein (RFP); the signal Ψ is the difference of the GFP and RFP signal in each pixel, normalized for the average RFP signal in the cytosol. (*B*) The Ras-GTP levels at the cell boundary starting at the arrow. Five patches are indicated in both images. (*C*) Scheme showing how pseudopods may start in the front of the cell in buffer or upon cyclic adenosine monophosphate (cAMP) stimulation. (*D*) Ras-GTP and F-actin levels were measured at the place of the future pseudopod; data show their onset relative to the start of the pseudopod. (*E*) Scheme showing how pseudopods may start at the side and in the rear of the cell, both in buffer or upon cAMP stimulation. The contractile cortex prevents the start of a pseudopod and is first degraded by the concerted activation of Ras, Rap, and myosin filament disassembly. Subsequently, Ras-induced formation of branched F-actin can push the membrane forward. (*F*) Ras-GTP, Rap1-GTP, myosin II, and F-actin levels were measured at the place of the future pseudopod. The data in *D* and *F* are the means and SEM of 10 measurements (panels based on data from Van Haastert et al. 2017 and Van Haastert 2021).

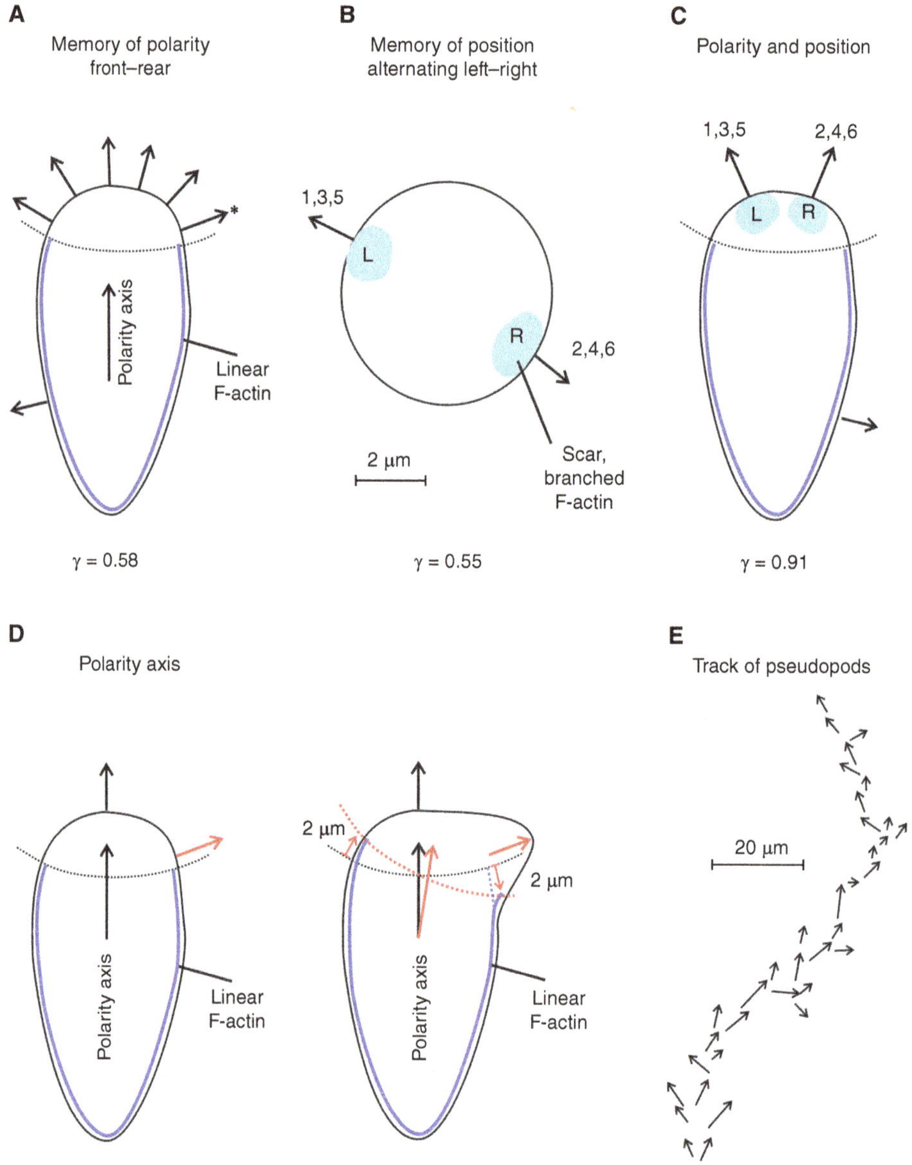

Figure 5. Memory. (*A*) Memory of polarity: The linear F-actin in the cortex induces a front–rear axis of polarity; the many pseudopods made in the front lead to persistence of movement with correlation factor $\gamma = 0.58$. (*B*) Memory of position: The activation of branched F-actin that is associated with a pseudopod leaves behind a modified molecule (presumably Scar) that strongly enhances the start of a future pseudopod at that place. Pseudopods are alternatingly *left* and *right*, which can be nearby, leading to persistence, or far away (as drawn here) leading to a wiggling cell. The correlation factor $\gamma = 0.55$. (*C*) The combination of memory of polarity and position leads to alternating *right–left* extension of pseudopods in a narrow front with very strong persistence; the measured correlation factor of *Dictyostelium* cells is $\gamma = 0.91$. (*D*) Change of polarity axis by a pseudopod at 70° (red arrow). Experiments have shown that the change of the polarity axis (y) depends on the direction of the new pseudopod (x) according to $y = 1.64 \times 10^{-5} \times x^3$ (Van Haastert 2021). For $x = 70°$, the change of the polarity axis is only 5.6°. Furthermore, the contractile cortex is degraded at the *right* side and grows at the *left* side, both more than a length of ~2 µm. (*E*) Pseudopods of a wild-type cell during movement of 9.5 min (panel based on data in Van Haastert 2021).

tyostelium mutants defective in cGMP signaling have strongly reduced polarity and extend many lateral pseudopods (Veltman and Van Haastert 2008).

Although the front–rear polarity axis is quite stable, its exact direction is adjusted by each new pseudopod as was analyzed in detail (Van Haastert 2021). Assume a pseudopod that starts at the right side in the front at an angle of 70° relative to the polarity axis (Fig. 5D). Such a pseudopod is still in the organelle-free front but near the contractile cortex. The pseudopod is associated with the local activation of Rap1 that will locally induce the disassembly of the nearby contractile cortex, and thereby shift the polarity axis to the right. Detailed quantitative analysis revealed two effects for a pseudopod extended at 70° to the right (Van Haastert 2021). First, the direction of the polarity axis changes by only ~6° to the right revealing the robustness of front–rear polarity axis. Secondly, at the right side of this pseudopod, linear F-actin is degraded and replaced by branched F-actin, while at the left side of the pseudopod, the reverse occurs: branched F-actin is replaced by linear F-actin, both over a distance of ~2 μm (Fig. 5D). Consequently, the boundary between branched and linear F-actin is not constant, but may change a few μm to the left or right at each new pseudopod. This changing boundary may lead to a weak cortex-membrane stability that could explain the observed very high frequency of bleb formation in the transition zone of front to rear of the cell (Fig. 1H).

POSITIONAL MEMORY WITH BRANCHED F-ACTIN IN THE FRONT

Dictyostelium cells extend pseudopods alternatingly to the right and left (Li et al. 2008; Bosgraaf and Van Haastert 2009b). All left pseudopods start at a similar position at the cell surface, and all right pseudopods start at their own position. In polarized cells, these right and left positions are nearby in the front. Initially we were hesitant to interpret these observations as positional memory. Since the position of the last pseudopod is refractory, the available space for the next pseudopod in the front is asymmetric and pseudopods must alternate to some extent in polarized cells (Van Haastert 2021). We measured the area in the front at the right and left side of the current pseudopod and calculated the expected fraction of alternating right–left pseudopods without memory, which is ~60% (random is 50%); the observed fraction is much higher: 77%. Secondly, vegetative cells do not have a restricted front but still have a high fraction of 74% alternating right–left. Third, polarized cells occasionally extend a pseudopod at the side of the cell. Now the entire front would be available for the next pseudopod, but we observed that the next pseudopod starts very frequently in the front at the position of the previous pseudopod (see Table 1, Fig. S3, and Fig. 8A in Van Haastert 2021). We concluded that cells have positional memory. In polarized cells these right and left positions are nearby in the front, but in unpolarized cells they can be at any position also at opposite sides of the cell. The following model has been proposed for this positional memory (Fig. 5B; Van Haastert 2021). When a first pseudopod stops, the second pseudopod cannot start at the same position but starts elsewhere in the cell, for example, to the right of the first pseudopod. Subsequently the branched F-actin machinery of the first pseudopod is degraded but leaves behind some memory molecules that can enhance the start of a new pseudopod at that position. Thus, when the second pseudopod stops, the third pseudopod is likely to start at the position of the memory molecule, that is, at the same position as the first pseudopod, to the left; and, subsequently, the fourth pseudopod at the position of the second pseudopod, to the right, etc. Detailed kinetic analysis revealed that the first left pseudopod must extend for at least 3.5 sec to build up memory, and the next left pseudopod must appear within 30 sec after the first left pseudopod, otherwise the memory is lost. This reveals that the right–left positional memory is short-lived and remembers only one pseudopod, in contrast to the long lived memory of polarity (Van Haastert 2021). In cells that have minimal front–rear polarity (e.g., vegetative *Dictyostelium* cells or myosin II–null cells), the persistence correlation factor of positional memory is $\gamma = 0.55$.

Of the ~50 *Dictyostelium* mutant strains analyzed, we have identified only one mutant that extends pseudopods not alternating right–left but randomly to the right or left (Van Haastert 2021). In this strain, the wild-type Scar protein is replaced by the Scar-S5D mutant in which five phosphorylatable serines are replaced by the phosphomimic aspartate (Ura et al. 2012). The Scar-activating Rac-GTP binds to a cleft in the Scar protein. This cleft is closed due to phosphorylation of the five serines. To activate Scar, these phosphates must be removed, followed by Rac-GTP binding and stimulation of the Scar complex (Davidson and Insall 2013). We envision that when the pseudopod stops and the Rac-GTP dissociates from the Scar complex, Scar may stay for some time in the unphosphorylated state. When the next pseudopod is induced, the area with unphosphorylated Scar is the preferred place for Rac-GTP to bind to Scar and activate Arp2/3, and thereby the new pseudopod starts at the same position as the previous pseudopod.

In summary, cells have two types of memory, a long-term memory of a front–rear polarity axis due to linear F-actin/myosin filaments in the rear and a short-term memory of pseudopod position due to branched F-actin in the front. Polarized wild-type cells have both polarity and positional memory. Pseudopods are extended in the front alternatingly to the right and to the left at a relatively small angle of ~55° (Fig. 5C); these cells have very strong persistence with $\gamma = 0.91$ (Fig. 5D). The opposing properties of linear and branched F-actin are highly synergistic for persistent amoeboid movement.

DIRECTIONAL MOVEMENT AND NAVIGATION

As described above, amoeboid movement in buffer is complex with sophisticated regulation of pseudopods. Directional cues add spatial information to bias pseudopod extension in the direction of the cue. We first consider cells that already move approximately in the direction of the chemotactic gradient. Polarity and positional memory stimulate the cell to continue to move in that direction, but persistence is not perfect ($\gamma = 0.9$). Therefore, after some pseudopods, the cell would no longer move in the direction of the gradient unless the gradient provides directional information. Chemotactic stimulation leads to the activation of Ras-GEFs in the part of the cell facing the highest concentration of chemoattractant (Fig. 6A). This results in an increase of active Ras-GTP at one side of the patch, thereby inducing a bias of local branched F-actin and pseudopods in the direction of the chemotactic gradient. Chemotactic stimulation also activates the cGMP pathway that induces the translocation of myosin filaments to the contractile cortex, leading to a further reduction of pseudopod formation in the rear of the cell. Therefore, cells moving approximately in the direction of the chemoattractant exhibit a small directional bias of branched F-actin and pseudopods toward the chemoattractant and have enhanced linear F-actin/myosin in the rear to suppress pseudopods that otherwise would be extended in a direction away from the chemoattractant.

Next, we consider cells that are chemotactically stimulated at the side of the cell perpendicular to the current direction of movement (Fig. 6B,C). The response is very different, depending on the steepness of the gradient. In shallow gradients (Fig. 6B), the activation of Ras-GTP at the side of the cell is not sufficient to break down the linear F-actin/myosin cortex needed to induce a local pseudopod at the side of the cell. Instead, the chemoattractant-induced activation of Ras-GTP combines with memory-derived activated Ras-GTP in the front of the cell thereby inducing a small bias of the pseudopod in the direction of gradient. Due to polarity and positional memory the subsequent pseudopod would be extended in this direction, but again with a small bias toward the chemoattractant, which ultimately leads to slow turning of the polarized cell in the direction of the shallow gradient of chemoattractant. In a steep gradient of chemoattractant (Fig. 6C), the strong stimulation of the Ras-GTP patch at the side of the cell may be sufficiently large to locally induce disassembly of the contractile linear F-actin cortex and to induce a branched F-actin pseudopod immediately in the direction of the gradient of chemo-

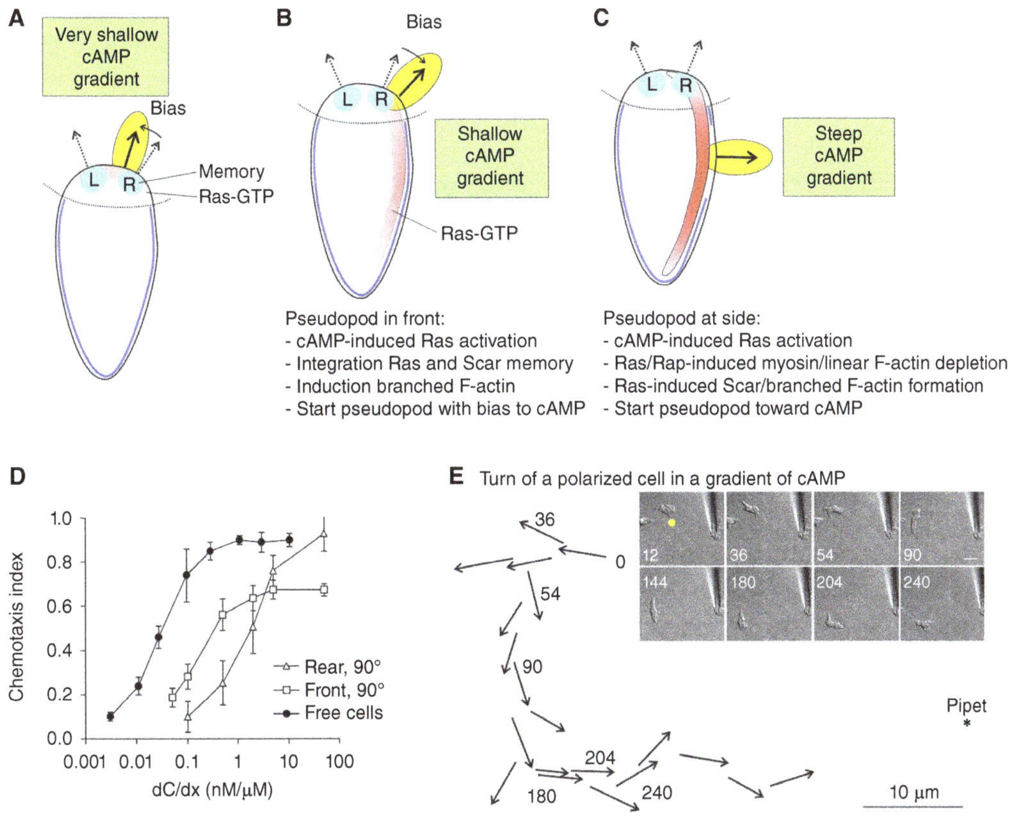

Figure 6. Chemotaxis to cyclic adenosine monophosphate (cAMP). (*A–C*) Schematics of chemotaxis. (*A*) Cell stimulated in the front. Pseudopod that would start at R receives a stronger Ras-GTP signal at the left and therefore starts with a small bias toward cAMP. (*B*) Cell stimulated at the side in a shallow cAMP gradient that does not induce a pseudopod at the side. Instead, the pseudopod that would start at R receives a stronger Ras-GTP signal at the right and therefore starts with a small bias toward cAMP. (*C*) Cell stimulated at the side in a steep cAMP gradient; Ras-GTP stimulation is sufficiently strong to locally degrade the contractile cortex (blue) and start a pseudopod in the direction of the gradient. (*D*) Dose–response curves for chemotaxis of "free" cells moving in a gradient for 15 min, and chemotaxis of a pseudopod with cAMP at the side (as in *B,C*). Data are the means and SEM from Van Haastert (2010b). (*E*) The turn of a cell that is initially stimulated from the back. The *insets* are stills from a phase contrast movie; the yellow dot indicates the cell analyzed. The main figure shows the pseudopods and how the cell turns toward the pipet.

attractant. Similar reactions may occur in neutrophils (Wang et al. 2014).

Dose–response curves for perpendicular stimulation reveal that very steep gradients are needed to induce a pseudopod at the side of the cell in the direction of the gradient; half-maximal chemotaxis requires a gradient of 1.4 nM/μm (Fig. 6D). A bias of pseudopods in the front ultimately leading to the slow turn of the cell in the direction of the gradient requires less steep gradients; half-maximal chemotaxis occurs in a gradient of 0.135 nM/μm. Cells that have turned toward the chemoattractant stay approximately on track by polarity and positional memory; the small corrections to precisely move in the direction of the gradient occur at very shallow gradients; half-maximal chemotaxis of cells occurs at a gradient of 0.022 nM/μm (Fig. 6D). Thus, positional memory of branched F-actin in the front and polarity of linear F-actin in the rear allow cells to chemotax in extremely shallow gradients. In these shallow gradients (0.022 nM/μm, midpoint 1 nM), only ~400 of the 40,000 cAMP receptors are occupied, and the half of the cell facing the highest cAMP concentration has only ~40 receptors more occupied than the other half of the cell. Theoretical considerations on the signal-noise-ratio reveal that this is statistically only possible with the combination of polarity and positional memory, which strongly improves the signal and reduces the noise (Van Haastert and Postma 2007; Bosgraaf and Van Haastert 2009a).

SUMMARY, COLLABORATION OF BRANCHED AND LINEAR F-ACTIN

Branched and linear F-actin are antipodes for pseudopod formation and pseudopod growth. Linear F-actin/myosin in the contractile cortex inhibits the formation of branched F-actin, and if some branched F-actin is formed, growth of the potentially emerging pseudopod is strongly inhibited due to the rigidity of the cortex and plasma membrane. On the other hand, myosin II filaments cannot easily translocate to areas with branched F-actin, which is preventing the formation of a strong contractile cortex in extending pseudopods. These opposite properties of branched and linear F-actin are highly cooperative in cell movement. First, the simultaneous stimulation of branched and linear F-actin provides a mechanism by which cells preferentially extend only one pseudopod at a time, not zero or more than one. Secondly, the two types of F-actin provide two memories: the linear F-actin-based front–rear polarity and the branched F-actin-based right–left positional memory, which together form a highly efficient mechanism for persistent cell movement. This persistence leads to the effective dispersal of cells, which is critical for food searching. Third, the memory also allows efficient chemotaxis in very shallow gradients, because cells suppress pseudopods at the side of the cell using linear F-actin and integrate directional information at the front using branched F-actin, by which cells move better toward the chemoattractant at each subsequent step.

The collaboration between branched and linear F-actin is likely to be important for any type of navigation such as electotaxis, thermotaxis, collective migration, haptotaxis, etc. The collaboration between branched and linear F-actin is also likely to operate in a very similar way in other amoeboid cells such as neutrophils.

ACKNOWLEDGMENTS

We thank Ineke Keizer-Gunnink for recording movies on *Dictyostelium* movement.

REFERENCES

Andrew N, Insall RH. 2007. Chemotaxis in shallow gradients is mediated independently of PtdIns 3-kinase by biased choices between random protrusions. *Nat Cell Biol* **9:** 193–200. doi:10.1038/ncb1536

Blagg SL, Insall RH. 2004. Control of SCAR activity in *Dictyostelium discoideum*. *Biochem Soc Trans* **32:** 1113–1114. doi:10.1042/BST0321113

Bosgraaf L, Van Haastert PJM. 2006. The regulation of myosin II in *Dictyostelium*. *Eur J Cell Biol* **85:** 969–979. doi:10.1016/j.ejcb.2006.04.004

Bosgraaf L, Van Haastert PJM. 2009a. Navigation of chemotactic cells by parallel signaling to pseudopod persistence and orientation. *PLoS One* **4:** e6842. doi:10.1371/journal.pone.0006842

Bosgraaf L, Van Haastert PJM. 2009b. The ordered extension of pseudopodia by amoeboid cells in the absence of external cues. *PLoS One* **4:** e5253. doi:10.1371/journal.pone.0005253

Bosgraaf L, Russcher H, Smith JL, Wessels D, Soll DR, Van Haastert PJM. 2002. A novel cGMP signalling pathway mediating myosin phosphorylation and chemotaxis in *Dictyostelium*. *EMBO J* **21:** 4560–4570. doi:10.1093/emboj/cdf438

Bovellan M, Romeo Y, Biro M, Boden A, Chugh P, Yonis A, Vaghela M, Fritzsche M, Moulding D, Thorogate R, et al. 2014. Cellular control of cortical actin nucleation. *Curr Biol* **24:** 1628–1635. doi:10.1016/j.cub.2014.05.069

Cao L, Ghasemi F, Way M, Jégou A, Romet-Lemonne G. 2023. Regulation of branched versus linear Arp2/3-generated actin filaments. *EMBO J* **42:** e113008. doi:10.15252/embj.2022113008

Chugh P, Paluch EK. 2018. The actin cortex at a glance. *J Cell Sci* **131:** jcs186254. doi:10.1242/jcs.186254

Clarke M, Baron A. 1987. Myosin filaments in cytoskeletons of *Dictyostelium* amoebae. *Cell Motil Cytoskeleton* **7:** 293–303. doi:10.1002/cm.970070402

Collier S, Paschke P, Kay RR, Bretschneider T. 2017. Image based modeling of bleb site selection. *Sci Rep* **7:** 6692. doi:10.1038/s41598-017-06875-9

Danuser G, Allard J, Mogilner A. 2013. Mathematical modeling of eukaryotic cell migration: insights beyond experiments. *Annu Rev Cell Dev Biol* **29:** 501–528. doi:10.1146/annurev-cellbio-101512-122308

Davidson AJ, Insall RH. 2013. SCAR/WAVE: a complex issue. *Commun Integr Biol* **6:** e27033. doi:10.4161/cib.27033

Diez S, Gerisch G, Anderson K, Müller-Taubenberger A, Bretschneider T. 2005. Subsecond reorganization of the actin network in cell motility and chemotaxis. *Proc Natl Acad Sci* **102:** 7601–7606. doi:10.1073/pnas.0408546102

Filić V, Marinović M, Faix J, Weber I. 2014. The IQGAP-related protein DGAP1 mediates signaling to the actin cytoskeleton as an effector and a sequestrator of Rac1 GTPases. *Cell Mol Life Sci* **71:** 2775–2785. doi:10.1007/s00018-014-1606-3

Fritz-Laylin LK, Riel-Mehan M, Chen BC, Lord SJ, Goddard TD, Ferrin TE, Nicholson-Dykstra SM, Higgs H, Johnson GT, Betzig E, et al. 2017. Actin-based protrusions of migrating neutrophils are intrinsically lamellar and facilitate direction changes. *eLife* **6:** e26990.

Gail MH, Boone C. 1970. The locomotion of mouse fibroblasts in tissue culture. *Biophys J* **10:** 980–993. doi:10.1016/S0006-3495(70)86347-0

García-Arcos JM, Jha A, Waterman CM, Piel M. 2024. Blebology: principles of bleb-based migration. *Trends Cell Biol* **34:** 838–853. doi:10.1016/j.tcb.2024.02.009

Gautreau AM, Fregoso FE, Simanov G, Dominguez R. 2022. Nucleation, stabilization, and disassembly of branched actin networks. *Trends Cell Biol* **32:** 421–432. doi:10.1016/j.tcb.2021.10.006

Gittes F, Mickey B, Nettleton J, Howard J. 1993. Flexural rigidity of microtubules and actin filaments measured from thermal fluctuations in shape. *J Cell Biol* **120:** 923–934. doi:10.1083/jcb.120.4.923

Hall RL. 1977. Amoeboid movement as a correlated walk. *J Math Biol* **4:** 327–335. doi:10.1007/BF00275081

Hill TL. 1981. Microfilament or microtubule assembly or disassembly against a force. *Proc Natl Acad Sci* **78:** 5613–5617. doi:10.1073/pnas.78.9.5613

Huang CH, Tang M, Shi C, Iglesias PA, Devreotes PN. 2013. An excitable signal integrator couples to an idling cytoskeletal oscillator to drive cell migration. *Nat Cell Biol* **15:** 1307–1316. doi:10.1038/ncb2859

Insall RH. 2010. Understanding eukaryotic chemotaxis: a pseudopod-centred view. *Nat Rev Mol Cell Biol* **11:** 453–458. doi:10.1038/nrm2905

Insall RH, Machesky LM. 2009. Actin dynamics at the leading edge: from simple machinery to complex networks. *Dev Cell* **17:** 310–322. doi:10.1016/j.devcel.2009.08.012

Janetopoulos C, Jin T, Devreotes P. 2001. Receptor-mediated activation of heterotrimeric G-proteins in living cells. *Science* **291:** 2408–2411. doi:10.1126/science.1055835

Jasnin M, Beck F, Ecke M, Fukuda Y, Martinez-Sanchez A, Baumeister W, Gerisch G. 2019. The architecture of traveling actin waves revealed by cryo-electron tomography. *Structure* **27:** 1211–1223.e5. doi:10.1016/j.str.2019.05.009

Kadzik RS, Homa KE, Kovar DR. 2020. F-Actin cytoskeleton network self-organization through competition and cooperation. *Annu Rev Cell Dev Biol* **36:** 35–60. doi:10.1146/annurev-cellbio-032320-094706

Kelkar M, Bohec P, Charras G. 2020. Mechanics of the cellular actin cortex: from signalling to shape change. *Curr Opin Cell Biol* **66:** 69–78. doi:10.1016/j.ceb.2020.05.008

Killich T, Plath PJ, Haß EC, Xiang W, Bultmann H, Rensing L, Vicker MG. 1994. Cell movement and shape are nonrandom and determined by intracellular, oscillatory rotating waves in *Dictyostelium* amoebae. *Biosystems* **33:** 75–87. doi:10.1016/0303-2647(94)90048-5

Kortholt A, Keizer-Gunnink I, Kataria R, Van Haastert PJM. 2013. Ras activation and symmetry breaking during *Dictyostelium* chemotaxis. *J Cell Sci* **126:** 4502–4513. doi:10.1242/jcs.132340

Kuhn J, Banerjee P, Haye A, Robinson DN, Iglesias PA, Devreotes PN. 2024. Complementary cytoskeletal feedback loops control signal transduction excitability and cell polarity. bioRxiv doi:10.1101/2024.02.13.580131

Li L, Nørrelykke SF, Cox EC. 2008. Persistent cell motion in the absence of external signals: a search strategy for eukaryotic cells. *PLoS One* **3:** e2093. doi:10.1371/journal.pone.0002093

Li TD, Bieling P, Weichsel J, Mullins RD, Fletcher DA. 2022. The molecular mechanism of load adaptation by branched actin networks. *eLife* **11:** e73145.

Litschko C, Linkner J, Brühmann S, Stradal TEB, Reinl T, Jänsch L, Rottner K, Faix J. 2017. Differential functions of WAVE regulatory complex subunits in the regulation of actin-driven processes. *Eur J Cell Biol* **96:** 715–727. doi:10.1016/j.ejcb.2017.08.003

Litschko C, Brühmann S, Csiszár A, Stephan T, Dimchev V, Damiano-Guercio J, Junemann A, Körber S, Winterhoff M, Nordholz B, et al. 2019. Functional integrity of the contractile actin cortex is safeguarded by multiple Diaphanous-related formins. *Proc Natl Acad Sci* **116:** 3594–3603. doi:10.1073/pnas.1821638116

Makarava N, Menz S, Theves M, Huisinga W, Beta C, Holschneider M. 2014. Quantifying the degree of persistence in random amoeboid motion based on the Hurst exponent of fractional Brownian motion. *Phys Rev E* **90:** 42703. doi:10.1103/PhysRevE.90.042703

McRobbie SJ, Newell PC. 1983. Changes in actin associated with the cytoskeleton following chemotactic stimulation of *Dictyostelium discoideum*. *Biochem Biophys Res Commun* **115:** 351–359. doi:10.1016/0006-291X(83)91011-2

Mogilner A, Oster G. 1996. Cell motility driven by actin polymerization. *Biophys J* **71:** 3030–3045. doi:10.1016/S0006-3495(96)79496-1

Mogilner A, Barnhart EL, Keren K. 2020. Experiment, theory, and the keratocyte: an ode to a simple model for cell motility. *Semin Cell Dev Biol* **100:** 143–151. doi:10.1016/j.semcdb.2019.10.019

Muresan CG, Sun ZG, Yadav V, Tabatabai AP, Lanier L, Kim JH, Kim T, Murrell MP. 2022. F-actin architecture determines constraints on myosin thick filament motion. *Nat Commun* **13:** 7008. doi:10.1038/s41467-022-34715-6

Noegel AA, Schleicher M. 2000. The actin cytoskeleton of *Dictyostelium*: a story told by mutants. *J Cell Sci* **113:** 759–766. doi:10.1242/jcs.113.5.759

Patlak CS. 1953. Random walk with persistence and external bias. *Bull Math Biophys* **15:** 311–338. doi:10.1007/BF02476407

Peskin CS, Odell GM, Oster GF. 1993. Cellular motions and thermal fluctuations: the Brownian ratchet. *Biophys J* **65:** 316–324. doi:10.1016/S0006-3495(93)81035-X

Potel MJ, Mackay SA. 1979. Preaggregative cell motion in *Dictyostelium*. *J Cell Sci* **36:** 281–309. doi:10.1242/jcs.36.1.281

Prahl LS, Odde DJ. 2018. Modeling cell migration mechanics. *Adv Exp Med Biol* **1092:** 159–187. doi:10.1007/978-3-319-95294-9_9

Ramalingam N, Franke C, Jaschinski E, Winterhoff M, Lu Y, Brühmann S, Junemann A, Meier H, Noegel AA, Weber I, et al. 2015. A resilient formin-derived cortical actin meshwork in the rear drives actomyosin-based motility in 2D confinement. *Nat Commun* **6:** 8496. doi:10.1038/ncomms9496

Ridley AJ. 2011. Life at the leading edge. *Cell* **145:** 1012–1022. doi:10.1016/j.cell.2011.06.010

Sasaki AT, Chun C, Takeda K, Firtel RA. 2004. Localized Ras signaling at the leading edge regulates PI3K, cell polarity, and directional cell movement. *J Cell Biol* **167:** 505–518. doi:10.1083/jcb.200406177

Skoge M, Wong E, Hamza B, Bae A, Martel J, Kataria R, Keizer-Gunnink I, Kortholt A, Van Haastert PJM, Charras G, et al. 2016. A worldwide competition to compare the speed and chemotactic accuracy of neutrophil-like cells. *PLoS One* **11:** e0154491. doi:10.1371/journal.pone.0154491

Šoštar M, Marinović M, Filić V, Pavin N, Weber I. 2024. Oscillatory dynamics of Rac1 activity in *Dictyostelium discoideum* amoebae. *PLoS Comput Biol* **20:** e1012025. doi:10.1371/journal.pcbi.1012025

Svitkina TM. 2020. Actin cell cortex: structure and molecular organization. *Trends Cell Biol* **30:** 556–565. doi:10.1016/j.tcb.2020.03.005

Tyson RA, Zatulovskiy E, Kay RR, Bretschneider T. 2014. How blebs and pseudopods cooperate during chemotaxis. *Proc Natl Acad Sci* **111:** 11703–11708. doi:10.1073/pnas.1322291111

Ura S, Pollitt AY, Veltman DM, Morrice NA, Machesky LM, Insall RH. 2012. Pseudopod growth and evolution during cell movement is controlled through SCAR/WAVE dephosphorylation. *Curr Biol* **22:** 553–561. doi:10.1016/j.cub.2012.02.020

Urban E, Jacob S, Nemethova M, Resch GP, Small JV. 2010. Electron tomography reveals unbranched networks of actin filaments in lamellipodia. *Nat Cell Biol* **12:** 429–435. doi:10.1038/ncb2044

Van Haastert PJM. 2010a. A model for a correlated random walk based on the ordered extension of pseudopodia. *PLoS Comput Biol* **6:** e1000874. doi:10.1371/journal.pcbi.1000874

Van Haastert PJM. 2010b. A stochastic model for chemotaxis based on the ordered extension of pseudopods. *Biophys J* **99:** 3345–3354. doi:10.1016/j.bpj.2010.09.042

Van Haastert PJM. 2020. Unified control of amoeboid pseudopod extension in multiple organisms by branched F-actin in the front and parallel F-actin/myosin in the cortex. *PLoS One* **15:** e0243442. doi:10.1371/journal.pone.0243442

Van Haastert PJM. 2021. Short- and long-term memory of moving amoeboid cells. *PLoS One* **16:** e0246345. doi:10.1371/journal.pone.0246345

Van Haastert PJM, Bosgraaf L. 2009a. Food searching strategy of amoeboid cells by starvation induced run length extension. *PLoS One* **4:** e6814. doi:10.1371/journal.pone.0006814

Van Haastert PJM, Bosgraaf L. 2009b. The local cell curvature guides pseudopodia towards chemoattractants. *HFSP J* **3:** 282–286. doi:10.2976/1.3185725

Van Haastert PJM, Postma M. 2007. Biased random walk by stochastic fluctuations of chemoattractant-receptor interactions at the lower limit of detection. *Biophys J* **93:** 1787–1796. doi:10.1529/biophysj.107.104356

Van Haastert PJM, Van Walsum H, Pasveer FJ. 1982. Nonequilibrium kinetics of a cGMP-binding protein in *Dictyostelium discoideum*. *J Cell Biol* **94:** 271–278. doi:10.1083/jcb.94.2.271

Van Haastert PJM, Keizer-Gunnink I, Kortholt A. 2017. Coupled excitable Ras and F-actin activation mediates spontaneous pseudopod formation and directed cell movement. *Mol Biol Cell* **28:** 922–934. doi:10.1091/mbc.e16-10-0733

Van Haastert PJM, Keizer-Gunnink I, Kortholt A. 2018. The cytoskeleton regulates symmetry transitions in moving amoeboid cells. *J Cell Sci* **131:** jcs208892. doi:10.1242/jcs.208892

Veltman DM, Van Haastert PJM. 2008. The role of cGMP and the rear of the cell in *Dictyostelium* chemotaxis and cell streaming. *J Cell Sci* **121:** 120–127. doi:10.1242/jcs.015602

Wang MJ, Artemenko Y, Cai WJ, Iglesias PA, Devreotes PN. 2014. The directional response of chemotactic cells depends on a balance between cytoskeletal architecture and the external gradient. *Cell Rep* **9:** 1110–1121. doi:10.1016/j.celrep.2014.09.047

Xiong Y, Huang CH, Iglesias PA, Devreotes PN. 2010. Cells navigate with a local-excitation, global-inhibition-biased excitable network. *Proc Natl Acad Sci* **107:** 17079–17086. doi:10.1073/pnas.1011271107

Yang C, Svitkina T. 2011. Visualizing branched actin filaments in lamellipodia by electron tomography. *Nat Cell Biol* **13:** 1012–1013. doi:10.1038/ncb2321

Ydenberg CA, Smith BA, Breitsprecher D, Gelles J, Goode BL. 2011. Cease-fire at the leading edge: new perspectives on actin filament branching, debranching, and cross-linking. *Cytoskeleton* **68:** 596–602. doi:10.1002/cm.20543

Zatulovskiy E, Tyson R, Bretschneider T, Kay RR. 2014. Bleb-driven chemotaxis of *Dictyostelium* cells. *J Cell Biol* **204:** 1027–1044. doi:10.1083/jcb.201306147

Leading the Way: Molecular Drivers of Single-Cell Migration

Dong Li,[1,2,5] Hui Tu,[1,3,5] and Huaqing Cai[1,4]

[1]National Laboratory of Biomacromolecules, CAS Center for Excellence in Biomacromolecules, Institute of Biophysics, Chinese Academy of Sciences, Beijing 100101, China

[2]Department of Cell Biology, School of Medicine, Johns Hopkins University, Baltimore, Maryland 21205, USA

[3]Institute of Systems Biomedicine, Beijing Key Laboratory of Tumor Systems Biology, School of Basic Medical Sciences, Peking University Health Science Center, Peking University, Beijing 100191, China

[4]College of Life Sciences, University of Chinese Academy of Sciences, Beijing 100049, China

Correspondence: huaqingcai@ibp.ac.cn

Cell migration plays a central role in a wide range of physiological, developmental, and disease-related processes. Studies using single-cell models, such as *Dictyostelium discoideum,* have provided important insights into the molecular principles underlying this process. Migrating cells exhibit a polarized morphology, with actin-rich protrusions at the leading edge driving forward motion and an actomyosin network at the trailing edge enabling retraction. While actin polymerization and direct cytoskeletal regulators are essential, a complex network of signaling molecules also play a critical role in cell migration. Initially viewed as part of the directional sensing machinery in guided migration, this signaling network is now also recognized as an integral component of the motility module itself. Its spontaneous activity coordinates with cytoskeletal reorganization, enabling cell migration even in the absence of external cues. This review highlights key cytoskeletal and signaling molecules involved in leading-edge protrusion formation, with an emphasis on findings from *Dictyostelium* studies. We also discuss recent advances in understanding how these cytoskeletal and signaling molecules organize into excitable networks to regulate cell motility.

Cell migration, the process by which cells move from one location to another, is a fundamental cellular activity that plays a critical role in numerous physiological processes, including gastrulation, wound healing, and immune responses (Scarpa and Mayor 2016; Stuelten et al. 2018; Yamada and Sixt 2019; Bodor et al. 2020). Dysregulated cell migration, however, contributes to the development and progression of various diseases, including cancer, atherosclerosis, and arthritis. Migrating cells adopt a polarized state in which key signaling and cytoskeletal components are spatiotemporally organized to form distinct leading and trailing edges. The leading edge, characterized by actin-rich protrusions, drives forward movement, whereas the

[5]These authors contributed equally to this work.

Cite this article as *Cold Spring Harb Perspect Biol* doi: 10.1101/cshperspect.a041748

trailing edge, containing an actomyosin network, facilitates retraction. In addition to random movement, cells can migrate directionally in response to external cues (Fig. 1), such as chemical gradients (chemotaxis or haptotaxis), changes in substrate stiffness (durotaxis), and electric fields (galvanotaxis). In these processes, cells sense environmental signals and transmit this information to the cytoskeletal machinery, which modulates protrusion dynamics to drive directed migration (Devreotes and Horwitz 2015; SenGupta et al. 2021).

The social amoeba *Dictyostelium discoideum* provides a valuable model for studying cell migration and other actin-driven processes (Cai and Devreotes 2011; Artemenko et al. 2014). This organism has a fascinating life cycle. Under nutrient-rich conditions, *Dictyostelium* exists as single cells that actively perform macropinocytosis and phagocytosis to acquire nutrients while migrating randomly. This phase is referred to as the vegetative stage. Once nutrients are depleted, the cells initiate a differentiation process that leads to the formation of multicellular structures, allowing them to survive in adverse conditions. The transition from the vegetative to the differentiating stage heavily relies on cyclic adenosine monophosphate (cAMP)-mediated chemotaxis (Tomchik and Devreotes 1981; Cai et al. 2014; Williams and Kay 2018; Hao et al. 2024). Numerous regulators of cell migration have been identified by leveraging these life cycle characteristics in conjunction with forward genetic screens. Many of these regulators have subsequently been found to perform conserved functions in higher eukaryotic cells. In this review, we focus primarily on the key cytoskeletal and signaling molecules involved in the formation of leading-edge protrusions in migrating cells, with an emphasis on insights gained from studies in *Dictyostelium* cells. We also highlight recent advances in understanding how these molecules integrate into excitable cytoskeletal and signaling networks to orchestrate cell motility.

CYTOSKELETAL NETWORK IN CELL MOTILITY

Migrating cells form actin-rich protrusions at the leading edge known as pseudopodia (*Dictyostelium* or mammalian cells in three-dimensional surroundings) or lamellipodia (mammalian cells in two-dimensional environments). Within these structures, actin polymerization generates a dendritic network that pushes against the plasma membrane, providing the driving force for migration. By coupling with the actomyosin retraction network at the trailing edge and their low-adhesion property, *Dictyostelium* cells can migrate at remarkable speeds, moving up to several times their own cell length in a minute.

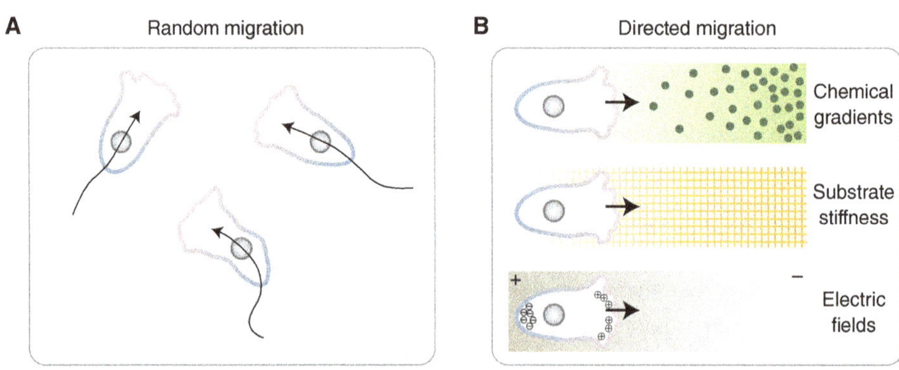

Figure 1. Modes of single-cell migration. (*A*) Cells can randomly migrate by spontaneously adopting a polarized state. In this mode, signaling and cytoskeletal elements self-organize into distinct leading and trailing edges. (*B*) Cells also migrate directionally in response to external cues, including chemical gradients, variations in substrate stiffness, and electric fields. These cues are sensed and transduced to the cytoskeletal machinery to drive directed migration.

Arp2/3 and Scar/WAVE Complexes

The actin-related protein 2/3 (Arp2/3) complex is the primary regulator of branched actin assembly in pseudopodia and the first major actin nucleator identified (Machesky et al. 1994). Instead of nucleating actin filaments de novo, this complex promotes the formation of new filaments by binding to the side of preexisting filaments, generating a Y-shaped branch at an angle of ~70° (Mullins et al. 1998). Both Arp2 and Arp3 resemble the actin fold and together mimic an actin dimer, providing a template for nucleation (Robinson et al. 2001). However, the Arp2/3 complex alone is an inefficient nucleator, and its ability to promote actin polymerization is greatly enhanced by the engagement of nucleation-promoting factors (NPFs) (Pollard and Borisy 2003). The main NPF for Arp2/3 in the formation of pseudopodia and lamellipodia is the suppressor of cAMP receptor/WASp family verprolin-homologous protein (Scar/WAVE) complex (Box 1), which consists of five subunits: Scar/WAVE, Abi/ABI, Nap1/NCKAP1, Pir121/CYFIP, and Hspc300/BRICK1 (Bear et al. 1998; Miki et al. 1998; Machesky et al. 1999; Chen et al. 2010). In *Dictyostelium*, each subunit is encoded by a single gene, and disruption of any of these genes impairs pseudopodia formation and reduces migration speed (Bear et al. 1998; Blagg et al. 2003; Ibarra et al. 2006; Pollitt and Insall 2008, 2009; Litschko et al. 2017). The Scar/WAVE subunit contains a carboxy-terminal verprolin central acidic (VCA) domain. When this domain is bound to Pir121, the complex remains inactive; however, upon release, the VCA domain interacts with both the Arp2/3 complex and actin, inducing conformational changes that activate Arp2/3 and initiate the nucleation of new actin filaments (Fig. 2A; Box 1; Robinson et al. 2001; Goley et al. 2004; Ti et al. 2011; Xu et al. 2012; Espinoza-Sanchez et al. 2018).

Regulation of Scar/WAVE complex activity remains a focus of ongoing research (Fig. 2A). The best-characterized activation pathway involves the small guanosine triphosphatase (GTPase) Rac1. GTP-bound Rac1 associates with Pir121, triggering the release of the VCA domain of Scar/WAVE, activating the complex (Miki et al. 1998; Chen et al. 2010; Steffen et al. 2013). Structural studies have identified two distinct Rac1-binding sites on Pir121—the A site and the D site (Chen et al. 2017). Functional analyses in *Dictyostelium* and melanoma cells, along with in vitro Arp2/3-mediated actin polymerization assays, indicate that Rac1 binding at the A site plays a key role in protrusion formation by directly triggering the allosteric activation of the Scar/WAVE complex. In contrast, binding at the D site appears to modulate protrusion morphology and function, possibly by enhancing the complex's membrane recruitment or increasing Rac1's affinity for the A site (Schaks et al. 2018; Ding et al. 2022). In addition to Rac1 binding, phosphorylation of Scar/WAVE and Abi has been shown to modulate the stability, lifetime, and recruitment of the complex (Danson et al. 2007; Lebensohn and Kirschner 2009; Mendoza et al. 2011; Singh et al. 2020, 2021). Furthermore, the propensity to associate with negatively curved membranes and self-organize into linear arrays (Pipathsouk et al. 2021; Wu et al. 2024) has been proposed to underpin the complex's role in generating sheet-like actin networks essential for protrusion formation and cell movement (Box 1).

Recent studies have also identified two negative regulators of the Scar/WAVE complex (Fig. 2A). CYFIP-related Rac interactor (CYRI) was initially discovered as a Nap1-interacting protein in *Dictyostelium*. It binds active Rac1 via a domain shared with Pir121/CYFIP and, therefore, acts as a competitive inhibitor of Rac1-mediated Scar/WAVE activation and actin polymerization. In CYRI-depleted *Dictyostelium* or mammalian cells, broad protrusions are enriched with Scar/WAVE, but protrusion retraction dynamics are reduced. Conversely, CYRI overexpression suppresses Scar/WAVE recruitment, resulting in unproductive protrusions (Fort et al. 2018; Yuki et al. 2019; Le et al. 2021; Yelland et al. 2021). A second negative regulator, leading edge enriched protein 1 (Leep1), was identified in *Dictyostelium* through a proteomics-based screen (Yang et al. 2021). Deletion of Leep1 results in more persistent localization of the Scar/WAVE complex at the leading edge, whereas its overexpression induces

BOX 1. STRUCTURE, ACTIVATION, AND SELF-ASSEMBLY OF THE Scar/WAVE COMPLEX

The suppressor of cyclic AMP receptor/WASp family verprolin-homologous protein (Scar/WAVE) complex is a heteropentameric assembly comprised of Scar/WAVE, Abi/ABI, Nap1/NCKAP1, Pir121/CYFIP, and Hspc300/BRICK1. It can be divided into two subcomplexes: a trimer formed by Scar, Abi, and Hspc300 and a dimer composed of Pir121 and Nap1, which serves as a platform for trimer binding (Chen et al. 2010). Structural analyses reveal that Hspc300 is sandwiched between Pir121 and the Abi-Scar unit, with Abi adopting an extended conformation that runs along the rim of a prominent cavity on Nap1. The meander region of Scar (residues 82–184) lies against a concave surface on Pir121, contributing to an interface that sequesters the verprolin central acidic (VCA) domain of Scar and maintains the complex in an autoinhibited conformation (Chen et al. 2010, 2017; Ding et al. 2022). Integration of signals through Rac1 binding, engagement of adaptor proteins, and phosphorylation triggers conformational changes that release the VCA domain and activate the complex. Furthermore, recent studies have begun to elucidate how the complex's abilities to sense membrane curvature and self-assemble into higher-order structures regulate its function in protrusion formation.

In migrating cells, the Scar/WAVE complex propagates as linear waves at the edges of pseudopodia and lamellipodia (Weiner et al. 2007; Veltman et al. 2012; Huang et al. 2013). These patterns likely emerge from an excitable feedback network involving both activation signals for the complex and delayed inhibition from negative regulators or actin polymerization (Weiner et al. 2007; Millius et al. 2012; Pipathsouk et al. 2021). Activation signals, such as Rac1, are broadly distributed at the leading edge, whereas the Scar/WAVE complex forms more structured patterns, indicating a self-organizing propensity (Weiner et al. 2007; Pipathsouk et al. 2021). Recent studies have provided new insights into how this complex achieves proper spatial organization. One study revealed that it forms actin-independent nanoscale rings around plasma membrane invaginations exhibiting saddle curvature—positive along one axis and negative along the other (Pipathsouk et al. 2021). Notably, the negative curvature in these regions is comparable to the curvature at the edges of protrusions. A subsequent study showed that the complex assembles into multilayered linear arrays, which are consistently associated with negative membrane curvature in the cellular context (Wu et al. 2024). Although the precise mechanism by which the complex recognizes membrane curvature remains unclear, it is reasonable to hypothesize that negative curvature at protrusion edges drives the formation of linear

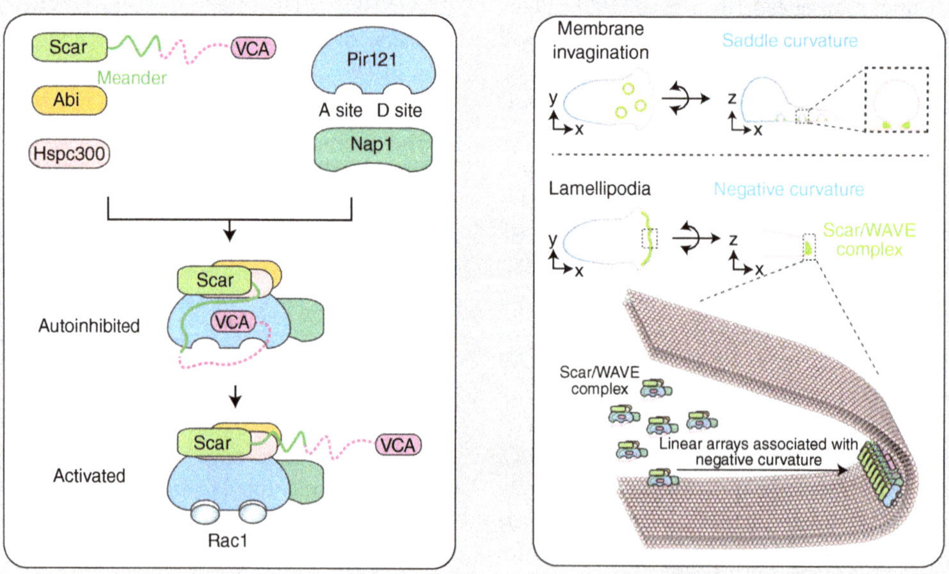

Continued

arrays of the complex. This organization, combined with the complex's ability to promote actin polymerization, serves as a template for generating sheet-like actin networks essential for protrusion formation. Furthermore, these studies demonstrated that the actin-related protein 2/3 (Arp2/3) complex is not required for the formation of these linear arrays. Together with findings showing that the Arp2/3 complex is not strictly needed for Scar/WAVE-mediated protrusion formation (Bieling et al. 2018; Buracco et al. 2024), these observations suggest that the Scar/WAVE complex contributes to protrusion formation beyond merely activating Arp2/3. Future research is needed to clarify the mechanisms underlying the curvature sensing of the Scar/WAVE complex, its full capacity for promoting actin polymerization, and how its spatial organization influences the morphology of actin networks and protrusions.

extensive filopodia formation, resembling the phenotype observed when the Scar/WAVE or Arp2/3 complex is disrupted (Steffen et al. 2006; Beli et al. 2008; Litschko et al. 2017; Schaks et al. 2018). Furthermore, Leep1 localizes to the edge of extending pseudopodia with a slight delay relative to the Scar/WAVE complex and frequently precedes pseudopodia retraction or splitting, consistent with its role in negatively regulating Scar/WAVE activity (Yang et al. 2021).

Loss of the Scar/WAVE complex in *Dictyostelium* results in relatively mild defects in migration, in contrast to the severe consequences observed in certain mammalian cells, such as neutrophils (Machesky and Insall 1998; Veltman et al. 2012; Leithner et al. 2016; Graziano et al. 2019; Tang et al. 2020). This milder phenotype in *Dictyostelium* is attributed to functional compensation by another NPF, Wiskott–Aldrich syndrome protein (WASP, *Dictyostelium* ortholog of N-WASP in vertebrates) (Fig. 2C). When the Scar/WAVE complex is intact, WASP is primarily localized to clathrin-coated pits, where it interacts with active Rac1 (Veltman et al. 2012; Li et al. 2023). This interaction facilitates the incorporation of Rac1 into endocytic vesicles, ensuring its removal from inappropriate locations and maintaining front–rear polarity (Amato et al. 2019). However, in the absence of Scar, WASP can relocate to the leading edge, where it functions similarly by activating the Arp2/3 complex and promoting the formation of morphologically normal pseudopodia (Veltman et al. 2012). *Dictyostelium* cells lacking both Scar and WASP fail to form pseudopodia (Davidson et al. 2018). The ability of WASP to compensate for the loss of Scar/WAVE has been observed to varying degrees in other organisms (Tang et al. 2013; Zhu et al. 2016). The mechanisms that determine the distinct subcellular localizations of different NPFs, the extent to which functional compensation occurs when one NPF is compromised, and the molecular basis of such compensation remain critical questions to be answered.

Formins

Another important actin nucleator at the leading edge is the formin family of proteins, which nucleate actin filaments by binding to their barbed ends. Unlike proteins that bind to barbed ends and inhibit filament elongation, formins act as "leaky cappers." They can cap the filament while still allowing actin monomers to be added, enabling continued filament growth (Pruyne et al. 2002; Shemesh and Kozlov 2007). Formins are crucial for the assembly of linear actin filaments and bundles, particularly in the formation of filopodia. *Dictyostelium* cells express 10 distinct formins (Rivero et al. 2005). Studies of ForH (also known as dDia2) have highlighted the role of formin in filopodia formation (Schirenbeck et al. 2005). ForH localizes to the tips of filopodia, where it promotes actin polymerization and competes with capping proteins. Deletion of *forH* results in shorter and fewer filopodia; however, it does not cause major defects in cell motility. Similarly, disruption of several other formin family members does not

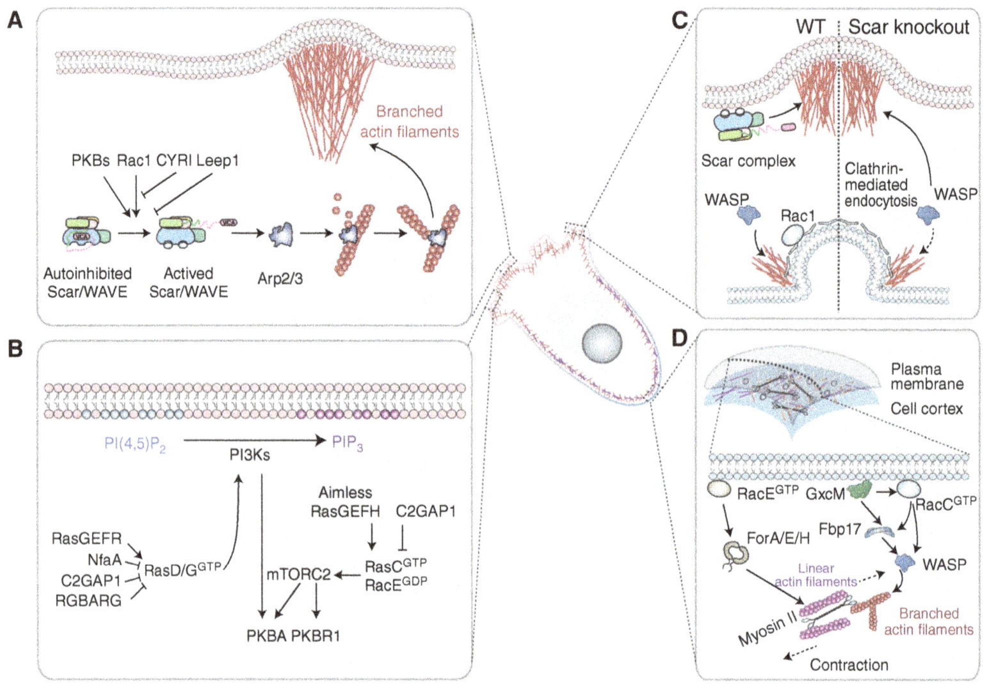

Figure 2. Key signaling and cytoskeletal events regulating cell migration. Migrating cells exhibit a polarized morphology, with actin-rich protrusions at the leading edge driving forward motion and an actomyosin network at the trailing edge facilitating retraction. (*A*) The suppressor of cyclic AMP receptor/WASp family verprolin-homologous protein (Scar/WAVE) complex is activated upon release of its verprolin central acidic (VCA) domain from Pir121. Once freed, the VCA domain interacts with the actin-related protein 2/3 (Arp2/3) complex and actin, promoting Arp2/3 activation and actin nucleation. The association of guanosine triphosphate (GTP)-bound Rac1 with Pir121 triggers the release of the VCA domain. CYRI and leading edge enriched protein 1 (Leep1) act as negative regulators of Scar/WAVE, whereas PKB kinases ensure proper Arp2/3 recruitment by Scar/WAVE. (*B*) In *Dictyostelium*, the phosphatidylinositol 3,4,5-trisphosphate (PIP_3) and Ras pathways are central to regulating protrusion dynamics. RasG, and possibly RasD, activates phosphoinositide 3-kinases (PI3Ks) to produce PIP_3. A hetero-oligomeric complex consisting of active RasC and phosphorylated guanosine diphosphate (GDP)-bound RacE activates mechanistic target of rapamycin complex 2 (mTORC2), which in turn phosphorylates PKBR1 and PKBA to modulate actin polymerization. Both RasC and RasG are regulated by multiples guanine nucleotide exchange factors (GEFs) and GTPase-activating proteins (GAPs), and PKBA is further recruited through its interaction with PIP_3. (*C*) In *Dictyostelium*, Wiskott–Aldrich syndrome protein (WASP) compensates for Scar/WAVE complex deficiencies. When Scar/WAVE is intact, WASP primarily localizes to clathrin-coated pits, where it sequesters active Rac1 into endocytic vesicles to help maintain front–rear polarity. In Scar knockout cells, however, WASP relocates to the leading edge, activating the Arp2/3 complex to promote pseudopodia formation. (WT) Wild-type. (*D*) The rear cortical actin network in *Dictyostelium* comprises distinct actin filament arrays. Active RacE recruits formins (ForA, ForE, and ForH) to assemble linear filaments, whereas the RhoGEF GxcM, together with F-BAR protein Fbp17, RacC, and WASP, activates Arp2/3 to promote branched actin assembly. These actin structures facilitate myosin II recruitment for rear retraction.

significantly impair cell migration (Junemann et al. 2016; Körber et al. 2023), suggesting that formins play a relatively minor role in *Dictyostelium* migration. These findings contrast with studies in other cell types. For example, in B16F1 melanoma cells, formins contribute prominently to migration by cooperating with the Arp2/3 complex to generate the actin network in lamellipodia. In these cells, formins localize not only at filopodia tips but also along the edges of lamellipodia (Yang et al. 2007; Kage et al. 2017). Disruption of formins, including mDia2,

FMNL2, and FMNL3, impairs lamellipodia formation, resulting in narrower, less dynamic protrusions with reduced actin filament networks (Yang et al. 2007; Kage et al. 2017). This leads to decreased migration efficiency and reduced force generation. In this context, formins have been proposed to function by protecting growing actin filaments from capping proteins and by elongating filaments nucleated by the Arp2/3 complex. As the Arp2/3 complex requires preexisting filaments to initiate branch formation, formins may also provide the necessary templates for Arp2/3 binding and catalysis.

Rho GTPases

Rho GTPases are key regulators of actin assembly and actin-based cellular processes (Nobes and Hall 1995). Like other small GTPases, they cycle between an active GTP-bound form and an inactive guanosine diphosphate (GDP)-bound form. This transition is controlled by guanine nucleotide exchange factors (GEFs), which facilitate the exchange of GDP for GTP, and GTPase-activating proteins (GAPs), which promote GTP hydrolysis. Rho GTPases exert their cellular functions through effector proteins, which preferentially bind to their GTP-bound state. Mammalian cells express 20 Rho GTPases, which are categorized into three subgroups: Rho, Rac, and Cdc42. *Dictyostelium* cells also express 20 Rho GTPases, but all of which are classified as Rac proteins. Notably, however, several *Dictyostelium* Rac proteins exhibit features similar to mammalian Cdc42 or Rho. The functions and regulation of these Rac proteins have been extensively reviewed elsewhere (Rivero and Xiong 2016; Filić et al. 2021). Here, we provide a brief overview of their roles in cell migration.

Several Rac GTPases, including Rac1 (Chung et al. 2000; Dumontier et al. 2000; Filić et al. 2012), RacB (Park et al. 2004), and RacE (Wang et al. 2013; Litschko et al. 2019; Senoo et al. 2019), have been implicated in the regulation of cell migration in *Dictyostelium*. At leading-edge protrusions, their primary role is to promote actin network assembly. For example, Rac1 interacts with the Scar/WAVE complex (as described in the previous section) and the actin-binding protein DdLIM (Prassler et al. 1998) to promote pseudopod formation. Rac GTPases also cooperate with other small GTPases or their regulators to fine-tune protrusion dynamics (Senoo et al. 2016, 2019, 2020). For example, RacE and the Ras family GTPase RasC form a hetero-oligomeric complex that activates the TOR-containing serine/threonine kinase complex mechanistic target of rapamycin complex 2 (mTORC2). In turn, mTORC2 phosphorylates and activates other serine/threonine kinases that regulate protrusion dynamics and cell migration (Kamimura et al. 2008; Senoo et al. 2019, 2020) as detailed below. Interestingly, RacE performs this function in its GDP-bound form and requires phosphorylation at Ser192, representing an unusual mechanism in which the function of a Rho GTPase is not mediated by its GTP-bound form.

Dictyostelium cells express a large number of RhoGEF and RhoGAP proteins, which are thought to regulate the spatiotemporal activity of Rac GTPases (Vlahou and Rivero 2006). However, only a few of these proteins have been thoroughly characterized. Among the 46 RhoGAP domain-containing proteins, DdRacGAP1, a putative GAP for Rac1B, has been shown to negatively regulate actin polymerization at the leading edge (Chung et al. 2000). In addition, 46 proteins are known to contain the conventional RhoGEF domain (Dbl-homology domain), including RacGEF1, and eight to contain the CZH2 RhoGEF domain, including ElmoE. Both RacGEF1 and ElmoE are putative GEFs for RacB (Park et al. 2004; Yan et al. 2012). These proteins respond to chemoattractant signals, and their disruption leads to reduced actin polymerization. ElmoE-deficient cells also exhibit reduced motility and impaired chemotaxis (Park et al. 2004; Yan et al. 2012). GxcT is a putative GEF for RacE and several other Rac GTPases. Cells lacking GxcT exhibit reduced motility and polarization, as well as abnormal production of lateral pseudopodia, a defect attributed to the failure to spatially restrict signaling events to the leading edge (Wang et al. 2013). RGBARG contains both a RhoGEF domain and a RhoGEF domain. Cells lacking RGBARG move at approximately half the speed of wild-type cells and display unusually broad and persistent leading edges during chemotaxis (Buckley et al. 2020). Despite these findings, the precise mechanisms by which *Dictyostelium* Rac proteins and their regulators coordinate the regulation of pseudopod morphology and dynamics are not fully understood and require further investigation.

Actin Assembly at the Rear Cortex

The actin network is also assembled at the rear cortex of migrating cells, where it contributes to the recruitment of myosin II. By coupling with myosin II activity, this network serves two main functions: facilitating cell retraction and maintaining cellular polarity by suppressing lateral and rear protrusions (Wessels et al. 1988; Jay et al. 1995; Heid et al. 2004; Bosgraaf and van Haastert 2006). The actomyosin network has also been shown to enable long-range coordination of actin network treadmilling in motile cells and promote protrusion persistence through mechanical feedback (Wilson et al. 2010; Tsai et al. 2019). Several actin assembly factors have been identified as key regulators of the rear cortical actin network in *Dictyostelium.* Among these, three formins (ForA, ForE, and ForH) are recruited and activated by RacE at the cell rear to safeguard cortical integrity (Fig. 2D; Litschko et al. 2019). Simultaneous deletion of all three formins or RacE severely impairs cortical rigidity and architecture. Under unconstrained conditions, this causes the formation of exaggerated protrusions in the mutant cells, whereas under two-dimensional confinement, the compromised rear cortex fails to recruit myosin II, resulting in reduced contractility and minimal movement. Similar defects in cell polarization and migration are observed upon disruption of the formins mDia1 and mDia3 in melanoma cells and fibroblasts (Litschko et al. 2019; Scholz et al. 2024). A recent study highlighted a role of Arp2/3 complex-mediated actin assembly at the rear cortex of *Dictyostelium* cells (Fig. 2D). The RhoGEF GxcM, localized specifically at the rear of migrating cells, functions together with F-BAR protein Fbp17, RacC, and WASP to activate the Arp2/3 complex and promote actin assembly. Overactivation of this pathway leads to excessive actin polymerization, whereas its disruption causes marked defects in cortical rigidity and functionality, both of which impair cell migration (Li et al. 2023). These findings suggest that the rear cortical actin network in rapidly migrating cells consists of structurally and biochemically distinct actin filament arrays, which may work together to shape its physical properties and influence cell migration behavior.

SIGNALING NETWORK IN CELL MOTILITY

In addition to the regulators directly involved in actin polymerization, a complex network of signaling molecules organized into parallel and interconnected pathways is crucial for cell motility (Devreotes et al. 2017). Much of our understanding of this network is derived from studies of *Dictyostelium* cells responding to global chemoattractant stimulation or migrating along chemoattractant gradients. As a result, the network was initially regarded as part of the directional sensing system of cells responsible for relaying extracellular chemical signals. However, many signaling events are also activated or inhibited within spontaneously extended protrusions, leading to cell migration in the absence of external cues (Sasaki et al. 2007; Bosgraaf and Van Haastert 2009; Arai et al. 2010; Huang et al. 2013). Moreover, migration defects in cells lacking some of the signaling molecules manifest as reduced motility rather than impaired directional sensing. These observations suggest that the signaling network is an integral component of the motility module, with its spontaneous activity orchestrating cytoskeletal reorganization to drive cell movement (Fig. 3A). Among the key pathways involved, the phosphatidylinositol 3,4,5-trisphosphate (PIP_3) and Ras pathways activated specifically at the leading edge of migrating cells play central roles in regulating cell motility.

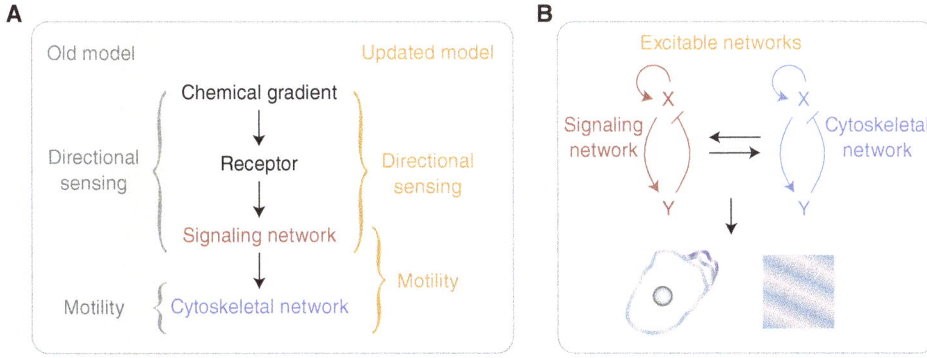

Figure 3. Cell motility arises from excitable signaling and cytoskeletal networks. (*A*) Traditionally, the signaling network was considered part of the directional sensing system responsible for relaying extracellular cues, whereas motility was attributed solely to the activity of the cytoskeletal network. However, signaling events are also detected within spontaneously extended protrusions, and cells lacking specific signaling molecules exhibit defects in random motility. These observations support an updated model in which the signaling network is an integral component of the motility module, with its spontaneous activity orchestrating cytoskeletal reorganization to drive cell movement. (*B*) Both the signaling and cytoskeletal networks exhibit excitability, driven by the interconnection of their components through positive and negative feedback loops. Their dynamic coupling generates propagating waves of signaling and cytoskeletal activities that promote protrusion formation and cell migration.

PIP_3 Signaling

Investigating the mechanisms underlying the subcellular localization of CRAC, a pleckstrin homology (PH) domain-containing protein, in migrating *Dictyostelium* cells, led to the identification of PIP_3 as the first signaling molecule at the leading edge (Parent et al. 1998). In *Dictyostelium*, PIP_3 levels are controlled by two types of enzymes: class I phosphoinositide 3-kinases (PI3Ks), which generate PIP_3 by phosphorylating phosphatidylinositol 4,5-bisphosphate (PI (4,5)P_2), and the phosphoinositide 3-phosphatase Pten, which catalyzes the reverse reaction. During migration, PI3Ks localize to the leading edge while Pten is confined to the trailing edge, creating a reciprocal distribution that enriches PIP_3 at the leading edge (Funamoto et al. 2002; Iijima and Devreotes 2002; Arai et al. 2010). Subsequent studies have shown that this internal PIP_3 gradient is a conserved feature across various migrating or polarized cell types (Servant et al. 2000; Nishio et al. 2007; Fivaz et al. 2008; Weiger et al. 2009).

A collection of data from *Dictyostelium* cells and neutrophils links PIP_3 production to actin polymerization and protrusion formation. First, genetic deletion of PI3Ks or pharmacological inhibition of their activity impairs cell motility (Funamoto et al. 2002; Loovers et al. 2006; Ferguson et al. 2007; Hoeller and Kay 2007). For example, *Dictyostelium* cells lacking class I PI3Ks retain the ability to orient and migrate in steep chemoattractant gradients but exhibit reduced migration speeds (Hoeller and Kay 2007). Their random motility is also severely compromised due to the formation of fewer and less stable protrusions (Hoeller and Kay 2007). Second, controlled increases in local PIP_3 activity, achieved through optogenetic tools in neutrophils, trigger pseudopod formation and initiate migration (Inoue and Meyer 2008; Graziano et al. 2017). Conversely, excess PIP_3 production, resulting from phosphatase deletion or mistargeting of PI3K, leads to uncontrolled pseudopod extension, loss of cell polarity, and migration defects in both *Dictyostelium* cells and neutrophils (Funamoto et al. 2002; Iijima and Devreotes 2002; Nishio et al. 2007). Third, PIP_3 effectors, such as MyoI family proteins, Rho GTPase regulators, and various PH domain-containing proteins, modulate protrusion dynamics (Funamoto et al. 2001; Zhang et al. 2010; Chen et al. 2012). In *Dictyostelium*, the PH domain-containing serine/threonine kinase PKBA binds PIP_3 at the leading edge (Fig. 2B), where it facilitates Arp2/3 complex activation and actin assembly (Funamoto et al. 2001; Singh et al. 2022). Disruption of *pkbA* reduces motility in wild-type cells but, intriguingly, partially rescues pseudopodial defects caused by elevated PIP_3 in *pten* knockout cells (Tang et al. 2011). Leep1, the aforementioned regulator of the Scar/WAVE complex, also contains a PH domain-like structure and interacts specifically with PIP_3 (Yang et al. 2021). In neutrophils, overexpression of P-Rex1, a PIP_3-dependent RacGEF, induces actin remodeling, whereas P-Rex1 deficiency impairs motility (Welch et al. 2002, 2005; Yoshizawa et al. 2005; Graziano et al. 2017). Finally, emerging evidence suggests bidirectional signaling between PIP_3 and F-actin, with components of the actin cytoskeletal network forming positive feedback loops that enhance PI3K localization and PIP_3 production at the leading edge (Sasaki et al. 2004; Inoue and Meyer 2008; Huang et al. 2013; Taniguchi et al. 2013; Kuhn et al. 2024).

Despite these findings, the role of PIP_3 in protrusion formation remains a subject of debate. The observation that *Dictyostelium* cells can migrate with minimal PIP_3 signaling has raised questions regarding its necessity. As mentioned previously, although PIP_3 is not strictly required, its absence significantly reduces both random and directional migration speeds (Hoeller and Kay 2007). This apparent dispensability likely arises from functional redundancy within the signaling network governing cell motility in which the loss of one pathway can often be compensated by other pathways. Indeed, multiple PIP_3-independent pathways have been identified, including those involving cyclic GMP (GMP), phospholipase A2 (PLA2), and the RasC/RacE-mTORC2 axis (discussed in the next section) (Chen et al. 2007; van Haastert et al. 2007; Kamimura et al. 2008; Veltman et al. 2008; Cai et al. 2010; Senoo et al. 2019, 2020). Importantly, inhibiting these pathways alongside PIP_3 signaling leads to more severe migration defects (Chen et al. 2007; Kamimura et al. 2008; Veltman et al. 2008), highlighting their collective contributions to cell motility. Studies also suggest that PIP_3 may inhibit chemotaxis, as axenic *Dictyostelium* cells, which exhibit elevated macropinocytosis—a process highly reliant on PIP_3 signaling—exhibit improved chemotaxis upon PI3K deletion (Hoeller et al. 2013; Veltman et al. 2014). This improvement may reflect an interplay between pseudopodia and macropinosome formation, which share many regulatory components and are likely driven by similar underlying mechanisms (Devreotes et al. 2017). Macropinocytosis may compete with migration for limited cytoskeletal resources, impairing migratory capacity; PI3K deletion abolishes macropinocytosis, alleviating this competition and hence improving migration. Supporting this concept, suppressing macropinocytosis in axenic cells using agarose overlay similarly enhances chemotaxis (Veltman et al. 2014). Thus, the inhibitory effect of PIP_3 on migration may be an indirect consequence of the trade-off between macropinocytosis and migration rather than direct suppression of cell motility. Alternatively, given the stronger PIP_3 signals at macropinocytic cups compared to pseudopodia, it is possible that pseudopodia require a moderate level of PIP_3, whereas an excess of PIP_3 favors macropinocytosis and interferes with migration.

Ras GTPases

Ras activation is another well-characterized signaling event that regulates protrusion formation. Among the 11 Ras proteins encoded by the *Dictyostelium* genome, RasG and RasC are the most extensively studied in the context of cell migration (Fig. 2B). RasG promotes protrusion formation primarily by activating PI3Ks (Tuxworth et al. 1997; Sasaki et al. 2004; Zhang et al. 2008). All class I PI3Ks in *Dictyostelium* contain a Ras-binding domain (RBD) that mediates their interaction with activated Ras proteins (Hoeller et al. 2013). Studies have shown that the RBDs of PI3K1 and PI3K2, two PI3Ks responsible for most PIP_3 production in cells, interact strongly with RasG (Hoeller et al. 2013). Mutations in the RBDs of PI3K1 and PI3K2 disrupt their kinase activity (Funamoto et al. 2002). Furthermore, chemoattractant-induced PI3K activation is diminished in *rasG* knockout cells, leading to reduced activation of PKBA, a key downstream effector of PIP_3 signaling (Sasaki et al. 2004; Zhang et al. 2008). In contrast, overexpression of constitutively active RasG results in prolonged PIP_3 accumulation and sustained PKBA activation (Sasaki et al. 2004; Zhang et al. 2008). Studies also suggest the existence of a positive feedback loop involving F-actin, PIP_3, and Ras. Inhibition of actin polymerization or PI3K activity reduces Ras activation at the leading edge, whereas *pten* deletion increases the number of Ras activity patches, leading to the extension of multiple pseudopodia (Sasaki et al. 2004, 2007; Huang et al. 2013). This feedback loop likely amplifies the initial Ras activation signal and stabilizes the leading edge. Disruption of *rasG* impairs cell motility, but the defect is only partial (Tuxworth et al. 1997). This is not surprising considering the presence of additional Ras family proteins and the partial motility defect seen in PI3K-deficient cells. At least two compensatory mechanisms contribute to this phenomenon. The first involves functional redundancy among Ras proteins. The expression of RasB and RasD, which share the same effector domain sequence as RasG, is upregulated in *rasG* knockout cells (Khosla et al. 2000). Our unpublished data reveal that simultaneous deletion of *rasD* and *rasG* results in more pronounced defects in cell motility and chemotaxis. The second mechanism involves the RasC/RacE-mTORC2 axis, as discussed in detail below.

The search for PIP_3-independent pathways regulating cell motility led to the discovery of a signaling cascade centered on RasC, RacE, and mTORC2 (Kamimura et al. 2008; Cai et al. 2010; Charest et al. 2010; Senoo et al. 2019, 2020). In *Dictyostelium* cells, chemoattractant stimulation activates RasC and induces phosphorylation of RacE at Ser192, promoting their assembly into a hetero-oligomeric complex that activates mTORC2 at the leading edge (Fig. 2B). mTORC2 phosphorylates and activates two PKB kinases, PKBR1 and PKBA, which then phosphorylate a range of substrates, including a cell adhesion molecule, a phosphoinositide kinase, and regulators of Ras and Rho family small GTPases, thereby modulating the protrusion dynamics and cell motility (Kamimura et al. 2008; Cai et al. 2010; Liao et al. 2010; Tang et al. 2011). Cells lacking *rasC* or expressing constitutively active RasC exhibit attenuated or prolonged mTORC2 signaling, respectively (Cai et al. 2010). These signaling events are triggered not only by chemoattractant stimulation, but also likely during spontaneous pseudopod formation. Consequently, deletion of pathway components impairs both chemotaxis and basal motility. While PKBA activation, as noted previously, requires additional interaction with PIP_3, PKBR1 activation is PIP_3-independent, enabling it to compensate for the loss of PIP_3 signaling and PKBA activity (Kamimura et al. 2008). Deletion of either *pkbA* or *pkbR1* causes partial motility defects, but their combined disruption nearly abolishes Arp2/3 complex recruitment to the leading edge, resulting in a near-complete loss of pseudopodia and severe motility impairment (Kamimura et al. 2008; Singh et al. 2022). In mammalian cells, mTORC2 has also been shown to regulate cytoskeleton-based processes such as chemotaxis (Liu et al. 2010; Fu and Hall 2020). In addition, a complex of phosphorylated GTP-bound KRAS4B and GDP-bound RHOA functions analogously to the RasC–RacE complex in activating mTORC2 (Senoo et al. 2021). However, whether this complex contributes to cell motility remains to be determined.

The activities of RasC and RasG in *Dictyostelium* are tightly regulated by GEFs and GAPs. A signaling complex comprising scaffold protein Sca1, RasGEFA, RasGEFH, and other components functions as a GEF for RasC (Charest et al. 2010). This complex localizes to protrusions in an F-actin and chemoattractant-dependent manner, providing spatiotemporal control of RasC activity. Deletion of this complex produces phenotypes similar to those of *rasC* knockout cells, including defects in chemotaxis and random motility (Insall et al. 1996; Lim et al. 2001; Charest et al. 2010). RasGEFR mediates RasG activation in response to chemoattractants, though its role in cell migration remains unclear (Kae et al. 2007). RasGEFX, which colocalizes with active Ras signals at protrusions, was recently identified as a key regulator of spontaneous Ras activation and random motility (Iwamoto et al. 2025). Whether it acts through RasG, RasC, or a combination of Ras proteins is yet to be determined. Several RasGAPs fine-tune Ras activity during migration. Loss of NfaA or C2GAP1, a protein containing C2 and RasGAP domains, elevates Ras activity, leading to excessive actin polymerization, impaired gradient sensing, and defective chemotaxis (Zhang et al. 2008; Xu et al. 2017). The phenotypes of *nfaA* knockout cells primarily arise from misregulation of RasG, whereas C2GAP1 interacts with both RasC and RasG. Interestingly, the migration defects in *c2gap1* knockout cells vary with chemoattractant concentration, with more severe impairments in high-concentration cAMP gradients and defect severity inversely correlating with distance from the cAMP source (Xu et al. 2017). Thus, C2GAP1 appears to facilitate Ras signaling adaptation, enabling cells to navigate across a wide range of gradient concentrations rather than directly regulating motility. Optical recruitment of C2GAP2, another C2 and RasGAP domain-containing protein, to the leading edge extinguishes existing protrusions (Lin et al. 2024). In contrast, uniform or rearward recruitment promotes cell polarization and migration through myosin II- and mTORC2-dependent mechanisms. Similarly, optical recruitment of a RasGAP protein, RASAL3, in neutrophils and macrophages produces comparable phenotypes (Lin et al. 2024). RGBARG, a protein with both RhoGEF and RasGAP domains, stimulates the GTPase activity of multiple Ras family proteins, including RasG (Buckley et al. 2020). As previously discussed, RGBARG-deficient cells migrate more slowly but form broad leading edges. However, as RGBARG is sparsely localized to pseudopodia, its specific role in regulating pseudopodia dynamics and the potential contribution of unregulated Ras activity to the observed phenotypes remain to be elucidated.

CELL MOTILITY ARISING FROM EXCITABLE CYTOSKELETAL AND SIGNALING NETWORKS

Recent studies have advanced our understanding of cell motility not only through the identification of key molecular components but also by unraveling the architecture of the cytoskeletal and signaling networks that orchestrate this complex process. Remarkably, both networks exhibit characteristics of excitability (Fig. 3B), including "all-or-nothing" responsiveness, oscillatory dynamics, and wave propagation (Devreotes et al. 2017). This excitability is particularly evident in the propagating waves of cytoskeletal and signaling components observed during cell migration. Actin waves, traveling along the basal surface of migrating cells, were initially described and characterized in *Dictyostelium* (Vicker 2002; Gerisch et al. 2004; Bretschneider et al. 2009). In addition, the recruitment of Scar/WAVE subunits, actin-binding proteins, and biosensors for Ras, PI3K, and Rac activation, along with the coordinated dissociation of PTEN and myosin, occurs in bursts or propagating waves (Bretschneider et al. 2004; Arai et al. 2010; Xiong et al. 2010; Gerisch et al. 2012; Veltman et al. 2012; Huang et al. 2013; van Haastert et al. 2017). Importantly, these waves can form autonomously, independent of external stimuli, suggesting that they are an intrinsic feature of motile cells. Theoretical models based on reaction–diffusion equations describing an activator–inhibitor excitable system have been developed to explain wave formation and propagation, with molecular components and feedback regulations described in the previous sections encoded into these equations (Meinhardt 1999; Vicker 2000; Xiong et al. 2010; Bement et al. 2015; Miao et al. 2019).

Further studies have revealed distinct dynamic patterns for cytoskeletal and signaling waves, indicating that the two excitable networks can operate independently (Huang et al. 2013; Gerhardt et al. 2014; van Haastert et al. 2017). However, when coupled, they can generate waves at the cell edge, driving protrusion formation and expanding the cell boundary (Weiger et al. 2009; Huang et al. 2013; Tang et al. 2014; Miao et al. 2019; Zhan et al. 2020). Altering the activation thresholds of these networks changes wave properties, influencing protrusion morphology and giving rise to distinct modes of migration (Miao et al. 2019, 2017). In contrast, in the absence of signal transduction, cytoskeletal components revert to rapid, nonpropagating flashes or oscillations, corresponding to diminished motility (Huang et al. 2013). Thus, cytoskeletal waves serve as dynamic readouts of propagating signaling waves. Computational modeling has demonstrated that the coupling of independent excitable modules is able to generate protrusions in unstimulated cells. Furthermore, directed cell migration can be conceptualized as a process in which guidance cues, such as chemoattractants, asymmetrically modulate excitability in the cell (Meinhardt 1999; Millius et al. 2009; Devreotes et al. 2017; Lin et al. 2024). Evidence of excitability and correlations between signaling/cytoskeletal waves and leading-edge protrusions have also been reported in various mammalian cell types, including neutrophils (Weiner et al. 2007; Millius et al. 2009), neurons (Winans et al. 2016), fibroblasts (Weiger et al. 2009; Case and Waterman 2011), and cancer cells (Yang et al. 2018; Zhan et al. 2020). Whether these dynamic behaviors represent a universal feature of migrating cells or have broader physiological implications remains to be fully elucidated in future research.

CONCLUDING REMARKS

Cell migration is a fundamental biological process with profound implications for both physiology and pathology. Extensive research has significantly advanced our understanding of the mechanisms underlying cell migration. Recent discoveries suggest that migration may arise from spontaneous activities within an excitable signaling network that regulates cytoskeletal dynamics. The excitable network hypothesis offers a novel framework for understanding both the migration process and accompanying morphological changes of the cell. Despite significant progress, many aspects of cell migration remain to be explored. Future research must tackle several critical challenges. First, while simplified two-dimensional systems and model organisms have revealed valuable insights, studying migration in more complex in vivo environments is essential. In these settings, migration differs considerably from behaviors observed on rigid planar substrates, necessitating the development of innovative tools and imaging techniques to capture these dynamic processes (Yamada and Sixt 2019). Second, although the core machinery driving migration has been identified, the field is just beginning to unravel how numerous cytoskeletal and signaling components act in concert. Recently developed tools, such as chemically or light-induced dimerization systems, offer promising potential to dissect these complex interactions with improved spatial and temporal precision (Kuhn et al. 2021). Third, how cells achieve the spatiotemporal enrichment of regulatory components to coordinate leading and trailing edge dynamics remains a major question. Emerging evidence suggests that the surface charge on the inner leaflet of the plasma membrane may play a key role in global molecular self-organization, supporting polarity establishment and migration (Banerjee et al. 2022). This represents a promising avenue for future investigation. Fourth, understanding how cells use similar molecular machinery to generate different types of protrusions and achieve diverse functional outcomes remains an active area of research. The topological connections between cortical actin waves and various protrusion types, such as filopodia, lamellipodia, and large-scale endocytic structures, are particularly intriguing (Devreotes et al. 2017; Miao et al. 2019). Continued advances in imaging technologies, genetic tools, and integrative modeling are crucial for unraveling these complexities and further enhancing our understanding of cell migration dynamics in both health and disease.

ACKNOWLEDGMENTS

We sincerely apologize to colleagues whose valuable work on this broad topic could not be cited due to space limitations. This work was supported by grants from the National Natural Science Foundation of China (32425018, 92254303, and 32170701 to H.C.), the Ministry of Science and Technology of China (2021YFA1300301 to H.C.), the Beijing Natural Science Foundation (5234029 to H.T.), and the National Laboratory of Biomacromolecules.

REFERENCES

Amato C, Thomason PA, Davidson AJ, Swaminathan K, Ismail S, Machesky LM, Insall RH. 2019. WASP restricts active Rac to maintain cells' front–rear polarization. *Curr Biol* **29:** 4169–4182.e4. doi:10.1016/j.cub.2019.10.036

Arai Y, Shibata T, Matsuoka S, Sato MJ, Yanagida T, Ueda M. 2010. Self-organization of the phosphatidylinositol lipids signaling system for random cell migration. *Proc Natl Acad Sci* **107:** 12399–12404. doi:10.1073/pnas.0908278107

Artemenko Y, Lampert TJ, Devreotes PN. 2014. Moving towards a paradigm: common mechanisms of chemotactic signaling in *Dictyostelium* and mammalian leukocytes. *Cell Mol Life Sci* **71:** 3711–3747. doi:10.1007/s00018-014-1638-8

Banerjee T, Biswas D, Pal DS, Miao Y, Iglesias PA, Devreotes PN. 2022. Spatiotemporal dynamics of membrane surface charge regulates cell polarity and migration. *Nat Cell Biol* **24:** 1499–1515. doi:10.1038/s41556-022-00997-7

Bear JE, Rawls JF, Saxe CL. 1998. SCAR, a WASP-related protein, isolated as a suppressor of receptor defects in late *Dictyostelium* development. *J Cell Biol* **142:** 1325–1335. doi:10.1083/jcb.142.5.1325

Beli P, Mascheroni D, Xu D, Innocenti M. 2008. WAVE and Arp2/3 jointly inhibit filopodium formation by entering into a complex with mDia2. *Nat Cell Biol* **10:** 849–857. doi:10.1038/ncb1745

Bement WM, Leda M, Moe AM, Kita AM, Larson ME, Golding AE, Pfeuti C, Su KC, Miller AL, Goryachev AB, et al. 2015. Activator–inhibitor coupling between Rho signalling and actin assembly makes the cell cortex an excitable medium. *Nat Cell Biol* **17:** 1471–1483. doi:10.1038/ncb3251

Bieling P, Hansen SD, Akin O, Li TD, Hayden CC, Fletcher DA, Mullins RD. 2018. WH2 and proline-rich domains of WASP-family proteins collaborate to accelerate actin filament elongation. *EMBO J* **37:** 102–121. doi:10.15252/embj.201797039

Blagg SL, Stewart M, Sambles C, Insall RH. 2003. PIR121 regulates pseudopod dynamics and SCAR activity in *Dictyostelium. Curr Biol* **13:** 1480–1487. doi:10.1016/S0960-9822(03)00580-3

Bodor DL, Pönisch W, Endres RG, Paluch EK. 2020. Of cell shapes and motion: the physical basis of animal cell migration. *Dev Cell* **52:** 550–562. doi:10.1016/j.devcel.2020.02.013

Bosgraaf L, van Haastert PJ. 2006. The regulation of myosin II in *Dictyostelium. Eur J Cell Biol* **85:** 969–979. doi:10.1016/j.ejcb.2006.04.004

Bosgraaf L, van Haastert PJ. 2009. The ordered extension of pseudopodia by amoeboid cells in the absence of external cues. *PLoS One* **4:** e5253. doi:10.1371/journal.pone.0005253

Bretschneider T, Diez S, Anderson K, Heuser J, Clarke M, Müller-Taubenberger A, Köhler J, Gerisch G. 2004. Dynamic actin patterns and Arp2/3 assembly at the substrate-attached surface of motile cells. *Curr Biol* **14:** 1–10. doi:10.1016/j.cub.2003.12.005

Bretschneider T, Anderson K, Ecke M, Müller-Taubenberger A, Schroth-Diez B, Ishikawa-Ankerhold HC, Gerisch G. 2009. The three-dimensional dynamics of actin waves, a model of cytoskeletal self-organization. *Biophys J* **96:** 2888–2900. doi:10.1016/j.bpj.2008.12.3942

Buckley CM, Pots H, Gueho A, Vines JH, Munn CJ, Phillips BA, Gilsbach B, Traynor D, Nikolaev A, Soldati T, et al. 2020. Coordinated Ras and Rac activity shapes macropinocytic cups and enables phagocytosis of geometrically diverse bacteria. *Curr Biol* **30:** 2912–2926.e5. doi:10.1016/j.cub.2020.05.049

Buracco S, Döring H, Engelbart S, Singh SP, Paschke P, Whitelaw J, Thomason PA, Paul NR, Tweedy L, Lilla S, et al. 2024. Scar/WAVE drives actin protrusions independently of its VCA domain using proline-rich domains. *Curr Biol* **34:** 4436–4451.e9. doi:10.1016/j.cub.2024.08.013

Cai H, Devreotes PN. 2011. Moving in the right direction: how eukaryotic cells migrate along chemical gradients. *Semin Cell Dev Biol* **22:** 834–841. doi:10.1016/j.semcdb.2011.07.020

Cai H, Das S, Kamimura Y, Long Y, Parent CA, Devreotes PN. 2010. Ras-mediated activation of the TORC2-PKB pathway is critical for chemotaxis. *J Cell Biol* **190:** 233–245. doi:10.1083/jcb.201001129

Cai H, Katoh-Kurasawa M, Muramoto T, Santhanam B, Long Y, Li L, Ueda M, Iglesias PA, Shaulsky G, Devreotes PN. 2014. Nucleocytoplasmic shuttling of a GATA transcription factor functions as a development timer. *Science* **343:** 1249531. doi:10.1126/science.1249531

Case LB, Waterman CM. 2011. Adhesive F-actin waves: a novel integrin-mediated adhesion complex coupled to ventral actin polymerization. *PLoS One* **6:** e26631. doi:10.1371/journal.pone.0026631

Charest PG, Shen Z, Lakoduk A, Sasaki AT, Briggs SP, Firtel RA. 2010. A Ras signaling complex controls the RasC-TORC2 pathway and directed cell migration. *Dev Cell* **18:** 737–749. doi:10.1016/j.devcel.2010.03.017

Chen L, Iijima M, Tang M, Landree MA, Huang YE, Xiong Y, Iglesias PA, Devreotes PN. 2007. PLA2 and PI3K/PTEN pathways act in parallel to mediate chemotaxis. *Dev Cell* **12:** 603–614. doi:10.1016/j.devcel.2007.03.005

Chen Z, Borek D, Padrick SB, Gomez TS, Metlagel Z, Ismail AM, Umetani J, Billadeau DD, Otwinowski Z, Rosen MK. 2010. Structure and control of the actin regulatory WAVE complex. *Nature* **468:** 533–538. doi:10.1038/nature09623

Chen CL, Wang Y, Sesaki H, Iijima M. 2012. Myosin I links PIP3 signaling to remodeling of the actin cytoskeleton in chemotaxis. *Sci Signal* **5:** ra10.

Chen B, Chou HT, Brautigam CA, Xing W, Yang S, Henry L, Doolittle LK, Walz T, Rosen MK. 2017. Rac1 GTPase activates the WAVE regulatory complex through two distinct binding sites. *eLife* **6:** e29795. doi:10.7554/eLife.29795

Chung CY, Lee S, Briscoe C, Ellsworth C, Firtel RA. 2000. Role of Rac in controlling the actin cytoskeleton and chemotaxis in motile cells. *Proc Natl Acad Sci* **97:** 5225–5230. doi:10.1073/pnas.97.10.5225

Danson CM, Pocha SM, Bloomberg GB, Cory GO. 2007. Phosphorylation of WAVE2 by MAP kinases regulates persistent cell migration and polarity. *J Cell Sci* **120:** 4144–4154. doi:10.1242/jcs.013714

Davidson AJ, Amato C, Thomason PA, Insall RH. 2018. WASP family proteins and formins compete in pseudopod- and bleb-based migration. *J Cell Biol* **217:** 701–714. doi:10.1083/jcb.201705160

Devreotes P, Horwitz AR. 2015. Signaling networks that regulate cell migration. *Cold Spring Harb Perspect Biol* **7:** a005959. doi:10.1101/cshperspect.a005959

Devreotes PN, Bhattacharya S, Edwards M, Iglesias PA, Lampert T, Miao Y. 2017. Excitable signal transduction networks in directed cell migration. *Annu Rev Cell Dev Biol* **33:** 103–125. doi:10.1146/annurev-cellbio-100616-060739

Ding B, Yang S, Schaks M, Liu Y, Brown AJ, Rottner K, Chowdhury S, Chen B. 2022. Structures reveal a key mechanism of WAVE regulatory complex activation by Rac1 GTPase. *Nat Commun* **13:** 5444. doi:10.1038/s41467-022-33174-3

Dumontier M, Höcht P, Mintert U, Faix J. 2000. Rac1 GTPases control filopodia formation, cell motility, endocytosis, cytokinesis and development in *Dictyostelium*. *J Cell Sci* **113:** 2253–2265. doi:10.1242/jcs.113.12.2253

Espinoza-Sanchez S, Metskas LA, Chou SZ, Rhoades E, Pollard TD. 2018. Conformational changes in Arp2/3 complex induced by ATP, WASp-VCA, and actin filaments. *Proc Natl Acad Sci* **115:** E8642–E8651. doi:10.1073/pnas.1717594115

Ferguson GJ, Milne L, Kulkarni S, Sasaki T, Walker S, Andrews S, Crabbe T, Finan P, Jones G, Jackson S, et al. 2007. PI(3)Kgamma has an important context-dependent role in neutrophil chemokinesis. *Nat Cell Biol* **9:** 86–91. doi:10.1038/ncb1517

Filić V, Marinovic M, Faix J, Weber I. 2012. A dual role for Rac1 GTPases in the regulation of cell motility. *J Cell Sci* **125:** 387–398. doi:10.1242/jcs.089680

Filić V, Mijanovic L, Putar D, Talajic A, Cetkovic H, Weber I. 2021. Regulation of the actin cytoskeleton via Rho GTPase signalling in *Dictyostelium* and mammalian cells: a parallel slalom. *Cells* **10:** 1592. doi:10.3390/cells10071592

Fivaz M, Bandara S, Inoue T, Meyer T. 2008. Robust neuronal symmetry breaking by Ras-triggered local positive feedback. *Curr Biol* **18:** 44–50. doi:10.1016/j.cub.2007.11.051

Fort L, Batista JM, Thomason PA, Spence HJ, Whitelaw JA, Tweedy L, Greaves J, Martin KJ, Anderson KI, Brown P, et al. 2018. Fam49/CYRI interacts with Rac1 and locally suppresses protrusions. *Nat Cell Biol* **20:** 1159–1171. doi:10.1038/s41556-018-0198-9

Fu W, Hall MN. 2020. Regulation of mTORC2 signaling. *Genes (Basel)* **11:** 1045. doi:10.3390/genes11091045

Funamoto S, Milan K, Meili R, Firtel RA. 2001. Role of phosphatidylinositol 3′ kinase and a downstream pleckstrin homology domain-containing protein in controlling chemotaxis in *Dictyostelium*. *J Cell Biol* **153:** 795–810. doi:10.1083/jcb.153.4.795

Funamoto S, Meili R, Lee S, Parry L, Firtel RA. 2002. Spatial and temporal regulation of 3-phosphoinositides by PI3-kinase and PTEN mediates chemotaxis. *Cell* **109:** 611–623. doi:10.1016/S0092-8674(02)00755-9

Gerhardt M, Ecke M, Walz M, Stengl A, Beta C, Gerisch G. 2014. Actin and PIP3 waves in giant cells reveal the inherent length scale of an excited state. *J Cell Sci* **127:** 4507–4517. doi:10.1242/jcs.156000

Gerisch G, Bretschneider T, Müller-Taubenberger A, Simmeth E, Ecke M, Diez S, Anderson K. 2004. Mobile actin clusters and traveling waves in cells recovering from actin depolymerization. *Biophys J* **87:** 3493–3503. doi:10.1529/biophysj.104.047589

Gerisch G, Schroth-Diez B, Müller-Taubenberger A, Ecke M. 2012. PIP_3 waves and PTEN dynamics in the emergence of cell polarity. *Biophys J* **103:** 1170–1178. doi:10.1016/j.bpj.2012.08.004

Goley ED, Rodenbusch SE, Martin AC, Welch MD. 2004. Critical conformational changes in the Arp2/3 complex are induced by nucleotide and nucleation promoting factor. *Mol Cell* **16:** 269–279. doi:10.1016/j.molcel.2004.09.018

Graziano BR, Gong D, Anderson KE, Pipathsouk A, Goldberg AR, Weiner OD. 2017. A module for Rac temporal signal integration revealed with optogenetics. *J Cell Biol* **216:** 2515–2531. doi:10.1083/jcb.201604113

Graziano BR, Town JP, Sitarska E, Nagy TL, Fošnarič M, Penič S, Iglič A, Kralj-Iglič V, Gov NS, Diz-Muñoz A, et al. 2019. Cell confinement reveals a branched-actin independent circuit for neutrophil polarity. *PLoS Biol* **17:** e3000457. doi:10.1371/journal.pbio.3000457

Hao Y, Yang Y, Tu H, Guo Z, Chen P, Chao X, Yuan Y, Wang Z, Miao X, Zou S, et al. 2024. A transcription factor complex in *Dictyostelium* enables adaptive changes in macropinocytosis during the growth-to-development transition. *Dev Cell* **59:** 645–660.e8. doi:10.1016/j.devcel.2024.01.012

Heid PJ, Wessels D, Daniels KJ, Gibson DP, Zhang H, Voss E, Soll DR. 2004. The role of myosin heavy chain phosphorylation in *Dictyostelium* motility, chemotaxis and F-actin localization. *J Cell Sci* **117:** 4819–4835. doi:10.1242/jcs.01358

Hoeller O, Kay RR. 2007. Chemotaxis in the absence of PIP_3 gradients. *Curr Biol* **17:** 813–817. doi:10.1016/j.cub.2007.04.004

Hoeller O, Bolourani P, Clark J, Stephens LR, Hawkins PT, Weiner OD, Weeks G, Kay RR. 2013. Two distinct functions for PI3-kinases in macropinocytosis. *J Cell Sci* **126:** 4296–4307. doi:10.1242/jcs.134015

Huang CH, Tang M, Shi C, Iglesias PA, Devreotes PN. 2013. An excitable signal integrator couples to an idling cytoskeletal oscillator to drive cell migration. *Nat Cell Biol* **15:** 1307–1316. doi:10.1038/ncb2859

Ibarra N, Blagg SL, Vazquez F, Insall RH. 2006. Nap1 regulates *Dictyostelium* cell motility and adhesion through SCAR-dependent and -independent pathways. *Curr Biol* **16:** 717–722. doi:10.1016/j.cub.2006.02.068

Iijima M, Devreotes P. 2002. Tumor suppressor PTEN mediates sensing of chemoattractant gradients. *Cell* **109:** 599–610. doi:10.1016/S0092-8674(02)00745-6

Inoue T, Meyer T. 2008. Synthetic activation of endogenous PI3K and Rac identifies an AND-gate switch for cell polarization and migration. *PLoS One* **3:** e3068. doi:10.1371/journal.pone.0003068

Insall RH, Borleis J, Devreotes PN. 1996. The *aimless* RasGEF is required for processing of chemotactic signals through G-protein-coupled receptors in *Dictyostelium*. *Curr Biol* **6:** 719–729. doi:10.1016/S0960-9822(09)00453-9

Iwamoto K, Matsuoka S, Ueda M. 2025. Excitable Ras dynamics-based screens reveal RasGEFX is required for macropinocytosis and random cell migration. *Nat Commun* **16:** 117. doi:10.1038/s41467-024-55389-2

Jay PY, Pham PA, Wong SA, Elson EL. 1995. A mechanical function of myosin II in cell motility. *J Cell Sci* **108**(Pt 1): 387–393. doi:10.1242/jcs.108.1.387

Junemann A, Filić V, Winterhoff M, Nordholz B, Litschko C, Schwellenbach H, Stephan T, Weber I, Faix J. 2016. A *Diaphanous*-related formin links Ras signaling directly to actin assembly in macropinocytosis and phagocytosis. *Proc Natl Acad Sci* **113:** E7464–E7473. doi:10.1073/pnas.1611024113

Kae H, Kortholt A, Rehmann H, Insall RH, Van Haastert PJ, Spiegelman GB, Weeks G. 2007. Cyclic AMP signalling in *Dictyostelium*: G-proteins activate separate Ras pathways using specific RasGEFs. *EMBO Rep* **8:** 477–482. doi:10.1038/sj.embor.7400936

Kage F, Winterhoff M, Dimchev V, Mueller J, Thalheim T, Freise A, Brühmann S, Kollasser J, Block J, Dimchev G, et al. 2017. FMNL formins boost lamellipodial force generation. *Nat Commun* **8:** 14832. doi:10.1038/ncomms14832

Kamimura Y, Xiong Y, Iglesias PA, Hoeller O, Bolourani P, Devreotes PN. 2008. PIP3-independent activation of TorC2 and PKB at the cell's leading edge mediates chemotaxis. *Curr Biol* **18:** 1034–1043. doi:10.1016/j.cub.2008.06.068

Khosla M, Spiegelman GB, Insall R, Weeks G. 2000. Functional overlap of the *Dictyostelium* RasG, RasD and RasB proteins. *J Cell Sci* **113:** 1427–1434. doi:10.1242/jcs.113.8.1427

Körber S, Junemann A, Litschko C, Winterhoff M, Faix J. 2023. Convergence of Ras- and Rac-regulated formin pathways is pivotal for phagosome formation and particle uptake in *Dictyostelium*. *Proc Natl Acad Sci* **120:** e2220825120. doi:10.1073/pnas.2220825120

Kuhn J, Lin Y, Devreotes PN. 2021. Using live-cell imaging and synthetic biology to probe directed migration in *Dictyostelium*. *Front Cell Dev Biol* **9:** 740205. doi:10.3389/fcell.2021.740205

Kuhn J, Banerjee P, Haye A, Robinson DN, Iglesias PA, Devreotes PN. 2024. Complementary cytoskeletal feedback loops control signal transduction excitability and cell polarity. bioRxiv doi:10.1101/2024.02.13.580131

Le AH, Yelland T, Paul NR, Fort L, Nikolaou S, Ismail S, Machesky LM. 2021. CYRI-A limits invasive migration through macropinosome formation and integrin uptake regulation. *J Cell Biol* **220:** e202009187. doi:10.1083/jcb.202012114

Lebensohn AM, Kirschner MW. 2009. Activation of the WAVE complex by coincident signals controls actin assembly. *Mol Cell* **36:** 512–524. doi:10.1016/j.molcel.2009.10.024

Leithner A, Eichner A, Müller J, Reversat A, Brown M, Schwarz J, Merrin J, de Gorter DJ, Schur F, Bayerl J, et al. 2016. Diversified actin protrusions promote environmental exploration but are dispensable for locomotion of leukocytes. *Nat Cell Biol* **18:** 1253–1259. doi:10.1038/ncb3426

Li D, Yang Y, Lv C, Wang Y, Chao X, Huang J, Singh SP, Yuan Y, Zhang C, Lou J, et al. 2023. GxcM-Fbp17/RacC-WASP signaling regulates polarized cortex assembly in migrating cells via Arp2/3. *J Cell Biol* **222:** e202208151. doi:10.1083/jcb.202208151

Liao XH, Buggey J, Kimmel AR. 2010. Chemotactic activation of *Dictyostelium* AGC-family kinases AKT and PKBR1 requires separate but coordinated functions of PDK1 and TORC2. *J Cell Sci* **123:** 983–992. doi:10.1242/jcs.064022

Lim CJ, Spiegelman GB, Weeks G. 2001. Rasc is required for optimal activation of adenylyl cyclase and Akt/PKB during aggregation. *EMBO J* **20:** 4490–4499. doi:10.1093/emboj/20.16.4490

Lin Y, Pal DS, Banerjee P, Banerjee T, Qin G, Deng Y, Borleis J, Iglesias PA, Devreotes PN. 2024. Ras suppression potentiates rear actomyosin contractility-driven cell polarization and migration. *Nat Cell Biol* **26:** 1062–1076. doi:10.1038/s41556-024-01453-4

Litschko C, Linkner J, Brühmann S, Stradal TE, Reinl T, Jänsch L, Rottner K, Faix J. 2017. Differential functions of WAVE regulatory complex subunits in the regulation of actin-driven processes. *Eur J Cell Biol* **96:** 715–727. doi:10.1016/j.ejcb.2017.08.003

Litschko C, Brühmann S, Csiszár A, Stephan T, Dimchev V, Damiano-Guercio J, Junemann A, Körber S, Winterhoff M, Nordholz B, et al. 2019. Functional integrity of the contractile actin cortex is safeguarded by multiple *Diaphanous*-related formins. *Proc Natl Acad Sci* **116:** 3594–3603. doi:10.1073/pnas.1821638116

Liu L, Das S, Losert W, Parent CA. 2010. mTORC2 regulates neutrophil chemotaxis in a cAMP- and RhoA-dependent fashion. *Dev Cell* **19:** 845–857. doi:10.1016/j.devcel.2010.11.004

Loovers HM, Postma M, Keizer-Gunnink I, Huang YE, Devreotes PN, van Haastert PJ. 2006. Distinct roles of PI(3,4,5)P_3 during chemoattractant signaling in *Dictyostelium*: a quantitative in vivo analysis by inhibition of PI3-kinase. *Mol Biol Cell* **17:** 1503–1513. doi:10.1091/mbc.e05-09-0825

Machesky LM, Insall RH. 1998. Scar1 and the related Wiskott–Aldrich syndrome protein, WASP, regulate the actin cytoskeleton through the Arp2/3 complex. *Curr Biol* **8:** 1347–1356. doi:10.1016/S0960-9822(98)00015-3

Machesky LM, Atkinson SJ, Ampe C, Vandekerckhove J, Pollard TD. 1994. Purification of a cortical complex containing two unconventional actins from *Acanthamoeba* by affinity chromatography on profilin-agarose. *J Cell Biol* **127:** 107–115. doi:10.1083/jcb.127.1.107

Machesky LM, Mullins RD, Higgs HN, Kaiser DA, Blanchoin L, May RC, Hall ME, Pollard TD. 1999. Scar, a WASp-related protein, activates nucleation of actin filaments by the Arp2/3 complex. *Proc Natl Acad Sci* **96:** 3739–3744. doi:10.1073/pnas.96.7.3739

Meinhardt H. 1999. Orientation of chemotactic cells and growth cones: models and mechanisms. *J Cell Sci* **112:** 2867–2874. doi:10.1242/jcs.112.17.2867

Mendoza MC, Er EE, Zhang W, Ballif BA, Elliott HL, Danuser G, Blenis J. 2011. ERK-MAPK drives lamellipodia protrusion by activating the WAVE2 regulatory complex. *Mol Cell* **41:** 661–671. doi:10.1016/j.molcel.2011.02.031

Miao Y, Bhattacharya S, Edwards M, Cai H, Inoue T, Iglesias PA, Devreotes PN. 2017. Altering the threshold of an excitable signal transduction network changes cell migratory modes. *Nat Cell Biol* **19:** 329–340. doi:10.1038/ncb3495

Miao Y, Bhattacharya S, Banerjee T, Abubaker-Sharif B, Long Y, Inoue T, Iglesias PA, Devreotes PN. 2019. Wave patterns organize cellular protrusions and control cortical dynamics. *Mol Syst Biol* **15:** e8585. doi:10.15252/msb.20188585

Miki H, Suetsugu S, Takenawa T. 1998. WAVE, a novel WASP-family protein involved in actin reorganization induced by Rac. *EMBO J* **17:** 6932–6941. doi:10.1093/emboj/17.23.6932

Millius A, Dandekar SN, Houk AR, Weiner OD. 2009. Neutrophils establish rapid and robust WAVE complex polarity in an actin-dependent fashion. *Curr Biol* **19:** 253–259. doi:10.1016/j.cub.2008.12.044

Millius A, Watanabe N, Weiner OD. 2012. Diffusion, capture and recycling of SCAR/WAVE and Arp2/3 complexes observed in cells by single-molecule imaging. *J Cell Sci* **125:** 1165–1176. doi:10.1242/jcs.091157

Mullins RD, Heuser JA, Pollard TD. 1998. The interaction of Arp2/3 complex with actin: nucleation, high affinity pointed end capping, and formation of branching networks of filaments. *Proc Natl Acad Sci* **95:** 6181–6186. doi:10.1073/pnas.95.11.6181

Nishio M, Watanabe K, Sasaki J, Taya C, Takasuga S, Iizuka R, Balla T, Yamazaki M, Watanabe H, Itoh R, et al. 2007. Control of cell polarity and motility by the PtdIns(3,4,5)P_3 phosphatase SHIP1. *Nat Cell Biol* **9:** 36–44. doi:10.1038/ncb1515

Nobes CD, Hall A. 1995. Rho, rac, and cdc42 GTPases regulate the assembly of multimolecular focal complexes associated with actin stress fibers, lamellipodia, and filopodia. *Cell* **81:** 53–62. doi:10.1016/0092-8674(95)90370-4

Parent CA, Blacklock BJ, Froehlich WM, Murphy DB, Devreotes PN. 1998. G protein signaling events are activated at the leading edge of chemotactic cells. *Cell* **95:** 81–91. doi:10.1016/S0092-8674(00)81784-5

Park KC, Rivero F, Meili R, Lee S, Apone F, Firtel RA. 2004. Rac regulation of chemotaxis and morphogenesis in *Dictyostelium*. *EMBO J* **23:** 4177–4189. doi:10.1038/sj.emboj.7600368

Pipathsouk A, Brunetti RM, Town JP, Graziano BR, Breuer A, Pellett PA, Marchuk K, Tran NT, Krummel MF, Stamou D, et al. 2021. The WAVE complex associates with sites of saddle membrane curvature. *J Cell Biol* **220:** e202003086. doi:10.1083/jcb.202003086

Pollard TD, Borisy GG. 2003. Cellular motility driven by assembly and disassembly of actin filaments. *Cell* **112:** 453–465. doi:10.1016/S0092-8674(03)00120-X

Pollitt AY, Insall RH. 2008. Abi mutants in *Dictyostelium* reveal specific roles for the SCAR/WAVE complex in cytokinesis. *Curr Biol* **18:** 203–210. doi:10.1016/j.cub.2008.01.026

Pollitt AY, Insall RH. 2009. Loss of *Dictyostelium* HSPC300 causes a scar-like phenotype and loss of SCAR protein. *BMC Cell Biol* **10:** 13. doi:10.1186/1471-2121-10-13

Prassler J, Murr A, Stocker S, Faix J, Murphy J, Marriott G. 1998. DdLIM is a cytoskeleton-associated protein involved in the protrusion of lamellipodia in *Dictyostelium*. *Mol Biol Cell* **9:** 545–559. doi:10.1091/mbc.9.3.545

Pruyne D, Evangelista M, Yang C, Bi E, Zigmond S, Bretscher A, Boone C. 2002. Role of formins in actin assembly: nucleation and barbed-end association. *Science* **297:** 612–615. doi:10.1126/science.1072309

Rivero F, Xiong H. 2016. Rho signaling in *Dictyostelium discoideum*. *Int Rev Cell Mol Biol* **322:** 61–181. doi:10.1016/bs.ircmb.2015.10.004

Rivero F, Muramoto T, Meyer AK, Urushihara H, Uyeda TQ, Kitayama C. 2005. A comparative sequence analysis reveals a common GBD/FH3-FH1-FH2-DAD architecture in formins from *Dictyostelium*, fungi and metazoa. *BMC Genomics* **6:** 28. doi:10.1186/1471-2164-6-28

Robinson RC, Turbedsky K, Kaiser DA, Marchand JB, Higgs HN, Choe S, Pollard TD. 2001. Crystal structure of Arp2/3 complex. *Science* **294:** 1679–1684. doi:10.1126/science.1066333

Sasaki AT, Chun C, Takeda K, Firtel RA. 2004. Localized Ras signaling at the leading edge regulates PI3K, cell polarity, and directional cell movement. *J Cell Biol* **167:** 505–518. doi:10.1083/jcb.200406177

Sasaki AT, Janetopoulos C, Lee S, Charest PG, Takeda K, Sundheimer LW, Meili R, Devreotes PN, Firtel RA. 2007. G protein-independent Ras/PI3K/F-actin circuit regulates basic cell motility. *J Cell Biol* **178:** 185–191. doi:10.1083/jcb.200611138

Scarpa E, Mayor R. 2016. Collective cell migration in development. *J Cell Biol* **212:** 143–155. doi:10.1083/jcb.201508047

Schaks M, Singh SP, Kage F, Thomason P, Klünemann T, Steffen A, Blankenfeldt W, Stradal TE, Insall RH, Rottner K. 2018. Distinct interaction sites of Rac GTPase with WAVE regulatory complex have non-redundant functions in vivo. *Curr Biol* **28:** 3674–3684.e6. doi:10.1016/j.cub.2018.10.002

Schirenbeck A, Bretschneider T, Arasada R, Schleicher M, Faix J. 2005. The diaphanous-related formin dDia2 is required for the formation and maintenance of filopodia. *Nat Cell Biol* **7:** 619–625. doi:10.1038/ncb1266

Scholz J, Stephan T, Pérez AG, Csiszár A, Hersch N, Fischer LS, Brühmann S, Körber S, Litschko C, Mijanovic L, et al. 2024. Decisive role of mDia-family formins in cell cortex function of highly adherent cells. *Sci Adv* **10:** eadp5929. doi:10.1126/sciadv.adp5929

SenGupta S, Parent CA, Bear JE. 2021. The principles of directed cell migration. *Nat Rev Mol Cell Biol* **22:** 529–547. doi:10.1038/s41580-021-00366-6

Senoo H, Cai H, Wang Y, Sesaki H, Iijima M. 2016. The novel RacE-binding protein GflB sharpens Ras activity at the leading edge of migrating cells. *Mol Biol Cell* **27:** 1596–1605. doi:10.1091/mbc.E15-11-0796

Senoo H, Kamimura Y, Kimura R, Nakajima A, Sawai S, Sesaki H, Iijima M. 2019. Phosphorylated Rho-GDP directly activates mTORC2 kinase towards AKT through dimerization with Ras-GTP to regulate cell migration. *Nat Cell Biol* **21:** 867–878. doi:10.1038/s41556-019-0348-8

Senoo H, Wai M, Matsubayashi HT, Sesaki H, Iijima M. 2020. Hetero-oligomerization of Rho and Ras GTPases connects GPCR activation to mTORC2-AKT signaling. *Cell Rep* **33:** 108427. doi:10.1016/j.celrep.2020.108427

Senoo H, Murata D, Wai M, Arai K, Iwata W, Sesaki H, Iijima M. 2021. KARATE: PKA-induced KRAS4B-

RHOA-mTORC2 supercomplex phosphorylates AKT in insulin signaling and glucose homeostasis. *Mol Cell* **81:** 4622–4634.e8. doi:10.1016/j.molcel.2021.09.001

Servant G, Weiner OD, Herzmark P, Balla T, Sedat JW, Bourne HR. 2000. Polarization of chemoattractant receptor signaling during neutrophil chemotaxis. *Science* **287:** 1037–1040. doi:10.1126/science.287.5455.1037

Shemesh T, Kozlov MM. 2007. Actin polymerization upon processive capping by formin: a model for slowing and acceleration. *Biophys J* **92:** 1512–1521. doi:10.1529/biophysj.106.098459

Singh SP, Thomason PA, Lilla S, Schaks M, Tang Q, Goode BL, Machesky LM, Rottner K, Insall RH. 2020. Cell-substrate adhesion drives Scar/WAVE activation and phosphorylation by a Ste20-family kinase, which controls pseudopod lifetime. *PLoS Biol* **18:** e3000774. doi:10.1371/journal.pbio.3000774

Singh SP, Thomason PA, Insall RH. 2021. Extracellular signalling modulates Scar/WAVE complex activity through Abi phosphorylation. *Cells* **10:** 3485. doi:10.3390/cells10123485

Singh SP, Paschke P, Tweedy L, Insall RH. 2022. AKT and SGK kinases regulate cell migration by altering Scar/WAVE complex activation and Arp2/3 complex recruitment. *Front Mol Biosci* **9:** 965921. doi:10.3389/fmolb.2022.965921

Steffen A, Faix J, Resch GP, Linkner J, Wehland J, Small JV, Rottner K, Stradal TE. 2006. Filopodia formation in the absence of functional WAVE- and Arp2/3-complexes. *Mol Biol Cell* **17:** 2581–2591. doi:10.1091/mbc.e05-11-1088

Steffen A, Ladwein M, Dimchev GA, Hein A, Schwenkmezger L, Arens S, Ladwein KI, Holleboom JM, Schur F, Small JV, et al. 2013. Rac function is crucial for cell migration but is not required for spreading and focal adhesion formation. *J Cell Sci* **126:** 4572–4588.

Stuelten CH, Parent CA, Montell DJ. 2018. Cell motility in cancer invasion and metastasis: insights from simple model organisms. *Nat Rev Cancer* **18:** 296–312. doi:10.1038/nrc.2018.15

Tang M, Iijima M, Kamimura Y, Chen L, Long Y, Devreotes P. 2011. Disruption of PKB signaling restores polarity to cells lacking tumor suppressor PTEN. *Mol Biol Cell* **22:** 437–447. doi:10.1091/mbc.e10-06-0522

Tang H, Li A, Bi J, Veltman DM, Zech T, Spence HJ, Yu X, Timpson P, Insall RH, Frame MC, et al. 2013. Loss of Scar/WAVE complex promotes N-WASP- and FAK-dependent invasion. *Curr Biol* **23:** 107–117. doi:10.1016/j.cub.2012.11.059

Tang M, Wang M, Shi C, Iglesias PA, Devreotes PN, Huang CH. 2014. Evolutionarily conserved coupling of adaptive and excitable networks mediates eukaryotic chemotaxis. *Nat Commun* **5:** 5175. doi:10.1038/ncomms6175

Tang Q, Schaks M, Koundinya N, Yang C, Pollard LW, Svitkina TM, Rottner K, Goode BL. 2020. WAVE1 and WAVE2 have distinct and overlapping roles in controlling actin assembly at the leading edge. *Mol Biol Cell* **31:** 2168–2178. doi:10.1091/mbc.E19-12-0705

Taniguchi D, Ishihara S, Oonuki T, Honda-Kitahara M, Kaneko K, Sawai S. 2013. Phase geometries of two-dimensional excitable waves govern self-organized morphodynamics of amoeboid cells. *Proc Natl Acad Sci* **110:** 5016–5021. doi:10.1073/pnas.1218025110

Ti SC, Jurgenson CT, Nolen BJ, Pollard TD. 2011. Structural and biochemical characterization of two binding sites for nucleation-promoting factor WASp-VCA on Arp2/3 complex. *Proc Natl Acad Sci* **108:** E463–E471. doi:10.1073/pnas.1100125108

Tomchik KJ, Devreotes PN. 1981. Adenosine 3′,5′-monophosphate waves in *Dictyostelium discoideum*: a demonstration by isotope dilution—fluorography. *Science* **212:** 443–446. doi:10.1126/science.6259734

Tsai TY, Collins SR, Chan CK, Hadjitheodorou A, Lam PY, Lou SS, Yang HW, Jorgensen J, Ellett F, Irimia D, et al. 2019. Efficient front–rear coupling in neutrophil chemotaxis by dynamic myosin II localization. *Dev Cell* **49:** 189–205.e6. doi:10.1016/j.devcel.2019.03.025

Tuxworth RI, Cheetham JL, Machesky LM, Spiegelmann GB, Weeks G, Insall RH. 1997. *Dictyostelium* RasG is required for normal motility and cytokinesis, but not growth. *J Cell Biol* **138:** 605–614. doi:10.1083/jcb.138.3.605

van Haastert PJ, Keizer-Gunnink I, Kortholt A. 2007. Essential role of PI3-kinase and phospholipase A2 in *Dictyostelium discoideum* chemotaxis. *J Cell Biol* **177:** 809–816. doi:10.1083/jcb.200701134

van Haastert PJ, Keizer-Gunnink I, Kortholt A. 2017. Coupled excitable Ras and F-actin activation mediates spontaneous pseudopod formation and directed cell movement. *Mol Biol Cell* **28:** 922–934. doi:10.1091/mbc.e16-10-0733

Veltman DM, Keizer-Gunnik I, van Haastert PJ. 2008. Four key signaling pathways mediating chemotaxis in *Dictyostelium discoideum*. *J Cell Biol* **180:** 747–753. doi:10.1083/jcb.200709180

Veltman DM, King JS, Machesky LM, Insall RH. 2012. SCAR knockouts in *Dictyostelium*: WASP assumes SCAR's position and upstream regulators in pseudopods. *J Cell Biol* **198:** 501–508. doi:10.1083/jcb.201205058

Veltman DM, Lemieux MG, Knecht DA, Insall RH. 2014. PIP_3-dependent macropinocytosis is incompatible with chemotaxis. *J Cell Biol* **204:** 497–505. doi:10.1083/jcb.201309081

Vicker MG. 2000. Reaction-diffusion waves of actin filament polymerization/depolymerization in *Dictyostelium* pseudopodium extension and cell locomotion. *Biophys Chem* **84:** 87–98. doi:10.1016/S0301-4622(99)00146-5

Vicker MG. 2002. F-actin assembly in *Dictyostelium* cell locomotion and shape oscillations propagates as a self-organized reaction-diffusion wave. *FEBS Lett* **510:** 5–9. doi:10.1016/S0014-5793(01)03207-0

Vlahou G, Rivero F. 2006. Rho GTPase signaling in *Dictyostelium discoideum*: insights from the genome. *Eur J Cell Biol* **85:** 947–959. doi:10.1016/j.ejcb.2006.04.011

Wang Y, Senoo H, Sesaki H, Iijima M. 2013. Rho GTPases orient directional sensing in chemotaxis. *Proc Natl Acad Sci* **110:** E4723–E4732. doi:10.1073/pnas.1312540110

Weiger MC, Wang CC, Krajcovic M, Melvin AT, Rhoden JJ, Haugh JM. 2009. Spontaneous phosphoinositide 3-kinase signaling dynamics drive spreading and random migration of fibroblasts. *J Cell Sci* **122:** 313–323. doi:10.1242/jcs.037564

Weiner OD, Marganski WA, Wu LF, Altschuler SJ, Kirschner MW. 2007. An actin-based wave generator organizes cell motility. *PLoS Biol* **5:** e221. doi:10.1371/journal.pbio.0050221

Welch HC, Coadwell WJ, Ellson CD, Ferguson GJ, Andrews SR, Erdjument-Bromage H, Tempst P, Hawkins PT, Stephens LR. 2002. P-Rex1, a PtdIns(3,4,5)P_3- and Gbetagamma-regulated guanine-nucleotide exchange factor for Rac. *Cell* **108:** 809–821. doi:10.1016/S0092-8674(02)00663-3

Welch HC, Condliffe AM, Milne LJ, Ferguson GJ, Hill K, Webb LM, Okkenhaug K, Coadwell WJ, Andrews SR, Thelen M, et al. 2005. P-Rex1 regulates neutrophil function. *Curr Biol* **15:** 1867–1873. doi:10.1016/j.cub.2005.09.050

Wessels D, Soll DR, Knecht D, Loomis WF, De Lozanne A, Spudich J. 1988. Cell motility and chemotaxis in *Dictyostelium* amebae lacking myosin heavy chain. *Dev Biol* **128:** 164–177. doi:10.1016/0012-1606(88)90279-5

Williams TD, Kay RR. 2018. The physiological regulation of macropinocytosis during *Dictyostelium* growth and development. *J Cell Sci* **131:** jcs213736. doi:10.1242/jcs.213736

Wilson CA, Tsuchida MA, Allen GM, Barnhart EL, Applegate KT, Yam PT, Ji L, Keren K, Danuser G, Theriot JA. 2010. Myosin II contributes to cell-scale actin network treadmilling through network disassembly. *Nature* **465:** 373–377. doi:10.1038/nature08994

Winans AM, Collins SR, Meyer T. 2016. Waves of actin and microtubule polymerization drive microtubule-based transport and neurite growth before single axon formation. *eLife* **5:** e12387. doi:10.7554/eLife.12387

Wu M, Marchando P, Meyer K, Tang Z, Woolfson DN, Weiner OD. 2024. The WAVE complex forms linear arrays at negative membrane curvature to instruct lamellipodia formation. bioRxiv doi:10.1101/2024.07.08.600855

Xiong Y, Huang CH, Iglesias PA, Devreotes PN. 2010. Cells navigate with a local-excitation, global-inhibition-biased excitable network. *Proc Natl Acad Sci* **107:** 17079–17086. doi:10.1073/pnas.1011271107

Xu XP, Rouiller I, Slaughter BD, Egile C, Kim E, Unruh JR, Fan X, Pollard TD, Li R, Hanein D, et al. 2012. Three-dimensional reconstructions of Arp2/3 complex with bound nucleation promoting factors. *EMBO J* **31:** 236–247. doi:10.1038/emboj.2011.343

Xu X, Wen X, Veltman DM, Keizer-Gunnink I, Pots H, Kortholt A, Jin T. 2017. GPCR-controlled membrane recruitment of negative regulator C2GAP1 locally inhibits Ras signaling for adaptation and long-range chemotaxis. *Proc Natl Acad Sci* **114:** E10092–E10101. doi:10.1073/pnas.1703208114

Yamada KM, Sixt M. 2019. Mechanisms of 3D cell migration. *Nat Rev Mol Cell Biol* **20:** 738–752. doi:10.1038/s41580-019-0172-9

Yan J, Mihaylov V, Xu X, Brzostowski JA, Li H, Liu L, Veenstra TD, Parent CA, Jin T. 2012. A Gbetagamma effector, ElmoE, transduces GPCR signaling to the actin network during chemotaxis. *Dev Cell* **22:** 92–103. doi:10.1016/j.devcel.2011.11.007

Yang C, Czech L, Gerboth S, Kojima S, Scita G, Svitkina T. 2007. Novel roles of formin mDia2 in lamellipodia and filopodia formation in motile cells. *PLoS Biol* **5:** e317. doi:10.1371/journal.pbio.0050317

Yang JM, Bhattacharya S, West-Foyle H, Hung CF, Wu TC, Iglesias PA, Huang CH. 2018. Integrating chemical and mechanical signals through dynamic coupling between cellular protrusions and pulsed ERK activation. *Nat Commun* **9:** 4673. doi:10.1038/s41467-018-07150-9

Yang Y, Li D, Chao X, Singh SP, Thomason P, Yan Y, Dong M, Li L, Insall RH, Cai H. 2021. Leep1 interacts with PIP_3 and the Scar/WAVE complex to regulate cell migration and macropinocytosis. *J Cell Biol* **220:** e202010096. doi:10.1083/jcb.202010096

Yelland T, Le AH, Nikolaou S, Insall R, Machesky L, Ismail S. 2021. Structural basis of CYRI-B direct competition with Scar/WAVE complex for Rac1. *Structure* **29:** 226–237.e4. doi:10.1016/j.str.2020.11.003

Yoshizawa M, Kawauchi T, Sone M, Nishimura YV, Terao M, Chihama K, Nabeshima Y, Hoshino M. 2005. Involvement of a Rac activator, P-Rex1, in neurotrophin-derived signaling and neuronal migration. *J Neurosci* **25:** 4406–4419. doi:10.1523/JNEUROSCI.4955-04.2005

Yuki KE, Marei H, Fiskin E, Eva MM, Gopal AA, Schwartzentruber JA, Majewski J, Cellier M, Mandl JN, et al. 2019. CYRI/FAM49B negatively regulates RAC1-driven cytoskeletal remodelling and protects against bacterial infection. *Nat Microbiol* **4:** 1516–1531. doi:10.1038/s41564-019-0484-8

Zhan H, Bhattacharya S, Cai H, Iglesias PA, Huang CH, Devreotes PN. 2020. An excitable Ras/PI3K/ERK signaling network controls migration and oncogenic transformation in epithelial cells. *Dev Cell* **54:** 608–623.e5. doi:10.1016/j.devcel.2020.08.001

Zhang S, Charest PG, Firtel RA. 2008. Spatiotemporal regulation of Ras activity provides directional sensing. *Curr Biol* **18:** 1587–1593. doi:10.1016/j.cub.2008.08.069

Zhang P, Wang Y, Sesaki H, Iijima M. 2010. Proteomic identification of phosphatidylinositol (3,4,5) triphosphate-binding proteins in *Dictyostelium discoideum*. *Proc Natl Acad Sci* **107:** 11829–11834. doi:10.1073/pnas.1006153107

Zhu Z, Chai Y, Jiang Y, Li W, Hu H, Li W, Wu JW, Wang ZX, Huang S, Ou G. 2016. Functional coordination of WAVE and WASP in *C. elegans* neuroblast migration. *Dev Cell* **39:** 224–238. doi:10.1016/j.devcel.2016.09.029

Pulling the Back to Steer Cell Migration: Mechanisms of Structural Polarity in Confining Signaling and Protrusion

Bo Gong[1,2] and Tobias Meyer[2]

[1]Department of Cell and Tissue Biology, University of California San Francisco, San Francisco, California 94143, USA

[2]Department of Cell and Developmental Biology, Weill Cornell Medicine, New York, New York 10065, USA

Correspondence: bo.gong@ucsf.edu, tom4003@med.cornell.edu

Directed cell migration is a coordinated process mediated by membrane protrusion at the front and contractile retraction at the back. This review compares a local excitation and global inhibition (LEGI) model and an emerging structural polarity of receptor activation (SPRA) model. The LEGI model posits that locally excitable signals generate membrane protrusions that become a single front through uncharacterized global inhibition that suppress competing protrusions. In contrast, the SPRA model proposes a central role of structural polarization, particularly membrane proximal F-actin, endoplasmic reticulum (ER)–plasma membrane (PM) contact sites, as well as actomyosin and nuclear positioning. In this model, the structural asymmetry ultimately sets up a stable front-to-back signal gradient, by enhancing receptor activity at the front and diminishing it at the back. Local differences in receptor sensitivity explains how uniform receptor stimuli initiate protrusions and trigger persistent cell migration. We discuss the mechanistic evidence for the two directed migration models and key open questions.

Cell migration is a fundamental cell behavior, playing essential roles in development, tissue maintenance, and inflammation. Understanding how cells migrate holds the potential to improve the treatment of various diseases linked to excessive migration as in cancer metastasis and autoimmune diseases and insufficient migration as in fibrosis (Mayor and Etienne-Manneville 2016; SenGupta et al. 2021). Cell migration is remarkably plastic and is frequently categorized into three modes: mesenchymal migration (Bear and Haugh 2014), characterized by stable adhesions, actin filament bundles, and actin-rich lamellipodia; amoeboid migration (Lämmermann and Sixt 2009), driven by low adhesion and dynamic pseudopodia extensions; and bleb-based migration (Fackler and Grosse 2008), where membrane protrusions form without actin support via contractility in the back. In addition, collectively migrating cells can use cell–cell or cell–matrix adhesion to undergo a type of migration that resembles mesenchymal migration (Mayor and Etienne-Manneville 2016).

Cite this article as *Cold Spring Harb Perspect Biol* doi: 10.1101/cshperspect.a041740

Cell migration requires precise coordination between leading-edge protrusion and rear-end retraction (Ridley et al. 2003; Devreotes and Horwitz 2015). To navigate, cells detect external cues, such as diffusible chemical gradients (chemotaxis), electric fields (galvanotaxis) and substrate mechanics and geometry (durotaxis and topotaxis). These cues are sensed via specific receptors, such as receptor tyrosine kinases (RTKs) and G-protein-coupled receptors (GPCRs) for chemotaxis, integrins and focal adhesion (FA) complexes for durotaxis, and cell–cell contacts in collective migration (Parent et al. 1998; Haugh et al. 2000; Servant et al. 2000; Duchek and Rorth 2001; Duchek et al. 2001; Funamoto et al. 2002; Iijima and Devreotes 2002; Yang et al. 2002; Jékely et al. 2005; Casaletto and McClatchey 2012; Plotnikov et al. 2012; Hayer et al. 2016; Chiasson-MacKenzie and McClatchey 2018). In response, stationary cells establish a front–rear polarity, initiate movement, and continuously orient their direction of migration.

A long-standing goal in cell migration is to develop a comprehensive model about how cells polarize and undergo directed migration. Such a model has to explain several puzzling observations: (1) uniform receptor stimuli generally suffice to initiate polarization and generate migration speeds comparable to gradients, a phenomenon termed chemokinesis (Gail and Boone 1970); (2) migrating cells exhibit a remarkable polarity persistence, retaining their polarization axis after receptor signaling is turned off (Gong et al. 2024); (3) migrating cells generally undergo biased random walks toward chemoattractants by turning their front toward the target rather than repeatedly losing and regaining polarity (Arrieumerlou and Meyer 2005); and (4) observations of various cells in vivo and in vitro suggest that cells can switch between migration modes, indicating that the bleb-based, mesenchymal, and amoeboid migration may share mechanistic features (SenGupta et al. 2021). Thus, models must explain how uniform external stimuli polarize internal cell signals, why cell polarization is more persistent than the signaling gradients themselves, how migrating cells suppress receptor signaling in the back, how they steer their front during biased random walks, and the identity of mechanisms that link the different migration modes.

Our review focuses on common underlying mechanisms that orchestrate cell migration, with particular attention to how external signals are integrated to establish front–back polarity and how cells steer their front. The canonical framework explaining symmetry breaking is a local excitation-global inhibition (LEGI) model (Kutscher et al. 2004; Devreotes et al. 2017; Li et al. 2020), wherein spontaneously excited local signals are self-amplified to promote protrusion, and global or long-range inhibitory signals maintain spatial asymmetry. Motivated by the alternative idea that asymmetric structures (Gomes et al. 2005; Renkawitz et al. 2019; Tsai et al. 2019; Bisaria et al. 2020), rather than asymmetric signaling, may be crucial for setting up the front–back polarity, we identified two critical structures: membrane proximal F-actin (Bisaria et al. 2020) and endoplasmic reticulum (ER)–plasma membrane (PM) contact sites (Gong et al. 2024), which can restrict receptor signaling to the front by suppressing it throughout the back during cell migration. We propose a "structural polarity of receptor activity model," in which global suppression of cortical receptor signaling and protrusion by local structures unifies how cells initiate membrane protrusions, sustain persistent migration, and steer the front. We summarize both models, highlight supporting evidence, and outline broader implications, limitations, and open questions.

KEY FRONT AND BACK SIGNALING

Extensive research over the past 50 years has focused on identifying and characterizing molecular factors, that are either enriched at the leading edge (front) or at the trailing edge (back) of the cell (Ridley et al. 2003; Devreotes and Horwitz 2015). "Front" signals predominantly promote F-actin polymerization to drive new membrane protrusions, while "back" signals were found to control the nonmuscle myosin II (NMII) activity to facilitate membrane contraction. Crucially, these "front" and "back" signals are thought to mutually inhibit each other, establishing a robust front–back polarity essential for direction migration (Fig. 1).

Genetic and pharmacological studies in *Dictyostelium*, leukocytes, mesenchymal cells, and epithelial cells have uncovered key signaling molecules that define front and back identities, many of which are evolutionarily conserved (Parent et al. 1998; Funamoto et al. 2002; Iijima and Devreotes 2002; Vitorino and Meyer 2008). At the front, signals such as phosphatidylinositol 3,4,5-triphosphate (PIP_3) and small GTPase activity including Ras, Rac, and Cdc42 dominate, whereas RhoA and Ca^{2+} signal are enriched at the rear (Tsai et al. 2014). Numerous regulators and effectors, such as Rho GTPase exchange factors (GEFs) and GTPase-activating proteins (GAPs), also exhibit polarized localization, as do downstream effectors, such as WASP family verprolin-homologous protein (WAVE), actin-related protein 2/3 complex (Arp2/3), Ste20-like kinase (SLK), Rho-associated coiled-coil containing protein kinase (ROCK), cofilin, ERM proteins (Ezrin, Radixin, Moesin family), and various PIP_3 effectors that jointly regulate actin polymerization, severing, and contraction

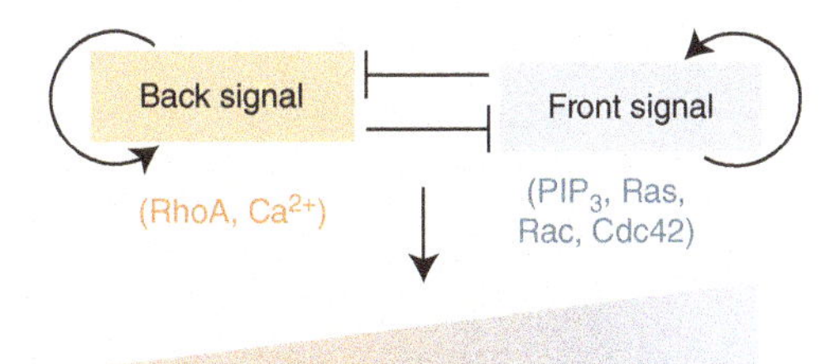

Figure 1. Positive–negative feedback loops in establishing cell polarity. Cells sense external cues in their environments through surface receptors, activating distinct signal modules at the front and back of cells. Here they are simplified as front and back signals respectively. These external gradients may originate from localized sources or be self-generated. Front signals are represented by PIP_3, Ras, Rac, and Cdc42 activity, in contrast to the back signals, including RhoA and Ca^{2+} activity. These front and back signals mutually activate and inhibit each other, forming interconnected feedback circuits that transform into a steep internal front–back polarity, thereby driving persistent cell migration.

(Park et al. 2008; Devreotes and Horwitz 2015). Live-cell imaging using fluorescence-based biosensors has provided invaluable insights into the spatiotemporal dynamics of these signaling events. For instance, pleckstrin homology (PH) domains from CRAC (Parent et al. 1998) or AKT protein (Watton and Downward 1999; Park et al. 2008) were fluorescently tagged and used to visualize PIP_3 accumulation at protrusive sites in *Dictyostelium* and mammalian cells. The Ras-binding domain (RBD) of Raf1 localizes to active Ras zones aligned with chemoattractant gradients, and optogenetic inhibition of Ras rapidly extinguishes existing protrusions and reverses migration (Lin et al. 2024). Fluorescence resonance energy transfer (FRET)-based sensors have revealed an active accumulation of Rac and Cdc42 in driving membrane protrusions, while Rho activity is generally absent when local protrusions are pushing outward but becomes activated to promote local membrane retractions (Pertz et al. 2006; Machacek et al. 2009; Yang et al. 2016; Marshall-Burghardt et al. 2024). Spatial and temporal correlation analyses between edge velocities and Rho GTPase signal dynamics were particularly informative regarding how front and back signals locally coordinate with oscillatory protrusion-retracting cycles at the leading edge.

Another key feature of Ras, PIP_3, and Rho GTPase signaling is their excitability, which manifests as rapid oscillations or traveling waves when observed using fluorescent sensors or dyes. In *Dictyostelium* cells, Ras (RBD) and PIP_3 (CRAC-PH) signals oscillate with a period of ~200 sec, much slower than the 10-sec oscillatory cytoskeleton network, revealed by SCAR/WAVE (suppressor of cAMP receptor, the *Dictyostelium* homolog of WAVE) and actin polymerization sensors, HSPC and LimE, respectively (Huang et al. 2013). These Ras and PIP_3 waves are self-sustaining and cytoskeleton-independent but are triggered by fluctuations in Ras activity or negative regulators such as PTEN (phosphatase and tensin homolog). In migrating dendritic cells, stochastic PIP_3 pulses at the membrane lead to transient protrusions and subtle reorientations in migration presumably due to a fluctuation of chemoattractant receptor occupancy (Arrieumerlou and Meyer 2005). Similar excitability has been observed for RhoA and Rac, the activity kinetics of which are generated and amplified by recruiting respective activators, for instance, the guanine nucleotide exchange factor GEF-H1 is indirectly activated by RhoA-GTP to further activate RhoA (Wittmann and Waterman-Storer 2001). RhoA and Rac can also mutually stimulate each other to potentiate the excitability. The Rac1 effector, Arhgef11/12, is identified as Rho GEFs to couple sequential Rac and Rho activation and synchronize cell protrusion–retraction cycles (Nanda et al. 2023), while the RhoA effector, diaphanous-related formin-1 (DIA), localizes to the membrane ruffles to stimulate Rac activity at the leading edge (Watanabe et al. 1997; Tkachenko et al. 2011). In addition, a reinforcing feedback loop among PI3K (phosphoinositide 3-kinase), Rac, and F-actin stabilizes PIP_3 polarity in neutrophils (Weiner et al. 2002). Together, these positive loops contribute to the excitability of Ras, PIP_3, and Rho GTPase signals (Fig. 2, right).

Equally important are local negative feedbacks at the front that spatially and temporally limit the positive feedbacks, preventing propagating waves from reaching backward or circling around the cell periphery, an event that would disrupt cell migration. The mutual inhibition between Rac1 and RhoA serves as a central mechanism for bistable polarity (Guilluy et al. 2011; Byrne et al. 2016; Salloum et al. 2020). Rac1 effectors such as p21-associated kinase (PAK) inhibit RhoA-specific GEFs (e.g., GEF-H1, Net1, p115-RhoGEF) (Nimnual et al. 2003; Barac et al. 2004; Alberts et al. 2005), while active Rac1 also activates p190RhoGAP to suppress RhoA (Bustos et al. 2008). Conversely, Rho-GTP activates Rho-associated kinase (ROCK), which enhances Rac-specific GAPs such as Fil-GAP and ArhGAP22, dampening Rac activity (Tsuji et al. 2002; Ohta et al. 2006; Sanz-Moreno et al. 2008). RhoA also inhibits itself via actomyosin-mediated recruitment of RhoGAPs such as Myo9b (Graessl et al. 2017), as shown by an increase of RhoA activity in cells treated with ROCK inhibitor, Y27632 or ML7 (Marshall-

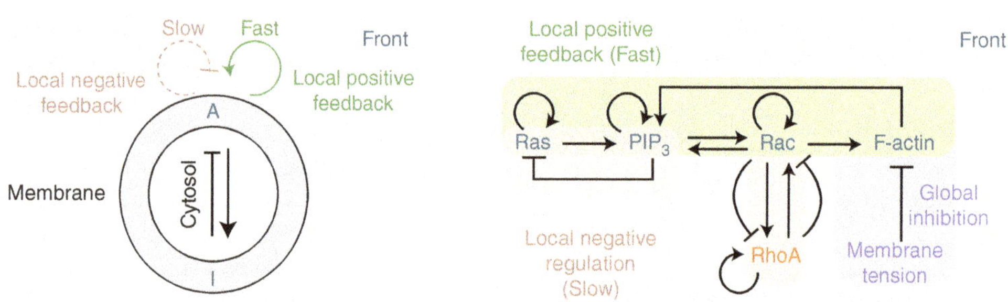

Figure 2. Molecular basis of the local excitation and global inhibition (LEGI) model for directed cell migration. (*Left*) Cell polarity is maintained through local excitation (A) coupled with global inhibition (I). Local excitation arises from a fast, local positive feedback plus slower, local negative feedback, producing spatially confined activity. This locally confined excitation signal induces a presumptive global inhibitory signal, which prevents additional activation elsewhere, thereby supporting persistent and directed cell migration. (*Right*) So far, self-sustaining feedback loops involve Ras, PIP_3, Rac, and RhoA, which recruit their respective activators, such as Rho GEFs, to reinforce signal localization (self-amplification mechanism). A well-characterized positive feedback loop among PI3K, Rac, and F-actin (green-shaded region) further stabilizes PIP_3-enriched domains. In parallel, local inhibitory circuits, such as mutual inhibition between Rac and RhoA via Rho GAPs, and Ras inhibition by the PIP_3-effector RASA2 (yellow-shaded region), contribute to spatially restricted and transient signaling activity. Superimposed on these local dynamics, an undefined global inhibitory mechanism (purple-shaded region; potentially membrane tension), is proposed to prevent front signal spreading to the back for additional membrane protrusion. Together, these mechanisms spatially coordinate front–back polarity to guide persistent cell migration.

Burghardt et al. 2024). Additionally, the PI3K effector RASA2 links PIP_3 signals to Ras inhibition. While its relevance in migrating cells remains to be evaluated, this negative feedback is crucial in neuronal differentiation (Chen et al. 2012). Overall, this feedback is important to generate oscillatory protrusion and retraction cycles, which strengthen integrin adhesions, supporting cell migration under conditions of low matrix adhesion (Tsai and Meyer 2012) and, by generating exploratory protrusions that help cells steer at the front to navigate complex matrix environments (Arrieumerlou and Meyer 2005).

LOCAL EXCITATION–GLOBAL INHIBITION (LEGI) MODEL

The finding of local positive feedbacks in the front of migrating cells led to the compelling model that forward protrusion is driven by an excitable signaling system that can locally activate PIP_3, Ras, and Rac to induce membrane protrusions (Kutscher et al. 2004; Devreotes et al. 2017; Li et al. 2020). However, the problem with having local positive feedback is that cells can trigger protrusion at multiple places and cells need therefore a mechanism whereby local protrusion at one site is suppressing protrusions at others. The LEGI model proposed that this second signaling mechanism may represent a slower diffusible factor that globally suppresses protrusions in cells, but such a diffusible factor has remained elusive (Fig. 2, left). The negative feedback discussed above could, in principle, contribute, but there is no experimental support that the components of these feedback loops diffuse to the back to suppress activity across the cell. Therefore, it remains an open question in the field what the identity and molecular basis is of this proposed diffusible long-range inhibitory factor.

The local and global feedback mechanisms represent the basis of the LEGI model, which can explain how cells may maintain stable front–back polarity exposed to gradients of chemoattractant. Notably, synthetic biology has validated related principles by reconstituting polarity circuits using mutual inhibition and feedback loops in yeast (Chau et al. 2012). One fundamental challenge with long-range inhibitory positive feedback is to make it function equally well in mammalian cells that can be short or long (Houk et al. 2012). For example, for a molecule to reach the back in a cell that is 100 µm instead of 10 µm long, it would take 100 times longer, since the diffusion timescales with the square root of the distance. Moreover, while the LEGI model includes many interesting concepts, it remains an open question how cells initiate symmetry breaking under uniform stimulus conditions and how they polarize receptor signaling activity and persistently migrate. Also, additional concepts are needed to explain how cells migrate in a biased random walk as observed in various experiments and how the same cell can switch between mesenchymal, amoeboid, and bleb-based cell migration. We believe that many of these conceptual challenges can be reconciled by the observation that upstream receptor signaling is stronger at the front than at the back, which we believe can be parsimoniously explained by structural suppression of signaling in the back.

STRUCTURAL POLARITY IN RECEPTOR ACTIVITY (SPRA) MODEL

In addition to the polarization of biochemical front–back signals, directed cell migration is also characterized by distinct and long-lasting morphological asymmetries (Devreotes and Horwitz 2015; Bodor et al. 2020). At the leading edge, actin polymerization is predominantly driven by the Arp2/3 complex, generating a dense, branched network of F-actin oriented perpendicular to the front PM. This structure supports the formation of lamellipodia and filopodia, which protrude the membrane and initiate forward movement (Insall and Machesky 2009). In contrast, the rear of the cell is marked by bundled F-actin arrays, such as stress fibers, that are cross-linked by NMII. These fibers run parallel to the PM and contribute to contractility and retraction of the trailing edge (Vicente-Manzanares et al. 2009). FAs further reflect this spatial distinction. Nascent adhesions, typically ∼0.25 µm in size, form rapidly at the leading edge within minutes. Some mature into focal complexes (∼0.5 μm) and subsequently into larger FAs (1–5 μm), which anchor the actin cytoskeleton to the extracellular matrix (ECM). Depending on cellular needs, FAs can either disassemble in the cell body within 10–20 min or mature into stable fibrillar adhesions (>5 μm) that participate in ECM remodeling (Smilenov et al. 1999; Laukaitis et al. 2001; Gupton and Waterman-Storer 2006). This spatiotemporal regulation of adhesion assembly and disassembly is essential for productive, directional migration.

Unlike the transient and oscillatory nature of biochemical signals such as Rho GTPases, which typically act over seconds, these morphological features provide longer-lived structural memory for maintaining front–back polarity. Indeed, studies have shown that molecular memory from receptor activation can be translated into morphological memory at the level of cell shape. In one-dimensional (1D) microchannels, chemotactic cells exhibit directional memory that depends on persistent polarized cytoskeletal structures, including moesin, actomyosin, and microtubules (Prentice-Mott et al. 2016; d'Alessandro et al. 2021). In a 1D assay under uniform serum conditions, cells transiently treated with a PI3K inhibitor (LY294002) resumed movement in the same direction upon drug washout (Gong et al. 2024), underscoring the importance of structural memory in sustaining direction migration.

So far, four internal structural modules have been proposed as key contributors to structural polarity in migrating cells: (1) actomyosin, (2) nuclear positioning, (3) membrane proximal F-actin, and (4) ER–PM contact sites. Together, these structures provide a physical framework that reinforces polarity and guides persistent migration, even in the absence of sustained external signals (Fig. 3).

ACTOMYOSIN

Actomyosin contractility plays a central role in initiating and maintaining structural polarity by coordinating protrusion, adhesion, and retraction across spatial and temporal scales. Locally, periodic lamellipodial contractions at the leading edge enable cells to sense and respond to matrix stiffness, regulating protrusion–retraction cycles during spreading and migration. Widely spread, actomyosin contractility defines the cell rear, generating sustained traction forces that balance front protrusion with back retraction to promote persistent migration (Vicente-Manzanares et al. 2009). Moreover, dynamic myosin II activity—manifested as rapid assembly and disassembly events or "flashes"—facilitates efficient turning during directional changes (Tsai et al. 2019). Under confinement or low-adhesion conditions, elevated cortical contractility mediated by NMII can drive mesenchymal-to-amoeboid transitions across diverse cell types, regardless of their fate or motility mode (Liu et al. 2015). These transitions rely on a positive feedback loop between cortical actin flows and contractility gradients. Intriguingly, in migrating cells, actomyosin also contributes to a cell-wide actin treadmilling mechanism by severing F-actin at the rear, enabling sustained motility (Wilson et al. 2010). These multifaceted roles underscore actomyosin's essential function as a structural driver of cell polarization and migration.

Structurally, actomyosin consists of bundled F-actin cross-linked by NMII, forming stress fibers. NMII is a hexametric motor protein composed of two heavy chains, two essential light chains (ELCs), and two regulatory light chains (RLCs). The heavy chain contains a coiled-coil rod domain that mediates dimerization, allowing NMII assembly into bipolar filaments. These filaments interact with actin via their actin-binding domains, enabling the formation of higher-order contractile structures. NMII filaments can assemble de novo or nucleate from preexisting structures, a process tightly regulated by receptor-mediated signaling cascades, particularly those involving Rac, calcium, and RhoA (Vicente-Manzanares et al. 2009). Rac, active at the leading edge, promotes actin polymerization and retrograde flow, which in turn transports F-actin-associated proteins, including NMII, toward the rear (Giannone et al. 2004; Maiuri et al. 2015). This flow-based redistribution is critical for rear contractility. In addition to localization,

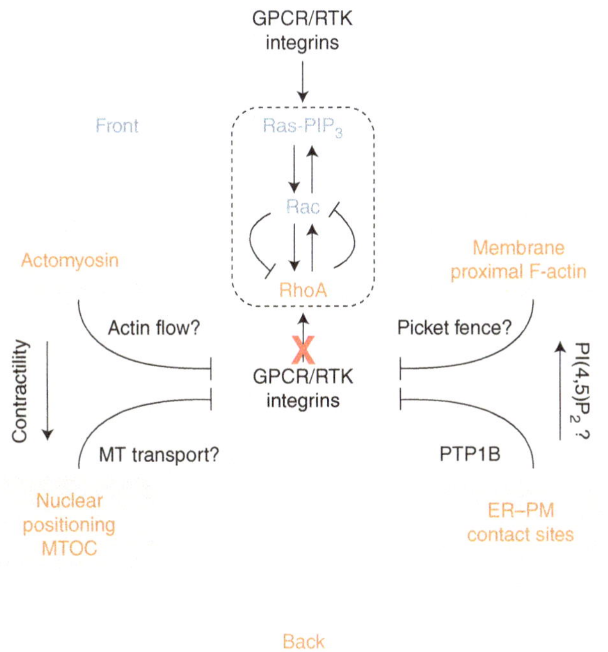

Figure 3. Molecular basis of the "structural polarity in receptor activity (SPRA)" model for directed cell migration. Structural asymmetries contribute to front–back polarity by modulating receptor sensitivity. Four polarized structures, including actomyosin, nuclear positioning, membrane proximal F-actin, and endoplasmic reticulum (ER)–plasma membrane (PM) contact sites, have been proposed to suppress receptor input at the cell rear, thereby biasing receptor activation toward the front. This structural repression promotes asymmetric initiation of signaling cascades, including local positive and negative feedback loops (dashed box region), as described in the local excitation–global inhibition (LEGI) model. ER–PM contact gradient promotes a gradient of ER-localized phosphatase PTP1B activity, which attenuates receptor tyrosine kinases (RTK) signaling at the rear PM. This polarized ER–PM contact gradient can also potentially provide PI(4,5)P$_2$ at the PM to promote F-actin organization. Membrane proximal F-actin may potentially serve as a physical barrier ("picket fence") to restrict receptor diffusion at the PM. Actomyosin driven rearward actin flow might interact with the endocytosed receptor trafficking pathway to regulate receptor sensitivity. Nuclear positioning coordinates with the microtubule-organizing center (MTOC) to organize MT distribution, which can shape the directional transport of receptors-containing vesicles. For clarity, potential feedback interactions between these structural elements and signaling circuits are not shown.

NMII must undergo a conformational switch from an assembly-incompetent folded state (10S) to an extended, assembly-competent state (6S) (Burgess et al. 2007). This switch is regulated by phosphorylation of Ser19 in the RLCs, primarily via two kinases: the Ca^{2+}/calmodulin-dependent myosin light chain kinase (MLCK) and the Rho-associated protein kinase (ROCK), a RhoA effector. MLCK directly phosphorylates RLCs, while ROCK both phosphorylates RLCs and inhibits myosin light chain phosphatase (MLCP), further enhancing NMII contractility (Totsukawa et al. 2000, 2004). Recent studies (Hadjitheodorou et al. 2023) employing optogenetic receptor systems and microfluidic confinement have shown that myosin II activity at the cell rear suppresses receptor input, creating a refractory zone that stabilizes cell polarity, although the underlying mechanisms remain to be elucidated.

Beyond contractility, actomyosin is also a key regulator of FA dynamics. While its roles in adhesion assembly and disassembly have been extensively reviewed elsewhere (Burridge and Chrzanowska-Wodnicka 1996; Gupton and Waterman-Storer 2006; Vicente-Manzanares et al. 2009), how actomyosin balances FA maturation versus turnover remains an open question. Additionally, actomyosin contributes to nuclear positioning (Gomes et al. 2005), retrograde flow of ER–PM contacts (Gong et al. 2024) and cortical actin (Theriot and Mitchison 1991; Yam et al. 2007), structural cues for cell polarity that will be further discussed in the following section. Additionally, it will be critical to determine whether actomyosin-mediated actin flow directly regulates the polarization of receptor activity during cell migration.

NUCLEAR POSITIONING

The nucleus, as the largest and stiffest intracellular organelle, plays a crucial role in regulating polarity across diverse cell types and environments, from planar two-dimensional (2D) substrates to complex three-dimensional (3D) tissue architectures (Gundersen and Worman 2013). In wound-healing assays, mesenchymal and epithelial cells exhibit coordinated front–rear polarity, wherein robust membrane protrusion at the leading edge and retrograde F-actin flow exert rearward forces that displace the nucleus, helping to define the cell rear (Gomes et al. 2005). In contrast, during immune surveillance, amoeboid cells exploit the nucleus as a mechanical gauge for dense tissue matrices. When the nucleus and its associated microtubule-organizing center (MTOC) encounter an optimally sized pore, competing cytoplasmic protrusions retract, and the entire cell reorients to follow the path cleared by nuclear translocation (Renkawitz et al. 2019). In polarized cells, the nucleus typically assumes an asymmetric position along the front–rear or apical–basal axis, a positioning tightly coupled to cytoskeletal forces and intracellular organization.

Nuclear positioning is mediated by dynamic linkages with the cytoskeleton, primarily via actin filaments and microtubules. Central is the linker of nucleoskeleton and cytoskeleton (LINC) complex, composed of outer nuclear membrane proteins such as Nesprin-2G and inner nuclear membrane proteins like SUN2, which together form transmembrane actin-associated nuclear (TAN) lines. SUN2 also interacts with A-type lamins and SAMP1, which are essential for coupling the nucleus to retrograde actin flow driven by the Cdc42–myosin II signaling axis (Luxton et al. 2010). The formin protein FHOD further enhances Nesprin-2G's affinity for F-actin, reinforcing these interactions (Antoku et al. 2015). Recently, the ER has been shown to spatially restrict nucleocytoskeletal interactions on the ventral side, favoring nuclear translocation via dorsal actin cables (Janota et al. 2022). In parallel, microtubule-dependent nuclear positioning involves a SUN1–Nesprin-2G–dynein-based LINC complex (Lele et al. 2018). In highly migratory systems such as developing neurons, nuclear translocation—termed interkinetic nuclear migration (INM)—relies on MT motor proteins including dynein, kinesin-1, dynactin, and the microtubule-associated protein Doublecortin (Francis et al. 1999; Gleeson et al. 1999; Schaar et al. 2004; Kosodo 2012), with additional contributions from actomyosin contractility (Schenk et al. 2009).

Despite growing evidence, how nuclear positioning modulates asymmetric polarity signaling remains incompletely understood. One plausible model suggests that nuclear–MTOC alignment may influence cell polarity via regulation of microtubule organization and integrin-based FA dynamics (Lele et al. 2018). Microtubules have been shown to interact directly with FAs, guiding the targeted delivery of metalloprotease-enriched vesicles to promote FA turnover (Stehbens et al. 2014). However, in confined 3D environments, where amoeboid cells migrate with limited dependence on FAs (Reversat et al. 2020), the nucleus often leads the MTOC (Renkawitz et al. 2019), indicating another polarity paradigm distinct from adhesion-coupled systems. It is also plausible that a main role of nuclear positioning in cell polarity is to direct microtubule organization and vesicular transport of various receptors. Overall, the contribution of nuclear positioning is context-dependent.

MEMBRANE PROXIMAL F-ACTIN

Remodeling of the actin cytoskeleton provides the fundamental driving force for cell migration, relying on continuous cycles of F-actin polymerization and depolymerization orchestrated by a complex network of regulatory proteins. Among the key actin nucleators, formins facilitate the assembly of linear actin filaments, while the Arp2/3 complex generates branched, dendritic networks essential for lamellipodial protrusions. In contrast, actin filament turnover is primarily mediated by severing proteins such as cofilin, which disassembles older filaments to recycle actin monomers (Bravo-Cordero et al. 2013). Migrating cells exhibit a polarized distribution of actin regulators, with enrichment of polymerization factors at the leading edge, resulting in higher total F-actin density at the front to drive membrane protrusion. Beyond filament turnover, the actin cytoskeleton is anchored to the PM by a set of tethering proteins—including the Ezrin, Radixin, and Moesin (ERM) family—forming the actin cortex (Chugh and Paluch 2018), a dense meshwork critical for controlling membrane tension and the spatial organization of surface receptors.

To probe the organization of F-actin near the PM, our group developed a membrane-proximal actin reporter, MPAct, by fusing a PM anchor (CAAX motif) to a fluorescent protein and the low-affinity F-actin-binding domain F-tractin. This probe enables quantitative imaging of the outermost cortex within ~10 nm of the membrane surface (membrane-proximal actin, or MPA) (Bisaria et al. 2020). We use the term MPA since the traditional term MCA generally refers to membrane cortical actin, which refers to the entire layer within ~200 nm of the membrane. Intriguingly, cultured epithelial and endothelial cells migrating in 1-, 2-, and 3D environments display stable MPA gradients, with a characteristically lower density at the leading edge. The steepness of the MPAct gradient correlates with migration velocity, suggesting that MPA organization serves as a persistence cue to regulate cell speed. Cross-correlation analyses between edge dynamics and MPA density in single-migrating cells revealed a local drop in MPA prior to membrane protrusion, supporting the idea that local cortex disassembly may permit forward advancement. Notably, elevated MPA levels appear to inhibit Rac activation, although the underlying mechanism remains elusive (Fig. 4). It is tempting to speculate that dense cortical actin restricts membrane receptor mobility via a "picket-fence" effect (Freeman et al. 2018), raising the possibility that diffusion-reaction dynamics of membrane-bound signaling components are constrained by the actin cortex during cell migration.

Another important role of MPA may serve as a physical basis of membrane tension. Tension can reflect "in-plane membrane stress," for example, during osmotic swelling, but often arises from the local attachment of cortical actin to the PM through PIP_2 lipids (Raucher et al. 2000). Several lines of evidence suggest a critical role of membrane tension in cell migration (Houk et al. 2012; Lieber et al. 2015; Bieling et al. 2016; Mueller et al. 2017; Hetmanski

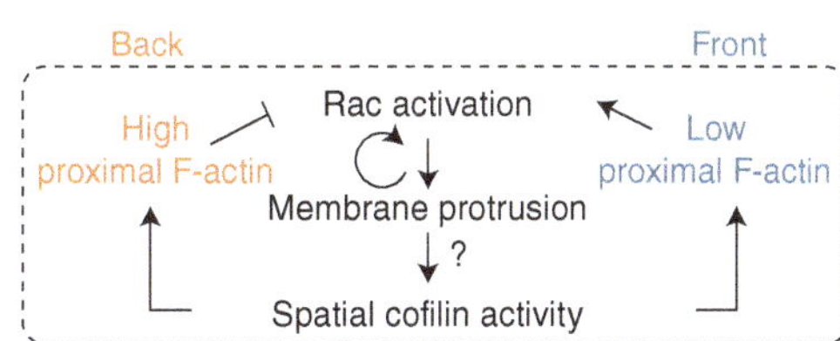

Figure 4. Roles of membrane-proximal F-actin gradients in cell polarization. Migratory cells have a gradient of membrane-proximal F-actin, with a higher density at the back. Regions enriched in membrane-proximal F-actin inhibit Rac activation, thereby restricting membrane protrusion to areas with lower membrane-proximal F-actin density. As protrusion progresses, cofilin activity becomes spatially distributed, with higher activity concentrated at the front, which promotes actin turnover at the front, reinforcing the gradient of membrane-proximal F-actin. This feedback loop consolidates front–back polarity by promoting membrane-proximal F-actin accumulation at the rear.

et al. 2019). Membrane tension has long been studied using membrane tethers (Dai and Sheetz 1995) and is thought to preserve membrane integrity by limiting actin- or bleb-based membrane protrusions before rupture. As membrane tension increases, an increased actin branch density is required for cells to keep protruding the membrane, suggesting that there is a general inverse link between actin-driven protrusion rate and membrane tension. Therefore, a critical role of membrane tension in cell migration is to slow or stop membrane protrusions at the front to protect the membrane integrity when the back retraction is slowing, or cells fail to retract their back, respectively. Elegant recent optogenetic experiments generating local membrane protrusions further showed that membrane tension rapidly increases and propagates across the cell (De Belly et al. 2023), highlighting the importance of membrane tension in cell migration. However, additional investigations are needed to determine how important membrane tension is in different contexts of cell migration. It is also not fully understood how local membrane tension is spatially and temporally linked to local membrane proximal F-actin.

ER–PM CONTACT SITES

In addition to the actin cortex, the ER forms another critical membrane-associated structure at the cell periphery—ER–PM contact sites (Porter and Palade 1957). The membranes in these specialized membrane junctions are typically spaced 10–25 nm apart and are stabilized by a variety of tethering and lipid transfer proteins, including members of the extended synaptotagmin (E-Syt) family (Takeshima et al. 2000; Manford et al. 2012; Chang et al. 2013; Giordano et al. 2013). ER–PM contact sites serve key functions in maintaining cellular homeostasis, such as replenishing ER calcium stores following depletion and mediating bidirectional lipid exchange between the ER and PM to maintain lipid homeostasis (Voeltz et al. 2024). To visualize these structures, a MAPPER (for membrane-attached peripheral ER) reporter was developed, featuring a single-pass ER-anchoring transmembrane domain and a polybasic PM-targeting motif that binds negatively charged lipids (Chang et al. 2013). Using MAPPER in combination with micropatterned substrates, we identified a striking gradient of ER–PM contacts in migrating cells: contact sites accumulate at the cell rear, both in individual and collectively migrating cells across diverse cell types, including cultured epithelial, endothelial, and fibroblasts (Gong et al. 2024). This rearward enrichment of ER–PM contacts correlates with increased cell migration speed in single-migrating cells, suggesting that ER–PM contact site asymmetry provides a structural basis for persistent directional migration.

Mechanistically, in single-migrating cells, this ER–PM contact gradient spatially restricts RTK signaling by confining the activity of the ER-localized phosphatase PTP1B to the rear of the cell. Because PTP1B can access PM-localized RTKs only at ER–PM junctions, receptor activation is suppressed at the back, thereby biasing signaling activity toward the front. This creates a permissive environment for new protrusion formation in regions with fewer ER–PM contacts, which can become the future leading edge. Initial protrusions reorganize the ER via microtubule-dependent transport, leading to an accumulation of tubular ER at the front and more sheet-like ER at the back. This in turn reinforces the polarized distribution of ER–PM contact sites, establishing a feedback loop that stabilizes front–rear polarity (Fig. 5). Together, these findings position the ER–PM contact gradients as a symmetry-breaking mechanism that underpins directional cell migration.

Our optogenetic manipulation in single-migrating cells revealed that locally increasing the densities of ER–PM contacts antagonize receptor signaling and membrane protrusions. This potentially explains how uniform receptor stimuli initiate stochastic protrusions by sampling local densities differences in these inhibitory structures in stationary cells. In turn, membrane protrusions initiate cell migration by locally increasing receptor signaling through strengthening ER–PM contacts gradients. Protrusions result in actomyosin-mediated backward flow of ER–PM contacts that further suppresses protrusion and promotes retraction of the back. This

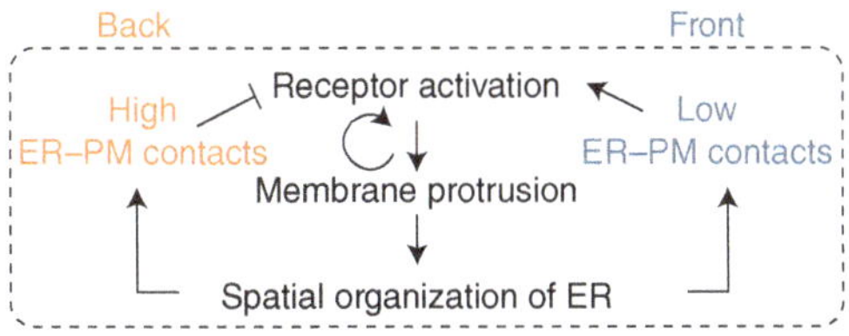

Figure 5. Roles of endoplasmic reticulum (ER)–plasma membrane (PM) contact gradients in symmetry breaking during cell migration. During the transition from an unpolarized to a polarized state, cells establish spatial asymmetry in ER–PM contact sites that contributes to directional migration. Regions with relatively sparse ER–PM contacts exhibit higher receptor activation, due to reduced inhibition by the ER-localized phosphatase PTP1B. This local receptor activation initiates downstream local feedback signaling, promoting membrane protrusion and designating the emerging front. As protrusion progresses, microtubule-mediated ER movements help reorganize the ER structure, leading to a polarized distribution of ER morphology, with more tubular ER concentrated at the front and sheet-like ER enriched at the back. This spatial segregation reinforces the gradient of ER–PM contacts, consolidating front–back polarity by accumulating ER–PM contacts at the rear. Such feedback between ER architecture and signaling ensures a robust and stable polarized state essential for persistent migration.

dynamic organization of ER–PM contacts is also applicable for cells when taking a turn. Upon their arrival to the end of a stripe, cells tend to stay in place to disassemble their ER–PM contacts at the former back before initiation of a new protrusion. Along with retrograde flow of the remaining ER–PM contacts toward the new back, a reversal gradient of ER–PM contacts forms, leading to a successful reversal of persistent cell migration. These observations highlight the importance of ER–PM contact gradients in cell migration and turning.

In addition to the receptor signaling regulation, given the roles of ER–PM contacts in calcium signaling and lipid regulation, future work is needed to determine how these gradients may further influence cell migration through localized modulation of Ca^{2+} and lipid signaling pathways, particularly $PI(4,5)P_2$, which is critical for both ER–PM contact and cortex actin organization.

CONCLUSIONS

In this review, we have summarized two conceptual models of cell migration, LEGI, and structural polarity of receptor activation (SPRA). Mechanistically, both converge on the regulation of Ras, PIP_3, and Rho GTPase signaling to coordinate membrane protrusion at the front and cell retraction at the rear, facilitating directed cell migration. However, the SPRA model highlights the regulation at the receptor level, providing a more mechanistic explanation of how cells interpret environmental cues, especially under uniform stimulation. In contrast, the LEGI model often invokes stochastic signaling oscillations to explain the initiation of symmetry breaking. Another key distinction between these models lies in their temporal dynamics. Temporally, the LEGI involves fast-acting molecular circuits that respond at the timescale of seconds, while the SPRA model replies on longer-lasting structural basis at the cell cortex, which usually operates on the scale of minutes. The longer timescale provides also a mechanism for migration persistence and directional memory during cell migration.

Despite recent progress, our understanding of the SPRA model remains at an early stage. To date, our mechanistic insights into receptor signaling by structural polarity have largely focused on RTK signaling regulated by the key phosphatase PTP1B. PTP1B, however, has numerous substrates, including the αIIbβ3 integrin/Src complex and $p130^{Cas}$, which could modulate the integrin and GPCR signaling important for cell migration. Moreover, PTP1B-independent mechanisms are likely to contribute to receptor signaling regulation. In addition, other key questions include disentangling actomyosin and nucleus's structural role from their signaling functions; identifying the molecular machinery underlying membrane-proximal F-actin and ER–PM contact sites, and linking them to the other two structural bases; defining the temporal sequence of polarization events during symmetry breaking; and elucidating cross talk between structural elements and biochemical networks and investigating roles of these structural polarity in collective cell migration. Addressing these issues will be crucial for a comprehensive and unified understanding of cell migration.

ACKNOWLEDGMENTS

We thank members of the Meyer and Teruel laboratories at Weill Cornell for ongoing discussions. Funding 5R35GM127026.

REFERENCES

Alberts AS, Qin H, Carr HS, Frost JA. 2005. PAK1 negatively regulates the activity of the Rho exchange factor NET1. *J Biol Chem* **280:** 12152–12161. doi:10.1074/jbc.M405073200

Antoku S, Zhu R, Kutscheidt S, Fackler OT, Gundersen GG. 2015. Reinforcing the LINC complex connection to actin filaments: the role of FHOD1 in TAN line formation and nuclear movement. *Cell Cycle* **14:** 2200–2205. doi:10.1080/15384101.2015.1053665

Arrieumerlou C, Meyer T. 2005. A local coupling model and compass parameter for eukaryotic chemotaxis. *Dev Cell* **8:** 215–227. doi:10.1016/j.devcel.2004.12.007

Barac A, Basile J, Vázquez-Prado J, Gao Y, Zheng Y, Gutkind JS. 2004. Direct interaction of p21-activated kinase 4 with PDZ-RhoGEF, a G protein-linked Rho guanine exchange factor. *J Biol Chem* **279:** 6182–6189. doi:10.1074/jbc.M309579200

Bear JE, Haugh JM. 2014. Directed migration of mesenchymal cells: where signaling and the cytoskeleton meet. *Curr Opin Cell Biol* **30:** 74–82. doi:10.1016/j.ceb.2014.06.005

Bieling P, Li TD, Weichsel J, McGorty R, Jreij P, Huang B, Fletcher DA, Mullins RD. 2016. Force feedback controls motor activity and mechanical properties of self-assembling branched actin networks. *Cell* **164:** 115–127. doi:10.1016/j.cell.2015.11.057

Bisaria A, Hayer A, Garbett D, Cohen D, Meyer T. 2020. Membrane-proximal F-actin restricts local membrane protrusions and directs cell migration. *Science* **368:** 1205–1210. doi:10.1126/science.aay7794

Bodor DL, Pönisch W, Endres RG, Paluch EK. 2020. Of cell shapes and motion: the physical basis of animal cell migration. *Dev Cell* **52:** 550–562. doi:10.1016/j.devcel.2020.02.013

Bravo-Cordero JJ, Magalhaes MA, Eddy RJ, Hodgson L, Condeelis J. 2013. Functions of cofilin in cell locomotion and invasion. *Nat Rev Mol Cell Biol* **14:** 405–415. doi:10.1038/nrm3609

Burgess SA, Yu S, Walker ML, Hawkins RJ, Chalovich JM, Knight PJ. 2007. Structures of smooth muscle myosin and heavy meromyosin in the folded, shutdown state. *J Mol Biol* **372:** 1165–1178. doi:10.1016/j.jmb.2007.07.014

Burridge K, Chrzanowska-Wodnicka M. 1996. Focal adhesions, contractility, and signaling. *Annu Rev Cell Dev Biol* **12:** 463–518. doi:10.1146/annurev.cellbio.12.1.463

Bustos RI, Forget MA, Settleman JE, Hansen SH. 2008. Coordination of Rho and Rac GTPase function via p190B RhoGAP. *Curr Biol* **18:** 1606–1611. doi:10.1016/j.cub.2008.09.019

Byrne KM, Monsefi N, Dawson JC, Degasperi A, Bukowski-Wills JC, Volinsky N, Dobrzyński M, Birtwistle MR, Tsyganov MA, Kiyatkin A, et al. 2016. Bistability in the Rac1, PAK, and RhoA signaling network drives actin cytoskeleton dynamics and cell motility switches. *Cell Syst* **2:** 38–48. doi:10.1016/j.cels.2016.01.003

Casaletto JB, McClatchey AI. 2012. Spatial regulation of receptor tyrosine kinases in development and cancer. *Nat Rev Cancer* **12:** 387–400. doi:10.1038/nrc3277

Chang CL, Hsieh TS, Yang TT, Rothberg KG, Azizoglu DB, Volk E, Liao JC, Liou J. 2013. Feedback regulation of receptor-induced Ca^{2+} signaling mediated by E-Syt1 and Nir2 at endoplasmic reticulum-plasma membrane junctions. *Cell Rep* **5:** 813–825. doi:10.1016/j.celrep.2013.09.038

Chau AH, Walter JM, Gerardin J, Tang C, Lim WA. 2012. Designing synthetic regulatory networks capable of self-organizing cell polarization. *Cell* **151:** 320–332. doi:10.1016/j.cell.2012.08.040

Chen JY, Lin JR, Cimprich KA, Meyer T. 2012. A two-dimensional ERK-AKT signaling code for an NGF-triggered cell-fate decision. *Mol Cell* **45:** 196–209. doi:10.1016/j.molcel.2011.11.023

Chiasson-MacKenzie C, McClatchey AI. 2018. Cell–Cell contact and receptor tyrosine kinase signaling. *Cold Spring Harb Perspect Biol* **10:** a029215. doi:10.1101/cshperspect.a029215

Chugh P, Paluch EK. 2018. The actin cortex at a glance. *J Cell Sci* **131:** jcs186254. doi:10.1242/jcs.186254

Dai J, Sheetz MP. 1995. Mechanical properties of neuronal growth cone membranes studied by tether formation with laser optical tweezers. *Biophys J* **68:** 988–996. doi:10.1016/S0006-3495(95)80274-2

d'Alessandro J, Barbier-Chebbah A, Cellerin V, Benichou O, Mège RM, Voituriez R, Ladoux B. 2021. Cell migration guided by long-lived spatial memory. *Nat Commun* **12:** 4118. doi:10.1038/s41467-021-24249-8

De Belly H, Yan S, Borja da Rocha H, Ichbiah S, Town JP, Zager PJ, Estrada DC, Meyer K, Turlier H, Bustamante C, et al. 2023. Cell protrusions and contractions generate long-range membrane tension propagation. *Cell* **186:** 3049–3061.e15. doi:10.1016/j.cell.2023.05.014

Devreotes P, Horwitz AR. 2015. Signaling networks that regulate cell migration. *Cold Spring Harb Perspect Biol* **7:** a005959. doi:10.1101/cshperspect.a005959

Devreotes PN, Bhattacharya S, Edwards M, Iglesias PA, Lampert T, Miao Y. 2017. Excitable signal transduction networks in directed cell migration. *Annu Rev Cell Dev Biol* **33:** 103–125. doi:10.1146/annurev-cellbio-100616-060739

Duchek P, Rorth P. 2001. Guidance of cell migration by EGF receptor signaling during *Drosophila* oogenesis. *Science* **291:** 131–133. doi:10.1126/science.291.5501.131

Duchek P, Somogyi K, Jékely G, Beccari S, Rørth P. 2001. Guidance of cell migration by the *Drosophila* PDGF/VEGF receptor. *Cell* **107:** 17–26. doi:10.1016/S0092-8674(01)00502-5

Fackler OT, Grosse R. 2008. Cell motility through plasma membrane blebbing. *J Cell Biol* **181:** 879–884. doi:10.1083/jcb.200802081

Francis F, Koulakoff A, Boucher D, Chafey P, Schaar B, Vinet MC, Friocourt G, McDonnell N, Reiner O, Kahn A, et al. 1999. Doublecortin is a developmentally regulated, microtubule-associated protein expressed in migrating and differentiating neurons. *Neuron* **23:** 247–256. doi:10.1016/S0896-6273(00)80777-1

Freeman SA, Vega A, Riedl M, Collins RF, Ostrowski PP, Woods EC, Bertozzi CR, Tammi MI, Lidke DS, Johnson P, et al. 2018. Transmembrane pickets connect cyto- and pericellular skeletons forming barriers to receptor engagement. *Cell* **172:** 305–317.e10. doi:10.1016/j.cell.2017.12.023

Funamoto S, Meili R, Lee S, Parry L, Firtel RA. 2002. Spatial and temporal regulation of 3-phosphoinositides by PI 3-kinase and PTEN mediates chemotaxis. *Cell* **109:** 611–623. doi:10.1016/S0092-8674(02)00755-9

Gail MH, Boone CW. 1970. The locomotion of mouse fibroblasts in tissue culture. *Biophys J* **10:** 980–993. doi:10.1016/S0006-3495(70)86347-0

Giannone G, Dubin-Thaler BJ, Döbereiner HG, Kieffer N, Bresnick AR, Sheetz MP. 2004. Periodic lamellipodial contractions correlate with rearward actin waves. *Cell* **116:** 431–443. doi:10.1016/S0092-8674(04)00058-3

Giordano F, Saheki Y, Idevall-Hagren O, Colombo SF, Pirruccello M, Milosevic I, Gracheva EO, Bagriantsev SN, Borgese N, De Camilli P. 2013. PI(4,5)P(2)-dependent and Ca^{2+}-regulated ER-PM interactions mediated by the extended synaptotagmins. *Cell* **153:** 1494–1509. doi:10.1016/j.cell.2013.05.026

Gleeson JG, Lin PT, Flanagan LA, Walsh CA. 1999. Doublecortin is a microtubule-associated protein and is expressed widely by migrating neurons. *Neuron* **23:** 257–271. doi:10.1016/S0896-6273(00)80778-3

Gomes ER, Jani S, Gundersen GG. 2005. Nuclear movement regulated by Cdc42, MRCK, myosin, and actin flow establishes MTOC polarization in migrating cells. *Cell* **121:** 451–463. doi:10.1016/j.cell.2005.02.022

Gong B, Johnston JD, Thiemicke A, de Marco A, Meyer T. 2024. Endoplasmic reticulum-plasma membrane contact gradients direct cell migration. *Nature* **631:** 415–423. doi:10.1038/s41586-024-07527-5

Graessl M, Koch J, Calderon A, Kamps D, Banerjee S, Mazel T, Schulze N, Jungkurth JK, Patwardhan R, Solouk D, et al. 2017. An excitable Rho GTPase signaling network generates dynamic subcellular contraction patterns. *J Cell Biol* **216:** 4271–4285. doi:10.1083/jcb.201706052

Guilluy C, Garcia-Mata R, Burridge K. 2011. Rho protein crosstalk: another social network? *Trends Cell Biol* **21:** 718–726. doi:10.1016/j.tcb.2011.08.002

Gundersen GG, Worman HJ. 2013. Nuclear positioning. *Cell* **152:** 1376–1389. doi:10.1016/j.cell.2013.02.031

Gupton SL, Waterman-Storer CM. 2006. Spatiotemporal feedback between actomyosin and focal-adhesion systems optimizes rapid cell migration. *Cell* **125:** 1361–1374. doi:10.1016/j.cell.2006.05.029

Hadjitheodorou A, Bell GRR, Ellett F, Irimia D, Tibshirani R, Collins SR, Theriot JA. 2023. Leading edge competition promotes context-dependent responses to receptor inputs to resolve directional dilemmas in neutrophil migration. *Cell Syst* **14:** 196–209.e6. doi:10.1016/j.cels.2023.02.001

Haugh JM, Codazzi F, Teruel M, Meyer T. 2000. Spatial sensing in fibroblasts mediated by 3′ phosphoinositides. *J Cell Biol* **151:** 1269–1280. doi:10.1083/jcb.151.6.1269

Hayer A, Shao L, Chung M, Joubert LM, Yang HW, Tsai FC, Bisaria A, Betzig E, Meyer T. 2016. Engulfed cadherin fingers are polarized junctional structures between collectively migrating endothelial cells. *Nat Cell Biol* **18:** 1311–1323. doi:10.1038/ncb3438

Hetmanski JHR, de Belly H, Busnelli I, Waring T, Nair RV, Sokleva V, Dobre O, Cameron A, Gauthier N, Lamaze C, et al. 2019. Membrane tension orchestrates rear retraction in matrix-directed cell migration. *Dev Cell* **51:** 460–475.e10. doi:10.1016/j.devcel.2019.09.006

Houk AR, Jilkine A, Mejean CO, Boltyanskiy R, Dufresne ER, Angenent SB, Altschuler SJ, Wu LF, Weiner OD. 2012. Membrane tension maintains cell polarity by confining signals to the leading edge during neutrophil migration. *Cell* **148:** 175–188. doi:10.1016/j.cell.2011.10.050

Huang CH, Tang M, Shi C, Iglesias PA, Devreotes PN. 2013. An excitable signal integrator couples to an idling cytoskeletal oscillator to drive cell migration. *Nat Cell Biol* **15:** 1307–1316. doi:10.1038/ncb2859

Iijima M, Devreotes P. 2002. Tumor suppressor PTEN mediates sensing of chemoattractant gradients. *Cell* **109:** 599–610. doi:10.1016/S0092-8674(02)00745-6

Insall RH, Machesky LM. 2009. Actin dynamics at the leading edge: from simple machinery to complex networks. *Dev Cell* **17:** 310–322. doi:10.1016/j.devcel.2009.08.012

Janota CS, Pinto A, Pezzarossa A, Machado P, Costa J, Campinho P, Franco CA, Gomes ER. 2022. Shielding of actin by the endoplasmic reticulum impacts nuclear positioning. *Nat Commun* **13:** 2763. doi:10.1038/s41467-022-30388-3

Jékely G, Sung HH, Luque CM, Rørth P. 2005. Regulators of endocytosis maintain localized receptor tyrosine kinase signaling in guided migration. *Dev Cell* **9:** 197–207. doi:10.1016/j.devcel.2005.06.004

Kosodo Y. 2012. Interkinetic nuclear migration: beyond a hallmark of neurogenesis. *Cell Mol Life Sci* **69:** 2727–2738. doi:10.1007/s00018-012-0952-2

Kutscher B, Devreotes P, Iglesias PA. 2004. Local excitation, global inhibition mechanism for gradient sensing: an interactive applet. *Sci STKE* **2004:** pl3. doi:10.1126/stke.2192004pl3

Lämmermann T, Sixt M. 2009. Mechanical modes of "amoeboid" cell migration. *Curr Opin Cell Biol* **21:** 636–644. doi:10.1016/j.ceb.2009.05.003

Laukaitis CM, Webb DJ, Donais K, Horwitz AF. 2001. Differential dynamics of alpha 5 integrin, paxillin, and alpha-actinin during formation and disassembly of adhesions in migrating cells. *J Cell Biol* **153:** 1427–1440. doi:10.1083/jcb.153.7.1427

Lele TP, Dickinson RB, Gundersen GG. 2018. Mechanical principles of nuclear shaping and positioning. *J Cell Biol* **217:** 3330–3342. doi:10.1083/jcb.201804052

Li X, Miao Y, Pal DS, Devreotes PN. 2020. Excitable networks controlling cell migration during development and disease. *Semin Cell Dev Biol* **100:** 133–142. doi:10.1016/j.semcdb.2019.11.001

Lieber AD, Schweitzer Y, Kozlov MM, Keren K. 2015. Front-to-rear membrane tension gradient in rapidly moving cells. *Biophys J* **108:** 1599–1603. doi:10.1016/j.bpj.2015.02.007

Lin Y, Pal DS, Banerjee P, Banerjee T, Qin G, Deng Y, Borleis J, Iglesias PA, Devreotes PN. 2024. Ras suppression potentiates rear actomyosin contractility-driven cell polarization and migration. *Nat Cell Biol* **26:** 1062–1076. doi:10.1038/s41556-024-01453-4

Liu YJ, Le Berre M, Lautenschlaeger F, Maiuri P, Callan-Jones A, Heuzé M, Takaki T, Voituriez R, Piel M. 2015. Confinement and low adhesion induce fast amoeboid migration of slow mesenchymal cells. *Cell* **160:** 659–672. doi:10.1016/j.cell.2015.01.007

Luxton GW, Gomes ER, Folker ES, Vintinner E, Gundersen GG. 2010. Linear arrays of nuclear envelope proteins harness retrograde actin flow for nuclear movement. *Science* **329:** 956–959. doi:10.1126/science.1189072

Machacek M, Hodgson L, Welch C, Elliott H, Pertz O, Nalbant P, Abell A, Johnson GL, Hahn KM, Danuser G. 2009. Coordination of Rho GTPase activities during cell protrusion. *Nature* **461:** 99–103. doi:10.1038/nature08242

Maiuri P, Rupprecht JF, Wieser S, Ruprecht V, Bénichou O, Carpi N, Coppey M, De Beco S, Gov N, Heisenberg CP, et al. 2015. Actin flows mediate a universal coupling between cell speed and cell persistence. *Cell* **161:** 374–386. doi:10.1016/j.cell.2015.01.056

Manford AG, Stefan CJ, Yuan HL, Macgurn JA, Emr SD. 2012. ER-to-plasma membrane tethering proteins regulate cell signaling and ER morphology. *Dev Cell* **23:** 1129–1140. doi:10.1016/j.devcel.2012.11.004

Marshall-Burghardt S, Migueles-Ramírez RA, Lin Q, El Baba N, Saada R, Umar M, Mavalwala K, Hayer A. 2024. Excitable Rho dynamics control cell shape and motility by sequentially activating ERM proteins and actomyosin contractility. *Sci Adv* **10:** eadn6858. doi:10.1126/sciadv.adn6858

Mayor R, Etienne-Manneville S. 2016. The front and rear of collective cell migration. *Nat Rev Mol Cell Biol* **17:** 97–109. doi:10.1038/nrm.2015.14

Mueller J, Szep G, Nemethova M, de Vries I, Lieber AD, Winkler C, Kruse K, Small JV, Schmeiser C, Keren K, et al. 2017. Load adaptation of lamellipodial actin networks. *Cell* **171:** 188–200.e16. doi:10.1016/j.cell.2017.07.051

Nanda S, Calderon A, Sachan A, Duong TT, Koch J, Xin X, Solouk-Stahlberg D, Wu YW, Nalbant P, Dehmelt L. 2023. Rho GTPase activity crosstalk mediated by Arhgef11 and Arhgef12 coordinates cell protrusion-retraction cycles. *Nat Commun* **14:** 8356. doi:10.1038/s41467-023-43875-y

Nimnual AS, Taylor LJ, Bar-Sagi D. 2003. Redox-dependent downregulation of Rho by Rac. *Nat Cell Biol* **5:** 236–241. doi:10.1038/ncb938

Ohta Y, Hartwig JH, Stossel TP. 2006. FilGAP, a Rho- and ROCK-regulated GAP for Rac binds filamin A to control actin remodelling. *Nat Cell Biol* **8:** 803–814. doi:10.1038/ncb1437

Parent CA, Blacklock BJ, Froehlich WM, Murphy DB, Devreotes PN. 1998. G protein signaling events are activated at the leading edge of chemotactic cells. *Cell* **95:** 81–91. doi:10.1016/S0092-8674(00)81784-5

Park WS, Heo WD, Whalen JH, O'Rourke NA, Bryan HM, Meyer T, Teruel MN. 2008. Comprehensive identification of PIP3-regulated PH domains from *C. elegans* to *H. sapiens* by model prediction and live imaging. *Mol Cell* **30:** 381–392. doi:10.1016/j.molcel.2008.04.008

Pertz O, Hodgson L, Klemke RL, Hahn KM. 2006. Spatiotemporal dynamics of RhoA activity in migrating cells. *Nature* **440:** 1069–1072. doi:10.1038/nature04665

Plotnikov SV, Pasapera AM, Sabass B, Waterman CM. 2012. Force fluctuations within focal adhesions mediate ECM-rigidity sensing to guide directed cell migration. *Cell* **151:** 1513–1527. doi:10.1016/j.cell.2012.11.034

Porter KR, Palade GE. 1957. Studies on the endoplasmic reticulum. III: Its form and distribution in striated muscle cells. *J Biophys Biochem Cytol* **3:** 269–300. doi:10.1083/jcb.3.2.269

Prentice-Mott HV, Meroz Y, Carlson A, Levine MA, Davidson MW, Irimia D, Charras GT, Mahadevan L, Shah JV. 2016. Directional memory arises from long-lived cytoskeletal asymmetries in polarized chemotactic cells. *Proc Natl Acad Sci* **113:** 1267–1272. doi:10.1073/pnas.1513289113

Raucher D, Stauffer T, Chen W, Shen K, Guo S, York JD, Sheetz MP, Meyer T. 2000. Phosphatidylinositol 4,5-bisphosphate functions as a second messenger that regulates cytoskeleton-plasma membrane adhesion. *Cell* **100:** 221–228. doi:10.1016/S0092-8674(00)81560-3

Renkawitz J, Kopf A, Stopp J, de Vries I, Driscoll MK, Merrin J, Hauschild R, Welf ES, Danuser G, Fiolka R, et al. 2019. Nuclear positioning facilitates amoeboid migration along the path of least resistance. *Nature* **568:** 546–550. doi:10.1038/s41586-019-1087-5

Reversat A, Gaertner F, Merrin J, Stopp J, Tasciyan S, Aguilera J, de Vries I, Hauschild R, Hons M, Piel M, et al. 2020. Cellular locomotion using environmental topography. *Nature* **582:** 582–585. doi:10.1038/s41586-020-2283-z

Ridley AJ, Schwartz MA, Burridge K, Firtel RA, Ginsberg MH, Borisy G, Parsons JT, Horwitz AR. 2003. Cell migration: integrating signals from front to back. *Science* **302:** 1704–1709. doi:10.1126/science.1092053

Salloum G, Jaafar L, El-Sibai M. 2020. Rho A and Rac1: antagonists moving forward. *Tissue Cell* **65:** 101364. doi:10.1016/j.tice.2020.101364

Sanz-Moreno V, Gadea G, Ahn J, Paterson H, Marra P, Pinner S, Sahai E, Marshall CJ. 2008. Rac activation and inactivation control plasticity of tumor cell movement. *Cell* **135:** 510–523. doi:10.1016/j.cell.2008.09.043

Schaar BT, Kinoshita K, McConnell SK. 2004. Doublecortin microtubule affinity is regulated by a balance of kinase and phosphatase activity at the leading edge of migrating neurons. *Neuron* **41:** 203–213. doi:10.1016/S0896-6273(03)00843-2

Schenk J, Wilsch-Bräuninger M, Calegari F, Huttner WB. 2009. Myosin II is required for interkinetic nuclear migration of neural progenitors. *Proc Natl Acad Sci* **106:** 16487–16492. doi:10.1073/pnas.0908928106

SenGupta S, Parent CA, Bear JE. 2021. The principles of directed cell migration. *Nat Rev Mol Cell Biol* **22:** 529–547. doi:10.1038/s41580-021-00366-6

Servant G, Weiner OD, Herzmark P, Balla T, Sedat JW, Bourne HR. 2000. Polarization of chemoattractant receptor signaling during neutrophil chemotaxis. *Science* **287:** 1037–1040. doi:10.1126/science.287.5455.1037

Smilenov LB, Mikhailov A, Pelham RJ, Marcantonio EE, Gundersen GG. 1999. Focal adhesion motility revealed in stationary fibroblasts. *Science* **286:** 1172–1174. doi:10.1126/science.286.5442.1172

Stehbens SJ, Paszek M, Pemble H, Ettinger A, Gierke S, Wittmann T. 2014. CLASPs link focal-adhesion-associated microtubule capture to localized exocytosis and adhesion site turnover. *Nat Cell Biol* **16:** 561–573. doi:10.1038/ncb2975

Takeshima H, Komazaki S, Nishi M, Iino M, Kangawa K. 2000. Junctophilins: a novel family of junctional membrane complex proteins. *Mol Cell* **6:** 11–22. doi:10.1016/s1097-2765(00)00003-4

Theriot JA, Mitchison TJ. 1991. Actin microfilament dynamics in locomoting cells. *Nature* **352:** 126–131. doi:10.1038/352126a0

Tkachenko E, Sabouri-Ghomi M, Pertz O, Kim C, Gutierrez E, Machacek M, Groisman A, Danuser G, Ginsberg MH. 2011. Protein kinase A governs a RhoA-RhoGDI protrusion-retraction pacemaker in migrating cells. *Nat Cell Biol* **13:** 660–667. doi:10.1038/ncb2231

Totsukawa G, Yamakita Y, Yamashiro S, Hartshorne DJ, Sasaki Y, Matsumura F. 2000. Distinct roles of ROCK (Rho-kinase) and MLCK in spatial regulation of MLC phosphorylation for assembly of stress fibers and focal adhesions in 3T3 fibroblasts. *J Cell Biol* **150:** 797–806. doi:10.1083/jcb.150.4.797

Totsukawa G, Wu Y, Sasaki Y, Hartshorne DJ, Yamakita Y, Yamashiro S, Matsumura F. 2004. Distinct roles of MLCK and ROCK in the regulation of membrane protrusions and focal adhesion dynamics during cell migration of fibroblasts. *J Cell Biol* **164:** 427–439. doi:10.1083/jcb.200306172

Tsai FC, Meyer T. 2012. Ca^{2+} pulses control local cycles of lamellipodia retraction and adhesion along the front of migrating cells. *Curr Biol* **22:** 837–842. doi:10.1016/j.cub.2012.03.037

Tsai FC, Seki A, Yang HW, Hayer A, Carrasco S, Malmersjö S, Meyer T. 2014. A polarized Ca^{2+}, diacylglycerol and STIM1 signalling system regulates directed cell migration. *Nat Cell Biol* **16:** 133–144. doi:10.1038/ncb2906

Tsai TY, Collins SR, Chan CK, Hadjitheodorou A, Lam PY, Lou SS, Yang HW, Jorgensen J, Ellett F, Irimia D, et al. 2019. Efficient front-rear coupling in neutrophil chemotaxis by dynamic myosin II localization. *Dev Cell* **49:** 189–205.e6. doi:10.1016/j.devcel.2019.03.025

Tsuji T, Ishizaki T, Okamoto M, Higashida C, Kimura K, Furuyashiki T, Arakawa Y, Birge RB, Nakamoto T, Hirai H, et al. 2002. ROCK and mDia1 antagonize in Rho-dependent Rac activation in Swiss 3T3 fibroblasts. *J Cell Biol* **157:** 819–830. doi:10.1083/jcb.200112107

Vicente-Manzanares M, Ma X, Adelstein RS, Horwitz AR. 2009. Non-muscle myosin II takes centre stage in cell adhesion and migration. *Nat Rev Mol Cell Biol* **10:** 778–790. doi:10.1038/nrm2786

Vitorino P, Meyer T. 2008. Modular control of endothelial sheet migration. *Genes Dev* **22:** 3268–3281. doi:10.1101/gad.1725808

Voeltz GK, Sawyer EM, Hajnóczky G, Prinz WA. 2024. Making the connection: how membrane contact sites have changed our view of organelle biology. *Cell* **187:** 257–270. doi:10.1016/j.cell.2023.11.040

Watanabe N, Madaule P, Reid T, Ishizaki T, Watanabe G, Kakizuka A, Saito Y, Nakao K, Jockusch BM, Narumiya S. 1997. P140mdia, a mammalian homolog of *Drosophila* diaphanous, is a target protein for Rho small GTPase and is a ligand for profilin. *EMBO J* **16:** 3044–3056. doi:10.1093/emboj/16.11.3044

Watton SJ, Downward J. 1999. Akt/PKB localisation and 3′ phosphoinositide generation at sites of epithelial cell–matrix and cell–cell interaction. *Curr Biol* **9:** 433–436. doi:10.1016/S0960-9822(99)80192-4

Weiner OD, Neilsen PO, Prestwich GD, Kirschner MW, Cantley LC, Bourne HR. 2002. A PtdInsP(3)- and Rho GTPase-mediated positive feedback loop regulates neutrophil polarity. *Nat Cell Biol* **4:** 509–513. doi:10.1038/ncb811

Wilson CA, Tsuchida MA, Allen GM, Barnhart EL, Applegate KT, Yam PT, Ji L, Keren K, Danuser G, Theriot JA. 2010. Myosin II contributes to cell-scale actin network treadmilling through network disassembly. *Nature* **465:** 373–377. doi:10.1038/nature08994

Wittmann T, Waterman-Storer CM. 2001. Cell motility: can Rho GTPases and microtubules point the way? *J Cell Sci* **114:** 3795–3803. doi:10.1242/jcs.114.21.3795

Yam PT, Wilson CA, Ji L, Hebert B, Barnhart EL, Dye NA, Wiseman PW, Danuser G, Theriot JA. 2007. Actin-myosin network reorganization breaks symmetry at the cell rear to spontaneously initiate polarized cell motility. *J Cell Biol* **178:** 1207–1221. doi:10.1083/jcb.200706012

Yang X, Dormann D, Münsterberg AE, Weijer CJ. 2002. Cell movement patterns during gastrulation in the chick are controlled by positive and negative chemotaxis mediated by FGF4 and FGF8. *Dev Cell* **3:** 425–437. doi:10.1016/S1534-5807(02)00256-3

Yang HW, Collins SR, Meyer T. 2016. Locally excitable Cdc42 signals steer cells during chemotaxis. *Nat Cell Biol* **18:** 191–201. doi:10.1038/ncb3292

Leukocyte Chemotaxis and the Elusive Negative Signals that Make It Work

Kwabena A. Badu-Nkansah and Sean R. Collins

Department of Microbiology and Molecular Genetics, University of California, Davis, Davis, California 95616, USA

Correspondence: srcollins@ucdavis.edu

The recruitment of leukocytes to target locations in the body is central to both effective immune responses and pathological inflammation. Leukocytes use chemotaxis to find their way, following shallow gradients of attractants through complex tissue environments. To accomplish this feat, the cells adapt to external attractant concentrations, sense spatial differences in attractant levels across the cell, and establish a stable but steerable cell polarity to move persistently while navigating around obstacles. These dynamic behaviors are critical for effective chemotaxis, and they emerge from the molecular wiring of the underlying signaling and cytoskeletal components. Positive and negative feedback, as well as cross talk connections between cell front and rear components, play central roles. However, our understanding of these connections is still limited, especially regarding the inhibitory connections necessary for information processing and spatial organization. Here, we highlight recent progress and remaining challenges in uncovering the key components and their regulation that could explain dynamic features of chemotaxis and may suggest new therapeutic approaches.

The migration of cells following gradients of chemical cues, known as chemotaxis, is fundamental for organisms across the tree of life. It is used by single-celled organisms to find food, during development for tissue patterning, to establish the wiring of the nervous system, and to focus immune responses within the body (Swaney et al. 2010; SenGupta et al. 2021). Chemotaxing cells can often sense very shallow gradients of attractants (Zigmond 1977), raising basic questions about how molecular pathways process spatial information. Chemotaxis is also highly relevant for disease, notably through its role in the immune system. Leukocytes use chemotaxis to travel from the vasculature through tissues to sites of injury and infection. Their rapid and accurate responses are critical for effective immune responses. However, the responses must also be properly limited to prevent excessive activity, which can result in inflammatory and autoimmune diseases (Wright et al. 2010; Kolaczkowska and Kubes 2013).

The chemotaxis of leukocytes, and chemotaxis more broadly, is characterized by dynamic features that allow the cells to track subtle gradients while migrating over large distances. Leukocytes can respond to very low concentrations of attractants but still adapt to navigate effectively when concentrations are 100-fold higher (Zigmond 1977; Herzmark et al. 2007). They move rapidly and accurately, even when directional gradients are very shallow or fluctuating,

Cite this article as *Cold Spring Harb Perspect Biol* doi: 10.1101/cshperspect.a041792

and they can change directions to find paths in complex environments with many obstacles. Even as they follow complicated paths, leukocytes move persistently, typically turning and adjusting their direction rather than stopping and reorienting. Underlying these features, parts of the chemotaxis signaling network exhibit emergent behaviors, including adaptation, oscillatory and wave dynamics, and spatial pattern formation.

These emergent properties are essential for chemotaxis, but their molecular mechanisms are incompletely understood. Their roles in mediating cell polarity and directional sensing are broadly conserved from human leukocytes to *Dictyostelium* amoeba, although some molecular details differ between cell types and organisms (Swaney et al. 2010; SenGupta et al. 2021). Importantly, while it is clear that coordinated positive and negative regulation are required, there are still major gaps in our understanding of how the signaling network generates its emergent properties. A central theme of this review is that there are especially large gaps related to the identities and roles of the negative regulators, leaving important questions about how leukocytes achieve their dynamic and effective chemotaxis.

GPCRs DRIVE CHEMOTAXIS

Immune cells use cell surface receptors to sense and follow gradients of chemical attractants. The majority of chemoattractant receptors in immune cells are G-protein-coupled receptors (GPCRs) that activate the Gαi family of heterotrimeric G-proteins (Neptune and Bourne 1997; Sun and Ye 2012; Hughes and Nibbs 2018). After binding attractant agonists, these receptors activate Gαi G-protein complexes and trigger their dissociation into Gαi and Gβγ subunits. To steer migration, these receptors need to communicate with a robust downstream polarity circuit that acts as the cell's internal compass (Fig. 1). By defining and reinforcing the cell's front–rear axis, the polarity circuit controls the cell's direction and promotes persistent migration (Weiner 2002). Activation of the G-protein complex connects receptor activity to the polarity circuit through a cascade of signaling events. This cascade amplifies signals, and it can trigger polarization and movement in a previously inactive cell or guide the migration of a cell moving in an attractant gradient field. Receptor activity can reorganize the cytoskeleton, and it promotes partitioning of the plasma membrane into front and rear domains that compartmentalize molecular activities (Xu et al. 2003; Wang 2009; Kamp et al. 2016; Banerjee et al. 2023). Although there is strong evidence for the importance of Gαi itself in chemotaxis (Kamakura et al. 2013; Collins et al. 2015; Surve et al. 2016; To and Smrcka 2018), the molecular mechanisms linking it to directed cell migration are only partly understood. More direct and detailed mechanisms have been identified, demonstrating a central role for effectors of its complementary Gβγ subunits (Stephens et al. 1994; Neptune and Bourne 1997; Li et al. 2003). There is also evidence that these receptors can engage divergent signaling through G12/G13 G-proteins that induce the formation of cell-rear-associated actin networks to complement Gαi signaling and promote migration (Xu et al. 2003).

In leukocytes, the association between Gαi-coupled GPCRs and chemotaxis is strong. Almost all chemokine receptors are Gαi-coupled GPCRs (Hughes and Nibbs 2018). Furthermore, exogenously expressed Gαi-coupled GPCRs are also able to drive directed migration in these cells (Karunarathne et al. 2013; Park et al. 2014; Bell et al. 2021). While it is not clear whether the exogenously expressed receptors are as effective as native chemoattractant receptors, this association suggests that Gαi G-proteins are central to directed migration in these cells. Even so, there are differences between chemokine GPCRs, and these differences fine-tune migration responses to help shape the complex trafficking of leukocytes in vivo (Campbell et al. 1997; Heit et al. 2002; Petrie Aronin et al. 2017; Lundgren et al. 2023).

Beyond leukocytes, chemokine GPCRs are used to guide migration in nonimmunological processes during development, and they are often co-opted in cancer metastasis (Bonecchi et al. 2009; Hughes and Nibbs 2018). Thus, the association between Gαi-coupled receptors and

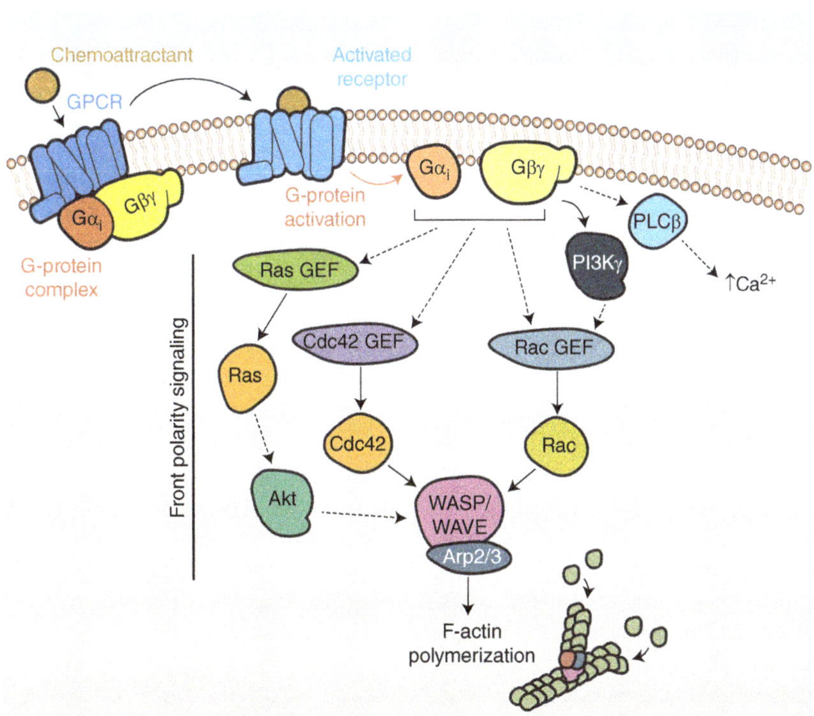

Figure 1. G-protein-coupled receptor (GPCR) signaling cascade during chemotaxis. During immune cell chemotaxis, GPCRs, activated by chemoattractant agonists, initiate a cascade of signaling events that induces spatiotemporally controlled actin polymerization. Bound receptors activate heterotrimeric G-protein complexes that contain Gαi and Gβγ subunits. Activated G-proteins bind and activate effectors that produce lipid and calcium (Ca^{2+}) signals through phosphoinositide 3-kinase γ (PI3Kγ) and phospholipase Cβ (PLCβ) and recruit guanine exchange factors (GEFs) to the plasma membrane where they activate Rac, Cdc42, and Ras guanosine triphosphatases (GTPases). Upon activation, Rac and Cdc42 recruit WASP-family verprolin-homologous protein (WAVE) and Wiskott–Aldrich syndrome protein (WASP) complexes that activate the Arp2/3 complex to initiate actin branching and polymerization. Ras activation has also been linked to F-actin polymerization through the downstream kinase Akt, although how these events engage actin nucleation complexes has not been fully elucidated. Solid arrows indicate direct interactions, while dashed arrows indicate multistep processes.

directed cell migration extends across cell types, including roles in many tissue morphogenetic processes.

POLARITY AND DIRECTIONAL SENSING

The signaling circuit controlling chemotaxis accomplishes two different but interconnected feats: It establishes cell polarity to allow the cell to move persistently, and it senses the direction of chemical gradients to steer migration. Polarity involves defining a cell front and rear by segregating molecular activities within the cell, and it is necessary for the cell to move. Directional sensing requires comparing receptor inputs across different parts of the cell. Information from sensing is then used to adjust the axis of polarity toward the direction of the strongest input signal.

In leukocyte polarity, the cell front is characterized by Rac, Cdc42, and Ras guanosine triphosphatase (GTPase) activities, production of phosphatidylinositol (3,4,5)-trisphosphate (PIP_3) lipids, and dynamic assembly of branched actin networks (Fig. 1). The rear is characterized by RhoA GTPase activity and contractile actomyosin assemblies (Xu et al. 2003; Wang 2009). To establish robust signaling domains, the front and rear subcircuits are each self-reinforcing through positive feedback loops. Separation of the two domains is thought to arise through mutual inhibition. Despite this simple conceptual framework, the molecular network mediating the positive and negative connections underlying polarity is complex and only partly understood.

Multiple positive feedback connections contribute to polarity, although molecular details are often missing. A positive feedback loop involving Rac, PIP_3, and polymerized actin helps reinforce the front (Wang et al. 2002; Weiner et al. 2002), although PIP_3 is not required for migration and chemotaxis (Afonso and Parent 2011). There is also evidence for actin-independent positive feedback involving Cdc42 (Li et al. 2003; Yang et al. 2016). Positive feedback involving RhoA and myosin helps reinforce the rear (Wong et al. 2007). Finally, both branched actin assembly and myosin-based contraction can be self-amplifying processes, as they generate actin structures compatible with further branching or myosin assembly. Actin assembled at the front flows rearward in the cell, accumulating myosin, which provides a positive link from the front to the rear (Fig. 2A; Tsai et al. 2019).

Similarly, mechanisms mediating antagonism between the front and rear domains are partially understood (Fig. 2A). Branched actin and contractile actomyosin networks are intrinsically antagonistic (Lomakin et al. 2015), and this is supported by multiple layers of cross talk between signaling components. Rac and RhoA have mutually antagonistic behavior in many contexts (Guilluy et al. 2011), and Cdc42 can also interact antagonistically with RhoA (Yang et al. 2016; Moe et al. 2021). Even so, the regulators mediating this cross talk in leukocytes are still largely unknown. Finally, mechanical interactions contribute to polarity through membrane and cortical tension that limit cell front activity to support two separate signaling domains (Houk et al. 2012; De Belly et al. 2023).

Directional sensing can be separated from polarity and migration. *Dictyostelium* amoeba are single-celled organisms with chemotaxis behaviors similar to those of leukocytes and underlying mechanisms that are evolutionarily related (Devreotes 2019). When these cells are treated with latrunculins to depolymerize actin, they are unable to move or morphologically polarize, but they can still orient PIP_3 signaling responses in the direction of an attractant source (Parent et al. 1998). Even so, during chemotaxis, directional sensing is tightly integrated with polarity and motility. Indeed, genetic screens for chemotaxis have identified very few components with specific roles in directional sensing that can be distinguished from roles in polarity and motility (Collins et al. 2015; Devreotes 2019; Belliveau et al. 2023). In particular, chemoattractant receptors and G-proteins are critical for directional sensing of attractant gradients, but they are not required for polarity or motility. Almost all other components identified in screens that affect chemotaxis are needed for efficient cell migration, regardless of directionality.

GEFs AND GAPs CONTRIBUTE TO COMPLEX SIGNAL WIRING

Direct regulators of the Rho and Ras family GTPases are likely mediators of cross talk and feedback in chemotaxis, and they can act as links between directional sensing and the polarity signaling network. These GTPases are activated by a family of guanine nucleotide exchange factors and inactivated by GTPase-activating proteins (GAPs). GEFs and GAPs are typically large multidomain proteins, each of which can have its own molecular regulation (Hodge and Ridley 2016). Adding to the complexity of this system, the expressions of GEFs and GAPs vary widely across cell types, which can lead to different wiring of the signaling network.

Multiple GEFs and GAPs contribute to the regulation of individual GTPases in leukocyte chemotaxis, likely playing specialized roles to help coordinate cellular behavior. DOCK2, PREX1, and VAV1 all contribute to Rac activation (Welch et al. 2002; Nishikimi et al. 2009; Phillipson et al. 2009). Similarly, multiple GEFs for RhoA, including PDZ-RhoGEF, GEF-H1, and p115-RhoGEF, contribute to migration and chemotaxis (Francis et al. 2006; Wong et al. 2007; Collins et al. 2015; Kopf et al. 2020; Joo and Olson 2021). While GEFs could be spe-

 Cite this article as *Cold Spring Harb Perspect Biol* doi: 10.1101/cshperspect.a041792

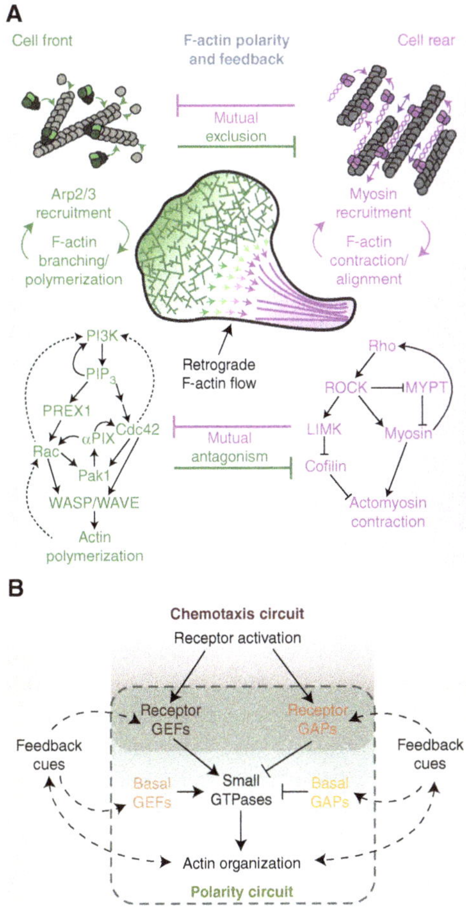

Figure 2. Feedback signals coordinate polarity signaling with actin networks. (*A*) Feedback signaling includes passive and active mechanisms. (*Top*) In polarized leukocytes, branched and contractile actin networks passively provide feedback and cross talk by producing actin structures suited for expansion of their own network type. At the cell front, the Arp2/3 complex nucleates new filaments that branch from existing filaments, which become substrates for further branching. During migration, actin assembled at the cell front moves retrograde in the cell frame of reference. As it approaches the cell rear, it accumulates nonmuscle myosin motors that break down branched structures into parallel bundles. Actomyosin contraction increases local actin density and alignment, thereby promoting further myosin recruitment and generating forces that propel the cell rear. (*Bottom*) Concurrently, active feedback mechanisms occur inside polarized compartments at the cell front and rear to complement actin network polarization. At the cell front, positive feedback drives polarized phosphoinositide 3-kinase (PI3K), Rac, and Cdc42 activation. p21-activated kinase (Pak1) is a central regulator of positive feedback to Rac and Cdc42 through the α-Pak1-interacting exchange factor (αPIX). At the cell rear, myosin filament organization may promote activation of RhoA through coordination with Rho guanine exchange factors (RhoGEFs). Downstream effectors of RhoA, including the kinases Rho-associated kinase (ROCK) and LIM kinase (LIMK), promote contraction by opposing myosin phosphatase (MYPT) and the actin filament severing protein cofilin. (*B*) For chemotaxis signaling to drive directional migration, receptor activity must seamlessly coordinate with polarity signaling. Receptor inputs locally integrate with self-organized polarity circuits through compartmentalized feedback cues. Spatiotemporal control of GEFs and GTPase-activating proteins (GAPs) for Rho family guanosine triphosphatases (GTPases) could be essential for this integration. Rho GEFs that drive polarity signaling could be categorized into two partially overlapping functional groups: (1) a "basal" group that is essential for self-organized polarity signaling and random migration independent of receptor signaling, and (2) a "receptor" group that is activated by receptor signals that steer cell polarity during chemotaxis. Although less well characterized, Rho GAPs may also fall into similar functional categories. Spatiotemporal organization of GEFs and GAPs by cytoskeletal cues could be an essential component of feedback signaling and provide secondary control of Rho GTPase activity. Investigating the feedback cues that organize GEFs and GAPs may offer important insights into how the chemotaxis and polarity circuits are compartmentalized.

cialized for transducing signals from receptors, mediating feedback connections, or maintaining basal regulation, there is evidence that the roles of individual GEFs span multiple of these categories (Fig. 2B). PREX1 can be activated downstream from GPCRs through Gβγ and PIP_3 (Welch et al. 2002), but it also participates in PIP_3-mediated feedback (Graziano et al. 2017). Similarly, α-Pak1-interacting exchange factor (α-PIX) transduces receptor inputs through Gβγ to Cdc42, but it also scaffolds positive feedback between Cdc42 and its effector Pak1 (Fig. 2A; Li et al. 2003). At the cell rear, PDZ-RhoGEF responds to Gα13 downstream from receptors, and it participates in positive feedback involving myosin II and RhoA (Wong et al. 2007). Indeed, PDZ-RhoGEF,

GEF-H1, and p115-RhoGEF all have motifs to bind active RhoA, suggesting roles in positive feedback (Medina et al. 2013).

While the molecular wiring of positive feedback in chemotaxis is only partially understood, our understanding of negative feedback and cross talk has larger holes. GAP proteins are likely to play key roles, but few have been directly implicated. Arhgap15 mediates negative feedback from PIP_3 to Rac, which contributes to signal processing, but it is not required for Rac signal attenuation after chemoattractant stimulation (Costa et al. 2007; Graziano et al. 2017). While other GAPs have been implicated in leukocyte chemotaxis, their roles in feedback or cross talk remain largely unclear (Gambardella et al. 2011; Csépányi-Kömi et al. 2016; Belliveau et al. 2023).

Determining which GEFs and GAPs mediate feedback and cross talk connections and how they mediate them could have a major impact on our understanding of signal processing and spatial pattern formation in chemotaxis. The gaps in our knowledge and potential for progress are especially large for the negative regulators that limit and segregate GTPase signals.

MECHANISMS OF DIRECTIONAL SENSING

A long-standing question in chemotaxis research is centered on directional sensing: How can a cell process information from receptors to detect a small difference in concentration across its surface? Early research by Sally Zigmond (1977) found that neutrophils are capable of sensing and following chemical gradients that differ by as little as 1% across the length of the cell. This suggests that the chemotaxis signaling system must use information very efficiently for directional sensing.

Theoretical models provide explanations for how such impressive directional sensing could be achieved. The most prominent model is the local excitation, global inhibition (LEGI) model. In this model, receptors send two signals, one that is positive and local, and another that is negative and spreads rapidly across the cell (Fig. 3; Iglesias and Devreotes 2008). The positive signal provides a measure of local receptor activity, while the negative signal reports the cell-wide average. By molecularly comparing or subtracting these two signals, the region of the cell closest to the gradient source could be determined. The positive signal promotes protrusion and cell front activities, which leads to movement toward the gradient source. The LEGI model also explains other empirically observed features of chemotaxis signaling. In response to a global increase in chemoattractant, cells typically respond with a pulse of cell front signaling activities, which then gradually attenuate, even as the stimulus persists. In the model, the positive signal is faster than the negative signal, which leads to a rapid response that attenuates as the negative signal catches up (Fig. 3). This same feature results in adaptation, which allows the system to function across a wide range of stimulus levels.

Despite its ability to explain spatial sensing, direct experimental support for the LEGI model is limited. There are extensive data supporting positive, protrusion- and front-promoting signals downstream from receptor activation. However, there is much less evidence for a negative signal generated by receptors. Experiments in *Dictyostelium* cells treated with latrunculin to block polarization and movement showed that PIP_3 lipids are locally enriched on the side of the cell closest to a gradient source (Parent et al. 1998). The fact that PIP_3 accumulation is limited to this side of the cell suggests that an opposing negative signal prevents its accumulation on the opposite side of the cell. This negative signal could be generated by receptors through a LEGI mechanism, but this result is also consistent with a downstream negative feedback mechanism.

A result in *Dictyostelium* cells lacking the Gβ G-protein subunit is perhaps the most compelling evidence for a negative signal generated by receptors in chemotaxis. These cells were stimulated with attractant, and actin assembly was measured. In these cells, the frequency of cortical actin patches decreased in response to stimulation, rather than increasing as they would in wild-type cells (Tang et al. 2014). This result suggests that there may be a G-protein-indepen-

 Cite this article as *Cold Spring Harb Perspect Biol* doi: 10.1101/cshperspect.a041792

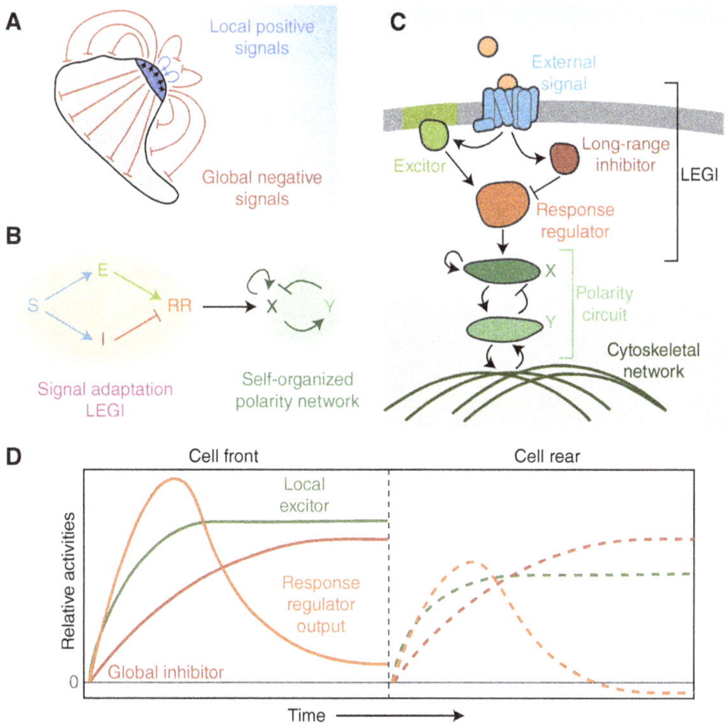

Figure 3. Local excitation global inhibition (LEGI) models provide a conceptual explanation for directional sensing. (*A*) LEGI models propose that directional sensing is achieved through integration of two signals generated by activated receptors, a local excitatory signal and an inhibitory signal that spreads rapidly across the cell. The excitatory signal promotes protrusion to direct migration, while the inhibitory signal provides a measure of the cell-wide average receptor input and suppresses protrusion in parts of the cell with lower inputs. (*B*) A LEGI circuit adapts to global changes in chemoattractant levels. The external input (S) drives both a rapid excitatory signal (E) and a slower, diffusive inhibitory signal (I) that collectively control a response regulator (RR). Signal integration by the RR provides adaptation upstream of signal transmission to the polarity circuit. The polarity circuit is represented by general front (X) and rear (Y) polarity domains that self-organize and can be triggered by noise fluctuations. (*C*) LEGI model components in a putative signaling schematic. Assigning LEGI model components to specific signaling regulators and events is still an ongoing effort. However, putative features of each component have been ascribed. External inputs are associated with receptor signals. Upstream excitatory and inhibitor signals are associated with slow and fast diffusing regulators, respectively. RRs directly integrate into the polarity circuit. (*D*) A simulated plot shows the activity of each LEGI component at the cell front (*left*) and rear (*right*) as cells adapt and sense an attractant gradient. Immediately after a chemoattractant gradient is applied, receptors activate excitatory responses at rates faster than inhibitors. At the side of the cell closest to the gradient source (cell front), the local excitation signals exceed the global inhibitory signals. In contrast, at the opposite side (cell rear), the inhibitory signals eventually dominate. As the system reaches a steady state, these differences result in net RR activity that is higher on the side of the cell with the highest external signal, leading to cell orientation and migration toward the gradient source.

dent negative signal that could serve as a global inhibitor. Even so, the mediators of this signal are not known. It is also unclear whether the same signal acts in leukocytes or whether it is strong enough to balance the positive signals in chemotaxis.

There is evidence that chemotactic GPCRs may generate negative or rear-promoting signals in human leukocytes as well, but the spatial scale of these signals is not clear, and they may be more local than global. Leukocyte cells treated with pertussis toxin, which inactivates the Gαi family of G-proteins, respond to a point source of attractant by assembling and orienting a cell rear toward the attractant source (Xu et al. 2003). This response is mediated by G12 and/or G13 G-proteins, which activate RhoA. Notably, however, the response is local, rather than global. It is also unclear whether G12/13 activity is needed for directional sensing.

There is also growing evidence that cells may use mechanisms other than LEGI to orient their migration during chemotaxis and that different receptors and/or cell types may use different mechanisms. For example, local inhibition can contribute to cell steering, leukocytes can rely on temporal sensing, and fibroblasts may use a global excitation-local inhibition (GELI) mechanism to follow some attractants, suggesting that directional sensing strategies may be context-specific (Petrie Aronin et al. 2017; Asokan et al. 2019; Town and Weiner 2023).

A LACK OF A GLOBAL INHIBITOR?

Recent work sought to test models of directional sensing by measuring spatial signal processing directly. Critical for this approach, light-controllable GPCRs that signal through Gαi drive-directed cell migration in leukocytes when these receptors are expressed exogenously (Karunarathne et al. 2013). In particular, parapinopsin receptors can be turned on and off with different wavelengths of light, providing a tool for highly precise control of receptor inputs. By combining this GPCR with a biosensor to measure downstream signaling, responses to subcellular receptor activation can be measured in single cells. In such experiments, a global negative signal would be expected to cause a negative response at locations distal to the stimulus. However, while these experiments revealed rapid local positive responses, there was no apparent negative response at a distance (Bell et al. 2021).

Furthermore, the same approach was used in cells migrating in branched microfluidic channels that cause their front to split into two (Hadjitheodorou et al. 2023). These fronts compete, and ultimately one wins to allow the cell to continue migrating. In this system, stimulating the light-sensitive receptor in one of the two fronts can bias the cell's decision, but there is no measurable effect on the opposite front until physical stretching mediates long-range communication.

These results indicate that Gαi G-proteins likely do not generate a long-range negative signal to balance their local protrusion-promoting effects. However, these experiments do not rule out possible contributions of G12/13 G-proteins to directional sensing, if their activity is less localized than the Gαi outputs, or a possible role for receptor-specific G-protein-independent negative signals. Finally, a LEGI-type signaling mechanism can process spatial information even if the inhibition is not global, so long as it spreads faster than the excitatory signal (Swaney et al. 2010). It is possible that local signal processing promotes directional sensing through unequal spread of the negative and positive signals.

POLARIZED SENSITIVITY

In actively migrating cells, inputs across the cell surface are not treated equally. Instead, these cells often exhibit "polarized sensitivity" in which responses to inputs near the front of the cell are stronger than responses to inputs of equal strength at the cell rear. When these cells encounter a change in gradient direction, they typically maintain their existing front and turn toward the new signal, rather than repolarizing to reorient directly (Gerisch and Keller 1981; Xu et al. 2003). This behavior allows the cells to keep moving, even as they struggle to track weak or fluctuating gradients, which may be important for migrating long distances efficiently to reach target locations.

The mechanisms underlying polarized sensitivity are incompletely understood, but the actin cytoskeleton appears to play an important role (Devreotes and Janetopoulos 2003; Wang et al. 2014; Hadjitheodorou et al. 2021). Treating cells with latrunculins that depolymerize actin blocks the ability of the cells to polarize or move, and it restores uniform sensitivity across the cell surface (Devreotes and Janetopoulos 2003). Thus, unequal sensitivity is strongly linked to polarity and polymerized actin. Supporting this idea, treating cells with a cocktail of actin-modulating drugs designed to freeze, but not depolymerize, the actin cytoskeleton results in extended periods of desensitization after cell stimulation (Dandekar et al. 2013). This result suggests that the fast actin remodeling at the front may reset sensitivity locally. Even so, a related study found that in such "frozen" cells, spatial differences in actin architecture still maintain polarized sensitivity (Wang et al. 2014).

Further experiments indicate that myosin acts to limit sensitivity at the cell rear. Leukocytes expressing a light-controllable receptor (described above) were stimulated specifically at the rear while migrating in narrow microfluidic channels that prevent them from turning. While some of the cells reverse their direction of migration, in others, polarity is too strong, and the cell continues forward. This insensitivity at the rear is strongly linked to the degree of rearward myosin accumulation. Cells with more myosin concentrated at the rear are less responsive, sometimes without even any apparent signaling response to the stimulus (Hadjitheodorou et al. 2021).

While it is not clear how sensitivity is modulated, the architecture of the actin cytoskeleton can have a large effect on the activation of key signaling nodes, both in response to receptor activity and independently. The connection between actin architecture and signaling activity extends across cell types and organisms. In epithelial cells, high levels of cortical actin, presumably including parallel microfilaments tethered to the membrane, limit the ability of a recruited Rac GEF to trigger protrusions (Bisaria et al. 2020), and in *Dicytostelium*, cortical actin antagonizes Ras activation (Kuhn et al. 2025). These observations suggest that actomyosin under the membrane may contribute to polarized sensitivity by limiting the activation of cell front signaling components. In contrast, positive feedback involving branched actin assembly can amplify cell front signaling activity (Wang et al. 2002; Weiner et al. 2002). Thus, polarized sensitivity may involve both tuning of receptor sensitivity and compartmentalization of downstream feedback by the actin cytoskeleton.

POLARIZED SENSITIVITY AS A TOOL FOR DIRECTIONAL SENSING AND CHEMOTAXIS

How does polarized sensitivity affect directional sensing? At first glance, it might appear to make directional sensing more difficult. Rather than integrating inputs from receptors across the entire cell surface, polarized sensitivity focuses the cell's measurement on a smaller region across its front. This reduces the effective number of receptors involved, and it reduces the distance across which a gradient must be measured. Thus, a smaller difference in attractant level must be measured with fewer sensors.

However, polarized sensitivity could also aid directional sensing in a different way—by serving as a mechanism for temporal averaging. Conceptually, polarity may act like a slow-moving compass needle responding to small pushes and pulls from a noisy environment. By maintaining polarity and making small sequential changes in direction, the cell integrates new information from receptors with its recent history (Fig. 4). At any moment in time, the direction of the cell's polarity axis is determined by the cumulative effects of recent small turns. Even if individual turns are noisy, this integration over time could allow the cell to bias its orientation and movement strongly toward the gradient source. Computational modeling of noisy but biased inputs that promote small turns in polarized cells supports the idea that this could be an accurate sensing strategy (Arrieumerlou and Meyer 2005). Furthermore, a study of in vivo imaging of neutrophils in zebrafish found that polymerized actin was necessary for the cells to orient signaling responses toward a gradient source, suggesting that actin, polarization, and movement may be critical for leukocytes to sense gradients in vivo (Georgantzoglou et al. 2022).

A LOCAL INHIBITOR PROMOTES TURNING

For sensing to work well in persistently polarized cells, the integration of existing polarity and receptor inputs must be balanced. Strong polarity could make it difficult for cells to track gra-

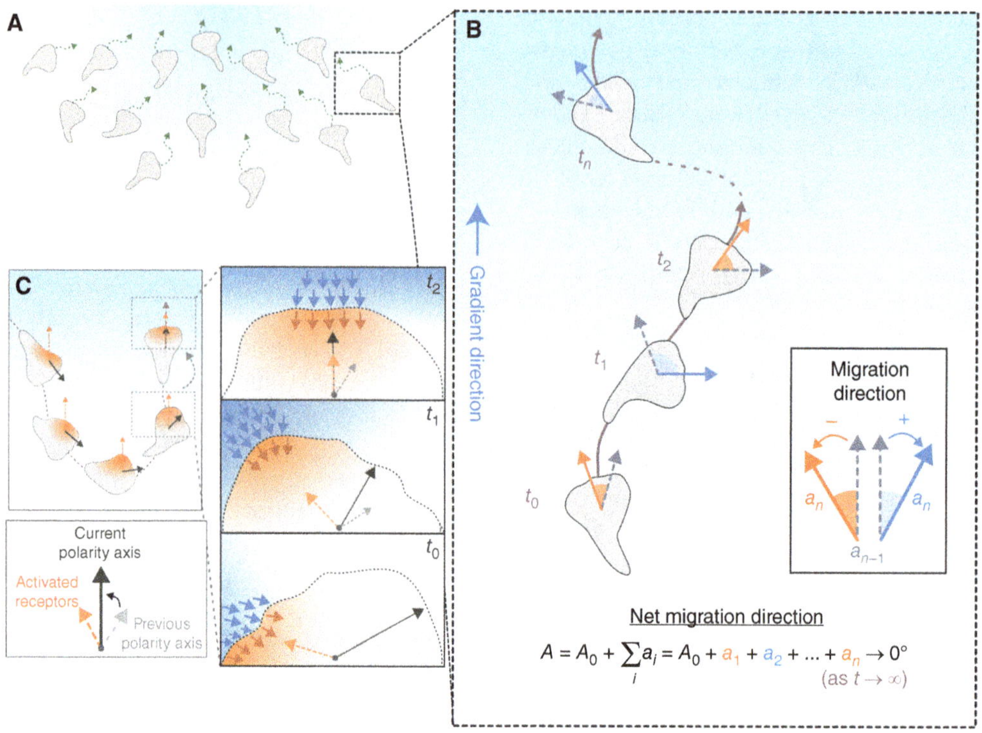

Figure 4. Stable polarity promotes temporal averaging to tune directionality as cells navigate in shallow chemoattractant gradients. (*A*) As cells navigate up chemoattractant gradients (blue), they often migrate in winding paths (green dotted arrows). (*B*) Cells tune their direction while maintaining polarity as they wander through chemoattractant fields. At short timescales, individual small turns (blue and orange angles) are noisy, and cell migration direction varies significantly. However, over longer periods of time, the accumulation of multiple small turns biased by receptor inputs allows cells to track the gradient direction. (*C*) (*Left*) As cells turn within chemoattractant gradient fields, receptors continuously receive directional inputs, which are integrated with the cell's polarity axis. Receptor inputs (orange) activate lamellipodial signaling machinery to drive protrusions, and this response is biased toward the cell front by the polarity circuit. (*Right*) New protrusions at the edge of the cell front cause the cell to turn toward receptor signals (orange dotted arrow), adjusting the polarity compass (from the gray dotted arrow to the black arrow).

dients if the polarity circuit overpowers inputs from receptors. It could also make it difficult for cells to navigate around obstacles that block their paths. Thus, polarity in chemotaxis needs to be robust, but still dynamic and "steerable." Cell polarity in leukocyte migration fits this description. These cells maintain a stable front, but their fronts wobble and frequently change direction during both random migration and chemotaxis in attractant gradients. Such a dynamic, steerable cell front could be achieved through local negative feedback. This feedback would act to destabilize an old cell front, promoting turning of the front and thus the cell.

Recent work identified a local, negative feedback pathway that mediates this effect (Town and Weiner 2023). The feedback was identified by manipulating PIP_3 signals at the cell front using optogenetics. Optogenetic tools based on protein recruitment to the plasma membrane can be used to activate Rho GTPase or PIP_3 lipid signaling in subcellular regions on very rapid timescales. Local activation of several different front-associated signals, including PIP_3, is sufficient to make cells turn in the direction of the stimulus (Yoo et al. 2010; O'Neill et al. 2016; Graziano et al. 2017; Pal et al. 2023). This strategy was used to test models of signal processing underlying cell steering (Fig. 5). First, a local stimulus on one side of the cell front induces a turn in that direction. This is then followed by a uniform stimulus across the entire cell. If the stimulation history had no impact, the cell would be expected to continue in a roughly straight path upon the global stimulation. If the prior inputs provided signaling "momentum" that built up during the local stimulation, cells would be expected to continue turning in the same direction. Instead, cells responded to the global stimulation by turning in the opposite direction (Fig. 5B,C; Town and Weiner 2023).

This turning result, combined with analysis of signaling reporters, revealed that an inhibitor of Rac accumulates locally through negative feedback. This limits responsiveness at sites of recent inputs and promotes turning of the cell front toward "newer" inputs. The negative feedback appears to act at the level of the Rac GTPase, as there was no apparent local negative feedback at the level of PIP_3. It is not yet clear whether the feedback acts directly on Rac or whether it acts on other cell front signaling activities as well.

A similar local negative signal contributes to chemotaxis in *Dictyostelium* amoeba (Xu et al. 2007). In this case, negative feedback appears to act at the level of PIP_3. When cells with depolymerized actin are exposed to a gradient of chemoattractant, they respond with a gradient of PIP_3 signaling oriented toward the gradient source. However, in response to a second spatially uniform attractant stimulus, the PIP_3 response is stronger in the direction opposite to that of the first stimulus. While these results could be explained by more than one model, they suggest that local negative feedback may

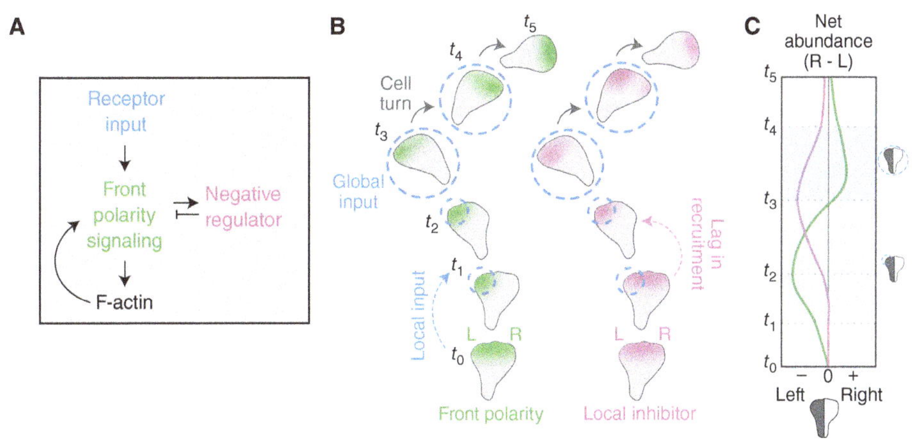

Figure 5. Local negative feedback promotes turning and helps cells interpret gradients. (*A*) A recent study (Town and Weiner 2023) identified a steering mechanism in which negative feedback activates a local inhibitor to promote turning toward newer inputs. The inhibitor is activated by front polarity signaling with a delay, and it provides spatiotemporal control of front activity, destabilizing older regions of the cell front and effectively increasing sensitivity of adjacent regions to promote turning. (*B*) In their experimental approach, local optogenetic activation of the front polarity signal PIP_3 was used to induce leftward turning of neutrophil-like cells (t_1 and t_2). Shortly afterward (t_3 and t_4), a global optogenetic input was applied, which induced cells to turn away from the previously stimulated side. These results imply the presence of inhibitory signals (pink) that accumulate locally in response to front polarity signaling (green). This negative feedback drives local decreases in sensitivity to receptor inputs, which helps determine the direction of new protrusions and subsequent shifts in cell polarity. (*C*) Example time course of front polarity (green) and local inhibitor levels (pink) during optogenetic cell steering and global activation. Here, migrating cells are steered by local stimulation first at a lateral edge followed by global activation. After global activation, lagging inhibitors at the stimulated *left* edge suppress increases of front polarity signals, while the *right* edge increases remain relatively unrestrained. (Figure based on data in Town and Weiner 2023.)

be a conserved feature of eukaryotic chemotaxis signaling. Such negative feedback may play a critical role in allowing polarity of these cells to be "steerable" while still being persistent.

EXCITABLE BEHAVIOR, LOCAL FEEDBACK, AND DYNAMIC POLARITY

A related conserved feature of chemotaxis is that front signaling components exhibit "excitable" behavior, including local pulses, traveling waves, and oscillations of activity (Nishikawa et al. 2014; Tang et al. 2014; Yang et al. 2016). This behavior is more obvious when actin is depolymerized, and it reflects a combination of strong positive and negative feedback regulation. The positive feedback causes rapid, local accumulation of signaling activity, and slower negative feedback drives attenuation, which can lead to wave movement or oscillation. This molecular wiring provides signal amplification, which allows cells to respond to very small numbers of activated receptors, and it drives spatial patterning of signaling to support polarity.

How do the positive and negative feedback connections underlying this excitable behavior relate to the connections described above underlying polarity, directional sensing, and "steerable" migration? It seems likely that each of the following is true: First, there are multiple positive and negative feedback connections that contribute to the dynamic signaling underlying chemotaxis and migration, and, second, the same feedback connections contribute to more than one of the emergent properties of chemotaxis signaling. Lastly, actin cytoskeletal networks have fundamental (but not fully understood) roles in spatially controlling many of these feedback connections. Positive feedback at the cell front involves signaling loops that depend on branched actin, as well as actin-independent loops. Although not yet elucidated, it seems likely that there are large overlaps between the molecular connections responsible for the actin-independent contributions to polarity and those that drive the excitable signaling behavior.

On the negative side, the signaling waves observed in the absence of actin must involve negative feedback that causes the waves to move. This feedback could be the same as the local negative feedback that destabilizes the front to promote cell turning and "steerability." Both promote the movement of signaling activity away from areas that have recently been elevated. Determining the molecular mediators should clarify this connection and shed light on how cell front activity is coordinated as cells make directional decisions.

TEMPORAL SENSING AND SPEED MODULATION

Effective chemotaxis can also be achieved without spatial comparisons of inputs. Indeed, bacteria chemotax using a temporal sensing mechanism (Sourjik and Wingreen 2012). Rather than sensing differences in attractant concentration across space, these cells measure whether attractant levels are increasing or decreasing. If levels are increasing, the cell continues moving ahead. If levels are decreasing, the cell stops, tumbles to randomize its direction, and starts moving again. By biasing "runs" of forward movement to be longer when cells experience increasing attractant levels, this mechanism biases migration toward the gradient source and results in efficient chemotaxis.

There is substantial evidence that leukocytes sense temporal changes in attractant concentrations. They respond to uniform increases in attractant concentration with signaling activation, actin polymerization, strengthened polarity, and increased migration speed (Wang 2009; Aranyosi et al. 2015). These cells also slow and stop in response to decreases in input signals (Ishida et al. 2025). The reduction in migration is associated with decreases in cell front signaling activities that can dip below basal or resting levels (Bell et al. 2021; Town and Weiner 2023; Algorta et al. 2025). This temporal sensing promotes migration when attractant signals appear, supports or accelerates migration when signals are increasing, and may limit wayward migration. In this way, it could support or synergize with spatial sensing for efficient chemotaxis.

While it has been appreciated that temporal sensing contributes to leukocyte chemotactic responses, recent evidence suggests that it may be

 Cite this article as *Cold Spring Harb Perspect Biol* doi: 10.1101/cshperspect.a041792

the primary sensing mechanism used in some contexts. Leukocytes can detect and follow gradients of multiple attractants using different attractant receptors, and they prioritize some attractants over others (Foxman et al. 1997; Heit et al. 2002). In particular, damage- or pathogen-associated signals including formyl peptides and the complement fragment C5a are prioritized over endogenously secreted inflammatory mediators such as leukotriene B4 (LTB4) and interleukin-8 (Heit et al. 2002). The signaling responses to high versus low priority attractant receptors are not the same (Lundgren et al. 2023), and the mechanisms they use to sense and follow gradients of the corresponding attractants appear to be different. A study using microfluidic devices to carefully control both spatial and temporal changes in attractant concentrations found that mouse dendritic cells and neutrophils effectively sense spatial gradients of high priority attractants during chemotaxis. However, their response to low priority attractants was quite different. Although they could initially orient using spatial cues, they were reliant on temporal sensing to continue migrating up these gradients (Petrie Aronin et al. 2017). Similarly, neutrophils migrating to wound sites in live zebrafish use an initial phase of spatial sensing to orient, but they rely heavily on temporal sensing to move persistently toward the gradient source (Georgantzoglou et al. 2022). This strategy may be advantageous for tracking shallow, long-range gradients.

How does temporal sensing contribute to leukocyte chemotaxis? It may contribute in multiple ways. For neutrophils moving toward a wound in zebrafish, cell speed is correlated with direction (Georgantzoglou et al. 2022). This suggests that temporal sensing modulates cell speed to promote chemotaxis. Temporal sensing may also promote accurate turning through a pilot pseudopod model (Gerisch et al. 1975; Town and Weiner 2023). In this case, negative feedback dampens signaling activities and opposes protrusions that experience decreasing input signals, whereas protrusions toward the gradient source would experience increasing input levels, which could outpace the negative feedback to promote continued protrusion.

Molecularly, negative feedback supporting temporal sensing may operate at multiple levels in the signaling cascade. In some cases, feedback appears to act at the level of receptors, desensitizing them to limit further responses. The study of Petrie Aronin et al. (2017) identified the GPCR kinase GRK3 as a mediator of adaptation for temporal sensing in mouse leukocytes responding to low priority attractants. In this case, GRK3 mediates receptor desensitization, which promotes temporal sensing while at the same time limiting spatial sensing. This likely represents a receptor-level negative feedback that is specifically important for this class of attractant receptors. Consistent with this observation, human leukocyte signaling responses attenuate much faster in response to low priority attractants than in response to high priority attractants (Lundgren et al. 2023). Although it is not clear which kinases control the response in human leukocytes, the rapid attenuation in response to the attractant LTB4 depends on two receptor phosphorylation sites. In each case, it is likely that receptor activation leads to activation of receptor kinases that desensitize the active receptors.

In contrast to this receptor-level regulation, other negative feedback mechanisms act farther downstream in the pathway. Negative feedback acts on Rac downstream from PIP_3 and actin (Costa et al. 2007; Graziano et al. 2017; Town and Weiner 2023). Negative feedback also tunes Cdc42 responses through both actin-dependent and actin-independent mechanisms (Bell et al. 2021). The leukocyte chemotaxis signaling network thus likely includes multiple negative feedback mechanisms acting in different contexts and on different components to promote both spatial and temporal signal processing.

In addition to negative feedback, temporal sensing could be achieved through an incoherent feedforward mechanism. Such a mechanism is central to the LEGI model, where receptor activation leads to parallel positive and negative outputs. The faster positive signal generates transient responses, which are attenuated by the slower-acting negative signal. A study combining kinetic measurements of signaling responses and mathematical modeling in *Dic-*

tyostelium found that the adaptation of Ras signaling has signatures of incoherent feedforward rather than negative feedback (Takeda et al. 2012). Thus, temporal sensing and adaptation may involve a combination of both negative feedback and incoherent feedforward mechanisms, which may depend on cell types and attractants.

IDENTIFYING THE NEGATIVE SIGNALS THAT GUIDE CHEMOTAXIS

Negative signals play critical roles in chemotaxis. They are necessary for cross talk in cell polarity to separate cell front and rear domains, for mediating the adaptation that underlies temporal sensing, and for balancing receptor inputs and excitable dynamics with existing polarity to mediate cell steering. They are also critical pieces of models for spatial sensing. Despite this great importance, many of the molecular identities and regulatory control mechanisms of negative signals in leukocyte chemotaxis are unknown.

During migration, negative signals including feedback and cross talk connections play central roles in separating and limiting the cell front and rear domains (Wang 2009). Importantly, each of these domains is reinforced by positive feedback loops, which need to be balanced by opposing mechanisms to keep one from spreading across the entire cell. In principle, this could be achieved without negative feedback through wave pinning or component depletion mechanisms (Altschuler et al. 2008; Mori et al. 2008). However, in practice, multiple negative signals appear to contribute to this balancing role, including cytoskeletal, mechanical, and purely biochemical mechanisms (Fig. 6). While these mechanisms overlap with the negative signals described in the sections above, we note that in each case the molecular mediators are not fully understood.

Membrane tension is the best-characterized mechanical signal contributing to polarity. Membrane tension describes lateral stretching forces within the membrane. The protrusion and membrane ruffling caused by branched actin assembly at the cell front can increase tension. Increased tension leads to inhibition of cell front activities including Rac signaling, completing a negative feedback loop (Houk et al. 2012). Changes in membrane tension can be rapidly transmitted across a leukocyte through interactions between the actin cortex and the plasma membrane, which allows membrane tension to act over long distances (De Belly et al. 2023). This long-range feedback helps organize polarity by limiting a cell to a single protrusive front. It could also promote directional migration by rapidly integrating cytoskeletal responses across the cell. Phospholipase D2 and the mammalian target of rapamycin complex 2 (mTORC2) have been identified as components needed for responding to changes in membrane tension, but it is still unclear how the effects are transmitted to GTPase signaling and actin assembly (Diz-Muñoz et al. 2016). Tension may act in part by directly modulating actin assembly, as branched actin assembly can be modulated by force (Bieling et al. 2016). Another component of the feedback appears to be biochemical, as mTORC2 can affect Rho GTPase signaling through both its kinase activity and through a kinase-independent activity (Liu et al. 2010; He et al. 2013; Saha et al. 2023).

More generally, chemotaxing leukocytes respond to physical obstacles in their path, and these encounters provide insights into the wiring of polarity. Waves of actin assembly can be extinguished upon encountering an obstacle (Weiner et al. 2007), supporting the idea that physical constraints can modulate migration by directly impacting actin assembly. The propagating wave behavior of both actin assembly and signaling activity may aid migration in this context by helping the cell find accessible paths adjacent to the obstacle. In some circumstances, obstacles can cause a single leading edge to split into two, perhaps by extinguishing waves at the center of the cell front (Lämmermann et al. 2013; Georgantzoglou et al. 2022). In experiments in branched microfluidic channels, the two separated edges initially both proceed with no apparent cross-inhibition until mechanical cues promote decision-making. In response to sharp obstacles, the nucleus acts as a mechanical gauge for comparing paths (Renkawitz et al.

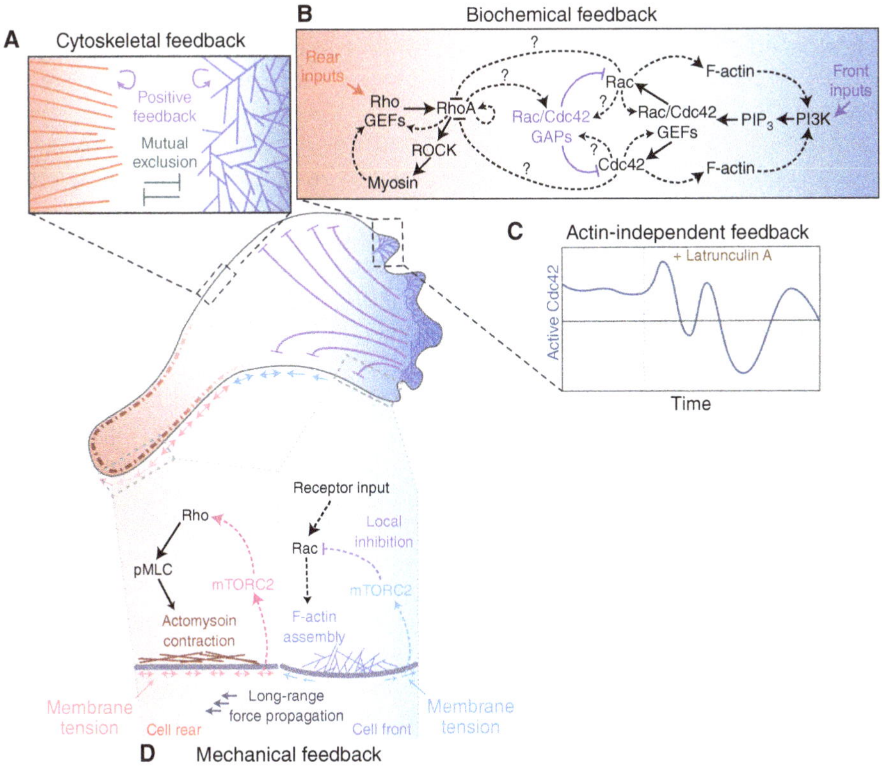

Figure 6. Multiple types of feedback networks collaborate to promote compartmentalized signaling and cytoskeletal activity in migrating leukocytes. (*A*) Cytoskeletal feedback promotes polarized compartments of branched and contractile actin networks through a combination of intrinsic antagonism and self-amplification between and within each domain. (*B*) Biochemical feedback signaling helps constrain front and rear domains through short- and long-range communication of diffusible components in a shared cytoplasmic space. Rho GTPase guanine exchange factors (GEFs) and GTPase-activating proteins (GAPs) are central to this network. Positive feedback loops may use Rho GEFs to reinforce inputs, while negative feedback loops incorporate Rho GAPs to help spatially shape and limit activity domains. Dotted arrows depict multistep pathways, while question marks indicate pathways whose components have not been fully elaborated. (*C*) Actin networks provide stabilizing control for biochemical feedback to Rho GTPases. Latrunculin A–induced depolymerization of actin in neutrophil-like cells can produce oscillations and activity waves of Cdc42, suggesting the existence of actin-independent biochemical feedback circuits, in addition to stabilizing functional links between actin polymerization and Rho GTPase signaling. (*D*) Mechanical feedback defines physical inputs into the polarity circuit. Membrane tension and cytoskeletal stresses generate mechanical feedback signals that are transmitted throughout the cell cortex. During chemotaxis, mechanical feedback from membrane tension drives RhoA activation, activating phosphorylation of myosin light chain (pMLC), and Rac inhibition through mTOR signaling. Importantly, these mechanical feedback connections propagate long-range communication and front–back coordination at rates faster than diffusion. (Figure 6C based on data in Yang et al. 2016.)

2019). In other contexts, cellular stretching mediates competition between the two fronts, reducing protrusion speed and Cdc42 signaling activity, and allowing one front to win (Hadjitheodorou et al. 2023). Membrane tension is a likely mediator of this mechanical competition. However, the fact that both fronts initially proceed raises the question of whether the tension feedback was engaged at this point or whether alternate signals were stabilizing polarity.

Indeed, there appear to be purely biochemical feedback mechanisms that complement mechanical feedback to maintain cell polarity in highly migratory cells. When either human leukocytes or *Dictyostelium* amoebae are treated with latrunculins to depolymerize the actin cytoskeleton, signaling waves and pulses spontaneously arise (Yang et al. 2016; Banerjee et al. 2023). The waves and pulses are highly dynamic and can travel rapidly across the cell membrane. The waves in leukocytes involve Cdc42 signaling activity, which self-organizes into dynamic regions of high and low activity within the plasma membrane (Yang et al. 2016). As described above, the wave and pulse behavior is characteristic of excitable systems and is indicative of a circuit containing both positive and negative feedback. Furthermore, since the actin cytoskeleton is depolymerized in these cells and the cell morphology is unchanging, these feedback loops are presumably independent of actin assembly, membrane tension, and other physical constraints, which contribute to polarity in migrating cells. While the negative regulator in this circuit may be the same as the local inhibitor contributing to steering, this remains to be verified.

Perhaps through related mechanisms, polarity can be restored in leukocytes lacking WASP-family verprolin-homologous protein (WAVE) complex–mediated branched actin assembly (Graziano et al. 2019). These cells have severe defects in actin assembly, polarization, and movement. However, when they are placed in an environment that physically compresses the cell, they are able to move persistently using bleb-like protrusions at the cell front. Rac signaling is strongly polarized in these cells, even though the positive feedback loop between Rac, branched actin, and PIP_3 is presumably broken.

A PATH FORWARD: FINDING THE GAPs THAT LEAD THE WAY

While Rho family GTPases are central regulators of polarity and migration in leukocytes, our understanding of the negative regulation that limits their activity in these contexts is still sparse. The negative feedback mechanism that drives waves and pulses of Cdc42 activity in latrunculin-treated cells remains unclear. Additionally, mechanisms that describe how membrane tension limits Rho GTPase activities and how compression restores polarity in cells lacking the WAVE complex are incomplete. Presumably, these regulatory connections involve GAP proteins, which are the canonical negative regulators of Rho GTPases. However, our understanding of their roles, and modes of regulation, lags far behind our understanding of their GEF counterparts. More than 20 GAPs that target Rho GTPases are expressed in leukocytes (Rincón et al. 2018), but how or whether they have specialized roles in basal and feedback-triggered signaling remains broadly unexplored.

Despite these significant gaps in knowledge, there are clues as to how GAPs could drive negative feedback in polarity signaling and chemotaxis. GAPs generally exhibit lower selectivity for specific Rho GTPases than their GEF counterparts (Amin et al. 2016). Akin to protein phosphatases, broader specificity is thought to promote control of global baseline activity of Rho GTPases. However, recent studies uncovered surprisingly specific and diverse localization dynamics of Rho family GAPs, including localization at membranous organelles and various cytoskeletal structures, pointing to the possibility that intricate recruitment dynamics of individual GAPs could be a key hidden feature of regulation (Galic et al. 2014; de Beco et al. 2018; Müller et al. 2020; Michaud et al. 2022). Additionally, recent evidence suggests that GAPs play central roles in directing negative feedback loops in polarity signaling in response to upstream inputs (Graziano et al. 2017; Town and Weiner 2023; Lin et al. 2024; Algorta et al. 2026). Even so, many fundamental questions remain about the regulatory mechanisms that allow GAPs to fill these roles, indicating broad unexplored areas for future investigation. Interestingly, many Rho GAPs display prominent localization to actin cytoskeletal structures, suggesting possible intimate connections between actin network states and GAP functions (Müller et al. 2020). Several studies have pointed to actin-mediated negative feedback being integral for tuning front polarity and receptor signals (Bell et al. 2021; Hadjitheodorou et al. 2021; Michaud et al. 2022; Kuhn et al. 2025), raising questions of whether and how GAP activity is coupled to local cytoskeletal states. Like GEFs, many Rho GAPs have numerous accessory domains presumed to be essential for controlling their localization dynamics and activity states (Amin et al. 2016). Yet, detailed characterization of these domains and their features is lacking. Understanding how these domains facilitate GAP recruitment and importantly how they spatially compartmentalize GAP activity can offer fundamental insights into how negative feedback is wired in leukocyte chemotaxis.

Although understanding how positive signals propagate in the polarity circuit has long been an area of extensive investigation, efforts to understand how negative signaling imparts precision and accuracy in leukocyte migration and chemotaxis are gradually narrowing the gap. To determine how negative signals are spatially controlled and how emergent behaviors arise from interconnected positive and negative feedback loops, systematic efforts to classify negative regulators at each level of the chemotactic cascade will be necessary. Together, these efforts will be valuable for updating the molecular roster of polarity signaling and advancing comprehensive models of leukocyte chemotaxis and recruitment. At the same time, such advances could provide new therapeutic targets and broaden the landscape of approaches for controlling inflammatory disease.

COMPETING INTEREST STATEMENT

The authors declare no competing interests.

ACKNOWLEDGMENTS

This work was supported by the National Institutes of Health (NIH) grant R01GM148769 to S.R.C. K.A.B.-N. was supported by the NIH/National Heart, Lung, and Blood Institute (NHLBI) training grant program T32 HL007013. We thank Cuauhtemoc Gonzalez and Laureline Lequint for critical feedback on the manuscript and members of the Collins laboratory for valuable discussions.

REFERENCES

Afonso PV, Parent CA. 2011. PI3K and chemotaxis: a priming issue? *Sci Signal* **4:** e22. doi:10.1126/scisignal.2002019

Algorta J, Town JP, Weiner OD, Edelstein-Keshet L. 2026. Investigating local negative feedback of Rac activity by mathematical models and cell motility simulations. *iScience* doi:10.1016/j.isci.2026.114641

Altschuler SJ, Angenent SB, Wang Y, Wu LF. 2008. On the spontaneous emergence of cell polarity. *Nature* **454:** 886–889. doi:10.1038/nature07119

Amin E, Jaiswal M, Derewenda U, Reis K, Nouri K, Koessmeier KT, Aspenström P, Somlyo AV, Dvorsky R, Ahmadian MR. 2016. Deciphering the molecular and functional basis of RHOGAP family proteins: a systematic approach toward selective inactivation of RHO family proteins. *J Biol Chem* **291:** 20353–20371. doi:10.1074/jbc.M116.736967

Aranyosi AJ, Wong EA, Irimia D. 2015. A neutrophil treadmill to decouple spatial and temporal signals during chemotaxis. *Lab Chip* **15:** 549–556. doi:10.1039/C4LC00970C

Arrieumerlou C, Meyer T. 2005. A local coupling model and compass parameter for eukaryotic chemotaxis. *Dev Cell* **8:** 215–227. doi:10.1016/j.devcel.2004.12.007

Asokan SB, Johnson HE, Sondek J, Shutova MS, Svitkina TM, Haugh JM, Bear JE. 2019. Lysophosphatidic acid provokes fibroblast chemotaxis through combinatorial regulation of myosin II. bioRxiv 355610. doi:10.1101/355610

Banerjee T, Matsuoka S, Biswas D, Miao Y, Pal DS, Kamimura Y, Ueda M, Devreotes PN, Iglesias PA. 2023. A dynamic partitioning mechanism polarizes membrane protein distribution. *Nat Commun* **14:** 7909. doi:10.1038/s41467-023-43615-2

Bell GRR, Rincón E, Akdoğan E, Collins SR. 2021. Optogenetic control of receptors reveals distinct roles for actin- and Cdc42-dependent negative signals in chemotactic signal processing. *Nat Commun* **12:** 6148. doi:10.1038/s41467-021-26371-z

Belliveau NM, Footer MJ, Akdoğan E, Van Loon AP, Collins SR, Theriot JA. 2023. Whole-genome screens reveal regulators of differentiation state and context-dependent migration in human neutrophils. *Nat Commun* **14:** 5770. doi:10.1038/s41467-023-41452-x

Bieling P, Li TD, Weichsel J, McGorty R, Jreij P, Huang B, Fletcher DA, Mullins RD. 2016. Force feedback controls motor activity and mechanical properties of self-assembling branched actin networks. *Cell* **164:** 115–127. doi:10.1016/j.cell.2015.11.057

Bisaria A, Hayer A, Garbett D, Cohen D, Meyer T. 2020. Membrane-proximal F-actin restricts local membrane protrusions and directs cell migration. *Science* **368:** 1205–1210. doi:10.1126/science.aay7794

Bonecchi R, Galliera E, Borroni EM, Corsi MM, Locati M, Mantovani A. 2009. Chemokines and chemokine receptors: an overview. *Front Biosci (Landmark Ed)* **14:** 540–551. doi:10.2741/3261

Campbell JJ, Foxman EF, Butcher EC. 1997. Chemoattractant receptor cross talk as a regulatory mechanism in leukocyte adhesion and migration. *Eur J Immunol* **27:** 2571–2578. doi:10.1002/eji.1830271016

Collins SR, Yang HW, Bonger KM, Guignet EG, Wandless TJ, Meyer T. 2015. Using light to shape chemical gradients for parallel and automated analysis of chemotaxis. *Mol Syst Biol* **11:** 804. doi:10.15252/msb.20156027

Costa C, Barberis L, Ambrogio C, Manazza AD, Patrucco E, Azzolino O, Neilsen PO, Ciraolo E, Altruda F, Prestwich GD, et al. 2007. Negative feedback regulation of Rac in leukocytes from mice expressing a constitutively active phosphatidylinositol 3-kinase γ. *Proc Natl Acad Sci* **104:** 14354–14359. doi:10.1073/pnas.0703175104

Csépányi-Kömi R, Wisniewski É, Bartos B, Lévai P, Németh T, Balázs B, Kurz ARM, Bierschenk S, Sperandio M, Ligeti E. 2016. Rac GTPase activating protein ARHGAP25 regulates leukocyte transendothelial migration in mice. *J Immunol* **197:** 2807–2815. doi:10.4049/jimmunol.1502342

Dandekar SN, Park JS, Peng GE, Onuffer JJ, Lim WA, Weiner OD. 2013. Actin dynamics rapidly reset chemoattractant receptor sensitivity following adaptation in neutrophils. *Philos Trans R Soc Lond B Biol Sci* **368:** 20130008. doi:10.1098/rstb.2013.0008

de Beco S, Vaidžiulytė K, Manzi J, Dalier F, di Federico F, Cornilleau G, Dahan M, Coppey M. 2018. Optogenetic dissection of Rac1 and Cdc42 gradient shaping. *Nat Commun* **9:** 4816. doi:10.1038/s41467-018-07286-8

De Belly H, Yan S, Borja da Rocha H, Ichbiah S, Town JP, Zager PJ, Estrada DC, Meyer K, Turlier H, Bustamante C, et al. 2023. Cell protrusions and contractions generate long-range membrane tension propagation. *Cell* **186:** 3049–3061.e15. doi:10.1016/j.cell.2023.05.014

Devreotes P. 2019. Moving toward molecular mechanisms for chemotaxis in eukaryotic cells. *Mol Biol Cell* **30:** 2873–2877. doi:10.1091/mbc.E19-07-0393

Devreotes P, Janetopoulos C. 2003. Eukaryotic chemotaxis: distinctions between directional sensing and polarization. *J Biol Chem* **278:** 20445–20448. doi:10.1074/jbc.R300010200

Diz-Muñoz A, Thurley K, Chintamen S, Altschuler SJ, Wu LF, Fletcher DA, Weiner OD. 2016. Membrane tension acts through PLD2 and mTORC2 to limit actin network assembly during neutrophil migration. *PLoS Biol* **14:** e1002474. doi:10.1371/journal.pbio.1002474

Foxman EF, Campbell JJ, Butcher EC. 1997. Multistep navigation and the combinatorial control of leukocyte chemotaxis. *J Cell Biol* **139:** 1349–1360. doi:10.1083/jcb.139.5.1349

Francis SA, Shen X, Young JB, Kaul P, Lerner DJ. 2006. Rho GEF Lsc is required for normal polarization, migration, and adhesion of formyl-peptide–stimulated neutrophils. *Blood* **107:** 1627–1635. doi:10.1182/blood-2005-03-1164

Galic M, Tsai FC, Collins SR, Matis M, Bandara S, Meyer T. 2014. Dynamic recruitment of the curvature-sensitive protein ArhGAP44 to nanoscale membrane deformations limits exploratory filopodia initiation in neurons. *eLife* **3:** e03116. doi:10.7554/eLife.03116

Gambardella L, Anderson KE, Nussbaum C, Segonds-Pichon A, Margarido T, Norton L, Ludwig T, Sperandio M, Hawkins PT, Stephens L, et al. 2011. The GTPase-activating protein ARAP3 regulates chemotaxis and adhesion-dependent processes in neutrophils. *Blood* **118:** 1087–1098. doi:10.1182/blood-2010-10-312959

Georgantzoglou A, Poplimont H, Walker HA, Lämmermann T, Sarris M. 2022. A two-step search and run response to gradients shapes leukocyte navigation in vivo. *J Cell Biol* **221:** e202103207. doi:10.1083/jcb.202103207

Gerisch G, Keller HU. 1981. Chemotactic reorientation of granulocytes stimulated with micropipettes containing fmet-leu-phe. *J Cell Sci* **52:** 1–10. doi:10.1242/jcs.52.1.1

Gerisch G, Hülser D, Malchow D, Wick U. 1975. Cell communication by periodic cyclic-AMP pulses. *Philos Trans R Soc Lond B Biol Sci* **272:** 181–192. doi:10.1098/rstb.1975.0080

Graziano BR, Gong D, Anderson KE, Pipathsouk A, Goldberg AR, Weiner OD. 2017. A module for Rac temporal signal integration revealed with optogenetics. *J Cell Biol* **216:** 2515–2531. doi:10.1083/jcb.201604113

Graziano BR, Town JP, Sitarska E, Nagy TL, Fošnarič M, Penič S, Iglič A, Kralj-Iglič V, Gov NS, Diz-Muñoz A, et al. 2019. Cell confinement reveals a branched-actin independent circuit for neutrophil polarity. *PLoS Biol* **17:** e3000457. doi:10.1371/journal.pbio.3000457

Guilluy C, Garcia-Mata R, Burridge K. 2011. Rho protein crosstalk: another social network? *Trends Cell Biol* **21:** 718–726. doi:10.1016/j.tcb.2011.08.002

Hadjitheodorou A, Bell GRR, Ellett F, Shastry S, Irimia D, Collins SR, Theriot JA. 2021. Directional reorientation of migrating neutrophils is limited by suppression of receptor input signaling at the cell rear through myosin II activity. *Nat Commun* **12:** 6619. doi:10.1038/s41467-021-26622-z

Hadjitheodorou A, Bell GRR, Ellett F, Irimia D, Tibshirani R, Collins SR, Theriot JA. 2023. Leading edge competition promotes context-dependent responses to receptor inputs to resolve directional dilemmas in neutrophil migration. *Cell Syst* **14:** 196–209.e6. doi:10.1016/j.cels.2023.02.001

He Y, Li D, Cook SL, Yoon MS, Kapoor A, Rao CV, Kenis PJA, Chen J, Wang F. 2013. Mammalian target of rapamycin and Rictor control neutrophil chemotaxis by regulating Rac/Cdc42 activity and the actin cytoskeleton. *Mol Biol Cell* **24:** 3369–3380. doi:10.1091/mbc.e13-07-0405

Heit B, Tavener S, Raharjo E, Kubes P. 2002. An intracellular signaling hierarchy determines direction of migration in opposing chemotactic gradients. *J Cell Biol* **159:** 91–102. doi:10.1083/jcb.200202114

Herzmark P, Campbell K, Wang F, Wong K, El-Samad H, Groisman A, Bourne HR. 2007. Bound attractant at the leading vs. the trailing edge determines chemotactic prowess. *Proc Natl Acad Sci* **104:** 13349–13354. doi:10.1073/pnas.0705889104

Hodge RG, Ridley AJ. 2016. Regulating Rho GTPases and their regulators. *Nat Rev Mol Cell Biol* **17:** 496–510. doi:10.1038/nrm.2016.67

Houk AR, Jilkine A, Mejean CO, Boltyanskiy R, Dufresne ER, Angenent SB, Altschuler SJ, Wu LF, Weiner OD. 2012. Membrane tension maintains cell polarity by confining signals to the leading edge during neutrophil migration. *Cell* **148:** 175–188. doi:10.1016/j.cell.2011.10.050

Hughes CE, Nibbs RJB. 2018. A guide to chemokines and their receptors. *FEBS J* **285:** 2944–2971. doi:10.1111/febs.14466

Iglesias PA, Devreotes PN. 2008. Navigating through models of chemotaxis. *Curr Opin Cell Biol* **20:** 35–40. doi:10.1016/j.ceb.2007.11.011

Ishida M, Uwamichi M, Nakajima A, Sawai S. 2025. Traveling-wave chemotaxis of neutrophil-like HL-60 cells. *Mol Biol Cell* **36:** ar17. doi:10.1091/mbc.E24-06-0245

Joo E, Olson MF. 2021. Regulation and functions of the RhoA regulatory guanine nucleotide exchange factor GEF-H1. *Small GTPases* **12:** 358–371. doi:10.1080/21541248.2020.1840889

Kamakura S, Nomura M, Hayase J, Iwakiri Y, Nishikimi A, Takayanagi R, Fukui Y, Sumimoto H. 2013. The cell polarity protein mInsc regulates neutrophil chemotaxis via a noncanonical G protein signaling pathway. *Dev Cell* **26:** 292–302. doi:10.1016/j.devcel.2013.06.008

Kamp ME, Liu Y, Kortholt A. 2016. Function and regulation of heterotrimeric G proteins during chemotaxis. *Int J Mol Sci* **17:** 90. doi:10.3390/ijms17010090

Karunarathne WKA, Giri L, Patel AK, Venkatesh KV, Gautam N. 2013. Optical control demonstrates switch-like PIP3 dynamics underlying the initiation of immune cell migration. *Proc Natl Acad Sci* **110:** E1575–E1583. doi:10.1073/pnas.1220755110

Kolaczkowska E, Kubes P. 2013. Neutrophil recruitment and function in health and inflammation. *Nat Rev Immunol* **13:** 159–175. doi:10.1038/nri3399

Kopf A, Renkawitz J, Hauschild R, Girkontaite I, Tedford K, Merrin J, Thorn-Seshold O, Trauner D, Häcker H, Fischer KD, et al. 2020. Microtubules control cellular shape and coherence in amoeboid migrating cells. *J Cell Biol* **219:** e201907154. doi:10.1083/jcb.201907154

Kuhn J, Banerjee P, Haye A, Robinson DN, Iglesias PA, Devreotes PN. 2025. Complementary cytoskeletal feedback loops control signal transduction excitability and cell polarity. *Nat Commun* **16:** 7482. doi:10.1038/s41467-025-62799-3

Lämmermann T, Afonso PV, Angermann BR, Wang JM, Kastenmüller W, Parent CA, Germain RN. 2013. Neutrophil swarms require LTB4 and integrins at sites of cell death in vivo. *Nature* **498:** 371–375. doi:10.1038/nature12175

Li Z, Hannigan M, Mo Z, Liu B, Lu W, Wu Y, Smrcka AV, Wu G, Li L, Liu M, et al. 2003. Directional sensing requires G beta gamma-mediated PAK1 and PIX alpha-dependent activation of Cdc42. *Cell* **114:** 215–227. doi:10.1016/S0092-8674(03)00559-2

Lin Y, Pal DS, Banerjee P, Banerjee T, Qin G, Deng Y, Borleis J, Iglesias PA, Devreotes PN. 2024. Ras suppression potentiates rear actomyosin contractility-driven cell polarization and migration. *Nat Cell Biol* **26:** 1062–1076. doi:10.1038/s41556-024-01453-4

Liu L, Das S, Losert W, Parent CA. 2010. mTORC2 regulates neutrophil chemotaxis in a cAMP- and RhoA-dependent fashion. *Dev Cell* **19:** 845–857. doi:10.1016/j.devcel.2010.11.004

Lomakin AJ, Lee KC, Han SJ, Bui DA, Davidson M, Mogilner A, Danuser G. 2015. Competition for actin between two distinct F-actin networks defines a bistable switch for cell polarization. *Nat Cell Biol* **17:** 1435–1445. doi:10.1038/ncb3246

Lundgren SM, Rocha-Gregg BL, Akdoğan E, Mysore MN, Hayes S, Collins SR. 2023. Signaling dynamics distinguish high- and low-priority neutrophil chemoattractant receptors. *Sci Signal* **16:** eadd1845. doi:10.1126/scisignal.add1845

Medina F, Carter AM, Dada O, Gutowski S, Hadas J, Chen Z, Sternweis PC. 2013. Activated RhoA is a positive feedback regulator of the Lbc family of Rho guanine nucleotide exchange factor proteins. *J Biol Chem* **288:** 11325–11333. doi:10.1074/jbc.M113.450056

Michaud A, Leda M, Swider ZT, Kim S, He J, Landino J, Valley JR, Huisken J, Goryachev AB, von Dassow G, et al. 2022. A versatile cortical pattern-forming circuit based on Rho, F-actin, Ect2, and RGA-3/4. *J Cell Biol* **221:** e202203017. doi:10.1083/jcb.202203017

Moe A, Holmes W, Golding AE, Zola J, Swider ZT, Edelstein-Keshet L, Bement W. 2021. Cross-talk-dependent cortical patterning of Rho GTPases during cell repair. *Mol Biol Cell* **32:** 1417–1432. doi:10.1091/mbc.E20-07-0481

Mori Y, Jilkine A, Edelstein-Keshet L. 2008. Wave-pinning and cell polarity from a bistable reaction-diffusion system. *Biophys J* **94:** 3684–3697. doi:10.1529/biophysj.107.12 0824

Müller PM, Rademacher J, Bagshaw RD, Wortmann C, Barth C, van Unen J, Alp KM, Giudice G, Eccles RL, Heinrich LE, et al. 2020. Systems analysis of RhoGEF and RhoGAP regulatory proteins reveals spatially organized RAC1 signalling from integrin adhesions. *Nat Cell Biol* **22:** 498–511. doi:10.1038/s41556-020-0488-x

Neptune ER, Bourne HR. 1997. Receptors induce chemotaxis by releasing the betagamma subunit of G_i, not by activating G_q or G_s. *Proc Natl Acad Sci* **94:** 14489–14494. doi:10.1073/pnas.94.26.14489

Nishikawa M, Hörning M, Ueda M, Shibata T. 2014. Excitable signal transduction induces both spontaneous and directional cell asymmetries in the phosphatidylinositol lipid signaling system for eukaryotic chemotaxis. *Biophys J* **106:** 723–734. doi:10.1016/j.bpj.2013.12.023

Nishikimi A, Fukuhara H, Su W, Hongu T, Takasuga S, Mihara H, Cao Q, Sanematsu F, Kanai M, Hasegawa H, et al. 2009. Sequential regulation of DOCK2 dynamics by two phospholipids during neutrophil chemotaxis. *Science* **324:** 384–387. doi:10.1126/science.1170179

O'Neill PR, Kalyanaraman V, Gautam N. 2016. Subcellular optogenetic activation of Cdc42 controls local and distal signaling to drive immune cell migration. *Mol Biol Cell* **27:** 1442–1450. doi:10.1091/mbc.E15-12-0832

Pal DS, Banerjee T, Lin Y, de Trogoff F, Borleis J, Iglesias PA, Devreotes PN. 2023. Actuation of single downstream nodes in growth factor network steers immune cell migration. *Dev Cell* **58:** 1170–1188.e7. doi:10.1016/j.devcel.2023.04.019

Parent CA, Blacklock BJ, Froehlich WM, Murphy DB, Devreotes PN. 1998. G protein signaling events are activated at the leading edge of chemotactic cells. *Cell* **95:** 81–91. doi:10.1016/S0092-8674(00)81784-5

Park JS, Rhau B, Hermann A, McNally KA, Zhou C, Gong D, Weiner OD, Conklin BR, Onuffer J, Lim WA. 2014. Synthetic control of mammalian-cell motility by engineering chemotaxis to an orthogonal bioinert chemical signal. *Proc Natl Acad Sci* **111:** 5896–5901. doi:10.1073/pnas.1402087111

Petrie Aronin CE, Zhao YM, Yoon JS, Morgan NY, Prüstel T, Germain RN, Meier-Schellersheim M. 2017. Migrating myeloid cells sense temporal dynamics of chemoattractant concentrations. *Immunity* **47:** 862–874.e3. doi:10.1016/j.immuni.2017.10.020

Phillipson M, Heit B, Parsons SA, Petri B, Mullaly SC, Colarusso P, Gower RM, Neely G, Simon SI, Kubes P.

2009. Vav1 is essential for mechanotactic crawling and migration of neutrophils out of the inflamed microvasculature. *J Immunol* **182:** 6870–6878. doi:10.4049/jimmunol.0803414

Renkawitz J, Kopf A, Stopp J, de Vries I, Driscoll MK, Merrin J, Hauschild R, Welf ES, Danuser G, Fiolka R, et al. 2019. Nuclear positioning facilitates amoeboid migration along the path of least resistance. *Nature* **568:** 546–550. doi:10.1038/s41586-019-1087-5

Rincón E, Rocha-Gregg BL, Collins SR. 2018. A map of gene expression in neutrophil-like cell lines. *BMC Genomics* **19:** 1–17. doi:10.1186/s12864-018-4957-6

Saha S, Town JP, Weiner OD. 2023. Mechanosensitive mTORC2 independently coordinates leading and trailing edge polarity programs during neutrophil migration. *Mol Biol Cell* **34:** ar35. doi:10.1091/mbc.E22-05-0191

SenGupta S, Parent CA, Bear JE. 2021. The principles of directed cell migration. *Nat Rev Mol Cell Biol* **22:** 529–547. doi:10.1038/s41580-021-00366-6

Sourjik V, Wingreen NS. 2012. Responding to chemical gradients: bacterial chemotaxis. *Curr Opin Cell Biol* **24:** 262–268. doi:10.1016/j.ceb.2011.11.008

Stephens L, Smrcka A, Cooke FT, Jackson TR, Sternweis PC, Hawkins PT. 1994. A novel phosphoinositide 3 kinase activity in myeloid-derived cells is activated by G protein beta gamma subunits. *Cell* **77:** 83–93. doi:10.1016/0092-8674(94)90237-2

Sun L, Ye RD. 2012. Role of G protein-coupled receptors in inflammation. *Acta Pharmacol Sin* **33:** 342–350. doi:10.1038/aps.2011.200

Surve CR, To JY, Malik S, Kim M, Smrcka AV. 2016. Dynamic regulation of neutrophil polarity and migration by the heterotrimeric G protein subunits $G\alpha_i$-GTP and $G\beta\gamma$. *Sci Signal* **9:** ra22. doi:10.1126/scisignal.aad8163

Swaney KF, Huang CH, Devreotes PN. 2010. Eukaryotic chemotaxis: a network of signaling pathways controls motility, directional sensing, and polarity. *Annu Rev Biophys* **39:** 265–289. doi:10.1146/annurev.biophys.093008.131228

Takeda K, Shao D, Adler M, Charest PG, Loomis WF, Levine H, Groisman A, Rappel WJ, Firtel RA. 2012. Incoherent feedforward control governs adaptation of activated Ras in a eukaryotic chemotaxis pathway. *Sci Signal* **5:** ra2. doi:10.1126/scisignal.2002413

Tang M, Wang M, Shi C, Iglesias PA, Devreotes PN, Huang CH. 2014. Evolutionarily conserved coupling of adaptive and excitable networks mediates eukaryotic chemotaxis. *Nat Commun* **5:** 5175. doi:10.1038/ncomms6175

To JY, Smrcka AV. 2018. Activated heterotrimeric G protein αi subunits inhibit Rap-dependent cell adhesion and promote cell migration. *J Biol Chem* **293:** 1570–1578. doi:10.1074/jbc.RA117.000964

Town JP, Weiner OD. 2023. Local negative feedback of Rac activity at the leading edge underlies a pilot pseudopod-like program for amoeboid cell guidance. *PLoS Biol* **21:** e3002307. doi:10.1371/journal.pbio.3002307

Tsai TYC, Collins SR, Chan CK, Hadjitheodorou A, Lam PY, Lou SS, Yang HW, Jorgensen J, Ellett F, Irimia D, et al. 2019. Efficient front-rear coupling in neutrophil chemotaxis by dynamic myosin II localization. *Dev Cell* **49:** 189–205.e6. doi:10.1016/j.devcel.2019.03.025

Wang F. 2009. The signaling mechanisms underlying cell polarity and chemotaxis. *Cold Spring Harb Perspect Biol* **1:** a002980. doi:10.1101/cshperspect.a002980

Wang F, Herzmark P, Weiner OD, Srinivasan S, Servant G, Bourne HR. 2002. Lipid products of PI(3)Ks maintain persistent cell polarity and directed motility in neutrophils. *Nat Cell Biol* **4:** 513–518. doi:10.1038/ncb810

Wang MJ, Artemenko Y, Cai WJ, Iglesias PA, Devreotes PN. 2014. The directional response of chemotactic cells depends on a balance between cytoskeletal architecture and the external gradient. *Cell Rep* **9:** 1110–1121. doi:10.1016/j.celrep.2014.09.047

Weiner OD. 2002. Regulation of cell polarity during eukaryotic chemotaxis: the chemotactic compass. *Curr Opin Cell Biol* **14:** 196–202. doi:10.1016/S0955-0674(02)00310-1

Weiner OD, Neilsen PO, Prestwich GD, Kirschner MW, Cantley LC, Bourne HR. 2002. A PtdInsP3- and Rho GTPase-mediated positive feedback loop regulates neutrophil polarity. *Nat Cell Biol* **4:** 509–513. doi:10.1038/ncb811

Weiner OD, Marganski WA, Wu LF, Altschuler SJ, Kirschner MW. 2007. An actin-based wave generator organizes cell motility. *PLoS Biol* **5:** e221. doi:10.1371/journal.pbio.0050221

Welch HCE, Coadwell WJ, Ellson CD, Ferguson GJ, Andrews SR, Erdjument-Bromage H, Tempst P, Hawkins PT, Stephens LR. 2002. P-Rex1, a PtdIns(3,4,5)P_3- and Gbetagamma-regulated guanine-nucleotide exchange factor for Rac. *Cell* **108:** 809–821. doi:10.1016/S0092-8674(02)00663-3

Wong K, Van Keymeulen A, Bourne HR. 2007. PDZRhoGEF and myosin II localize RhoA activity to the back of polarizing neutrophil-like cells. *J Cell Biol* **179:** 1141–1148. doi:10.1083/jcb.200706167

Wright HL, Moots RJ, Bucknall RC, Edwards SW. 2010. Neutrophil function in inflammation and inflammatory diseases. *Rheumatology* **49:** 1618–1631. doi:10.1093/rheumatology/keq045

Xu J, Wang F, Van Keymeulen A, Herzmark P, Straight A, Kelly K, Takuwa Y, Sugimoto N, Mitchison T, Bourne HR. 2003. Divergent signals and cytoskeletal assemblies regulate self-organizing polarity in neutrophils. *Cell* **114:** 201–214. doi:10.1016/S0092-8674(03)00555-5

Xu X, Meier-Schellersheim M, Yan J, Jin T. 2007. Locally controlled inhibitory mechanisms are involved in eukaryotic GPCR-mediated chemosensing. *J Cell Biol* **178:** 141–153. doi:10.1083/jcb.200611096

Yang HW, Collins SR, Meyer T. 2016. Locally excitable Cdc42 signals steer cells during chemotaxis. *Nat Cell Biol* **18:** 191–201. doi:10.1038/ncb3292

Yoo SK, Deng Q, Cavnar PJ, Wu YI, Hahn KM, Huttenlocher A. 2010. Differential regulation of protrusion and polarity by PI3K during neutrophil motility in live zebrafish. *Dev Cell* **18:** 226–236. doi:10.1016/j.devcel.2009.11.015

Zigmond SH. 1977. Ability of polymorphonuclear leukocytes to orient in gradients of chemotactic factors. *J Cell Biol* **75:** 606–616. doi:10.1083/jcb.75.2.606

Neutrophil Migration: Using All the Tricks of Mesenchymal and Amoeboid Cells to Navigate Many Environments

Martina Lerche[1,2] and Clare M. Waterman[1]

[1]Cell and Developmental Biology Center, National Heart Lung and Blood Institute, National Institutes of Health, Bethesda, Maryland 20892, USA

[2]Division of Cell Matrix Biology and Regenerative Medicine, School of Biological Sciences, Faculty of Biology, Medicine and Health, The University of Manchester, Manchester M13 9NT, United Kingdom

Correspondence: watermancm@nhlbi.nih.gov; martina.lerche@manchester.ac.uk

Humans encounter numerous pathogens every day and therefore rely heavily on the innate immune system that acts as the first line of defense against infectious agents. Neutrophils make up more than half of the circulating leukocytes in humans and are well equipped to neutralize invaders and thus are a crucial part of our defense system [Friedl and Weigelin (2008). *Nat Immunol* **9:** 960; Rankin (2010). *J Leukoc Biol* **88:** 241]. To get from the bloodstream to sites of insult, neutrophils often have to migrate through various tissues and cover distances much further than many other cells. Thus, it is not surprising that they use elements from both amoeboid and mesenchymal migration, allowing them to migrate efficiently through various environments as they navigate in the body. Here we will discuss the cellular properties of neutrophils that make them exceptional migrators in a variety of physiological contexts, along with the various signaling pathways that guide neutrophil migration.

Neutrophil migration is guided by both physical cues of the environment and chemical signaling molecules, known as chemoattractants, during the various stages of their life, ranging from mobilization from the bone marrow into the blood circulation, crawling along and through the endothelium as well as during migration into and out of the tissue at the site of inflammation. During inflammation, chemoattractants are released by invading microorganisms, damaged and dead cells, or tissue-resident immune cells that sense these and secrete additional inflammatory chemoattractants (Tang et al. 2012; Sokol and Luster 2015). Chemoattractants bind to G-protein-coupled receptors (GPCRs) and receptor tyrosine kinases (RTKs) expressed on the cell surface to initiate downstream signaling processes that mediate directed cell migration toward the chemotactic stimulus in a process known as chemotaxis (Petri and Sanz 2018).

A prerequisite for cell migration is establishment of cell polarization relative to the chemoattractant, a process where cells obtain and sustain an asymmetric organization of subcellular components and asymmetric shape with a clear protruding front and a retracting rear (Van Keymeulen et al. 2006). This polarized shape is gen-

Cite this article as *Cold Spring Harb Perspect Biol* doi: 10.1101/cshperspect.a041749

erated by the cytoskeleton, which exerts intracellular forces to mediate displacement of the cell body. Signaling through GPCRs and RTKs promote local regulation of the actomyosin cytoskeleton and cell adhesion receptors to generate protrusion of the cell in the direction of migration, traction on the environment, and pulling forward of the cell body to advance the cell (Lauffenburger and Horwitz 1996; Sánchez-Madrid and Del Pozo 1999; Morin et al. 2008). Neutrophils use hundreds of accessory proteins to orchestrate the cytoskeletal remodeling during migration and the activity of many of these are regulated by Rho GTPases that signal downstream from GPCRs (Xu et al. 2003; Swaney et al. 2010; McCormick et al. 2017). Three members of the Rho family of small GTPases—Rac, Rho, and Cdc42—have been extensively studied for their role in organizing the actin cytoskeleton and generating specific structures required for cell migration. Rac GTP induces actin polymerization and lamellipodia formation and Cdc42 GTP mediates filopodia like protrusions, while Rho GTP stimulates cell contractility and stress fiber formation (Nobes and Hall 1995). The activity of small GTPases is regulated by activating guanine nucleotide exchange factors (GEFs) that catalyze the exchange of guanosine diphosphate (GDP) for guanosine triphosphate (GTP), and GTPase-activating proteins (GAPs) that induce GTP hydrolysis. GEFs and GAPs also play an important part in several feedback loops to regulate timely activation of GTPases (McCormick et al. 2017; Michael and Vermeren 2019). Thus, intricate signaling pathways, induced by extracellular cues, guide cytoskeletal changes to mediate timely migration of the neutrophil toward the site of inflammation.

Here we will describe how the above general principles of cell migration differ in two distinct modes of single cell migration, mesenchymal and amoeboid, and discuss how neutrophils use components from both of these modes to move forward depending on the physiological context in which they need to move. In addition, we will discuss how both the physical properties of the neutrophil and their environment impact migration and how various signaling cues initiate downstream signaling to regulate cell polarization and guide-directed migration of neutrophils.

TWO MAJOR MODES OF SINGLE-CELL MIGRATION

The mechanisms of directed cell movement can roughly be divided into mesenchymal and amoeboid modes of migration. The mode of migration used depends both on cell-intrinsic properties, such as expression of adhesion receptors and signaling state, and on environmental properties such as confinement, guidance cues, and biochemical and biomechanical properties of the extracellular matrix (ECM) (Roussel and Gingras 1997; Friedl et al. 1998; Dangerfield et al. 2002; Lämmermann and Germain 2014).

Mesenchymal migration is classically used by undifferentiated cells in the mesenchyme in early development as well as by fibroblasts and some cancer cells that express high levels of ECM-binding integrin receptors (Ichida et al. 2011; McAndrews et al. 2015; Yang et al. 2020; Doyle et al. 2021). A key feature of mesenchymal migration is that it is largely a form of single-cell migration where cells such as fibroblasts and certain cells during development lack, or in the case of cancer cells, lose their tight cell–cell contacts, in contrast to collective cell migration of, for example, epithelial sheets that depend on tight cell–cell interactions (Yonemura et al. 1995; Angres et al. 1996; Cano et al. 2000; Vedula et al. 2012). In mesenchymal migration, the cell attaches and moves along ECM fibers via transmembrane integrin receptors that are coupled to the contractile actin cytoskeleton by intracellular adaptor proteins (Gardel et al. 2010; Schiller et al. 2013; Chastney et al. 2021). Mesenchymal cells are usually elongated, highly adherent, and capable of remodeling the ECM they migrate along by strong contraction and local secretion of proteolytic enzymes (Larsen et al. 2006; Bear and Haugh 2014; Yamada and Sixt 2019).

In contrast to mesenchymal migration, amoeboid migration operates with low amounts of engaged adhesion receptors or no specific adhesion receptors, generates very little traction, does generally not involve proteolytic degradation, and is used mainly by primordial germ cells and immune cells (Lämmermann et al. 2008; Renkawitz et al. 2019; George et al. 2023). As this mode of migration is similar to the migration of lower eukaryotic single-celled amoebae, germ cells and immune cell migration in tissue is usually classified as amoeboid. Due to lack of stable adhesions, amoeboid migration is generally much faster than mesenchymal migration as it does not require assembly and disassembly of organized adhesion complexes (Friedl et al. 1998). Instead, amoeboid cells rapidly change their morphology by retracting and extending protrusions that can be generated by both internal hydrostatic pressure and actin polymerization (Yamada and Sixt 2019; García-Arcos et al. 2024).

While neutrophils express high amounts of β2 integrins that they use for adhesion to endothelial cells, they express a lower amount of ECM-binding integrins and do not form prominent focal adhesions as compared to mesenchymal cells, and thus are generally considered to primarily use amoeboid migration (Smith et al. 1989; Lämmermann et al. 2013; Barros-Becker et al. 2017; Cunin et al. 2019). However, as professional migrators, neutrophils actually exploit elements from both migration modes when navigating through the body. Like mesenchymal cells, neutrophils lack cell–cell junctions and migrate in a single-cell manner, we will therefore focus on the mechanism of single-cell migration in this section. However, neutrophils do signal extensively to each other and migrate collectively, in a swarming manner, which will be described later. We will begin by introducing the classical steps of migration and how neutrophils migrate directionally, and in the following section we will discuss how neutrophils employ elements from both mesenchymal and amoeboid migration during various stages of their recruitment to, and from, the site of inflammation.

Chemotaxis and Establishment of Cell Polarity

Directed migration takes place through a mechanism called chemotaxis, a process where cells polarize and migrate in the direction of a chemoattractant gradient. Chemotaxis is essential in inflammation, as it guides both amoeboid immune cells and mesenchymal cells to the site of inflammation to mediate an inflammatory response and tissue repair (Ross and Benditt 1961; Seppä et al. 1982; McDonald et al. 2010). Once a cell senses a directional chemotactic cue, it responds by establishing spatially segregated regions of specific signaling cascades within the cell that drive an asymmetry in shape and organization of internal organelles. Upon asymmetric chemoattractant binding on the cell surface, GPCRs and RTKs activate, resulting in formation or stabilization of local protrusions (Hawkins et al. 1992; Stephens et al. 1994; Klarlund et al. 1997; Vanhaesebroeck et al. 2012; Chen et al. 2024).

Polarized mesenchymal cells such as fibroblasts usually display a broad, fanlike lamella in the cell front, with thin actomyosin filaments mediating contraction behind the leading edge and strong stress fibers coupled to mature adhesion complexes that extend into a narrow tail in the cell rear (Omelchenko et al. 2002; Asokan et al. 2014). The Golgi apparatus and microtubule organizing center (MTOC) are positioned in front of the nucleus with microtubules radiating toward the leading edge, which allows for rapid delivery of Golgi-derived vesicles along the microtubules into the protruding lamella (Ridley et al. 2003). Furthermore, these cells usually display several competing lamellipodia and mainly respond to chemoattractants such as growth factors via RTKs that signal via Rac and phosphoinositide 3-kinase (PI3K) (Ridley et al. 1992; Pertz et al. 2006; Melvin et al. 2011; Vorotnikov and Tyurin-Kuzmin 2014). Activation and localization of Rac and PI3K to lamellipodia promotes lateral spreading and stabilization. Thus, only protrusions that experience high enough concentrations of the chemoattractant will prevail, and the competing protrusions and the cell will gradually reorient toward the chemoattractant (Welf et al. 2012). In addition, mesenchymal cell polarization lacks the global inhibitory signal that is present in amoeboid cells (discussed below), and thus fibroblasts require much steeper chemotactic gradients (10% steepness across the cell) to polarize, compared to human neutrophils (1% steepness across the cell) (Zigmond 1977; Schneider and Haugh 2005; Melvin et al. 2011).

Polarized amoeboid cells display a clear protruding pseudopod at the front and a highly contractile cell rear with a lifted cylindrical uropod (Hind et al. 2016). In contrast to mesenchymal cells, microtubules radiate toward the posterior uropod of neutrophils (Eddy et al. 2002; Kroll et al. 2023). Neutrophils display a highly chemotactic behavior that is very similar to the amoeba *Dictyostelium discoideum* toward a wide range of chemoattractants (Devreotes and Zigmond 1988). These attractants are released by both microbes and host cells and can be classified into four separate subgroups based on their biochemical properties: formyl peptides (e.g., *N*-formyl-methionyl-leucyl-phenylalanine [fMLP]), chemokines (e.g., stromal cell–derived factor 1 [SDF-1]), chemotactic lipids (e.g., leukotriene B4 [LTB4]), and complement anaphylatoxins (e.g., C5a). All these attractants regulate cell behavior by binding to GPCRs, the main receptor family responding to these molecules, and couple to heterotrimeric G proteins to initiate downstream signaling events (Metzemaekers et al. 2020). GPCR binding resulting in local activation of PI3K at the leading edge drives regional activation of the small GTPase Rac to promote actin polymerization and protrusion of the pseudopod of amoebae and neutrophils (Hawkins et al. 1992; Stephens et al. 1994; Klarlund et al. 1997; Van Haastert and Devreotes 2004; Vanhaesebroeck et al. 2012; Chen et al. 2024).

As GPCRs are globally distributed along the cell membrane, it is fascinating how well neutrophils and amoeboid cells can create and maintain polarity in chemotactic gradients with as low as 1% concentration differences between the front and the back of a cell that is only ~10 µm in length (Zigmond 1977; Servant et al. 1999; Onsum and Rao 2007). The local-excitation global-inhibition model (LEGI) has been used to explain this phenomenon (Parent and Devreotes 1999). The model suggests that binding of a chemoattractant induces a local self-enhancement of the signal at the cell front and a slower, diffusible, global inhibition of the signal. While the inhibition signal suppresses the back of the cell, it can be overridden by the positive feedback loop at the front allowing for polarized signaling activity (Parent and Devreotes 1999; Surve et al. 2016).

In neutrophils, long-range activation of Rho GTPases cooperates with local GPCR signaling to regulate cell polarity. Using optogenetic tools, a recent study from the Weiner group showed that local ectopic activation of Rac not only creates a protrusive front in unpolarized neutrophil-like HL-60 cells, but also activates Rho and actomyosin contraction at the opposite pole of the cell. Similarly local Rho-based contraction in the back can induce activation of Rac at the cell front (De Belly et al. 2025). The signaling networks regulating cell polarity and migration has been covered extensively elsewhere and readers are referred to excellent reviews for more in-depth details of this process (Devreotes and Horwitz 2015; Rappel and Edelstein-Keshet 2017).

Thus, chemoattractant binding to cell surface receptors initiates signaling cascades that, via PI3K and small GTPases, induce intracellular cytoskeletal changes to mediate cell polarization with a clear contractile back and a protruding front facing the chemoattractant.

Extension of the Leading Edge

Once a cell has established an internal polarity, the first physical step of migration is membrane protrusion at the leading edge. In mesenchymal migration, forward protrusion of the cell edge is generated by a broad, 1–2 µm-thick flap of branched actin filaments covered by a membrane that is known as a lamellipodium (Abercrombie et al. 1970; Ponti et al. 2004; Chhabra and Higgs 2007). Addition of actin monomers to the barbed ends of actin filaments that are facing the leading edge of the cell generates a pushing force on the membrane (Pollard and Borisy 2003). After growth factors induce Rac activation at the plasma membrane, this causes activation of the WAVE regulatory complex which in turn activates the Arp2/3 complex that nucleates branched actin assembly (Ridley et al. 1992; Mehidi et al. 2019). This causes the cell membrane to be pushed forward while simultaneously pushing the growing actin network backward, causing actin retrograde flow, during which filamentous actin (F-actin) is severed and depolymerized (Iwasa and Mullins 2007). This assembly and disassembly of actin causes actin treadmilling within the protruding cell front (Mitchison and Kirschner 1988).

In contrast to mesenchymal migration mediated by lamellipodial protrusion, in amoeboid migration, membrane protrusion takes place either through actin polymerization in cylindrical pseudopods or membrane blebbing at the cell front (Stockem et al. 1982; Condeelis et al. 1988; Blaser et al. 2006). Upon GPCR activation, PI3K converts phosphatidylinositol 4,5-bisphosphate (PIP_2) to phosphatidylinositol 3,4,5-trisphosphate (PIP_3) at the cell membrane that causes Rac and Cdc42 activation to drive pseudopod formation (Stephens et al. 1994; Benard et al. 1999; Wang et al. 2002; Chen et al. 2024). Pseudopods are rich in branched actin assembly activators and consist of dense actin networks and lack cortical actin (Fritz-Laylin et al. 2017a,b). The key difference between Arp2/3 driving lamellipodial versus pseudopodial protrusion is that lamellipodia are physically coupled to their environment via adhesion receptors, as described later, while pseudopods are not. In contrast to pseudopods, membrane blebs are spherical membrane protrusions detached from the cortical actin cytoskeleton consisting of cytosol and devoid of polymerized actin and myosin II during their initial stages (Cunningham 1995; Charras et al. 2006). Under confinement, cell protrusion by blebbing is dependent on myosin II–mediated contraction of the cell cortex that generates fluid pressure to create a protruding membrane bubble that detaches from the underlying actin cortex (Blaser et al. 2006; Ibo et al. 2016). In amoeba, stimulation with the chemoattractant cyclic adenosine monophosphate (AMP) induces phosphorylation of myosin II, which drives membrane blebbing to drive bleb protrusions (Zatulovskiy et al. 2014).

Neutrophils have been observed to extend lamellar structures, pseudopods and membrane blebs, and are hence able to form all these protruding structures (Zhelev et al. 2004; Fritz-Laylin et al. 2017b; Schrope et al. 2025; Wu et al. 2025). In addition, neutrophil-like HL-60 cells have also been observed to protrude pseudopod sheets and rosettes that, in contrast to mesenchymal lamellipodial protrusions, are not in constant contact with the substrate (Fritz-Laylin et al. 2017b). Interestingly, stimulating primary human neutrophils and HL-60 cells with a chemoattractant induces downstream signaling that activates chloride-bicarbonate exchanger 2 (AE2) and sodium-proton antiporter (NHE1), causing an influx of water and cell swelling that correlates with enhanced cell migration (Nagy et al. 2024). The precise mechanism of how chemoattractant-induced cell swelling mediates neutrophil migration is yet to be determined, but as cell swelling increases hydrostatic pressure it is possible that this takes place by inducing membrane bleb-based cell protrusions.

Thus, extracellular stimuli can activate intracellular signaling pathways that induce various membrane protrusions. These cell protrusions are all highly dynamic and need to interact with the cell environment to mediate forward movement of the cell.

Attachment of Protrusion

For a protrusion to lead to productive cell movement, it must somehow attach to its environment and generate a force against it. In mesenchymal migration, nascent adhesions physically couple the force generated by actin treadmilling in lamellipodial protrusions to the underlying substrate via integrin adhesion receptors, while the lack of adhesion results in nonproductive actin retrograde flow and/or membrane ruffling (Borm et al. 2005; Choi et al. 2008; Bachir et al. 2014). Integrin receptors bind to ECM ligands on the outside of the cell and the actin cytoskeleton, via adaptor proteins such as talin and vinculin, on the inside of the cell, causing the actin retrograde flow to become coupled to the extracellular environment and to thus stall its movement (Horwitz et al. 1986; Alexandrova et al. 2008; Chastney et al. 2021; Lu et al. 2022). As actin still polymerizes at the leading edge, but is now coupled to the substrate, the force of further polymerization causes the leading edge to move forward, while the stabilized and growing adhesions exert rearward pushing traction on the underlying substrate to drive the cell forward in the direction of the protrusion (Gardel et al. 2008, 2010).

In contrast to mesenchymal motility, amoeboid migration does not involve formation of adhesion complexes at the leading edge (O'Neill et al. 2018). Instead, amoeboid cells move forward by actin retrograde flow generating nonspecific friction against the environment through any transmembrane protein that is coupled to the cytoskeleton (Malawista and De Boisfleury Chevance 1997; Hawkins et al. 2009; Lämmermann and Sixt 2009; Bergert et al. 2015; Aoun et al. 2020). In confined environments, friction is enhanced by cells exerting forces perpendicular to the surface to chimney forward without using adhesion receptors (Malawista et al. 2000; Charras and Paluch 2008). In this mode, bleb protrusions at the cell front causes shear stress in the direction of the expanding bleb while normal stresses are generated onto the confining substrate to anchor the cell in place. Simultaneously, actomyosin contraction at the back of the cell both causes the bleb to grow and the actin cortex behind the bleb to be pulled rearward, allowing the cell to squeeze itself forward and generate rearward friction onto the confining environment. This enables the cell to migrate forward and wedge itself where the initial bleb was located (Yip et al. 2015).

While amoeboid pseudopods are nonadherent, pseudopods of neutrophils occasionally contain integrins that mediate adhesion and aid in migration on ECM (Lawson and Maxfield 1995). In addition, in response to inflammation, primary human neutrophils redistribute adhesion molecules such as L-selectin and platelet endothelial cell adhesion molecule (PECAM) to pseudopods to mediate initial adhesion to the endothelium (Liu et al. 1998; Fu et al. 2023). Furthermore, adhesion molecules are also present in the uropod, at the back of human and mouse neutrophils, where they mediate adhesion to the substrate (Szczur et al. 2009; Hind et al. 2016; Andreata et al. 2023). Thus, neutrophils can interact with their environment both by adhering their pseudopods and uropods and by generating nonspecific friction against the environment when blebbing.

Cell Advance and Retraction of Rear

The final step of cell migration consists of forward movement of the cell body and retraction and de-adhesion of the cell rear to allow the contractile forces to pull the back of the cell forward and recycle actin monomers and myosin filaments to the front of the cell. Following adhesion of the leading edge protrusion, the rearward force of retrograde flow pushes backward against stiff ECM and generates rearward traction force that results in net forward movement of the cell body (DiMilla et al. 1991; Jay et al. 1995; Lauffenburger and Horwitz 1996; Bangasser et al. 2013). Mesenchymal cells use anterior actomyosin contractions, localized between the nucleus and the leading edge, and integrin ligation to generate traction on the ECM to push the cell forward and also locally remodel the ECM (Doyle et al. 2021; Kanchanawong and Calderwood 2023). In mesenchymal cells, de-adhesion can be induced purely by the force of contraction, in which case the linkage between the actin cytoskeleton and integrin receptors is physically broken and integrins are left on the substrate as a trail (Regen and Horwitz 1992). Adhesions can also be disassembled in a regulated manner, driven by intracellular signaling inducing cleavage and internalization of adhesion adaptor proteins (Ridley et al. 2003; Wehrle-Haller and Imhof 2003).

In amoeboid migration that lacks stable adhesions, cell retraction is driven by high Rho signaling that via its downstream effector Rho-associated protein kinase (ROCK) induces actomyosin contraction (George et al. 2023). The contraction zone of amoeboid cells, which is localized behind the protruding zone, consists of a thick cortical layer with both thin actin filaments and actomyosin bundles (Stockem et al. 1982). These structures are in contact with the plasma membrane and generate local contraction of the actin network at the back of the cell causing forward flow of the cytoplasm, squeezing the nucleus and cell body forward

 Cite this article as *Cold Spring Harb Perspect Biol* doi: 10.1101/cshperspect.a041749

(Stockem et al. 1982; Lämmermann et al. 2008). Actomyosin contraction allows for a quick retraction of the most posterior zone of amoeboid cells (Uchida et al. 2003; Delanoë-Ayari et al. 2008). This region is characterized by dense cortical actin and thick actomyosin bundles along with single myosin filaments in the cytoplasm, indicating that actomyosin disassembly take place in this region, allowing for continued streaming of monomers from the cell rear back to the regenerative front where the migration cycle can start again (Stockem et al. 1982).

Chemotaxing primary human neutrophils have been observed to generate traction forces that, like mesenchymal cells, are counteropposed to the direction of cell movement. In stark contrast to forces in mesenchymal cells, these traction forces are generated almost exclusively by the contractile uropod while no traction forces are observed at the lamellipodia. These uropod generated traction forces are thought to break the adhesions in the uropod while also squeezing the cell interior forward in an amoeboid-like manner (Smith et al. 2007). In addition, inhibiting integrin internalization from the uropod promotes frustrated migration where primary mouse and human neutrophils have been seen to become elongated as they are unable to detach the uropod while attempting to move forward, suggesting that regulated disassembly of uropod adhesions also play a role in rear retraction of neutrophils (Cera et al. 2009).

Thus, neutrophils use elements of both mesenchymal and amoeboid cells as they advance forward and retract their rear. We next discuss the physiological settings in which neutrophils migrate, and, where it is known, how neutrophils use different aspects of mesenchymal and amoeboid migration in these different microenvironments.

NEUTROPHIL RECRUITMENT INTO AND OUT OF THE CIRCULATORY SYSTEM

Upon inflammation, the number of white blood cells swiftly increases in the blood circulation due to their release from the bone marrow (Furze and Rankin 2008). From the circulation, they are captured by and roll along the endothelium until they arrest and form strong adhesions to migrate along the endothelial wall until finding a suitable location for extravasation in a process known as the leukocyte adhesion cascade. Once adhered, they exit the bloodstream and migrate through tissue to the site of inflammation, where they do their job of neutralizing invaders and clearing dead cells and debris, after which they finally return to the circulation (Ley et al. 2007). During these processes, neutrophils are required to squeeze through extremely narrow spaces and they also experience exceptionally high shear flow forces in the bloodstream that they have to overcome to firmly adhere and migrate (Atherton and Born 1972; Lawrence et al. 1997; Kim and Sarelius 2004; Voisin et al. 2009; Nourshargh et al. 2010; Mukhopadhyay et al. 2024). Here we will begin by discussing neutrophil release from the bone marrow into the bloodstream. We will then explore the adhesion cascade and how neutrophils extravasate from the blood circulation and navigate to and from the inflamed tissue under these challenging conditions.

Neutrophil Maturation and Release from the Bone Marrow

On a daily basis in humans as many as 10^{11} mature neutrophils are released from the hematopoietic compartment of the bone marrow into the circulation via transcellular migration through the sinusoidal endothelium (Burdon et al. 2008; Furze and Rankin 2008). As neutrophils are produced and matured in the bone marrow their nuclei undergo significant changes becoming both lobulated and softer compared to other cell types (Lammerding et al. 2004, 2006; Cohen et al. 2008; Lee et al. 2011; Manley et al. 2018; Wang et al. 2021). These physical properties of the nuclei are thought to facilitate the migration through constricted spaces, for instance when neutrophils migrate out from the bone marrow stroma and squeeze themselves between and through endothelial cells into the blood circulation (Nourshargh et al. 2010; Rowat et al. 2013; Shen et al. 2023). The number of circulating neutrophils can be significantly elevated upon inflammation as more neutrophils are recruited from the bone marrow reserve (Wengner et al. 2008).

The decision to stay in or leave the bone marrow is regulated by the balance between different chemoattractants in the marrow versus in the circulating plasma (Furze and Rankin 2008). Stromal cells in the bone marrow constitutively express the chemoattractant stromal cell–derived factor 1 (SDF1, Cxcl12) that binds and retains neutrophils in the bone marrow via the chemokine receptor CXCR4 (Martin et al. 2003). The granulocyte-colony stimulating factor (G-CSF) is the major regulator of this interaction and is upregulated by several inflammatory stimuli in response to infection (e.g., interleukin-1β, lipopolysaccharide, and tumor necrosis factor α) (Martins et al. 2010). An increase in both local and circulating G-CSF causes disruption of the SDF1–CXCR4 interaction and thus interrupts the bone marrow retention of neutrophils (Lévesque et al. 2003; Hyun et al. 2006; Panopoulos and Watowich 2008; De Filippo and Rankin 2018). G-CSF also induces increased expression of the chemokine receptor CXCR2 in murine neutrophils, whose ligands CXCL1 and CXCL2 are constitutively expressed on the bone marrow endothelium, thus causing a mobilization into the vascular space of the bone marrow (Eash et al. 2010). Macrophages also produce the inflammatory protein-2 (MIP-2) in response to infection, which has been shown to increase the expression of the adhesion molecule integrin α4 on murine neutrophils and their mobilization from the bone marrow (Burdon et al. 2005). In human neutrophils, integrin α4 is able to bind vascular cell adhesion molecule-1 (VCAM-1), which is expressed both in the bone marrow stroma and on sinusoidal endothelium, suggesting that upregulation of this adhesion molecule allows for adhesion to the sinusoidal endothelium to facilitate neutrophil migration into the circulatory system (Simmons et al. 1992; Burdon et al. 2005). Supporting this, blocking integrin α4 significantly reduces MIP-2-induced mobilization of murine neutrophils into the bone marrow, suggesting that integrin α4 mediates adhesion-dependent migration of neutrophils out of the bone marrow into the circulatory system (Burdon et al. 2005).

Live cell imaging of neutrophil mobilization from the bone marrow has proven to be difficult and therefore the migration mode of neutrophils exiting the bone marrow has not to our knowledge been characterized in detail yet. It is likely that they migrate in a mostly amoeboid fashion, squeezing themselves into the blood circulation while also using elements of adhesion-dependent mesenchymal migration as inflammatory signaling upregulates the expression of the adhesion receptor integrin α4 to induce neutrophil mobilization from the bone marrow.

Neutrophil Adhesion to and Crawling along the Endothelium

In addition to mobilizing neutrophils, inflammatory stimuli activate endothelial cells close to the site of infection to upregulate the expression of selectin adhesion molecules. These endothelial selectins bind to the constitutively expressed counter selectin ligands on neutrophils and capture them from the bloodstream (Wyble et al. 1997; Fassbender et al. 1998; Miyamoto et al. 1998). The transient selectin–selectin interactions allow neutrophils to roll on the endothelium due to the shear stress they experience from the blood flow (Damiano et al. 1996; Sundd et al. 2010). During this rolling, neutrophils are exposed to inflammatory cues presented by the endothelium, which induce activation of leukocyte-specific β2 integrin adhesion receptors that mediate firm neutrophil adhesion and arrest (Wen et al. 2022a,b; Singh et al. 2023).

Once firm adhesion is established, neutrophils start crawling along the endothelium allowing them to probe their surfaces for optimal sites of transmigration to extravasate. This crawling is guided by gradients of chemoattractants and chemokines presented by endothelial cells to guide neutrophils toward suitable extravasation sites (Grönloh et al. 2021). Neutrophil crawling is mediated by integrin adhesions to intercellular adhesion molecule (ICAM)-1 expressed on the surface of the endothelial cells (Gorina et al. 2014). While some redundancy between the leukocyte-specific integrins exists, as they all are capable of mediating endothelial adhesion, the integrin αLβ2 (also knowns as lymphocyte function-associated antigen 1 [LFA-1]) is recognized as the main receptor mediating firm adhesion while integrin αMβ2 (macrophage-1 antigen [Mac-1]) is crucial for neutrophil crawling (Phillipson et al. 2006). This difference in function is most likely due to differences in the kinetics and bond strengths of these two integrins, with αLβ2 providing stronger interactions to ICAM optimal for firm adhesion while the weaker bond strength of αMβ2 is better suited for migration (Li et al. 2013, 2018). Similarly, the ligands ICAM-1 and -2 play various roles during neutrophil crawling. Human neutrophils under shear flow are able to migrate equally against and in the direction of flow as they search for transmigration locations on the endothelium. However, migration against the flow is dependent on ICAM-2 interactions (Grönloh et al. 2023a). While ICAM-2 is steadily expressed on the endothelium, ICAM-1 expression is upregulated in response to inflammatory signals and distributed to endothelial cell–cell junctions where it serves as adhesion and transmigration hotspots for neutrophils (Grönloh et al. 2023b,c).

Thus, during arrest and migration on the endothelial surface, neutrophils use a migration mode mostly resembling mesenchymal migration that depends clearly on integrin adhesion complexes that allow for firm attachment to, and migration along, the endothelium, while withstanding the high forces caused by the blood flow.

Neutrophil Extravasation

After migration along the endothelium, neutrophils pick a site to exit the bloodstream by crawling through the blood vessel wall. Once migration on the endothelium is stopped, neutrophils adhere to the endothelium via β2-integrins and display microvillar contacts and small invaginations into the endothelium (Cinamon et al. 2004). In response to shear flow, human neutrophils extend protrusions that invade into the endothelial cell surface and endothelial cell–cell junctions and begin the transmigration process in an integrin-dependent manner, usually within a cell-lengths distance from where they adhered (Cinamon et al. 2004). Exactly where neutrophils decide to transmigrate is an interesting question. Migrating murine leukocytes have been observed to extend 400–800 nm long, spherically tipped podosome-like protrusions into the endothelium and the cell–cell junctions, as if to probe the endothelial surface stiffness while they migrate (Martinelli et al. 2014). Many cell types, including neurons and cancer cells, use tiny finger-like protrusions known as filopodia to probe their environment during cell migration (Chan and Odde 2008; Jacquemet et al. 2016). These filopodial protrusions contain growth factor receptors, cell adhesion receptors, and mechanosensitive adhesion proteins and are specialized in sensing ECM-derived cues (Jacquemet et al. 2019). It is still not clear whether the protrusions of neutrophils can act as mechanosensors, and, if so, which proteins would be involved in this process. However, the efficiency of diapedesis inversely correlates with F-actin density and stiffness, suggesting that leukocyte protrusions probe the endothelium for softer transmigration sites (Martinelli et al. 2014).

Neutrophils transmigrate either in a paracellular manner, where they migrate between the cell–cell junctions of two or more cells, or a transcellular manner where the migration takes place through one endothelial cell (Mamdouh et al. 2009; Yeh et al. 2018). During transendothelial migration, the endothelial cells actively participate in the transmigration process (Mooren et al. 2014). In paracellular transmigration, neutrophils crawl along endothelial cell–cell junctions and prefer to exit at dynamic junctional membrane protrusions that functions as transendothelial hotspots with an apical lamella that guides neutrophils to efficiently traverse the inflamed endothelia (Arts et al. 2021). These protrusions are dynamic and vary highly in their morphology appearance, with some resembling wide flaps while others are more finger-like (Arts et al. 2021). Once a neutrophil exits at a junctional membrane protrusion, the junctional membrane protrusion becomes even more active, which seems to reinforce the site for transmigration as other neutrophils follow the same path to cross the endothelium (Hyun et al. 2019; Arts et al. 2021). The endothelium has also been shown to form protrusive cup-like structures with vertical projections surrounding the transmigrating leukocytes and guiding the extravasation process both in para- and transcellular migration (Carman and Springer 2004). During transcellular migration, human leukocytes extend protrusions that probe the endothelium for permissive transmigration sites. This probing also induces transcellular endothelial invaginations and pores that rapidly widen to allow transmigration and then constrict and reseal once the neutrophil has transmigrated (Carman et al. 2007). While it is still somewhat unclear what regulates the route of choice for neutrophils during transmigration, they seem to prefer the path of least resistance, as only ~10% of human leukocyte diapedesis take place via transcellular migration while the remaining portion extravasate at cell–cell junctions (Carman and Springer 2004). Reducing endothelial cell–cell junction integrity increases the fraction of murine leukocytes undergoing paracellular migration, whereas increasing junction integrity increases the fraction of transcellular migration and reduces paracellular migration (Martinelli et al. 2014). Furthermore, treating endothelial cells with cytochalasin D for a short time to decrease the actin filament density throughout the cells, without disrupting the cortical actin at cell junctions, increases leukocyte transcellular migration (Martinelli et al. 2014). These results suggest that in cases where the cell–cell junctions are too tight for paracellular migration, leukocytes use transcellular migration to breach the endothelium (Martinelli et al. 2014).

Once neutrophils have transmigrated the endothelium they still need to migrate through the pericyte layer embedded in the venular basement membrane. Pericytes are long cells that wrap around the endothelial cell layer in a non-confluent manner, displaying gaps between adjacent cells, and they play a crucial role in both maintaining vascular integrity and generating the venular basement membrane (Bergers and Song 2005). These cells are able to secrete chemokines and upregulate their expression of integrin ligands in response to proinflammatory signals, thereby taking part in leukocyte recruitment to site of inflammation (Navarro et al. 2016). Confocal intravital microscopy imaging of mouse cremasteric venules has shown that following endothelial transmigration, a process that takes ~4 min, neutrophils undergo abluminal crawling for ~20 min between the endothelial and pericyte layer before exiting the pericyte sheet (Proebstl et al. 2012). During abluminal crawling, neutrophils migrate in an ICAM-, Mac-1-, and LFA-1-dependent manner on top of the pericytes, suggesting that they use adhesions in a mesenchymal manner (Proebstl et al. 2012). However, the compression of the endothelial layer makes both human and murine neutrophils take on a flattened phenotype with reduced elongation and their migration resembles amoeboid cells with the nucleus being squeezed forward by the cell rear (Proebstl et al. 2012; Grönloh et al. 2023b). Unlike transendothelial migration, neutrophils do not undergo transcellular pericyte migration but instead exit through enlarged pericyte paracellular gaps (Wang et al. 2006). However, similar to transendothelial migration, neutrophils seem to follow each other and use the same pericyte gaps as neutrophils before them when extravasating (Proebstl et al. 2012).

Thus, neutrophils adhere and crawl on top of the endothelium in a mesenchymal, integrin-dependent manner while probing the endothelium for suitable transmigration sites. As neutrophils begin their transmigration, they extend long protrusions that invade into the endothelium either in a trans- or paracellular process. While integrin adhesions allow neutrophils to adhere to the endothelium during this process, neutrophils squeeze themselves through the endothelial layer in an amoeboid manner. During abluminal crawling, neutrophils migrate on top of the pericyte layer in an integrin-dependent manner, but with an amoeboid-like morphology as they are squeezed between the two cell layers. Next, we will discuss how neutrophils migrate in the tissue toward the site of inflammation.

 Cite this article as *Cold Spring Harb Perspect Biol* doi: 10.1101/cshperspect.a041749

NEUTROPHIL MIGRATION TOWARD SITE OF INFLAMMATION

Once neutrophils leave the circulatory system, they must then begin their journey through tissue from the vascular exit point to the site of insult. As neutrophils migrate in the tissue toward the site of inflammation, the biochemical and physical properties of the extracellular environment affect neutrophil migration as they maze their way through various ECM structures and between cells. Here we will discuss both how physical properties of the ECM affects cell migration, and how neutrophils themselves amplify inflammatory signals to enhance neutrophil recruitment to the site of infection.

Neutrophil Migration in Tissue

As neutrophils migrate in tissue, they are surrounded by tightly packed cells and the ECM, a heterogenous network of macromolecules creating fibers of various stiffness and orientation. These both provide ligands for neutrophil adhesion but also create physical barriers for their migration. Neutrophil migration was demonstrated to resemble amoeboid more than mesenchymal migration in the 1980s when a study showed that rabbit neutrophils adhere and migrate poorly on collagen in two-dimensional (2D) but are capable of migrating efficiently in three-dimensional (3D) collagen matrices (Brown 1982). Since then, adhesion-independent migration of leukocytes has been observed both in vitro and in vivo. Blocking integrin activity with ethylenediaminetetraacetic acid or function-blocking antibodies does not inhibit directional migration of human neutrophils in vitro, and knockdown of all integrins in murine dendritic cells, another subtype of leukocytes, likewise does not inhibit migration in the ear dermis in vivo (Malawista and De Boisfleury Chevance 1997; Lämmermann et al. 2008). These results suggest that leukocytes can migrate in an integrin adhesion-independent manner by actin-based protrusion of the front and myosin generated contraction of the rear that increases the intracellular pressure and thus friction against the environment to drive cell migration in tissue (Malawista et al. 2000; Lämmermann et al. 2008; Lämmermann and Germain 2014).

Although neutrophils can migrate in tissues independent of integrins, they are capable of forming integrin-ECM adhesions, as immunostaining of human neutrophils migrating in 3D ECM of the amnion reveal the presence of ECM-binding integrins, together with F-actin and actin cross-linkers within pseudopods at the leading edge (Mandeville et al. 1997). These pseudopod structures extend laterally and become stationary in close association with ECM fibers as neutrophils migrate, suggesting that pseudopods can anchor the neutrophil to the ECM as they migrate in 3D (Mandeville et al. 1997). Measuring adhesion forces generated by neutrophils in 3D and in vivo is technically challenging, and it is thus hard to determine to what extent they use integrin adhesions to pull themselves forward and mechanically deform the ECM. However, in a mouse study examining neutrophil recruitment to a sterile dermal wound, neutrophils were observed to locally clear the dense collagen from the wound center by physically remodeling the ECM network that they then clustered to (Lämmermann et al. 2013). This suggests that neutrophils are able to form strong enough integrin adhesions to rearrange the ECM even if these adhesions are dispensable for neutrophil migration. While the mouse study did not examine the precise role of integrin adhesions in ECM remodeling, it found that neutrophils lacking the integrin-actin adaptor molecule talin were capable of chemotaxing toward the wound but not able to migrate into the ECM-cleared center (Lämmermann et al. 2013). These results suggests that integrin adhesions are dispensable when neutrophils are surrounded by dense ECM but indispensable for adhesion and migration in areas free from collagen fibers.

For continued migration, neutrophils can either degrade the barrier, mechanically move the barrier or navigate around the barrier. Even if neutrophils express a large number of proteinases, they do not extensively degrade matrix fibers while migrating (Mandeville et al. 1997). Instead, they tend to migrate through existing gaps. Human neutrophils extend balloon-like protrusions resembling blebs that grow as the cell squeezes itself forward through a stationary structure resembling an actomyosin constriction ring produced by narrow openings in the ECM (Mandeville et al. 1997). As the neutrophil squeezes itself through such openings, it mechanically stretches the ECM in a reversible manner (Mandeville et al. 1997). While small pores may restrict neutrophils from migrating through, they also can serve as footholds for neutrophils. Pseudopods rich in actin and integrins have been observed extending from neutrophils into small pores, being used to migrate across the pore instead of through it (Mandeville et al. 1997). Neutrophils also use these dynamic pseudopod extensions to probe ECM fibers that they wrap around and pull themselves forward. Interestingly, neutrophils tend to choose more stable and stiffer fibers when migrating, suggesting that in addition to migrating up chemical gradients, they also migrate up ECM stiffness gradients (Mandeville et al. 1997).

Thus, as neutrophils migrate in tissue, they are capable of using elements from both mesenchymal and amoeboid migration. While their pseudopods display integrins, actin and actin cross-linkers that they use to migrate along ECM fibers in a mesenchymal manner, these adhesions are dispensable for interstitial migration. In the absence of integrin adhesions, neutrophils migrate in an amoeboid-like manner by generating friction against the surrounding ECM, induced by rear contraction and increased intracellular pressure. It is therefore likely that neutrophils use a combination of both of these modes, using both the provided ECM ligands to migrate along in an adhesion-dependent manner and the compression of the ECM and other cells for friction-based chimneying.

Amplification of Inflammatory Signals and Neutrophil Swarming

Neutrophils migrate in a highly coordinated fashion toward the source of inflammation in tissue. In addition to primary chemoattractants inducing cell-intrinsic feedforward loops mediating cell polarity and migration, secondary cell extrinsic signals are also important in reinforcing neutrophil migration to the right place (Parent and Devreotes 1999; Chou et al. 2010; Afonso et al. 2012; Surve et al. 2016). Migrating human and murine neutrophils that are initially recruited to inflammation sites by primary chemoattractants subsequently secrete secondary chemoattractants such as LTB4 that act as signal-relay molecules to amplify chemotaxis of neutrophils (Dahinden et al. 1988; Afonso et al. 2012). This signal amplification allows for a biphasic neutrophil response, with an initial recruitment of neutrophils close to the damage site taking place within the first 15 min (Ng et al. 2011). The local recruitment is followed by a secondary recruitment of neutrophils from more distal sites in the tissue during which the number of recruited neutrophils substantially increases for an additional 15 min before it plateaus (Chtanova et al. 2008; Ng et al. 2011). This secondary recruitment displays a swarming behavior, compared to a bee swarm, as large amounts of neutrophils, located further away from the wound, migrate both faster and more directionally than the first responders and form a large cluster at the site of inflammation (Chtanova et al. 2008; Lämmermann et al. 2013). Neutrophil swarming is LTB4-dependent, as the secondary recruitment of neutrophils from distal sites but not the initial local recruitment is nearly abolished in neutrophils lacking the LTB4 receptor (Lämmermann et al. 2013).

LTB4 is produced and released by neutrophils in a concentration-dependent manner in response to primary chemoattractants (Afonso et al. 2012). Both LTB4 and its synthesizing enzymes are packaged into nuclear envelope–derived multivesicular bodies that fuse with the plasma membrane and are secreted as exosomes (Majumdar et al. 2021; Arya et al. 2022). These LTB4 exosomes induce neutrophil chemotaxis in both an auto- and paracrine manner. LTB4 enhances chemoattractant-induced myosin phosphorylation and cell polarization, and also works as an attractant for neutrophils that are further away and unable to sense the primary chemoattractant, and thus functions as a signaling relay molecule (Afonso et al. 2012; Majumdar et al. 2016; Subramanian et al. 2020). Interestingly, a recent preprint from the Parent laboratory (Arya et al. 2025b) found that the nuclear constriction that neutrophils often are subjected to as they migrate through confining tissue microenvironments increases LTB4 synthesis and secretion in neutrophil-like HL-60 cells. In addition, constricted HL-60 cells also undergo a shape change, displaying a wider and more lateral lamellipodium instead of the more amoeboid-like pseudopod, suggesting that the migration mode might switch from amoeboid to an intermediate mode using elements from both adhesive mesenchymal and amoeboid migration. These results suggest that the mechanical constriction neutrophils experienced by the ECM or other cells, induces increased LTB4 signaling to promote efficient chemotaxis of neutrophils in tissue (Arya et al. 2025b).

Thus, as neutrophils migrate in tissue, they are not only guided by primary chemoattractants released at the site of inflammation, but neutrophils themselves also secrete secondary chemoattractants. These secondary attractant molecules influence both cell polarity and chemotaxis in an autocrine and a paracrine manner to guide abundant neutrophil recruitment to the site of inflammation. The extremely rapid and directional migration displayed by swarming neutrophils suggest that their migration takes place in an amoeboid manner. Next, we will discuss how neutrophils migrate away from the site of inflammation and back to the circulation.

REVERSE MIGRATION AND RESOLUTION

Once the infection is resolved, infiltration of additional neutrophils is ceased and residing neutrophils are either cleared by other immune cells or undergo reverse migration back to the bloodstream or the lymphatic system, from where they return to the lung, spleen, liver, and bone marrow for clearance (Furze and Rankin 2008). In sterile injury in mice, neutrophils migrate to the site of injury within 4 h and their presence peaks at 12 h postinjury, where after the amount progressively reduces until returning to nearly basal levels 48 h postinjury (Arya et al. 2025a). This is important, as excess neutrophil infiltration can cause chronic inflammation (Elgazar-Carmon et al. 2008; Herrero-Cervera et al. 2022). While it was long thought that neutrophil clearance take place via programmed cell death at the site of inflammation, it is now known that only ~10% of neutrophils undergo apoptosis while the majority return to the bone marrow (Mathias et al. 2006; Wang et al. 2017).

Clearance of damaged or dead neutrophils at the site of inflammation takes place through efferocytosis, a noninflammatory phagocytic process where macrophages and neutrophils clear apoptotic cells (Ge et al. 2022; Ramos and Oehler 2024). In contrast to phagocytosis of pathogens, that often induce a proinflammatory response in the phagocytosing cell, efferocytosis activates an anti-inflammatory response to induce resolution (Ramos and Oehler 2024). Some mediators of resolution dampen the recruitment of neutrophils by downregulating the expression of chemokine receptors, thus making neutrophils insensitive to chemoattractants (Schett and Neurath 2018). Efferocytosis can also activate anti-inflammatory transcription factors such as members of the peroxisome proliferator-activated receptors (PPARs) family and their downstream signaling pathways in phagocytosing cells to upregulate the production and secretion of anti-inflammatory cytokines that aid in the resolution of inflammation (Ramos and Oehler 2024). Recent work from the Parent group displayed how this is mediated by migrating human neutrophils secreting LTB4-containing nonconventional exosomes together with sticky DNA fragments allowing for local spots with high concentrations of LTB4. While low concentration of LTB4 activates the GPCR leukotriene B4 receptor (BLT1) to promote chemotaxis, high concentration of LTB4 activates the BLT2 GPCR, which activates PPARα to induce reverse migration of neutrophils and resolution of inflammation (Arya et al. 2025a).

Human neutrophils migrating away from a highly concentrated chemoattractant display a similar morphology to neutrophils migrating toward chemoattractants, but with their leading edge facing away from the chemoattractant source, instead of toward (Tharp et al. 2006). In a mouse study where neutrophils were photoactivated to allow tracking of their migration from the site of inflammation, it was revealed that they either migrated into healthy tissue or reverse-migrated back into the vasculature. Most of these photoactivated neutrophils were found in the bone marrow 24 h later, while a small population was observed in the lung where they shortly resided before migrating to the bone marrow (Wang et al. 2017). Furthermore, it was found that CXCR4 drives this injury-derived neutrophil trafficking from lung to bone marrow, as treatment with the antagonist, AMD3100, significantly reduced the number of neutrophils that trafficked to the bone marrow (Wang et al. 2017). CXCR4 is also involved in regulating the return of aging neutrophils back to the bone marrow (Martin et al. 2003). The estimated half-life of human neutrophils in the circulatory system under homeostatic conditions was long thought to only be a few hours, but more recent measurements point to half-lives as long as a few days, in contrast to murine neutrophils that are still thought to have a life span of less than a day (Mauer et al. 1960; Pillay et al. 2010). However, pathogen encounter can either prolong or reduce the life span of neutrophils, thus altering the resolution of inflammation (McCracken and Allen 2014). Thus, the precise half-life of neutrophils is still debated. Interestingly, as neutrophils age (6 and 3 h onward in murine and human neutrophils, respectively), they display an increase in surface expression of CXCR4 (Martin et al. 2003; Möhle et al. 2009). In addition, they display an increased ability to migrate toward the CXCR4 ligand SDF-1 (also known as CXCL12) that is highly expressed by stromal cells in the bone marrow (Martin et al. 2003). It is therefore likely that CXCR4 expression on neutrophils and SDF-1 expression in the bone marrow regulate the return of aged neutrophils back to the bone marrow for clearance (Martin et al. 2003). Interestingly, neutrophils from aged mice and elderly human patients display reduced sensitivity and migration toward chemokines, suggesting that not only the age of neutrophils themselves, but also the age of the organism, affects the neutrophil phenotype (Fulop et al. 2004; Brubaker et al. 2013).

Hence, while damaged or dead neutrophils are phagocytosed in the tissue, healthy neutrophils return to the bone marrow through reverse migration once the inflammation has been resolved. Both phagocytic activity and high local secretion of LTB4 by neutrophils can activate transcription factors such as PPARα to upregulate the production and release of anti-inflammatory cytokines, and thus induce neutrophil reverse migration out of the tissue. While high-resolution imaging of reverse-migrating neutrophils to our knowledge has not been performed, neutrophils exhibit similar morphologies migrating into and out of tissue, suggesting that they use primarily an amoeboid-like migration in the dense ECM, possibly also using integrin-adhesion-dependent modes when possible.

CONCLUSIONS AND PERSPECTIVES

Neutrophil migration is often classified as amoeboid migration. While their characteristic highly polarized morphology, with a contractile uropod and a pseudopod most often displaying low adhesion and traction, is in line with an amoeboid mode of migration, they certainly use elements of mesenchymal migration during their journey from the bone marrow to the site of inflammation.

Integrin expression is upregulated during neutrophil mobilization from the bone marrow to support migration into the circulatory system. Likewise, as neutrophils roll along the blood vessels, selectin engagement and chemokine exposure induces integrin activation to mediate firm adhesion and cell arrest to the endothelium. The adhered neutrophils then migrate in an integrin-dependent and mesenchymal manner along the substrate, probing the endothelial layer for optimal extravasation sites. For this process, they extend protrusions that are yet to be fully characterized, but it is possible that these protrusions contain mechanosensitive integrin adhesion complexes as neutrophils prefer to extravasate at softer sites. During extravasation, neutrophils squeeze themselves through the endothelial layer, a process that is facilitated by their comparatively soft and lobulated nuclei. This process is followed by abluminal crawling where neutrophils display a highly amoeboid morphology with a clear leading edge and contractile uropod pushing the cell and nuclei forward. However, this crawling is integrin-dependent, suggesting that neutrophils use a combination of mesenchymal and amoeboid migration. In tissue, neutrophils are capable of migrating in a completely integrin-independent and amoeboid manner as long as they are surrounded by ECM to push against and chimney through. Nevertheless, pseudopods rich in both integrins and actin regulatory proteins have been observed to interact with ECM fibers that neutrophils migrate along.

To what extent neutrophils push forward by generating friction against their environment, versus move forward by exerting traction forces during interstitial migration remains unknown, but using elements from both of these modes is likely what make them extraordinarily efficient and fast migrators. Due to neutrophils being short-lived and challenging to genetically modify, few studies have addressed the nature and composition of the adhesive and frictional interactions between neutrophils and their environment in vivo. As new force measuring tools and signaling probes are developed and improved, it will be interesting to see how much integrin adhesion signaling cross talks with intracellular signaling induced by chemotactic molecules to dictate neutrophil polarization and directed migration in vivo.

ACKNOWLEDGMENTS

The authors thank Denisa Wagner for comments on the manuscript, as well as for her and her laboratory members' Viktoria Jung and Camila M.S. Silva inspiration and knowledge of neutrophil biology that led us to work on these fabulous little cellular speedsters. In addition, we are grateful to Tim Mitchison and the members of the Waterman Laboratory for insightful discussions and feedback. We also thank the Marine Biological Laboratory in Woods Hole, MA, where the Waterman and Wagner laboratories collaborated together for 10 years. This work was supported by the Division of Intramural Research of the National Heart, Lung, and Blood Institute and the National Institutes of Health.

REFERENCES

Abercrombie M, Joan E, Heaysman M, Pegrum SM. 1970. The locomotion of fibroblasts in culture. II: "Ruffling." *Exp Cell Res* **60:** 437–444. doi:10.1016/0014-4827(70)90537-9

Afonso PV, Janka-Junttila M, Lee YJ, McCann CP, Oliver CM, Aamer KA, Losert W, Cicerone MT, Parent CA. 2012. LTB4 is a signal-relay molecule during neutrophil chemotaxis. *Dev Cell* **22:** 1079–1091. doi:10.1016/j.devcel.2012.02.003

Alexandrova AY, Arnold K, Schaub S, Vasiliev JM, Meister JJ, Bershadsky AD, Verkhovsky AB. 2008. Comparative dynamics of retrograde actin flow and focal adhesions: formation of nascent adhesions triggers transition from fast to slow flow. *PLoS One* **3:** e3234. doi:10.1371/journal.pone.0003234

Andreata F, Clément M, Benson RA, Hadchouel J, Procopio E, Even G, Vorbe J, Benadda S, Ollivier V, Ho-Tin-Noe B, et al. 2023. CD31 signaling promotes the detachment at the uropod of extravasating neutrophils allowing their migration to sites of inflammation. *eLife* **12:** e84752. doi:10.7554/eLife.84752

Angres B, Barth A, Nelson WJ. 1996. Mechanism for transition from initial to stable cell–cell adhesion: kinetic analysis of E-cadherin-mediated adhesion using a quantitative adhesion assay. *J Cell Biol* **134:** 549–557. doi:10.1083/jcb.134.2.549

Aoun L, Farutin A, Garcia-Seyda N, Nègre P, Rizvi MS, Tlili S, Song S, Luo X, Biarnes-Pelicot M, Galland R, et al. 2020. Amoeboid swimming is propelled by molecular paddling in lymphocytes. *Biophys J* **119:** 1157–1177. doi:10.1016/j.bpj.2020.07.033

Arts JJG, Mahlandt EK, Grönloh MLB, Schimmel L, Noordstra I, Gordon E, van Steen ACI, Tol S, Walzog B, van Rijssel J, et al. 2021. Endothelial junctional membrane protrusions serve as hotspots for neutrophil transmigration. *eLife* **10:** e66074. doi:10.7554/eLife.66074

Arya SB, Chen S, Jordan-Javed F, Parent CA. 2022. Ceramide-rich microdomains facilitate nuclear envelope budding for non-conventional exosome formation. *Nat Cell Biol* **24:** 1019–1028. doi:10.1038/s41556-022-00934-8

Arya SB, Collie SP, Xu Y, Fernandez M, Sexton JZ, Mosalaganti S, Coulombe PA, Parent CA. 2025a. Neutrophils secrete exosome-associated DNA to resolve sterile acute inflammation. *Nat Cell Biol* **27:** 931–947. doi:10.1038/s41556-025-01671-4

Arya SB, Jordan-Javed F, Loesel K, Choi Y, Collie SP, Hein LE, Baker BM, Yoon E, Parent CA. 2025b. cPLA$_2\alpha$ targeting to exosomes connects nuclear deformation to LTB$_4$-signaling during neutrophil chemotaxis. bioRxiv doi:10.1101/2025.07.02.662855

Asokan SB, Johnson HE, Rahman A, King SJ, Rotty JD, Lebedeva IP, Haugh JM, Bear JE. 2014. Mesenchymal chemotaxis requires selective inactivation of myosin II at the leading edge via a non-canonical PLCγ/PKCα pathway. *Dev Cell* **31:** 747–760. doi:10.1016/j.devcel.2014.10.024

Atherton A, Born GVR. 1972. Quantitative investigations of the adhesiveness of circulating polymorphonuclear leucocytes to blood vessel walls. *J Physiol* **222:** 447–474. doi:10.1113/jphysiol.1972.sp009808

Bachir AI, Zareno J, Moissoglu K, Plow EF, Gratton E, Horwitz AR. 2014. Integrin-associated complexes form hierarchically with variable stoichiometry in nascent adhesions. *Curr Biol* **24:** 1845–1853. doi:10.1016/j.cub.2014.07.011

Bangasser BL, Rosenfeld SS, Odde DJ. 2013. Determinants of maximal force transmission in a motor-clutch model of cell traction in a compliant microenvironment. *Biophys J* **105:** 581–592. doi:10.1016/j.bpj.2013.06.027

Barros-Becker F, Lam PY, Fisher R, Huttenlocher A. 2017. Live imaging reveals distinct modes of neutrophil and macrophage migration within interstitial tissues. *J Cell Sci* **130:** 3801–3808. doi:10.1242/jcs.206128

Bear JE, Haugh JM. 2014. Directed migration of mesenchymal cells: where signaling and the cytoskeleton meet. *Curr Opin Cell Biol* **30:** 74–82. doi:10.1016/j.ceb.2014.06.005

Benard V, Bohl BP, Bokoch GM. 1999. Characterization of Rac and Cdc42 activation in chemoattractant-stimulated human neutrophils using a novel assay for active GTPases. *J Biol Chem* **274:** 13198–13204. doi:10.1074/jbc.274.19.13198

Bergers G, Song S. 2005. The role of pericytes in blood-vessel formation and maintenance. *Neuro Oncol* **7:** 452–464. doi:10.1215/S1152851705000232

Bergert M, Erzberger A, Desai RA, Aspalter IM, Oates AC, Charras G, Salbreux G, Paluch EK. 2015. Force transmission during adhesion-independent migration. *Nat Cell Biol* **17:** 524–529. doi:10.1038/ncb3134

Blaser H, Reichman-Fried M, Castanon I, Dumstrei K, Marlow FLL, Kawakami K, Solnica-Krezel L, Heisenberg CP, Raz E. 2006. Migration of zebrafish primordial germ cells: a role for myosin contraction and cytoplasmic flow. *Dev Cell* **11:** 613–627. doi:10.1016/j.devcel.2006.09.023

Borm B, Requardt RP, Herzog V, Kirfel G. 2005. Membrane ruffles in cell migration: indicators of inefficient lamellipodia adhesion and compartments of actin filament reorganization. *Exp Cell Res* **302:** 83–95. doi:10.1016/j.yexcr.2004.08.034

Brown AF. 1982. Neutrophil granulocytes: adhesion and locomotion on collagen substrata and in collagen matrices. *J Cell Sci* **58:** 455–467. doi:10.1242/jcs.58.1.455

Brubaker AL, Rendon JL, Ramirez L, Choudhry MA, Kovacs EJ. 2013. Reduced neutrophil chemotaxis and infiltration contributes to delayed resolution of cutaneous wound infection with advanced age. *J Immunol* **190:** 1746–1757. doi:10.4049/jimmunol.1201213

Burdon PCE, Martin C, Rankin SM. 2005. The CXC chemokine MIP-2 stimulates neutrophil mobilization from the rat bone marrow in a CD49d-dependent manner. *Blood* **105:** 2543–2548. doi:10.1182/blood-2004-08-3193

Burdon PCE, Martin C, Rankin SM. 2008. Migration across the sinusoidal endothelium regulates neutrophil mobilization in response to ELR^+ CXC chemokines. *Br J Haematol* **142:** 100–108. doi:10.1111/j.1365-2141.2008.07018.x

Cano A, Pérez-Moreno MA, Rodrigo I, Locascio A, Blanco MJ, Del Barrio MG, Portillo F, Nieto MA. 2000. The transcription factor Snail controls epithelial–mesenchymal transitions by repressing E-cadherin expression. *Nat Cell Biol* **2:** 76–83. doi:10.1038/35000025

Carman CV, Springer TA. 2004. A transmigratory cup in leukocyte diapedesis both through individual vascular endothelial cells and between them. *J Cell Biol* **167:** 377–388. doi:10.1083/jcb.200404129

Carman CV, Sage PT, Sciuto TE, de la Fuente MA, Geha RS, Ochs HDD, Dvorak HF, Dvorak AM, Springer TA. 2007. Transcellular diapedesis is initiated by invasive podosomes. *Immunity* **26:** 784–797. doi:10.1016/j.immuni.2007.04.015

Cera MR, Fabbri M, Molendini C, Corada M, Orsenigo F, Rehberg M, Reichel CA, Krombach F, Pardi R, Dejana E. 2009. JAM-A promotes neutrophil chemotaxis by controlling integrin internalization and recycling. *J Cell Sci* **122:** 268–277. doi:10.1242/jcs.037127

Chan CE, Odde DJ. 2008. Traction dynamics of filopodia on compliant substrates. *Science* **322:** 1687–1691. doi:10.1126/science.1163595

Charras G, Paluch E. 2008. Blebs lead the way: how to migrate without lamellipodia. *Nat Rev Mol Cell Biol* **9:** 730–736. doi:10.1038/nrm2453

Charras GT, Hu CK, Coughlin M, Mitchison TJ. 2006. Reassembly of contractile actin cortex in cell blebs. *J Cell Biol* **175:** 477–490. doi:10.1083/jcb.200602085

Chastney MR, Conway JRW, Ivaska J. 2021. Integrin adhesion complexes. *Curr Biol* **31:** R536–R542. doi:10.1016/j.cub.2021.01.038

Chen CL, Syahirah R, Ravala SK, Yen YC, Klose T, Deng Q, Tesmer JJG. 2024. Molecular basis for Gβγ-mediated activation of phosphoinositide 3-kinase γ. *Nat Struct Mol Biol* **31:** 1198–1207. doi:10.1038/s41594-024-01265-y

Chhabra ES, Higgs HN. 2007. The many faces of actin: matching assembly factors with cellular structures. *Nat Cell Biol* **9:** 1110–1121. doi:10.1038/ncb1007-1110

Choi CK, Vicente-Manzanares M, Zareno J, Whitmore LA, Mogilner A, Horwitz AR. 2008. Actin and α-actinin orchestrate the assembly and maturation of nascent adhesions in a myosin II motor-independent manner. *Nat Cell Biol* **10:** 1039–1050. doi:10.1038/ncb1763

Chou RC, Kim ND, Sadik CD, Seung E, Lan Y, Byrne MH, Haribabu B, Iwakura Y, Luster AD. 2010. Lipid–cytokine–chemokine cascade drives neutrophil recruitment in a murine model of inflammatory arthritis. *Immunity* **33:** 266–278. doi:10.1016/j.immuni.2010.07.018

Chtanova T, Schaeffer M, Han SJ, van Dooren GG, Nollmann M, Herzmark P, Chan SW, Satija H, Camfield K, Aaron H, et al. 2008. Dynamics of neutrophil migration in lymph nodes during infection. *Immunity* **29:** 487–496. doi:10.1016/j.immuni.2008.07.012

Cinamon G, Shinder V, Shamri R, Alon R. 2004. Chemoattractant signals and β2 integrin occupancy at apical endothelial contacts combine with shear stress signals to promote transendothelial neutrophil migration. *J Immunol* **173:** 7282–7291. doi:10.4049/jimmunol.173.12.7282

Cohen TV, Klarmann KD, Sakchaisri K, Cooper JP, Kuhns D, Anver M, Johnson PF, Williams SC, Keller JR, Stewart CL. 2008. The lamin B receptor under transcriptional control of C/EBPε is required for morphological but not functional maturation of neutrophils. *Hum Mol Genet* **17:** 2921–2933. doi:10.1093/hmg/ddn191

Condeelis J, Hall A, Bresnick A, Warren V, Hock R, Bennett H, Ogihara S. 1988. Actin polymerization and pseudopod extension during amoeboid chemotaxis. *Cell Motil Cytoskeleton* **10:** 77–90. doi:10.1002/cm.970100113

Cunin P, Lee PY, Kim E, Schmider AB, Cloutier N, Pare A, Gunzer M, Soberman RJ, Lacroix S, Boilard E, et al. 2019. Differential attenuation of β2 integrin–dependent and –independent neutrophil migration by Ly6G ligation. *Blood Adv* **3:** 256–267. doi:10.1182/bloodadvances.2018026732

Cunningham CC. 1995. Actin polymerization and intracellular solvent flow in cell surface blebbing. *J Cell Biol* **129:** 1589–1599. doi:10.1083/jcb.129.6.1589

Dahinden CA, Zingg J, Maly FE, De Weck AL. 1988. Leukotriene production in human neutrophils primed by recombinant human granulocyte/macrophage colony-stimulating factor and stimulated with the complement component C5a and fMLP as second signals. *J Exp Med* **167:** 1281–1295. doi:10.1084/jem.167.4.1281

Damiano ER, Westheider J, Tözeren A, Ley K. 1996. Variation in the velocity, deformation, and adhesion energy density of leukocytes rolling within venules. *Circ Res* **79:** 1122–1130. doi:10.1161/01.RES.79.6.1122

Dangerfield J, Larbi KY, Huang MT, Dewar A, Nourshargh S. 2002. PECAM-1 (CD31) homophilic interaction up-regulates α6β1 on transmigrated neutrophils in vivo and plays a functional role in the ability of α6 integrins to mediate leukocyte migration through the perivascular basement membrane. *J Exp Med* **196:** 1201–1212. doi:10.1084/jem.20020324

De Belly H, Gallen AF, Strickland E, Estrada DC, Zager PJ, Nagy TL, Burkhardt JK, Turlier H, Weiner OD. 2025. Long range mutual activation establishes Rho and Rac polarity during cell migration. bioRxiv doi:10.1101/2024.10.01.616161

De Filippo K, Rankin SM. 2018. CXCR4, the master regulator of neutrophil trafficking in homeostasis and disease. *Eur J Clin Invest* **48:** e12949. doi:10.1111/eci.12949

Delanoë-Ayari H, Iwaya S, Maeda YT, Inose J, Rivière C, Sano M, Rieu JP. 2008. Changes in the magnitude and distribution of forces at different Dictyostelium developmental stages. *Cell Motil Cytoskeleton* **65:** 314–331. doi:10.1002/cm.20262

Devreotes P, Horwitz AR. 2015. Signaling networks that regulate cell migration. *Cold Spring Harb Perspect Biol* **7:** a005959. doi:10.1101/cshperspect.a005959

Devreotes PN, Zigmond SH. 1988. Chemotaxis in eukaryotic cells: a focus on leukocytes and *Dictyostelium*. *Annu Rev Cell Biol* **4:** 649–686. doi:10.1146/annurev.cb.04.110188.003245

DiMilla PA, Barbee K, Lauffenburger DA. 1991. Mathematical model for the effects of adhesion and mechanics on cell migration speed. *Biophys J* **60:** 15–37. doi:10.1016/S0006-3495(91)82027-6

Doyle AD, Sykora DJ, Pacheco GG, Kutys ML, Yamada KM. 2021. 3D mesenchymal cell migration is driven by anterior cellular contraction that generates an extracellular matrix prestrain. *Dev Cell* **56:** 826–841.e4. doi:10.1016/j.devcel.2021.02.017

Eash KJ, Greenbaum AM, Gopalan PK, Link DC. 2010. CXCR2 and CXCR4 antagonistically regulate neutrophil trafficking from murine bone marrow. *J Clin Invest* **120:** 2423–2431. doi:10.1172/JCI41649

Eddy RJ, Pierini LM, Maxfield FR. 2002. Microtubule asymmetry during neutrophil polarization and migration. *Mol Biol Cell* **13:** 4470–4483. doi:10.1091/mbc.e02-04-0241

Elgazar-Carmon V, Rudich A, Hadad N, Levy R. 2008. Neutrophils transiently infiltrate intra-abdominal fat early in the course of high-fat feeding. *J Lipid Res* **49:** 1894–1903. doi:10.1194/jlr.M800132-JLR200

Fassbender K, Kaptur S, Becker P, Gröschl J, Hennerici M. 1998. Adhesion molecules in tissue injury: kinetics of expression and shedding and association with cytokine release in humans. *Clin Immunol Immunopathol* **89:** 54–60. doi:10.1006/clin.1998.4583

Friedl P, Weigelin B. 2008. Interstitial leukocyte migration and immune function. *Nat Immunol* **9:** 960–969. doi:10.1038/ni.f.212

Friedl P, Zänker KS, Zänker Z, Bröcker EB. 1998. Cell migration strategies in 3-D extracellular matrix: differences in morphology, cell matrix interactions, and integrin function. *Microsc Res Tech* **43:** 369–378. doi:10.1002/(SICI)1097-0029(19981201)43:5<369::AID-JEMT3>3.0.CO;2-6

Fritz-Laylin LK, Lord SJ, Mullins RD. 2017a. WASP and SCAR are evolutionarily conserved in actin-filled pseudopod-based motility. *J Cell Biol* **216:** 1673–1688. doi:10.1083/jcb.201701074

Fritz-Laylin LK, Riel-Mehan M, Chen BC, Lord SJ, Goddard TD, Ferrin TE, Nicholson-Dykstra SM, Higgs H, Johnson GT, Betzig E, et al. 2017b. Actin-based protrusions of migrating neutrophils are intrinsically lamellar and facilitate direction changes. *eLife* **6:** e26990. doi:10.7554/eLife.26990

Fu T, Sullivan DP, Gonzalez AM, Haynes ME, Dalal PJ, Rutledge NS, Tierney AL, Yescas JA, Weber EW, Muller WA. 2023. Mechanotransduction via endothelial adhesion molecule CD31 initiates transmigration and reveals a role for VEGFR2 in diapedesis. *Immunity* **56:** 2311–2324.e6. doi:10.1016/j.immuni.2023.08.001

Fulop T, Larbi A, Douziech N, Fortin C, Guérard KP, Lesur O, Khalil A, Dupuis G. 2004. Signal transduction and functional changes in neutrophils with aging. *Aging Cell* **3:** 217–226. doi:10.1111/j.1474-9728.2004.00110.x

Furze RC, Rankin SM. 2008. Neutrophil mobilization and clearance in the bone marrow. *Immunology* **125:** 281–288. doi:10.1111/j.1365-2567.2008.02950.x

García-Arcos JM, Jha A, Waterman CM, Piel M. 2024. Blebology: principles of bleb-based migration. *Trends Cell Biol* **34:** 838–853. doi:10.1016/j.tcb.2024.02.009

Gardel ML, Sabass B, Ji L, Danuser G, Schwarz US, Waterman CM. 2008. Traction stress in focal adhesions correlates biphasically with actin retrograde flow speed. *J Cell Biol* **183:** 999–1005. doi:10.1083/jcb.200810060

Gardel ML, Schneider IC, Aratyn-Schaus Y, Waterman CM. 2010. Mechanical integration of actin and adhesion dynamics in cell migration. *Annu Rev Cell Dev Biol* **26:** 315–333. doi:10.1146/annurev.cellbio.011209.122036

Ge Y, Huang M, Yao YM. 2022. Efferocytosis and its role in inflammatory disorders. *Front Cell Dev Biol* **10:** 839248. doi:10.3389/fcell.2022.839248

George S, Martin JAJ, Graziani V, Sanz-Moreno V. 2023. Amoeboid migration in health and disease: immune responses versus cancer dissemination. *Front Cell Dev Biol* **10:** 1091801. doi:10.3389/fcell.2022.1091801

Gorina R, Lyck R, Vestweber D, Engelhardt B. 2014. B2 integrin–mediated crawling on endothelial ICAM-1 and ICAM-2 is a prerequisite for transcellular neutrophil diapedesis across the inflamed blood–brain barrier. *J Immunol* **192:** 324–337. doi:10.4049/jimmunol.1300858

Grönloh MLB, Arts JJG, van Buul JD. 2021. Neutrophil transendothelial migration hotspots—mechanisms and implications. *J Cell Sci* **134:** jcs255653. doi:10.1242/jcs.255653

Grönloh MLB, Tebbens ME, Kotsi M, Arts JJG, van Buul JD. 2023a. Intercellular adhesion molecule 2 regulates diapedesis hotspots by allowing neutrophil crawling against the direction of flow. *Vasc Biol* **5:** e230005. doi:10.1530/VB-23-0005

Grönloh MLB, Arts JJG, Mahlandt EK, Nolte MA, Goedhart J, van Buul JD. 2023b. Primary adhered neutrophils increase JNK1-MARCKSL1-mediated filopodia to promote secondary neutrophil transmigration. *iScience* **26:** 107406. doi:10.1016/j.isci.2023.107406

Grönloh MLB, Arts JJG, Palacios Martínez S, van der Veen AA, Kempers L, van Steen ACI, Roelofs JJTH, Nolte MA, Goedhart J, van Buul JD. 2023c. Endothelial transmigration hotspots limit vascular leakage through heterogeneous expression of ICAM-1. *EMBO Rep* **24:** e55483. doi:10.15252/embr.202255483

Hawkins PT, Jackson TR, Stephens LR. 1992. Platelet-derived growth factor stimulates synthesis of Ptdlns(3,4,5)P_3 by activating a Ptdlns(4,5)P_2 3-OH kinase. *Nature* **358:** 157–159. doi:10.1038/358157a0

Hawkins RJ, Piel M, Faure-Andre G, Lennon-Dumenil AM, Joanny JF, Prost J, Voituriez R. 2009. Pushing off the walls: a mechanism of cell motility in confinement. *Phys Rev Lett* **102:** 058103. doi:10.1103/PhysRevLett.102.058103

Herrero-Cervera A, Soehnlein O, Kenne E. 2022. Neutrophils in chronic inflammatory diseases. *Cell Mol Immunol* **19:** 177–191. doi:10.1038/s41423-021-00832-3

Hind LE, Vincent WJB, Huttenlocher A. 2016. Leading from the back: the role of the uropod in neutrophil polarization and migration. *Dev Cell* **38:** 161–169. doi:10.1016/j.devcel.2016.06.031

Horwitz A, Duggan K, Buck C, Beckerle MC, Burridge K. 1986. Interaction of plasma membrane fibronectin receptor with talin—a transmembrane linkage. *Nature* **320:** 531–533. doi:10.1038/320531a0

Hyun KK, De La Luz Sierra M, Williams CK, Gulino AV, Tosato G. 2006. G-CSF down-regulation of CXCR4 expression identified as a mechanism for mobilization of myeloid cells. *Blood* **108:** 812–820. doi:10.1182/blood-2005-10-4162

Hyun YM, Choe YH, Park SA, Kim M. 2019. LFA-1 (CD11a/CD18) and Mac-1 (CD11b/CD18) distinctly regulate neutrophil extravasation through hotspots I and II. *Exp Mol Med* **51:** 1–13. doi:10.1038/s12276-019-0227-1

Ibo M, Srivastava V, Robinson DN, Gagnon ZR. 2016. Cell blebbing in confined microfluidic environments. *PLoS One* **11:** e0163866. doi:10.1371/journal.pone.0163866

Ichida M, Yui Y, Yoshioka K, Tanaka T, Wakamatsu T, Yoshikawa H, Itoh K. 2011. Changes in cell migration of mesenchymal cells during osteogenic differentiation. *FEBS Lett* **585:** 4018–4024. doi:10.1016/j.febslet.2011.11.014

Iwasa JH, Mullins RD. 2007. Spatial and temporal relationships between actin-filament nucleation, capping, and disassembly. *Curr Biol* **17:** 395–406. doi:10.1016/j.cub.2007.02.012

Jacquemet G, Baghirov H, Georgiadou M, Sihto H, Peuhu E, Cettour-Janet P, He T, Perälä M, Kronqvist P, Joensuu H, et al. 2016. L-type calcium channels regulate filopodia stability and cancer cell invasion downstream of integrin signalling. *Nat Commun* **7:** 13297. doi:10.1038/ncomms13297

Jacquemet G, Stubb A, Saup R, Miihkinen M, Kremneva E, Hamidi H, Ivaska J. 2019. Filopodome mapping identifies p130Cas as a mechanosensitive regulator of filopodia stability. *Curr Biol* **29:** 202–216.e7. doi:10.1016/j.cub.2018.11.053

Jay PY, Pham PA, Wong SA, Elson EL. 1995. A mechanical function of myosin II in cell motility. *J Cell Sci* **108:** 387–393. doi:10.1242/jcs.108.1.387

Kanchanawong P, Calderwood DA. 2023. Organization, dynamics and mechanoregulation of integrin-mediated cell–ECM adhesions. *Nat Rev Mol Cell Biol* **24:** 142–161. doi:10.1038/s41580-022-00531-5

Kim MB, Sarelius IH. 2004. Role of shear forces and adhesion molecule distribution on P-selectin–mediated leukocyte rolling in postcapillary venules. *Am J Physiol Heart Circ Physiol* **287:** H2705–H2711. doi:10.1152/ajpheart.00448.2004

Klarlund JK, Guilherme A, Holik JJ, Virbasius JV, Chawla A, Czech MP. 1997. Signaling by phosphoinositide-3,4,5-trisphosphate through proteins containing pleckstrin and Sec7 homology domains. *Science* **275:** 1927–1930. doi:10.1126/science.275.5308.1927

Kroll J, Hauschild R, Kuznetcov A, Stefanowski K, Hermann MD, Merrin J, Shafeek L, Müller-Taubenberger A, Renkawitz J. 2023. Adaptive pathfinding by nucleokinesis during amoeboid migration. *EMBO J* **42:** e114557. doi:10.15252/embj.2023114557

Lammerding J, Schulze PC, Takahashi T, Kozlov S, Sullivan T, Kamm RD, Stewart CL, Lee RT. 2004. Lamin A/C deficiency causes defective nuclear mechanics and mechanotransduction. *J Clin Invest* **113:** 370–378. doi:10.1172/JCI200419670

Lammerding J, Fong LG, Ji JY, Reue K, Stewart CL, Young SG, Lee RT. 2006. Lamins A and C but not lamin B1 regulate nuclear mechanics. *J Biol Chem* **281:** 25768–25780. doi:10.1074/jbc.M513511200

Lämmermann T, Germain RN. 2014. The multiple faces of leukocyte interstitial migration. *Semin Immunopathol* **36:** 227–251. doi:10.1007/s00281-014-0418-8

Lämmermann T, Sixt M. 2009. Mechanical modes of "amoeboid" cell migration. *Curr Opin Cell Biol* **21:** 636–644. doi:10.1016/j.ceb.2009.05.003

Lämmermann T, Bader BL, Monkley SJ, Worbs T, Wedlich-Söldner R, Hirsch K, Keller M, Förster R, Critchley DR, Fässler R, et al. 2008. Rapid leukocyte migration by integrin-independent flowing and squeezing. *Nature* **453:** 51–55. doi:10.1038/nature06887

Lämmermann T, Afonso PV, Angermann BR, Wang JM, Kastenmüller W, Parent CA, Germain RN. 2013. Neutrophil swarms require LTB4 and integrins at sites of cell death in vivo. *Nature* **498:** 371–375. doi:10.1038/nature12175

Larsen M, Artym VV, Green JA, Yamada KM. 2006. The matrix reorganized: extracellular matrix remodeling and integrin signaling. *Curr Opin Cell Biol* **18:** 463–471. doi:10.1016/j.ceb.2006.08.009

Lauffenburger DA, Horwitz AF. 1996. Cell migration: a physically integrated molecular process. *Cell* **84:** 359–369. doi:10.1016/S0092-8674(00)81280-5

Lawrence MB, Kansas GS, Kunkel EJ, Ley K. 1997. Threshold levels of fluid shear promote leukocyte adhesion through selectins (CD62L,P,E). *J Cell Biol* **136:** 717–727. doi:10.1083/jcb.136.3.717

Lawson MA, Maxfield FR. 1995. Ca^{2+}- and calcineurin-dependent recycling of an integrin to the front of migrating neutrophils. *Nature* **377:** 75–79. doi:10.1038/377075a0

Lee YJ, Patel D, Park S. 2011. Local rheology of human neutrophils investigated using atomic force microscopy. *Int J Biol Sci* **7:** 102–111. doi:10.7150/ijbs.7.102

Lévesque JP, Hendy J, Takamatsu Y, Simmons PJ, Bendall LJ. 2003. Disruption of the CXCR4/CXCL12 chemotactic interaction during hematopoietic stem cell mobilization induced by GCSF or cyclophosphamide. *J Clin Invest* **111:** 187–196. doi:10.1172/JCI15994

Ley K, Laudanna C, Cybulsky MI, Nourshargh S. 2007. Getting to the site of inflammation: the leukocyte adhesion cascade updated. *Nat Rev Immunol* **7:** 678–689. doi:10.1038/nri2156

Li N, Mao D, Lü S, Tong C, Zhang Y, Long M. 2013. Distinct binding affinities of Mac-1 and LFA-1 in neutrophil activation. *J Immunol* **190:** 4371–4381. doi:10.4049/jimmunol.1201374

Li N, Yang H, Wang M, Lü S, Zhang Y, Long M. 2018. Ligand-specific binding forces of LFA-1 and Mac-1 in neutrophil adhesion and crawling. *Mol Biol Cell* **29:** 408–418. doi:10.1091/mbc.E16-12-0827

Liu Q, Kishimoto TK, Mainolfi E, Deleon RP, Myers C, Moretz RC. 1998. Dynamic expression of L-selectin in cell-to-cell interactions between neutrophils and endothelial cells in vitro. *Exp Cell Res* **243:** 87–93. doi:10.1006/excr.1998.4139

Lu F, Zhu L, Bromberger T, Yang J, Yang Q, Liu J, Plow EF, Moser M, Qin J. 2022. Mechanism of integrin activation by talin and its cooperation with kindlin. *Nat Commun* **13:** 1–19. doi:10.1038/s41467-022-30117-w

Majumdar R, Tavakoli Tameh A, Parent CA. 2016. Exosomes mediate LTB4 release during neutrophil chemotaxis. *PLoS Biol* **14:** e1002336. doi:10.1371/journal.pbio.1002336

Majumdar R, Tameh AT, Arya SB, Parent CA. 2021. Exosomes mediate LTB4 release during neutrophil chemotaxis. *PLoS Biol* **19:** e3001271. doi:10.1371/journal.pbio.3001271

Malawista SE, De Boisfleury Chevance A. 1997. Random locomotion and chemotaxis of human blood polymorphonuclear leukocytes (PMN) in the presence of EDTA: PMN in close quarters require neither leukocyte integrins nor external divalent cations. *Proc Natl Acad Sci* **94:** 11577–11582. doi:10.1073/pnas.94.21.11577

Malawista SE, De Boisfleury Chevance A, Boxer LA. 2000. Random locomotion and chemotaxis of human blood polymorphonuclear leukocytes from a patient with leukocyte adhesion deficiency-1: normal displacement in close quarters via chimneying. *Cell Motil Cytoskeleton* **46:** 183–189. doi:10.1002/1097-0169(200007)46:3<183::AID-CM3>3.0.CO;2-2

Mamdouh Z, Mikhailov A, Muller WA. 2009. Transcellular migration of leukocytes is mediated by the endothelial lateral border recycling compartment. *J Exp Med* **206:** 2795–2808. doi:10.1084/jem.20082745

Mandeville JTH, Lawson MA, Maxfield FR. 1997. Dynamic imaging of neutrophil migration in three dimensions: mechanical interactions between cells and matrix. *J Leukoc Biol* **61:** 188–200. doi:10.1002/jlb.61.2.188

Manley HR, Keightley MC, Lieschke GJ. 2018. The neutrophil nucleus: an important influence on neutrophil migration and function. *Front Immunol* **9:** 2867. doi:10.3389/fimmu.2018.02867

Martin C, Burdon PCE, Bridger G, Gutierrez-Ramos JC, Williams TJ, Rankin SM. 2003. Chemokines acting via CXCR2 and CXCR4 control the release of neutrophils from the bone marrow and their return following senescence. *Immunity* **19:** 583–593. doi:10.1016/S1074-7613(03)00263-2

Martinelli R, Zeiger AS, Whitfield M, Sciuto TE, Dvorak A, van Vliet KJ, Greenwood J, Carman CV. 2014. Probing the biomechanical contribution of the endothelium to lymphocyte migration: diapedesis by the path of least resistance. *J Cell Sci* **127:** 3720–3734. doi:10.1242/jcs.148619

Martins A, Han J, Kim SO. 2010. The multifaceted effects of granulocyte colony-stimulating factor in immunomodulation and potential roles in intestinal immune homeostasis. *IUBMB Life* **62:** 611–617. doi:10.1002/iub.361

Mathias JR, Perrin BJ, Liu TX, Kanki J, Look AT, Huttenlocher A. 2006. Resolution of inflammation by retrograde chemotaxis of neutrophils in transgenic zebrafish. *J Leukoc Biol* **80:** 1281–1288. doi:10.1189/jlb.0506346

Mauer AM, Athens JW, Ashenbrucker H, Cartwright GE, Wintrobe MM. 1960. Leukokinetic studies. II: A method for labeling granulocytes in vitro with radioactive diisopropylfluorophosphate (DFP32). *J Clin Invest* **39:** 1481–1486. doi:10.1172/JCI104167

McAndrews KM, McGrail DJ, Ravikumar N, Dawson MR. 2015. Mesenchymal stem cells induce directional migration of invasive breast cancer cells through TGF-β. *Sci Rep* **5:** 1–13. doi:10.1038/srep16941

McCormick B, Chu JY, Vermeren S. 2017. Cross-talk between Rho GTPases and PI3K in the neutrophil. *Small GTPases* **10:** 187–195. doi:10.1080/21541248.2017.1304855

McCracken JM, Allen LAH. 2014. Regulation of human neutrophil apoptosis and lifespan in health and disease. *J Cell Death* **7:** 15–23. doi:10.4137/JCD.S11038

McDonald B, Pittman K, Menezes GB, Hirota SA, Slaba I, Waterhouse CCM, Beck PL, Muruve DA, Kubes P. 2010. Intravascular danger signals guide neutrophils to sites of sterile inflammation. *Science* **330:** 362–366. doi:10.1126/science.1195491

Mehidi A, Rossier O, Schaks M, Chazeau A, Binamé F, Remorino A, Coppey M, Karatas Z, Sibarita JB, Rottner K, et al. 2019. Transient activations of Rac1 at the lamellipodium tip trigger membrane protrusion. *Curr Biol* **29:** 2852–2866.e5. doi:10.1016/j.cub.2019.07.035

Melvin AT, Welf ES, Wang Y, Irvine DJ, Haugh JM. 2011. In chemotaxing fibroblasts, both high-fidelity and weakly biased cell movements track the localization of PI3K signaling. *Biophys J* **100:** 1893–1901. doi:10.1016/j.bpj.2011.02.047

Metzemaekers M, Gouwy M, Proost P. 2020. Neutrophil chemoattractant receptors in health and disease: double-edged swords. *Cell Mol Immunol* **17:** 433–450. doi:10.1038/s41423-020-0412-0

Michael M, Vermeren S. 2019. A neutrophil-centric view of chemotaxis. *Essays Biochem* **63:** 607–618. doi:10.1042/EBC20190011

Mitchison T, Kirschner M. 1988. Cytoskeletal dynamics and nerve growth. *Neuron* **1:** 761–772. doi:10.1016/0896-6273(88)90124-9

Miyamoto K, Ogura Y, Hamada M, Nishiwaki H, Hiroshiba N, Tsujikawa A, Mandai M, Suzuma K, Tojo SJ, Honda Y. 1998. In vivo neutralization of P-selectin inhibits leukocyte–endothelial interactions in retinal microcirculation during ocular inflammation. *Microvasc Res* **55:** 230–240. doi:10.1006/mvre.1998.2084

Möhle R, Weisel KC, Bautz F, Seitz G, Yildirim S, Kanz L. 2009. Modulation of CXC chemokine receptor expression and function in human neutrophils during aging in vitro suggests a role in their clearance from circulation. *Mediators Inflamm* **2009:** 790174. doi:10.1155/2009/790174

Mooren OL, Li J, Nawas J, Cooper JA. 2014. Endothelial cells use dynamic actin to facilitate lymphocyte transendothelial migration and maintain the monolayer barrier. *Mol Biol Cell* **25:** 4115–4129. doi:10.1091/mbc.e14-05-0976

Morin NA, Oakes PW, Hyun YM, Lee D, Chin EY, King MR, Springer TA, Shimaoka M, Tang JX, Reichner JS, et al. 2008. Nonmuscle myosin heavy chain IIA mediates integrin LFA-1 de-adhesion during T lymphocyte migration. *J Exp Med* **205:** 195–205. doi:10.1084/jem.20071543

Mukhopadhyay A, Tsukasaki Y, Chan WC, Le JP, Kwok ML, Zhou J, Natarajan V, Mostafazadeh N, Maienschein-Cline M, Papautsky I, et al. 2024. *trans*-Endothelial neutrophil migration activates bactericidal function via Piezo1 mechanosensing. *Immunity* **57:** 52–67.e10. doi:10.1016/j.immuni.2023.11.007

Nagy TL, Strickland J, Weiner OD. 2024. Neutrophils actively swell to potentiate rapid migration. *eLife* **12:** RP90551. doi:10.7554/eLife.90551

Navarro R, Compte M, Álvarez-Vallina L, Sanz L. 2016. Immune regulation by pericytes: modulating innate and adaptive immunity. *Front Immunol* **7:** 480. doi:10.3389/fimmu.2016.00480

Ng LG, Qin JS, Roediger B, Wang Y, Jain R, Cavanagh LL, Smith AL, Jones CA, De Veer M, Grimbaldeston MA, et al. 2011. Visualizing the neutrophil response to sterile tissue injury in mouse dermis reveals a three-phase cascade of events. *J Invest Dermatol* **131:** 2058–2068. doi:10.1038/jid.2011.179

Nobes CD, Hall A. 1995. Rho, Rac, and Cdc42 GTPases regulate the assembly of multimolecular focal complexes associated with actin stress fibers, lamellipodia, and filopodia. *Cell* **81:** 53–62. doi:10.1016/0092-8674(95)90370-4

Nourshargh S, Hordijk PL, Sixt M. 2010. Breaching multiple barriers: leukocyte motility through venular walls and the interstitium. *Nat Rev Mol Cell Biol* **11:** 366–378. doi:10.1038/nrm2889

Omelchenko T, Vasiliev JM, Gelfand IM, Feder HH, Bonder EM. 2002. Mechanisms of polarization of the shape of fibroblasts and epitheliocytes: separation of the roles of microtubules and Rho-dependent actin–myosin contractility. *Proc Natl Acad Sci* **99:** 10452–10457. doi:10.1073/pnas.152339899

O'Neill PR, Castillo-Badillo JA, Meshik X, Kalyanaraman V, Melgarejo K, Gautam N. 2018. Membrane flow drives an adhesion-independent amoeboid cell migration mode. *Dev Cell* **46:** 9–22.e4. doi:10.1016/j.devcel.2018.05.029

Onsum M, Rao CV. 2007. A mathematical model for neutrophil gradient sensing and polarization. *PLoS Comput Biol* **3:** e36. doi:10.1371/journal.pcbi.0030036

Panopoulos AD, Watowich SS. 2008. Granulocyte colony-stimulating factor: molecular mechanisms of action during steady state and "emergency" hematopoiesis. *Cytokine* **42:** 277–288. doi:10.1016/j.cyto.2008.03.002

Parent CA, Devreotes PN. 1999. A cell's sense of direction. *Science* **284:** 765–770. doi:10.1126/science.284.5415.765

Pertz O, Hodgson L, Klemke RL, Hahn KM. 2006. Spatiotemporal dynamics of RhoA activity in migrating cells. *Nature* **440:** 1069–1072. doi:10.1038/nature04665

Petri B, Sanz MJ. 2018. Neutrophil chemotaxis. *Cell Tissue Res* **371:** 425–436. doi:10.1007/s00441-017-2776-8

Phillipson M, Heit B, Colarusso P, Liu L, Ballantyne CM, Kubes P. 2006. Intraluminal crawling of neutrophils to emigration sites: a molecularly distinct process from adhesion in the recruitment cascade. *J Exp Med* **203:** 2569–2575. doi:10.1084/jem.20060925

Pillay J, Den Braber I, Vrisekoop N, Kwast LM, De Boer RJ, Borghans JAM, Tesselaar K, Koenderman L. 2010. In vivo labeling with 2H2O reveals a human neutrophil lifespan of 5.4 days. *Blood* **116:** 625–627. doi:10.1182/blood-2010-01-259028

Pollard TD, Borisy GG. 2003. Cellular motility driven by assembly and disassembly of actin filaments. *Cell* **112:** 453–465. doi:10.1016/S0092-8674(03)00120-X

Ponti A, Machacek M, Gupton SL, Waterman-Storer CM, Danuser G. 2004. Two distinct actin networks drive the protrusion of migrating cells. *Science* **305:** 1782–1786. doi:10.1126/science.1100533

Proebstl D, Voisin MB, Woodfin A, Whiteford J, D'Acquisto F, Jones GE, Rowe D, Nourshargh S. 2012. Pericytes support neutrophil subendothelial cell crawling and breaching of venular walls in vivo. *J Exp Med* **209:** 1219–1234. doi:10.1084/jem.20111622

Ramos C, Oehler R. 2024. Clearance of apoptotic cells by neutrophils in inflammation and cancer. *Cell Death Discov* **10:** 1–8. doi:10.1038/s41420-024-01809-7

Rankin SM. 2010. The bone marrow: a site of neutrophil clearance. *J Leukoc Biol* **88:** 241–251. doi:10.1189/jlb.0210112

Rappel WJ, Edelstein-Keshet L. 2017. Mechanisms of cell polarization. *Curr Opin Syst Biol* **3:** 43–53. doi:10.1016/j.coisb.2017.03.005

Regen CM, Horwitz AF. 1992. Dynamics of beta 1 integrin-mediated adhesive contacts in motile fibroblasts. *J Cell Biol* **119:** 1347–1359. doi:10.1083/jcb.119.5.1347

Renkawitz J, Kopf A, Stopp J, de Vries I, Driscoll MK, Merrin J, Hauschild R, Welf ES, Danuser G, Fiolka R, et al. 2019. Nuclear positioning facilitates amoeboid migration along the path of least resistance. *Nature* **568:** 546–550. doi:10.1038/s41586-019-1087-5

Ridley AJ, Paterson HF, Johnston CL, Diekmann D, Hall A. 1992. The small GTP-binding protein Rac regulates growth factor-induced membrane ruffling. *Cell* **70:** 401–410. doi:10.1016/0092-8674(92)90164-8

Ridley AJ, Schwartz MA, Burridge K, Firtel RA, Ginsberg MH, Borisy G, Parsons JT, Horwitz AR. 2003. Cell migration: integrating signals from front to back. *Science* **302:** 1704–1709. doi:10.1126/science.1092053

Ross R, Benditt EP. 1961. Wound healing and collagen formation. I: Sequential changes in components of guinea pig skin wounds observed in the electron microscope. *J Biophys Biochem Cytol* **11:** 677–700. doi:10.1083/jcb.11.3.677

Roussel E, Gingras MC. 1997. Transendothelial migration induces rapid expression on neutrophils of granule-release VLA6 used for tissue infiltration. *J Leukoc Biol* **62:** 356–362. doi:10.1002/jlb.62.3.356

Rowat AC, Jaalouk DE, Zwerger M, Ung WL, Eydelnant IA, Olins DE, Olins AL, Herrmann H, Weitz DA, Lammerding J. 2013. Nuclear envelope composition determines the ability of neutrophil-type cells to passage through micron-scale constrictions. *J Biol Chem* **288:** 8610–8618. doi:10.1074/jbc.M112.441535

Sánchez-Madrid F, Del Pozo MA. 1999. Leukocyte polarization in cell migration and immune interactions. *EMBO J* **18:** 501–511. doi:10.1093/emboj/18.3.501

Schett G, Neurath MF. 2018. Resolution of chronic inflammatory disease: universal and tissue-specific concepts. *Nat Commun* **9:** 1–8. doi:10.1038/s41467-018-05800-6

Schiller HB, Hermann MR, Polleux J, Vignaud T, Zanivan S, Friedel CC, Sun Z, Raducanu A, Gottschalk KE, Théry M, et al. 2013. β1- and αv-class integrins cooperate to regulate myosin II during rigidity sensing of fibronectin-based microenvironments. *Nat Cell Biol* **15:** 625–636. doi:10.1038/ncb2747

Schneider IC, Haugh JM. 2005. Quantitative elucidation of a distinct spatial gradient-sensing mechanism in fibroblasts. *J Cell Biol* **171:** 883–892. doi:10.1083/jcb.200509028

Schrope JH, Horn A, Lazorchak K, Tinnen CW, Stevens JJ, Farooqui M, Robertson T, Li J, Bennin D, Juang T, et al. 2025. Confinement by liquid–liquid interface replicates in vivo neutrophil deformations and elicits bleb-based migration. *Adv Sci* **12:** 2414024. doi:10.1002/advs.202414024

Seppä H, Grotendorst G, Seppä S, Schiffmann E, Martin GR. 1982. Platelet-derived growth factor in chemotactic for fibroblasts. *J Cell Biol* **92:** 584–588. doi:10.1083/jcb.92.2.584

Servant G, Weiner OD, Neptune ER, Sedat JW, Bourne HR. 1999. Dynamics of a chemoattractant receptor in living neutrophils during chemotaxis. *Mol Biol Cell* **10:** 1163–1178. doi:10.1091/mbc.10.4.1163

Shen C, Mulder E, Buitenwerf W, Postat J, Jansen A, Kox M, Mandl JN, Vrisekoop N. 2023. Nuclear segmentation facilitates neutrophil migration. *J Cell Sci* **136:** jcs260546. doi:10.1242/jcs.260768

Simmons P, Masinovsky B, Longenecker B, Berenson R, Torok-Storb B, Gallatin W. 1992. Vascular cell adhesion molecule-1 expressed by bone marrow stromal cells mediates the binding of hematopoietic progenitor cells. *Blood* **80:** 388–395. doi:10.1182/blood.V80.2.388.bloodjournal802388

Singh V, Kaur R, Kumari P, Pasricha C, Singh R. 2023. ICAM-1 and VCAM-1: gatekeepers in various inflammatory and cardiovascular disorders. *Clin Chim Acta* **548:** 117487. doi:10.1016/j.cca.2023.117487

Smith CW, Marlin SD, Rothlein R, Toman C, Anderson DC. 1989. Cooperative interactions of LFA-1 and Mac-1 with intercellular adhesion molecule-1 in facilitating adherence and transendothelial migration of human neutrophils in vitro. *J Clin Invest* **83:** 2008–2017. doi:10.1172/JCI114111

Smith LA, Aranda-Espinoza H, Haun JB, Dembo M, Hammer DA. 2007. Neutrophil traction stresses are concentrated in the uropod during migration. *Biophys J* **92:** L58–L60. doi:10.1529/biophysj.106.102822

Sokol CL, Luster AD. 2015. The chemokine system in innate immunity. *Cold Spring Harb Perspect Biol* **7:** a016303. doi:10.1101/cshperspect.a016303

Stephens L, Smrcka A, Cooke FT, Jackson TR, Sternweis PC, Hawkins PT. 1994. A novel phosphoinositide 3 kinase activity in myeloid-derived cells is activated by G protein βγ subunits. *Cell* **77:** 83–93. doi:10.1016/0092-8674(94)90237-2

Stockem W, Hoffmann HU, Gawlitta W. 1982. Spatial organization and fine structure of the cortical filament layer in normal locomoting *Amoeba proteus*. *Cell Tissue Res* **221:** 505–519. doi:10.1007/BF00215699

Subramanian BC, Melis N, Chen D, Wang W, Gallardo D, Weigert R, Parent CA. 2020. The LTB4–BLT1 axis regulates actomyosin and β2-integrin dynamics during neutrophil extravasation. *J Cell Biol* **219:** e201910215. doi:10.1083/jcb.201910215

Sundd P, Gutierrez E, Pospieszalska MK, Zhang H, Groisman A, Ley K. 2010. Quantitative dynamic footprinting microscopy reveals mechanisms of neutrophil rolling. *Nat Methods* **7:** 821–824. doi:10.1038/nmeth.1508

Surve CR, To JY, Malik S, Kim M, Smrcka AV. 2016. Dynamic regulation of neutrophil polarity and migration by the heterotrimeric G protein subunits $G\alpha_i$-GTP and Gβγ. *Sci Signal* **9:** ra22. doi:10.1126/scisignal.aad8163

Swaney KF, Huang CH, Devreotes PN. 2010. Eukaryotic chemotaxis: a network of signaling pathways controls motility, directional sensing, and polarity. *Annu Rev Biophys* **39:** 265–289. doi:10.1146/annurev.biophys.093008.131228

Szczur K, Zheng Y, Filippi MD. 2009. The small Rho GTPase Cdc42 regulates neutrophil polarity via CD11b integrin signaling. *Blood* **114:** 4527–4537. doi:10.1182/blood-2008-12-195164

Tang D, Kang R, Coyne CB, Zeh HJ, Lotze MT. 2012. PAMPs and DAMPs: signal 0s that spur autophagy and immunity. *Immunol Rev* **249:** 158–175. doi:10.1111/j.1600-065X.2012.01146.x

Tharp WG, Yadav R, Irimia D, Upadhyaya A, Samadani A, Hurtado O, Liu SY, Munisamy S, Brainard DM, Mahon MJ, et al. 2006. Neutrophil chemorepulsion in defined interleukin-8 gradients in vitro and in vivo. *J Leukoc Biol* **79:** 539–554. doi:10.1189/jlb.0905516

Uchida KSK, Kitanishi-Yumura T, Yumura S. 2003. Myosin II contributes to the posterior contraction and the anterior extension during the retraction phase in migrating *Dictyostelium* cells. *J Cell Sci* **116:** 51–60. doi:10.1242/jcs.00195

Van Haastert P, Devreotes P. 2004. Chemotaxis: signalling the way forward. *Nat Rev Mol Cell Biol* **5:** 626–634. doi:10.1038/nrm1435

Vanhaesebroeck B, Stephens L, Hawkins P. 2012. PI3K signalling: the path to discovery and understanding. *Nat Rev Mol Cell Biol* **13:** 195–203. doi:10.1038/nrm3290

Van Keymeulen A, Wong K, Knight ZA, Govaerts C, Hahn KM, Shokat KM, Bourne HR. 2006. To stabilize neutrophil polarity, PIP_3 and Cdc42 augment RhoA activity at the back as well as signals at the front. *J Cell Biol* **174:** 437–445. doi:10.1083/jcb.200604113

Vedula SRK, Leong MC, Lai TL, Hersen P, Kabla AJ, Lim CT, Ladoux B. 2012. Emerging modes of collective cell migration induced by geometrical constraints. *Proc Natl Acad Sci* **109:** 12974–12979. doi:10.1073/pnas.1119313109

Voisin MB, Woodfin A, Nourshargh S. 2009. Monocytes and neutrophils exhibit both distinct and common mechanisms in penetrating the vascular basement membrane in vivo. *Arterioscler Thromb Vasc Biol* **29:** 1193–1199. doi:10.1161/ATVBAHA.109.187450

Vorotnikov AV, Tyurin-Kuzmin PA. 2014. Chemotactic signaling in mesenchymal cells compared to amoeboid cells. *Genes Dis* **1:** 162–173. doi:10.1016/j.gendis.2014.09.006

Wang F, Herzmark P, Weiner OD, Srinivasan S, Servant G, Bourne HR. 2002. Lipid products of PI(3)Ks maintain persistent cell polarity and directed motility in neutrophils. *Nat Cell Biol* **4:** 513–518. doi:10.1038/ncb810

Wang S, Voisin MB, Larbi KY, Dangerfield J, Scheiermann C, Tran M, Maxwell PH, Sorokin L, Nourshargh S. 2006. Venular basement membranes contain specific matrix protein low expression regions that act as exit points for emigrating neutrophils. *J Exp Med* **203:** 1519–1532. doi:10.1084/jem.20051210

Wang J, Hossain M, Thanabalasuriar A, Gunzer M, Meininger C, Kubes P. 2017. Visualizing the function and fate of neutrophils in sterile injury and repair. *Science* **358:** 111–116. doi:10.1126/science.aam9690

Wang K, Qin Y, Chen Y. 2021. In situ AFM detection of the stiffness of the in situ exposed cell nucleus. *Biochim Biophys Acta Mol Cell Res* **1868:** 118985. doi:10.1016/j.bbamcr.2021.118985

Wehrle-Haller B, Imhof BA. 2003. Actin, microtubules and focal adhesion dynamics during cell migration. *Int J Biochem Cell Biol* **35:** 39–50. doi:10.1016/S1357-2725(02)00071-7

Welf ES, Ahmed S, Johnson HE, Melvin AT, Haugh JM. 2012. Migrating fibroblasts reorient directionality by a metastable, PI3K-dependent mechanism. *J Cell Biol* **197:** 105–114. doi:10.1083/jcb.201108152

Wen L, Marki A, Wang Z, Kiosses WB, Mikulski Z, Ley K, Orecchioni M, Makings J, Billitti M, Wang E, et al. 2022a. A humanized β2 integrin knockin mouse reveals localized intra- and extravascular neutrophil integrin activation in vivo. *Cell Rep* **39:** 110876. doi:10.1016/j.celrep.2022.110876

Wen L, Moser M, Ley K. 2022b. Molecular mechanisms of leukocyte β2 integrin activation. *Blood* **139:** 3480–3492. doi:10.1182/blood.2021013500

Wengner AM, Pitchford SC, Furze RC, Rankin SM. 2008. The coordinated action of G-CSF and ELR^+ CXC chemokines in neutrophil mobilization during acute inflammation. *Blood* **111:** 42–49. doi:10.1182/blood-2007-07-099648

Wu M, Sadhu RK, Meyer K, Tang Z, Marchando P, Woolfson DN, Gov NS, Weiner OD. 2025. WAVE complex forms linear arrays at negative membrane curvature to instruct lamellipodia formation. *J Cell Biol* **224:** e202410098. doi:10.1083/jcb.202410098

Wyble CW, Hynes KL, Kuchibhotla J, Marcus BC, Hallahan D, Gewertz BL. 1997. TNF-α and IL-1 upregulate membrane-bound and soluble E-selectin through a common pathway. *J Surg Res* **73:** 107–112. doi:10.1006/jsre.1997.5207

Xu J, Wang F, Van Keymeulen A, Herzmark P, Straight A, Kelly K, Takuwa Y, Sugimoto N, Mitchison T, Bourne HR. 2003. Divergent signals and cytoskeletal assemblies regulate self-organizing polarity in neutrophils. *Cell* **114:** 201–214. doi:10.1016/S0092-8674(03)00555-5

Yamada KM, Sixt M. 2019. Mechanisms of 3D cell migration. *Nat Rev Mol Cell Biol* **20:** 738–752. doi:10.1038/s41580-019-0172-9

Yang J, Antin P, Berx G, Blanpain C, Brabletz T, Bronner M, Campbell K, Cano A, Casanova J, Christofori G, et al. 2020. Guidelines and definitions for research on epithelial–mesenchymal transition. *Nat Rev Mol Cell Biol* **21:** 341–352. doi:10.1038/s41580-020-0237-9

Yeh YT, Serrano R, François J, Chiu JJ, Li YSJ, Del Álamo JC, Chien S, Lasheras JC. 2018. Three-dimensional forces exerted by leukocytes and vascular endothelial cells dynamically facilitate diapedesis. *Proc Natl Acad Sci* **115:** 133–138. doi:10.1073/pnas.1717489115

Yip AK, Chiam KH, Matsudaira P. 2015. Traction stress analysis and modeling reveal that amoeboid migration in confined spaces is accompanied by expansive forces and requires the structural integrity of the membrane–cortex interactions. *Integr Biol* **7:** 1196–1211. doi:10.1039/C4IB00245H

Yonemura S, Itoh M, Nagafuchi A, Tsukita S. 1995. Cell-to-cell adherens junction formation and actin filament organization: similarities and differences between non-polarized fibroblasts and polarized epithelial cells. *J Cell Sci* **108:** 127–142. doi:10.1242/jcs.108.1.127

Zatulovskiy E, Tyson R, Bretschneider T, Kay RR. 2014. Bleb-driven chemotaxis of *Dictyostelium* cells. *J Cell Biol* **204:** 1027–1044. doi:10.1083/jcb.201306147

Zhelev DV, Alteraifi AM, Chodniewicz D. 2004. Controlled pseudopod extension of human neutrophils stimulated with different chemoattractants. *Biophys J* **87:** 688–695. doi:10.1529/biophysj.103.036699

Zigmond SH. 1977. Ability of polymorphonuclear leukocytes to orient in gradients of chemotactic factors. *J Cell Biol* **75:** 606–616. doi:10.1083/jcb.75.2.606

Many Cells Make Light Work: Self-Generated Gradients Organize *Dictyostelium* Aggregates and Neutrophil Swarms

Evelyn Strickland,[1] John F. Durel,[2] Allyson E. Sgro,[2] and Orion D. Weiner[1]

[1]Department of Biochemistry and Biophysics; Cardiovascular Research Institute, University of California, San Francisco, San Francisco, California 94158, USA

[2]Janelia Research Campus, Howard Hughes Medical Institute, Ashburn, Virginia 20147, USA

Correspondence: sgroa@janelia.hhmi.org; orion.weiner@ucsf.edu

Cells work together to accomplish complex tasks. For example, both neutrophils and Dictyostelid collectives use self-generated multicellular signaling gradients to coordinate aggregation over large areas through local interactions. However, these aggregation programs occur for different reasons that necessitate different implementations. Dictyostelids are soil-dwelling amoeba that aggregate when starving to facilitate dispersal to new locations. These aggregates do not require specific locations or group sizes. In contrast, neutrophils are innate immune cells that collectively migrate to sites of injury and infection. These swarms need to occur in specific locations and must be constrained in size to avoid collateral damage to the host. Here, we review how these evolutionarily divergent systems sculpt long-range gradients at the molecular and cellular levels, discussing their similarities and differences in light of their distinctive goals. Convergence on self-generated gradients for aggregation despite different goals suggests that it is an optimal strategy to bring individuals together in complex environments.

Across biological scales, individuals come together into groups to collectively solve problems that are beyond the ability of individuals acting alone. At the organismal scale, ants communicate with one another to forage for food (Chandra et al. 2021), and starlings flock to evade predators (King and Sumpter 2012). At the cellular scale, immune cells called neutrophils collectively migrate to distant sites of injury and infection to rapidly limit infection and heal tissues, and the cellular slime mold, *Dictyostelium discoideum*, aggregates together during starvation to form a multicellular dispersal structure. Despite being separated by over a billion years of evolution, both cell types use evolutionarily conserved approaches to sense chemoattractants, polarize, and move during aggregation (Devreotes and Zigmond 1988; Parent 2004; Lawal et al. 2020). However, their collective gradient sculpting circuits are quite distinct, with the differences in single-celled behaviors and signaling machinery reflecting the distinctive tasks the cellular collectives are trying to solve. Neutrophils act together as a part of host defense, which requires them to recruit many other neutrophils to damage and

Cite this article as *Cold Spring Harb Perspect Biol* doi: 10.1101/cshperspect.a041739

infection sites without triggering a runaway reaction that would lead to unchecked inflammation. Controlling this immune response requires cells to both aggregate to a specific location and manage the number of cells converging on this location (Siwicki and Kubes 2023). In contrast, *Dictyostelium* is a soil-dwelling microbe that aggregates as part of a developmental program for survival, which requires cells to come together to create a fruiting body structure where stalk cells hold a collection of spore cells up in the air for dispersal (Kessin 2001). Both the size and location of this structure are flexible, with the exact location of the initial aggregation being an emergent property of the environment and the spatial distribution of the cells (Kessin 2001; Eckstein et al. 2020; Ford et al. 2023). While many reviews cover swarming and aggregation and the underlying mechanisms in either neutrophils (Kienle and Lämmermann 2016; Brown and Yipp 2023; Song et al. 2023) or *Dictyostelium* (Aubry and Firtel 1999; Weijer 2004; Loomis 2014; Consalvo et al. 2019; Kawabe et al. 2019) from multiple perspectives, this review will focus on juxtaposing these two systems to inform our understanding of how cells working together might generally solve shortcomings of chemotactic cells acting alone. To show this, we discuss the highly disparate goals of these two model systems, how differences in implementation support these different goals, and how the conceptual similarities between them enable them to solve the common challenge presented by bringing large numbers of individuals together. Both systems have converged on the general approach of relaying and reshaping long-range guidance cues, suggesting that collective chemotactic control strategies represent an optimal approach for bringing cells together while simultaneously modulating group properties to serve a wide variety of biological needs.

WHAT MIGRATION TASKS ARE NEUTROPHIL AND *DICTYOSTELIUM* SWARMS TRYING TO SOLVE?

Both neutrophils and *Dictyostelium* are model systems that biologists have turned to for decades to better understand behaviors such as cellular motility, micropinocytosis, and bacterial predation in eukaryotes. While both neutrophils and *Dictyostelium* undergo collective aggregation (Fig. 1), this behavior is driven by very different goals. In both cell types, the collective aggregation programs use self-generated chemotactic gradients for guidance, with activated cells release signals that trigger their neighbors in an excitable manner. For both cell types under common laboratory conditions, this process of cell-to-cell signal regeneration can generate propagating waves of chemoattractant that move outward, away from the original source, and instruct cells to move inward toward a central area. Despite these conceptual similarities, there are fundamental differences in how these systems each create and shape their chemotactic gradients. Appreciating why these differences exist and why we should not expect identical molecular circuitry or aggregation strategies requires understanding the end tasks that *Dictyostelium* and neutrophils are working to solve.

Dictyostelium is a single-celled eukaryotic microbe that lives in the topsoil and leaf-litter layers and consumes bacteria (Douglas et al. 2013). When available bacteria are depleted, transcriptional changes in the amoeba initiate a developmental program that enables cells to find and migrate toward one another to build a dispersal structure (Kessin 2001; Loomis 2015). To accomplish this task, *D. discoideum* and several related species use cyclic AMP (cAMP) as both the chemoattractant to locate each other and the signal to synchronize gene expression as part of the developmental program (Bonner 1947; Konijn et al. 1967; Barkley 1969; Loomis 2014; Du et al. 2015). Microbial developmental programs have different design constraints than metazoan developmental programs. They are part of stress responses and need the flexibility to tolerate a wide range of environmental conditions, group sizes, and genetic heterogeneity in the population. As a result, the signaling program coordinating development needs to function across many different population sizes and spatial geometries without reliance on a fixed center of aggregation (Ponte et al. 1998; Brock and Gomer 1999; Medina et al. 2019; Ricci-Tam et al. 2023).

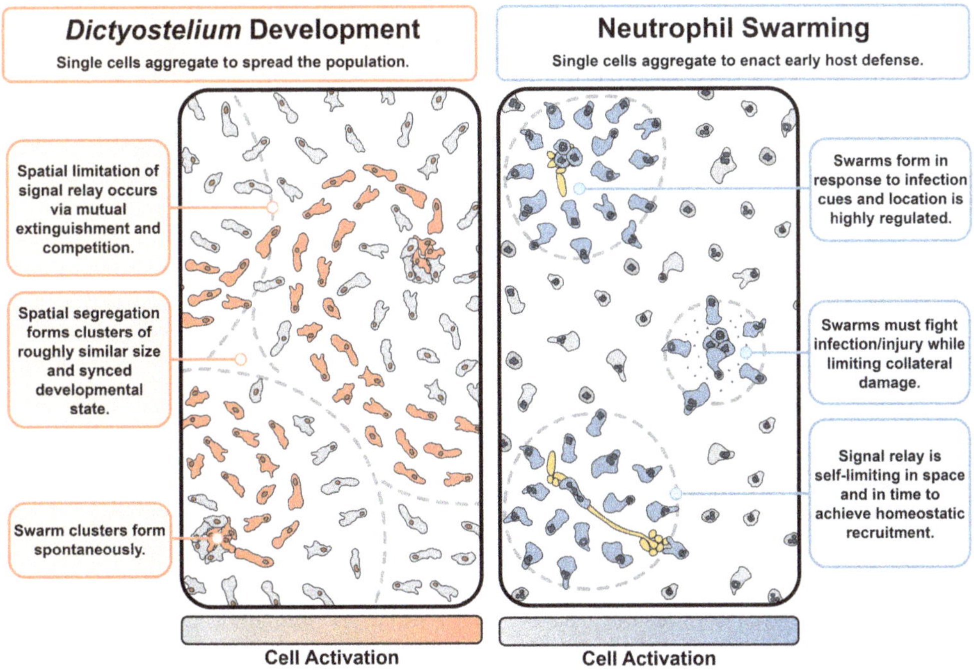

Figure 1. Migration patterns of *Dictyostelium* aggregation and neutrophil swarming. *Dictyostelium discoideum* is a single-celled amoeba that aggregates under starvation conditions to develop into a multicellular structure to support dispersal of cells to more favorable locations. Cell movement is controlled through self-generated gradients of the chemoattractant cyclic AMP (cAMP) (here red indicates cells in a high cAMP state) that are relayed from cell to cell. As development progresses, the cAMP waves generate target and spiral patterns that sweep outward and partition the field through mutual extinction into adjacent aggregation sites. The goal of this process is to form distinct clusters for multicellular morphogenesis, but the exact number of cells and the precise location of each aggregation site are not important. Neutrophils are innate immune cells that collectively migrate to sites of injury and infection, enabling them to rapidly accumulate from distances beyond the reach of primary injury/infection cues. The dynamics of the self-generated gradients of the leukotriene B_4 (LTB_4) chemoattractant can be visualized by cell biosensors (drawn as blue cells having a high calcium state). Unlike in *Dictyostelium*, the specific location of the swarming sites is essential, and the range of the self-generated gradients is tuned by a self-extinguishing relay rather than competition among aggregation sites.

In contrast, neutrophils are the first responders of the immune system, specialized to move fast, sense a wide array of damage and pathogen-associated cues, and escalate the local inflammation response. To carry out these functions, they self-amplify their recruitment to sites of infected tissue where they trap and contain invading pathogens (Chtanova et al. 2008; Peters et al. 2008; Li et al. 2024) and to sterile injuries to break up tissue damage (McDonald et al. 2010; Ng et al. 2011; Lämmermann et al. 2013; Poplimont et al. 2020). This collective signal amplification is essential because most neutrophils are too far away to sense the primary cues of injury and infection directly. Specifically, swarming relies on neutrophils' innate ability to synthesize the lipid chemoattractant leukotriene B_4 (LTB_4), which strengthens primary chemoattractant gradients (Afonso et al. 2012; Lämmermann et al. 2013; Reátegui et al. 2017; Kienle et al. 2021; Strickland et al. 2024). Given neutrophils' capacity for cell and tissue destruction, swarm sites must be chosen carefully, and the number of cells arriving at any given site must be limited and proportional to the insult to minimize collateral damage to the host. To enable robust recruitment over a wide range of starting conditions, recent work has shown how these homeostatic behaviors are baked into this self-amplification process at the neutrophil level (Kienle et al. 2021; Strickland et al. 2024). To best defend the host, neutrophils must work among themselves and with surrounding cells to organize swarms that typically do not involve every cell in the vicinity, instead "choosing" to only recruit cells in proportion to the magnitude of the insult.

While both systems converged on self-generated chemotactic gradients as an effective strategy for gathering widely dispersed cells in complex natural environments, they evolved in distinctive niches under markedly dissimilar selective pressures. In cellular slime molds, aggregation is a proactive first step in a starvation-driven development program that is required for survival and occurs in the soil. In neutrophils, aggregation is a key defensive mechanism to protect the host from illness and occurs inside of an organism. To better appreciate whether there are common rules underlying the convergent approach of using self-generated chemotactic gradients to drive collective migration represent convergent approaches, it is necessary to dive into the details of each of these cellular collectives in their respective contexts while appreciating where they have independently innovated to facilitate the specific task at hand.

WHAT SIMILARITIES EXIST IN NEUTROPHIL AND *DICTYOSTELIUM* CHEMOTACTIC GRADIENT GENERATION?

Both of these aggregating multicellular systems leverage self-generated chemotactic gradients that are relayed and amplified by multiple cells to overcome the physical limitations of local gradient detection and permit aggregation over distances up to the centimeter scale in *Dictyostelium* (Alcantara and Monk 1974; Tomchik and Devreotes 1981), and up to the millimeter scale in neutrophils (Strickland et al. 2024), which are 2–3 orders of magnitude larger than can be interpreted by a single cell. While there are multiple ways cells can create chemical gradients to drive their own motility (Insall et al. 2022), signal relays where each cell triggers its neighbor to produce the cue provide a number of practical advantages. Specifically, relays propagate information more robustly over long distances by regenerating a dissipating signal at each cell "node" and creating a traveling wave, analogous to how nodes of Ranvier regenerate the action potential traveling down myelinated neurons (Arancibia-Carcamo and Attwell 2014). In comparison to a single, long-distance gradient, relaying the signal is also more energy efficient (Kessin 2001; Qiu et al. 2017), as well as more adaptable to changes in the environment. For a single long-range gradient, small differences in receptor occupancy across the cell limit the distance over which the direction of the source can be inferred (Devreotes and Zigmond 1988; Tweedy et al. 2016b; Insall et al. 2022). In contrast, the combination of spatial and temporal chemoattractant changes in a self-generated wave enables more efficient guidance over longer distances than static or slowly developing diffusive gradients (Vicker et al. 1986; Albrecht and Petty 1998; Geiger et al. 2003).

Traveling chemoattractant waves present two potential problems that cells need to solve: (1) The cells need to avoid reversing direction as the wave passes over them so they only move toward the aggregation center, registering the "back of the wave" as part of the same stimulus, and (2) they also need to reset so subsequent new waves passing over the cell register as new stimuli that they should move toward. Two studies using microfluidic devices to create traveling waves of chemoattractant to probe how *Dictyostelium* and neutrophils respond suggest that these systems have wave properties that are optimized for their specific molecular motility machinery (Skoge et al. 2014; Ishida et al. 2025). Specifically, waves that pass over a cell too quickly do not allow the cell to discern the direction of the wave origin. Conversely, waves that pass too slowly over a cell result in cells following the passing wave away from the wave source, showing the "back-of-the-wave" problem (Goldstein 1996; Skoge et al. 2014; Ishida et al. 2025). Waves of an intermediate speed and periodicity, however, yield an optimum where cells unidirectionally migrate toward the source of the wave. In the neutrophil case, this optimum speed (95–170 μm/min) brackets the wave speeds observed in neutrophil swarming (Strickland et al. 2024; Ishida et al. 2025). Similarly, the optimum wave parameters at which *Dictyostelium* migrate toward the source are comparable to those endogenously generated by cells at the onset of starvation (~200–500 µm/min wave speed and 5–8 min period depending on the strain and density) (Alcantara and Monk 1974; Devreotes et al. 1983; Siegert and Weijer 1989; Postma and van Haastert 2009; Skoge et al. 2014). Together, these studies suggest that both neutrophils and *Dictyostelium* have optimized the parameters of their signal relay systems to match their molecular motility machinery.

These similarities in implementation speak to the similarities in the problems both organisms face during aggregation: Both systems need to spread signals over populations orders of magnitude larger than a single cell and operate in complex and potentially noisy environments. This self-generated signal relay with a built-in optimal frequency enables both systems to more effectively and robustly coordinate over large distances.

HOW DO NEUTROPHIL AND *DICTYOSTELIUM* SIGNAL RELAYS DIFFER?

Despite sharing conserved migratory machinery and having converged on signal relay as an efficient strategy for multicellular aggregation, neutrophils and *Dictyostelium* have distinctive implementations of signal relay. We will focus on what these differences are and how they support each organism's aggregation goals. For example, *Dictyostelium* signal relay circuitry should be expected to favor an active state since their survival depends on aggregating. Accordingly, they continually secrete prestarvation factors that trigger the relay without negative feedback from the presence of food (Rathi and Clarke 1992; Yuen et al. 1995). Neutrophil circuitry, conversely, should favor an inactive state because host survival depends on minimizing collateral damage (Brown and Yipp 2023). Accordingly, these cells require the permissive input of an insult to trigger their aggregation program. Similar constraints affect how both cellular swarms regulate their group size, with *Dictyostelium* using an indirect secretory method to regulate group size, while in neutrophils, group size regulation is directly baked into the relay circuitry (Brock and Gomer 1999; Kienle et al. 2021; Strickland et al. 2024). These different approaches reflect their needs for approximate and tight size control, respectively.

Dictyostelium needs to create a fruiting body that functions as a dispersal structure where cells are held up off the ground. Ideally, this structure is large but not so large as to render it structurally unstable, but the developmental program is flexible and can handle lower cell numbers that yield smaller aggregates (Bonner and Hoffman 1963). This final fruiting body does not necessarily have to form where the cells originally aggregated, so unlike in neutrophils, the aggregation location is unimportant. A combination of experimental data and theoretical models suggests that cell density in combination with properties of the signaling circuit sets the frequency of stochastic cAMP signaling centers that begin the aggregation process (Hashimoto et al. 1975; Sawai et al. 2005; Gregor et al. 2010). The excitable cells propagate the cAMP waves through the population from the centers until they run into one another and coextinguish (Durston 1973; Goldbeter and Erneux 1978; Sgro et al. 2015). The sensed intensity of the cAMP wave is modulated by macromolecular "counting factors" that are secreted as the cells begin to stream into a signaling center. These additional extracellular signals build up in the environment over time and can cause streams to break up, limiting the final size of the aggregation center (Brock and Gomer 1999). The relay also functions as an oscillatory signal that transcriptionally synchronizes cells, so they are primed to move to the next stage of development together (Cai et al. 2014). In the aggregate, cells use allorecognition mechanisms to sort out related and unrelated cells such that they create fruiting bodies primarily with close kin, and then the developmental process proceeds with cells beginning to express prespore and prestalk genes (Hirose et al. 2015; Strassmann 2016).

Local attractant breakdown is also an essential feature of *Dictyostelium* chemotaxis. During development, cells produce both a membrane-bound and a secreted extracellular phosphodiesterase (PDE) that degrades cAMP into AMP in the immediate vicinity (Chang 1968). These mitigate chemoattractant buildup and improve chemotactic precision by further steepening the passing waves (Chang 1968; Malchow et al. 1972; Pannbacker and Bravard 1972; Kessin 2001; Masaki et al. 2013; Tweedy et al. 2020). Rates of cAMP degradation are further fine-tuned by the production and secretion of a phosphodiesterase inhibitor (PDI) (Pálsson and Cox 1996). Interestingly, because the release of PDE is insensitive to cAMP receptor adaptation while the cAMP sensing and production are adaptation-sensitive, a sustained cAMP signal will decrease its own sensing and production but not its depletion (Kessin 2001). This degradation can even overwhelm production of the cAMP signal and result in a "repulsive" cAMP gradient (Karmakar et al. 2021). It remains unclear whether such tightly controlled local degradation mechanisms play similarly important roles in neutrophil chemotaxis.

In comparison, neutrophil swarming is an inherently self-limiting amplification process. Unlike *Dictyostelium*, neutrophils have an innate ability to limit the amplification of their chemoattractant cues in space and time. Originally, neutrophils were known to climb simple diffusive gradients of *N*-formyl-methionyl-leucyl-phenylalanine (fMLP) better when they could reinforce the gradient with a secondary cue, LTB_4 (Afonso et al. 2012). This gradient amplification mechanism was soon tied to neutrophil swarming (Lämmermann et al. 2013) and has recently been shown to exhibit striking self-limiting relay-like wave propagation in a simplified ex vivo context (Strickland et al. 2024). These chemoattractant waves do not have a fixed radius of propagation. Instead, cells adjust the radius of recruitment as a function of environmental cues. Neutrophils generate a small radius of response for high neutrophil density and a large radius of response for low neutrophil density, thereby recruiting similar numbers of cells to the target over a wide range of initial cell densities (Strickland et al. 2024). This ability to tune and self-limit relay is an unusual and unexpected feature of neutrophil collective migration (Dieterle et al. 2020), as most other excitable systems continue to propagate indefinitely in an otherwise permissive field or extinguish as a result of interference between multiple aggregation centers (as in *Dictyostelium).* Beyond their ability to spatially restrict the spatial extent of swarming, other recent work has shown how neutrophils temporally limit their swarming response by desensitization to repeat exposures of LTB_4 (Kienle et al. 2021). This desensitization enables cells to selectively reorganize their clusters, facilitating persistent responses and pathogen sequestration by preventing cells from moving to the most recent wave source. While it is unclear whether a cell–cell relay is used in all contexts, neutrophils exhibit an extraordinary ability to shape and respond to their own self-generated gradients in space and time.

Neutrophils also cluster around select sites of swarm initiation. While *Dictyostelium* swarm centers stochastically arise or are promoted at certain locations by environmental heterogeneity, neutrophil clusters form in response to and position themselves around infection and injury cues. While the exact logic of neutrophil swarm initiation remains poorly understood, neutrophils appear highly resistant to spurious activation. When swarms do form, swarm size appears directly correlated with infection size, as shown by an ex vivo assay where the swarm size was tuned by the number of foreign particles (Reátegui et al. 2017). This work also showed that a single yeast does not trigger swarming, indicating that simple receptor engagement is not sufficient and cells must sense a certain threshold amount of a danger or damage cue to activate a collective response, possibly reflecting the threshold of an excitable system. Furthermore, multiple signals, both physical and chemical, appear to be needed to trigger neutrophil swarming to sites of injury in animal studies. In mice, physical contact with an injury site appears necessary to initiate swarming, and this requirement is exploited by other cell types to prevent swarming altogether in certain contexts (Lämmermann et al. 2013; Uderhardt et al. 2019). Beyond contact alone, in zebrafish, neutrophils appear to locally reinforce calcium "alarm signals" via purinergic signaling that could strengthen the pioneer neutrophil's ability to create and release LTB_4 (Poplimont et al. 2020). While these observations show how careful neutrophils are in choosing swarm initiation sites, they also present a challenge for the field. Experimental differences in the sufficient or necessary features of swarm initiation may differ between different model systems—not only between the distantly related neutrophils and *Dictyostelium* but even between neutrophils in different animal species. Having outlined the broad commonalities and differences in end goals and signaling mechanisms, we are better positioned to explore which peculiarities of the underlying molecular mechanisms give rise to their different context-specific collective behaviors (Fig. 2).

WHAT ARE THE UNDERLYING CIRCUITS DRIVING SIGNAL RELAY IN THESE SYSTEMS?

Dictyostelium's signaling system is very well studied, both due to its role as a model organism and the availability of technology facilitating mutagenesis and phenotypic screening. The idea that a chemical attractant, rather than some other mechanism, helps cells find one another and aggregate was confirmed in 1947 by John Bonner (1947). This "acrasin" was later determined in *D. discoideum* to be the small molecule cAMP (Konijn et al. 1967). Over time, the receptors and production enzyme for cAMP were identified, as well as the molecules that degrade cAMP both externally and internally (Chang 1968; Malchow et al. 1972; Pannbacker and Bravard 1972; Green and Newell 1975; Klein 1976; Klein et al. 1988; Saxe et al. 1991). Many other signaling components were identified using restriction enzyme-mediated integration, a technique that facilitates the generation of libraries of strains with a single site interrupted in the genome to link specific genes to aberrant signaling and developmental phenotypes (Kuspa and Loomis 1992). Over time,

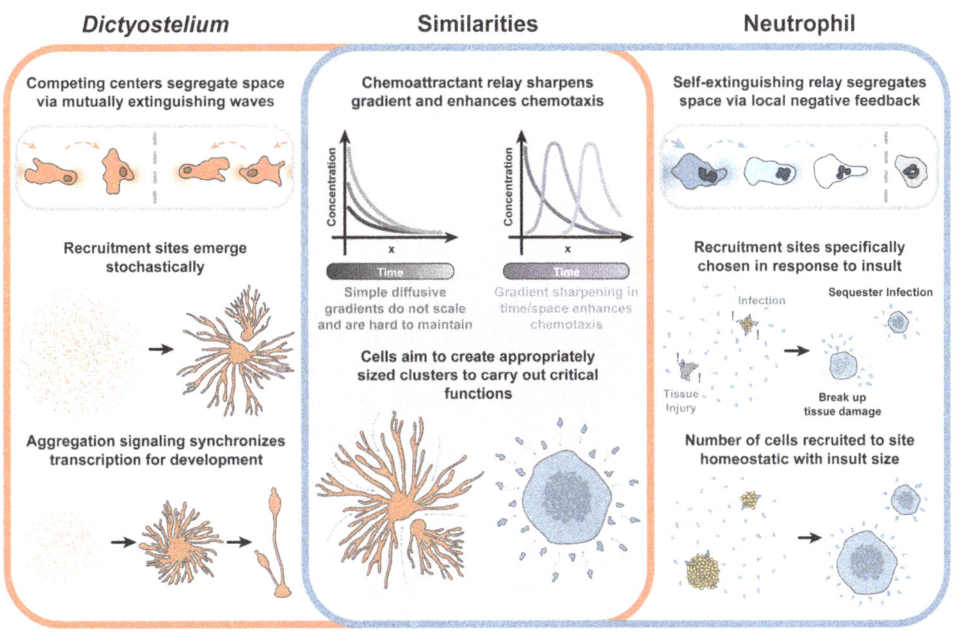

Figure 2. Key similarities and differences. Neutrophil swarming and *Dictyostelium* aggregation both employ dynamic self-generated gradients to enable directional guidance over longer distances than are possible via static spatial gradients. Both cell types employ self-generated relays to generate propagating waves of chemoattractant that guide cell movement toward aggregation sites. Beyond these key similarities, these systems have several important differences: (1) the mechanism by which relay is limited (*top*), (2) how the location of aggregation sites emerges (*middle*), and (3) the purpose of cell aggregation (*bottom*).

through a combination of theoretical modeling and experimental interrogation, a rough circuit for signal relay in *Dictyostelium* has emerged, but there is not yet complete agreement on its architecture (Fig. 3 shows a few key players and features; Martiel and Goldbeter 1987; Laub and Loomis 1998; Maeda et al. 2004; Sawai et al. 2005; Sgro et al. 2015; Jaiswal et al. 2024). Despite this lack of agreement on how the circuit is wired, key functional features include excitability and fold change detection (Sgro et al. 2015; Kamino et al. 2017; Huyan et al. 2021). Excitability enables cells to amplify changes they detect in the environment, and in *Dictyostelium*, there is a refractory period following an excitable stimulus where cells cannot effectively amplify any external changes. As a result of this refractory period, cells that have recently been stimulated by another signaling center cannot fully respond to a signal from another potential aggregation center (Goldbeter and Erneux 1978; Sgro et al. 2015). This refractory period is important because it helps set the size of the area that will aggregate: If a cell has recently experienced a stimulus from one signaling center, it will not respond to another, and the area where those signals meet becomes the edge of the aggregation territory. Similarly, fold-change detection allows cells to respond only to large changes in signals in the environment, avoiding responses to small noisy fluctuations and potentially enabling better coordination across a wide range of group sizes (Sgro et al. 2015; Kamino et al. 2017; Huyan et al. 2021).

Upstream of the cAMP production and relay circuit is the part of the network that interprets extracellular cAMP signals and translates them into chemotactic responses. In contrast to the cAMP production circuit, it is well agreed upon that the chemotactic circuitry has an incoherent feedforward architecture (Takeda et al. 2012; Nakajima et al. 2014; Skoge et al. 2014). These different parts of the intracellular signaling network have different functions that these different architectures enable: The chemotaxis network is designed to climb gradients, whereas the signal relay network is designed to amplify and pass along signals both to drive chemotaxis and to synchronize gene expression. These behaviors also happen on different timescales, with movement responses over tens of seconds and a full cycle of signal relay happening over minutes. This is different from neutrophils, where it is unknown whether the primary swarm signal carrier, LTB_4, drives transcriptional changes in the cells or serves other functions that would necessitate different cellular subcircuits.

It is also interesting to note that there is even diversity in signal relays at the molecular level among species closely related to *D. discoideum*, as these other organisms are performing very similar behaviors with the same final goal. Only a handful of Dictyostelids, such as *D. discoideum*, *Dictyostelium purpureum*, and *Dictyostelium mucoroides*, are known to use cAMP as their primary aggregation cue in a signal relay (Shaffer 1956; Bonner et al. 1969). Other species rely on extracellular cAMP only in later periods of development or use other molecules entirely: *Dictyostelium minutum* sends out aggregation signals using a derivative of folic acid, *Dictyostelium lacteum* uses a derivative of pterin, and the more distantly related *Polysphondylium* uses glorin (Shimomura et al. 1982; Van Haastert et al. 1982; de Wit and Konijn 1983). While many of these aggregating amoeba lineages are well-characterized both morphologically and phylogenetically, it remains to be seen whether they use similar molecules and have similar relay circuit structures to those in *D. discoideum* (Brown and Silberman 2013). Given that even closely related species implement signal relay mechanisms differently at the molecular level, it is evident that identifying the molecular players in one aggregating cell type is insufficient to base an understanding of specifics even in qualitatively similar systems.

Unlike *Dictyostelium*, the neutrophil swarming relay circuit structure is not well understood. While its key molecular player LTB_4 has long been implicated in autocrine and paracrine amplification of chemotactic signaling (Afonso et al. 2012; Lämmermann et al. 2013), its role in cell-to-cell relay has only recently been observed (Dieterle et al. 2020; Strickland et al. 2024). While cAMP exposure causes the creation of cAMP and its release in *Dictyostelium*, it

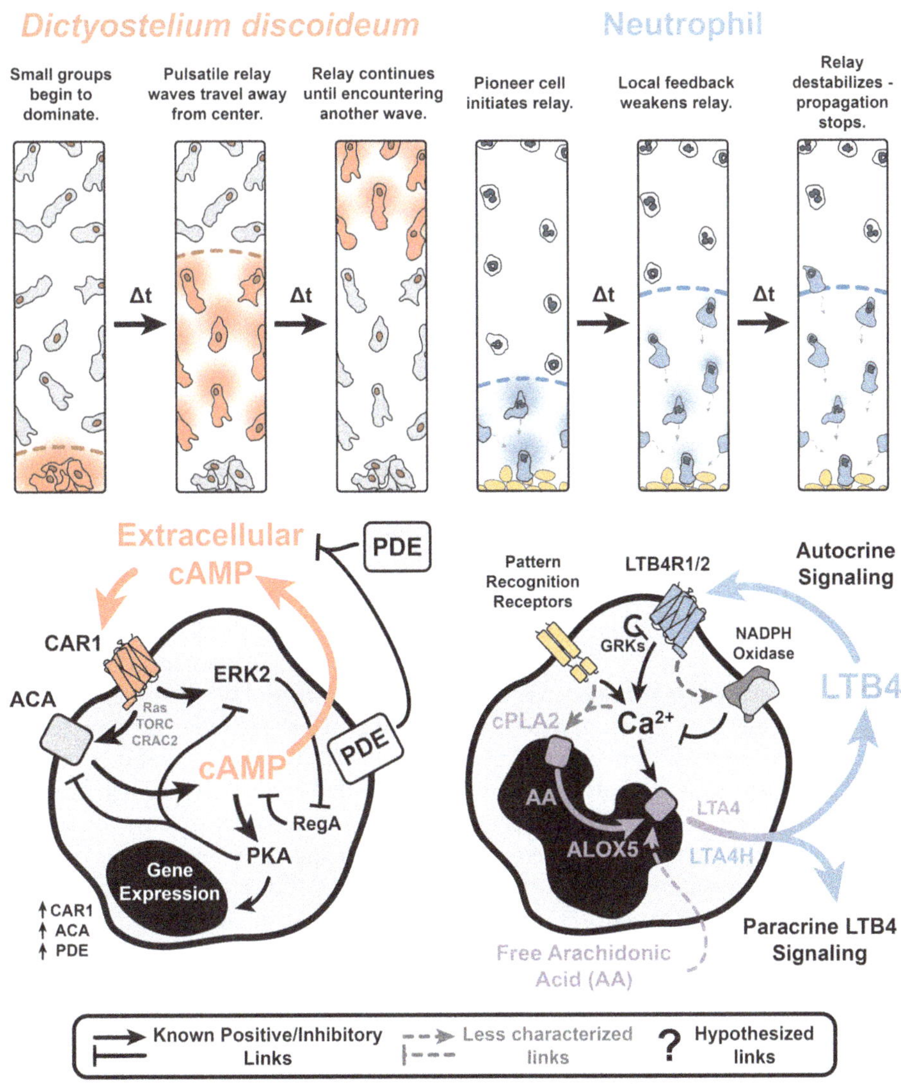

Figure 3. Cell circuits driving collective motility through gradient generation. *Dictyostelium discoideum* cyclic AMP (cAMP) signaling begins as cells starve and stochastically release pulses of cAMP. These pulses draw in neighboring cells, creating stochastically generated small clusters by driving cell movement toward aggregation sites. After cells are stimulated by a pulse, they have a refractory period during which they cannot respond either by making additional cAMP or by moving toward the new pulse, and this refractory period means waves mutually extinguish one another and partition the cells into different groups based on which wave stimulates the cell to move toward it. Neutrophil signal relay differs from *D. discoideum* relay in two key respects. First, neutrophils require distinct initiation signals (diffusive damage-associated molecular pattern [DAMP]/pathogen-associated molecular patterns [PAMPs] or contact-based bacterial/fungal cues) to initiate a swarming response—these control the specific sites of cell aggregation unlike the stochastically determined sites of aggregation for *D. discoideum*. Second, after initiation, neutrophils generate propagating waves of leukotriene B_4 (LTB_4) that are self-extinguishing, even for a single site of initiation. Note that this figure shows a few key players and some of their known and potential links and is not meant to be exhaustive.

does not appear that LTB_4 exposure is sufficient to cause LTB_4 release as part of the neutrophil signal relay. Furthering the mystery, recent work has shown that swarm signals show relay-like travel before being dampened in a self-extinguishing manner further from the initiation site (Strickland et al. 2024). While a simple circuit to account for this self-extinguishing relay has been proposed, this understanding has yet to be integrated into a larger network, and it is not yet understood how neutrophils tune their excitability across a population of cells (Fig. 3). Just as in the study of *Dictyostelium*, a combination of theory with quantitative single-cell and population-wide experiments has just begun to reveal important information processing features of the signal relay network. While the full LTB_4 relay circuit is not clearly understood, we will detail the better-known aspects of this system and its multiple layers of regulation.

LTB_4 synthesis and release is more complicated than a binary on or off switch, and different inputs trigger different bulk amounts of output (Ford-Hutchinson et al. 1980; Palmer and Salmon 1983; Afonso et al. 2012). While calcium signaling has been shown to be a central node regulating LTB_4 synthesis (Ford-Hutchinson et al. 1980; Palmer and Salmon 1983; Khazen et al. 2022), bypassing other regulatory layers and plugging directly into calcium potentially obscures the true complexity of the circuit. For example, LTB_4 pre-exposure can both enhance (Naccache et al. 1989; McDonald et al. 1994; Serio et al. 1997; Surette et al. 1999) and limit LTB_4 output (Fiedler et al. 1998) depending on the exact timing and presentation of the stimulus, as well as the presence of other stimuli. These data perhaps should be seen in the light of highly regulated LTB_4 receptors, where phosphorylation state changes in a stepwise, concentration-dependent manner (Nakanishi et al. 2018; Tatsumi et al. 2023). These features enable the receptor to have different downstream signaling kinetics and internalization at even a single exposure to physiological concentrations of cue (Kienle et al. 2021; Lundgren et al. 2023), allowing for a complex layer of positive and negative feedback at the level of the receptor itself.

Furthermore, LTB_4 synthesis is also strongly limited by the activation state and activity of the neutrophil nicotinamide adenine dinucleotide phosphate (NADPH) oxidase complex (Song et al. 2023). While links between the LTB_4 receptor and the activation of the NADPH oxidase complex exist (Lindsay et al. 1998; Serezani et al. 2005), it is not a direct interaction and likely involves currently unidentified intermediates (Wu et al. 2025). However, NADPH oxidase has been shown to regulate the membrane polarization state to directly limit calcium influx (Tintinger et al. 2001; Song et al. 2020) and thus puts a "ceiling" on LTB_4 output in a cell-intrinsic fashion that is critical for the spatial limitation of relay (Strickland et al. 2024). Importantly, perturbations to this circuit result in disease. For example, in the human autoinflammatory condition, chronic granulomatous disease, neutrophils overproduce LTB_4 due to defects in the NADPH oxidase complex (Hamasaki et al. 1989; Tintinger et al. 2001; Song et al. 2020), causing an overrecruitment of myeloid cells to sites of infection and aberrant granuloma formation that can be devastating when left unmanaged (Dinauer 2019).

While the above self-limiting features of the LTB_4 relay circuit have recently captured significant attention, the importance of its upstream precursor, arachidonic acid (AA), and how it is regulated have perhaps been neglected. AA is predominately liberated from membrane phospholipids by the cytosolic phospholipase A2 (cPLA2), and this release serves as the necessary substrate during LTB_4 synthesis in response to stimuli such as fMLP, platelet-activating factor (PAF), or zymosan (Claesson et al. 1981; Palmer and Salmon 1983; Haurand and Flohé 1989; Burke and Dennis 2009). In direct challenge to existing relay models, however, exogenous LTB_4 exposure alone does not lead to detectable amounts of endogenous AA mobilization (or subsequent LTB_4 synthesis) (Prescott et al. 1984; Haurand and Flohé 1989; McDonald et al. 1992). While this paradox remains unaddressed within the larger context of neutrophil signal relay, how neutrophils interact with and process exogenous sources of AA might provide insight into how AA regulates LTB_4 generation broadly. In particular, the work that originally characterized LTB_4 showed that exogenous AA is readily taken up by neutrophils and converted into leukotrienes (Borgeat and Samuelsson 1979). Since then, many others have detailed how exogenous AA might provide a copermissive input into the amplification of LTB_4 synthesis (Naccache et al. 1989; Krump and Borgeat 1999; Flamand et al. 2006). Perhaps the most illuminating finding is that detection of LTB_4 in the environment enhances LTB_4 synthesis from exogenous sources of AA (McDonald et al. 1992, 1994; Serio et al. 1997; Surette et al. 1999), indicating a strong positive feedback loop limited by AA availability.

While these circuits have been studied in isolation, LTB_4 synthesis is a multilayered regulatory circuit that will require a much more thorough, quantitative treatment to reach the understanding we have of *Dictyostelium* signaling. However, even a partial enumeration of the molecular circuits makes it evident that there is minimal overlap between the molecular means used to accomplish the aggregation task in these systems. Despite these molecular differences, it is remarkable that both systems have mechanistically converged on signal relay, suggesting that this strategy may be optimal for long-range guidance. These observations suggest that signal relay is not highly constrained from an evolvability perspective, as a wide range of molecules have been co-opted into serving in these and similar relays. This flexibility may allow signal relay circuits to accommodate system-specific needs, such as those of aggregation positioning and group size.

WHY IS IT IMPORTANT TO UNDERSTAND DIFFERENCES IN COLLECTIVE AGGREGATION BEHAVIORS?

Historically, we have focused on a few model systems in biology with the hope that by studying a single example of a phenomenon, we would understand how it might be implemented in other systems. In the case of aggregation behavior, as we have outlined here, even in closely related systems such as different Dictyostelids, different molecules are used to accomplish this task. While the more distantly related systems such as *Dictyostelium* and neutrophils do not share a molecular mechanism for swarming behavior, studying both of these implementations of aggregation has helped the field to identify key principles that could broadly underlie multicellular coordination. Furthermore, the shared methodological approaches and conceptual frameworks developed to understand these two systems have benefitted one another. By continuing to work on these different models in tandem, we will identify approaches to help us tackle the remaining overlapping challenges facing the field of collective cellular migration.

While it appears that signal relay has many advantages for bringing together many cells across large areas, it remains unclear to what extent this mechanism is leveraged in situ. This reflects one of the major quandaries of laboratory science: Does what we have learned in the laboratory reflect how these systems behave in nature? We can make reasonable assumptions about how what we observe in the laboratory may apply to native environments, but behavior is a product of interactions between individuals and the environment, and different environments may elicit or demand different signaling programs to coordinate behaviors effectively within them. For example, the aggregation phenomena we are observing evolved in complex three-dimensional (3D) environments, yet we often rely on two-dimensional (2D) or oversimplified 3D environments in the laboratory setting. The greater level of control in laboratory assays enables us to more effectively disentangle molecular mechanisms, and they are more amenable to observation and quantification than working in vivo or in situ. However, we may not be seeing the full behavioral repertoire of these systems without a complex environment to induce different behaviors. Additionally, simplified assays limit our understanding of the pressures the native environment places on how aggregation can be coordinated. For example, both *Dictyostelium* and neutrophils use diffusive signals that are impacted by both the cell medium and the geometry of their environment. Indeed, theory has shown that dimensionality is important for understanding signal relay wave dynamics (Dieterle et al. 2020). Furthermore, the environment should have substantial effects on cellular motility as it will not only provide different types of surfaces for ligand diffusion but also cellular modes of migration. While it seems likely that recreating the signal at each individual makes signal relay robust to these types of environmental perturbations, further experiments in more complex environments are important to test this intuition.

The other major component missing from laboratory assays is the broad array of diffusive cues that cells interpret in their native environments, both in terms of their spatiotemporal presentation and their biochemical diversity. While early studies of chemoattraction were done with static gradients, cells can actively shape their own gradients, and it is important to understand how they are shaping their gradients if we want to link gradient shape to behavior (Tweedy et al. 2016a, 2020; Tweedy and Insall 2020). Indeed, it is known that in addition to signal relay, *Dictyostelium* actively degrades external cAMP during starvation. While chemoattractant breakdown enhances its pathfinding abilities, how this impacts its ability to aggregate in natural conditions is unclear (Tweedy et al. 2020). Furthermore, while in *Dictyostelium* and neutrophils these self-generated chemoattractant gradients propagate through pulsatile relays, in other models of collective migration such as lateral line migration during zebrafish development and cancer cell migration, chemoattractant sequestration/degradation also shapes the signal gradient (Donà et al. 2013; Venkiteswaran et al. 2013; Susanto et al. 2017). These gradient-shaping mechanisms can enable cells to learn about their environments, such as the ability to determine different path lengths and dead-ends in a maze (Tweedy et al. 2020). In addition to chemoattractants such as cAMP and LTB_4, there are also chemorepellents that play a role in other types of cell migration, and chemorepellents may be important for regulating other signal relay processes (Phillips and Gomer 2012; El-Sobky et al. 2025). Together, these examples underscore the importance of considering the active role cells play in their own directed migration through creating highly complex biochemical environments even in relatively simple physical environments.

In addition to characterizing the natural state of these systems, a synthetic biology approach where simple circuits are built from the ground up and used to understand what features different types of circuitry impart could yield deep insights into how self-generated gradients for collective migration are best regulated (see Youk and Lim 2014 for an example of this approach to understanding autocrine signaling). Continuing to dive into a diverse range of systems that create their own chemotactic gradients to regulate multicellular aggregation will help us identify the full toolbox biology has at its disposal to regulate these types of complex behaviors in situ. As we have outlined in this review, even very closely related cells can have different implementations of a mechanism; through examining both their similarities and differences, we can create a framework to identify what features of a mechanism are important for function. That both *Dictyostelium* and neutrophils use signal relay to bring cells together suggests that this mechanism may be optimal for long-range coordination of many individuals. By regenerating the signal at each cell, there is no single point of failure, and even those far away from the wave center receive a strong signal. Furthermore, in comparison to a long-range gradient, the dynamic nature of the wave makes it easy for cells to follow it as a guidance cue both near and far from the source. Future work comparing these systems and others, both experimentally and in silico, will help us to identify how they function in nature and to compare alternative signaling mechanisms. This joint experiment–theory approach, which has already been so impactful for understanding collective cell motility, will demonstrate whether signal relay is the optimal way of coordinating multicellular aggregation in complex environments or merely one context-dependent solution to a critical task that recurs across the tree of life.

ACKNOWLEDGMENTS

We thank members of the Weiner and Sgro laboratories, and Ariel Amir for helpful discussion and text review. This work was supported by

R35GM118167 (O.D.W.) and the Howard Hughes Medical Institute (A.E.S. and J.F.D.).

REFERENCES

Afonso PV, Janka-Junttila M, Lee YJ, McCann CP, Oliver CM, Aamer KA, Losert W, Cicerone MT, Parent CA. 2012. LTB_4 is a signal-relay molecule during neutrophil chemotaxis. *Dev Cell* **22:** 1079–1091. doi:10.1016/j.devcel.2012.02.003

Albrecht E, Petty HR. 1998. Cellular memory: neutrophil orientation reverses during temporally decreasing chemoattractant concentrations. *Proc Natl Acad Sci* **95:** 5039–5044. doi:10.1073/pnas.95.9.5039

Alcantara F, Monk M. 1974. Signal propagation during aggregation in the slime mould *Dictyostelium discoideum*. *J Gen Microbiol* **85:** 321–334. doi:10.1099/00221287-85-2-321

Arancibia-Carcamo I, Attwell D. 2014. The node of Ranvier in CNS pathology. *Acta Neuropathol* **128:** 161–175. doi:10.1007/s00401-014-1305-z

Aubry L, Firtel R. 1999. Integration of signaling networks that regulate *Dictyostelium* differentiation. *Annu Rev Cell Dev Biol* **15:** 469–517. doi:10.1146/annurev.cellbio.15.1.469

Barkley DS. 1969. Adenosine-3′,5′-phosphate: identification as acrasin in a species of cellular slime mold. *Science* **165:** 1133–1134. doi:10.1126/science.165.3898.1133

Bonner JT. 1947. Evidence for the formation of cell aggregates by chemotaxis in the development of the slime mold *Dictyostelium discoideum*. *J Exp Zool* **106:** 1–26. doi:10.1002/jez.1401060102

Bonner JT, Hoffman ME. 1963. Evidence for a substance responsible for the spacing pattern of aggregation and fruiting in the cellular slime molds. *J Embryol Exp Morphol* **11:** 571–589. doi:10.1242/dev.11.3.571

Bonner JT, Barkley DS, Hall EM, Konijn TM, Mason JW, O'Keefe G III, Wolfe PB. 1969. Acrasin, acrasinase, and the sensitivity to acrasin in *Dictyostelium discoideum*. *Dev Biol* **20:** 72–87. doi:10.1016/0012-1606(69)90005-0

Borgeat P, Samuelsson B. 1979. Transformation of arachidonic acid by rabbit polymorphonuclear leukocytes. Formation of a novel dihydroxyeicosatetraenoic acid. *J Biol Chem* **254:** 2643–2646. doi:10.1016/S0021-9258(17)30120-5

Brock DA, Gomer RH. 1999. A cell-counting factor regulating structure size in *Dictyostelium*. *Genes Dev* **13:** 1960–1969. doi:10.1101/gad.13.15.1960

Brown MW, Silberman JD. 2013. The non–*Dictyostelid sorocarpic* amoebae. In *Dictyostelids*, pp. 219–242. Springer, Berlin.

Brown L, Yipp BG. 2023. Neutrophil swarming: is a good offense the best defense? *iScience* **26:** 107655. doi:10.1016/j.isci.2023.107655

Burke JE, Dennis EA. 2009. Phospholipase A2 structure/function, mechanism, and signaling. *J Lipid Res* **50** (Suppl.)**:** S237–S242. doi:10.1194/jlr.R800033-JLR200

Cai H, Katoh-Kurasawa M, Muramoto T, Santhanam B, Long Y, Li L, Ueda M, Iglesias PA, Shaulsky G, Devreotes PN. 2014. Nucleocytoplasmic shuttling of a GATA transcription factor functions as a development timer. *Science* **343:** 1249531. doi:10.1126/science.1249531

Chandra V, Gal A, Kronauer DJC. 2021. Colony expansions underlie the evolution of army ant mass raiding. *Proc Natl Acad Sci* **118:** e2026534118. doi:10.1073/pnas.2026534118

Chang YY. 1968. Cyclic 3′,5′-adenosine monophosphate phosphodiesterase produced by the slime mold *Dictyostelium discoideum*. *Science* **161:** 57–59. doi:10.1126/science.161.3836.57

Chtanova T, Schaeffer M, Han S-J, van Dooren GG, Nollmann M, Herzmark P, Chan SW, Satija H, Camfield K, Aaron H, et al. 2008. Dynamics of neutrophil migration in lymph nodes during infection. *Immunity* **29:** 487–496. doi:10.1016/j.immuni.2008.07.012

Claesson HE, Lundberg U, Malmsten C. 1981. Serum-coated zymosan stimulates the synthesis of leukotriene B_4 in human polymorphonuclear leukocytes. Inhibition by cyclic AMP. *Biochem Biophys Res Commun* **99:** 1230–1237. doi:10.1016/0006-291X(81)90751-8

Consalvo KM, Rijal R, Tang Y, Kirolos SA, Smith MR, Gomer RH. 2019. Extracellular signaling in *Dictyostelium*. *Int J Dev Biol* **63:** 395–405. doi:10.1387/ijdb.190259rg

Devreotes PN, Zigmond SH. 1988. Chemotaxis in eukaryotic cells: a focus on leukocytes and *Dictyostelium*. *Annu Rev Cell Biol* **4:** 649–686. doi:10.1146/annurev.cb.04.110188.003245

Devreotes PN, Potel MJ, MacKay SA. 1983. Quantitative analysis of cyclic AMP waves mediating aggregation in *Dictyostelium discoideum*. *Dev Biol* **96:** 405–415. doi:10.1016/0012-1606(83)90178-1

de Wit RJ, Konijn TM. 1983. Identification of the acrasin of *Dictyostelium minutum* as a derivative of folic acid. *Cell Differ* **12:** 205–210. doi:10.1016/0045-6039(83)90029-5

Dieterle PB, Min J, Irimia D, Amir A. 2020. Dynamics of diffusive cell signaling relays. *eLife* **9:** e61771. doi:10.7554/eLife.61771

Dinauer MC. 2019. Inflammatory consequences of inherited disorders affecting neutrophil function. *Blood* **133:** 2130–2139. doi:10.1182/blood-2018-11-844563

Donà E, Barry JD, Valentin G, Quirin C, Khmelinskii A, Kunze A, Durdu S, Newton LR, Fernandez-Minan A, Huber W, et al. 2013. Directional tissue migration through a self-generated chemokine gradient. *Nature* **503:** 285–289. doi:10.1038/nature12635

Douglas TE, Brock DA, Adu-Oppong B, Queller DC, Strassmann JE. 2013. Collection and cultivation of Dictyostelids from the wild. *Methods Mol Biol* **983:** 113–124. doi:10.1007/978-1-62703-302-2_6

Du Q, Kawabe Y, Schilde C, Chen Z-H, Schaap P. 2015. The evolution of aggregative multicellularity and cell–cell communication in the Dictyostelia. *J Mol Biol* **427:** 3722–3733. doi:10.1016/j.jmb.2015.08.008

Durston AJ. 1973. *Dictyostelium discoideum* aggregation fields as excitable media. *J Theor Biol* **42:** 483–504. doi:10.1016/0022-5193(73)90242-7

Eckstein TF, Vidal-Henriquez E, Bae AJ, Gholami A. 2020. Spatial heterogeneities shape the collective behavior of signaling amoeboid cells. *Sci Signal* **13:** eaaz3975. doi:10.1126/scisignal.aaz3975

El-Sobky MH, Rijal R, Gomer RH. 2025. Two endogenous *Dictyostelium discoideum* chemorepellents use different mechanisms to induce repulsion. *Proc Natl Acad Sci* **122:** e2503168122. doi:10.1073/pnas.2503168122

Fiedler J, Wheelan P, Henson PM, Murphy RC. 1998. Exogenous leukotriene B_4 (LTB_4) inhibits human neutrophil generation of LTB_4 from endogenous arachidonic acid during opsonized zymosan phagocytosis. *J Pharmacol Exp Ther* **287:** 150–156. doi:10.1016/S0022-3565(24)37775-4

Flamand N, Picard S, Lemieux L, Pouliot M, Bourgoin SG, Borgeat P. 2006. Effects of pyrrophenone, an inhibitor of group IVA phospholipase A_2, on eicosanoid and PAF biosynthesis in human neutrophils: pyrrophenone inhibits its lipid mediator biosynthesis. *Br J Pharmacol* **149:** 385–392. doi:10.1038/sj.bjp.0706879

Ford HZ, Manhart A, Chubb JR. 2023. Controlling periodic long-range signalling to drive a morphogenetic transition. *eLife* **12:** e83796. doi:10.7554/eLife.83796

Ford-Hutchinson AW, Bray MA, Doig MV, Shipley ME, Smith MJ. 1980. Leukotriene B, a potent chemokinetic and aggregating substance released from polymorphonuclear leukocytes. *Nature* **286:** 264–265. doi:10.1038/286264a0

Geiger J, Wessels D, Soll DR. 2003. Human polymorphonuclear leukocytes respond to waves of chemoattractant, like *Dictyostelium*. *Cell Motil Cytoskeleton* **56:** 27–44. doi:10.1002/cm.10133

Goldbeter A, Erneux T. 1978. Excitability in the adenylate cyclase reaction in *Dictyostelium discoideum*. *FEBS Lett* **89:** 237–241. doi:10.1016/0014-5793(78)80226-9

Goldstein RE. 1996. Traveling-wave chemotaxis. *Phys Rev Lett* **77:** 775–778. doi:10.1103/PhysRevLett.77.775

Green AA, Newell PC. 1975. Evidence for the existence of two types of cAMP binding sites in aggregating cells of *Dictyostelium discoideum*. *Cell* **6:** 129–136. doi:10.1016/0092-8674(75)90003-3

Gregor T, Fujimoto K, Masaki N, Sawai S. 2010. The onset of collective behavior in social amoebae. *Science* **328:** 1021–1025. doi:10.1126/science.1183415

Hamasaki T, Sakano T, Kobayashi M, Sakura N, Ueda K, Usui T. 1989. Leukotriene B_4 metabolism in neutrophils of patients with chronic granulomatous disease: phorbol myristate acetate decreases endogenous leukotriene B_4 via NADPH oxidase-dependent mechanism. *Eur J Clin Invest* **19:** 404–411. doi:10.1111/j.1365-2362.1989.tb00249.x

Hashimoto Y, Cohen MH, Robertson A. 1975. Cell density dependence of the aggregation characteristics of the cellular slime mould *Dictyostelium discoideum*. *J Cell Sci* **19:** 215–229. doi:10.1242/jcs.19.1.215

Haurand M, Flohé L. 1989. Leukotriene formation by human polymorphonuclear leukocytes from endogenous arachidonate. Physiological triggers and modulation by prostanoids. *Biochem Pharmacol* **38:** 2129–2137. doi:10.1016/0006-2952(89)90067-1

Hirose S, Santhanam B, Katoh-Kurosawa M, Shaulsky G, Kuspa A. 2015. Allorecognition, via TgrB1 and TgrC1, mediates the transition from unicellularity to multicellularity in the social amoeba *Dictyostelium discoideum*. *Development* **142:** 3561–3570. doi:10.1242/dev.123281

Huyan C, Golden A, Zhu X, Mehta P, Sgro AE. 2021. Robust coordination of collective oscillatory signaling requires single-cell excitability and fold-change detection. bioRxiv doi:10.1101/2021.09.02.457527

Insall RH, Paschke P, Tweedy L. 2022. Steering yourself by the bootstraps: how cells create their own gradients for chemotaxis. *Trends Cell Biol* **32:** 585–596. doi:10.1016/j.tcb.2022.02.007

Ishida M, Uwamichi M, Nakajima A, Sawai S. 2025. Traveling-wave chemotaxis of neutrophil-like HL-60 cells. *Mol Biol Cell* **36:** ar17. doi:10.1091/mbc.E24-06-0245

Jaiswal P, Meena NP, Chang F-S, Liao X-H, Kim L, Kimmel AR. 2024. An integrated, cross-regulation pathway model involving activating/adaptive and feed-forward/feedback loops for directed oscillatory cAMP signal-relay/response during the development of *Dictyostelium*. *Front Cell Dev Biol* **11:** 1263316. doi:10.3389/fcell.2023.1263316

Kamino K, Kondo Y, Nakajima A, Honda-Kitahara M, Kaneko K, Sawai S. 2017. Fold-change detection and scale invariance of cell-cell signaling in social amoeba. *Proc Natl Acad Sci* **114:** E4149–E4157. doi:10.1073/pnas.1702181114

Karmakar R, Tyree T, Gomer RH, Rappel W-J. 2021. Cell dispersal by localized degradation of a chemoattractant. *Proc Natl Acad Sci* **118:** e2008126118. doi:10.1073/pnas.2008126118

Kawabe Y, Du Q, Schilde C, Schaap P. 2019. Evolution of multicellularity in Dictyostelia. *Int J Dev Biol* **63:** 359–369. doi:10.1387/ijdb.190108ps

Kessin RH. 2001. *Dictyostelium: evolution, cell biology, and the development of multicellularity (developmental and cell biology series, series number 38: evolution)*. Cambridge University Press, Cambridge.

Khazen R, Corre B, Garcia Z, Lemaître F, Bachellier-Bassi S, d'Enfert C, Bousso P. 2022. Spatiotemporal dynamics of calcium signals during neutrophil cluster formation. *Proc Natl Acad Sci* **119:** e2203855119. doi:10.1073/pnas.2203855119

Kienle K, Lämmermann T. 2016. Neutrophil swarming: an essential process of the neutrophil tissue response. *Immunol Rev* **273:** 76–93. doi:10.1111/imr.12458

Kienle K, Glaser KM, Eickhoff S, Mihlan M, Knöpper K, Reátegui E, Epple MW, Gunzer M, Baumeister R, Tarrant TK, et al. 2021. Neutrophils self-limit swarming to contain bacterial growth in vivo. *Science* **372:** eabe7729. doi:10.1126/science.abe7729

King AJ, Sumpter DJT. 2012. Murmurations. *Curr Biol* **22:** R112–R114. doi:10.1016/j.cub.2011.11.033

Klein C. 1976. Adenylate cyclase activity in *Dictyostelium discoideum* amoebae and its changes during differentiation. *FEBS Lett* **68:** 125–128. doi:10.1016/0014-5793(76)80419-X

Klein PS, Sun TJ, Saxe CL III, Kimmel AR, Johnson RL, Devreotes PN. 1988. A chemoattractant receptor controls development in *Dictyostelium discoideum*. *Science* **241:** 1467–1472. doi:10.1126/science.3047871

Konijn TM, Van De Meene JG, Bonner JT, Barkley DS. 1967. The acrasin activity of adenosine-3′,5′-cyclic phosphate. *Proc Natl Acad Sci* **58:** 1152–1154. doi:10.1073/pnas.58.3.1152

Krump E, Borgeat P. 1999. Adenosine: an endogenous inhibitor of arachidonic acid release and leukotriene biosynthesis in human neutrophils. *Adv Exp Med Biol* **447:** 107–115. doi:10.1007/978-1-4615-4861-4_10

Kuspa A, Loomis WF. 1992. Tagging developmental genes in *Dictyostelium* by restriction enzyme-mediated integration of plasmid DNA. *Proc Natl Acad Sci* **89:** 8803–8807. doi:10.1073/pnas.89.18.8803

Lämmermann T, Afonso PV, Angermann BR, Wang JM, Kastenmüller W, Parent CA, Germain RN. 2013. Neutrophil swarms require LTB_4 and integrins at sites of cell death in vivo. *Nature* **498:** 371–375. doi:10.1038/nature12175

Laub MT, Loomis WF. 1998. A molecular network that produces spontaneous oscillations in excitable cells of *Dictyostelium*. *Mol Biol Cell* **9:** 3521–3532. doi:10.1091/mbc.9.12.3521

Lawal HM, Schilde C, Kin K, Brown MW, James J, Prescott AR, Schaap P. 2020. Cold climate adaptation is a plausible cause for evolution of multicellular sporulation in Dictyostelia. *Sci Rep* **10:** 8797. doi:10.1038/s41598-020-65709-3

Li S, Yao Q, Li J, Yang H, Qian R, Zheng M, Wu N, Jiang H, Li L, Zeng Z. 2024. Inhibition of neutrophil swarming by type I interferon promotes intracellular bacterial evasion. *Nat Commun* **15:** 8663. doi:10.1038/s41467-024-53060-4

Lindsay MA, Perkins RS, Barnes PJ, Giembycz MA. 1998. Leukotriene B_4 activates the NADPH oxidase in eosinophils by a pertussis toxin-sensitive mechanism that is largely independent of arachidonic acid mobilization. *J Immunol* **160:** 4526–4534. doi:10.4049/jimmunol.160.9.4526

Loomis WF. 2014. Cell signaling during development of *Dictyostelium*. *Dev Biol* **391:** 1–16. doi:10.1016/j.ydbio.2014.04.001

Loomis WF. 2015. Genetic control of morphogenesis in *Dictyostelium*. *Dev Biol* **402:** 146–161. doi:10.1016/j.ydbio.2015.03.016

Lundgren SM, Rocha-Gregg BL, Akdoğan E, Mysore MN, Hayes S, Collins SR. 2023. Signaling dynamics distinguish high- and low-priority neutrophil chemoattractant receptors. *Sci Signal* **16:** eadd1845. doi:10.1126/scisignal.add1845

Maeda M, Lu S, Shaulsky G, Miyazaki Y, Kuwayama H, Tanaka Y, Kuspa A, Loomis WF. 2004. Periodic signaling controlled by an oscillatory circuit that includes protein kinases ERK2 and PKA. *Science* **304:** 875–878. doi:10.1126/science.1094647

Malchow D, Nägele B, Schwarz H, Gerisch G. 1972. Membrane-bound cyclic AMP phosphodiesterase in chemotactically responding cells of *Dictyostelium discoideum*. *Eur J Biochem* **28:** 136–142. doi:10.1111/j.1432-1033.1972.tb01894.x

Martiel J-L, Goldbeter A. 1987. A model based on receptor desensitization for cyclic AMP signaling in *Dictyostelium* cells. *Biophys J* **52:** 807–828. doi:10.1016/S0006-3495(87)83275-7

Masaki N, Fujimoto K, Honda-Kitahara M, Hada E, Sawai S. 2013. Robustness of self-organizing chemoattractant field arising from precise pulse induction of its breakdown enzyme: a single-cell level analysis of PDE expression in *Dictyostelium*. *Biophys J* **104:** 1191–1202. doi:10.1016/j.bpj.2013.01.023

McDonald PP, McColl SR, Naccache PH, Borgeat P. 1992. Activation of the human neutrophil 5-lipoxygenase by leukotriene B_4. *Br J Pharmacol* **107:** 226–232. doi:10.1111/j.1476-5381.1992.tb14491.x

McDonald PP, McColl SR, Braquet P, Borgeat P. 1994. Autocrine enhancement of leukotriene synthesis by endogenous leukotriene B_4 and platelet-activating factor in human neutrophils. *Br J Pharmacol* **111:** 852–860. doi:10.1111/j.1476-5381.1994.tb14816.x

McDonald B, Pittman K, Menezes GB, Hirota SA, Slaba I, Waterhouse CCM, Beck PL, Muruve DA, Kubes P. 2010. Intravascular danger signals guide neutrophils to sites of sterile inflammation. *Science* **330:** 362–366. doi:10.1126/science.1195491

Medina JM, Shreenidhi PM, Larsen TJ, Queller DC, Strassmann JE. 2019. Cooperation and conflict in the social amoeba *Dictyostelium discoideum*. *Int J Dev Biol* **63:** 371–382. doi:10.1387/ijdb.190158jm

Naccache PH, McColl SR, Caon AC, Borgeat P. 1989. Arachidonic acid-induced mobilization of calcium in human neutrophils: evidence for a multicomponent mechanism of action. *Br J Pharmacol* **97:** 461–468. doi:10.1111/j.1476-5381.1989.tb11973.x

Nakajima A, Ishihara S, Imoto D, Sawai S. 2014. Rectified directional sensing in long-range cell migration. *Nat Commun* **5:** 5367. doi:10.1038/ncomms6367

Nakanishi Y, Tan M, Ichiki T, Inoue A, Yoshihara JI, Maekawa N, Takenoshita I, Yanagida K, Yamahira S, Yamaguchi S, et al. 2018. Stepwise phosphorylation of leukotriene B_4 receptor 1 defines cellular responses to leukotriene B_4. *Sci Signal* **11:** eaao5390. doi:10.1126/scisignal.aao5390

Ng LG, Qin JS, Roediger B, Wang Y, Jain R, Cavanagh LL, Smith AL, Jones CA, de Veer M, Grimbaldeston MA, et al. 2011. Visualizing the neutrophil response to sterile tissue injury in mouse dermis reveals a three-phase cascade of events. *J Invest Dermatol* **131:** 2058–2068. doi:10.1038/jid.2011.179

Palmer RM, Salmon JA. 1983. Release of leukotriene B_4 from human neutrophils and its relationship to degranulation induced by *N*-formyl-methionyl-leucyl-phenylalanine, serum-treated zymosan and the ionophore A23187. *Immunology* **50:** 65–73.

Pálsson E, Cox CE. 1996. Origin and evolution of circular waves and spirals in *Dictyostelium discoideum* territories. *Proc Natl Acad Sci* **93:** 1151–1155. doi:10.1073/pnas.93.3.1151

Pannbacker RG, Bravard LJ. 1972. Phosphodiesterase in *Dictyostelium discoideum* and the chemotactic response to cyclic adenosine monophosphate. *Science* **175:** 1014–1015. doi:10.1126/science.175.4025.1014

Parent CA. 2004. Making all the right moves: chemotaxis in neutrophils and *Dictyostelium*. *Curr Opin Cell Biol* **16:** 4–13. doi:10.1016/j.ceb.2003.11.008

Peters NC, Egen JG, Secundino N, Debrabant A, Kimblin N, Kamhawi S, Lawyer P, Fay MP, Germain RN, Sacks D. 2008. In vivo imaging reveals an essential role for neutrophils in leishmaniasis transmitted by sand flies. *Science* **321:** 970–974. doi:10.1126/science.1159194

Phillips JE, Gomer RH. 2012. A secreted protein is an endogenous chemorepellant in *Dictyostelium discoideum*. *Proc Natl Acad Sci* **109:** 10990–10995. doi:10.1073/pnas.1206350109

Ponte E, Bracco E, Faix J, Bozzaro S. 1998. Detection of subtle phenotypes: the case of the cell adhesion molecule CsA in *Dictyostelium*. *Proc Natl Acad Sci* **95:** 9360–9365. doi:10.1073/pnas.95.16.9360

Poplimont H, Georgantzoglou A, Boulch M, Walker HA, Coombs C, Papaleonidopoulou F, Sarris M. 2020. Neutrophil swarming in damaged tissue is orchestrated by connexins and cooperative calcium alarm signals. *Curr Biol* **30:** 2761–2776.e7. doi:10.1016/j.cub.2020.05.030

Postma M, van Haastert PJM. 2009. Mathematics of experimentally generated chemoattractant gradients. *Methods Mol Biol* **571:** 473–488. doi:10.1007/978-1-60761-198-1_31

Prescott SM, Zimmerman GA, Seeger AR. 1984. Leukotriene B_4 is an incomplete agonist for the activation of human neutrophils. *Biochem Biophys Res Commun* **122:** 535–541. doi:10.1016/S0006-291X(84)80066-2

Qiu S, Haselmayr W, Li B, Zhao C, Guo W. 2017. Bacterial relay for energy-efficient molecular communications. *IEEE Trans Nanobioscience* **16:** 555–562. doi:10.1109/TNB.2017.2741669

Rathi A, Clarke M. 1992. Expression of early developmental genes in *Dictyostelium discoideum* is initiated during exponential growth by an autocrine-dependent mechanism. *Mech Dev* **36:** 173–182. doi:10.1016/0925-4773(92)90068-U

Reátegui E, Jalali F, Khankhel AH, Wong E, Cho H, Lee J, Serhan CN, Dalli J, Elliott H, Irimia D. 2017. Microscale arrays for the profiling of start and stop signals coordinating human-neutrophil swarming. *Nat Biomed Eng* **1:** 0094. doi:10.1038/s41551-017-0094

Ricci-Tam C, Kuipa S, Kostman MP, Aronson MS, Sgro AE. 2023. Microbial models of development: inspiration for engineering self-assembled synthetic multicellularity. *Semin Cell Dev Biol* **141:** 50–62. doi:10.1016/j.semcdb.2022.04.014

Sawai S, Thomason PA, Cox EC. 2005. An autoregulatory circuit for long-range self-organization in *Dictyostelium* cell populations. *Nature* **433:** 323–326. doi:10.1038/nature03228

Saxe CL III, Johnson R, Devreotes PN, Kimmel AR. 1991. Multiple genes for cell surface cAMP receptors in *Dictyostelium discoideum*. *Dev Genet* **12:** 6–13. doi:10.1002/dvg.1020120104

Serezani CH, Aronoff DM, Jancar S, Peters-Golden M. 2005. Leukotriene B_4 mediates p47phox phosphorylation and membrane translocation in polyunsaturated fatty acid-stimulated neutrophils. *J Leukoc Biol* **78:** 976–984. doi:10.1189/jlb.1004587

Serio KJ, Baker JR, Ring WL, Riddick CA, Bigby TD. 1997. Leukotriene B_4 costimulates 5-lipoxygenase activity in neutrophils via increased 5-lipoxygenase translocation. *Am J Physiol Cell Physiol* **272:** C1329–C1334. doi:10.1152/ajpcell.1997.272.4.C1329

Sgro AE, Schwab DJ, Noorbakhsh J, Mestler T, Mehta P, Gregor T. 2015. From intracellular signaling to population oscillations: bridging size- and time-scales in collective behavior. *Mol Syst Biol* **11:** 779. doi:10.15252/msb.20145352

Shaffer BM. 1956. Acrasin, the chemotactic agent in cellular slime moulds. *J Exp Biol* **33:** 645–657. doi:10.1242/jeb.33.4.645

Shimomura O, Suthers HL, Bonner JT. 1982. Chemical identity of the acrasin of the cellular slime mold *Polysphondylium violaceum*. *Proc Natl Acad Sci* **79:** 7376–7379. doi:10.1073/pnas.79.23.7376

Siegert F, Weijer C. 1989. Digital image processing of optical density wave propagation in *Dictyostelium discoideum* and analysis of the effects of caffeine and ammonia. *J Cell Sci* **93:** 325–335. doi:10.1242/jcs.93.2.325

Siwicki M, Kubes P. 2023. Neutrophils in host defense, healing, and hypersensitivity: dynamic cells within a dynamic host. *J Allergy Clin Immunol* **151:** 634–655. doi:10.1016/j.jaci.2022.12.004

Skoge M, Yue H, Erickstad M, Bae A, Levine H, Groisman A, Loomis WF, Rappel W-J. 2014. Cellular memory in eukaryotic chemotaxis. *Proc Natl Acad Sci* **111:** 14448–14453. doi:10.1073/pnas.1412197111

Song Z, Huang G, Paracatu LC, Grimes D, Gu J, Luke CJ, Clemens RA, Dinauer MC. 2020. NADPH oxidase controls pulmonary neutrophil infiltration in the response to fungal cell walls by limiting LTB_4. *Blood* **135:** 891–903. doi:10.1182/blood.2019003525

Song Z, Bhattacharya S, Clemens RA, Dinauer MC. 2023. Molecular regulation of neutrophil swarming in health and disease: lessons from the phagocyte oxidase. *iScience* **26:** 108034. doi:10.1016/j.isci.2023.108034

Strassmann JE. 2016. Kin discrimination in *Dictyostelium* social amoebae. *J Eukaryot Microbiol* **63:** 378–383. doi:10.1111/jeu.12307

Strickland E, Pan D, Godfrey C, Kim JS, Hopke A, Ji W, Degrange M, Villavicencio B, Mansour MK, Zerbe CS, et al. 2024. Self-extinguishing relay waves enable homeostatic control of human neutrophil swarming. *Dev Cell* **59:** 2659–2671.e4. doi:10.1016/j.devcel.2024.06.003

Surette ME, Krump E, Picard S, Borgeat P. 1999. Activation of leukotriene synthesis in human neutrophils by exogenous arachidonic acid: inhibition by adenosine A_{2a} receptor agonists and crucial role of autocrine activation by leukotriene B_4. *Mol Pharmacol* **56:** 1055–1062. doi:10.1124/mol.56.5.1055

Susanto O, Koh YWH, Morrice N, Tumanov S, Thomason PA, Nielson M, Tweedy L, Muinonen-Martin AJ, Kamphorst JJ, Mackay GM, et al. 2017. LPP3 mediates self-generation of chemotactic LPA gradients by melanoma cells. *J Cell Sci* **130:** 3455–3466. doi:10.1242/jcs.207514

Takeda K, Shao D, Adler M, Charest PG, Loomis WF, Levine H, Groisman A, Rappel W-J, Firtel RA. 2012. Incoherent feedforward control governs adaptation of activated Ras in a eukaryotic chemotaxis pathway. *Sci Signal* **5:** ra2. doi:10.1126/scisignal.2002413

Tatsumi R, Aihara S, Matsune S, Aoki J, Inoue A, Shimizu T, Nakamura M. 2023. Stepwise phosphorylation of BLT1 defines complex assemblies with β-arrestin serving distinct functions. *FASEB J* **37:** e23213. doi:10.1096/fj.202301440R

Tintinger GR, Theron AJ, Steel HC, Anderson R. 2001. Accelerated calcium influx and hyperactivation of neutrophils in chronic granulomatous disease. *Clin Exp Immunol* **123:** 254–263. doi:10.1046/j.1365-2249.2001.01447.x

Tomchik KJ, Devreotes PN. 1981. Adenosine 3′,5′-monophosphate waves in *Dictyostelium discoideum*: a demonstration by isotope dilution—fluorography. *Science* **212:** 443–446. doi:10.1126/science.6259734

Tweedy L, Insall RH. 2020. Self-generated gradients yield exceptionally robust steering cues. *Front Cell Dev Biol* **8:** 133. doi:10.3389/fcell.2020.00133

Tweedy L, Knecht DA, Mackay GM, Insall RH. 2016a. Self-generated chemoattractant gradients: attractant depletion extends the range and robustness of chemotaxis. *PLoS Biol* **14:** e1002404. doi:10.1371/journal.pbio.1002404

Tweedy L, Susanto O, Insall RH. 2016b. Self-generated chemotactic gradients—cells steering themselves. *Curr Opin Cell Biol* **42:** 46–51. doi:10.1016/j.ceb.2016.04.003

Tweedy L, Thomason PA, Paschke PI, Martin K, Machesky LM, Zagnoni M, Insall RH. 2020. Seeing around corners: cells solve mazes and respond at a distance using attractant breakdown. *Science* **369:** eaay9792. doi:10.1126/science.aay9792

Uderhardt S, Martins AJ, Tsang JS, Lämmermann T, Germain RN. 2019. Resident macrophages cloak tissue microlesions to prevent neutrophil-driven inflammatory damage. *Cell* **177:** 541–555.e17. doi:10.1016/j.cell.2019.02.028

Van Haastert PJ, De Wit RJ, Grijpma Y, Konijn TM. 1982. Identification of a pterin as the acrasin of the cellular slime mold *Dictyostelium lacteum*. *Proc Natl Acad Sci* **79:** 6270–6274. doi:10.1073/pnas.79.20.6270

Venkiteswaran G, Lewellis SW, Wang J, Reynolds E, Nicholson C, Knaut H. 2013. Generation and dynamics of an endogenous, self-generated signaling gradient across a migrating tissue. *Cell* **155:** 674–687. doi:10.1016/j.cell.2013.09.046

Vicker MG, Lackie JM, Schill W. 1986. Neutrophil leucocyte chemotaxis is not induced by a spatial gradient of chemoattractant. *J Cell Sci* **84:** 263–280. doi:10.1242/jcs.84.1.263

Weijer CJ. 2004. *Dictyostelium* morphogenesis. *Curr Opin Genet Dev* **14:** 392–398. doi:10.1016/j.gde.2004.06.006

Wu Y, Dahlgren C, Forsman H, Sundqvist M. 2025. LTB_4 is converted into a potent human neutrophil NADPH oxidase activator via a receptor transactivation mechanism in which the BLT_1 receptor activates the free fatty acid receptor 2. *Prostaglandins Leukot Essent Fatty Acids* **205:** 102680. doi:10.1016/j.plefa.2025.102680

Youk H, Lim WA. 2014. Secreting and sensing the same molecule allows cells to achieve versatile social behaviors. *Science* **343:** 1242782. doi:10.1126/science.1242782

Yuen IS, Jain R, Bishop JD, Lindsey DF, Deery WJ, Van Haastert PJ, Gomer RH. 1995. A density-sensing factor regulates signal transduction in *Dictyostelium*. *J Cell Biol* **129:** 1251–1262. doi:10.1083/jcb.129.5.1251

Collective Mesoderm and Endoderm Cell Migration during Vertebrate Gastrulation

Cornelis J. Weijer

Division of Molecular Cell and Developmental Biology, School of Life Sciences, University of Dundee, Dundee DD1 5EH, United Kingdom

Correspondence: c.j.weijer@dundee.ac.uk

Collective migration is a key mechanism during embryonic development, particularly during gastrulation when the basic three-dimensional body plan is established. During this critical phase, mesoderm and endoderm tissues internalize from the surface to form the embryo's inner structures through species-specific processes. This internalization involves extensive collective migration of large cohorts of mesenchymal cells along characteristic routes inside the embryo. Despite decades of research, the mechanisms that control and guide the movement of these cells to their destinations remain largely unresolved. Understanding these complex interactions remains a central challenge in developmental biology. This review examines complementary insights obtained through the study of multiple vertebrate model systems—frogs, fish, chick, mouse—as well as embryonic stem cell–derived gastruloids. The evidence points to key roles for chemotactic movement and guidance mechanisms operating in concert with dynamic changes in cell–cell and cell–substrate adhesion. Recent studies have increasingly revealed critical roles for cell- and tissue-generated mechanical stresses and mechanosensing in executing and coordinating motion. Furthermore, dynamic feedback between signaling and motion generates emergent properties that enable large-scale coordination of collective migration.

Gastrulation constitutes an essential phase in development of many invertebrates and all vertebrates. This process arranges the three primary germ layers—ectoderm, mesoderm, and endoderm—in their correct topological positions within the developing embryo. The ectoderm, which forms the skin and nervous system, occupies the outer layer. The mesoderm, forming the muscles, skeleton, circulatory system, and many major organs, forms the middle layer. The endoderm, which becomes the lining of the digestive tract and associated glands, constitutes the innermost layer.

Gastrulation has been extensively studied in a variety of model systems (Solnica-Krezel and Sepich 2012). The primary vertebrate model systems include amphibians (frogs), fish, and amniotes such as the chick and mice (Schauer and Heisenberg 2021; Sheng et al. 2021). More recently, embryonic stem (ES) cell–derived organoid systems have been developed to recapitulate key aspects of the early gastrulation

Cite this article as *Cold Spring Harb Perspect Biol* doi: 10.1101/cshperspect.a041793

in vitro (van den Brink and van Oudenaarden 2021).

THE CHALLENGE OF COLLECTIVE MIGRATION

Most embryos begin development as simple epithelial spheres or sheets of cells. Following the induction of mesoderm and endoderm precursors in characteristic localized regions of the embryo, these sheets undergo extensive deformation and rearrangement to internalize the mesoderm and endoderm, driven by cell divisions, shape changes, intercalations, and ingressions. Mesendoderm precursors undergo complete or partial epithelial-to-mesenchymal transition (EMT), internalize, and form mesenchymal cells that collectively migrate to establish the internal mesoderm and endoderm layers.

This process presents remarkable challenges. Mesoderm and endoderm cells typically migrate considerable distances to reach their destinations, requiring highly integrated collective migration of thousands to tens of thousands of cells. The mechanisms underlying this collective migration, the nature of signals coordinating cellular behaviors, and the instructive signals guiding cells to specific destinations represent areas of intensive ongoing research.

Key Questions

Several fundamental questions guide current research efforts:

1. What mechanisms drive individual and collective cell movement?
2. Which substrates do migrating cells use, and how do they generate traction?
3. Which signals guide and coordinate movement?
4. How does cell movement provide feedback to signaling dynamics?

This review compares complementary insights into collective migration mechanisms and coordination obtained through studying gastrulation across various model organisms, while highlighting several critical open questions.

Model System Comparisons

Here we summarize the species-specific tissue dynamics and currently known underlying cellular mechanisms driving mesoderm and endoderm internalization and dispersal inside the embryo. We highlight emerging insights in the control and guidance mechanisms and their molecular basis.

COLLECTIVE CELL BEHAVIORS DRIVING GASTRULATION IN *XENOPUS*

Amphibians have been instrumental in elucidating cell behaviors underlying the large-scale tissue movements driving the involution and positioning of the mesoderm and endoderm layers both in vivo and in tissue explants (Keller 2005; Keller and Sutherland 2020).

Tissue Dynamics

After induction of the mesoderm by a combination of transforming growth factor (TGF)-β and Wnt signals from the endoderm (Kimelman 2006), the blastula stage embryo consists of a vegetal located solid mass of endodermal cells and a central several cell thick layer of mesodermal cells, surrounding the blastocoel, capped by a several-cell-layer-thick mass of ectodermal cells. During gastrulation, the mesoderm and endoderm involute through a circular blastopore and migrate toward the animal pole, closing the blastocoel and opening the gastrocoel, the cavity that is going to form the lumen of the digestive tract. During all stages of gastrulation, mesoderm and endoderm cells stay connected in compact tissues, in which cells can locally rearrange through cell-on-cell movements. During involution of the endoderm and mesoderm, the blastopore closes, moving to the ventral pole, while the ectoderm of the animal cap thins and expands in a process known as epiboly (Fig. 1A).

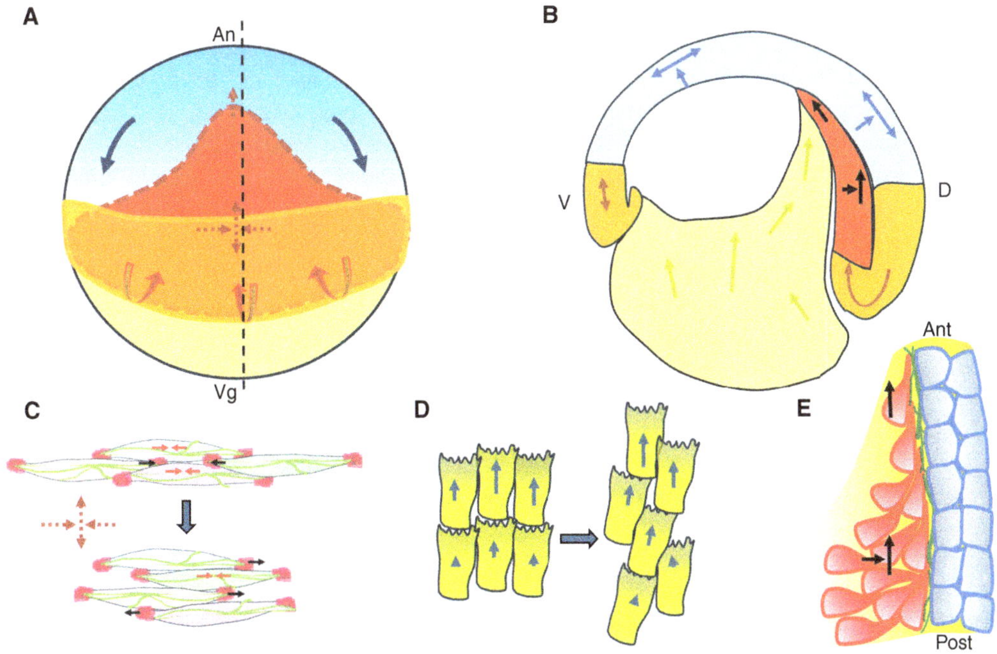

Figure 1. *Xenopus* gastrulation. (*A*) Dorsal surface view of a midgastrulation-stage *Xenopus* embryo. The ectoderm (blue) expands to cover the embryo during the involution (brown/red arrows) of the mesoderm (brown surface, red involuted) and endoderm (yellow) through the blastopore (brown stippled line). The mesoderm shows mediolateral intercalation indicated by the stippled red arrows, resulting in convergence toward the dorsal midline and extension toward the animal pole. (*B*) Dorsal ventral cross section through the embryo at the location of the stippled line in *A*. Dorsal is on the *right* and ventral on the *left*. The ectoderm undergoes a process of radial intercalation (blue arrows). Ingression of the mesoderm through the blastopore is driven by directional migration and thinning of the leading-edge mesoderm (red) as indicated by the black arrows (*B*,*E*), followed by axial mesoderm (brown), extending through mediolateral intercalation (shown in *A*). The endoderm (yellow) ingresses through vegetal rotation movement, resulting in endoderm thinning and extension driven by ingression-type cell-on-cell migration (*D*). (*C*) Schematic of bipolar intercalating mesoderm cells. Intercalation is driven via alternating cycles of contraction (red arrows) of the internal actin–myosin cable network and cell-on-cell migration driven by bipolar extensile protrusive activity (blue arrows). (*D*) Schematic of endoderm cell-on-cell migration at differential velocity (blue arrows) resulting in expansion and thinning of the endoderm. (*E*) Schematic of leading-edge mesoderm cell (red) migration on the extracellular matrix (green fibers) separating the overlaying ectoderm cells (blue). The mesoderm cells move in response to signals such as platelet-derived growth factor A (PDGFA) and stromal cell–derived factor 1 (SDF1, yellow background) secreted by the ectoderm. Directional intercalation of the mesoderm toward the ectoderm surface toward these signals contributes to the expansion and thinning of the leading-edge mesoderm. (An) Animal, (Vg) vegetal, (D) dorsal, (V) ventral, (Ant) anterior, (Post) posterior. (Figures based on data in Huang and Winklbauer 2018; Keller and Sutherland 2020; Sutherland et al. 2020).

Cellular Mechanisms

Blastopore Formation, Involution of Endoderm, and Epiboly

The mesoderm and endoderm involution process starts on the dorsal side of the embryo visible through the formation of the blastopore, which rapidly spreads to form a circumferential blastoporal ring. Blastopore formation is initiated by the apical contraction of endoderm-derived bottle cells, which is coupled to a simultaneous elongation and internalization of the cell body (Hardin and Keller 1988). Sustained internalization of the endoderm and its expansion inside the embryo requires an active movement, vegetal rotation of endoderm cells (Fig. 1B; Winklbauer and Schürfeld 1999). Vegetal rotation is driven by apical migration of unipolar endoderm amoeboid cells via cell-on-cell movement. Apically located endoderm cells move faster than basal (vegetal) cells creating a velocity gradient, generating an extension and thinning of the endoderm (Fig. 1D). During the contraction/closure of the blastopore, the involuting mesoderm locally thickens, convergent thickening, generating a circumferential tensile force helping to constrict and close the blastopore (Shook et al. 2018). It is driven by an affinity change between the surface layer of epithelial cells of the involuting marginal zone and the deeper layer of mesodermal cells (Shook et al. 2022). During the internalization of the mesoderm (involuting marginal zone), the mesoderm cells execute mediolateral intercalation, which contributes additional circumferential tension, contributing to closure of the blastoporal ring (Fig. 1A,C; Shook et al. 2018). The process of mesoderm and endoderm internalization is coupled to the epiboly of the animal cap tissue (Fig 1A). During epiboly, the animal cap ectoderm engulfs the vegetal endoderm, and the blastoporal ring contracts and closes. Epiboly is driven by radial intercalation of cells from the deeper layers of the animal cap into the layer of epithelial surface cells, expanding its area. This intercalation process involves a chemotactic response to complement 3a produced by the external tissue layer (Szabó et al. 2016). Altogether, bottle cells constriction, endoderm vegetal rotation, convergent thickening, and mediolateral intercalation of the involuting mesoderm coupled with ectodermal epiboly result in contraction of the blastopore and internalization of the mesoderm and endoderm.

Internal Expansion and Migration of Mesoderm

The cell behaviors that drive the internal anterior expansion of the mesoderm have been characterized in considerable detail. The anterior-most Hex and Goosecoid-expressing leading-edge mesoderm (LEM), forming the prechordal plate (PPL), actively move directionally toward the animal pole (Fig. 1B). They move via extension of lamellipodia along the ECM lining the blastocoel roof. Their movement appears to be guided by a long signal that is deposited in the ECM (Nagel et al. 2004). Follower cells extend lamellipodia under the cell bodies of the cells in front resulting in a shingle-like packing. Initially, the mesoderm layer behind the leading edge increases to several cells thick. These cells show a protrusive activity toward the directed blastocoel roof, driving intercalation of cells, resulting in thinning and expansion of the mesoderm tissue (Fig. 1E). The LEM migration also generates a pulling force, aiding following axial mesoderm to elongate and intercalate independent of the planar cell polarity (PCP) pathway (Hara et al. 2013).

The involuting mesoderm tissue shows extensive convergence flows from lateral to medial coupled to extensive flows toward the animal pole. This process is driven by directional intercalations, of elongated bipolar cells, oriented perpendicular to the central midline. These intercalating cells go through alternating cycles of contraction of medial contraction and lateral extension driven by a dynamic node and cable internal actin–myosin network, combined with making and breaking of adhesive cell–cell contacts (Fig. 1C; Pfister et al. 2016; Weng et al. 2025b).

Autonomous behaviors such as convergent extension of mesoderm and neural plate, as well as convergent thickening and vegetal rotation, have been studied in detail in tissue explants of the mesoderm marginal zone from early gastrulation stages (Keller and Sutherland 2020). Furthermore, animal cap isolates, in which mesoderm can be induced through a variety of signaling molecules, Activin/Nodal, fibroblast growth factors (FGFs), Wnts, and or combinations thereof (Kimelman 2006), can be used to study convergent extension in a more synthetic environment (Barrett et al. 2025).

Molecular Mechanisms

Endoderm Cell Dynamics

Noncanonical Wnt signaling through Wnt5 has been proposed to trigger release of the polarity protein lethal giant larva (LGL), thereby triggering apical constriction of bottle cells (Choi and Sokol 2009). Apical contraction is myosin II–driven and requires endocytosis of the apical surface (Lee and Harland 2010). It is furthermore dependent on the apical localized PCP pathway component Vangl2 and its downstream effectors Daam1 and Rho kinase as well as the Rho-GEF Plekhg5 (Ossipova et al. 2015; Popov et al. 2018; Baldwin et al. 2023). The amoeboid cell-on-cell migration underlying vegetal rotation requires C-cadherin-mediated cell–cell contact, fibronectin, and trailing end retraction through ephrinB1-dependent macropinocytosis (Wen and Winklbauer 2017; Huang and Winklbauer 2018). The signals controlling cell polarity and movement speed remain to be determined.

Migration of Leading-Edge Mesoderm

LEM cells can move collectively following directional information encoded in the ECM replica of the inner part of the blastocoel roof on a glass slide in a vegetal-to-animal direction. The guidance signal is a long matrix-bound isoform of platelet-derived growth factor (PDGF), signaling to the cytoskeleton in a phosphatidyl inositide 3 (PI3) kinase-dependent manner (Nagel et al. 2004; Smith et al. 2009). This directional orientation of protrusive activity is dependent on a short form diffusible PDGF signal released by the ectodermal blastocoel roof and detected by PDGF receptor α (PDGFRα) on the mesoderm cells (Damm and Winklbauer 2011). This signaling system interacts in parallel with stromal-derived factor 1 secreted by the blastocoel roof, detected by the C-X-C chemokine receptor type 4 (CXCR4), and self-generated gradient formation is possibly helped by CXCR7 decoy receptors in the LEM (Fig. 1E; Fukui et al. 2007). The resulting intercalation process is driven by cell-on-cell migration, requiring a combination of adhesive C-cadherin and integrin/fibronectin distributed on the cell surfaces (Huang and Winklbauer 2018). Isolation experiments on conditioned substrate have shown that masses of LEM cells not only migrate anteriorly in a collective manner, but that migrating cohorts of cells have an internal front-to-back polarity. Xfrz7 acts upstream of Pak1 to control lamellipodia stability, and Ephrin B1 and Dishevelled act in front-to-back fusion. These control modules, together with substratum-bound guidance cues, combine preferential front–back adhesion and diminished lateral adhesion of cells to promote collective directional mesoderm migration (Nagel and Winklbauer 2023).

Mesoderm Intercalation

Both cell polarity and convergent extension behavior can be modulated by altering the activity and or localization of components of the PCP pathway (Wallingford et al. 2000; Butler and Wallingford 2017; Keller and Sutherland 2020). Cell dissociation and reaggregation experiments with anterior and posterior-derived chorda-mesoderm precursor cells have demonstrated that the direction of convergent extension is under the control of the anterior–posterior (A/P) signaling system (Ninomiya et al. 2004), likely controlled by opposing Nodal and bone morphogenetic protein (BMP) gradients that control mesoderm differentiation and patterning (Shen 2007; Niehrs 2010). How the PCP pathway interacts with the A/P system is a focus of investigation. BMP could inhibit convergent extension through inhibition of Xarhgef3.2 expression that interacts with Dishevelled and Daam1 (Yoon et al. 2022). Vangl2 and Prickle, key components of the PCP pathway, are found enriched on the anterior membranes of intercalating cells (Huebner and Wallingford 2018). This asymmetric localization could shape the A/P organization of the actin–myosin network and contractile mechanics with the anterior junction being under higher tension (Weng et al. 2025a,b). How the cellular A/P membrane localization of PCP components is achieved and how it orchestrates directional intercalation across aligned cells remain to be determined.

Interestingly, recent studies with Activin-treated animal caps have shown that the epithelial surface ectoderm cells partially engulf the underlying mesenchymal cells. The protruding deeper mesenchymal cells orient and elongate along this epithelial–mesenchymal interface and intercalate, which then propagates outward resulting in the extension of the structure (Barrett et al. 2025). The mechanisms that trigger this initial alignment remain unresolved, but could depend on surface tension differences between the surface and deeper cells similar to the proposed mechanism for convergent thickening (Shook et al. 2022).

Collective Cell Migration during Zebrafish Gastrulation

A combination of detailed imaging studies and molecular perturbation studies in zebrafish has elucidated interesting principles and mechanisms underlying the collective migration of mesoderm and endoderm cells. They show that it involves a combination of chemical and mechanical signaling that needs to be integrated at the cellular and tissue levels. Fish embryos are highly transparent, which enables detailed imaging studies of various cell behaviors in vivo. Light sheet microscopy studies of global migration of mesoderm and endoderm cells are now being undertaken (Shah et al. 2019; Kondow et al. 2020). When these studies are linked to single-cell and spatial transcriptomics, they can give novel insights linking the dynamics of collective cell migration and gene expression (Liberali and Schier 2024; Wan et al. 2024).

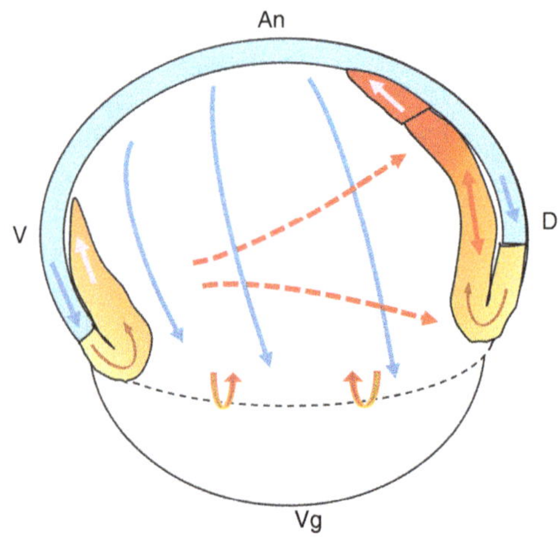

Figure 2. Animal vegetal cross section of midgastrulation stage zebrafish embryo (dorsal *right*, ventral *left*). Mesendoderm cells ingress at the blastoderm margin. Their migration is guided by pole cells (red), followed by axial mesoderm and endoderm (brown). The epiblast extends through epiboly (blue arrows), and the mesoderm and endoderm cells converge to and extend the midline (red arrows). (An) Animal, (Vg) vegetal, (D) dorsal, (V) ventral. (Figure based on data in Roszko et al. 2015.)

Tissue Dynamics

In zebrafish embryos located on the animal pole of the yolk, the cells divide rapidly and start to spread over the yolk by epiboly (Fig. 2). During epiboly, the surface of the yolk is progressively covered by the blastoderm, which is overlaid by the epithelial enveloping layer (EVL). The EVL is connected by tight junctions to the yolk cell membrane (Carvalho and Heisenberg 2010). At the midblastula stage, the yolk syncytial layer (YSL) starts to form by the release of nuclei from some of the marginal blastomeres into the yolk. There, they undergo three rounds of division, stopping at the sphere stage. During epiboly, many nuclei stay located at the contracting germband edge, while the others move upward in animal direction during the following epiboly (D'Amico and Cooper 2001). At the start of gastrulation, at 50% epiboly, an initially broad actin–myosin meshwork is formed under the YSL membrane at the EVL boundary. Lateral and circumferential contraction of this network to a thinner cable results in a vegetal-directed friction-mediated pulling force that drives epiboly of the EVL (Behrndt et al. 2012). During the initial stages of epiboly, Nodal signaling from the YSL induces endoderm and mesoderm in the cells of the blastoderm margin in a dose-dependent manner (Hill 2018; Cheng et al. 2023). At the start of gastrulation, mesendoderm cells all around the blastoderm ring start to ingress and migrate anteriorly. Initially, both the mesoderm and endoderm cells migrate toward the animal pole, but when they are halfway on their migration toward the animal pole, they change direction of migration to converge toward the dorsal midline showing extensive intercalation (Sepich et al. 2005).

Cellular Mechanisms

The first cells to ingress on the dorsal side are the Goosecoid prechordal plate (PPL) cells. They have been shown to move collectively at uniform speed and require contact for directional movement (Dumortier et al. 2012). Further experiments suggest that the anterior migration of the very first ingressing PPL cells (polster cells) is dependent on contact from follower axial mesoderm. This effect is thought to be mediated through tension generated by follower cells pulling on the leader cells (Boutillon et al. 2022, 2024).

During the early stages of gastrulation, there is no significant extracellular matrix for the cells to get traction from, and the mesoderm cells migrate predominantly on overlying ectoderm, while endoderm cell migration requires contact with both the ectodermal and YSL layers. It has been suggested that BMP signaling inhibits the cohesiveness/viscosity of the ectodermal layer through modulation of E-cadherin expression, making areas of high BMP expression areas of low cohesiveness and viscosity, providing loss of traction and a stop migration when the PPL cells reach the animal pole (Tavano et al. 2025).

Endoderm cells are also formed in response to Nodal signaling (Economou et al. 2022), and they undergo a series of characteristic movements. Initially, there is significant random motility, which then converts into anterior-directed motility (Pézeron et al. 2008). The random motility is associated with extension and retraction of randomly oriented filopodia driven by heightened actin polymerization. Dispersal may be facilitated through contact inhibition of locomotion mechanism (LaBelle et al. 2025). During the onset of convergent extension, the endoderm cells align their motion with the overlying mesoderm cells.

Molecular Mechanisms

Ingression and Migration of Mesoderm

On the dorsal side, ingression is controlled by high Nodal signaling in the blastoderm edge cells, which results in increased protrusion formation and movement in the direction of the YSL (Pinheiro et al. 2022). This increase in protrusive activity requires Rho kinase and Rac1 activation. Nodal signaling also controls a gradient of E-cadherin expression in the ingressing blastoderm cells, with the first cells to ingress expressing the highest levels of E-cadherin, allowing them to act as leaders for the following cells. It is yet unknown which signals control the ingression of more lateral and ventral blastoderm ring cells, where nodal signaling is thought to be very much weaker or absent. Movement of the PPL cells requires signaling through PI3 kinase and target of rapamycin complex 2 (TORC2) (Montero et al. 2003; Dumortier and David 2015), while Wnt-PCP signaling is permissive (Čapek et al. 2019). Coordinated anterior migration of the most anterior polster cells appears to be controlled by pulling of follower cells on polster cells. This results in activation of a mechanotransduction process in the leader cells, requiring E-cadherin and α-actinin. Mechanotransduction results in vinculin recruitment to α-catenin in polster cells and activation of PI3 kinase signaling and extension of directed protrusions at the leading edge, via an as-yet unresolved mechanism (Boutillon et al. 2022, 2024). The ingressing cells all around the blastoderm start to move in apical direction. This movement is driven by the ubiquitously expressed peptide Toddler/Apelin, which is detected through the G-protein-coupled Apelin receptor (APR) (Pauli et al. 2014). Toddler acts downstream from Nodal signaling (Norris et al. 2017). It has been suggested that the migrating mesoderm cells follow a self-generated Toddler gradient created through receptor-mediated removal of Toddler through internalization and inactivation of the Toddler–APR complex (Stock et al. 2022).

The later convergence and extension movements are modulated by noncanonical Wnt signaling through Wnt 11 and Wnt 5 (Heisenberg et al. 2000; Kilian et al. 2003), and migration is sensitive to the alteration of components of the PCP pathway. However, noncanonical Wnt signaling through Frz 7 is thought to be permissive for this dorsal migration (Čapek et al. 2019). The convergence movements are inhibited ventrally by BMP signaling (Myers et al. 2002), which functions through negative modulation of calcium/cadherin-dependent cell adhesion necessary for cell migration and intercalation (von der Hardt et al. 2007). Although FGF signaling plays a major role in mesoderm differentiation, its role in the guidance of mesoderm or endoderm cells is not yet very clear due to the complexity and redundancy of the FGF signaling pathway components (Leerberg et al. 2019).

Endoderm Cell Migration

It has been shown that this is a RAC-dependent process that is enhanced by the RAC guanine nucleotide exchange factor (GEF) Prex1, a Nodal transcriptional target (Woo et al. 2012). There is strong evidence that directional migration of the endoderm is controlled by diffusible signaling molecules SDF1 and CXCL12B secreted by mesoderm cells and detected by CXCR4 receptors on the endoderm (Mizoguchi et al. 2008; Nair and Schilling 2008). CXCL12b detection has been shown to result in integrin expression by endoderm cells binding mesoderm-associated fibronectin, thus tethering the endoderm to mesoderm cells. It is thought that through this tether, the Toddler that controls mesoderm migration also indirectly contributes to the guidance of endoderm cell migration (Nair and Schilling 2008). However, in Toddler CXCR4 double mutants, endoderm cell migration is rather normal, possibly pointing to as-yet undiscovered guidance cues (Stock et al. 2022). The precise coordination of mesoderm and endoderm is still under active investigation.

COLLECTIVE MIGRATION DURING CHICK GASTRULATION

Tissue Dynamics

Chick embryos at the time of laying consist of two concentric discs of cells; the central single-cell thick layer of epiblast cells will form the embryo, while the surrounding several-cell-thick area opaca will form the extraembryonic tissues. The embryo at this stage contains around 50,000 cells, and an A/P polarity has already been established (Serrano Nájera and Weijer 2020). Scattered cells throughout the epiblast apically contract and ingress to contribute to the forming hypoblast that forms in a posterior-to-anterior direction (Stern and Downs 2012). Mesendoderm precursor cells are induced in a posteriorly located sickle-shaped region of the epiblast at the boundary between the embryonic and extraembryonic tissue by signals from the extraembryonic tissue and hypoblast involving long-range gradients of Wnt and BMP signaling (Lee et al. 2024). Following the induction of the mesendoderm precursor cells, the sickle-shaped region elongates the cells start to elongate and align along the embryonic extraembryonic direction and directionally intercalate toward the central midline (Voiculescu et al. 2007; Rozbicki et al. 2015). Already during the displacement of the sickle cells toward the central midline, a fraction of these epiblast cells apically contract and ingress. They integrate into and displace the posterior hypoblast layer to start to form the endoderm (Lawson and Schoenwolf 2003). Cells in the center of the forming streak start to contract, and cells will individually go through an EMT process and ingress into the streak to form the mesoderm (Nakaya et al. 2008). While this is happening, there is a considerable remodeling of the basal lamina that takes place, allowing the cells to ingress into the streak. It remains to be established what role the remodeling and breakdown of the ECM plays in the control of ingression (Hamidi et al. 2020). Once the cells have ingressed, they start to move away as a dense mass of individual cells in the space between the epiblast and the hypoblast. The migration pattern and fate of these ingressing mesoderm cells are dependent on the region of ingression into the embryo (Fig. 3A). The cells that ingress from the extending tip of the streak will contribute to the prechordal mesoderm, while the cells at ingress from the anterior streak will form somites. The cells that ingress from the middle of the extending streak will form the lateral plate mesoderm, and the cells that ingress from the posterior streak will form the embryonic and extraembryonic mesoderm that will give rise to the embryonic and extraembryonic components of the circulatory system and the heart (Fig. 3B, C; Psychoyos and Stern 1996; Yang et al. 2002).

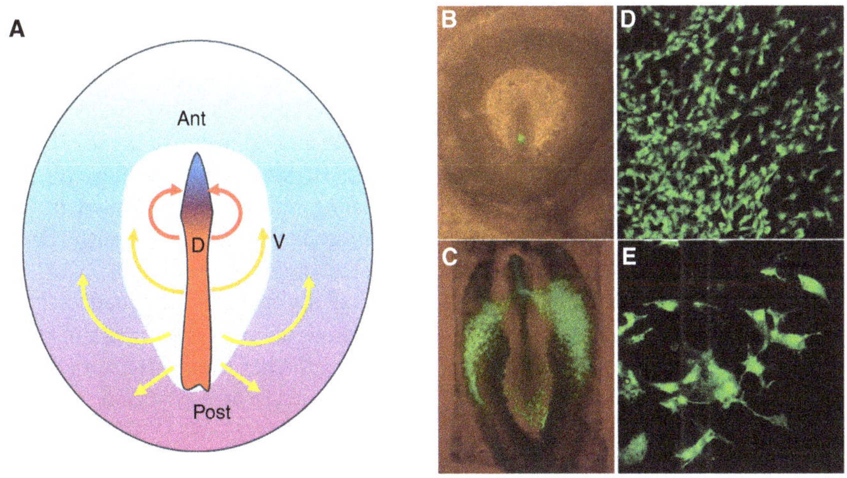

Figure 3. Schematic of cell migration patterns during chick gastrulation. (*A*) Mesoderm cell migration routes after their ingression through the primitive streak. Cells all along the streak migrate out in response to fibroblast growth factor (FGF)8 (red). Anterior cells, forming somites, migrate back toward the midline in response to FGF4 (blue). Middle streak, cells form lateral plate mesoderm. Cells in the back of the streak migrate in response to VEGF in the extraembryonic region (purple). Effective migration of all mesoderm cells is dependent on expression of platelet-derived growth factor (PDGF) in the epiblast (light blue). (*B*) Image of Hamburger and Hamilton (HH) 3+ streak stage, with green fluorescent protein (GFP)-expressing cells in the planted posterior streak. (*C*) Same embryo as *B* after further development to stage HH9; the labeled cells have migrated into the embryonic and extraembryonic to form the vasculature. (*D,E*) Intermediate and high-magnification images of the migrating cells showing extensive protrusions making cell–cell contacts of a group of cells. (Ant) Anterior, (Post) posterior, (D) dorsal, (V) ventral. (Figure data courtesy of Dr. Manli Chuai.)

Cellular Mechanisms

Directional Epiblast Cell Intercalation and Ingression

The large-scale tissue flows associated with streak formation that extends anteriorly along the central midline of the embryo are driven by mediolateral intercalation of mesendoderm cells (Voiculescu et al. 2007). This intercalation of the mesendoderm precursors in the epiblast is accompanied and dependent on junctional myosin II accumulation in aligned junctions in neighboring cells, giving the appearance of supercellular myosin cables. These cables align with the direction of intercalations toward the central midline of the embryo (Rozbicki et al. 2015; Saadaoui et al. 2020; Serra et al. 2023; Sknepnek et al. 2023). During streak formation, individual cells undergo extensive apical contractions and EMT and ingress into the deeper layers of the embryo. Ingression into the forming streak has been suggested to be a Nodal-dependent self-amplifying process resulting in an accumulation of individual ingressions (Voiculescu et al. 2014). Once the cells have ingressed as individual cells, they migrate away in a collective manner (Fig. 3B–E).

Signals Guiding the Migration of Mesoderm Cells

Experiments using a combination of ectopic sources and tissue transplantation experiments have shown that cells in the anterior and middle streaks are highly responsive to FGF signals. FGF4 from an ectopic source acts as a very strong chemoattractant for middle and anterior streak cells, while FGF8 acts from an ectopic source as a strong repellent. Blocking FGF receptors in responding cells via a small molecule inhibitor or expression of a dominant-negative FGFR1c receptor blocks these effects, suggesting that the attractive and repellent effects of FGF4 and FGF8 are direct (Yang et al. 2002). Based on these observations and FGF expression patterns, it was hypothesized that mesoderm cells migrate away from the streak due to the repulsive effect of FGF8 produced by cells in the streak, while the cells from the anterior and middle streak are attracted back toward the midline in response to the secretion of FGF4 secreted by the axial mesoderm formed after the regression of the node (Fig 3A). Exposure of streak stage embryos to an FGF receptor inhibitor blocks ingression of mesoderm precursor cells into the streak, without affecting SNAI2 expression; however, it affects expression of many other key components controlling migration, such as PDGFRA and FGF4, but not FGF8 (Hardy et al. 2011). During body axis elongation and somitogenesis, FGF signaling continues to control mesoderm cell motility. The migration speed of cells in the presomitic mesoderm is controlled by a gradient of FGF signaling, with high FGF and high speed in the posterior and low FGF resulting in low speed in the anterior (Bénazéraf et al. 2010). This gradient of cell motility has been suggested to generate extensile forces driving the posterior expansion of the primitive streak through a compression resulting in polarized intercalations of the presomitic mesoderm (Xiong et al. 2020). The FGF-dependent production of hyaluronic acid and swelling of this polymer further contribute a mechanical driving force for axis elongation (Michaut et al. 2025).

PDGFA is expressed in the epiblast, and its receptor PDGFRA is expressed in the migrating mesoderm (Yang et al. 2008). Experiments with localized sources of PDGF failed to show a strong chemotactic effect, but inhibition of PDGF signaling resulted in loss of mesoderm migration toward ectopic FGF4 sources. PDGF signaling is required for the expression of N-cadherin protein on the surface of mesoderm cells (Yang et al. 2008). N-cadherin is enriched at the tips of long filopodia through which mesoderm cells contact each other and may provide traction. PI3 kinase signaling is not required for EMT but is required for directional migration of mesoderm cells away from the streak (Leslie et al. 2007). Cells in the posterior streak express vascular endothelial growth factor receptor 2 (VEGFR2) receptors and respond highly to VEGF signals in the posterior epiblast (Fig. 3A; Eichmann et al. 1997, 1998). Inhibition of VEGF receptor signaling results in reduced ingression and extension of the streak (Chuai et al. 2023). Besides FGF signaling, Wnt and BMP signaling have been implicated in the control of mesoderm cell movements. Ectopic local sources of Wnt (Sweetman et al. 2008) and BMP (Song et al. 2014) can deflect the migration of mesoderm cells from grafts in the posterior streak. Whether these effects are direct or indirect remains to be established.

COLLECTIVE MIGRATION DURING MOUSE GASTRULATION

The mammalian system used most extensively to study gastrulation is the mouse embryo. This system has been and is instrumental in the study of relevant signaling pathways during gastrulation through genetic manipulation (Ramkumar and Anderson 2011; Morgani and Hadjantonakis 2020). Recent advances in imaging and culture technology have improved the ability to make live observations of early development, including gastrulation (Ichikawa et al. 2013; McDole et al. 2018).

Tissue-Level Shape Changes

At the start of gastrulation, after implantation in the uterine wall, the cup-shaped embryo consists of a few hundred cells. The single-cell-thick epiblast epithelium is surrounded by a single-cell-thick visceral endoderm layer (hypoblast). In preparation for streak formation, the most distal located hypoblast cells, the distal visceral endoderm (DVE) move anterior to form the anterior visceral endoderm (AVE) (Fig. 4A). Gastrulation starts with the formation of the streak that as in chick extend from posterior to anterior; however, formation does not involve large-scale epiblast cell movements, and the streak forms by in situ ingression of mesoderm cells (Williams et al. 2012). The cells undergo a full EMT and migrate away as individual cells between the epiblast and visceral endoderm layer. Throughout this period, all the cells in vari-

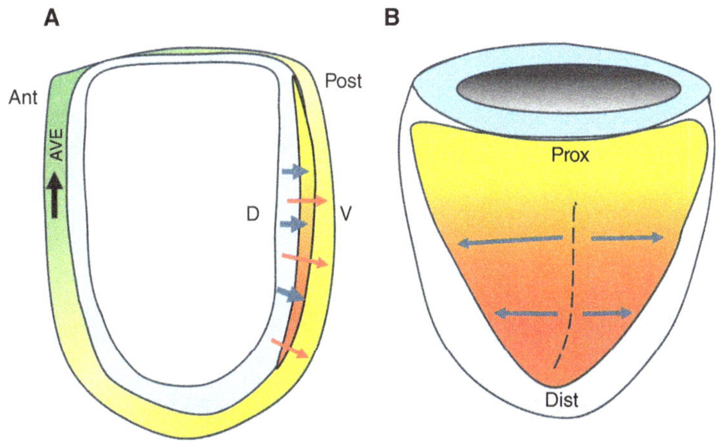

Figure 4. Schematic of extended primitive streak stage mouse embryo. (*A*) Cross-section of embryo. The anterior visceral endoderm cells (green) have migrated from distal to proximal. Epiblast cells (blue) after undergoing epithelial-to-mesenchymal transition (EMT) migrate out (blue arrows) to form the mesoderm (red), while endoderm precursors integrate (red arrows) into the visceral endoderm (yellow) to form the definitive endoderm. (*B*) Dorsal surface view of outward migration of wings of mesoderm cells, posterior (yellow) and anterior (red). Stippled line indicates the location of the primitive streak, and blue arrows indicate the direction of mesoderm migration away from the streak. (Ant) Anterior, (Post) posterior, (D) dorsal, (V) ventral, (Prox) proximal, (Dist) distal, (AVE) anterior visceral endoderm. (Figure based on data in Saykali et al. 2019.)

ous tissues of the embryo continuously divide, and the embryo significantly increases in size to reach several tens of thousands of cells by the end of gastrulation, resulting in considerable expansion of the embryo.

Cellular Mechanisms

Cell-tracking experiments have indicated that during AVE formation, the cells actively migrate, sending out many protrusions as in the chick embryo. The trajectories show that they migrate to the proximal embryonic–extraembryonic boundary in a meandering manner (Srinivas et al. 2004). Collective migration of AVE cells is dependent on the expression of β-Pix, regulating cell division cycle 42 (CDC42) and Pak activity (Omelchenko et al. 2014). Further analysis has shown that the rearrangement of the cells involves the formation and resolution of multicellular rosettes (Trichas et al. 2012). It has been established that AVE migration is not driven by Sdf1 and CXCR4 signaling (Miller et al. 2008). It has been suggested that migration requires active cell division and that tissue expansion generated by cell divisions in the posterior visceral endoderm cells may contribute to a mechanical displacement (Rodriguez et al. 2005). Breakdown of the basement membrane is required for AVE migration, and a slight gradient in membrane perforations might bias the anterior migration of the DVE (Chen et al. 2025).

Ingression of the mesoderm and endoderm cells involves apical contraction of cells and degradation of the ECM as in the case of the chick embryos (Hashimoto et al. 1987; Chen et al. 2025). Apical contraction occurs through a ratchet type mechanism (Francou et al. 2023) and dissolution of junctions involving dynamic interactions between Crumbs and Myosin 2 localization (Ramkumar et al. 2016).

Ingressing endoderm cells insert into the overlying hypoblast to form part of the definitive endoderm, while pushing the hypoblast toward the anterior (Fig. 4B). Ingressing mesoderm cells stay in the space between the epiblast and the hypoblast and migrate to their destinations. Initial light sheet microscopy-based investigations suggested that individual mesodermal cells migrate not as a sheet but as single cells without extensive coordination (Ichikawa et al. 2013). More localized conventional imaging at higher magnification of groups of labeled cells has shown that the mesoderm cells extend many long and dynamic processes and may contact each other as in the chick embryo. The tracks of individual cells show that they migrate slowly and directionally, but extremely meanderingly (Saykali et al. 2019). Interestingly, it has been suggested that daughter cells tend to stay close to each other during their migration, possibly close enough to stay in contact through longer cellular extensions (Saykali et al. 2019). There is no clear correlation between the time of ingression and the position of the cells away from the streak, suggesting that cells may overtake each other in the migration process. More recent whole embryo light sheet microscopy analysis of the behavior of all cells in gastrulating embryos with labeled nuclei has shown that there is a clear overall migration pattern of the mesoderm cells; however, the tracking and detailed analysis of individual trajectories at the full embryo scale is still extremely challenging (McDole et al. 2018).

MOLECULAR MECHANISMS OF MIGRATION AND GUIDANCE

Mesoderm cell migration is characterized by directional extension of filopodia and lamellipodia. Many components of the cytoskeletal machinery that produce these protrusions have been identified in mutant studies that have different gastrulation defects. Genetic perturbation studies have shown that effective mesoderm migration is dependent on key components of the actin–myosin cytoskeleton and cell adhesion, such as Rac1, stripac activity, myosin, and Scribble (Bazzi et al. 2017; Omelchenko et al. 2020). Major components are the small guanosine triphosphatases (GTPases) Rac1, CDC42, RhoA, as well as some of their GEFs and GTPase-activating proteins (GAPs), and components that affect the actin–myosin (Omelchenko 2022). There is less evidence for a major role of the PCP pathway in the control of mesoderm cell migration during gastrulation; however, there is substantial evidence that it has a major role in extension of the body axis in later development. Vangl2–Cofilin double mutants show extensive axis extension defects underlining the important role of PCP components in axis elongation and rearrangements of epithelial cells (Mahaffey et al. 2013).

Signals that affect the migration of mesoderm cells have been identified through mutational analysis and highlighted key roles for FGF and VEGF signaling. *Fgf8* and *Fgf4* are coexpressed in the primitive streak of the gastrulating mouse embryo. *Fgf8* deletion results in a loss of *Fgf4* expression, and the absence of both *Fgf8* and *Fgf4*, epiblast cells can undergo an EMT and move into the streak, but then fail to move away from the streak. As a consequence, no embryonic mesoderm- or endoderm-derived tissues develop, although extraembryonic tissues form (Sun et al. 1999). The FGFR1 receptor is a major transducer of FGF signals during gastrulation. Early studies found that in the absence of *Fgfr1*, embryos displayed early growth defects; however, they remained capable of gastrulating and generating some mesoderm, pointing toward redundancy in FGF signal detection (Zhang et al. 2006). The nascent mesoderm of *Fgfr1* homozygous mutant embryos differentiated into diverse mesodermal subtypes, but mesodermal patterning was aberrant in *Fgfr1* knockout (KO) mice (Yamaguchi et al. 1994). Further analysis showed that deletion of the *Fgfr1* receptor does not inhibit basal cell migration of mesoderm cells in vitro. However, null mutants were shown to display a reduction in *Snail* expression and downregulation of E-cadherin (Ciruna and Rossant 2001). However, to date, few experiments have been performed to elucidate possible instructive roles of FGF signaling for mesoderm migration in mouse embryos.

A *Pdgfrα* KO has defects in neural crest and somite structure (Soriano 1997). The role of *Pdgfrα* showed that defects occur mostly after gastrulation, the embryos are small, and partial mutants have shown PI3 kinase to be a major upstream component (Klinghoffer et al. 2002; Betsholtz 2004). The precise role of PDGF signaling at the level of single-cell migration and mesoderm migration has not yet been analyzed. It has been well established that cells in the posterior streak move in response to VEGF signals. The cells in the posterior streak express the *Vegfr2* receptor, and these cells migrate from the streak outward to form the extraembryonic and embryonic primitive vasculature (Fong et al. 1995).

The regulatory role of the Aplnr pathway in the differentiation of mesoderm and mesoderm migration remains unclear. Activating and suppressing Aplnr signaling pathway via peptides, small molecules, and genetic modifications, including small interfering RNA (siRNA)- and short hairpin RNA (shRNA)-mediated knockdown and CRISPR-Cas9-mediated KO, revealed that Aplnr signaling not only induces migration of cells during germ layer formation but also enhances mesoderm differentiation through the FGF/mitogen-activated protein kinase (MAPK) pathway. Antibody array and liquid chromatography-mass spectrometry (LC/MS) protein profiling data demonstrated that Apelin-13 treatment enhanced cell cycle, endothelial growth factor receptor (EGFR), FGF, Wnt, and integrin signaling pathway proteins (Şişli et al. 2023).

GASTRULOIDS

In recent years, there has been a tremendous increase in the ability to produce two- and three-dimensional gastruloid model systems from mouse and human stem cells. Development of gastrulation-stage 2d cultures results in the formation of concentric regions of extraembryonic and embryonic tissues (Warmflash et al. 2014). These concentric structures, under some conditions, start to show evidence for processes like EMT and mesendoderm cell migration directed from the periphery of the colony toward the center (Martyn et al. 2019). Mesoderm cell migration appears to require nodal signaling, since it can be inhibited by treatment with small molecule inhibitors targeting nodal signaling (Martyn et al. 2019). These 2D systems may open up a convenient way to study the signals that control mesendoderm migration. Recently, it has been shown that cadherin switching and EMT are modulated independently of cell fate commitment in mesendodermal differentiation (Zhu and Warmflash 2025).

ES cell–derived three-dimensional gastruloids form structures that can polarize and develop defined AP axes with a regionalized expression of Brachyury apparently through a process of cell sorting (Schauer and Heisenberg 2021; van den Brink and van Oudenaarden 2021; Turner and Martinez Arias 2024). The evidence so far indicates that these mesoderm cells sort to the posterior pole of the gastruloid through a process that involves downregulation of cadherin and sorting of cadherin cells into regions that will form endoderm. There is emerging evidence for a role of Wnt5a possibly acting upstream of PCP signaling in A/P axis elongation (Fiuza et al. 2025). Chimeric analysis of wild-type and mutants should help unravel the mechanisms that guide mesoderm cell movement in these conditions. Current ES cell-only derived gastruloids do not yet undergo a complete gastrulation program, since they do not form a morphological and functional primitive streak. However, gastruloids derived by combining mouse trophoblast stem (TS) cells, ES cells exposed to a pulse of Gata4 expression to drive extraembryonic endoderm fate and ES cells can undergo morphogenesis to a mid-gastrulation stage equivalent. These gastruloids form an anterior signaling center that migrates unilaterally, thus breaking symmetry and forming a primitive streak, embryonic and extraembryonic mesoderm, as well as definitive endoderm (Amadei et al. 2021). All these different gastruloids will undoubtedly be useful in the exploration of the range of behaviors that cells can execute under specific experimentally controlled external conditions, not necessarily the same as those found in vivo, revealing the full plasticity of the system (Villaronga-Luque et al. 2025).

CONCLUSIONS AND OUTLOOK

Gastrulation is the process of transforming a sheet or ball of cells into a three-layered structure with a central digestive tract from mouth to anus. This involves internalization of mesoderm and endoderm tissue, a process requiring a partial or complete EMT. After internalization of the mesoderm and endoderm, these tissues need to disperse to line the inside of the embryo to form the digestive tract from mouth to anus, surrounding skeleton and musculature, and internal organs. Much progress has been made in describing the often-complex cell behaviors that cells display both in vivo and in vitro in the various model organisms as outlined above. Through forward and reverse genetics, many of the molecular components of the cellular machinery essential for completing some of the critical cell motility behaviors have been identified. Considerably less progress has been made in the elucidation of the molecular basis of the mechanisms guiding and coordinating these motility behaviors and their coordination with the control of gene expression.

Migration of mesenchymal mesoderm shows interesting similarities and differences in packing density and migration between lower vertebrates (fish and frogs) and amniotes (chick and mouse). In fish and frogs, mesodermal cells appear to be tightly packed and as a result rely on a high degree of cell-on-cell migration. This has led to the identification of distinct mechanisms for the directional collective migration of the PPL cells requiring monopolar cells and axial and paraxial mesoderm intercalation movements requiring the extensive formation of bipolar cells. The high packing density also results in highly confined directed movement trajectories. Mesoderm cells in amniotes show a lower packing density and migrate in a space that has an extensive diffuse ECM, from which cells likely can get some traction. Migrating cells extend many long filopodial protrusions that contact neighboring cells in N-cadherin-mediated cell–cell interactions, which are also required for effective migration. Individual cells follow more meandering tracks, but collectively their movement is highly coordinated.

Migration patterns of endoderm precursor cells appear more variable across organisms. In frogs, endoderm cells migrate through a cell-on-cell ingression-type migration mechanism, where the cells get traction to move from neighbors. Velocity gradients result in stretching and thinning of the initially large and solid cell mass, converting it into an epithelial sheet that is folded inside the embryo. In fish, however, endoderm cells form a dispersed monolayer of individual cells between the yolk cell membrane and closely packed overlying mesoderm and move through protrusion of filopodia and lamellipodia. After mesoderm-guided dorsal convergence, they fuse to form epithelial structures. In amniotes, endoderm cells arise in the epithelial epiblast sheet, undergo EMT and ingress, but immediately insert a hypoblast epithelial layer again to form a mosaic tissue that then collectively displaces toward anterior.

The signals guiding these different cell behaviors have only been partly resolved. There appear to be two major mechanisms. Chemotactic movement plays a key part in the migration of the LEM in frogs, radial intercalation of cells in the blastocoel roof, the migration of mesoderm and endoderm cells in fish, and the migration of mesoderm cells in the chick and possibly in the mouse. In some cases, there is strong evidence for self-generated gradients resulting in effective cell dispersal of mesoderm cells in many instances.

The second mechanism that has been proposed to give rise to directional migration cues is the viscosity of the substrate in the case of cell-on-cell migration. This has been proposed for the anterior migration of LEM cells in fish. This is an interesting concept and should be further investigated in other systems, where substrate cell density varies across migration territories.

Thirdly, it is well established that in many embryos, cell intercalation is the mechanism that drives part of the elongation of the body axis. For this to happen, the cells need to be closely packed and intercalate collectively in the same direction perpendicular to the A/P axis. There is extensive evidence that PCP signaling couples to the organization of the actin–myosin cytoskeleton and affects the autonomous polarization of cells. However, the molecular mechanisms underlying the global intercalation alignment perpendicular to the A/P axis still need to be resolved. Two competing ideas have been proposed, one is that it is due to a long-range chemical gradient such as noncanonical Wnt5 signaling. Another competing explanation of this spatial organization of intercalations is an emergent self-organizing property involving mechanical feedback (Plum et al. 2025).

To further progress understanding, the feedback between signaling and cell behaviors will be essential to visualize signaling dynamics in vivo. Being able to visualize and quantify the signal dynamics signals and the cellular responses such as movement gene expression will allow building up of a complete picture of key morphogenetic events and feedback between signaling and movement at the scale of the embryo. This is well illustrated in the case of simple systems such as the multicellular development and morphogenesis of *Dictyostelium discoideum* (Ford et al. 2023; Brimson et al. 2025). However, to date, there are surprisingly few studies that show the feasibility of these approaches during gastrulation in embryos.

It is becoming increasingly evident that mechanical stresses and mechanotransduction play essential roles in the coordination of collective cell behaviors during development. So far, cell shape and deformation dynamics and cytoskeletal dynamics, especially of actin and myosin, are often used as indicators of these processes. However, more quantitative methods to measure stresses and their dynamics at the cell-to-tissue scale are urgently needed.

ACKNOWLEDGMENTS

Dr. Manli Chuai kindly provided the data used in Figure 3. C.J.W. acknowledges support from the BBSRC (BB/T006781/1) and EPSRC EP/W023946/1.

REFERENCES

Amadei G, Lau KYC, De Jonghe J, Gantner CW, Sozen B, Chan C, Zhu M, Kyprianou C, Hollfelder F, Zernicka-Goetz M. 2021. Inducible stem-cell-derived embryos capture mouse morphogenetic events in vitro. *Dev Cell* **56:** 366–382.e9. doi:10.1016/j.devcel.2020.12.004

Baldwin A, Popov IK, Keller R, Wallingford J, Chang C. 2023. The RhoGEF protein Plekhg5 regulates medioapical and junctional actomyosin dynamics of apical constriction during *Xenopus* gastrulation. *Mol Biol Cell* **34:** ar64. doi:10.1091/mbc.E22-09-0411

Barrett K, Anand S, Thome V, Kodjabachian L, Merkel M, Lenne PF. 2025. Epithelial–mesenchymal interface guides cell shapes and axis elongation in embryonic explants. *Development* **152:** dev204344. doi:10.1242/dev.204344

Bazzi H, Soroka E, Alcorn HL, Anderson KV. 2017. STRIP1, a core component of STRIPAK complexes, is essential for normal mesoderm migration in the mouse embryo. *Proc Natl Acad Sci* **114:** E10928–E10936. doi:10.1073/pnas.1713535114

Behrndt M, Salbreux G, Campinho P, Hauschild R, Oswald F, Roensch J, Grill SW, Heisenberg CP. 2012. Forces driving epithelial spreading in zebrafish gastrulation. *Science* **338:** 257–260. doi:10.1126/science.1224143

Bénazéraf B, Francois P, Baker RE, Denans N, Little CD, Pourquié O. 2010. A random cell motility gradient downstream of FGF controls elongation of an amniote embryo. *Nature* **466:** 248–252. doi:10.1038/nature09151

Betsholtz C. 2004. Insight into the physiological functions of PDGF through genetic studies in mice. *Cytokine Growth Factor Rev* **15:** 215–228. doi:10.1016/j.cytogfr.2004.03.005

Boutillon A, Escot S, Elouin A, Jahn D, González-Tirado S, Starruß J, Brusch L, David NB. 2022. Guidance by followers ensures long-range coordination of cell migration through alpha-catenin mechanoperception. *Dev Cell* **57:** 1529–1544.e5. doi:10.1016/j.devcel.2022.05.001

Boutillon A, Banavar SP, Campàs O. 2024. Conserved physical mechanisms of cell and tissue elongation. *Development* **151:** dev202687. doi:10.1242/dev.202687

Brimson CA, Baines R, Sams-Dodd E, Stefanescu I, Evans B, Kuwana S, Hashimura H, Sawai S, Thompson CRL. 2025. Collective oscillatory signaling in *Dictyostelium discoideum* acts as a developmental timer initiated by weak coupling of a noisy pulsatile signal. *Dev Cell* **60:** 918–933.e4. doi:10.1016/j.devcel.2024.11.016

Butler MT, Wallingford JB. 2017. Planar cell polarity in development and disease. *Nat Rev Mol Cell Biol* **18:** 375–388. doi:10.1038/nrm.2017.11

Čapek D, Smutny M, Tichy AM, Morri M, Janovjak H, Heisenberg CP. 2019. Light-activated Frizzled7 reveals a permissive role of non-canonical wnt signaling in mesendoderm cell migration. *eLife* **8:** e42093. doi:10.7554/eLife.42093

Carvalho L, Heisenberg CP. 2010. The yolk syncytial layer in early zebrafish development. *Trends Cell Biol* **20:** 586–592. doi:10.1016/j.tcb.2010.06.009

Chen DY, Claussen NH, Titus S, Hu W, Weatherbee BAT, Mandelbaum RS, Scott RT Jr, Seli E, Streichan SJ, Zernicka-Goetz M. 2025. Basement membrane perforations guide anterior-posterior axis formation. *Nat Commun* **16:** 6763. doi:10.1038/s41467-025-61441-6

Cheng T, Xing YY, Liu C, Li YF, Huang Y, Liu X, Zhang YJ, Zhao GQ, Dong Y, Fu XX, et al. 2023. Nodal coordinates the anterior–posterior patterning of germ layers and induces head formation in zebrafish explants. *Cell Rep* **42:** 112351. doi:10.1016/j.celrep.2023.112351

Choi SC, Sokol SY. 2009. The involvement of lethal giant larvae and Wnt signaling in bottle cell formation in *Xenopus* embryos. *Dev Biol* **336:** 68–75. doi:10.1016/j.ydbio.2009.09.033

Chuai M, Serrano Nájera G, Serra M, Mahadevan L, Weijer CJ. 2023. Reconstruction of distinct vertebrate gastrulation modes via modulation of key cell behaviors in the chick embryo. *Sci Adv* **9:** eabn5429. doi:10.1126/sciadv.abn5429

Ciruna B, Rossant J. 2001. FGF signaling regulates mesoderm cell fate specification and morphogenetic movement at the primitive streak. *Dev Cell* **1:** 37–49. doi:10.1016/S1534-5807(01)00017-X

D'Amico LA, Cooper MS. 2001. Morphogenetic domains in the yolk syncytial layer of axiating zebrafish embryos. *Dev Dyn* **222:** 611–624. doi:10.1002/dvdy.1216

Damm EW, Winklbauer R. 2011. PDGF-A controls mesoderm cell orientation and radial intercalation during *Xenopus* gastrulation. *Development* **138:** 565–575. doi:10.1242/dev.056903

Dumortier JG, David NB. 2015. The TORC2 component, Sin1, controls migration of anterior mesendoderm during zebrafish gastrulation. *PLoS One* **10:** e0118474. doi:10.1371/journal.pone.0118474

Dumortier JG, Martin S, Meyer D, Rosa FM, David NB. 2012. Collective mesendoderm migration relies on an intrinsic directionality signal transmitted through cell contacts. *Proc Natl Acad Sci* **109:** 16945–16950. doi:10.1073/pnas.1205870109

Economou AD, Guglielmi L, East P, Hill CS. 2022. Nodal signaling establishes a competency window for stochastic cell fate switching. *Dev Cell* **57:** 2604–2622.e5. doi:10.1016/j.devcel.2022.11.008

Eichmann A, Corbel C, Nataf V, Vaigot P, Bréant C, Douarin NML. 1997. Ligand-dependent development of the endothelial and hemopoietic lineages from embryonic mesodermal cells expressing vascular endothelial growth factor receptor 2. *Proc Natl Acad Sci* **94:** 5141–5146. doi:10.1073/pnas.94.10.5141

Eichmann A, Corbel C, Jaffredo T, Bréant C, Joukov V, Kumar V, Alitalo K, le Douarin NM. 1998. Avian VEGF-C: cloning, embryonic expression pattern and stimulation of the differentiation of VEGFR2-expressing endothelial cell precursors. *Development* **125:** 743–752. doi:10.1242/dev.125.4.743

Fiuza UM, Bonavia S, Pascual-Mas P, Torregrosa-Cortés G, Casaní-Galdón P, Robertson G, Dias A, Martinez Arias A. 2025. Morphogenetic constraints in the development of gastruloids: implications for mouse gastrulation. *Cells Dev* **183:** 204043. doi:10.1016/j.cdev.2025.204043

Fong GH, Rossant J, Gertsenstein M, Breltman ML. 1995. Failure of blood-island formation and vasculogenesis in Flk-1-deficient mice. *Nature* **376:** 62–66. doi:10.1038/376062a0

Ford HZ, Manhart A, Chubb JR. 2023. Controlling periodic long-range signalling to drive a morphogenetic transition. *eLife* **12:** e83796. doi:10.7554/eLife.83796

Francou A, Anderson KV, Hadjantonakis AK. 2023. A ratchet-like apical constriction drives cell ingression during the mouse gastrulation EMT. *eLife* **12:** e84019. doi:10.7554/eLife.84019

Fukui A, Goto T, Kitamoto J, Homma M, Asashima M. 2007. SDF-1 alpha regulates mesendodermal cell migration during frog gastrulation. *Biochem Biophys Res Commun* **354:** 472–477. doi:10.1016/j.bbrc.2007.01.007

Hamidi S, Nakaya Y, Nagai H, Alev C, Kasukawa T, Chhabra S, Lee R, Niwa H, Warmflash A, Shibata T, et al. 2020. Mesenchymal-epithelial transition regulates initiation of pluripotency exit before gastrulation. *Development* **147:** dev184960. doi:10.1242/dev.184960

Hara Y, Nagayama K, Yamamoto TS, Matsumoto T, Suzuki M, Ueno N. 2013. Directional migration of leading-edge mesoderm generates physical forces: implication in *Xenopus* notochord formation during gastrulation. *Dev Biol* **382:** 482–495. doi:10.1016/j.ydbio.2013.07.023

Hardin J, Keller R. 1988. The behaviour and function of bottle cells during gastrulation of *Xenopus laevis*. *Development* **103:** 211–230. doi:10.1242/dev.103.1.211

Hardy KM, Yatskievych TA, Konieczka J, Bobbs AS, Antin PB. 2011. FGF signalling through RAS/MAPK and PI3K pathways regulates cell movement and gene expression in the chicken primitive streak without affecting E-cadherin expression. *BMC Dev Biol* **11:** 1–17. doi:10.1186/1471-213X-11-20

Hashimoto K, Fujimoto H, Nakatsuji N. 1987. An ECM substratum allows mouse mesodermal cells isolated from the primitive streak to exhibit motility similar to that inside the embryo and reveals a deficiency in the *T*/*T* mutant cells. *Development* **100:** 587–598. doi:10.1242/dev.100.4.587

Heisenberg CP, Tada M, Rauch GJ, Saúde L, Concha ML, Geisler R, Stemple DL, Smith JC, Wilson SW. 2000. Silberblick/Wnt11 mediates convergent extension movements during zebrafish gastrulation. *Nature* **405:** 76–81. doi:10.1038/35011068

Hill CS. 2018. Spatial and temporal control of NODAL signaling. *Curr Opin Cell Biol* **51:** 50–57. doi:10.1016/j.ceb.2017.10.005

Huang Y, Winklbauer R. 2018. Cell migration in the *Xenopus* gastrula. *Wiley Interdiscip Rev Dev Biol* **7:** e325. doi:10.1002/wdev.325

Huebner RJ, Wallingford JB. 2018. Coming to consensus: a unifying model emerges for convergent extension. *Dev Cell* **46:** 389–396. doi:10.1016/j.devcel.2018.08.003

Ichikawa T, Nakazato K, Keller PJ, Kajiura-Kobayashi H, Stelzer EH, Mochizuki A, Nonaka S. 2013. Live imaging of whole mouse embryos during gastrulation: migration analyses of epiblast and mesodermal cells. *PLoS One* **8:** e64506. doi:10.1371/journal.pone.0064506

Keller R. 2005. Cell migration during gastrulation. *Curr Opin Cell Biol* **17:** 533–541. doi:10.1016/j.ceb.2005.08.006

Keller R, Sutherland A. 2020. Convergent extension in the amphibian, *Xenopus laevis*. *Curr Top Dev Biol* **136:** 271–317. doi:10.1016/bs.ctdb.2019.11.013

Kilian B, Mansukoski H, Barbosa FC, Ulrich F, Tada M, Heisenberg CP. 2003. The role of Ppt/Wnt5 in regulating cell shape and movement during zebrafish gastrulation. *Mech Dev* **120:** 467–476. doi:10.1016/S0925-4773(03)00004-2

Kimelman D. 2006. Mesoderm induction: from caps to chips. *Nat Rev Genet* **7:** 360–372. doi:10.1038/nrg1837

Klinghoffer RA, Hamilton TG, Hoch R, Soriano P. 2002. An allelic series at the PDGFαR locus indicates unequal contributions of distinct signaling pathways during development. *Dev Cell* **2:** 103–113. doi:10.1016/S1534-5807(01)00103-4

Kondow A, Ohnuma K, Kamei Y, Taniguchi A, Bise R, Sato Y, Yamaguchi H, Nonaka S, Hashimoto K. 2020. Light-sheet microscopy-based 3D single-cell tracking reveals a correlation between cell cycle and the start of endoderm cell internalization in early zebrafish development. *Dev Growth Differ* **62:** 495–502. doi:10.1111/dgd.12695

LaBelle J, Wyatt T, Ramirez J, Woo S. 2025. Endodermal cells use contact inhibition of locomotion to achieve uniform cell dispersal during zebrafish gastrulation. *iScience* **28:** 114144. doi:10.1016/j.isci.2025.114144

Lawson A, Schoenwolf GC. 2003. Epiblast and primitive-streak origins of the endoderm in the gastrulating chick embryo. *Development* **130:** 3491–3501. doi:10.1242/dev.00579

Lee JY, Harland RM. 2010. Endocytosis is required for efficient apical constriction during *Xenopus* gastrulation. *Curr Biol* **20:** 253–258. doi:10.1016/j.cub.2009.12.021

Lee HC, Oliveira NMM, Hastings C, Baillie-Benson P, Moverley AA, Lu HC, Zheng Y, Wilby EL, Weil TT, Page KM, et al. 2024. Regulation of long-range BMP gradients and embryonic polarity by propagation of local calcium-firing activity. *Nat Commun* **15:** 1463. doi:10.1038/s41467-024-45772-4

Leerberg DM, Hopton RE, Draper BW. 2019. Fibroblast growth factor receptors function redundantly during zebrafish embryonic development. *Genetics* **212:** 1301–1319. doi:10.1534/genetics.119.302345

Leslie NR, Yang X, Downes CP, Weijer CJ. 2007. PtdIns(3,4,5)P_3-dependent and -independent roles for PTEN in the control of cell migration. *Curr Biol* **17:** 115–125. doi:10.1016/j.cub.2006.12.026

Liberali P, Schier AF. 2024. The evolution of developmental biology through conceptual and technological revolutions. *Cell* **187:** 3461–3495. doi:10.1016/j.cell.2024.05.053

Mahaffey JP, Grego-Bessa J, Liem KF Jr, Anderson KV. 2013. Cofilin and Vangl2 cooperate in the initiation of planar cell polarity in the mouse embryo. *Development* **140:** 1262–1271. doi:10.1242/dev.085316

Martyn I, Siggia ED, Brivanlou AH. 2019. Mapping cell migrations and fates in a gastruloid model to the human primitive streak. *Development* **146:** dev179564. doi:10.1242/dev.179564

McDole K, Guignard L, Amat F, Berger A, Malandain G, Royer LA, Turaga SC, Branson K, Keller PJ. 2018. In toto imaging and reconstruction of post-implantation mouse development at the single-cell level. *Cell* **175:** 859–876.e33. doi:10.1016/j.cell.2018.09.031

Michaut A, Mongera A, Gupta A, Tarazona OA, Serra M, Kefala GM, Rigoni P, Lee JG, Rivas F, Hall AR, et al. 2025. Extracellular volume expansion drives vertebrate axis elongation. *Curr Biol* **35:** 843–853.e6. doi:10.1016/j.cub.2024.12.051

Miller RJ, Banisadr G, Bhattacharyya BJ. 2008. CXCR4 signaling in the regulation of stem cell migration and development. *J Neuroimmunol* **198:** 31–38. doi:10.1016/j.jneuroim.2008.04.008

Mizoguchi T, Verkade H, Heath JK, Kuroiwa A, Kikuchi Y. 2008. Sdf1/Cxcr4 signaling controls the dorsal migration of endodermal cells during zebrafish gastrulation. *Development* **135:** 2521–2529. doi:10.1242/dev.020107

Montero JA, Kilian B, Chan J, Bayliss PE, Heisenberg CP. 2003. Phosphoinositide 3-kinase is required for process outgrowth and cell polarization of gastrulating mesendodermal cells. *Curr Biol* **13:** 1279–1289. doi:10.1016/S0960-9822(03)00505-0

Morgani SM, Hadjantonakis AK. 2020. Signaling regulation during gastrulation: insights from mouse embryos and in vitro systems. In *Current topics in developmental biology* (ed. Small S, Briscoe J), pp. 391–431. Academic, Cambridge, MA.

Myers DC, Sepich DS, Solnica-Krezel L. 2002. Bmp activity gradient regulates convergent extension during zebrafish gastrulation. *Dev Biol* **243:** 81–98. doi:10.1006/dbio.2001.0523

Nagel M, Winklbauer R. 2023. Polarized contact behavior in directionally migrating *Xenopus* gastrula mesendoderm. *Int J Dev Biol* **67:** 79–90. doi:10.1387/ijdb.230123rw

Nagel M, Tahinci E, Symes K, Winklbauer R. 2004. Guidance of mesoderm cell migration in the *Xenopus* gastrula requires PDGF signaling. *Development* **131:** 2727–2736. doi:10.1242/dev.01141

Nair S, Schilling TF. 2008. Chemokine signaling controls endodermal migration during zebrafish gastrulation. *Science* **322:** 89–92. doi:10.1126/science.1160038

Nakaya Y, Sukowati EW, Wu Y, Sheng G. 2008. RhoA and microtubule dynamics control cell-basement membrane interaction in EMT during gastrulation. *Nat Cell Biol* **10:** 765–775. doi:10.1038/ncb1739

Niehrs C. 2010. On growth and form: a Cartesian coordinate system of Wnt and BMP signaling specifies bilaterian body axes. *Development* **137:** 845–857. doi:10.1242/dev.039651

Ninomiya H, Elinson RP, Winklbauer R. 2004. Antero-posterior tissue polarity links mesoderm convergent extension to axial patterning. *Nature* **430:** 364–367. doi:10.1038/nature02620

Norris ML, Pauli A, Gagnon JA, Lord ND, Rogers KW, Mosimann C, Zon LI, Schier AF. 2017. Toddler signaling regulates mesodermal cell migration downstream of Nodal signaling. *eLife* **6:** e22626. doi:10.7554/eLife.22626

Omelchenko T. 2022. Cellular protrusions in 3D: orchestrating early mouse embryogenesis. *Semin Cell Dev Biol* **129:** 63–74. doi:10.1016/j.semcdb.2022.05.005

Omelchenko T, Rabadan MA, Hernández-Martínez R, Grego-Bessa J, Anderson KV, Hall A. 2014. Beta-Pix directs collective migration of anterior visceral endoderm cells in the early mouse embryo. *Genes Dev* **28:** 2764–2777. doi:10.1101/gad.251371.114

Omelchenko T, Hall A, Anderson KV. 2020. Beta-Pix-dependent cellular protrusions propel collective mesoderm migration in the mouse embryo. *Nat Commun* **11:** 6066. doi:10.1038/s41467-020-19889-1

Ossipova O, Chuykin I, Chu CW, Sokol SY. 2015. Vangl2 cooperates with Rab11 and Myosin V to regulate apical constriction during vertebrate gastrulation. *Development* **142:** 99–107. doi:10.1242/dev.111161

Pauli A, Norris ML, Valen E, Chew GL, Gagnon JA, Zimmerman S, Mitchell A, Ma J, Dubrulle J, Reyon D, et al. 2014. Toddler: an embryonic signal that promotes cell movement via apelin receptors. *Science* **343:** 1248636. doi:10.1126/science.1248636

Pézeron G, Mourrain P, Courty S, Ghislain J, Becker TS, Rosa FM, David NB. 2008. Live analysis of endodermal layer formation identifies random walk as a novel gastrulation movement. *Curr Biol* **18:** 276–281. doi:10.1016/j.cub.2008.01.028

Pfister K, Shook DR, Chang C, Keller R, Skoglund P. 2016. Molecular model for force production and transmission during vertebrate gastrulation. *Development* **143:** 715–727. doi:10.1242/dev.128090

Pinheiro D, Kardos R, Hannezo É, Heisenberg CP. 2022. Morphogen gradient orchestrates pattern-preserving tissue morphogenesis via motility-driven unjamming. *Nat Phys* **18:** 1482–1493. doi:10.1038/s41567-022-01787-6

Plum AM, Serra M, Weijer CJ. 2025. Advances in mechanochemical modelling of vertebrate gastrulation. *Biochem Soc Trans* **53:** 871–880. doi:10.1042/BST20240469

Popov IK, Ray HJ, Skoglund P, Keller R, Chang C. 2018. The RhoGEF protein Plekhg5 regulates apical constriction of bottle cells during gastrulation. *Development* **145:** dev168922. doi:10.1242/dev.168922

Psychoyos D, Stern CD. 1996. Restoration of the organizer after radical ablation of Hensen's node and the anterior primitive streak in the chick embryo. *Development* **122:** 3263–3273. doi:10.1242/dev.122.10.3263

Ramkumar N, Anderson KV. 2011. Snapshot: mouse primitive streak. *Cell* **146:** 488–488.e2. doi:10.1016/j.cell.2011.07.028

Ramkumar N, Omelchenko T, Silva-Gagliardi NF, McGlade CJ, Wijnholds J, Anderson KV. 2016. Crumbs2 promotes cell ingression during the epithelial-to-mesenchymal transition at gastrulation. *Nat Cell Biol* **18:** 1281–1291. doi:10.1038/ncb3442

Rodriguez TA, Srinivas S, Clements MP, Smith JC, Beddington RSP. 2005. Induction and migration of the anterior visceral endoderm is regulated by the extra-embryonic ectoderm. *Development* **132:** 2513–2520. doi:10.1242/dev.01847

Roszko I, S. Sepich D, Jessen JR, Chandrasekhar A, Solnica-Krezel L. 2015. A dynamic intracellular distribution of Vangl2 accompanies cell polarization during zebrafish gastrulation. *Development* **142:** 2508–2520. doi:10.1242/dev.119032

Rozbicki E, Chuai M, Karjalainen AI, Song F, Sang HM, Martin R, Knölker HJ, MacDonald MP, Weijer CJ. 2015. Myosin-II-mediated cell shape changes and cell intercalation contribute to primitive streak formation. *Nat Cell Biol* **17:** 397–408. doi:10.1038/ncb3138

Saadaoui M, Rocancourt D, Roussel J, Corson F, Gros J. 2020. A tensile ring drives tissue flows to shape the gastrulating amniote embryo. *Science* **367:** 453–458. doi:10.1126/science.aaw1965

Saykali B, Mathiah N, Nahaboo W, Racu ML, Hammou L, Defrance M, Migeotte I. 2019. Distinct mesoderm migration phenotypes in extra-embryonic and embryonic regions of the early mouse embryo. *eLife* **8:** e42434. doi:10.7554/eLife.42434

Schauer A, Heisenberg CP. 2021. Reassembling gastrulation. *Dev Biol* **474:** 71–81. doi:10.1016/j.ydbio.2020.12.014

Sepich DS, Calmelet C, Kiskowski M, Solnica-Krezel L. 2005. Initiation of convergence and extension movements of lateral mesoderm during zebrafish gastrulation. *Dev Dyn* **234:** 279–292. doi:10.1002/dvdy.20507

Serra M, Serrano Nájera G, Chuai M, Plum AM, Santhosh S, Spandan V, Weijer CJ, Mahadevan L. 2023. A mechanochemical model recapitulates distinct vertebrate gastrulation modes. *Sci Adv* **9:** eadh8152. doi:10.1126/sciadv.adh8152

Serrano Nájera G, Weijer CJ. 2020. Cellular processes driving gastrulation in the avian embryo. *Mech Dev* **163:** 103624. doi:10.1016/j.mod.2020.103624

Shah G, Thierbach K, Schmid B, Waschke J, Reade A, Hlawitschka M, Roeder I, Scherf N, Huisken J. 2019. Multiscale imaging and analysis identify pan-embryo cell dynamics of germlayer formation in zebrafish. *Nat Commun* **10:** 5753. doi:10.1038/s41467-019-13625-0

Shen MM. 2007. Nodal signaling: developmental roles and regulation. *Development* **134:** 1023–1034. doi:10.1242/dev.000166

Sheng G, Martinez Arias A, Sutherland A. 2021. The primitive streak and cellular principles of building an amniote body through gastrulation. *Science* **374:** abg1727. doi:10.1126/science.abg1727

Shook DR, Kasprowicz EM, Davidson LA, Keller R. 2018. Large, long range tensile forces drive convergence during *Xenopus* blastopore closure and body axis elongation. *eLife* **7:** e26944. doi:10.7554/eLife.26944

Shook DR, Wen JWH, Rolo A, O'Hanlon M, Francica B, Dobbins D, Skoglund P, DeSimone DW, Winklbauer R, Keller RE. 2022. Characterization of convergent thickening, a major convergence force producing morphogenic movement in amphibians. *eLife* **11:** e57642. doi:10.7554/eLife.57642

Şişli HB, Şenkal S, Hayal TB, Bulut E, Doğan A. 2023. Regulatory role of apelin receptor signaling in migration and differentiation of mouse embryonic stem cell-derived mesoderm cells and mesenchymal stem/stromal cells. *Hum Cell* **36:** 612–630. doi:10.1007/s13577-023-00861-2

Sknepnek R, Djafer-Cherif I, Chuai M, Weijer C, Henkes S. 2023. Generating active T1 transitions through mechanochemical feedback. *eLife* **12:** e79862. doi:10.7554/eLife.79862

Smith EM, Mitsia M, Nugenta MA, Symes K. 2009. PDGF-A interactions with fibronectin reveal a critical role for heparan sulfate in directed cell migration during *Xenopus* gastrulation. *Proc Natl Acad Sci* **106:** 21683–21688. doi:10.1073/pnas.0902510106

Solnica-Krezel L, Sepich DS. 2012. Gastrulation: making and shaping germ layers. *Annu Rev Cell Dev Biol* **28:** 687–717. doi:10.1146/annurev-cellbio-092910-154043

Song J, McColl J, Camp E, Kennerley N, Mok GF, McCormick D, Grocott T, Wheeler GN, Münsterberg AE. 2014. Smad1 transcription factor integrates BMP2 and Wnt3a signals in migrating cardiac progenitor cells. *Proc Natl Acad Sci* **111:** 7337–7342. doi:10.1073/pnas.1321764111

Soriano P. 1997. The PDGFα receptor is required for neural crest cell development and for normal patterning of the somites. *Development* **124:** 2691–2700. doi:10.1242/dev.124.14.2691

Srinivas S, Rodriguez T, Clements M, Smith JC, Beddington RS. 2004. Active cell migration drives the unilateral movements of the anterior visceral endoderm. *Development* **131:** 1157–1164. doi:10.1242/dev.01005

Stern CD, Downs KM. 2012. The hypoblast (visceral endoderm): an evo-devo perspective. *Development* **139:** 1059–1069. doi:10.1242/dev.070730

Stock J, Kazmar T, Schlumm F, Hannezo E, Pauli A. 2022. A self-generated Toddler gradient guides mesodermal cell migration. *Sci Adv* **8:** eadd2488. doi:10.1126/sciadv.add2488

Sun X, Meyers EN, Lewandoski M, Martin GR. 1999. Targeted disruption of Fgf8 causes failure of cell migration in the gastrulating mouse embryo. *Genes Dev* **13:** 1834–1846. doi:10.1101/gad.13.14.1834

Sutherland A, Keller R, Lesko A. 2020. Convergent extension in mammalian morphogenesis. *Semin Cell Dev Biol* **100:** 199–211. doi:10.1016/j.semcdb.2019.11.002

Sweetman D, Wagstaff L, Cooper O, Weijer C, Münsterberg A. 2008. The migration of paraxial and lateral plate mesoderm cells emerging from the late primitive streak is controlled by different Wnt signals. *BMC Dev Biol* **8:** 63. doi:10.1186/1471-213X-8-63

Szabó A, Cobo I, Omara S, McLachlan S, Keller R, Mayor R. 2016. The molecular basis of radial intercalation during tissue spreading in early development. *Dev Cell* **37:** 213–225. doi:10.1016/j.devcel.2016.04.008

Tavano S, Brückner DB, Tasciyan S, Tong X, Kardos R, Schauer A, Hauschild R, Heisenberg CP. 2025. BMP-dependent patterning of ectoderm tissue material properties modulates lateral mesendoderm cell migration during early zebrafish gastrulation. *Cell Rep* **44:** 115387. doi:10.1016/j.celrep.2025.115387

Trichas G, Smith AM, White N, Wilkins V, Watanabe T, Moore A, Joyce B, Sugnaseelan J, Rodriguez TA, Kay D, et al. 2012. Multi-cellular rosettes in the mouse visceral endoderm facilitate the ordered migration of anterior visceral endoderm cells. *PLoS Biol* **10:** e1001256. doi:10.1371/journal.pbio.1001256

Turner DA, Martinez Arias A. 2024. Three-dimensional stem cell models of mammalian gastrulation. *Bioessays* **46:** e2400123. doi:10.1002/bies.202400123

van den Brink SC, van Oudenaarden A. 2021. 3D gastruloids: a novel frontier in stem cell-based in vitro modeling of mammalian gastrulation. *Trends Cell Biol* **31:** 747–759. doi:10.1016/j.tcb.2021.06.007

Villaronga-Luque A, Savill RG, López-Anguita N, Bolondi A, Garai S, Gassaloglu SI, Rouatbi R, Schmeisser K, Poddar A, Bauer L, et al. 2025. Integrated molecular-phenotypic profiling reveals metabolic control of morphological variation in a stem-cell-based embryo model. *Cell Stem Cell* **32:** 759–777.e13. doi:10.1016/j.stem.2025.03.012

Voiculescu O, Bertocchini F, Wolpert L, Keller RE, Stern CD. 2007. The amniote primitive streak is defined by epithelial cell intercalation before gastrulation. *Nature* **449:** 1049–1052. doi:10.1038/nature06211

Voiculescu O, Bodenstein L, Lau IJ, Stern CD. 2014. Local cell interactions and self-amplifying individual cell ingression drive amniote gastrulation. *eLife* **3:** e01817. doi:10.7554/eLife.01817

von der Hardt S, Bakkers J, Inbal A, Carvalho L, Solnica-Krezel L, Heisenberg CP, Hammerschmidt M. 2007. The Bmp gradient of the zebrafish gastrula guides migrating lateral cells by regulating cell-cell adhesion. *Curr Biol* **17:** 475–487. doi:10.1016/j.cub.2007.02.013

Wallingford JB, Rowning BA, Vogeli KM, Rothbächer U, Fraser SE, Harland RM. 2000. Dishevelled controls cell polarity during *Xenopus* gastrulation. *Nature* **405:** 81–85. doi:10.1038/35011077

Wan Y, El Kholtei J, Jenie I, Colomer-Rosell M, Liu J, Acedo JN, Du LY, Codina-Tobias M, Wang M, Sawh A, et al. 2024. Whole-embryo spatial transcriptomics at subcellular resolution from gastrulation to organogenesis. bioRxiv doi:10.1101/2024.08.27.609868

Warmflash A, Sorre B, Etoc F, Siggia ED, Brivanlou AH. 2014. A method to recapitulate early embryonic spatial patterning in human embryonic stem cells. *Nat Methods* **11:** 847–854. doi:10.1038/nmeth.3016

Wen JW, Winklbauer R. 2017. Ingression-type cell migration drives vegetal endoderm internalisation in the *Xenopus* gastrula. *eLife* **6:** e27190. doi:10.7554/eLife.27190

Weng S, Devitt CC, Nyaoga BM, Alvarado J, Wallingford JB. 2025a. PCP-dependent polarized mechanics in the cortex of individual cells during convergent extension. *Dev Biol* **523:** 59–67. doi:10.1016/j.ydbio.2025.04.007

Weng S, Hayashi M, Inoue Y, Wallingford JB. 2025b. Planar polarized force propagation integrates cell behavior with tissue shaping during convergent extension. *Curr Biol* **35:** 1–10.e3. doi:10.1016/j.cub.2024.10.070

Williams M, Burdsal C, Periasamy A, Lewandoski M, Sutherland A. 2012. Mouse primitive streak forms in situ by initiation of epithelial to mesenchymal transition without migration of a cell population. *Dev Dyn* **241:** 270–283. doi:10.1002/dvdy.23711

Winklbauer R, Schürfeld M. 1999. Vegetal rotation, a new gastrulation movement involved in the internalization of the mesoderm and endoderm in *Xenopus*. *Development* **126:** 3703–3713. doi:10.1242/dev.126.16.3703

Woo S, Housley MP, Weiner OD, Stainier DY. 2012. Nodal signaling regulates endodermal cell motility and actin dynamics via Rac1 and Prex1. *J Cell Biol* **198:** 941–952. doi:10.1083/jcb.201203012

Xiong F, Ma W, Bénazéraf B, Mahadevan L, Pourquié O. 2020. Mechanical coupling coordinates the co-elongation of axial and paraxial tissues in avian embryos. *Dev Cell* **55:** 354–366.e5. doi:10.1016/j.devcel.2020.08.007

Yamaguchi TP, Harpal K, Henkemeyer M, Rossant J. 1994. fgfr-1 is required for embryonic growth and mesodermal patterning during mouse gastrulation. *Genes Dev* **8:** 3032–3044. doi:10.1101/gad.8.24.3032

Yang X, Dormann D, Münsterberg AE, Weijer CJ. 2002. Cell movement patterns during gastrulation in the chick are controlled by positive and negative chemotaxis mediated by FGF4 and FGF8. *Dev Cell* **3:** 425–437. doi:10.1016/S1534-5807(02)00256-3

Yang X, Chrisman H, Weijer CJ. 2008. PDGF signalling controls the migration of mesoderm cells during chick gastrulation by regulating N-cadherin expression. *Development* **135:** 3521–3530. doi:10.1242/dev.023416

Yoon J, Kumar V, Goutam RS, Kim SC, Park S, Lee U, Kim J. 2022. Bmp signal gradient modulates convergent cell movement via Xarhgef3.2 during gastrulation of *Xenopus* embryos. *Cells* **11:** 44. doi:10.3390/cells11010044

Zhang X, Ibrahimi OA, Olsen SK, Umemori H, Mohammadi M, Ornitz DM. 2006. Receptor specificity of the fibroblast growth factor family. The complete mammalian FGF family. *J Biol Chem* **281:** 15694–15700. doi:10.1074/jbc.M601252200

Zhu Y, Warmflash A. 2025. Dependence of cell fate potential and cadherin switching on primitive streak coordinate during differentiation of human pluripotent stem cells. bioRxiv doi:10.1101/2025.01.31.635963

Chemical and Mechanical Regulation of Leukocyte Migration

Jonathan H. Schrope,[1,2,3,4] **Tanner F. Robertson,**[2] **Milka Sarris,**[5] **and Anna Huttenlocher**[2,6]

[1]Department of Biomedical Engineering, University of Wisconsin-Madison, Madison, Wisconsin 53706, USA

[2]Department of Medical Microbiology and Immunology, University of Wisconsin-Madison, Madison, Wisconsin 53706, USA

[3]Department of Pathology and Laboratory Medicine, University of Wisconsin-Madison, Madison, Wisconsin 53705, USA

[4]Medical Scientist Training Program, University of Wisconsin-Madison, Madison, Wisconsin 53792, USA

[5]Department of Physiology, Development and Neuroscience, University of Cambridge, Cambridge CB2 3EL, United Kingdom

[6]Department of Pediatrics, University of Wisconsin-Madison, Madison, Wisconsin 53792, USA

Correspondence: ms543@cam.ac.uk; huttenlocher@wisc.edu

Directed leukocyte motility is essential for immunity and host defense. Dysregulated leukocyte migration is implicated in clinical immunodeficiency and hyperinflammatory conditions. Leukocytes sense both chemical and physical cues within the environment to regulate internal migration machinery and thus coordinate the immune response and its resolution. In response to environmental cues, leukocytes cater migration strategies to both exert forces on surrounding tissues and alter the chemical environment through self-generated gradients. Here, we synthesize recent advances in our understanding of how chemical and physical cues within the tissue environment regulate leukocyte motility, with implications to develop therapeutic strategies to modulate the immune response in human disease.

IMPORTANCE OF LEUKOCYTE MIGRATION

Cell motility plays a central role in tissue formation during development and subsequent protection through regulation of the immune response. In the context of inflammation, innate immune cells (neutrophils and macrophages) are recruited as first defenders to sites of tissue injury or infection and subsequently coordinate an adaptive immune response mediated by T and B cells. The functional purpose of different leukocyte cell types necessitates motility within distinct tissue environments. Neutrophils and monocytes are recruited through the vascular endothelial lining and then epithelial tissues by chemical gradients arising from focal tissue damage or infection. Dendritic cells and tissue resident macrophages survey peripheral tissues before migrating to lymphoid organs to present antigen to B and T cells, which migrate interstitially within lymph nodes. Fundamental to both innate and adaptive immunity is a requirement for precise regulation of leukocyte motility. Dysregulated migration manifests in

Cite this article as *Cold Spring Harb Perspect Biol* doi: 10.1101/cshperspect.a041752

human disease—both immunodeficiency and inflammatory conditions. Deficits in innate cell motility result in recurrent infections. On the other hand, excessive leukocyte recruitment is implicated in chronic inflammation and autoimmune conditions. Therefore, understanding the molecular mechanisms that regulate leukocyte motility holds potential for developing therapeutic strategies to treat human diseases. Decades of research has uncovered the molecular machinery that leukocytes use to migrate within tissues (Friedl and Weigelin 2008; Weninger et al. 2014). This has led to the development of Food and Drug Administration (FDA)-approved drugs targeting leukocyte motility to treat inflammatory conditions (Mackay 2008; Arseneau and Cominelli 2015; Chandar et al. 2015). While much focus has been placed on the chemical signaling networks that regulate this machinery, emerging evidence has identified a significant role for mechanical cues in regulating leukocyte migration. Such physical cues vary spatially within distinct organs and temporally over the course of disease. Here we outline recent advances in our understanding of how leukocytes integrate both chemical and physical cues to regulate migration with focus on the importance of understanding how leukocytes tailor migration strategies within distinct tissue contexts.

BASICS OF LEUKOCYTE MOTILITY: FORCE GENERATION AND POLARITY

At a fundamental level, cell migration is dependent on spatiotemporal regulation of substrate adhesion and intracellular force generation to propel the cell forward. Adhesion is mediated by membrane-spanning integrins, which recognize ligands present within the environment. Integrins are linked to the actin cytoskeleton intracellularly via recruitment of supporting proteins (i.e., focal adhesion kinase, vinculin, talin, paxillin, etc.) (Huttenlocher et al. 1995; Parsons et al. 2010; Chastney et al. 2021; Yamaguchi and Knaut 2022) to form mature complexes known as focal adhesions. In tandem, cells generate forces to extend forward protrusions and contract the cell rear, so that forward migration is obtained through dynamic coupling of adhesion with differential force generation across the cell. Adhesion maturation and disassembly is regulated by the push and pull forces generated through cytoskeletal dynamics (Geiger et al. 2009; Gardel et al. 2010; Parsons et al. 2010; Plotnikov et al. 2012), which are dependent on cells responding to both chemical and physical cues within the tissue environment. Therefore, while spatiotemporal cues at the tissue scale direct leukocyte motility in the context of inflammation (i.e., endogenous gradients and tissue mechanical properties), at the single-cell level, spatiotemporal regulation of intracellular signaling networks and force generation act as the basis for cell motility. While leukocytes can fit within this canonical view, they present a unique malleability in their migration strategies, including adhesion-independent mechanisms (Lämmermann et al. 2008; Pinner and Sahai 2009; Schmidt and Friedl 2010; Paluch and Raz 2013), often dependent on environmental mechanical cues. While physical and chemical regulation of cell adhesion has been extensively reviewed (Geiger et al. 2009; Gardel et al. 2010; Parsons et al. 2010; Plotnikov et al. 2012), in this section we focus on regulation of cytoskeletal dynamics. We first outline the mechanisms by which leukocytes generate protrusive forces (i.e., regulation of actin polymerization) and organize those forces over the length of the cell (i.e., polarity). Next, we discuss how the environmental context regulates the organization of such machinery (i.e., migration modes).

Actin as a Central Mediator of Cell Motility

Intracellular force generation is a fundamental requirement for motility and is achieved through the regulation of actin cytoskeletal dynamics. Actin polymerization at the leading edge generates expansive forces, contrasted by actomyosin-mediated contractile forces at the cell rear. Intracellular actin consists of monomers at very high concentration (100 μM) (Pollard et al. 2000); however, forward rate constants driving polymerization into filaments are slow. This enables precise spatiotemporal regulation by a number of actin regulatory proteins, most notably actin regulatory binding protein 2/3 (ARP2/3), which binds to existing actin filaments to nucleate branched actin networks (Goley and Welch 2006), and formins, which drive formation of linear stress fibers that link focal adhesion complexes between the front and rear of the cell (Burridge and Guilluy 2016). Since its discovery in 1994 (Machesky et al. 1994), ARP2/3 has become a central player in understanding the regulation of cell protrusions. However, multiple mechanisms exist to generate forward protrusions through actin polymerization, therefore requiring mechanisms to differentially regulate ARP2/3. This is largely accomplished through nucleation-promoting factors (NPFs) such as Wiskott–Aldrich syndrome protein (WASP, named after the immunodeficiency syndrome associated with its dysfunction) and the WASP family verprolin-homologous protein (WAVE) (Fig. 1). These NPFs link upstream signaling events to control spatiotemporal activation of ARP2/3. In general, activation of WASP drives formation of needle-like filopodia (Miki et al. 1998) and pseudopodia (Fritz-Laylin et al. 2017; Davidson et al. 2018), while WAVE triggers formation of membrane ruffles and sheet-like lamellipodia (Miki et al. 1998; Nakagawa et al. 2001). NPFs are activated by various chemical or mechanical stimuli to therefore regulate types of actin-driven protrusions. Activation can occur downstream from chemokine sensing through membrane-bound G-protein-coupled receptors (GPCRs). However, purely physical cues also regulate actin cytoskeletal dynamics. For example, membrane deformations from application of local mechanical force induce the recruitment of WASP to drive local actin polymerization (Brunetti et al. 2022) and facilitate dendritic cell migration through dense tissues (Gaertner et al. 2022). In dense tissue environments, such expansive forces must overcome confining forces from the environment. Recent work has demonstrated that under confinement, leukocytes redistribute actin to form a centralized actin "cloud" to exert pushing forces on the environment, dependent on dedicator of cytokinesis protein 8 (DOCK8)-mediated activation of Cdc42 (Reis-Rodrigues et al. 2025; Shen et al. 2025). Taken together, these studies suggest that the mechanical properties of the surrounding tissue environment regulate migration machinery. However, cytoskeletal regulation is not limited to external physical cues alone. Cell-intrinsic chemical and physical cues also regulate spatial organization of actin dynamics over the length of the cell, which we discuss in the next section.

Leukocyte Polarity: Chemical and Mechanical Regulation

Cell motility requires asymmetries in force generation across the front and rear to establish a net driving force, dependent on spatial coordination of actin dynamics within a single cell. The concept of establishing distinct processes across the length of a cell is known as polarity, which can be described at multiple levels: morphologically, driven by differential cytoskeletal processes altering membrane shape, and/or chemically, through spatial asymmetries in signaling networks that regulate the cytoskeletal dynamics at the cell front and rear. These signaling networks are coordinated by spatially variant activation of Rho-guanosine triphosphatases (GTPases), including Rac and Cdc42 at the front and Rho at the rear (Fig. 1; Sánchez-Madrid and Angel del Pozo 1999; Raftopoulou and Hall 2004; Ridley 2015; Lawson and Ridley 2018). An early step in leukocyte polarization downstream from chemoattractant signaling through GPCRs is the asymmetric localization of key signaling molecules like phosphoinositide 3-kinase (PI3K). Production of the PI3K product phosphatidylinositol (3,4,5)-trisphosphate (PtdIns(3,4,5)P_3) at the front of polarized neutrophils is a hallmark of gradient sensing induced by chemoattractants that recruit specific adaptor proteins and guanine nucleotide-exchange factors that lead to the activation of Rac at the leading edge. Rac and Cdc42 activation at the front drives ARP2/3-mediated branched-actin polymerization through the NPFs WAVE or WASP, respectively. In contrast, Rho activation at the rear drives actomyosin contractility and membrane flow to generate forward propulsion.

Although it is well established that activation of Rac and Cdc42 at the front is driven by GPCR

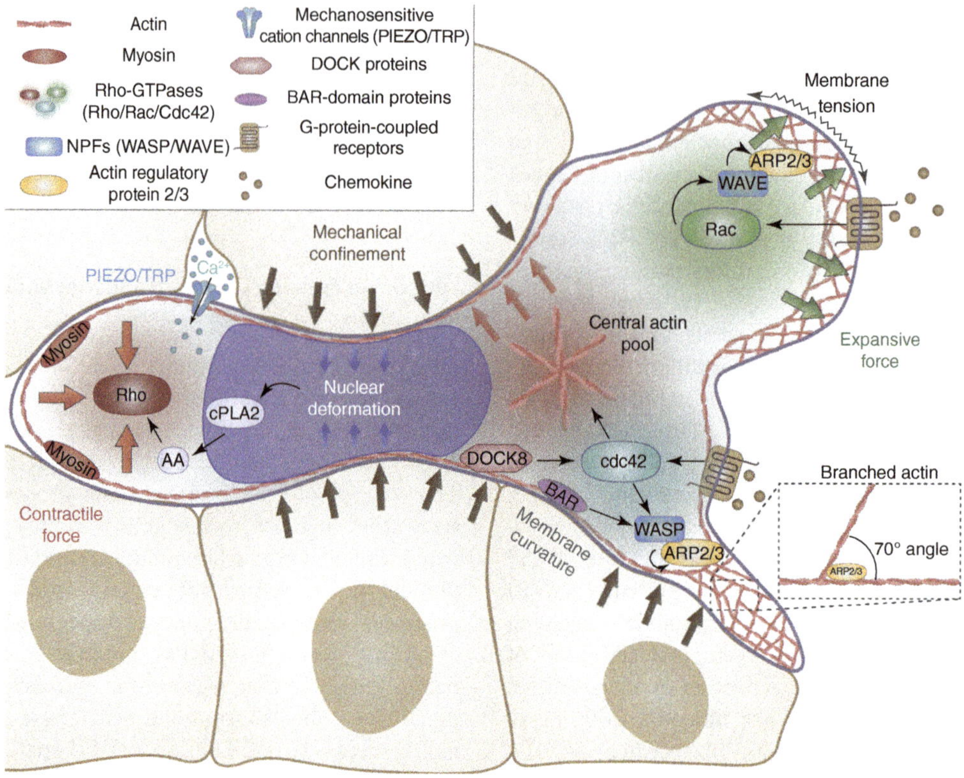

Figure 1. Summary of chemical and mechanical regulation of cell migration. Differential activation of Rho-guanosine triphosphatases (GTPases) (Rho/Rac/Cdc42) establishes polarity to generate expansive force at the front and contraction at the rear. Rac and Cdc42 at the front activate actin nucleation-promoting factors (NPFs) WASP family verprolin-homologous protein (WAVE) or Wiskott–Aldrich syndrome protein (WASP) to initiate formation of actin regulatory binding protein 2/3 (ARP2/3)-mediated branched actin networks. These pathways are regulated by G-protein-coupled receptor (GPCR) binding of chemokines at the front, or by a number of either external or internal mechanical cues (i.e., mechanical confinement, membrane curvature, and/or tension). Expansive forces increase membrane tension which in turn regulates polarity signaling. Leukocytes use multiple mechanisms to sense external physical forces. Mechanical confinement drives centralized actin polymerization (labeled central actin cloud) through dedicator of cytokinesis 8 (DOCK8)-mediated activation of Cdc42 to exert transverse pushing forces on the environment. Mechanosensitive cation channels such as PIEZO or transient receptor potentials (TRPs) open under physical confinement, driving influx of calcium from the extracellular space. BAR-domain proteins sense alterations in membrane curvature. Nuclear deformations lead to the release of arachidonic acid (AA) through cytosolic phospholipase A2 (cPLA2) activation to regulate Rho-mediated actomyosin contractility. (Figure generated with BioRender; https://biorender.com.)

recognition of chemokines, the feedback mechanisms that act to limit Rho-GTPase activation at the front, and activate Rho at the rear, continue to be a challenging question. A longstanding hypothesis was that purely chemical, often diffusion-based, mechanisms provided such feedback across the length of a cell (Jilkine et al. 2007; Mori et al. 2008; Fritz and Pertz 2016; Yang et al. 2016; Deng et al. 2024). Indeed, Ras levels can control front- and back-promoting feedback loops to affect actomyosin contractility at the rear during chemotaxis (Lin et al. 2024). However, other studies have identified a role of cell-intrinsic mechanical properties in regulating polarity during leukocyte migration. For example, membrane tension, defined as the force per unit length acting on a cross section of the membrane, regulates polarity by confining signals to the leading edge of the cell (Houk et al. 2012). In recent work, De Belly and colleagues showed that cell protrusions and contractions generate long-range membrane tension propagation (De Belly et al. 2023) and such tension regulates Rho activation at the rear (Belly et al. 2024). From an evolutionary perspective, these results provoke an appreciation for the complexity of how cells integrate both chemical and mechanical mechanisms to regulate intracellular migration machinery. While cells sense physical forces in the environment, they also sense cell-intrinsic forces to organize polarity signaling. On one hand, one might intuit that selection of such a mechanism is rooted in the rapid timescale by which mechanical cues can propagate within a cell compared to diffusion-based mechanisms of long-range inhibition. That said, Shi and colleagues found that fluctuations in membrane tension propagate in a diffusive manner (Shi et al. 2018), drawing skepticism to this rationale. In any case, these data indicate the importance of exploring how leukocytes integrate both chemical and physical signals to regulate polarity and thus motility. It also provokes the question of how cells sense intrinsic mechanical properties such as membrane tension, which we discuss in a later section.

How the Physical Environment Regulates Leukocyte Migration Strategies

In contrast to developmental processes, which require cell motility through a dynamic yet pre-established set of tissue environments, leukocytes must traverse spatiotemporally variant tissue environments throughout the body depending on the site and type of inflammation. To confront this challenge, leukocytes cater their migration strategies depending on the properties of the physical environment. Such strategies are increasingly described as either "mesenchymal," characterized by high adhesion and a dominant actin-driven mechanism of protrusion formation, versus "amoeboid," characterized by low adhesion and multiple distinct mechanisms of protrusion generation (Fig. 2). However, these terms likely represent a continuum. In general, leukocyte motility is considered amoeboid in nature, albeit cell-specific differences exist. In zebrafish, macrophages exhibit morphology typical of mesenchymal migration that is dependent on proteases (Barros-Becker et al. 2017), distinct from neutrophils, which act as a classic model of amoeboid motility. Within the amoeboid regime (i.e., neutrophils, T cells, dendritic cells), leukocytes can adopt distinct mechanisms of motility, largely in response to physical cues arising from the surrounding environment (Fig. 2). In dense tissues inducing mechanical confinement, leukocytes can migrate independent of adhesion (Lämmermann et al. 2008, Pinner and Sahai 2009; Schmidt and Friedl 2010). Following this seminal finding, it was initially unclear how cells could couple internal forces to the environment to generate forward propulsion in the absence of adhesion to a substrate. Recently, studies have demonstrated that in the absence of adhesion, cells use aberrations in environmental topologies (Reversat et al. 2020) and friction forces at the cell–substrate interface (Bergert et al. 2015) to couple intracellular forces to the environment and enable forward motility.

Environmental cues also impact mechanisms of protrusion generation within the amoeboid regime. Under confinement, neutrophils can migrate independent of ARP2/3-mediated protrusion formation (Graziano et al. 2019), transitioning to a "bleb"-based migration mode where increases in intracellular hydraulic pressure through rear contractility induce rapid extensions of cell membrane initially devoid of actin (Fackler and Grosse 2008; Bergert et al. 2012; Tyson et al. 2014; Zatulovskiy et al. 2014). As the actin cortex reforms, it serves as a scaffold for the next bleb, allowing cells to migrate through successive intermittent blebbing (Fackler and Grosse 2008; Zatulovskiy et al. 2014; Schick and Raz 2022). While such blebbing in leukocytes has traditionally been observed within in vitro models providing confinement by rigid materials, our laboratory has recently shown that physical interaction with a soft, deformable barrier like that of surrounding cells in vivo is sufficient to induce a transition to bleb-based motility in neutrophils (Schrope et al. 2025). The emerging picture is that environmental cues, specifically high confine-

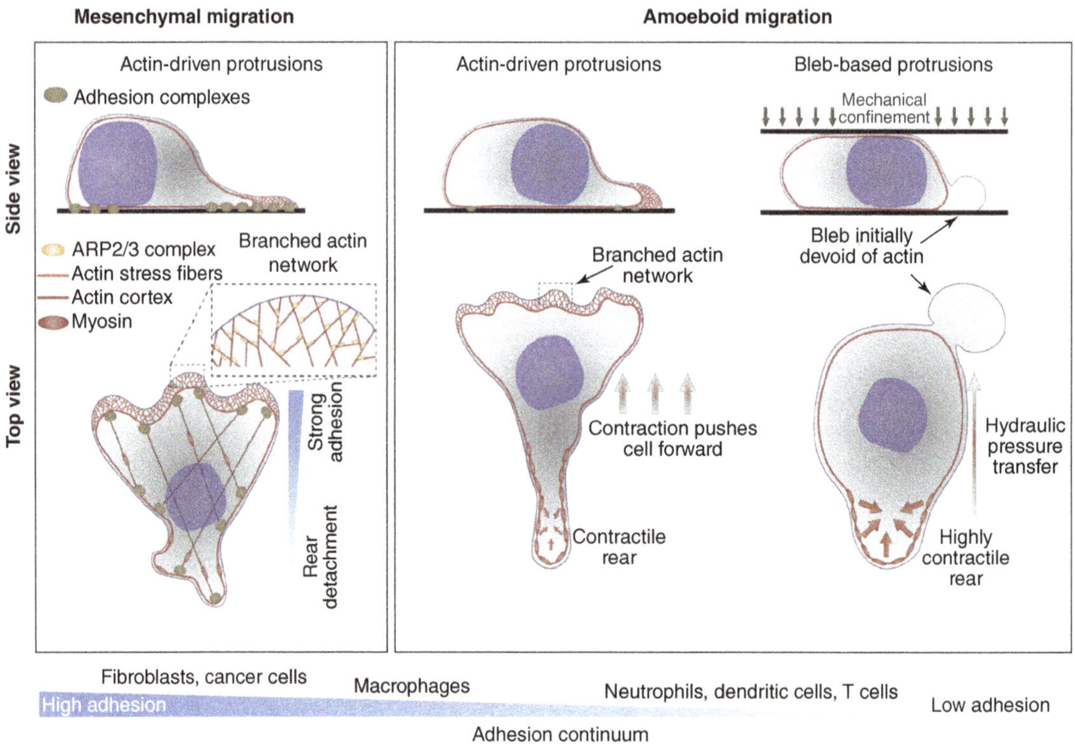

Figure 2. Spectrum of cell migration modes. Mesenchymal migration is dependent on high adhesion at the front and coordinated detachment at the rear (*left* panel). Amoeboid migration typical of most leukocytes is characterized by low adhesion and high rear contractile forces pushing the cell forward. Within the amoeboid spectrum, leukocytes can adopt different mechanisms of protrusion formation, dependent on the external environment. Under confinement, leukocytes can transition to a bleb-based mechanism of motility, whereby rear contractility provokes detachment of membrane from the actin cortex at the cell front, forming rounded protrusions initially devoid of actin.

ment and low adhesion (Liu et al. 2015), induce a transition to bleb-based motility. With that in mind, the lipid chemoattractant sphingosine-1-phosphate (S1P), which prompts T-cell egress from lymphoid tissues in the blood and lymph, elicits intermittent blebbing in naive T cells even in the absence of confinement, suggesting that both chemical and mechanical signals can regulate the choice between using F-actin-rich leading edges versus blebs (Robertson et al. 2021). Thus, the mechanisms by which leukocytes transduce environmental cues to regulate this transition between migration modes remain an intriguing question of interest.

The physiological relevance of bleb-based leukocyte motility in living organisms remains an active area of research, as observations of intermittent blebbing in vivo have largely been limited to zebrafish primordial germ cell migration (Blaser et al. 2006; Kardash et al. 2010; Goudarzi et al. 2017; Grimaldi et al. 2020). Recently, Obeidy et al. (2020) found that partial loss of ARP2/3 induced intermittent blebbing of mouse T cells transplanted into zebrafish; however, the migration modes used by leukocytes to traverse various interstitial tissue environments remain an active question. Under planar confinement in vitro, cells can also use "stable blebs" characterized by a large cytoplasmic balloon-like leading edge devoid of F-actin and maintained by retrograde actin flow and actomyosin contraction (Bergert et al. 2015;

Liu et al. 2015; Logue et al. 2015; Ruprecht et al. 2015). Historically observed in vitro, our laboratory recently found that T cells can migrate via stable blebs under planar-like confinement between keratinocyte layers in the zebrafish epidermis (Robertson et al. 2025). In all, these studies provoke further investigation of the migration strategies used by leukocytes within various tissue contexts and disease states.

In the previous sections, we have outlined emerging mechanisms by which physical cues regulate actin polymerization and thus protrusion formation at the cell front, with less focus on how external physical forces regulate Rho-mediated rear contractility, which is critical to bleb-based migration. While established that differential adhesion across the cell length plays a role in regulating actomyosin contractility at the rear (Ren et al. 1999; Parsons et al. 2010; Weißenbruch and Mayor 2024), a recent study has suggested that leukocytes sense mechanical confinement through the mechanosensitive cation channel PIEZO1 to transition to bleb-based motility (Srivastava et al. 2020). Additional evidence suggests that PIEZO-mediated calcium influx drives Rho-mediated contractility at the cell rear in *Drosophila* nephrocytes (Schulz et al. 2024), cancer cells (Pardo-Pastor et al. 2018), and human T cells (Liu et al. 2024). That said, a comprehensive understanding of how cells transduce external physical cues to regulate polarity and migration strategies remains an area of active research. In the next section, we review recent advances in our understanding of how leukocytes sense physical cues, to provide a framework for future studies to link these mechanisms of mechanosensing to internal signaling networks.

HOW LEUKOCYTES SENSE THE EXTERNAL PHYSICAL ENVIRONMENT

As discussed, leukocytes adopt different migration strategies depending on the properties of the physical environment. This prompts the question of how leukocytes sense external physical forces. Over the last decade, seminal advances have been made in our understanding of how cells sense substrate stiffness (Janmey et al. 2020; Yi et al. 2021), mainly in mesenchymal cell types. In this context, mechanical resistance applied by the environment regulates adhesion maturation, which in turn coordinates intracellular migration machinery. That said, less is understood regarding amoeboid migration typical of leukocytes. Here we discuss emerging mechanisms enabling leukocyte sensing of the physical environment.

Mechanosensitive Cation Channels

Recent work has elucidated how membrane-bound mechanosensitive cation channels enable cells to sense external forces. Leukocytes express a number of these mechanosensitive cation channels, namely, transient receptor potential (TRP) and PIEZO channels (Fig. 1). Both reside in the membrane and open in response to physical membrane perturbations to allow cation influx (primarily calcium), which regulates downstream signaling pathways and thus mechanisms of motility. TRPM7 has been implicated in directional decision-making in cancer cells (Zhao et al. 2019). Furthermore, neutrophils lacking TRPC6 exhibit impaired chemotaxis (Damann et al. 2009; Lindemann et al. 2013), and mice lacking TRPC6 exhibit an impaired allergic response (Sel et al. 2008). The discovery of PIEZO channels resulted in the 2021 Nobel Prize in Physiology or Medicine and represents an emerging target in understanding how leukocytes transduce environmental cues to regulate migration machinery. As mentioned above, PIEZO1-mediated mechanosensing regulates integrin-dependent chemotaxis in T cells (Liu et al. 2024) and the transition to bleb-based motility in the leukocyte-like amoeba *Dictyostelium* (Srivastava et al. 2020). In addition to regulating leukocyte-intrinsic motility pathways, PIEZO activity within surrounding tissues can also influence immune cell motility. For example, PIEZO-mediated mechanosensing in endothelial cells is required for neutrophil vascular egress, termed transendothelial migration (TEM) (Wang et al. 2022). In turn, PIEZO-mediated mechanosensing in neutrophils during TEM enhances bactericidal function (Mukhopadhyay et al. 2024). Following TEM, leukocytes also mechanically interact with surrounding epithelial tissues during interstitial migration (Schrope et al. 2025). Therefore, future work might focus on how physical cell–cell interactions with epithelial tissues regulate leukocyte motility and therefore how changes in tissue mechanics throughout the course of disease regulate the immune response.

Membrane Tension and Curvature Sensing

Both intra- and extracellular forces modulate membrane shape during migration. Specifically, confining forces arising from the surrounding tissue architecture resist intracellular expansion forces to define an equilibrium state that determines cell shape and membrane tension. While established that membrane tension regulates polarity, the mechanisms by which leukocytes sense this physical cue is an active area of investigation. Recent work has established that membrane tension sensing is mediated through phospholipase D2 (PLD2) and the mammalian target of rapamycin complex 2 (mTORC2) in neutrophils (Diz-Muñoz et al. 2016). Membrane tension is intricately related to alterations in membrane curvature. Leukocytes use a number of mechanisms to sense changes in local membrane curvature, including PIEZO and BAR domain proteins. The latter are characterized by a banana-shaped BAR (Bin/Amphiphysin/Rvs) domain that binds to the cell membrane. These proteins become activated by alterations in membrane curvature and act on effectors WASP and WAVE to modulate actin assembly (Fig. 1; Stanishneva-Konovalova et al. 2016; Almeida-Souza et al. 2018; Bieling et al. 2018). In leukocytes, BAR-domain proteins enable cells to sense curvature imposed by physical obstacles within the environment to regulate decisions of migration path (Sitarska et al. 2023). The F-BAR protein proline-serine-threonine phosphatase interacting protein 1 (PSTPIP1) has been shown to control matrix degradation and filopodia formation in macrophages (Starnes et al. 2014); however, it is unclear how the tissue context or intrinsic cues such as membrane tension activate PSTPIP1. Thus, the role of BAR-domain proteins in regulating tissue-specific motility in vivo remains an important area of investigation.

The Cell Nucleus as a Mechanosensor

The cell nucleus is often thought to play a passive, and often limiting, role in cell motility, as its rigidity limits migration within confined environments (Rowat et al. 2013; Shin et al. 2013; Yadav et al. 2018). That said, emerging evidence has identified a mechanosensory role of the cell nucleus to actively regulate cytoskeletal dynamics within mechanically complex environments. In addition to its role in protrusion formation at the cell front, perinuclear ARP2/3 also enables nuclear deformation (Thiam et al. 2016). DOCK8 mediates central organization of actin, which protects the nucleus in confined environments (Reis-Rodrigues et al. 2025; Shen et al. 2025). Leukocytes use forward positioning of the nucleus as a mechanosensor to determine the path of least mechanical resistance at branch points within microfluidic channels (Renkawitz et al. 2019). While much of this work has been in vitro, recent evidence in *Drosophila* suggests that leukocytes adapt their nuclear lamina in response to confining forces, which enables its deformation and preserves nuclear membrane integrity (Karling and Weavers 2025). In the event that extreme external pressures lead to nuclear rupture, endosomal sorting complexes required for transport III protein (ESCRT III) enable rapid nuclear repair mechanisms during the course of migration (Raab et al. 2016).

A key mechanism by which nuclear deformations are transduced to chemical signals has recently been identified in both cancer cells and leukocytes. Nuclear deformation induces activation of the calcium-dependent phospholipase cytosolic phospholipase A2 (cPLA2), which produces arachidonic acid (AA) to induce actomyosin contractility (Fig. 1; Lomakin et al. 2020; Venturini et al. 2020). In a zebrafish wound model, osmotic swelling of wounded epithelial cells induced activation of cPLA2, which was critical in generating chemokine signals to organize the immune response (Enyedi et al. 2016). Recently, leukocytes have been shown to use a similar mechanism, whereby cPLA2 and ARP2/3-dependent shape sensing in dendritic cells regulated C–C chemokine receptor type 7 (CCR7) surface expression and therefore homing to lymphoid tissues (Alraies et al. 2024). As noted above, leukocytes are constantly interacting with surrounding cells and matrix during interstitial migration; therefore, it will be important to understand how nuclear deformations regulate leukocyte motility through tissues in vivo.

HOW LEUKOCYTES SENSE AND GENERATE CHEMICAL GRADIENTS

The Role of Endogenous Gradients in Directing Leukocyte Migration In Vivo

While clear that the physical tissue context regulates cell motility, it has long been understood that leukocyte trafficking in contexts of both homeostasis and inflammatory states is directed by the formation of endogenous chemical gradients that direct motility. The role for such gradients has been demonstrated extensively in situ. Neutrophils sense the bacterial derived peptide *N*-formyl-methionyl-leucyl-phenylalanine (fMLP) to migrate to sites of infection (Yang et al. 2012) and migrate up gradients of CXCL8 (Sarris et al. 2012) and the lipid mediator leukotriene B_4 (LTB_4) (Afonso et al. 2012; Lämmermann et al. 2013) in the context of tissue damage. Gradients of CCR7 ligands (i.e., CCL19 and CCL21) direct dendritic cell homing to lymph nodes to present antigen to lymphocytes to link innate and adaptive processes (Braun et al. 2011; Seth et al. 2011; Weber et al. 2013; Worbs et al. 2017). Within the adaptive immune system, memory T cells also use gradients of CCR7 ligands to exit peripheral tissues via the lymphatics (Debes et al. 2005). High-to-low gradients of S1P between the circulatory fluids and the interstitial space of lymphoid tissues drive T-cell egress into the blood and lymph (Schwab and Cyster 2007). Similarly, high-to-low gradients of lipoprotein A between the high endothelial venule and lymph node interstitial space are thought to promote T-cell entry into lymph nodes (Knowlden et al. 2014). Distinct gradients of ligands for CCR7 (CCL19 and CCL21) and CXCR5 (CXCL13) are critical for the spatial segregation of T and B cells within lymphoid tissues (Ngo et al. 1998; Förster et al. 1999; Ansel et al. 2000). The dynamic regulation of CCR7 and CXCR5 enables the movement of activated T and B cells during germinal center formation (Ansel et al. 2000; Reif et al. 2002; Okada et al. 2005). Chemokine gradients are similarly thought to be critical for the infiltration of leukocytes into solid tumors (Kohli et al. 2022). Nonchemoattractant gradients can also influence leukocyte motility. For example, gradients of H_2O_2 generated in the context of focal tissue injury regulate neutrophil recruitment in zebrafish (Niethammer et al. 2009; Yoo et al. 2011). While much of these findings are based in genetic knockout of chemokine ligands or receptors, recent work has introduced genetically encoded reporters to directly visualize chemoattractant gradients, such as LTB_4 (Tamás et al. 2023) and S1P (Ramos-Perez et al. 2015), in live tissues.

How Cells Sense Endogenous Gradients

Following the well-known video capture of neutrophil migration toward bacterial cells by David Rogers in the 1950s, the mechanisms by which leukocytes sense chemoattractant gradients to direct motility have been a major focus of research. The primary effect of chemoattractant sensing is establishing cell polarity and regulation of actin dynamics. Differential signal activation at the front versus rear regulates cytoskeletal dynamics to establish distinct mechanisms of force generation across the length of a cell. The mechanisms by which leukocytes transduce chemical gradients to regulate actin dynamics have been extensively reviewed (Snyderman and Goetzl 1981; Zigmond et al. 1981; Iijima et al. 2002; Swaney et al. 2010). Briefly, membrane-spanning GPCRs recognize chemoattractant ligands to regulate PI3K, Rac, and Cdc42 at the leading edge to generate protrusions toward regions of greatest ligand concentration. In vivo, neutrophils resolve gradients to direct migration path through a two-step search and run process (Georgantzoglou et al. 2022). That said, the generation of cytoskeletal polarity can be independent of local gradient sensing (i.e., in the case of uniform chemoattractant), through spontaneous fluctuations (Wedlich-Soldner and Li 2003), or in response to mechanical cues (i.e., confinement, membrane tension). In this case, decisions of migration path can be distinct from polarity machinery. For example, generating local gradients at the cell rear of already polarized cells by micropipette results in cell turning rather than reversing polarity (Zigmond et al. 1981; Swaney et al. 2010). In zebrafish, photoactivation of Rac can steer neutrophil directionality in cells with already established polarity (Yoo et al. 2010). Blocking prior polarity and actin dynamics prevents gradient responses in zebrafish neutrophils (Georgantzoglou et al. 2022). This evidence suggests that gradients can act as modulators of a pre-established polarity circuit, which is relevant to situations where cells are already migratory when encountering an established gradient in vivo.

To put the complexities of leukocyte gradient sensing into perspective, it is useful to compare to strategies used by distant prokaryotic ancestors, per Sarris and Sixt (2015). Devoid of a dynamic actin cytoskeleton, many bacterial species use a flagellar mechanism of motility. In the presence of a gradient, redirection of flagellar orientation drives motility toward regions of greatest ligand concentration. Lack of integration between chemokine sensing and cytoskeletal machinery, in addition to the small size of prokaryotes, limits their ability to resolve gradients across the length of a single cell, therefore limiting the spatial resolution of turning capacity. In contrast, reliance on cytoskeletal machinery for protrusion formation in eukaryotes enables leukocytes to resolve gradients at very local regions across the cell membrane and extend multiple protrusions in the face of a single chemical gradient. This is particularly useful for circumventing physical obstacles present in vivo (i.e., surrounding cells). We have previously discussed the mechanisms by which leukocytes probe the physical properties of distinct migration paths, through BAR-domain proteins (Sitarska et al. 2023) and the cell nucleus (Renkawitz et al. 2019). It is likely that evolution of such a mechanism of eukaryotic gradient sensing was partially driven by selection pressures to enable sensing physical properties of surrounding tissues. An emerging question is how cells integrate competition between chemical (i.e., GPCR activation) and mechanical cues (i.e., resistance and/or confinement) to resolve directional decisions. For example, how much receptor activation is required to overcome a path of incrementally greater mechanical resistance? Answers to questions like this would not only advance our understanding of the fundamental mechanisms that regulate immunity but might also confer therapeutic potential. For example, while the development of CAR-T cells has been revolutionary in treating hematological malignancies, success in solid tumors remains limited. This is partially explained by an inability to penetrate dense tissue environments (Lanitis et al. 2017; Golo et al. 2024). While T cells elect the path of least mechanical resistance (Martinelli et al. 2014; Renkawitz et al. 2019), it is likely that enhanced GPCR sensitivity to endogenous gradients emanating from tumors might override such mechanosensing mechanisms to enable deeper penetration into solid tumors.

Self-Generated Leukocyte Gradients

Just as leukocytes possess strategies to deform and degrade surrounding physical barriers, they also possess mechanisms to alter the endogenous chemical environment by generating their own gradients. Self-generated gradients can either amplify recruitment of surrounding leukocytes in a paracrine manner (i.e., neutrophil swarming) or steer the cell itself through generation of a local gradient by degradation of a uniformly distributed global chemoattractant (Fig. 3); here we review recent advances in both of these mechanisms.

Neutrophil swarming refers to a paracrine mechanism by which motile neutrophils secrete chemoattractants to direct the migration of surrounding neutrophils (Fig. 3A; Chtanova et al. 2008; Ng et al. 2011). Among neutrophil chemoattractants, the lipid mediator LTB_4 has the most prominent role in long-range recruitment (Afonso et al. 2012; Lämmermann et al. 2013); however, CXCR1/2 ligands also play a role in this process (Lämmermann et al. 2013; Reátegui et al. 2017). This self-amplifying recruitment results in convergence of an immense number of leukocytes on rapid timescales to mediate bacterial and fungal clearance (Lämmermann et al. 2013; Hopke et al. 2020, 2022). Studies in

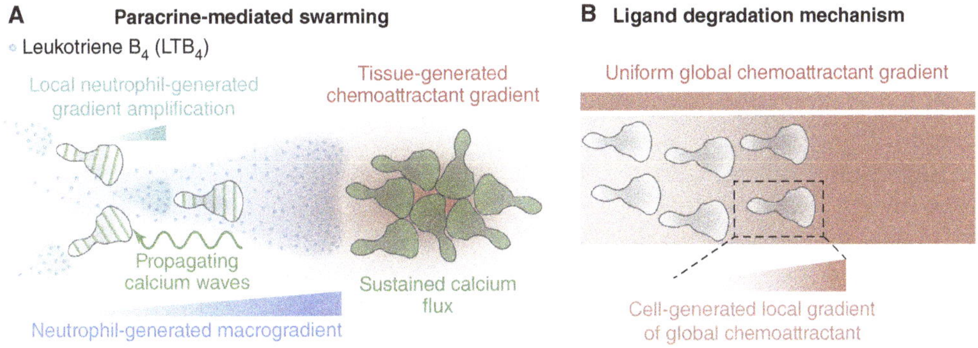

Figure 3. Leukocyte self-generated gradients. (*A*) Neutrophil swarming describes the paracrine-mediated process whereby neutrophil-generated gradients enable long-range signal amplification and thus rapid recruitment to sites of inflammation. Sustained calcium flux (green) in arrested cells promotes activation of 5-lipoxygenase (5-LPO) to produce leukotriene B_4 (LTB_4), corresponding with generation of central macrogradients (blue) of LTB_4 or other 5-LPO products and derivatives. In addition, migrating cells exhibit intermittent calcium fluxes of variable frequency that propagate as waves traveling radially from the central site of inflammation, consistent with intercellular relay through production of local gradients (cyan). (*B*) In a uniform chemoattractant gradient, the leukocyte-like soil amoeba *Dictyostelium* can degrade chemoattractant at the cell front to establish a local gradient across the cell front to enable chemotaxis.

zebrafish (Poplimont et al. 2020) and ex vivo models of human neutrophils (Khazen et al. 2022) have demonstrated that chemoattractant dynamics are governed by two distinct calcium signaling patterns in pioneer cells. (1) Coordinated and relatively sustained calcium fluxes have been observed in the arrested clustering cells, which promote activation of nuclear compartments of 5-lipoxygenase (5-LPO), the rate-limiting enzyme of LTB_4 production (Poplimont et al. 2020). These calcium signals appear consistent with the generation of central chemoattractant macrogradients by the core-clustering cells. (2) Propagating calcium waves emanating across the migratory cells have been additionally observed with a frequency that varies with neutrophil density and target size (Strickland et al. 2024). These waves are consistent with local attractant gradients relayed among the responding migrating cells and appear to be self-limited through reactive oxygen species (ROS) signaling and GPCR desensitization to limit excessive recruitment (Kienle et al. 2021; Strickland et al. 2024). The emerging picture is that spatial gradients from the core-clustering cells and attractant relays may cooperate to shape neutrophil swarms: While spatial macrogradients provide tissue-scale directional information, signal relays enable long-range recruitment through gradient amplification. It remains unclear how chemoattractant release can be coordinated in these distinct fashions. One possibility is that the core cells secrete specific subpools of LTB_4 (originating from nuclear 5-LPO compartments) along with other ligands, such arachidonic acid or CXCR1/2 ligands, providing a basal cocktail of tissue-scale spatial gradients. Overlaid to this, propagating relays might originate from alternative cellular pools of LTB_4. Interestingly, LTB_4 biosynthetic machinery has been shown to be packaged in nuclear-derived exosomes within multivesicular bodies (Arya et al. 2022). It remains to be explored whether these subcellular pools could provide stores of soluble LTB_4 that could be released rapidly upon calcium activation and diffuse in timescales compatible with the observed calcium waves (Strickland et al. 2024). Aside from the coordination of attractant synthesis, an additional level of cooperation lies in the transcellular supply of LTB_4 precursors. Specifically, mouse neutrophils deficient in 5-LPO or LTA_4H are deficient in swarming but can complement each other when combined, suggesting production of these mediators can be shared across the population (Hopke et al. 2022).

The triggers of neutrophil swarming are also not fully understood. Observations in zebrafish have demonstrated that pioneer neutrophils (the first to recognize an endogenous gradient) release chromatin reminiscent of neutrophil extracellular trap (NET) release (Isles et al. 2021), a neutrophil-specific effector function, which might explain cell-type specificity. Furthermore, it is unclear whether physical cues also play a role in neutrophil swarming. Given the role of calcium signaling in LTB_4 production, which increases through PIEZO and TRP channel opening under mechanical stress, future work is needed to understand how tissue physical properties might regulate swarming. It may be that multiple types of chemical and physical triggers converge to generate neutrophil swarms. The broader relevance of these cooperative guidance mechanisms will be interesting to explore in other leukocyte types. Cytotoxic T cells swarm via homotypic chemokine secretion, reminiscent of the CXCL1/2/8-driven neutrophil swarms (Galeano Niño et al. 2020). T cells can also move in a cooperative and collective fashion through a T-cell network in zebrafish that is likely mediated by both mechanical and chemical mechanisms (Robertson et al. 2023). In addition, neutrophil-derived attractants can seed monocyte swarming, suggesting further levels of cooperation across cell types (Walters et al. 2021).

Distinct from the concept of paracrine-mediated gradient generation, whereby leukocytes themselves secrete chemoattractant to direct the motility of nearby leukocytes as above, cells can also locally alter global gradients to enhance chemotactic ability. This addresses a central problem in chemotaxis; as gradients are applied over longer distances, the chemokine concentration difference experienced by the front and rear of the cell becomes smaller and limits effective chemotaxis. In response to this limitation, a number of cell types have evolved mechanisms to alter and sharpen exogenous chemokine gradients. Seminal studies in the larval zebrafish discovered that neural progenitor cells within the lateral line primordium generate local gradients through the breakdown of chemoattractant CXCL12a at the collective front (Donà et al. 2013) and sequestration of stromal-derived factor 1 (SDF1) (Venkiteswaran et al. 2013) at the collective rear to enable directed motility throughout the organism in early development. A similar mechanism has been observed in melanoma cells, whereby local degradation of the chemoattractant lysophosphatidic acid (LPA) within a globally uniform gradient enables forward chemotaxis in vitro (Muinonen-Martin et al. 2014). Critically, this has recently been demonstrated in the leukocyte-like amoeba *Dictyostelium*, which can locally degrade a chemoattractant within a uniform gradient to efficiently migrate through complex microfluidic mazes (Fig. 3B; Tweedy et al. 2020). Intriguingly, this mechanism of local degradation-driven chemotaxis is more robust than that within externally established exogenous gradients (Tweedy et al. 2016; Tweedy and Insall 2020). Dendritic cells can similarly use CCR7 to both sense and "sink" (internalize) CCL19 to self-shape exogenous gradients and migrate over considerable distances (Alanko et al. 2023). Altogether, these works have identified a fundamentally new mechanism of how leukocytes might orchestrate path-finding in the absence of a guiding exogenous gradient. It is commonly thought that saturation of GPCR binding in the context of high chemoattractant concentration limits recruitment or can induce reverse migration away from wounds in the context of focal inflammation (reviewed in the next section). However, this provides a mechanism that might drive continued (and possibly excessive) leukocyte recruitment by degradation of chemoattractant at the cell front. Further work is required to establish which leukocytes might use this strategy and whether it is relevant to regulating migration in vivo.

THE RESOLUTION OF INFLAMMATION: REVERSE MIGRATION

The Role of Reverse Migration in the Resolution of Inflammation

Thus far, we have discussed how leukocytes sense chemical and physical cues to regulate motility, largely in the context of migration toward a site of inflammation. However, these concepts also hold in regard to migration away from injury or infection, implicated in the resolution of inflammation (de Oliveira et al. 2016; Rocha-Gregg and Huttenlocher 2021). Leukocytes recruited to sites of inflammation establish tissue-scale systems to initiate repair and pathogen clearance. However, excessive recruitment and accumulation of innate immune cells such as neutrophils can drive superfluous damage through production of ROS and formation of NETs (termed NETosis), both implicated in chronic wound healing and autoimmune disease. Under the longstanding traditional view, neutrophil clearance at the site of inflammation is mediated through apoptosis and phagocytosis by other immune cells (i.e., macrophages) (Silva 2011; Buckley et al. 2013). However, over the last 20 years, a novel mechanism of resolution has emerged, whereby neutrophils mediate their own clearance by migrating away from regions of focal inflammation in a process termed "reverse migration" (Fig. 4). This process was first suggested by work from Hughes and colleagues who tracked the fate of radiolabeled neutrophils in mice and found that >70% of neutrophils that entered inflamed glomerular capillaries returned to the circulation (Hughes et al. 1997). Neutrophil migration away from a wound was first directly visualized in zebrafish (Mathias et al. 2006), with subsequent studies in the same model establishing that neutrophils can reenter the vasculature and traffic to other organs (Hall et al. 2009; Yoo and Huttenlocher 2011). Parallel studies provided evidence for human neutrophil reverse transmigration into blood vessels in vitro (Buckley et al. 2006), and in a murine model of muscle inflammation (Woodfin et al. 2011). Subsequent photoconversion experiments in mice provided further evidence that neutrophils can migrate from inflamed liver injury sites to other tissues such as the lung and bone marrow (Wang et al. 2017).

Regarding the physiological implications of this phenomenon, studies have suggested that reverse migration not only functions to reduce neutrophil accumulation at the site of inflammation but can impact adaptive immunity. Duffy and colleagues identified that reverse migration from the dermis to the bone marrow initiates a source of memory $CD8^{+}$ T cells (Duffy et al. 2012). Recent evidence suggests that this mechanism is not confined to neutrophils alone, but also used by macrophages; mouse Kupffer cells in the liver reverse migrate into liver sinusoids in neonatal sepsis and meningitis (Araujo David et al. 2024). Notably, emerging evidence suggests that reverse migration might also have a detrimental effect by disseminating inflammation (Ji et al. 2024). Thus, while reverse migration has been observed a number of times within zebrafish (Elks et al. 2011; Robertson et al. 2014; Tauzin et al. 2014; Ellett et al. 2015), and mice

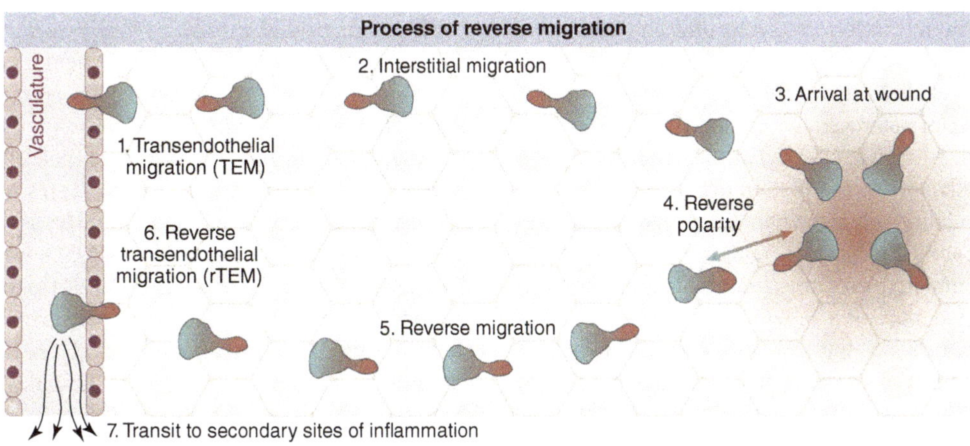

Figure 4. Process of reverse migration. Following transendothelial migration (TEM) and recruitment to a site of tissue injury through interstitial tissues, neutrophils can reverse migrate away from the wound and enter the vasculature for transit to secondary sites. This process has been implicated in the resolution of inflammation.

(Woodfin et al. 2011; Duffy et al. 2012; Wang et al. 2017), identifying the precise physical and molecular mechanisms that regulate this process remains an active area of research. Identifying such mechanisms might enable novel therapeutic strategies to regulate inflammation and its resolution in human disease.

Potential Mechanisms of Reverse Migration

While studies have identified characteristics of reverse migrated neutrophils, the precise mechanisms that regulate this process remain poorly understood. In mice, reduced endothelial cell expression of the junctional adhesion molecule JAM-C facilitates neutrophil reverse transmigration into the vasculature (Woodfin et al. 2011). Such reduction of junctional adhesion molecule-C (JAM-C) occurs through LTB_4-stimulated proteolytic cleavage by neutrophil elastase (NE) (Colom et al. 2015). Reverse migration of Kupffer cell macrophages is dependent on macrophage inhibitory factor (MIF), its cell surface receptor CD74, and the adhesion molecule CD44 (Araujo David et al. 2024). In the zebrafish, activation of hypoxia-inducible factor-1α (HIF-1α) delayed the resolution of inflammation by reducing apoptosis and reverse migration (Elks et al. 2011). More recent work has identified a role for myeloid-derived growth factor (MYDGF) and its interaction with HIF-1α in resolution of neutrophil inflammation through effects on reverse migration (Houseright et al. 2021). In the wake of emerging evidence of heterogeneity in innate leukocytes (Ng et al. 2019; Xie et al. 2020), it remains unclear whether reverse migration occurs within a subpopulation of leukocytes preprogrammed for this ability, or whether temporal changes in environmental cues reprogram polarity to initiate reverse migration.

One possibility is that the establishment of a secondary, competing gradient emanating from the vasculature recruits leukocytes away from the wound. However, this would not explain how "naive" neutrophils encountering both gradients for the first time continue to arrive at the wound while others have reverse migrated. Rather, this would suggest dynamic reprogramming of the leukocyte-intrinsic response to an established gradient emanating from a site of tissue injury. In vitro studies have found that both neutrophils and T cells can exhibit retrotaxis, that is, the counterintuitive migration toward lower concentrations of an established chemoattractant gradient (Boneschansker et al. 2014). This is typically observed at high chemoattractant concentrations, particularly of complement component 5a (c5a) and CXCL8 (Tharp et al. 2006; Boneschansker et al. 2014). Retrotaxis was enhanced by the presence of lipoxin A4 (LXA4), an established mediator of inflammation resolution (Hamza et al. 2014). These data suggest that increased chemoattractant production at a wound site might trigger a reverse in cell polarity and counterintuitively drive chemotaxis away from the wound. That said, retrotaxis down an established gradient is yet to be confirmed in vivo, with some studies suggesting that neutrophils exhibit more random (i.e., chemokinetic) motility when undergoing reverse migration in zebrafish (Holmes et al. 2012). Wang et al. (2017) found that neutrophils exhibit random, chemokinetic, motility within the wound, but then transition to directional motility at the wound edge (Hind and Huttenlocher 2018), suggesting a combination of these phenomena.

A large body of evidence suggests that receptor desensitization and/or internalization plays a role in neutrophil-intrinsic reprogramming, which might drive a transition from chemotaxis to chemokinetic or retrotaxic behavior. Human neutrophils undergoing reverse transendothelial migration through an in vitro endothelial monolayer exhibit decreased surface expression of the CXCL8 receptor CXCR1 (Buckley et al. 2006). Both CXCR2 and zebrafish CXCL8 were demonstrated to regulate neutrophil reverse migration in zebrafish by promoting neutrophil chemokinetic motility (Powell et al. 2017). Coombs et al. demonstrated that CXCR2 is able to promote bidirectional motion in response to CXCL8b through the ability of CXCR2 to remain sustained at the cell surface following ligand exposure (Coombs et al. 2019). Consistent with this, in an acute lung injury mouse model, reverse migrated neutrophils exhibited increased expression of C–C motif chemokine receptor-like 2 (CCRL2) (Ji et al. 2024), which is required for CXCR2-mediated neutrophil migration in mouse models of chronic inflammation (Del Prete et al. 2017). On the other hand, other receptors (CXCR1) promoting recruitment to alternative inflammatory chemokine ligands (CXCL8a) are downregulated at the cell surface, and blocking their desensitization and internalization led to unidirectional motion and enhanced clustering (Coombs et al. 2019). This suggests that reverse migration requires certain subsets of chemoattractant receptors to shut down signaling and others to sustain signaling. The reasons why certain receptors might support bidirectional motion and other persistent motion might have to do with the distributions of the corresponding ligands or differences in signaling downstream from the receptors. Intriguingly, Dowdell and colleagues recently demonstrated a counterintuitive mechanism by which gradients of competing chemoattractants arising from the same source and recognized by the same receptor can induce reverse chemotaxis in *Dictyostelium* cells, mediated by differential degrees of ligand degradation (Dowdell et al. 2023). It is also possible that interaction with other immune cells reprograms neutrophils to initiate reverse motility. Physical contact with macrophages has been found to trigger motility away from wound sites in zebrafish (Tauzin et al. 2014; Miskolci et al. 2021). Macrophages have also been found to "cloak" wounds thereby limiting chemoattractant signaling and neutrophil inflammation in mice (Uderhardt et al. 2019).

Notably, these theories are by no means mutually exclusive, and it is likely that a comprehensive mechanistic understanding of reverse migration will contain components of each. Given the abundance of evidence that tissue mechanical properties influence leukocyte polarity and motility mechanisms, reviewed above, it is also possible that alterations in tissue physical properties at sites of tissue injury or infection might contribute to reverse migration. It is known that tissue architecture changes following focal injury, including initial epithelial cell separation or "fracking," followed by contraction (Casares et al. 2015; Dumortier et al. 2019; Kennard et al. 2023). In light of the fact that physical cues such as progressive mechanical confinement induces neutrophil retrotaxis (Wang et al. 2018), it is possible that epithelial tissue dynamics play a role in reverse migration. Further identification of the chemical and physical cues that regulate leukocyte migration might enable the development of therapeutics to modulate the resolution of inflammation in the context of human disease.

IMPLICATIONS TO HUMAN DISEASE: THERAPEUTICS TARGETING LEUKOCYTE MOTILITY

Thus far, we have discussed the mechanisms by which leukocytes sense and generate both physical forces and chemokine gradients to regulate motility. Deficiencies in either of these processes manifest clinically as primary immunodeficiency syndromes. These are characterized by high morbidity and mortality, and often curable only by bone marrow transplant (BMT). Mutations in genes that regulate leukocyte actin dynamics are termed "actinopathies," and lead to recurrent infections in childhood, resulting from impaired leukocyte recruitment (Kamnev et al. 2021). For example, deficiencies in WASP result in Wiskott–Aldrich syndrome (Massaad et al. 2013), manifesting as recurrent infections early in childhood and limiting life expectancy in the absence of BMT. Loss of DOCK8, a regulator of Cdc42, manifests in a similar manner with high mortality, curable only by BMT. Mutations in moesin, which links the actin cortex to the plasma membrane, results in X-linked moesin-associated immunodeficiency (XMAID), a primary immunodeficiency characterized by lymphopenia and defective T-cell migration (Lagresle-Peyrou et al. 2016; Bradshaw et al. 2018; Avery et al. 2022). Other mutations regulate leukocyte chemokine sensing. For example, loss of the chemokine receptor CXCR4 leads to abnormal leukocyte trafficking resulting in warts, hypogammaglobulinemia, infections, and myelokathexis (WHIM) syndrome (Kawai and Malech 2009).

Conversely, hyperactivation and excessive leukocyte accumulation is a defining feature of autoimmune diseases such as rheumatoid arthritis, multiple sclerosis, systemic lupus erythematosus, and inflammatory bowel disease (i.e., Crohn's and ulcerative colitis). Emerging evidence suggests that dysregulation of leukocyte motility is also associated with chronic obstructive pulmonary disease and atherosclerosis (Baker and Donnelly 2021; Brassington et al. 2022; Mauersberger et al. 2022). Efforts to elucidate the fundamental mechanisms that regulate leukocyte motility have resulted in several FDA-approved pharmaceutical agents to treat autoimmune disorders. These largely target mechanisms of integrin-mediated adhesion. Early examples include inhibition of CD11a (integrin αL) for the treatment of psoriasis (Dedrick et al. 2002) albeit its use has largely been replaced by tumor necrosis factor (TNF)-α inhibitors. More recently, inhibitors of α4 integrin through the monoclonal antibodies vedolizumab and natalizumab have demonstrated clinical efficacy for treatment of Crohn's disease, ulcerative colitis, and multiple sclerosis (Chandar et al. 2015). The S1P receptor modulator fingolimod has been used for the treatment of multiple sclerosis to decrease T-cell recruitment to the brain (Brinkmann et al. 2010). Other agents remain under investigation, including those targeting β7 integrins through etrolizumab, CCR9 and CXCL10 with vercirnon and BMS-936557, and S1P with RPC 1063. Such translational work represents accomplishments in understanding mechanisms of leukocyte migration through basic science endeavors and motivates further understanding of how leukocytes regulate motility within various tissue contexts.

DISCUSSION/CONCLUSION

Here, we have outlined recent advances in our understanding of how leukocytes integrate both chemical and physical cues to regulate motility. Until recently, our understanding of leukocyte migration has largely been constrained to a chemical context. The impact of physical cues in regulating motility has been a relatively recent focus within the field. Within any disease context, changes in the physical tissue environment occur. The relevance of these changes has become increasingly emphasized in the field of oncology. Decades of work have identified the role of physical changes that occur in the context of malignancy in regulating motility of metastatic cells, including reorganizations in extracellular matrix (ECM) fiber alignment (Ray and Provenzano 2021) and generalized tissue stiffening (Zhang and Reinhart-King 2020). Extensive in vivo validation of these concepts has resulted in a myriad of pharmaceutical agents that target physical tissue properties (Lampi and Reinhart-King 2018). While the role of the tissue environment in regulating leukocyte motility has largely been confined to in vitro studies, these advances in oncology provide a clear pathway forward. Success will likely hinge on the development of in vitro models that better recapitulate physical properties in vivo (Park et al. 2019; Virumbrales-Muñoz et al. 2020; Schrope et al. 2025) and use of animal models that enable characterization and modulation of tissue properties in vivo (Barriga et al. 2018; Marchant et al. 2022). In comparison to cancer metastasis, capture of leukocyte migration in vivo is hampered by the rapid timescale of motility. Therefore, strategies to improve visualization within mammalian models (Patel et al. 2022; Ou et al. 2024) and use of other model systems that offer greater optical access (i.e., zebrafish, *Drosophila*, and *Xenopus*) and genetic tractability will likely prove critical to this effort. Understanding how tissue context regulates leukocyte motility will enable development of therapeutic strategies to better control the immune response in the context of human disease.

ACKNOWLEDGMENTS

J.H.S. was supported by a National Institutes of Health (NIH) Ruth L. Kirschstein National Research Service Award (NRSA) Fellowship 1F30HL174128. T.F.R. was supported by an NIH Ruth L. Kirschstein Postdoctoral Individual National Research Service Award F32GM146398. M.S. was supported by the Leverhulme Trust (grant RPG-2021-226) and the European Research Council (ERC) under the Horizon 2020 program and UKRI, grant agreement No. EP/Y02799X/1. A.H. was supported by an NIH MIRA R35GM118027.

REFERENCES

Afonso PV, Janka-Junttila M, Lee YJ, McCann CP, Oliver CM, Aamer KA, Losert W, Cicerone MT, Parent CA. 2012. LTB4 is a signal-relay molecule during neutrophil chemotaxis. *Dev Cell* **22:** 1079–1091. doi:10.1016/j.devcel.2012.02.003

Alanko J, Uçar MC, Canigova N, Stopp J, Schwarz J, Merrin J, Hannezo E, Sixt M. 2023. CCR7 acts as both a sensor and a sink for CCL19 to coordinate collective leukocyte migration. *Sci Immunol* **8:** eadc9584. doi:10.1126/sciimmunol.adc9584

Almeida-Souza L, Frank RAW, García-Nafría J, Colussi A, Gunawardana N, Johnson CM, Yu M, Howard G, Andrews B, Vallis Y, et al. 2018. A flat BAR protein promotes actin polymerization at the base of clathrin-coated pits. *Cell* **174:** 325–337.e14. doi:10.1016/j.cell.2018.05.020

Alraies Z, Rivera CA, Delgado M-G, Sanséau D, Maurin M, Amadio R, Maria Piperno G, Dunsmore G, Yatim A, Lacerda Mariano L, et al. 2024. Cell shape sensing licenses dendritic cells for homeostatic migration to lymph nodes. *Nat Immunol* **25:** 1193–1206. doi:10.1038/s41590-024-01856-3

Ansel KM, Ngo VN, Hyman PL, Luther SA, Förster R, Sedgwick JD, Browning JL, Lipp M, Cyster JG. 2000. A chemokine-driven positive feedback loop organizes lymphoid follicles. *Nature* **406:** 309–314. doi:10.1038/35018581

Araujo David B, Atif J, Vargas E Silva Castanheira F, Yasmin T, Guillot A, Ait Ahmed Y, Peiseler M, Hommes JW, Salm L, Brundler MA, et al. 2024. Kupffer cell reverse migration into the liver sinusoids mitigates neonatal sepsis and meningitis. *Sci Immunol* **9:** eadq9704. doi:10.1126/sciimmunol.adq9704

Arseneau K, Cominelli F. 2015. Targeting leukocyte trafficking for the treatment of inflammatory bowel disease. *Clin Pharmacol Ther* **97:** 22–28. doi:10.1002/cpt.6

Arya SB, Chen S, Jordan-Javed F, Parent CA. 2022. Ceramide-rich microdomains facilitate nuclear envelope budding for non-conventional exosome formation. *Nat Cell Biol* **24:** 1019–1028. doi:10.1038/s41556-022-00934-8

Avery L, Robertson TF, Wu CF, Roy NH, Chauvin SD, Perkey E, Vanderbeck A, Maillard I, Burkhardt JK. 2022. A murine model of X-linked moesin-associated immunodeficiency (X-MAID) reveals defects in T cell homeostasis and migration. *Front Immunol* **12:** 726406. doi:10.3389/fimmu.2021.726406

Baker JR, Donnelly LE. 2021. Leukocyte function in COPD: clinical relevance and potential for drug therapy. *Int J Chron Obstruct Pulmon Dis* **16:** 2227–2242. doi:10.2147/COPD.S266394

Barriga EH, Franze K, Charras G, Mayor R. 2018. Tissue stiffening coordinates morphogenesis by triggering collective cell migration in vivo. *Nature* **554:** 523–527. doi:10.1038/nature25742

Barros-Becker F, Lam PY, Fisher R, Huttenlocher A. 2017. Live imaging reveals distinct modes of neutrophil and macrophage migration within interstitial tissues. *J Cell Sci* **130:** 3801–3808. doi:10.1242/jcs.206128

Belly HD, Gallen AF, Strickland E, Estrada DC, Zager PJ, Burkhardt JK, Turlier H, Weiner OD. 2024. Long range mutual activation establishes Rho and Rac polarity during cell migration. bioRxiv doi:10.1101/2024.10.01.616161v1

Bergert M, Chandradoss SD, Desai RA, Paluch E. 2012. Cell mechanics control rapid transitions between blebs and lamellipodia during migration. *Proc Natl Acad Sci* **109:** 14434–14439. doi:10.1073/pnas.1207968109

Bergert M, Erzberger A, Desai RA, Aspalter IM, Oates AC, Charras G, Salbreux G, Paluch EK. 2015. Force transmission during adhesion-independent migration. *Nat Cell Biol* **17:** 524–529. doi:10.1038/ncb3134

Bieling P, Hansen SD, Akin O, Li TD, Hayden CC, Fletcher DA, Mullins RD. 2018. WH2 and proline-rich domains of WASP-family proteins collaborate to accelerate actin filament elongation. *EMBO J* **37:** 102–121. doi:10.15252/embj.201797039

Blaser H, Reichman-Fried M, Castanon I, Dumstrei K, Marlow FL, Kawakami K, Solnica-Krezel L, Heisenberg CP, Raz E. 2006. Migration of zebrafish primordial germ cells: a role for myosin contraction and cytoplasmic flow. *Dev Cell* **11:** 613–627. doi:10.1016/j.devcel.2006.09.023

Boneschansker L, Yan J, Wong E, Briscoe DM, Irimia D. 2014. Microfluidic platform for the quantitative analysis of leukocyte migration signatures. *Nat Commun* **5:** 4787. doi:10.1038/ncomms5787

Bradshaw G, Lualhati RR, Albury CL, Maksemous N, Roos-Araujo D, Smith RA, Benton MC, Eccles DA, Lea RA, Sutherland HG, et al. 2018. Exome sequencing diagnoses X-linked moesin-associated immunodeficiency in a primary immunodeficiency case. *Front Immunol* **9:** 420. doi:10.3389/fimmu.2018.00420

Brassington K, Selemidis S, Bozinovski S, Vlahos R. 2022. Chronic obstructive pulmonary disease and atherosclerosis: common mechanisms and novel therapeutics. *Clin Sci (Lond)* **136:** 405–423. doi:10.1042/CS20210835

Braun A, Worbs T, Moschovakis GL, Halle S, Hoffmann K, Bölter J, Münk A, Förster R. 2011. Afferent lymph–derived T cells and DCs use different chemokine receptor CCR7–dependent routes for entry into the lymph node and intranodal migration. *Nat Immunol* **12:** 879–887. doi:10.1038/ni.2085

Brinkmann V, Billich A, Baumruker T, Heining P, Schmouder R, Francis G, Aradhye S, Burtin P. 2010. Fingolimod (FTY720): discovery and development of an oral drug to treat multiple sclerosis. *Nat Rev Drug Discov* **9:** 883–897. doi:10.1038/nrd3248

Brunetti RM, Kockelkoren G, Raghavan P, Bell GRR, Britain D, Puri N, Collins SR, Leonetti MD, Stamou D, Weiner OD. 2022. WASP integrates substrate topology and cell polarity to guide neutrophil migration. *J Cell Biol* **221:** e202104046. doi:10.1083/jcb.202104046

Buckley CD, Ross EA, McGettrick HM, Osborne CE, Haworth O, Schmutz C, Stone PCW, Salmon M, Matharu NM, Vohra RK, et al. 2006. Identification of a phenotypically and functionally distinct population of long-lived neutrophils in a model of reverse endothelial migration. *J Leukoc Biol* **79:** 303–311. doi:10.1189/jlb.0905496

Buckley CD, Gilroy DW, Serhan CN, Stockinger B, Tak PP. 2013. The resolution of inflammation. *Nat Rev Immunol* **13:** 59–66. doi:10.1038/nri3362

Burridge K, Guilluy C. 2016. Focal adhesions, stress fibers and mechanical tension. *Exp Cell Res* **343:** 14–20. doi:10.1016/j.yexcr.2015.10.029

Casares L, Vincent R, Zalvidea D, Campillo N, Navajas D, Arroyo M, Trepat X. 2015. Hydraulic fracture during epithelial stretching. *Nat Mater* **14:** 343–351. doi:10.1038/nmat4206

Chandar AK, Singh S, Murad MH, Peyrin-Biroulet L, Loftus EV. 2015. Efficacy and safety of natalizumab and vedolizumab for the management of Crohn's disease: a systematic review and meta-analysis. *Inflamm Bowel Dis* **21:** 1695–1708. doi:10.1097/MIB.0000000000000373

Chastney MR, Conway JRW, Ivaska J. 2021. Integrin adhesion complexes. *Curr Biol* **31:** R536–R542. doi:10.1016/j.cub.2021.01.038

Chtanova T, Schaeffer M, Han S-J, van Dooren GG, Nollmann M, Herzmark P, Chan SW, Satija H, Camfield K, Aaron H, et al. 2008. Dynamics of neutrophil migration in lymph nodes during infection. *Immunity* **29:** 487–496. doi:10.1016/j.immuni.2008.07.012

Colom B, Bodkin JV, Beyrau M, Woodfin A, Ody C, Rourke C, Chavakis T, Brohi K, Imhof BA, Nourshargh S. 2015. Leukotriene B4-neutrophil elastase axis drives neutrophil reverse transendothelial cell migration in vivo. *Immunity* **42:** 1075–1086. doi:10.1016/j.immuni.2015.05.010

Coombs C, Georgantzoglou A, Walker HA, Patt J, Merten N, Poplimont H, Busch-Nentwich EM, Williams S, Kotsi C, Kostenis E, et al. 2019. Chemokine receptor trafficking coordinates neutrophil clustering and dispersal at wounds in zebrafish. *Nat Commun* **10:** 5166. doi:10.1038/s41467-019-13107-3

Damann N, Owsianik G, Li S, Poll C, Nilius B. 2009. The calcium-conducting ion channel transient receptor potential canonical 6 is involved in macrophage inflammatory protein-2-induced migration of mouse neutrophils. *Acta Physiol (Oxf)* **195:** 3–11. doi:10.1111/j.1748-1716.2008.01918.x

Davidson AJ, Amato C, Thomason PA, Insall RH. 2018. WASP family proteins and formins compete in pseudopod- and bleb-based migration. *J Cell Biol* **217:** 701–714. doi:10.1083/jcb.201705160

De Belly H, Yan S, Borja da Rocha H, Ichbiah S, Town JP, Zager PJ, Estrada DC, Meyer K, Turlier H, Bustamante C, et al. 2023. Cell protrusions and contractions generate long-range membrane tension propagation. *Cell* **186:** 3049–3061.e15. doi:10.1016/j.cell.2023.05.014

Debes GF, Arnold CN, Young AJ, Krautwald S, Lipp M, Hay JB, Butcher EC. 2005. Chemokine receptor CCR7 required for T lymphocyte exit from peripheral tissues. *Nat Immunol* **6:** 889–894. doi:10.1038/ni1238

Dedrick RL, Walicke P, Garovoy M. 2002. Anti-adhesion antibodies: efalizumab, a humanized anti-CD11a monoclonal antibody. *Transpl Immunol* **9:** 181–186. doi:10.1016/S0966-3274(02)00029-1

Del Prete A, Martínez-Muñoz L, Mazzon C, Toffali L, Sozio F, Za L, Bosisio D, Gazzurelli L, Salvi V, Tiberio L, et al. 2017. The atypical receptor CCRL2 is required for CXCR2-dependent neutrophil recruitment and tissue damage. *Blood* **130:** 1223–1234. doi:10.1182/blood-2017-04-777680

Deng Y, Banerjee T, Pal DS, Banerjee P, Zhan H, Borleis J, Igleias PA, Devreotes PN. 2024. PIP5K-Ras bistability initiates plasma membrane symmetry breaking to regulate cell polarity and migration. bioRxiv doi:10.1101/2024.09.15.613115

de Oliveira S, Rosowski EE, Huttenlocher A. 2016. Neutrophil migration in infection and wound repair: going forward in reverse. *Nat Rev Immunol* **16:** 378–391. doi:10.1038/nri.2016.49

Diz-Muñoz A, Thurley K, Chintamen S, Altschuler SJ, Wu LF, Fletcher DA, Weiner OD. 2016. Membrane tension acts through PLD2 and mTORC2 to limit actin network assembly during neutrophil migration. *PLoS Biol* **14:** e1002474. doi:10.1371/journal.pbio.1002474

Donà E, Barry JD, Valentin G, Quirin C, Khmelinskii A, Kunze A, Durdu S, Newton LR, Fernandez-Minan A, Huber W, et al. 2013. Directional tissue migration through a self-generated chemokine gradient. *Nature* **503:** 285–289. doi:10.1038/nature12635

Dowdell A, Paschke PI, Thomason PA, Tweedy L, Insall RH. 2023. Competition between chemoattractants causes unexpected complexity and can explain negative chemotaxis. *Curr Biol* **33:** 1704–1715.e3. doi:10.1016/j.cub.2023.03.006

Duffy D, Perrin H, Abadie V, Benhabiles N, Boissonnas A, Liard C, Descours B, Reboulleau D, Bonduelle O, Verrier B, et al. 2012. Neutrophils transport antigen from the dermis to the bone marrow, initiating a source of memory $CD8^+$ T cells. *Immunity* **37:** 917–929. doi:10.1016/j.immuni.2012.07.015

Dumortier JG, Le Verge-Serandour M, Tortorelli AF, Mielke A, de Plater L, Turlier H, Maître JL. 2019. Hydraulic fracturing and active coarsening position the lumen of the mouse blastocyst. *Science* **365:** 465–468. doi:10.1126/science.aaw7709

Elks PM, van Eeden FJ, Dixon G, Wang X, Reyes-Aldasoro CC, Ingham PW, Whyte MKB, Walmsley SR, Renshaw SA. 2011. Activation of hypoxia-inducible factor-1α (Hif-1α) delays inflammation resolution by reducing neutrophil apoptosis and reverse migration in a zebrafish inflammation model. *Blood* **118:** 712–722. doi:10.1182/blood-2010-12-324186

Ellett F, Elks PM, Robertson AL, Ogryzko NV, Renshaw SA. 2015. Defining the phenotype of neutrophils following reverse migration in zebrafish. *J Leukoc Biol* **98:** 975–981. doi:10.1189/jlb.3MA0315-105R

Enyedi B, Jelcic M, Niethammer P. 2016. The cell nucleus serves as a mechanotransducer of tissue damage-induced inflammation. *Cell* **165:** 1160–1170. doi:10.1016/j.cell.2016.04.016

Fackler OT, Grosse R. 2008. Cell motility through plasma membrane blebbing. *J Cell Biol* **181:** 879–884. doi:10.1083/jcb.200802081

Förster R, Schubel A, Breitfeld D, Kremmer E, Renner-Müller I, Wolf E, Lipp M. 1999. CCR7 coordinates the primary immune response by establishing functional microenvironments in secondary lymphoid organs. *Cell* **99:** 23–33. doi:10.1016/S0092-8674(00)80059-8

Friedl P, Weigelin B. 2008. Interstitial leukocyte migration and immune function. *Nat Immunol* **9:** 960–969. doi:10.1038/ni.f.212

Fritz RD, Pertz O. 2016. The dynamics of spatio-temporal Rho GTPase signaling: formation of signaling patterns. *F1000Res* **5:** F1000 Faculty Rev-749. doi:10.12688/f1000research.7370.1

Fritz-Laylin LK, Lord SJ, Mullins RD. 2017. WASP and SCAR are evolutionarily conserved in actin-filled pseudopod-based motility. *J Cell Biol* **216:** 1673–1688. doi:10.1083/jcb.201701074

Gaertner F, Reis-Rodrigues P, de Vries I, Hons M, Aguilera J, Riedl M, Leithner A, Tasciyan S, Kopf A, Merrin J, et al. 2022. WASp triggers mechanosensitive actin patches to facilitate immune cell migration in dense tissues. *Dev Cell* **57:** 47–62.e9. doi:10.1016/j.devcel.2021.11.024

Galeano Niño JL, Pageon SV, Tay SS, Colakoglu F, Kempe D, Hywood J, Mazalo JK, Cremasco J, Govendir MA, Dagley LF, et al. 2020. Cytotoxic T cells swarm by homotypic chemokine signalling. *eLife* **9:** e56554. doi:10.7554/eLife.56554

Gardel ML, Schneider IC, Aratyn-Schaus Y, Waterman CM. 2010. Mechanical integration of actin and adhesion dynamics in cell migration. *Annu Rev Cell Dev Biol* **26:** 315–333. doi:10.1146/annurev.cellbio.011209.122036

Geiger B, Spatz JP, Bershadsky AD. 2009. Environmental sensing through focal adhesions. *Nat Rev Mol Cell Biol* **10:** 21–33. doi:10.1038/nrm2593

Georgantzoglou A, Poplimont H, Walker HA, Lämmermann T, Sarris M. 2022. A two-step search and run response to gradients shapes leukocyte navigation in vivo. *J Cell Biol* **221:** e202103207. doi:10.1083/jcb.202103207

Goley ED, Welch MD. 2006. The ARP2/3 complex: an actin nucleator comes of age. *Nat Rev Mol Cell Biol* **7:** 713–726. doi:10.1038/nrm2026

Golo M, Newman PLH, Kempe D, Biro M. 2024. Mechanoimmunology in the solid tumor microenvironment. *Biochem Soc Trans* **52:** 1489–1502. doi:10.1042/BST20231427

Goudarzi M, Tarbashevich K, Mildner K, Begemann I, Garcia J, Paksa A, Reichman-Fried M, Mahabaleshwar H, Blaser H, Hartwig J, et al. 2017. Bleb expansion in migrating cells depends on supply of membrane from cell surface invaginations. *Dev Cell* **43:** 577–587.e5. doi:10.1016/j.devcel.2017.10.030

Graziano BR, Town JP, Sitarska E, Nagy TL, Fošnarič M, Penič S, Iglič A, Kralj-Iglič V, Gov NS, Diz-Muñoz A, et al. 2019. Cell confinement reveals a branched-actin independent circuit for neutrophil polarity. *PLoS Biol* **17:** e3000457. doi:10.1371/journal.pbio.3000457

Grimaldi C, Schumacher I, Boquet-Pujadas A, Tarbashevich K, Vos BE, Bandemer J, Schick J, Aalto A, Olivo-Marin JC, Betz T, et al. 2020. E-cadherin focuses protrusion formation at the front of migrating cells by impeding actin flow. *Nat Commun* **11:** 5397. doi:10.1038/s41467-020-19114-z

Hall C, Flores MV, Chien A, Davidson A, Crosier K, Crosier P. 2009. Transgenic zebrafish reporter lines reveal conserved Toll-like receptor signaling potential in embryonic myeloid leukocytes and adult immune cell lineages. *J Leukoc Biol* **85:** 751–765. doi:10.1189/jlb.0708405

Hamza B, Wong E, Patel S, Cho H, Martel J, Irimia D. 2014. Retrotaxis of human neutrophils during mechanical confinement inside microfluidic channels. *Integr Biol (Camb)* **6:** 175–183. doi:10.1039/C3IB40175H

Hind LE, Huttenlocher A. 2018. Neutrophil reverse migration and a chemokinetic resolution. *Dev Cell* **47:** 404–405. doi:10.1016/j.devcel.2018.11.004

Holmes GR, Anderson SR, Dixon G, Robertson AL, Reyes-Aldasoro CC, Billings SA, Renshaw SA, Kadirkamanathan V. 2012. Repelled from the wound, or randomly dispersed? Reverse migration behaviour of neutrophils characterized by dynamic modelling. *J R Soc Interface* **9:** 3229–3239. doi:10.1098/rsif.2012.0542

Hopke A, Scherer A, Kreuzburg S, Abers MS, Zerbe CS, Dinauer MC, Mansour MK, Irimia D. 2020. Neutrophil swarming delays the growth of clusters of pathogenic fungi. *Nat Commun* **11:** 2031. doi:10.1038/s41467-020-15834-4

Hopke A, Lin T, Scherer AK, Shay AE, Timmer KD, Wilson-Mifsud B, Mansour MK, Serhan CN, Irimia D, Hurley BP. 2022. Transcellular biosynthesis of leukotriene B4 orchestrates neutrophil swarming to fungi. *iScience* **25:** 105226. doi:10.1016/j.isci.2022.105226

Houk AR, Jilkine A, Mejean CO, Boltyanskiy R, Dufresne ER, Angenent SB, Altschuler SJ, Wu LF, Weiner OD. 2012. Membrane tension maintains cell polarity by confining signals to the leading edge during neutrophil migration. *Cell* **148:** 175–188. doi:10.1016/j.cell.2011.10.050

Houseright RA, Miskolci V, Mulvaney O, Bortnov V, Mosher DF, Rindy J, Bennin DA, Huttenlocher A. 2021. Myeloid-derived growth factor regulates neutrophil motility in interstitial tissue damage. *J Cell Biol* **220:** e202103054. doi:10.1083/jcb.202103054

Hughes J, Johnson RJ, Mooney A, Hugo C, Gordon K, Savill J. 1997. Neutrophil fate in experimental glomerular capillary injury in the rat. Emigration exceeds in situ clearance by apoptosis. *Am J Pathol* **150:** 223–234.

Huttenlocher A, Sandborg RR, Horwitz AF. 1995. Adhesion in cell migration. *Curr Opin Cell Biol* **7:** 697–706. doi:10.1016/0955-0674(95)80112-X

Iijima M, Huang YE, Devreotes P. 2002. Temporal and spatial regulation of chemotaxis. *Dev Cell* **3:** 469–478. doi:10.1016/S1534-5807(02)00292-7

Isles HM, Loynes CA, Alasmari S, Kon FC, Henry KM, Kadochnikova A, Hales J, Muir CF, Keightley MC, Kadirkamanathan V, et al. 2021. Pioneer neutrophils release chromatin within in vivo swarms. *eLife* **10:** e68755. doi:10.7554/eLife.68755

Janmey PA, Fletcher DA, Reinhart-King CA. 2020. Stiffness sensing by cells. *Physiol Rev* **100:** 695–724. doi:10.1152/physrev.00013.2019

Ji J, Zhong H, Wang Y, Liu J, Tang J, Liu Z. 2024. Chemerin attracts neutrophil reverse migration by interacting with C–C motif chemokine receptor-like 2. *Cell Death Dis* **15:** 1–10. doi:10.1038/s41419-024-06820-5

Jilkine A, Marée AFM, Edelstein-Keshet L. 2007. Mathematical model for spatial segregation of the Rho-family GTPases based on inhibitory crosstalk. *Bull Math Biol* **69:** 1943–1978. doi:10.1007/s11538-007-9200-6

Kamnev A, Lacouture C, Fusaro M, Dupré L. 2021. Molecular tuning of actin dynamics in leukocyte migration as revealed by immune-related actinopathies. *Front Immunol* **12:** 750537. doi:10.3389/fimmu.2021.750537

Kardash E, Reichman-Fried M, Maître JL, Boldajipour B, Papusheva E, Messerschmidt EM, Heisenberg CP, Raz E. 2010. A role for Rho GTPases and cell-cell adhesion in single-cell motility in vivo. *Nat Cell Biol* **12:** 47–53. doi:10.1038/ncb2003

Karling T, Weavers H. 2025. Immune cells adapt to confined environments in vivo to optimise nuclear plasticity for migration. *EMBO Rep* **26:** 1238–1268. doi:10.1038/s44319-025-00381-0

Kawai T, Malech HL. 2009. WHIM syndrome: congenital immune deficiency disease. *Curr Opin Hematol* **16:** 20–26. doi:10.1097/MOH.0b013e32831ac557

Kennard AS, Sathe M, Labuz EC, Prinz CK, Theriot JA. 2023. Post-injury hydraulic fracturing drives fissure formation in the zebrafish basal epidermal cell layer. *Curr Biol* **33:** 2616–2631.e5. doi:10.1016/j.cub.2023.05.021

Khazen R, Corre B, Garcia Z, Lemaître F, Bachellier-Bassi S, d'Enfert C, Bousso P. 2022. Spatiotemporal dynamics of calcium signals during neutrophil cluster formation. *Proc Natl Acad Sci* **119:** e2203855119. doi:10.1073/pnas.2203855119

Kienle K, Glaser KM, Eickhoff S, Mihlan M, Knöpper K, Reátegui E, Epple MW, Gunzer M, Baumeister R, Tarrant TK, et al. 2021. Neutrophils self-limit swarming to contain bacterial growth in vivo. *Science* **372:** eabe7729. doi:10.1126/science.abe7729

Knowlden SA, Capece T, Popovic M, Chapman TJ, Rezaee F, Kim M, Georas SN. 2014. Regulation of T cell motility in vitro and in vivo by LPA and LPA2. *PLoS One* **9:** e101655. doi:10.1371/journal.pone.0101655

Kohli K, Pillarisetty VG, Kim TS. 2022. Key chemokines direct migration of immune cells in solid tumors. *Cancer Gene Ther* **29:** 10–21. doi:10.1038/s41417-021-00303-x

Lagresle-Peyrou C, Luce S, Ouchani F, Soheili TS, Sadek H, Chouteau M, Durand A, Pic I, Majewski J, Brouzes C, et al. 2016. X-linked primary immunodeficiency associated with hemizygous mutations in the moesin (MSN) gene. *J Allergy Clin Immunol* **138:** 1681–1689.e8. doi:10.1016/j.jaci.2016.04.032

Lämmermann T, Bader BL, Monkley SJ, Worbs T, Wedlich-Söldner R, Hirsch K, Keller M, Förster R, Critchley DR, Fässler R, et al. 2008. Rapid leukocyte migration by integrin-independent flowing and squeezing. *Nature* **453:** 51–55. doi:10.1038/nature06887

Lämmermann T, Afonso PV, Angermann BR, Wang JM, Kastenmüller W, Parent CA, Germain RN. 2013. Neutrophil swarms require LTB4 and integrins at sites of cell death in vivo. *Nature* **498:** 371–375. doi:10.1038/nature12175

Lampi MC, Reinhart-King CA. 2018. Targeting extracellular matrix stiffness to attenuate disease: from molecular mechanisms to clinical trials. *Sci Transl Med* **10:** eaao0475. doi:10.1126/scitranslmed.aao0475

Lanitis E, Dangaj D, Irving M, Coukos G. 2017. Mechanisms regulating T-cell infiltration and activity in solid tumors. *Ann Oncol* **28:** xii18–xii32. doi:10.1093/annonc/mdx238

Lawson CD, Ridley AJ. 2018. Rho GTPase signaling complexes in cell migration and invasion. *J Cell Biol* **217:** 447–457. doi:10.1083/jcb.201612069

Lin Y, Sankar Pal D, Banerjee P, Banerjee T, Qin G, Deng Y, Borleis J, Iglesias PA, Devreotes PN. 2024. Ras suppression potentiates rear actomyosin contractility-driven cell polarization and migration. *Nat Cell Biol* **26:** 1062–1076. doi:10.1038/s41556-024-01453-4

Lindemann O, Umlauf D, Frank S, Schimmelpfennig S, Bertrand J, Pap T, Hanley PJ, Fabian A, Dietrich A, Schwab A. 2013. TRPC6 regulates CXCR2-mediated chemotaxis of murine neutrophils. *J Immunol* **190:** 5496–5505. doi:10.4049/jimmunol.1201502

Liu YJ, Le Berre M, Lautenschlaeger F, Maiuri P, Callan-Jones A, Heuzé M, Takaki T, Voituriez R, Piel M. 2015. Confinement and low adhesion induce fast amoeboid migration of slow mesenchymal cells. *Cell* **160:** 659–672. doi:10.1016/j.cell.2015.01.007

Liu CSC, Mandal T, Biswas P, Hoque MA, Bandopadhyay P, Sinha BP, Sarif J, D'Rozario R, Sinha DK, Sinha B, et al. 2024. Piezo1 mechanosensing regulates integrin-dependent chemotactic migration in human T cells. *eLife* **12:** RP91903. doi:10.7554/eLife.91903

Logue JS, Cartagena-Rivera AX, Baird MA, Davidson MW, Chadwick RS, Waterman CM. 2015. Erk regulation of actin capping and bundling by Eps8 promotes cortex tension and leader bleb-based migration. *eLife* **4:** e08314. doi:10.7554/eLife.08314

Lomakin AJ, Cattin CJ, Cuvelier D, Alraies Z, Molina M, Nader GPF, Srivastava N, Sáez PJ, Garcia-Arcos JM, Zhitnyak IY, et al. 2020. The nucleus acts as a ruler tailoring cell responses to spatial constraints. *Science* **370:** eaba2894. doi:10.1126/science.aba2894

Machesky LM, Atkinson SJ, Ampe C, Vandekerckhove J, Pollard TD. 1994. Purification of a cortical complex containing two unconventional actins from Acanthamoeba by affinity chromatography on profilin-agarose. *J Cell Biol* **127:** 107–115. doi:10.1083/jcb.127.1.107

Mackay CR. 2008. Moving targets: cell migration inhibitors as new anti-inflammatory therapies. *Nat Immunol* **9:** 988–998. doi:10.1038/ni.f.210

Marchant CL, Malmi-Kakkada AN, Espina JA, Barriga EH. 2022. Cell clusters softening triggers collective cell migration in vivo. *Nat Mater* **21:** 1314–1323. doi:10.1038/s41563-022-01323-0

Martinelli R, Zeiger AS, Whitfield M, Sciuto TE, Dvorak A, Van Vliet KJ, Greenwood J, Carman CV. 2014. Probing the biomechanical contribution of the endothelium to lymphocyte migration: diapedesis by the path of least resistance. *J Cell Sci* **127:** 3720–3734. doi:10.1242/jcs.148619

Massaad MJ, Ramesh N, Geha RS. 2013. Wiskott–Aldrich syndrome: a comprehensive review. *Ann NY Acad Sci* **1285:** 26–43. doi:10.1111/nyas.12049

Mathias JR, Perrin BJ, Liu TX, Kanki J, Look AT, Huttenlocher A. 2006. Resolution of inflammation by retrograde chemotaxis of neutrophils in transgenic zebrafish. *J Leukoc Biol* **80:** 1281–1288. doi:10.1189/jlb.0506346

Mauersberger C, Hinterdobler J, Schunkert H, Kessler T, Sager HB. 2022. Where the action is—leukocyte recruitment in atherosclerosis. *Front Cardiovasc Med* **8:** 813984. doi:10.3389/fcvm.2021.813984

Miki H, Suetsugu S, Takenawa T. 1998. WAVE, a novel WASP-family protein involved in actin reorganization induced by Rac. *EMBO J* **17:** 6932–6941. doi:10.1093/emboj/17.23.6932

Miskolci V, Klemm LC, Huttenlocher A. 2021. Cell migration guided by cell-cell contacts in innate immunity. *Trends Cell Biol* **31:** 86–94. doi:10.1016/j.tcb.2020.11.002

Mori Y, Jilkine A, Edelstein-Keshet L. 2008. Wave-pinning and cell polarity from a bistable reaction-diffusion system. *Biophys J* **94:** 3684–3697. doi:10.1529/biophysj.107.120824

Muinonen-Martin AJ, Susanto O, Zhang Q, Smethurst E, Faller WJ, Veltman DM, Kalna G, Lindsay C, Bennett DC, Sansom OJ, et al. 2014. Melanoma cells break down LPA to establish local gradients that drive chemotactic dispersal. *PLoS Biol* **12:** e1001966. doi:10.1371/journal.pbio.1001966

Mukhopadhyay A, Tsukasaki Y, Chan WC, Le JP, Kwok ML, Zhou J, Natarajan V, Mostafazadeh N, Maienschein-Cline M, Papautsky I, et al. 2024. *trans*-Endothelial neutrophil migration activates bactericidal function via Piezo1 mechanosensing. *Immunity* **57:** 52–67.e10. doi:10.1016/j.immuni.2023.11.007

Nakagawa M, Fukata M, Yamaga M, Itoh N, Kaibuchi K. 2001. Recruitment and activation of Rac1 by the formation of E-cadherin-mediated cell-cell adhesion sites. *J Cell Sci* **114:** 1829–1838. doi:10.1242/jcs.114.10.1829

Ng LG, Ostuni R, Hidalgo A. 2019. Heterogeneity of neutrophils. *Nat Rev Immunol* **19:** 255–265. doi:10.1038/s41577-019-0141-8

Ng LG, Qin JS, Roediger B, Wang Y, Jain R, Cavanagh LL, Smith AL, Jones CA, de Veer M, Grimbaldeston MA, et al. 2011. Visualizing the neutrophil response to sterile tissue injury in mouse dermis reveals a three-phase cascade of events. *J Invest Dermatol* **131:** 2058–2068. doi:10.1038/jid.2011.179

Ngo VN, Lucy Tang H, Cyster JG. 1998. Epstein–Barr virus-induced molecule 1 ligand chemokine is expressed by dendritic cells in lymphoid tissues and strongly attracts naive T cells and activated B cells. *J Exp Med* **188:** 181–191. doi:10.1084/jem.188.1.181

Niethammer P, Grabher C, Look AT, Mitchison TJ. 2009. A tissue-scale gradient of hydrogen peroxide mediates rapid wound detection in zebrafish. *Nature* **459:** 996–999. doi:10.1038/nature08119

Obeidy P, Ju LA, Oehlers SH, Zulkhernain NS, Lee Q, Galeano Niño JL, Kwan RY, Tikoo S, Cavanagh LL, Mrass P, et al. 2020. Partial loss of actin nucleator actin-related protein 2/3 activity triggers blebbing in primary T lymphocytes. *Immunol Cell Biol* **98:** 93–113. doi:10.1111/imcb.12304

Okada T, Miller MJ, Parker I, Krummel MF, Neighbors M, Hartley SB, O'Garra A, Cahalan MD, Cyster JG. 2005. Antigen-engaged B cells undergo chemotaxis toward the T zone and form motile conjugates with helper T cells. *PLoS Biol* **3:** e150. doi:10.1371/journal.pbio.0030150

Ou Z, Duh YS, Rommelfanger NJ, Keck CHC, Jiang S, Brinson K, Zhao S, Schmidt EL, Wu X, Yang F, et al. 2024. Achieving optical transparency in live animals with absorbing molecules. *Science* **385:** eadm6869. doi:10.1126/science.adm6869

Paluch EK, Raz E. 2013. The role and regulation of blebs in cell migration. *Curr Opin Cell Biol* **25:** 582–590. doi:10.1016/j.ceb.2013.05.005

Pardo-Pastor C, Rubio-Moscardo F, Vogel-González M, Serra SA, Afthinos A, Mrkonjic S, Destaing O, Abenza JF, Fernández-Fernández JM, Trepat X, et al. 2018. Piezo2 channel regulates RhoA and actin cytoskeleton to promote cell mechanobiological responses. *Proc Natl Acad Sci* **115:** 1925–1930. doi:10.1073/pnas.1718177115

Park SM, Eom S, Hong H, Yoon J, Lee SJ, Kim BC, Kim HW, Kim DS. 2019. Reconstruction of in vivo–like in vitro model: enabling technologies of microfluidic systems for dynamic biochemical/mechanical stimuli. *Microelectron Eng* **203–204:** 6–24. doi:10.1016/j.mee.2018.10.010

Parsons JT, Horwitz AR, Schwartz MA. 2010. Cell adhesion: integrating cytoskeletal dynamics and cellular tension. *Nat Rev Mol Cell Biol* **11:** 633–643. doi:10.1038/nrm2957

Patel KB, Liang W, Casper MJ, Voleti V, Li W, Yagielski AJ, Zhao HT, Perez Campos C, Lee GS, Liu JM, et al. 2022. High-speed light-sheet microscopy for the in-situ acquisition of volumetric histological images of living tissue. *Nat Biomed Eng* **6:** 569–583. doi:10.1038/s41551-022-00849-7

Pinner and Sahai. 2009. Integrin-independent movement of immune cells. *F1000 Biol Rep* **8:** 67. doi:10.3410/B1-67

Plotnikov SV, Pasapera AM, Sabass B, Waterman CM. 2012. Force fluctuations within focal adhesions mediate ECM-rigidity sensing to guide directed cell migration. *Cell* **151:** 1513–1527. doi:10.1016/j.cell.2012.11.034

Pollard TD, Blanchoin L, Mullins RD. 2000. Molecular mechanisms controlling actin filament dynamics in nonmuscle cells. *Annu Rev Biophys Biomol Struct* **29:** 545–576. doi:10.1146/annurev.biophys.29.1.545

Poplimont H, Georgantzoglou A, Boulch M, Walker HA, Coombs C, Papaleonidopoulou F, Sarris M. 2020. Neutrophil swarming in damaged tissue is orchestrated by connexins and cooperative calcium alarm signals. *Curr Biol* **30:** 2761–2776.e7. doi:10.1016/j.cub.2020.05.030

Powell D, Tauzin S, Hind LE, Deng Q, Beebe DJ, Huttenlocher A. 2017. Chemokine signaling and the regulation of bidirectional leukocyte migration in interstitial tissues. *Cell Rep* **19:** 1572–1585. doi:10.1016/j.celrep.2017.04.078

Raab M, Gentili M, de Belly H, Thiam HR, Vargas P, Jimenez AJ, Lautenschlaeger F, Voituriez R, Lennon-Duménil AM, Manel N, et al. 2016. ESCRT III repairs nuclear envelope ruptures during cell migration to limit DNA damage and cell death. *Science* **352:** 359–362. doi:10.1126/science.aad7611

Raftopoulou M, Hall A. 2004. Cell migration: Rho GTPases lead the way. *Dev Biol* **265:** 23–32. doi:10.1016/j.ydbio.2003.06.003

Ramos-Perez WD, Fang V, Escalante-Alcalde D, Cammer M, Schwab SR. 2015. A map of sphingosine 1-phosphate distribution in the spleen. *Nat Immunol* **16:** 1245–1252. doi:10.1038/ni.3296

Ray A, Provenzano PP. 2021. Aligned forces: origins and mechanisms of cancer dissemination guided by extracellular matrix architecture. *Curr Opin Cell Biol* **72:** 63–71. doi:10.1016/j.ceb.2021.05.004

Reátegui E, Jalali F, Khankhel AH, Wong E, Cho H, Lee J, Serhan CN, Dalli J, Elliott H, Irimia D. 2017. Microscale arrays for the profiling of start and stop signals coordinating human-neutrophil swarming. *Nat Biomed Eng* **1:** 0094. doi:10.1038/s41551-017-0094

Reif K, Ekland EH, Ohl L, Nakano H, Lipp M, Förster R, Cyster JG. 2002. Balanced responsiveness to chemoattractants from adjacent zones determines B-cell position. *Nature* **416:** 94–99. doi:10.1038/416094a

Reis-Rodrigues P, Avellaneda MJ, Canigova N, Gaertner F, Vaahtomeri K, Riedl M, de Vries I, Merrin J, Hauschild R, Fukui Y, et al. 2025. Migrating immune cells globally coordinate protrusive forces. *Nat Immunol* **26:** 1258–1266. doi:10.1038/s41590-025-02211-w

Ren X, Kiosses WB, Alexander Schwartz M. 1999. Regulation of the small GTP-binding protein Rho by cell adhesion and the cytoskeleton. *EMBO J* **18:** 578–585. doi:10.1093/emboj/18.3.578

Renkawitz J, Kopf A, Stopp J, de Vries I, Driscoll MK, Merrin J, Hauschild R, Welf ES, Danuser G, Fiolka R, et al. 2019. Nuclear positioning facilitates amoeboid migration along the path of least resistance. *Nature* **568:** 546–550. doi:10.1038/s41586-019-1087-5

Reversat A, Gaertner F, Merrin J, Stopp J, Tasciyan S, Aguilera J, de Vries I, Hauschild R, Hons M, Piel M, et al. 2020. Cellular locomotion using environmental topography. *Nature* **582:** 582–585. doi:10.1038/s41586-020-2283-z

Ridley AJ. 2015. Rho GTPase signalling in cell migration. *Curr Opin Cell Biol* **36:** 103–112. doi:10.1016/j.ceb.2015.08.005

Robertson AL, Holmes GR, Bojarczuk AN, Burgon J, Loynes CA, Chimen M, Sawtell AK, Hamza B, Willson J, Walmsley SR, et al. 2014. A zebrafish compound screen reveals modulation of neutrophil reverse migration as an anti-inflammatory mechanism. *Sci Transl Med* **6:** 225ra29. doi:10.1126/scitranslmed.3007672

Robertson TF, Chengappa P, Gomez Atria D, Wu CF, Avery L, Roy NH, Maillard I, Petrie RJ, Burkhardt JK. 2021. Lymphocyte egress signal sphingosine-1-phosphate promotes ERM-guided, bleb-based migration. *J Cell Biol* **220:** e202007182. doi:10.1083/jcb.202007182

Robertson TF, Hou Y, Schrope J, Shen S, Rindy J, Sauer JD, Dinh HQ, Huttenlocher A. 2023. A tessellated lymphoid network provides whole-body T cell surveillance in zebrafish. *Proc Natl Acad Sci* **120:** e2301137120. doi:10.1073/pnas.2301137120

Robertson TF, Schrope J, Zwick Z, Rindy J, Horn A, Hou Y, Huttenlocher A. 2025. Live imaging in zebrafish reveals tissue-specific strategies for amoeboid migration. *Development* **152:** dev204351. doi:10.1242/dev.204351

Rocha-Gregg B, Huttenlocher A. 2021. Signal integration in forward and reverse neutrophil migration: fundamentals and emerging mechanisms. *Curr Opin Cell Biol* **72:** 124–130. doi:10.1016/j.ceb.2021.07.002

Rowat AC, Jaalouk DE, Zwerger M, Ung WL, Eydelnant IA, Olins DE, Olins AL, Herrmann H, Weitz DA, Lammerding J. 2013. Nuclear envelope composition determines the ability of neutrophil-type cells to passage through micron-scale constrictions. *J Biol Chem* **288:** 8610–8618. doi:10.1074/jbc.M112.441535

Ruprecht V, Wieser S, Callan-Jones A, Smutny M, Morita H, Sako K, Barone V, Ritsch-Marte M, Sixt M, Voituriez R, et al. 2015. Cortical contractility triggers a stochastic switch to fast amoeboid cell motility. *Cell* **160:** 673–685. doi:10.1016/j.cell.2015.01.008

Sánchez-Madrid F, Angel del Pozo M. 1999. Leukocyte polarization in cell migration and immune interactions. *EMBO J* **18:** 501–511. doi:10.1093/emboj/18.3.501

Sarris M, Sixt M. 2015. Navigating in tissue mazes: chemoattractant interpretation in complex environments. *Curr Opin Cell Biol* **36:** 93–102. doi:10.1016/j.ceb.2015.08.001

Sarris M, Masson JB, Maurin D, Van der Aa LM, Boudinot P, Lortat-Jacob H, Herbomel P. 2012. Inflammatory chemokines direct and restrict leukocyte migration within live tissues as glycan-bound gradients. *Curr Biol* **22:** 2375–2382. doi:10.1016/j.cub.2012.11.018

Schick J, Raz E. 2022. Blebs—formation, regulation, positioning, and role in amoeboid cell migration. *Front Cell Dev Biol* **10:** 926394. doi:10.3389/fcell.2022.926394

Schmidt S, Friedl P. 2010. Interstitial cell migration: integrin-dependent and alternative adhesion mechanisms. *Cell Tissue Res* **339:** 83–92. doi:10.1007/s00441-009-0892-9

Schrope JH, Horn A, Lazorchak K, Tinnen CW, Stevens JJ, Farooqui M, Robertson T, Li J, Bennin D, Juang T, et al. 2025. Confinement by liquid–liquid interface replicates in vivo neutrophil deformations and elicits bleb based migration. *Adv Sci* **12:** e2414024. doi:10.1002/advs.202414024

Schulz K, Hazelton-Cavill P, Alornyo KK, Edenhofer I, Lindenmeyer M, Lohr C, Huber TB, Denholm B, Koehler S. 2024. Piezo activity levels need to be tightly regulated to maintain normal morphology and function in pericardial nephrocytes. *Sci Rep* **14:** 28254. doi:10.1038/s41598-024-79352-9

Schwab SR, Cyster JG. 2007. Finding a way out: lymphocyte egress from lymphoid organs. *Nat Immunol* **8:** 1295–1301. doi:10.1038/ni1545

Sel S, Rost BR, Yildirim AÖ, Sel B, Kalwa H, Fehrenbach H, Renz H, Gudermann T, Dietrich A. 2008. Loss of classical transient receptor potential 6 channel reduces allergic airway response. *Clin Exp Allergy* **38:** 1548–1558. doi:10.1111/j.1365-2222.2008.03043.x

Seth S, Oberdörfer L, Hyde R, Hoff K, Thies V, Worbs T, Schmitz S, Förster R. 2011. CCR7 essentially contributes to the homing of plasmacytoid dendritic cells to lymph nodes under steady-state as well as inflammatory conditions. *J Immunol* **186:** 3364–3372. doi:10.4049/jimmunol.1002598

Shen C, Cerf A, Postat J, Bhagrath A, Merino M, Mingarelli A, Barnes G, Patel D, Rogers D, Luo VM, et al. 2025. A Dock8-dependent mechanosensitive central actin pool maintains T cell shape and protects the nucleus during migration. *Sci Immunol* **25:** eadt9239. doi:10.1126/sciimmunol.adt9239

Shi Z, Graber ZT, Baumgart T, Stone HA, Cohen AE. 2018. Cell membranes resist flow. *Cell* **175:** 1769–1779.e13. doi:10.1016/j.cell.2018.09.054

Shin JW, Spinler KR, Swift J, Chasis JA, Mohandas N, Discher DE. 2013. Lamins regulate cell trafficking and lineage maturation of adult human hematopoietic cells. *Proc Natl Acad Sci* **110:** 18892–18897. doi:10.1073/pnas.1304996110

Silva MT. 2011. Macrophage phagocytosis of neutrophils at inflammatory/infectious foci: a cooperative mechanism in the control of infection and infectious inflammation. *J Leukoc Biol* **89:** 675–683. doi:10.1189/jlb.0910536

Sitarska E, Almeida SD, Beckwith MS, Stopp J, Czuchnowski J, Siggel M, Roessner R, Tschanz A, Ejsing C, Schwab Y, et al. 2023. Sensing their plasma membrane curvature allows migrating cells to circumvent obstacles. *Nat Commun* **14:** 5644. doi:10.1038/s41467-023-41173-1

Snyderman R, Goetzl EJ. 1981. Molecular and cellular mechanisms of leukocyte chemotaxis. *Science* **213:** 830–837. doi:10.1126/science.6266014

Srivastava N, Traynor D, Piel M, Kabla AJ, Kay RR. 2020. Pressure sensing through Piezo channels controls whether cells migrate with blebs or pseudopods. *Proc Natl Acad Sci* **117:** 2506–2512. doi:10.1073/pnas.1905730117

Stanishneva-Konovalova TB, Kelley CF, Eskin TL, Messelaar EM, Wasserman SA, Sokolova OS, Rodal AA. 2016. Coordinated autoinhibition of F-BAR domain membrane binding and WASp activation by Nervous Wreck. *Proc Natl Acad Sci* **113:** E5552–E5561. doi:10.1073/pnas.1524412113

Starnes TW, Bennin DA, Bing X, Eickhoff JC, Grahf DC, Bellak JM, Seroogy CM, Ferguson PJ, Huttenlocher A. 2014. The F-BAR protein PSTPIP1 controls extracellular matrix degradation and filopodia formation in macrophages. *Blood* **123:** 2703–2714. doi:10.1182/blood-2013-07-516948

Strickland E, Pan D, Godfrey C, Kim JS, Hopke A, Ji W, Degrange M, Villavicencio B, Mansour MK, Zerbe CS, et al. 2024. Self-extinguishing relay waves enable homeostatic control of human neutrophil swarming. *Dev Cell* **59:** 2659–2671.e4. doi:10.1016/j.devcel.2024.06.003

Swaney KF, Huang CH, Devreotes PN. 2010. Eukaryotic chemotaxis: a network of signaling pathways controls motility, directional sensing, and polarity. *Annu Rev Biophys* **39:** 265–289. doi:10.1146/annurev.biophys.093008.131228

Tamás SX, Roux BT, Vámosi B, Dehne FG, Török A, Fazekas L, Enyedi B. 2023. A genetically encoded sensor for visualizing leukotriene B4 gradients in vivo. *Nat Commun* **14:** 4610. doi:10.1038/s41467-023-40326-6

Tauzin S, Starnes TW, Becker FB, Lam P, Huttenlocher A. 2014. Redox and Src family kinase signaling control leukocyte wound attraction and neutrophil reverse migration. *J Cell Biol* **207:** 589–598. doi:10.1083/jcb.201408090

Tharp WG, Yadav R, Irimia D, Upadhyaya A, Samadani A, Hurtado O, Liu S-Y, Munisamy S, Brainard DM, Mahon MJ, et al. 2006. Neutrophil chemorepulsion in defined interleukin-8 gradients in vitro and in vivo. *J Leukoc Biol* **79:** 539–554. doi:10.1189/jlb.0905516

Thiam HR, Vargas P, Carpi N, Crespo CL, Raab M, Terriac E, King MC, Jacobelli J, Alberts AS, Stradal T, et al. 2016. Perinuclear Arp2/3-driven actin polymerization enables nuclear deformation to facilitate cell migration through complex environments. *Nat Commun* **7:** 10997. doi:10.1038/ncomms10997

Tweedy L, Insall RH. 2020. Self-generated gradients yield exceptionally robust steering cues. *Front Cell Dev Biol* **8:** 133. doi:10.3389/fcell.2020.00133

Tweedy L, Knecht DA, Mackay GM, Insall RH. 2016. Self-generated chemoattractant gradients: attractant depletion extends the range and robustness of chemotaxis. *PLoS Biol* **14:** e1002404. doi:10.1371/journal.pbio.1002404

Tweedy L, Thomason PA, Paschke PI, Martin K, Machesky LM, Zagnoni M, Insall RH. 2020. Seeing around corners: cells solve mazes and respond at a distance using attractant breakdown. *Science* **369:** eaay9792. doi:10.1126/science.aay9792

Tyson RA, Zatulovskiy E, Kay RR, Bretschneider T. 2014. How blebs and pseudopods cooperate during chemotaxis. *Proc Natl Acad Sci* **111:** 11703–11708. doi:10.1073/pnas.1322291111

Uderhardt S, Martins AJ, Tsang JS, Lämmermann T, Germain RN. 2019. Resident macrophages cloak tissue microlesions to prevent neutrophil-driven inflammatory damage. *Cell* **177:** 541–555.e17. doi:10.1016/j.cell.2019.02.028

Venkiteswaran G, Lewellis SW, Wang J, Reynolds E, Nicholson C, Knaut H. 2013. Generation and dynamics of an endogenous, self-generated signaling gradient across a migrating tissue. *Cell* **155:** 674–687. doi:10.1016/j.cell.2013.09.046

Venturini V, Pezzano F, Català Castro F, Häkkinen HM, Jiménez-Delgado S, Colomer-Rosell M, Marro M, Tolosa-Ramon Q, Paz-López S, Valverde MA, et al. 2020. The nucleus measures shape changes for cellular proprioception to control dynamic cell behavior. *Science* **370:** eaba2644. doi:10.1126/science.aba2644

Virumbrales-Muñoz M, Ayuso JM, Gong MM, Humayun M, Livingston MK, Lugo-Cintrón KM, McMinn P, Álvarez-García YR, Beebe DJ. 2020. Microfluidic lumen-based systems for advancing tubular organ modeling. *Chem Soc Rev* **49:** 6402–6442. doi:10.1039/D0CS00705F

Walters N, Zhang J, Rima XY, Nguyen LTH, Germain RN, Lämmermann T, Reátegui E. 2021. Analyzing inter-leukocyte communication and migration in vitro: neutrophils play an essential role in monocyte activation during swarming. *Front Immunol* **12:** 671546. doi:10.3389/fimmu.2021.671546

Wang J, Hossain M, Thanabalasuriar A, Gunzer M, Meininger C, Kubes P. 2017. Visualizing the function and fate of neutrophils in sterile injury and repair. *Science* **358:** 111–116. doi:10.1126/science.aam9690

Wang X, Jodoin E, Jorgensen J, Lee J, Markmann JJ, Cataltepe S, Irimia D. 2018. Progressive mechanical confinement of chemotactic neutrophils induces arrest, oscillations, and retrotaxis. *J Leukoc Biol* **104:** 1253–1261. doi:10.1002/JLB.5TA0318-110RRR

Wang S, Wang B, Shi Y, Möller T, Stegmeyer RI, Strilic B, Li T, Yuan Z, Wang C, Wettschureck N, et al. 2022. Mechanosensation by endothelial PIEZO1 is required for leukocyte diapedesis. *Blood* **140:** 171–183. doi:10.1182/blood.2021014614

Weber M, Hauschild R, Schwarz J, Moussion C, de Vries I, Legler DF, Luther SA, Bollenbach T, Sixt M. 2013. Interstitial dendritic cell guidance by haptotactic chemokine gradients. *Science* **339:** 328–332. doi:10.1126/science.1228456

Wedlich-Soldner R, Li R. 2003. Spontaneous cell polarization: undermining determinism. *Nat Cell Biol* **5:** 267–270. doi:10.1038/ncb0403-267

Weißenbruch K, Mayor R. 2024. Actomyosin forces in cell migration: moving beyond cell body retraction. *Bioessays* **46:** 2400055. doi:10.1002/bies.202400055

Weninger W, Biro M, Jain R. 2014. Leukocyte migration in the interstitial space of non-lymphoid organs. *Nat Rev Immunol* **14:** 232–246. doi:10.1038/nri3641

Woodfin A, Voisin MB, Beyrau M, Colom B, Caille D, Diapouli FM, Nash GB, Chavakis T, Albelda SM, Rainger GE, et al. 2011. The junctional adhesion molecule JAM-C regulates polarized transendothelial migration of neutrophils in vivo. *Nat Immunol* **12:** 761–769. doi:10.1038/ni.2062

Worbs T, Hammerschmidt SI, Förster R. 2017. Dendritic cell migration in health and disease. *Nat Rev Immunol* **17:** 30–48. doi:10.1038/nri.2016.116

Xie X, Shi Q, Wu P, Zhang X, Kambara H, Su J, Yu H, Park SY, Guo R, Ren Q, et al. 2020. Single-cell transcriptome profiling reveals neutrophil heterogeneity in homeostasis and infection. *Nat Immunol* **21:** 1119–1133. doi:10.1038/s41590-020-0736-z

Yadav SK, Feigelson SW, Roncato F, Antman-Passig M, Shefi O, Lammerding J, Alon R. 2018. Frontline Science: elevated nuclear lamin A is permissive for granulocyte transendothelial migration but not for motility through collagen I barriers. *J Leukoc Biol* **104:** 239–251. doi:10.1002/JLB.3HI1217-488R

Yamaguchi N, Knaut H. 2022. Focal adhesion-mediated cell anchoring and migration: from in vitro to in vivo. *Development* **149:** dev200647. doi:10.1242/dev.200647

Yang CT, Cambier CJ, Davis JM, Hall CJ, Crosier PS, Ramakrishnan L. 2012. Neutrophils exert protection in the early tuberculous granuloma by oxidative killing of mycobacteria phagocytosed from infected macrophages. *Cell Host Microbe* **12:** 301–312. doi:10.1016/j.chom.2012.07.009

Yang HW, Collins SR, Meyer T. 2016. Locally excitable Cdc42 signals steer cells during chemotaxis. *Nat Cell Biol* **18:** 191–201. doi:10.1038/ncb3292

Yi B, Xu Q, Liu W. 2021. An overview of substrate stiffness guided cellular response and its applications in tissue regeneration. *Bioact Mater* **15:** 82–102. doi:10.1016/j.bioactmat.2021.12.005

Yoo SK, Huttenlocher A. 2011. Spatiotemporal photolabeling of neutrophil trafficking during inflammation in live zebrafish. *J Leukoc Biol* **89:** 661–667. doi:10.1189/jlb.1010567

Yoo SK, Deng Q, Cavnar PJ, Wu YI, Hahn KM, Huttenlocher A. 2010. Differential regulation of protrusion and polarity by PI3K during neutrophil motility in live zebrafish. *Dev Cell* **18:** 226–236. doi:10.1016/j.devcel.2009.11.015

Yoo SK, Starnes TW, Deng Q, Huttenlocher A. 2011. Lyn is a redox sensor that mediates leukocyte wound attraction in vivo. *Nature* **480:** 109–112. doi:10.1038/nature10632

Zatulovskiy E, Tyson R, Bretschneider T, Kay RR. 2014. Bleb-driven chemotaxis of *Dictyostelium* cells. *J Cell Biol* **204:** 1027–1044. doi:10.1083/jcb.201306147

Zhang J, Reinhart-King CA. 2020. Targeting tissue stiffness in metastasis: mechanomedicine improves cancer therapy. *Cancer Cell* **37:** 754–755. doi:10.1016/j.ccell.2020.05.011

Zhao R, Afthinos A, Zhu T, Mistriotis P, Li Y, Serra SA, Zhang Y, Yankaskas CL, He S, Valverde MA, et al. 2019. Cell sensing and decision-making in confinement: the role of TRPM7 in a tug of war between hydraulic pressure and cross-sectional area. *Sci Adv* **5:** eaaw7243. doi:10.1126/sciadv.aaw7243

Zigmond SH, Levitsky HI, Kreel BJ. 1981. Cell polarity: an examination of its behavioral expression and its consequences for polymorphonuclear leukocyte chemotaxis. *J Cell Biol* **89:** 585–592. doi:10.1083/jcb.89.3.585

Harnessing the Power of *Drosophila* Primordial Germ Cell Migration

Lacy J. Barton[1] and Benjamin Lin[2]

[1]Department of Neuroscience, Developmental and Regenerative Biology, The University of Texas at San Antonio, San Antonio, Texas 78249, USA

[2]Department of Biochemistry and Cell Biology, Stony Brook University, Stony Brook, New York 11794, USA

Correspondence: lacy.barton@utsa.edu; benjamin.c.lin@stonybrook.edu

Primordial germ cells (PGCs) are embryonic germline precursors of sperm and egg that connect one generation to the next. In many animals, maintaining this intergenerational connection requires long-distance PGC migration across varied tissues during embryonic development. How PGCs traverse these dynamic cellular terrains and respond to a select number of cues while ignoring others remains a mystery. This review describes the molecular mechanisms underlying PGC migration in *Drosophila melanogaster*, focusing on cell motility modes and the myriad of signals and cells PGCs interact with during their journey. We describe the tools and techniques available in *Drosophila* to assess and manipulate migrating PGCs and the surrounding cells they encounter. As PGC migration entails wide-ranging cell behaviors from transepithelial migration to directed cell motility across divergent tissues, harnessing the power of *Drosophila* can provide fundamental insights into how cell movement in vivo drives health and disease.

The migration of primordial germ cells (PGCs) through the embryo to the developing somatic gonad is a shared feature of reproductive development. Indeed, in humans, mice, fish, and fruit flies, PGCs actively migrate rather than being passive carriers of the next generation's genome (Barton et al. 2024a). This journey is often necessary because in many species, PGCs are specified early in embryogenesis outside the embryo, long before and far away from their destination, the developing somatic gonad (Hansen and Pelegri 2021). To decrease this travel distance, PGCs opportunistically use developmental morphogenesis to translocate passively into the embryo interior. Once within the embryo, PGCs migrate during a challenging period when morphogenetic movements coordinated by a myriad of signaling molecules dynamically shape the surrounding environment. How PGCs robustly arrive at the right place at the right time is a fascinating question for cell migration researchers.

PGC migration has several desirable features to push the frontier of three-dimensional (3D) cell migration forward. PGCs in many species undergo transepithelial migration (TEM) and exhibit changes in adhesion and migration modes as they migrate through several tissues to reach the gonad. Notably, PGCs in several model organisms, including flies, fish, and frogs,

Cite this article as *Cold Spring Harb Perspect Biol* doi: 10.1101/cshperspect.a041800

use amoeboid migration to move between other cells and the extracellular matrix (ECM) (Blaser et al. 2006; Terayama et al. 2013; Lin et al. 2022; Morita et al. 2023). Amoeboid migration broadly refers to cells migrating with constant shape changes (Paluch et al. 2016). In this review, we use a definition centered on observations of PGC migration, referring to amoeboid migration as low-adhesion, rapid migration driven by blebs (hydrostatic pressure–driven extrusions of the plasma membrane from the actin cortex) or cortical actin flow or both. It is increasingly appreciated that cells migrating in 3D are generally confined and can move without firm adhesion to the external environment (Paluch et al. 2016; Yamada and Sixt 2019), which diversifies 3D cell migration strategies from canonical mesenchymal migration. With diverse tissue navigation, adhesive dynamics, and migration modes, PGCs provide a rich platform to inform us about emerging noncanonical cell migration strategies observed in cancer (Tozluoğlu et al. 2013).

There are many practical advantages to using PGCs as an in vivo model for cell migration. Continued improvements in light microscopy have made it feasible to study solitary migrating cells in their native context. Indeed, imaging locomoting cells in 3D environments has revealed new migration modes (Petrie et al. 2014). In many model organisms, PGCs reproducibly migrate in transparent, optically accessible ex utero embryos. A wealth of germline biology knowledge and facile genetic manipulations enable targeted expression in either the PGCs (Van Doren et al. 1998b; Köprunner et al. 2001; Knaut et al. 2002; Sano et al. 2005) or the surrounding somatic cells they migrate through (Brand and Perrimon 1993). The genetic toolkit is expanding to include the latest-generation biosensors (Zhang et al. 2023) and optogenetic molecular actuators, which permit light-inducible changes in cellular signaling (Izquierdo et al. 2018). While available in several classic multicellular model organisms, coordinated efforts in the *Drosophila melanogaster* (hereafter flies) community have allowed for unmatched ease to tag all protein-coding genes (Bellen et al. 2011) for live cell imaging and protein manipulation (Caussinus et al. 2012; Harmansa et al. 2017).

Here, we focus on the path and mechanisms of PGC migration in flies. We discuss the assays, tools, challenges, and opportunities of using this model organism to gain a deeper understanding of cell migration in a complex ecosystem (see Box 1 for special considerations). Because the core steps of *Drosophila* PGC migration are best understood with live imaging, we present unpublished movies from dorsal and lateral perspectives, along with a comprehensive description of essential milestones in PGC migration (Fig. 1; Supplemental Movies S1 and S2). Together, this review will cover how PGC migration in flies is poised to advance our understanding of the molecular logic enabling cells to reconcile migratory cues in complex in vivo environments.

PGC INTERNALIZATION

Fly PGCs are specified through preformation. The posterior pole of young fly embryos is enriched in maternally deposited germplasm, consisting of ribonucleoprotein condensates called germ granules, organelles, proteins, and messenger RNAs (mRNAs). Germ granules are a defining feature of PGCs across the animal kingdom and were historically used to trace the developmental path of PGCs in *Drosophila* (Counce 1963; Mahowald 1968). Germplasm instructs an alternative, earlier cell formation program and is necessary and sufficient for PGC formation (Ephrussi and Lehmann 1992). Accordingly, PGCs are defined as individual cells ~50 min before all other cells (referred to as somatic cells hereafter) in the syncytial blastoderm embryo at 25°C (Foe et al. 1993). Approximately 30–40 PGCs form at the embryo posterior and reside outside the embryo on the apical surface of epithelial cells destined to become the posterior endoderm, also known as the posterior midgut. After PGC formation at the end of stage 3, somatic cells cellularize at stage 5, and gastrulation commences (Fig. 1A; Foe et al. 1993).

During *Drosophila* gastrulation, PGCs internalize from their original location outside the

BOX 1. SPECIAL CONSIDERATIONS FOR INVESTIGATIONS OF PGC MIGRATION

Challenges and solutions for primordial germ cell (PGC)-specific manipulations: *Drosophila* embryogenesis has a high reliance on maternal RNAs and proteins due to a delayed maternal-to-zygotic transition and PGC preformation based on messenger RNA (mRNA) entrapment in germ granules rather than signal induction. Thus, maternal GAL4 drivers are required to express transgenes in early PGCs; yet, many drivers, including *maternal α-tubulin-GAL4* and some *nanos-GAL4*, are loaded into the entire oocyte, impacting somatic cells and PGCs of the early embryo (Table 1). Overcoming this challenge is of particular importance given that many common genes are required for cellularization, gastrulation, mesodermal patterning, and PGC migration, which occur in rapid succession in the developing embryo (Moore et al. 1998; Jemc 2011). To enrich for PGC-specific expression, most genetic PGC tools append the 3′ untranslated region (UTR) of *nanos* mRNA to a foreign transcript to enrich the mRNA in germ granules. The translated protein products are concentrated in PGCs through two mechanisms: (1) germ granule–localized translation (Gavis and Lehmann 1994; Chen et al. 2024b), and (2) translational silencing outside germplasm mediated by the RNA-binding protein Smaug (Dahanukar et al. 1999). Reporter constructs can be constitutively expressed in PGCs using the *nanos* promoter (Sano et al. 2005), while conditional expression of potentially toxic constructs, such as constitutively active and dominant-negative proteins, can be achieved using a nanos-driven Gal4-VP16 and a cognate upstream activation sequence (UAS) construct (Van Doren et al. 1998b). Adding the 3′ UTR from other germ granule–localized mRNAs, such as *pgc* and *gcl*, is also sufficient to enrich a user-defined mRNA in PGCs (Rangan et al. 2009; Eagle et al. 2018; Lin et al. 2020). The combined targeting of maternal GAL4 to the posterior pole using the 3′ UTR of *nanos* with UASt responders improves PGC specificity, but GAL4 is still active in the embryonic endoderm underlying the newly formed PGCs in many embryos (Masukawa et al. 2021). Thus, care must be taken to choose the best drivers and responders and to validate restricted expression. Another way to genetically manipulate only PGCs to test autonomy is to generate mutant clone oocytes and rescue embryonic development with a paternally provided wild-type allele or transgenes using an early somatic cell–specific GAL4 (Kunwar et al. 2008). Additional approaches to manipulate only PGCs include the use of PGC transplantation into host embryos or optogenetics for in vivo assessments, as well as ex vivo migration or live imaging experiments using mass-isolated or fluorescence-activated cell sorting (FACS)-enriched PGCs (Lin et al. 2022; Barton et al. 2024b). An ideal workflow to test for autonomous PGCs requirements would include at least one of the above approaches and manipulating the factor-of-interest in somatic cells.

Challenges and solutions for identification of front–back polarity: PGCs alternate between phases of directed and random migration. PGCs first migrate directionally through the posterior endoderm and then randomly migrate on the dorsal endoderm surface, before once again moving directionally toward the somatic gonadal precursors (SGPs) in the lateral mesoderm (Fig. 1D,E; Supplemental Movie S1). Once in the mesoderm, front–back polarity is complicated by germband retraction, in which the position of PGCs and SGPs flip relative to the whole embryo (Figs. 1 and 2). After arriving at the SGPs, PGCs begin to randomly migrate a second time to keep their position (Fig. 1F; Supplemental Movie S2). These alternating migratory phases make it challenging to assess candidate proteins involved in establishing PGC front–back polarity in fixed embryos. We suggest using live imaging whenever feasible during directed PGC movements (Fig. 1E; Supplemental Movie S1). In fixed embryos, we suggest identifying directionally migrating PGCs based on morphology (teardrop shape in which the bottom of the teardrop is the front) or with a known internal PGC polarity marker, such as myosin II. F-actin can be enriched at both the PGC front and back (Supplemental Movies S1 and S2).

Challenge and potential solutions for overcoming the robustness of PGC migration: Like most animals, PGC migration to the gonad is essential to fertility and thus the survival of the species. Given this importance, it is perhaps not surprising that genetic redundancy is common among axes required for PGC migration (Zhang et al. 1997; Sano et al. 2005; Barton et al. 2024b). Moreover, multiple

Continued

attractive and repulsive cues involving Wunens, Hedgehog (HH), and Juvenile hormones ensure that at least some PGCs reach the gonad. Indeed, to our knowledge, there is not a single perturbation in mice or flies in which no PGCs colonize the gonad. In flies, one PGC is all it takes to yield a fertile adult due to a compensatory mechanism that increases PGC proliferation rates when the number of colonizing or surviving PGCs is reduced (Gilboa and Lehmann 2006). The practical implications of this compensation and redundancy are that even severely compromised PGC migration does not cause infertility in flies. Although this robustness benefits the species and the laboratory maintenance of mutant strains, unbiased forward genetic screens are more cumbersome and less fruitful. Advances in our ability to generate animals mutant for many genes at the same time or new approaches to detect migration failures in vivo in a high-throughput manner may help the field overcome this challenge.

Potential challenges and solutions for overcoming environmental contributions to PGC migration: While external embryogenesis makes flies a great model for live imaging studies, it also necessitates a high degree of environmental control. Indeed, recent work has revealed that many factors required for fly PGC migration are influenced by the environment. Factors required for PGC migration are influenced by temperature, nutritional status, parental age, and circadian rhythm (Sarov-Blat et al. 2000; Gnocchi et al. 2015; Kurogi et al. 2021; Amatobi et al. 2023; Kim et al. 2023). While the impact of these parameters on PGC migration has yet to be directly tested, these findings suggest that tightly controlling environmental conditions, parental age, and population density may result in cleaner and more reproducible studies.

embryo. Massed PGCs adhere to each other and the underlying epithelium while being drawn into the embryo interior (Fig. 1B). Some PGCs can prematurely enter the embryo during cellularization (Counce 1963), but it remains unclear whether these PGCs migrate through or are carried in by attachment to the invaginating membrane furrows, also called cleavage furrows, enclosing the underlying somatic cells. The adhesive transmembrane protein E-cadherin and associated junctional proteins are enriched on descending cleavage furrow tips and later in apicolateral regions beneath PGCs, forming spot junctions between the epithelium (Hunter and Wieschaus 2000). PGCs also express E-cadherin (Kunwar et al. 2008), suggesting a potential competition between PGC–PGC and PGC–cleavage furrow adhesion. Live imaging qualitatively indicates the following: (1) the speed of membrane furrow descent and precocious PGC entrance are correlated, (2) the membranes of a subset of PGCs are pulled downward with furrows and retract back, and (3) perturbing E-cadherin increases the propensity for early PGC crossing (Kunwar et al. 2008; DeGennaro et al. 2011). Past and more recent live imaging studies also corroborate that PGCs are stationary during transit into the embryo (Kunwar et al. 2008; also see Bailles et al. 2019 and supplemental video 17 therein). Future live imaging experiments, where PGC migration and adhesion are individually manipulated, will be needed to clarify whether signaling perturbations that increase premature PGC entrance are driven by migration or defective cell adhesion (Li et al. 2003a; DeGennaro et al. 2011; Kim et al. 2021).

ESCAPING THE ENDODERM

PGC internalization into the developing gut is a common phenomenon. How PGCs enter the gut and their behavior within the gut appear to be species-specific. Fly PGCs inside the posterior midgut cavity at stage 9 of embryogenesis remain clustered and assume an intriguing rosette conformation. Each PGC adopts a "teardrop" shape and orients the broad portion of the teardrop to the midgut interface (Fig. 1C; Kunwar et al. 2008). This morphological change arises from PGCs establishing a front–back migration axis, characterized by enhanced actomyosin contractility at the cell rear through autonomous signaling from the orphan G-protein-coupled receptor (GPCR), Tre1 (Kunwar et al. 2003, 2008; LeBlanc and Lehmann 2017; Lin et al. 2020). Tre1 signaling directs PGC migration outward from the cluster against the confines of the midgut. Conversely, *Tre1* mutant

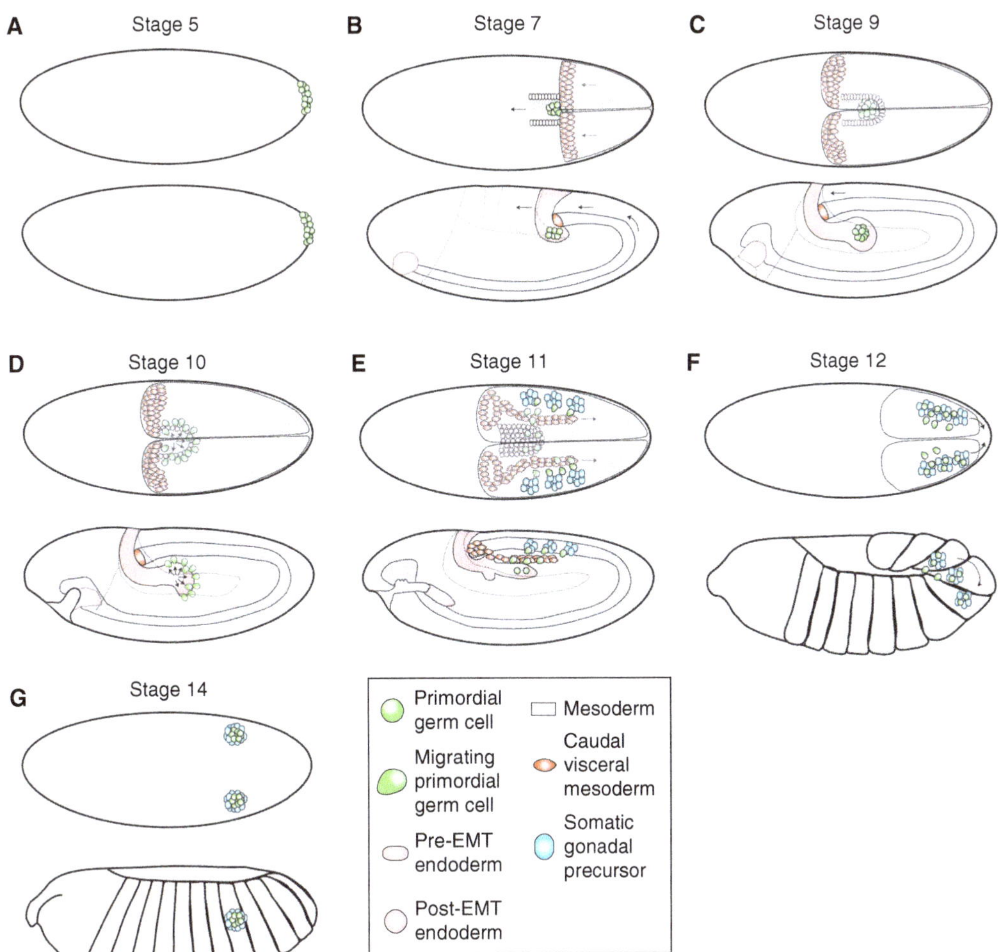

Figure 1. Major steps in *Drosophila* primordial germ cell (PGC) migration. Each panel shows dorsal (*top*) and lateral (*bottom*) embryo illustrations with relevant cell types, tissues, and morphological changes. In the dorsal and lateral view embryo illustrations, the anterior–posterior axis is *left* to *right*. The dorsoventral axis is *top* to *bottom* in the lateral view embryo illustrations. Arrows indicate direction of movement. (*A*) PGCs form at the posterior pole before all other cells (referred to as somatic cells) at the end of stage 3. Somatic cells first sit in a common cytoplasm before becoming individualized by membranes during a process called cellularization during stage 5. (*B*) PGCs are cupped by the posterior endoderm and internalized during gastrulation. The germ band, which includes the mesoderm and ectoderm, undergoes convergent extension and elongates from the ventral surface to wrap around the dorsal surface, while simultaneously the posterior endoderm containing the PGCs moves in an anterior direction. (*C*) The germ band continues to extend as PGCs are carried further into the embryo. As a group, PGCs form a rosette structure by individually polarizing and establishing a front–back migration axis toward the endoderm. (*D*) The posterior endoderm undergoes a developmentally regulated epithelial-to-mesenchymal transition (EMT), providing space for PGCs to transmigrate through the posterior endoderm. This step in PGC migration is called transepithelial migration (TEM). Arrows indicate the direction of PGC migration. (*E*) Now on the exterior of the posterior endoderm, PGCs first move toward the dorsal side of the posterior endoderm and then split into two groups while migrating toward the somatic gonadal precursors (SGPs) specified in the lateral mesoderm. Shown here are the three bilateral SGP clusters within females. The caudal visceral mesoderm cells collectively migrate posteriorly and are frequently used by PGCs as intermediaries on the way to the SGPs. The germ band has reached its full extension. (*F*) Germ band retraction occurs, and the germ band is pulled back toward the ventral side of the embryo. The PGCs and SGPs are embedded within the germ band and are dramatically pulled along during this movement. (*G*) The PGCs and SGPs compact into a single spherical gonad on each side of the embryo.

Table 1. Two-component systems for *Drosophila* PGC migration

Line	Details	Stocks/references
GAL4/LexA drivers active in migrating PGCs		
nos-GAL4::VP16	*nanos* promoter/5′UTR and 3′UTR; allows zygotic expression of UAS transgenes in PGCs, PE	BL4937; BL7303 (Van Doren et al. 1998b)
GAL4/LexA drivers active in the maternal germline for loading RNAs into eggs		
NGT40-GAL4	*nanos* promoter/5′UTR and *alpha-tubulin 84b* 3′UTR, PE	BL4442; BL25394; BL32563 (Li and Gergen 1999)
mat-alpha-GAL4::VP16	*alpha-tubulin 67c* promoter/5′UTR and *alpha-tubulin 84b* 3′UTR, PE	BL7062; BL7063; BL80361 (Häcker and Perrimon 1998)
mat-alpha-GAL4	*alpha-tubulin 67c* promoter/5′UTR and *alpha-tubulin 84b* 3′UTR; weaker than Mat-alpha-GAL4::VP16, PI (attp2, attp40)	Lin et al. 2020
mat-alpha-LexA	*alpha-tubulin 67c* promoter/5′UTR and *alpha-tubulin 84b* 3′UTR, PI (attp2, attp40)	Lin et al. 2022
vasa-GAL4	*vasa* promoter/5′UTR and *hsp70* 3′UTR, PE	Zhao et al. 2013
osk-GAL4::VP16	*oskar* promoter/5′UTR, 24 nucleotides of *oskar* coding sequence, *alpha-tubulin 84b* and *hsp70* 3′UTR, PE	BL44241; BL44242 (Telley et al. 2012)
otu-GAL4::VP16	*ovarian tumor* promoter/5′UTR and K10 3′UTR, PE	BL58424 (Rørth 1998)
MTD-GAL4	Combination of otu-GAL4-VP16, NGT-40, and Nos-GAL4::VP16, PE	BL31777 (Mazzalupo and Cooley 2006)
GAL4/LexA drivers active in relevant embryonic somatic cells		
nullo-GAL4	*nullo*; earliest zygotic soma-specific driver; St 5; PE	BL26875; BL38417 (Kunwar et al. 2003)
hkb-GAL4	*huckebein*; early endoderm driver, anterior and posterior endoderm; St 9 (Lin laboratory, unpubl.); PE	Kusch et al. 2002
48Y-GAL4	Cytological position 26C; early endoderm driver; predominantly anterior and posterior endoderm but lowly in lateral visceral mesoderm; St 10; PE	BL4935 (Martin-Bermudo et al. 1997)
byn-GAL4	*brachyenteron*; early endoderm driver, hindgut, midgut; St 8; PE	Iwaki and Lengyel 2002
twist-GAL4	*twist*; broad mesoderm driver; St 8; PE	BL914; BL2517; BL58804; BL80574 (Greig and Akam 1993; Jiang and Levine 1993; Staehling-Hampton et al. 1994)
24B-GAL4	*how*; maternally deposited, zygotically broad mesoderm driver; ST 6; PE	BL1767 (Fyrberg et al. 1997)
elav-GAL4	*embryonic lethal abnormal vision*; neuronal cells; St 12; PE	BL458; BL8765; BL602358 (Lin and Goodman 1994; Luo et al. 1994)
sim-GAL4	*single-minded*; midline mesectoderm; St 5; PE	Xiao et al. 1996
G447-GAL4	Unknown; caudal visceral mesoderm; St 11; PE	Georgias et al. 1997
h-GAL4	*hairy*; odd parasegments, some overlap with even parasegments; St 10; PE	Brand and Perrimon 1993
en-GAL4	*engrailed*; anterior region of parasegments; St 10; PE	Fietz et al. 1995
NP5141-GAL4	*ken and barbie*; parasegments 2 and 14; St 5; PE	Kenwrick et al. 2019
eve-GAL4	*even-skipped*; parasegments 5 and 13; PE	Fujioka et al. 2003
eya-GAL4	*eyes absent*; predicted mesoderm; PE	Leatherman and DiNardo 2008

Continued

Table 1. *Continued*

Line	Details	Stocks/references
six4-GAL4, six4-GAL4::VP16	*six4*; initially pan-mesodermal, becoming restricted to somatic gonadal precursors; St 9; PE	Anllo et al. 2019; Warder et al. 2024
adbA-GAL4	*abdominal A*; abdominal segments 2–7; PE	Wang et al. 2013
tj-GAL4	*traffic jam*; somatic gonadal precursors; St 12; PE; note: not expressed in male-specific gonadal precursors	Tanentzapf et al. 2007; Panchal et al. 2017
btl-GAL4	*breathless*; early endoderm, trachea; St 10; PE	Shiga et al. 1996
UAS behaviors in PGCs when used with nos-GAL4::VP16		
UASt	5xUAS, *hsp70* promoter and 5′UTR, SV40 terminator; active in PGCs beginning at St 9 and some expression in underlying endoderm, male bias after St 14	See Masukawa et al. (2021) for a description of activity in PGCs
UASp	14xUAS, P-element promoter, K10 terminator; active in PGCs beginning at St 11; weaker than UASt; male bias after St 12	Masukawa et al. 2021
UASz	10xUAS, *hsp70* promoter with piRNA targets removed from *hsp70* 5′UTR, active in PGCs beginning at St 9 with immediate male bias	Masukawa et al. 2021

(St) Stage of embryogenesis, (PE) P-element insertion, (PI) PhiC integrase, (BL) Bloomington Stock Center, (PGC) primordial germ cell, (UAS) upstream activation sequence, (*how*) *held out wings*, (MTD) maternal triple driver, (UTR) untranslated region.

PGCs aimlessly migrate within the PGC cluster and frequently exchange neighbors. As detailed below, Tre1 has multiple roles in *Drosophila* PGC migration, but the primary one is to orient PGCs in the endoderm for the next step in their migration.

To reach the gonad, PGCs must eventually escape the gut. Theoretically, escape could be accomplished through PGC invasion or endodermal dynamics. In flies, PGCs escape the posterior midgut epithelium and enter the embryo interior after a developmentally programmed posterior midgut epithelial-to-mesenchymal transition (EMT) (Fig. 1D; Warrior 1994; Jaglarz and Howard 1995; Seifert and Lehmann 2012; Lin et al. 2020). Adherens junctions linking the posterior midgut epithelium are dissolved during EMT, providing space for PGC TEM. The posterior midgut EMT is the rate-limiting step for PGC separation and embryo entry, as premature junction breakdown leads to early PGC TEM (Seifert and Lehmann 2012). Following midgut junction dissolution, Tre1-directed PGCs continue their outward migration and break away from each other during TEM. *Tre1* mutant PGCs, on the other hand, continue to randomly migrate within the PGC cluster and remain confined within the developing gut, despite the absence of midgut restraints. Thus, directed migration drives PGC separation; PGCs migrate away from each other to break E-cadherin-dependent PGC–PGC adhesion. There are no detectable changes in plasma membrane E-cadherin levels preceding PGC separation (Lin et al. 2020), in contrast to the reduction of E-cadherin observed prior to zebrafish PGC separation (Blaser et al. 2005). Migration-mediated cell–cell separation may preserve cadherin function for subsequent cell behaviors, as suggested for hepatocyte growth factor–stimulated myogenic cell release (Brand-Saberi et al. 1996).

Escaping the narrow passages through the endoderm is inherently risky. PGC TEM is accompanied by transient cell and potential nuclear deformation (Kunwar et al. 2008; Seifert and Lehmann 2012; Lin et al. 2020) and has been linked to PGC death (Renault et al. 2010; Tarayrah-Ibraheim et al. 2021). Of the 30–40 PGCs born at stage 5 of embryogenesis, ~35%–45% of PGCs perish postmidgut egress (Yamada et al. 2008; Slaidina and Lehmann 2017; Tarayrah-

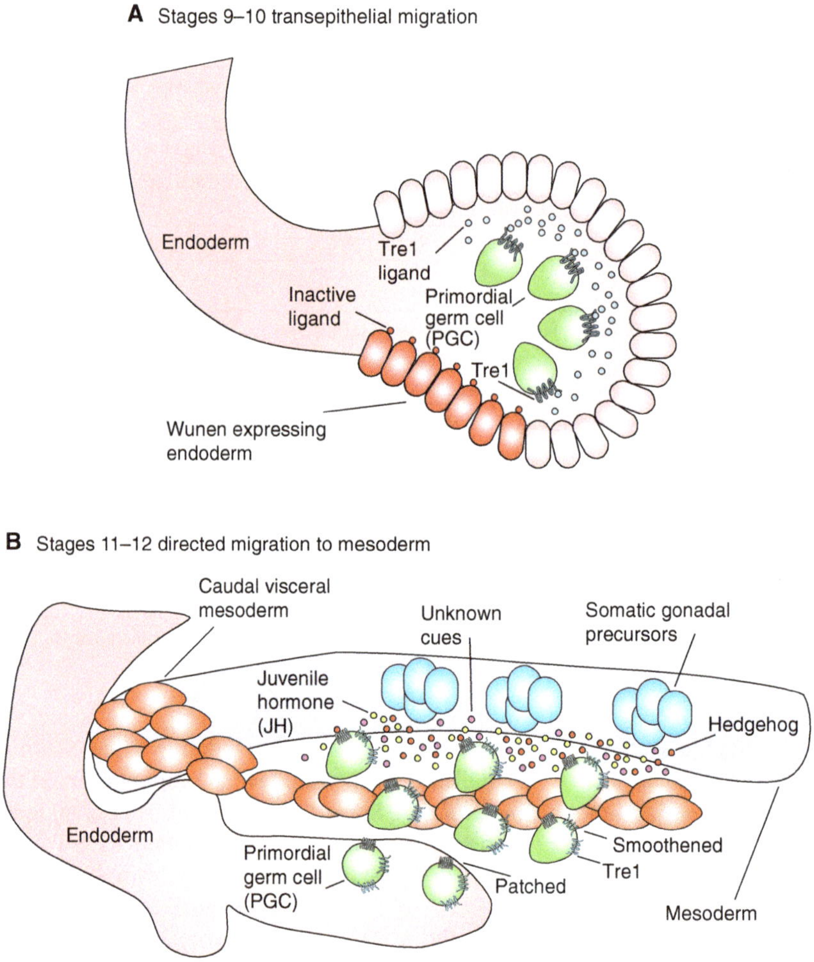

Figure 2. Extracellular signaling guiding primordial germ cell (PGC) migration. Lateral view illustrations of directed PGC (green cells) migration. Anterior is to the *left* and posterior is to the *right*. Dorsal is *up* and ventral is *down*. (*A*) Illustration of directed PGC migration out of the posterior endoderm. PGCs use the G-protein-coupled receptor (GPCR) Tre1 to sense an unknown guidance cue presumably produced by the endoderm (peach) to orient their migration toward the posterior endoderm. Wunens expressed by the ventral posterior endoderm are thought to degrade the Tre1 ligand. (*B*) Illustration of directed PGC migration from the endoderm to the somatic gonadal precursors (SGPs) (blue cells) situated in the lateral mesoderm, which produce chemical guidance cues, including Hedgehog and Juvenile hormones, and also physically contact PGCs. PGCs express several guidance receptors, including the hedgehog receptor Patched and the GPCRs, Tre1 and Smoothened. Live imaging indicates that many PGCs use the caudal visceral mesoderm (red oval cells) as a bridge to reach the lateral mesoderm.

Ibraheim et al. 2021). Similarly, endoderm exit has recently been connected to PGC fragmentation and death in mice (Goodwin et al. 2025). While PGC death is mediated by apoptosis in mice (Stallock et al. 2003), fly PGCs perish through an alternative parthanatos-like pathway. Mechanistically, DNase II translocates from the lysosomes to the nuclei of dying PGCs, causing DNA damage that triggers the activation of poly(ADP-ribose) polymerase 1 (PARP-1) and the DNA damage response pathways (Tarayrah-Ibraheim et al. 2021). Whether

PGCs in other organisms use similar mechanisms to cull PGCs remains an open question. As to why a subset of PGCs die, developmental PGC migration has been likened to a selection process, where only the "fittest" and perhaps least damaged PGCs survive to pass on their genomes to the next generation (Cantú and Laird 2017). Understanding how PGCs survive or die during their escape from the endoderm will provide insights into cell invasion and the emerging field of nuclear mechanotransduction during cell motility (Yamada and Sixt 2019).

As fly PGCs escape the gut, they individualize and then sort into two bilateral populations that will eventually colonize the two gonads (Fig. 1E; Supplemental Movie S1). PGC individualization requires Tre1, while bilateral sorting requires the integral membrane lipid phosphate phosphatases, Wunen and Wunen 2 (Zhang et al. 1997; Starz-Gaiano et al. 2001). *Wunens* are expressed in PGCs and various somatic tissues. During PGC cluster dispersal and TEM, somatic *wunens* are expressed in regions that PGCs avoid, the midgut ventral surface and hindgut (Fig. 2A; Zhang et al. 1996). In embryos lacking zygotic *wunens* (affecting somatic tissues), PGCs ectopically migrate toward these endoderm regions, suggesting somatic *wunens* create repulsive zones for PGC migration. Intriguingly, *wunens* and *Tre1* share PGC migration phenotypes such that *wunen* mutant PGCs migrate randomly and remain trapped in clusters (Renault et al. 2010), mimicking *Tre1* mutant PGCs. *Tre1 wunen* double mutant PGCs likewise have disrupted migratory polarity within the endoderm (LeBlanc and Lehmann 2017). The concerted actions of Wunens in PGCs and somatic tissues presumably shape external lipid phosphate distributions for Tre1-mediated PGC guidance, suggesting the Tre1 ligand is a lipid phosphate (LeBlanc and Lehmann 2017). However, the identity of the Tre1 ligand remains elusive. Wunens dephosphorylate extracellular lipid phosphates; some, such as lysophosphatidic acid and sphingosine-1-phosphate, are known chemoattractants (Kassmer et al. 2015), suggesting the Tre1 ligand could be one of these lipid phosphates (Renault et al. 2010; LeBlanc and Lehmann 2017). We anticipate that (1) ex vivo experiments observing changes in PGC cytoskeletal dynamics upon stimulation with candidate lipid phosphates, and (2) studying Tre1 signaling in other cell contexts, such as astrocyte growth (Chen et al. 2024a), immune cell TEM (Thuma et al. 2018), and asymmetric neuroblast division (Yoshiura et al. 2012), will provide clarity on the Tre1 ligand.

GPCR CIRCUITRY IN PGC MIGRATION

GPCR signaling plays a central role in directing PGC migration across species. Below, we discuss what is known about GPCR and guidance signaling from the perspective of the GPCR, Tre1, but the downstream molecular mechanisms are relevant for subsequent steps in fly PGC migration in the mesoderm (Fig. 3). As a canonical class A GPCR, Tre1 activates heterotrimeric G-proteins; Tre1 catalyzes guanosine diphosphate (GDP)/guanosine triphosphate (GTP) exchange on Gα subunits, causing dissociation from Gβγ heterodimers. Each liberated signaling entity, Gα-GTP and Gβγ, can then engage independent signaling cascades. In vitro assays in insect cells and pull-down experiments suggest Tre1 binds Gαo but not the related Gα subunit, Gαi (Yoshiura et al. 2012). Gαo is an inhibitory Gα subunit that lowers cyclic AMP (cAMP) levels indirectly through modulation of adenylyl cyclases (Birnbaumer 2007). Whether Tre1 associates with Gαo in PGCs requires further validation, as removing the maternal contribution of Gαo is complicated by the requirements for Gαo in oogenesis (Katanaev et al. 2005). Overexpressing a dominant-negative Gαo (constitutively bound to GDP) in PGCs causes migration defects (LeBlanc and Lehmann 2017), but whether this perturbation leads to navigation defects within or outside the midgut requires further live imaging validation. A dominant-negative Gαo protein will also deplete available Gβγ heterodimers required for other GPCRs (yet to be determined) involved in guiding PGCs after midgut egress. Future pull-down experiments using PGC lysates will be needed to assess Tre1-Gα coupling specificity. Another avenue to assess Gαo signaling in PGCs is to overexpress pertussis toxin, which specifically inactivates Gαo in *Drosophila* be-

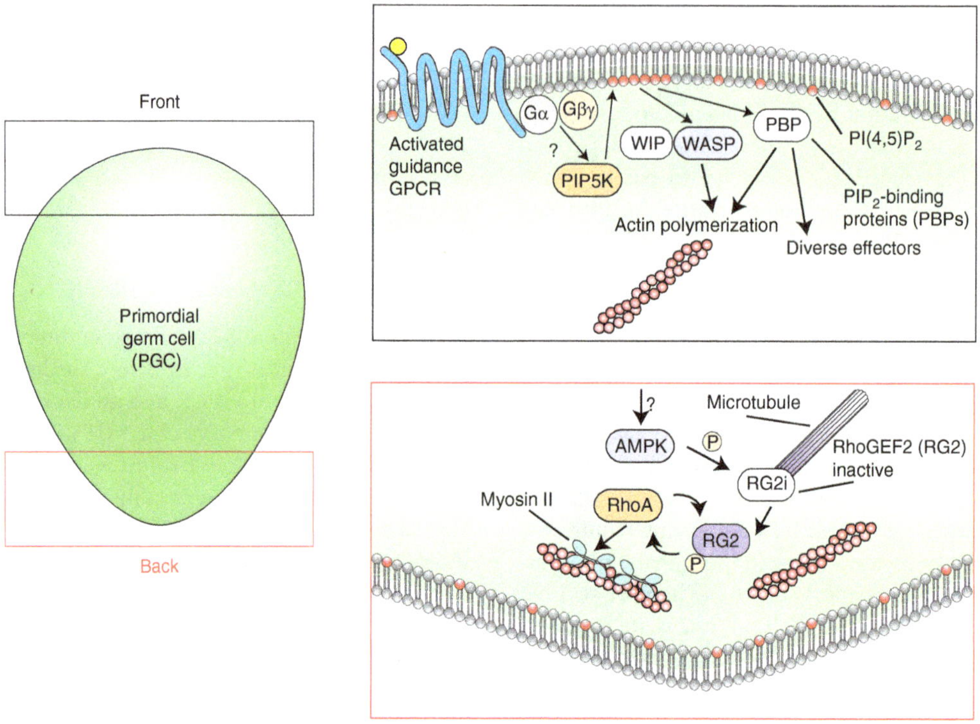

Figure 3. Internal signaling pathways guiding primordial germ cell (PGC) migration. (*Left*) Zoomed-out illustration of front (*top*) and back (*bottom*) signaling pathways in directed PGC migration. (*Right*) Zoomed-in illustration of the front and back of a migrating PGC. At the *front* (black box), activation of a guidance receptor (using a G-protein-coupled receptor [GPCR] as an example) leads to the local recruitment of phosphatidylinositol-4-phosphate 5-kinase (PIP5K), producing phosphatidylinositol 4,5-bisphosphate (PI(4,5)P_2). PI(4,5)P_2-binding proteins, such as Wiskott–Aldrich syndrome protein (WASp) and its regulator WASp-interacting protein (WIP), are then enriched and activated at these plasma membrane regions with elevated PI(4,5)P_2. WASp promotes F-actin assembly by activating the Arp2/3 complex. Many proteins bind PI(4,5)P_2 and may be recruited to the front to trigger other signaling pathways. At the back (red box), an unknown upstream regulator activates AMP-activated protein kinase (AMPK), which then phosphorylates inactive Rho guanine nucleotide exchange factor 2 (RhoGEF2) present on the growing ends of microtubules. Phosphorylation releases and activates RhoGEF2, which subsequently catalyzes guanosine diphosphate (GDP)/guanosine triphosphate (GTP) exchange on RhoA. The now activated RhoA (RhoA-GTP) potentiates the activity of myosin II to drive actomyosin contractility. Both inactive and active RhoA can bind RhoGEF2, forming a localization-based positive feedback loop.

cause the fly Gαi does not have a properly positioned carboxy-terminal cysteine residue (Katanaev et al. 2005). More generally, comparing GPCR signaling in migrating cells in vivo versus ex vivo will be informative in discerning environmental influences on guidance signaling.

In contrast to the ambiguity surrounding Tre1-Gα coupling, the identity of the Gβγ dimer involved in Tre1 signaling is clearer. Only a single Gβ (Gβ13f) and Gγ (Gγ1) subunit are maternally deposited into embryos (Kunwar et al. 2008). Removing the maternal contribution of either protein while rescuing expression in somatic cells with the upstream activation sequence UAS/GAL4 system phenocopies the *Tre1* mutant phenotype, suggesting Tre1 couples to these Gβγ subunits in PGCs (Kunwar et al. 2008). Several downstream proteins have been linked to Tre1 signaling in establishing PGC front–back polarity, but none have been definitively connected to either Gα and/or Gβγ (Fig. 3). These include proteins implicated in establishing the rear, Rho1 (*Drosophila* RhoA homolog), the Rho1-specific Rho guanine exchange factor (GEF), RhoGEF2, and AMP-activated protein kinase (AMPK), and the front, the phosphatidylinositol-4-phosphate 5-kinase (PIP5K), dPIP5K, dWIP (WASp-interacting protein), and Wiskott–Aldrich syndrome protein (WASp) (Kunwar et al. 2003, 2008; LeBlanc and Lehmann 2017; Lin et al. 2020, 2022; Kim et al. 2021). At the rear, AMPK phosphorylates RhoGEF2, freeing RhoGEF2 from inhibitory microtubule interactions to activate Rho1, which drives rear-localized actomyosin contractility (Lin et al. 2022). At the front, dPIP5K creates phosphatidylinositol 4,5-bisphosphate ($PI(4,5)P_2$), recruiting WASp and generating branched actin polymerization through the Arp2/3 complex (Kim et al. 2021). The same signaling axis, $PI(4,5)P_2$-WASp-Arp2/3, is also involved in the formation of PGCs (Kilwein et al. 2025). Given that diverse effectors bind $PI(4,5)P_2$ (Fig. 3; Balakrishnan et al. 2015), assessing $PI(4,5)P_2$-dependent signaling events outside the WASp signaling axis will be an interesting area for future exploration.

Several open questions remain concerning how GPCR signaling guides PGCs. (1) In addition to identifying Gα effectors, what proteins associate with free Gβγ heterodimers after GPCR activation? Gβγ heterodimers do not have intrinsic catalytic activity and bind a variety of effectors upon dissociation from Gα subunits (Smrcka 2008). Clarity may come from pulling down biotinylated Gβγ from PGC lysates and identifying interactors using mass spectrometry or employing a candidate approach based on the known binding of pleckstrin homology domains to Gβγ (Touhara et al. 1994). (2) How does AMPK, a metabolic sensor, integrate inputs from metabolic and GPCR signaling? AMPK activity is modulated by AMP, ADP, and ATP levels, phosphorylation of its activation loop via the upstream kinases LKB1 and CAMKK2, and phosphorylation of the serine/threonine-rich (ST-loop) domain in the α subunit (Garcia and Shaw 2017). AMPK biosensors will be needed to determine the distribution of AMPK activity and changes in activity upon perturbing upstream regulators and metabolic state. (3) How is WASp activated at the front of migrating PGCs? WASp is autoinhibited and canonically activated by the binding of the small Rho-GTPase, Cdc42, to its GTPase-binding domain (GBD) (Higgs and Pollard 2000). However, Cdc42 does not appear to be essential for PGC migration; overexpressing a dominant-negative Cdc42 does not perturb PGC migration (Kunwar et al. 2003) and a fluorescently tagged WASp Rho-binding domain (RBD), which binds active Cdc42 (Cdc42-GTP), is not enriched at the front of polarized PGCs during cluster dispersal (Lin et al. 2020). Purified WASp is activated by $PI(4,5)P_2$ alone, suggesting a similar mechanism may underlie WASp activation in PGCs (Higgs and Pollard 2000). While we currently appreciate that many cells use amoeboid migration for motility in vivo, the underlying intracellular signaling pathways that impinge on the cytoskeleton to guide movement remain unclear. Dissecting guidance signaling pathways in fly PGCs will provide one avenue to understand how amoeboid migrating cells are directed in vivo.

NAVIGATING THE MESODERM

PGCs begin active migration within the endoderm just prior to entering the mesoderm. Once in the mesoderm, PGCs are exposed to a multitude of signals and cell types, using some to navigate to the gonad, while ignoring others. In this section, we discuss critical cell types and signals that fly PGCs encounter once they are no longer shielded by the endodermal midgut.

PGCs are significantly influenced by the somatic gonadal precursors (SGPs) and the dorsolateral mesoderm from which SGPs are derived. Unlike in fish and mice, the fly gonad develops from three distinct SGP clusters in females and four clusters in males (DeFalco et al. 2003). SGP development and dynamics have been reviewed in detail (Jemc 2011). Briefly, SGPs are specified at stage 11 within the lateral *even skipped* (*eve*) domain, which is characterized by expression of *eve*. While much of the *eve* domain becomes fat body, SGPs develop within the dorsal portion of parasegments 10–12 in females and 10–13 in males. Many genes are required for germ band patterning and the subsequent specification and maintenance of SGP fate, including *abdominal A*, *abdominal B*, *eyes absent (eya) (also known as clift)*, *Dwnt2*, *fushi tarazu (ftz)*, *hedgehog (hh)*, *Notch*, *odd-paired*, *serrate*, *tinman*, *trithorax*, and *zinc finger homeodomain protein 1 (zfh1)* (reviewed in Jemc 2011). Interestingly, factors expressed in SGPs, like *six4* and the retrotransposon 412, are initially broadly expressed in the dorsolateral mesoderm and eventually become SGP-specific (Brookman et al. 1992; Kirby et al. 2001), highlighting the progressive patterning of the dorsolateral mesoderm that occurs simultaneously with PGC migration. Loss of factors important for mesoderm specification and patterning, such as *twist*, *snail*, *tinman*, *hh*, *six4*, or *eya*, compromises PGC migration (Warrior 1994; Boyle et al. 1997; Broihier et al. 1998; Moore et al. 1998; Kirby et al. 2001). However, the mesoderm and SGPs are not required for PGC TEM, as PGCs still cross the midgut epithelia even in animals lacking mesoderm entirely (Warrior 1994). Together, these findings reveal the importance of a properly patterned mesoderm and SGP specification for PGC migration.

The mesoderm and SGPs facilitate proper PGC migration in multiple ways, which include providing chemical guidance cues. Two such cues in flies are Juvenile hormones (JHs) and Hh. Both JHs and Hh are necessary and sufficient for directed PGC migration (Moore et al. 1998; Deshpande et al. 2001; Barton et al. 2024b), but how these signals facilitate PGC migration is not fully understood, and Hh's role has long been debated. At the center of the debate is whether Hh signaling is required directly and autonomously within PGCs (Deshpande et al. 2001, 2023; Deshpande and Schedl 2005; Kim et al. 2021) or whether Hh signaling is required indirectly for PGC migration through its role in patterning the germ band (Moore et al. 1998; Renault et al. 2009; Kenwrick et al. 2019). To distinguish between these two possibilities and circumvent patterning defects associated with Hh loss in somatic cells, two groups leveraged the delayed maternal-to-zygotic transition in PGCs, using germline clones to generate embryos without the maternal contributions of various Hh signaling factors (see Box 1 for rationale). One study found that maternal loss of Hh signaling factors, *patched* and the catalytic subunit of PKA, compromised PGC migration, while any defects in PGC migration upon maternal loss of *smoothened* were resolved by stage 15 (Deshpande et al. 2001). A follow-up study was unable to make *patched* and *PKA* mutant oocytes, consistent with previously published essential functions of both genes in oogenesis (Renault et al. 2009). However, they were able to generate *smoothened* mutant oocytes and expressed dominant-negative forms of Hh signaling components in PGCs, such as a truncated *Cubitus interruptus* (Renault et al. 2009). In their hands, these manipulations did not impair PGC migration (Renault et al. 2009). Since one caveat to both studies is that zygotic expression of these Hh components in migrating PGCs can potentially rescue PGC migration defects, Hh signaling was also manipulated by RNAi and expression of dominant mutants. These manipulations led to PGC migration defects in 30%–60% of embryos, impairing PGC internalization prior to active migration in the mesoderm (Kim et al. 2021; Deshpande et al. 2023).

JH and Hh signaling share a requirement for 3-hydroxy-3-methylglutaryl-coenzyme A reductase (Hmgcr) but in different ways. As the rate-limiting step in the mevalonate pathway, Hmgcr drives the production of farnesyl-pyrophosphate, which is the precursor of JHs. Thus, Hmgcr is directly required for JH synthesis (Wen et al. 2015). The role for Hmgcr in Hh signaling is less direct, with Hmgcr being required for Hh secretion and transmission (Deshpande and Schedl 2005), through prenylation of Gγ1 (Deshpande et al. 2009) and the secreted WNT-inhibitory factor 1 ortholog, Shifted (Deshpande et al. 2013). During PGC migration, Hmgcr expression is first high in mesoderm and subsequently becomes restricted to SGPs (Van Doren et al. 1998a). Like JHs and Hh, Hmgcr is also sufficient for attracting PGCs. Indeed, ectopic expression of Hmgcr can misdirect PGCs up to 50 µm away (Van Doren et al.

1998a; Santos and Lehmann 2004; Kenwrick et al. 2019).

The requirements of JH and Hh in Hmgcr-dependent PGC migration have been assessed in various ways. Hmgcr and a JH-specific synthesis enzyme, Jhamt, genetically interact to support PGC migration (Barton et al. 2024a). Loss of Jhamt or JH transcription factors suppresses PGC mislocalization near ectopic Hmgcr expression (Barton et al. 2024a). Likewise, some studies have shown that loss of Hmgcr suppresses dominant Hh mutations (Deshpande and Schedl 2005) and knockdown of Hh components suppresses Hmgcr-mediated PGC mismigration in a portion of embryos (Deshpande et al. 2023). Other studies have shown that full loss of Hh has no effect on Hmgcr-dependent attraction (Renault et al. 2009; Kenwrick et al. 2019), and Hmgcr can misdirect PGCs even when expressed in cells that do not typically express Hh (Van Doren et al. 1998a; Renault et al. 2009; Deshpande et al. 2023). Together, these findings suggest that more work is needed to fully understand how Hh and JH impact PGC migration. Given the role of Hh in developmental patterning, it would be important to systematically determine whether ectopic Hmgcr or Hh expression leads to ectopic expression of SGP-enriched genes such as *eya* or *Six4*. Since questions surrounding the autonomy of Hh signaling requirements are linked to inherent difficulties in genetically manipulating PGCs in early embryogenesis (Box 1), PGC transplantation experiments or in vitro migration assays with purified Hh ligand (Chen et al. 1999) would clarify the role of Hh in *Drosophila* PGC migration. Nonetheless, many studies have explored Hh's role in PGC migration. In contrast, much remains to be understood about JHs, including where and how JHs support PGC migration.

SGPs also support PGC incorporation through adhesion and ensheathment. SGPs first appear to contact PGCs, while the germ band is fully extended at stage 11 (Sano et al. 2012) and intermingle with PGCs during stage 12 (Fig. 1E, F; Boyle and DiNardo 1995). Following the dramatic morphogenic movement called germ band retraction, SGPs ensheath PGCs at stage 13 (Jenkins et al. 2003). SGP compaction begins at the end of stage 13 (Fig. 1G), and the impact of this process on PGCs will be discussed later in this review. Together, SGPs facilitate PGC homing through cue production and physical contact. Interestingly, these two mechanisms are uncoupled in *six4* mutants, which have reduced *Hmgcr* expression in SGPs and thus do not attract PGCs over long distances but still contact nearby PGCs (Clark et al. 2007). Exploiting separation-of-function mutants, such as *six4* may help untangle the multitude of ways SGPs support PGC migration.

PGCs encounter many other cells as they migrate to the gonad. The caudal visceral mesoderm (CVM) collectively migrates from the posterior germ band past migratory PGCs (Figs. 1E and 2B). The CVM and PGCs influence each other's trajectory while crossing paths, such that PGC migration is compromised when CVM migration is compromised and vice versa (Broihier et al. 1998; Stepanik et al. 2016). Observations from fixed embryos indicate PGCs are often in contact with CVM cells and reach the lateral mesoderm after CVM cells have arrived to bridge the gap between endoderm and lateral mesoderm. This suggests PGCs use the CVM as a stepping-stone to the lateral mesoderm (Broihier et al. 1998). Shared genetic requirements may also contribute to this interdependence, as the CVM and PGCs require Wunens and fibroblast growth factor (FGF) signaling for their migration (Kadam et al. 2012; Pares and Ricardo 2016; Stepanik et al. 2016). Insights may come from recent single-cell RNA-sequencing studies that have uncovered differences in the transcriptomes of fly CVM cells relative to their position along the anterior–posterior axis, which mirror recent findings in mouse PGCs (Sun et al. 2023; Jaszczak et al. 2025). SGPs and CVMs are certainly not the only cells PGCs encounter, as neighbors change quickly in early embryogenesis. Fly PGCs can encounter and theoretically be influenced by many other cell types while migrating in the mesoderm, including hemocytes, neurons, and developing fat body. Despite a wealth of genetic requirements, the influence of cell types on PGC migration remains poorly resolved. Nonetheless, this constellation of encounters suggests that PGCs are exposed to many different signals.

How PGCs receive and respond to some cues while ignoring others is an open area of investigation. Several studies have demonstrated through in vivo and ex vivo perturbations that PGCs appear capable of responding to appropriate mesodermal cues even before they exit the midgut (Li et al. 2003a; Ricardo and Lehmann 2009; Kim et al. 2021; Lin et al. 2022; Barton et al. 2024b), suggesting they are physically shielded from mesodermal signals within the gut. Young fly PGCs employ several other strategies to limit signal responsiveness, including degrading signal transduction factors (Pae et al. 2017), blocking the import of factors into the nucleus (Asaoka et al. 2019), or globally inhibiting transcription (Hanyu-Nakamura et al. 2008). However, there are likely additional ways PGCs are shielded from inductive somatic signals waiting to be discovered.

PGC MOTILITY

The somatic cells and the signals they produce influence how PGCs move. Recent work in zebrafish demonstrated that PGCs can switch protrusion types based on their surroundings (Truszkowski et al. 2023). In flies, early studies using phalloidin staining hinted at how PGCs migrate to the gonads. The Dallai and Howard laboratories observed a strong cortical F-actin staining in PGCs relative to somatic cells and a rounded cell shape, with variable regions of F-actin aggregation and small pseudopod extensions (Callaini et al. 1995; Jaglarz and Howard 1995). Electron microscopy uncovered that contact points between PGCs and other cells lacked specialized structures, suggesting these contacts are transient (Jaglarz and Howard 1995). Live imaging of fly PGC migration solidified these observations but lagged behind live PGC observations in mice and fish by several years (Molyneaux et al. 2001; Knaut et al. 2002; Wolke et al. 2002). This is likely because of the technical difficulty of imaging fly PGCs relatively deep within the embryo. Accordingly, two-photon microscopy was required to resolve subcellular fly PGC dynamics (Sano et al. 2005). From this seminal work, two observations suggest PGCs use amoeboid cell migration: (1) PGCs retain a rounded cell shape during migration with a uropod-like tail, and (2) although not appreciated at the time of this study, F-actin dense regions at the cortex travel along the periphery of PGCs, reflecting cortical actin flow, the translocation of cortical F-actin, and associated proteins along the inner surface of the plasma membrane (Supplemental Movies S1–S3). Cortical actin flows are driven toward regions of high actomyosin contractility and are seen during various cell behaviors, including cell division and migration (Bray and White 1988).

Two-photon microscopy of the dorsal PGC cortex recently provided concrete evidence that PGCs use cortical flows for motility. Lin et al. (2022) observed a retrograde flow of cortical actin foci (presumed to be embedded in the actin cortex) in directionally migrating PGCs in vivo. Perturbing actin cortex organization by overexpressing a constitutively active or dominant-negative RhoA halted cortical flows and PGC migration, indicating that cortical flows are causal for migration. Notably, overexpressing constitutively active but not dominant-negative Rac1 perturbs PGC migration (Renault et al. 2010), suggesting branched actin polymerization, a downstream output from Rac1 signaling, is dispensable for fly PGC migration and antagonistic when elevated. The diffuse localization of a Rac1 biosensor in migrating PGCs further supports this notion (Lin et al. 2020). Although Rac1 appears to play a minor role in fly and frog PGCs (Terayama et al. 2013), it is essential for fish PGC migration (Grimaldi et al. 2020).

Other reviews in this collection provide excellent mechanistic details underlying mesenchymal migration. Therefore, we focus on describing the cortical flow-driven amoeboid cell migration mode employed by fly PGCs (Lin et al. 2022), commonly referred to as fast amoeboid migration (FAM) (Bergert et al. 2015; Liu et al. 2015; Logue et al. 2015; Ruprecht et al. 2015). Cells using FAM have high levels of actomyosin contractility, retain a rounded cell shape during propulsion, are driven by retrograde (front to rear) cortical actin flow, and are minimally adhered to the external environment. In most cases, FAM is characterized by an invariant leading bleb. However, although isolated fly PGCs will occasionally display blebs ex vivo, there have been no concrete observations of blebbing in migrating PGCs in vivo. The absence of a leading bleb in fly PGCs compared to other cells using FAM may arise from the slower cortical flow speeds observed in fly PGCs (~3 µm/min in fly PGCs vs. 14 µm/min at the front of HeLa cells) (Liu et al. 2015; Lin et al. 2022), which may not deplete F-actin from the front fast enough. To our knowledge, MDA-MB-231 metastatic human breast cancer cells migrating in Matrigel were the first cells described using what is now termed FAM in 3D (Poincloux et al. 2011), while Walker carcinosarcoma cells treated with colchicine and latrunculin A were observed migrating with hallmarks of FAM as early as 2002 (Keller et al. 2002). More recently, several groups established that FAM is not limited to specific cell types (Bergert et al. 2015; Liu et al. 2015; Logue et al. 2015; Ruprecht et al. 2015). Liu et al. (2015) convincingly showed that various cells placed in a low-adhesion environment under geometric confinement would spontaneously adopt FAM, while Ruprecht et al. (2015) observed embryonic zebrafish cells using FAM in vivo after wounding. Larval zebrafish T cells also adopt FAM when migrating between confining skin cells (Robertson et al. 2025), and rounded *Drosophila* larval fat body cells are propelled by cortical actin flows (Andrieu et al. 2025). Although FAM appears to be labile and readily reversed by changes in mechanical confinement or contractile stimuli (Liu et al. 2015; Ruprecht et al. 2015), cultured fly PGCs outside the confining embryo environment and without known actomyosin contractility-stimulating factors maintain several hallmarks of FAM, including a rounded cell shape and cortical flows (Lin et al. 2022). Thus, fly PGCs autonomously employ FAM and are a model system to uncover the molecular requirements of FAM in vivo.

THE END OF THE JOURNEY

PGCs contact, or recontact, with SGPs after germ band retraction. At the end of stage 13, SGPs begin the process of compaction, in which the connected but elongated SGP- and PGC-containing gonads compact into a sphere (Boyle and DiNardo 1995). Curiously, loss of many factors required for SGP compaction causes PGCs to fail to incorporate into the gonad (Van Doren et al. 2003; Sano et al. 2012; Tripathy et al. 2014; Silva et al. 2016). How proper gonad compaction prevents extragonadal PGCs is not completely understood but likely involves E-cadherin-based adhesion (Jenkins et al. 2003). However, recent live imaging has revealed that PGCs are able to migrate toward SGPs and colonize the gonad later than previously appreciated. Indeed, PGCs are observed actively migrating to the gonad even as SGPs have started compaction at the end of stage 13 (Fig. 1G; Supplemental Movies S2 and S3). Practically, this observation means that investigators conducting end point analysis of mis-migration should only quantify PGCs outside of the gonad after SGP compaction, or stage 14 and beyond, as PGCs may still be in the process of migrating. This new insight should also be considered when interpreting previous results in which PGC migration was assessed prior to gonad compaction. After SGP compaction, PGCs appear no longer able to migrate toward SGPs despite continued cortical flow (Supplemental Movie S3). What prevents continued migration toward SGPs after gonad compaction remains unclear. Promigratory signals may cease to be made or released after the SGP compaction has started. PGCs may no longer be able to respond to such cues due to an intrinsic developmental clock. Indeed, other aspects of PGC development occur simultaneously with their migration across the animal kingdom. In flies, the germline-specific organelle called the fusome forms while PGCs migrate in the mesoderm (Lin and Spradling 1997), although this structure develops within PGCs in the absence of SGPs or mesoderm (Wawersik and Van Doren 2005). Alternatively, the ability of PGCs to migrate toward SGPs may be regulated by the appearance or disappearance of systemic cues. Indeed, the major insect hormone, Ecdysone, first becomes active at stage 13 when PGC migration is complete (Kozlova and Thummel 2003). Whether systemic, local, intrinsic mechanisms or all three prevent further migration

Table 2. One-component systems for *Drosophila* primordial germ cell (PGC) migration

Line	Details	Stocks/references
One-component systems to visualize PGCs		
vasa::eGFP, vasa::mCherry, vasa::mApple, vasa::tdTomato, vasa::mEos3.2	*vasa*; St 5, to mark PGCs; all are endogenous knock-in	Kina et al. 2019
eGFP::aub	*aubergine*; St 5, to mark PGCs; endogenous knock-in	Kina et al. 2019
nos-Moesin-ABD::GFP nos-Moesin-ABD::RFP	*nanos* promoter/5′UTR and *nanos* 3′UTR; St 5 onward; monitor F-actin in PGCs; PE	Sano et al. 2005; Renault et al. 2010
nos-lifeact-tdTomato-p2a-tdKatushka2-CAAX	*nanos* promoter/5′UTR and *nanos* 3′UTR; St 5 onward; monitor F-actin dynamics and plasma membrane; PI (attp8, attp40, attp2)	Lin et al. 2020
nos-lifeact-tdStaygold	*nanos* promoter/5′UTR and *nanos* 3′UTR; St 5 onward; monitor F-actin in PGCs; PI (attp2 and attp40)	Lin laboratory, unpubl.
nos-squash-tdTomato-p2a-tdKatushka2-CAAX	*nanos* promoter/5′UTR and *nanos* 3′UTR; St 5 onward; monitor myosin II and plasma membrane; PI (attp2 and attp40)	Lin et al. 2022
faf::lacZ	*fat facets*; St 5 onward; mark PGCs; PI	Fischer-Vize et al. 1992
One-component systems to visualize SGPs		
six4-nls-eGFP	*six4* promoter/5′UTR and SV40 3′UTR; broad mesoderm resolving to SGPs; St 11; PE	Clark et al. 2006
six4-eGFP::moesin-ABD	*six4* promoter/5′UTR and SV40 3′UTR with fusion to actin-binding domain of moesin; broad mesoderm resolving to SGPs; St 11; PE	Sano et al. 2012
six4-nlsRFP	*six4* promoter/5′UTR and SV40 3′UTR, broad mesoderm resolving to SGPs; St 11; PI (attp1 and attp2)	Rhymer and Anllo 2025
six4-GFP	*six4* promoter/5′UTR and *six4* 3′UTR; predicted broad mesoderm resolving to SGPs; St 11; PI (attp40)	BL67733 (Kudron et al. 2018)
tj-GFP	*traffic jam* promoter/5′UTR and 3′UTR; predicted SGPs, St 12; PI (VK00033)	BL66391 (Kudron et al. 2018)
68-77-lacZ	trap that reflects Ubx; mesoderm resolving to SGPs; St 11; PE	Boyle and DiNardo 1995
Lines to assess signaling cascades relevant to PGC migration		
nos-GCaMP3	*nanos* promoter/5′UTR and *nanos* 3′UTR; St 5, monitor calcium dynamics in PGCs; PI (attp2 and attp5)	Kaneuchi et al. 2015
nos-GCaMP6m	*nanos* promoter/5′UTR and *nanos* 3′UTR; St 5 onward, monitor calcium dynamics in PGCs; PI (attp2 and attp40)	Hu and Wolfner 2019
nos-mCherry-Atg8a	*nanos* promoter/5′UTR and *nanos* 3′UTR; predicted St 5, monitor autophagy in PGCs; PE	Nilangekar et al. 2019
nos-GFP-Ref(2)P	*nanos* promoter/5′UTR and *nanos* 3′UTR; predicted St 5, monitor autophagy in PGCs; PE	Nilangekar et al. 2019

Continued

Table 2. *Continued*

Line	Details	Stocks/references
nos-roGFP2-Orp1	*nanos* promoter/5′UTR and *nanos* 3′UTR; predicted St 5; mitochondrial H_2O_2 sensor using bioluminescence in PGCs; PE	Nilangekar et al. 2019
nos-roGFP2-Grxl	*nanos* promoter/5′UTR and *nanos* 3′UTR; predicted St 5; mitochondrial glutathione redox state in PGCs; PE	Nilangekar and Shravage 2019
nos-nanos-lantern-cAMP	*nanos* promoter/5′UTR and *nanos* 3′UTR; predicted St 5; monitor cAMP levels using bioluminescence in PGCs; PI (attp40)	Tu et al. 2023
nos-G-Flamp1	*nanos* promoter/5′UTR and *nanos* 3′UTR; predicted St 5; monitor cAMP levels in PGCs; PI (attp2)	Tu et al. 2023
nos-Anillin-RBD-p2a-tdKatushka2-CAAX	*nanos* promoter/5′UTR and *nanos* 3′UTR; St 5; monitor Rho1-GTP and plasma membrane dynamics in PGCs with some signal in soma; PI (attp2 and attp40)	Kong et al. 2019
UAS-PLCdelta-PH::eGFP	Fluorescently tagged pleckstrin homology domain from human PLCD1 under UAS control, St 5, monitor localization of phosphatidylinositol 4,5-bisphosphate ($PI(4,5)P_2$); PE	BL39693
UASp-PLCdelta-PH::mCerulean	Fluorescently tagged pleckstrin homology domain from human PLCD1 under UASp control, St 5, monitor localization of $PI(4,5)P_2$; PE	BL30895; BL31421
UAS-4C-PLCdelta-PH::mcherry	Fluorescently tagged pleckstrin homology domain from human PLCD1 under UAS control, St 5, monitor localization of $PI(4,5)P_2$; PE	BL51658
JHRE-GFP	Early trypsin gene from *Aedes aegypti*; St 10 onward, monitor JH III signaling broadly; PI (attp2)	Barton et al. 2024b
JHRE-GFP and JHRE-dsRed	early trypsin gene from *A. aegypti*; St 10 onward, monitor JH III signaling broadly; PE	Barton laboratory, unpubl.
ptc-lacZ (H84 and PG)	*patched*; monitor HH signaling broadly; PE	Ingham et al. 1991; Sanicola et al. 1995

(PGC) Primordial germ cell, (SGP) somatic gonadal precursor, (RBD) Rho1-binding domain, ($PI(4,5)P_2$) phosphatidylinositol 4,5-bisphosphate, (HH) Hedgehog, (JH) Juvenile hormone, (St) stage of embryogenesis, (PE) P-element insertion, (PI) PhiC integrase, (BL) Bloomington Stock Center, (ABD) actin-binding domain.

toward SGPs will be an interesting area of future exploration.

THE POWER OF *DROSOPHILA* PGCs AS AN IN VIVO MODEL FOR CELL MIGRATION

The migration of PGCs through the embryo has recaptured the interest of reproductive and cell biologists in recent years (Cantú and Laird 2017; Roelen and Chuva de Sousa Lopes 2022). Indeed, this journey is a rich platform to understand how cells move through many dynamic tissues with high fidelity. We are only just beginning to understand the many signals and cells required for this journey and appreciate the varied ways the environmental context can influence PGCs, the risks involved in the journey, and developmental programs that can mitigate such risks. Capitalizing on the strengths of various model organisms will bring us closer to understanding how this epic journey is accomplished and shed light on general principles underlying the various facets of in vivo cell migration.

Table 3. Antibodies and probes to identify *Drosophila* PGCs, SGPs, and other relevant cell types

Antibodies for immunohistochemistry		
Species and target	**Cell type**	**Source**
Rabbit anti-Vasa	PGCs all stages, also in SGPs after St 14	Lehmann laboratory (Fischer-Vize et al. 1992)
Mouse anti-Vasa	PGCs all stages, perhaps also in SGPs after St 14	DSHB[a] 46F11
Rat anti-Vasa	PGCs all stages, also in SGPs after St 14	DSHB Anti-vasa
Rabbit anti-Vasa	PGCs all stages, perhaps also in SGPs after St 14	DSHB Anti-vasa (Rb-IgG)
Guinea pig anti-Vasa	PGCs all stages, perhaps also in SGPs after St 14	DSHB Anti-vasa (Gp-IgG)
Mouse anti-Vasa	PGCs all stages, perhaps also in SGPs after St 14	Santa Cruz Sc-514249
Mouse anti-Eya	Initially in mesoderm, narrowing to SGPs, and then present in posterior somatic gonad, along with nervous system	DSHB Eya10H6
Rabbit anti-Sox100b	Male-specific SGPs	Jin et al. 2020
Mouse anti-AdbA	Germ band and anterior somatic gonad	DSHB FP6.87
Mouse anti-AdbB	Germ band but not in anterior somatic gonad	DSHB 1A2E9
Rabbit and rat anti-Escargot	Anterior somatic gonad	Sanchez-Díaz et al. 2015; Konstantinides et al. 2022
Mouse anti-Zfh-1.d	Pan-mesoderm and then hemocytes and caudal visceral mesoderm	Lai et al. 1991
Mouse anti-Even-skipped	Odd parasegments	DSHB 3c10
Mouse anti-Hindsight	Endoderm	DSHB 1G9
Rat anti-Traffic jam	SGPs	Li et al. 2003b
Mouse anti-Mef2	Pan-mesoderm	DSHB Mef2

RNAs for in situ hybridization		
RNA	**Cell types and notes**	**Source**
Bluetail transposon	Present before gonad compaction, posterior somatic gonad	Galloni et al. 1993
dWnt-2	Present before gonad compaction, posterior somatic gonad	DeFalco et al. 2003
412 retrotransposon	Broad mesoderm that narrows to SGPs	Brookman et al. 1992

[a]DSHB—Developmental Studies Hybridoma Bank, created by the NICHD of the NIH and maintained at Department of Biology, The University of Iowa, Iowa City, IA 52242.

(PGCs) Primordial germ cells, (SGPs) somatic gonadal precursors.

Flies are an excellent model to explore the intricate interplay between migrating cells and their natural environment. As described above, PGCs must (1) adhere to the endoderm and themselves for efficient internalization during gastrulation, (2) break adhesion to escape from the endoderm, (3) polarize at the right time to capitalize on gut EMT, (4) escape the endoderm and enter the mesoderm, (5) respond to migratory cues from the mesoderm while ignoring the many other developmental cues, (6) migrate through multiple cell types, and (7) incorporate into the SGPs (Fig. 1). In flies, these and many other processes occur within a few hours in an externally fertilized embryo, which is ideal for live imaging. External fertilization also aids pharmacological perturbation, either through injection or permeabilization. With high fecundity, *Drosophila* PGCs are easily isolated in large quantities for ex vivo migration assays or biochemistry. The time difference between PGC and somatic cell formation conveniently allows for mass PGC isolation with or without fluorescence-activated cell sorting (FACS) (Allis et al.

1977; Ricardo and Lehmann 2009; Lin et al. 2022; Barton et al. 2024b). Adding confidence to ex vivo studies, cultured PGCs follow aspects of the in vivo development program and can differentiate into functional germ cells in heterochronic embryo transplants (Allis et al. 1979). Thus, there are not only many biological aspects of cell migration that can be examined using fly PGCs, but there is also a wide diversity of approaches that can be employed to increase rigor.

Ultimately, the most significant advantage of flies is the vast and growing genetic toolkit. For decades, researchers have been generating and sharing mutants, antibodies, and transgenic tools to visualize and untangle spatiotemporal requirements (Tables 1–3). Commercial injection services, well-characterized genomic landing sites, transgenic germline-restricted integrase lines, and a fast life cycle all contribute to the rapidly expanding genetic toolkit. The most widely used in vivo expression system in flies is the two-component GAL4/UAS system, although the LexA/LexAop and QF/QUAS systems are gaining popularity (Zirin et al. 2024). Indeed, there are many driver and responder lines that can be used for manipulating PGCs, SGPs, or other somatic cells (Tables 1 and 2). As leaders in the fly community continue to generate new strain collections for GAL4 gene traps, protein traps, and fluorescent tagging (Diao et al. 2015; Sarov et al. 2016; Lee et al. 2018), the genetic tools available to visualize protein and cell dynamics as PGCs migrate are sure to increase.

With the versatility of available techniques, the vast genetic toolkit, and the well-characterized and rapid progression of PGC migration, flies are well positioned to answer emerging questions in how cells navigate in vivo. To fully capitalize on these many strengths, the development of transgenic tools with increased PGC specificity and an expansion of ex vivo models would refine our understanding of the complex process of PGC migration. Despite the demonstrated utility of ex vivo PGC experimentation, PGC migration has not been fully recapitulated in this setting. Future ex vivo studies could optimize the spatial distribution of chemoattractants and place PGCs in more physiological environments, either on or between other cells. The recent development of immortalized *Drosophila* cell lines of epithelial and mesodermal origin can facilitate this effort (Coleman-Gosser et al. 2023). Advances in ex vivo systems can help untangle the requirements of signaling networks that are spatiotemporally difficult to separate in vivo. Indeed, many PGC signaling axes are genetically interlinked either for PGC migration or in other contexts, such as Tre1 with Wunens, Wunens with Hmgcr, Hmgcr with JHs and HH, and HH with Tre1 (Deshpande and Schedl 2005; Deshpande et al. 2009, 2023; LeBlanc and Lehmann 2017; Kenwrick et al. 2019; Kim et al. 2021; Barton et al. 2024b). Ordering these requirements will be challenging but important and will provide a template for those interested in similarly complex mammalian PGC migration or, more generally, cell migration in vivo.

ACKNOWLEDGMENTS

B.L. is supported by the NIH's National Institute of General Medical Sciences (1R35GM157022). L.J.B. is supported by the NIH's Eunice Kennedy Shriver National Institute for Child Health and Development (R00 HD097306). This was written while L.J.B. was an AFAR Grant for Junior Faculty awardee. L.J.B. holds a Voelcker Fund Young Investigator Award from the Max and Minnie Tomerlin Voelcker Fund. We thank Creehan Healy from the Barton Laboratory for input on the tables.

REFERENCES

Allis CD, Waring GL, Mahowald AP. 1977. Mass isolation of pole cells from *Drosophila melanogaster*. *Dev Biol* **56:** 372–381. doi:10.1016/0012-1606(77)90277-9

Allis CD, Underwood EM, Caulton JH, Mahowald AP. 1979. Pole cells of *Drosophila melanogaster* in culture: normal metabolism, ultrastructure, and functional capabilities. *Dev Biol* **69:** 451–465. doi:10.1016/0012-1606(79)90304-X

Amatobi KM, Ozbek-Unal AG, Schäbler S, Deppisch P, Helfrich-Förster C, Mueller MJ, Wegener C, Fekete A. 2023. The circadian clock is required for rhythmic lipid transport in *Drosophila* in interaction with diet and photic condition. *J Lipid Res* **64:** 100417. doi:10.1016/j.jlr.2023.100417

Andrieu C, Lee BH, Franz A. 2025. Cell deformations generated by stochastic actomyosin waves drive in vivo random-walk swimming migration. *J Cell Sci* **138:** jcs263787. doi:10.1242/jcs.263787

Anllo L, Plasschaert LW, Sui J, DiNardo S. 2019. Live imaging reveals hub cell assembly and compaction dynamics during morphogenesis of the *Drosophila* testis niche. *Dev Biol* **446:** 102–118. doi:10.1016/j.ydbio.2018.12.014

Asaoka M, Hanyu-Nakamura K, Nakamura A, Kobayashi S. 2019. Maternal Nanos inhibits Importin-alpha2/pendulin-dependent nuclear import to prevent somatic gene expression in the *Drosophila* germline. *PLoS Genet* **15:** e1008090. doi:10.1371/journal.pgen.1008090

Bailles A, Collinet C, Philippe JM, Lenne PF, Munro E, Lecuit T. 2019. Genetic induction and mechanochemical propagation of a morphogenetic wave. *Nature* **572:** 467–473. doi:10.1038/s41586-019-1492-9

Balakrishnan SS, Basu U, Raghu P. 2015. Phosphoinositide signalling in *Drosophila*. *Biochim Biophys Acta* **1851:** 770–784. doi:10.1016/j.bbalip.2014.10.010

Barton LJ, Roa-de la Cruz L, Lehmann R, Lin B. 2024a. The journey of a generation: advances and promises in the study of primordial germ cell migration. *Development* **151:** dev201102. doi:10.1242/dev.201102

Barton LJ, Sanny J, Packard Dawson E, Nouzova M, Noriega FG, Stadtfeld M, Lehmann R. 2024b. Juvenile hormones direct primordial germ cell migration to the embryonic gonad. *Curr Biol* **34:** 505–518.e6. doi:10.1016/j.cub.2023.12.033

Bellen HJ, Levis RW, He Y, Carlson JW, Evans-Holm M, Bae E, Kim J, Metaxakis A, Savakis C, Schulze KL, et al. 2011. The *Drosophila* gene disruption project: progress using transposons with distinctive site specificities. *Genetics* **188:** 731–743. doi:10.1534/genetics.111.126995

Bergert M, Erzberger A, Desai RA, Aspalter IM, Oates AC, Charras G, Salbreux G, Paluch EK. 2015. Force transmission during adhesion-independent migration. *Nat Cell Biol* **17:** 524–529. doi:10.1038/ncb3134

Birnbaumer L. 2007. Expansion of signal transduction by G proteins: the second 15 years or so: from 3 to 16 α subunits plus βγ dimers. *Biochim Biophys Acta* **1768:** 772–793. doi:10.1016/j.bbamem.2006.12.002

Blaser H, Eisenbeiss S, Neumann M, Reichman-Fried M, Thisse B, Thisse C, Raz E. 2005. Transition from non-motile behaviour to directed migration during early PGC development in zebrafish. *J Cell Sci* **118:** 4027–4038. doi:10.1242/jcs.02522

Blaser H, Reichman-Fried M, Castanon I, Dumstrei K, Marlow Florence L, Kawakami K, Solnica-Krezel L, Heisenberg CP, Raz E. 2006. Migration of zebrafish primordial germ cells: a role for myosin contraction and cytoplasmic flow. *Dev Cell* **11:** 613–627. doi:10.1016/j.devcel.2006.09.023

Boyle M, DiNardo S. 1995. Specification, migration and assembly of the somatic cells of the *Drosophila* gonad. *Development* **121:** 1815–1825. doi:10.1242/dev.121.6.1815

Boyle M, Bonini N, DiNardo S. 1997. Expression and function of *clift* in the development of somatic gonadal precursors within the *Drosophila* mesoderm. *Development* **124:** 971–982. doi:10.1242/dev.124.5.971

Brand AH, Perrimon N. 1993. Targeted gene expression as a means of altering cell fates and generating dominant phenotypes. *Development* **118:** 401–415. doi:10.1242/dev.118.2.401

Brand-Saberi B, Müller TS, Wilting J, Christ B, Birchmeier C. 1996. Scatter factor/hepatocyte growth factor (SF/HGF) induces emigration of myogenic cells at interlimb level in vivo. *Dev Biol* **179:** 303–308. doi:10.1006/dbio.1996.0260

Bray D, White JG. 1988. Cortical flow in animal cells. *Science* **239:** 883–888. doi:10.1126/science.3277283

Broihier HT, Moore LA, Van Doren M, Newman S, Lehmann R. 1998. *zfh-1* is required for germ cell migration and gonadal mesoderm development in *Drosophila*. *Development* **125:** 655–666. doi:10.1242/dev.125.4.655

Brookman JJ, Toosy AT, Shashidhara LS, White RA. 1992. The 412 retrotransposon and the development of gonadal mesoderm in *Drosophila*. *Development* **116:** 1185–1192. doi:10.1242/dev.116.4.1185

Callaini G, Riparbelli MG, Dallai R. 1995. Pole cell migration through the gut wall of the *Drosophila* embryo: analysis of cell interactions. *Dev Biol* **170:** 365–375. doi:10.1006/dbio.1995.1222

Cantú AV, Laird DJ. 2017. A pilgrim's progress: seeking meaning in primordial germ cell migration. *Stem Cell Res* **24:** 181–187. doi:10.1016/j.scr.2017.07.017

Caussinus E, Kanca O, Affolter M. 2012. Fluorescent fusion protein knockout mediated by anti-GFP nanobody. *Nat Struct Mol Biol* **19:** 117–121. doi:10.1038/nsmb.2180

Chen CH, von Kessler DP, Park W, Wang B, Ma Y, Beachy PA. 1999. Nuclear trafficking of cubitus interruptus in the transcriptional regulation of hedgehog target gene expression. *Cell* **98:** 305–316. doi:10.1016/S0092-8674(00)81960-1

Chen J, Stork T, Kang Y, Nardone KAM, Auer F, Farrell RJ, Jay TR, Heo D, Sheehan A, Paton C, et al. 2024a. Astrocyte growth is driven by the Tre1/S1pr1 phospholipid-binding G protein-coupled receptor. *Neuron* **112:** 93–112.e10. doi:10.1016/j.neuron.2023.11.008

Chen R, Stainier W, Dufourt J, Lagha M, Lehmann R. 2024b. Direct observation of translational activation by a ribonucleoprotein granule. *Nat Cell Biol* **26:** 1322–1335. doi:10.1038/s41556-024-01452-5

Clark IBN, Boyd J, Hamilton G, Finnegan DJ, Jarman AP. 2006. D-six4 plays a key role in patterning cell identities deriving from the *Drosophila* mesoderm. *Dev Biol* **294:** 220–231. doi:10.1016/j.ydbio.2006.02.044

Clark IB, Jarman AP, Finnegan DJ. 2007. Live imaging of *Drosophila* gonad formation reveals roles for Six4 in regulating germline and somatic cell migration. *BMC Dev Biol* **7:** 52. doi:10.1186/1471-213X-7-52

Coleman-Gosser N, Hu Y, Raghuvanshi S, Stitzinger S, Chen W, Luhur A, Mariyappa D, Josifov M, Zelhof A, Mohr SE, et al. 2023. Continuous muscle, glial, epithelial, neuronal, and hemocyte cell lines for *Drosophila* research. *eLife* **12:** e85814. doi:10.7554/eLife.85814

Counce SJ. 1963. Developmental morphology of polar granules in *Drosophila*. Including observations on pole cell behavior and distribution during embryogenesis. *J Morphol* **112:** 129–145. doi:10.1002/jmor.1051120203

Dahanukar A, Walker JA, Wharton RP. 1999. Smaug, a novel RNA-binding protein that operates a translational switch in *Drosophila*. *Mol Cell* **4:** 209–218. doi:10.1016/S1097-2765(00)80368-8

DeFalco TJ, Verney G, Jenkins AB, McCaffery JM, Russell S, Van Doren M. 2003. Sex-specific apoptosis regulates sexual dimorphism in the *Drosophila* embryonic gonad. *Dev Cell* **5:** 205–216. doi:10.1016/S1534-5807(03)00204-1

DeGennaro M, Hurd TR, Siekhaus DE, Biteau B, Jasper H, Lehmann R. 2011. Peroxiredoxin stabilization of DE-cadherin promotes primordial germ cell adhesion. *Dev Cell* **20:** 233–243. doi:10.1016/j.devcel.2010.12.007

Deshpande G, Schedl P. 2005. HMGCoa reductase potentiates hedgehog signaling in *Drosophila melanogaster*. *Dev Cell* **9:** 629–638. doi:10.1016/j.devcel.2005.09.014

Deshpande G, Swanhart L, Chiang P, Schedl P. 2001. Hedgehog signaling in germ cell migration. *Cell* **106:** 759–769. doi:10.1016/S0092-8674(01)00488-3

Deshpande G, Godishala A, Schedl P. 2009. Ggamma1, a downstream target for the *hmgcr*-isoprenoid biosynthetic pathway, is required for releasing the Hedgehog ligand and directing germ cell migration. *PLoS Genet* **5:** e1000333. doi:10.1371/journal.pgen.1000333

Deshpande G, Zhou K, Wan JY, Friedrich J, Jourjine N, Smith D, Schedl P. 2013. The hedgehog pathway gene shifted functions together with the *hmgcr*-dependent isoprenoid biosynthetic pathway to orchestrate germ cell migration. *PLoS Genet* **9:** e1003720. doi:10.1371/journal.pgen.1003720

Deshpande G, Ng C, Jourjine N, Chiew JW, Dasilva J, Schedl P. 2023. Hedgehog signaling guides migration of primordial germ cells to the *Drosophila* somatic gonad. *Genetics* **225:** iyad165. doi:10.1093/genetics/iyad165

Diao F, Ironfield H, Luan H, Diao F, Shropshire WC, Ewer J, Marr E, Potter CJ, Landgraf M, White BH. 2015. Plug-and-play genetic access to *Drosophila* cell types using exchangeable exon cassettes. *Cell Rep* **10:** 1410–1421. doi:10.1016/j.celrep.2015.01.059

Eagle WVI, Yeboah-Kordieh DK, Niepielko MG, Gavis ER. 2018. Distinct *cis*-acting elements mediate targeting and clustering of *Drosophila* polar granule mRNAs. *Development* **145:** dev164657. doi:10.1242/dev.164657

Ephrussi A, Lehmann R. 1992. Induction of germ cell formation by oskar. *Nature* **358:** 387–392. doi:10.1038/358387a0

Fietz MJ, Jacinto A, Taylor AM, Alexandre C, Ingham PW. 1995. Secretion of the amino-terminal fragment of the Hedgehog protein is necessary and sufficient for hedgehog signalling in *Drosophila*. *Curr Biol* **5:** 643–650. doi:10.1016/S0960-9822(95)00129-1

Fischer-Vize JA, Rubin GM, Lehmann R. 1992. The *fat facets* gene is required for *Drosophila* eye and embryo development. *Development* **116:** 985–1000. doi:10.1242/dev.116.4.985

Foe VE, Odell GM, Edgar A. 1993. Mitosis and morphogenesis in the *Drosophila* embryo: point and counterpoint. In *The development of Drosophila melanogaster* (ed. Bate M, Arias AM), pp. 149–300. Cold Spring Harbor Laboratory Press, New York.

Fujioka M, Lear BC, Landgraf M, Yusibova GL, Zhou J, Riley KM, Patel NH, Jaynes JB. 2003. Even-skipped, acting as a repressor, regulates axonal projections in *Drosophila*. *Development* **130:** 5385–5400. doi:10.1242/dev.00770

Fyrberg C, Becker J, Barthmaier P, Mahaffey J, Fyrberg E. 1997. A *Drosophila* muscle-specific gene related to the mouse quaking locus. *Gene* **197:** 315–323. doi:10.1016/S0378-1119(97)00278-3

Galloni M, Gyurkovics H, Schedl P, Karch F. 1993. The bluetail transposon: evidence for independent *cis*-regulatory domains and domain boundaries in the bithorax complex. *EMBO J* **12:** 1087–1097. doi:10.1002/j.1460-2075.1993.tb05750.x

Garcia D, Shaw RJ. 2017. AMPK: mechanisms of cellular energy sensing and restoration of metabolic balance. *Mol Cell* **66:** 789–800. doi:10.1016/j.molcel.2017.05.032

Gavis ER, Lehmann R. 1994. Translational regulation of nanos by RNA localization. *Nature* **369:** 315–318. doi:10.1038/369315a0

Georgias C, Wasser M, Hinz U. 1997. A basic-helix-loop-helix protein expressed in precursors of *Drosophila* longitudinal visceral muscles. *Mech Dev* **69:** 115–124. doi:10.1016/S0925-4773(97)00169-X

Gilboa L, Lehmann R. 2006. Soma-germline interactions coordinate homeostasis and growth in the *Drosophila* gonad. *Nature* **443:** 97–100. doi:10.1038/nature05068

Gnocchi D, Pedrelli M, Hurt-Camejo E, Parini P. 2015. Lipids around the clock: focus on circadian rhythms and lipid metabolism. *Biology (Basel)* **4:** 104–132. doi:10.3390/biology4010104

Goodwin K, Emrich TA, Arnold S, McDole K. 2025. Primordial germ cells experience increasing physical confinement and DNA damage during migration in the mouse embryo. bioRxiv doi:10.1101/2025.03.03.641275

Greig S, Akam M. 1993. Homeotic genes autonomously specify one aspect of pattern in the *Drosophila* mesoderm. *Nature* **362:** 630–632. doi:10.1038/362630a0

Grimaldi C, Schumacher I, Boquet-Pujadas A, Tarbashevich K, Vos BE, Bandemer J, Schick J, Aalto A, Olivo-Marin JC, Betz T, et al. 2020. E-cadherin focuses protrusion formation at the front of migrating cells by impeding actin flow. *Nat Commun* **11:** 5397. doi:10.1038/s41467-020-19114-z

Häcker U, Perrimon N. 1998. *DRhoGEF2* encodes a member of the Dbl family of oncogenes and controls cell shape changes during gastrulation in *Drosophila*. *Genes Dev* **12:** 274–284. doi:10.1101/gad.12.2.274

Hansen CL, Pelegri F. 2021. Primordial germ cell specification in vertebrate embryos: phylogenetic distribution and conserved molecular features of preformation and induction. *Front Cell Dev Biol* **9:** 730332. doi:10.3389/fcell.2021.730332

Hanyu-Nakamura K, Sonobe-Nojima H, Tanigawa A, Lasko P, Nakamura A. 2008. *Drosophila* Pgc protein inhibits P-TEFb recruitment to chromatin in primordial germ cells. *Nature* **451:** 730–733. doi:10.1038/nature06498

Harmansa S, Alborelli I, Bieli D, Caussinus E, Affolter M. 2017. A nanobody-based toolset to investigate the role of protein localization and dispersal in *Drosophila*. *eLife* **6:** e22549. doi:10.7554/eLife.22549

Higgs HN, Pollard TD. 2000. Activation by Cdc42 and Pip2 of Wiskott–Aldrich syndrome protein (Wasp) stimulates actin nucleation by Arp2/3 complex. *J Cell Biol* **150:** 1311–1320. doi:10.1083/jcb.150.6.1311

Hu Q, Wolfner MF. 2019. The *Drosophila* Trpm channel mediates calcium influx during egg activation. *Proc Natl Acad Sci* **116:** 18994–19000. doi:10.1073/pnas.1906967116

Hunter C, Wieschaus E. 2000. Regulated expression of *nullo* is required for the formation of distinct apical and basal adherens junctions in the *Drosophila* blastoderm. *J Cell Biol* **150:** 391–402. doi:10.1083/jcb.150.2.391

Ingham PW, Taylor AM, Nakano Y. 1991. Role of the *Drosophila* patched gene in positional signalling. *Nature* **353:** 184–187. doi:10.1038/353184a0

Iwaki DD, Lengyel JA. 2002. A delta–notch signaling border regulated by engrailed/invected repression specifies boundary cells in the *Drosophila* hindgut. *Mech Dev* **114:** 71–84. doi:10.1016/S0925-4773(02)00061-8

Izquierdo E, Quinkler T, De Renzis S. 2018. Guided morphogenesis through optogenetic activation of Rho signalling during early *Drosophila* embryogenesis. *Nat Commun* **9:** 2366. doi:10.1038/s41467-018-04754-z

Jaglarz MK, Howard KR. 1995. The active migration of *Drosophila* primordial germ cells. *Development* **121:** 3495–3503. doi:10.1242/dev.121.11.3495

Jaszczak RG, Zussman JW, Wagner DE, Laird DJ. 2025. Comprehensive profiling of migratory primordial germ cells reveals niche-specific differences in non-canonical Wnt and Nodal-Lefty signaling in anterior vs posterior migrants. *eLife* **14:** RP103074. doi:10.7554/eLife.103074.3

Jemc JC. 2011. Somatic gonadal cells: the supporting cast for the germline. *Genesis* **49:** 753–775. doi:10.1002/dvg.20784

Jenkins AB, McCaffery JM, Van Doren M. 2003. *Drosophila* E-cadherin is essential for proper germ cell-soma interaction during gonad morphogenesis. *Development* **130:** 4417–4426. doi:10.1242/dev.00639

Jiang J, Levine M. 1993. Binding affinities and cooperative interactions with bHLH activators delimit threshold responses to the dorsal gradient morphogen. *Cell* **72:** 741–752. doi:10.1016/0092-8674(93)90402-C

Jin Z, Chen J, Huang H, Wang J, Lv J, Yu M, Guo X, Zhang Y, Cai T, Xi R. 2020. The *Drosophila* ortholog of mammalian transcription factor Sox9 regulates intestinal homeostasis and regeneration at an appropriate level. *Cell Rep* **31:** 107683. doi:10.1016/j.celrep.2020.107683

Kadam S, Ghosh S, Stathopoulos A. 2012. Synchronous and symmetric migration of *Drosophila* caudal visceral mesoderm cells requires dual input by two FGF ligands. *Development* **139:** 699–708. doi:10.1242/dev.068791

Kaneuchi T, Sartain CV, Takeo S, Horner VL, Buehner NA, Aigaki T, Wolfner MF. 2015. Calcium waves occur as *Drosophila* oocytes activate. *Proc Natl Acad Sci* **112:** 791–796. doi:10.1073/pnas.1420589112

Kassmer SH, Rodriguez D, Langenbacher AD, Bui C, De Tomaso AW. 2015. Migration of germline progenitor cells is directed by sphingosine-1-phosphate signalling in a basal chordate. *Nat Commun* **6:** 8565. doi:10.1038/ncomms9565

Katanaev VL, Ponzielli R, Sémériva M, Tomlinson A. 2005. Trimeric G protein-dependent frizzled signaling in *Drosophila*. *Cell* **120:** 111–122. doi:10.1016/j.cell.2004.11.014

Keller H, Zadeh AD, Eggli P. 2002. Localised depletion of polymerised actin at the front of Walker carcinosarcoma cells increases the speed of locomotion. *Cell Motil* **53:** 189–202. doi:10.1002/cm.10063

Kenwrick K, Mukherjee A, Renault AD. 2019. Hmgcr promotes a long-range signal to attract *Drosophila* germ cells independently of Hedgehog. *J Cell Sci* **132:** jcs232637. doi:10.1242/jcs.232637

Kilwein MD, Miller P, Lee KY, Osterfield M, Mogilner A, Shvartsman SY, Gavis ER. 2025. Formation of *Drosophila* germ cells requires spatial patterning of phospholipids. *Curr Biol* **35:** 1612–1621.e3. doi:10.1016/j.cub.2025.01.071

Kim JH, Hanlon CD, Vohra S, Devreotes PN, Andrew DJ. 2021. Hedgehog signaling and Tre1 regulate actin dynamics through PI(4,5)P_2 to direct migration of *Drosophila* embryonic germ cells. *Cell Rep* **34:** 108799. doi:10.1016/j.celrep.2021.108799

Kim HS, Parker DJ, Hardiman MM, Munkácsy E, Jiang N, Rogers AN, Bai Y, Brent C, Mobley JA, Austad SN, et al. 2023. Early-adulthood spike in protein translation drives aging via juvenile hormone/germline signaling. *Nat Commun* **14:** 5021. doi:10.1038/s41467-023-40618-x

Kina H, Yoshitani T, Hanyu-Nakamura K, Nakamura A. 2019. Rapid and efficient generation of GFP-knocked-in *Drosophila* by the CRISPR-Cas9-mediated genome editing. *Dev Growth Differ* **61:** 265–275. doi:10.1111/dgd.12607

Kirby RJ, Hamilton GM, Finnegan DJ, Johnson KJ, Jarman AP. 2001. *Drosophila* homolog of the myotonic dystrophy-associated gene, SIX5, is required for muscle and gonad development. *Curr Biol* **11:** 1044–1049. doi:10.1016/S0960-9822(01)00319-0

Knaut H, Steinbeisser H, Schwarz H, Nüsslein-Volhard C. 2002. An evolutionary conserved region in the vasa 3′UTR targets RNA translation to the germ cells in the zebrafish. *Curr Biol* **12:** 454–466. doi:10.1016/S0960-9822(02)00723-6

Kong D, Lv Z, Häring M, Lin B, Wolf F, Großhans J. 2019. In vivo optochemical control of cell contractility at single cell resolution. *EMBO Rep* **20:** e47755. doi:10.15252/embr.201947755

Konstantinides N, Holguera I, Rossi AM, Escobar A, Dudragne L, Chen YC, Tran TN, Martínez Jaimes AM, Özel MN, Simon F, et al. 2022. A complete temporal transcription factor series in the fly visual system. *Nature* **604:** 316–322. doi:10.1038/s41586-022-04564-w

Köprunner M, Thisse C, Thisse B, Raz E. 2001. A zebrafish *nanos*-related gene is essential for the development of primordial germ cells. *Genes Dev* **15:** 2877–2885. doi:10.1101/gad.212401

Kozlova T, Thummel CS. 2003. Essential roles for ecdysone signaling during *Drosophila* mid-embryonic development. *Science* **301:** 1911–1914. doi:10.1126/science.1087419

Kudron MM, Victorsen A, Gevirtzman L, Hillier LW, Fisher WW, Vafeados D, Kirkey M, Hammonds AS, Gersch J, Ammouri H, et al. 2018. The ModERN resource: genome-wide binding profiles for hundreds of *Drosophila* and *Caenorhabditis elegans* transcription factors. *Genetics* **208:** 937–949. doi:10.1534/genetics.117.300657

Kunwar PS, Starz-Gaiano M, Bainton RJ, Heberlein U, Lehmann R. 2003. Tre1, a G protein-coupled receptor, directs transepithelial migration of *Drosophila* germ cells. *PLoS Biol* **1:** e80. doi:10.1371/journal.pbio.0000080

Kunwar PS, Sano H, Renault AD, Barbosa V, Fuse N, Lehmann R. 2008. Tre1 GPCR initiates germ cell transepithelial migration by regulating *Drosophila melanogaster* E-cadherin. *J Cell Biol* **183:** 157–168. doi:10.1083/jcb.200807049

Kurogi Y, Mizuno Y, Imura E, Niwa R. 2021. Neuroendocrine regulation of reproductive dormancy in the fruit fly *Drosophila melanogaster*: a review of juvenile hormone-dependent regulation. *Front Ecol Evol* **9:** 715029. doi:10.3389/fevo.2021.715029

Kusch T, Storck T, Walldorf U, Reuter R. 2002. Brachyury proteins regulate target genes through modular binding sites in a cooperative fashion. *Genes Dev* **16:** 518–529. doi:10.1101/gad.213002

Lai Z, Fortini ME, Rubin GM. 1991. The embryonic expression patterns of zfh-1 and zfh-2, two *Drosophila* genes encoding novel zinc-finger homeodomain proteins. *Mech Dev* **34:** 123–134. doi:10.1016/0925-4773(91)90049-C

Leatherman JL, DiNardo S. 2008. Zfh-1 controls somatic stem cell self-renewal in the *Drosophila* testis and non-autonomously influences germline stem cell self-renewal. *Cell Stem Cell* **3:** 44–54. doi:10.1016/j.stem.2008.05.001

LeBlanc MG, Lehmann R. 2017. Domain-specific control of germ cell polarity and migration by multifunction Tre1 GPCR. *J Cell Biol* **216:** 2945–2958. doi:10.1083/jcb.201612053

Lee PT, Zirin J, Kanca O, Lin WW, Schulze KL, Li-Kroeger D, Tao R, Devereaux C, Hu Y, Chung V, et al. 2018. A gene-specific T2A-GAL4 library for *Drosophila*. *eLife* **7:** e35574. doi:10.7554/eLife.35574

Li LH, Gergen JP. 1999. Differential interactions between Brother proteins and Runt domain proteins in the *Drosophila* embryo and eye. *Development* **126:** 3313–3322. doi:10.1242/dev.126.15.3313

Li J, Xia F, Li WX. 2003a. Coactivation of STAT and Ras is required for germ cell proliferation and invasive migration in *Drosophila*. *Dev Cell* **5:** 787–798. doi:10.1016/S1534-5807(03)00328-9

Li MA, Alls JD, Avancini RM, Koo K, Godt D. 2003b. The large Maf factor traffic jam controls gonad morphogenesis in *Drosophila*. *Nat Cell Biol* **5:** 994–1000. doi:10.1038/ncb1058

Lin DM, Goodman CS. 1994. Ectopic and increased expression of fasciclin II alters motoneuron growth cone guidance. *Neuron* **13:** 507–523. doi:10.1016/0896-6273(94)90022-1

Lin H, Spradling AC. 1997. A novel group of *pumilio* mutations affects the asymmetric division of germline stem cells in the *Drosophila* ovary. *Development* **124:** 2463–2476. doi:10.1242/dev.124.12.2463

Lin B, Luo J, Lehmann R. 2020. Collectively stabilizing and orienting posterior migratory forces disperses cell clusters in vivo. *Nat Commun* **11:** 4477. doi:10.1038/s41467-020-18185-2

Lin B, Luo J, Lehmann R. 2022. An AMPK phosphoregulated RhoGEF feedback loop tunes cortical flow–driven amoeboid migration in vivo. *Sci Adv* **8:** eabo0323. doi:10.1126/sciadv.abo0323

Liu YJ, Le Berre M, Lautenschlaeger F, Maiuri P, Callan-Jones A, Heuzé M, Takaki T, Voituriez R, Piel M. 2015. Confinement and low adhesion induce fast amoeboid migration of slow mesenchymal cells. *Cell* **160:** 659–672. doi:10.1016/j.cell.2015.01.007

Logue JS, Cartagena-Rivera AX, Baird MA, Davidson MW, Chadwick RS, Waterman CM. 2015. Erk regulation of actin capping and bundling by Eps8 promotes cortex tension and leader bleb-based migration. *eLife* **4:** e08314. doi:10.7554/eLife.08314

Luo L, Liao YJ, Jan LY, Jan YN. 1994. Distinct morphogenetic functions of similar small GTPases: *Drosophila* Drac1 is involved in axonal outgrowth and myoblast fusion. *Genes Dev* **8:** 1787–1802. doi:10.1101/gad.8.15.1787

Mahowald AP. 1968. Polar granules of *Drosophila*. II: Ultrastructural changes during early embryogenesis. *J Exp Zool* **167:** 237–261. doi:10.1002/jez.1401670211

Martin-Bermudo MD, Dunin-Borkowski OM, Brown NH. 1997. Specificity of PS integrin function during embryogenesis resides in the α subunit extracellular domain. *EMBO J* **16:** 4184–4193. doi:10.1093/emboj/16.14.4184

Masukawa M, Ishizaki Y, Miura H, Hayashi M, Ota R, Kobayashi S. 2021. Male-biased protein expression in primordial germ cells, identified through a comparative study of UAS vectors in *Drosophila*. *Sci Rep* **11:** 21482. doi:10.1038/s41598-021-00729-1

Mazzalupo S, Cooley L. 2006. Illuminating the role of caspases during *Drosophila* oogenesis. *Cell Death Differ* **13:** 1950–1959. doi:10.1038/sj.cdd.4401892

Molyneaux KA, Stallock J, Schaible K, Wylie C. 2001. Time-lapse analysis of living mouse germ cell migration. *Dev Biol* **240:** 488–498. doi:10.1006/dbio.2001.0436

Moore LA, Broihier HT, Van Doren M, Lunsford LB, Lehmann R. 1998. Identification of genes controlling germ cell migration and embryonic gonad formation in *Drosophila*. *Development* **125:** 667–678. doi:10.1242/dev.125.4.667

Morita M, Morimoto M, Teramoto T, Ikenouchi J, Atsuta Y, Saito D. 2023. The SOCE system is critical for membrane bleb formation to drive avian primordial germ cell migration. bioRxiv doi:10.1101/2023.06.12.544577

Nilangekar KS, Shravage BV. 2019. Mitochondrial redox sensor for *Drosophila* female germline stem cells. In *Autophagy in differentiation and tissue maintenance: methods and protocols* (ed. Turksen K), pp. 13–20. Springer, New York.

Nilangekar K, Murmu N, Sahu G, Shravage BV. 2019. Generation and characterization of germline-specific autophagy and mitochondrial reactive oxygen species reporters in *Drosophila*. *Front Cell Dev Biol* **7:** 47. doi:10.3389/fcell.2019.00047

Pae J, Cinalli RM, Marzio A, Pagano M, Lehmann R. 2017. GCL and CUL3 control the switch between cell lineages by mediating localized degradation of an RTK. *Dev Cell* **42:** 130–142.e7. doi:10.1016/j.devcel.2017.06.022

Paluch EK, Aspalter IM, Sixt M. 2016. Focal adhesion–independent cell migration. *Annu Rev Cell Dev Biol* **32:** 469–490. doi:10.1146/annurev-cellbio-111315-125341

Panchal T, Chen X, Alchits E, Oh Y, Poon J, Kouptsova J, Laski FA, Godt D. 2017. Specification and spatial arrangement of cells in the germline stem cell niche of the *Drosophila* ovary depend on the Maf transcription factor

Traffic jam. *PLoS Genet* **13:** e1006790. doi:10.1371/journal.pgen.1006790

Pares G, Ricardo S. 2016. FGF control of E-cadherin targeting in the *Drosophila* midgut impacts on primordial germ cell motility. *J Cell Sci* **129:** 354–366. doi:10.1242/jcs.174284

Petrie RJ, Koo H, Yamada KM. 2014. Generation of compartmentalized pressure by a nuclear piston governs cell motility in a 3D matrix. *Science* **345:** 1062–1065. doi:10.1126/science.1256965

Poincloux R, Collin O, Lizárraga F, Romao M, Debray M, Piel M, Chavrier P. 2011. Contractility of the cell rear drives invasion of breast tumor cells in 3D Matrigel. *Proc Natl Acad Sci* **108:** 1943–1948. doi:10.1073/pnas.1010396108

Rangan P, DeGennaro M, Jaime-Bustamante K, Coux RX, Martinho RG, Lehmann R. 2009. Temporal and spatial control of germ-plasm RNAs. *Curr Biol* **19:** 72–77. doi:10.1016/j.cub.2008.11.066

Renault AD, Ricardo S, Kunwar PS, Santos A, Starz-Gaiano M, Stein JA, Lehmann R. 2009. Hedgehog does not guide migrating *Drosophila* germ cells. *Dev Biol* **328:** 355–362. doi:10.1016/j.ydbio.2009.01.042

Renault AD, Kunwar PS, Lehmann R. 2010. Lipid phosphate phosphatase activity regulates dispersal and bilateral sorting of embryonic germ cells in *Drosophila. Development* **137:** 1815–1823. doi:10.1242/dev.046110

Rhymer E, Anllo L. 2025. Generation of *six4*-nlsRFP: a red somatic gonadal nuclear marker for live imaging *Drosophila* gonadogenesis. *MicroPubl Biol* doi:10.17912/micropub.biology.001652

Ricardo S, Lehmann R. 2009. An ABC transporter controls export of a *Drosophila* germ cell attractant. *Science* **323:** 943–946. doi:10.1126/science.1166239

Robertson TF, Schrope J, Zwick Z, Rindy J, Horn A, Hou Y, Huttenlocher A. 2025. Live imaging in zebrafish reveals tissue-specific strategies for amoeboid migration. *Development* **152:** dev204351. doi:10.1242/dev.204351

Roelen BAJ, Chuva de Sousa Lopes SM. 2022. Stay on the road: from germ cell specification to gonadal colonization in mammals. *Philos Trans R Soc Lond B Biol Sci* **377:** 20210259. doi:10.1098/rstb.2021.0259

Rørth P. 1998. Gal4 in the *Drosophila* female germline. *Mech Dev* **78:** 113–118. doi:10.1016/S0925-4773(98)00157-9

Ruprecht V, Wieser S, Callan-Jones A, Smutny M, Morita H, Sako K, Barone V, Ritsch-Marte M, Sixt M, Voituriez R, et al. 2015. Cortical contractility triggers a stochastic switch to fast amoeboid cell motility. *Cell* **160:** 673–685. doi:10.1016/j.cell.2015.01.008

Sanchez-Díaz I, Rosales-Bravo F, Reyes-Taboada JL, Covarrubias AA, Narvaez-Padilla V, Reynaud E. 2015. The Esg gene is involved in nicotine sensitivity in *Drosophila melanogaster. PLoS One* **10:** e0133956. doi:10.1371/journal.pone.0133956

Sanicola M, Sekelsky J, Elson S, Gelbart WM. 1995. Drawing a stripe in *Drosophila* imaginal disks: negative regulation of decapentaplegic and patched expression by engrailed. *Genetics* **139:** 745–756. doi:10.1093/genetics/139.2.745

Sano H, Renault AD, Lehmann R. 2005. Control of lateral migration and germ cell elimination by the *Drosophila melanogaster* lipid phosphate phosphatases Wunen and Wunen 2. *J Cell Biol* **171:** 675–683. doi:10.1083/jcb.200506038

Sano H, Kunwar PS, Renault AD, Barbosa V, Clark IB, Ishihara S, Sugimura K, Lehmann R. 2012. The *Drosophila* actin regulator ENABLED regulates cell shape and orientation during gonad morphogenesis. *PLoS One* **7:** e52649. doi:10.1371/journal.pone.0052649

Santos AC, Lehmann R. 2004. Isoprenoids control germ cell migration downstream of HMGCoA reductase. *Dev Cell* **6:** 283–293. doi:10.1016/S1534-5807(04)00023-1

Sarov M, Barz C, Jambor H, Hein MY, Schmied C, Suchold D, Stender B, Janosch S, KJ VV, Krishnan RT, et al. 2016. A genome-wide resource for the analysis of protein localisation in *Drosophila. eLife* **5:** e12068. doi:10.7554/eLife.12068

Sarov-Blat L, So WV, Liu L, Rosbash M. 2000. The *Drosophila* takeout gene is a novel molecular link between circadian rhythms and feeding behavior. *Cell* **101:** 647–656. doi:10.1016/S0092-8674(00)80876-4

Seifert JRK, Lehmann R. 2012. *Drosophila* primordial germ cell migration requires epithelial remodeling of the endoderm. *Development* **139:** 2101–2106. doi:10.1242/dev.078949

Shiga Y, Tanaka-Matakatsu M, Hayashi S. 1996. A nuclear GFP/β-galactosidase fusion protein as a marker for morphogenesis in living *Drosophila. Dev Growth Differ* **38:** 99–106. doi:10.1046/j.1440-169X.1996.00012.x

Silva D, Olsen KW, Bednarz MN, Droste A, Lenkeit CP, Chaharbakhshi E, Temple-Wood ER, Jemc JC. 2016. Regulation of gonad morphogenesis in *Drosophila* melanogaster by BTB family transcription factors. *PLoS One* **11:** e0167283. doi:10.1371/journal.pone.0167283

Slaidina M, Lehmann R. 2017. Quantitative differences in a single maternal factor determine survival probabilities among *Drosophila* germ cells. *Curr Biol* **27:** 291–297. doi:10.1016/j.cub.2016.11.048

Smrcka AV. 2008. G protein βγ subunits: central mediators of G protein-coupled receptor signaling. *Cell Mol Life Sci* **65:** 2191–2214. doi:10.1007/s00018-008-8006-5

Staehling-Hampton K, Hoffmann FM, Baylies MK, Rushtont E, Bate M. 1994. Dpp induces mesodermal gene expression in *Drosophila. Nature* **372:** 783–786. doi:10.1038/372783a0

Stallock J, Molyneaux K, Schaible K, Knudson CM, Wylie C. 2003. The pro-apoptotic gene Bax is required for the death of ectopic primordial germ cells during their migration in the mouse embryo. *Development* **130:** 6589–6597. doi:10.1242/dev.00898

Starz-Gaiano M, Cho NK, Forbes A, Lehmann R. 2001. Spatially restricted activity of a *Drosophila* lipid phosphatase guides migrating germ cells. *Development* **128:** 983–991. doi:10.1242/dev.128.6.983

Stepanik V, Dunipace L, Bae YK, Macabenta F, Sun J, Trisnadi N, Stathopoulos A. 2016. The migrations of *Drosophila* muscle founders and primordial germ cells are interdependent. *Development* **143:** 3206–3215. doi:10.1242/dev.134346

Sun J, Zhang C, Gao F, Stathopoulos A. 2023. Single-cell transcriptomics illuminates regulatory steps driving anterior-posterior patterning of *Drosophila* embryonic mesoderm. *Cell Rep* **42:** 113289. doi:10.1016/j.celrep.2023.113289

Tanentzapf G, Devenport D, Godt D, Brown NH. 2007. Integrin-dependent anchoring of a stem-cell niche. *Nat Cell Biol* **9:** 1413–1418. doi:10.1038/ncb1660

Tarayrah-Ibraheim L, Maurice EC, Hadary G, Ben-Hur S, Kolpakova A, Braun T, Peleg Y, Yacobi-Sharon K, Arama E. 2021. DNase II mediates a parthanatos-like developmental cell death pathway in *Drosophila* primordial germ cells. *Nat Commun* **12:** 2285. doi:10.1038/s41467-021-22622-1

Telley IA, Gáspár I, Ephrussi A, Surrey T. 2012. Aster migration determines the length scale of nuclear separation in the *Drosophila* syncytial embryo. *J Cell Biol* **197:** 887–895. doi:10.1083/jcb.201204019

Terayama K, Kataoka K, Morichika K, Orii H, Watanabe K, Mochii M. 2013. Developmental regulation of locomotive activity in *Xenopus* primordial germ cells. *Dev Growth Differ* **55:** 217–228. doi:10.1111/dgd.12018

Thuma L, Carter D, Weavers H, Martin P. 2018. *Drosophila* immune cells extravasate from vessels to wounds using Tre1 GPCR and Rho signaling. *J Cell Biol* **217:** 3045–3056. doi:10.1083/jcb.201801013

Touhara K, Inglese J, Pitcher JA, Shaw G, Lefkowitz RJ. 1994. Binding of G protein beta gamma-subunits to pleckstrin homology domains. *J Biol Chem* **269:** 10217–10220. doi:10.1016/S0021-9258(17)34048-6

Tozluoğlu M, Tournier AL, Jenkins RP, Hooper S, Bates PA, Sahai E. 2013. Matrix geometry determines optimal cancer cell migration strategy and modulates response to interventions. *Nat Cell Biol* **15:** 751–762. doi:10.1038/ncb2775

Tripathy R, Kunwar PS, Sano H, Renault AD. 2014. Transcriptional regulation of *Drosophila* gonad formation. *Dev Biol* **392:** 193–208. doi:10.1016/j.ydbio.2014.05.026

Truszkowski L, Batur D, Long H, Tarbashevich K, Vos BE, Trappmann B, Raz E. 2023. Primordial germ cells adjust their protrusion type while migrating in different tissue contexts in vivo. *Development* **150:** dev200603. doi:10.1242/dev.200603

Tu R, Tang XA, Xu R, Ping Z, Yu Z, Xie T. 2023. Gap junction–transported cAMP from the niche controls stem cell progeny differentiation. *Proc Natl Acad Sci* **120:** e2304168120. doi:10.1073/pnas.2304168120

Van Doren M, Broihier HT, Moore LA, Lehmann R. 1998a. HMG-CoA reductase guides migrating primordial germ cells. *Nature* **396:** 466–469. doi:10.1038/24871

Van Doren M, Williamson AL, Lehmann R. 1998b. Regulation of zygotic gene expression in *Drosophila* primordial germ cells. *Curr Biol* **8:** 243–246. doi:10.1016/S0960-9822(98)70091-0

Van Doren M, Mathews WR, Samuels M, Moore LA, Broihier HT, Lehmann R. 2003. Fear of intimacy encodes a novel transmembrane protein required for gonad morphogenesis in *Drosophila*. *Development* **130:** 2355–2364. doi:10.1242/dev.00454

Wang W, Tindell N, Yan S, Yoder JH. 2013. Homeotic functions of the Teashirt transcription factor during adult *Drosophila* development. *Biol Open* **2:** 18–29. doi:10.1242/bio.20122915

Warder BN, Nelson KA, Sui J, Anllo L, DiNardo S. 2024. An actomyosin network organizes niche morphology and responds to feedback from recruited stem cells. *Curr Biol* **34:** 3917–3930.e6. doi:10.1016/j.cub.2024.07.041

Warrior R. 1994. Primordial germ cell migration and the assembly of the *Drosophila* embryonic gonad. *Dev Biol* **166:** 180–194. doi:10.1006/dbio.1994.1306

Wawersik M, Van Doren M. 2005. *Nanos* is required for formation of the spectrosome, a germ cell-specific organelle. *Dev Dyn* **234:** 22–27. doi:10.1002/dvdy.20503

Wen D, Rivera-Perez C, Abdou M, Jia Q, He Q, Liu X, Zyaan O, Xu J, Bendena WG, Tobe SS, et al. 2015. Methyl farnesoate plays a dual role in regulating *Drosophila* metamorphosis. *PLoS Genet* **11:** e1005038. doi:10.1371/journal.pgen.1006559

Wolke U, Weidinger G, Köprunner M, Raz E. 2002. Multiple levels of posttranscriptional control lead to germ line-specific gene expression in the zebrafish. *Curr Biol* **12:** 289–294. doi:10.1016/S0960-9822(02)00679-6

Xiao H, Hrdlicka LA, Nambu JR. 1996. Alternate functions of the single-minded and rhomboid genes in development of the *Drosophila* ventral neuroectoderm. *Mech Dev* **58:** 65–74. doi:10.1016/S0925-4773(96)00559-X

Yamada KM, Sixt M. 2019. Mechanisms of 3D cell migration. *Nat Rev Mol Cell Biol* **20:** 738–752. doi:10.1038/s41580-019-0172-9

Yamada Y, Davis KD, Coffman CR. 2008. Programmed cell death of primordial germ cells in *Drosophila* is regulated by p53 and the Outsiders monocarboxylate transporter. *Development* **135:** 207–216. doi:10.1242/dev.010389

Yoshiura S, Ohta N, Matsuzaki F. 2012. Tre1 GPCR signaling orients stem cell divisions in the *Drosophila* central nervous system. *Dev Cell* **22:** 79–91. doi:10.1016/j.devcel.2011.10.027

Zhang N, Zhang J, Cheng Y, Howard K. 1996. Identification and genetic analysis of wunen, a gene guiding *Drosophila melanogaster* germ cell migration. *Genetics* **143:** 1231–1241. doi:10.1093/genetics/143.3.1231

Zhang N, Zhang J, Purcell KJ, Cheng Y, Howard K. 1997. The *Drosophila* protein Wunen repels migrating germ cells. *Nature* **385:** 64–67. doi:10.1038/385064a0

Zhang Y, Rózsa M, Liang Y, Bushey D, Wei Z, Zheng J, Reep D, Broussard GJ, Tsang A, Tsegaye G, et al. 2023. Fast and sensitive GCaMP calcium indicators for imaging neural populations. *Nature* **615:** 884–891. doi:10.1038/s41586-023-05828-9

Zhao S, Chen D, Geng Q, Wang Z. 2013. The highly conserved LAMMER/CLK2 protein kinases prevent germ cell overproliferation in *Drosophila*. *Dev Biol* **376:** 163–170. doi:10.1016/j.ydbio.2013.01.023

Zirin J, Jusiak B, Lopes R, Ewen-Campen B, Bosch JA, Risbeck A, Forman C, Villalta C, Hu Y, Perrimon N. 2024. Expanding the *Drosophila* toolkit for dual control of gene expression. *eLife* **12:** RP94073. doi:10.7554/eLife.94073

Neighboring Cells as Living Substrates for Guiding Collective Cell Migration during Development

Hoang Anh Le[1] and Roberto Mayor[1,2]

[1]Department of Cell and Developmental Biology, Anatomy Building, University College London, London WC1E 6BT, United Kingdom

[2]Center for Integrative Biology, Faculty of Sciences, Universidad Mayor, 7500994 Providencia, Santiago, Chile

Correspondence: anh.le@ucl.ac.uk; r.mayor@ucl.ac.uk

As cells migrate inside the body, they encounter various biochemical and physical cues that provide them with directional guidance. In the past 20 years or so, there has been a significant shift in the effort to understand how physical factors contribute to cellular behaviors. Nevertheless, much of the research has been focused on the interactions between migrating cells and the extracellular matrix in vitro as these are simpler and more accessible models, while neglecting the importance of the cellular environment, which often requires in vivo model systems. With the development of new technology along with the appropriate choice of model organisms, the interesting topic of cell-on-cell interaction during migration is beginning to unravel. In this review, we will take a deep dive into some of the recent results that demonstrate how the biophysics of the cellular environment can impact cell migration, with a strong focus on the use of in vivo model systems, naming the *Drosophila* border cells, the *Xenopus* cephalic neural crest, and the zebrafish posterior lateral line primordium.

Historically, cell migration research has been largely focused on cells migrating on plastic surfaces coated with some matrix proteins. The easy accessibility of this approach has given us a deep knowledge of the molecular mechanisms and behaviors of migrating cells. However, approaches like this largely ignore a very fundamental fact that within a living organism, cells do not exist in isolation but are surrounded by matrix networks and other cells in a three-dimensional space. Therefore, understanding how cells behave and interact in their native histological environment remains the ultimate goal of migration research.

Over the years, the clever use of various animal model systems combined with the development of imaging techniques has helped us approach this question in a more physiologically relevant manner. One interesting aspect that we soon realized is that apart from chemistry, physical cues were a major force at play in guiding cell migration in an organism. As cells migrate through tissue, they encounter various pathways and obstacles of different physical properties

Cite this article as *Cold Spring Harb Perspect Biol* doi: 10.1101/cshperspect.a041741

created by the extracellular matrix (ECM)—a network of fibrous proteins that fills the interstitial space and the surrounding neighbor cells. The role of cell–matrix interactions has been extensively studied and reviewed in other articles (Humphrey et al. 2014; Yamada et al. 2022; Saraswathibhatla et al. 2023), yet our understanding of the contribution of cells that act as substrates to guide migrating cells remains significantly lacking.

In this review, we focus exclusively on the roles of the physical interactions between collective migrating cells and their surrounding cellular environment in vivo. We will discuss the various physical factors, such as stiffness, shape, and topography of the cellular environment and their impact on migrating cell clusters. While the typical cell–matrix mechanical interactions require almost exclusively integrins (Lo et al. 2000; Elosegui-Artola et al. 2018; Winkler et al. 2020) as a hub for force transmission, the mechanics of cell–cell interactions rely more on the function of cadherins (Leckband and de Rooij 2014; Das et al. 2015; Yap et al. 2018; Mishra et al. 2019; Angulo-Urarte et al. 2020). We will draw on examples from three different biological systems: border cell migration in the *Drosophila* egg chamber, neural crest cells in the *Xenopus* embryos, and the posterior lateral line primordium (pLLP) in the zebrafish embryos. We hope that this review helps to demonstrate the importance and abundance of cell-on-cell interactions, particularly in a physiologically relevant context, and bring some of the focus and effort in the mechanobiology field to better understand this lesser-known aspect in modulating collective cell migration.

Drosophila melanogaster BORDER CELL CLUSTER

Perhaps one of the most prominent examples to study the biophysics of cell–cell interactions in vivo is the migration of border cell clusters during *Drosophila*'s egg chamber development (Montell et al. 2012; Mishra et al. 2019). Anatomically speaking, the egg chamber at stage 8, which is the onset of the border cells' migration journey, is composed of three parts (Fig. 1A): the innermost layer is composed of 15 nurse cells clustered on the anterior, and one oocyte resides on the posterior of the chamber; surrounding this is a layer of ~650 epithelial follicle cells; and on the outermost is an ECM layer where laminin is one of the major proteins (Gutzeit et al. 1991; Montell et al. 2012). At the initiation of the migration process, two specific follicle epithelial cells that have been predesignated in previous developmental stages, called polar cells, release a cytokine called unpaired (UPD). In coordination with the hormone ecdysone, UPD triggers the expression of many migration-related genes in ~4–8 surrounding epithelial cells, which subsequently delaminate from the epithelium and become border cells. The two polar cells are positioned at the center of the border cell cluster and are immotile on their own, while the border cells organize around to form a moving chamber escorting these cells through the tightly packed space between nurse cells toward the oocyte. The border cell–polar cell cluster, once it reaches the oocyte, is involved in the formation of a structure known as micropyle, which is where sperm can enter for fertilization of the oocyte (Horne-Badovinac 2020).

The interior of the egg chamber contains little to no matrix proteins (Montell et al. 2012); therefore, the border cell cluster truly is a cell-on-cell migrating system and relies on cadherin interactions. Within the cluster, border cells retain their epithelial characteristics by maintaining DE-cadherin junctions with each other and with the polar cell pair. In contrast, a significantly lower DE-cadherin signal was detected at the interface between border cells and nurse cells (Cai et al. 2014), yet this remains vital for the migration of the cluster (Fig. 1A). Interestingly, in DE-cadherin null mutants or when DE-cadherin was abolished in all polar, border, and nurse cells, cells within the cluster remained together (Niewiadomska et al. 1999). However, the cluster would split apart if the junctions between border cells and polar cells were weakened while those between border cells and nurse cells remained intact (Cai et al. 2014). These results suggest that apart from structural functions, DE-cadherin also plays a role in modulating mi-

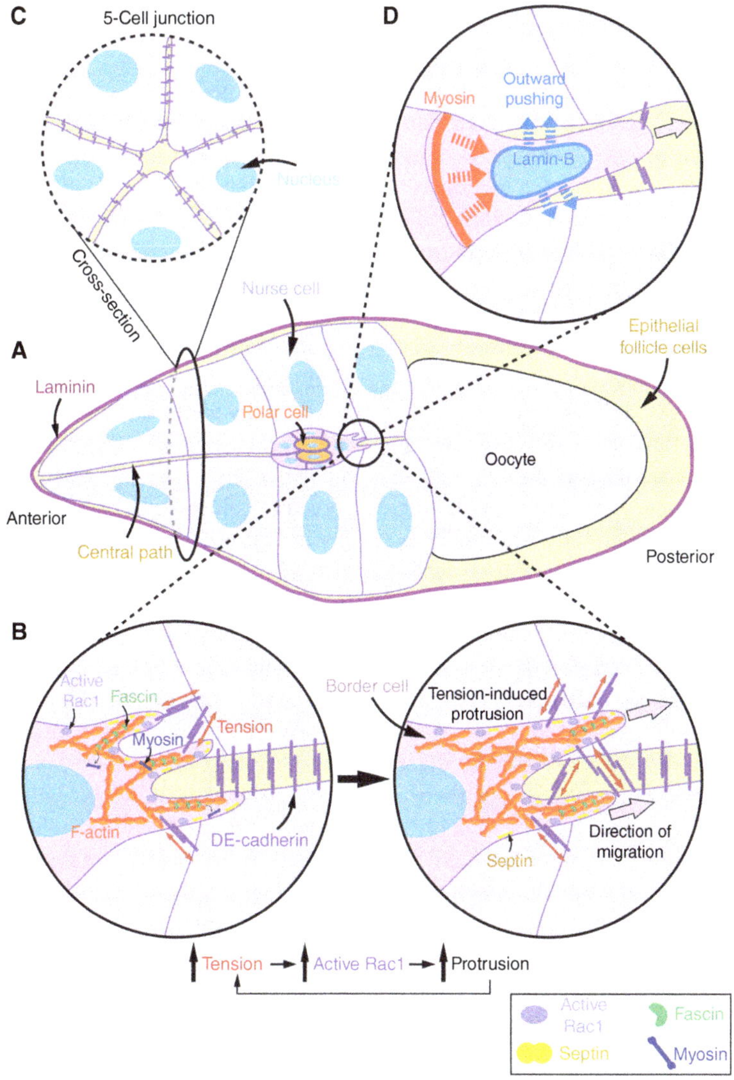

Figure 1. The *Drosophila* border cell cluster uses cadherin interactions with nurse cells and a nuclear wedging mechanism to guide their collective migration. (*A*) The general anatomy of a stage 9 *Drosophila*'s egg chamber. The border cell cluster migrates from the anterior side toward the posterior side, where the oocyte resides. DE-cadherin (dark purple) is at a higher level between polar cells (light yellow) and border cells (light pink) (thicker purple lines) than between polar cells and nurse cells (purple). (*B*) The positive feedback loop between border cells and nurse cells: Border cells squeeze between nurse cells and form homotypic DE-cadherin adhesions with the surrounding nurse cells to pull themselves forward. These DE-cadherin interactions create tensions (red double-headed arrows), which trigger the recruitment of active Rac1 (light purple) to the cell's leading edge, driving actin polymerization (orange), which in turn allows for more DE-cadherin interactions and more protrusions to form. Fascin (green) bundles actin filaments together and displaces myosin (dark blue), therefore reducing actomyosin contractility and cellular stiffness. The level of septin (yellow) also controls the shape of border cells. Too high levels cause the border cell cluster to become more rugged in texture, while too low levels cause the cluster to appear smoother. Both impair migration. (*C*) A cross-section through the egg chamber shows the 5-cell junction forming the central path for the border cell cluster to move through. (*D*) Actomyosin contractility (orange) from the back pushes the nucleus of the leading border cell forward to wedge open a path for the cluster to move through. This function of the nucleus is dependent on the level of lamin-B (dark blue).

gration. This was further demonstrated when DE-cadherin was depleted from the outer border cells, and the cluster's migration persistence was significantly decreased. When DE-cadherin was abolished from nurse cells, this caused the border cell cluster to wander and diverge from their usual path, even when natural chemoattractants were still present (Cai et al. 2014; Dai et al. 2020). In contrast, when DE-cadherin was overexpressed homogenously in nurse cells, the cluster's speed decreased. These data strongly suggest that the traction generated between the border cell cluster and its surrounding nurse cells via cadherins is one of the major regulators of their migration. Although the literature tends to refer to this migration as being frictionally guided, whether one can truly refer to this as such remains debatable. Friction is nonspecific by nature, meaning any irregularity or texture on the interacting surfaces, and, in this case, any transmembrane protein, should be able to generate such traction and therefore should affect the border cell cluster's migration. This theory can easily be tested by overexpressing or depleting any random transmembrane protein in border cells or nurse cells, but this has yet to be done. Furthermore, when DE-cadherin was overexpressed in half of the nurse cells on one side of the border cell cluster, effectively creating higher friction on one side relative to the other side, no significant change in the border cell cluster's migration was observed (Dai et al. 2020). This suggests that perhaps it is more about the specific optimal number of functioning cadherin–cadherin interactions that is important rather than any nonspecific friction force. A recent study also identified septin—a GTP-binding protein that is capable of self-oligomerization to form filaments (Mostowy and Cossart 2012)—to be a controller of surface texture for the border cell cluster (Fig. 1B). Too low a level of septin caused the cluster to be smoother, while too high a level caused the cluster to be more rugged in texture (Gabbert et al. 2023). In both cases, the cluster failed to migrate. This further favors the idea that it is the specific DE-cadherin–cadherin interaction rather than any nonspecific friction force that provides the traction needed to drag the cluster forward. This was confirmed using a fluorescence resonance energy transfer-based cadherin tension sensor where higher tension was detected at the front of the border cell cluster (Fig. 1B; Cai et al. 2014). It is believed that the strong cadherin–cadherin interaction at the front promotes protrusion formation possibly through the recruitment of the active form of the small GTPase Rac1—a major regulator of the actin cytoskeleton (Nobes and Hall 1995), which then allows for more cadherin interactions with the nurse cells, and drives more protrusion formation, effectively forming a positive feedback loop. Although Rac1 was observed to colocalize with E-cadherin in Madin–Darby canine kidney (MDCK) cells (Nakagawa et al. 2001), a precise mechanism in vivo is yet to be determined.

The significance of the physical cellular environment is highlighted even further when we consider the fact that border cell clusters can, in theory, choose any of the 40 or so different pathways that exist between nurse cells to get to the oocyte (Dai et al. 2020), yet only the path at the center of the egg chamber is chosen. It was then discovered that the way cells puzzle up together can influence other cells' behavior. For example, 2-cell junctions are much harder to squeeze in between compared to 3-cell junctions since the space between two side-by-side cells is essentially nonexistent. Inside the egg chamber, the central path is formed by a 5-cell junction (Fig. 1C), which allows for protrusions from the leading border cells to enter through, essentially choosing the path of least resistance. The preference for high-cell-number junctions also exists in other systems, such as when multiciliated cells probe for 3-cell junctions to intercalate into the ectoderm during *Xenopus* embryo development (Ventura et al. 2022), suggesting this is a general biological phenomenon. As the protrusion is threaded through the central gap, the border cell's nucleus is also pushed forward by myosin contractility from behind to act as a wedge to widen the gap for the entire cluster to slowly move through (Fig. 1D; Penfield and Montell 2023). Nuclear deformation is most prominent in the leader cell. It is also important to note that this is not the same as the nuclear piston model used by cancer cells (Petrie et al. 2014, 2017; Lee et al. 2021). In this model, the propelled nucleus formed a sealed cork separating the front and the back of the cell, leading to an increase in the intracellular pressure at the front protrusion to push the cell forward (Le and Mayor 2023). In the case of border cell migration, no such intracellular separation was detected as measured using fluorescence photoactivation (Penfield and Montell 2023), and instead, the nucleus quite literally acted as a physical stopper to prevent the central gap from closing up. Interestingly, this function of the nucleus is dependent on lamin-B since depleting it causes the nucleus to slide back toward the cell body and causes myosin to accumulate at the front. Lamin is a family of structural proteins that support the mechanical properties of the nucleus, with lamin-A tending to be more prevalent in stiffer tissue, while lamin-B is more prevalent in softer ones (Penfield and Montell 2023). Although the literature on the role of different lamin isoforms in nuclear stiffness is complicated, a few studies have indicated that lamin-B acts to soften and make the nucleus more pliable (Ferrera et al. 2014; Kim et al. 2021). Therefore, it is tempting to speculate that when lamin-B is depleted from border cells, their nuclei become stiffer and less deformable, causing it to slide back upon the constriction recoil from the 5-way cell junction, similar to how a wet bar of soap slips out of your hand when you try to squeeze on it. The importance of the topology of the cellular environment was also demonstrated further when it was recently shown that reducing the overall compaction of the egg chamber by reducing the level of laminin in the outer basement membrane allowed for the border cell cluster to migrate faster (Molina López et al. 2023). In laminin-knockdown egg chambers, the border cell cluster formed more protrusions, increased their rotation speed, and overall migrated faster, which was likely due to more space being available and a lower level of crowding from the surrounding nurse cells. It would also be interesting to test whether the nuclear wedging mechanism would still be at play in this scenario.

Another factor that is intricately linked with nurse cells' topology is stiffness. Recent evidence pointed out that the migrating border cell cluster could actively influence the stiffness of their surrounding tissue, and vice versa (Lamb et al. 2021). Fascin is an actin-bundling protein (Jansen et al. 2011) and is thought to negatively regulate nonmuscle myosin activity (Fig. 1B; Elkhatib et al. 2014). When the level of fascin is high, actin filaments are tightly packed, which effectively blocks myosin molecules from accessing binding sites on the filaments, therefore reducing the cellular stiffness. When fascin was depleted from nurse cells, this increased their apparent stiffness and triggered myosin accumulation in border cells (Lamb et al. 2021), an observation that agrees with a previous finding (Aranjuez et al. 2016). When fascin was depleted from border cells, the stiffness of nurse cells also increased as measured by atomic force microscopy (AFM). In both cases, this impeded the border cell cluster's migration. How this stiffness is relayed across different cells remains an interesting question.

Overall, border cell migration has proven to be an immensely powerful system for studying cell-on-cell interactions in vivo. However, direct mechanical manipulations and validations in *Drosophila* tissue in vivo can be challenging. This disadvantage can be overcome by utilizing another model organism, the *Xenopus* embryo. The most direct evidence of reciprocal interactions between a moving cell cluster and the cellular environment during development has been demonstrated better using this system, which we will be discussing in the next segment.

XENOPUS EMBRYO CEPHALIC NEURAL CREST CLUSTER

Possibly one of the most classical models of collective cell migration, the neural crest has given us a significantly better understanding of the biology of cell–tissue interactions. The neural crest is a group of migratory multipotent stem cells that will eventually give rise to various cell types such as neurons, glial cells, and melanocytes (Mayor and Theveneau 2013; Szabó and Mayor 2018). In *Xenopus* embryos, the neural crest population is first specified at neurulation on the dorsal side at around stage 13. Similar to how border cells delaminate from the epithelium in *Drosophila* egg chambers, the neural crest cells also enter a delamination process to leave the neuroepithelium layer, which is then followed by a series of characteristic molecular changes known as an epithelial-to-mesenchymal transition (EMT) (Theveneau and Mayor 2012). Depending on the anatomical location of these neural crest populations in an embryo, they can be classified as either cephalic (head), vagal, or trunk neural crest. Arguably, the most studied population is the cephalic neural crest, which will be the focus of this section. At the mid-neurula stage (stage 16), the delamination process begins. As the cell cluster migrates dorsoventrally, the neural crest cells start to separate into distinct streams (Fig. 2A) and progressively gain more mesenchymal markers, one of which is the switch from the cell–cell junction molecule E-cadherin to N-cadherin via a complex gene regulatory network, which includes a Twist-induced HIF1α stabilization signaling axis (Mayor and Theveneau 2013; Scarpa et al. 2015). This also makes the neural crest a popular model system for studying cancer cell biology, as many of the cellular activities and molecular signals are shared between migrating neural crest and metastatic cancer cells (Powell et al. 2013).

Inside an embryo, neural crest cells are sandwiched between two layers of fibronectin, one is deposited underneath the ectoderm and the other is on top of the mesoderm, forming a thin cage that can be slowly opened as the cell streams move (Fig. 2B). This layer of fibronectin provides the substrate that neural crest cells use to adhere to albeit does not contribute much to the mechanical cue that neural crest cells sense, but rather it is the cellular component of the surrounding tissue that does (Barriga et al. 2018). Neural crest cells migrate on top of the mesoderm as their direct substrate. Using AFM, it was shown that as the embryos got older, the head mesoderm gradually became stiffer, from ~50 to ~150 Pa. Grafting a nonmigrating population of neural crest cells onto an older premigratory embryo triggered premature migration. However, when a migratory neural crest was grafted onto early, nonmigratory embryos, their migration was inhibited. This evidence strongly suggests a correlation between increasing substrate stiffness and the propensity to migrate of the neural crest. Directly decreasing the stiffness of the mesoderm layer by either overexpressing constitutively active myosin phosphatase, depleting myosin contractility with morpholino, or mechanically ablating the mesoderm impaired the migration of neural crest cells in vivo. The reverse was also true since increasing the stiffness of the mesoderm by overexpressing the constitutively active myosin light chain or by pressing on the embryo with a cantilever enhanced neural crest migration. Knocking down the integrin–vinculin axis impaired neural crest cell migration in vivo without affecting their adhesion to fibronectin ex vivo, suggesting that this pathway was responsible for the mechanosensing of the neural crest. However, in wild-type embryos, staining of phosphor-myosin showed no detectable differences in signal intensity in the head mesoderm between nonmigratory versus premigratory embryos. This suggested that myosin contractility was not the main controller of the natural stiffening of the cephalic mesoderm. Interestingly, it was shown that the increase in mesoderm cell density due to convergent extension was one of the main factors driving this stiffening. Inhibiting convergent extension by blocking the planar cell polarity pathway inhibited neural crest streaming, which was rescued by restoring the stiffness levels. It is interesting to note that for a cell to sense an increase in cell density, it should be able to sense the combined mechanical property of the tissue. The exact mechanism remains unexplored, although a similar mechanism of cell density-dependent stiffening has been shown during axonal growth (Koser et al. 2016). Nevertheless, this work on neural crest migration was the first to demonstrate the existence of stiffness-induced migration in vivo.

Durotaxis is a phenomenon where cells preferentially migrate toward stiffer substrates (Shellard and Mayor 2021b). Although stiffness-induced migration is a hallmark of durotaxis, the existence of a stiffness gradient, much like a chemoattractant gradient in chemotaxis, was still elusive. It was previously found that one of the major driving mechanisms for the directed mi-

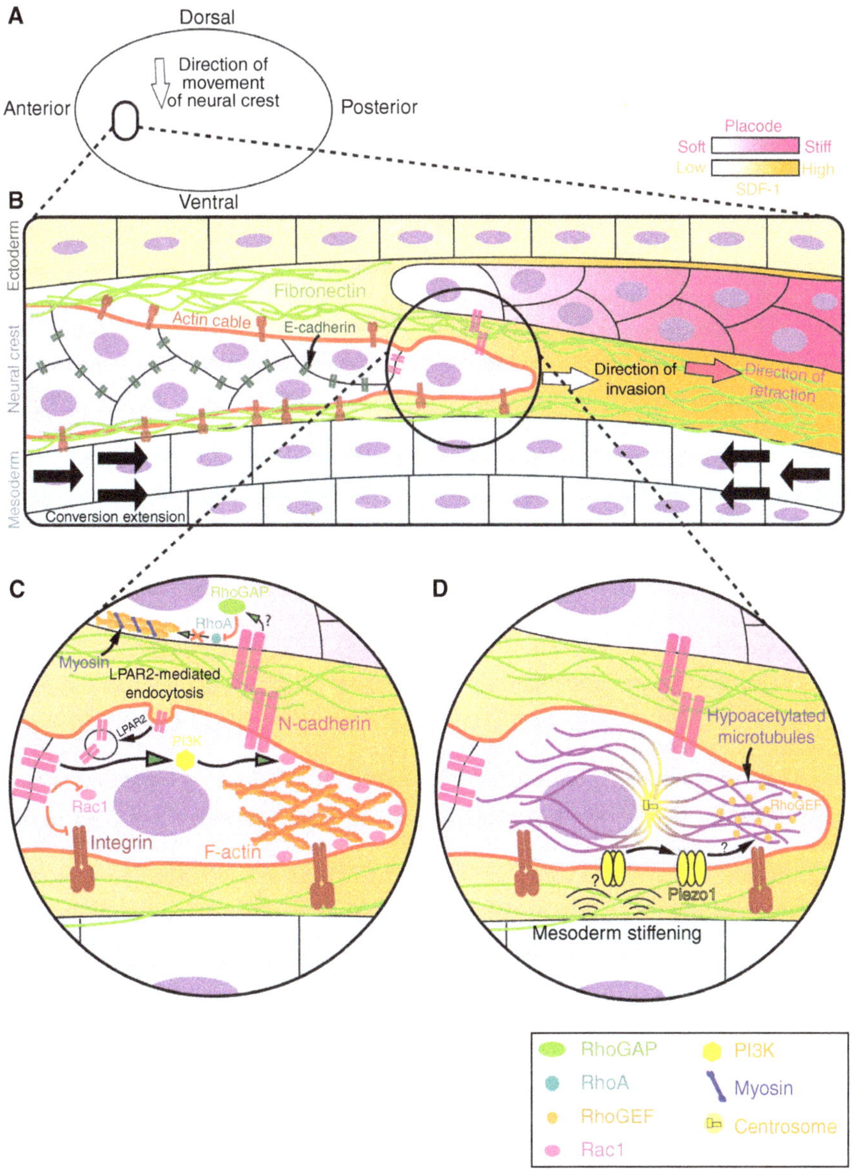

Figure 2. The role of N-cadherin and tissue stiffness in regulating the migration of neural crest cells in *Xenopus laevis* embryos. (*A*) The cephalic neural crest forms three distinct streams (light purple) and migrates dorsoventrally in a *Xenopus* embryo. (*B*) The stiffening of the mesoderm triggers the onset of neural crest migration. There is a gradual switch from E-cadherin (dark green) to N-cadherin (dark pink) as the cluster enters epithelial-to-mesenchymal transition. The neural crest interacts with the placode and follows a chase-and-run directed migration. The placode secretes SDF1 (CXCL12) (shade of orange), which attracts the neural crest. As the neural crest cells make contact with the placode through N-cadherin–N-cadherin interactions, this triggers the placode to retreat. The placode continues to release SDF1, which then attracts the neural crest further, and the cycle continues. The neural crest cluster has an actin cable (red) that mechanically connects each cell together, forming a supracell. (*C*) N-cadherin–N-cadherin interactions between neural crest cells suppress Rac1 (pink) and focal adhesion (represented with integrin) formation at the cell–cell junction but promote Rac1 activation at the leading front through phosphoinositide-3 kinase (yellow). This polarizes the leading cell to move forward. The surface receptor protein LPAR2 mediates N-cadherin internalization to maintain the optimal amount of surface N-cadherin to allow for the fluidization of the neural crest cluster as it migrates. N-cadherin–N-cadherin interactions between neural crest cells and the placode led to a potential activation of Rho-GTPase-activating protein, which inhibits RhoA signaling and reduces cellular contractility and placode stiffness. (*D*) Mesoderm stiffening is sensed by the neural crest via a Piezo1 (yellow)-dependent mechanism. This leads to an increase in hypoacetylated microtubules (dark purple) in neural crest cells. This increases RhoGEF (orange) sequestration, therefore lowering the level of activated Rho signaling, and thus reducing cellular contractility and stiffness.

gration of neural crest cells relies on a rather curious chase-and-run interaction between the neural crest cells and the placode (Fig. 2B; Theveneau et al. 2013). The placode releases the chemoattractant SDF1 (CXCL12), which attracts the neural crest both in vitro and in vivo. The neural crest cluster, despite being composed of hundreds of individually connected cells, migrates toward the SDF1 in a manner that resembles that of a single cell, which is appropriately termed a supracell (Shellard et al. 2018). The leading edge of the cluster is closer to the chemoattractant source, and thus Rac1 is activated and triggers protrusion formation. The rear of the cluster is further away, therefore creating a bias toward the activation of another small GTPase protein, RhoA, which is a major regulator of myosin-induced contractility. The contraction, analogous to an amoeboid cell, squeezes the cluster forward. As the neural crest migrates toward the placodal tissue, it makes brief contact through N-cadherin–N-cadherin homotypic interactions, and this causes the placode to retract and move away (Theveneau et al. 2013). As SDF1 is still being released, the neural crest cells then follow until they contact the placode again, and the cycle of chase-and-run repeats. This interaction increases the directionality of the placode as evidenced by traction force microscopy showing the repolarization of the placode's traction force vectors upon interacting with the neural crest. It was also shown that homotypic N-cadherin interactions between neural crest cells themselves inhibit Rac1 and reduce cell–matrix adhesion at the cell–cell junction via the Src pathway while promoting the activity of phosphoinositide-3 kinase (PI3K) at the free edge (Theveneau et al. 2010; Ouyang et al. 2013; Scarpa et al. 2015; Roycroft et al. 2018). PI3K is a known activator of cell migration by catalyzing the formation of phosphatidylinositol-3,4,5-triphosphate (PIP_3) that activates Rac1 and the actin polymerization machinery (Campa et al. 2015; Le et al. 2021). This causes the cells to repolarize and move away from each other, a phenomenon known as contact inhibition of locomotion. It would be reasonable to assume that the same mechanism might be at play at the interface of neural crest cells and the placode, and perhaps, a similar supracellular migration mode might also govern the retreating rear of the placodal tissue. When the stiffness of the placode was measured, a gradient was observed dorsoventrally, and importantly, this was established coincidentally with the initiation of neural crest migration (Shellard and Mayor 2021a). More intriguingly, this gradient appears to be moving toward the ventral region as the embryos mature. Combined with previous knowledge of the previously described chase-and-run mechanism, it was hypothesized that this stiffness gradient is generated by the neural crest cells as they migrate. When grafting migratory neural crest cells to the ventral side of the placode, a stiffness gradient was observed but with a reverse orientation. This suggests that neural crest cells were able to induce a stiffness gradient in the placode. Mechanically straining the tissue above the neural crest to generate an ectopic stiffness gradient is sufficient to trigger ectopic neural crest migration. Mechanical ablation of the ectoderm above the neural crest, in contrast, completely abrogated the placode's stiffness gradient, and this resulted in the complete inhibition of the migration of the neural crest. Subsequent experiments revealed that N-cadherin–N-cadherin homotypic interactions between the neural crest cells and the placodal tissue were responsible for this stiffness gradient (Fig. 2C). Plating the placode on the fibronectin-coated substrate supplemented with N-cadherin disrupted their actin cytoskeleton organization and decreased their apparent stiffness. The exact mechanism of how N-cadherin reduces the placodal stiffness in *Xenopus* is unclear since different model systems seem to provide contradictory results. In cultured Chinese Hamster Ovary cells, increased myosin activity was seen near N-cadherin (Ouyang et al. 2013), while the model of neural tube closure in *Ciona robusta* suggested the sequestration of Rho-GTPase-activating protein to the homotypic junction and thus reduced cortical RhoA activity (Hashimoto and Munro 2019) and presumably cellular stiffness (Fig. 2C). However, cellular stiffness in most cells depends on cortical actomyosin, which is consistent with the observation that neural crest disassembles the placode's actomyosin cortex in an N-cadherin-dependent manner, leading to a reduction in its stiffness. Altogether, these data suggest that neural crest cells rely on a self-generated durotactic gradient of the placode for their migration in vivo. Despite this, many questions remain unanswered. For example, if the placode is continuously retreating away from the neural crest, and the two cell clusters only make contact temporarily each time, then how is the temporal interaction between the two cell populations precisely maintained? If the placode retreats away too fast, the neural crest will not be able to have sufficient contacts to create and sense the stiffness gradient. If the placode retreats too slowly, the directional migration of neural crest cells can be impaired. Therefore, precise control of the relative speed between the retreating placode and the advancing neural crest must exist, yet this question remains challenging to explore in vivo; therefore, perhaps more sophisticated approaches will be required.

The effect of mesoderm stiffening is also shown to be a two-way street. A jamming-to-unjamming transition occurring at the onset of neural crest migration is mediated by the activity of the membrane receptor LPAR2 (Kuriyama et al. 2014). LPAR2 is a G-protein-coupled receptor and is a member of the lysophosphatidic acid (LPA) receptor family. This protein family binds and responds to LPA signals and can induce a plethora of cellular responses (Geraldo et al. 2021). In this case, LPAR2 mediates the internalization of N-cadherin on the surface of neural crest cells (Fig. 2C). When LPAR2 is depleted, the high abundance of surface N-cadherin holds the cells together and causes the cluster to lose its fluidity and, therefore, be unable to migrate efficiently. As it turns out, the mechanical property of the mesoderm on which neural crest cells migrate also regulates this fluidity transition. Despite recent evidence from cultured cells showing that the cortical stiffness is independent of the stiffness of the substrate (Rheinlaender et al. 2020), neural crest clusters appear to be mechanoresponsive. While the mesoderm gets stiffer as the embryo matures, the neural crest gets softer, decreasing its apparent stiffness from ~300 Pa premigratory to ~200 Pa during the migratory phase (Marchant et al. 2022). This reduction was achieved through a Piezo1-mediated microtubule deacetylation mechanism (Fig. 2D). When plating neural crest explants on soft or stiff polyacrylamide gel, their microtubule acetylation status increases or decreases, respectively. When the level of microtubule acetylation was artificially increased, this increased the stiffness of the neural crest cluster and reduced its migration in vivo. In contrast, introducing hypoacetylated microtubules induced premature migration. Inhibiting Piezo1 with morpholinos or inhibitors increased microtubule acetylation and blocked their migration. Mechanistically, microtubule acetylation could affect cellular contractility by controlling the availability and the activation of RhoGEFs, which can activate RhoA and cellular contraction (Fig. 2D; Schmidt and Stehbens 2024) and, therefore, cellular stiffness. Perhaps in an in vivo context where neural crest cells must migrate in a tight space between the ectoderm and the mesoderm, a more flexible phenotype provided by microtubule deacetylation will allow them to easily squeeze in between these tissues (Collins et al. 2021). However, counterintuitively, inhibiting Piezo1 ex vivo increased neural crest cell dispersion through an overactive Rac1 mechanism (Canales Coutiño and Mayor 2021). However, this phenotype did not translate to a better invasion in vivo, suggesting that within an intact animal, a fine-tuned signaling pathway between Piezo1, microtubules, and Rac1 and by extension, the actin cytoskeleton, is crucial for normal neural crest migration. Overall, these molecular mechanisms serve as a demonstration of the importance of cell–tissue interactions in modulating the migration behavior of neural crests in vivo.

ZEBRAFISH POSTERIOR LATERAL LINE PRIMORDIUM

The pLLP is a group of ~100–120 cells that are derived from the posterior lateral line placode epithelium (Chitnis et al. 2012; Karaiwa et al. 2020), which is reminiscent of border cells delaminating from the follicle epithelium in the *Drosophila* egg chamber. This group of cells migrates along the horizontal myoseptum, starting from just behind the otic vesicle toward the tip of the tail, and depositing small clusters of ~20 cells called neuromasts along the way, which will develop into water current–sensing organs (Fig. 3A). The pLLP can be subdivided into two main parts: a leading portion that contains flatter and more mesenchymal cells and a follower portion that contains more columnar epithelial-like cells that still retain their apicobasal polarity (Chitnis et al. 2012). Similar to the neural crest, leader cells in pLLP also have a higher level of N-cadherin relative to E-cadherin, while the reverse is true for follower cells that are required to be more cohesive (Chitnis et al. 2012; Olson and Nechiporuk 2018).

The biochemical pathway that governs the directed migration of pLLP is relatively well understood. The chemokine CXCL12a (also known as SDF-1), which is involved in neural crest chemotaxis, is responsible for directing pLLP migration (David et al. 2002; Donà et al. 2013; Dalle Nogare et al. 2014). It is homogenously distributed throughout the length of the myoseptum and is secreted by the muscle pioneer (MP) cells and the muscle fast fiber (MFF) cells (Karaiwa et al. 2020). It is believed that the pLLP self-generates its own chemical gradient by differentially expressing two CXCL12a receptors—CXCR4b and CXCR7b. While CXCR4b is predominantly expressed by the leader cells and can trigger protrusion formation upon ligand binding, CXCR7b is predominantly expressed in the trailing edge and binds and internalizes CXCL12a but does not trigger any downstream activity (David et al. 2002; Wong et al. 2020). This effectively creates a chemical gradient of CXCL12a driving pLLP's forward migration (Insall et al. 2022).

However, recent work has started to explore other factors that may also contribute to the migration of the pLLP, one of which is the architecture of the tissue that pLLP cells interact with. Anatomically, the groove where the pLLP cluster migrates is situated between the dorsal and the ventral myotomes, the muscle segments that assemble in a chevron pattern along the rostral–caudal body axis of the fish embryo. The above-mentioned MP and MFF cells contribute to the formation of the myotome groove itself through a Sonic hedgehog (Shh)-dependent differentiation pathway (Fig. 3B; Blagden et al. 1997; Karaiwa et al. 2020). Without proper differentiation by these two cell types, the dorsal and ventral myotomes fuse together, and the groove disappears (Karaiwa et al. 2020). When zebrafish embryos at 15 hours postfertilization (hpf) were treated with cyclopamine (an Shh inhibitor), this led to a consistent loss of MP and MFF cells and a fusion of the furrow from the 16th myotome onward. The pLLP cluster migrated normally up until the 16th myotome and then lost its directionality. This change of migration was not due to the loss of CXCL12a because at 32 hpf when the myotome groove had already formed, depletion of MP and MFF cells no longer affected pLLP's migration trajectory. It could be argued that other sources for CXCL12a might exist, or that CXCL12a might have diffused from the rostral MP and MFF cells. Nevertheless, this evidence suggests that it was likely the physical presence of an anatomical pathway, the myotome furrow, that allowed for the normal passage of the pLLP cluster.

Detailed microscopy analysis of the architecture and the use of mosaic labeling of the pLLP also reveal the presence of both filopodia-like (finger-like protrusions made up of mostly parallel actin filaments) and lamellipodia-like (thin sheet-like protrusions composed of branched actin filaments) (Mattila and Lappalainen 2008; Krause and Gautreau 2014) protrusions throughout the cluster, both on the basal side where the cells make contact with the basement membrane and on the lateral and apical side where they make contacts with the overlying skin cells (Fig. 3C; Dalle Nogare et al. 2020; Karaiwa et al. 2020; Olson and Nechiporuk 2021). These protrusions align in the same traveling direction with the pLLP and are decorated with numerous focal adhesions (adhesive structures that allow cells to attach to their surrounding) as visualized using phosphor-paxillin. To address whether traction force is generated as the pLLP migrates, the fluorescently labeled basement membrane was bleached in circular regions in a hexagonal pattern to create local landmarks (Yamaguchi et al. 2022). Shift-

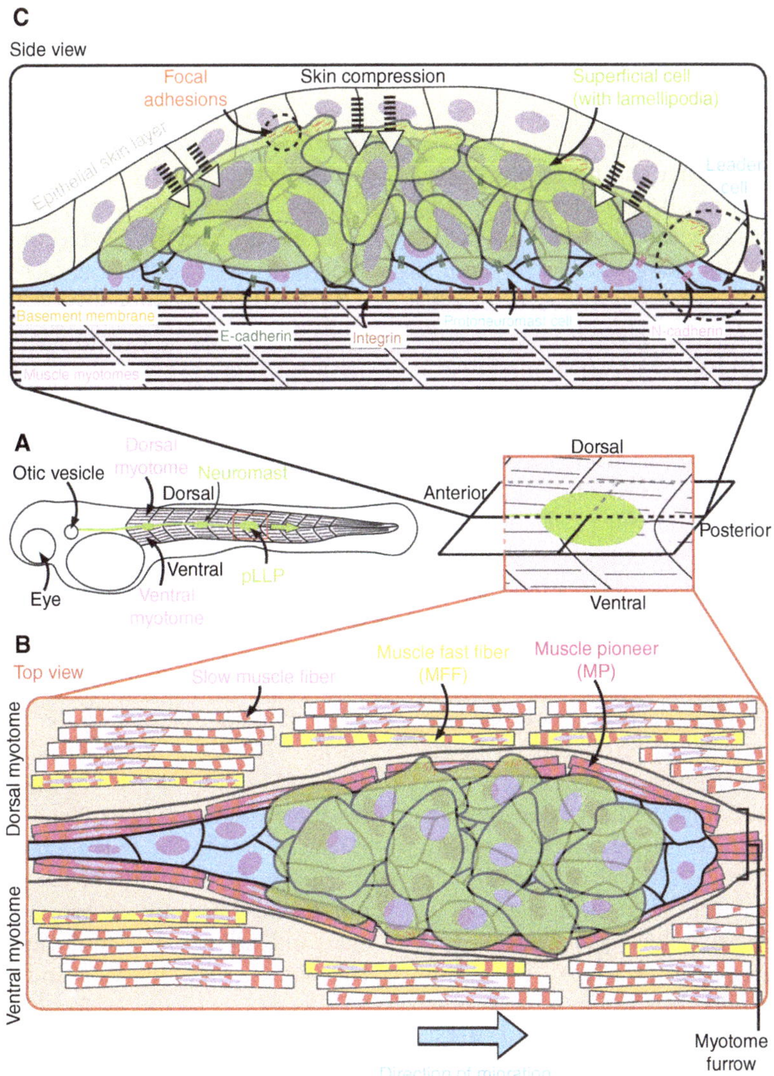

Figure 3. The role of tissue compression and topology in regulating the migration of the posterior lateral line primordium in zebrafish embryos. (*A*) The anatomy of an early zebrafish embryo. The posterior lateral line primordium (pLLP) (green) migrates from behind the otic vesicle toward the tip of the tail. The cluster migrates in a groove between the dorsal and ventral myotomes (light pink). As it migrates, clumps of cells called neuromasts are regularly deposited. (*B*, *top*) View of the pLLP (the skin layer is not drawn but assumed to be present). The pLLP migrates within the myotome furrow. The myotome furrow is formed by the differentiation of the muscle pioneer (MP) cells (dark pink) and the muscle fast fiber cells (yellow) that are decorated along the front–rear body axis of the embryo. The basement membrane (light orange) separates the pLLP from the muscle layer underneath. (*C*) *Side* view of the pLLP. The cells of the pLLP can be divided into two main groups: the leader cells that migrate at the front of the cluster and the follower cells. The follower cells can then be subdivided into two groups: the superficial cells (green) that make extensive contact with the epithelial skin layer, and the protoneuromast cells (light blue) that sit at the basal side. Transparency indicates the cells are lying *above*. There is a higher level of N-cadherin (pink) near the front of the cluster, while there is a higher level of E-cadherin (dark green) near the trailing end. The pLLP cluster forms numerous focal adhesions (orange) and integrin adhesions (red) with the epithelial skin layer (beige) and the basement membrane made of laminin (light orange). The epithelial skin layer provides some degree of compression to the migrating pLLP.

ing the position of these darkened areas, particularly in the middle and the rear region, to the pLLP cluster showed that traction force was being generated. Although not directly observed, the traction force is also likely to be present between the pLLP cluster and the skin epithelial layer due to this being the largest contact surface area with the pLLP cluster (~2350 μm^2) (Dalle Nogare et al. 2020) along with the presence of numerous focal adhesions. When talin—a scaffolding protein that connects and transmits forces between integrins and the actin cytoskeleton was knocked down using morpholino—cells within the pLLP cluster lost their protrusions and assumed a round morphology, and their migration was impaired (Olson and Nechiporuk 2021). It is important to note that since focal adhesions are present between cells of the pLLP and the skin epithelial layer, there is most likely a thin layer of ECM sandwiched between these two cell populations, albeit difficult to detect. This matrix layer may also be mechanically linked to the epithelial layer and assist in transducing any mechanical forces exerted by the migrating pLLP (Humphrey et al. 2014).

As the pLLP migrates, it wedges open the space between the epithelial skin layer and the basement membrane. The cells of the pLLP cluster that make direct contact with the skin are named superficial cells (Dalle Nogare et al. 2020), and they have the largest surface area (~51 μm^2 per cell) than any other cell within the cluster. This suggests that physical contact between the pLLP and the skin may be important. To test this hypothesis, the skin above the pLLP was removed. Upon skin removal, the pLLP completely lost its ability to migrate, while simultaneously becoming taller (Dalle Nogare et al. 2020; Olson and Nechiporuk 2021). The cells also lost their lamellipodia and formed random bleb protrusions instead. This suggests that the skin provided some degree of compression to the pLLP cells and that even when adhesions to the basement membrane are still present, contact with the skin remains essential to pLLP's migration and morphology (Fig. 3C). Artificially compressing the deskinned embryo by encasing it in Matrigel or agarose only partially rescued the morphology of the superficial cells but not the migration speed or directionality of the pLLP (Dalle Nogare et al. 2020). Only when the skin regrew and covered the pLLP, was the migration rescued. This suggests that focal adhesions and compression are important for the maintenance of the pLLP's morphology, but its migration behavior may likely rely more on other forms of cell–cell interactions between the superficial cells and the skin, such as cadherins. This was partially supported by the fact that when α-catenin, the protein that is responsible for mediating force transmission between cadherin and the actin cytoskeleton, was depleted, the pLLP cluster lost its directionality, albeit with no change in its migration speed (Yamaguchi et al. 2022). This minor discrepancy hints at other factors that these experimental setups might not have taken into consideration, such as the difference in stiffness between the skin and Matrigel/agarose, how strong the natural compression from the skin versus the Matrigel, or perhaps other functions of α-catenin that we might not have accounted for (Kobielak and Fuchs 2004; Erez et al. 2005; Schlegelmilch et al. 2011). It would also be interesting to explore more specifically what and how cadherin interactions with the skin epithelial layer contribute to the migration of the pLLP and the differential contributions between focal adhesions and cadherins in transducing the mechanical forces between the pLLP and the skin.

PERSPECTIVES AND CONCLUSIONS

Cells are constantly receiving and responding to signals from the tissue. Chemical signals aside, the physical interactions between cells and their surrounding environment play an important role in affecting their behaviors. Within a tissue, cells make contact with both the ECM and other cells that make up that tissue. However, it is the latter that remains less explored (Le and Mayor 2023). In this review, we have discussed in depth three different model systems of cell–cell interactions across insects and vertebrates: the border cells in *Drosophila melanogaster*, the neural crest in *Xenopus laevis* and, finally, the pLLP of the zebrafish. Although each of these systems is unique in its specific biology, they share many commonalities and principles that can be applied to each other and other systems. In all three systems, the surrounding tissue actively plays a role in guiding the migrating cell population through direct interactions. As these clusters migrate, they rely on the formation of some physical bonds with the surrounding cells, whether it is cadherin or focal adhesions, to generate traction forces pulling them forward. Being able to make direct cell–cell interactions also implies that these migrating cells exist within a densely packed tissue environment, which means there is a certain degree of confinement imposed on the cell cluster in all three systems. It turns out that this natural compression helps cells move through the environment more effectively, either by triggering the nuclear piston model (Petrie et al. 2017; Lee et al. 2021), the nuclear ruler mechanism (Lomakin et al. 2020), the nuclear wedge mechanism (Penfield and Montell 2023), or simply by providing a substrate for protrusion spreading. Cells are considered active matter (Needleman and Dogic 2017), meaning a substrate made from living cells can actively react to an interaction with another cell. This is an important distinction with the ECM, which is mainly fibrous proteins and is largely passive. This is evident in the example of nurse cell stiffening in response to fascin depletion in border cells, or in neural crest cells that induce the retraction and softening of the placode. This action–reaction relationship between migrating cells and their cellular substrate can result in a feedback mechanism that can sustain itself, such as in the case of the chase-and-run mechanism in *Xenopus*'s neural crest.

Although not discussed in this review, apart from developmental processes, it is important to note that cell-on-cell migration also exists under normal physiological conditions in adult organisms and with single-cell migration. The $CD8^+$ T cells of the gut mucosa have been observed to adopt an elongated shape in a densely packed environment, such as between the longitudinal muscle cells of the small intestine, while having a more amoeboid shape in sparse areas (Thompson et al. 2019). Tissue-resident T cells in the salivary gland use the topology of macrophages to help them find their way around and increase their migration and patrolling capability (Stolp et al. 2020). The concept of cell–cell interactions can be further applied in a disease context like cancer. For example, fibroblasts have been observed to make extensive physical contact with cancer cells through heterotypic cadherin interactions and physically pull streams of cancer cells outwards, assisting their metastasis (Gaiser et al. 2012). The degree of immune infiltration into a tumor also naturally depends on how densely packed or how protected the tumor is through the formation of a surrounding matrix barrier (Grantab et al. 2006; Kuczek et al. 2019), which can have direct therapeutic implications.

In conclusion, we hope that this review has brought some attention to the interesting biology of cell-on-cell migration. Within a complex tissue where both cells and matrix proteins are intertwined, it is expected that both factors will play important regulatory functions. However, we hope that the few examples discussed here have demonstrated the diversity as well as the importance of the cellular environment in modulating migratory behaviors of collectively migrating cells. Within a multicellular organism, cell–cell interactions are likely to be just as widespread as cell–matrix interactions. Therefore, a better understanding of this phenomenon will prove to be vital in better understanding the biology of cell migration in vivo in both normal and pathological processes, with a promising yet untapped therapeutic potential.

COMPETING INTEREST STATEMENT

The authors declare no competing interests.

ACKNOWLEDGMENTS

Work in the R.M. laboratory is supported by grants from the Medical Research Council (MR/S007792/1), Biotechnology and Biological Sciences Research Council (M008517 and BB/T013044), and Wellcome Trust (102489/Z/13/Z). H.A.L. is supported by the Sir Henry Wellcome fellowship (224063/Z/21/Z).

REFERENCES

Angulo-Urarte A, van der Wal T, Huveneers S. 2020. Cell–cell junctions as sensors and transducers of mechanical forces. *Biochim Biophys Acta Biomembr* **1862:** 183316. doi:10.1016/j.bbamem.2020.183316

Aranjuez G, Burtscher A, Sawant K, Majumder P, McDonald JA. 2016. Dynamic myosin activation promotes collective morphology and migration by locally balancing oppositional forces from surrounding tissue. *Mol Biol Cell* **27:** 1898–1910. doi:10.1091/mbc.e15-10-0744

Barriga EH, Franze K, Charras G, Mayor R. 2018. Tissue stiffening coordinates morphogenesis by triggering collective cell migration in vivo. *Nature* **554:** 523–527. doi:10.1038/nature25742

Blagden CS, Currie PD, Ingham PW, Hughes SM. 1997. Notochord induction of zebrafish slow muscle mediated by Sonic hedgehog. *Genes Dev* **11:** 2163–2175. doi:10.1101/gad.11.17.2163

Cai D, Chen SC, Prasad M, He L, Wang X, Choesmel-Cadamuro V, Sawyer JK, Danuser G, Montell DJ. 2014. Mechanical feedback through E-cadherin promotes direction sensing during collective cell migration. *Cell* **157:** 1146–1159. doi:10.1016/j.cell.2014.03.045

Campa CC, Ciraolo E, Ghigo A, Germena G, Hirsch E. 2015. Crossroads of PI3K and Rac pathways. *Small GTPases* **6:** 71–80. doi:10.4161/21541248.2014.989789

Canales Coutiño B, Mayor R. 2021. The mechanosensitive channel Piezo1 cooperates with semaphorins to control neural crest migration. *Development* **148:** dev200001. doi:10.1242/dev.200001

Chitnis AB, Dalle Nogare D, Matsuda M. 2012. Building the posterior lateral line system in zebrafish. *Dev Neurobiol* **72:** 234–255. doi:10.1002/dneu.20962

Collins C, Kim SK, Ventrella R, Carruzzo HM, Wortman JC, Han H, Suva EE, Mitchell JW, Yu CC, Mitchell BJ. 2021. Tubulin acetylation promotes penetrative capacity of cells undergoing radial intercalation. *Cell Rep* **36:** 109556. doi:10.1016/j.celrep.2021.109556

Dai W, Guo X, Cao Y, Mondo JA, Campanale JP, Montell BJ, Burrous H, Streichan S, Gov N, Rappel WJ, et al. 2020. Tissue topography steers migrating *Drosophila* border cells. *Science* **370:** 987–990. doi:10.1126/science.aaz4741

Dalle Nogare D, Somers K, Rao S, Matsuda M, Reichman-Fried M, Raz E, Chitnis AB. 2014. Leading and trailing cells cooperate in collective migration of the zebrafish posterior lateral line primordium. *Development* **141:** 3188–3196. doi:10.1242/dev.106690

Dalle Nogare DE, Natesh N, Vishwasrao HD, Shroff H, Chitnis AB. 2020. Zebrafish Posterior Lateral Line primordium migration requires interactions between a superficial sheath of motile cells and the skin. *eLife* **9:** e58251. doi:10.7554/eLife.58251

Das T, Safferling K, Rausch S, Grabe N, Boehm H, Spatz JP. 2015. A molecular mechanotransduction pathway regulates collective migration of epithelial cells. *Nat Cell Biol* **17:** 276–287. doi:10.1038/ncb3115

David NB, Sapède D, Saint-Etienne L, Thisse C, Thisse B, Dambly-Chaudière C, Rosa FM, Ghysen A. 2002. Molecular basis of cell migration in the fish lateral line: role of the chemokine receptor CXCR4 and of its ligand, SDF1. *Proc Natl Acad Sci* **99:** 16297–16302. doi:10.1073/pnas.252339399

Donà E, Barry JD, Valentin G, Quirin C, Khmelinskii A, Kunze A, Durdu S, Newton LR, Fernandez-Minan A, Huber W, et al. 2013. Directional tissue migration through a self-generated chemokine gradient. *Nature* **503:** 285–289. doi:10.1038/nature12635

Elkhatib N, Neu MB, Zensen C, Schmoller KM, Louvard D, Bausch AR, Betz T, Vignjevic DM. 2014. Fascin plays a role in stress fiber organization and focal adhesion disassembly. *Curr Biol* **24:** 1492–1499. doi:10.1016/j.cub.2014.05.023

Elosegui-Artola A, Trepat X, Roca-Cusachs P. 2018. Control of mechanotransduction by molecular clutch dynamics. *Trends Cell Biol* **28:** 356–367. doi:10.1016/j.tcb.2018.01.008

Erez N, Bershadsky A, Geiger B. 2005. Signaling from adherens-type junctions. *Eur J Cell Biol* **84:** 235–244. doi:10.1016/j.ejcb.2004.12.007

Ferrera D, Canale C, Marotta R, Mazzaro N, Gritti M, Mazzanti M, Capellari S, Cortelli P, Gasparini L. 2014. Lamin B1 overexpression increases nuclear rigidity in autosomal dominant leukodystrophy fibroblasts. *FASEB J* **28:** 3906–3918. doi:10.1096/fj.13-247635

Gabbert AM, Campanale JP, Mondo JA, Mitchell NP, Myers A, Streichan SJ, Miolane N, Montell DJ. 2023. Septins regulate border cell surface geometry, shape, and motility downstream of Rho in *Drosophila*. *Dev Cell* **58:** 1399–1413.e5. doi:10.1016/j.devcel.2023.05.017

Gaiser MR, Lämmermann T, Feng X, Igyarto BZ, Kaplan DH, Tessarollo L, Germain RN, Udey MC. 2012. Cancer-associated epithelial cell adhesion molecule (EpCAM; CD326) enables epidermal Langerhans cell motility and migration in vivo. *Proc Natl Acad Sci* **109:** E889–E897. doi:10.1073/pnas.1117674109

Geraldo LHM, Spohr TCLS, Amaral RFD, Fonseca ACCD, Garcia C, Mendes FA, Freitas C, dosSantos MF, Lima FRS. 2021. Role of lysophosphatidic acid and its receptors in health and disease: novel therapeutic strategies. *Sig Transduct Target Ther* **6:** 1–18. doi:10.1038/s41392-020-00451-w

Grantab R, Sivananthan S, Tannock IF. 2006. The penetration of anticancer drugs through tumor tissue as a function of cellular adhesion and packing density of tumor cells. *Cancer Res* **66:** 1033–1039. doi:10.1158/0008-5472.CAN-05-3077

Gutzeit HO, Eberhardt W, Gratwohl E. 1991. Laminin and basement membrane-associated microfilaments in wild-type and mutant *Drosophila* ovarian follicles. *J Cell Sci* **100:** 781–788. doi:10.1242/jcs.100.4.781

Hashimoto H, Munro E. 2019. Differential expression of a classic cadherin directs tissue-level contractile asymmetry during neural tube closure. *Dev Cell* **51:** 158–172.e4. doi:10.1016/j.devcel.2019.10.001

Horne-Badovinac S. 2020. The *Drosophila* micropyle as a system to study how epithelia build complex extracellular structures. *Philos Trans R Soc Lond B Biol Sci* **375:** 20190561. doi:10.1098/rstb.2019.0561

Humphrey JD, Dufresne ER, Schwartz MA. 2014. Mechanotransduction and extracellular matrix homeostasis. *Nat Rev Mol Cell Biol* **15:** 802–812. doi:10.1038/nrm3896

Insall RH, Paschke P, Tweedy L. 2022. Steering yourself by the bootstraps: how cells create their own gradients for chemotaxis. *Trends Cell Biol* **32:** 585–596. doi:10.1016/j.tcb.2022.02.007

Jansen S, Collins A, Yang C, Rebowski G, Svitkina T, Dominguez R. 2011. Mechanism of actin filament bundling by fascin. *J Biol Chem* **286:** 30087–30096. doi:10.1074/jbc.M111.251439

Karaiwa A, Yamada S, Yamamoto H, Wakasa M, Ishijima H, Akiyama R, Hosokawa Y, Bessho Y, Matsui T. 2020. Relationship between surrounding tissue morphology and directional collective migration of the posterior lateral line primordium in zebrafish. *Genes Cells* **25:** 582–592. doi:10.1111/gtc.12793

Kim PH, Chen NY, Heizer PJ, Tu Y, Weston TA, Fong JLC, Gill NK, Rowat AC, Young SG, Fong LG. 2021. Nuclear membrane ruptures underlie the vascular pathology in a mouse model of Hutchinson–Gilford progeria syndrome. *JCI Insight* **6:** e151515. doi:10.1172/jci.insight.151515

Kobielak A, Fuchs E. 2004. α-Catenin: at the junction of intercellular adhesion and actin dynamics. *Nat Rev Mol Cell Biol* **5:** 614–625. doi:10.1038/nrm1433

Koser DE, Thompson AJ, Foster SK, Dwivedy A, Pillai EK, Sheridan GK, Svoboda H, Viana M, Costa LDF, Guck J, et al. 2016. Mechanosensing is critical for axon growth in the developing brain. *Nat Neurosci* **19:** 1592–1598. doi:10.1038/nn.4394

Krause M, Gautreau A. 2014. Steering cell migration: lamellipodium dynamics and the regulation of directional persistence. *Nat Rev Mol Cell Biol* **15:** 577–590. doi:10.1038/nrm3861

Kuczek DE, Larsen AMH, Thorseth ML, Carretta M, Kalvisa A, Siersbæk MS, Simões AMC, Roslind A, Engelholm LH, Noessner E, et al. 2019. Collagen density regulates the activity of tumor-infiltrating T cells. *J Immunother Cancer* **7:** 68. doi:10.1186/s40425-019-0556-6

Kuriyama S, Theveneau E, Benedetto A, Parsons M, Tanaka M, Charras G, Kabla A, Mayor R. 2014. In vivo collective cell migration requires an LPAR2-dependent increase in tissue fluidity. *J Cell Biol* **206:** 113–127. doi:10.1083/jcb.201402093

Lamb MC, Kaluarachchi CP, Lansakara TI, Mellentine SQ, Lan Y, Tivanski AV, Tootle TL. 2021. Fascin limits Myosin activity within *Drosophila* border cells to control substrate stiffness and promote migration. *eLife* **10:** e69836. doi:10.7554/eLife.69836

Le HA, Mayor R. 2023. Cell–matrix and cell–cell interaction mechanics in guiding migration. *Biochem Soc Trans* **51:** 1733–1745. doi:10.1042/BST20230211

Le AH, Yelland T, Paul NR, Fort L, Nikolaou S, Ismail S, Machesky LM. 2021. CYRI-A limits invasive migration through macropinosome formation and integrin uptake regulation. *J Cell Biol* **220:** e202012114. doi:10.1083/jcb.202012114

Leckband DE, de Rooij J. 2014. Cadherin adhesion and mechanotransduction. *Annu Rev Cell Dev Biol* **30:** 291–315. doi:10.1146/annurev-cellbio-100913-013212

Lee HP, Alisafaei F, Adebawale K, Chang J, Shenoy VB, Chaudhuri O. 2021. The nuclear piston activates mechanosensitive ion channels to generate cell migration paths in confining microenvironments. *Sci Adv* **7:** eabd4058. doi:10.1126/sciadv.abd4058

Lo CM, Wang HB, Dembo M, Wang YL. 2000. Cell movement is guided by the rigidity of the substrate. *Biophys J* **79:** 144–152. doi:10.1016/S0006-3495(00)76279-5

Lomakin AJ, Cattin CJ, Cuvelier D, Alraies Z, Molina M, Nader GPF, Srivastava N, Sáez PJ, Garcia-Arcos JM, Zhitnyak IY, et al. 2020. The nucleus acts as a ruler tailoring cell responses to spatial constraints. *Science* **370:** eaba2894. doi:10.1126/science.aba2894

Marchant CL, Malmi-Kakkada AN, Espina JA, Barriga EH. 2022. Cell clusters softening triggers collective cell migration in vivo. *Nat Mater* **21:** 1314–1323. doi:10.1038/s41563-022-01323-0

Mattila PK, Lappalainen P. 2008. Filopodia: molecular architecture and cellular functions. *Nat Rev Mol Cell Biol* **9:** 446–454. doi:10.1038/nrm2406

Mayor R, Theveneau E. 2013. The neural crest. *Development* **140:** 2247–2251. doi:10.1242/dev.091751

Mishra AK, Campanale JP, Mondo JA, Montell DJ. 2019. Cell interactions in collective cell migration. *Development* **146:** dev172056. doi:10.1242/dev.172056

Molina López E, Kabanova A, Winkel A, Franze K, Palacios IM, Martín-Bermudo MD. 2023. Constriction imposed by basement membrane regulates developmental cell migration. *PLoS Biol* **21:** e3002172. doi:10.1371/journal.pbio.3002172

Montell DJ, Yoon WH, Starz-Gaiano M. 2012. Group choreography: mechanisms orchestrating the collective movement of border cells. *Nat Rev Mol Cell Biol* **13:** 631–645. doi:10.1038/nrm3433

Mostowy S, Cossart P. 2012. Septins: the fourth component of the cytoskeleton. *Nat Rev Mol Cell Biol* **13:** 183–194. doi:10.1038/nrm3284

Nakagawa M, Fukata M, Yamaga M, Itoh N, Kaibuchi K. 2001. Recruitment and activation of Rac1 by the formation of E-cadherin-mediated cell–cell adhesion sites. *J Cell Sci* **114:** 1829–1838. doi:10.1242/jcs.114.10.1829

Needleman D, Dogic Z. 2017. Active matter at the interface between materials science and cell biology. *Nat Rev Mater* **2:** 1–14. doi:10.1038/natrevmats.2017.48

Niewiadomska P, Godt D, Tepass U. 1999. *DE*-cadherin is required for intercellular motility during *Drosophila* oogenesis. *J Cell Biol* **144:** 533–547. doi:10.1083/jcb.144.3.533

Nobes CD, Hall A. 1995. Rho, Rac, and Cdc42 GTPases regulate the assembly of multimolecular focal complexes associated with actin stress fibers, lamellipodia, and filopodia. *Cell* **81:** 53–62. doi:10.1016/0092-8674(95)90370-4

Olson HM, Nechiporuk AV. 2018. Using zebrafish to study collective cell migration in development and disease. *Front Cell Dev Biol* **6:** 83. doi:10.3389/fcell.2018.00083

Olson HM, Nechiporuk AV. 2021. Lamellipodia-like protrusions and focal adhesions contribute to collective cell migration in zebrafish. *Dev Biol* **469:** 125–134. doi:10.1016/j.ydbio.2020.10.007

Ouyang M, Lu S, Kim T, Chen CE, Seong J, Leckband DE, Wang F, Reynolds AB, Schwartz MA, Wang Y. 2013. N-cadherin regulates spatially polarized signals through distinct p120ctn and β-catenin-dependent signaling pathways. *Nat Commun* **4:** 1589. doi:10.1038/ncomms2560

Penfield L, Montell DJ. 2023. Nuclear lamin facilitates collective border cell invasion into confined spaces in vivo. *J Cell Biol* **222:** e202212101. doi:10.1083/jcb.202212101

Petrie RJ, Koo H, Yamada KM. 2014. Generation of compartmentalized pressure by a nuclear piston governs cell motility in a 3D matrix. *Science* **345:** 1062–1065. doi:10.1126/science.1256965

Petrie RJ, Harlin HM, Korsak LIT, Yamada KM. 2017. Activating the nuclear piston mechanism of 3D migration in tumor cells. *J Cell Biol* **216:** 93–100. doi:10.1083/jcb.201605097

Powell DR, Blasky AJ, Britt SG, Artinger KB. 2013. Riding the crest of the wave: parallels between the neural crest and cancer in epithelial-to-mesenchymal transition and migration. *Wiley Interdiscip Rev Syst Biol Med* **5:** 511–522. doi:10.1002/wsbm.1224

Rheinlaender J, Dimitracopoulos A, Wallmeyer B, Kronenberg NM, Chalut KJ, Gather MC, Betz T, Charras G, Franze K. 2020. Cortical cell stiffness is independent of substrate mechanics. *Nat Mater* **19:** 1019–1025. doi:10.1038/s41563-020-0684-x

Roycroft A, Szabó A, Bahm I, Daly L, Charras G, Parsons M, Mayor R. 2018. Redistribution of adhesive forces through Src/FAK drives contact inhibition of locomotion in neural crest. *Dev Cell* **45:** 565–579.e3. doi:10.1016/j.devcel.2018.05.003

Saraswathibhatla A, Indana D, Chaudhuri O. 2023. Cell–extracellular matrix mechanotransduction in 3D. *Nat Rev Mol Cell Biol* **24:** 495–516. doi:10.1038/s41580-023-00583-1

Scarpa E, Szabó A, Bibonne A, Theveneau E, Parsons M, Mayor R. 2015. Cadherin switch during EMT in neural crest cells leads to contact inhibition of locomotion via repolarization of forces. *Dev Cell* **34:** 421–434. doi:10.1016/j.devcel.2015.06.012

Schlegelmilch K, Mohseni M, Kirak O, Pruszak J, Rodriguez JR, Zhou D, Kreger BT, Vasioukhin V, Avruch J, Brummelkamp TR, et al. 2011. Yap1 acts downstream of α-catenin to control epidermal proliferation. *Cell* **144:** 782–795. doi:10.1016/j.cell.2011.02.031

Schmidt CJ, Stehbens SJ. 2024. Microtubule control of migration: coordination in confinement. *Curr Opin Cell Biol* **86:** 102289. doi:10.1016/j.ceb.2023.102289

Shellard A, Mayor R. 2021a. Collective durotaxis along a self-generated stiffness gradient in vivo. *Nature* **600:** 690–694. doi:10.1038/s41586-021-04210-x

Shellard A, Mayor R. 2021b. Durotaxis: the hard path from in vitro to in vivo. *Dev Cell* **56:** 227–239. doi:10.1016/j.devcel.2020.11.019

Shellard A, Szabó A, Trepat X, Mayor R. 2018. Supracellular contraction at the rear of neural crest cell groups drives collective chemotaxis. *Science* **362:** 339–343. doi:10.1126/science.aau3301

Stolp B, Thelen F, Ficht X, Altenburger LM, Ruef N, Inavalli VVGK, Germann P, Page N, Moalli F, Raimondi A, et al. 2020. Salivary gland macrophages and tissue-resident $CD8^+$ T cells cooperate for homeostatic organ surveillance. *Sci Immunol* **5:** eaaz4371. doi:10.1126/sciimmunol.aaz4371

Szabó A, Mayor R. 2018. Mechanisms of neural crest migration. *Annu Rev Genet* **52:** 43–63. doi:10.1146/annurev-genet-120417-031559

Theveneau E, Mayor R. 2012. Neural crest delamination and migration: from epithelium-to-mesenchyme transition to collective cell migration. *Dev Biol* **366:** 34–54. doi:10.1016/j.ydbio.2011.12.041

Theveneau E, Marchant L, Kuriyama S, Gull M, Moepps B, Parsons M, Mayor R. 2010. Collective chemotaxis requires contact-dependent cell polarity. *Dev Cell* **19:** 39–53. doi:10.1016/j.devcel.2010.06.012

Theveneau E, Steventon B, Scarpa E, Garcia S, Trepat X, Streit A, Mayor R. 2013. Chase-and-run between adjacent cell populations promotes directional collective migration. *Nat Cell Biol* **15:** 763–772. doi:10.1038/ncb2772

Thompson EA, Mitchell JS, Beura LK, Torres DJ, Mrass P, Pierson MJ, Cannon JL, Masopust D, Fife BT, Vezys V. 2019. Interstitial migration of CD8αβ T cells in the small intestine is dynamic and is dictated by environmental cues. *Cell Rep* **26:** 2859–2867.e4. doi:10.1016/j.celrep.2019.02.034

Ventura G, Amiri A, Thiagarajan R, Tolonen M, Doostmohammadi A, Sedzinski J. 2022. Multiciliated cells use filopodia to probe tissue mechanics during epithelial integration in vivo. *Nat Commun* **13:** 6423. doi:10.1038/s41467-022-34165-0

Winkler J, Abisoye-Ogunniyan A, Metcalf KJ, Werb Z. 2020. Concepts of extracellular matrix remodelling in tumour progression and metastasis. *Nat Commun* **11:** 5120. doi:10.1038/s41467-020-18794-x

Wong M, Newton LR, Hartmann J, Hennrich ML, Wachsmuth M, Ronchi P, Guzmán-Herrera A, Schwab Y, Gavin AC, Gilmour D. 2020. Dynamic buffering of extracellular chemokine by a dedicated scavenger pathway enables robust adaptation during directed tissue migration. *Dev Cell* **52:** 492–508.e10. doi:10.1016/j.devcel.2020.01.013

Yamada KM, Doyle AD, Lu J. 2022. Cell–3D matrix interactions: recent advances and opportunities. *Trends Cell Biol* **32:** 883–895. doi:10.1016/j.tcb.2022.03.002

Yamaguchi N, Zhang Z, Schneider T, Wang B, Panozzo D, Knaut H. 2022. Rear traction forces drive adherent tissue migration in vivo. *Nat Cell Biol* **24:** 194–204. doi:10.1038/s41556-022-00844-9

Yap AS, Duszyc K, Viasnoff V. 2018. Mechanosensing and mechanotransduction at cell–cell junctions. *Cold Spring Harb Perspect Biol* **10:** a028761. doi:10.1101/cshperspect.a028761

Traveling in a Pack: The Border Cells' Collective Journey Guided by Landscapes and Chemical Cues

Alexander George[1] and Michelle Starz-Gaiano[2]

[1]Dartmouth Cancer Center, One Medical Center Drive, Lebanon, New Hampshire 03756, USA
[2]Department of Biological Sciences, University of Maryland, Baltimore, Maryland 21250, USA
Correspondence: starz@umbc.edu

Many cell types migrate collectively, a process critical for animal development and co-opted in some medical disorders. Thus, uncovering the molecular regulation of collective cell migration is of broad interest, yet this process remains understudied as compared to individual cell motility. Both collective and individual cell movements rely on similar mechanisms to change cell shapes and adhesiveness. Although grouped cells face the added challenge of maintaining coordination and communication, they can then leverage group-level advantages, like animals in a pack. How motile cells work together to accomplish these feats is an active area of study. The border cells of the *Drosophila* ovary provide an ideal case for investigating collective cell migration, because they can be imaged within their native tissue and are amenable to genetic manipulation. Here, we discuss how border cell movement is controlled genetically, including recent insights into group guidance, how these cells interact with their surroundings, and how they divide up functions and coordinate behaviors.

From herds of elephants to colonies of army ants, many animals travel in packs, which can provide significant selective advantages, including creating layers of protection, facilitating broader surveillance of the environment, and the ability to share labor. There is safety and strength in numbers—especially if the group communicates well. Thus, it should not be surprising that on the microscopic level, many cell types also travel in packs. Historically, research has predominantly focused on individual cell migration. Groups of cells rely on many of the same molecular mechanisms that individual cells do, so past studies have been extremely informative for both types of migrations. However, groups of motile cells must additionally stay coordinated and communicate, which inspires additional, interesting research questions. Cells that migrate as a collective can divide up roles, and the group can leverage physical properties that individual cells may not possess, but how this occurs is only beginning to be understood. Because collective cell migrations are critical for healthy animal development and can be aberrantly induced to trigger medical problems, like autoimmune disorders and metastatic cancer, it is essential to uncover the mechanisms that regulate these processes.

Cite this article as *Cold Spring Harb Perspect Biol* doi: 10.1101/cshperspect.a041884

The border cells of the *Drosophila* (fruit fly) ovary provide an ideal case for investigating how cells can travel in a pack. Researchers have focused on these cells because they can be observed within their native context—the developing egg—and because they are amenable to many types of genetic manipulations engineered for use in flies. As a result, a lot is known about how border cell migration is controlled, including how the collective divides up tasks and coordinates behaviors within the group while being responsive to its surroundings. In this review, we describe the molecular and biophysical mechanisms of collective cell migration that have been characterized by studying border cells and what open questions are still being explored.

TRACING THE STEPS

Tracking an individual animal's behavior can provide insight into group dynamics; similarly, the basic steps governing cell motility have been inferred from studies of individually migrating cells, and this framework applies broadly to most cell types (Tschumperlin 2013; De Pascalis and Etienne-Manneville 2017; SenGupta et al. 2021). Migration is often directed by external cues from the environment. In response, a cell polarizes through dynamic cytoskeletal reorganization, where regulatory proteins drive actin polymerization and branching at the front, creating a membrane protrusion toward the direction of migration (Le Clainche and Carlier 2008; Insall and Machesky 2009; Svitkina 2018). (Although we note that some cell types rely instead on blebbing, which has a different mechanism; Fackler and Grosse 2008.) To gain traction and translocate, the membrane at the protruding front makes contact and adheres with a substrate, and adhesion proteins connected to the internal cytoskeleton exert force. The molecular motor protein, myosin II, powers contractility and rearrangements by pulling on actin filaments, and these forces translocate the cell forward while the back of the cell releases its adherence to the substrate and retracts (Ridley et al. 2003). Myosin activity is tightly and locally controlled through phosphorylation of its regulatory light chain subunits (Kimura et al. 1996; Jordan and Karess 1997; Vicente-Manzanares et al. 2009).

Collectively migrating cells, including the border cells, follow this basic routine and have similar mechanistic players as in individual cells, but these alone are not sufficient to explain the observed group dynamics (Montell et al. 2012; Olson and Nechiporuk 2021; Roberto and Emery 2022). (An overview of border cell migration is shown in Fig. 1 and Supplemental Movie 1.) Working with a group always presents challenges for cooperation and communication. In this review, we explore how cells overcome these challenges and the advantages that they gain from migrating collectively.

SELECTING THE GROUP AND DIVIDING THE LABOR

Within a traveling pack, individuals can divide up tasks for efficiency. Within the *Drosophila* ovary, many cell types arise with different functions to enable successful egg production. The migrating border cell cluster itself contains two clearly differentiated cell types with different roles: the polar cells, which are centrally located and nonmotile, and the outer border cells, which tightly adhere to the polar cells (Han et al. 2000; Grammont and Irvine 2002). These two cell types arise as neighbors within the monolayer of the follicular epithelium, which surrounds germline cells as they develop (Fig. 1A). Each egg forms from an "egg chamber" that contains one oocyte and 15 sister cells—the nurse cells—which accumulate biomolecules to feed the oocyte as it goes from stem cell to a mature egg over about a week (Horne-Badovinac and Bilder 2005; He et al. 2011). The follicle cells help pattern the oocyte and form the needed eggshell. While many interesting processes occur during egg development (reviewed in McLaughlin and Bratu 2015; Duhart et al. 2017; Roberto and Emery 2022), we focus here on the middle stages. By stage 8 (Fig. 1B), most follicle cells appear identical, except for two cells on either end of the anterior–posterior (AP) axis—the polar cells. Polar cells activate numerous signaling pathways, including Notch and Hippo, differentiating them from their neighbors

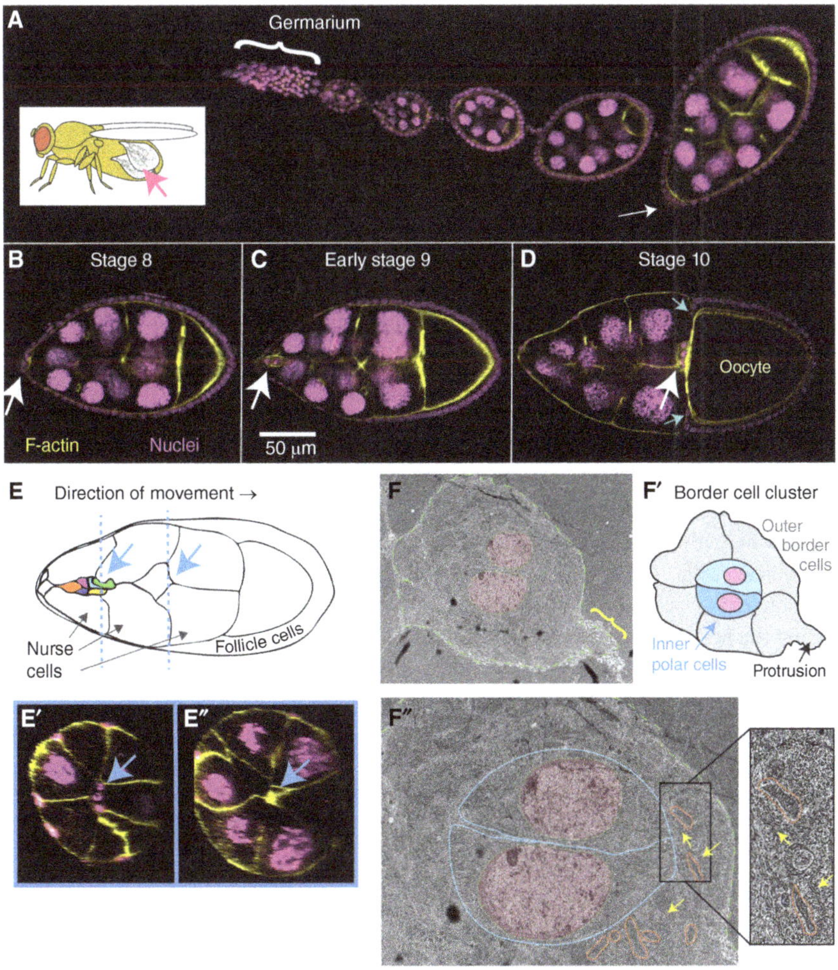

Figure 1. Border cell specification and migration stages during *Drosophila* oogenesis. (*A–D*) *Drosophila* egg chambers stained for F-actin (yellow) and nuclei (magenta). Anterior is to the *left* and posterior is to the *right*. Large white arrows denote the border cell clusters. (*A*) Egg chambers are formed in the adult female fly ovary (*inset*, pink arrow). A fluorescently labeled ovariole shows the germarium (bracket), which houses the germline and follicle stem cells, and progressively developing egg chambers. In this example, the *rightmost* egg chamber has developed to stage 7; the medium white arrow indicates anterior polar cells. (*B–D*) Border cells (white arrow) are specified from the *anteriormost* follicle cells in stage 8 egg chambers (*B*), detach and migrate posteriorly toward the oocyte in stage 9 (*C*), and then reach and dock with the oocyte in stage 10, where they aggregate with the invading centripetal cells (blue arrows) (*D*) to complete their migration. (*E*) In early stage 9, the border cell cluster (rainbow-colored in *E*) rounds up and extends forward protrusions, pinches off from the epithelial neighbors, and invades into the domain of large substrate germline cells, the nurse cells (illustration courtesy of Mackenzie Primrose). Along their route, the border cells encounter prominent multicellular junctures between substrate nurse cells (blue arrows). Blue dashed lines represent egg chamber cross-sections highlighted in *E′–E″*. (*E′–E″*) X–Z cross-sections of egg chambers through multicellular junctures. The border cell cluster (small nuclei) is entering into a juncture surrounded by many nurse cells in *E′*. Another prominent multicellular juncture exists ahead of the cluster more posteriorly in *E″*. (*F*) Transmission electron micrograph of a border cell cluster mid-migration. Anterior–posterior axis is *left* to *right*. The cluster is comprised of border cells (green outline) and the immotile polar cells in the center (red pseudocolored nuclei). The posterior (migration front) of the cluster is to the *right*, and the cluster is surrounded by the larger nurse cells (darker, outer regions). Yellow bracket annotates a forward protrusion. (*F′*) A schematic to show border cell cluster organization. (*F″*) Higher magnification image of *F*. Polar cell membranes are highlighted in blue ellipses. Yellow arrows denote the endoplasmic reticulum and orange ellipses encircle the mitochondria, both of which may be under regulation/stress during migration (image courtesy of Erika Geisbrecht and Yungui Guo).

(Wang et al. 2007; Montell et al. 2012; Lin et al. 2014). Importantly, a key task for the polar cells is to induce border cell fate and motility by releasing a cytokine-like protein, Unpaired (Upd), which activates the Janus kinase and Signal transducer and activator of transcription (Jak/STAT) pathway in nearby epithelial cells (Fig. 2A; Silver and Montell 2001; Ghiglione et al. 2002; McGregor et al. 2002).

Jak/STAT signaling first determines the migratory cells—the outer border cells that surround the polar cells—and later is both required and sufficient for cell motility. Thus, this signaling pathway must be tightly regulated. Overexpression of Upd or constitutively active Jak expressed in follicle cells can induce ectopic numbers of migratory cells, evidence for pathway activation being the key migratory induction signal (Silver and Montell 2001; Ghiglione et al. 2002; Silver et al. 2005). Genetic and physical mechanisms have been identified that normally control the switch to induce the motile cell fate (reviewed in Saadin and Starz-Gaiano 2016a; Peercy and Starz-Gaiano 2020). Controlled Upd release, its graded extracellular distribution, the nuclear import of STAT in receiving cells, and a regulatory feedback loop limit high STAT activation levels to only six to eight follicle cells, enabling their motility (Xi et al. 2003; Starz-Gaiano et al. 2009; Van de Bor et al. 2011; Hayashi et al. 2012; Saadin and Starz-Gaiano 2018; Wu et al. 2022). Downstream, Jak/STAT signaling activates the CCAAT enhancer-binding protein (C/EBP), Slow border cells (Slbo), a required transcriptional activator (Montell et al. 1992; Beccari et al. 2002). Slbo activates expression of the homophilic adhesion molecule E-cadherin (E-cad, encoded by *shotgun*) (Niewiadomska et al. 1999) and multiple other targets, including many cytoskeletal and adhesion regulators (Borghese et al. 2006; Wang et al. 2006; Saadin and Starz-Gaiano 2016b; Burghardt et al. 2023). Expression of these components causes the outer border cells to coalesce around the polar cells and enables the cluster to begin migrating, starting with formation of actin-rich membrane protrusions (Inaki et al. 2022).

In contrast to the dynamic outer border cells, the polar cells have a different job: They do not become motile, but instead get carried in the center of the cluster throughout migration (Prasad and Montell 2007; Cliffe et al. 2017). Surprisingly, the polar cells still have critical roles in the collective migration. The central position of the polar cells is important as they maintain STAT activation by continuous cytokine secretion, which is necessary to keep the group moving (Silver et al. 2005). Polar cells also organize and maintain partial apical-basal polarity in the border cells along their juxtaposed surfaces, and this is required for efficient cohort movement (Fig. 3; Pinheiro and Montell 2004; Wang et al. 2018; Alhadyian et al. 2021; Campanale et al. 2022; Miao et al. 2022; Bhattacharya and Starz-Gaiano 2024). While this axis is independent of the front–back axis of migration, it is not understood how these axes are integrated. However, it is clear that the cell cluster needs polar cells to supervise signaling and outer border cells to translate signals into movement. Thus, polar cells act much like the elders in the animal pack—less active but important for organizing the group.

COORDINATING A DEPARTURE TIME

As for birds or butterflies migrating seasonally, cell migration timing is critical, and this information is imparted by the local environment. How do border cells know when to make their move? Temporal regulation comes from properly titrated ecdysone steroid hormone levels that signal through conserved nuclear hormone receptor complexes (Bai et al. 2000; Cherbas et al. 2003; Hackney et al. 2007; Jang et al. 2009; Domanitskaya et al. 2014). These levels peak at stage 9 of oogenesis (Riddiford 1993), and the hormone is necessary for promoting oocyte growth (Schwartz et al. 1989; Buszczak et al. 1999). All follicle cells appear to be competent to produce the hormone (Buszczak et al. 1999; Domanitskaya et al. 2014); however, only border cells respond to ecdysone by moving. This occurs in response to transcriptional activation via the Ecdysone receptor (EcR) and Ultraspiracle heterodimer, along with the Taiman coactivator (orthologous to human steroid receptor coactivator-3) (Fig. 2A; Bai et al. 2000; McDonald et al. 2003;

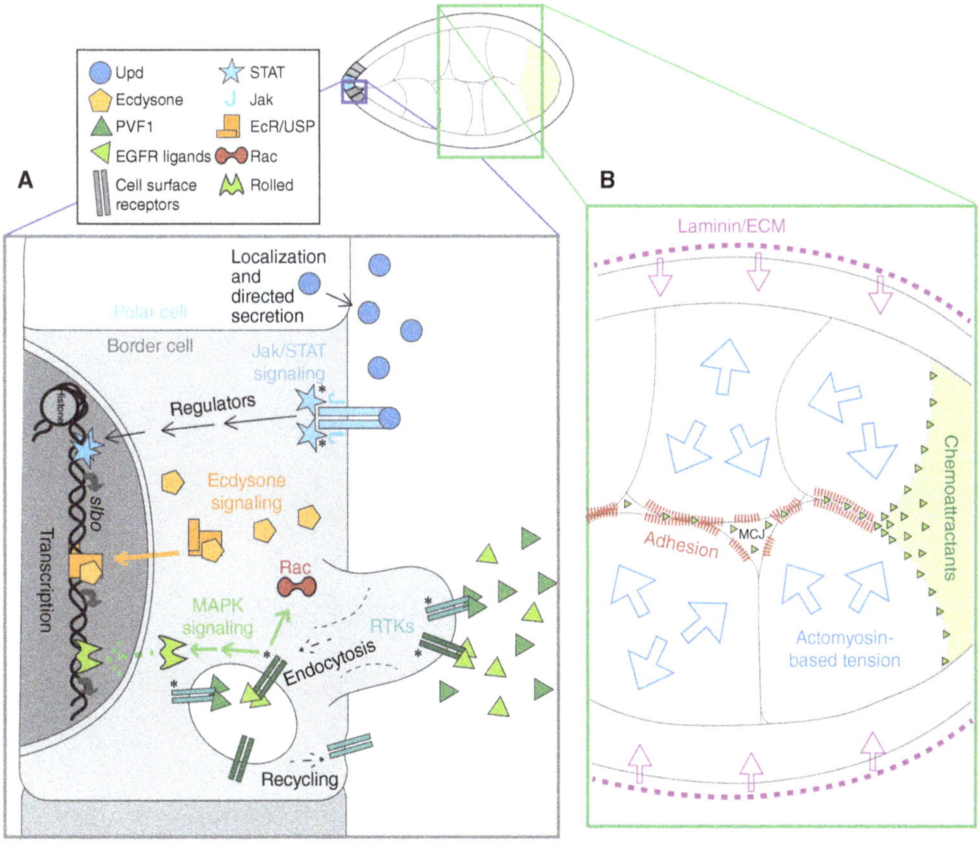

Figure 2. Signaling pathways that specify border cells promote motility and influence the migration path between germline cells. (*Top center*) Cartoon depicts a stage 8 egg chamber with border cells being specified in the follicular epithelium. Boxed regions are enlarged *below*. (*A*) Unpaired (Upd) is secreted by polar cells, which binds to and activates a cell surface receptor in nearby epithelial cells to induce Janus kinase and Signal transducer and activator of transcription (Jak/STAT) activation. Phospho-STAT is shuttled to the nucleus. STAT signaling levels depend on many regulators. High levels of activated STAT in the nucleus lead to expression of numerous targets, including *slbo* and *shg (E-cad)*, to specify border cell fate at stage 8 of egg development. At the start of stage 9, high levels of the ecdysone steroid hormone, produced within follicle cells, activate the EcR/Usp heterodimer and cofactors to promote expression of downstream targets and license the cells to move. Chemoattractants Platelet-derived growth factor/Vascular endothelial growth factor–related Factor 1 (PVF1) and Epidermal growth factor (EGF)-like ligands (green triangles) are expressed from the oocyte in these stages and bind and activate receptor tyrosine kinases (RTKs) at the cell surface. Activated RTKs are endocytosed into signaling endosomes via regulation by ArfGAP1 and then locally activate Rac and trigger mitogen-activated protein (MAP) kinase signaling to enable directional movement. Receptors can then be recycled. (*B*) Forces from the outer basement membrane (purple arrows), including contributions from Laminin, compress the center of the egg chamber, while actomyosin in the nurse cells provides stiffness and maintains shape as cells resist compression and balance intracellular forces (large blue arrows). Homophilic E-cadherin (E-cad) binding, represented by the hatched red lines, keeps nurse cells in close contact except at multicellular junctures (MCJs). Chemoattractants (green triangles) are secreted from the oocyte and form an irregular concentration gradient due to the geometry of large MCJs—the largest MCJs are also called "intersections." (See text for details and Table 1 for recently defined regulators in these pathways.)

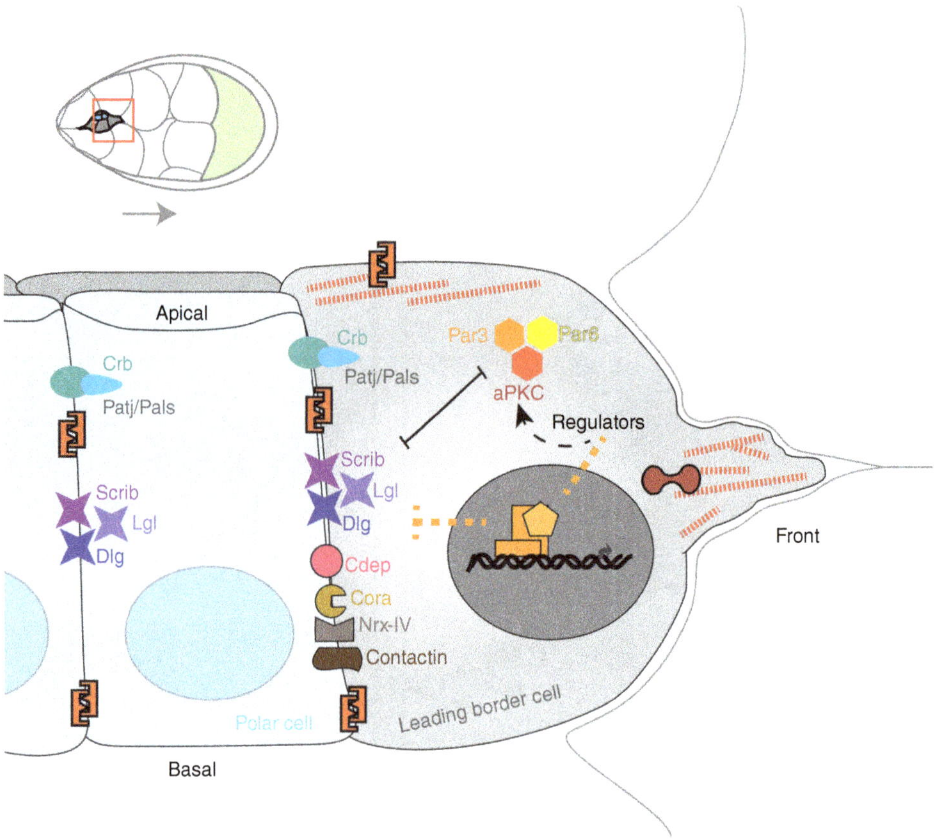

Figure 3. Apical-basal polarity promotes efficient clustered migration. (*Top inset*) Depicts a stage 9 egg chamber with border cells migrating to the *right*. Boxed region is enlarged *below*. Separate from the front/back direction of migration, the cell cluster maintains an apical-basal polarity evident by the specific localization of polarity proteins, especially along the polar cell/polar cell (blue) and polar cell/border cell interfaces (blue/gray). Multiple regulators are needed to maintain this organization, including the polarity proteins themselves, the aPKC/Par-3/Par-6 complex, and steroid hormone signaling (from the nucleus). Actin-based protrusions are enhanced at the front of the cell (*right*) in clusters that have proper apical-basal polarity. (See text for details and Table 1 for recently defined regulators involved in these pathways.)

Jang et al. 2009). The key transcriptional target(s) of steroid hormone signaling that triggers movement remains elusive, despite multiple transcriptomic studies in border cells (Borghese et al. 2006; Wang et al. 2006; Manning et al. 2017; Burghardt et al. 2023). However, a cascade of nuclear hormone receptor activation occurs downstream from EcR, and activation of other transcription factors and a chromatin remodeling factor are needed for a proper response (see Table 1; Saadin and Starz-Gaiano 2016b; Manning et al. 2017; Wang et al. 2020b; Xia et al. 2024). EcR signaling is necessary for apical-basal polarity, cluster shape changes, and adhesion, suggesting it can license the border cells to move by coordinating different aspects of the cells' dynamic properties (Bhattacharya and Starz-Gaiano 2024). Thus, it seems the steroid hormone exerts its effects through many targets, some of which remain unidentified.

Insulin signaling also promotes movement (Sharma et al. 2018), but whether other cues work in parallel to ecdysone to control timing remains to be seen. Border cells migrating prematurely provide some clues. Knockdown of spectraplakin protein induced a phenotype reminiscent of precocious migration (Zhao et al. 2022), although whether this is an effector of ecdysone signaling or an independent timing cue is untested. Interestingly, genetic reduction of Laminin, a protein thought to constrain egg chamber shape at the peripheral extracellular matrix (ECM), resulted in premature completion of migration, suggesting there may be a mechanical component to the timing signals (Molina López et al. 2023). More work is needed to determine the links between movement timing and developmental progression.

Once border cells sense that the time is ripe and are competent to begin migration, they are ready to go. However, like grounded birds preparing to flock, additional preparation is required.

PACKING UP AND LEAVING HOME

When it is time to enact a seasonal migration, animals must pack up and leave the nest. For migrating cell collectives, leaving requires a physical untethering from their origin tissue. Upon specification, border cells transform, rounding up in morphology, detaching from the ECM, and breaking adhesive tethers and syncytial contacts with their neighbors (Szafranski and Goode 2004, 2007; Medioni and Noselli 2005; McDonald et al. 2008; Airoldi et al. 2011; Montell et al. 2012; Fox et al. 2020). While the outer border cells actively extend many actin-rich protrusions at this time to pull the cluster forward (Fulga and Rørth 2002; Plutoni et al. 2019), full detachment from the anterior epithelium can take hours (Prasad and Montell 2007; Inaki et al. 2022). Despite detaching themselves from cells in the epithelium, border and polar cells stay tightly attached to neighbors within the motile cluster (Cai et al. 2014).

Live-imaging studies demonstrate that detachment-specific phenotypes are distinct from motility phenotypes and may be controlled separately. Indeed, aberrant levels of septins, GTP-binding proteins that act as noncanonical cytoskeletal scaffolding proteins, result in clusters that do not detach from the epithelium (Gabbert et al. 2023). Similarly, clusters mutant for the motility signal *slbo* exhibit small changes in position without detaching or exhibiting invasive protrusions (Inaki et al. 2022). The intracellular force transduction protein, myosin II, needs to be regulated at the site of detachment to help "pinch off" the cluster from the rest of the epithelium (Fig. 4; Majumder et al. 2012; Plutoni et al. 2019; Zeledon et al. 2019). Misspecification of "extra" border cells also results in poor detachment, and motile clusters often appear tethered to the anterior by a subset of trailing cells (Silver and Montell 2001; Silver et al. 2005; Starz-Gaiano et al. 2008; Monahan and Starz-Gaiano 2015). In addition, proper Par-1 signaling in both border cells and epithelial neighbors (McDonald et al. 2008; Majumder et al. 2012) and regulated Notch signaling in the motile cells (Prasad et al. 2007; Wang et al. 2007) are all required for proper border cell exit from the epithelium.

For the cluster to leave its place of origin, it must become invasive—the ability of cells to force their way into the adjacent tissue and blaze a trail forward. Unlike some invasive cells, border cells do not have to power through a dense ECM—instead, they squeeze between the large nurse cells. Multiple molecules contribute to initial invasion. Jak/STAT signaling is known to be a driving factor since abnormally high signaling induces ectopic invasion of epithelial cells (Silver and Montell 2001; Beccari et al. 2002; Ghiglione et al. 2002; Silver et al. 2005). E-cad is required but also must be controlled properly and localized correctly—high or low levels block invasion (Niewiadomska et al. 1999; Cai et al. 2014; Dai et al. 2020). Other contributors are being identified. For example, the structural protein nuclear Lamin facilitates invasion of the collective (Fig. 5; Penfield and Montell 2023). In egg chambers mutant for generating prostaglandins, a type of lipid-signaling molecule, ectopic cells become invasive but always trail behind the main cluster, possibly by being pulled in (Fox et al. 2020). Border cells mutant for other signaling pathways fail to invade due to downregulated protrusions, cytoskeletal misregulation, and abnormal polarity (see newly identified regulators in Table 1, and previously reviewed ones in Montell et al. 2012; Veeman and McDonald 2016).

Table 1. Recently discovered regulators of border cell migration

Regulators—gene name(s)	Human ortholog (symbols)	Category	References
Vinculin	VCL	Actomyosin/intercellular communication	Khaitan et al. 2023
Misshapen	MINK1	Actomyosin/intercellular communication	Plutoni et al. 2019
Rap1 GTPase	RAP1A	Actomyosin/intercellular communication	Chang et al. 2018; Sawant et al. 2018
Protein phosphatase 1α at 96A, protein phosphatase 1 at 87B, protein phosphatase 1 at 13C	PPP gene group	Actomyosin/intercellular communication	Chen et al. 2020
Arf79F, Arf102F, Gartenzwarg, Drongo	ARF1, ARF4, GBF1, AGFG1	Actomyosin/intercellular communication	Zeledon et al. 2019
Tao, Ras-associated protein 2-like	TAO, RAP2B	Actomyosin/intercellular communication	Roberto et al. 2025
Insulin-like receptor, *Chico	IGF1R, IRS1	Actomyosin/intercellular communication	*Ghiglione et al. 2018; Sharma et al. 2018
Fascin	FSCN2	Actomyosin/intercellular communication	Lamb et al. 2021
Cell division cycle 42	CDC42	Actomyosin/intercellular communication	Colombié et al. 2017
Ubiquitin-specific protease 16/45, Strawberry notch	USP16, SBNO1	Actomyosin/intercellular communication	Messer et al. 2025
Tropomyosin 1	TPM1	Actomyosin/intercellular communication	Cho et al. 2016
Ras association domain–containing protein 8	RASSF8	Actomyosin/intercellular communication	Höhne et al. 2025
microRNA-34, Eip74EF	MIR34A, ELF2	Ecdysone/timing	Xia et al. 2024
Short stop	DST	Ecdysone/timing	Zhao et al. 2022
Laminin B, Viking	LAMB2, COL4A1	Ecdysone/timing	Molina López et al. 2023
Ecdysone-induced protein 75B, *hormone receptor 3, *ftz transcription factor 1	NR1D2, RORB, NR5A1	Ecdysone/timing	Manning et al. 2017; *Wang et al. 2020
Spindly	SPDL1	Ecdysone/timing	Clemente et al. 2018

Continued

Table 1. *Continued*

Regulators—gene name(s)	Human ortholog (symbols)	Category	References
Protein phosphatase 1 regulatory subunit 15	PPP1R15A	Organelles/stress	Chen and McDonald 2024
Catecholamines up	SLC39A7	Organelles/stress	Guo et al. 2024
Lamin	LMNB2	Organelles/stress	Penfield and Montell 2023
Dynamin-related protein 1	DNM1L	Organelles/stress	Qu et al. 2022
Cactin	CACTIN	Polarity/junctions/adhesion	Miao et al. 2022
Pak3	PAK3	Polarity/junctions/adhesion	Felix et al. 2015
Nonstop	USP22	Polarity/junctions/adhesion	Badmos et al. 2021
Macroglobulin complement-related, Coracle, Neurexin-IV, Contactin	CD109, EPB41L1, CNTNAP2, CNTN2	Polarity/junctions/adhesion	Alhadyian et al. 2021
Mind bomb 2; Rho GTPase-activating protein at 19D; Supervillin	MIB2, ARHGAP23, SVIL	Polarity/junctions/adhesion	Trivedi et al. 2022
Patj, Stardust, atypical protein kinase C	PATJ, PALS1, PRKCI	Polarity/junctions/adhesion	Wang et al. 2018
Chondrocyte-derived ezrin-like domain containing protein	FARP2	Rac regulation	Campanale et al. 2022
Rho GTPase activating protein at 18B	ARHGAP12	Rac regulation	Lei et al. 2023
CYFIP-related Rac1 interactor	CYRIA	Rac regulation	Rötte et al. 2024
CTP synthase	CTPS1	RTK-endosomes	Wang et al. 2022
ADP-ribosylation factor GTPase activating protein 1	ARFGAP1	RTK-endosomes	Boutet et al. 2023
Innexin-2	N/A	STAT pathway regulation	Sahu et al. 2017
Dysfusion	NPAS4	STAT pathway regulation	Wu et al. 2022
A-soluble NSF attachment protein, Syntaxin 1A	NAPA, STX1A	STAT pathway regulation	Saadin and Starz-Gaiano 2018
Trithorax-like	N/A	STAT pathway regulation	Ogienko et al. 2020
Carboxy-terminal Src kinase, Src oncogene at 42A	CSK, FRK	Miscellaneous	Torres et al. 2023
Rickets	LGR4	Miscellaneous	Anllo and Schüpbach 2016
Cytosolic prostaglandin E synthase, Akr1B	PTGES3	Miscellaneous	Mellentine et al. 2024
Roughex	N/A	Miscellaneous	Herriage and Calvi 2024
Septin 1, Septin 2, Peanut	SEPTIN2, SEPTIN6, SEPTIN7	Miscellaneous	Gabbert et al. 2023
Peroxinectin-like	N/A	Miscellaneous	Fox et al. 2020
Blistery	TNS1	Miscellaneous	Cha et al. 2017

Drosophila genes with published roles in border cell migration over the last decade are organized by broad cellular processes (category). The predicted human orthologs are listed with their HGNC symbols in the order of their corresponding *Drosophila* genes in the same row. (RTK) Receptor tyrosine kinase, (STAT) signal transducer and activator of transcription.

As they detach and invade, the border cells enter the domain of nurse cells—their migratory substrate—and begin their posterior migration. However, like salmon leaving the ocean for the great migration, the journey is long, arduous, and surprisingly stereotypic. The border cells' destination is the oocyte, where they will dock and form the micropyle—a required structure to create a fertilizable egg (for review, see Horne-Badovinac 2020)—but how do they get there? What directional information do border cells have, and how do they interpret it to reach their target?

Finding and Staying on the Right Path

While animals use odors to track food and follow landscapes to guide their migration, cells monitor chemical attractants (chemoattractants) and use topographical information within a tissue to determine which way to go. We will first consider chemoattractant signaling. Prevailing chemotaxis models suggest that a single cell locally senses the levels of chemoattractant, stabilizing protrusions in the direction of the highest relative levels while suppressing extensions elsewhere via subcellular signaling mechanisms (reviewed in Iglesias and

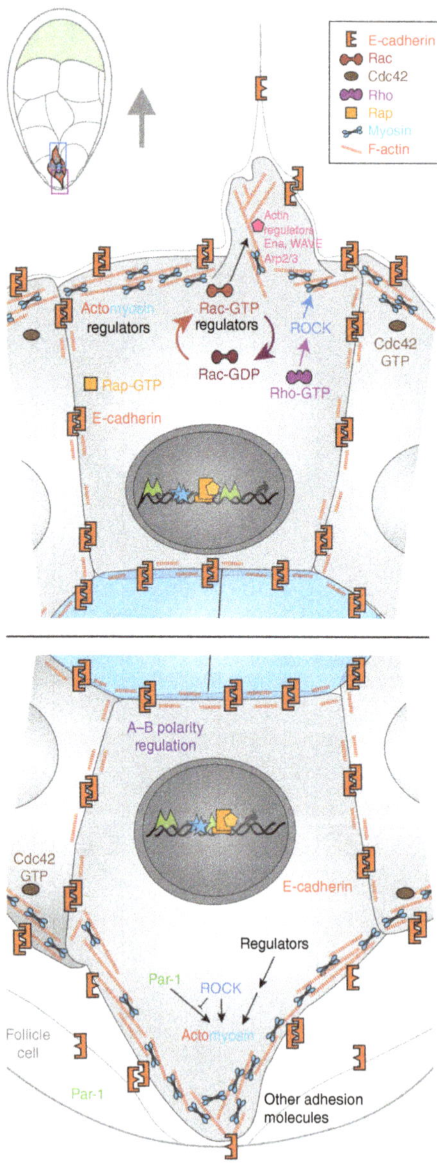

Figure 4. Adhesion and cytoskeletal regulation govern border cell migration. (*Top inset*) Depicts a stage 9 egg chamber with border cells migrating toward the oocyte (green at the *top*), and the blue and purple boxes are the areas represented below in detail. Actomyosin cables exist at the cluster periphery. (*Top*) A leading cell organizing its E-cad on the surface of border cells (gray) and nurse cells (white) provides adhesion and connects to the actin cytoskeleton to generate traction. E-cad is differentially regulated by Rap1 and other regulators on different border cell surfaces and is found at intermediate levels at border cell/border cell sites and highest at the border cell interface with polar cells (blue). Rac and the other small GTPases are regulated by specific exchange factors and activating proteins (not shown). Actin is extended in protrusions through nucleation, polymerization, and branching protein complexes downstream from activated Rac including Ena, WASP and Verprolin-homologous protein (WAVE), and Arp2/3 (*top*). Downstream from Rho and Rho-associated kinase (ROCK), myosin is activated and pulls on actin at the base of a protrusion. An actomyosin supracellular cable pulls the border cell cluster together at the outer surfaces, influenced by Cdc42, and provides a compact shape to the cluster. As the cells exit from the follicular epithelium, several regulators target myosin II for activation along the rear cells as they retract (*bottom*). Apical-basal polarity runs above and below the plane of the schematic, near the polar cells (light blue). Proper polarity and reorganization of adhesion molecules are needed as border cells detach from their neighbors. (See text for details and Table 1 for recently defined regulators involved in these pathways.)

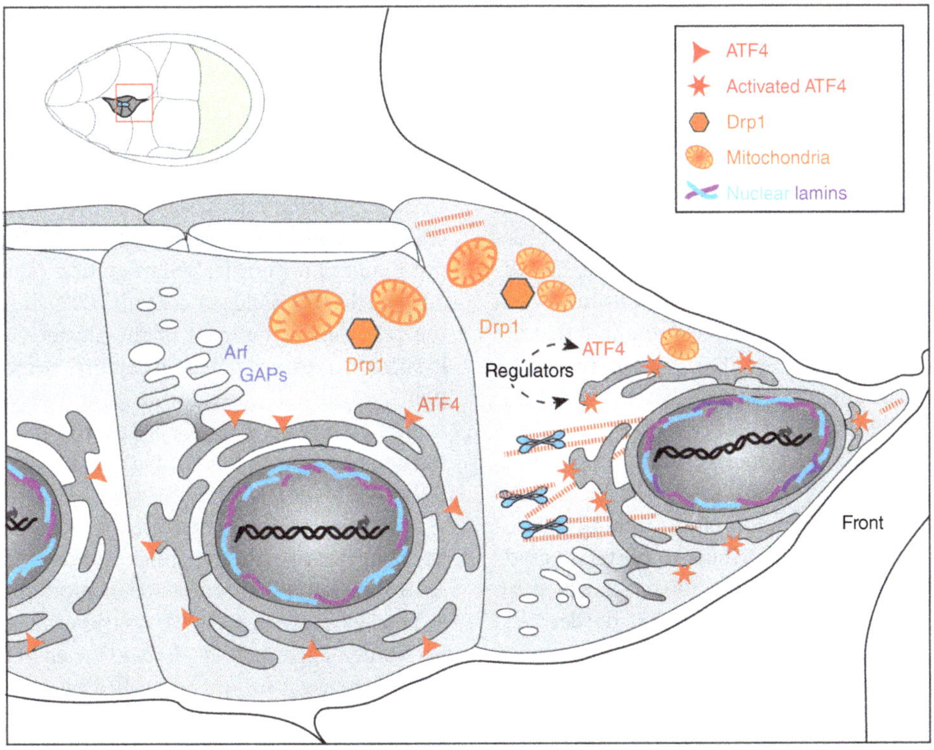

Figure 5. Organelles respond to their surroundings and promote effective border cell migration. (*Top inset*) Depicts a stage 9 egg chamber with border cells migrating to the *right*. Boxed region is enlarged *below*. As the cluster squeezes between tightly packed nurse cells (white), the leading border cell (*right*) forces its nucleus forward via actomyosin changes and resists compression through upregulation of certain nuclear lamin proteins (blue and purple), opening up a less restricted path. Migration activates the Protein kinase R-like ER kinase (PERK)/activating transcription factor 4 (ATF4)-mediated endoplasmic reticulum (ER) stress response via ATF4 activation on the ER. Mitochondrial fission increases during cluster movement via Drp1 downstream from Rho and ROCK, while ArfGAPs regulate the Golgi organization, contributing to pleiotropic effects on migration (see text for details).

Devreotes 2008; Stephens et al. 2008; SenGupta et al. 2021). Border cells provide an interesting case, since this model would require comparisons and communication between cells to sense differences and achieve directional movement for the whole cluster, and it is not yet clear how this works.

The border cell cluster migrates toward several chemoattractants. These signals are ligands for two types of receptor tyrosine kinases (RTKs): the *Drosophila* homolog of the vertebrate Epidermal growth factor receptor (EGFR) and one similar to both Platelet-derived growth factor (PDGF) and Vascular endothelial growth factor (VEGF) receptor (called PVR) (Fig. 2). Ligand binding induces receptor dimerization and phosphorylation, recruiting factors that trigger downstream signaling. EGFR and PVR are each required in border cells but are functionally redundant for posterior movements (Duchek et al. 2001; McDonald et al. 2006; Poukkula et al. 2011). Notably, expressing both dominant negative EGFR and PVR concurrently results in clusters that protrude erratically, suggesting that functional, forward-oriented protrusions are stabilized downstream from RTK signaling (Prasad and Montell 2007). Conversely, constitutively active PVR or EGFR impedes border cell migration, indicating that RTK activity needs to be properly regulated, not just stimulated, to confer guidance (Duchek and Rørth 2001; Duchek et al. 2001).

Several relevant RTK ligands are expressed at the oocyte, the border cells' destination, and misexpression can attract border cells elsewhere (Duchek and Rørth 2001; Duchek et al. 2001; McDonald et al. 2003, 2006; Dai et al. 2020). These results indicate that these chemoattractant ligands are sufficient for guidance, are diffusible in extracellular space, and normally act by establishing a concentration gradient that the border cells move up. The known posterior-guiding chemoattractants are PDGF/VEGF-related factor 1 (PVF1) and two EGFR ligands—Spitz and Keren (Duchek and Rørth 2001; McDonald et al. 2006). Other EGFR ligands guide a dorsal migration after reaching the oocyte and may also contribute to posterior guidance (Duchek and Rørth 2001; Bianco et al. 2007). Live-imaging revealed that border cells exhibit variations in posterior migration speed that are dependent on chemoattractant signaling and may be due to variations in chemoattractant levels (Bianco et al. 2007; Poukkula et al. 2011; George et al. 2025). Intriguingly, the collectiveness of the cluster is predicted to confer advantages in interpreting the chemoattractant gradient compared to an individual border cell. Larger collectives may offset the effects of variations in concentration along the migration path because they can sample the environment over a larger domain, which would not be possible in individual cells (George et al. 2025).

Protrusions may function as a searching structure prior to their stabilization by a perceived chemoattractant gradient, as seen in other systems (Van Haastert 2010). The border cell cluster protrudes even in the absence of RTK signaling, although the protrusions are smaller, less stabilized, and directionally more random (Prasad and Montell 2007; Inaki et al. 2022). While border cell protrusions initially sample the environment (Prasad et al. 2007) and increasingly do so after migration is permitted, they become stabilized when activation of high levels of chemoattractant receptors at the front of the cluster triggers reorganization of actin filaments (Fulga and Rørth 2002; Poukkula et al. 2011). Activation of guidance RTKs in border cells leads to activation of Rho GTPases, predominantly Rac, as well as increased levels of mitogen-activated protein kinase (MAPK) signaling in the border cells, consistent with other cell types (Fig. 2A; Duchek and Rørth 2001; Duchek et al. 2001; Bianco et al. 2007). Unlike some other eukaryotic systems, phospholipase C (PLC) and phosphoinositol-3 kinase (PI3K) are not required for border cell migration (Duchek and Rørth 2001; Fulga and Rørth 2002). Instead, the cells define the front of the cluster via the localization of activated receptors, which is tightly regulated via endocytosis and recycling factors (Jékely et al. 2005; Assaker et al. 2010; Janssens et al. 2010; Ramel et al. 2013; Wan et al. 2013; Boutet et al. 2023).

Downstream from activated RTKs, Rac is a key regulator of actin dynamics and cytoskeletal reorganization that promotes translocation (Fig. 4; Duchek and Rørth 2001; Wang et al. 2010; Fernández-Espartero et al. 2013). Rac activity in border cells is locally controlled by exchange factors and other regulators (Duchek and Rørth 2001; Bianco et al. 2007; Fernández-Espartero et al. 2013; Chang et al. 2018; Lei et al. 2023; Rötte et al. 2024). Rac is not just required: experiments with photoactivatable Rac (PA-Rac) indicate that its activation in one cell is sufficient to designate that cell as the leader and to direct the other cells of the cluster to follow, even if in the wrong direction (Wang et al. 2010; Inaki et al. 2012). However, controlled Rac activity is essential, as constitutively high Rac signaling can be deleterious to border cells and other follicle cells, blocking migration (Geisbrecht and Montell 2004) or even inducing striking phagocytic activity in follicle cells (Mishra et al. 2023). Under normal conditions, though, RTK activation at the leading edge recruits active Rac from an existing pool that resides at supracellular actomyosin cables, to polymerize actin specifically at protrusion tips (Zhou et al. 2022).

Many components respond to Rac activation and control productive cytoskeletal dynamics in the border cell cluster. Some well-conserved regulators determine where and when actin filaments are nucleated, how far they are extended, and whether they are branched, bundled, or severed (reviewed in Montell et al. 2012;

 Cite this article as *Cold Spring Harb Perspect Biol* doi: 10.1101/cshperspect.a041884

Mayor and Etienne-Manneville 2016). Recently, additional components have been newly shown to fit into these actin regulatory networks to control efficient border cell migration, cohesion, and polarity (see Table 1; Law et al. 2013; Felix et al. 2015; Campanale et al. 2022; Rötte et al. 2024). In addition, myosin II activation is needed at the base of protrusions and at the back of the cluster to promote efficient forward movements (Majumder et al. 2012; Aranjuez et al. 2016; Mishra et al. 2019; Plutoni et al. 2019), although how front–back information is transduced through the cluster is not entirely clear. Interestingly, the actin-bundling protein, Fascin, also contributes to timely migration, both autonomously and nonautonomously by regulating myosin activity (Lamb et al. 2021). Thus, the border cell system remains a strong platform for discovery of new molecular mechanisms.

Once oriented toward the trail of chemoattractants and poised by cytoskeletal reorganization, the border cells can "follow their noses" toward the oocyte. But, environmental factors still play a role.

Pathfinding by Following the Natural Landscape

In addition to following chemical factors, cells follow the physical terrain of their environment and, like animals, often prefer the path of least resistance. Border cells directly contact their migratory substrate cells through homophilic E-cad attachments, and E-cad is required in both cell types for directional movement (Niewiadomska et al. 1999; Cai et al. 2014; Dai et al. 2020). E-cad adhesions are less concentrated at multicellular junctures between nurse cells (Dai et al. 2020), and this can create small pockets of extracellular space (see Figs. 1E′,E″,2B). Border cells frequently navigate into these spaces, guiding the border cells along the central path rather than on the side. Remarkably, even when chemoattractant or adhesion information is genetically perturbed, border cells reliably travel centrally, albeit poorly posteriorly (Dai et al. 2020). However, since abnormally high RTK activation in a single cell and/or PA-Rac can drive migration of the whole cluster in directions independent of the central path (Wang et al. 2010; Inaki et al. 2012), there may be some degree of prioritization for chemoattractant guidance over physical guidance.

Several aspects of migratory behaviors might be explained by the irregular landscape of the large nurse cells. Models have considered the drag and osmotic pressure exerted on the border cells from their surroundings. One possibility is that the cluster gets trapped within certain junctures, hindering its speed and advancement, and potentially contributing to a tumbling motion of the cluster (Stonko et al. 2015; Cai et al. 2016). In addition, our recent work showed that migration speed is tied to changes in the chemoattractant concentration gradients along the migration path where germline cells meet, as these multicellular junctures create "intersections" that force border cells to evaluate different directions to go and may impact the levels of diffusible signals. Mathematical modeling corroborated the idea that the slower speeds in these regions are due to flattened chemoattractant gradients (George et al. 2025). This does not rule out an additional role for physical effects in these variable behaviors. Given the demonstrated force dynamics between the nurse cells and border cells throughout migration (Aranjuez et al. 2016; Lamb et al. 2021; Mellentine et al. 2024), it will be interesting to uncover how much of clustered cell migration behavior is attributable to physical versus chemical guidance.

Just as herds of animals rely on various cues like scent, terrain, and each other to navigate their seasonal migrations, border cells are guided by an array of chemical and biophysical inputs—chemoattractants, adhesion molecules, and tissue topography—all contributing to their journey. But, more questions arise: what keeps the cluster moving, intact, and well-coordinated on its way to the oocyte?

Banding Together to Advance

Like the muscles that allow animals to run or fly or swim, mechanical systems are engaged for cells to translocate. After invasion begins, cell–cell adhesion and cytoskeletal dynamics are altered to adjust mechanical properties and communicate within the migrating cohort, so no cell gets left behind. Although flies have multiple cadherins, E-cad seems to predominate in border cells in the formation of transient attachments to nurse cells and for traction (Niewiadomska et al. 1999; Cai et al. 2014). E-cad is subcellularly regulated within each cell of the cluster: It is most highly localized at the surfaces between the polar cells and border cells, it is high between neighboring border cells to provide cohesion and coordination within the group, and it is at lower levels at the periphery facing nurse cells, but still required there for traction (Fig. 4). How this subcellular regulation is controlled is an open question, although steroid hormone signaling, ubiquitin-pathway proteins, and Rap1 GTPase are involved (Bai et al. 2000; Jang et al. 2009; Chang et al. 2018; Sawant et al. 2018; Trivedi and Starz-Gaiano 2018; Bhattacharya and Starz-Gaiano 2024). E-cad-dependent migration is linked to the cortical actin cytoskeleton by adapters including α- and β-catenins (Niewiadomska et al. 1999; Desai et al. 2013), and the cluster's stiffness and shape can be mediated through myosin II (Majumder et al. 2012; Aranjuez et al. 2016; Mishra et al. 2019; Lamb et al. 2021).

In many migrating cells, force transduction comes not just from binding of similar adhesion molecules but also from interactions between ECM and integrin-dependent focal adhesions (De Pascalis and Etienne-Manneville 2017). Border cells express integrins, which are important for intercluster adhesion; however, a requirement for integrins in border cell migration has not been shown (Dinkins et al. 2008; Llense and Martín-Blanco 2008; Kókai et al. 2012; reviewed in Montell et al. 2012). Currently, any ECM components that may localize along the nurse cell migratory domain and their potential effects on border cell migration remain uncharacterized.

Throughout migration, the cluster must balance its internal forces with the forces exerted on it from surrounding nurse cells (Saadin and Starz-Gaiano 2016a). An actomyosin network outlines the cluster periphery, but the overall shape of the cluster varies. To explore how this is controlled, Aranjuez et al. (2016) altered myosin II regulation in and around the border cells. These and subsequent experiments showed that compressive forces from the surrounding substrate cells are opposed by internal myosin-dependent forces to maintain overall cluster shape (Aranjuez et al. 2016; Lamb et al. 2021). Complementarily, signals from a myosin II activity reporter coincided with cluster extension and retraction events throughout migration (Aranjuez et al. 2016; Mishra et al. 2019; Lamb et al. 2021), demonstrating a cluster-wide coordination of force transduction. These external forces have notable effects on live behaviors. Changes in basement membrane stiffness, for instance, modulate migration speed, suggesting there is an optimal tissue rigidity for efficient movement (Molina López et al. 2023). Mathematical models propose that a critical size threshold for border cell clusters aids in overcoming these forces and adhesion considerations in the heterogenous migration domain (Stonko et al. 2015; Cai et al. 2016).

The cluster of border cells band together during migration and distribute external forces through an intercellular network of cortical actomyosin (Fig. 4). This peripheral network mediates communication within the cluster, to bias protrusions frontward, and it is impacted through a combination of confinement and tension (reviewed in Roberto and Emery 2022, and see Wang et al. 2020a; Zhou et al. 2022). The network is controlled in part through the Ste20-like kinase Misshapen (Msn) (Plutoni et al. 2019), Cdc42 (Colombié et al. 2017), Jun-kinase signaling (Llense and Martín-Blanco 2008), and ArfGAP (Zeledon et al. 2019; Boutet et al. 2023). The levels and localization of these regulators are, themselves, tightly regulated. As RTK signaling establishes the front of the cluster, localized activation of myosin (via phosphorylation) at the base of protrusions and at the sides and back of the cluster pulls the back of the cluster forward, restricts protrusions to the front (Mishra et al. 2019; Wang et al. 2020a), and may also contribute to rotational movement (Combedazou et al. 2017). While ArfGAP increases myosin activation in the back of the cluster, it also facilitates chemoattractant receptor recycling to the front, thus refining the forward–backward response (Zeledon et al. 2019; Boutet et al. 2023). Interestingly, Msn is regulated by the Tao kinase (Plutoni et al. 2019), but the much stronger loss-of-function phenotype of *Tao* compared to *msn* (Roberto et al. 2025) suggests that Tao likely has additional targets, potentially Par-1, which is known to be required for border cell detachment and to restrict protrusions (McDonald et al. 2008; Majumder et al. 2012). Protein phosphatase 1 (Pp1) may also be important for transducing information between cells and maintaining collectiveness, acting on myosin and adhesion complexes (Chen et al. 2020). Thus, in response to directional cues, border cells stabilize actin-filled protrusions through adhesion, relaying force and tension through a supracellular communication network to reinforce forward-facing protrusions.

The intracellular and intercellular mechanics described here keep the cluster moving and coordinated. But who is in charge? Will one individual lead, while others merely follow, or will each individual carry its own weight?

All for One and One for All?

Some packs of animals designate one individual as a leader, while the rest are followers. In other cases, leaders are switched out to avoid burnout or exhaustion. Border cells may follow the latter situation, with outer border cells observed to exchange positions with neighbors in the cohort and taking turns as the lead cell (Prasad and Montell 2007; Cliffe et al. 2017). The "decision" to switch is proposed to depend on both biochemical and biophysical factors. One possibility is that when the chemoattractant receptors in the leading cell are saturated, this pauses signaling, and an adjacent cell takes over (proposed in Bianco et al. 2007). Indeed, border cells rotate more frequently as they get closer to the oocyte, where presumably the highest concentrations of chemoattractant are encountered (Bianco et al. 2007; Poukkula et al. 2011; Cai et al. 2016). Alternatively, physical factors such as gaps between substrate cells may limit the ability of the leader to pull forward, forcing another cell to take the lead and the cluster to rotate (Stonko et al. 2015; Cai et al. 2016). Another intriguing possibility is that dips in the chemoattractant gradient at multicellular junctions temporarily disrupt guidance information for the leader (George et al. 2025), allowing an adjacent cell still capable of detecting a concentration difference to move to the front. This concept may provide new insight into how cells behave in complex environments.

Recent evidence suggests that a leader cell can take on specific roles, potentially sacrificing itself for the good of the group. For instance, Penfield and Montell (2023) demonstrated that as a leader cell protrudes forward, its nucleus wedges itself ahead reinforced by Lamin and exerts a physical force mediated by actomyosin to squeeze between nurse cells. This clears a path, making it easier for other cells to follow (Fig. 5). This finding is consistent with studies in the fly embryo (Deshpande and Telley 2021) and in individual cells like leukocytes (reviewed in Li et al. 2022), in which nuclear tensile strength and repositioning play important roles in movement within confined domains. The other cell nuclei shapes remain largely unchanged, which is probably beneficial given the potentially damaging effects of nuclear deformation on the genome and the cell generally (Calero-Cuenca et al. 2018; Shah et al. 2021). It is therefore interesting to wonder whether leaders change throughout migration to confer a selective advantage to the group, so that no one cell bears the entire physical burden.

Interestingly, like ailing members of an animal group, weaker cells in a migratory group sometimes fall behind but are assisted by the rest. This is illustrated in the case of the border cells when mutant cells remain in the cluster but consistently reside at the back of the group (Bianco et al. 2007; Wan et al. 2013). These mosaic clusters often migrate more slowly than normal. Other times, mutant cells may be left behind, as is the case with cells defective for E-cad or STAT signaling (Ghiglione et al. 2002; Silver et al. 2005; Cai et al. 2014; Campanale et al. 2022). These experiments exemplify the need for each cell to individually maintain sufficient ability for motility but also reveal that individual defective cells can be carried by the rest of the cluster.

Different models for the role of follower cell behaviors in migratory cohorts have been proposed. Emerging paradigms emphasize the contributing roles of followers to the collective migration (Theveneau and Linker 2017; Qin et al. 2021). In border cells, leader cells have defined roles at the front of the cluster, being the first to invade and promoting forward movements of the group. Meanwhile, follower cells are surprisingly active at the back of the cluster, where they crawl over other cells and still protrude actively in a Rac-dependent process that allows them to "keep up with the leaders" (Campanale et al. 2022). These data suggest that, normally, follower cells in the cluster are not passive. Tracking the contributing factors in individual cells throughout migration to predict cell switching behaviors will be an interesting avenue of future study.

Besides the signaling and mechanics that need to be individually maintained and coordinated through the cohort, individual cells must also maintain proper organelle structure and function, an intriguing aspect of cell migration that we are only beginning to appreciate (Fig. 5). Given high energetic demands of molecular machinery required for motility (Mosier et al. 2021; Parlani et al. 2023), it makes sense that mitochondrial dynamics play a role in migration capacity (Qu et al. 2022), and this opens new lines of investigation into fueling cell movement. Consistent with the idea that cells undergo physical exertion, endoplasmic reticulum (ER) stress seems to be triggered during border cell migration (Chen and McDonald 2024). Because of this, border cells require a key regulatory subunit called dPPP1R15, which opposes the Activating transcription factor 4 (ATF4/Cryptocephalic)-dependent integrated stress response and permits efficient migration. In fact, border cells lacking *dPPP1R15* do not invade well between germline cells, and loss of its target, *Pp1*, results in cluster cohesion defects. Avoidance of the ER stress response in border cells also requires the Zinc transporter Zrt-Irt-like protein7 (Zip7/Catsup) (Guo et al. 2024). Additionally, proper organization of the Golgi in border cells through Arf GTPases is tied to proper function of the secretory pathway in regulating key migration components, like E-Cad (Zeledon et al. 2019). It will be interesting in the future to understand how cells distinguish between damaging stress versus temporary stress due to movement and respond appropriately, and more generally, how organelles contribute to collective cell migration.

Arrival—and What Is on the Horizon

Like the animal herd settling down after an arduous journey, border cells complete their migration by docking with the oocyte (Miao et al. 2020). The docked border cells are surrounded by another interesting, invasive epithelial cell type, the centripetal cells, which move as a sheet to enclose the oocyte face (Parsons et al. 2023). With their migration complete, the border cells undergo further development into the micropyle, which is required for fertilization (reviewed in Horne-Badovinac 2020).

Although the border cells' journey is complete once it reaches the oocyte, the journey to understand it endures. To date, the strengths in genetic, molecular, and imaging tools in *Drosophila* have made it possible to elucidate detailed mechanisms governing the border cells' collective migration—from the physical and local factors to the intracellular and intercellular signaling pathways that allow them to reach their destination. Excitingly, newly available genetic tools and pioneering techniques promise to advance the field even further. As with those who study annual animal migrations with new tracking technologies, border cell researchers can leverage advances in experimental tools to make new discoveries.

Recent advancements in "-omics" methods applied to egg chambers have revealed a plethora of information about the genes and proteins that regulate different phases of border cell migration. A single-cell RNA-seq study conducted on the whole ovary, for instance, found that the expression profile of border cells is like that of migratory centripetal cells (Jevitt et al. 2020). Additional profiling revealed novel genes that overlap between these cell types (Rust et al. 2020; Slaidina et al. 2021). These studies identified conserved and distinct genetic mechanisms involved in different types of cell motility and invasion. Another study isolated border cells in premigratory, migratory, and postmigratory phases, and differences in their mRNA expression revealed transcriptional switches throughout (Burghardt et al. 2023). Although some cell types only require posttranslational mechanisms to move, many transcription factors activated at the start of border cell migration, for example, STAT, EcR, and Slbo, among others (Saadin and Starz-Gaiano 2016a), suggest that transcriptional responses are beneficial for regulating migration over long time frames of development. These studies introduce a host of potential candidate genes involved in collective cell migration, which will be interesting to validate and explore mechanistically—especially considering that a fifth of genes in the fly genome remain uncharacterized and at least half are considered functionally understudied (Mohr et al. 2023; Rocha et al. 2023).

Complementing these ʻomics approaches, innovative tools are increasingly available for rapid, efficient genetic manipulation to study protein function and cell dynamics, enabling new lines of inquiry. For instance, clustered regularly interspaced short palindromic repeats (CRISPR)-based strategies for altering gene expression in specific cell types and developmental times are now common, with commercially available materials facilitating access (Gratz et al. 2015; Lin et al. 2015; Bier et al. 2018; Zirin et al. 2022). In border cells, precise genetic modifications have yielded epitope-tagged ligands for RTKs (Dai et al. 2020), protein trapping strategies to monitor protein expression (Li-Kroeger et al. 2018), and targeted guide RNAs to generate cell-specific disruptions (Port et al. 2014; Port and Bullock 2016; Badmos et al. 2021). The advent of additional tools allows for exquisite control of protein function, such as with optogenetics, which allows for spatially precise photoactivatable (PA) control over protein activity. For example, PA-Rac biases border cell cluster polarization and drives migration (Wang et al. 2010), while PA-GFP-tagged tubulin allowed measurement of intracellular diffusion rates (Penfield and Montell 2023). Additionally, nanobody-based strategies can drive proteins to abnormal subcellular domains in border cells or nurse cells to assay the functional relevance of their precise localization (Harmansa et al. 2017; Campanale et al. 2022; Miao et al. 2022; George et al. 2025).

A long-standing strength of the border cell system is the ability to image cells in their native context, and technical improvements in live microscopy enable visualization of collective cell behaviors in unprecedented detail, including variations in migration speeds, protrusion dynamics, and leader and follower cell changes (Dai and Montell 2016; Dai et al. 2020; Campanale et al. 2022; Inaki et al. 2022; George et al. 2025). Better fluorescent protein expression and gentler imaging systems have uncovered intracellular signaling mechanisms within the cluster. For example, use of a LifeAct marker expressed in the border cells (Cai et al. 2014) enabled visualization of the intracellular mechanisms of a supracellular actomyosin network (Mishra et al. 2019; Wang et al. 2020a; Zhou et al. 2022) and identified stereotypic locations of high myosin II activity within the cluster that correlated with behaviors, like rear retraction and protrusion extension (Aranjuez et al. 2016; Mishra et al. 2019; Lamb et al. 2021). However, imaging at high resolution in space and time poses phototoxicity challenges (Icha et al. 2017); thus, experiments to ascertain the dynamics of intracellular proteins using traditional confocal microscopy have been limited. Recent advances in microscopy combine high spatial and temporal resolution with less light exposure (Planchon et al. 2011; Liu et al. 2018), which should dramatically improve characterizations of the dynamics of receptor signaling, cytoskeletal dynamics, and cell–cell communication in the border cells.

An exciting and complementary approach for understanding collective cell migration lies in leveraging innovative interdisciplinary strategies. Mathematical models have focused on discrete, agent-based strategies, and these have added insight to several aspects of border cell movements and generated testable predictions (Mekus et al. 2018; Peercy and Starz-Gaiano 2020). Newer models have informed key parameters in border cell specification, chemotaxis, and physical path preference and have been especially helpful for predicting governing parameters for cases in which biological data are incomplete (Mekus et al. 2018; Dai et al. 2020; Gabbert et al. 2023; George et al. 2025; reviewed in Peercy and Starz-Gaiano 2020). Other migration models have a continuum of cell density and of chemoattractant allowing observation of macroscopic characteristics such as pattern formation (Keller and Segel 1971; Painter 2019); such approaches could be useful for border cells as well. Improved modeling will take advantage of our increasing ability to measure biological changes at molecular and biophysical levels, in part through use of improved super-resolution techniques for image capture and analysis (Amat et al. 2015; Schroeder et al. 2021; Lin et al. 2022; Melanthota et al. 2022; Chen et al. 2023). Different types of representation, such as continuum models, may provide additional insights. As experimental technologies continue to advance, so do databases of physiological, cellular information that can apply to properties of the border cells. Thus, more biologically defined parameters can be incorporated into computational models to capture increasingly accurate and generalizable emergent properties of clustered cell migration.

Border cell migration has been an informative platform for revealing fundamental molecular mechanisms of collective cell behaviors, and the cell migration field's potential in understanding disease continues to deepen. Border cells have long been a canonical model system for cancer metastasis and invasion (Stuelten et al. 2018). However, the high conservation of gene expression and mechanistic regulation observed in other systems, such as glioblastoma, hypoxia, and immunotherapy, have also provided valuable insights into disease states (Volovetz et al. 2020; Mishra et al. 2023). Recent work assaying activating Rac mutations (characterized earlier in border cells; Murphy and Montell 1996) has led to innovative ideas for potential anticancer therapies (Mishra et al. 2023). Therefore, by taking advantage of diverse, improved experimental methods and integrating them with evolving genetic strategies, researchers can continue to use border cell migration to elucidate critical mechanisms that control collective migrations in development and disease.

ACKNOWLEDGMENTS

We acknowledge support from the National Science Foundation grant IOS #2303587. We thank Jocelyn McDonald, Bradford Peercy, Elana Frazier, Andrew Opincar, and Christopher Welsh for critical evaluation of the manuscript and helpful suggestions. We appreciate the work of Yungui Guo, Erika Geisbrecht, and Mackenzie Primrose and their contributions of figure panels. We are grateful for the wealth of updated data cataloged in FlyBase (Öztürk-Çolak et al. 2024). We thank the participants of the Collective Behaviour Workshop 2023 at the Isaac Newton Institute at Cambridge University for inspiration toward our theme. Finally, we acknowledge the many individuals who have contributed to our understanding of border cell migration whose work we did not discuss due to limitations on the length and scope of this paper.

REFERENCES

Airoldi SJ, McLean PF, Shimada Y, Cooley L. 2011. Intercellular protein movement in syncytial *Drosophila* follicle cells. *J Cell Sci* **124:** 4077–4086. doi:10.1242/jcs.090456

Alhadyian H, Shoaib D, Ward RE. 2021. Septate junction proteins are required for egg elongation and border cell migration during oogenesis in *Drosophila. G3 (Bethesda)* **11:** jkab127. doi:10.1093/g3journal/jkab127

Amat F, Höckendorf B, Wan Y, Lemon WC, McDole K, Keller PJ. 2015. Efficient processing and analysis of large-scale light-sheet microscopy data. *Nat Protoc* **10:** 1679–1696. doi:10.1038/nprot.2015.111

Anllo L, Schüpbach T. 2016. Signaling through the G-protein-coupled receptor Rickets is important for polarity, detachment, and migration of the border cells in *Drosophila. Dev Biol* **414:** 193–206. doi:10.1016/j.ydbio.2016.04.017

Aranjuez G, Burtscher A, Sawant K, Majumder P, McDonald JA. 2016. Dynamic myosin activation promotes collective morphology and migration by locally balancing oppositional forces from surrounding tissue. *Mol Biol Cell* **27:** 1898–1910. doi:10.1091/mbc.e15-10-0744

Assaker G, Ramel D, Wculek SK, González-Gaitán M, Emery G. 2010. Spatial restriction of receptor tyrosine kinase activity through a polarized endocytic cycle controls border cell migration. *Proc Natl Acad Sci* **107:** 22558–22563. doi:10.1073/pnas.1010795108

Badmos H, Cobbe N, Campbell A, Jackson R, Bennett D. 2021. *Drosophila* USP22/nonstop polarizes the actin cytoskeleton during collective border cell migration. *J Cell Biol* **220:** e202007005. doi:10.1083/jcb.202007005

Bai J, Uehara Y, Montell DJ. 2000. Regulation of invasive cell behavior by taiman, a *Drosophila* protein related to AIB1, a steroid receptor coactivator amplified in breast cancer. *Cell* **103:** 1047–1058. doi:10.1016/S0092-8674(00)00208-7

Beccari S, Teixeira L, Rørth P. 2002. The JAK/STAT pathway is required for border cell migration during *Drosophila* oogenesis. *Mech Dev* **111:** 115–123. doi:10.1016/S0925-4773(01)00615-3

Bhattacharya M, Starz-Gaiano M. 2024. Steroid hormone signaling synchronizes cell migration machinery, adhesion and polarity to direct collective movement. *J Cell Sci* **137:** jcs261164. doi:10.1242/jcs.261164

Bianco A, Poukkula M, Cliffe A, Mathieu J, Luque CM, Fulga TA, Rørth P. 2007. Two distinct modes of guidance signalling during collective migration of border cells. *Nature* **448:** 362–365. doi:10.1038/nature05965

Bier E, Harrison MM, O'Connor-Giles KM, Wildonger J. 2018. Advances in engineering the fly genome with the CRISPR-Cas system. *Genetics* **208:** 1–18. doi:10.1534/genetics.117.1113

Borghese L, Fletcher G, Mathieu J, Atzberger A, Eades WC, Cagan RL, Rørth P. 2006. Systematic analysis of the transcriptional switch inducing migration of border cells. *Dev Cell* **10:** 497–508. doi:10.1016/j.devcel.2006.02.004

Boutet A, Zeledon C, Emery G. 2023. ArfGAP1 regulates the endosomal sorting of guidance receptors to promote directed collective cell migration. *iScience* **26:** 107467. doi:10.1016/j.isci.2023.107467

Burghardt E, Rakijas J, Tyagi A, Majumder P, Olson BJSC, McDonald JA. 2023. Transcriptome analysis reveals temporally regulated genetic networks during *Drosophila* border cell collective migration. *BMC Genomics* **24:** 728. doi:10.1186/s12864-023-09839-8

Buszczak M, Freeman MR, Carlson JR, Bender M, Cooley L, Segraves WA. 1999. Ecdysone response genes govern egg chamber development during mid-oogenesis in *Drosophila*. *Development* **126:** 4581–4589. doi:10.1242/dev.126.20.4581

Cai D, Chen SC, Prasad M, He L, Wang X, Choesmel-Cadamuro V, Sawyer JK, Danuser G, Montell DJ. 2014. Mechanical feedback through E-cadherin promotes direction sensing during collective cell migration. *Cell* **157:** 1146–1159. doi:10.1016/j.cell.2014.03.045

Cai D, Dai W, Prasad M, Luo J, Gov NS, Montell DJ. 2016. Modeling and analysis of collective cell migration in an in vivo three-dimensional environment. *Proc Natl Acad Sci* **113:** E2134–E2141. doi:10.1073/pnas.1522656113

Calero-Cuenca FJ, Janota CS, Gomes ER. 2018. Dealing with the nucleus during cell migration. *Curr Opin Cell Biol* **50:** 35–41. doi:10.1016/j.ceb.2018.01.014

Campanale JP, Mondo JA, Montell DJ. 2022. A Scribble/Cdep/Rac pathway controls follower-cell crawling and cluster cohesion during collective border–cell migration. *Dev Cell* **57:** 2483–2496.e4. doi:10.1016/j.devcel.2022.10.004

Cha IJ, Lee JH, Cho KS, Lee SB. 2017. *Drosophila* tensin plays an essential role in cell migration and planar polarity formation during oogenesis by mediating integrin-dependent extracellular signals to actin organization. *Biochem Biophys Res Commun* **484:** 702–709. doi:10.1016/j.bbrc.2017.01.183

Chang YC, Wu JW, Hsieh YC, Huang TH, Liao ZM, Huang YS, Mondo JA, Montell D, Jang AC. 2018. Rap1 negatively regulates the hippo pathway to polarize directional protrusions in collective cell migration. *Cell Rep* **22:** 2160–2175. doi:10.1016/j.celrep.2018.01.080

Chen Y, McDonald JA. 2024. Collective cell migration relies on PPP1R15-mediated regulation of the endoplasmic reticulum stress response. *Curr Biol* **34:** 1390–1402.e4. doi:10.1016/j.cub.2024.02.014

Chen Y, Kotian N, Aranjuez G, Chen L, Messer CL, Burtscher A, Sawant K, Ramel D, Wang X, McDonald JA. 2020. Protein phosphatase 1 activity controls a balance between collective and single cell modes of migration. *eLife* **9:** e52979. doi:10.7554/eLife.52979

Chen Y, Kotian N, McDonald JA. 2023. Quantitative image analysis of dynamic cell behaviors during border cell migration. *Methods Mol Biol* **2626:** 193–217. doi:10.1007/978-1-0716-2970-3_10

Cherbas L, Hu X, Zhimulev I, Belyaeva E, Cherbas P. 2003. EcR isoforms in *Drosophila*: testing tissue-specific requirements by targeted blockade and rescue. *Development* **130:** 271–284. doi:10.1242/dev.00205

Cho A, Kato M, Whitwam T, Kim JH, Montell DJ. 2016. An atypical tropomyosin in *Drosophila* with intermediate filament-like properties. *Cell Rep* **16:** 928–938. doi:10.1016/j.celrep.2016.06.054

Clemente GD, Hannaford MR, Beati H, Kapp K, Januschke J, Griffis ER, Müller HJ. 2018. Requirement of the dynein-adaptor spindly for mitotic and post-mitotic functions in *Drosophila*. *J Dev Biol* **6:** 9. doi:10.3390/jdb6020009

Cliffe A, Doupé DP, Sung H, Lim IK, Ong KH, Cheng L, Yu W. 2017. Quantitative 3D analysis of complex single border cell behaviors in coordinated collective cell migration. *Nat Commun* **8:** 14905. doi:10.1038/ncomms14905

Colombié N, Choesmel-Cadamuro V, Series J, Emery G, Wang X, Ramel D. 2017. Non-autonomous role of Cdc42 in cell–cell communication during collective migration. *Dev Biol* **423:** 12–18. doi:10.1016/j.ydbio.2017.01.018

Combedazou A, Choesmel-Cadamuro V, Gay G, Liu J, Dupré L, Ramel D, Wang X. 2017. Myosin II governs collective cell migration behaviour downstream of guidance receptor signalling. *J Cell Sci* **130:** 9–103. doi:10.1242/jcs.179952

Dai W, Montell DJ. 2016. Live imaging of border cell migration in *Drosophila*. *Methods Mol Biol* **1407:** 153–168. doi:10.1007/978-1-4939-3480-5_12

Dai W, Guo X, Cao Y, Mondo JA, Campanale JP, Montell BJ, Burrous H, Streichan S, Gov N, Rappel WJ, et al. 2020. Tissue topography steers migrating *Drosophila* border cells. *Science* **370:** 987–990. doi:10.1126/science.aaz4741

De Pascalis C, Etienne-Manneville S. 2017. Single and collective cell migration: the mechanics of adhesions. *Mol Biol Cell* **28:** 1833–1846. doi:10.1091/mbc.e17-03-0134

Desai R, Sarpal R, Ishiyama N, Pellikka M, Ikura M, Tepass U. 2013. Monomeric α-catenin links cadherin to the actin cytoskeleton. *Nat Cell Biol* **15:** 261–273. doi:10.1038/ncb2685

Deshpande O, Telley IA. 2021. Nuclear positioning during development: pushing, pulling and flowing. *Semin Cell Dev Biol* **120:** 10–21. doi:10.1016/j.semcdb.2021.09.020

Dinkins MB, Fratto VM, Lemosy EK. 2008. Integrin alpha chains exhibit distinct temporal and spatial localization patterns in epithelial cells of the *Drosophila* ovary. *Dev Dyn* **237:** 3927–3939. doi:10.1002/dvdy.21802

Domanitskaya E, Anllo L, Schüpbach T. 2014. Phantom, a cytochrome P450 enzyme essential for ecdysone biosynthesis, plays a critical role in the control of border cell migration in *Drosophila. Dev Biol* **386:** 408–418. doi:10.1016/j.ydbio.2013.12.013

Duchek P, Rørth P. 2001. Guidance of cell migration by EGF receptor signaling during *Drosophila* oogenesis. *Science* **291:** 131–133. doi:10.1126/science.291.5501.131

Duchek P, Somogyi K, Jékely G, Beccari S, Rørth P. 2001. Guidance of cell migration by the *Drosophila* PDGF/VEGF receptor. *Cell* **107:** 17–26. doi:10.1016/S0092-8674(01)00502-5

Duhart JC, Parsons TT, Raftery LA. 2017. The repertoire of epithelial morphogenesis on display: progressive elaboration of *Drosophila* egg structure. *Mech Dev* **148:** 18–39. doi:10.1016/j.mod.2017.04.002

Fackler OT, Grosse R. 2008. Cell motility through plasma membrane blebbing. *J Cell Biol* **181:** 879–884. doi:10.1083/jcb.200802081

Felix M, Chayengia M, Ghosh R, Sharma A, Prasad M. 2015. Pak3 regulates apical-basal polarity in migrating border cells during *Drosophila* oogenesis. *Development* **142:** 3692–3703. doi:10.1242/dev.125682

Fernández-Espartero CH, Ramel D, Farago M, Malartre M, Luque CM, Limanovich S, Katzav S, Emery G, Martín-Bermudo MD. 2013. GTP exchange factor Vav regulates guided cell migration by coupling guidance receptor signalling to local Rac activation. *J Cell Sci* **126:** 2285–2293. doi:10.1242/jcs.124438

Fox EF, Lamb MC, Mellentine SQ, Tootle TL. 2020. Prostaglandins regulate invasive, collective border cell migration. *Mol Biol Cell* **31:** 1584–1594. doi:10.1091/mbc.E19-10-0578

Fulga TA, Rørth P. 2002. Invasive cell migration is initiated by guided growth of long cellular extensions. *Nat Cell Biol* **4:** 715–719. doi:10.1038/ncb848

Gabbert AM, Campanale JP, Mondo JA, Mitchell NP, Myers A, Streichan SJ, Miolane N, Montell DJ. 2023. Septins regulate border cell surface geometry, shape, and motility downstream of Rho in *Drosophila. Dev Cell* **58:** 1399–1413.e5. doi:10.1016/j.devcel.2023.05.017

Geisbrecht ER, Montell DJ. 2004. A role for *Drosophila* IAP1-mediated caspase inhibition in Rac-dependent cell migration. *Cell* **118:** 111–125. doi:10.1016/j.cell.2004.06.020

George A, Akhavan N, Peercy B, Starz-Gaiano M. 2025. Chemotaxis of *Drosophila* border cells is modulated by tissue geometry through dispersion of chemoattractants. *iScience* **28:** 111959. doi:10.1016/j.isci.2025.111959

Ghiglione C, Devergne O, Georgenthum E, Carballès F, Médioni C, Cerezo D, Noselli S. 2002. The *Drosophila* cytokine receptor Domeless controls border cell migration and epithelial polarization during oogenesis. *Development* **129:** 5437–5447. doi:10.1242/dev.00116

Ghiglione C, Jouandin P, Cérézo D, Noselli S. 2018. The *Drosophila* insulin pathway controls *Profilin* expression and dynamic actin-rich protrusions during collective cell migration. *Development* **145:** dev161117. doi:10.1242/dev.161117

Grammont M, Irvine KD. 2002. Organizer activity of the polar cells during *Drosophila* oogenesis. *Development* **129:** 5131–5140. doi:10.1242/dev.129.22.5131

Gratz SJ, Rubinstein CD, Harrison MM, Wildonger J, O'Connor-Giles KM. 2015. CRISPR-Cas9 genome editing in *Drosophila. Curr Protoc Mol Biol* **111:** 31.32.31–31.32.20. doi:10.1002/0471142727.mb3102s111

Guo X, Mutch M, Torres AY, Nano M, Rauth N, Harwood J, McDonald D, Chen Z, Montell C, Dai W, et al. 2024. The Zn^{2+} transporter ZIP7 enhances endoplasmic-reticulum-associated protein degradation and prevents neurodegeneration in *Drosophila. Dev Cell* **59:** 1655–1667.e6. doi:10.1016/j.devcel.2024.04.003

Hackney JF, Pucci C, Naes E, Dobens L. 2007. Ras signaling modulates activity of the ecdysone receptor EcR during cell migration in the *Drosophila* ovary. *Dev Dyn* **236:** 1213–1226. doi:10.1002/dvdy.21140

Han DD, Stein D, Stevens LM. 2000. Investigating the function of follicular subpopulations during *Drosophila* oogenesis through hormone-dependent enhancer-targeted cell ablation. *Development* **127:** 573–583. doi:10.1242/dev.127.3.573

Harmansa S, Alborelli I, Bieli D, Caussinus E, Affolter M. 2017. A nanobody-based toolset to investigate the role of protein localization and dispersal in *Drosophila. eLife* **6:** e22549. doi:10.7554/eLife.22549

Hayashi Y, Sexton TR, Dejima K, Perry DW, Takemura M, Kobayashi S, Nakato H, Harrison DA. 2012. Glypicans regulate JAK/STAT signaling and distribution of the Unpaired morphogen. *Development* **139:** 4162–4171. doi:10.1242/dev.078055

He L, Wang X, Montell DJ. 2011. Shining light on *Drosophila* oogenesis: live imaging of egg development. *Curr Opin Genet Dev* **21:** 612–619. doi:10.1016/j.gde.2011.08.011

Herriage HC, Calvi BR. 2024. Premature endocycling of *Drosophila* follicle cells causes pleiotropic defects in oogenesis. *Genetics* **226:** iyae009. doi:10.1093/genetics/iyae009

Höhne MY, Milani W, Ramlow K, Bogdan S. 2025. The tumor suppressor RASSF8: a WAVE interaction partner controlling migration and cohesion of invasive border cells in *Drosophila. Proc Natl Acad Sci* **122:** e2426702122. doi:10.1073/pnas.2426702122

Horne-Badovinac S. 2020. The *Drosophila* micropyle as a system to study how epithelia build complex extracellular structures. *Philos Trans R Soc Lond B Biol Sci* **375:** 20190561. doi:10.1098/rstb.2019.0561

Horne-Badovinac S, Bilder D. 2005. Mass transit: epithelial morphogenesis in the *Drosophila* egg chamber. *Dev Dyn* **232:** 559–574. doi:10.1002/dvdy.20286

Icha J, Weber M, Waters JC, Norden C. 2017. Phototoxicity in live fluorescence microscopy, and how to avoid it. *Bioessays* **39:** 1700003. doi:10.1002/bies.201700003

Iglesias PA, Devreotes PN. 2008. Navigating through models of chemotaxis. *Curr Opin Cell Biol* **20:** 35–40. doi:10.1016/j.ceb.2007.11.011

Inaki M, Vishnu S, Cliffe A, Rørth P. 2012. Effective guidance of collective migration based on differences in cell states. *Proc Natl Acad Sci* **109:** 2027–2032. doi:10.1073/pnas.1115260109

Inaki M, Vishnu S, Matsuno K. 2022. Live imaging of delamination in *Drosophila* shows epithelial cell motility and invasiveness are independently regulated. *Sci Rep* **12:** 16210. doi:10.1038/s41598-022-20492-1

Insall RH, Machesky LM. 2009. Actin dynamics at the leading edge: from simple machinery to complex networks. *Dev Cell* **17:** 310–322. doi:10.1016/j.devcel.2009.08.012

Jang AC, Chang YC, Bai J, Montell D. 2009. Border-cell migration requires integration of spatial and temporal signals by the BTB protein Abrupt. *Nat Cell Biol* **11:** 569–579. doi:10.1038/ncb1863

Janssens K, Sung HH, Rørth P. 2010. Direct detection of guidance receptor activity during border cell migration. *Proc Natl Acad Sci* **107:** 7323–7328. doi:10.1073/pnas.0915075107

Jékely G, Sung HH, Luque CM, Rørth P. 2005. Regulators of endocytosis maintain localized receptor tyrosine kinase signaling in guided migration. *Dev Cell* **9:** 197–207. doi:10.1016/j.devcel.2005.06.004

Jevitt A, Chatterjee D, Xie G, Wang XF, Otwell T, Huang YC, Deng WM. 2020. A single-cell atlas of adult *Drosophila* ovary identifies transcriptional programs and somatic cell lineage regulating oogenesis. *PLoS Biol* **18:** e3000538. doi:10.1371/journal.pbio.3000538

Jordan P, Karess R. 1997. Myosin light chain-activating phosphorylation sites are required for oogenesis in *Drosophila*. *J Cell Biol* **139:** 1805–1819. doi:10.1083/jcb.139.7.1805

Keller EF, Segel LA. 1971. Model for chemotaxis. *J Theor Biol* **30:** 225–234. doi:10.1016/0022-5193(71)90050-6

Khaitan V, Shill K, Chatterjee P, Mukherjee S, Majumder P. 2023. Singed and vinculin play redundant roles in cell migration by regulating F-actin. *Dev Dyn* **252:** 986–1008. doi:10.1002/dvdy.585

Kimura K, Ito M, Amano M, Chihara K, Fukata Y, Nakafuku M, Yamamori B, Feng J, Nakano T, Okawa K, et al. 1996. Regulation of myosin phosphatase by Rho and Rho-associated kinase (Rho-kinase). *Science* **273:** 245–248. doi:10.1126/science.273.5272.245

Kókai E, Páldy FS, Somogyi K, Chougule A, Pál M, Kerekes É, Deák P, Friedrich P, Dombrádi V, Ádám G. 2012. Calpb modulates border cell migration in *Drosophila* egg chambers. *BMC Dev Biol* **12:** 20. doi:10.1186/1471-213X-12-20

Lamb MC, Kaluarachchi CP, Lansakara TI, Mellentine SQ, Lan Y, Tivanski AV, Tootle TL. 2021. Fascin limits Myosin activity within *Drosophila* border cells to control substrate stiffness and promote migration. *eLife* **10:** e69836. doi:10.7554/eLife.69836

Law AL, Vehlow A, Kotini M, Dodgson L, Soong D, Theveneau E, Bodo C, Taylor E, Navarro C, Perera U, et al. 2013. Lamellipodin and the Scar/WAVE complex cooperate to promote cell migration in vivo. *J Cell Biol* **203:** 673–689. doi:10.1083/jcb.201304051

Le Clainche C, Carlier MF. 2008. Regulation of actin assembly associated with protrusion and adhesion in cell migration. *Physiol Rev* **88:** 489–513. doi:10.1152/physrev.00021.2007

Lei F, Xu X, Huang J, Su D, Wan P. 2023. *Drosophila* RhoGAP18B regulates actin cytoskeleton during border cell migration. *PLoS One* **18:** e0280652. doi:10.1371/journal.pone.0280652

Li Y, Chen M, Chang W. 2022. Roles of the nucleus in leukocyte migration. *J Leukoc Biol* **112:** 771–783. doi:10.1002/JLB.1MR0622-473RR

Li-Kroeger D, Kanca O, Lee PT, Cowan S, Lee MT, Jaiswal M, Salazar JL, He Y, Zuo Z, Bellen HJ. 2018. An expanded toolkit for gene tagging based on MiMIC and scarless CRISPR tagging in *Drosophila*. *eLife* **7:** e38709. doi:10.7554/eLife.38709

Lin TH, Yeh TH, Wang TW, Yu JY. 2014. The Hippo pathway controls border cell migration through distinct mechanisms in outer border cells and polar cells of the *Drosophila* ovary. *Genetics* **198:** 1087–1099. doi:10.1534/genetics.114.167346

Lin S, Ewen-Campen B, Ni X, Housden BE, Perrimon N. 2015. In vivo transcriptional activation using CRISPR/Cas9 in *Drosophila*. *Genetics* **201:** 433–442. doi:10.1534/genetics.115.181065

Lin CH, Lin TY, Hsu SC, Hsu HJ. 2022. Expansion microscopy-based imaging for visualization of mitochondria in *Drosophila* ovarian germline stem cells. *FEBS Open Bio* **12:** 2102–2110. doi:10.1002/2211-5463.13506

Liu TL, Upadhyayula S, Milkie DE, Singh V, Wang K, Swinburne IA, Mosaliganti KR, Collins ZM, Hiscock TW, Shea J, et al. 2018. Observing the cell in its native state: imaging subcellular dynamics in multicellular organisms. *Science* **360:** eaaq1392. doi:10.1126/science.aaq1392

Llense F, Martín-Blanco E. 2008. JNK signaling controls border cell cluster integrity and collective cell migration. *Curr Biol* **18:** 538–544. doi:10.1016/j.cub.2008.03.029

Majumder P, Aranjuez G, Amick J, McDonald JA. 2012. Par-1 controls myosin-II activity through myosin phosphatase to regulate border cell migration. *Curr Biol* **22:** 363–372. doi:10.1016/j.cub.2012.01.037

Manning L, Sheth J, Bridges S, Saadin A, Odinammadu K, Andrew D, Spencer S, Montell D, Starz-Gaiano M. 2017. A hormonal cue promotes timely follicle cell migration by modulating transcription profiles. *Mech Dev* **148:** 56–68. doi:10.1016/j.mod.2017.06.003

Mayor R, Etienne-Manneville S. 2016. The front and rear of collective cell migration. *Nat Rev Mol Cell Biol* **17:** 97–109. doi:10.1038/nrm.2015.14

McDonald JA, Pinheiro EM, Montell DJ. 2003. PVF1, a PDGF/VEGF homolog, is sufficient to guide border cells and interacts genetically with Taiman. *Development* **130:** 3469–3478. doi:10.1242/dev.00574

McDonald JA, Pinheiro EM, Kadlec L, Schupbach T, Montell DJ. 2006. Multiple EGFR ligands participate in guiding migrating border cells. *Dev Biol* **296:** 94–103. doi:10.1016/j.ydbio.2006.04.438

McDonald JA, Khodyakova A, Aranjuez G, Dudley C, Montell DJ. 2008. PAR-1 kinase regulates epithelial detachment and directional protrusion of migrating border cells. *Curr Biol* **18:** 1659–1667. doi:10.1016/j.cub.2008.09.041

McGregor JR, Xi R, Harrison DA. 2002. JAK signaling is somatically required for follicle cell differentiation in *Drosophila*. *Development* **129:** 705–717. doi:10.1242/dev.129.3.705

McLaughlin JM, Bratu DP. 2015. *Drosophila melanogaster* oogenesis: an overview. *Methods Mol Biol* **1328:** 1–20. doi:10.1007/978-1-4939-2851-4_1

Medioni C, Noselli S. 2005. Dynamics of the basement membrane in invasive epithelial clusters in *Drosophila*. *Development* **132:** 3069–3077. doi:10.1242/dev.01886

Mekus Z, Cooley J, George A, Sabo V, Strzegowski M, Starz-Gaiano M, Peercy BE. 2018. effects of cell packing on chemoattractant distribution within a tissue. *AIMS Biophys* **5:** 1–21. doi:10.3934/biophy.2018.1.1

Melanthota SK, Gopal D, Chakrabarti S, Kashyap AA, Radhakrishnan R, Mazumder N. 2022. Deep learning-based image processing in optical microscopy. *Biophys Rev* **14:** 463–481. doi:10.1007/s12551-022-00949-3

Mellentine SQ, Brown HN, Ramsey AS, Li J, Tootle TL. 2024. Specific prostaglandins are produced in the migratory cells and the surrounding substrate to promote *Drosophila* border cell migration. *Front Cell Dev Biol* **11:** 1257751. doi:10.3389/fcell.2023.1257751

Messer CL, Burghardt E, McDonald JA. 2025. A deficiency screen of the *X* chromosome for Rap1 GTPase dominant interacting genes in *Drosophila* border cell migration. *G3 (Bethesda)* **15:** jkaf040. doi:10.1093/g3journal/jkaf040

Miao G, Godt D, Montell DJ. 2020. Integration of migratory cells into a new site in vivo requires channel-independent functions of innexins on microtubules. *Dev Cell* **54:** 501–515.e9. doi:10.1016/j.devcel.2020.06.024

Miao G, Guo L, Montell DJ. 2022. Border cell polarity and collective migration require the spliceosome component Cactin. *J Cell Biol* **221:** e202202146. doi:10.1083/jcb.202202146

Mishra AK, Mondo JA, Campanale JP, Montell DJ. 2019. Coordination of protrusion dynamics within and between collectively migrating border cells by myosin II. *Mol Biol Cell* **30:** 2490–2502. doi:10.1091/mbc.E19-02-0124

Mishra AK, Rodriguez M, Torres AY, Smith M, Rodriguez A, Bond A, Morrissey MA, Montell DJ. 2023. Hyperactive Rac stimulates cannibalism of living target cells and enhances CAR-M-mediated cancer cell killing. *Proc Natl Acad Sci* **120:** e2310221120. doi:10.1073/pnas.2310221120

Mohr SE, Kim AR, Hu Y, Perrimon N. 2023. Finding information about uncharacterized *Drosophila melanogaster* genes. *Genetics* **225:** iyad187. doi:10.1093/genetics/iyad187

Molina López E, Kabanova A, Winkel A, Franze K, Palacios IM, Martín-Bermudo MD. 2023. Constriction imposed by basement membrane regulates developmental cell migration. *PLoS Biol* **21:** e3002172. doi:10.1371/journal.pbio.3002172

Monahan AJ, Starz-Gaiano M. 2015. Socs36E limits STAT signaling via Cullin2 and a SOCS-box independent mechanism in the *Drosophila* egg chamber. *Mech Dev* **138:** 313–327. doi:10.1016/j.mod.2015.08.003

Montell DJ, Rorth P, Spradling AC. 1992. Slow border cells, a locus required for a developmentally regulated cell migration during oogenesis, encodes *Drosophila* C/EBP. *Cell* **71:** 51–62. doi:10.1016/0092-8674(92)90265-E

Montell D, Yoon W, Starz-Gaiano M. 2012. Group choreography: mechanisms orchestrating the collective movement of border cells. *Nat Rev Mol Cell Biol* **13:** 631–645. doi:10.1038/nrm3433

Mosier JA, Wu Y, Reinhart-King CA. 2021. Recent advances in understanding the role of metabolic heterogeneities in cell migration. *Fac Rev* **10:** 8. doi:10.12703/r/10-8

Murphy AM, Montell DJ. 1996. Cell type-specific roles for Cdc42, Rac, and RhoL in *Drosophila* oogenesis. *J Cell Biol* **133:** 617–630. doi:10.1083/jcb.133.3.617

Niewiadomska P, Godt D, Tepass U. 1999. DE-cadherin is required for intercellular motility during *Drosophila* oogenesis. *J Cell Biol* **144:** 533–547. doi:10.1083/jcb.144.3.533

Ogienko AA, Yarinich LA, Fedorova EV, Dorogova NV, Bayborodin SI, Baricheva EM, Pindyurin AV. 2020. GAGA regulates border cell migration in *Drosophila*. *Int J Mol Sci* **21:** 7468. doi:10.3390/ijms21207468

Olson HM, Nechiporuk AV. 2021. Lamellipodia-like protrusions and focal adhesions contribute to collective cell migration in zebrafish. *Dev Biol* **469:** 125–134. doi:10.1016/j.ydbio.2020.10.007

Öztürk-Çolak A, Marygold SJ, Antonazzo G, Attrill H, Goutte-Gattat D, Jenkins VK, Matthews BB, Millburn G, Dos Santos G, Tabone CJ, et al. 2024. FlyBase: updates to the *Drosophila* genes and genomes database. *Genetics* **227:** iyad211. doi:10.1093/genetics/iyad211

Painter KJ. 2019. Mathematical models for chemotaxis and their applications in self-organisation phenomena. *J Theor Biol* **481:** 162–182. doi:10.1016/j.jtbi.2018.06.019

Parlani M, Jorgez C, Friedl P. 2023. Plasticity of cancer invasion and energy metabolism. *Trends Cell Biol* **33:** 388–402. doi:10.1016/j.tcb.2022.09.009

Parsons TT, Mosallaei S, Raftery LA. 2023. Two phases for centripetal migration of *Drosophila melanogaster* follicle cells: initial ingression followed by epithelial migration. *Development* **150:** dev200492. doi:10.1242/dev.200492

Peercy BE, Starz-Gaiano M. 2020. Clustered cell migration: modeling the model system of *Drosophila* border cells. *Semin Cell Dev Biol* **100:** 167–176. doi:10.1016/j.semcdb.2019.11.010

Penfield L, Montell DJ. 2023. Nuclear lamin facilitates collective border cell invasion into confined spaces in vivo. *J Cell Biol* **222:** e202212101. doi:10.1083/jcb.202212101

Pinheiro EM, Montell DJ. 2004. Requirement for Par-6 and Bazooka in *Drosophila* border cell migration. *Development* **131:** 5243–5251. doi:10.1242/dev.01412

Planchon TA, Gao L, Milkie DE, Davidson MW, Galbraith JA, Galbraith CG, Betzig E. 2011. Rapid three-dimensional isotropic imaging of living cells using Bessel beam plane illumination. *Nat Methods* **8:** 417–423. doi:10.1038/nmeth.1586

Plutoni C, Keil S, Zeledon C, Delsin LEA, Decelle B, Roux PP, Carréno S, Emery G. 2019. Misshapen coordinates protrusion restriction and actomyosin contractility during collective cell migration. *Nat Commun* **10:** 3940. doi:10.1038/s41467-019-11963-7

Port F, Bullock SL. 2016. Augmenting CRISPR applications in *Drosophila* with tRNA-flanked sgRNAs. *Nat Methods* **13:** 852–854. doi:10.1038/nmeth.3972

Port F, Chen HM, Lee T, Bullock SL. 2014. Optimized CRISPR/Cas tools for efficient germline and somatic genome engineering in *Drosophila*. *Proc Natl Acad Sci* **111:** E2967–E2976. doi:10.1073/pnas.1405500111

Poukkula M, Cliffe A, Changede R, Rørth P. 2011. Cell behaviors regulated by guidance cues in collective migration of border cells. *J Cell Biol* **192:** 513–524. doi:10.1083/jcb.201010003

Prasad M, Montell DJ. 2007. Cellular and molecular mechanisms of border cell migration analyzed using time-lapse live-cell imaging. *Dev Cell* **12:** 997–1005. doi:10.1016/j.devcel.2007.03.021

Prasad M, Jang AC, Starz-Gaiano M, Melani M, Montell DJ. 2007. A protocol for culturing *Drosophila* melanogaster stage 9 egg chambers for live imaging. *Nat Protoc* **2:** 2467–2473. doi:10.1038/nprot.2007.363

Qin L, Yang D, Yi W, Cao H, Xiao G. 2021. Roles of leader and follower cells in collective cell migration. *Mol Biol Cell* **32:** 1267–1272. doi:10.1091/mbc.E20-10-0681

Qu C, Yang W, Kan Y, Zuo H, Wu M, Zhang Q, Wang H, Wang D, Chen J. 2022. RhoA/ROCK signaling regulates Drp1-mediated mitochondrial fission during collective cell migration. *Front Cell Dev Biol* **10:** 882581. doi:10.3389/fcell.2022.882581

Ramel D, Wang X, Laflamme C, Montell DJ, Emery G. 2013. Rab11 regulates cell-cell communication during collective cell movements. *Nat Cell Biol* **15:** 317–324. doi:10.1038/ncb2681

Riddiford L. 1993. Hormones and *Drosophila* development. In *The development of Drosophila melanogaster* (ed. Bate M, Arias AM), pp. 899–939. Cold Spring Harbor Laboratory Press, Cold Spring Harbor, NY.

Ridley AJ, Schwartz MA, Burridge K, Firtel RA, Ginsberg MH, Borisy G, Parsons JT, Horwitz AR. 2003. Cell migration: integrating signals from front to back. *Science* **302:** 1704–1709. doi:10.1126/science.1092053

Roberto GM, Emery G. 2022. Directing with restraint: mechanisms of protrusion restriction in collective cell migrations. *Semin Cell Dev Biol* **129:** 75–81. doi:10.1016/j.semcdb.2022.03.037

Roberto GM, Boutet A, Keil S, Del Guidice E, Duramé E, Tremblay MG, Moss T, Therrien M, Emery G. 2025. Tao and Rap2l ensure proper Misshapen activation and levels during *Drosophila* border cell migration. *Dev Cell* **60:** 119–132.e6. doi:10.1016/j.devcel.2024.09.014

Rocha JJ, Jayaram SA, Stevens TJ, Muschalik N, Shah RD, Emran S, Robles C, Freeman M, Munro S. 2023. Functional unknomics: systematic screening of conserved genes of unknown function. *PLoS Biol* **21:** e3002222. doi:10.1371/journal.pbio.3002222

Rötte M, Höhne MY, Klug D, Ramlow K, Zedler C, Lehne F, Schneider M, Bischoff MC, Bogdan S. 2024. CYRI controls epidermal wound closure and cohesion of invasive border cell cluster in *Drosophila*. *J Cell Biol* **223:** e202310153. doi:10.1083/jcb.202310153

Rust K, Byrnes LE, Yu KS, Park JS, Sneddon JB, Tward AD, Nystul TG. 2020. A single-cell atlas and lineage analysis of the adult *Drosophila* ovary. *Nat Commun* **11:** 5628. doi:10.1038/s41467-020-19361-0

Saadin A, Starz-Gaiano M. 2016a. Circuitous genetic regulation governs a straightforward cell migration. *Trends Genet* **32:** 660–673. doi:10.1016/j.tig.2016.08.001

Saadin A, Starz-Gaiano M. 2016b. Identification of novel regulators of the JAK/STAT signaling pathway that control border cell migration in the *Drosophila* ovary. *G3 (Bethesda)* **6:** 1991–2002. doi:10.1534/g3.116.028100

Saadin A, Starz-Gaiano M. 2018. Cytokine exocytosis and JAK/STAT activation in the *Drosophila* ovary requires the vesicle trafficking regulator α-Snap. *J Cell Sci* **131:** jcs217638. doi:10.1242/jcs.217638

Sahu A, Ghosh R, Deshpande G, Prasad M. 2017. A gap junction protein, Inx2, modulates calcium flux to specify border cell fate during *Drosophila* oogenesis. *PLoS Genet* **13:** e1006542. doi:10.1371/journal.pgen.1006542

Sawant K, Chen Y, Kotian N, Preuss KM, McDonald JA. 2018. Rap1 GTPase promotes coordinated collective cell migration in vivo. *Mol Biol Cell* **29:** 2656–2673. doi:10.1091/mbc.E17-12-0752

Schroeder AB, Dobson ETA, Rueden CT, Tomancak P, Jug F, Eliceiri KW. 2021. The ImageJ ecosystem: open-source software for image visualization, processing, and analysis. *Protein Sci* **30:** 234–249. doi:10.1002/pro.3993

Schwartz M, Kelly T, Woods C, Imberski R. 1989. Ecdysteroid fluctuations in adult *Drosophila* melanogaster caused by elimination of pupal reserves and synthesis by early vitellogenic ovarian follicles. *Insect Biochem* **19:** 243–249. doi:10.1016/0020-1790(89)90068-1

SenGupta S, Parent CA, Bear JE. 2021. The principles of directed cell migration. *Nat Rev Mol Cell Biol* **22:** 529–547. doi:10.1038/s41580-021-00366-6

Shah P, Hobson CM, Cheng S, Colville MJ, Paszek MJ, Superfine R, Lammerding J. 2021. Nuclear deformation causes DNA damage by increasing replication stress. *Curr Biol* **31:** 753–765.e6. doi:10.1016/j.cub.2020.11.037

Sharma A, Halder S, Felix M, Nisaa K, Deshpande G, Prasad M. 2018. Insulin signaling modulates border cell movement in *Drosophila* oogenesis. *Development* **145:** dev166165. doi:10.1242/dev.166165

Silver DL, Montell DJ. 2001. Paracrine signaling through the JAK/STAT pathway activates invasive behavior of ovarian epithelial cells in *Drosophila*. *Cell* **107:** 831–841. doi:10.1016/S0092-8674(01)00607-9

Silver DL, Geisbrecht ER, Montell DJ. 2005. Requirement for JAK/STAT signaling throughout border cell migration in *Drosophila*. *Development* **132:** 3483–3492. doi:10.1242/dev.01910

Slaidina M, Gupta S, Banisch TU, Lehmann R. 2021. A single-cell atlas reveals unanticipated cell type complexity in *Drosophila* ovaries. *Genome Res* **31:** 1938–1951. doi:10.1101/gr.274340.120

Starz-Gaiano M, Melani M, Wang X, Meinhardt H, Montell DJ. 2008. Feedback inhibition of Jak/STAT signaling by apontic is required to limit an invasive cell population. *Dev Cell* **14:** 726–738. doi:10.1016/j.devcel.2008.03.005

Starz-Gaiano M, Melani M, Meinhardt H, Montell D. 2009. Interpretation of the UPD/JAK/STAT morphogen gradient in *Drosophila* follicle cells. *Cell Cycle* **8:** 2918–2926. doi:10.4161/cc.8.18.9547

Stephens L, Milne L, Hawkins P. 2008. Moving towards a better understanding of chemotaxis. *Curr Biol* **18:** R485–R494. doi:10.1016/j.cub.2008.04.048

Stonko DP, Manning L, Starz-Gaiano M, Peercy BE. 2015. A mathematical model of collective cell migration in a three-dimensional, heterogeneous environment. *PLoS One* **10:** e0122799. doi:10.1371/journal.pone.0122799

Stuelten CH, Parent CA, Montell DJ. 2018. Cell motility in cancer invasion and metastasis: insights from simple model organisms. *Nat Rev Cancer* **18:** 296–312. doi:10.1038/nrc.2018.15

Svitkina T. 2018. The actin cytoskeleton and actin-based motility. *Cold Spring Harb Perspect Biol* **10:** a018267. doi:10.1101/cshperspect.a018267

Szafranski P, Goode S. 2004. A Fasciclin 2 morphogenetic switch organizes epithelial cell cluster polarity and motility. *Development* **131:** 2023–2036. doi:10.1242/dev.01097

Szafranski P, Goode S. 2007. Basolateral junctions are sufficient to suppress epithelial invasion during *Drosophila* oogenesis. *Dev Dyn* **236:** 364–373. doi:10.1002/dvdy.21020

Theveneau E, Linker C. 2017. Leaders in collective migration: are front cells really endowed with a particular set of skills? *F1000Res* **6:** 1899. doi:10.12688/f1000research.11889.1

Torres AY, Nano M, Campanale JP, Deak S, Montell DJ. 2023. Activated Src kinase promotes cell cannibalism in *Drosophila*. *J Cell Biol* **222:** e202302076. doi:10.1083/jcb.202302076

Trivedi S, Starz-Gaiano M. 2018. *Drosophila* Jak/STAT signaling: regulation and relevance in human cancer and metastasis. *Int J Mol Sci* **19:** 4056. doi:10.3390/ijms19124056

Trivedi S, Bhattacharya M, Starz-Gaiano M. 2022. Mind bomb 2 promotes cell migration and epithelial structure by regulating adhesion complexes and the actin cytoskeleton. *Dev Biol* **491:** 94–104. doi:10.1016/j.ydbio.2022.08.009

Tschumperlin DJ. 2013. Fibroblasts and the ground they walk on. *Physiology* **28:** 380–390. doi:10.1152/physiol.00024.2013

Van de Bor V, Zimniak G, Cérézo D, Schaub S, Noselli S. 2011. Asymmetric localisation of cytokine mRNA is essential for JAK/STAT activation during cell invasiveness. *Development* **138:** 1383–1393. doi:10.1242/dev.056184

Van Haastert PJ. 2010. Chemotaxis: insights from the extending pseudopod. *J Cell Sci* **123:** 3031–3037. doi:10.1242/jcs.071118

Veeman MT, McDonald JA. 2016. Dynamics of cell polarity in tissue morphogenesis: a comparative view from *Drosophila* and *Ciona*. *F1000Res* **5:** 1084. doi:10.12688/f1000research.8011.1

Vicente-Manzanares M, Ma X, Adelstein RS, Horwitz AR. 2009. Non-muscle myosin II takes centre stage in cell adhesion and migration. *Nat Rev Mol Cell Biol* **10:** 778–790. doi:10.1038/nrm2786

Volovetz J, Berezovsky AD, Alban T, Chen Y, Lauko A, Aranjuez GF, Burtscher A, Shibuya K, Silver DJ, Peterson J, et al. 2020. Identifying conserved molecular targets required for cell migration of glioblastoma cancer stem cells. *Cell Death Dis* **11:** 152. doi:10.1038/s41419-020-2342-2

Wan P, Wang D, Luo J, Chu D, Wang H, Zhang L, Chen J. 2013. Guidance receptor promotes the asymmetric distribution of exocyst and recycling endosome during collective cell migration. *Development* **140:** 4797–4806. doi:10.1242/dev.094979

Wang X, Bo J, Bridges T, Dugan KD, Pan TC, Chodosh LA, Montell DJ. 2006. Analysis of cell migration using whole-genome expression profiling of migratory cells in the *Drosophila* ovary. *Dev Cell* **10:** 483–495. doi:10.1016/j.devcel.2006.02.003

Wang X, Adam JC, Montell D. 2007. Spatially localized Kuzbanian required for specific activation of Notch during border cell migration. *Dev Biol* **301:** 532–540. doi:10.1016/j.ydbio.2006.08.031

Wang X, He L, Wu YI, Hahn KM, Montell DJ. 2010. Light-mediated activation reveals a key role for Rac in collective guidance of cell movement in vivo. *Nat Cell Biol* **12:** 591–597. doi:10.1038/ncb2061

Wang H, Qiu Z, Xu Z, Chen SJ, Luo J, Wang X, Chen J. 2018. aPKC is a key polarity determinant in coordinating the function of three distinct cell polarities during collective migration. *Development* **145:** dev158444. doi:10.1242/dev.158444

Wang H, Guo X, Wang X, Wang X, Chen J. 2020a. Supracellular actomyosin mediates cell-cell communication and shapes collective migratory morphology. *iScience* **23:** 101204. doi:10.1016/j.isci.2020.101204

Wang X, Wang H, Liu L, Li S, Emery G, Chen J. 2020b. Temporal coordination of collective migration and lumen formation by antagonism between two nuclear receptors. *iScience* **23:** 101335. doi:10.1016/j.isci.2020.101335

Wang PY, Chakraborty A, Ma HJ, Wu JW, Jang AC, Lin WC, Pi HW, Yeh CT, Cheng ML, Yu JS, et al. 2022. *Drosophila* CTP synthase regulates collective cell migration by controlling the polarized endocytic cycle. *Development* **149:** dev200190. doi:10.1242/dev.200190

Wu JW, Wang CW, Chen RY, Hung LY, Tsai YC, Chan YT, Chang YC, Jang AC. 2022. Spatiotemporal gating of stat nuclear influx by *Drosophila* Npas4 in collective cell migration. *Sci Adv* **8:** eabm2411. doi:10.1126/sciadv.abm 2411

Xi R, McGregor JR, Harrison DA. 2003. A gradient of JAK pathway activity patterns the anterior–posterior axis of the follicular epithelium. *Dev Cell* **4:** 167–177. doi:10.1016/S1534-5807(02)00412-4

Xia J, Wang L, Lei F, Pan L, Liu L, Wan P. 2024. MicroRNA-34 disrupts border cell migration by targeting Eip74EF in *Drosophila* melanogaster. *J Insect Physiol* **159:** 104724. doi:10.1016/j.jinsphys.2024.104724

Zeledon C, Sun X, Plutoni C, Emery G. 2019. The ArfGAP drongo promotes actomyosin contractility during collective cell migration by releasing myosin phosphatase from the trailing edge. *Cell Rep* **28:** 3238–3248.e3. doi:10.1016/j.celrep.2019.08.044

Zhao AJ, Montes-Laing J, Perry WMG, Shiratori M, Merfeld E, Rogers SL, Applewhite DA. 2022. The *Drosophila* spectraplakin Short stop regulates focal adhesion dynamics by cross-linking microtubules and actin. *Mol Biol Cell* **33:** ar19. doi:10.1091/mbc.E21-09-0434

Zhou S, Li P, Liu J, Liao J, Li H, Chen L, Li Z, Guo Q, Belguise K, Yi B, et al. 2022. Two Rac1 pools integrate the direction and coordination of collective cell migration. *Nat Commun* **13:** 6014. doi:10.1038/s41467-022-33727-6

Zirin J, Bosch J, Viswanatha R, Mohr SE, Perrimon N. 2022. State-of-the-art CRISPR for in vivo and cell-based studies in *Drosophila*. *Trends Genet* **38:** 437–453. doi:10.1016/j.tig.2021.11.006

Collective Cell Migration Strategies: Patterning, Motility, and Directionality of the Posterior Lateral Line Primordium in Zebrafish

Alex V. Nechiporuk[1] and Holger Knaut[2]

[1]Department of Cell, Developmental & Cancer Biology, Oregon Health & Science University, The Knight Cancer Institute, Portland, Oregon 97239, USA

[2]Department of Cell Biology, New York University Grossman School of Medicine, New York, New York 10016, USA

Correspondence: nechipor@ohsu.edu; holger.knaut@nyulangone.org

During development and homeostasis, tissues move and rearrange to form organs, seal wounds, or—in the case of cancer—spread in the body. To accomplish this, cells in tissues need to communicate with each other, generate force to push themselves forward, and know where to go to—all of this with little to no error. Here, we discuss how a migrating tissue—the zebrafish posterior lateral line primordium—solves these challenges. We focus on the strategies that ensure signaling within the tissue, enable the tissue to generate and transmit force to its substrate for propulsion, and allow robust directional sensing and migration by the tissue. These strategies include facilitated diffusion and ligand trapping for focal signaling, a self-generated attractant gradient for long-distance migration, clamping of the attractant concentration to the attractant receptor's K_d for most sensitive signaling, mechanical coupling among cells for averaging directional sensing in a tissue, and large rear traction stresses to propel the tissue forward. Many of these strategies likely apply to collectively migrating cells in other contexts and should thus provide insights with direct relevance to human health.

The lateral line is a mechanosensory system in aquatic vertebrates that detects water movement and modulates swimming behavior. The peripheral aspect of this sensory system consists of sensory organs on the animal's surface, called neuromasts, and the sensory neurons that innervate them. In zebrafish, the neuromasts of the embryonic or primary lateral line in the trunk, referred to as the posterior lateral line (pLL), are established during the first 2 days of development through collective cell migration. Both the pLL sensory organs and the associated neurons originate from the pLL placode, a transient thickening of the embryonic ectoderm. This placode, located just posterior to the developing ear, contains mixed progenitors for both cell types at 14 hours postfertilization (hpf) (Fig. 1A).

By 22 hpf, a subset of sensory progenitors, known as the pLL primordium (primordium), initiates caudal migration along the trunk. Concurrently, the pLL neurons extend axons that

Cite this article as *Cold Spring Harb Perspect Biol* doi: 10.1101/cshperspect.a041751

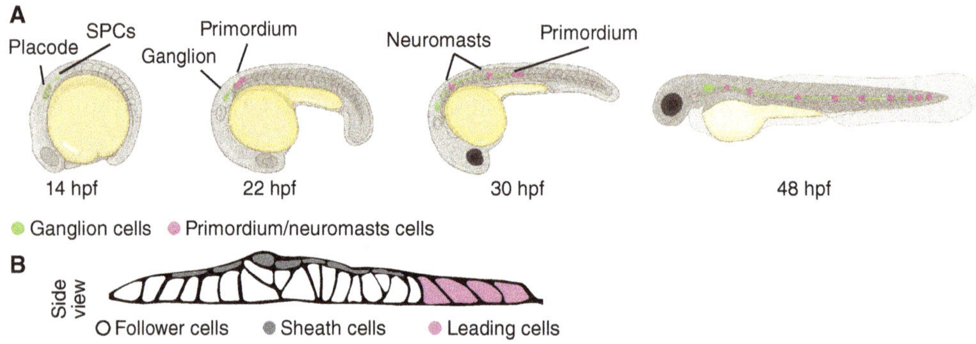

Figure 1. Development of the posterior lateral line sensory system. (*A*) The posterior lateral sensory system originates from the posterior lateral line placode. The sensory neurons and the primordium cells are initially mixed in the placode with a population of primordium cells: the separate primordium cells (SPCs) are born posterior to the main placode. Upon unmixing, the sensory neurons coalesce into the posterior lateral line ganglion, and the primordium cells coalesce and start migrating along the body of the embryo. The axons of the sensory neurons track the primordium closely and start innervating the neuromasts that are deposited from the back of the migrating primordium. (*B*) The primordium consists of leading cells, follower cells, and sheath cells. The leading and follower cells form a single layered epithelium that is more mesenchymal-like in the front and more epithelia-like in the back. The back two-thirds of this cell sheet are covered by sheath cells that separate the inner primordium cells from the embryonic skin. (hpf) Hours postfertilization.

closely follow the migrating primordium. As the primordium moves, it periodically deposits clusters of 20–30 cells from its back, which differentiate into neuromasts containing sensory hair cells that transduce water currents into electrochemical signals. By 48 hpf, the primordium reaches the end of the tail, where it forms a terminal cluster of two to three neuromasts (Fig. 1A).

The primordium serves as an exceptional model for studying collective cell migration because it can be imaged live at high resolution and easily manipulated genetically. This model thus provides a powerful framework for investigating the coordination of cell movement, leader–follower dynamics, and interactions with the surrounding environment in an in vivo setting.

BRIEF HISTORY

The pLL emerged as a model system for studying collective cell migration during the 1980s and 1990s, but its migratory behavior had been documented much earlier. In 1904, Harrison demonstrated that the pLL develops through active migration using chimeric frogs (Harrison 1903). He fused the head of a black tadpole to the tail of a brown tadpole and observed a dark streak (black cells) forming on the brown tadpole's tail. This streak eventually separated into pigmented dots, leading Harrison to conclude that the lateral line forms via the coordinated migration of a group of cells down the trunk.

Later, Stone identified the origin of these migratory cells (Stone 1933). By staining donor embryos with Nile blue sulfate dye and grafting labeled cells from the postauditory region into unstained hosts, he observed labeled cells migrating along the flank of the embryo. These migrating cells deposited small clusters of blue-stained cells (presumptive neuromasts), which eventually differentiated into lateral line organs. This work established that the postauditory region, later identified as the pLL placode, serves as the origin of the pLL.

EARLY PRIMORDIUM DEVELOPMENT

The pLL, including both the sensory organs and the neurons that innervate them, derives from the pLL placode, a thickening of the embryonic ectoderm. Like other cranial placodes, the pLL

Table 1. Summary of protein abbreviation and function

Molecule	Function
Eye1, Atoh1a, Atoh1b	Transcription factors, specify neuronal fate such as the hair cell fate in neuromasts
RA	Retinoic acid, a signaling molecule that often patterns tissues in a concentration-dependent manner (morphogen)
Fgf3, Fgf10a	Fibroblast growth factors, signaling molecules that bind Fgf receptors and promote neuromast differentiation in the primordium
Fgfr1a	Fgf receptor, signals through the Ras-Mapk pathway upon activation by Fgf ligands
Wnt	Signaling molecule that signals to induce cell fate and cell proliferation in the primordium
Sef/il17rd	Similar expression to Fgf genes/interleukin 12 receptor, transmembrane protein that reduces Fgf signaling in the front of the primordium
Dkk1	Dickkopf, a secreted protein that inhibits Wnt signaling and restricts the Wnt expression domain to the front of the primordium
Dusp6	Dual specificity protein phosphatase, dephosphorylates Fgf signaling components and dampens Fgf signaling, reduces Fgf signaling in the front of the primordium
Anosmin1	Secreted protein, mutated in humans with Kallmann syndrome, binds and shuttles Fgf10a across the primordium to increase Fgf10a's signaling range
Mcf2lb	A RhoGEF, activates RhoA by facilitating the exchange of GDP for GTP at the apical constrictions in forming neuromasts
Rock2a	Rho-associated kinase, activated by active RhoA, phosphorylates and activates Myosin II
Myosin II	Motor protein that binds and constricts F-actin network
Shroom3	Scaffold protein that binds F-actin and recruits Rho-associated kinase (Rock) to apical junctions, leading to phosphorylation of Myosin II, and apical constriction
EpCAM	Epithelial cell adhesion molecule, mediates cell–cell adhesion
Cxcl12	A chemokine, provides directionality by attracting cells to places with higher chemokine levels
Cxcr4	A chemokine receptor, binds the Cxcl12 and mediates Cxcl12 guidance
Ackr3/Cxcr7	An atypical chemokine receptor, does not signal upon Cxcl12 binding but instead internalizes Cxcl12 and targets it for degradation

placode originates from the pan-placodal ectoderm (Piotrowski and Baker 2014), a transient embryonic structure formed at the end of epiboly and the beginning of somitogenesis (10 hpf in zebrafish). The pan-placodal ectoderm expresses several transcription factors, including *eye1*. Loss of *eye1* results in a smaller pLL placode due to increased cell death and fewer hair cells in both the ear and neuromasts (Whitfield et al. 1996; Kozlowski et al. 2005). This highlights the conserved mechanisms in the early development of both mechanosensory systems, including the hair cells of the ear and the hair cells of the neuromast.

The pLL placode is induced between 10 and 14 hpf. Work by the Piotrowski laboratory demonstrated that activation of retinoic acid (RA) signaling, coupled with inhibition of Fgf signaling, is essential for pLL placode induction (see Table 1 for protein abbreviations and functions; Sarrazin et al. 2010; Nikaido et al. 2017). Inhibition of RA signaling completely abolishes placode formation, although it remains unclear whether RA acts directly within the placode and/or in the surrounding tissues.

Neural and sensory progenitors initially intermingle within the pLL placode (Mizoguchi et al. 2011). Time-lapse imaging of fluorescently labeled pLL neuronal and primordium progenitors revealed a "salt-and-pepper" distribution of these two cell types at 14–15 hpf (Fig. 2; Supplemental Movie S1). Within a few hours, pLL neuroblasts delaminate, migrate anterodorsally, and aggregate to form the pLL ganglion, while the remaining cells organize in the pLL primordium. Shortly after separating from the pLL ganglion, the cells in the back/anterior region of the primordium arrange into rosettes. Initial rosette formation is likely mediated by Fgf signaling, which is upregulated in the back/anterior region of the primordium (Nechiporuk and Raible 2008; Breau et al. 2012). At about 22 hpf, the primordium starts migrating, and the pLL neurons extend axons that track the primordium's rear, comigrating with the primordium throughout its journey. When the primordium is misrouted, the axons follow the primordium to ectopic sites (Gilmour et al. 2004). However, the primordium does not require the axons for migration; genetic ablation of the pLL ganglion does not disrupt primordium migration or neuromast deposition (Gilmour et al. 2004; Grant et al. 2005). The towing of the axons by the primordium was suggested to depend on glial cell line–derived neurotrophic factor (GDNF), a signaling molecule expressed by the primordium that potentially could attract and "tow" the axons along (Schuster et al. 2010), although the requirement of GDNF in axon guidance was not observed in another study (Shepherd et al. 2004). However, neurotrophic factor receptor Ret is required for axons to track the primordium—in its absence the axon follows the primordium track only halfway (Tuttle et al. 2019), supporting the idea that neurotrophic factors secreted by the primordium recruit the axons.

Notch signaling regulates the balance between neural/ganglion and sensory/primordium progenitors within the pLL placode. Global

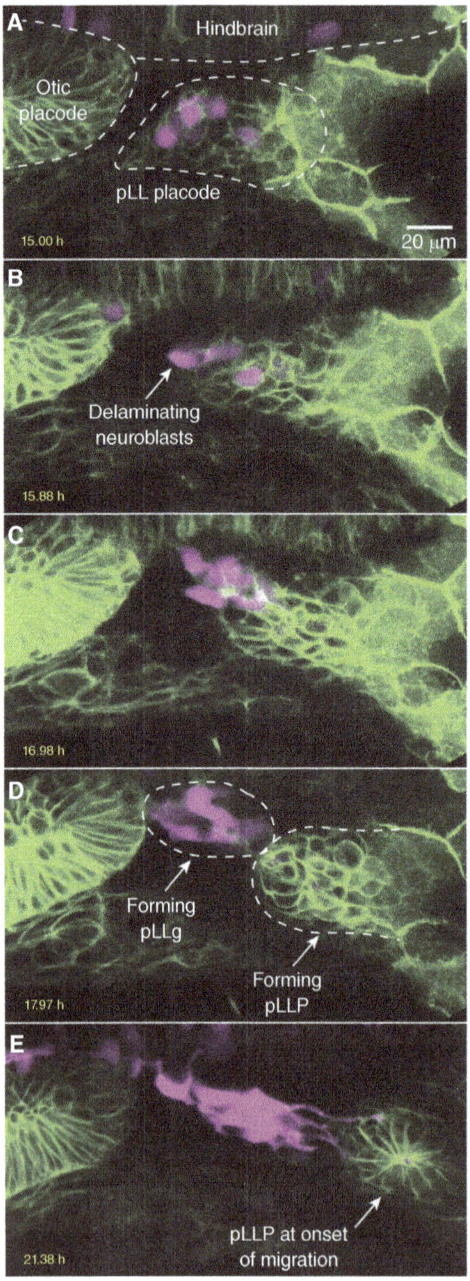

Figure 2. Unmixing of the neurons from the primordium cells in the posterior lateral line (pLL) placode. Stills from Supplemental Movie S1 visualizing the formation of the pLL system between 15 and 23 hours postfertilization (hpf). The pLL placode is marked by the (*−8.0 cldnB:lyn$_2$-eGFP*) transgene, and the pLL progenitors undergoing neurogenesis are visualized by the (*−5 kb neurod1:Zebrabow*) transgene. The pLL placode, otic placode, and the neural tube are outlined by dashed lines. (*A*) At the onset of pLL neurogenesis (15 hpf), neural and sensory progenitors are intermixed within the pLL placode. (*B*,*C*) Neural and sensory progenitors are in a process of separating into the pLL ganglion (pLLg) and primordium, respectively. (*D*) Neural and sensory progenitors separate at 18 hpf. (*E*) Posterior lateral line primordium (pLLP) migration begins at ~22 hpf; pLLg axons will extend with the migrating pLLP. Time stamp indicates hpf. Scale bar, 20 μm.

inhibition of Notch signaling increases the number of neural progenitors, whereas conditional activation of Notch signaling starting at 12 hpf reduces the number of neural progenitors. Delta ligands (Delta A, B, and D) act redundantly to limit neurogenesis by signaling through Notch1a and Notch3. Simultaneous knockdown of these ligands leads to an increase in pLL neuroblasts (Mizoguchi et al. 2011; Kozlovskaja-Gumbrienė et al. 2017). Thus, RA and Fgf signaling induce placode formation, while Notch signaling ensures proper progenitor allocation within the placode.

CELLULAR ORGANIZATION OF THE PRIMORDIUM

The primordium consists of three primary cell types: leading cells, follower cells, and sheath cells (Fig. 1B). The leading region of the primordium comprises two cell types: flat mesenchymal tip cells and epithelial columnar cells that are not yet arranged into polarized rosettes. Leading cells exhibit active protrusive behavior, including long filopodia that explore the environment (Haas and Gilmour 2006), likely providing directional guidance. Immediately posterior to the tip region, columnar epithelial cells line the midline of the primordium (Harding and Nechiporuk 2012). These leading cells change their front-back polarity as they constrict to form rosettes, which will mature into neuromasts.

The follower region contains two to three polarized rosettes surrounded by less organized epithelial cells. As in many other systems (Harding et al. 2014), rosettes in the primordium are transient structures that will eventually form neuromasts after deposition. Each rosette consists of apicobasally polarized epithelial cells, whose apical membranes are constricted; an actomyosin ring and tight junctions are present at the apical region of each rosette (Ernst et al. 2012; Harding and Nechiporuk 2012). Follower cells also proliferate as they form new rosettes. This partly replenishes the cells in the primordium after neuromast deposition (Aman et al. 2011; McGraw et al. 2011, 2014; Valdivia et al. 2011).

Sheath cells are flat cells that cover the apical side of the rear two-thirds of the primordium, separating it from the overlying skin (Nogare et al. 2020; Olson and Nechiporuk 2021). Similar to the other cells in the primordium, the sheath cells extend protrusions and comigrate with the "inner" cells of the primordium.

SIGNALING PATHWAYS THAT ORGANIZE THE MIGRATING PRIMORDIUM

The migrating primordium is patterned by the Wnt, Fgf, and Notch signaling pathways. Wnt ligands are expressed in front of the primordium and induce the *fgf10a* and *fgf3* expression in the front (Lecaudey et al. 2008; Nechiporuk and Raible 2008; Aman et al. 2011). The Fgf receptor *fgfr1a* is expressed throughout the primordium (Lecaudey et al. 2008; Nechiporuk and Raible 2008). However, Fgf signaling is restricted to the back of the primordium. This compartmentalization of signaling is due to several mechanisms. First, Wnt signaling not only induces *fgf* expression in the front but also the expression of the Fgf inhibitors *dusp6* and *il17rd/sef*. The phosphatase Dusp6 dampens Fgf signaling by dephosphorylating and thus inactivating extracellular signal-regulated kinase 2 (ERK2), an intracellular downstream signaling transducer (Tsang et al. 2004; Li et al. 2007), and the interleukin receptor Il17rd downregulates FGF signaling at the level of signal transduction by preventing ERK phosphorylation (Fürthauer et al. 2002; Tsang et al. 2002). These inhibitors block Fgf signaling in the front and restrict it to the primordium's back (Aman and Piotrowski 2008; Matsuda et al. 2013). Second, signaling from the front to the back is facilitated by Anosmin1, an extracellular protein that is secreted by all the primordium cells and increases the diffusion of Fgf10a across the primordium. Anosmin1 is thought to increase Fgf10a diffusion by liberating Fgf10a from the extracellular matrix and shuttling it across the primordium, effectively increasing Fgf10a's ability to diffuse and thus its signaling range (Wang et al. 2018). Third, Fgf signaling from the front to the back induces the expression of the Wnt antagonist *dkk1* in the back of the primordium (Aman and Piotrowski 2008; Venero Galanternik et al. 2015). Dkk1 is secreted by the cells in the primordium's back, binds to and induces the internalization of a Wnt receptor complex component (Xue et al. 2025). This blocks Wnt signaling in the primordium's back (Fig. 3).

Within the follower region, Fgf signaling is required for rosette formation and maturation. Fgf signaling activates the Fgfr-Ras-MAPK pathway, leading to the assembly of polarized rosettes (Harding and Nechiporuk 2012). MAPK pathway activation promotes apical constriction of rosette cells through a cascade involving expression of Shroom3, a scaffold protein that localizes Rock2a, a downstream effector of RhoA, to the apical membrane (Ernst et al. 2012; Harding and Nechiporuk 2012). Mcf2lb, a Rho guanine nucleotide exchange factor, activates RhoA that in turn stimulates Rock2a. Rock2a then phosphorylates nonmuscle Myosin II (referred to as Myosin II), which induces contraction of the apical actomyosin network and apical constriction (Olson et al. 2024).

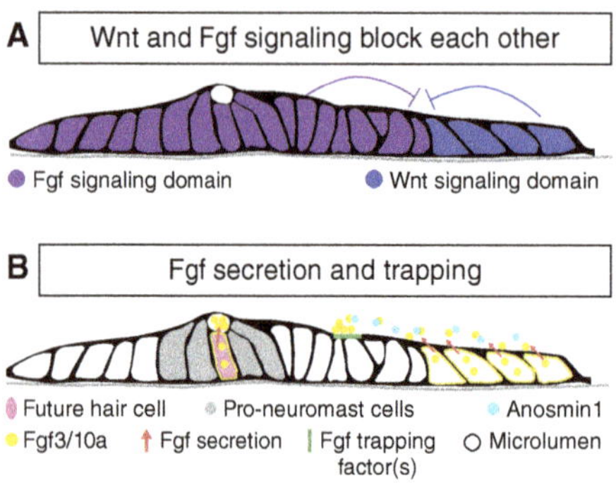

Figure 3. Fgf and Wnt signaling in the primordium. (*A*) The Wnt signaling pathway is active in the cells in the front of the primordium, while the Fgf signaling pathway is active in the back of the primordium. Wnt signaling induces the expression of two Fgf signaling pathway antagonists, Il17rd/Sef and Dusp6, in front cells. This renders the primordium's front unresponsive to Fgf. Fgf signaling induces the secretion of the Wnt receptor antagonist Dkk1 by the cells in the back of the primordium. This renders the primordium's back unresponsive to Wnt. This mutual antagonism restricts the activity of the two pathways to the front and back of the primordium. (*B*) Fgf3/10a are secreted by the front cells. Both growth factors bind tightly to the extracellular matrix. Anosmin1 is secreted by the primordium and binds to Fgf10a (and likely Fgf3). Anosmin1-bound Fgf10a diffuses farther because it is not retained by binding to the extracellular matrix. In this manner, Anosmin1 serves as a shuttle for Fgf10a. Above forming neuromasts, Fgf3/10a are locally trapped by unknown factors. This increases Fgf signaling locally and induces the future hair cell that starts to secrete Fgf10a. Continued local Fgf signaling around the future hair cell causes the apical sides of the adjacent primordium cells to form an extracellular microlumen. The hair cell continues to secrete Fgf10a into the microlumen and thus sustain high levels of Fgf signaling in the forming neuromasts. Note that for clarity Anosmin1 is only depicted in the front although it is secreted throughout by the primordium.

Notably, discrepancies exist regarding the role of Shroom3 in this process. Kozlovskaja-Gumbrienė et al. (2017) reported that dominant-negative Shroom3 did not affect rosette formation, suggesting the presence of alternative scaffolds that localize Rock2a. Furthermore, Olson et al. demonstrated that rosettes could form even in the absence of RhoA activity, albeit with defects, implicating a RhoA-independent mechanism in rosette assembly (Olson et al. 2024).

The forming rosettes harbor one to two hair cell precursors that are surrounded by support cells. Fgf signaling orchestrates rosette differentiation by inducing the expression of *atoh1a*, encoding a proneural factor and master regulator of hair cell fate, in a small subset of the rosette cells (Sarrazin et al. 2006; Nechiporuk and Raible 2008). Atoh1a, in turn, promotes the expression of DeltaA and DeltaD that activate the Delta receptor Notch3 on neighboring cells. Notch activation in the neighboring cells inhibits *atoh1* expression such that *atoh1* expression becomes restricted to one to two cells. In this manner, lateral inhibition ensures that each rosette is composed of one to two hair cell precursor and neighboring cells destined to adopt supporting cell fates (Itoh and Chitnis 2001; Matsuda and Chitnis 2010). In addition, Atoh1a induces the expression of *fgf10a* and *atoh1b* in the hair cell precursor. The expression of *fgf10a* in the hair cell precursors establishes a new Fgf signaling center in each rosette in the back of the primordium that promotes rosette maturation into neuromasts while Atoh1b maintains *atoh1a* expression (Matsuda and Chitnis 2010). Maturation causes the apical domains of the rosette cells to condense into a cup-like structure that traps Fgf10a secreted by the front cells through an unknown mechanism (Wang et al. 2018). This cup-like structure then matures into a soccer ball–like microlumen with the soccer ball patches formed by the apical sides of the rosette cells (Durdu et al. 2014). The *fgf10a*-expressing hair cell precursors secrete Fgf10a directly into this microlumen such that the Fgf10a concentration increases about 50-fold (Wang et al. 2018). Disruption of microlumen formation reduces Fgf signaling and slows neuromast maturation (Durdu et al. 2014), highlighting the role of microlumina in focusing high levels of Fgf activity to the center of the rosettes in the back of the primordium (Fig. 3B).

Finally, Notch signaling complements Fgf signaling in organizing the primordium by maintaining epithelial rosettes. Intracellular Notch activation induces apical localization of Rock2a and phosphorylated Myosin II in a Shroom3-independent manner, promoting apical constriction (Kozlovskaja-Gumbrienė et al. 2017). Inhibition of Notch signaling disrupts apical constriction and cell–cell adhesion, likely due to Notch regulation of adhesion molecules such as epithelial cell adhesion molecule (EpCAM).

In summary, the organization of the primordium into distinct front and back regions, mediated by Wnt, Fgf, and Notch signaling, is critical for its function. This spatial compartmentalization ensures the dynamic coordination of polarized rosette formation, neuromast maturation, and hair cell differentiation, all of which are essential for the primordium to deposit functional neuromasts along the embryonic body axis. Disruption of these signaling pathways or their interactions compromises the structural integrity and cellular specialization within the primordium, leading to defects in sensory organ development. Thus, the precise regulation of primordium organization is fundamental to its migratory capacity and the establishment of the lateral line sensory system.

PRIMORDIUM MIGRATION

Initiation of Migration

Shortly after the primordium is born at ~18 hpf, it starts to express the G-protein-coupled chemokine receptor Cxcr4b and initiates its migration along the body of the embryo to the tip of the tail. It moves on top of a stripe of cells, the muscle pioneer cells of the horizontal myoseptum, which express the chemokine Cxcl12a (Fig. 4A; Gompel et al. 2001; David et al. 2002). Since Cxcl12 signals through Cxcr4 and guides migrating cells in many contexts (Wang and Knaut 2014), this suggests that chemokine signaling also guides the migrating primordium. Indeed, morpholino-mediated reduction of Cxcl12a or Cxcr4b function blocked or severely slowed migration of the primordium, respectively (David et al. 2002; Li et al. 2004). These defects were also confirmed in *cxcl12a* and *cxcr4b* mutants (Haas and Gilmour 2006; Valentin et al. 2007). The primordium migrates sometimes slowly in *cxcr4b* mutant embryos because the primordium also expresses the *cxcr4b* paralog *cxcr4a*. In *cxcr4a, cxcr4b* double mutant embryos, the primordium fails to migrate, phenocopying the migration defect observed in *cxcl12a* mutant embryos (Colak-Champollion et al. 2019). Importantly, live imaging showed that the primordium cells still move, displaying a tumbling behavior, in *cxcr4b* mutant embryos, indicating that Cxcr4b-mediated signaling is not essential for the motility of the primordium cells (Haas and Gilmour 2006). Also, global overexpression of the Cxcl12a paralog Cxcl12b or the induction of Cxcl12a expression in a few cells along the migratory route stalled the primordium (Li et al. 2004) or recruited it to the ectopic chemokine source (Venkiteswaran et al. 2013), respectively. This indicates that Cxcr4b-mediated Cxcl12a signaling is required and sufficient for primordium migration.

Live imaging showed that the primordium is initially split into a larger and a smaller cluster. The smaller cluster of cells, called the separate primordium cells (SPCs), migrates rostrally toward the larger cluster at ~20–22 hpf, fuses with it, and constitutes the source of the leading cells of the primordium (Fig. 1A). The recruitment of

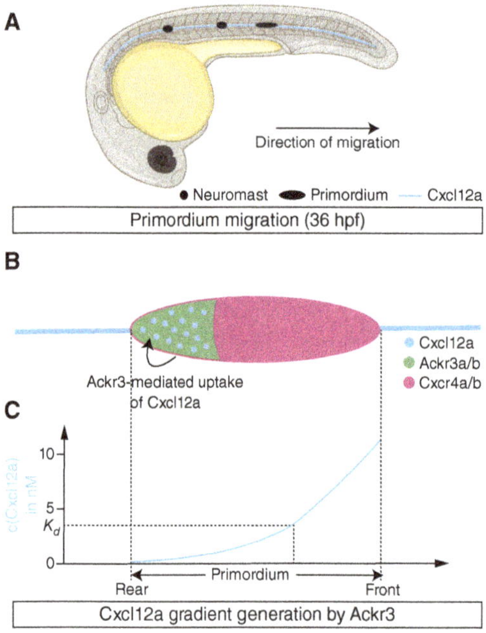

Figure 4. A self-generated chemokine attractant gradient guides the primordium. (*A*) A stripe of Cxcl12a-producing cells demarcates the migratory route of the primordium from behind the ear to the tip of the tail. (*B*) The primordium expresses the Cxcl12a guidance receptors Cxcr4a and Cxcr4b in all the cells and the Cxcl12a clearance receptors Ackr3a and Ackr3b only in the cells in its back. The restricted expression of Ackr3 constitutes a local chemokine sink in the back of the primordium that clears Cxcl12a and thus generates a chemokine attractant gradient across the primordium. (*C*) The chemokine gradient across the primordium ranges from 12 nM in the front to close to 0 nM in the back of signaling-available Cxcl12a. The Cxcl12a concentration for most sensitive signaling through Cxcr4, the K_d of Cxcl12a for Cxcr4b, is located about one-third into the primordium.

these cells to the main primordium requires Fgf signaling. Fgf3 and Fgf10a are expressed by the main primordium, while the small cell cluster expresses Fgfr1, suggesting that the main primordium attracts the small cell cluster (Breau et al. 2012). Intriguingly, this Fgf-mediated attraction seems to persist in the migrating primordium—although reversed; the front cells express Fgf3 and Fgf10a and the rear cells express Fgfr1, and when separated by a laser cut, the rear cells require Fgf signaling to fuse with the front cells again (Nogare et al. 2014). Whether the cells of the smaller cluster are required for primordium migration is unclear. However, the leading cells of the primordium express *snail1b*, a zinc-finger transcription factor, that is induced by Cxcl12a and required for timely onset of migration (Neelathi et al. 2018). Since the SPCs form the leading cells, they might also form the *snail1b*-expressing cluster and could thus contribute to the timely initiation of migration.

A Self-Generated Attractant Gradient Guides the Primordium

The initial migration of the primordium toward the stripe of chemokine-expressing cells is consistent with the idea that the primordium uses a graded distribution of Cxcl12a for its navigation—akin to the classical idea of guidance by a local attractant source (Crick 1970). However, once the primordium reaches the stripe of chemokine-expressing cells and migrates along it, the primordium should encounter the same Cxcl12a levels in its front and in its back. Yet, the primordium follows the stripe of Cxcl12a-expressing cells without ever reversing its direction—although when faced with an abrogated chemokine stripe, the primordium will make a U-turn and follow the chemokine stripe in the opposite direction (Haas and Gilmour 2006), indicating that the chemokine stripe can serve as a two-way street. A few models were proposed to explain how the primordium extracts directional information from an attractant stripe source: Antagonistic interactions between Ackr3b (Cxcr7b), an alternate Cxcl12a receptor expressed in the rear of the primordium only, and Cxcr4b (Dambly-Chaudière et al. 2007), or activation of Ackr3b in the rear and Cxcr4b in the front (Valentin et al. 2007); both scenarios could endow the primordium with a polarized response to the Cxcl12a stripe source. However, quantification of Cxcl12a-induced Cxcr4b turnover and internalization showed that the primordium is exposed to a graded chemokine distribution with high Cxcl12a levels in the front and low levels in the rear (Donà et al. 2013; Venkiteswaran et al. 2013). This chemokine gradient is generated by two Cxcl12a scavenger receptors, Ackr3a and Ackr3b (Miyasaka et al.

 Cite this article as *Cold Spring Harb Perspect Biol* doi: 10.1101/cshperspect.a041751

2007; Boldajipour et al. 2008; Lewellis et al. 2013). Both scavenger receptors are expressed in the rear of the primordium where they take up Cxcl12a for intracellular degradation (Fig. 4B). Ackr3a clears 31% and Ackr3b 40% of Cxcl12a from underneath the primordium. Consistent with these clearance activities, the primordium migration is not affected in embryos with reduced *ackr3a* activity and 70% slowed in *ackr3b* mutant embryos. Combined inactivation of Ackr3a and Ackr3b abrogates the Cxcl12a gradient across the primordium and completely stalls primordium migration. Ackr3 clears Cxcl12a rapidly: When flooded with chemokine by globally overexpressing Cxcl12a for a short period, Ackr3 reestablishes the chemokine gradient in 2 h (Venkiteswaran et al. 2013). Thus, this local and efficient chemokine sink generates a Cxcl12a gradient across the primordium, which then provides the collective with directional information.

Robustness of Migration

The primordium migrates for 20 h over a distance of 1.2 mm without deviating from its path (Nogare et al. 2017; Colak-Champollion et al. 2019). To attain such error-free migration, the embryo must provide precise guidance while safeguarding from perturbations in directional sensing.

A Small, Locally Restricted Chemokine Pool Guides the Primordium

The primordium needs a narrow Cxcl12a trail to carve out a gradient that reliably directs it forward (Fig. 4A). However, the stripe-source expresses the chemokine continuously for 30 h, and during this time freely diffusing Cxcl12a should fill the extracellular space around the chemokine-producing cells and create a wide Cxcl12a trail. This should confuse the primordium. Since the primordium migrates highly directionally, Cxcl12a is unlikely to diffuse far from its stripe source. Indeed, GFP-tagged Cxcl12a localizes tightly to the chemokine-producing cells and seems to be hindered in its diffusion; based on the dynamics of the recovery of the chemokine gradient upon abolishing the gradient by global overexpression of Cxcl12a, Cxcl12a diffuses more than four times slower than it diffuses in water (Venkiteswaran et al. 2013). Together, this suggests that the chemokine is binding tightly to the extracellular matrix. This notion is supported by the observation that only a small fraction of the total Cxcl12a pool is available for signaling. Ackr3-mediated chemokine uptake in the rear of the primordium reduces Cxcl12a to levels too low to induce Cxcr4b internalization and, presumably, signaling. This is based on the observation that the degree of Cxcr4b internalization in the rear is similar to the degree of Cxcr4b internalization in *cxcl12a* mutant embryos and indicates that Ackr3 removes all signaling-available Cxcl12a. Yet, total Cxcl12a protein levels are only reduced by 1%. This is based on the GFP-tagged Cxcl12a uptake in the back of the primordium compared to the total Cxcl12a-GFP produced along the migratory route (Venkiteswaran et al. 2013). Thus, the signaling-available Cxcl12a pool—the pool that is depleted in the back of the primordium by Ackr3—constitutes only ~1% of the total Cxcl12a pool and 99% of the total Cxcl12a pool are not available for signaling through Cxcr4b. Together, this suggests that most of the chemokine is sequestered at its source by the extracellular matrix and likely constitutes a reservoir that feeds and potentially buffers a small pool of signaling-available Cxcl12a to provide guidance to the primordium (Fig. 5A).

The Primordium Buffers Chemokine Levels for Optimal Directional Sensing

To extract accurate directional information, the primordium needs to generate a chemokine gradient across itself with a slope that is steep enough for the cells in the collective to sense the Cxcl12a gradient between their fronts and rears and with Cxcl12a concentrations around the K_d of Cxcl12a for Cxcr4b—at these concentrations small changes in Cxcl12a levels cause maximal changes in the ligated receptor concentrations and thus allow for most sensitive signal-

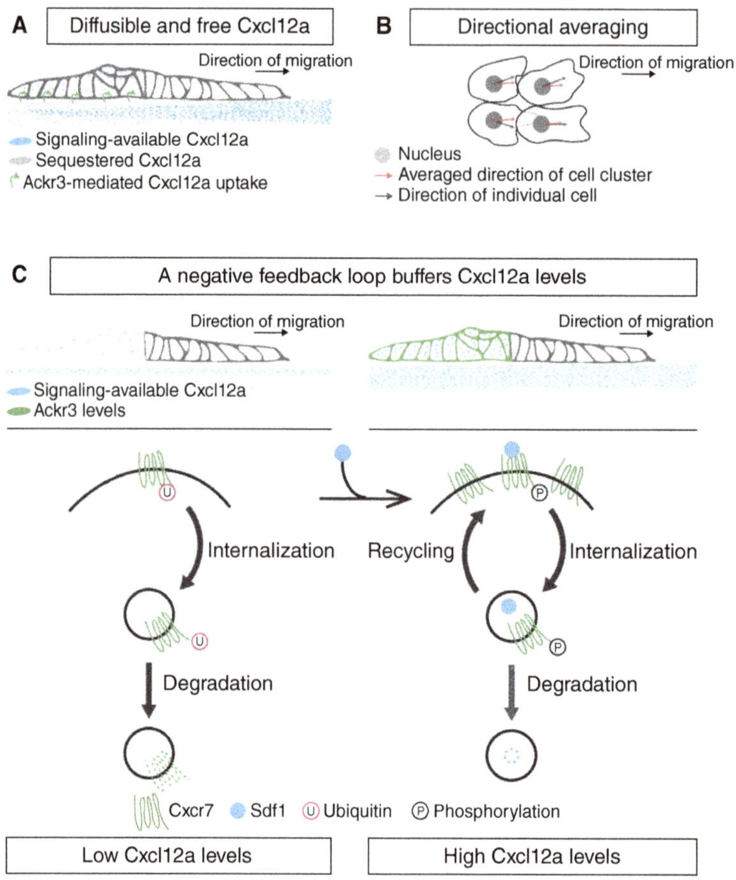

Figure 5. Fail-safe mechanism for robust directional primordium migration. (*A*) The primordium migrates across a stripe of secreted Cxcl12a. Most Cxcl12a is sequestered, probably by the extracellular matrix, and only a small Cxcl12a fraction is available for signaling to the primordium. (*B*) The primordium cells adhere to each other through cadherin-mediated cell–cell interactions. Mistakes in directional sensing by individual cells are reduced by directional averaging over all the cells in the collective such that the net movement of the collective is affected little by directional sensing mistakes of a few cells. (*C*) The primordium buffers chemokine levels. At low chemokine levels, Ackr3b is not protected from degradation such that its levels are low and few Cxcl12a molecules are internalized and degraded (*left*). At high chemokine levels, Ackr3b is phosphorylated and protected from degradation such that its levels are high and many Cxcl12a molecules are internalized and degraded (*right*). This negative feedback loop maintains the Cxcl12a concentration around the concentration of Cxcl12a's K_d for Cxcr4b and thus ensures most sensitive chemokine signaling through Cxcr4b.

ing (Herzmark et al. 2007; Ueda and Shibata 2007; Van Haastert and Postma 2007; Rappel and Levine 2008; Fuller et al. 2010). These ideas were tested by measuring the endogenous Cxcl12a concentrations available to the cells in the primordium using a ratiometric chemokine sensor. The chemokine sensor expresses Cxcr4b fused to a red fluorescent protein (RFP) and a membrane-tethered green fluorescent protein (GFP) from the same transcript through an internal ribosomal entry site. The membrane-tethered GFP serves as a measure for the total Cxcr4b-RFP produced. Because Cxcl12a-bound Cxcr4b is internalized and degraded, the ratio of the red to green fluorescence on the membrane is a readout of the levels of Cxcl12a available to the cells—more internalized Cxcr4b-RFP indicates more Cxcl12a exposure. This sensor was calibrated with purified Cxcl12a in cell culture and then expressed at endogenous levels in the primordium to measure the signaling-available chemokine concentration (Lewellis et al. 2013; Venkiteswaran et al. 2013). This analysis showed that the combined activities of Ackr3a and Ackr3b carve out a gradient across the primordium with 12 nM signaling-available Cxcl12a in the front and 0 nM signaling-available Cxcl12a in the back (Fig. 4C; Lau et al. 2020). The chemokine gradient translates to about a 7% or 0.8 nM difference in Cxcl12a between the fronts and rears of individual cells in the collective. Lowering the difference in Cxcl12a across cells to 5% by removing Ackr3a-mediated Cxcl12a clearance does not impede primordium migration but lowering the gradient further to 4% across cells by removing Ackr3b-mediated Cxcl12a clearance results in slowed and stalled primordium migration. Importantly, these gradient measurements are averages over the entire primordium and not across individual primordium cells (Venkiteswaran et al. 2013). Together, this indicates that the primordium generates a Cxcl12a gradient that is steeper than it needs for directed migration, endowing the primordium with robustness to perturbations in the chemokine gradient.

The concentration of signaling-available Cxcl12a is 12 nM in the front falls to 0 nM in the back of the primordium, and the Cxcl12a's K_d for Cxcr4b is 3.4 nM (Lau et al. 2020). The Cxcl12a gradient reaches this concentration 60 µm/one-third into primordium from the front (Fig. 4C). This exposes much of the primordium to optimal Cxcl12a concentrations and thus optimal attractant gradient sensing; mismatching the ligand concentration and the receptor K_d by changing Cxcr4b's affinity for Cxcl12a results in less directional primordium migration (Lau et al. 2020). The primordium actively maintains the Cxcl12a concentration around Cxcl12a's K_d for Cxcr4b and thus the steepness of the chemokine gradient. The primordium achieves this through a negative-feedback loop between Cxcl12a and Ackr3b; increased or decreased Cxcl12a levels result in higher or lower Ackr3b levels, respectively. Molecularly, it is thought that Ackr3b is rapidly degraded in the absence of Cxcl12a. Upon chemokine binding, Ackr3b is phosphorylated on its cytoplasmic tail, protected from degradation, and recycled back to the cell membrane (Fig. 5C). In support of this model, mutating the phosphorylation sites in Ackr3b's cytoplasmic tail blocks its upregulation in response to elevated Cxcl12a levels and thus breaks the negative-feedback loop (Lau et al. 2020; Wong et al. 2020). Importantly, blocking Ackr3b phosphorylation also results in less directional primordium migration (Lau et al. 2020). Thus, "clamping" the concentration of Cxcl12a close to its K_d for Cxcr4b through a negative-feedback loop—at the energetic cost of constantly turning over Ackr3b—allows the primordium to extract reliably directional information from a shallow and noisy attractant gradient.

Cell–Cell Adhesion Averages Directional Sensing by the Cells across the Primordium

The self-generated Cxcl12a gradient stretches across most of the primordium such that all but the cells in the rear of the collective are exposed to the chemokine gradient and can extract directional information. This suggests that most primordium cells—both in the front and the back—could contribute to directional sensing and thus overall speed of the collective. Consistent with this idea, *cxcr4a*$^{-/+}$; *cxcr4b*$^{-/-}$ primordia that are almost completely insensitive to Cxcl12a and rarely move along the chemokine stripe will migrate short distances if they harbor wild-type cells. The distance that such chimeric primordia travel increases linearly with the number of wild-type cells in the *cxcr4a*$^{-/+}$; *cxcr4b*$^{-/-}$ primordia (Colak-Champollion et al. 2019). This indicates that all cells, including the leading cells in the front and the follower cells in the rear, need to sense the chemokine guidance cue and contribute equally to the migration of the primordium. Importantly though, such chimeric primordia only migrate productively if the front cells are wild-type and can sense Cxcl12a (Haas and Gilmour 2006; Colak-Champollion et al. 2019). This is likely because Cxcl12a-blind cells at the front block those behind them—and there are no Cxcl12a-responsive cells in front of them to pull them along, causing the primordium to jam, rather than the front cells having a distinct role in leading the collective. Together, these observations suggest that wild-type cells can pull Cxcl12a-blind cells along, but that optimal migration requires all the cells in the primordium to be able to sense the Cxcl12a gradient.

The primordium cells adhere tightly to each other. They express the two cell–cell adhesion receptors E-cadherin/cadherin-1 (Cdh1) and N-cadherin/cadherin-2 (Cdh2) (Liu et al. 2003; Matsuda and Chitnis 2010; Revenu et al. 2014; Kozlovskaja-Gumbrienė et al. 2017; Colak-Champollion et al. 2019). Although the primordium cells express other adhesion molecules (Sur et al. 2023), reducing cadherin-mediated adhesion in a few cells (*cdh1*$^{-/-}$; *cdh2*$^{-/+}$ or *cdh1*$^{-/+}$; *cdh2*$^{-/-}$ cells) in an otherwise wild-type primordium causes these cells to detach from the collective and fall behind. Similarly, depleting αE-catenin (Ctnna1), a linker protein that connects cadherins to the actin network inside the cells, in the primordium causes cells to shed from the primordium's rear but does not affect overall migration. Intriguingly, in both scenarios the cells migrate less directionally, and, in the case of *cdh1* or *cxcr4b* mutant cells in wild-type primordia, the directionality of the neighboring wild-type cells is also affected (Colak-Champollion et al. 2019). This suggests two mechanisms that maintain and refine directional migration. Frist, mechanical coupling through cell–cell adhesions averages directional sensing across the primordium. Second, and in line with observations in other settings (Weber et al. 2012; Cai et al. 2014; Das et al. 2015; Hayer et al. 2016), cells reinforce directional information by pulling on the cadherins of their neighbors (Fig. 5B).

Together, these observations point to several mechanisms that the primordium uses to attain error-free migration. The primordium generates a Cxcl12a gradient that is almost twice as steep as is needed for directional sensing and maintains the chemokine concentration close to its optimum for most sensitive signaling through Cxcr4. Both of these are achieved through a negative feedback loop between Cxcl12a and its clearance receptor Ackr3b (likely Ackr3a as well) that "clamps" the chemokine concentration close to zero underneath the back of the primordium. Importantly, the Cxcl12a gradient is sensed by most of the cells in the primordium and through cadherin-mediated cell–cell coupling the derived directional information is averaged across the cells in the collective. Collectively, these fail-safe mechanisms endow the primordium with robustness against perturbations and highly directional migration.

PRIMORDIUM MOTILITY

To propel themselves forward, cells generate force internally and transmit it externally. Studies of singly migrating cells have detailed fundamental modes of locomotion. In lamellipodia-driven or mesenchymal migration, actin polymerization beneath the plasma membrane pushes the plasma membrane outward while driving the actin network backward. The balance between membrane protrusion and actin retrograde flow depends on the membrane's resistance: Greater resistance increases backward actin flow relative to protrusion (Bodor et al. 2020). Often, the retrograde flow is aided by Myosin II–mediated pulling on the actin network in the rear of the cells (Lin et al. 1996; Ponti et al. 2004; Wilson et al. 2010; Yolland et al. 2019). In bleb-driven cell migration, hydrostatic pressure separates the plasma membrane from the underlying cortical actin network and pushes out the plasma membrane purely through fluid influx as blebs (Paluch and Raz 2013). These blebs are devoid of F-actin. Rather than retrograde actin flow in the protrusions of lamellipodia-driven migrating cells, the cortical actomyosin underneath the cell membrane flows toward the rear in bleb-driven migrating cells. This is because Myosin II contracts and pulls on the cortical actomyosin in the rear of the cells. In both cases, the actin flow in the protrusions or underneath the cell cortex needs to be coupled to the outside to propel the cell forward. In principle, this can be achieved through specific adhesion to the substrate such as integrin-mediated binding to the extracellular matrix or unspecific adhesion of surface molecules to the outside substrate (Paluch et al. 2016). Migrating cells using lamellipodia for locomotion mainly rely on specific adhesion while migrating cells using blebs are thought to rely on unspecific adhesion.

In the primordium, the cells use lamellipodia-driven migration. Cells in the front and rear of the primordium extend protrusions on the basal cell sides along and on top of the basement membrane, and the primordium's superficial cells extend protrusions between the primordium's apical side and the overlying skin (Haas and Gilmour 2006; Nogare et al. 2020; Olson and Nechiporuk 2021). Although protrusions extend in all directions, they are biased in the direction of migration (Colak-Champollion et al. 2019; Nogare et al. 2020; Olson and Nechiporuk 2021). Intriguingly, when the skin above the primordium is surgically removed, the primordium stalls and the primordium cells cease to form lamellipodia and extend bleb-like protrusions instead. Replacing the removed skin with Matrigel does not restore primordium migration but rescues lamellipodia-like protrusions. This is probably because the skin compresses the primordium. Skin removal causes the primordium to rise in height by 18%. Adding Matrigel is thought to restore compression but might lack factors required for productive lamellipodia-driven translocation (Nogare et al. 2020).

As expected for lamellipodia-driven cell migration, the actin network in the protrusions flows toward the center of the cells (Nogare et al. 2020; Yamaguchi et al. 2022; Qian et al. 2024). This requires Arp2/3 activity and RhoA-mediated signaling. In the absence of Arp2/3 activity, the lamellipodia-like protrusions are mostly lost (Olson and Nechiporuk 2021). Similarly, blocking RhoA signaling also results in protrusion loss. Intriguingly, increasing RhoA signal by overexpressing a RhoGEF increases protrusive activity, resulting in shorter, more frequent protrusions and, similarly to loss of RhoA activity, stalled primordium migration (Qian et al. 2024). As observed in other contexts (Levayer and Lecuit 2012), RhoA signaling pulses in the cells of the primordium. Although the mechanisms that turn on and off RhoA in the primordium cells are unclear, depletion of F-actin increases and inhibition of Myosin II blocks RhoA activation, suggesting that low levels of F-actin initiate RhoA activation and active Myosin II recruits active RhoA into clusters. In the primordium, pulses of active RhoA promote protrusion formation in two ways. At the base of protrusions, RhoA is activated, probably by local depletion of F-actin, and activates Rock. Active Rock then phosphorylates and activates non-muscle Myosin II, which clusters at the base of the protrusion, pulls on the F-actin network, and hauls in the protrusion. At the tip of the protrusions, RhoA is activated and promotes actin polymerization through Rock and unidentified actin polymerization factors (Qian et al. 2024). Together, these two RhoA activities generate rearward-directed actin flow (Fig. 6A).

The retrograde actin flow generated in the primordium cells needs to be transmitted to the outside to propel the cells forward. The superficial cells extend their protrusion along the skin, and the skin and the primordium cells both ex-

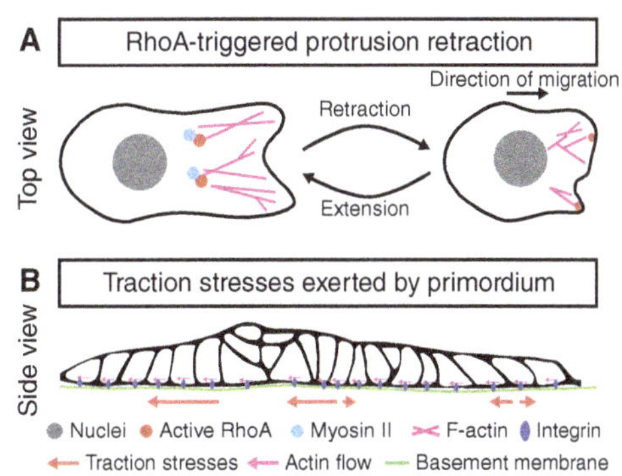

Figure 6. RhoA-triggered actin flow in the primordium cells is coupled through integrins to the basement membrane. (*A*) Pulses of active RhoA at the bases of protrusions activate Myosin II to pull on the F-actin network and haul in cell protrusions. At the tip of protrusions, active RhoA induces F-actin formation. (*B*) Actin flow on the basal sides of the cells in the primordium is coupled through integrins—and likely unspecific adhesion—to the basement membrane and generates traction stresses. The traction stresses are small and less oriented underneath the front of the primordium and increase more than threefold and point mostly toward the rear underneath the back of the primordium.

press the cell–cell adhesion receptor E-cadherin/cadherin-1 (Cdh1) (Colak-Champollion et al. 2019; Nogare et al. 2020). This suggests that Cdh1 could transmit the actin flow in the basal cells to the overlying skin. However, degron-mediated depletion αE-catenin (Ctnna1), a linker between cadherins and the actin network, does not affect the migration speed of the primordium cells (Yamaguchi et al. 2022), indicating that retrograde actin flow in the superficial cells is likely not transmitted to the skin through Cdh1 but possibly through other adhesion receptors or unspecific adhesion.

The inner cells of the primordium (the cells underneath the superficial primordium cells) extend protrusion along the basement membrane and express the extracellular matrix adhesion receptor β-integrin-1b (Itgb1b). Thus, integrin-mediated adhesion could couple actin flow in the primordium cells to the basement membrane. Consistent with this idea, impairing the basement membrane blocks primordium migration. Also, reducing β-integrin-1 function in the primordium slows the migrating primordium by 30% and increases actin flow threefold in the cells (Yamaguchi et al. 2022). It is likely that the increased actin flow, together with unspecific adhesions, partly compensates for the loss of integrin-mediated adhesion to sustain migration albeit at a reduced speed.

Although the primordium relies on integrins for efficient migration, Itgb1b and its cytoplasmic adaptor Talin-1 do not aggregate into classical focal adhesions but form nascent adhesion-like structures that turn over in <2 min. Importantly, when placed in culture, primordium cells do form focal adhesion-like structures similarly to those observed in migrating cells in culture (Yamaguchi et al. 2022), indicating that the primordium cells can from long-lived cell–substrate interactions but rely on transient cell–substrate interactions for movement. This was also observed for migrating melanoma cells; in vitro, the cells formed prominent focal adhesion-like structures but only formed fewer and smaller ones in vivo (Xue et al. 2023).

Force transmission from inside the cell to the outside of the cell should deform the cell's substrate—if the substrate is sufficiently pliable. How much a cell is pushing or pulling on its substrate is measured as the force per area or stress. The magnitude, direction, and distribution of the stresses that a cell or a group of cells exert on their substrate yield information about how they push and/or pull themselves forward. In vitro, stresses exerted by cells on their substrate are often back calculated from the displacement of fluorescent beads in a gel of known stiffness (Cheung et al. 2024). In vivo, this can be replicated by labeling the basement membrane fluorescently, bleaching a dot pattern on it, and recording the displacement of the dots as cells move across the bleach pattern. Additionally, one needs to measure or approximate the physical properties of the substrate—the stiffness and the Poisson ratio of the basement membrane (Yamaguchi et al. 2022). The stiffness is the ratio of the stress applied to a material and the resulting deformation or strain, and the Poisson ratio is the ratio between transverse to axial contraction (or expansion) or strain. Using this approach, it was shown that the front cells of the primordium push and pull with low stresses (~20 Pa) in all directions on the basement membrane. In contrast, the cells on the sides and in the rear of the primordium exert threefold higher stresses (~65 Pa peaking at 500–600 Pa) that are mostly directed toward the rear of the cell collective. This indicates that the front cells are more exploratory, while the cells at the sides and in the rear of the primordium push more strongly and more directionally on the basement membrane to propel the cell collective forward (Fig. 6B; Yamaguchi et al. 2022). Since the primordium moves at constant speed but exerts transient stresses on the basement membrane, its uniform motion is likely due to the averaging of the fluctuating stresses across the cell collective. On a tissue level, the primordium moves in a continuous breaststroke-like manner, with its front pushing the basement membrane in random directions while its sides and rear strongly push the basement membrane sideways and backward. This stress distribution is reminiscent of the stress distribution in singly migrating cells and entire animals. Locomoting keratocytes in culture generate forward-directed thrust in their backs, while the leading lamellae do not show detectable propulsive traction (Oliver et al. 1995, 1999). Similarly, gliding snails and slugs push themselves forward with their rears (Lai et al. 2010). This suggests that a rear-engine design can drive locomotion across very different length scales.

ACKNOWLEDGMENTS

We thank the members of the Knaut and Nechiporuk laboratories for critical comments and insightful suggestions. We apologize to authors whose work we were unable to cite due to space constraints. Parts of the figures were created in BioRender (https://biorender.com). This work was supported by grants from the National Institutes of Health (R01NS119449 to H.K. and R01GM130868 to A.V.N.).

REFERENCES

Aman A, Piotrowski T. 2008. Wnt/beta-catenin and Fgf signaling control collective cell migration by restricting chemokine receptor expression. *Dev Cell* **15:** 749–761. doi:10.1016/j.devcel.2008.10.002

Aman A, Nguyen M, Piotrowski T. 2011. Wnt/β-catenin dependent cell proliferation underlies segmented lateral line morphogenesis. *Dev Biol* **349:** 470–482. doi:10.1016/j.ydbio.2010.10.022

Bodor DL, Pönisch W, Endres RG, Paluch EK. 2020. Of cell shapes and motion: the physical basis of animal cell migration. *Dev Cell* **52:** 550–562. doi:10.1016/j.devcel.2020.02.013

Boldajipour B, Mahabaleshwar H, Kardash E, Reichman-Fried M, Blaser H, Minina S, Wilson D, Xu Q, Raz E. 2008. Control of chemokine-guided cell migration by ligand sequestration. *Cell* **132:** 463–473. doi:10.1016/j.cell.2007.12.034

Breau MA, Wilson D, Wilkinson DG, Xu Q. 2012. Chemokine and Fgf signalling act as opposing guidance cues in formation of the lateral line primordium. *Development* **139:** 2246–2253. doi:10.1242/dev.080275

Brockway NL, Cook ZT, O'Gallagher MJ, Tobias ZJC, Gedi M, Carey KM, Unni VK, Pan YA, Metz MR, Weissman TA. 2019. Multicolor lineage tracing using in vivo time-lapse imaging reveals coordinated death of clonally related cells in the developing vertebrate brain. *Dev Biol* **453:** 130–140. doi:10.1016/j.ydbio.2019.05.006

Cai D, Chen SC, Prasad M, He L, Wang X, Choesmel-Cadamuro V, Sawyer JK, Danuser G, Montell DJ. 2014. Mechanical feedback through E-cadherin promotes direction sensing during collective cell migration. *Cell* **157:** 1146–1159. doi:10.1016/j.cell.2014.03.045

Cheung BCH, Abbed RJ, Wu M, Leggett SE. 2024. 3D traction force microscopy in biological gels: from single cells to multicellular spheroids. *Annu Rev Biomed Eng* **26:** 93–118. doi:10.1146/annurev-bioeng-103122-031130

Colak-Champollion T, Lan L, Jadhav AR, Yamaguchi N, Venkiteswaran G, Patel H, Cammer M, Meier-Schellersheim M, Knaut H. 2019. Cadherin-mediated cell coupling coordinates chemokine sensing across collectively migrating cells. *Curr Biol* **29:** 2570–2579.e7. doi:10.1016/j.cub.2019.06.061

Crick F. 1970. Diffusion in embryogenesis. *Nature* **225:** 420–422. doi:10.1038/225420a0

Dambly-Chaudière C, Cubedo N, Ghysen A. 2007. Control of cell migration in the development of the posterior lateral line: antagonistic interactions between the chemokine receptors CXCR4 and CXCR7/RDC1. *BMC Dev Biol* **7:** 23. doi:10.1186/1471-213X-7-23

Das T, Safferling K, Rausch S, Grabe N, Boehm H, Spatz JP. 2015. A molecular mechanotransduction pathway regulates collective migration of epithelial cells. *Nat Cell Biol* **17:** 276–287. doi:10.1038/ncb3115

David NB, Sapède D, Saint-Etienne L, Thisse C, Thisse B, Dambly-Chaudière C, Rosa FM, Ghysen A. 2002. Molecular basis of cell migration in the fish lateral line: role of the chemokine receptor CXCR4 and of its ligand, SDF1. *Proc Natl Acad Sci* **99:** 16297–16302. doi:10.1073/pnas.252339399

Donà E, Barry JD, Valentin G, Quirin C, Khmelinskii A, Kunze A, Durdu S, Newton LR, Fernandez-Minan A, Huber W, et al. 2013. Directional tissue migration through a self-generated chemokine gradient. *Nature* **503:** 285–289. doi:10.1038/nature12635

Durdu S, Iskar M, Revenu C, Schieber N, Kunze A, Bork P, Schwab Y, Gilmour D. 2014. Luminal signalling links cell communication to tissue architecture during organogenesis. *Nature* **515:** 120–124. doi:10.1038/nature13852

Ernst S, Liu K, Agarwala S, Moratscheck N, Avci ME, Nogare DD, Chitnis AB, Ronneberger O, Lecaudey V. 2012. Shroom3 is required downstream of FGF signalling to mediate proneuromast assembly in zebrafish. *Development* **139:** 4571–4581. doi:10.1242/dev.083253

Fuller D, Chen W, Adler M, Groisman A, Levine H, Rappel W-J, Loomis WF. 2010. External and internal constraints on eukaryotic chemotaxis. *Proc National Acad Sci* **107:** 9656–9659. doi:10.1073/pnas.0911178107

Fürthauer M, Lin W, Ang SL, Thisse B, Thisse C. 2002. Sef is a feedback-induced antagonist of Ras/MAPK-mediated FGF signalling. *Nat Cell Biol* **4:** 170–174. doi:10.1038/ncb750

Gilmour D, Knaut H, Maischein HM, Nüsslein-Volhard C. 2004. Towing of sensory axons by their migrating target cells in vivo. *Nat Neurosci* **7:** 491–492. doi:10.1038/nn1235

Gompel N, Cubedo N, Thisse C, Thisse B, Dambly-Chaudière C, Ghysen A. 2001. Pattern formation in the lateral line of zebrafish. *Mech Dev* **105:** 69–77. doi:10.1016/S0925-4773(01)00382-3

Grant KA, Raible DW, Piotrowski T. 2005. Regulation of latent sensory hair cell precursors by glia in the zebrafish lateral line. *Neuron* **45:** 69–80. doi:10.1016/j.neuron.2004.12.020

Haas P, Gilmour D. 2006. Chemokine signaling mediates self-organizing tissue migration in the zebrafish lateral line. *Dev Cell* **10:** 673–680. doi:10.1016/j.devcel.2006.02.019

Harding MJ, Nechiporuk AV. 2012. Fgfr-Ras-MAPK signaling is required for apical constriction via apical positioning of Rho-associated kinase during mechanosensory organ formation. *Development* **139:** 3467–3467. doi:10.1242/dev.087999

Harding MJ, McGraw HF, Nechiporuk A. 2014. The roles and regulation of multicellular rosette structures during morphogenesis. *Development* **141:** 2549–2558. doi:10.1242/dev.101444

Harrison RG. 1903. Experimentelle Untersuchungen Über die Entwicklung der Sinnesorgane der Seitenlinie bei den Ampkibien. *Arch für Mikrosk Anat* **63:** 35–149. doi:10.1007/BF02978174

Hayer A, Shao L, Chung M, Joubert LM, Yang HW, Tsai FC, Bisaria A, Betzig E, Meyer T. 2016. Engulfed cadherin fingers are polarized junctional structures between collectively migrating endothelial cells. *Nat Cell Biol* **18:** 1311–1323. doi:10.1038/ncb3438

Herzmark P, Campbell K, Wang F, Wong K, El-Samad H, Groisman A, Bourne HR. 2007. Bound attractant at the leading vs. the trailing edge determines chemotactic prowess. *Proc Natl Acad Sci* **104:** 13349–13354. doi:10.1073/pnas.0705889104

Itoh M, Chitnis AB. 2001. Expression of proneural and neurogenic genes in the zebrafish lateral line primordium correlates with selection of hair cell fate in neuromasts. *Mech Dev* **102:** 263–266. doi:10.1016/S0925-4773(01)00308-2

Kozlovskaja-Gumbrienė A, Yi R, Alexander R, Aman A, Jiskra R, Nagelberg D, Knaut H, McClain M, Piotrowski T. 2017. Proliferation-independent regulation of organ size by Fgf/Notch signaling. *eLife* **6:** e21049. doi:10.7554/eLife.21049

Kozlowski DJ, Whitfield TT, Hukriede NA, Lam WK, Weinberg ES. 2005. The zebrafish dog-eared mutation disrupts eya1, a gene required for cell survival and differentiation in the inner ear and lateral line. *Dev Biol* **277:** 27–41. doi:10.1016/j.ydbio.2004.08.033

Lai JH, del Alamo JC, Rodríguez-Rodríguez J, Lasheras JC. 2010. The mechanics of the adhesive locomotion of terrestrial gastropods. *J Exp Biol* **213:** 3920–3933. doi:10.1242/jeb.046706

Lau S, Feitzinger A, Venkiteswaran G, Wang J, Lewellis SW, Koplinski CA, Peterson FC, Volkman BF, Meier-Schellersheim M, Knaut H. 2020. A negative-feedback loop maintains optimal chemokine concentrations for directional cell migration. *Nat Cell Biol* **22:** 266–273. doi:10.1038/s41556-020-0465-4

Lecaudey V, Cakan-Akdogan G, Norton WHJ, Gilmour D. 2008. Dynamic Fgf signaling couples morphogenesis and migration in the zebrafish lateral line primordium. *Development* **135:** 2695–2705. doi:10.1242/dev.025981

Levayer R, Lecuit T. 2012. Biomechanical regulation of contractility: spatial control and dynamics. *Trends Cell Biol* **22:** 61–81. doi:10.1016/j.tcb.2011.10.001

Lewellis SW, Nagelberg D, Subedi A, Staton A, LeBlanc M, Giraldez A, Knaut H. 2013. Precise SDF1-mediated cell guidance is achieved through ligand clearance and microRNA-mediated decay. *J Cell Biol* **200:** 337–355. doi:10.1083/jcb.201207099

Li Q, Shirabe K, Kuwada JY. 2004. Chemokine signaling regulates sensory cell migration in zebrafish. *Dev Biol* **269:** 123–136. doi:10.1016/j.ydbio.2004.01.020

Li C, Scott DA, Hatch E, Tian X, Mansour SL. 2007. Dusp6 (*Mkp3*) is a negative feedback regulator of FGF-stimulated ERK signaling during mouse development. *Development* **134:** 167–176. doi:10.1242/dev.02701

Lin CH, Espreafico EM, Mooseker MS, Forscher P. 1996. Myosin drives retrograde F-Actin flow in neuronal growth cones. *Neuron* **16:** 769–782. doi:10.1016/S0896-6273(00)80097-5

Liu Q, Ensign RD, Azodi E. 2003. Cadherin-1, -2 and -4 expression in the cranial ganglia and lateral line system of developing zebrafish. *Gene Expr Patterns* **3:** 653–658. doi:10.1016/S1567-133X(03)00109-1

Matsuda M, Chitnis AB. 2010. Atoh1a expression must be restricted by Notch signaling for effective morphogenesis of the posterior lateral line primordium in zebrafish. *Development* **137:** 3477–3487. doi:10.1242/dev.052761

Matsuda M, Nogare DD, Somers K, Martin K, Wang C, Chitnis AB. 2013. Lef1 regulates Dusp6 to influence neuromast formation and spacing in the zebrafish posterior lateral line primordium. *Development* **140:** 2387–2397. doi:10.1242/dev.091348

McGraw HF, Drerup CM, Culbertson MD, Linbo T, Raible DW, Nechiporuk AV. 2011. Lef1 is required for progenitor cell identity in the zebrafish lateral line primordium. *Development* **138:** 3921–3930. doi:10.1242/dev.062554

McGraw HF, Culbertson MD, Nechiporuk AV. 2014. Kremen1 restricts Dkk activity during posterior lateral line development in zebrafish. *Development* **141:** 3212–3221. doi:10.1242/dev.102541

Miyasaka N, Knaut H, Yoshihara Y. 2007. Cxcl12/Cxcr4 chemokine signaling is required for placode assembly and sensory axon pathfinding in the zebrafish olfactory system. *Development* **134:** 2459–2468. doi:10.1242/dev.001958

Mizoguchi T, Togawa S, Kawakami K, Itoh M. 2011. Neuron and sensory epithelial cell fate is sequentially determined by notch signaling in zebrafish lateral line development. *J Neurosci* **31:** 15522–15530. doi:10.1523/JNEUROSCI.3948-11.2011

Nechiporuk A, Raible DW. 2008. FGF-dependent mechanosensory organ patterning in zebrafish. *Science* **320:** 1774–1777. doi:10.1126/science.1156547

Neelathi UM, Nogare DD, Chitnis AB. 2018. Cxcl12a induces *snail1b* expression to initiate collective migration and sequential Fgf-dependent neuromast formation in the zebrafish posterior lateral line primordium. *Development* **145:** dev162453. doi:10.1242/dev.162453

Nikaido M, Acedo JN, Hatta K, Piotrowski T. 2017. Retinoic acid is required and Fgf, Wnt, and Bmp signaling inhibit posterior lateral line placode induction in zebrafish. *Dev Biol* **431:** 215–225. doi:10.1016/j.ydbio.2017.09.017

Nogare DD, Somers K, Rao S, Matsuda M, Reichman-Fried M, Raz E, Chitnis AB. 2014. Leading and trailing cells cooperate in collective migration of the zebrafish posterior lateral line primordium. *Development* **141:** 3188–3196. doi:10.1242/dev.106690

Nogare DD, Nikaido M, Somers K, Head J, Piotrowski T, Chitnis AB. 2017. In toto imaging of the migrating zebrafish lateral line primordium at single cell resolution. *Dev Biol* **422:** 14–23. doi:10.1016/j.ydbio.2016.12.015

Nogare DED, Natesh N, Vishwasrao HD, Shroff H, Chitnis AB. 2020. Zebrafish posterior lateral line primordium migration requires interactions between a superficial sheath of motile cells and the skin. *eLife* **9:** e58251. doi:10.7554/eLife.58251

Oliver T, Jacobson K, Dembo M. 1995. Traction forces in locomoting cells. *Cell Motil Cytoskelet* **31:** 225–240. doi:10.1002/cm.970310306

Oliver T, Dembo M, Jacobson K. 1999. Separation of propulsive and adhesive traction stresses in locomoting keratocytes. *J Cell Biol* **145:** 589–604. doi:10.1083/jcb.145.3.589

Olson HM, Nechiporuk AV. 2021. Lamellipodia-like protrusions and focal adhesions contribute to collective cell migration in zebrafish. *Dev Biol* **469:** 125–134. doi:10.1016/j.ydbio.2020.10.007

Olson HM, Maxfield A, Calistri NL, Heiser LM, Qian W, Knaut H, Nechiporuk AV. 2024. Rhoa GEF Mcf2lb regulates rosette integrity during collective cell migration. *Development* **151:** dev201898. doi:10.1242/dev.201898

Paluch EK, Raz E. 2013. The role and regulation of blebs in cell migration. *Curr Opin Cell Biol* **25:** 582–590. doi:10.1016/j.ceb.2013.05.005

Paluch EK, Aspalter IM, Sixt M. 2016. Focal adhesion–independent cell migration. *Annu Rev Cell Dev Biol* **32:** 469–490.

Piotrowski T, Baker CVH. 2014. The development of lateral line placodes: taking a broader view. *Dev Biol* **389:** 68–81. doi:10.1016/j.ydbio.2014.02.016

Ponti A, Machacek M, Gupton SL, Waterman-Storer CM, Danuser G. 2004. Two distinct actin networks drive the protrusion of migrating cells. *Science* **305:** 1782–1786. doi:10.1126/science.1100533

Qian W, Yamaguchi N, Lis P, Cammer M, Knaut H. 2024. Pulses of RhoA signaling stimulate actin polymerization and flow in protrusions to drive collective cell migration. *Curr Biol* **34:** 245–259.e8. doi:10.1016/j.cub.2023.11.044

Rappel WJ, Levine H. 2008. Receptor noise limitations on chemotactic sensing. *Proc Natl Acad Sci* **105:** 19270–19275. doi:10.1073/pnas.0804702105

Revenu C, Streichan S, Donà E, Lecaudey V, Hufnagel L, Gilmour D. 2014. Quantitative cell polarity imaging defines leader-to-follower transitions during collective migration and the key role of microtubule-dependent adherens junction formation. *Development* **141:** 1282–1291. doi:10.1242/dev.101675

Sarrazin AF, Villablanca EJ, Nuñez VA, Sandoval PC, Ghysen A, Allende ML. 2006. Proneural gene requirement for hair cell differentiation in the zebrafish lateral line. *Dev Biol* **295:** 534–545. doi:10.1016/j.ydbio.2006.03.037

Sarrazin AF, Nunez VA, Sapede D, Tassin V, Dambly-Chaudière C, Ghysen A. 2010. Origin and early development of the posterior lateral line system of zebrafish. *J Neurosci* **30:** 8234–8244. doi:10.1523/JNEUROSCI.5137-09.2010

Schuster K, Dambly-Chaudière C, Ghysen A. 2010. Glial cell line-derived neurotrophic factor defines the path of developing and regenerating axons in the lateral line system of zebrafish. *Proc Natl Acad Sci* **107:** 19531–19536. doi:10.1073/pnas.1002171107

Shepherd IT, Pietsch J, Elworthy S, Kelsh RN, Raible DW. 2004. Roles for GFRα1 receptors in zebrafish enteric nervous system development. *Development* **131:** 241–249. doi:10.1242/dev.00912

Stone LS. 1933. The development of lateral-line sense organs in amphibians observed in living and vital-stained preparations. *J Comp Neurol* **57:** 507–540. doi:10.1002/cne.900570307

Sur A, Wang Y, Capar P, Margolin G, Prochaska MK, Farrell JA. 2023. Single-cell analysis of shared signatures and transcriptional diversity during zebrafish development. *Dev Cell* **58:** 3028–3047.e12. doi:10.1016/j.devcel.2023.11.001

Tsang M, Friesel R, Kudoh T, Dawid IB. 2002. Identification of Sef, a novel modulator of FGF signalling. *Nat Cell Biol* **4:** 165–169. doi:10.1038/ncb749

Tsang M, Maegawa S, Kiang A, Habas R, Weinberg E, Dawid IB. 2004. A role for MKP3 in axial patterning of the zebrafish embryo. *Development* **131:** 2769–2779. doi:10.1242/dev.01157

Tuttle A, Drerup CM, Marra M, McGraw H, Nechiporuk AV. 2019. Retrograde Ret signaling controls sensory pioneer axon outgrowth. *eLife* **8:** e46092. doi:10.7554/eLife.46092

Ueda M, Shibata T. 2007. Stochastic signal processing and transduction in chemotactic response of eukaryotic cells. *Biophys J* **93:** 11–20. doi:10.1529/biophysj.106.100263

Valdivia LE, Young RM, Hawkins TA, Stickney HL, Cavodeassi F, Schwarz Q, Pullin LM, Villegas R, Moro E, Argenton F, et al. 2011. Lef1-dependent Wnt/β-catenin signalling drives the proliferative engine that maintains tissue homeostasis during lateral line development. *Development* **138:** 3931–3941. doi:10.1242/dev.062695

Valentin G, Haas P, Gilmour D. 2007. The chemokine SDF1a coordinates tissue migration through the spatially restricted activation of Cxcr7 and Cxcr4b. *Curr Biol* **17:** 1026–1031. doi:10.1016/j.cub.2007.05.020

Van Haastert PJM, Postma M. 2007. Biased random walk by stochastic fluctuations of chemoattractant-receptor interactions at the lower limit of detection. *Biophys J* **93:** 1787–1796. doi:10.1529/biophysj.107.104356

Venero Galanternik M, Kramer KL, Piotrowski T. 2015. Heparan sulfate proteoglycans regulate Fgf signaling and cell polarity during collective cell migration. *Cell Rep* **10:** 414–428. doi:10.1016/j.celrep.2014.12.043

Venkiteswaran G, Lewellis SW, Wang J, Reynolds E, Nicholson C, Knaut H. 2013. Generation and dynamics of an endogenous, self-generated signaling gradient across a migrating tissue. *Cell* **155:** 674–687. doi:10.1016/j.cell.2013.09.046

Wang J, Knaut H. 2014. Chemokine signaling in development and disease. *Development* **141:** 4199–4205. doi:10.1242/dev.101071

Wang J, Yin Y, Lau S, Sankaran J, Rothenberg E, Wohland T, Meier-Schellersheim M, Knaut H. 2018. Anosmin1 shuttles Fgf to facilitate its diffusion, increase its local concentration, and induce sensory organs. *Dev Cell* **46:** 751–766.e12. doi:10.1016/j.devcel.2018.07.015

Weber GF, Bjerke MA, DeSimone DW. 2012. A mechanoresponsive cadherin-keratin complex directs polarized protrusive behavior and collective cell migration. *Dev Cell* **22:** 104–115. doi:10.1016/j.devcel.2011.10.013

Whitfield TT, Granato M, van Eeden FJM, Schach U, Brand M, Furutani-Seiki M, Haffter P, Hammerschmidt M, Heisenberg CP, Jiang YJ, et al. 1996. Mutations affecting development of the zebrafish inner ear and lateral line. *Development* **123:** 241–254. doi:10.1242/dev.123.1.241

Wilson CA, Tsuchida MA, Allen GM, Barnhart EL, Applegate KT, Yam PT, Ji L, Keren K, Danuser G, Theriot JA. 2010. Myosin II contributes to cell-scale actin network treadmilling via network disassembly. *Nature* **465:** 373–377. doi:10.1038/nature08994

Wong M, Newton LR, Hartmann J, Hennrich ML, Wachsmuth M, Ronchi P, Guzmán-Herrera A, Schwab Y, Gavin AC, Gilmour D. 2020. Dynamic buffering of extracellular chemokine by a dedicated scavenger pathway enables robust adaptation during directed tissue migration. *Dev Cell* **52:** 492–508.e10. doi:10.1016/j.devcel.2020.01.013

Xue Q, Varady SRS, Waddell TQA, Roman MR, Carrington J, Roh-Johnson M. 2023. Lack of Paxillin phosphorylation promotes single-cell migration in vivo. *J Cell Biol* **222:** e202206078. doi:10.1083/jcb.202206078

Xue C, Chu Q, Shi Q, Zeng Y, Lu J, Li L. 2025. Wnt signaling pathways in biology and disease: mechanisms and therapeutic advances. *Signal Transduct Target Ther* **10:** 106. doi:10.1038/s41392-025-02142-w

Yamaguchi N, Zhang Z, Schneider T, Wang B, Panozzo D, Knaut H. 2022. Rear traction forces drive adherent tissue migration in vivo. *Nat Cell Biol* **24:** 194–204. doi:10.1038/s41556-022-00844-9

Yolland L, Burki M, Marcotti S, Luchici A, Kenny FN, Davis JR, Serna-Morales E, Müller J, Sixt M, Davidson A, et al. 2019. Persistent and polarized global actin flow is essential for directionality during cell migration. *Nat Cell Biol* **21:** 1370–1381. doi:10.1038/s41556-019-0411-5

Running to Their Fates: Neural Crest and Placode Migratory Behavior and Cell Fate Decisions

Junpeng Gao, Hugo A. Urrutia, and Marianne E. Bronner

Division of Biology and Biological Engineering, California Institute of Technology, Pasadena, California 91125, USA

Correspondence: mbronner@caltech.edu

Neural crest cells are among the most migratory cells in the vertebrate embryo. They initially arise within the forming neural tube, the precursor to the brain and spinal cord, but then migrate away by undergoing an epithelial-to-mesenchymal transition (EMT). Different neural crest populations exist along the body axis that differ in their flavor of EMT, migratory pathways, and cell types into which they differentiate. In the head region, neural crest cells interact with ectodermal placodal cells, another migratory cell type unique to vertebrates, that form sensory structures of the head. Here, we focus on select neural crest and placodal populations, highlighting how their region-specific migrations may be linked to cell fate choices and discuss how neural crest–placode interactions influence each other's behavior and derivative formation.

The process of embryogenesis, whereby a single cell develops into a complex, multicellular organism, involves not only extensive cell proliferation and differentiation but also a series of morphogenetic events that assemble one or more cell populations into distinct territories and anatomical structures. A prime example occurs at the neural plate border (NPB) region of developing vertebrate embryos. The NPB lies between the neural plate (presumptive central nervous system) and nonneuronal ectoderm (presumptive epidermis) of developing gastrula stage embryos. This territory gives rise to two embryonic cell types, the neural crest (NC) and the cranial ectodermal placodes, which undergo extensive movements and migrations during early development and are both unique to vertebrates.

The NC and cranial ectodermal placodes together give rise to sensory structures of the head (Baker and Bronner-Fraser 1997, 2001; Edens and Bronner 2025). Both cell populations are born within the NPB region during gastrulation, where they are induced by a battery of signaling factors, including bone morphogenetic protein (BMP), fibroblast growth factor (FGF), Notch, retinoic acid, and WNT (Thawani and Groves 2020; Thiery et al. 2020; Stundl and Bronner 2025). After induction, NC cells are regionalized along the anteroposterior axis into several distinct subpopulations, referred to as cranial, vagal, trunk, and lumbosacral populations (Fig. 1A). NC cells have broad developmental potential to differentiate into a wide variety of distinct cell types, including bone and cartilage of the head, neurons, and glia of the

Cite this article as *Cold Spring Harb Perspect Biol* doi: 10.1101/cshperspect.a041753

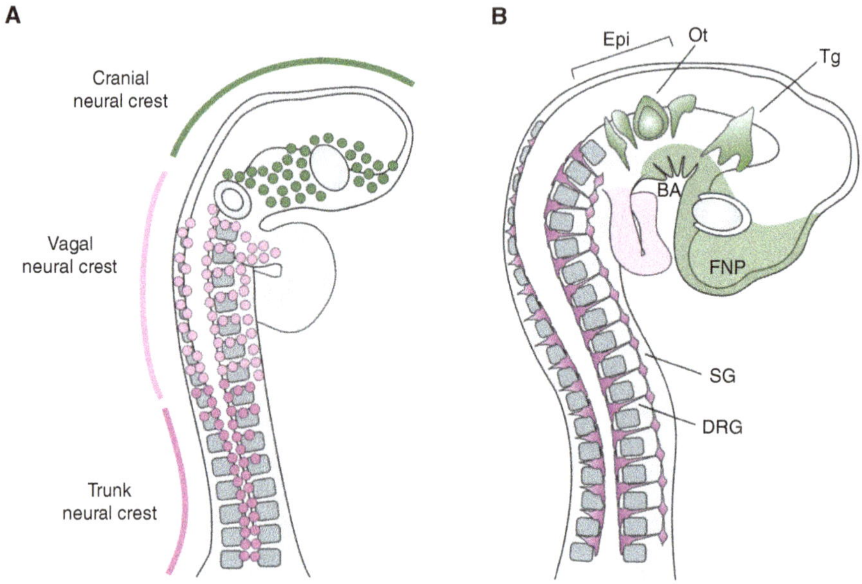

Figure 1. Overview of neural crest subpopulations and their derivatives. (*A*) The neural crest is divided along the anteroposterior axis into the cranial (green), vagal (pink), trunk (purple), and lumbosacral (not shown) populations. The cranial neural crest cells form between the mid-forebrain and posterior hindbrain and migrate as a collective sheet across the head region. The vagal neural crest cells arise from the level of somite 1–7 and migrate toward the heart or caudally into and along the gut. Finally, the trunk neural crest cells are derived from somite 8 and beyond. Diagram is a representation of HH13 chick embryo. (*B*) The neural crest gives rise to a number of derivatives that span the entire embryo. The cranial neural crest cells (green) contribute to the frontal nasal process (FNP) and branchial arches (BAs) that will form the bone, cartilage, and connective tissues of the head, along with the cranial peripheral nervous system, which includes the trigeminal (Tg) and epibranchial (Epi) ganglion and the acoustic ganglion that associates with the otic (Ot) placode/vesicle. The vagal neural crest cells (pink) will contribute to the septum of the heart as well as contributing to the enteric nervous system. Lastly, the trunk neural crest cells will form the sympathetic ganglia (SG) and the dorsal root ganglia (DRG). Other neural crest derivatives not shown include the smooth muscle cells of the vessels, secretory cells of the adrenal gland, pigment cells across the entire body, etc. Diagram is a representation of the HH18 chick embryo. (HH) Hamburger and Hamilton stages.

peripheral nervous system and melanocytes (Fig. 1B; Sauka-Spengler and Bronner-Fraser 2008; Rothstein et al. 2018). In contrast to NC cells which extend along the entire body axis, ectodermal placode cells arise only in the head. They segregate into subgroups that occupy stereotypical locations, forming the olfactory, lens, otic, and lateral line (only in aquatic vertebrates) placodes that contribute to the paired sense organs and cranial sensory neurons of the trigeminal and epibranchial ganglia (Fig. 2; Schlosser 2006; Saint-Jeannet and Moody 2014).

Given that they are among the most migratory of embryonic cell types, the mechanisms underlying NC and placode cell migration have been studied extensively. Classical hypotheses of epithelial-to-mesenchymal transitions (EMTs) and collective movements have been generated using these cell types as models (Duband 2010; Breau and Schneider-Maunoury 2015; Mayor and Etienne-Manneville 2016; Szabó and Mayor 2018; Leathers and Rogers 2022). Furthermore, both populations give rise to diverse and important structures, such that a wealth of literature over the past decades have created an increasingly detailed molecular network underlying their specification and differentiation processes (Baker and Bronner-Fraser

 Cite this article as *Cold Spring Harb Perspect Biol* doi: 10.1101/cshperspect.a041753

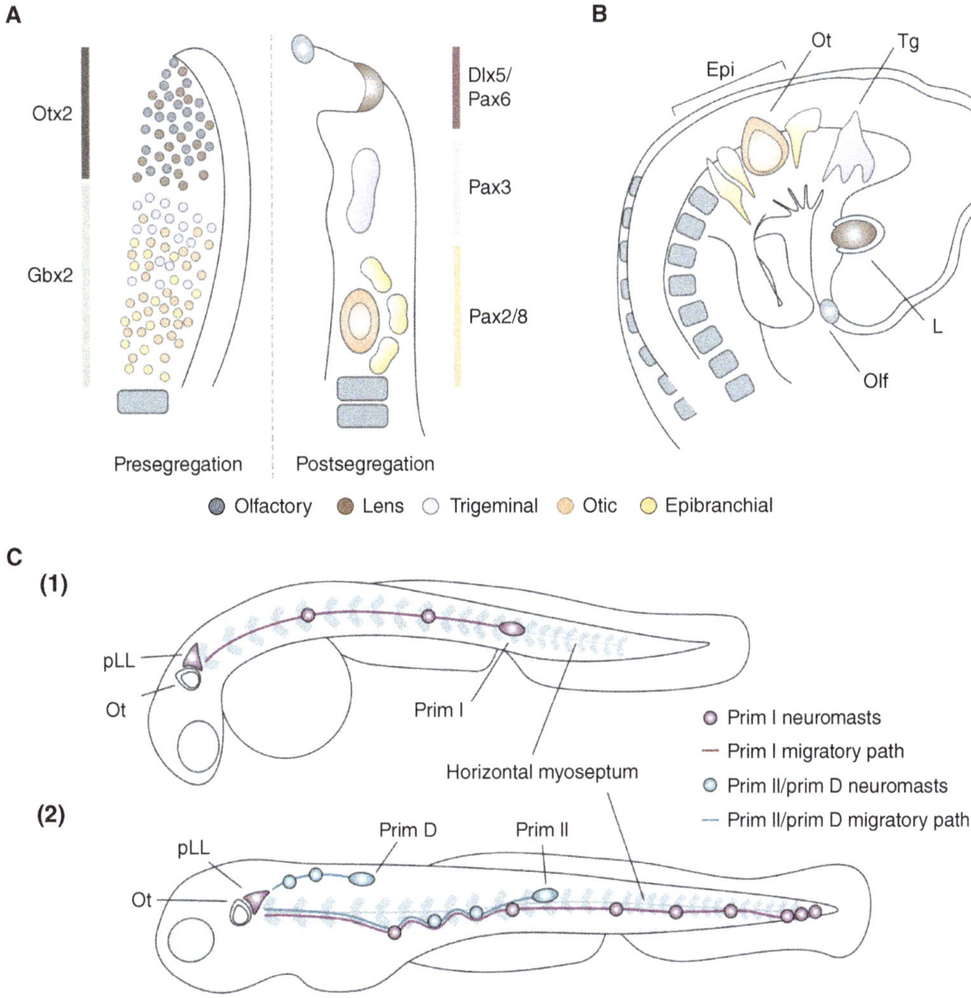

Figure 2. Overview of placode subpopulations and their derivatives. (*A*) Individual placode populations undergo progressive segregation to occupy stereotypical sites within the cranial ectoderm. They are initially intermixed and only broadly separated into anterior and posterior groups, linked to Otx2 versus Gbx2 expression (*left*, presegregation). Later, placodes are distinctly regionalized into the olfactory (Olf, gray), lens (bronze), trigeminal (Tg) (heather), otic (Ot) (orange), and epibranchial (Epi) (yellow) populations, where each population is marked by a combination of Pax and Dlx genes (*right*, postsegregation). Diagrams reflect HH8 (presegregation) and HH12 (postsegregation) chick embryos. (*B*) Placodes eventually contribute to the paired sense organs, including the Olf, lens (L), and Ot and the Epi and Tg ganglia. Diagram is a representation of HH18 chick embryo. (*C*) Aquatic vertebrates additionally possess the lateral line placodes. The posterior lateral line (pLL) placode of zebrafish is the best studied. (1) The pLL arises from a region posterior to the Ot, and initially generates a primordium (prim I, purple), that migrates along the horizontal myoseptum and deposits mechanosensory organs called neuromasts at approximately every fifth somite (gray rectangles). (2) Later, a second primordium is generated from the same pLL field and further splits into two migrating primordia (teal). One is designated prim D and migrates dorsally, whereas the second one is named prim II, and travels the same path as prim I, along the horizontal myoseptum. Prim II begins migration when prim I has already reached the tail, where it further dissociates into three neuromast structures. Due to a global, ventral movement of the lateral line system at this time point, the migratory path of prim II is slightly distorted. Prim II migrates slower than prim I but deposits neuromasts more frequently, at around approximately every second somite. Diagrams are representations of 32 h postfertilization (hpf) (1) and 3 days postfertilization (dpf) (2) zebrafish embryos.

2001; Schlosser 2006; Sauka-Spengler and Bronner-Fraser 2008; Martik and Bronner 2017; Thiery et al. 2020; Erickson et al. 2023). However, the migratory behavior and differentiation of these cells are often discussed as separate topics such that their connections are often missed. To help better understand the coupling between cell motility and cell fate decisions, here we provide a review of insights into the mechanisms of NC and cranial placode migration, their specification and differentiation programs, while specifically focusing on how these two processes are intertwined.

NEURAL CREST SPECIFICATION AND EMT

After initial induction at the NPB, premigratory NC cells come to lie in the dorsal region of the closing neural tube prior to undergoing EMT to become migratory (Theveneau and Mayor 2012). During EMT, cell–cell connections within the epithelium are disrupted, accompanied by downregulation of adhesion molecules like E-cadherin and cadherin-6B, as well as tight junction molecules including claudins and occludins (Nakagawa and Takeichi 1995; Aaku-Saraste et al. 1996; Coles et al. 2007). Expression of other adhesion molecules like cadherin-7 and -11, often associated with mesenchymal state and motility, is upregulated in a manner that differs between species (Fig. 3A; Nakagawa and Takeichi 1995; Hadeball et al. 1998). This "cadherin switch" has been viewed as a classical feature of EMT in other migratory cells including cancer cells undergoing metastasis (Kuriyama and Mayor 2008; Kerosuo and Bronner-Fraser 2012).

Intriguingly, the acquisition of migratory potential goes hand in hand with specification of NC identity. NC specification involves a series of events that progressively restrict cell fate, with the acquisition of migratory potential as an important criterion for defining bona fide NC cells (Sauka-Spengler and Bronner-Fraser 2008; Bronner 2012). Several NC specifier genes, such as Snail1/2 and FoxD3, are known to directly regulate EMT processes. In both cultured cells and in the cranial neural tube of chick embryos, Snail2 represses E-cadherin, cadherin-6B, and tight junction proteins (Batlle et al. 2000; Cano et al. 2000; Bolós et al. 2003; Ikenouchi et al. 2003; Taneyhill et al. 2007). Furthermore, FoxD3 has also been shown to repress expression of N-cadherin (Cheung et al. 2005) and Tspan18, a scaffolding protein for cadherin-6B (Fairchild and Gammill 2013), while upregulating cadherin-7 (Dottori et al. 2001). Finally, genes that are not restricted to but are highly expressed in premigratory NC cells, such as c-Myc, AP-2, and Twist, can also directly repress E-cadherin expression, thereby enabling EMT (Fig. 3A; Batsché and Crémisi 1999; Barriga et al. 2013).

On the other hand, the classical view of an EMT program under direct regulation of NC specifying factors has been challenged. First, not all cells expressing Snail2 become migratory NC cells (Linker et al. 2000). More recent research has also identified a plethora of layers by which NC EMT can be regulated, including epigenetic modifications (Liu et al. 2013; Hu et al. 2014; Lee et al. 2019; Gandhi et al. 2020), posttranscriptional and posttranslational control (Liu et al. 2013; Mathavan et al. 2017; Sánchez-Vásquez et al. 2019; Galton et al. 2022), and even shifts in lipid content (Piacentino et al. 2022), although most of these regulators still affect EMT indirectly via known NC specifier genes. Finally, increased interest in the role of tissue biophysics in development has highlighted the importance of physical stiffness of the surrounding environment as a determinant of NC EMT (Fig. 3A; Barriga et al. 2018).

To further complicate the role of NC specification in regulating EMT, early transplantation experiments have hinted that NC migration can in turn affect cell identity. In a striking example, placing an already migratory NC cell back into the neural tube was shown to reprogram the cell toward a neural tube fate (Ruffins et al. 1998). Such observations suggest that even established NC identity is amenable to changes depending on the external environment. This is not surprising given that there is strong evidence that cells of the dorsal neural folds possess high-fate flexibility. Early tracking experiments showed that individual cells of this region can give rise not only to migratory NC cells but also to neural tube and epidermal cells (Bronner-Fraser and

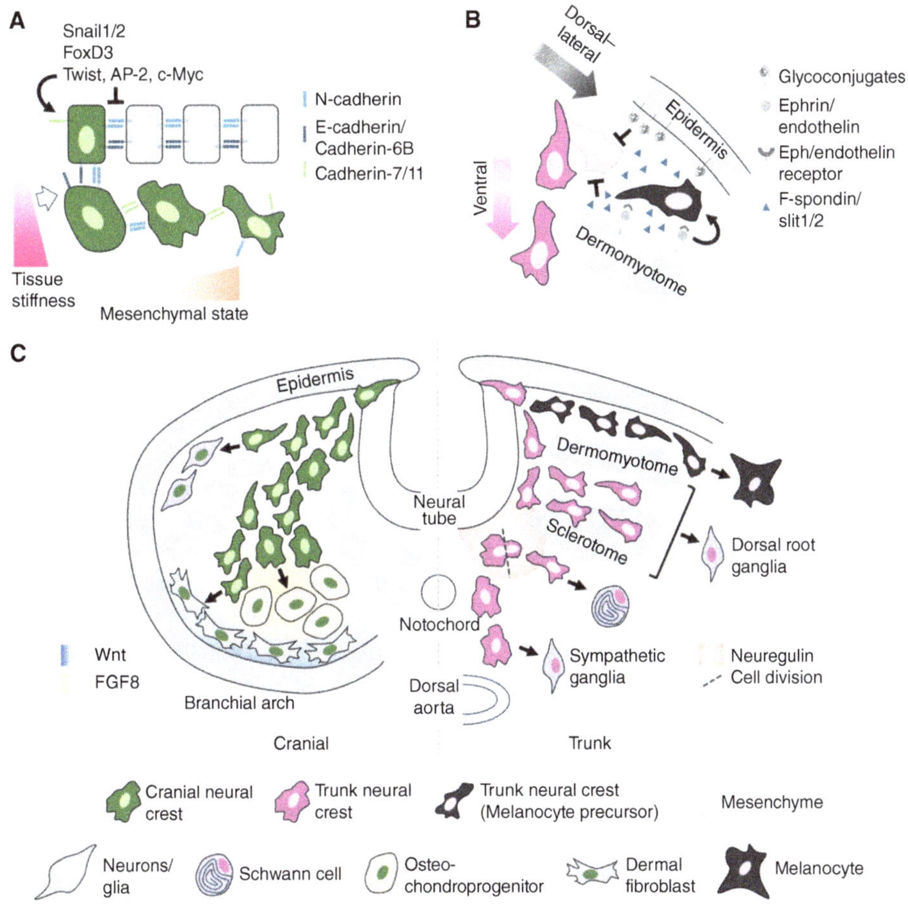

Figure 3. Neural crest specification and cell fate decisions are tightly coupled to their migratory behaviors. (*A*) Epithelium-to-mesenchymal transition (EMT) of neural crest is regulated by a battery of specification factors including Snail1/2, FoxD3, Twist, AP-2, and c-Myc. These transcription factors orchestrate a "cadherin switch," whereby the delaminating neural crest cells downregulate E-cadherin along with cadherin-6B and upregulate cadherin-7 and -11. N-cadherin is also reported to be downregulated, but this varies between axial level and species, and migratory neural crest cells still retain some N-cadherin expression. Regardless, the resulted change in cell adhesion properties, together with environmental influences such as tissue stiffness, allows neural crest cells to leave the epithelium (gray) and assume a mesenchymal state, thereby becoming migratory cells (green). (*B*) The trunk neural crest cells migrate along two pathways depending on their fate: a dorsal–lateral route reserved for melanocyte precursors (black) and a ventral route for neurons, glia, and adrenomedullary cells (purple). Inhibitory signals expressed by the dorsal dermomyotome and the epidermis, including glycoconjugates, F-spondin, and Slit1/2, redirect nonmelanocyte precursors away from this route (mirrored cells with shadow), whereas endothelin and ephrin signalings attract neural crest cells that are prespecified for melanocytic fate. (*C*) Schematic diagram of transverse section showing migration of cranial (*left*) and trunk (*right*) neural crest cells. (*Left*) Cranial neural crest cells (green) migrate as a collective group through a rather loose mesenchymal tissue environment (gray, round cells). Cells migrate toward and populate the branchial arch (BA), where local FGF8 signals (wheat) promote their differentiation into osteochondroprogenitors (wheat). Simultaneously, a high level of Wnt expression locally in the epidermis (blue) is proposed to redirect these cells to become dermal fibroblasts (gray, spiky). In contrast, cells that do not migrate into the BA adopt neuronal/glial fates (heather). (*Right*) Trunk neural crest cells (purple) can migrate either between the somite and the epidermis (dorsal–lateral route) and differentiate into melanocytes (black) or invade the somite or between the somite and the neural tube (ventral route). Along this ventral pathway, cells mainly migrate as a single chain. A leader cell that migrates first to the level of the ventral neural tube receives local signals such as neuregulins (bronze) and halts. It then undergoes an asymmetric cell division (dashed line) to produce one progeny that further migrates to the level of dorsal aorta and contributes to the sympathetic ganglia (SG) and another progeny that is retained locally and is directed toward Schwann cell fate. Finally, all remaining cells contribute to neurons and glia of the dorsal root ganglia.

Fraser 1988; Collazo et al. 1993; Serbedzija et al. 1994; Selleck and Bronner-Fraser 1995), and more recent analysis of neural plate and NPB genes revealed significant overlap in expression domains, with many cells coexpressing multiple markers that were thought to be lineage-specific (Roellig et al. 2017). Therefore, the transplantation experiment discussed above raises the intriguing possibility that NC cells may only maintain their identity when migrating along proper pathways but retain fate flexibility to respond to alternate environments. It would thus be interesting to see whether directly manipulating the migratory behavior of NC by physically blocking NC delamination or emigration can subsequently rewire its identity into other ectodermal cell types. Taken together, our current understanding has suggested that there is a highly intertwined coupling between cell fate choice and migration even at early stages of NC cell formation.

NEURAL CREST MIGRATION

Depending upon axial level of origin, NC cells undergo EMT that is either collective (e.g., in the cranial region) or individual (e.g., in the trunk region). After leaving the neural tube, NC cells migrate extensively along well-defined pathways characteristic of the axial level of origin. By responding to local cues along these pathways, they invade distinct regions and tissues to differentiate into a diverse range of cell types (Minoux and Rijli 2010; Theveneau and Mayor 2012).

Detailed analyses of the NC gene regulatory network have delineated the specification and differentiation of NC cells across axial levels with focus on formation of several derivative cell types (Sauka-Spengler and Bronner-Fraser 2008; Martik and Bronner 2017). Separately, there exists a rich literature describing the route, cellular behavior, and molecular mechanisms of NC migration (Szabó and Mayor 2018). However, there are fewer studies bridging the acquisition of NC fate to their migratory routes and target sites. In the sections below, we separately focus on the cranial and trunk NC populations as these are the most well studied. We will briefly summarize current knowledge of their differentiation programs and migratory behaviors, while highlighting links between migration and cell fate decisions.

Cranial Neural Crest Migration

The cranial NC population arises from the mid-forebrain to the rostral hindbrain adjacent to the forming otic placode (Fig. 1A). These cells migrate in a sheetlike fashion that becomes subdivided into distinct streams. The first stream emerges from the forebrain, midbrain, and rostral hindbrain, spreading like a cobra's hood to populate the frontal nasal process and the first branchial arch. The second and third NC streams from rhombomeres (r) 4 and 6 of the hindbrain migrate into branchial arches 2 and 3–6. In particular, the third stream reflects the "cardiac" NC, which contributes to the outflow tract of the heart and other cardiovascular structures (Minoux and Rijli 2010).

A rich literature has demonstrated that NC axial level identities regulate their migratory paths. The formation of segregated migratory streams is mediated by the combinatorial activation of neuropilin/semaphorin (Nrp/Sema) and ephrin receptor/ephrin (Eph/Ephrin) signaling pathways within different NC populations and in regions lateral to NC-free rhombomeres (Fig. 4B; Minoux and Rijli 2010). For example, expression of Nrp2-only or Nrp1 and 2 in the first and second streams prevents their intermixing across a Sema3a positive region (Gammill et al. 2007; Schwarz et al. 2008), whereas expression of EphrinB2 in the second stream deters the EphB1-positive cells of the third NC stream (Smith et al. 1997; Adams et al. 2001). On top of these external cues, NC–NC interactions further refine stream boundaries, mostly via contact inhibition of locomotion (Fig. 4A). This is demonstrated both ex and in vivo, where NC cells retract their filopodia and turn against each other upon contact, thereby preventing intermingling between streams (Teddy and Kulesa 2004; Carmona-Fontaine et al. 2008).

Whether NC migratory behaviors and trajectories feed back into regulating their identity is a long-term debate of "nature" versus "nur-

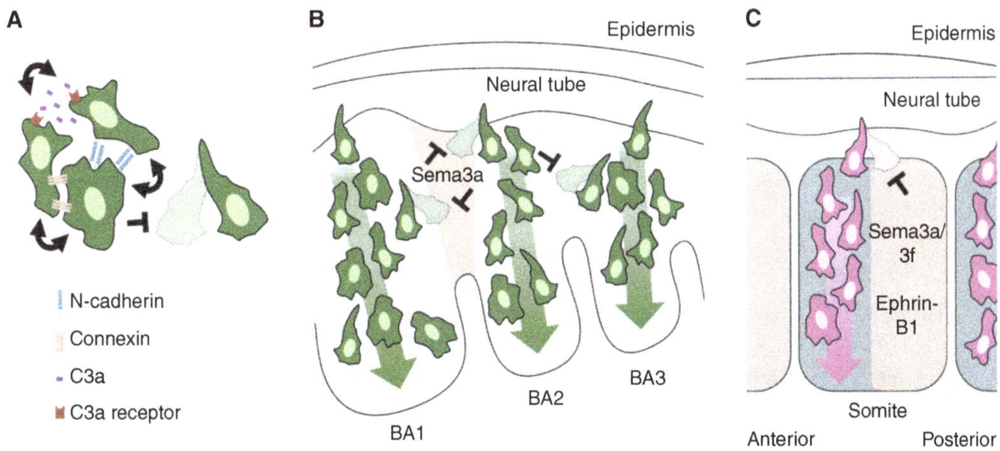

Figure 4. Migrating neural crest streams are regulated by both internal and external signalings. (*A*) Neural crest cells display a number of intercellular interactions, best studied in the cranial crest population. Cranial neural crest cells are connected and attracted to each other via cell adhesion factors such as N-cadherin, gap junction proteins such as connexin, and chemotactic cues such as C3a. The cells further display contact inhibition where neural crest cells that touch each other would reorient and turn away. (*B*) Segmentation of cranial neural crest streams (green) invading the branchial arches (BAs) is mediated by both inhibitory, Sema3a-expressing region (peach) between BA1 and BA2, and contact inhibition between the BA2 and BA3 streams. (*C*) Patterning of the trunk neural crest streams (purple) is generated by a repulsive territory (peach) at the posterior fragment of each somite, where semaphorin (Sema3a) and ephrin (Ephrin-B1) signalings prevent cells from migrating into the region. In all of the above, cells that have reoriented their migratory direction are depicted as mirrored cells with shadow.

ture." Attempts to reconcile this debate have been expertly reviewed elsewhere (Dawes and Kelsh 2021; Erickson et al. 2023). Current understanding suggests that while intrinsic biases are often introduced early and have clear influences on final fate, they are neither strictly complete nor immutable. Here, we particularly focus on a few key examples to address the coupling between NC migration and fate decision. These examples show that what NC cells encounter and interact with during migration and localization has a pivotal influence on the cell's eventual fate. In a number of cases, a NC cell's initial default developmental trajectory may be altered by migration to a different environment that adjusts the cell's differentiation in response to the local flavor.

Two major fates of cranial NC cells are neurogenic/gliogenic or chondrogenic. The early migrating NC cells travel long distances to populate the facial prominence and branchial arches, mostly contributing to skeletal and cartilaginous tissues, whereas the late-migrating population remains close to the neural tube and mainly forms cranial ganglia (Baker and Bronner-Fraser 1997). It has then been proposed that NC cells undergo a Sox-switch: NC cells all express Sox10 when they initially emigrate from the neural tube. Retention of Sox10 allows cells to proceed to neuronal/glial fates. In contrast, turning on Sox9 expression at the expense of Sox10 redirects cells toward osteochondrogenic fates (Haldin and LaBonne 2010; Schock and LaBonne 2020). Interestingly, this switch was shown to be under regulation of environmental cues. FGF8 is locally expressed in the pharyngeal arches where it signals NC cells in the vicinity to upregulate Sox9, thereby directing these cells toward cartilage development. On the other hand, NC cells that do not enter the arch, or are experimentally manipulated to be insensitive to FGF8 signaling pathways, retain Sox10 expression and become neurogenic (Fig. 3C; Blentic et al. 2008). In addition to osteoblasts and chondrocytes, cranial NCs are also capable of forming connective tissue such as dermal fibroblasts. It was proposed that a WNT gradient exists near the cranial ectoderm, between the WNT-expressing superficial epidermis and internal mesenchyme. NC cells that migrate close to the epidermis therefore receive more WNT signals, which inhibits osteoblast and chondrocyte fates (Hartmann 2006) and specifies the cells to become dermal fibroblasts, whereas those localized internally receive less WNT signals and become osteoblast progenitors (Fig. 3C; Goodnough et al. 2012, 2014). The above signaling molecules, together with other known ligands such as transforming growth factor β (TGF-β) (John et al. 2011), therefore play pivotal roles in generating local niches/environments that shape the cell fate decisions of NC cells that have entered the territory (Santagati and Rijli 2003; Dash and Trainor 2020).

Within the cranial NC population that contributes to bone and cartilage, it has been shown that axial level identities, mainly influenced by a colinear Hox code, are imparted onto the NC prior to emigration and bias their eventual fate (Hunt and Krumlauf 1991; Hunt et al. 1991). It has been proposed that the Hox negative NC that originates from forebrain to anterior hindbrain may represent a "ground" state, where they migrate and occupy the face and the first pharyngeal arch, while expression of Hox genes in more caudal regions redirects the cells to colonize the other branchial arches (Dash and Trainor 2020). Indeed, loss or overexpression of Hox genes such as Hoxa2 results in homeotic transformations of the second, third, or fourth arch into the first arch, or vice versa (Rijli et al. 1993; Grammatopoulos et al. 2000; Minoux et al. 2009).

Interestingly, the maintenance of such Hox identity in NC streams may be linked to their migratory behavior. Numerous studies have shown that cranial NC cells travel in dense, continuous groups with constant intercellular interactions, a behavior termed collective migration. This is afforded by a combination of mechanisms including contact inhibition of locomotion and coattraction (Fig. 4A; Szabó and Mayor 2018). The latter is achieved through both cell adhesion/junction molecules such as N-cadherin and connexins (Xu et al. 2001, 2006) and chemoattractants such as the complement component factor C3a (Carmona-Fontaine et al. 2011). Together, these mechanisms allow NC cells to maintain cohesiveness and prevent straying cells (Thomas and Yamada 1992). The collective migration of NC cells then allows for a community effect, such that cells in groups can be more readily induced and better maintain their identity than individual cells during tissue development (Gurdon 1988). Intriguingly, this was indeed demonstrated in transplantation studies of hindbrain NC cells: Large transplants with closely clustered NC cells maintain their initial Hox identity, whereas transplanted single cells, or individual cells that have disconnected from large transplants, are much more plastic and switch their Hox expression and identity according to new axial level (Trainor and Krumlauf 2000; Schilling et al. 2001). The molecular mechanisms underlying such community effects remain to be elucidated. Meanwhile, existing observations have hinted that gap junction—based cellular connections that exist among collectively migrating NC cells—can facilitate signal transduction within the group (Wang et al. 2010). This provides a plausible mechanism for the synchronization of cellular identities.

In another example, NC cells that migrate into the pharyngeal arches are further patterned along the proximal-distal axis by a *Dlx* code. In the first arch, NC cells that migrate to the proximal portion turn on *Dlx1/2* and give rise to the upper jaw or maxilla, while those that migrate to the distal portion additionally turn on *Dlx5/6*, which specifies them to generate the lower jaw or mandible (Depew et al. 1999; Shimizu et al. 2018). It has been shown that this *Dlx* code is partially created by the distally restricted expression of endothelin-1. Endothelin-1 deletion results in loss of Dlx5 and transformation of the mandible into the maxilla, whereas overexpression of the ligand results in the opposite (Kurihara et al. 1994; Clouthier et al. 1998; Ozeki et al. 2004; Sato et al. 2008). Therefore, NC cells that migrate into the first arch are capable of forming both maxillary and mandibular structures, but their final fate is strongly influenced by their physical end point along the proximal-distal axis.

Trunk Neural Crest Migration

Trunk NC cells emerge from the posterior part of the embryo (between somite 7 and 28 in chick) and adopt migratory routes and behaviors that are different from their cranial counterparts (Fig. 1A). Migrating through a much denser environment than the head, the trunk NC cells undertake two main routes: (1) a ventral path where the cells traverse between the somite and the neural tube or through the somite, and (2) a dorsolateral path where cells migrate between the somite and the overlaying ectoderm. There are also noted differences in the timing and mode of EMT between trunk NC populations. Emigration away from the neural tube proceeds in an anterior–posterior manner, with anterior cells migrating out earlier. Furthermore, in organisms including chick and zebrafish, migration through the ventral pathway initiates earlier than that through the dorsolateral route, although in mouse, NC cells enter both pathways simultaneously (Vega-Lopez et al. 2017; Häkkinen et al. 2025).

There are striking correlations between the migratory route of the trunk NCs and their fate: Trunk NC migration through the somite is patterned, such that they travel through the anterior half of each somitic sclerotome but are excluded from the posterior half due to repulsion via Nrp/Sema and Eph/Ephrin signaling pathways (Fig. 4C). This segmental migration of NC streams foreshadows the later patterning of the dorsal root ganglia into which NC cells condense. In addition, cells that migrate using the ventral pathway differentiate into a variety of cell types including neurons and glia of the sensory and sympathetic ganglia, as well as chromaffin cells of the adrenal medulla, and Schwann cells, whereas those that follow the dorsolateral stream exclusively give rise to pigment cells (Vega-Lopez et al. 2017; Häkkinen et al. 2025).

Particularly for trunk NC, the literature has debated whether these cells emigrate as multipotent progenitors and differentiate according to local environments, or are progressively fate restricted, with some lineage decisions made as early as cells are delaminating from the neural tube (Dawes and Kelsh 2021; Häkkinen et al. 2025). Regardless of the extent of their potency, there are at least strong early biases in cell fate that influence the NC migratory path. For example, the restriction of dorsolateral migratory path to pigment cell progenitors was shown to be mediated via a combination of repulsion and attraction mechanisms (Fig. 3B). A number of inhibitory cues, ranging from F-spondin (Debby-Brafman et al. 1999), Slit1/2 (Jia et al. 2005), to glycoconjugates such as chondroitin-6-sulfate and peanut agglutinin (PNA) lectin–binding molecules (Oakley et al. 1994), are expressed under the epidermis, which can prevent nonmelanocyte progenitors from entering. On the other hand, melanocyte specifier genes, such as Mitf (Dorsky et al. 2000; Steingrímsson et al. 2004; Greenhill et al. 2011), can directly turn on endothelin and ephrin receptors that attract pigment cell progenitors to migrate into the dorsolateral space (Santiago and Erickson 2002; Harris et al. 2008; Kawasaki-Nishihara et al. 2011).

NC populations that enter the ventral pathway also display varied developmental potential depending on their timing of emigration (Artinger and Bronner-Fraser 1992; Raible and Eisen 1994). Cells that migrate early adopt neurogenic fate, whereas those that migrate later do not. While such differences may be intrinsic, transplantation experiments have shown that early migrating NC populations will no longer adopt neuronal fates when put into older hosts, thereby suggesting that temporal differences in the tissue environment may play important roles in sustaining the potential of migrating NC cells (Raible and Eisen 1996). Furthermore, ablating the early migrating cells allows the late-migrating populations to take on neurogenic fate, implying that interactions between the two migrating populations may have inhibitory effects on cell fate choice (Raible and Eisen 1996).

Indeed, interactions within a single migrating trunk NC stream were shown to generate an intricate spatial map that is crucial for the specification of cell fates. Live imaging has shown that trunk NC cells that take the ventral route migrate in single chains with dynamic but constant interactions between the front (or leader) and rear (or follower) cells, while removing such interactions can pause migration (Richardson et al. 2016). Moreover, cells strictly follow their relative positions and migrate in a temporally synchronized manner: The leader cell of the group will thus always end up at a more ventral position than the followers. Within a single migrating chain, a cell's position along the dorsoventral axis has been shown to accurately predict its fates (Fig. 3C; Alhashem et al. 2022a). When in the appropriate position, the leader cell would go through an asymmetric cell division, yielding one progeny that inherits high Notch activity, activates the sympathetic fate specifier Phox2bb (Pattyn et al. 1999), and further migrates ventrally to give rise to the sympathetic ganglion (Alhashem et al. 2022a,b). On the other hand, the low-Notch activity daughter cell, and in some cases also its immediate follower, will halt migration and differentiate into Schwann cells, a type of glia. Finally, the remaining dorsal follower cells all contribute to dorsal root ganglion. In this scheme, the position at which the leader cell halts to undergo division and specification is likely determined by the tissue environment. It was shown that the ligands Neuregulin1/2 are expressed by the somites and ventral spinal cord neurons and signal to the receptors ErbB2 and ErbB3 expressed by trunk NC cells. This signal is crucial for pausing the cells at the correct location to form sympathetic and dorsal root ganglion (Honjo et al. 2008). Together, these observations suggest a model in which intercellular interactions within the migrating NC chain and responses to environmental cues direct cellular specification with high spatial resolution.

PLACODE SPECIFICATION AND A CELL SORTING MODEL

Akin to the NC, the cranial placodes represent another population of cells from the NPB region that can undergo movement and migration. However, NC and placode cells also display important differences even at early stages of specification. The NC cells, regardless of their axial level, all reside in the NPB and subsequently the dorsal neural fold before being fully specified and initiating migration. On the other hand, placode cells initially arise from a single primordium termed the pan-placodal region, a horseshoe-shaped domain abutting the anterior NPB. Later, the primordium subdivides and is specified into individually distinct populations (namely lens, olfactory, trigeminal, otic, epibranchial, and lateral line) that occupy diverse locations within the head ectoderm (Fig. 2A; Baker and Bronner-Fraser 2001; Schlosser 2006; Streit 2025).

The pan-placodal primordium is comprised either of multipotent cells or a heterogeneous mixture of fate-determined progenitors. Grafting experiments coupled to analysis of expression patterns for a number of transcription factors, notably in the Six/Eya/Dash network, have identified the pan-placodal primordium as a common ground from which all placodes arise (Baker and Bronner-Fraser 2001; Schlosser 2006). Fate mapping studies and examination of known placodal markers revealed that cells with different developmental trajectories are spatially arranged in a salt-and-pepper pattern, therefore suggesting intermingling of cells with varied identities (Kozlowski et al. 1997; Whitlock and Westerfield 2000; Streit 2002; Bhattacharyya et al. 2004; Xu et al. 2008). On the other hand, more recent analysis of the transcription factor signatures, along with single-cell transcriptomics, of the NPB region has further highlighted that cell fates are biased but flexible until advanced stages, even beyond neural tube closure (Roellig et al. 2017; Thiery et al. 2023).

How the pan-placodal region segregates into discrete placode domains is a subject of much debate. Two major models have been put forward (Fig. 5A; Schlosser 2006; Breau and Schneider-Maunoury 2014). In the first, local signals are proposed to activate different gene regulatory networks within multipotent placode progenitors to impart them with regional identities. In the alternative view, segregation of placode domains is attributed to cell sorting of fate-biased populations with differential expression

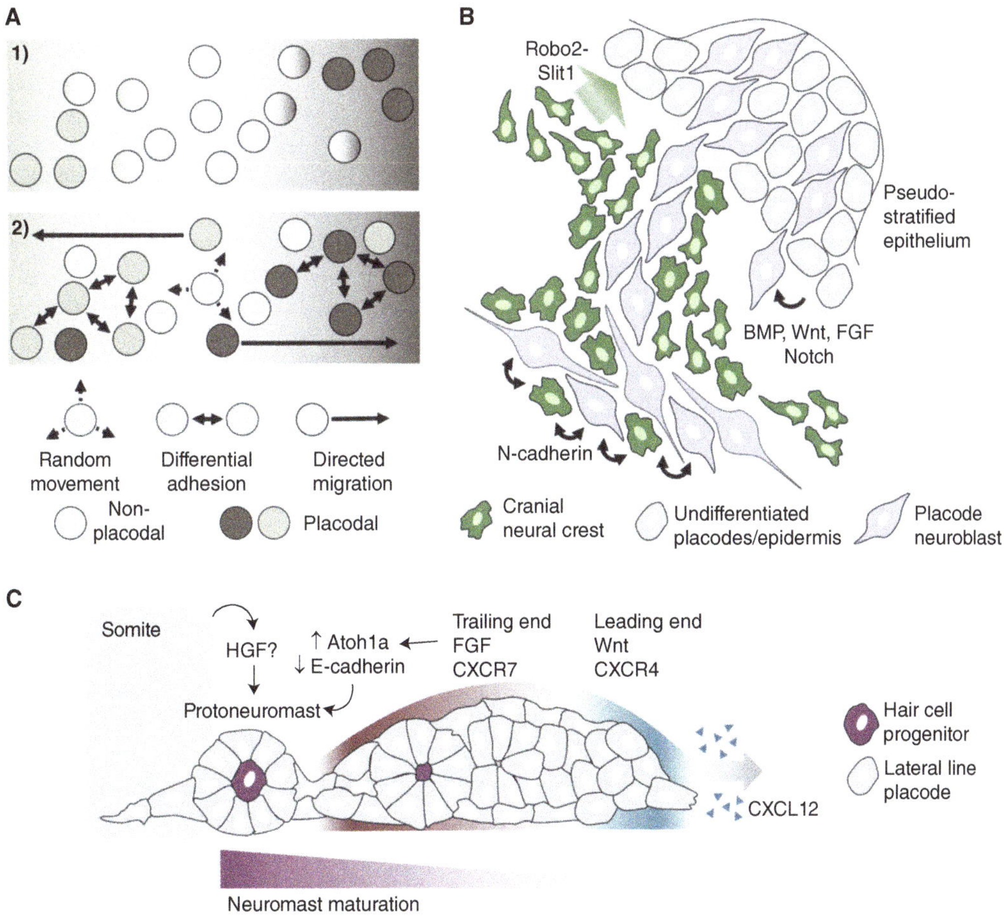

Figure 5. Placode specification and cell fate decisions are linked to their morphogenesis. (*A*) Two mechanisms have been proposed for placode segregation: (1) different placode identities arise from transcription factor interactions within gene regulatory networks that gradually direct the cells to different lineages (generically illustrated here as brown and beige cells), and (2) active movement of prespecified (brown and beige) placodes to their designated territory via differential adhesion-mediated neighborhood exchange or directional migration, and their motion is distinct from random movement of nonplacodal cells. (*B*) Neurogenic placodes such as the trigeminal and epibranchial placodes ingress as neuroblast cells (heather) from a pseudo-stratified epithelium. Signaling molecules such as BMP, Wnt, FGF, and Notch are known to regulate both neurogenesis and delamination. Numerous placode–neural crest (NC) interactions also underlie this process. The cranial NC stream (green cells) migrates underneath the epidermis and can signal to the placodes via the Robo2-Slit1 pathway to promote ingression. NC cells further form a corridor-like structure that tunnels the placodes as they migrate. In return, placode cells possibly act as an anchor to retain the migrating NC stream via cell adhesion proteins such as N-cadherin. (*C*) The migrating lateral line primordium is a polarized structure, with a leading end (teal) that expresses Wnt and CXCR4 and a trailing end (red) that expresses FGF and CXCR7. The coordination between CXCR4 and CXCR7 allows the primordium to migrate along a path of CXCL12 ligands. Polarization of the primordium further directs the formation, maturation, and deposition of protoneuromasts. Specifically, the FGF signaling pathway initiates rosette formation and activates the hair cell specifier Atoh1a to generate hair cell progenitors (purple) at the center. As the protoneuromast progressively matures, increased Atoh1a causes a decrease in E-cadherin, which, together with hepatocyte growth factors (HGFs) secreted from the somitic region prepares the protoneuromast for deposition.

of adhesion molecules. A putative gene regulatory network has been assembled that underlies the process of placode segregation and diversification, with *Six/Eya* genes specifying general placode identities, while transcription factors such as Otx2 and Gbx2, along with Pax, Dlx, and Irx genes, gradually segment the placode primordium along the anterior–posterior axis (Fig. 2A). This occurs under local FGF, BMP, and WNT signaling inputs (Schlosser 2006; Streit 2008). However, the existence of a gene regulatory network does not preclude the possibility of cell sorting, and the establishment of sharp domain boundaries is more likely than not to involve the concerted action of transcription factor–based feedback loops and cell movements. While the latter mechanism has not been formally tested, live tracking experiments using DiI labeling on neural plate stage chicken embryos support a role for cell sorting in placode segregation (Streit 2002; Bhattacharyya et al. 2004). Groups of cells in the pan-placodal region were labeled and exhibited directional movements toward future olfactory, lens, or otic territories, different from cellular motions associated with normal embryonic growth. On the other hand, contradictory observations were made in *Xenopus* where high-throughput DiI labeling failed to show movements beyond random cellular shuffling within the ectoderm (Pieper et al. 2011). More recently, direct testing of the cell sorting hypothesis was performed by transplanting and live tracking fluorescently labeled placodal and adjacent nonplacodal *Xenopus* ectoderm. It was shown that transplanted placode cells indeed undergo more directional migration than nonplacodal counterparts (Steventon et al. 2016). Furthermore, the transcription factors Otx2 and Gbx2, known for patterning the anterior–posterior axis of the neuroectoderm, were found to be involved: Overexpressing the genes sharpened the boundary between Otx2 and Gbx2 expressing domains, while directly repressing their downstream genes within transplanted placodes decreased their directionality of migration (Steventon et al. 2012, 2016). Corroborating these observations, Gbx2 and Otx2 have been shown to directly mediate cell sorting at the midbrain–hindbrain boundary via differential expressions of N-cadherin and Eph/Ephrin (Kesavan et al. 2020). In addition to the hypothesized broad segregation of the pan-placodal region via Gbx2 and Otx2, cell sorting was also proposed to allow the separation of Pax6-expressing lens and Dlx5-expressing olfactory cells within the anterior placodes (Bhattacharyya et al. 2004), and between otic versus epibranchial placodes in the posterior, in a Pax2/8 dosage-dependent manner (Bhat and Riley 2011; McCarroll et al. 2012). While the exact molecular factors such as cell adhesion proteins have yet to be determined, the above observations taken together favor a role for cell sorting in placode segregation and specification.

PLACODE INGRESSION AND MIGRATION

After individual placodes have adopted their stereotypical locations within the epithelium, they begin local differentiation processes that are accompanied by extensive morphogenesis. Within the epithelium, the olfactory, hypophyseal, lens, and otic placodes all contribute to forming the paired sense organs by undergoing epithelial invagination, where a cohesive sheet of cells bend into cup-like structures. In addition, all except the lens placodes are capable of neurogenesis, in which case cells delaminate from the epithelium and invade the underlying mesenchyme to later contribute to the cranial ganglia and sensory neurons (Baker and Bronner-Fraser 2001; Schlosser 2006). Our current understandings of placode invagination has been reviewed elsewhere (Breau and Schneider-Maunoury 2015). In the following sections, we focus on placode ingression and migratory behavior. Here, in contrast to the highly migratory NC, placode cells migrate, but to lesser and varied extents. Among the well-studied populations, the trigeminal and epibranchial placodes display short-range migration, termed ingression, away from the epithelium, whereas the lateral-line placodes in aquatic vertebrates present a classical model for long-range collective migration, traversing nearly the entire length of the embryo (Breau and Schneider-Maunoury 2015). Below, we separately discuss these placode populations, again highlighting how their mobility is coupled to their differentiation processes.

Trigeminal and Epibranchial Placode Ingression and Migration

The trigeminal placodes originate lateral to the midbrain on both sides of the embryo in a speckly manner and contribute to both the ophthalmic and maxillomandibular lobes of the ganglion. The epibranchial placodes arise at caudal hindbrain level, from a territory surrounding the future otic placodes, and generate the geniculate, petrosal, and nodose ganglia (Fig. 2A; Baker and Bronner-Fraser 2001). In contrast to NC cells that delaminate as multipotent progenitors, the trigeminal and epibranchial placodes already adopt neuronal fates and express neuronal markers while still within the ectodermal epithelium (Fig. 5B) and exist as a mixture of mitotically active and terminally differentiated cells (D'amico-Martel 1982; Begbie 2002; McCabe et al. 2009). Fixed and live imaging of transverse tissue sections at the level of trigeminal and epibranchial placodes has revealed a number of interesting migratory features (Fig. 5B; Graham et al. 2007; Shiau et al. 2008, 2011). Both placode populations exit from a focally pseudo-stratified epithelium, locally breach the basement membrane, and migrate out as neuroblasts with round bodies and short neuronal processes, in stark contrast to the mesenchymal cell shape of migrating NC cells. During migration, placodes also exhibit ample cell–cell contacts and often organize as single chains of cells. Once they reach their destination, placode cells begin to coalesce and develop longer axonal protrusions. The above observations show that placodes migrate quite differently from NC cells, hinting at inherent differences in migratory behaviors between multipotent cells and fate-committed neuronal cells.

Despite these clear morphological changes, there remain two key questions that are only partially resolved on the molecular level: (1) what are the signaling events that direct placode ingression, and (2) what are the molecular shifts that need to occur for placode delamination. A number of studies have revealed a role for BMP (Begbie et al. 1999), Wnt (Canning et al. 2008), FGF (Lassiter et al. 2009), and Notch (Lassiter et al. 2010) signaling pathways in regulating neuronal differentiation in the trigeminal and epibranchial placodes (Fig. 5B). Interestingly, in all of these cases, gain or loss of functions of these factors concomitantly altered the extent of neurogenesis and cell delamination (Lassiter et al. 2014). Furthermore, directly manipulating the expression of neuronal fate specifiers within the placodes also disrupts their ingression (Fode et al. 1998; Bina et al. 2023). These data coupled with the observation that all migrating trigeminal and epibranchial placodes exhibit neuroblast morphology and express neuronal markers may suggest that neurogenesis is a prerequisite for placode delamination (Lassiter et al. 2014).

Even less is known about the molecular events that transition the trigeminal and epibranchial placodes from stationary residents of the epithelium to migrating cells. In the epibranchial placodes, molecular factors associated with NC EMT, such as RhoB and Snail2, are not expressed in placodes (Graham et al. 2007). Furthermore, cell-adhesion molecules such as glypican-1 and N-cadherin are found on ingressing trigeminal placodes, whereas they are absent or actively turned off in delaminating and migrating NC cells (Shiau and Bronner-Fraser 2009; Shiau et al. 2010). Together, these observations suggest that placode ingression appears to involve a set of molecular programs that are distinct from the well-studied NC.

Lateral Line Placode Migration

The lateral line placode originates from a domain surrounding the otic region of aquatic vertebrates and generates a set of mechanosensory and electrosensory organs along the body of the animal to guide them during swimming and for detecting predators and prey (Ghysen and Dambly-Chaudière 2007). Different species possess different lateral line populations, and up to six placodal populations are proposed to exist in the primitive lateral line of ancient gnathostomes (Northcutt 1997). The posterior lateral line of zebrafish is by far the best studied system (Piotrowski and Baker 2014) and thus our subject of discussion here.

In zebrafish, the posterior lateral line placode arises rostral to the otic placodes and generates both the posterior lateral line ganglion and a highly migratory primordium, which travels as a compact group of cells along the horizontal myoseptum (Fig. 2C) on each side of the embryo. As it migrates, the primordium deposits neuromasts, which contain hair cells that are morphologically similar to those of the ear and will later mature into mechanosensory organs (Gompel et al. 2001; Dalle Nogare and Chitnis 2017). Early transplantation experiments have shown that the placode primordium can migrate cell-autonomously (Stone 1928). Later studies have established a classical model involving the CXCR4/CXCL12 (Sdf1) axis to explain this collective and cell-autonomous migratory behavior (Fig. 5C; Piotrowski and Baker 2014; Dalle Nogare and Chitnis 2017). Briefly, the posterior lateral line primordium is divided into two compartments: a leading end that activates the Wnt/β-catenin signaling pathway and expresses CXCR4 and a trailing end that upregulates the FGF signaling pathway and expresses CXCR7 (Aman and Piotrowski 2008; Nechiporuk and Raible 2008). Both receptors bind to the chemokine CXCL12, which is expressed along the migration route of the primordium (David et al. 2002; Haas and Gilmour 2006; Valentin et al. 2007). However, the receptors react with the chemokine differently: CXCR7 competes with CXCR4 for CXCL12 binding but does not initiate downstream signaling events, thereby acting as a sink for the ligand, and generates an internal gradient along the primordium (Dambly-Chaudière et al. 2007; Venkiteswaran et al. 2013). This gradient promotes the autonomous migration of the posterior lateral line placode. It has been further hypothesized that migration of the primordium likely terminates via the oversaturation of its internal gradient by a high level of CXCL12 at the tail of the embryo, akin to the case for primordial germ cells (Doitsidou et al. 2002). Alternatively, other studies have postulated an autonomous mechanism where the primordium has an intrinsic specification regarding the number of neuromasts it can deposit (Stone 1937; Piotrowski and Baker 2014).

An important derivative of the migrating primordium is the mechanosensory neuromast. As the primordium travels, a portion of the trailing cells begin to undergo apical constriction and form rosette structures called the protoneuromast, which gradually loses speed and detaches from the more migratory primordium. The protoneuromast finally halts at stereotypical locations and further organizes and differentiates into sensory neuromast structures (Gompel et al. 2001; Venero Galanternik et al. 2016; Nogare et al. 2017). Development of the protoneuromasts is highly coupled to primordium migration (Fig. 5C). First, in the trailing end, high FGF expression "polarizes" the primordium to initiate rosette structure formation and activate the hair cell specifier Atoh1a and the Notch-Delta signaling pathway, thereby patterning a protoneuromast with a hair cell progenitor at the center and support cells in the periphery (Itoh and Chitnis 2001; Lecaudey et al. 2008; Nechiporuk and Raible 2008; Kozlovskaja-Gumbrienė et al. 2017). As the protoneuromast matures, the central hair cell progenitor starts to express FGF ligands and becomes a new source of signals, allowing it to reorganize the protoneuromast around itself. Furthermore, Atoh1a represses E-cadherin, which is hypothesized to allow the destabilization of the protoneuromast and its detachment from the migrating primordium (Matsuda and Chitnis 2010). Finally, the deposition of neuromasts occurs periodically at every fifth somite. A gradient of expression of the EMT factor Met occurs across the migrating primordium, while its ligand hepatocyte growth factor (HGF) is expressed at somite boundaries. Thus, the migrating primordium appears to undergo gradual de-epithelialization as it passes through successive somite boundaries and eventually deposits the trailing protoneuromast (Haines et al. 2004). Together, these observations suggest an intricate system where the internal "polarization" of the lateral line primordium not only drives its own migration but also organizes the specification of sensory cells and the deposition of maturing neuromast structures at spatially correct locations (Ghysen and Dambly-Chaudière 2007; Aman and Piotrowski 2011; Dalle Nogare and Chitnis 2017).

The posterior lateral line placode was also shown to generate two additional primordia (named primII and primD) during a second wave of development (Fig. 2C), only hours after the first primordium (primI) appears. Both new primordia arise from the same place as primI, and one of them, primII, follows the same migratory route as primI and deposits additional neuromasts along the way (Sapède et al. 2002; Nuñez et al. 2009; Sarrazin et al. 2010). Interestingly, a major difference between neuromasts laid by primI and primII is the orientation of their hair cells: primI-derived neuromasts all harbor hair bundles that are oriented parallel to the embryo's anteroposterior axis, whereas those derived from primII are perpendicular, or dorsoventral, in orientation (López-Schier et al. 2004). It was then shown that hair cell polarity is correlated with the migratory direction of the primordium when the neuromast was deposited. Indeed, mutations that randomized primordium migration trajectories resulted in scrambled hair bundle orientations. Thus, it was proposed that anteroposteriorly polarized hair cells arose because primI migrated and deposited neuromasts in a horizontal fashion. In contrast, neuromasts deposited by primII were affected by a global ventral migration of the entire lateral line system at the time primII was traveling (Fig. 2C), thereby resulting in dorsoventrally polarized hair cells (López-Schier et al. 2004). However, this scenario has been challenged, and more recent studies have suggested that the difference in hair bundle polarity of primI and primII derivatives might be due to intrinsic differences between either the hair cells or support cells of the two populations (Navajas Acedo et al. 2019; Kindt et al. 2021). It remains unclear how such intrinsic biases are imparted into each primordium to yield correlated migration trajectory and cell polarity. Regardless, the above example provides another interesting case where placode migration may possibly be linked to cellular differentiation.

INTERACTIONS BETWEEN MIGRATORY NEURAL CREST AND PLACODES MUTUALLY INFLUENCE THEIR FATES

In addition to giving rise to distinct structures of the cranial facial region, NC and placode cells also interact extensively to generate important components of the peripheral nervous system (Steventon et al. 2014). For example, the otic placodes are known to act as physical barriers that split NC streams (Baker and Bronner-Fraser 2001). Recent advances have shed light on additional ways in which NC–placode interactions not only influence migratory behaviors but also affect their respective development.

A number of studies have suggested an early role for NC cells in mediating placode segregation (Fleenor and Begbie 2014). For example, it was proposed that the lens placode represents a "ground state," and the remaining placodes arise through active repression of lens identity (Bailey et al. 2006). NC cells were shown to mediate this repression, as they migrate underneath the cranial epidermis but spare the lens domain due to physical blockage by the optic vesicle, thereby maintaining lens identity in the anterior while permitting the further specification of other placode populations (Bailey et al. 2006). In addition, migrating NC cells can physically separate placodes into discrete populations. Originating from a common precursor domain (Ladher et al. 2010), the individual epibranchial placodes are segmented by a chase-and-run interaction with the NC, mediated by both the CXCR4/CXCL12 signaling pathway and contact inhibition of locomotion (Theveneau et al. 2013).

Beyond early specification, NC and placode interactions also mutually regulate each other to allow proper patterning and formation of the cranial ganglia. The trigeminal ganglion represents a prime example (Fig. 5B), in which experimental ablation of either population severely disrupts gangliogenesis (Hamburger 1961; Shiau et al. 2008). When placode cells are absent, the NC cells become heavily dispersed and fail to condense into the bilobed structure of the trigeminal ganglion. This may suggest a role for trigeminal placodes in halting and retaining the NC cells as they migrate beneath the epidermis. Indeed, a number of molecules expressed by the placodes, including the cell adhesion protein N-cadherin (Shiau and Bronner-Fraser 2009), are required for the assembly of the condensed trigeminal ganglion. At the same time, placode delamination might be temporally controlled by the migrating NC cells: Migratory NC cells express the ligand Slit1, while the placodes express its receptor Robo2, and disruption of this signaling pathway partially blocks placode ingression (Shiau et al. 2008). A similar mechanism may also be at play in the epibranchial placodes, where the NC streams are shaped into "corridors" that ensheathe the ingressed placodes and act as substrate for their migration (Freter et al. 2013). Together, these observations support an intriguing model where migratory NC cells and placodes provide mutual attraction/retention cues to anchor each other at the correct location for proper structure formation.

A third example of NC–placode interactions takes place within the lateral line system. In addition to depositing neuromasts, the migrating posterior lateral line primordium also drops off interneuromast cells that are reserved as progenitors for the postembryonic development of so-called intercalary neuromasts (Gompel et al. 2001; Grant et al. 2005). What allows for the delayed differentiation of interneuromast cells to postembryonic stages? It was hypothesized that NC-derived glial cells that associate with the lateral line ganglia may interact with and inhibit the differentiation of intercalary neuromasts cells (Grant et al. 2005; López-Schier and Hudspeth 2005). However, such dormancy may be disrupted later when the interneuromast cells undergo an overall ventral migration, thereby removing the inhibitory contacts (Ghysen and Dambly-Chaudière 2007).

PERSPECTIVE/CONCLUSION

Development requires the coordination of cell proliferation, differentiation, and spatial rearrangements. NC and ectodermal placodes are two transient embryonic cell populations that undergo all these processes, rendering them great models for not only understanding aspects of their individual development but also how interactions between them guide development of the other population. In the present review, we highlight the idea that cellular motility is not just a consequence of cellular identity, but rather actively engages in the shaping of cell fate. Unlike stationary cells that can only interact via local cues, migration affords cells the ability to leave their place of birth, explore diverse external environments, and adjust their fates accordingly, thereby greatly expanding the repertoire of possible fates beyond intrinsic biases and prepatterning. In some sense, migration is nature's own transplantation experiment. Still, our understanding of the intertwined relationship between cell migration and fate decision remains far from complete. With the advent of new molecular tools, including single-cell and spatial transcriptomics, genetic manipulations, and advanced live imaging, this area of research is a ripe arena for future investigations at the cellular and molecular level.

COMPETING INTEREST STATEMENT

The authors declare no competing interests.

ACKNOWLEDGMENTS

This work was partially supported by National Institutes of Health (NIH) R01DE027538 and NIH R01DE024157 to M.E.B.

REFERENCES

Aaku-Saraste E, Hellwig A, Huttner WB. 1996. Loss of occludin and functional tight junctions, but not ZO-1, during neural tube closure—remodeling of the neuroepithelium prior to neurogenesis. *Dev Biol* **180:** 664–679. doi:10.1006/dbio.1996.0336

Adams RH, Diella F, Hennig S, Helmbacher F, Deutsch U, Klein R. 2001. The cytoplasmic domain of the ligand ephrinB2 is required for vascular morphogenesis but not cranial neural crest migration. *Cell* **104:** 57–69. doi:10.1016/S0092-8674(01)00191-X

Alhashem Z, Camargo-Sosa K, Kelsh RN, Linker C. 2022a. Trunk neural crest migratory position and asymmetric division predict terminal differentiation. *Front Cell Dev Biol* **10:** 887393. doi:10.3389/fcell.2022.887393

Alhashem Z, Feldner-Busztin D, Revell C, Alvarez-Garcillan Portillo M, Camargo-Sosa K, Richardson J, Rocha M, Gauert A, Corbeaux T, Milanetto M, et al. 2022b. Notch controls the cell cycle to define leader versus follower identities during collective cell migration. *eLife* **11:** e73550. doi:10.7554/eLife.73550

Aman A, Piotrowski T. 2008. Wnt/β-catenin and Fgf signaling control collective cell migration by restricting chemokine receptor expression. *Dev Cell* **15:** 749–761. doi:10.1016/j.devcel.2008.10.002

Aman A, Piotrowski T. 2011. Cell-cell signaling interactions coordinate multiple cell behaviors that drive morphogenesis of the lateral line. *Cell Adh Migr* **5:** 499–508. doi:10.4161/cam.5.6.19113

Artinger KB, Bronner-Fraser M. 1992. Partial restriction in the developmental potential of late emigrating avian neural crest cells. *Dev Biol* **149:** 149–157. doi:10.1016/0012-1606(92)90271-H

Bailey AP, Bhattacharyya S, Bronner-Fraser M, Streit A. 2006. Lens specification is the ground state of all sensory placodes, from which FGF promotes olfactory identity. *Dev Cell* **11:** 505–517. doi:10.1016/j.devcel.2006.08.009

Baker CVH, Bronner-Fraser M. 1997. The origins of the neural crest. Part I: embryonic induction. *Mech Dev* **69:** 3–11. doi:10.1016/S0925-4773(97)00132-9

Baker CVH, Bronner-Fraser M. 2001. Vertebrate cranial placodes I. Embryonic induction. *Dev Biol* **232:** 1–61. doi:10.1006/dbio.2001.0156

Barriga EH, Maxwell PH, Reyes AE, Mayor R. 2013. The hypoxia factor Hif-1α controls neural crest chemotaxis and epithelial to mesenchymal transition. *J Cell Biol* **201:** 759–776. doi:10.1083/jcb.201212100

Barriga EH, Franze K, Charras G, Mayor R. 2018. Tissue stiffening coordinates morphogenesis by triggering collective cell migration in vivo. *Nature* **554:** 523–527. doi:10.1038/nature25742

Batlle E, Sancho E, Francí C, Domínguez D, Monfar M, Baulida J, García De Herreros A. 2000. The transcription factor snail is a repressor of E-cadherin gene expression in epithelial tumour cells. *Nat Cell Biol* **2:** 84–89. doi:10.1038/35000034

Batsché E, Crémisi C. 1999. Opposite transcriptional activity between the wild type c-*myc* gene coding for c-Myc1 and c-Myc2 proteins and c-Myc1 and c-Myc2 separately. *Oncogene* **18:** 5662–5671. doi:10.1038/sj.onc.1202927

Begbie J. 2002. Early steps in the production of sensory neurons by the neurogenic placodes. *Mol Cell Neurosci* **21:** 502–511. doi:10.1006/mcne.2002.1197

Begbie J, Brunet JF, Rubenstein JLR, Graham A. 1999. Induction of the epibranchial placodes. *Development* **126:** 895–902. doi:10.1242/dev.126.5.895

Bhat N, Riley BB. 2011. Integrin-α5 coordinates assembly of posterior cranial placodes in zebrafish and enhances Fgf-dependent regulation of otic/epibranchial cells. *PLoS One* **6:** e27778. doi:10.1371/journal.pone.0027778

Bhattacharyya S, Bailey AP, Bronner-Fraser M, Streit A. 2004. Segregation of lens and olfactory precursors from a common territory: cell sorting and reciprocity of Dlx5 and Pax6 expression. *Dev Biol* **271:** 403–414. doi:10.1016/j.ydbio.2004.04.010

Bina P, Hines MA, Sanyal J, Taneyhill LA. 2023. Neurogenin 2 and neuronal differentiation 1 control proper development of the chick trigeminal ganglion and its nerve branches. *J Dev Biol* **11:** 8. doi:10.3390/jdb11010008

Blentic A, Tandon P, Payton S, Walshe J, Carney T, Kelsh RN, Mason I, Graham A. 2008. The emergence of ectomesenchyme. *Dev Dyn* **237:** 592–601. doi:10.1002/dvdy.21439

Bolós V, Peinado H, Pérez-Moreno MA, Fraga MF, Esteller M, Cano A. 2003. The transcription factor Slug represses *E-cadherin* expression and induces epithelial to mesenchymal transitions: a comparison with Snail and E47 repressors. *J Cell Sci* **116:** 499–511. doi:10.1242/jcs.00224

Breau MA, Schneider-Maunoury S. 2014. Mechanisms of cranial placode assembly. *Int J Dev Biol* **58:** 9–19. doi:10.1387/ijdb.130351mb

Breau MA, Schneider-Maunoury S. 2015. Cranial placodes: models for exploring the multi-facets of cell adhesion in epithelial rearrangement, collective migration and neuronal movements. *Dev Biol* **401:** 25–36. doi:10.1016/j.ydbio.2014.12.012

Bronner ME. 2012. Formation and migration of neural crest cells in the vertebrate embryo. *Histochem Cell Biol* **138:** 179–186. doi:10.1007/s00418-012-0999-z

Bronner-Fraser M, Fraser SE. 1988. Cell lineage analysis reveals multipotency of some avian neural crest cells. *Nature* **335:** 161–164. doi:10.1038/335161a0

Canning CA, Lee L, Luo SX, Graham A, Jones CM. 2008. Neural tube derived Wnt signals cooperate with FGF signaling in the formation and differentiation of the trigeminal placodes. *Neural Dev* **3:** 35. doi:10.1186/1749-8104-3-35

Cano A, Pérez-Moreno MA, Rodrigo I, Locascio A, Blanco MJ, Del Barrio MG, Portillo F, Nieto MA. 2000. The transcription factor Snail controls epithelial–mesenchymal transitions by repressing E-cadherin expression. *Nat Cell Biol* **2:** 76–83. doi:10.1038/35000025

Carmona-Fontaine C, Matthews HK, Kuriyama S, Moreno M, Dunn GA, Parsons M, Stern CD, Mayor R. 2008. Contact inhibition of locomotion in vivo controls neural crest directional migration. *Nature* **456:** 957–961. doi:10.1038/nature07441

Carmona-Fontaine C, Theveneau E, Tzekou A, Tada M, Woods M, Page KM, Parsons M, Lambris JD, Mayor R. 2011. Complement fragment C3a controls mutual cell attraction during collective cell migration. *Dev Cell* **21:** 1026–1037. doi:10.1016/j.devcel.2011.10.012

Cheung M, Chaboissier MC, Mynett A, Hirst E, Schedl A, Briscoe J. 2005. The transcriptional control of trunk neural crest induction, survival, and delamination. *Dev Cell* **8:** 179–192. doi:10.1016/j.devcel.2004.12.010

Clouthier DE, Hosoda K, Richardson JA, Clay Williams S, Yanagisawa H, Kuwaki T, Kumada M, Hammer RE, Yanagisawa M. 1998. Cranial and cardiac neural crest defects in endothelin-A receptor-deficient mice. *Development* **125:** 813–824. doi:10.1242/dev.125.5.813

Coles EG, Taneyhill LA, Bronner-Fraser M. 2007. A critical role for cadherin-6B in regulating avian neural crest emigration. *Dev Biol* **312:** 533–544. doi:10.1016/j.ydbio.2007.09.056

Collazo A, Bronner-Fraser M, Fraser SE. 1993. Vital dye labelling of *Xenopus laevis* trunk neural crest reveals multipotency and novel pathways of migration. *Development* **118:** 363–376. doi:10.1242/dev.118.2.363

Dalle Nogare D, Chitnis AB. 2017. A framework for understanding morphogenesis and migration of the zebrafish posterior lateral line primordium. *Mech Dev* **148:** 69–78. doi:10.1016/j.mod.2017.04.005

Dambly-Chaudière C, Cubedo N, Ghysen A. 2007. Control of cell migration in the development of the posterior lateral line: antagonistic interactions between the chemokine receptors CXCR4 and CXCR7/RDC1. *BMC Dev Biol* **7:** 23. doi:10.1186/1471-213X-7-23

D'amico-Martel A. 1982. Temporal patterns of neurogenesis in avian cranial sensory and autonomic ganglia. *Am J Anat* **163:** 351–372. doi:10.1002/aja.1001630407

Dash S, Trainor PA. 2020. The development, patterning and evolution of neural crest cell differentiation into cartilage and bone. *Bone* **137:** 115409. doi:10.1016/j.bone.2020.115409

David NB, Sapède D, Saint-Etienne L, Thisse C, Thisse B, Dambly-Chaudière C, Rosa FM, Ghysen A. 2002. Molecular basis of cell migration in the fish lateral line: role of the chemokine receptor CXCR4 and of its ligand, SDF1. *Proc Natl Acad Sci* **99:** 16297–16302. doi:10.1073/pnas.252339399

Dawes JHP, Kelsh RN. 2021. Cell fate decisions in the neural crest, from pigment cell to neural development. *Int J Mol Sci* **22:** 13531. doi:10.3390/ijms222413531

Debby-Brafman A, Burstyn-Cohen T, Klar A, Kalcheim C. 1999. F-Spondin, expressed in somite regions avoided by neural crest cells, mediates inhibition of distinct somite domains to neural crest migration. *Neuron* **22:** 475–488. doi:10.1016/S0896-6273(00)80703-5

Depew MJ, Liu JK, Long JE, Presley R, Meneses JJ, Pedersen RA, Rubenstein JLR. 1999. *Dlx5* regulates regional development of the branchial arches and sensory capsules. *Development* **126:** 3831–3846. doi:10.1242/dev.126.17.3831

Doitsidou M, Reichman-Fried M, Stebler J, Köprunner M, Dörries J, Meyer D, Esguerra CV, Leung T, Raz E. 2002. Guidance of primordial germ cell migration by the chemokine SDF-1. *Cell* **111:** 647–659. doi:10.1016/S0092-8674(02)01135-2

Dorsky RI, Raible DW, Moon RT. 2000. Direct regulation of *nacre*, a zebrafish *MITF* homolog required for pigment cell formation, by the Wnt pathway. *Genes Dev* **14:** 158–162. doi:10.1101/gad.14.2.158

Dottori M, Gross MK, Labosky P, Goulding M. 2001. The winged-helix transcription factor Foxd3 suppresses interneuron differentiation and promotes neural crest cell fate. *Development* **128:** 4127–4138. doi:10.1242/dev.128.21.4127

Duband JL. 2010. Diversity in the molecular and cellular strategies of epithelium-to-mesenchyme transitions: insights from the neural crest. *Cell Adh Migr* **4:** 458–482. doi:10.4161/cam.4.3.12501

Edens BM, Bronner ME. 2025. Making sense of vertebrate senses from a neural crest and cranial placode evo-devo perspective. *Trends Neurosci* **48:** 213–226. doi:10.1016/j.tins.2024.12.008

Erickson AG, Kameneva P, Adameyko I. 2023. The transcriptional portraits of the neural crest at the individual cell level. *Semin Cell Dev Biol* **138:** 68–80. doi:10.1016/j.semcdb.2022.02.017

Fairchild CL, Gammill LS. 2013. Tetraspanin18 is a FoxD3-responsive antagonist of cranial neural crest epithelial to mesenchymal transition that maintains cadherin-6B protein. *J Cell Sci* **126:** 1464–1476. doi:10.1242/jcs.120915

Fleenor SJ, Begbie J. 2014. Neural crest cell and placode interactions in cranial PNS development. *Neural Crest Cells* **2014:** 153–165. doi:10.1016/B978-0-12-401730-6.00009-0

Fode C, Gradwohl G, Morin X, Dierich A, LeMeur M, Goridis C, Guillemot F. 1998. The bHLH protein NEUROGENIN 2 is a determination factor for epibranchial placode–derived sensory neurons. *Neuron* **20:** 483–494. doi:10.1016/S0896-6273(00)80989-7

Freter S, Fleenor SJ, Freter R, Liu KJ, Begbie J. 2013. Cranial neural crest cells form corridors prefiguring sensory neuroblast migration. *Development* **140:** 3595–3600. doi:10.1242/dev.091033

Galton R, Fejes-Toth K, Bronner ME. 2022. Co-option of the piRNA pathway to regulate neural crest specification. *Sci Adv* **8:** eabn1441. doi:10.1126/sciadv.abn1441

Gammill LS, Gonzalez C, Bronner-Fraser M. 2007. Neuropilin 2/semaphorin 3F signaling is essential for cranial neural crest migration and trigeminal ganglion condensation. *Dev Neurobiol* **67:** 47–56. doi:10.1002/dneu.20326

Gandhi S, Hutchins EJ, Maruszko K, Park JH, Thomson M, Bronner ME. 2020. Bimodal function of chromatin remodeler Hmga1 in neural crest induction and Wnt-dependent emigration. *eLife* **9:** e57779. doi:10.7554/eLife.57779

Ghysen A, Dambly-Chaudière C. 2007. The lateral line microcosmos. *Genes Dev* **21:** 2118–2130. doi:10.1101/gad.1568407

Gompel N, Cubedo N, Thisse C, Thisse B, Dambly-Chaudière C, Ghysen A. 2001. Pattern formation in the lateral line of zebrafish. *Mech Dev* **105:** 69–77. doi:10.1016/S0925-4773(01)00382-3

Goodnough LH, Chang AT, Treloar C, Yang J, Scacheri PC, Atit RP. 2012. Twist1 mediates repression of chondrogenesis by β-catenin to promote cranial bone progenitor specification. *Development* **139:** 4428–4438. doi:10.1242/dev.081679

Goodnough LH, DiNuoscio GJ, Ferguson JW, Williams T, Lang RA, Atit RP. 2014. Distinct requirements for cranial ectoderm and mesenchyme-derived Wnts in specification and differentiation of osteoblast and dermal progenitors. *PLoS Genet* **10:** e1004152. doi:10.1371/journal.pgen.1004152

Graham A, Blentic A, Duque S, Begbie J. 2007. Delamination of cells from neurogenic placodes does not involve an epithelial-to-mesenchymal transition. *Development* **134:** 4141–4145. doi:10.1242/dev.02886

Grammatopoulos GA, Bell E, Toole L, Lumsden A, Tucker AS. 2000. Homeotic transformation of branchial arch identity after *Hoxa2* overexpression. *Development* **127:** 5355–5365. doi:10.1242/dev.127.24.5355

Grant KA, Raible DW, Piotrowski T. 2005. Regulation of latent sensory hair cell precursors by glia in the zebrafish lateral line. *Neuron* **45:** 69–80. doi:10.1016/j.neuron.2004.12.020

Greenhill ER, Rocco A, Vibert L, Nikaido M, Kelsh RN. 2011. An iterative genetic and dynamical modelling approach identifies novel features of the gene regulatory network underlying melanocyte development. *PLoS Genet* **7:** e1002265. doi:10.1371/journal.pgen.1002265

Gurdon JB. 1988. A community effect in animal development. *Nature* **336:** 772–774. doi:10.1038/336772a0

Haas P, Gilmour D. 2006. Chemokine signaling mediates self-organizing tissue migration in the zebrafish lateral line. *Dev Cell* **10:** 673–680. doi:10.1016/j.devcel.2006.02.019

Hadeball B, Borchers A, Wedlich D. 1998. *Xenopus* cadherin-11 (Xcadherin-11) expression requires the Wg/Wnt signal. *Mech Dev* **72:** 101–113. doi:10.1016/S0925-4773(98)00022-7

Haines L, Neyt C, Gautier P, Keenan DG, Bryson-Richardson RJ, Hollway GE, Cole NJ, Currie PD. 2004. Met and Hgf signaling controls hypaxial muscle and lateral line development in the zebrafish. *Development* **131:** 4857–4869. doi:10.1242/dev.01374

Häkkinen HM, Villaseca S, De La Burgade M, Alhashem Z, Scarpa E. 2025. Collective migration in complex environments: the case of the trunk neural crest. *Curr Top Dev Biol* **164:** 109–151. doi:10.1016/bs.ctdb.2025.01.004

Haldin CE, LaBonne C. 2010. Soxe factors as multifunctional neural crest regulatory factors. *Int J Biochem Cell Biol* **42:** 441–444. doi:10.1016/j.biocel.2009.11.014

Hamburger V. 1961. Experimental analysis of the dual origin of the trigeminal ganglion in the chick embryo. *J Exp Zool* **148:** 91–123. doi:10.1002/jez.1401480202

Harris ML, Hall R, Erickson CA. 2008. Directing pathfinding along the dorsolateral path—the role of EDNRB2 and EphB2 in overcoming inhibition. *Development* **135:** 4113–4122. doi:10.1242/dev.023119

Hartmann C. 2006. A Wnt canon orchestrating osteoblastogenesis. *Trends Cell Biol* **16:** 151–158. doi:10.1016/j.tcb.2006.01.001

Honjo Y, Kniss J, Eisen JS. 2008. Neuregulin-mediated ErbB3 signaling is required for formation of zebrafish dorsal root ganglion neurons. *Development* **135:** 2615–2625. doi:10.1242/dev.022178

Hu N, Strobl-Mazzulla PH, Simoes-Costa M, Sánchez-Vásquez E, Bronner ME. 2014. DNA methyltransferase 3B regulates duration of neural crest production via repression of *Sox10*. *Proc Natl Acad Sci* **111:** 17911–17916. doi:10.1073/pnas.1318408111

Hunt P, Krumlauf R. 1991. Deciphering the Hox code: clues to patterning branchial regions of the head. *Cell* **66:** 1075–1078. doi:10.1016/0092-8674(91)90029-X

Hunt P, Gulisano M, Cook M, Sham M-H, Faiella A, Wilkinson D, Boncinelli E, Krumlauf R. 1991. A distinct Hox code for the branchial region of the vertebrate head. *Nature* **353:** 861–864. doi:10.1038/353861a0

Ikenouchi J, Matsuda M, Furuse M, Tsukita S. 2003. Regulation of tight junctions during the epithelium-mesenchyme transition: direct repression of the gene expression of claudins/occludin by Snail. *J Cell Sci* **116:** 1959–1967. doi:10.1242/jcs.00389

Itoh M, Chitnis AB. 2001. Expression of proneural and neurogenic genes in the zebrafish lateral line primordium correlates with selection of hair cell fate in neuromasts. *Mech Dev* **102:** 263–266. doi:10.1016/S0925-4773(01)00308-2

Jia L, Cheng L, Raper J. 2005. Slit/Robo signaling is necessary to confine early neural crest cells to the ventral migratory pathway in the trunk. *Dev Biol* **282:** 411–421. doi:10.1016/j.ydbio.2005.03.021

John N, Cinelli P, Wegner M, Sommer L. 2011. Transforming growth factor β-mediated Sox10 suppression controls mesenchymal progenitor generation in neural crest stem cells. *Stem Cells* **29:** 689–699. doi:10.1002/stem.607

Kawasaki-Nishihara A, Nishihara D, Nakamura H, Yamamoto H. 2011. ET3/Ednrb2 signaling is critically involved in regulating melanophore migration in *Xenopus*. *Dev Dyn* **240:** 1454–1466. doi:10.1002/dvdy.22649

Kerosuo L, Bronner-Fraser M. 2012. What is bad in cancer is good in the embryo: importance of EMT in neural crest development. *Semin Cell Dev Biol* **23:** 320–332. doi:10.1016/j.semcdb.2012.03.010

Kesavan G, Machate A, Hans S, Brand M. 2020. Cell-fate plasticity, adhesion and cell sorting complementarily establish a sharp midbrain–hindbrain boundary. *Development* **147:** dev186882. doi:10.1242/dev.186882

Kindt KS, Akturk A, Jarysta A, Day M, Beirl A, Flonard M, Tarchini B. 2021. EMX2-GPR156-Gαi reverses hair cell orientation in mechanosensory epithelia. *Nat Commun* **12:** 2861. doi:10.1038/s41467-021-22997-1

Kozlovskaja-Gumbrienė A, Yi R, Alexander R, Aman A, Jiskra R, Nagelberg D, Knaut H, McClain M, Piotrowski T. 2017. Proliferation-independent regulation of organ size by Fgf/Notch signaling. *eLife* **6:** e21049. doi:10.7554/eLife.21049

Kozlowski DJ, Murakami T, Ho RK, Weinberg ES. 1997. Regional cell movement and tissue patterning in the zebrafish embryo revealed by fate mapping with caged fluorescein. *Biochem Cell Biol* **75:** 551–562.

Kurihara Y, Kurihara H, Suzukf H, Kodama T, Maemura K, Nagaf R, Oda H, Kuwaki T, Cao H, Kamadat N, et al. 1994. Elevated blood pressure and craniofacial abnormalities in mice deficient in endothelin-l. *Nature* **368:** 703–710. doi:10.1038/368703a0

Kuriyama S, Mayor R. 2008. Molecular analysis of neural crest migration. *Phil Trans R Soc B* **363:** 1349–1362. doi:10.1098/rstb.2007.2252

Ladher RK, O'Neill P, Begbie J. 2010. From shared lineage to distinct functions: the development of the inner ear and epibranchial placodes. *Development* **137:** 1777–1785. doi:10.1242/dev.040055

Lassiter RNT, Reynolds SB, Marin KD, Mayo TF, Stark MR. 2009. FGF signaling is essential for ophthalmic trigeminal placode cell delamination and differentiation. *Dev Dyn* **238:** 1073–1082. doi:10.1002/dvdy.21949

Lassiter RNT, Ball MK, Adams JS, Wright BT, Stark MR. 2010. Sensory neuron differentiation is regulated by notch signaling in the trigeminal placode. *Dev Biol* **344:** 836–848. doi:10.1016/j.ydbio.2010.05.514

Lassiter RNT, Stark MR, Zhao T, Zhou CJ. 2014. Signaling mechanisms controlling cranial placode neurogenesis and delamination. *Dev Biol* **389:** 39–49. doi:10.1016/j.ydbio.2013.11.025

Leathers TA, Rogers CD. 2022. Time to go: neural crest cell epithelial-to-mesenchymal transition. *Development* **149:** dev200712. doi:10.1242/dev.200712

Lecaudey V, Cakan-Akdogan G, Norton WHJ, Gilmour D. 2008. Dynamic Fgf signaling couples morphogenesis and migration in the zebrafish lateral line primordium. *Development* **135:** 2695–2705. doi:10.1242/dev.025981

Lee H, Ismail T, Kim C, Kim Y, Park J, Kwon O, Kang B, Lee D, Kwon T, Park T, et al. 2019. Lysine demethylase 3a in craniofacial and neural development during *Xenopus* embryogenesis. *Int J Mol Med* **43:** 1105–1113. doi:10.3892/ijmm.2018.402

Linker C, Bronner-Fraser M, Mayor R. 2000. Relationship between gene expression domains of Xsnail, Xslug, and Xtwist and cell movement in the prospective neural crest of *Xenopus*. *Dev Biol* **224:** 215–225. doi:10.1006/dbio.2000.9723

Liu JAJ, Wu MH, Yan CH, Chau BKH, So H, Ng A, Chan A, Cheah KSE, Briscoe J, Cheung M. 2013. Phosphorylation of *Sox9* is required for neural crest delamination and is regulated downstream of BMP and canonical Wnt signaling. *Proc Natl Acad Sci* **110:** 2882–2887. doi:10.1073/pnas.1211747110

López-Schier H, Hudspeth AJ. 2005. Supernumerary neuromasts in the posterior lateral line of zebrafish lacking peripheral glia. *Proc Natl Acad Sci* **102:** 1496–1501. doi:10.1073/pnas.0409361102

López-Schier H, Starr CJ, Kappler JA, Kollmar R, Hudspeth AJ. 2004. Directional cell migration establishes the axes of planar polarity in the posterior lateral-line organ of the zebrafish. *Dev Cell* **7:** 401–412. doi:10.1016/j.devcel.2004.07.018

Martik ML, Bronner ME. 2017. Regulatory logic underlying diversification of the neural crest. *Trends Genet* **33:** 715–727. doi:10.1016/j.tig.2017.07.015

Mathavan K, Khedgikar V, Bartolo V, Alfandari D. 2017. The ectodomain of cadherin-11 binds to erbB2 and stimulates Akt phosphorylation to promote cranial neural crest cell migration. *PLoS One* **12:** e0188963. doi:10.1371/journal.pone.0188963

Matsuda M, Chitnis AB. 2010. Atoh1a expression must be restricted by Notch signaling for effective morphogenesis of the posterior lateral line primordium in zebrafish. *Development* **137:** 3477–3487. doi:10.1242/dev.052761

Mayor R, Etienne-Manneville S. 2016. The front and rear of collective cell migration. *Nat Rev Mol Cell Biol* **17:** 97–109. doi:10.1038/nrm.2015.14

McCabe KL, Sechrist JW, Bronner-Fraser M. 2009. Birth of ophthalmic trigeminal neurons initiates early in the placodal ectoderm. *J Comp Neurol* **514:** 161–173. doi:10.1002/cne.22004

McCarroll MN, Lewis ZR, Culbertson MD, Martin BL, Kimelman D, Nechiporuk AV. 2012. Graded levels of Pax2a and Pax8 regulate cell differentiation during sensory placode formation. *Development* **139:** 2740–2750. doi:10.1242/dev.076075

Minoux M, Rijli FM. 2010. Molecular mechanisms of cranial neural crest cell migration and patterning in craniofacial development. *Development* **137:** 2605–2621. doi:10.1242/dev.040048

Minoux M, Antonarakis GS, Kmita M, Duboule D, Rijli FM. 2009. Rostral and caudal pharyngeal arches share a common neural crest ground pattern. *Development* **136:** 637–645. doi:10.1242/dev.028621

Nakagawa S, Takeichi M. 1995. Neural crest cell–cell adhesion controlled by sequential and subpopulation-specific expression of novel cadherins. *Development* **121:** 1321–1332. doi:10.1242/dev.121.5.1321

Navajas Acedo J, Voas MG, Alexander R, Woolley T, Unruh JR, Li H, Moens C, Piotrowski T. 2019. PCP and Wnt pathway components act in parallel during zebrafish mechanosensory hair cell orientation. *Nat Commun* **10:** 3993. doi:10.1038/s41467-019-12005-y

Nechiporuk A, Raible DW. 2008. FGF-Dependent mechanosensory organ patterning in zebrafish. *Science* **320:** 1774–1777. doi:10.1126/science.1156547

Nogare DD, Nikaido M, Somers K, Head J, Piotrowski T, Chitnis AB. 2017. In toto imaging of the migrating Zebrafish lateral line primordium at single cell resolution. *Dev Biol* **422:** 14–23. doi:10.1016/j.ydbio.2016.12.015

Northcutt RG. 1997. Evolution of gnathostome lateral line ontogenies. *Brain Behav Evol* **50:** 25–37. doi:10.1159/000113319

Nuñez VA, Sarrazin AF, Cubedo N, Allende ML, Dambly-Chaudière C, Ghysen A. 2009. Postembryonic development of the posterior lateral line in the zebrafish. *Evol Dev* **11:** 391–404. doi:10.1111/j.1525-142X.2009.00346.x

Oakley RA, Lasky CJ, Erickson CA, Tosney KW. 1994. Glycoconjugates mark a transient barrier to neural crest migration in the chicken embryo. *Development* **120:** 103–114. doi:10.1242/dev.120.1.103

Ozeki H, Kurihara Y, Tonami K, Watatani S, Kurihara H. 2004. Endothelin-1 regulates the dorsoventral branchial arch patterning in mice. *Mech Dev* **121:** 387–395. doi:10.1016/j.mod.2004.02.002

Pattyn A, Morin X, Cremer H, Goridis C, Brunet JF. 1999. The homeobox gene Phox2b is essential for the development of autonomic neural crest derivatives. *Nature* **399:** 366–370. doi:10.1038/20700

Piacentino ML, Hutchins EJ, Andrews CJ, Bronner ME. 2022. Temporal changes in plasma membrane lipid content induce endocytosis to regulate developmental epithelial-to-mesenchymal transition. *Proc Natl Acad Sci* **119:** e2212879119. doi:10.1073/pnas.2212879119

Pieper M, Eagleson GW, Wosniok W, Schlosser G. 2011. Origin and segregation of cranial placodes in *Xenopus laevis*. *Dev Biol* **360:** 257–275. doi:10.1016/j.ydbio.2011.09.024

Piotrowski T, Baker CVH. 2014. The development of lateral line placodes: taking a broader view. *Dev Biol* **389:** 68–81. doi:10.1016/j.ydbio.2014.02.016

Raible DW, Eisen JS. 1994. Restriction of neural crest cell fate in the trunk of the embryonic zebrafish. *Development* **120:** 495–503. doi:10.1242/dev.120.3.495

Raible DW, Eisen JS. 1996. Regulative interactions in zebrafish neural crest. *Development* **122:** 501–507. doi:10.1242/dev.122.2.501

Richardson J, Gauert A, Briones Montecinos L, Fanlo L, Alhashem ZM, Assar R, Marti E, Kabla A, Härtel S, Linker C. 2016. Leader cells define directionality of trunk, but not cranial, neural crest cell migration. *Cell Rep* **15:** 2076–2088. doi:10.1016/j.celrep.2016.04.067

Rijli FM, Mark M, Lakkaraju S, Dierich A, Dollé P, Chambon P. 1993. A homeotic transformation is generated in the rostral branchial region of the head by disruption of Hoxa-2, which acts as a selector gene. *Cell* **75:** 1333–1349. doi:10.1016/0092-8674(93)90620-6

Roellig D, Tan-Cabugao J, Esaian S, Bronner ME. 2017. Dynamic transcriptional signature and cell fate analysis reveals plasticity of individual neural plate border cells. *eLife* **6:** e21620. doi:10.7554/eLife.21620

Rothstein M, Bhattacharya D, Simoes-Costa M. 2018. The molecular basis of neural crest axial identity. *Dev Biol* **444:** S170–S180. doi:10.1016/j.ydbio.2018.07.026

Ruffins S, Bruk Artinger K, Bronner-Fraser M. 1998. Early migrating neural crest cells can form ventral neural tube derivatives when challenged by transplantation. *Dev Biol* **203:** 295–304. doi:10.1006/dbio.1998.8973

Saint-Jeannet JP, Moody SA. 2014. Establishing the pre-placodal region and breaking it into placodes with distinct identities. *Dev Biol* **389:** 13–27. doi:10.1016/j.ydbio.2014.02.011

Sánchez-Vásquez E, Bronner ME, Strobl-Mazzulla PH. 2019. Epigenetic inactivation of miR-203 as a key step in neural crest epithelial-to-mesenchymal transition. *Development* **146:** dev171017. doi:10.1242/dev.171017

Santagati F, Rijli FM. 2003. Cranial neural crest and the building of the vertebrate head. *Nat Rev Neurosci* **4:** 806–818. doi:10.1038/nrn1221

Santiago A, Erickson CA. 2002. Ephrin-B ligands play a dual role in the control of neural crest cell migration. *Development* **129:** 3621–3632. doi:10.1242/dev.129.15.3621

Sapède D, Gompel N, Dambly-Chaudière C, Ghysen A. 2002. Cell migration in the postembryonic development of the fish lateral line. *Development* **129:** 605–615. doi:10.1242/dev.129.3.605

Sarrazin AF, Nunez VA, Sapede D, Tassin V, Dambly-Chaudiere C, Ghysen A. 2010. Origin and early development of the posterior lateral line system of zebrafish. *J Neurosci* **30:** 8234–8244. doi:10.1523/JNEUROSCI.5137-09.2010

Sato T, Kurihara Y, Asai R, Kawamura Y, Tonami K, Uchijima Y, Heude E, Ekker M, Levi G, Kurihara H. 2008. An endothelin-1 switch specifies maxillomandibular identity. *Proc Natl Acad Sci* **105:** 18806–18811. doi:10.1073/pnas.0807345105

Sauka-Spengler T, Bronner-Fraser M. 2008. A gene regulatory network orchestrates neural crest formation. *Nat Rev Mol Cell Biol* **9:** 557–568. doi:10.1038/nrm2428

Schilling TF, Prince V, Ingham PW. 2001. Plasticity in zebrafish hox expression in the hindbrain and cranial neural crest. *Dev Biol* **231:** 201–216. doi:10.1006/dbio.2000.9997

Schlosser G. 2006. Induction and specification of cranial placodes. *Dev Biol* **294:** 303–351. doi:10.1016/j.ydbio.2006.03.009

Schock EN, LaBonne C. 2020. Sorting Sox: diverse roles for Sox transcription factors during neural crest and craniofacial development. *Front Physiol* **11:** 606889. doi:10.3389/fphys.2020.606889

Schwarz Q, Vieira JM, Howard B, Eickholt BJ, Ruhrberg C. 2008. Neuropilin 1 and 2 control cranial gangliogenesis and axon guidance through neural crest cells. *Development* **135:** 1605–1613. doi:10.1242/dev.015412

Selleck MAJ, Bronner-Fraser M. 1995. Origins of the avian neural crest: the role of neural plate-epidermal interactions. *Development* **121:** 525–538. doi:10.1242/dev.121.2.525

Serbedzija GN, Bronner-Fraser M, Fraser SE. 1994. Developmental potential of trunk neural crest cells in the mouse. *Development* **120:** 1709–1718. doi:10.1242/dev.120.7.1709

Shiau CE, Bronner-Fraser M. 2009. N-cadherin acts in concert with Slit1-Robo2 signaling in regulating aggregation of placode-derived cranial sensory neurons. *Development* **136:** 4155–4164. doi:10.1242/dev.034355

Shiau CE, Lwigale PY, Das RM, Wilson SA, Bronner-Fraser M. 2008. Robo2-Slit1 dependent cell–cell interactions mediate assembly of the trigeminal ganglion. *Nat Neurosci* **11:** 269–276. doi:10.1038/nn2051

Shiau CE, Hu N, Bronner-Fraser M. 2010. Altering Glypican-1 levels modulates canonical Wnt signaling during trigeminal placode development. *Dev Biol* **348:** 107–118. doi:10.1016/j.ydbio.2010.09.017

Shiau CE, Das RM, Storey KG. 2011. An effective assay for high cellular resolution time-lapse imaging of sensory placode formation and morphogenesis. *BMC Neurosci* **12:** 37. doi:10.1186/1471-2202-12-37

Shimizu M, Narboux-Nême N, Gitton Y, De Lombares C, Fontaine A, Alfama G, Kitazawa T, Kawamura Y, Heude E, Marshall L, et al. 2018. Probing the origin of matching functional jaws: roles of Dlx5/6 in cranial neural crest cells. *Sci Rep* **8:** 14975. doi:10.1038/s41598-018-33207-2

Smith A, Robinson V, Patel K, Wilkinson DG. 1997. The EphA4 and EphB1 receptor tyrosine kinases and ephrin-B2 ligand regulate targeted migration of branchial neural crest cells. *Curr Biol* **7:** 561–570. doi:10.1016/S0960-9822(06)00255-7

Steingrímsson E, Copeland NG, Jenkins NA. 2004. Melanocytes and the *Microphthalmia* transcription factor network. *Annu Rev Genet* **38:** 365–411. doi:10.1146/annurev.genet.38.072902.092717

Steventon B, Mayor R, Streit A. 2012. Mutual repression between Gbx2 and Otx2 in sensory placodes reveals a general mechanism for ectodermal patterning. *Dev Biol* **367:** 55–65. doi:10.1016/j.ydbio.2012.04.025

Steventon B, Mayor R, Streit A. 2014. Neural crest and placode interaction during the development of the cranial sensory system. *Dev Biol* **389:** 28–38. doi:10.1016/j.ydbio.2014.01.021

Steventon B, Mayor R, Streit A. 2016. Directional cell movements downstream of Gbx2 and Otx2 control the assembly of sensory placodes. *Biol Open* **5:** 1620–1624. doi:10.1242/bio.020966

Stone LS. 1928. Experiments on the transplantation of placodes of the cranial ganglia in the amphibian embryo. III: Preauditory and postauditory placodal materials interchanged. *J Comp Neurol* **47:** 117–154. doi:10.1002/cne.900470107

Stone LS. 1937. Further experimental studies of the development of lateral-line sense organs in amphibians observed in living preparations. *J Comp Neurol* **68:** 83–115. doi:10.1002/cne.900680105

Streit A. 2002. Extensive cell movements accompany formation of the otic placode. *Dev Biol* **249:** 237–254. doi:10.1006/dbio.2002.0739

Streit A. 2008. The cranial sensory nervous system: specification of sensory progenitors and placodes. In *StemBook*. Harvard Stem Cell Institute, Cambridge. https://www.ncbi.nlm.nih.gov/books/NBK27038/. doi: 10.3824/stembook.1.31.1

Streit A. 2025. The neural plate border: multipotent progenitors or cells of mixed identity? *Dev Biol* **523:** 51–58. doi:10.1016/j.ydbio.2025.04.002

Stundl J, Bronner ME. 2025. Tempting fate: neural crest induction along the body axis. *Cells Dev* **2025:** 204000. doi:10.1016/j.cdev.2025.204000

Szabó A, Mayor R. 2018. Mechanisms of neural crest migration. *Annu Rev Genet* **52:** 43–63. doi:10.1146/annurev-genet-120417-031559

Taneyhill LA, Coles EG, Bronner-Fraser M. 2007. Snail2 directly represses cadherin-6B during epithelial-to-mesenchymal transitions of the neural crest. *Development* **134:** 1481–1490. doi:10.1242/dev.02834

Teddy JM, Kulesa PM. 2004. In vivo evidence for short- and long-range cell communication in cranial neural crest cells. *Development* **131:** 6141–6151. doi:10.1242/dev.01534

Thawani A, Groves AK. 2020. Building the border: development of the chordate neural plate border region and its derivatives. *Front Physiol* **11:** 608880. doi:10.3389/fphys.2020.608880

Theveneau E, Mayor R. 2012. Neural crest delamination and migration: from epithelium-to-mesenchyme transition to collective cell migration. *Dev Biol* **366:** 34–54. doi:10.1016/j.ydbio.2011.12.041

Theveneau E, Steventon B, Scarpa E, Garcia S, Trepat X, Streit A, Mayor R. 2013. Chase-and-run between adjacent cell populations promotes directional collective migration. *Nat Cell Biol* **15:** 763–772. doi:10.1038/ncb2772

Thiery A, Buzzi AL, Streit A. 2020. Cell fate decisions during the development of the peripheral nervous system in the vertebrate head. *Curr Top Dev Biol* **139:** 127–167. doi:10.1016/bs.ctdb.2020.04.002

Thiery AP, Buzzi AL, Hamrud E, Cheshire C, Luscombe NM, Briscoe J, Streit A. 2023. scRNA-sequencing in chick suggests a probabilistic model for cell fate allocation at the neural plate border. *eLife* **12:** e82717. doi:10.7554/eLife.82717

Thomas LA, Yamada KM. 1992. Contact stimulation of cell migration. *J Cell Sci* **103:** 1211–1214. doi:10.1242/jcs.103.4.1211

Trainor P, Krumlauf R. 2000. Plasticity in mouse neural crest cells reveals a new patterning role for cranial mesoderm. *Nat Cell Biol* **2:** 96–102. doi:10.1038/35000051

Valentin G, Haas P, Gilmour D. 2007. The chemokine SDF1a coordinates tissue migration through the spatially restricted activation of Cxcr7 and Cxcr4b. *Curr Biol* **17:** 1026–1031. doi:10.1016/j.cub.2007.05.020

Vega-Lopez GA, Cerrizuela S, Aybar MJ. 2017. Trunk neural crest cells: formation, migration and beyond. *Int J Dev Biol* **61:** 5–15. doi:10.1387/ijdb.160408gv

Venero Galanternik M, Navajas Acedo J, Romero-Carvajal A, Piotrowski T. 2016. Imaging collective cell migration and hair cell regeneration in the sensory lateral line. *Methods Cell Biol* **134:** 211–256. doi:10.1016/bs.mcb.2016.01.004

Venkiteswaran G, Lewellis SW, Wang J, Reynolds E, Nicholson C, Knaut H. 2013. Generation and dynamics of an endogenous, self-generated signaling gradient across a migrating tissue. *Cell* **155:** 674–687. doi:10.1016/j.cell.2013.09.046

Wang X, Veruki ML, Bukoreshtliev NV, Hartveit E, Gerdes HH. 2010. Animal cells connected by nanotubes can be electrically coupled through interposed gap-junction channels. *Proc Natl Acad Sci* **107:** 17194–17199. doi:10.1073/pnas.1006785107

Whitlock KE, Westerfield M. 2000. The olfactory placodes of the zebrafish form by convergence of cellular fields at the edge of the neural plate. *Development* **127:** 3645–3653. doi:10.1242/dev.127.17.3645

Xu X, Li WEI, Huang GY, Meyer R, Chen T, Luo Y, Thomas MP, Radice GL, Lo CW. 2001. Modulation of mouse neural crest cell motility by N-cadherin and connexin 43 gap junctions. *J Cell Biol* **154:** 217–230. doi:10.1083/jcb.200105047

Xu X, Francis R, Wei CJ, Linask KL, Lo CW. 2006. Connexin 43-mediated modulation of polarized cell movement and the directional migration of cardiac neural crest cells. *Development* **133:** 3629–3639. doi:10.1242/dev.02543

Xu H, Dude CM, Baker CVH. 2008. Fine-grained fate maps for the ophthalmic and maxillomandibular trigeminal placodes in the chick embryo. *Dev Biol* **317:** 174–186. doi:10.1016/j.ydbio.2008.02.012

Illuminating Mammalian Cell Migration in Native Tissues at Subcellular and Molecular Resolution

Bhagawat C. Subramanian,[1] Marco Heydecker,[1] Yeap Ng,[1] Nicolas Melis,[1,2] and Roberto Weigert[1]

[1]Laboratory of Cellular and Molecular Biology, Center for Cancer Research, National Cancer Institute, Bethesda, Maryland 20814, USA

[2]Université de Poitiers, Poitiers, Nouvelle-Aquitaine 86000, France

Correspondence: weigertr@mail.nih.gov; subramanianb@nih.gov

Although cell migration has been extensively investigated using in vitro model systems, the mechanisms underlying mammalian cell migration in native tissue environments remain underexplored. Moreover, efforts to directly manipulate and visualize molecular regulators in live mammalian tissues have been scarce. In this article, we first review the current insights into various single-cell migration phenomena, including stem cell types, observed in mammalian tissues under homeostatic and pathophysiological conditions. Thereafter, we discuss intravital subcellular microscopy (ISMic) as a tool to unravel membrane remodeling mechanisms underlying cell migration in live animal tissues. Lastly, we emphasize the need for innovative microscopy and complementary advanced approaches to achieve a deeper fundamental understanding of cell migration modalities and their impact on mammalian tissue in homeostasis and pathophysiology.

Cell migration is a fundamental and complex process observed across phyla that is essential for development, growth, survival, and propagation of multicellular organisms. Cells integrate diverse biochemical and biophysical cues from their tissue microenvironment to migrate toward or away from these signals. In mammals, such migratory responses are crucial for physiological homeostasis and pathological tissue responses (SenGupta et al. 2021). At the cellular level, cues from the immediate microenvironment are sensed via cognate receptors on the plasma membrane, which transmit downstream signals leading to the organization of asymmetric force-generating machineries based on the actin cytoskeleton (Devreotes and Horwitz 2015; SenGupta et al. 2021). Such force-generating cytoskeletal machineries in turn remodel cellular membrane and drive subcellular dynamics (Ebrahim and Weigert 2019; Chakrabarti et al. 2021; Quintanilla et al. 2023; De Belly and Weiner 2024). Cell migration therefore requires asymmetric remodeling of the plasma membrane by force-generating cytoskeletal machineries (collectively referred to as "membrane remodeling mechanisms" hereafter), which couple protrusive leading edge with the contractile rear of a cell (Stock and Pauli 2021; De Belly and Weiner 2024).

Cite this article as *Cold Spring Harb Perspect Biol* doi: 10.1101/cshperspect.a041742

The cell migration phenomenon of "diapedesis" was first documented in frogs by Elie Metchnikoff in 1891 using a simple light microscope, making it an early example of intravital microscopy (IVM) (Metchnikoff 1893; Secklehner et al. 2017). Since then, various modes and forms of mammalian cell migration have been studied in high spatiotemporal and molecular detail, including mesenchymal and amoeboid modes during random and directed (chemo-, hapto-, duro-, topo-, galvanotaxis, etc.) migration. To this end, cells plated on a coated surface (two-dimensional [2D]) or in 3D environments, and cell-free molecular assemblies have been widely used to uncover how membrane remodeling mechanisms control cell migration response (Petrie et al. 2009; Bugyi and Carlier 2010; SenGupta et al. 2021; De Belly and Weiner 2024). While these simplified model systems continue to provide fundamental insights into individual cell migration modalities in vitro, mammalian cells exhibit remarkable plasticity in adopting multiple modes of migration in their complex native tissue microenvironments (Te Boekhorst et al. 2016; Yamada and Sixt 2019; Pourjafar and Tiwari 2024). This realization has driven the development of approaches to directly investigate cell migration phenomena at subcellular and molecular resolution within native tissue microenvironments using IVM, which could then be complemented using ex vivo and 3D models (Frattolin et al. 2021; Mehta et al. 2022; Scheele et al. 2022; Yamada et al. 2022).

In this article, we focus on in vivo studies that have used IVM to advance our understanding of single-cell migration responses in adult mammalian tissues. We examine and contrast single-cell migration phenomena in both homeostatic and pathological conditions, highlighting their role in tissue physiology. In doing so, we identify the gaps in our understanding of the molecular mechanisms behind these phenomena and emphasize the need for intravital subcellular microscopy (ISMic) to gain fundamental insights. ISMic, a technique pioneered by our group, combines multiphoton IVM with genetic mouse reporters to achieve subcellular and molecular resolution, enabling the study of membrane remodeling mechanisms in live mammalian tissues (Masedunskas et al. 2012a; Amornphimoltham et al. 2013; Weigert 2014; Ebrahim and Weigert 2019; Ebrahim et al. 2019; Heydecker et al. 2024). We provide specific examples of how ISMic enables the observation of subcellular dynamics at high spatiotemporal resolution in migrating cells within live mammalian tissues. Lastly, we discuss the need for innovative and combinatorial approaches to unravel the molecular mechanisms underlying various cell migration phenomena in tissue pathophysiology. Due to space constraints, we focus our discussion on select key IVM studies and regret the omission of many other important contributions. For additional insights into cell migration during mammalian development and tumor progression, topics beyond the scope of this article, we encourage readers to consult other papers.

CELL MIGRATION IN TISSUE HOMEOSTASIS

Cell migration is crucial for maintaining tissue integrity by continuously serving to replace degenerated cells as well as enabling immune surveillance. Although many mature cells in adult tissues remain largely static, migratory activity of certain stem and immune cell types is essential for tissue homeostasis. Here, we discuss examples of single-cell migration phenomena observed in adult tissues during homeostasis.

Stem cells within adult tissues play an integral role in replenishing differentiated cell populations. Over time, resident cells degenerate and are replaced by stem cells differentiating either within the same or adjacent tissues harboring specific stem cell niches (Beumer and Clevers 2024). In many tissues, stem cells migrate from their niches, replenishing cells lost to normal turnover or injury (Laird et al. 2008; Iismaa et al. 2018). For instance, hematopoietic stem cells (HSCs), primarily located in the bone marrow (BM), interact with stromal cells within specialized perivascular niches. Recent IVM studies have revealed that HSCs alternate between slow, directed migration and brief periods of confined movement (Upadhaya et al. 2020). Interestingly, a small fraction of HSCs periodically exit the BM into systemic circulation, potentially migrating to other tissues, where they might differentiate into tissue-resident immune cells (Wright et al. 2001; de Lucas et al. 2018). The circulating HSCs eventually home back to the BM niche (Fig. 1A), a process involving rolling and arrest interactions requiring vascular E- and P-selectins and vascular cell adhesion molecule-1 (VCAM-1) on BM venules (Mazo et al. 1998; Massberg et al. 2007; Laird et al. 2008).

Another intriguing example is that of stem cell migration within the intestinal epithelium, a tissue continually subjected to mechanical and biological stresses. To maintain barrier integrity and nutrient absorption efficiency, the intestinal epithelium is entirely renewed approximately every 4–5 days (Beumer and Clevers 2016; Iismaa et al. 2018). Two principal stem cell populations drive this renewal: crypt base columnar (CBC) cells (marked by $Lgr5^+$) and the +4 cells located in adjacent transit-amplifying regions (Barker et al. 2007; Tian et al. 2011; Beumer and Clevers 2016; Capdevila et al. 2024). Recent IVM studies suggest a "conveyor belt-like" mechanism (Fig. 1B), where proliferating CBC cells at the crypt base displace cells upward progressively migrating from the crypt toward the villus tip (Ritsma et al. 2014). It is unclear whether crypt stem cell migration is purely passive or whether there are active mechanisms to directionally orient daughter stem cells to migrate specifically toward the villus tip. Substantial knowledge gaps persist concerning the precise molecular and subcellular mechanisms that warrant further investigation of crypt stem cell migration phenomena (Azkanaz et al. 2022).

Unlike the stem cells, immune cells actively patrol and maintain tissue homeostasis. Immune cells collectively represent the most diverse and dynamic cell types, adopting complex migration patterns in a tissue-specific manner (Germain et al. 2012; Fan and Rudensky 2016; Mihlan et al. 2022). For example, nonclassical monocytes, distinct from inflammatory monocytes, actively patrol the vascular endothelium under steady-state conditions (Fig. 1C). This surveillance mechanism involves crawling migration mediated by β_2-integrin lymphocyte function-associated antigen 1 (LFA-1), allowing nonclassical monocytes to survey, sense, and remove debris or microparticles along the endothelial surfaces (Carlin et al. 2013; Rua and McGavern 2015). Neutrophils, typically known for their rapid response to inflammatory stimuli, also exist and migrate within lymphoid tissues under homeostasis. Their entry and exit from lymph nodes, for example, involves specific molecular interactions such as endothelial tethering via L-selectin and lymph node exit via sphingosine-1-phosphate (S1P) signaling (Waite et al. 2011; Bogoslowski et al. 2020).

In the perivascular and interstitial space, resident macrophages establish an extensive network that actively monitors and rapidly responds to perturbations in various tissues (Fig. 1D; Ginhoux and Jung 2014; Uderhardt et al. 2019; Mihlan et al. 2022). For example, mammary ductal macrophages form an extensive network and are necessary for postlactation epithelial remodeling (Dawson et al. 2020). In the skin, a subset of resident macrophages migrate slowly and bidirectionally to patrol sensory nerves, although the precise migration mechanism remains unclear (Kolter et al. 2019). Lung-resident alveolar macrophages, conversely, exhibit highly amoeboid movement, actively squeezing through alveolar pores to patrol lung tissue (Neupane et al. 2020).

Dendritic cells (DCs), another pivotal immune cell type, similarly contribute to immune surveillance. While appearing stationary, DCs extend and retract numerous spiny protrusive processes, presumably filamentous actin (F-actin)-rich, that dynamically sample their surrounding tissue environment under steady-state conditions (Fig. 1E; Lindquist et al. 2004; Ng et al. 2008; Waite et al. 2011). Certain DC populations, particularly those located beneath epithelial barriers, exhibit enhanced amoeboid migration modality, requiring Gαi signaling and RhoA-activated kinase (ROCK)-dependent activity to respond quickly to environmental signals (Ng et al. 2008; Nitschké et al. 2012).

Various tissue-resident lymphoid cells also patrol tissue compartments, maintaining local immune surveillance (Fig. 1E). For instance, invariant natural killer T (iNKT) cells patrol liver sinusoids, effectively interacting with DCs and

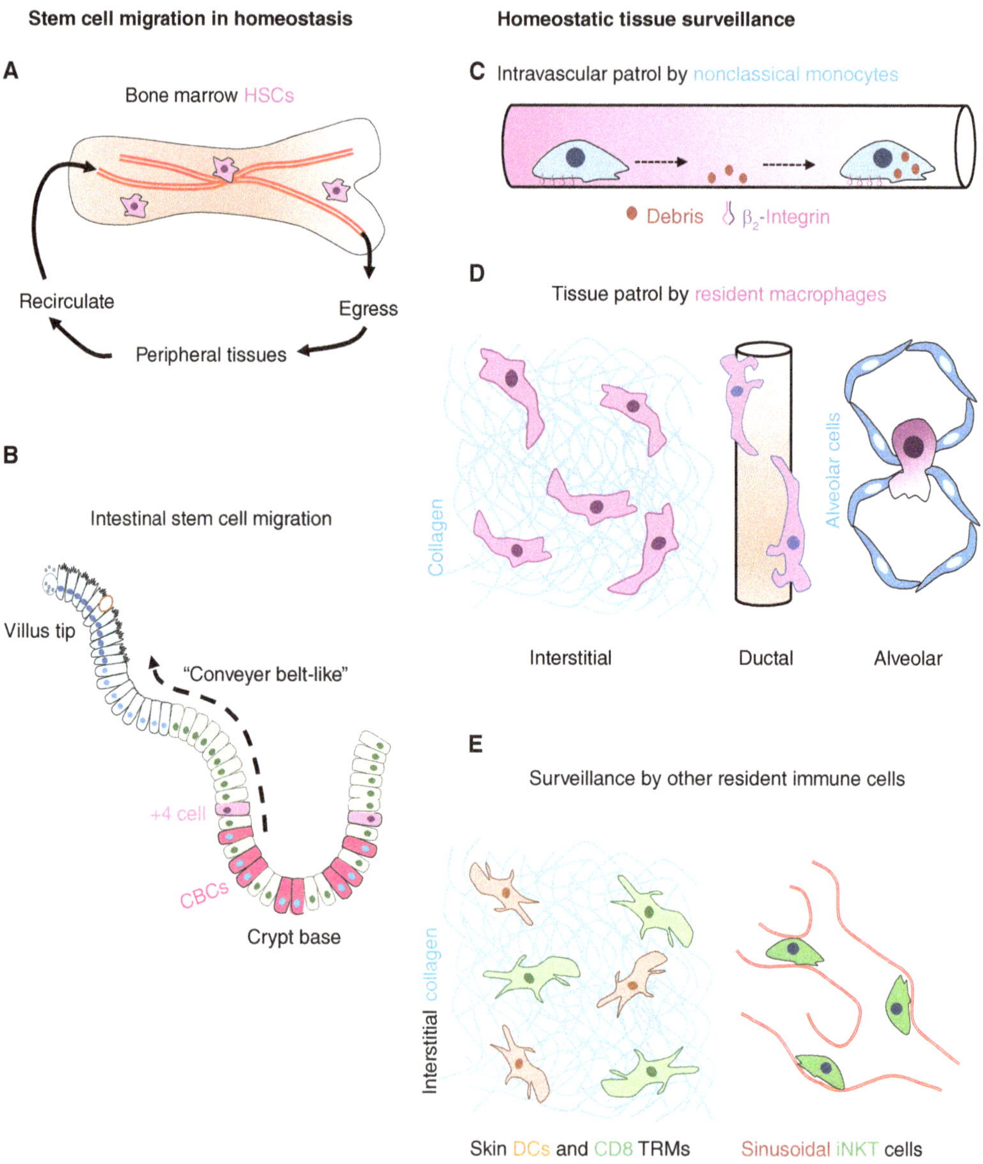

Figure 1. Examples of single-cell migration observed in mammalian tissues under homeostatic conditions. Stem cell migration: (*A*) Hematopoietic stem cells (HSCs) migrate within bone marrow (BM) as well as egress to peripheral tissues via vascular sinusoids (red), before circulating back to the BM. (*B*) Intestinal stem cells in crypt base migrate using a "conveyer belt–like" mechanism to maintain intestinal homeostasis. Immune cell migration: (*C*) Nonclassical monocytes clear particles or debris from the vascular lumen. (*D*) Resident macrophages of the skin, mammary duct, and alveoli patrol the respective tissues. (*E*) Tissue surveillance by resident immune cell types such as dendritic cells (DCs) and CD8 tissue-resident memory T cells (TRMs) in the skin, as well as invariant natural killer T (iNKT) cells in the liver sinusoids, is depicted.

 Cite this article as *Cold Spring Harb Perspect Biol* doi: 10.1101/cshperspect.a041742

T cells (Geissmann et al. 2005). Innate lymphoid cells (ILC3) dynamically migrate within intestinal cryptopatches (Pearson et al. 2016), while tissue-resident memory CD8$^+$ T cells (TRMs) in the skin exhibit continuous, spiny protrusions and movements, crucial for surveillance against previously encountered antigens (Zaid et al. 2014).

These diverse examples underscore the fundamental role of cell migration in maintaining mammalian tissue homeostasis (Fig. 1), emphasizing the need to elucidate the underlying molecular mechanisms using advanced imaging methods, such as ISMic, for a more comprehensive understanding and targeting of such phenomena.

CELL MIGRATION IN TISSUE PATHOPHYSIOLOGY

Cell migration is essential not only for maintaining tissue homeostasis but also for enabling effective responses to tissue injury, infection, and inflammation. Various cell types, including stem cells and immune cells, demonstrate dynamic migratory behaviors crucial for tissue regeneration and defense. Here, we present key examples illustrating single-cell migration events across different tissues under pathological conditions, primarily captured through IVM.

Beyond routine maintenance, stem cells also play a pivotal role in tissue regeneration following injury. For instance, satellite cells (SCs) in skeletal muscle, normally quiescent and immobile along the basal lamina, become activated upon muscle damage (Wang and Rudnicki 2012; Fumagalli et al. 2020). The activated SCs proliferate and exhibit bidirectional migration along myofiber remnants, actively sampling their microenvironment (Fig. 2A). Interactions with infiltrating immune cells, such as monocytes and neutrophils, further enhance SC proliferation and differentiation, accelerating the muscle regeneration process (Webster et al. 2016; He et al. 2023).

Similarly, bone regeneration following injury involves tightly coordinated migration of osteoclasts (BM-derived monocytes) and osteoblasts (mesenchymal cells) (de Lucas et al. 2018; Kim and Bixel 2020). Under homeostatic conditions, these cells occupy discrete regions within bone tissue. Upon injury, osteoclasts initiate slow, dynamic migration and closely interact with osteoblasts, balancing bone resorption and formation processes (Fig. 2B). Specific subsets of osteoclasts regulated by the cytokine receptor activator of nuclear factor-κB ligand (RANKL)-expressing Th17 cells remain static, while migratory osteoclasts interact with osteoblasts to potentially modulate bone resorption (Kikuta et al. 2013; Furuya et al. 2018). Osteoblast migration and differentiation following injury further contribute to bone remodeling, although the molecular details of these interactions remain incompletely understood (Park et al. 2012; Furuya et al. 2018).

The highly vascularized lung alveoli, responsible for gas exchange, frequently undergo tissue perturbation. Alveolar type 2 (AT2) cells, a rare stem cell population, regenerate alveolar type 1 cells after lung injury. Recent studies have shown AT2 cells to migrate dynamically within and between alveoli through branched actin-driven protrusions (Fig. 2C), emphasizing their migration capacity in mediating postinflammation alveolar repair (Desai et al. 2014; Chioccioli et al. 2024).

Immune cell migration during inflammation and infection has been extensively studied via IVM, yielding significant insights. Monocytes, for example, undergo complex intravascular interactions—rolling, crawling, adhesion, and extravasation—to reach inflammation sites (Schmidt et al. 2013; Rua and McGavern 2015). These intravascular leukocyte responses are coordinated by selectins and β_2-integrins (Von Andrian et al. 1993; von Brühl et al. 2012). Once extravasated, monocytes adopt amoeboid migration mode, which occurs independent of branched actin formation, actively migrating toward inflammation site where they clear apoptotic cells and support tissue repair (Fig. 2D; Lämmermann et al. 2013; Dal-Secco et al. 2015; Rotty et al. 2017).

Tissue-resident macrophages likewise show dynamic responses to infection and inflammation. Kupffer cells (KCs) migrate from liver pa-

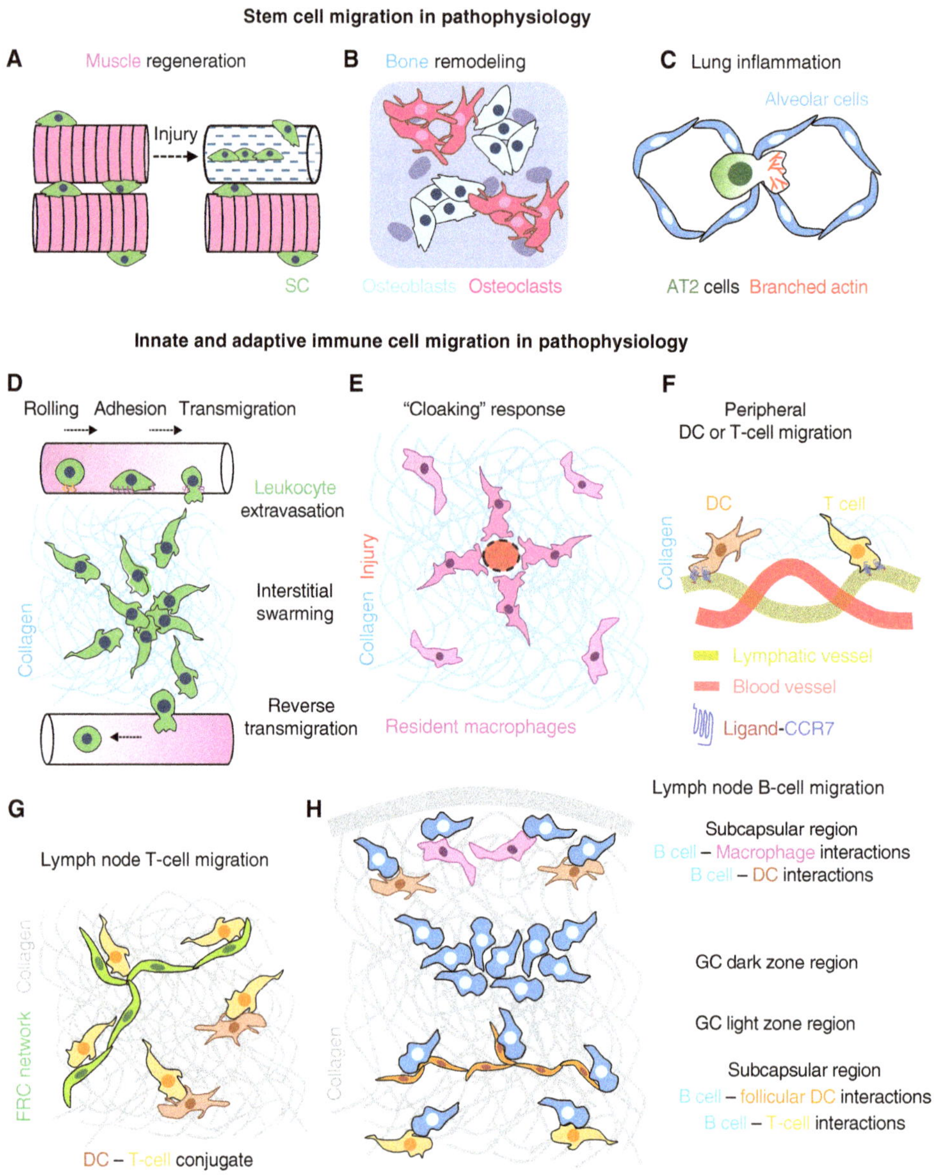

Figure 2. Examples of single-cell migration documented in mammalian tissues under pathophysiological conditions. Stem cells in tissue regeneration and remodeling: (*A*) Satellite cells (SCs) migrate along the remnants of an injured myofiber to seed and regenerate the tissue. (*B*) In the bones, osteoclasts and osteoblasts segregate to distinct pools but interact with each other to remodel the bone under inflammatory conditions. (*C*) Upon lung inflammation and infection, alveolar type 2 (AT2) stem cells migrate through alveolar pores using branched actin to promote regeneration of damaged tissue. Immune cell migration in tissue pathophysiology: (*D*) The intravascular, interstitial, and reverse transmigration migration phases of leukocytes, such as monocytes and neutrophils, are depicted. (*E*) The resident macrophage cloaking response to wounding observed in tissues such as brain, peritoneal serosa, muscle etc., is shown. (*F*) Dendritic cells (DCs) and T cells upon stimulation migrate and drain into the lymphatics, a process requiring the haptotactic CCR7 signaling is showcased. (*G*) In the lymph node parenchyma, T cells migrating along the follicular reticular cell (FRC) network and engaging with the antigen-presenting DCs to form stable conjugates are displayed. (*H*) B cells within the lymph node engage with multiple cell types to undergo maturation and expansion in the germinal center (GC).

renchyma into sinusoids, capturing pathogens via β_1-integrin, CD44, and macrophage migration inhibitory factor signaling (Araujo David et al. 2024). Gata6$^+$ peritoneal macrophages traverse mesothelial layers to remove necrotic cells and facilitate liver recovery, while alveolar macrophages navigate alveolar pores to intercept pathogens, thereby limiting neutrophil-induced lung damage (Wang and Kubes 2016; Neupane et al. 2020). Brain-resident microglia dynamically respond to injury through rapid extension of protrusions, regulated by purinergic receptors and connexin channels (Fig. 2E; Davalos et al. 2005; Nimmerjahn et al. 2005; d'Errico et al. 2022). Using dynamic cellular protrusions, interstitial macrophages such as microglia effectively sample for cellular debris and apoptotic cells. Upon encountering dead cells or debris in the injured interstitium, protrusions from resident macrophages emanate toward the injured site, a response termed "cloaking," that relies on alarmin sensing from apoptotic cells and β_1-integrin–actomyosin engagement in resident macrophages, to restrict the damage from recruiting circulating leukocytes to the injury site (Davalos et al. 2005; Uderhardt et al. 2019; Paterson and Lämmermann 2022).

DC migration underlies antigen-sampling and antigen-presenting functions for a productive immune response. For example, intestinal DCs respond to pathogens by moving into epithelial layers via $G\alpha_i$ and Toll-like receptor (TLR) signaling, whereas ocular surface DCs and skin Langerhans cells become more motile under dry eye conditions and exposure to tumor necrosis factor α, respectively (Nishibu et al. 2006; Farache et al. 2013; Jamali et al. 2020). Upon activation, DCs enter the lymphatics using CCR7-dependent haptotaxis (Fig. 2F) and thereafter migrate in an integrin-independent amoeboid mode within lymph node parenchyma (Lämmermann et al. 2008; Weber et al. 2013; Worbs et al. 2017). In the lymph node, migrating antigen-presenting DCs interact with multiple T cells over time, a crucial factor in the fine-tuning of specific T- and B-cell responses during adaptive immunity (Fig. 2G,H; Mempel et al. 2004; Castellino et al. 2006).

Neutrophils are rapid first responders to inflammation and are a prototype for amoeboid migration modality. Their migration encompasses endothelial interactions, extravasation, and directed interstitial movement toward inflammation sites (Fig. 2D). IVM studies have identified critical and diverse molecular cues, such as lipids, cytokines, and chemokine signaling pathways underlying neutrophil extravasation and interstitial migration responses (McDonald et al. 2008, 2010; Chou et al. 2010; Nourshargh and Alon 2014; Miyabe et al. 2017; Girbl et al. 2018; Choudhury et al. 2019; Giovanni et al. 2022). Notably, while engagement of integrin or adhesion receptor is necessary for neutrophil extravasation response or migration within the vascular lumen to inflammation, neutrophil swarming response in the interstitial space occurs in an integrin-independent manner (Von Andrian et al. 1993; McDonald et al. 2010; Lämmermann et al. 2013; Nourshargh and Alon 2014; Glaser et al. 2021). Upon resolution of inflamed site, neutrophils could reverse transmigration back into the vasculature and end up in distant organs or undergo terminal clearance in the BM (Colom et al. 2015; Wang et al. 2017). The above IVM-based studies demonstrate the complex, coordinated, and intricate mechanisms required to mount a robust neutrophil response to infection and inflammation.

T cells mature in the thymus and subsequently migrate to peripheral lymphoid organs, displaying varied migratory behaviors depending on tissue context and immune state. Within lymph nodes, naive T cells undergo phases of low and high motility along the follicular reticular cell network, regulated by CCR7 (Miller et al. 2003; Bajénoff et al. 2006; Worbs et al. 2007). Upon antigen encounter, T cells proceed to form stable conjugates with DCs for activation (Fig. 2G) and thereafter resume rapid migration to eventually exit lymph nodes via cortical sinuses guided by S1P signaling (Mempel et al. 2004; Grigorova et al. 2009). Alternately, T cells also interact and form stable conjugates with B cells within the lymph node (Fig. 2H; Qi et al. 2008). Detailed IVM analyses across pathological states, such as infection, graft rejection, etc., have revealed precise dynamics of antigen-specific T-cell responses in peripheral tissues (Egen et al. 2008; Chtanova et al. 2009; Kim et al. 2009; Celli et al. 2011; Gebhardt et al. 2011; Groom et al. 2012; Murooka et al. 2012; Sung et al. 2012).

B-cell migration within BM and in lymphoid germinal centers (GCs) is also dynamic. Within the BM, immature B cells migrate using $\alpha_4\beta_1$-integrin and in response to CXCL4/CXCL12, before they egress from the BM by reverse transmigration (Beck et al. 2014). Once in the lymph node, B cells interact with subcapsular macrophages and antigen-bearing DCs in the extrafollicular areas and thereafter localize to lymphoid GCs (Qi et al. 2006; Junt et al. 2007; Phan et al. 2007). Within GCs, B cells migrate along the follicular DC network, transitioning between light and dark zones, critical for affinity maturation and clonal expansion (Bajénoff et al. 2006; Allen et al. 2007; Hauser et al. 2007; Schwickert et al. 2007). Interactions with T cells in the GC are essential for B-cell activation and proliferation (Fig. 2H; Allen et al. 2007; Qi et al. 2008; Victora et al. 2010).

Besides specific immune cell types mentioned above, IVM studies have also shed light on the migration behavior of lesser studied leukocyte subtypes to infection and inflammation. These include the following: platelet migration to thrombus requiring nonmuscle myosin IIA (NMIIA) (Gaertner et al. 2017); iNKT cells in the lymph node, liver, and joints (Kastenmüller et al. 2012; Lee et al. 2014; Liew et al. 2017); tissue-resident eosinophils of the lymph node and lungs (Chojnacki et al. 2019); tissue-resident γδT cells in barrier tissue epithelium (Chennupati et al. 2010; Edelblum et al. 2012; Wang et al. 2014; Wilharm et al. 2019; Fischer et al. 2020); and ILC2 and ILC3 cells within the inflamed lymph node and intestine (Pearson et al. 2016; Lok et al. 2021).

Collectively, the above examples emphasize the impact of IVM as the key approach in elucidating the complex and highly adaptable migratory responses of stem and immune cell types to tissue injury and inflammation. Although these studies have laid foundational framework to our understanding of the upstream signaling cues and downstream effector functions, the precise membrane remodeling mechanisms controlling single-cell migration plasticity observed in vivo remain poorly understood. This necessitates the application of advanced imaging techniques such as ISMic, as well as complimentary innovative approaches, to further unravel the molecular interplay dictating membrane remodeling mechanisms during cell migration in vivo.

ISMic AND IN VIVO CELL MIGRATION DYNAMICS AT A MOLECULAR LEVEL

Several studies have documented subcellular events that occur in cells that migrate. For example, numerous immune cell types are known to internalize antigens as well as microbial pathogens in vivo (Phan et al. 2007; Ng et al. 2008; Gonzalez et al. 2010; Uderhardt et al. 2019; Wang et al. 2021). Similarly, whole-cell calcium fluxes during leukocyte migration have been visualized using IVM of genetically engineered calcium-sensitive reporters, contributing to intracellular signaling events that are crucial for cell activation and related responses (Yoshikawa et al. 2016; Kyratsous et al. 2017; Bohineust et al. 2020; Ulbricht et al. 2021; Mehari et al. 2022; Bauer et al. 2023; Mukhopadhyay et al. 2024). Although these studies provide invaluable insights into the downstream signaling and phagocytosis response within cells in vivo, they do not directly address the membrane remodeling mechanisms that underlie cell migration observed under specified contexts.

ISMic, an advanced imaging approach developed by our group, bridges this gap by providing high-resolution insights into membrane remodeling mechanisms in live mammalian tissues. ISMic uses state-of-the-art IVM instrumentation—including optimized objective lenses, sensitive detectors, and high-speed cameras—to visualize the molecular dynamics of fluorescently tagged proteins or antibody conjugates within the native tissue (Masedunskas et al. 2012a; Weigert 2014; Ebrahim and Weigert 2019). Our pioneering efforts have leveraged ISMic to elucidate cytoskeletal dynamics and membrane remodeling mechanisms during endocytic and regulated exocytic processes in live animal models (Masedunskas et al. 2011, 2012b, 2017; Milberg et al. 2017; Ebrahim et al. 2019; Shitara et al. 2019; Heydecker et al. 2024).

We further adapted ISMic to study actomyosin-driven membrane remodeling mechanisms during both intravascular and interstitial leukocyte migration, particularly in inflamed skin contexts (Subramanian et al. 2020; Melis et al. 2022). Through a combination of genetic reporters and antibody labeling, we recently visualized and precisely characterized the dynamics and functional roles of NMIIA and β_2-integrin during neutrophil extravasation, exemplifying ISMic's potential to reveal molecular details otherwise unattainable with traditional imaging approaches (Subramanian et al. 2020). Indeed, while NMIIA redistributed to the cell periphery, antibody-labeled β2-integrin gets enriched on the membrane interacting with endothelium, both mechanisms critical for neutrophils to undergo arrest and extravasation from inflamed vessels (Fig. 3A). Such a redistribution of NMIIA and β_2-integrin in extravasating neutrophils requires extracellular vesicle-mediated leukotriene B4 signaling (Subramanian et al. 2020). Likewise, a combination of fluorescent probe and targeted chemical perturbations along with ISMic led to the revelation of a striking F-actin enrichment at sites within T cells that coincides with the pore formed in the endothelium during transmigration (Fig. 3B; Yan et al. 2019). While actomyosin is necessary for T-cell arrest and extravasation from high endothelial venules (HEVs), branched or linear F-actin appear individually dispensable for arrest but not transmigration response (Yan et al. 2019). ISMic has also been combined with fluorescence resonance energy transfer (FRET) sensors to successfully document oscillatory patterns of key regulatory protein activity such as RhoA during neutrophil interstitial migration (Nobis et al. 2017).

To further show the extent to which ISMic can be exploited to resolve the membrane remodeling mechanisms in cell migration, we provide two examples from our ongoing investigations. In the first example, ISMic enabled high-resolution visualization of a migrating DC in the salivary gland, where the green fluorescent protein (GFP)–tagged NMIIA can be visualized as bundles that undergo retrograde flow to form a branching mesh-like structure underneath the protrusive front, offering the first glimpse of NMIIA dynamics in a resident immune cell type within a native tissue (Fig. 3C; Supplemental Movie S1). Second, we used ISMic to visualize the dynamics of active mitochondria in conjunction with GFP-NMIIA in extravasating neutrophils. Active mitochondria cluster into large strand-like structures during transmigration step, which coincides with dense NMIIA accumulation to the rear of cells. These arrangements are dissolved once neutrophil enter the perivascular area (Fig. 3D; Supplemental Movie S2). This example sheds light on the potential coupling of actomyosin machinery with mitochondrial dynamics that occurs in directed leukocyte migration response in vivo.

Together, the above examples underscore the power and versatility of ISMic in providing unprecedented subcellular and molecular insights into the dynamic mechanisms behind single-cell migration responses within live mammalian tissues. Future investigations incorporating ISMic will be instrumental in bridging the gap between the known molecular cues that promote cell migration with that of functional consequence to tissue pathophysiology, by providing a comprehensive mechanistic understanding of the underlying membrane remodeling mechanisms.

SUMMARY AND OUTLOOK

While reductionist model systems have historically provided foundational insights into cell migration, recent technological advances emphasize the superiority of in vivo imaging approaches such as ISMic. Real-time, 4D subcellular imaging captures biological processes within their natural tissue microenvironments, where crucial cell–cell and cell–matrix interactions occur, factors that simplified in vitro systems fail to replicate adequately.

Advanced imaging technologies, including lattice light sheet and adaptive optics microscopy, now offer sufficient spatiotemporal resolution to monitor dynamic subcellular

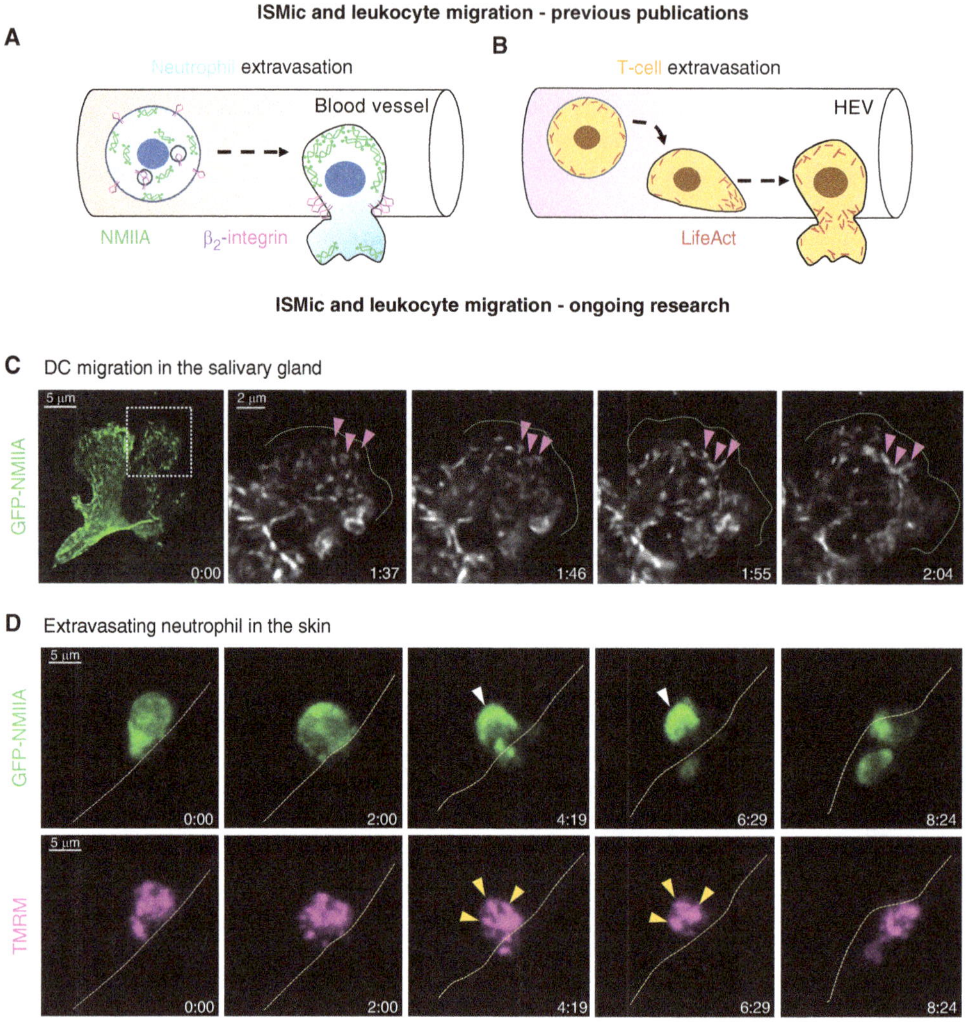

Figure 3. Intravital subcellular microscopy (ISMic) in single-cell migration. The results from previously published ISMic studies have been presented as schematics for neutrophils (*A*) (Subramanian et al. 2020) and T-cell (*B*) (Yan et al. 2019) extravasation from respective vessels. A couple of examples from our ongoing research involving the roles of membrane remodeling mechanisms in leukocyte migration are presented. A dendritic cell (DC) is imaged using spinning disc microscopy in the salivary gland of a green fluorescent protein (GFP)-nonmuscle myosin IIA (NMIIA) mouse (*C*). Area within the dashed white box is represented as grayscale images. Dashed green line and magenta arrowheads mark the protrusive edge and retrograde movement of NMIIA bundles, respectively. Scale bars, 5 µm (whole cell); 2 µm (enlarged area). Time stamp is presented as min:sec. (*D*) A two-photon IVM of a recipient mouse footpad is used to visualize the dynamics of GFP-NMIIA (green) within donor neutrophils costained with mitochondria potential dye tetramethylrhodamine, methyl ester (TMRM; magenta). White and orange arrowheads mark the accumulating GFP-NMIIA in the cell rear and clustering mitochondria, respectively. Dashed yellow line represents the endothelial lining at each time point. Scale bars, 5 µm, and the time stamp is presented as min:sec.

events during cell migration within live tissues. Additionally, genetically engineered mouse models as well as specific multispectral probes being actively developed enable precise molecular perturbations specifically within migrating cell populations, allowing investigators to dissect the roles of individual proteins and pathways under physiological and pathological conditions.

Furthermore, the integration of machine learning approaches in microscopy has substantially improved the quantitative analysis of imaging data, enabling researchers to perform sophisticated cell segmentation, tracking, and morphological analysis in complex 3D tissue architectures over time. These capabilities facilitate comprehensive, unbiased analyses of cellular dynamics, thus significantly enhancing the biological relevance, interpretability, and reproducibility of in vivo studies.

Ultimately, some critical biological processes driving cell migration, such as immune surveillance, tissue regeneration, and vascular remodeling, simply cannot be replicated entirely through reductionist approaches. Thus, advanced in vivo imaging methods, represented by ISMic, have become indispensable tools, providing unparalleled insights into complex biological processes that maintain tissue homeostasis and contribute to pathological conditions. For additional information, see supplemental material.

COMPETING INTEREST STATEMENT

The authors declare no competing interests.

ACKNOWLEDGMENTS

We thank all the Weigert group members for their suggestions toward this manuscript. This research was supported by the Intramural Research Program of the National Institutes of Health (NIH). The contributions of the NIH authors were made as part of their official duties as NIH federal employees, are in compliance with agency policy requirements, and are considered Works of the United States Government. However, the findings and conclusions presented in this manuscript are those of the authors and do not necessarily reflect the views of the NIH or the U.S. Department of Health and Human Services.

REFERENCES

Allen CDC, Okada T, Tang HL, Cyster JG. 2007. Imaging of germinal center selection events during affinity maturation. *Science* **315:** 528–531. doi:10.1126/science.1136736

Amornphimoltham P, Rechache K, Thompson J, Masedunskas A, Leelahavanichkul K, Patel V, Molinolo A, Gutkind JS, Weigert R. 2013. Rab25 regulates invasion and metastasis in head and neck cancer. *Clin Cancer Res* **19:** 1375–1388. doi:10.1158/1078-0432.CCR-12-2858

Araujo David B, Atif J, Vargas e Silva Castanheira F, Yasmin T, Guillot A, Ait Ahmed Y, Peiseler M, Hommes JW, Salm L, Brundler MA, et al. 2024. Kupffer cell reverse migration into the liver sinusoids mitigates neonatal sepsis and meningitis. *Sci Immunol* **9:** eadq9704. doi:10.1126/sciimmunol.adq9704

Azkanaz M, Corominas-Murtra B, Ellenbroek SIJ, Bruens L, Webb AT, Laskaris D, Oost KC, Lafirenze SJA, Annusver K, Messal HA, et al. 2022. Retrograde movements determine effective stem cell numbers in the intestine. *Nature* **607:** 548–554. doi:10.1038/s41586-022-04962-0

Bajénoff M, Egen JG, Koo LY, Laugier JP, Brau F, Glaichenhaus N, Germain RN. 2006. Stromal cell networks regulate lymphocyte entry, migration, and territoriality in lymph nodes. *Immunity* **25:** 989–1001. doi:10.1016/j.immuni.2006.10.011

Barker N, van Es JH, Kuipers J, Kujala P, van den Born M, Cozijnsen M, Haegebarth A, Korving J, Begthel H, Peters PJ, et al. 2007. Identification of stem cells in small intestine and colon by marker gene Lgr5. *Nature* **449:** 1003–1007. doi:10.1038/nature06196

Bauer IJ, Fang P, Lämmle KF, Tyystjärvi S, Alterauge D, Baumjohann D, Yoon H, Korn T, Wekerle H, Kawakami N. 2023. Visualizing the activation of encephalitogenic T cells in the ileal lamina propria by in vivo two-photon imaging. *Proc Natl Acad Sci* **120:** e2302697120. doi:10.1073/pnas.2302697120

Beck TC, Gomes AC, Cyster JG, Pereira JP. 2014. CXCR4 and a cell-extrinsic mechanism control immature B lymphocyte egress from bone marrow. *J Exp Med* **211:** 2567–2581. doi:10.1084/jem.20140457

Beumer J, Clevers H. 2016. Regulation and plasticity of intestinal stem cells during homeostasis and regeneration. *Development* **143:** 3639–3649. doi:10.1242/dev.133132

Beumer J, Clevers H. 2024. Hallmarks of stemness in mammalian tissues. *Cell Stem Cell* **31:** 7–24. doi:10.1016/j.stem.2023.12.006

Bogoslowski A, Wijeyesinghe S, Lee WY, Chen CS, Alanani S, Jenne C, Steeber DA, Scheiermann C, Butcher EC, Masopust D, et al. 2020. Neutrophils recirculate through lymph nodes to survey tissues for pathogens. *J Immunol* **204:** 2552–2561. doi:10.4049/jimmunol.2000022

Bohineust A, Garcia Z, Corre B, Lemaître F, Bousso P. 2020. Optogenetic manipulation of calcium signals in single T cells in vivo. *Nat Commun* **11:** 1143. doi:10.1038/s41467-020-14810-2

Bugyi B, Carlier MF. 2010. Control of actin filament treadmilling in cell motility. *Annu Rev Biophys* **39:** 449–470. doi:10.1146/annurev-biophys-051309-103849

Capdevila C, Miller J, Cheng L, Kornberg A, George JJ, Lee H, Botella T, Moon CS, Murray JW, Lam S, et al. 2024. Time-resolved fate mapping identifies the intestinal upper crypt zone as an origin of Lgr5$^+$ crypt base columnar cells. *Cell* **187:** 3039–3055.e14. doi:10.1016/j.cell.2024.05.001

Carlin LM, Stamatiades EG, Auffray C, Hanna RN, Glover L, Vizcay-Barrena G, Hedrick CC, Cook HT, Diebold S, Geissmann F. 2013. Nr4a1-dependent Ly6C(low) monocytes monitor endothelial cells and orchestrate their disposal. *Cell* **153:** 362–375. doi:10.1016/j.cell.2013.03.010

Castellino F, Huang AY, Altan-Bonnet G, Stoll S, Scheinecker C, Germain RN. 2006. Chemokines enhance immunity by guiding naive CD8$^+$ T cells to sites of CD4$^+$ T cell–dendritic cell interaction. *Nature* **440:** 890–895. doi:10.1038/nature04651

Celli S, Albert ML, Bousso P. 2011. Visualizing the innate and adaptive immune responses underlying allograft rejection by two-photon microscopy. *Nat Med* **17:** 744–749. doi:10.1038/nm.2376

Chakrabarti R, Lee M, Higgs HN. 2021. Multiple roles for actin in secretory and endocytic pathways. *Curr Biol* **31:** R603–R618. doi:10.1016/j.cub.2021.03.038

Chennupati V, Worbs T, Liu X, Malinarich FH, Schmitz S, Haas JD, Malissen B, Förster R, Prinz I. 2010. Intra- and intercompartmental movement of γδ T cells: intestinal intraepithelial and peripheral γδ T cells represent exclusive nonoverlapping populations with distinct migration characteristics. *J Immunol* **185:** 5160–5168. doi:10.4049/jimmunol.1001652

Chioccioli M, Liu S, Magruder S, Tata A, Borriello L, McDonough JE, Konkimalla A, Kim SH, Nouws J, Gonzalez DG, et al. 2024. Stem cell migration drives lung repair in living mice. *Dev Cell* **59:** 830–840.e4. doi:10.1016/j.devcel.2024.02.003

Chojnacki A, Wojcik K, Petri B, Aulakh G, Jacobsen EA, LeSuer WE, Colarusso P, Patel KD. 2019. Intravital imaging allows real-time characterization of tissue resident eosinophils. *Commun Biol* **2:** 1–12. doi:10.1038/s42003-019-0425-3

Chou RC, Kim ND, Sadik CD, Seung E, Lan Y, Byrne MH, Haribabu B, Iwakura Y, Luster AD. 2010. Lipid-cytokine-chemokine cascade drives neutrophil recruitment in a murine model of inflammatory arthritis. *Immunity* **33:** 266–278. doi:10.1016/j.immuni.2010.07.018

Choudhury SR, Babes L, Rahn JJ, Ahn BY, Goring KAR, King JC, Lau A, Petri B, Hao X, Chojnacki AK, et al. 2019. Dipeptidase-1 is an adhesion receptor for neutrophil recruitment in lungs and liver. *Cell* **178:** 1205–1221.e17. doi:10.1016/j.cell.2019.07.017

Chtanova T, Han SJ, Schaeffer M, van Dooren GG, Herzmark P, Striepen B, Robey EA. 2009. Dynamics of T cell, antigen-presenting cell, and pathogen interactions during recall responses in the lymph node. *Immunity* **31:** 342–355. doi:10.1016/j.immuni.2009.06.023

Colom B, Bodkin JV, Beyrau M, Woodfin A, Ody C, Rourke C, Chavakis T, Brohi K, Imhof BA, Nourshargh S. 2015. Leukotriene B4-neutrophil elastase axis drives neutrophil reverse transendothelial cell migration in vivo. *Immunity* **42:** 1075–1086. doi:10.1016/j.immuni.2015.05.010

Dal-Secco D, Wang J, Zeng Z, Kolaczkowska E, Wong CHY, Petri B, Ransohoff RM, Charo IF, Jenne CN, Kubes P. 2015. A dynamic spectrum of monocytes arising from the in situ reprogramming of CCR2$^+$ monocytes at a site of sterile injury. *J Exp Med* **212:** 447–456. doi:10.1084/jem.20141539

Davalos D, Grutzendler J, Yang G, Kim JV, Zuo Y, Jung S, Littman DR, Dustin ML, Gan WB. 2005. ATP mediates rapid microglial response to local brain injury in vivo. *Nat Neurosci* **8:** 752–758. doi:10.1038/nn1472

Dawson CA, Pal B, Vaillant F, Gandolfo LC, Liu Z, Bleriot C, Ginhoux F, Smyth GK, Lindeman GJ, Mueller SN, et al. 2020. Tissue-resident ductal macrophages survey the mammary epithelium and facilitate tissue remodelling. *Nat Cell Biol* **22:** 546–558. doi:10.1038/s41556-020-0505-0

De Belly H, Weiner OD. 2024. Follow the flow: actin and membrane act as an integrated system to globally coordinate cell shape and movement. *Curr Opin Cell Biol* **89:** 102392. doi:10.1016/j.ceb.2024.102392

de Lucas B, Pérez LM, Gálvez BG. 2018. Importance and regulation of adult stem cell migration. *J Cell Mol Med* **22:** 746–754. doi:10.1111/jcmm.13422

d'Errico P, Ziegler-Waldkirch S, Aires V, Hoffmann P, Mezö C, Erny D, Monasor LS, Liebscher S, Ravi VM, Joseph K, et al. 2022. Microglia contribute to the propagation of Aβ into unaffected brain tissue. *Nat Neurosci* **25:** 20–25. doi:10.1038/s41593-021-00951-0

Desai TJ, Brownfield DG, Krasnow MA. 2014. Alveolar progenitor and stem cells in lung development, renewal and cancer. *Nature* **507:** 190–194. doi:10.1038/nature12930

Devreotes P, Horwitz AR. 2015. Signaling networks that regulate cell migration. *Cold Spring Harb Perspect Biol* **7:** a005959. doi:10.1101/cshperspect.a005959

Ebrahim S, Weigert R. 2019. Intravital microscopy in mammalian multicellular organisms. *Curr Opin Cell Biol* **59:** 97–103. doi:10.1016/j.ceb.2019.03.015

Ebrahim S, Chen D, Weiss M, Malec L, Ng Y, Rebustini I, Krystofiak E, Hu L, Liu J, Masedunskas A, et al. 2019. Dynamic polyhedral actomyosin lattices remodel micron-scale curved membranes during exocytosis in live mice. *Nat Cell Biol* **21:** 933–939. doi:10.1038/s41556-019-0365-7

Edelblum KL, Shen L, Weber CR, Marchiando AM, Clay BS, Wang Y, Prinz I, Malissen B, Sperling AI, Turner JR. 2012. Dynamic migration of γδ intraepithelial lymphocytes requires occludin. *Proc Natl Acad Sci* **109:** 7097–7102. doi:10.1073/pnas.1112519109

Egen JG, Rothfuchs AG, Feng CG, Winter N, Sher A, Germain RN. 2008. Macrophage and T cell dynamics during the development and disintegration of mycobacterial granulomas. *Immunity* **28:** 271–284. doi:10.1016/j.immuni.2007.12.010

Fan X, Rudensky AY. 2016. Hallmarks of tissue-resident lymphocytes. *Cell* **164:** 1198–1211. doi:10.1016/j.cell.2016.02.048

Farache J, Koren I, Milo I, Gurevich I, Kim KW, Zigmond E, Furtado GC, Lira SA, Shakhar G. 2013. Luminal bacteria recruit CD103$^+$ dendritic cells into the intestinal epithelium to sample bacterial antigens for presentation. *Immunity* **38:** 581–595. doi:10.1016/j.immuni.2013.01.009

Fischer MA, Golovchenko NB, Edelblum KL. 2020. Γδ T cell migration: separating trafficking from surveillance behaviors at barrier surfaces. *Immunol Rev* **298:** 165–180. doi:10.1111/imr.12915

Frattolin J, Watson DJ, Bonneuil WV, Russell MJ, Fasanella Masci F, Bandara M, Brook BS, Nibbs RJB, Moore JE. 2021. The critical importance of spatial and temporal scales in designing and interpreting immune cell migration assays. *Cells* **10:** 3439. doi:10.3390/cells10123439

Fumagalli A, Bruens L, Scheele CLGJ, van Rheenen J. 2020. Capturing stem cell behavior using intravital and live cell microscopy. *Cold Spring Harb Perspect Biol* **12:** a035949. doi:10.1101/cshperspect.a035949

Furuya M, Kikuta J, Fujimori S, Seno S, Maeda H, Shirazaki M, Uenaka M, Mizuno H, Iwamoto Y, Morimoto A, et al. 2018. Direct cell–cell contact between mature osteoblasts and osteoclasts dynamically controls their functions in vivo. *Nat Commun* **9:** 300. doi:10.1038/s41467-017-02541-w

Gaertner F, Ahmad Z, Rosenberger G, Fan S, Nicolai L, Busch B, Yavuz G, Luckner M, Ishikawa-Ankerhold H, Hennel R, et al. 2017. Migrating platelets are mechano-scavengers that collect and bundle bacteria. *Cell* **171:** 1368–1382.e23. doi:10.1016/j.cell.2017.11.001

Gebhardt T, Whitney PG, Zaid A, Mackay LK, Brooks AG, Heath WR, Carbone FR, Mueller SN. 2011. Different patterns of peripheral migration by memory $CD4^+$ and $CD8^+$ T cells. *Nature* **477:** 216–219. doi:10.1038/nature10339

Geissmann F, Cameron TO, Sidobre S, Manlongat N, Kronenberg M, Briskin MJ, Dustin ML, Littman DR. 2005. Intravascular immune surveillance by $CXCR6^+$ NKT cells patrolling liver sinusoids. *PLoS Biol* **3:** e113. doi:10.1371/journal.pbio.0030113

Germain RN, Robey EA, Cahalan MD. 2012. A decade of imaging cellular motility and interaction dynamics in the immune system. *Science* **336:** 1676–1681. doi:10.1126/science.1221063

Ginhoux F, Jung S. 2014. Monocytes and macrophages: developmental pathways and tissue homeostasis. *Nat Rev Immunol* **14:** 392–404. doi:10.1038/nri3671

Giovanni MD, Tam H, Valet C, Xu Y, Looney MR, Cyster JG. 2022. GPR35 promotes neutrophil recruitment in response to serotonin metabolite 5-HIAA. *Cell* **185:** 815–830.e19. doi:10.1016/j.cell.2022.01.010

Girbl T, Lenn T, Perez L, Rolas L, Barkaway A, Thiriot A, Del Fresno C, Lynam E, Hub E, Thelen M, et al. 2018. Distinct compartmentalization of the chemokines CXCL1 and CXCL2 and the atypical receptor ACKR1 determine discrete stages of neutrophil diapedesis. *Immunity* **49:** 1062–1076.e6. doi:10.1016/j.immuni.2018.09.018

Glaser KM, Mihlan M, Lämmermann T. 2021. Positive feedback amplification in swarming immune cell populations. *Curr Opin Cell Biol* **72:** 156–162. doi:10.1016/j.ceb.2021.07.009

Gonzalez SF, Lukacs-Kornek V, Kuligowski MP, Pitcher LA, Degn SE, Kim YA, Cloninger MJ, Martinez-Pomares L, Gordon S, Turley SJ, et al. 2010. Capture of influenza by medullary dendritic cells via SIGN-R1 is essential for humoral immunity in draining lymph nodes. *Nat Immunol* **11:** 427–434. doi:10.1038/ni.1856

Grigorova IL, Schwab SR, Phan TG, Pham THM, Okada T, Cyster JG. 2009. Cortical sinus probing, S1P1-dependent entry and flow-based capture of egressing T cells. *Nat Immunol* **10:** 58–65. doi:10.1038/ni.1682

Groom JR, Richmond J, Murooka TT, Sorensen EW, Sung JH, Bankert K, von Andrian UH, Moon JJ, Mempel TR, Luster AD. 2012. CXCR3 chemokine receptor-ligand interactions in the lymph node optimize $CD4^+$ T helper 1 cell differentiation. *Immunity* **37:** 1091–1103. doi:10.1016/j.immuni.2012.08.016

Hauser AE, Junt T, Mempel TR, Sneddon MW, Kleinstein SH, Henrickson SE, von Andrian UH, Shlomchik MJ, Haberman AM. 2007. Definition of germinal-center B cell migration in vivo reveals predominant intrazonal circulation patterns. *Immunity* **26:** 655–667. doi:10.1016/j.immuni.2007.04.008

He Y, Heng Y, Qin Z, Wei X, Wu Z, Qu J. 2023. Intravital microscopy of satellite cell dynamics and their interaction with myeloid cells during skeletal muscle regeneration. *Sci Adv* **9:** eadi1891. doi:10.1126/sciadv.adi1891

Heydecker M, Shitara A, Chen D, Tran DT, Masedunskas A, Tora MS, Ebrahim S, Appaduray MA, Galeano Niño JL, Bhardwaj A, et al. 2024. Coordination of force-generating actin-based modules stabilizes and remodels membranes in vivo. *J Cell Biol* **223:** e202401091. doi:10.1083/jcb.202401091

Iismaa SE, Kaidonis X, Nicks AM, Bogush N, Kikuchi K, Naqvi N, Harvey RP, Husain A, Graham RM. 2018. Comparative regenerative mechanisms across different mammalian tissues. *NPJ Regen Med* **3:** 1–20. doi:10.1038/s41536-018-0044-5

Jamali A, Seyed-Razavi Y, Chao C, Ortiz G, Kenyon B, Blanco T, Harris DL, Hamrah P. 2020. Intravital multiphoton microscopy of the ocular surface: alterations in conventional dendritic cell morphology and kinetics in dry eye disease. *Front Immunol* **11:** 742. doi:10.3389/fimmu.2020.00742

Junt T, Moseman EA, Iannacone M, Massberg S, Lang PA, Boes M, Fink K, Henrickson SE, Shayakhmetov DM, Di Paolo NC, et al. 2007. Subcapsular sinus macrophages in lymph nodes clear lymph-borne viruses and present them to antiviral B cells. *Nature* **450:** 110–114. doi:10.1038/nature06287

Kastenmüller W, Torabi-Parizi P, Subramanian N, Lämmermann T, Germain RN. 2012. A spatially-organized multicellular innate immune response in lymph nodes limits systemic pathogen spread. *Cell* **150:** 1235–1248. doi:10.1016/j.cell.2012.07.021

Kikuta J, Wada Y, Kowada T, Wang Z, Sun-Wada GH, Nishiyama I, Mizukami S, Maiya N, Yasuda H, Kumanogoh A, et al. 2013. Dynamic visualization of RANKL and Th17-mediated osteoclast function. *J Clin Invest* **123:** 866–873. doi:10.1172/JCI65054

Kim J, Bixel MG. 2020. Intravital multiphoton imaging of the bone and bone marrow environment. *Cytometry A* **97:** 496–503. doi:10.1002/cyto.a.23937

Kim JV, Kang SS, Dustin ML, McGavern DB. 2009. Myelomonocytic cell recruitment causes fatal CNS vascular injury during acute viral meningitis. *Nature* **457:** 191–195. doi:10.1038/nature07591

Kolter J, Feuerstein R, Zeis P, Hagemeyer N, Paterson N, d'Errico P, Baasch S, Amann L, Masuda T, Lösslein A, et al. 2019. A subset of skin macrophages contributes to the surveillance and regeneration of local nerves. *Immunity* **50:** 1482–1497.e7. doi:10.1016/j.immuni.2019.05.009

Kyratsous NI, Bauer IJ, Zhang G, Pesic M, Bartholomäus I, Mues M, Fang P, Wörner M, Everts S, Ellwart JW, et al. 2017. Visualizing context-dependent calcium signaling in encephalitogenic T cells in vivo by two-photon microscopy. *Proc Natl Acad Sci* **114:** E6381–E6389. doi:10.1073/pnas.1701806114

Laird DJ, von Andrian UH, Wagers AJ. 2008. Stem cell trafficking in tissue development, growth, and disease. *Cell* **132:** 612–630. doi:10.1016/j.cell.2008.01.041

Lämmermann T, Bader BL, Monkley SJ, Worbs T, Wedlich-Söldner R, Hirsch K, Keller M, Förster R, Critchley DR, Fässler R, et al. 2008. Rapid leukocyte migration by integrin-independent flowing and squeezing. *Nature* **453:** 51–55. doi:10.1038/nature06887

Lämmermann T, Afonso PV, Angermann BR, Wang JM, Kastenmüller W, Parent CA, Germain RN. 2013. Neutrophil swarms require LTB4 and integrins at sites of cell death in vivo. *Nature* **498:** 371–375. doi:10.1038/nature12175

Lee WY, Sanz MJ, Wong CHY, Hardy PO, Salman-Dilgimen A, Moriarty TJ, Chaconas G, Marques A, Krawetz R, Mody CH, et al. 2014. Invariant natural killer T cells act as an extravascular cytotoxic barrier for joint-invading Lyme *Borrelia*. *Proc Natl Acad Sci* **111:** 13936–13941. doi:10.1073/pnas.1404769111

Liew PX, Lee WY, Kubes P. 2017. iNKT cells orchestrate a switch from inflammation to resolution of sterile liver injury. *Immunity* **47:** 752–765.e5. doi:10.1016/j.immuni.2017.09.016

Lindquist RL, Shakhar G, Dudziak D, Wardemann H, Eisenreich T, Dustin ML, Nussenzweig MC. 2004. Visualizing dendritic cell networks in vivo. *Nat Immunol* **5:** 1243–1250. doi:10.1038/ni1139

Lok LSC, Walker JA, Jolin HE, Scanlon ST, Ishii M, Fallon PG, McKenzie ANJ, Clatworthy MR. 2021. Group 2 innate lymphoid cells exhibit tissue-specific dynamic behaviour during type 2 immune responses. *Front Immunol* **12:** 711907. doi:10.3389/fimmu.2021.711907

Masedunskas A, Sramkova M, Parente L, Sales KU, Amornphimoltham P, Bugge TH, Weigert R. 2011. Role for the actomyosin complex in regulated exocytosis revealed by intravital microscopy. *Proc Natl Acad Sci* **108:** 13552–13557. doi:10.1073/pnas.1016778108

Masedunskas A, Milberg O, Porat-Shliom N, Sramkova M, Wigand T, Amornphimoltham P, Weigert R. 2012a. Intravital microscopy: a practical guide on imaging intracellular structures in live animals. *Bioarchitecture* **2:** 143–157. doi:10.4161/bioa.21758

Masedunskas A, Porat-Shliom N, Rechache K, Aye MP, Weigert R. 2012b. Intravital microscopy reveals differences in the kinetics of endocytic pathways between cell cultures and live animals. *Cells* **1:** 1121–1132. doi:10.3390/cells1041121

Masedunskas A, Chen Y, Stussman R, Weigert R, Mather IH. 2017. Kinetics of milk lipid droplet transport, growth, and secretion revealed by intravital imaging: lipid droplet release is intermittently stimulated by oxytocin. *Mol Biol Cell* **28:** 935–946. doi:10.1091/mbc.e16-11-0776

Massberg S, Schaerli P, Knezevic-Maramica I, Köllnberger M, Tubo N, Moseman EA, Huff IV, Junt T, Wagers AJ, Mazo IB, et al. 2007. Immunosurveillance by hematopoietic progenitor cells trafficking through blood, lymph, and peripheral tissues. *Cell* **131:** 994–1008. doi:10.1016/j.cell.2007.09.047

Mazo IB, Gutierrez-Ramos JC, Frenette PS, Hynes RO, Wagner DD, von Andrian UH. 1998. Hematopoietic progenitor cell rolling in bone marrow microvessels: parallel contributions by endothelial selectins and vascular cell adhesion molecule 1. *J Exp Med* **188:** 465–474. doi:10.1084/jem.188.3.465

McDonald B, McAvoy EF, Lam F, Gill V, de la Motte C, Savani RC, Kubes P. 2008. Interaction of CD44 and hyaluronan is the dominant mechanism for neutrophil sequestration in inflamed liver sinusoids. *J Exp Med* **205:** 915–927. doi:10.1084/jem.20071765

McDonald B, Pittman K, Menezes GB, Hirota SA, Slaba I, Waterhouse CCM, Beck PL, Muruve DA, Kubes P. 2010. Intravascular danger signals guide neutrophils to sites of sterile inflammation. *Science* **330:** 362–366. doi:10.1126/science.1195491

Mehari FT, Miller M, Pick R, Bader A, Pekayvaz K, Napoli M, Uhl B, Reichel CA, Sperandio M, Walzog B, et al. 2022. Intravital calcium imaging in myeloid leukocytes identifies calcium frequency spectra as indicators of functional states. *Sci Signal* **15:** eabe6909. doi:10.1126/scisignal.abe6909

Mehta P, Rahman Z, ten Dijke P, Boukany PE. 2022. Microfluidics meets 3D cancer cell migration. *Trends Cancer* **8:** 683–697. doi:10.1016/j.trecan.2022.03.006

Melis N, Subramanian B, Chen D, Weigert R. 2022. Imaging neutrophil migration in the mouse skin to investigate subcellular membrane remodeling under physiological conditions. *J Vis Exp* 10.3791/63581. doi:10.3791/63581

Mempel TR, Henrickson SE, von Andrian UH. 2004. T-cell priming by dendritic cells in lymph nodes occurs in three distinct phases. *Nature* **427:** 154–159. doi:10.1038/nature02238

Metchnikoff E. 1893. *Lectures on the comparative pathology of inflammation delivered at the Pasteur Institute in 1891*. Kegan Paul, Trench, Trübner, London.

Mihlan M, Safaiyan S, Stecher M, Paterson N, Lämmermann T. 2022. Surprises from intravital imaging of the innate immune response. *Annu Rev Cell Dev Biol* **38:** 467–489. doi:10.1146/annurev-cellbio-120420-112849

Milberg O, Shitara A, Ebrahim S, Masedunskas A, Tora M, Tran DT, Chen Y, Conti MA, Adelstein RS, Ten Hagen KG, et al. 2017. Concerted actions of distinct nonmuscle myosin II isoforms drive intracellular membrane remodeling in live animals. *J Cell Biol* **216:** 1925–1936. doi:10.1083/jcb.201612126

Miller MJ, Wei SH, Cahalan MD, Parker I. 2003. Autonomous T cell trafficking examined in vivo with intravital two-photon microscopy. *Proc Natl Acad Sci* **100:** 2604–2609. doi:10.1073/pnas.2628040100

Miyabe Y, Miyabe C, Murooka TT, Kim EY, Newton GA, Kim ND, Haribabu B, Luscinskas FW, Mempel TR, Luster AD. 2017. Complement C5a receptor is the key initiator of neutrophil adhesion igniting immune complex–induced arthritis. *Sci Immunol* **2:** eaaj2195. doi:10.1126/sciimmunol.aaj2195

Mukhopadhyay A, Tsukasaki Y, Chan WC, Le JP, Kwok ML, Zhou J, Natarajan V, Mostafazadeh N, Maienschein-Cline M, Papautsky I, et al. 2024. *trans*-Endothelial neutrophil migration activates bactericidal function via Piezo1 mechanosensing. *Immunity* **57:** 52–67.e10. doi:10.1016/j.immuni.2023.11.007

Murooka TT, Deruaz M, Marangoni F, Vrbanac VD, Seung E, von Andrian UH, Tager AM, Luster AD, Mempel TR. 2012. HIV-infected T cells are migratory vehicles for viral dissemination. *Nature* **490:** 283–287. doi:10.1038/nature11398

Neupane AS, Willson M, Chojnacki AK, Castanheira FVES, Morehouse C, Carestia A, Keller AE, Peiseler M, DiGiandomenico A, Kelly MM, et al. 2020. Patrolling alveolar macrophages conceal bacteria from the immune system to maintain homeostasis. *Cell* **183:** 110–125.e11. doi:10.1016/j.cell.2020.08.020

Ng LG, Hsu A, Mandell MA, Roediger B, Hoeller C, Mrass P, Iparraguirre A, Cavanagh LL, Triccas JA, Beverley SM, et al. 2008. Migratory dermal dendritic cells act as rapid sensors of protozoan parasites. *PLoS Pathog* **4:** e1000222. doi:10.1371/journal.ppat.1000222

Nimmerjahn A, Kirchhoff F, Helmchen F. 2005. Resting microglial cells are highly dynamic surveillants of brain parenchyma in vivo. *Science* **308:** 1314–1318. doi:10.1126/science.1110647

Nishibu A, Ward BR, Jester JV, Ploegh HL, Boes M, Takashima A. 2006. Behavioral responses of epidermal Langerhans cells in situ to local pathological stimuli. *J Invest Dermatol* **126:** 787–796. doi:10.1038/sj.jid.5700107

Nitschké M, Aebischer D, Abadier M, Haener S, Lucic M, Vigl B, Luche H, Fehling HJ, Biehlmaier O, Lyck R, et al. 2012. Differential requirement for ROCK in dendritic cell migration within lymphatic capillaries in steady-state and inflammation. *Blood* **120:** 2249–2258. doi:10.1182/blood-2012-03-417923

Nobis M, Herrmann D, Warren SC, Kadir S, Leung W, Killen M, Magenau A, Stevenson D, Lucas MC, Reischmann N, et al. 2017. A RhoA-FRET biosensor mouse for intravital imaging in normal tissue homeostasis and disease contexts. *Cell Rep* **21:** 274–288. doi:10.1016/j.celrep.2017.09.022

Nourshargh S, Alon R. 2014. Leukocyte migration into inflamed tissues. *Immunity* **41:** 694–707. doi:10.1016/j.immuni.2014.10.008

Park D, Spencer JA, Koh BI, Kobayashi T, Fujisaki J, Clemens TL, Lin CP, Kronenberg HM, Scadden DT. 2012. Endogenous bone marrow MSCs are dynamic, fate-restricted participants in bone maintenance and regeneration. *Cell Stem Cell* **10:** 259–272. doi:10.1016/j.stem.2012.02.003

Paterson N, Lämmermann T. 2022. Macrophage network dynamics depend on haptokinesis for optimal local surveillance. *eLife* **11:** e75354. doi:10.7554/eLife.75354

Pearson C, Thornton EE, McKenzie B, Schaupp A-L, Huskens N, Griseri T, West N, Tung S, Seddon BP, Uhlig HH, et al. 2016. ILC3 GM-CSF production and mobilisation orchestrate acute intestinal inflammation. *eLife* **5:** e10066. doi:10.7554/eLife.10066

Petrie RJ, Doyle AD, Yamada KM. 2009. Random versus directionally persistent cell migration. *Nat Rev Mol Cell Biol* **10:** 538–549. doi:10.1038/nrm2729

Phan TG, Grigorova I, Okada T, Cyster JG. 2007. Subcapsular encounter and complement-dependent transport of immune complexes by lymph node B cells. *Nat Immunol* **8:** 992–1000. doi:10.1038/ni1494

Pourjafar M, Tiwari VK. 2024. Plasticity in cell migration modes across development, physiology, and disease. *Front Cell Dev Biol* **12:** 1363361. doi:10.3389/fcell.2024.1363361

Qi H, Egen JG, Huang AYC, Germain RN. 2006. Extrafollicular activation of lymph node B cells by antigen-bearing dendritic cells. *Science* **312:** 1672–1676. doi:10.1126/science.1125703

Qi H, Cannons JL, Klauschen F, Schwartzberg PL, Germain RN. 2008. SAP-controlled T–B cell interactions underlie germinal centre formation. *Nature* **455:** 764–769. doi:10.1038/nature07345

Quintanilla MA, Hammer JA, Beach JR. 2023. Non-muscle myosin 2 at a glance. *J Cell Sci* **136:** jcs260890. doi:10.1242/jcs.260890

Ritsma L, Ellenbroek SIJ, Zomer A, Snippert HJ, de Sauvage FJ, Simons BD, Clevers H, van Rheenen J. 2014. Intestinal crypt homeostasis revealed at single-stem-cell level by in vivo live imaging. *Nature* **507:** 362–365. doi:10.1038/nature12972

Rotty JD, Brighton HE, Craig SL, Asokan SB, Cheng N, Ting JP, Bear JE. 2017. Arp2/3 complex is required for macrophage integrin functions but is dispensable for FcR phagocytosis and in vivo motility. *Dev Cell* **42:** 498–513.e6. doi:10.1016/j.devcel.2017.08.003

Rua R, McGavern DB. 2015. Elucidation of monocyte/macrophage dynamics and function by intravital imaging. *J Leukoc Biol* **98:** 319–332. doi:10.1189/jlb.4RI0115-006RR

Scheele CLGJ, Herrmann D, Yamashita E, Lo Celso C, Jenne CN, Oktay MH, Entenberg D, Friedl P, Weigert R, Meijboom FLB, et al. 2022. Multiphoton intravital microscopy of rodents. *Nat Rev Methods Primers* **2:** 1–26. doi:10.1038/s43586-021-00091-6

Schmidt S, Moser M, Sperandio M. 2013. The molecular basis of leukocyte recruitment and its deficiencies. *Mol Immunol* **55:** 49–58. doi:10.1016/j.molimm.2012.11.006

Schwickert TA, Lindquist RL, Shakhar G, Livshits G, Skokos D, Kosco-Vilbois MH, Dustin ML, Nussenzweig MC. 2007. In vivo imaging of germinal centres reveals a dynamic open structure. *Nature* **446:** 83–87. doi:10.1038/nature05573

Secklehner J, Lo Celso C, Carlin LM. 2017. Intravital microscopy in historic and contemporary immunology. *Immunol Cell Biol* **95:** 506–513. doi:10.1038/icb.2017.25

SenGupta S, Parent CA, Bear JE. 2021. The principles of directed cell migration. *Nat Rev Mol Cell Biol* **22:** 529–547. doi:10.1038/s41580-021-00366-6

Shitara A, Malec L, Ebrahim S, Chen D, Bleck C, Hoffman MP, Weigert R. 2019. Cdc42 negatively regulates endocytosis during apical membrane maintenance in live animals. *Mol Biol Cell* **30:** 324–332. doi:10.1091/mbc.E18-10-0615

Stock J, Pauli A. 2021. Self-organized cell migration across scales—from single cell movement to tissue formation. *Development* **148:** dev191767. doi:10.1242/dev.191767

Subramanian BC, Melis N, Chen D, Wang W, Gallardo D, Weigert R, Parent CA. 2020. The LTB4–BLT1 axis regulates actomyosin and β2-integrin dynamics during neutrophil extravasation. *J Cell Biol* **219:** e201910215. doi:10.1083/jcb.201910215

Sung JH, Zhang H, Moseman EA, Alvarez D, Iannacone M, Henrickson SE, de la Torre JC, Groom JR, Luster AD, von Andrian UH. 2012. Chemokine guidance of central memory T cells is critical for antiviral recall responses in lymph nodes. *Cell* **150:** 1249–1263. doi:10.1016/j.cell.2012.08.015

Te Boekhorst V, Preziosi L, Friedl P. 2016. Plasticity of cell migration in vivo and in silico. *Annu Rev Cell Dev Biol* **32:** 491–526. doi:10.1146/annurev-cellbio-111315-125201

Tian H, Biehs B, Warming S, Leong KG, Rangell L, Klein OD, de Sauvage FJ. 2011. A reserve stem cell population in small intestine renders Lgr5-positive cells dispensable. *Nature* **478:** 255–259. doi:10.1038/nature10408

Uderhardt S, Martins AJ, Tsang JS, Lämmermann T, Germain RN. 2019. Resident macrophages cloak tissue microlesions to prevent neutrophil-driven inflammatory damage. *Cell* **177:** 541–555.e17. doi:10.1016/j.cell.2019.02.028

Ulbricht C, Leben R, Rakhymzhan A, Kirchhoff F, Nitschke L, Radbruch H, Niesner RA, Hauser AE. 2021. Intravital quantification reveals dynamic calcium concentration changes across B cell differentiation stages. *eLife* **10:** e56020. doi:10.7554/eLife.56020

Upadhaya S, Krichevsky O, Akhmetzyanova I, Sawai CM, Fooksman DR, Reizis B. 2020. Intravital imaging reveals motility of adult hematopoietic stem cells in the bone marrow niche. *Cell Stem Cell* **27:** 336–345.e4. doi:10.1016/j.stem.2020.06.003

Victora GD, Schwickert TA, Fooksman DR, Kamphorst AO, Meyer-Hermann M, Dustin ML, Nussenzweig MC. 2010. Germinal center dynamics revealed by multiphoton microscopy with a photoactivatable fluorescent reporter. *Cell* **143:** 592–605. doi:10.1016/j.cell.2010.10.032

Von Andrian UH, Berger EM, Ramezani L, Chambers JD, Ochs HD, Harlan JM, Paulson JC, Etzioni A, Arfors KE. 1993. In vivo behavior of neutrophils from two patients with distinct inherited leukocyte adhesion deficiency syndromes. *J Clin Invest* **91:** 2893–2897. doi:10.1172/JCI116535

von Brühl ML, Stark K, Steinhart A, Chandraratne S, Konrad I, Lorenz M, Khandoga A, Tirniceriu A, Coletti R, Köllnberger M, et al. 2012. Monocytes, neutrophils, and platelets cooperate to initiate and propagate venous thrombosis in mice in vivo. *J Exp Med* **209:** 819–835. doi:10.1084/jem.20112322

Waite JC, Leiner I, Lauer P, Rae CS, Barbet G, Zheng H, Portnoy DA, Pamer EG, Dustin ML. 2011. Dynamic imaging of the effector immune response to listeria infection in vivo. *PLoS Pathog* **7:** e1001326. doi:10.1371/journal.ppat.1001326

Wang J, Kubes P. 2016. A reservoir of mature cavity macrophages that can rapidly invade visceral organs to affect tissue repair. *Cell* **165:** 668–678. doi:10.1016/j.cell.2016.03.009

Wang YX, Rudnicki MA. 2012. Satellite cells, the engines of muscle repair. *Nat Rev Mol Cell Biol* **13:** 127–133. doi:10.1038/nrm3265

Wang X, Sumida H, Cyster JG. 2014. GPR18 is required for a normal CD8αα intestinal intraepithelial lymphocyte compartment. *J Exp Med* **211:** 2351–2359. doi:10.1084/jem.20140646

Wang J, Hossain M, Thanabalasuriar A, Gunzer M, Meininger C, Kubes P. 2017. Visualizing the function and fate of neutrophils in sterile injury and repair. *Science* **358:** 111–116. doi:10.1126/science.aam9690

Wang H, Li L, Li Y, Li Y, Sha Y, Wen S, You Q, Liu L, Shi M, Zhou H. 2021. Intravital imaging of interactions between iNKT and Kupffer cells to clear free lipids during steatohepatitis. *Theranostics* **11:** 2149–2169. doi:10.7150/thno.51369

Weber M, Hauschild R, Schwarz J, Moussion C, de Vries I, Legler DF, Luther SA, Bollenbach T, Sixt M. 2013. Interstitial dendritic cell guidance by haptotactic chemokine gradients. *Science* **339:** 328–332. doi:10.1126/science.1228456

Webster MT, Manor U, Lippincott-Schwartz J, Fan CM. 2016. Intravital imaging reveals ghost fibers as architectural units guiding myogenic progenitors during regeneration. *Cell Stem Cell* **18:** 243–252. doi:10.1016/j.stem.2015.11.005

Weigert R. 2014. Imaging the dynamics of endocytosis in live mammalian tissues. *Cold Spring Harb Perspect Biol* **6:** a017012. doi:10.1101/cshperspect.a017012

Wilharm A, Tabib Y, Nassar M, Reinhardt A, Mizraji G, Sandrock I, Heyman O, Barros-Martins J, Aizenbud Y, Khalaileh A, et al. 2019. Mutual interplay between IL-17–producing γδT cells and microbiota orchestrates oral mucosal homeostasis. *Proc Natl Acad Sci* **116:** 2652–2661. doi:10.1073/pnas.1818812116

Worbs T, Mempel TR, Bölter J, von Andrian UH, Förster R. 2007. CCR7 ligands stimulate the intranodal motility of T lymphocytes in vivo. *J Exp Med* **204:** 489–495. doi:10.1084/jem.20061706

Worbs T, Hammerschmidt SI, Förster R. 2017. Dendritic cell migration in health and disease. *Nat Rev Immunol* **17:** 30–48. doi:10.1038/nri.2016.116

Wright DE, Wagers AJ, Gulati AP, Johnson FL, Weissman IL. 2001. Physiological migration of hematopoietic stem and progenitor cells. *Science* **294:** 1933–1936. doi:10.1126/science.1064081

Yamada KM, Sixt M. 2019. Mechanisms of 3D cell migration. *Nat Rev Mol Cell Biol* **20:** 738–752. doi:10.1038/s41580-019-0172-9

Yamada KM, Doyle AD, Lu J. 2022. Cell–3D matrix interactions: recent advances and opportunities. *Trends Cell Biol* **32:** 883–895. doi:10.1016/j.tcb.2022.03.002

Yan SLS, Hwang IY, Kamenyeva O, Kehrl JH. 2019. In vivo F-actin filament organization during lymphocyte transendothelial and interstitial migration revealed by intravital microscopy. *iScience* **16:** 283–297. doi:10.1016/j.isci.2019.05.040

Yoshikawa S, Usami T, Kikuta J, Ishii M, Sasano T, Sugiyama K, Furukawa T, Nakasho E, Takayanagi H, Tedder TF, et al. 2016. Intravital imaging of Ca^{2+} signals in lymphocytes of Ca^{2+} biosensor transgenic mice: indication of autoimmune diseases before the pathological onset. *Sci Rep* **6:** 18738. doi:10.1038/srep18738

Zaid A, Mackay LK, Rahimpour A, Braun A, Veldhoen M, Carbone FR, Manton JH, Heath WR, Mueller SN. 2014. Persistence of skin-resident memory T cells within an epidermal niche. *Proc Natl Acad Sci* **111:** 5307–5312. doi:10.1073/pnas.1322292111

Soft Matter Physics Meets Cell Biology: Transitions of Collective Cell Migration in 3D Environments

Mirjam M. Zegers,[1] Pablo Gottheil,[2] Josef Käs,[2] and Peter Friedl[1]

[1]Department of Medical Biosciences, Radboud University Medical Centre, 6525 GA Nijmegen, Netherlands
[2]Peter Debye Institute for Soft Matter Physics, Leipzig University, 04103 Leipzig, Germany
Correspondence: peter.friedl@radboudumc.nl; jkaes@uni-leipzig.de

Plasticity of cell migration is a hallmark of cell movement during morphogenesis, tissue repair, and cancer metastasis. Interconversions of migration modes are tissue context–dependent and range from (1) collective migration of cohesive cells, migrating as epithelial sheets and strands; (2) multicellular networks of individualized cells moving while maintaining short-lived interactions; and (3) fully individualized cells moving by mesenchymal or amoeboid migration. Modes of cell migration, which are controlled by cell–cell adhesion, cell density, and active forces, can also be represented by physics-derived parameters, including temperature, applied stress, and volume fraction in classical passive jamming phase diagrams. Cell-packing density, cell–cell adhesion strength, and intrinsic migratory capacity have been defined as the key parameters driving jamming transitions in 2D sheet models, where extracellular matrix (ECM) is typically not considered. Here, we review how plasticity of cell migration programs intersects with jamming/unjamming principles and specifically focus on the impact of ECM architectures. In three-dimensional (3D) tissues, additional spatial parameters determine cell density and cell–cell interactions, including the degree of confinement forcing cells together versus the availability of free space. Integrating mechanisms of jamming/unjamming with actin-based active movement of cells in a 3D environment, similar to the motion of active nematic droplets in a passive nematic matrix, will enable building realistic models to predict cell behaviors in physiological and pathological contexts, including cancer metastasis.

The ability to migrate is intrinsic to nearly all metazoan cell types and biological processes, including development, wound healing, immune surveillance, and tissue homeostasis. For cells to change position and build tissues, specific evolutionarily conserved modes of migration have emerged in different cell types and environments. In mammalian tissues, cells migrate using mesenchymal, amoeboid, or collective modes (Te Boekhorst et al. 2016; Yamada and Sixt 2019). Mesenchymal migration is characterized by an elongated, spindle-like cell shape, which is maintained by frontal protrusion driven by actin polymerization and actomyosin-dependent rear retraction. Mesenchymal migration depends on strong adhesion to

Cite this article as *Cold Spring Harb Perspect Biol* doi: 10.1101/cshperspect.a041794

extracellular matrix (ECM) via integrin adhesion receptors and the cell's ability to remodel ECM both by traction forces and by proteolytic degradation to create the migration path (Wolf and Friedl 2011). Amoeboid migration occurs by a rounded cell shape with occasional elongation, mediated by actin-rich pseudopodal or blebbing protrusions in combination with cortical actomyosin contractions at the rear, which do not generate strong forces toward the ECM but suffice to mediate cell squeezing through spaces (Te Boekhorst et al. 2016; Yamada and Sixt 2019). Collective migration involves two or more cells connected by cadherin-mediated and/or other cell–cell interactions of variable strength that maintain multicellular coordination and directional persistence (Friedl and Mayor 2017; Giavazzi et al. 2018). Each migration mode is involved in particular processes, such as amoeboid leukocyte migration in inflammation, mesenchymal migration of fibroblasts upon wound healing, and collective migration of sprouting endothelial or epithelial cells to build vessels, ducts, glands, and epithelial barriers.

Migration modes are adaptive to various triggers, allowing cells to transition between migration strategies. Transitions between migration modes include collective-to-single-cell transition, mesenchymal-to-amoeboid transition, and interconversion between cohesive-to-network-like and swarming collective movement (Te Boekhorst and Friedl 2016; Deutsch et al. 2020). Transitions between migration modes are mediated by adaptation of cell–cell and cell–matrix interactions, and changes in cytoskeletal organization (Friedl and Wolf 2010) and regulated upstream by genetically determined transcriptional programs during morphogenesis and/or by extracellular signals including cytokine signaling, growth factor activation, or ECM stiffness (Clark and Vignjevic 2015). For example, epithelial cells maintain nearly stationary behavior when forming barriers or ducts, as multicellular sheets during tissue homeostasis, but also adopt highly motile behaviors during cancer invasion, where they may form moving multicellular sheets, loosely connected groups, or become completely individualized (Te Boekhorst and Friedl 2016). Stromal fibroblasts and leukocytes, which typically migrate as individual cells through three-dimensional (3D) tissues, can establish cell–cell interactions and move as multicellular swarms or networks when fibroblasts are activated by injury (Jiang et al. 2020) or when leukocytes encounter a collagen IV–rich ECM (Rey-Barroso et al. 2018). Each migration mode can be defined using molecular and physical metrics, such as actomyosin contractility, force generation, and the extent of ECM degradation, including features unique to each mode, which provides a foundation for standardizing migration research and distinguishing general principles from mode-specific mechanisms (Hu et al. 2023).

As a concurrent development to live-cell-based analyses of cell migration, transitions between migration modes have recently been explored through theoretical approaches using principles of colloidal physics as inspiration (Table 1; Sadati et al. 2013; Blauth et al. 2021). Inert materials, such as granular matter, foams, and colloids, were used as a basis for exploring their analogies in biological tissues (Lawson-Keister and Manning 2021; Fredberg 2022). Accordingly, solid, liquid–crystal, liquid, or gas states of inert materials depend on the interactions and organization of their constituent particles. In solid-like, crystalline, or jammed states, particles are tightly packed and positionally frozen by their constraints. Conversely, liquid–crystal states allow streaming motion, and fluid-like, unjammed states allow for neighbor exchanges of particles. These transitions depend on state variables such as packing densities, shear stresses, and viscoelasticity and contributions from thermal or active energy (Liu and Nagel 2023). Due to its intuitive appeal, "unjamming" has established itself as the overarching term for the mechanical concept of cellular motility transitions (Bi et al. 2016; Atia et al. 2021; Lawson-Keister and Manning 2021). Unjamming was originally coined for colloidal systems and emphasizes interactions through steric effects, such as Onsager repulsion (Onsager 1949). The Onsager repulsion predicts that elongated particles spontaneously align at high densities because parallel alignment minimizes the volume each particle excludes from others,

Table 1. Soft matter physics meets cell biology: terms and definitions

State[a]	Biological features	Physical concept	Example	Key molecular elements	References
1. Crystalline	Nonmotile cells with high cell–cell adhesion at high-cell density with stable anchorage to the substrate, hexagonal order, transition to jammed	Solid-like ordered	Adult nonregenerating epithelia (glands, ducts), epidermis	AJ, TJ, integrin, and hemidesmosome junctions to basement membrane, apicobasal polarity, cell and tissue stiffness	Classen et al. 2005; Armengol-Collado et al. 2023
2. Jammed	Nonmotile cells, confluent clusters (100% cell packing fraction), high number densities, amorphous order, transition to sheet motion	Solid-like unordered	Jammed cancer clusters, bronchial epithelial cell layers, early dysplasia	High cell number densities, roundish cells, cell and tissue stiffness	Atia et al. 2018; Gottheil et al. 2023
3. Sheet motion of a solid	Cells migrate as coordinated cohorts with correlated speed and direction, no neighbor exchange, stable cell–cell adhesion, high cell density, movement along ordered continuous substrate, transition to nematic flocking	Solid-like, unordered, jammed sheet motion	Regenerating epithelia during wound healing, gut epithelia moving up the villi, rotating egg chamber	AJ, TJ, apicobasal polarity, basal dynamic anchorage to ECM (e.g., basement membrane) through integrins	Giavazzi et al. 2018; Krndija et al. 2019; Williams and Horne-Badovinac 2024; Shen et al. 2025
4. Polar nematic flocking in cell clusters	Unidirectional streams of cells with neighbor exchanges, transition to unjammed motion	Cells can only move in one direction of cell alignment and are jammed in all other directions	Tumor cell clusters of rhabdomyosarcoma, melanoma, and breast and cervix carcinomas, *Drosophila* border cell migration	AJ, traction forces to enable movement depend on integrin–ECM adhesion	Friedl et al. 1995; Hegerfeldt et al. 2002; Peercy and Starz-Gaiano 2020; Fuhs et al. 2022; Shen et al. 2025
5. Active nematic motion in cell clusters	Cells are aligned as in streams, but they can move in both directions of the alignment	In the direction of alignment, cells behave like a fluid and can move in both directions of the alignment; in all other directions, cells are jammed	Cell monolayers of stem cells, epithelial cells, cancer cells, adipocytes; morphogenesis in muscles	Dynamic AJs, cell–ECM traction forces	Malinverno et al. 2017; Saw et al. 2017; Blanch-Mercader et al. 2018, 2021; Guillamat et al. 2022

Continued

Table 1. *Continued*

State[a]	Biological features	Physical concept	Example	Key molecular elements	References
6. Strand invasion by polar nematic strands	Independently moving cells with maintained cell–cell adhesion and correlated direction, but some neighbor exchange; cells depend on cell–ECM and cell–cell adhesion for forward motion	Active, polar nematic motion in strands (anisotropic fluid-like); viscous motion with instabilities	Leader/tip cell–guided collective migration, endothelia, invading carcinoma strands	AJ, mechanocoupling, traction forces to enable movement depend on integrin–ECM adhesion	Xi et al. 2017; Chepizhko et al. 2018; Ilina et al. 2020; Kato et al. 2023
7. Strand invasion by active nematic motion	Cells with low cell–cell adhesion, confined together within narrow ordered (ECM) spaces at high cell density; parallel and antiparallel motion with frequent neighbor exchange; cells rely mostly on cell–ECM adhesion for self-propulsion	Active nematic motion in single-file geometries (anisotropic fluid-like)	Mesenchymal cells in ECM channels; oncostreams in glioblastoma and breast tumors	No AJ, confining ordered ECM, cell–ECM adhesion prevails over cell–cell adhesion	Comba et al. 2022; Law et al. 2023; Barberis et al. 2024; Argento et al. 2025; Chepizhko et al. 2025
8. Unjammed, active fluid with retained cell–cell contact	Cells migrate independently without directional coordination with random neighbor exchanges; traction through intercellular contacts	Fluid-like, noisy unjammed motion with diffusion-like displacements	Cancer cells in the tumor core	Low-affinity cell–cell interactions with high turnover (high cadherin dynamics, ALCAM)	Park et al. 2015; Atia et al. 2018; Staneva et al. 2019; Grosser et al. 2021; Gottheil et al. 2023
9. Unjammed in 3D confinement	Cells migrate independently without directional coordination with random neighbor exchanges; cells are forced together by extracellular confinement	Fluid-like unjammed motion with variable persistence	Cancer cells after complete EMT, experimental depletion of AJs	Extremely weak cell–cell cohesion, migration mediated through ECM interactions	Rey-Barroso et al. 2018; Ilina et al. 2020; Kang et al. 2021

Continued

Table 1. *Continued*

State[a]	Biological features	Physical concept	Example	Key molecular elements	References
10. Gas-like individual cell motion (mesenchymal)	Cells without cell–cell interaction at low cell density in low-density ECM	Gas-like random single-cell motion	Immune cells in interstitial matrix, single mesenchymal cells during development or cancer invasion	No AJ, integrin-dependent or -independent interaction to fibrillar ECM with pores; interaction with isotropically ordered ECM prevails over cell–cell adhesion	Haeger et al. 2014; Zanotelli et al. 2024
11. Gas-like individual cell motion (amoeboid)	Amoeboid cells in ECM or within cell aggregates	Gas-like random single-cell motion, plasma fluctuations	Amoeboid moving leukocytes	Interactions are solely generated by friction, no adhesion	Friedl et al. 1998; Lämmermann et al. 2008

(AJ) Adherens junction, (TJ) tight junction, (ECM) extracellular matrix, ALCAM, activated leukocyte cell adhesion molecule; (EMT) epithelial-to-mesenchymal transition, (3D) three dimensional.

[a]Numbers refer to states depicted in Figure 1B.

allowing for more efficient packing. In a cell and tissue context, the term unjamming has been used to describe fluid-like tissue states with cells changing positions relative to neighbor cells. Unjamming can be induced by weakening of cell–cell interactions and/or by increased cell–ECM traction forces that maintain or even strengthen cell–cell adhesions while supporting active motion. Fluid-like motion lacks positional order and thereby differs from, for example, anisotropic, nematic fluid-like states and directed collective states. In this context, "nematic" describes elongated cells with aligned principal axes within a shared neighborhood, typically corresponding to the migration direction. The key distinction between nematic and fully fluid phases lies in the directional degrees of freedom available to motile cells. A nematic phase represents a liquid crystalline state where cells can move fluidly along the alignment direction but remain jammed in all other directions. In contrast, a fully fluid (isotropically unjammed) state allows unrestricted movement in all directions. Within nematic phases, "active nematic" permits movement in both directions along the alignment axis, while "polar nematic" restricts motion to a single direction along that axis. These nematic states represent "ordered" motion since cell orientation is nonrandom within local neighborhoods, positioning them as intermediate states between fully jammed (nonmotile) and completely isotropically unjammed conditions.

By connecting immobile, solid-like behavior with mobile, fluid-like behaviors, state or phase diagrams have helped to better understand the (active) material properties of biological tissues (Fig. 1). These diagrams distinguish the following states: (1) largely immobile positioning, defined as solid-like crystalline or solid-like jammed (amorphous) state; (2) ordered motion, including anisotropic states of sheet-like motion, solid flocking, and active polar nematic motion. Solid flocking refers to the process where complete strands or sheets of cells move, but the cells retain positional stability to their neighbors; thus, only the "flock" is motile. Strand invasion is not a technical term, yet it aims to describe the morphology of cell collectives exploring the microenvironment in a strand-like manner; (3) unjammed random motion as isotropic fluid-like cell groups; and (4) gas-like states of moving single cells (Table 1). This nomenclature can help to gain mechanistic insights and predict dynamic cell behaviors and their transitions in response to changing conditions (Angelini et al. 2011; Park et al. 2015; Mongera et al. 2018; Palamidessi et al. 2019; Ilina et al. 2020; Gottheil et al. 2023, 2024).

In this review, we explore wet-laboratory and theoretical approaches describing the adaptive responses of cell migration in the light of the jamming/unjamming paradigm, side-by-side. We will highlight how the nonbinary, gradual, or stepwise plasticity of cell migration can be represented by concepts of passive or active soft matter physics to describe essential mechanisms of adaptive cell responses to environmental triggers. This will allow us to classify and ultimately to quantify different types, transitions, and instabilities of collective cell migration modes, including cohesive, coordinated migration reshaping the tissue and active fluids flowing along tissue boundaries into interstitial space.

TRANSITIONS OF CELL MIGRATION AND INTERMEDIATE STATES

The adaptive responses to cell-intrinsic or extracellular stimuli enable migrating cells to adapt cell–cell and cell–matrix adhesion, cell polarity, and cytoskeletal dynamics.

Cadherin-based adherens junctions (AJs) are important regulators of the transitions in collective behaviors. Cadherins regulate cell–cell interactions between neighboring cells via homophilic *trans*-interactions of their extracellular domains and through their intracellular domains connecting to cortical actin and actin cables to regulate active force transmission, traction, and tension across multiple cells (Hidalgo-Carcedo et al. 2011). E-cadherin provides the highest stability of cell–cell interactions in epithelial tissues (Campàs et al. 2024). Other cadherins, including P-cadherin, VE-cadherin, and cadherin-6, contribute to AJs of variable stability and dynamics (Halbleib and Nelson 2006; Vitorino and Meyer 2008; Garcia et al. 2015; Plutoni et al. 2016). The stability of cell–cell interactions is regulated by intracellular adaptors, including catenins, and/or transcriptional programs, which regulate the expression of cadherins and other adhesion molecules connecting cells to form multicellular groups, strands, and sheets (Friedl and Mayor 2017). The stability of AJs is reinforced by mechanical tension, which promotes cadherin signaling, actin polymerization, and increases adhesion strength (Warmt et al. 2021; Campàs et al. 2024). AJs stabilize the position of cells relative to their neighbors by junctional tension, which aligns cells along the main migration direction (Trepat et al. 2009; Cai et al. 2014; Ozawa et al. 2020). Thereby, AJs enhance cell–cell coordination between neighboring cells, typically in concert with signaling from integrin–ECM interaction and actomyosin cables (Mui et al. 2016; Hadjisavva et al. 2022), thereby securing leader–follower cell coordination (Reffay et al. 2014) and preventing detachment of individual cells.

The roles of tight junctions (TJs) in transitions in collective behaviors remain poorly defined. TJs consist of transmembrane proteins including junctional adhesion molecules (JAMs), claudins, and occludin. In epithelia, TJs form mechanosensitive complexes, which develop alongside and cooperate with AJs (Campbell et al. 2017) to maintain apicobasolateral cell polarization and epithelial barrier integrity (Citi et al. 2024). By connecting with the actomyosin cytoskeleton, particularly through ZO1 and ZO2, TJs regulate the balance between apical and basolateral contractility, cell–cell adhesion strength, and cell elongation. These functions enable the directional, active polar nematic motion of stably connected cell sheets, where cells exhibit orientational order by aligning their elongated shapes with their migration direction (Skamrahl et al. 2021; Jipp et al. 2024). Additionally, contractile forces acting on TJs stimulate leading edge dynamics through Rac1 activation, which contributes to the polarity and nematic movement of cohesive groups (Das et al. 2015). Taken together, TJs are important regulators of cell shape and cellular tension during sheet motion and polar nematic migration. However, to what extent TJs control, or are abandoned during polar and active nematic motion, when TJs no longer act as barriers, remains to be established.

In epithelial cell sheets, tricellular AJs (tAJs) and tricellular TJs (tTJs) form at the sites where three cells meet and where junctional forces are high (Higashi and Miller 2017; Bosveld et al. 2018). Tricellular junctions act as mechanical stress coordinators that prevent intercellular gaps where tensile or compressive forces are highest and thereby maintain integrity of cell sheets (Kim et al. 2021; Cho et al. 2022). For example, the tAJ-specific adhesion protein Sidekick prevents intercellular gap formation during cellular nematic rearrangements in embryonic *Drosophila* epithelia (Finegan et al. 2019; Letizia et al. 2019). Beyond their structural role, the ability of tTJs and tAJs to regulate tension and cell elongation has been proposed to play key roles in the fluidization of epithelial sheets (Campàs et al. 2024). Tricellular junctions further contribute to collective cell motion by planar cell polarity signaling and Wiskott–Aldrich syndrome protein family verprolin homologous (WAVE)-dependent basal actin polymerization (Squarr et al. 2016). The importance of tricellular junctions in collective dynamics is corroborated by mathematical modeling, which identified tTJ maturation state as a key biological descriptor for jamming transitions in epithelial monolayers (Lohmann et al. 2020).

ADAPTIVE RESPONSE OF CELL–CELL INTERACTIONS

Consistent with their function in cell–cell coordination, downregulation of AJs or TJs promotes a transition from cohesive, jammed state toward fluid-like unjammed motion and further facilitates single-cell detachment toward a gas-like state (Fig. 1; Das et al. 2015; Ilina et al. 2020). The stability of AJs is primarily controlled by endocytosis and recycling of E-cadherin (Katsuno-Kambe and Yap 2020; Sigismund et al. 2021). This process can be triggered by phosphorylation of its stabilizing binding partner p120-catenin (Fukumoto et al. 2008), by actomyosin tension (Iyer et al. 2019), or through the activation of endocytic regulators such as Rab5 (Malinverno et al. 2017) and Rab11 (Aiello et al.

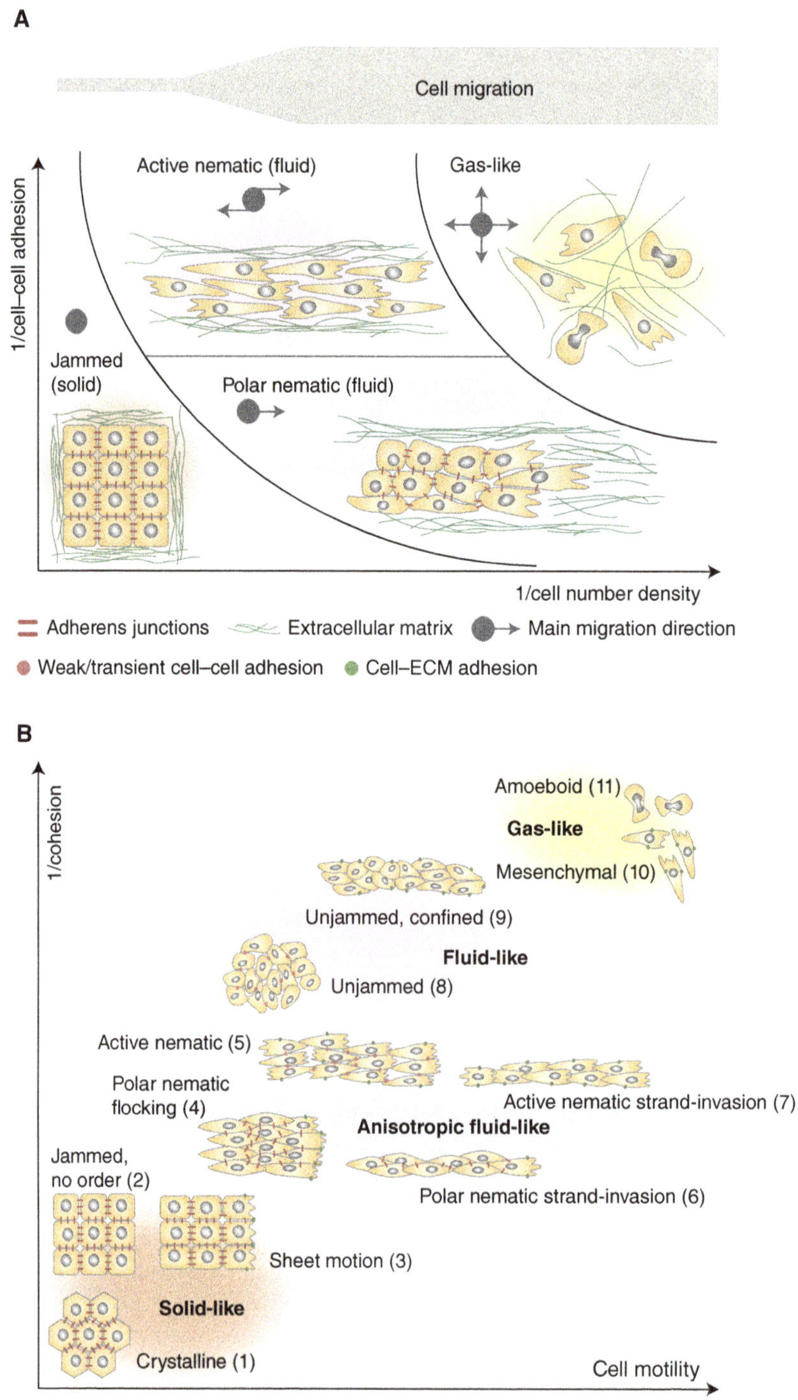

Figure 1. (*See following page for legend.*)

2018). AJs are further downregulated by proteolytic cleavage of E-cadherin by extracellular proteases (Hu et al. 2016), which additionally results in competition and autoinhibition of homotypic E-cadherin interactions by cleaved E-cadherin ectodomains (Venhuizen et al. 2020). Soluble cadherin ectodomains and full-length E-cadherin can activate receptor tyrosine kinases (RTKs), including epidermal growth factor receptor (EGFR), by acting as ligands or by transactivation, respectively, and thereby enhance front–rear cell polarization and directional migration (Brouxhon et al. 2014; Fu et al. 2024). As a result of a decreased AJ stability, many pathways, including Wingless-related integration site (Wnt)/β-catenin, transforming growth factor β (TGF-β), RhoGTPases, phosphoinositide 3-kinase (PI3K)/AKT serine/threonine kinase/protein kinase B (AKT), mitogen-activated protein kinase (MAPK)/extracellular signal–regulated kinase (ERK), Hippo/Yes-associated protein (YAP)/transcriptional coactivator with PDZ-binding motif (TAZ), and focal adhesion kinase (FAK)/Src are activated, which directly or indirectly favor motility (Venhuizen et al. 2020).

In morphogenesis and cancer, the downregulation of AJs and TJs is a hallmark of the epithelial-to-mesenchymal transition (EMT). EMT can be induced by various signaling pathways, including cytokines activating EMT-related transcription (e.g., TGF-β, Wnt, epidermal growth factor [EGF]), environmental triggers including hypoxia, or altered topology or mechanics of the extracellular environment, such as increased substrate stiffness resulting from, for example, fibrosis, allowing cells to adapt to varying microenvironmental cues (Haerinck et al. 2023). EMT causes epithelial cells to reduce cell–cell adhesion, change cell polarity from apicobasal to front–rear orientation, and gain cell–ECM interactions to develop into mesenchymally migrating cells (Pastushenko and Blanpain 2019). Depending on the strength of stimuli, EMT can be partial or complete along the epithelial-to-mesenchymal spectrum, leading to cell migration with variably mature cell–cell interactions or complete cell individualization.

In EMT-dependent and other types of AJ deregulation, the responses are gradual and probabilistic rather than deterministic in activating collective or single-cell migration: Cells may (1) retain cohesion and develop high persistence along shared trajectories, (2) increase dynamics of cell–cell interactions and exhibit more frequent neighbor exchanges, (3) exhibit rotational or swirling patterns (vorticity), (4) display parallel streaming with noise, or (5) undergo cell detachment and individualization (Fig. 2). These behaviors represent a spectrum from cohesive to dispersive states. The intermediate patterns, particularly rotational movements and streaming behaviors, often indicate proximity to jamming/unjamming transitions, where the correlation

Figure 1. Modes of collective and single-cell migration and nomenclature developed by jamming–unjamming concepts. (*A*) Hypothetical jamming phase diagram for cell invasion into three-dimensional (3D) tissue. Cells maintaining stable cell–cell junctions in tissue conditions of high confinement are immobile (jammed, red shading). Decreasing extracellular matrix (ECM) density and reduced stability of cell–cell contacts result in polar nematic migration of cohesive groups (blue shading) or fluid-like movement of weakly connected cell groups (magenta shading). When cell–cell contacts are very weak and ECM conditions are porous, cell individualization corresponds to gas-like behavior (yellow shading). (*B*) Expanded nomenclature of cell migration modes and transitions based on cell cohesion states and jamming transitions (as defined in Table 1). Movement of solid-like cell clusters occurs as sheet migration without neighbor exchange due to strong cell–cell adhesions by adherens junctions (AJs). Polar nematic flocking and strand invasion are distinguished from motion by active nematic and active nematic strand invasion by the level of cell–cell interactions, which enable unidirectional motion in polar nematic migration types. Fluid-like, uncoordinated motion can occur in cell clusters, with very low cellular interactions to enable cell motion, whereas in confined 3D ECM spaces, cellular interactions are absent, and cell motion is driven by cell–ECM interactions. Shading colors correspond to the main categories indicated in *A*. Colored dots indicate the main source of traction force to enable cell migration. Numbers in parentheses refer to the classification in Table 1.

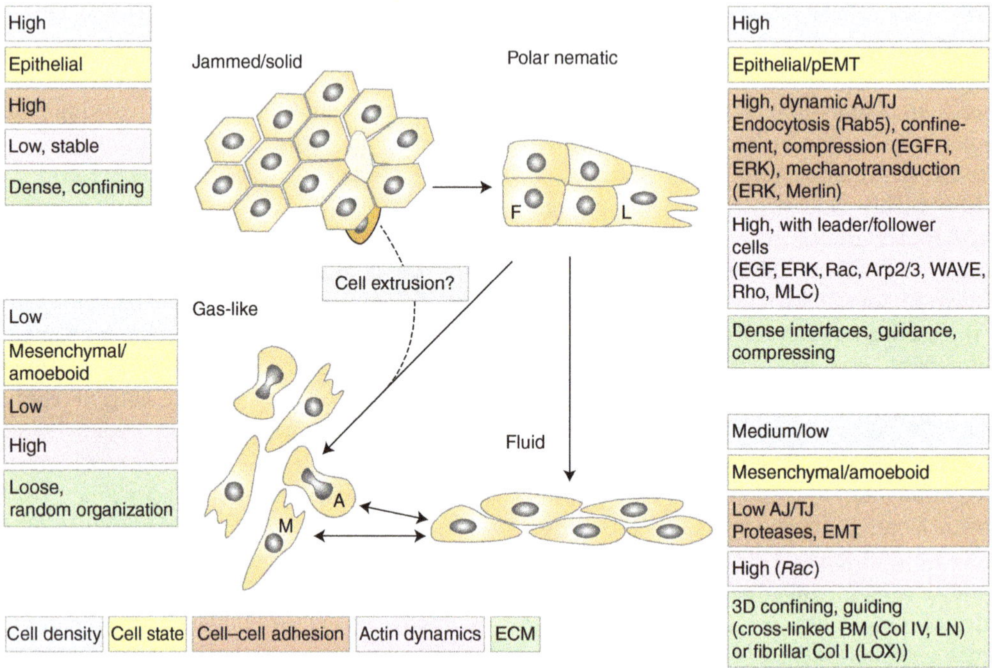

Figure 2. Integration of jamming/unjamming states of cell migration and underlying molecular programs (boxes). (A) Amoeboid, (AJ) adherens junction, (BM) basement membrane, (Col) collagen, (F) follower cells, (L) leader cell, (LOX) lysyl oxidase, (M) mesenchymal, (pEMT) partial epithelial-to-mesenchymal transition, (TJ) tight junction.

length (the distance over which cells coordinate migration) increases significantly. The concept of "noise" in computational analysis of collective movements refers to stochastic fluctuations that deviate from the main migration orientation, arising from both cell-intrinsic processes and correlated interactions that reduce coordination between neighboring cells. Consequently, all forms of correlated multicellular migration, including parallel streaming, which exhibit varying levels of cell–cell coordination, comprise varying levels of noise. This overall progression from strongly adherent, solid-like epithelia to more loosely connected anisotropic and fluid-like tissue behaviors can be understood as the "fluidization of tissues."

GENERAL MECHANISMS OF UNJAMMING TRANSITIONS

Based on physical concepts, experimental data, and computational modeling, motility transitions in cell groups are defined by multiple concurrent morphological parameters, including changes of cell shape, nucleus morphology, cell volume fraction, and, hence, the packing density of nuclei in the group. Jamming transitions are further defined by changes in cell–cell adhesion creating active pulling and pushing stresses generated between cells via the cytoskeleton (Fig. 2). Whether a cell collective is in a more solid-like or liquid-like state is effectively determined by the combined metrics of cell–cell adhesion, cell and nucleus density, and cell and nuclear shape (Table 1). Fluid-like behavior is associated with prominent cell elongation and a lower number of cells per unit volume or area (cell number density), because individual cells exploit available space and respond to neighbor dynamics by cytoskeletal rearrangement (Grosser et al. 2021; Gottheil et al. 2023). Elongated cell shapes are closely linked to nuclear elongation, with round-to-elongated transitions in both nuclei and cells correlating with increased fluid-like motion and bulk tissue fluidity in cancer cell models (Grosser et al. 2021; Cai et al. 2022; Gottheil et al. 2023; Sauer et al. 2023).

With high cell–cell interaction strength, in the solid-like, jammed state, tissues exhibit high cell number densities, high apicobasal polarity, and high lateral anchorage paired with low-to-absent basal traction forces toward the substrate, as found in ordered epithelia confined by dense, hexagonal-like, immobile packing (Bi et al. 2016; Atia et al. 2021; Lawson-Keister and Manning 2021). At intermediate cell–cell interaction strength, a liquid crystalline phase enables active nematic behavior, with cells exhibiting directionally constrained movement. This nematic migration is characterized by increased dynamics of cell–ECM contacts, transitioning from positionally stable adhesions to coordinated polarized traction forces. While cells align along preferred directions, cell–cell contacts by AJs remain intact and keep cells physically connected and spatially crowded, so that neighboring cells limit each other's movement through excluded volume effects. In sheet-like motion, cells remain connected and move as a coherent unit, while in a polar active nematic state, neighbor exchanges become possible primarily along the direction of motion. In 2D epithelial sheets, an increase in cell shape elongation beyond a critical threshold correlates with increased cell movement and more fluid-like behavior, as observed in bronchial epithelial cells (Tse et al. 2012; Park et al. 2015; De Marzio et al. 2021).

At medium-to-low levels of cell–cell adhesion, independently migrating cells can move by solid-flocking migration, in which complete strands or sheets of cells move, without neighbor exchange; thus, only the "flock" is motile (Shen et al. 2025). Likewise, a solid-flocking state with ballistic movement is achieved by heterogeneous levels of cell–cell adhesion and migration; groups of solid, stably connected cells without neighbor exchange migrate together in the same direction as a unit in an otherwise solid epithelium (Malinverno et al. 2017; Giavazzi et al. 2018).

At very low levels of cell–cell adhesion, when both cell density and migration activity are high, cells exhibit a fluid-like unjammed state (active fluid-like). Here cells migrate predominantly individually in an uncorrelated, random manner. Cell migration is either enabled by weak intercellular interactions that provide sufficient traction force for neighbor exchanges. Alternatively, unjammed fluid-like motion can occur in the absence of intercellular interaction, when cells apply traction forces toward the ECM and simultaneously are pushed together in narrow, cleft-like ECM spaces. Finally, when cell–cell adhesion is low compared to the forces generated by cells toward the ECM, cells detach, cell number density drops below a collective behavior threshold, causing cells to move independently, analogous to a gas-like state (Lenne and Trivedi 2022). In this state, cells move without cell–cell interaction and directional coordination between cells, and traction for motility is solely provided by the ECM.

Multiple upstream mechanisms have been identified to trigger changes of collective cell dynamics and jamming transitions, including (1) mechanical feedback from cell–cell adhesion controlling adhesion strength and turnover; (2) stiffness differences in ECM, which alter the strength of traction forces toward cell–cell junctions and promigratory signaling pathways; (3) cell proliferation and cell death, which impact the local cell packing density and locally disrupt cell–cell adhesion; (4) changes in front–rear cell polarization, including chemotactic responses and guidance by tissue structures; and (5) changes of cell and nuclear softness and deformability, which promote fluidization by lowering the energy needed for cells to squeeze by each other (Fig. 2; Fuhs et al. 2022). The transition from immobile, solid state to cohesive sheet movement occurs in wound healing models in vitro. This large-scale unjamming transition is initiated at the outward edge, leading to sheet-like collective migration of cohesive polarized cells into the free space (Trepat et al. 2009; Angelini et al. 2011; Reffay et al. 2014; Garcia et al. 2015; Aoki et al. 2017). Similar solid-to-fluid-like nematic transitions can occur in the absence of free space, manifesting as swirling and rotational patterns (vorticity and active turbulence) (Lång et al. 2022). Unjamming transitions start from solid-like monolayers when cell motility is activated (e.g., by increasing turnover of cell–cell adhesions induced by Rab5a-dependent endocytosis) (Malinverno et al. 2017).

JAMMING TRANSITIONS IN 2D TISSUES

Unjamming transitions of cell sheets without downregulation of AJs can be induced by tissue compression. Primary human bronchial epithelial cells cultured at an air–liquid interface proliferate to reach a solid-like jammed state, whereas cells from asthmatic donors exhibit fluid-like unjammed motion, even at high confluency (Tse et al. 2012; Park et al. 2015; De Marzio et al. 2021). This fluid-like unjammed state can be induced in solid-like jammed healthy bronchial epithelial cells (Tse et al. 2012; Park et al. 2015; De Marzio et al. 2021) and in normal and transformed mammary epithelial cells (Cai et al. 2022; Lee et al. 2023) when mechanical compressive stress exceeds the tissue's yield stress, a critical mechanical threshold above which the epithelial collective begins to flow rather than merely deform elastically. This yield stress behavior is fundamental to the jamming–unjamming transition, as cells rearrange when sufficient mechanical force overrides their initially stable organization, causing cells to change shape and move.

Compression-induced unjammed cells maintain cell–cell adhesion by AJs and TJs and are characterized by an elongated cell shape, aligned basal stress fibers, and increased traction forces toward the ECM (Saraswathibhatla and Notbohm 2020; De Marzio et al. 2021; Phung et al. 2022). Mechanical compression leads to the activation of mechanosensitive ERK activation (De Marzio et al. 2021) through multiple pathways (Crozet and Levayer 2023). As an example, compression opens stretch-activated Piezo1 channels causing calcium influx that activates ERK (Gudipaty et al. 2017). In *Drosophila* wings during embryonic development, mechanosensitive pathways enhance E-cadherin endocytosis and turnover and promote epithelial fluidization (Iyer et al. 2019). Compaction further increases local growth factor concentrations by decreasing the size of intercellular lumens, thereby activating EGFR (Tschumperlin et al. 2004). As a consequence, ERK signaling drives cytoskeletal remodeling and changes in adhesion dynamics that facilitate unjamming transitions and coordinated migration of cell sheets (Hino et al. 2020).

Growth factor signaling, in particular activation of EGFR, promotes unjamming of 2D cell sheets by initiating WAVE-dependent lamellipodia formation and directed cell motility, typically with the maintenance of AJs (Tschumperlin et al. 2004; Aoki et al. 2013, 2017; Malinverno et al. 2017; Palamidessi et al. 2019; Deguchi et al. 2024). As examples, ligands for TGF-β and EGFR family receptors, including Nodal, drive unjamming in zebrafish gastrula, primarily by mechanisms that rely on increased protrusion and motility (Pinheiro et al. 2022).

Lastly, unjamming of cell sheets may be induced by local triggers. Mitotic events in the blastoderm of an early zebrafish embryo are associated with locally decreased cell–cell adhesion and cell rounding, which drives tissue fluidization (Petridou et al. 2019). In polarized epithelium, reduction of the apical cell diameter by apical constriction induces unjamming by local deformation of neighbor cells (Atia et al. 2018). Local constriction can induce rotational motion of circular *Drosophila* photoreceptor clusters in otherwise solid epithelia (Founounou et al. 2021).

JAMMING TRANSITIONS IN 3D TISSUES

Only a few studies have focused on jamming transitions in 3D environments. In 2D monolayer culture, migration relies on traction forces against flat, inert, uniform, and often stiff substrate. In 3D tissue environments, cell groups are surrounded by other cells or ECM, which provides a dynamic, heterogeneous viscoelastic 3D microenvironment that can either promote or restrict movement. Physical confinement is caused by other cells, channel-like ECM interfaces, or ECM barriers that guide cells along specific paths but restrict other directions. The organization of the ECM varies depending on its geometrical structure, its composition, and the relative contributions of ECM proteins (collagen, fibronectin, and laminin), glycosaminoglycans, and the extent of collagen cross-linking (Naba 2024). Furthermore, in 3D tissues, migrating cells interact with deformable and plastic ECM, which can be restructured during cell migration and, in turn, influence the migration and jamming state of follower cells (Wolf et al. 2007; Haeger et al. 2014). The 3D ECM also presents biochemical cues, such as adhesion ligands and bound cytokines as well as topological features, including curvature that can induce, direct, or delay cell migration (Pieuchot et al. 2018). Therefore, the 3D organization of the ECM limits the cellular degrees of freedom and controls unjamming transitions by mechanical confinement in three dimensions (Yamada and Sixt 2019).

Commonly studied ECM conditions that influence jamming transitions and cellular confinement in 3D multicellular cohorts include (1) highly ordered, high-density ECM creating tissue interfaces (e.g., basement membrane, collagen bundles); (2) high-density, low porosity ECM without guiding interfaces, for example, reconstituted high-density collagen scaffolds; and (3) low-density randomly structured ECM containing fibrillar networks with pores that can accommodate moving cells (Wolf et al. 2009; van Helvert et al. 2018). These ECM conditions differ in density, stiffness, and topology and determine the degree of cellular confinement, cell–cell adhesion, and migration strategies.

JAMMING BETWEEN 3D TISSUE INTERFACES AND ALIGNED FIBROTIC ECM

Basement membranes along epithelia, blood, and lymph vessels, peripheral nerves, and myofibers form 2D surfaces within 3D interstitial tissue, which support cell motion along the interface but also limit migration in a wall-like manner (Weigelin et al. 2012; Naba 2024). Likewise, collagen bundles form curved surfaces and provide 2D-like interfaces within 3D tissue. In normal connective tissue, but also in the fibrotic stroma surrounding tumors, tissue clefts between collagen bundles consisting of primarily cross-linked collagen I form narrow (1–10 µm) 3D interfaces, which serve as guiding structures for directed single-cell and collective migration along the interface and into the free space (Weigelin et al. 2012; Carey et al. 2015; Benias et al. 2018; Cenaj et al. 2021; Chepizhko et al. 2025). Similar alignment of collective invasion along ECM interfaces is present in clinical samples of invasive carcinoma (Khalil et al. 2017; Ilina et al. 2018, 2020). As a result, epithelial tumors invade tissue as multicellular fingerlike protrusions, which align along tissue interfaces (Provenzano et al. 2008; Weigelin et al. 2012; Wolf et al. 2013; Ilina et al. 2018; Kato et al. 2023).

ECM-guided migration of cell groups may develop nematic flow patterns, in which cell–cell adhesion is maintained while AJs exhibit dynamic turnover. For example, rotational collective migration has been observed in MCF10A 3D spheroids in vitro (Wang et al. 2013) and in the *Drosophila* egg chamber in vivo (Squarr et al. 2016; Palamidessi et al. 2019; Williams and Horne-Badovinac 2024). In these systems, noninvasive rotational collective migration guided by basement membranes relies on endocytosis- and WAVE-dependent mechanisms that produce highly correlated nematic motion patterns.

Similarly, MDCK cells migrating into narrow tubular polydimethylsiloxane (PDMS) tracks reduce migration speed and undergo a jamming transition toward active polar nematic motion, possibly by mechanical confinement and reinforced cell–cell adhesion (Xi et al. 2017). In a similar experimental setup using MCF10A cells, dense cell populations entering tubular confinement undergo partial EMT, increased cell elongation and motility, and transit from fluid-like to active nematic motion (Nasrollahi and Pathak 2016; Lin et al. 2025). Differences between cell models may depend upon the quality of cell–cell interactions. MCF10A cells have less developed cell–cell adhesion as compared to MDCK cells, lack TJs (Plachot et al. 2009), and therefore may be more susceptible to confinement-related pathways that downregulate AJs.

When cell–cell adhesion is low, cell–ECM interactions and the degree of confinement by ECM jointly control the fluidization of cell groups along extracellular tissue structures. Hence, whereas tightly connected E-cadherin-positive mouse mammary tumor cells exhibit a polar nematic state and invade the ECM by polar strand invasion, experimental downregulation of E-cadherin or p120-catenin causes either active fluid-like unjammed motion when cells are confined by ECM and pushed together to high cell density or gas-like behavior when confinement by ECM is low, cell density decreases, and cells move individually, in vitro and in vivo (Ilina et al. 2020). The collective yet unjammed migration in 3D tissue, which laterally confines the cells, may also explain the multicellular invasive behavior of invasive lobular carcinomas, which lack AJs but are typically found as chain-like patterns along and in between highly ordered collagen bundles in patient samples (Khalil et al. 2017; Batra et al. 2023). Likewise, mesenchymal cells can maintain active nematic motion. The collective movements of melanoma or fibrosarcoma cells along paths of least resistance promote streaming patterns of cell cohorts with locally high density in vitro (Haeger et al. 2014; Nam et al. 2016; Beunk et al. 2023), and in vivo (Weigelin et al. 2012; Law et al. 2023). Likewise, invasion of mesenchymal cancer cells along collagen bundles of the mouse dermis can be as nematic ordered fluid migration (Chepizhko et al. 2025). The collective forward and backward axial motion of groups of mesenchymal cells in high-grade human gliomas and mouse glioma models, named glioma networks or oncostreams, depends on the secretion of α1-type I collagen and develops fluid-like nematic motion along collagen I scaffolds (Comba et al. 2022; Barberis et al. 2024; Argento et al. 2025). In summary, 3D confinement promotes nematic motion, where stronger cell–cell adhesion favors coordinated active polar motion and weaker cell–cell adhesion allows more fluid active nematic behavior (Fig. 1).

TRANSITIONS IN HIGH-DENSITY ECM WITHOUT INTERFACES

The majority of studies analyzing in vitro or ex vivo cell migration in 3D culture relies upon cell spheroids or tumor organoids embedded in homogeneous reconstituted ECMs, including randomly polymerized fibrillar collagen or reconstituted basement membrane, which both lack significant predefined structures for directed cell guidance. Under these conditions, when spaces are limited, cells can overcome ECM barriers using cell deformation, proteolytic remodeling of the ECM, or both. In 3D ECM conditions, thick squamous cell strands and invading breast cancer cohorts retain AJs and depend on MMP14-mediated proteolytic activity for 3D tissue invasion (Wolf and Friedl 2011; Wolf et al. 2013; Marchant et al. 2022; Pogoda et al. 2022; Kato et al. 2023).

Invasion patterns in randomly organized 3D collagen in vitro can range from thin, single-file strands to broad, pushing masses and correlate with similar patterns in vivo (Kato et al. 2023). In high-density ECM, proteolytic remodeling results in 3D ECM tunnels, which confine cell migration laterally and favor fluid-like active nematic motion as cell strands, whereas in low-density fibrillar 3D ECM, cells move as a loosely connected network or individually (Haeger et al. 2014; Kato et al. 2023).

Deformation of ECM and cell deformation occur in conjunction, resulting in oscillating ECM deformation or mechanical ECM compression and ECM densification (Ilina et al. 2011; Chepizhko et al. 2018). Cell deformation is generally mediated by changes in actomyosin-dependent cortical stiffness, with additional roles for intermediate filaments and microtubules (Marchant et al. 2022; Pogoda et al. 2022), as well as shape adaptation of the stiffest organelle, the nucleus (Wolf et al. 2013; Fuhs et al. 2022). Mechanical pushing forces contribute to the ability of tumor cohorts to widen migration tracks (Ilina et al. 2011). The formation of supracellular actomyosin cables at the cluster periphery of cells that retain AJs aids the pushing forces of the invading cohort by maintaining cluster integrity (Hidalgo-Carcedo et al. 2011; Blauth et al. 2021) along with additional AJ-independent outward forces that operate in cell cohorts with low cell–cell adhesions (Chepizhko et al. 2025). Thus, irrespective of cell–cell interaction strength, cell-generated ECM tracks combined with cell deformation build and exploit the migration space required for confined fluid-like active nematic cell motion.

GAS-LIKE BEHAVIOR IN 3D ENVIRONMENTS

During the transition from solid, immobile, or fluid-like motion to gas-like motion, single cells detach from neighbor cells and migrate individually. Detachment into the ECM depends on downregulation of receptors mediating cell–cell adhesion together with active cytoskeletal dynamics and interaction with the ECM substrate (Clark and Vignjevic 2015; Aiello et al. 2018; Kato et al. 2023). Importantly, in 3D ECM, individual cell dissemination depends upon unconfined extracellular space, such as pores or openings large enough to accommodate the cell body and the nucleus (Wolf et al. 2013; McGregor et al. 2016). Depending on the degree of cell–ECM interaction and traction force generation, the mode of migration is mesenchymal, with strong engagement toward the ECM, amoeboid, with low adhesion and force generation toward the ECM, or alternating when cell–ECM interactions are of variable strength (Fig. 2). In addition, cells may deploy different levels of migration persistence, depending on intrinsic polarity, durotactic response to stiffness gradients (durotaxis), guidance by extracellular chemical cues, for example, chemokines, as well as guidance by random or ordered ECM organization (van Helvert et al. 2018). Mesenchymal single cells, which develop high cell-intrinsic polarity and further pull and stiffen ECM at the cell edges, develop directionally persistent migration along the cell-generated corridor of aligned collagen fibers (van Helvert and Friedl 2016; Zanotelli et al. 2024). Thus, conditions of low cell–cell adhesion and low-density ECM structures typically favor gas-like migration.

The known examples of gas-like migration in 3D ECM depend on downregulation of cell–cell adhesion in epithelial cells or intrinsic absence of AJs. The best-understood mechanism of cell individualization is complete EMT. Here, cells in near-solid or fluid states transcriptionally downregulate cell–cell adhesion molecules while upregulating mesenchymal features, such as increased traction forces on ECM fibers and enhanced ECM proteolysis, thus facilitating the shift from fluid-like to gas-like migration (Giampieri et al. 2009). Mesenchymal fibrosarcoma or melanoma cells with intrinsically low cell–cell adhesion or epithelial breast or lung cancer cells with experimentally downregulated AJs can transition from an active fluid migration in confining ECM, to a gas-like migration when encountering low-density ECM conditions (Haeger et al. 2014; Plou et al. 2018; Ilina et al. 2020; Kang et al. 2021). When ECM pores are large, cells with intrinsically low cell–cell adhesion detach from neighbor cells and move individually (Haeger et al. 2014; van der Net et al. 2024). Amoeboid cells, including leukocytes or amoeboid tumor cells, develop gas-like movement as a result of intrinsically low-avidity, short-lived cell–cell interaction (George et al. 2023).

Solid-to-gas transition may arise in immobile cells passively, including by extrusion from their neighbors in response to nematic ordering defects of the epithelium (Saw et al. 2017), oncogenic transformation (Tanimura and Fujita 2020), or apoptosis causing local unjamming (Andrade and Rosenblatt 2011). Cell extrusion of dying or defective cells in epithelial sheets occurs when cellular crowding, ERK signaling, and cortical actomyosin-mediated forces generated by directly neighboring cells along their cell–cell junctions push the defective cell toward the basal side and into the matrix (Fig. 2). This process is favored by soft ECM (Balasubramaniam et al. 2025) or FAK activation through integrin–ECM contacts (Ray and Provenzano 2021). As a related two-step mechanism to individualization, heterogeneous levels of cell–cell adhesion within confined spheroids can lead to sorting of less adhesive cells to the outside, followed by gas-like dissemination of the less adhesive cells from the spheroid edge (Pawlizak et al. 2015; Cai et al. 2024).

Strong confinement within dense cell populations can induce heterogeneity in intrinsic motility, cell–cell adhesion, and mechanical properties, resulting in unjamming responses and induce gas-like states. While it was shown that mechanical confinement and high cell number densities can lead to solid-like jammed behaviors (Grosser et al. 2021; Gottheil et al. 2023), mechanical compression and associated high densities can initiate key migration-related signaling pathways, including TGF-β, β-catenin, Rho-ROCK), and ERK (Linke et al. 2024). This may result in variable compression-induced fluid-like behaviors. Collectively migrating epidermal A431 and mammary MCF7 tumor cells entering narrow PDMS-based channels destabilize E-cadherin at cell–cell junctions, which favors dissociation of leader cells due to their high integrin-dependent traction and faster migration (Law et al. 2023). Likewise, single melanoma cells unjam and detach from the tip of collective invasion zones followed by fast, linear, nearly ballistic migration paths (Lévy-like migration patterns) along ECM interfaces of the mouse dermis in vivo (Huda et al. 2018). Confinement-induced leader cell dissociation involves changes in microtubule polymerization and downstream activation of the RhoA/ROCK/myosin II pathway causing the mechanical separation of leader and follower cells by a mechanism similar to cytokinesis (Law et al. 2023). As an upstream mechanism, strong confinement favors nuclear anillin, a Rho activator, to translocate to the plasma membrane where it activates Rho/ROCK signaling and actomyosin contractility (Tran et al. 2025). Strong confinement can further induce EMT in epithelial cells independent of ECM stiffness or cell–ECM adhesion (Nasrollahi and Pathak 2016), which likely contributes to the gas-like behavior observed in cadherin-low tumor cells escaping the dense environment of a primary tumor (Haeger et al. 2014; Huda et al. 2018; Ilina et al. 2020). In summary, gas-like individualization of cells is induced by a complex cooperation of cell junction regulation, available space in 3D tissues, and signals induced by strong confinement.

Limitations of the Jamming/Unjamming Paradigm

While jamming phase diagrams can plausibly describe transitions between immobile solids and active fluid behaviors of homogeneous cohorts, several dynamic cell behaviors remain to be integrated into the framework. These include shape-induced versus density-induced unjamming, solid-to-gas transitions at the free edge of cell groups, and the individual-to-collective transition imposed by 3D confinement. Additionally, the framework does not fully account for cell dynamics and migration in dense tumor cores with no available free space, local heterogeneity of molecular and cytoskeletal states that may lead to jammed cell clusters moving within an unjammed environment of soft cells (Fuhs et al. 2022), and the combination of active and passive transport of cells in fluids. An example of this complexity involves leader cells at the front of migrating epithelial cohorts, which typically develop partial mesenchymal traits compared to follower cells and exhibit increased cell–ECM forces to pull follower cells forward, thereby overriding jamming constraints present in the core of the group (Vishwakarma et al. 2018; Wang et al. 2025b).

Spatial and molecular heterogeneity, along with the effects of cellular packing in 3D confinement, remain significant challenges to be integrated within theoretical frameworks of unjamming transitions. In dense 3D cell cohorts, this complexity is compounded by the repositioning of AJs and TJs as compared to 2D sheets, which leads to poorly defined junction-associated cell–cell interactions that extend beyond current jamming paradigms. Consequently, applying jamming paradigms in 3D tissues requires consideration of additional factors, including cell heterogeneity and plasticity, the mechanics of the nucleus, noisy migration behaviors induced by stochastic fluctuations of intracellular signaling, turnover of cell adhesions, cell proliferation, and the specific state and architecture of the ECM. These complexities are further amplified in tumor microenvironments, where the heterogeneity of cell–cell and cell–ECM interactions is enhanced in tumor microenvironments in response to localized gradients of oxygen, nutrients, and mechanical stresses that differentially activate transcriptional programs and migration-enhancing signaling pathways (Te Boekhorst et al. 2022). Moreover, while shape-induced unjamming transitions provide quantitative predictions in 2D systems, cell motion in 3D tumor environments introduces additional complexity through cell death upon nutrient deprivation and therapy stress, changes in nuclear stiffness regulation, and spatial constraints that contribute to jamming behavior (Gottheil et al. 2023).

The unjamming processes we describe here have been predominately modeled by three approaches: vertex models that describe shape-induced unjamming (Park et al. 2015), (nemato) hydrodynamic modeling of cells and ECM as passive and active nematic phases (Gottheil et al. 2024), and approaches (e.g., hydrodynamics-based or vertex-based) to describe flocking (Giavazzi et al. 2018; Shen et al. 2025). The vertex approaches focus on the dense packing at high cellular volume fractions, nematohydrodynamics modeling is well suited to describe collective cell motility in dense microenvironments on mesoscales, and flocking methods consider cellular cross-talk in collective motion. More intricate effects such as the contribution of the nucleus to jamming may require Potts models (Braat et al. 2025). Thus, there is currently no universal physical description of the unjamming phenomena in tissues; rather, which description to consider depends on the time and length scales. For instance, cancer modeling on the mesoscale (~10–100 µm) over large time spans (~years) can be carried out by hydrodynamic approaches (Gottheil et al. 2024), while phenomena like cell neighbor exchanges that occur within hours on the length scales of cells (~1–10 µm) can be described by approaches resolving individual cells (Bi et al. 2016; Fuhs et al. 2022).

OUTLOOK

Extending the jamming paradigm to the third dimension and including confined space will be required to realistically capture 3D invasion processes, including branching morphogenesis, angiogenesis, and collective cancer invasion. Additional metrics to be considered for 3D jamming phase diagrams may include pressure to appreciate passive cell–cell interactions when active adhesion is low or even repulsive behaviors are present (Pajic-Lijakovic et al. 2024). To investigate fluid-to-gas-like behaviors, combining 3D confinement with more open 3D tissue space will be required. To accommodate active motion despite cell volume fractions of 100% (confluency) and concurrent cell–cell junction stability, additional parameters describing local fluid-like behavior, detachment, and evasion explaining nematic migration need to be considered. Such additional parameters may include cell-intrinsic actin dynamics, nuclear mechanics, dynamic turnover of cell–cell and cell–ECM interactions, and subcellular localized cues, which may disrupt cell–cell interaction stability; mitosis or cell death and tissue-derived signals activating migration properties, such as hypoxia and/or compression-induced pathways (Crozet and Levayer 2023; Wang et al. 2025a).

Many systemic and targeted antitumor therapies, including irradiation, microtubule-targeting agents, anthracyclines, and EGFR inhibitors, not only reduce cell density by inducing cell death but also profoundly impact key parameters of jamming transitions, such as cell migration, cell size, cell deformability, and cell and nuclear shape. These transitions are characterized by distinct morphological changes that can be observed in diagnostic histopathological specimens from cancer patients. Therefore, therapy-induced alterations in cell density and nuclear shape may emerge as valuable biomarkers for predicting jamming transitions in tumor samples.

Transitions of nonepithelial cells with intrinsically low adhesion at low density require additional considerations. Extreme examples on either side of the spectrum include low-density leukocytes in loose ECM, which exhibit uncorrelated gas-like motion; cells without active cytoskeleton such as erythrocytes propelled by fluid flows; and low-density chondrocytes or osteocytes, which, at comparable cell density, are fully jammed due to their entrapment by solidified ECM matrices of the cartilage or bone. Therefore, additional parameters on extracellular confinement, including order, porosity, compressibility, and susceptibility to remodeling, will need to be considered. By broadening the parameter space and biological context, expanded concepts describing solid-like (jammed) to fluid-like (unjammed) tissue behaviors will continue to deliver computational and wet-laboratory insights into the cell-intrinsic and extracellular key components controlling these transitions of multicellular migration in diverse biological contexts.

ACKNOWLEDGMENTS

The original work covered in this article was supported by grants from the European Research Council (ERC-AdG-2021-101054921), National Institutes of Health (NIH) (U54 CA210184-01 and U54 CA261694-01), The Netherlands Organisation for Scientific Research (NWO) (IMAGINE! NWO 024.005 .009 and NWO OCENW.GROOT.2019.022), Worldwide Cancer Research (WWCR 23-0177), and the European Union (ITN-Inflanet EU-H2020-ITN 955576).

REFERENCES

Aiello NM, Maddipati R, Norgard RJ, Balli D, Li J, Yuan S, Yamazoe T, Black T, Sahmoud A, Furth EE, et al. 2018. EMT subtype influences epithelial plasticity and mode of cell migration. *Dev Cell* **45:** 681–695.e4. doi:10.1016/j.devcel.2018.05.027

Andrade D, Rosenblatt J. 2011. Apoptotic regulation of epithelial cellular extrusion. *Apoptosis* **16:** 491–501. doi:10.1007/s10495-011-0587-z

Angelini TE, Hannezo E, Trepat X, Marquez M, Fredberg JJ, Weitz DA. 2011. Glass-like dynamics of collective cell migration. *Proc Natl Acad Sci* **108:** 4714–4719. doi:10.1073/pnas.1010059108

Aoki K, Kumagai Y, Sakurai A, Komatsu N, Fujita Y, Shionyu C, Matsuda M. 2013. Stochastic ERK activation induced by noise and cell-to-cell propagation regulates cell density-dependent proliferation. *Mol Cell* **52:** 529–540. doi:10.1016/j.molcel.2013.09.015

Aoki K, Kondo Y, Naoki H, Hiratsuka T, Itoh RE, Matsuda M. 2017. Propagating wave of ERK activation orients collective cell migration. *Dev Cell* **43:** 305–317.e5. doi:10.1016/j.devcel.2017.10.016

Argento AE, Varela ML, Singh G, Visnuk DP, Jacobovitz B, Rutherford ME, Edwards MB, Chaboche Q, Orringer DA, Heth JA, et al. 2025. Three-dimensional topological defects and quasi-long-range order in biological liquid crystals. bioRxiv doi:10.1101/2025.04.14.648711

Armengol-Collado JM, Carenza LN, Eckert J, Krommydas D, Giomi L. 2023. Epithelia are multiscale active liquid crystals. *Nat Phys* **19:** 1773–1779. doi:10.1038/s41567-023-02179-0

Atia L, Bi D, Sharma Y, Mitchel JA, Gweon B, Koehler S, DeCamp SJ, Lan B, Kim JH, Hirsch R, et al. 2018. Geometric constraints during epithelial jamming. *Nat Phys* **14:** 613–620. doi:10.1038/s41567-018-0089-9

Atia L, Fredberg JJ, Gov NS, Pegoraro AF. 2021. Are cell jamming and unjamming essential in tissue development? *Cells Dev* **168:** 203727. doi:10.1016/j.cdev.2021.203727

Balasubramaniam L, Monfared S, Ardaševa A, Rosse C, Schoenit A, Dang T, Maric C, Hautefeuille M, Kocgozlu L, Chilupuri R, et al. 2025. Dynamic forces shape the survival fate of eliminated cells. *Nat Phys* **21:** 269–278. doi:10.1038/s41567-024-02716-5

Barberis L, Condat CA, Faisal SM, Lowenstein PR. 2024. The self-organized structure of glioma oncostreams and the disruptive role of passive cells. *Sci Rep* **14:** 25435. doi:10.1038/s41598-024-74823-5

Batra H, Mouabbi JA, Ding Q, Sahin AA, Raso MG. 2023. Lobular carcinoma of the breast: a comprehensive review with translational insights. *Cancers (Basel)* **15:** 5491. doi:10.3390/cancers15225491

Benias PC, Wells RG, Sackey-Aboagye B, Klavan H, Reidy J, Buonocore D, Miranda M, Kornacki S, Wayne M, Carr-Locke DL, et al. 2018. Structure and distribution of an unrecognized interstitium in human tissues. *Sci Rep* **8:** 4947. doi:10.1038/s41598-018-23062-6

Beunk L, Wen N, van Helvert S, Bekker B, Ran L, Kang R, Paulat T, Syga S, Deutsch A, Friedl P, et al. 2023. Cell jamming in a collagen-based interface assay is tuned by collagen density and proteolysis. *J Cell Sci* **136:** jcs260207. doi:10.1242/jcs.260207

Bi D, Yang X, Marchetti MC, Manning ML. 2016. Motility-driven glass and jamming transitions in biological tissues. *Phys Rev X* **6:** 021011. doi:10.1103/PhysRevX.6.021011

Blanch-Mercader C, Yashunsky V, Garcia S, Duclos G, Giomi L, Silberzan P. 2018. Turbulent dynamics of epithelial cell cultures. *Phys Rev Lett* **120:** 208101. doi:10.1103/PhysRevLett.120.208101

Blanch-Mercader C, Guillamat P, Roux A, Kruse K. 2021. Integer topological defects of cell monolayers: mechanics and flows. *Phys Rev E* **103:** 012405. doi:10.1103/PhysRevE.103.012405

Blauth E, Kubitschke H, Gottheil P, Grosser S, Käs JA. 2021. Jamming in embryogenesis and cancer progression. *Front Phys* **9:** 666709. doi:10.3389/fphy.2021.666709

Bosveld F, Wang Z, Bellaïche Y. 2018. Tricellular junctions: a hot corner of epithelial biology. *Curr Opin Cell Biol* **54:** 80–88. doi:10.1016/j.ceb.2018.05.002

Braat QJS, Janzen G, Jansen BC, Debets VE, Ciarella S, Janssen LMC. 2025. Shape matters: inferring the motility of confluent cells from static images. *Soft Matter* **21:** 6504–6515. doi:10.1039/D5SM00222B

Brouxhon SM, Kyrkanides S, Teng X, Athar M, Ghazizadeh S, Simon M, O'Banion MK, Ma L. 2014. Soluble E-cadherin: a critical oncogene modulating receptor tyrosine kinases, MAPK and PI3K/Akt/mTOR signaling. *Oncogene* **33:** 225–235. doi:10.1038/onc.2012.563

Cai D, Chen SC, Prasad M, He L, Wang X, Choesmel-Cadamuro V, Sawyer JK, Danuser G, Montell DJ. 2014. Mechanical feedback through E-cadherin promotes direction sensing during collective cell migration. *Cell* **157:** 1146–1159. doi:10.1016/j.cell.2014.03.045

Cai G, Nguyen A, Bashirzadeh Y, Lin SS, Bi D, Liu AP. 2022. Compressive stress drives adhesion-dependent unjamming transitions in breast cancer cell migration. *Front Cell Dev Biol* **10:** 933042. doi:10.3389/fcell.2022.933042

Cai G, Li X, Lin SS, Chen SJ, Rodgers NC, Koning KM, Bi D, Liu AP. 2024. Matrix confinement modulates 3D spheroid sorting and burst-like collective migration. *Acta Biomater* **179:** 192–206. doi:10.1016/j.actbio.2024.03.007

Campàs O, Noordstra I, Yap AS. 2024. Adherens junctions as molecular regulators of emergent tissue mechanics.

Nat Rev Mol Cell Biol **25:** 252–269. doi:10.1038/s41580-023-00688-7

Campbell HK, Maiers JL, DeMali KA. 2017. Interplay between tight junctions and adherens junctions. *Exp Cell Res* **358:** 39–44. doi:10.1016/j.yexcr.2017.03.061

Carey SP, Rahman A, Kraning-Rush CM, Romero B, Somasegar S, Torre OM, Williams RM, Reinhart-King CA. 2015. Comparative mechanisms of cancer cell migration through 3D matrix and physiological microtracks. *Am J Physiol Cell Physiol* **308:** C436–C447. doi:10.1152/ajpcell.00225.2014

Cenaj O, Allison DHR, Imam R, Zeck B, Drohan LM, Chiriboga L, Llewellyn J, Liu CZ, Park YN, Wells RG, et al. 2021. Evidence for continuity of interstitial spaces across tissue and organ boundaries in humans. *Commun Biol* **4:** 436. doi:10.1038/s42003-021-01962-0

Chepizhko O, Lionetti MC, Malinverno C, Giampietro C, Scita G, Zapperi S, La Porta CAM. 2018. From jamming to collective cell migration through a boundary induced transition. *Soft Matter* **14:** 3774–3782. doi:10.1039/C8SM00128F

Chepizhko O, Armengol-Collado JM, Alexander S, Wagena E, Weigelin B, Giomi L, Friedl P, Zapperi S, La Porta CAM. 2025. Confined cell migration along extracellular matrix space in vivo. *Proc Natl Acad Sci* **122:** e2414009121. doi:10.1073/pnas.2414009121

Cho Y, Haraguchi D, Shigetomi K, Matsuzawa K, Uchida S, Ikenouchi J. 2022. Tricellulin secures the epithelial barrier at tricellular junctions by interacting with actomyosin. *J Cell Biol* **221:** e202009037. doi:10.1083/jcb.202009037

Citi S, Fromm M, Furuse M, González-Mariscal L, Nusrat A, Tsukita S, Turner JR. 2024. A short guide to the tight junction. *J Cell Sci* **137:** jcs261776. doi:10.1242/jcs.261776

Clark AG, Vignjevic DM. 2015. Modes of cancer cell invasion and the role of the microenvironment. *Curr Opin Cell Biol* **36:** 13–22. doi:10.1016/j.ceb.2015.06.004

Classen AK, Anderson KI, Marois E, Eaton S. 2005. Hexagonal packing of *Drosophila* wing epithelial cells by the planar cell polarity pathway. *Dev Cell* **9:** 805–817. doi:10.1016/j.devcel.2005.10.016

Comba A, Faisal SM, Dunn PJ, Argento AE, Hollon TC, Al-Holou WN, Varela ML, Zamler DB, Quass GL, Apostolides PF, et al. 2022. Spatiotemporal analysis of glioma heterogeneity reveals COL1A1 as an actionable target to disrupt tumor progression. *Nat Commun* **13:** 3606. doi:10.1038/s41467-022-31340-1

Crozet F, Levayer R. 2023. Emerging roles and mechanisms of ERK pathway mechanosensing. *Cell Mol Life Sci* **80:** 355. doi:10.1007/s00018-023-05007-z

Das T, Safferling K, Rausch S, Grabe N, Boehm H, Spatz JP. 2015. A molecular mechanotransduction pathway regulates collective migration of epithelial cells. *Nat Cell Biol* **17:** 276–287. doi:10.1038/ncb3115

Deguchi E, Lin S, Hirayama D, Matsuda K, Tanave A, Sumiyama K, Tsukiji S, Otani T, Furuse M, Sorkin A, et al. 2024. Low-affinity ligands of the epidermal growth factor receptor are long-range signal transmitters in collective cell migration of epithelial cells. *Cell Rep* **43:** 114986. doi:10.1016/j.celrep.2024.114986

De Marzio M, Kiliç A, Maiorino E, Mitchel JA, Mwase C, O'Sullivan MJ, McGill M, Chase R, Fredberg JJ, Park JA, et al. 2021. Genomic signatures of the unjamming transition in compressed human bronchial epithelial cells. *Sci Adv* **7:** eabf1088. doi:10.1126/sciadv.abf1088

Deutsch A, Friedl P, Preziosi L, Theraulaz G. 2020. Multiscale analysis and modelling of collective migration in biological systems. *Philos Trans R Soc Lond B Biol Sci* **375:** 20190377. doi:10.1098/rstb.2019.0377

Finegan TM, Hervieux N, Nestor-Bergmann A, Fletcher AG, Blanchard GB, Sanson B. 2019. The tricellular vertex-specific adhesion molecule Sidekick facilitates polarised cell intercalation during *Drosophila* axis extension. *PLoS Biol* **17:** e3000522. doi:10.1371/journal.pbio.3000522

Founounou N, Farhadifar R, Collu GM, Weber U, Shelley MJ, Mlodzik M. 2021. Tissue fluidity mediated by adherens junction dynamics promotes planar cell polarity-driven ommatidial rotation. *Nat Commun* **12:** 6974. doi:10.1038/s41467-021-27253-0

Fredberg JJ. 2022. On the origins of order. *Soft Matter* **18:** 2346–2353. doi:10.1039/D1SM01716K

Friedl P, Mayor R. 2017. Tuning collective cell migration by cell-cell junction regulation. *Cold Spring Harb Perspect Biol* **9:** a029199. doi: 10.1101/cshperspect.a029199

Friedl P, Wolf K. 2010. Plasticity of cell migration: a multiscale tuning model. *J Cell Biol* **188:** 11–19. doi:10.1083/jcb.200909003

Friedl P, Noble PB, Walton PA, Laird DW, Chauvin PJ, Tabah RJ, Black M, Zanker KS. 1995. Migration of coordinated cell clusters in mesenchymal and epithelial cancer explants in vitro. *Cancer Res* **55:** 4557–4560.

Friedl P, Entschladen F, Conrad C, Niggemann B, Zänker KS. 1998. $CD4^+$ T lymphocytes migrating in three-dimensional collagen lattices lack focal adhesions and utilize beta1 integrin-independent strategies for polarization, interaction with collagen fibers and locomotion. *Eur J Immunol* **28:** 2331–2343. doi:10.1002/(SICI)1521-4141(199808)28:08<2331::AID-IMMU2331>3.0.CO;2-C

Fu C, Dilasser F, Lin SZ, Karnat M, Arora A, Rajendiran H, Ong HT, Mui Hoon Brenda N, Phow SW, Hirashima T, et al. 2024. Regulation of intercellular viscosity by E-cadherin-dependent phosphorylation of EGFR in collective cell migration. *Proc Natl Acad Sci* **121:** e2405560121. doi:10.1073/pnas.2405560121

Fuhs T, Wetzel F, Fritsch AW, Li X, Stange R, Pawlizak S, Kießling TR, Morawetz E, Grosser S, Sauer F, et al. 2022. Rigid tumours contain soft cancer cells. *Nat Phys* **18:** 1510–1519. doi:10.1038/s41567-022-01755-0

Fukumoto Y, Shintani Y, Reynolds AB, Johnson KR, Wheelock MJ. 2008. The regulatory or phosphorylation domain of p120 catenin controls E-cadherin dynamics at the plasma membrane. *Exp Cell Res* **314:** 52–67. doi:10.1016/j.yexcr.2007.07.024

Garcia S, Hannezo E, Elgeti J, Joanny JF, Silberzan P, Gov NS. 2015. Physics of active jamming during collective cellular motion in a monolayer. *Proc Natl Acad Sci* **112:** 15314–15319. doi:10.1073/pnas.1510973112

George S, Martin JAJ, Graziani V, Sanz-Moreno V. 2023. Amoeboid migration in health and disease: immune responses versus cancer dissemination. *Front Cell Dev Biol* **10:** 1091801. doi:10.3389/fcell.2022.1091801

Giampieri S, Manning C, Hooper S, Jones L, Hill CS, Sahai E. 2009. Localized and reversible TGFbeta signalling switches breast cancer cells from cohesive to single cell motility. *Nat Cell Biol* **11:** 1287–1296. doi:10.1038/ncb1973

Giavazzi F, Paoluzzi M, Macchi M, Bi D, Scita G, Manning ML, Cerbino R, Marchetti MC. 2018. Flocking transitions in confluent tissues. *Soft Matter* **14:** 3471–3477. doi:10.1039/C8SM00126J

Gottheil P, Lippoldt J, Grosser S, Renner F, Saibah M, Tschodu D, Poßögel AK, Wegscheider AS, Ulm B, Friedrichs K, et al. 2023. State of cell unjamming correlates with distant metastasis in cancer patients. *Phys Rev X* **13:** 031003. doi:10.1103/PhysRevX.13.031003

Gottheil P, Bhattacharyya S, Lettl K, Friedrich P, Roth K, Rivera-Moreno S, Merkel M, Aktas B, Sauer I, Daneshgar A, et al. 2024. Self organisation of invasive breast cancer driven by the interplay of active and passive nematic dynamics. arXiv doi:10.48550/arXiv.2412.01285

Grosser S, Lippoldt J, Oswald L, Merkel M, Sussman DM, Renner F, Gottheil P, Morawetz EW, Fuhs T, Xie X, et al. 2021. Cell and nucleus shape as an indicator of tissue fluidity in carcinoma. *Phys Rev X* **11:** 011033. doi:10.1103/PhysRevX.11.011033

Gudipaty SA, Lindblom J, Loftus PD, Redd MJ, Edes K, Davey CF, Krishnegowda V, Rosenblatt J. 2017. Mechanical stretch triggers rapid epithelial cell division through Piezo1. *Nature* **543:** 118–121. doi:10.1038/nature21407

Guillamat P, Blanch-Mercader C, Pernollet G, Kruse K, Roux A. 2022. Integer topological defects organize stresses driving tissue morphogenesis. *Nat Mater* **21:** 588–597. doi:10.1038/s41563-022-01194-5

Hadjisavva R, Anastasiou O, Ioannou PS, Zheltkova M, Skourides PA. 2022. Adherens junctions stimulate and spatially guide integrin activation and extracellular matrix deposition. *Cell Rep* **40:** 111091. doi:10.1016/j.celrep.2022.111091

Haeger A, Krause M, Wolf K, Friedl P. 2014. Cell jamming: collective invasion of mesenchymal tumor cells imposed by tissue confinement. *Biochim Biophys Acta* **1840:** 2386–2395. doi:10.1016/j.bbagen.2014.03.020

Haerinck J, Goossens S, Berx G. 2023. The epithelial-mesenchymal plasticity landscape: principles of design and mechanisms of regulation. *Nat Rev Genet* **24:** 590–609. doi:10.1038/s41576-023-00601-0

Halbleib JM, Nelson WJ. 2006. Cadherins in development: cell adhesion, sorting, and tissue morphogenesis. *Genes Dev* **20:** 3199–3214. doi:10.1101/gad.1486806

Hegerfeldt Y, Tusch M, Brocker EB, Friedl P. 2002. Collective cell movement in primary melanoma explants: plasticity of cell-cell interaction, beta1-integrin function, and migration strategies. *Cancer Res* **62:** 2125–2130.

Hidalgo-Carcedo C, Hooper S, Chaudhry SI, Williamson P, Harrington K, Leitinger B, Sahai E. 2011. Collective cell migration requires suppression of actomyosin at cell-cell contacts mediated by DDR1 and the cell polarity regulators Par3 and Par6. *Nat Cell Biol* **13:** 49–59. doi:10.1038/ncb2133

Higashi T, Miller AL. 2017. Tricellular junctions: how to build junctions at the TRICkiest points of epithelial cells. *Mol Biol Cell* **28:** 2023–2034. doi:10.1091/mbc.e16-10-0697

Hino N, Rossetti L, Marín-Llauradó A, Aoki K, Trepat X, Matsuda M, Hirashima T. 2020. ERK-Mediated mechanochemical waves direct collective cell polarization. *Dev Cell* **53:** 646–660.e8. doi:10.1016/j.devcel.2020.05.011

Hu QP, Kuang JY, Yang QK, Bian XW, Yu SC. 2016. Beyond a tumor suppressor: soluble E-cadherin promotes the progression of cancer. *Int J Cancer* **138:** 2804–2812. doi:10.1002/ijc.29982

Hu J, Serra-Picamal X, Bakker GJ, Van Troys M, Winograd-Katz S, Ege N, Gong X, Didan Y, Grosheva I, Polansky O, et al. 2023. Multisite assessment of reproducibility in high-content cell migration imaging data. *Mol Syst Biol* **19:** e11490. doi:10.15252/msb.202211490

Huda S, Weigelin B, Wolf K, Tretiakov KV, Polev K, Wilk G, Iwasa M, Emami FS, Narojczyk JW, Banaszak M, et al. 2018. Levy-like movement patterns of metastatic cancer cells revealed in microfabricated systems and implicated in vivo. *Nat Commun* **9:** 4539. doi:10.1038/s41467-018-06563-w

Ilina O, Bakker GJ, Vasaturo A, Hofmann RM, Friedl P. 2011. Two-photon laser-generated microtracks in 3D collagen lattices: principles of MMP-dependent and -independent collective cancer cell invasion. *Phys Biol* **8:** 015010. doi:10.1088/1478-3975/8/1/015010

Ilina O, Campanello L, Gritsenko PG, Vullings M, Wang C, Bult P, Losert W, Friedl P. 2018. Intravital microscopy of collective invasion plasticity in breast cancer. *DisModel Mech* **11:** dmm034330. doi:10.1242/dmm.034330

Ilina O, Gritsenko PG, Syga S, Lippoldt J, La Porta CAM, Chepizhko O, Grosser S, Vullings M, Bakker GJ, Starruß J, et al. 2020. Cell-cell adhesion and 3D matrix confinement determine jamming transitions in breast cancer invasion. *Nat Cell Biol* **22:** 1103–1115. doi:10.1038/s41556-020-0552-6

Iyer KV, Piscitello-Gómez R, Paijmans J, Jülicher F, Eaton S. 2019. Epithelial viscoelasticity is regulated by mechanosensitive E-cadherin turnover. *Curr Biol* **29:** 578–591.e5. doi:10.1016/j.cub.2019.01.021

Jiang D, Christ S, Correa-Gallegos D, Ramesh P, Kalgudde Gopal S, Wannemacher J, Mayr CH, Lupperger V, Yu Q, Ye H, et al. 2020. Injury triggers fascia fibroblast collective cell migration to drive scar formation through N-cadherin. *Nat Commun* **11:** 5653. doi:10.1038/s41467-020-19425-1

Jipp M, Wagner BD, Egbringhoff L, Teichmann A, Rübeling A, Nieschwitz P, Honigmann A, Chizhik A, Oswald TA, Janshoff A. 2024. Cell-substrate distance fluctuations of confluent cells enable fast and coherent collective migration. *Cell Rep* **43:** 114553. doi:10.1016/j.celrep.2024.114553

Kang W, Ferruzzi J, Spatarelu CP, Han YL, Sharma Y, Koehler SA, Mitchel JA, Khan A, Butler JP, Roblyer D, et al. 2021. A novel jamming phase diagram links tumor invasion to non-equilibrium phase separation. *iScience* **24:** 103252. doi:10.1016/j.isci.2021.103252

Kato T, Jenkins RP, Derzsi S, Tozluoglu M, Rullan A, Hooper S, Chaleil RAG, Joyce H, Fu X, Thavaraj S, et al. 2023. Interplay of adherens junctions and matrix proteolysis determines the invasive pattern and growth of squamous cell carcinoma. *eLife* **12:** e76520. doi:10.7554/eLife.76520

Katsuno-Kambe H, Yap AS. 2020. Endocytosis, cadherins and tissue dynamics. *Traffic* **21:** 268–273. doi:10.1111/tra.12721

Khalil AA, Ilina O, Gritsenko PG, Bult P, Span PN, Friedl P. 2017. Collective invasion in ductal and lobular breast cancer associates with distant metastasis. *Clin Exp Metastasis* **34:** 421–429. doi:10.1007/s10585-017-9858-6

Kim S, Pochitaloff M, Stooke-Vaughan GA, Campàs O. 2021. Embryonic tissues as active foams. *Nat Phys* **17:** 859–866. doi:10.1038/s41567-021-01215-1

Krndija D, El Marjou F, Guirao B, Richon S, Leroy O, Bellaiche Y, Hannezo E, Matic Vignjevic D. 2019. Active cell migration is critical for steady-state epithelial turnover in the gut. *Science* **365:** 705–710. doi:10.1126/science.aau3429

Lämmermann T, Bader BL, Monkley SJ, Worbs T, Wedlich-Söldner R, Hirsch K, Keller M, Förster R, Critchley DR, Fässler R, et al. 2008. Rapid leukocyte migration by integrin-independent flowing and squeezing. *Nature* **453:** 51–55. doi:10.1038/nature06887

Lång E, Pedersen C, Lång A, Blicher P, Klungland A, Carlson A, Bøe SO. 2022. Mechanical coupling of supracellular stress amplification and tissue fluidization during exit from quiescence. *Proc Natl Acad Sci* **119:** e2201328119. doi:10.1073/pnas.2201328119

Law RA, Kiepas A, Desta HE, Perez Ipiña E, Parlani M, Lee SJ, Yankaskas CL, Zhao R, Mistriotis P, Wang N, et al. 2023. Cytokinesis machinery promotes cell dissociation from collectively migrating strands in confinement. *Sci Adv* **9:** eabq6480. doi:10.1126/sciadv.abq6480

Lawson-Keister E, Manning ML. 2021. Jamming and arrest of cell motion in biological tissues. *Curr Opin Cell Biol* **72:** 146–155. doi:10.1016/j.ceb.2021.07.011

Lee YL, Mathur J, Walter C, Zmuda H, Pathak A. 2023. Matrix obstructions cause multiscale disruption in collective epithelial migration by suppressing leader cell function. *Mol Biol Cell* **34:** ar94. doi:10.1091/mbc.E22-06-0226

Lenne PF, Trivedi V. 2022. Sculpting tissues by phase transitions. *Nat Commun* **13:** 664. doi:10.1038/s41467-022-28151-9

Letizia A, He D, Astigarraga S, Colombelli J, Hatini V, Llimargas M, Treisman JE. 2019. Sidekick is a key component of tricellular adherens junctions that acts to resolve cell rearrangements. *Dev Cell* **50:** 313–326.e5. doi:10.1016/j.devcel.2019.07.007

Lin WJ, Yu H, Pathak A. 2025. Gradients in cell density and shape transitions drive collective cell migration into confining environments. *Soft Matter* **21:** 719–728. doi:10.1039/D3SM01240A

Linke JA, Munn LL, Jain RK. 2024. Compressive stresses in cancer: characterization and implications for tumour progression and treatment. *Nat Rev Cancer* **24:** 768–791. doi:10.1038/s41568-024-00745-z

Liu AJ, Nagel SR. 2023. Twenty-five years of the jamming phase diagram. *Nat Rev Phys* **5:** 630–631. doi:10.1038/s42254-023-00661-5

Lohmann S, Giampietro C, Pramotton FM, Al-Nuaimi D, Poli A, Maiuri P, Poulikakos D, Ferrari A. 2020. The role of tricellulin in epithelial jamming and unjamming via segmentation of tricellular junctions. *Adv Sci (Weinh)* **7:** 2001213. doi:10.1002/advs.202001213

Malinverno C, Corallino S, Giavazzi F, Bergert M, Li Q, Leoni M, Disanza A, Frittoli E, Oldani A, Martini E, et al. 2017. Endocytic reawakening of motility in jammed epithelia. *Nat Mater* **16:** 587–596. doi:10.1038/nmat4848

Marchant CL, Malmi-Kakkada AN, Espina JA, Barriga EH. 2022. Cell clusters softening triggers collective cell migration in vivo. *Nat Mater* **21:** 1314–1323. doi:10.1038/s41563-022-01323-0

McGregor AL, Hsia CR, Lammerding J. 2016. Squish and squeeze—the nucleus as a physical barrier during migration in confined environments. *Curr Opin Cell Biol* **40:** 32–40. doi:10.1016/j.ceb.2016.01.011

Mongera A, Rowghanian P, Gustafson HJ, Shelton E, Kealhofer DA, Carn EK, Serwane F, Lucio AA, Giammona J, Campàs O. 2018. A fluid-to-solid jamming transition underlies vertebrate body axis elongation. *Nature* **561:** 401–405. doi:10.1038/s41586-018-0479-2

Mui KL, Chen CS, Assoian RK. 2016. The mechanical regulation of integrin-cadherin crosstalk organizes cells, signaling and forces. *J Cell Sci* **129:** 1093–1100. doi:10.1242/jcs.183699

Naba A. 2024. Mechanisms of assembly and remodelling of the extracellular matrix. *Nat Rev Mol Cell Biol* **25:** 865–885. doi:10.1038/s41580-024-00767-3

Nam KH, Kim P, Wood DK, Kwon S, Provenzano PP, Kim DH. 2016. Multiscale cues drive collective cell migration. *Sci Rep* **6:** 29749. doi:10.1038/srep29749

Nasrollahi S, Pathak A. 2016. Topographic confinement of epithelial clusters induces epithelial-to-mesenchymal transition in compliant matrices. *Sci Rep* **6:** 18831. doi:10.1038/srep18831

Onsager L. 1949. The effects of shape on the interaction of colloidal particles. *Ann NY Acad Sci* **51:** 627–659. doi:10.1111/j.1749-6632.1949.tb27296.x

Ozawa M, Hiver S, Yamamoto T, Shibata T, Upadhyayula S, Mimori-Kiyosue Y, Takeichi M. 2020. Adherens junction regulates cryptic lamellipodia formation for epithelial cell migration. *J Cell Biol* **219:** e202006196. doi:10.1083/jcb.202006196

Pajic-Lijakovic I, Milivojevic M, McClintock PVE. 2024. Epithelial cell–cell interactions in an overcrowded environment: jamming or live cell extrusion. *J Biol Eng* **18:** 47. doi:10.1186/s13036-024-00442-3

Palamidessi A, Malinverno C, Frittoli E, Corallino S, Barbieri E, Sigismund S, Beznoussenko GV, Martini E, Garre M, Ferrara I, et al. 2019. Unjamming overcomes kinetic and proliferation arrest in terminally differentiated cells and promotes collective motility of carcinoma. *Nat Mater* **18:** 1252–1263. doi:10.1038/s41563-019-0425-1

Park JA, Kim JH, Bi D, Mitchel JA, Qazvini NT, Tantisira K, Park CY, McGill M, Kim SH, Gweon B, et al. 2015. Unjamming and cell shape in the asthmatic airway epithelium. *Nat Mater* **14:** 1040–1048. doi:10.1038/nmat4357

Pastushenko I, Blanpain C. 2019. EMT transition states during tumor progression and metastasis. *Trends Cell Biol* **29:** 212–226. doi:10.1016/j.tcb.2018.12.001

Pawlizak S, Fritsch AW, Grosser S, Ahrens D, Thalheim T, Riedel S, Kießling TR, Oswald L, Zink M, Manning ML, et al. 2015. Testing the differential adhesion hypothesis across the epithelial−mesenchymal transition. *New J Phys* **17:** 083049. doi:10.1088/1367-2630/17/8/083049

Peercy BE, Starz-Gaiano M. 2020. Clustered cell migration: modeling the model system of *Drosophila* border cells.

Semin Cell Dev Biol **100:** 167–176. doi:10.1016/j.semcdb.2019.11.010

Petridou NI, Grigolon S, Salbreux G, Hannezo E, Heisenberg CP. 2019. Fluidization-mediated tissue spreading by mitotic cell rounding and non-canonical Wnt signalling. *Nat Cell Biol* **21:** 169–178. doi:10.1038/s41556-018-0247-4

Phung TKN, Mitchel JA, O'Sullivan MJ, Park JA. 2022. In airway epithelium, basal stem cells and their stress fibers remodel during the unjamming transition. bioRxiv doi:10.1101/2022.08.18.504453

Pieuchot L, Marteau J, Guignandon A, Dos Santos T, Brigaud I, Chauvy PF, Cloatre T, Ponche A, Petithory T, Rougerie P, et al. 2018. Curvotaxis directs cell migration through cell-scale curvature landscapes. *Nat Commun* **9:** 3995. doi:10.1038/s41467-018-06494-6

Pinheiro D, Kardos R, Hannezo É, Heisenberg CP. 2022. Morphogen gradient orchestrates pattern-preserving tissue morphogenesis via motility-driven unjamming. *Nat Phys* **18:** 1482–1493. doi:10.1038/s41567-022-01787-6

Plachot C, Chaboub LS, Adissu HA, Wang L, Urazaev A, Sturgis J, Asem EK, Lelièvre SA. 2009. Factors necessary to produce basoapical polarity in human glandular epithelium formed in conventional and high-throughput three-dimensional culture: example of the breast epithelium. *BMC Biol* **7:** 77. doi:10.1186/1741-7007-7-77

Plou J, Juste-Lanas Y, Olivares V, Del Amo C, Borau C, García-Aznar JM. 2018. From individual to collective 3D cancer dissemination: roles of collagen concentration and TGF-beta. *Sci Rep* **8:** 12723. doi:10.1038/s41598-018-30683-4

Plutoni C, Bazellieres E, Le Borgne-Rochet M, Comunale F, Brugues A, Séveno M, Planchon D, Thuault S, Morin N, Bodin S, et al. 2016. P-cadherin promotes collective cell migration via a Cdc42-mediated increase in mechanical forces. *J Cell Biol* **212:** 199–217. doi:10.1083/jcb.201505105

Pogoda K, Byfield F, Deptula P, Cieśluk M, Suprewicz L, Skłodowski K, Shivers JL, van Oosten A, Cruz K, Tarasovetc E, et al. 2022. Unique role of vimentin networks in compression stiffening of cells and protection of nuclei from compressive stress. *Nano Lett* **22:** 4725–4732. doi:10.1021/acs.nanolett.2c00736

Provenzano PP, Inman DR, Eliceiri KW, Knittel JG, Yan L, Rueden CT, White JG, Keely PJ. 2008. Collagen density promotes mammary tumor initiation and progression. *BMC Med* **6:** 11. doi:10.1186/1741-7015-6-11

Ray A, Provenzano PP. 2021. Aligned forces: origins and mechanisms of cancer dissemination guided by extracellular matrix architecture. *Curr Opin Cell Biol* **72:** 63–71. doi:10.1016/j.ceb.2021.05.004

Reffay M, Parrini MC, Cochet-Escartin O, Ladoux B, Buguin A, Coscoy S, Amblard F, Camonis J, Silberzan P. 2014. Interplay of RhoA and mechanical forces in collective cell migration driven by leader cells. *Nat Cell Biol* **16:** 217–223. doi:10.1038/ncb2917

Rey-Barroso J, Calovi DS, Combe M, German Y, Moreau M, Canivet A, Wang X, Sire C, Theraulaz G, Dupré L. 2018. Switching between individual and collective motility in B lymphocytes is controlled by cell-matrix adhesion and inter-cellular interactions. *Sci Rep* **8:** 5800. doi:10.1038/s41598-018-24222-4

Sadati M, Taheri Qazvini N, Krishnan R, Park CY, Fredberg JJ. 2013. Collective migration and cell jamming. *Differentiation* **86:** 121–125. doi:10.1016/j.diff.2013.02.005

Saraswathibhatla A, Notbohm J. 2020. Tractions and stress fibers control cell shape and rearrangements in collective cell migration. *Phys Rev X* **10:** 011016. doi:10.1103/PhysRevX.10.011016

Sauer F, Grosser S, Shahryari M, Hayn A, Guo J, Braun J, Briest S, Wolf B, Aktas B, Horn LC, et al. 2023. Changes in tissue fluidity predict tumor aggressiveness in vivo. *Adv Sci (Weinh)* **10:** e2303523. doi:10.1002/advs.202303523

Saw TB, Doostmohammadi A, Nier V, Kocgozlu L, Thampi S, Toyama Y, Marcq P, Lim CT, Yeomans JM, Ladoux B. 2017. Topological defects in epithelia govern cell death and extrusion. *Nature* **544:** 212–216. doi:10.1038/nature21718

Shen Y, O'Byrne J, Schoenit A, Maitra A, Mège RM, Voituriez R, Ladoux B. 2025. Flocking and giant fluctuations in epithelial active solids. *Proc Natl Acad Sci* **122:** e2421327122. doi:10.1073/pnas.2421327122

Sigismund S, Lanzetti L, Scita G, Di Fiore PP. 2021. Endocytosis in the context-dependent regulation of individual and collective cell properties. *Nat Rev Mol Cell Biol* **22:** 625–643. doi:10.1038/s41580-021-00375-5

Skamrahl M, Pang H, Ferle M, Gottwald J, Rübeling A, Maraspini R, Honigmann A, Oswald TA, Janshoff A. 2021. Tight junction ZO proteins maintain tissue fluidity, ensuring efficient collective cell migration. *Adv Sci (Weinh)* **8:** e2100478. doi:10.1002/advs.202100478

Squarr AJ, Brinkmann K, Chen B, Steinbacher T, Ebnet K, Rosen MK, Bogdan S. 2016. Fat2 acts through the WAVE regulatory complex to drive collective cell migration during tissue rotation. *J Cell Biol* **212:** 591–603. doi:10.1083/jcb.201508081

Staneva R, El Marjou F, Barbazan J, Krndija D, Richon S, Clark AG, Vignjevic DM. 2019. Cancer cells in the tumor core exhibit spatially coordinated migration patterns. *J Cell Sci* **132:** jcs220277. doi:10.1242/jcs.220277

Tanimura N, Fujita Y. 2020. Epithelial defense against cancer (EDAC). *Semin Cancer Biol* **63:** 44–48. doi:10.1016/j.semcancer.2019.05.011

Te Boekhorst V, Friedl P. 2016. Plasticity of cancer cell invasion-mechanisms and implications for therapy. *Adv Cancer Res* **132:** 209–264. doi:10.1016/bs.acr.2016.07.005

Te Boekhorst V, Preziosi L, Friedl P. 2016. Plasticity of cell migration in vivo and in silico. *Annu Rev Cell Dev Biol* **32:** 491–526. doi:10.1146/annurev-cellbio-111315-125201

Te Boekhorst V, Jiang L, Mählen M, Meerlo M, Dunkel G, Durst FC, Yang Y, Levine H, Burgering BMT, Friedl P. 2022. Calpain-2 regulates hypoxia/HIF-induced plasticity toward amoeboid cancer cell migration and metastasis. *Curr Biol* **32:** 412–427.e8. doi:10.1016/j.cub.2021.11.040

Tran AT, Wisniewski EO, Mistriotis P, Stoletov K, Parlani M, Amitrano A, Ifemembi B, Lee SJ, Bera K, Zhang Y, et al. 2025. Cytoplasmic anillin and Ect2 promote RhoA/myosin II-dependent confined migration and invasion. *Nat Mater* **24:** 1476–1488. doi:10.1038/s41563-025-02269-9

Trepat X, Wasserman MR, Angelini TE, Millet E, Weitz DA, Butler JP, Fredberg JJ. 2009. Physical forces during collective cell migration. *Nat Phys* **5:** 426–430. doi:10.1038/nphys1269

Tschumperlin DJ, Dai G, Maly IV, Kikuchi T, Laiho LH, McVittie AK, Haley KJ, Lilly CM, So PT, Lauffenburger DA, et al. 2004. Mechanotransduction through growth-factor shedding into the extracellular space. *Nature* **429:** 83–86. doi:10.1038/nature02543

Tse JM, Cheng G, Tyrrell JA, Wilcox-Adelman SA, Boucher Y, Jain RK, Munn LL. 2012. Mechanical compression drives cancer cells toward invasive phenotype. *Proc Natl Acad Sci* **109:** 911–916. doi:10.1073/pnas.1118910109

van der Net A, Rahman Z, Bordoloi AD, Muntz I, Ten Dijke P, Boukany PE, Koenderink GH. 2024. EMT-related cell–matrix interactions are linked to states of cell unjamming in cancer spheroid invasion. *iScience* **27:** 111424. doi:10.1016/j.isci.2024.111424

van Helvert S, Friedl P. 2016. Strain stiffening of fibrillar collagen during individual and collective cell migration identified by AFM nanoindentation. *ACS Appl Mater Interfaces* **8:** 21946–21955. doi:10.1021/acsami.6b01755

van Helvert S, Storm C, Friedl P. 2018. Mechanoreciprocity in cell migration. *Nat Cell Biol* **20:** 8–20. doi:10.1038/s41556-017-0012-0

Venhuizen JH, Jacobs FJC, Span PN, Zegers MM. 2020. P120 and E-cadherin: double-edged swords in tumor metastasis. *Semin Cancer Biol* **60:** 107–120. doi:10.1016/j.semcancer.2019.07.020

Vishwakarma M, Di Russo J, Probst D, Schwarz US, Das T, Spatz JP. 2018. Mechanical interactions among followers determine the emergence of leaders in migrating epithelial cell collectives. *Nat Commun* **9:** 3469. doi:10.1038/s41467-018-05927-6

Vitorino P, Meyer T. 2008. Modular control of endothelial sheet migration. *Genes Dev* **22:** 3268–3281. doi:10.1101/gad.1725808

Wang H, Lacoche S, Huang L, Xue B, Muthuswamy SK. 2013. Rotational motion during three-dimensional morphogenesis of mammary epithelial acini relates to laminin matrix assembly. *Proc Natl Acad Sci* **110:** 163–168. doi:10.1073/pnas.1201141110

Wang H, Ardila C, Jindal A, Aggarwal V, Wang W, Vande Geest J, Jiang Y, Xing J, Sant S. 2025a. Protrusion force and cell-cell adhesion-induced polarity alignment govern collective migration modes. *Biophys J* **124:** 1674–1692. doi:10.1016/j.bpj.2025.04.010

Wang Y, Wang Y, Zhu Y, Yu P, Zhou F, Zhang A, Gu Y, Jin R, Li J, Zheng F, et al. 2025b. Angiomotin cleavage promotes leader formation and collective cell migration. *Dev Cell* **60:** 101–118.e7. doi:10.1016/j.devcel.2024.09.013

Warmt E, Grosser S, Blauth E, Xie X, Kubitschke H, Stange R, Sauer F, Schnauß J, Tomm JM, von Bergen M, et al. 2021. Differences in cortical contractile properties between healthy epithelial and cancerous mesenchymal breast cells. *New J Phys* **23:** 103020. doi:10.1088/1367-2630/ac254e

Weigelin B, Bakker GJ, Friedl P. 2012. Intravital third harmonic generation microscopy of collective melanoma cell invasion: principles of interface guidance and microvesicle dynamics. *Intravital* **1:** 32–43. doi:10.4161/intv.21223

Williams AM, Horne-Badovinac S. 2024. Fat2 polarizes Lar and Sema5c to coordinate the motility of collectively migrating epithelial cells. *J Cell Sci* **137:** jcs261173. doi:10.1242/jcs.261173

Wolf K, Friedl P. 2011. Extracellular matrix determinants of proteolytic and non-proteolytic cell migration. *Trends Cell Biol* **21:** 736–744. doi:10.1016/j.tcb.2011.09.006

Wolf K, Wu YI, Liu Y, Geiger J, Tam E, Overall C, Stack MS, Friedl P. 2007. Multi-step pericellular proteolysis controls the transition from individual to collective cancer cell invasion. *Nat Cell Biol* **9:** 893–904. doi:10.1038/ncb1616

Wolf K, Alexander S, Schacht V, Coussens LM, von Andrian UH, van Rheenen J, Deryugina E, Friedl P. 2009. Collagen-based cell migration models in vitro and in vivo. *Semin Cell Dev Biol* **20:** 931–941. doi:10.1016/j.semcdb.2009.08.005

Wolf K, Te Lindert M, Krause M, Alexander S, Te Riet J, Willis AL, Hoffman RM, Figdor CG, Weiss SJ, Friedl P. 2013. Physical limits of cell migration: control by ECM space and nuclear deformation and tuning by proteolysis and traction force. *J Cell Biol* **201:** 1069–1084. doi:10.1083/jcb.201210152

Xi W, Sonam S, Beng Saw T, Ladoux B, Teck Lim C. 2017. Emergent patterns of collective cell migration under tubular confinement. *Nat Commun* **8:** 1517. doi:10.1038/s41467-017-01390-x

Yamada KM, Sixt M. 2019. Mechanisms of 3D cell migration. *Nat Rev Mol Cell Biol* **20:** 738–752. doi:10.1038/s41580-019-0172-9

Zanotelli MR, Miller JP, Wang W, Ortiz I, Tahon E, Bordeleau F, Reinhart-King CA. 2024. Tension directs cancer cell migration over fiber alignment through energy minimization. *Biomaterials* **311:** 122682. doi:10.1016/j.biomaterials.2024.122682

Engineered In Vitro Platforms for Mechanochemical Control of Cell Migration

Bishwa Ranjan Si,[1,2] Qinling Yuan,[1,2] Sanjiban Nath,[1,2] Bhawana Agarwal,[1,2] Alice Amitrano,[1,2] Debanik Choudhury,[1,2] and Konstantinos Konstantopoulos[1,2,3,4]

[1]Department of Chemical and Biomolecular Engineering, The Johns Hopkins University, Baltimore, Maryland 21218, USA

[2]Institute for NanoBioTechnology, The Johns Hopkins University, Baltimore, Maryland 21218, USA

[3]Department of Biomedical Engineering, The Johns Hopkins University, Baltimore, Maryland 21218, USA

[4]Department of Oncology, Sidney Kimmel Comprehensive Cancer Center, Baltimore, Maryland 21218, USA

Correspondence: konstant@jhu.edu

Cell migration in confined environments follows distinct mechanisms compared to conventional 2D migration. By using in vitro models and incorporating extracellular cues from the tissue microenvironment, we can gain deeper insights into the complexities of cell migration. In this work, we explore various engineered in vitro models to study cell migration. We delve into biophysical tools, such as traction force microscopy, to understand how cells generate forces in response to their surroundings. We highlight the use of novel optogenetic tools for precise, spatiotemporal control of protein expression at the cellular level. Lastly, we examine emerging therapeutic strategies designed to target abnormal cell migration.

Cell migration is a fundamental process essential for diverse physiological phenomena, including embryonic development where it drives tissue morphogenesis (Herbert and Stainier 2011) and shapes the structure and function of the developing organism. Cell migration also plays a vital role in wound healing (Aragona et al. 2017), tissue regeneration, and immune surveillance (Luster et al. 2005), enabling immune cells to navigate to infection sites and initiate defense mechanisms. Impairment of cell migration can have profound consequences, including neurological disorders (Sukumaran et al. 2022) and immunodeficiencies (Janssen and Geha 2019), while aberrant cell migration contributes to pathological conditions, such as cancer metastasis (Friedl and Alexander 2011), autoimmune diseases (García-Cuesta et al. 2019), and fibrosis (Glass et al. 2020), exacerbating disease progression. Thus, gaining deeper mechanistic insights into the regulation of cell migration is crucial for developing effective therapeutic strategies, which necessitates the use of engineered in vitro model systems mimicking key aspects of the in vivo microenvironment along with animal models. In vitro models provide flexibility in experimental design, allowing precise control over parameters like substrate stiffness and pore size, while enabling high-throughput experimentation. They are valuable for studying molecular mechanisms and signaling pathways in cell migration by not only mim-

Cite this article as *Cold Spring Harb Perspect Biol* doi: 10.1101/cshperspect.a041795

icking the in vivo environment but also isolating parameters, such as extracellular matrix (ECM) composition, microenvironment stiffness, fluid viscosity, and chemical gradients (Paul et al. 2016a). However, conventional 2D cell migration models, although commonly used (Ridley et al. 2003), fail to capture the complexity of the in vivo environment, where cells must navigate confining spaces and interact with diverse microenvironments (Friedl and Alexander 2011). Confined spaces can arise from pores within the ECM or from ECM remodeling driven by cell–matrix interactions involving cancer cells, macrophages, and cancer-associated fibroblasts, which result in formation of tunnel-like tracks. These tracks represent the highways for cancer cells to efficiently move throughout the body and establish new metastatic colonies. Such confined pathways are commonly found within organ vasculature, inside blood vessels, and between nerve fibers (Winkler et al. 2020), emphasizing the importance of studying cell migration in confined environments to better reflect physiological conditions. This article covers engineered in vitro models of cell migration, along with sophisticated biophysical and optogenetic tools, to understand how microenvironmental cues regulate cell function. Moreover, we discuss current therapeutic strategies aimed at targeting aberrant cell migration for disease management.

IN VITRO PLATFORMS OF CELL MIGRATION

Microfluidic devices are widely used to study cell migration. These platforms enable comprehensive investigations of migration modes and mechanisms in response to pathophysiologically relevant mechanochemical cues present in the tissue microenvironment. Box 1 summarizes currently used microfluidic devices to study cell migration with special emphasis on confining systems (Fig. 1; Paul et al. 2016a, 2017).

As an illustrative example of how different track geometries influence migration, prior studies have shown that cancer cells exhibit distinct behaviors when confined along either their dorsoventral (top-to-bottom) axis or its orthogonal (side-to-side) direction. When the microchannel height is limited to 3 µm, thus compressing the nucleus in the dorsoventral direction, the setup is classified as vertically confined (Mistriotis et al. 2019; Wisniewski et al. 2020). In the reverse configuration where the microchannel width is restricted to 3 µm, the setup is considered laterally confined (Balzer et al. 2012; Wisniewski et al. 2020). Applying vertical confinement to HT1080 fibrosarcoma cells induces myosin II–dependent nuclear tension and stiffening, resulting in slower bleb-based cell migration (Wisniewski et al. 2020). In contrast, lateral confinement promotes faster mesenchymal mode of migration. This highlights how different migration track geometries can lead to distinct cellular properties and migration modes.

MODES AND MECHANISMS OF CELL MIGRATION

Cells navigate the intricate 3D architecture of living tissues using various migration strategies, including single-cell amoeboid, mesenchymal, and lobopodial motility, as well as collective migration (Yamada and Sixt 2019). In mesenchymal migration, cells adopt a spindle-shaped morphology, forming actin stress fibers linked to the extracellular substrate through integrin-based focal adhesions (FAs), which generate traction forces (Yamada and Sixt 2019). Confined amoeboid migration involves pill-shaped cells using cyclic expansion and contraction to traverse narrow ECM spaces. The amoeboid mode in confinement, characterized by RhoA-mediated membrane blebs, exhibits polarized distribution of actin, ion channels, and other molecules between the leading and trailing edges (Stroka et al. 2014; Zhao et al. 2021). In contrast, mesenchymal cells display protrusions and a more uniform actin network (Zhao et al. 2021). Under low-adhesion conditions, confined mesenchymal cells transition to one of two rapid amoeboid phenotypes. The A1 phenotype, defined by a rounded cell body and minimal leading edge, occurs in low-confinement, low-contractility environments (Liu et al. 2015; Ruprecht et al. 2015), while the A2 phenotype,

BOX 1. ENGINEERED IN VITRO PLATFORMS OF CELL MOTILITY

3D Microchannel Arrays—These devices comprise parallel microchannels with rectangular cross-sections, typically ranging from 3 µm—limit of soft lithography—to 50 µm in height and width. They are fabricated using a patterned silicon wafer as a template, followed by micromolding with materials, such as polydimethylsiloxane (PDMS) (Heuzé et al. 2011; Balzer et al. 2012; Tong et al. 2012; Irimia and Ellett 2016), polyacrylamide (PA) (Afthinos et al. 2022), or collagen (Kraning-Rush et al. 2013).

Microcontact Printed Arrays or Islands—Microcontact printing uses a PDMS stamp, designed with precise shapes and dimensions through soft lithography, which is coated with an extracellular matrix (ECM)-like fibronectin or collagen (Ghosh et al. 2008). The stamp is pressed onto glass or another chosen substrate, followed by washing with a non-adhesive reagent like poly(ethylene glycol) (PLL-PEG) to ensure cell adhesion is restricted to the ECM-coated regions (Ghosh et al. 2008; Théry and Piel 2009). Additionally, micropatterning techniques have been used to create rigid islands, made of cross-linked photoresist, in various configurations on non-adhesive PA hydrogels with specific stiffnesses (Wong et al. 2014).

Vertical Confinement—This model uses compressive forces to confine cells on 2D substrates, with device height typically varying between 3 µm (high confinement) and 5 µm (low confinement) using PDMS pillar structures, along with the application of negative pressure to control the confinement height (Liu et al. 2015). In some cases, cells can be sandwiched between two rigid substrates, such as PDMS or glass, or between a rigid and a soft substrate, such as agarose (Srivastava et al. 2017). Certain hybrid designs combine vertical confinement (e.g., agarose gel) with lateral confinement using micropillars, allowing the study of directional migration under combined constraints (Elpers et al. 2023).

3D Encapsulated Hydrogels—Cells are embedded within 3D hydrogels, where pore size and fiber arrangement can be precisely controlled through the hydrogel's composition and polymerization conditions (Wolf et al. 2013). This approach simulates cell migration through fibrillar collagen. However, precise control of length and width of migration tracks is relatively limited. Cell-derived matrices (CDMs) from fibroblasts are generally linearly elastic (Petrie et al. 2014). Alginate-based hydrogels with tunable stress relaxation properties have been developed to capture the time-dependent viscoelastic behavior of tissues (Chaudhuri et al. 2016; Blache et al. 2022).

In addition to studying cell–matrix interactions, these models also provide a valuable platform to study cell–cell communication and migration in complex environments. Studying the interactions between immune cells, like T cells or macrophages, with cancer cells, can offer insights into immune response and facilitate drug screening applications (Zhou et al. 2023; Alieva et al. 2024).

defined by an elongated ellipsoidal shape and a pronounced uropod, arises under high confinement and contractility (Ruprecht et al. 2015).

In 3D elastic matrices, RhoA activation switches fibroblasts from lamellipodial (mesenchymal) migration to either amoeboid migration or lobopodial migration under low or high matrix adhesion, respectively (Yamada and Sixt 2019). Lobopodial migration, also observed in cancer cells upon matrix metalloproteinase inhibition, involves an elongated cylindrical protrusion (Petrie et al. 2017). This mode combines characteristics of mesenchymal and amoeboid migration, by using asymmetric intracellular pressure for generating a bleb-like protrusion at the leading edge, while anchoring and pulling against the substrate with strong adhesion (Yamada and Sixt 2019). In addition to single-cell migration, cells often migrate as multicellular groups, displaying front–rear polarity, and coordinated mechanical and signaling interactions at cell–cell junctions (Yamada and Sixt 2019; Law et al. 2023). In migrating epithelial and endothelial cell sheets, cadherin adhesion receptors form the mechanical link between neighboring cells, coordinating actomyosin contractility and maintaining collective polarity (Yamada and Sixt 2019). Confinement promotes the transition from collectively migrating sheets to finger-like strands and cell individualization (Yamada and

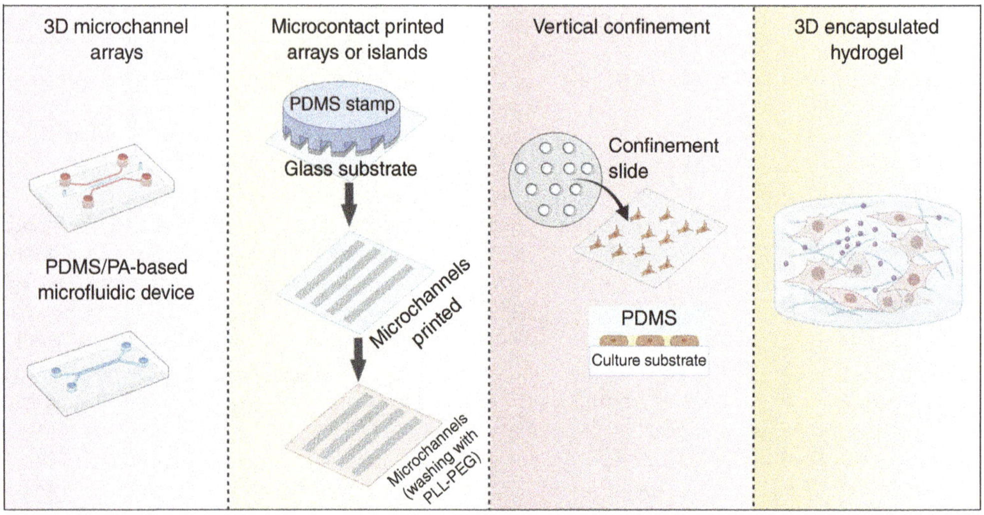

Figure 1. In vitro platforms of confined cell migration. (PDMS) Polydimethylsiloxane, (PA) polyacrylamide, (PLL-PEG) poly(ethylene glycol). (Figure generated with BioRender; https://biorender.com.)

Sixt 2019). This process is driven by myosin IIA (MIIA) contractility, which dismantles E-cadherin junctions between leader and follower cells (Law et al. 2023). Cell dissociation can also occur through downregulation of E-cadherin at cell–cell junctions (Friedl and Gilmour 2009; Ilina et al. 2020).

Two-Dimensional versus Confined Migration

In conventional 2D cell migration, cycles of actin polymerization-driven protrusions, transient integrin-mediated substrate adhesions, and myosin II contractility coordinate cell body retraction and forward movement. In 2D migration, different types of cells sense and adapt to their surroundings by generating traction through a "motor-clutch" model (Chan and Odde 2008). This mechanism uses molecular clutches to link F-actin and the extracellular substrate, creating a frictional slippage interface. This results in traction force transmission and reduces F-actin retrograde flow, enabling actin polymerization to propel the leading edge forward. On stiff substrates, "frictional slippage" occurs due to the limited compliance of the substrate, which causes rapid tension buildup within engaged clutches, shortened clutch interaction times, and elevated retrograde flow. On soft substrates, slower tension buildup prolongs the clutch engagement, reducing retrograde flow. Both extremes impair motility, while optimal motility is achieved when F-actin–clutch connections persist long enough to generate maximal traction (Chan and Odde 2008). The "motor-clutch" model explains actin-based motility in 2D environments.

While the nucleus is not essential for migration in 2D environments, it plays a crucial role in 3D and confined settings. In cells with a lobopodial phenotype, RhoA–myosin II contractility, in coordination with the nucleoskeleton-intermediate filament linker protein nesprin-3 and integrin-mediated adhesion, propels the nucleus in a piston-like motion (Petrie et al. 2014). In contrast, blebbing cells migrating under vertical confinement primarily use rear-generated forces to propel the nucleus forward (Mistriotis et al. 2019). Vertical confinement in this context induces myosin II–driven nuclear stiffening, which activates RhoA at the cell poles. The polarized activation of RhoA, combined with the linker of nucleoskeleton and cytoskeleton (LINC) complex's role in positioning the nucleus at the posterior of the cell, increases local cytoplasmic pressure, effectively pushing the nucleus forward and directing cell migration (Wisniewski et al. 2020).

The Osmotic Engine Model (OEM) of Cell Migration

In lateral confinement—directed perpendicularly to the dorsoventral polarity axis—cancer cells can migrate even when actin polymerization is completely disrupted, or myosin II contractility and β1 integrins are blocked, as explained by OEM. OEM operates through the polarization of specific ion transporters, ion channels, and aquaporins (AQPs) that regulate cell volume and facilitate cell motility (Stroka et al. 2014). According to OEM, cancer cell movement in confined spaces is driven by coordinated cycles of isosmotic swelling at the leading edge and shrinkage at the trailing edge of the cell (Stroka et al. 2014; Zhang et al. 2022). During migration through rigid, confining microchannels, the Na^+/H^+ exchanger 1 (NHE1) and AQP5 polarize at the leading edge of breast cancer cells, facilitating isosmotic cell swelling. On the other hand, SWELL1 (LRRC8A) accumulates at the trailing edge along with AQP4, mediating isosmotic shrinkage (Zhang et al. 2022). SWELL1 polarization at the cell rear determines migration directionality, and is regulated by RhoA activity, as demonstrated via optogenetic tools. Optimal cell migration depends on maintaining front-to-rear polarity, making efficient cell reversal contingent upon simultaneous repolarization of both NHE1 and SWELL1 mediated by cell division control protein 42 homolog (Cdc42) and RhoA, respectively (Zhang et al. 2022). The role of OEM in cell migration was recently extended to immune cells, including neutrophils and T cells (de Boer et al. 2023; Nagy et al. 2024; Amitrano et al. 2025a).

Macropinocytosis and Cell Migration

The polarized distribution of signaling and membrane components at the cell front and rear is essential for establishing directional migration. Macropinocytosis contributes to this process by regulating membrane trafficking and spatial organization of signaling molecules. In *Dictyostelium* amoeba, activated Ras and phosphatidylinositol-3,4,5-triphosphate (PIP_3) localize at the front, driving membrane extension, whereas PTEN, myosin II, phosphatidylinositol-3,4-biphosphate $PI(3,4)P_2$, and GTPase-activating proteins (GAPs), localize at the back. $PI(3,4)P_2$ gradually degrades as it diffuses toward the cell front, and disappears when actin-rich protrusions form at the leading edge but re-emerges as these protrusions retract. $PI(3,4)P_2$ is also internalized within nascent macropinosomes, which subsequently fuse with the plasma membrane at the cell rear. This sequence of events is analogous to a "reverse-fountain flow" of $PI(3,4)P_2$, creating a back-to-front gradient essential for polarity (Li et al. 2021). Further research is needed to elucidate how $PI(3,4)P_2$ cross talks with ion channels and other membrane components.

EFFECTS OF LOCAL MICROENVIRONMENTAL CUES ON CELL MIGRATION

Cells respond to various mechanochemical cues that guide their migration, directing movement either toward or away from the stimulus. These cues, encountered in vivo (Fig. 2), trigger signal transduction pathways that convert mechanical, chemical, or electrical signals into cellular responses (SenGupta et al. 2021). Incorporating such cues into in vitro models enables researchers to better mimic in vivo conditions and understand the molecular mechanisms underlying migration in the presence of such cues.

Substrate Stiffness and Viscoelasticity

Substrate stiffness is a key physical cue influencing cell morphology and function (Amitrano et al. 2025b). Most microfluidic devices used to study cell migration use polydimethylsiloxane (PDMS) whose stiffness is in the range of MPa. However, the stiffness of native tissues typically varies from a few hundred Pa to a few kPa (Butcher et al. 2009). Although the stiffness of PDMS can be reduced to the kPa range, its lack of water permeability limits its physiological relevance. To overcome these limitations,

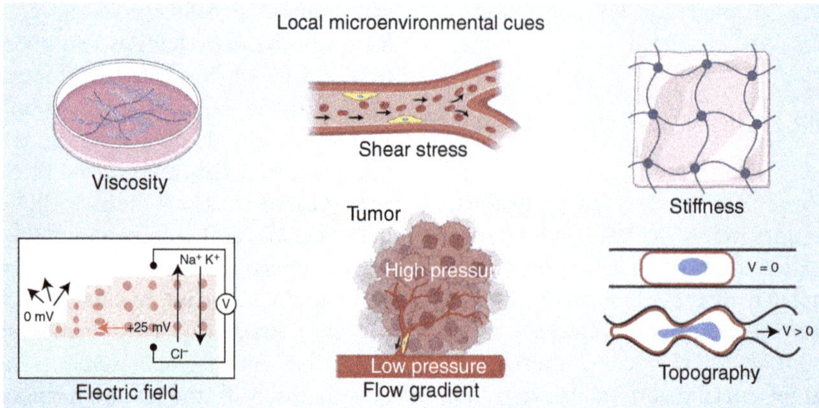

Figure 2. Extracellular microenvironmental cues encountered by cells during confined migration. (Figure generated with BioRender; https://biorender.com.)

water-permeable polyacrylamide (PA) gels have been adopted.

Unlike synthetic substrates, such as PDMS and PA, biological tissues and the ECM are not purely elastic but exhibit viscoelasticity, a time-dependent response to material deformation, as well as mechanical plasticity, which is irreversible material deformation in response to applied force. Viscoelasticity allows tissues to gradually relax stress over time when subjected to strain. Stress relaxation tests have shown that soft tissues, including the liver, breast, muscle, skin, and adipose tissue, can recover from deformation over timescales ranging from tens to hundreds of seconds (Chaudhuri et al. 2020). Stiffer tissues, like cartilage and bone, also display viscoelasticity (Chaudhuri et al. 2020). In interpenetrating polymer network (IPN) hydrogels whose plasticity can be tuned independent of stiffness, cells can traverse pores <3 µm without any protease dependence (Wisdom et al. 2018). Due to mechanical plasticity of the hydrogel, cell invadopodia protrusions can mechanically create tissue tracks, signifying the importance of both the elastic and viscous elements of a material.

Extracellular Fluid Viscosity

Viscosity, which is a measure of fluid resistance to flow, has emerged as a key physical cue influencing cell migration in vitro and cancer cell extravasation in vivo (Bera et al. 2022). To date, most in vitro cell functional assays are conducted in media with a viscosity close to that of water (0.77 cP at 37°C). However, interstitial fluid viscosity can reach up to 3.5 cP, while whole blood viscosity typically ranges from 4 to 6 cP and may exceed 8 cP under pathological conditions. Elevated fluid viscosity counterintuitively increases cell migration in 2D, 3D, and confined microenvironments in vitro (Bera et al. 2022; Maity et al. 2022). Cells subjected to elevated viscosity respond by forming an actin-related protein 2/3 (Arp2/3) complex–dependent dense actin network to counteract the increased viscous forces. This viscosity-induced actin reorganization recruits NHE1 to the cell leading edge via its actin binding partner ezrin and drives water influx, which increases membrane tension (Bera et al. 2022). This elevated tension activates the mechanosensitive ion channel (MIC) called transient receptor potential cation vanilloid 4 (TRPV4), resulting in calcium influx and heightened RhoA-mediated contractility at the cell front (Bera et al. 2022). This cascade collectively enables cells to achieve faster migration in environments with elevated fluid viscosity. Notably, cells not only sense and respond to viscosity but also have the ability to develop and retain memory of their previous exposure to viscous forces via a TRPV4-dependent pathway. This memory was shown using breast cancer cells, where their preconditioning at elevated viscosities enhances their propensity to colonize mouse lungs in vivo compared to those exposed to basal viscosity (Bera et al. 2022).

Shear Stress and Pressure Gradients

Shear stresses are present in vivo due to interstitial fluid flow (Munson and Shieh 2014) and blood flow in capillaries. Low levels of shear stress facilitate cancer cell intravasation (Yankaskas et al. 2021). Thus, cancer cells preferentially enter lymphatic and venous vessels over the arterial circulation due to the lower prevailing shear stresses. Microfluidic platforms have been used to simulate aspects of cell intravasation, wherein cells migrating in confining microchannels encounter a larger orthogonal channel with physiologically relevant shear stresses (Yankaskas et al. 2021). Non-cancerous cells, like human dermal fibroblasts, exhibit high sensitivity to shear stress due to elevated expression levels of the MIC transient receptor potential melastatin 7 (TRPM7), which triggers the reversal in their migration direction to avoid shear stress exposure. In contrast, cancerous cells, like HT1080, which display lower levels of TRPM7, remain relatively unaffected by the shear stress and are able to successfully intravasate both in vivo and in vitro. This study highlights the important role of TRMP7 as a shear stress sensor, which induces Ca^{2+} influx and RhoA/myosin II contractility, promoting cell reversal through polarity switching.

During pathological conditions, such as cancer, fluid flow can increase three to five times in magnitude, resulting from heightened pressure gradients, between the intratumoral space and the surrounding healthy tissue. This pressure differential is characterized by outward convective fluid flows, facilitating tumor cell dissemination and subsequent metastasis (Boucher et al. 1990). However, in vertically confined microfluidic devices, cells exhibit seemingly counterintuitive behavior. Upon application of a pressure differential at the ends of the microchannels, cancer cells tend to migrate preferentially toward regions of higher pressure (Amiri et al. 2024). This cell reversal toward regions of higher pressure is driven by myosin IIA polarization at the new cell rear, actin polymerization, and NHE1 activation at the new cell front. Myosin IIA contractility is activated by increased influx of calcium through the fluid force sensor TRPM7 in response to a pressure differential. Interestingly, knockdown of Lamin A/C disrupts this cell reversal due to reduced calcium influx and myosin IIA repolarization (Amiri et al. 2024), highlighting the crucial role of nucleus in cytoskeletal rearrangement and in ion channel mechanosensation (Mistriotis et al. 2024). These findings may help explain the observed directional response in vitro, particularly when considering tumor heterogeneity. While the majority of confined cells respond to fluid forces by reversing direction, a more aggressive subpopulation, characterized by low lamin A/C expression and reduced sensitivity to mechanical cues, may retain the ability to migrate downstream. This directional insensitivity could facilitate their escape from high-pressure tumor regions and contribute to metastatic progression.

Topography

Contact guidance in 3D refers to the phenomenon where the substrate or channel geometry provides directional cues that guide cell movement through the anisotropic features of the microenvironment. In highly aligned collagen gels, tumor explants and MDA-MB-231 cells invade more deeply than in low-alignment matrices (Provenzano et al. 2008). This process is Rho/ROCK-dependent, as contractility aids in matrix alignment; however, migration becomes independent of contractility when the matrix is prealigned. In another study using a Y-shaped microchannel device, cells develop contact guidance when the feeder channel is wider than the cell body (Paul et al. 2016b). In this case, cells elongate along one of the side channel walls and are guided by contact, steering them toward the adjacent branch channel, regardless of branch width. Notably, this contact guidance is suppressed when integrins or both myosin II isoforms (IIA and IIB) are knocked down. Intriguingly, contact guidance-driven decision-making is enhanced upon knockdown or inhibition of Cdc42 (Paul et al. 2016b). These findings underscore the critical role of contact guidance in cell migration under different microtopologies.

Cells typically migrate by linking the force generated from integrin-based adhesion receptors to their surrounding microenvironment. However, amoeboid cells, such as leukocytes, can migrate very fast even on weakly adhering surfaces, as they can exploit the topographical features of their surroundings to propel themselves forward (Yamada and Sixt 2019). Specifically, T cells can generate migratory forces by coupling actin retrograde flow to the local geometric irregularities of the substrate, even in the absence of traditional adhesion complex components like talin (Reversat et al. 2020). Using serrated confining microchannels with periodic patterns ranging from 6 to 24 µm, wild-type T cells migrate across all topographies, including smooth channels. In contrast, talin-knockout cells migrate only within serrated channels of 6 or 12 µm periodicity, but not 24 µm, as this exceeds the average length scale of the actin retrograde flow. This study reveals that channel irregularities generate non-vanishing normal forces, which promote cortical actin flow and enable migration even in the absence of adhesive forces, highlighting the critical role of topographical cues in directing migration.

Voltage Gradients

Cells respond to endogenous electric fields (EFs) in vivo, playing crucial roles in physiological processes, such as wound healing (Zhao et al. 2006) and embryogenesis, as well as pathological conditions like cancer metastasis and bacterial infections (Sun et al. 2024). On 2D substrates, cells exposed to a direct current (DC) EF ranging from 0.5 to 5 V/cm using Ag/AgCl electrodes orient themselves perpendicularly to the EF direction (Huang et al. 2013). However, inside moderately confining microchannels (width $[W] = 20$ µm), cells are unable to reorient in a perpendicular direction in response to EF. Yet, EF enhances the migration of NIH 3T3 cells compared to their movement on 2D substrates, irrespective of the EF strength (Huang et al. 2013). The combined effects of chemotaxis and electrotaxis on dHL-60 cells, a neutrophil-like cell line, were recently examined in an Electrotaxis-On-Chip (ETOC) device (Moarefian et al. 2021). This device consists of microchannels ($W \times H = 10 \times 10$ µm) with applied DC EFs of 0.4–1 V/cm. When the chemoattractant *N*-formylmethionine-leucyl-phenylalanine (fMLP) was introduced parallel to the EF, migration velocities increased sixfold compared to fMLP alone. Further, the addition of an inflammatory signal, leukotriene B4 (LTB4), perpendicular to the field direction resulted in a significant 2.9-fold decrease in migration compared to LTB4 alone. These findings emphasize the intricate interplay between EFs, chemical cues, and spatial constraints in regulating cell motility.

THE ROLE OF TRACTION FORCES IN CELL MIGRATION

The generation of traction forces is fundamental for cell migration and ECM probing (Roca-Cusachs et al. 2017). However, due to their micron-scale dimensions, cells exert forces in the piconewton to nanonewton range, making accurate measurements challenging. Reliable quantification of these minute forces demands highly sensitive techniques and precise calibration (Polacheck and Chen 2016). Early evidence of cellular traction was observed as wrinkles formed on silicone rubber substrates during cell migration (Harris et al. 1980). This led to the development of the traction force microscopy (TFM) technique, which uses fiducial markers, fluorescent beads, embedded in deformable substrates, such as PA, enabling precise mapping and quantification of traction forces at the cell–substrate interface (Dembo and Wang 1999).

Traction Force Microscopy (TFM)

TFM enables precise measurement of fluorescent bead displacement within substrates and the calculation of traction forces from displacement vectors, a process critically dependent on the known material properties of the substrate (Polacheck and Chen 2016). However, from a mathematical perspective, measuring traction forces presents the challenge of being an ill-posed problem, where even minor errors in bead displacement measurements can result in large inaccuracies in force computations (Schwarz and Soiné 2015). To overcome this, TFM has undergone significant advancements over time, including improvements in computational algorithm and enhanced imaging methodologies, such as total internal reflection fluorescence (TIRF) microscopy (Gutierrez et al. 2011), two-photon microscopy, superresolved stimulated emission depletion (STED) microscopy (Colin-York et al. 2016), structured illumination microscopy (SIM) (Colin-York et al. 2019), or combinations of these techniques (Barbieri et al. 2021). These innovations have enabled the resolution of forces on 2D substrates with spatial precision as fine as 100 nm and temporal resolution as high as 1 sec. In addition to TFM, other external deformation-based techniques have emerged including micropillar-based TFM (Tan et al. 2003), deformable hydrogel microparticles (Vorselen et al. 2020), and oil droplets systems (Campàs et al. 2014), offering alternative platforms for studying cellular traction forces. Moreover, the recent development of molecular biosensors (Glazier et al. 2019) has facilitated the measurements of forces exerted by specific molecules both in vitro and in vivo, broadening our understanding of cellular mechanics.

Mechanisms of Generation of Forces

Early studies using 2D TFM measurements revealed that fibroblasts, which display a mesenchymal mode of migration, generate in-plane contractile tractions on the substrate (Beningo et al. 2001). These forces arise from a coordinated sequence of events, beginning with the formation of a protrusion at the cell's leading edge, followed by the establishment of integrin-based adhesions with the ECM, forward transport of the cell body and nucleus, and disassembly of adhesions at the rear. Protrusions at the leading edge are driven by rapid polarized actin polymerization, where filament elongation exerts pushing forces on the plasma membrane. If the actin filaments are not anchored to FAs, polymerization drives retrograde flow, causing a backward movement of the actin network (Wilson et al. 2010). However, if the plasma membrane is anchored to the underlying substrate ECM, the retrograde actin flow slows down and is converted into rearward contractile tractions that propel the cell forward. This anchoring occurs via integrin-mediated adhesions, which couple actin filaments to the ECM through FAs. Force generated by actin retrograde flow alone is sufficient to produce traction forces on the substrate (Gardel et al. 2008). This force-generating machinery is further enhanced by the activity of myosin II, a molecular motor protein that binds to actin filaments to create pulling forces, amplifying retrograde flow and traction generation (Quintanilla et al. 2023). Away from the cell periphery, F-actin filaments are bundled by proteins, such as α-actinin, to form long cable-like structures known as stress fibers. These stress fibers are interlaced with contractile myosin filaments, which enable the generation of tensile forces at a larger scale across the cell. Disrupting stress fibers reduces the traction forces exerted by adherent cells on the substrate, highlighting their critical role in force generation during cell migration (Kumar et al. 2006). Recent advancements using optogenetic tools to locally increase myosin II activity have revealed that tension generated within stress fibers can enhance traction forces at distant sites (Oakes et al. 2017) (discussed in more detail in the optogenetics section). This finding highlights the role of myosin motors in generating tension on the actin scaffold, which is transmitted across the cell via stress fibers. Additional studies have also shed light on the fact that membrane tension itself can drive contractile forces as demonstrated in actomyosin-deficient liposome systems (Murrell et al. 2014). Furthermore, the actin cortex, located just beneath the cell membrane, has emerged as a critical component for tension transmission across the cell, functioning in concert with stress fibers to regulate cellular forces effectively (Vignaud et al. 2021).

Transmission of Cellular Traction Force

The "motor-clutch" model describes how contractile forces mediated by the actomyosin cytoskeleton are transmitted to the substrate via integrins using a series of adaptor proteins that link integrins to the cytoskeleton. These protein–protein interactions collectively form the molecular clutch, enabling effective force transmission. One of the primary molecular clutches connecting actin to the integrins is talin (Critchley 2009). Knockdown of talin using siRNA leads to faster retrograde flow and lower traction forces, a consequence of actin filaments losing their anchorage to the substrate. Similarly, disruption of vinculin, which reinforces the actin–talin connection, results in accelerated retrograde flow and diminished traction stresses (Plotnikov et al. 2012).

Traction Forces in Various Physiological Settings

Traction forces generated by mesenchymal fibroblasts are crucial for single-cell and collective cell migration during wound healing, facilitating efficient tissue repair. Collective migration involves the coordinated movement of cells, facilitated by two primary mechanisms: the formation of a supracellular actomyosin ring at the wound margin, which contracts to draw the wound edges together like a purse-string, and the collective migration of marginal and submarginal cells guided by lamellipodial and filopodial extensions (Anon et al. 2012). A traction dipole, inward-directed tractions (away from the wound) at the leading edge, followed by outward-directed tractions (toward the wound) near the actomyosin ring, is observed (Brugués et al. 2014) resembling single-cell migration, where cells in wound healing processes generate contractile forces at the leading edge to pull themselves forward, whereas the actomyosin ring produces tractions perpendicular to the wound, deforming the underlying substrate and aiding wound closure.

Increasing substrate stiffness in elastic gels enhances cellular traction (Lo et al. 2000), with cells migrating preferentially toward regions of greater stiffness (Isenberg et al. 2009; Plotnikov et al. 2012). Interestingly, while most adherent cells generate higher traction forces on stiffer substrates, certain cells such as U251 glioma cells exhibit maximal traction on substrates with intermediate stiffness (Bangasser et al. 2017). These cells migrate from regions of higher stiffness where traction forces are reduced, to areas of lower stiffness where they can exert greater forces (Isomursu et al. 2022), further reinforcing the idea that mechanical forces govern migration directionality. In viscoelastic alginate gels, cells migrated more rapidly in fast-relaxing gels compared to slow-relaxing ones, which correlates with higher maximal traction forces exerted on the substrate. Higher traction forces in fast-relaxing gels are attributed to increased nascent adhesions and prolonged lifetime of integrin–ECM bonds (Adebowale et al. 2021). Recent work has shed light on the role of extracellular fluid viscosity in regulating cellular migration demonstrating that increasing viscosity significantly enhances migration in 2D, 3D, and confined environments (Bera et al. 2022). When cells were exposed to elevated viscosity, migration rates increase concurrently with a doubling of traction forces due to increased nascent adhesions and actomyosin reinforcement (Pittman et al. 2022).

In confined 3D microenvironments, such as microchannels, cells exhibit traction patterns that are distinct from those in unconfined 2D settings. Using PDMS-based micropillars, it was observed that both 3T3 fibroblasts and human osteosarcoma (HOS) cells generated lower forces in vertically confined channels compared to unconfined regions (Raman et al. 2013). Interestingly, many of the forces exerted in these channels were directed normal to the direction of migration. Similar findings were reported in laterally confined PA-based microchannel systems, where MDA-MB-231 cells primarily applied perpendicular forces during migration (Afthinos et al. 2022). In a similar light, cells in confined microwells generated outward-directed pushing forces (Faure et al. 2024), highlighting a new and exciting area of research. These extensile pushing forces, unique to confined microenvironments, open avenues for exploring not only their role in facilitating migration and adaptation but also their underlying origins. A deeper understanding of how these forces are generated and regulated could provide crucial insights into cellular mechanics and behavior in constrained spaces.

OPTOGENETIC CONTROL OF CELL DYNAMICS AND SIGNALING PATHWAYS IN CELL MIGRATION

Cell migration is a dynamic process governed by the spatiotemporal regulation of cytoskeletal and membrane components along with associated signaling pathways. Although traditional genetic approaches, such as gene overexpression, knockdown or knockout, have advanced our understanding of cell migration mechanisms, they often fail to capture dynamic events, such as subcellular protein localization and activation status that drive cell migration. Optogenetic systems, which incorporate light-responsive molecules, overcome these limitations by enabling precise spatiotemporal control of signaling and protein dynamics. These tools have proven invaluable for studying cell migration where polarized signaling molecules play a critical role. Using live-cell imaging, optogenetic tools allow researchers to toggle transcription, target molecules to specific subcellular compartments, and modulate protein activity in real time, thereby providing deeper mechanistic insights into migration processes. There are five principal categories of optogenetic tools extensively used in research. The first category involves light-sensitive ion channels, which regulate ion flux across cellular membranes and have been extensively used to study neuronal function and neurodegenerative mechanisms. The second approach, protein caging and uncaging, involves light-responsive groups attached to proteins or small molecules; specific wavelengths of light activate these molecules by detaching inhibitory groups. The third approach centers on protein dimerization, leveraging light-induced interactions between two proteins or protein domains, where light-sensitive domains undergo conformational changes to initiate binding. The fourth approach involves dimerization via protein caging, using light-triggered release of restrained interaction domains, enabling protein interaction upon illumination. Last, protein oligomerization harnesses light to orchestrate the assembly of proteins into larger complexes. As the mechanisms underlying these optogenetic systems have been thoroughly explored in existing literature (Tischer and Weiner 2014; Wittmann et al. 2020; Manoilov et al. 2021), our discussion will focus on their applications, particularly in the context of cell migration.

Optogenetics in Ion Channels/Ion Transporters

Polarized ion transporters and ion channels are critical to controlling the direction and efficiency of cell migration. SWELL1, a key component of the volume-regulated anion channel (VRAC), predominantly localizes at the rear of migrating cells in confined environments. The role of SWELL1 in cell migration direction was established through an optogenetic strategy, wherein SWELL1 was fused to Cry2-mCherry, and a green fluorescent protein (GFP)-labeled amino-terminal portion of cryptochrome-interacting basic helix–loop–helix 1 amino-terminal (CIBN) dimerization partner was engineered to anchor it to the plasma membrane via a CAAX motif (Zhang et al. 2022). This configuration enabled precise enrichment of SWELL1 at the cell leading edge following light stimulation, effectively redefining the "old" leading edge as the "new" cell rear, thereby resulting in the reversal of migration direction.

Optogenetics in PI3K/Akt Signaling Pathway

Phosphoinositide 3-kinase (PI3K)/Akt is a major signal transduction pathway that plays a critical role in regulating cell migration. Activation of this pathway begins with PI3K phosphorylating the lipid substrate phosphatidylinositol 4,5-bisphosphate (PIP_2) to generate PIP_3 on the cell membrane. This lipid transformation creates a docking site for Akt, recruiting it to the membrane where it is subsequently phosphorylated and activated. Once activated, Akt facilitates cell motility by coordinating a range of cellular processes including cytoskeletal remodeling, membrane rearrangements, cell polarity establishment, and adhesion site turnover. By integrating these mechanisms and interacting with other signaling cascades, the PI3K/Akt pathway ensures efficient and directed cell migration (Cain and Ridley 2009; Karar and Maity 2011; Martini et al. 2014). In highly invasive breast cancer cells, activation of the PI3K/Akt signaling pathway correlates with reduced levels of Lamin A/C, resulting in increased nuclear deformability. This enhanced deformability facilitates the migration of cells through confined microenvironments that mimic the interstitial spaces encountered during metastasis (Wisniewski et al. 2020).

In an opto-PI3K system, Cry2 is fused to the iSH2 domain of the PI3K regulatory subunit, thereby enabling it to bind to the endogenous catalytic subunit of PI3K (Zhou et al. 2018). Upon blue-light stimulation, this construct is recruited to the cell surface by CIBN-CAAX, promoting the conversion of PIP_2 to PIP_3 and subsequently activating Akt. Optogenetic activation of PI3K leads to light dose-dependent downregulation of E-cadherin expression, simulating aspects of epithelial–mesenchymal transition (EMT). This activation enhanced migration of A549 lung adenocarcinoma cells, as demonstrated by the accelerated wound closure in scratch assays compared to nonactivated controls. Similarly, an opto-iSH2 system was used to achieve localized PIP_3 production (Idevall-Hagren et al. 2012), likely activating downstream effectors, such as Rac1, which controls actin nucleation to drive membrane ruffling. As membrane ruffling precedes lamellipodia formation and initiates migration (Mahankali et al. 2011), this finding highlights the importance of PI3K signaling in dynamic cytoskeletal reorganization, essential for directional migration. Building on this, an improved light-induced dimer (iLID) system was developed (Town and Weiner 2023), where a membrane-bound anchor peptide (SsrA) exposed to blue light recruits SspB fused to the iSH2 domain to the plasma membrane, initiating localized PI3K signaling. Precise modulation of PI3K activity at the leading edge of neutrophil-like HL-60 cells induces highly directed movement, whereas activation at the rear causes the cells to either perform a "U-turn" or lose their polarity. These findings collectively emphasize the critical role of spatially regulated PI3K signaling in maintaining cell polarity and controlling the directionality of cell migration.

Optogenetics in Rho GTPase Modulation

Small Rho GTPases are key regulators of intracellular signals and processes that are important to cell migration, including actin polymerization, cell rear retraction, and the establishment of cell polarity. Among these, Rho, Rac, and Cdc42 orchestrate complex signaling cascades that mediate dynamic cytoskeletal rearrangements required for cell motility. Optogenetic approaches have emerged as powerful tools to dissect the precise roles and mechanisms of these GTPases, providing unprecedented resolution and specificity (Wittmann et al. 2020).

A sophisticated optogenetic tool was developed to enable precise control over subcellular RhoA activation, by fusing the DHPH domain of ARHGEF11, a RhoA-specific guanine nucleotide exchange factor (GEF), to Cry2-mCherry, termed optoGEF-RhoA (Fig. 3; Valon et al. 2017). This construct was targeted to distinct subcellular locations, either the plasma membrane or mitochondria, via CIBN. Recruitment of optoGEF-RhoA to the plasma membrane results in elevated RhoA activity, which in turn enhances cortical contractility, traction forces, intercellular tension, and nuclear localization of Yes-associated protein (YAP). In contrast, targeting optoGEF-RhoA to the mitochondria leads to a significant decrease in RhoA activity, resulting in diminished contractility and reduced nuclear YAP accumulation. Using this optogenetic system, it was demonstrated that light-induced RhoA activation at the leading edge of migrating MDA-MB-231 breast cancer cells in confining channels promotes local enrichment of SWELL1 localization (Zhang et al. 2022), which results in the reversal of the migration direction. Conversely, light-induced reduction of RhoA activity at the trailing edge of confined cells disrupts local SWELL1 enrichment, resulting in decreased migration velocities. These findings illustrate the role of spatially reg-

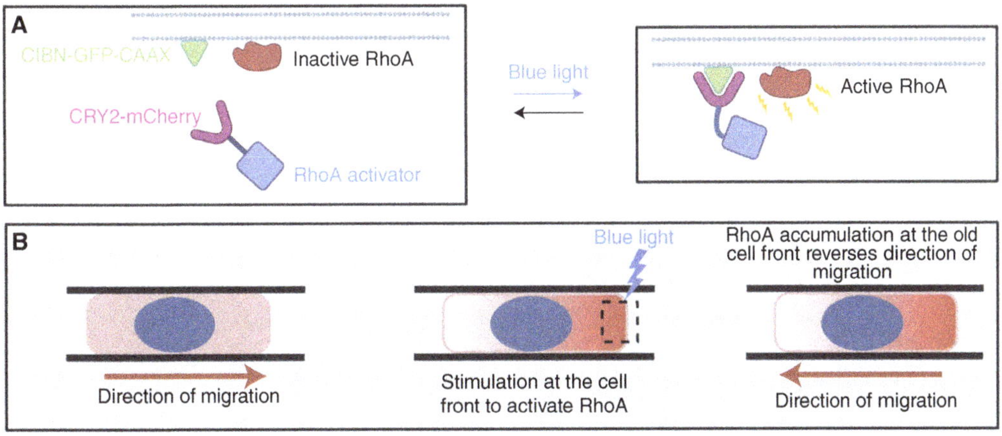

Figure 3. (*A*) Mechanism of optoGEF-RhoA upon stimulation with blue light. As CIBN is anchored to the membrane via the CAAX motif, the cytosolic optoGEF can be recruited to the stimulated membrane region through light-induced interaction between CRY2 and CIBN. This localized recruitment enables spatial activation of RhoA. (*B*) Use of the optoGEF-RhoA expressing cells in confined channels and spatial stimulation of cell with blue light. Targeted activation of RhoA at the "old" leading edge triggers reversal of migration direction. (Figure generated with BioRender; https://biorender.com.)

ulated RhoA activity in controlling SWELL1 polarization, thereby governing both direction and efficiency of cell migration. In a separate study, the iLID optogenetic system was used to achieve localized activation of RhoA in RAW 264.7 macrophage cells (O'Neill et al. 2018). RhoA-stimulated actomyosin contractility at the cell rear generated a rearward plasma membrane flow, leading to uropod formation and persistent migration in the direction opposite to the activation site.

A migration mode termed "collective amoeboid" was introduced to describe clusters of cancer cells navigating through nonadhesive microchannels, independent of FAs, relying on rear actomyosin contractility for migration (Pagès et al. 2022). Using CRY2–CIBN dimerizer system to photoactivate RhoA at the anterior of an HT29-MTX colon cancer cell cluster resulted in the reversal of migration direction, suggesting that polarized contractility at the collective level dictates the migration directionality. Actomyosin-driven force dynamics was investigated by combining optogenetic activation of RhoA, live-cell imaging, and TFM (Oakes et al. 2017). By locally activating RhoA through the recruitment of DH domain of LARG, a specific GEF, actin and myosin accumulation and localized stress fiber contraction were induced. This enhanced contractility propels myosin and α-actinin toward the stimulated area, increasing strain at both the stress fiber–adhesion interface and the activation site, thereby recruiting the mechanosensitive protein zyxin. Upon deactivation, relaxation of the system, mediated by the release of elastic energy, resulted in a temporary reverse cytoskeletal flow. Through this optogenetic manipulation, the authors demonstrated the elastic nature of stress fibers, highlighting their role in regulating force transmission and cytoskeletal organization (Fig. 4).

Unlike RhoA, which regulates actomyosin contractility primarily at the rear of motile cells, Cdc42 and Rac1 exhibit an activity gradient that emanates from the leading edge, promoting actin polymerization and formation of filopodial and lamellipodial protrusions (Raftopoulou and Hall 2004; de Beco et al. 2018). In the realm of optogenetics targeting Cdc42 and Rac, a common strategy for modulating their activity, similar to that used for Rho, involves the manipulation of their GEFs or GAPs. An opto-Rac1/CIBN system was developed by fusing T-cell lymphoma invasion and metastasis-inducing protein 1 (TIAM),

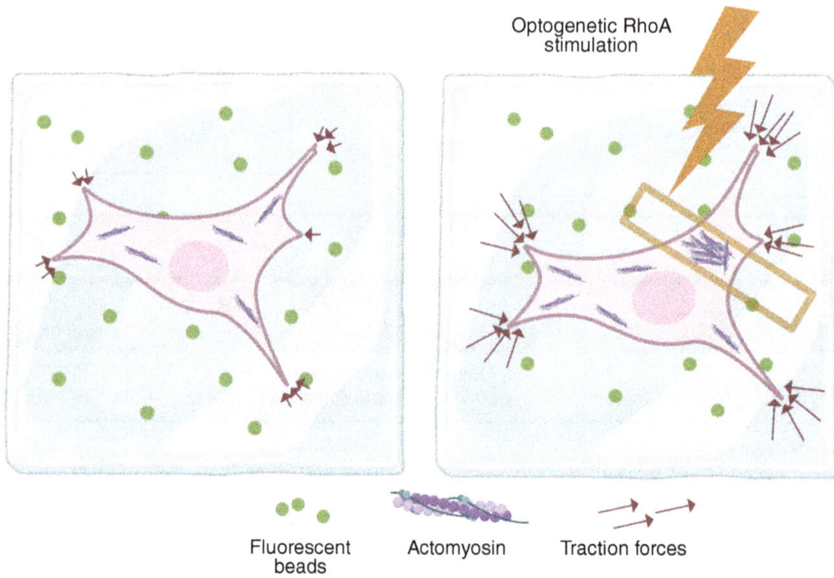

Figure 4. Optogenetic RhoA stimulation locally increases myosin II contractility, leading to increased traction forces even at sites away from the area of optical stimulation. (Figure generated with BioRender; https://biorender.com.)

a Rac1 activator, to CRY2-mCherry (Rossetti et al. 2024). Upon stimulation of the leading edge of Madin–Darby canine kidney (MDCK) kidney cells, this system effectively induced lamellipodium formation and FA assembly, thereby generating leader cells necessary for collective migration. Interestingly, their findings reveal that a single cell activated by this method could effectively lead to only one follower cell but not larger groups, attributed to the dependence of coherent collective migration on asymmetric traction forces and tension fields within the cell cluster. A similar system was used to examine how Rac1 and Cdc42 gradients are shaped during cell migration (de Beco et al. 2018). This study revealed that activating these Rho GTPases at the cell leading edge and regulating the formation of their spatial gradients are essential for promoting cell migration. Specifically, it was found that a steep gradient of Cdc42 enhances directional movement, whereas a wider gradient of Rac1, facilitated by the activation of its GAP, β2-chimaerin, increases speed. This GAP acts downstream from Cdc42 and Rac1 to locally inhibit Rac1 at the very front of the cell. These findings highlight the crucial role of precise gradient control in directing cellular migration.

Other Optogenetics Applications

Beyond targeting ion channels, PI3K/Akt signaling pathways, and GTPase activity, a number of optogenetic constructs have been developed to explore other critical components of cell migration. In *Dictyostelium discoideum*, CIBN–CRY2-based actuators carrying positive charges targeted to plasma membrane could effectively initiate de novo protrusion formation and disrupt pre-existing cell polarity by locally activating the Ras/PI3K/Akt/F-actin signaling cascade (Banerjee et al. 2022), highlighting the importance of membrane surface charge dynamics in cell migration. In differentiated human HL-60 neutrophils and murine RAW 264.7 macrophages, global optogenetic recruitment of active Ras enhanced cell spreading and random motility, while local Ras activation at the cell rear induced the formation of new protrusions, and redirected migration (Pal et al. 2023). Additionally, the application of a RasGAP at the cell front suppressed protrusions and altered migration direction.

Optogenetics has quickly evolved into a powerful tool for spatiotemporal control of cellular dynamics, offering deeper insights into processes like cell migration. However, challenges remain, including insufficient system characterization, limited spatiotemporal resolution, and difficulties in translating findings from in vitro to in vivo. These challenges highlight the ongoing opportunities for innovation and refinement in optogenetics, paving the way for its full integration into broader biological research and clinical applications.

THERAPEUTIC APPROACHES TARGETING CELL MIGRATION

Dysregulated cell migration, whether excessive or impaired, plays a central role in the development and progression of various diseases. Fibrosis involves abnormal fibroblast migration and ECM deposition, with therapies targeting excessive migration showing potential for diseases like liver fibrosis and systemic sclerosis. In neurological disorders, and autoimmune diseases, such as rheumatoid arthritis and multiple sclerosis, modulating cell migration may aid tissue repair and disease management. In cardiovascular diseases, targeting smooth muscle cell migration can stabilize plaques, while enhancing endothelial migration supports vessel repair. Modulating immune cell migration, such as neutrophils and macrophages, can improve infection responses. Therefore, therapies disrupting migration-related pathways, including integrin inhibitors and actomyosin modulators, offer promising precision medicine approaches to manage various diseases.

Cancer

The cytoskeleton, primarily composed of actin filaments, microtubules, and intermediate filaments, regulates cell shape, polarity, and motility through complex structural and signaling networks. Several compounds targeting microtubule filament stabilization or inhibition are in clinical trials to disrupt cellular functions (Negi et al. 2015). Histone deacetylase 6 (HDAC6) has emerged as a key regulator in microtubule organization, facilitating cell migration with significant implications for cancer progression and neurological diseases. ACY-1215, an oral selective HDAC6 inhibitor, alongside some other previously identified inhibitors, is under investigation for multiple cancer types and cardiovascular diseases (Li et al. 2022). Unlike conventional microtubule-disrupting agents, like vinblastine and taxol, ACY-1215 indirectly stabilizes microtubules by enhancing α-tubulin acetylation, reducing F-actin-dependent motility without inducing mitotic arrest (Li et al. 2022). Actin filaments are known to play an important role in cancer cell invasiveness and migratory potential, particularly in 2D microenvironments. Multiple inhibitors of actin polymerization have been approved by the Food and Drug Administration (FDA) for disease treatment including innovative nanoparticle-based therapeutics (Xu et al. 2023). Raltegravir, a previously FDA-approved drug for its antiviral properties, has recently entered phase 0 clinical trials for head and neck cancer due to its proven inhibitory effect on Fascin1, which is an actin-bundling protein (Alburquerque-Gonzalez et al. 2021).

EMT has been reported to drive tumor progression, metastasis, and therapy resistance. Rho GTPases regulate actin dynamics, FAs, and matrix remodeling to facilitate cell–ECM interactions as well as promote cell–cell junction disassembly and invasive protrusions for endothelial barrier breach. Rac1, a key GTPase, influences cytoskeletal dynamics and motility, making it a promising therapeutic target for inhibiting metastasis. NSC23766, Rac1 inhibitor, is currently in an early clinical trial for hepatocellular carcinoma (Ma et al. 2023). Similarly, integrins, key mediators of cell–ECM interactions and tumor survival, are potential therapeutic targets to improve cancer treatment outcomes (Chen et al. 2022). Dasatinib, a tyrosine kinase inhibitor, suppresses collective cell migration in colon cancer by disrupting the coordination between FAs and E-cadherin-mediated cell–cell junctions. It exerts a dual inhibitory effect by suppressing FA kinase (FAK) signaling, thereby destabilizing cell–matrix adhesion,

Table 1. List of drugs: molecular targets, mechanisms of action, and clinical trial levels related to cancer

Antagonists/ activators	Target	Mechanism of action	Clinical trial	References
ACY1215	HDAC6	Disrupts microtubule organization	Phase I and phase II	Li et al. 2022
Raltegravir	Fascin1	Inhibits actin bundling via fascin1	Phase 0	Alburquerque-Gonzalez et al. 2021
NSC23766	Rac1	Disrupts cytoskeleton remodeling	Preclinical stage	Jiang et al. 2020
Dasatinib	Tyrosine kinase	Disrupts coordination between FAs and E-cadherin-mediated cell–cell junctions	Approved, marketed as Sprycel	Lu et al. 2024
uPAcyclin	α_v-Integrin	Inhibits integrin-mediated interactions with the ECM	Preclinical stage	Franco et al. 2023
ISO1	MIF	Inhibits MIF cytokine	Preclinical stage	Cheng et al. 2020
Maraviroc	CCR5	Combined with PD1 inhibitors for reduced metastasis	Phase I	Karampoor et al. 2020

(CCR5) C-C chemokine receptor type 5, (ECM) extracellular matrix, (MIF) macrophage migration inhibitory factor, (PD1) programmed death protein 1.

while also impairing E-cadherin function, weakening cell–cell cohesion (Lu et al. 2024). The uPAcyclin decapeptide inhibits glioblastoma progression by targeting α_v-integrin-dependent pathways, effectively reducing cell migration, invasion, and vasculogenic mimicry (Franco et al. 2023).

Stromal components and immune cells in the tumor microenvironment regulate cancer cell invasion, survival, and colonization, making them key therapeutic targets to prevent cancer spread. Strategies include disrupting ECM remodeling, modulating immune responses, and inhibiting premetastatic niche formation (Neophytou et al. 2021). ECM proteins, with cytokine-like domains, attract immune cells to disease sites, while ECM density can either impede or facilitate their migration (Bonnans et al. 2014). Alternatively, researchers have also targeted cancer progression via modulating immune cell migration into the tumor microenvironment. In this regard, some current studies target integrins, such as $\alpha_v\beta_3$-integrin, showing reduced malignant T-cell migration in hematological cancers (Huang et al. 2023). Additionally, immune cells constitute a major portion of the tumor environment and release cytokines, such as macrophage migration inhibitory factor (MIF), which enhance pancreatic cancer cell proliferation, migration, and invasion. Recent studies have demonstrated that ISO1 effectively inhibits MIF-mediated pancreatic cell migration and proliferation in vitro and in xenograft models, showing its promising role as a therapeutic agent (Cheng et al. 2020).

The list of drugs along with their molecular targets and underlying mechanisms of action is provided in Table 1. Therapies targeting cell migration offer promise for treating diseases from cancer to chronic inflammation. However, challenges like off-target effects, drug resistance, and limited efficacy in complex in vivo settings hinder clinical translation. Moreover, because migration plays roles in both health and disease, strategies must avoid disrupting essential functions like immune trafficking and tissue repair. Continued research is crucial to address these challenges and realize the full potential of migration-targeted treatments.

CONCLUDING REMARKS

In conclusion, engineered in vitro platforms enable us to dissect the unique mechanisms of confined cell migration, offering insights beyond conventional 2D systems. Integrated with microenvironmental cues, these models are indispensable for studying cellular mechanotransduction. Coupled with biophysical tools, such as TFM and cutting-edge optogenetic systems, they enable precise analysis of force generation and spatiotemporal signaling dynamics. As these platforms evolve, they will become increasingly vital for identifying therapeutic targets and advancing treatments for diseases driven by aberrant cell migration.

ACKNOWLEDGMENTS

This work was supported, in part, by R01 CA 254193, R01 CA257647, and R35 GM156305.

REFERENCES

Adebowale K, Gong Z, Hou JC, Wisdom KM, Garbett D, Lee HP, Nam S, Meyer T, Odde DJ, Shenoy VB, et al. 2021. Enhanced substrate stress relaxation promotes filopodia-mediated cell migration. *Nat Mater* **20:** 1290–1299. doi:10.1038/s41563-021-00981-w

Afthinos A, Bera K, Chen J, Ozcelikkale A, Amitrano A, Choudhury MI, Huang R, Pachidis P, Mistriotis P, Chen Y, et al. 2022. Migration and 3D traction force measurements inside compliant microchannels. *Nano Lett* **22:** 7318–7327. doi:10.1021/acs.nanolett.2c01261

Alburquerque-Gonzalez B, Bernabe-Garcia A, Bernabe-Garcia M, Ruiz-Sanz J, Lopez-Calderon FF, Gonnelli L, Banci L, Pena-Garcia J, Luque I, Nicolas FJ, et al. 2021. The FDA-approved antiviral raltegravir inhibits Fascin1-dependent invasion of colorectal tumor cells in vitro and in vivo. *Cancers (Basel)* **13:** 861. doi:10.3390/cancers13040861

Alieva M, Barrera Román M, de Blank S, Petcu D, Zeeman AL, Dautzenberg NMM, Cornel AM, van de Ven C, Pieters R, den Boer ML, et al. 2024. BEHAV3D: a 3D live imaging platform for comprehensive analysis of engineered T cell behavior and tumor response. *Nat Protoc* **19:** 2052–2084. doi:10.1038/s41596-024-00972-6

Amiri F, Akinpelu AA, Keith WC, Hemmati F, Vaghasiya RS, Bowen D, Waliagha RS, Wang C, Chen P, Mitra AK, et al. 2024. Confinement controls the directional cell responses to fluid forces. *Cell Rep* **43:** 114692. doi:10.1016/j.celrep.2024.114692

Amitrano A, Choudhury D, Konstantopoulos K. 2025a. Navigating confinement: mechanotransduction and metabolic adaptation. *Curr Opin Cell Biol* **94:** 102487. doi:10.1016/j.ceb.2025.102487

Amitrano A, Yuan Q, Agarwal B, Sen A, Dance YW, Zuo Y, Phillip JM, Gu L, Konstantopoulos K. 2025b. Extracellular fluid viscosity regulates human mesenchymal stem cell lineage and function. *Sci Adv* **11:** eadr5023. doi:10.1126/sciadv.adr5023

Anon E, Serra-Picamal X, Hersen P, Gauthier NC, Sheetz MP, Trepat X, Ladoux B. 2012. Cell crawling mediates collective cell migration to close undamaged epithelial gaps. *Proc Natl Acad Sci* **109:** 10891–10896. doi:10.1073/pnas.1117814109

Aragona M, Dekoninck S, Rulands S, Lenglez S, Mascré G, Simons BD, Blanpain C. 2017. Defining stem cell dynamics and migration during wound healing in mouse skin epidermis. *Nat Commun* **8:** 14684. doi:10.1038/ncomms14684

Balzer EM, Tong Z, Paul CD, Hung WC, Stroka KM, Boggs AE, Martin SS, Konstantopoulos K. 2012. Physical confinement alters tumor cell adhesion and migration phenotypes. *FASEB J* **26:** 4045–4056. doi:10.1096/fj.12-211441

Banerjee T, Biswas D, Pal DS, Miao Y, Iglesias PA, Devreotes PN. 2022. Spatiotemporal dynamics of membrane surface charge regulates cell polarity and migration. *Nat Cell Biol* **24:** 1499–1515. doi:10.1038/s41556-022-00997-7

Bangasser BL, Shamsan GA, Chan CE, Opoku KN, Tüzel E, Schlichtmann BW, Kasim JA, Fuller BJ, McCullough BR, Rosenfeld SS, et al. 2017. Shifting the optimal stiffness for cell migration. *Nat Commun* **8:** 15313. doi:10.1038/ncomms15313

Barbieri L, Colin-York H, Korobchevskaya K, Li D, Wolfson DL, Karedla N, Schneider F, Ahluwalia BS, Seternes T, Dalmo RA, et al. 2021. Two-dimensional TIRF-SIM-traction force microscopy (2D TIRF-SIM-TFM). *Nat Commun* **12:** 2169. doi:10.1038/s41467-021-22377-9

Beningo KA, Dembo M, Kaverina I, Small JV, Wang YL. 2001. Nascent focal adhesions are responsible for the generation of strong propulsive forces in migrating fibroblasts. *J Cell Biol* **153:** 881–888. doi:10.1083/jcb.153.4.881

Bera K, Kiepas A, Godet I, Li Y, Mehta P, Ifemembi B, Paul CD, Sen A, Serra SA, Stoletov K, et al. 2022. Extracellular fluid viscosity enhances cell migration and cancer dissemination. *Nature* **611:** 365–373. doi:10.1038/s41586-022-05394-6

Blache U, Ford EM, Ha B, Rijns L, Chaudhuri O, Dankers PYW, Kloxin AM, Snedeker JG, Gentleman E. 2022. Engineered hydrogels for mechanobiology. *Nat Rev Methods Primers* **2:** 98. doi:10.1038/s43586-022-00179-7

Bonnans C, Chou J, Werb Z. 2014. Remodelling the extracellular matrix in development and disease. *Nat Rev Mol Cell Biol* **15:** 786–801. doi:10.1038/nrm3904

Boucher Y, Baxter LT, Jain RK. 1990. Interstitial pressure gradients in tissue-isolated and subcutaneous tumors: implications for therapy. *Cancer Res* **50:** 4478–4484.

Brugués A, Anon E, Conte V, Veldhuis JH, Gupta M, Colombelli J, Muñoz JJ, Brodland GW, Ladoux B, Trepat X. 2014. Forces driving epithelial wound healing. *Nat Phys* **10:** 683–690. doi:10.1038/nphys3040

Butcher DT, Alliston T, Weaver VM. 2009. A tense situation: forcing tumour progression. *Nat Rev Cancer* **9:** 108–122. doi:10.1038/nrc2544

Cain RJ, Ridley AJ. 2009. Phosphoinositide 3-kinases in cell migration. *Biol Cell* **101:** 13–29. doi:10.1042/BC20080079

Campàs O, Mammoto T, Hasso S, Sperling RA, O'Connell D, Bischof AG, Maas R, Weitz DA, Mahadevan L, Ingber DE. 2014. Quantifying cell-generated mechanical forces within living embryonic tissues. *Nat Methods* **11:** 183–189. doi:10.1038/nmeth.2761

Chan CE, Odde DJ. 2008. Traction dynamics of filopodia on compliant substrates. *Science* **322:** 1687–1691. doi:10.1126/science.1163595

Chaudhuri O, Gu L, Klumpers D, Darnell M, Bencherif SA, Weaver JC, Huebsch N, Lee HP, Lippens E, Duda GN, et al. 2016. Hydrogels with tunable stress relaxation regulate stem cell fate and activity. *Nat Mater* **15:** 326–334. doi:10.1038/nmat4489

Chaudhuri O, Cooper-White J, Janmey PA, Mooney DJ, Shenoy VB. 2020. Effects of extracellular matrix viscoelasticity on cellular behaviour. *Nature* **584:** 535–546. doi:10.1038/s41586-020-2612-2

Chen JR, Zhao JT, Xie ZZ. 2022. Integrin-mediated cancer progression as a specific target in clinical therapy. *Biomed Pharmacother* **155:** 113745. doi:10.1016/j.biopha.2022.113745

Cheng B, Wang Q, Song Y, Liu Y, Liu Y, Yang S, Li D, Zhang Y, Zhu C. 2020. MIF inhibitor, ISO-1, attenuates human pancreatic cancer cell proliferation, migration and invasion in vitro, and suppresses xenograft tumour growth in vivo. *Sci Rep* **10:** 6741. doi:10.1038/s41598-020-63778-y

Colin-York H, Shrestha D, Felce JH, Waithe D, Moeendarbary E, Davis SJ, Eggeling C, Fritzsche M. 2016. Super-resolved traction force microscopy (STFM). *Nano Lett* **16:** 2633–2638. doi:10.1021/acs.nanolett.6b00273

Colin-York H, Javanmardi Y, Barbieri L, Li D, Korobchevskaya K, Guo Y, Hall C, Taylor A, Khuon S, Sheridan GK, et al. 2019. Spatiotemporally super-resolved volumetric traction force microscopy. *Nano Lett* **19:** 4427–4434. doi:10.1021/acs.nanolett.9b01196

Critchley DR. 2009. Biochemical and structural properties of the integrin-associated cytoskeletal protein talin. *Annu Rev Biophys* **38:** 235–254. doi:10.1146/annurev.biophys.050708.133744

de Beco S, Vaidžiulytė K, Manzi J, Dalier F, di Federico F, Cornilleau G, Dahan M, Coppey M. 2018. Optogenetic dissection of Rac1 and Cdc42 gradient shaping. *Nat Commun* **9:** 4816. doi:10.1038/s41467-018-07286-8

de Boer LL, Vanes L, Melgrati S, Biggs O'May J, Hayward D, Driscoll PC, Day J, Griffiths A, Magueta R, Morrell A, et al. 2023. T cell migration requires ion and water influx to regulate actin polymerization. *Nat Commun* **14:** 7844. doi:10.1038/s41467-023-43423-8

Dembo M, Wang YL. 1999. Stresses at the cell-to-substrate interface during locomotion of fibroblasts. *Biophys J* **76:** 2307–2316. doi:10.1016/S0006-3495(99)77386-8

Elpers MA, Varlet AA, Agrawal R, Lammerding J. 2023. Agarose-based 3D cell confinement assay to study nuclear mechanobiology. *Curr Protoc* **3:** e847. doi:10.1002/cpz1.847

Faure LM, Gómez-González M, Baguer O, Comelles J, Martínez E, Arroyo M, Trepat X, Roca-Cusachs P. 2024. 3D micropatterned traction force microscopy: a technique to control 3D cell shape while measuring cell-substrate force transmission. *Adv Sci (Weinh)* **11:** e2406932. doi:10.1002/advs.202406932

Franco P, Camerino I, Merlino F, D'Angelo M, Cimmino A, Carotenuto A, Colucci-D'Amato L, Stoppelli MP. 2023. αV-Integrin-dependent inhibition of glioblastoma cell migration, invasion and vasculogenic mimicry by the uPAcyclin decapeptide. *Cancers (Basel)* **15:** 4775. doi:10.3390/cancers15194775

Friedl P, Alexander S. 2011. Cancer invasion and the microenvironment: plasticity and reciprocity. *Cell* **147:** 992–1009. doi:10.1016/j.cell.2011.11.016

Friedl P, Gilmour D. 2009. Collective cell migration in morphogenesis, regeneration and cancer. *Nat Rev Mol Cell Biol* **10:** 445–457. doi:10.1038/nrm2720

García-Cuesta EM, Santiago CA, Vallejo-Díaz J, Juarranz Y, Rodríguez-Frade JM, Mellado M. 2019. The role of the CXCL12/CXCR4/ACKR3 axis in autoimmune diseases. *Front Endocrinol (Lausanne)* **10:** 585. doi:10.3389/fendo.2019.00585

Gardel ML, Sabass B, Ji L, Danuser G, Schwarz US, Waterman CM. 2008. Traction stress in focal adhesions correlates biphasically with actin retrograde flow speed. *J Cell Biol* **183:** 999–1005. doi:10.1083/jcb.200810060

Ghosh M, Alves C, Tong Z, Tettey K, Konstantopoulos K, Stebe KJ. 2008. Multifunctional surfaces with discrete functionalized regions for biological applications. *Langmuir* **24:** 8134–8142. doi:10.1021/la8006525

Glass DS, Grossfeld D, Renna HA, Agarwala P, Spiegler P, Kasselman LJ, Glass AD, DeLeon J, Reiss AB. 2020. Idiopathic pulmonary fibrosis: molecular mechanisms and potential treatment approaches. *Respir Investig* **58:** 320–335. doi:10.1016/j.resinv.2020.04.002

Glazier R, Brockman JM, Bartle E, Mattheyses AL, Destaing O, Salaita K. 2019. DNA mechanotechnology reveals that integrin receptors apply pN forces in podosomes on fluid substrates. *Nat Commun* **10:** 4507. doi:10.1038/s41467-019-12304-4

Gutierrez E, Tkachenko E, Besser A, Sundd P, Ley K, Danuser G, Ginsberg MH, Groisman A. 2011. High refractive index silicone gels for simultaneous total internal reflection fluorescence and traction force microscopy of adherent cells. *PLoS One* **6:** e23807. doi:10.1371/journal.pone.0023807

Harris AK, Wild P, Stopak D. 1980. Silicone rubber substrata: a new wrinkle in the study of cell locomotion. *Science* **208:** 177–179. doi:10.1126/science.6987736

Herbert SP, Stainier DY. 2011. Molecular control of endothelial cell behaviour during blood vessel morphogenesis. *Nat Rev Mol Cell Biol* **12:** 551–564. doi:10.1038/nrm3176

Heuzé ML, Collin O, Terriac E, Lennon-Duménil AM, Piel M. 2011. Cell migration in confinement: a micro-channel-based assay. *Methods Mol Biol* **769:** 415–434. doi:10.1007/978-1-61779-207-6_28

Huang YJ, Samorajski J, Kreimer R, Searson PC. 2013. The influence of electric field and confinement on cell motility. *PLoS One* **8:** e59447. doi:10.1371/journal.pone.0059447

Huang L, Zhu Y, Kong Q, Guan X, Lei X, Zhang L, Yang H, Yao X, Liang S, An X, et al. 2023. Inhibition of integrin alpha(v)beta(3)-FAK-MAPK signaling constrains the invasion of T-ALL cells. *Cell Adh Migr* **17:** 1–14.

Idevall-Hagren O, Dickson EJ, Hille B, Toomre DK, De Camilli P. 2012. Optogenetic control of phosphoinositide metabolism. *Proc Natl Acad Sci* **109:** E2316–E2323. doi:10.1073/pnas.1211305109

Ilina O, Gritsenko PG, Syga S, Lippoldt J, La Porta CAM, Chepizhko O, Grosser S, Vullings M, Bakker GJ, Starruß J, et al. 2020. Cell-cell adhesion and 3D matrix confinement determine jamming transitions in breast cancer invasion. *Nat Cell Biol* **22:** 1103–1115. doi:10.1038/s41556-020-0552-6

Irimia D, Ellett F. 2016. Big insights from small volumes: deciphering complex leukocyte behaviors using microfluidics. *J Leukoc Biol* **100:** 291–304. doi:10.1189/jlb.5RU0216-056R

Isenberg BC, Dimilla PA, Walker M, Kim S, Wong JY. 2009. Vascular smooth muscle cell durotaxis depends on substrate stiffness gradient strength. *Biophys J* **97:** 1313–1322. doi:10.1016/j.bpj.2009.06.021

Isomursu A, Park KY, Hou J, Cheng B, Mathieu M, Shamsan GA, Fuller B, Kasim J, Mahmoodi MM, Lu TJ, et al. 2022. Directed cell migration towards softer environments. *Nat Mater* **21:** 1081–1090. doi:10.1038/s41563-022-01294-2

Janssen E, Geha RS. 2019. Primary immunodeficiencies caused by mutations in actin regulatory proteins. *Immunol Rev* **287:** 121–134. doi:10.1111/imr.12716

Jiang C, Sun ZM, Zhu DC, Guo Q, Xu JJ, Lin JH, Chen ZX, Wu YS. 2020. Inhibition of Rac1 activity by NSC23766 prevents cartilage endplate degeneration via Wnt/beta-catenin pathway. *J Cell Mol Med* **24:** 3582–3592. doi:10.1111/jcmm.15049

Karampoor S, Zahednasab H, Amini R, Esghaei M, Sholeh M, Keyvani H. 2020. Maraviroc attenuates the pathogenesis of experimental autoimmune encephalitis. *Int Immunopharmacol* **80:** 106138. doi:10.1016/j.intimp.2019.106138

Karar J, Maity A. 2011. PI3K/AKT/mTOR pathway in angiogenesis. *Front Mol Neurosci* **4:** 51. doi:10.3389/fnmol.2011.00051

Kraning-Rush CM, Carey SP, Lampi MC, Reinhart-King CA. 2013. Microfabricated collagen tracks facilitate single cell metastatic invasion in 3D. *Integr Biol (Camb)* **5:** 606–616. doi:10.1039/c3ib20196a

Kumar S, Maxwell IZ, Heisterkamp A, Polte TR, Lele TP, Salanga M, Mazur E, Ingber DE. 2006. Viscoelastic retraction of single living stress fibers and its impact on cell shape, cytoskeletal organization, and extracellular matrix mechanics. *Biophys J* **90:** 3762–3773. doi:10.1529/biophysj.105.071506

Law RA, Kiepas A, Desta HE, Perez Ipiña E, Parlani M, Lee SJ, Yankaskas CL, Zhao R, Mistriotis P, Wang N, et al. 2023. Cytokinesis machinery promotes cell dissociation from collectively migrating strands in confinement. *Sci Adv* **9:** eabq6480. doi:10.1126/sciadv.abq6480

Li X, Pal DS, Biswas D, Iglesias PA, Devreotes PN. 2021. Reverse fountain flow of phosphatidylinositol-3,4-bisphosphate polarizes migrating cells. *EMBO J* **40:** e105094. doi:10.15252/embj.2020105094

Li J, Yu M, Fu S, Liu D, Tan Y. 2022. Role of selective histone deacetylase 6 inhibitor ACY-1215 in cancer and other human diseases. *Front Pharmacol* **13:** 907981. doi:10.3389/fphar.2022.907981

Liu YJ, Le Berre M, Lautenschlaeger F, Maiuri P, Callan-Jones A, Heuze M, Takaki T, Voituriez R, Piel M. 2015. Confinement and low adhesion induce fast amoeboid migration of slow mesenchymal cells. *Cell* **160:** 659–672. doi:10.1016/j.cell.2015.01.007

Lo CM, Wang HB, Dembo M, Wang YL. 2000. Cell movement is guided by the rigidity of the substrate. *Biophys J* **79:** 144–152. doi:10.1016/S0006-3495(00)76279-5

Lu YW, Hou XL, Koo HM, Chao WT. 2024. Dasatinib suppresses collective cell migration through the coordination of focal adhesion and E-cadherin in colon cancer cells. *Heliyon* **10:** e23501. doi:10.1016/j.heliyon.2023.e23501

Luster AD, Alon R, von Andrian UH. 2005. Immune cell migration in inflammation: present and future therapeutic targets. *Nat Immunol* **6:** 1182–1190. doi:10.1038/ni1275

Ma N, Xu E, Luo Q, Song G. 2023. Rac1: a regulator of cell migration and a potential target for cancer therapy. *Molecules* **28:** 2976. doi:10.3390/molecules28072976

Mahankali M, Peng HJ, Cox D, Gomez-Cambronero J. 2011. The mechanism of cell membrane ruffling relies on a phospholipase D2 (PLD2), Grb2 and Rac2 association. *Cell Signal* **23:** 1291–1298. doi:10.1016/j.cellsig.2011.03.010

Maity D, Bera K, Li Y, Ge Z, Ni Q, Konstantopoulos K, Sun SX. 2022. Extracellular hydraulic resistance enhances cell migration. *Adv Sci (Weinh)* **9:** e2200927. doi:10.1002/advs.202200927

Manoilov KY, Verkhusha VV, Shcherbakova DM. 2021. A guide to the optogenetic regulation of endogenous molecules. *Nat Methods* **18:** 1027–1037. doi:10.1038/s41592-021-01240-1

Martini M, De Santis MC, Braccini L, Gulluni F, Hirsch E. 2014. PI3K/AKT signaling pathway and cancer: an updated review. *Ann Med* **46:** 372–383. doi:10.3109/07853890.2014.912836

Mistriotis P, Wisniewski EO, Bera K, Keys J, Li Y, Tuntithavornwat S, Law RA, Perez-Gonzalez NA, Erdogmus E, Zhang Y, et al. 2019. Confinement hinders motility by inducing RhoA-mediated nuclear influx, volume expansion, and blebbing. *J Cell Biol* **218:** 4093–4111. doi:10.1083/jcb.201902057

Mistriotis P, Wisniewski EO, Si BR, Kalab P, Konstantopoulos K. 2024. Coordinated in confined migration: crosstalk between the nucleus and ion channel-mediated mechanosensation. *Trends Cell Biol* **34:** 809–825. doi:10.1016/j.tcb.2024.01.001

Moarefian M, Davalos RV, Burton MD, Jones CN. 2021. Electrotaxis-on-Chip to quantify neutrophil migration towards electrochemical gradients. *Front Immunol* **12:** 674727. doi:10.3389/fimmu.2021.674727

Munson JM, Shieh AC. 2014. Interstitial fluid flow in cancer: implications for disease progression and treatment. *Cancer Manag Res* **6:** 317–328. doi:10.2147/CMAR.S65444

Murrell MP, Voituriez R, Joanny JF, Nassoy P, Sykes C, Gardel ML. 2014. Liposome adhesion generates traction stress. *Nat Phys* **10:** 163–169. doi:10.1038/nphys2855

Nagy TL, Strickland E, Weiner OD. 2024. Neutrophils actively swell to potentiate rapid migration. *eLife* **12:** RP90551. doi:10.7554/eLife.90551

Negi AS, Gautam Y, Alam S, Chanda D, Luqman S, Sarkar J, Khan F, Konwar R. 2015. Natural antitubulin agents: importance of 3,4,5-trimethoxyphenyl fragment. *Bioorg Med Chem* **23:** 373–389. doi:10.1016/j.bmc.2014.12.027

Neophytou CM, Panagi M, Stylianopoulos T, Papageorgis P. 2021. The role of tumor microenvironment in cancer metastasis: molecular mechanisms and therapeutic opportunities. *Cancers (Basel)* **13:** 2053. doi:10.3390/cancers13092053

Oakes PW, Wagner E, Brand CA, Probst D, Linke M, Schwarz US, Glotzer M, Gardel ML. 2017. Optogenetic control of RhoA reveals zyxin-mediated elasticity of stress fibres. *Nat Commun* **8:** 15817. doi:10.1038/ncomms15817

O'Neill PR, Castillo-Badillo JA, Meshik X, Kalyanaraman V, Melgarejo K, Gautam N. 2018. Membrane flow drives an adhesion-independent amoeboid cell migration mode. *Dev Cell* **46:** 9–22.e4. doi:10.1016/j.devcel.2018.05.029

Pagès DL, Dornier E, de Seze J, Gontran E, Maitra A, Maciejewski A, Wang L, Luan R, Cartry J, Canet-Jourdan C, et al. 2022. Cell clusters adopt a collective amoeboid mode of migration in confined nonadhesive environments. *Sci Adv* **8:** eabp8416. doi:10.1126/sciadv.abp8416

Pal DS, Banerjee T, Lin Y, de Trogoff F, Borleis J, Iglesias PA, Devreotes PN. 2023. Actuation of single downstream nodes in growth factor network steers immune cell migration. *Dev Cell* **58:** 1170–1188.e7. doi:10.1016/j.devcel.2023.04.019

Paul CD, Hung WC, Wirtz D, Konstantopoulos K. 2016a. Engineered models of confined cell migration. *Annu Rev Biomed Eng* **18:** 159–180. doi:10.1146/annurev-bioeng-071114-040654

Paul CD, Shea DJ, Mahoney MR, Chai A, Laney V, Hung WC, Konstantopoulos K. 2016b. Interplay of the physical microenvironment, contact guidance, and intracellular signaling in cell decision making. *FASEB J* **30:** 2161–2170. doi:10.1096/fj.201500199R

Paul CD, Mistriotis P, Konstantopoulos K. 2017. Cancer cell motility: lessons from migration in confined spaces. *Nat Rev Cancer* **17:** 131–140. doi:10.1038/nrc.2016.123

Petrie RJ, Koo H, Yamada KM. 2014. Generation of compartmentalized pressure by a nuclear piston governs cell motility in a 3D matrix. *Science* **345:** 1062–1065. doi:10.1126/science.1256965

Petrie RJ, Harlin HM, Korsak LI, Yamada KM. 2017. Activating the nuclear piston mechanism of 3D migration in tumor cells. *J Cell Biol* **216:** 93–100. doi:10.1083/jcb.201605097

Pittman M, Iu E, Li K, Wang M, Chen J, Taneja N, Jo MH, Park S, Jung WH, Liang L, et al. 2022. Membrane ruffling is a mechanosensor of extracellular fluid viscosity. *Nat Phys* **18:** 1112–1121. doi:10.1038/s41567-022-01676-y

Plotnikov SV, Pasapera AM, Sabass B, Waterman CM. 2012. Force fluctuations within focal adhesions mediate ECM-rigidity sensing to guide directed cell migration. *Cell* **151:** 1513–1527. doi:10.1016/j.cell.2012.11.034

Polacheck WJ, Chen CS. 2016. Measuring cell-generated forces: a guide to the available tools. *Nat Methods* **13:** 415–423. doi:10.1038/nmeth.3834

Provenzano PP, Inman DR, Eliceiri KW, Trier SM, Keely PJ. 2008. Contact guidance mediated three-dimensional cell migration is regulated by Rho/ROCK-dependent matrix reorganization. *Biophys J* **95:** 5374–5384. doi:10.1529/biophysj.108.133116

Quintanilla MA, Hammer JA, Beach JR. 2023. Non-muscle myosin 2 at a glance. *J Cell Sci* **136:** jcs260890. doi:10.1242/jcs.260890

Raftopoulou M, Hall A. 2004. Cell migration: Rho GTPases lead the way. *Dev Biol* **265:** 23–32. doi:10.1016/j.ydbio.2003.06.003

Raman PS, Paul CD, Stroka KM, Konstantopoulos K. 2013. Probing cell traction forces in confined microenvironments. *Lab Chip* **13:** 4599–4607. doi:10.1039/c3lc50802a

Reversat A, Gaertner F, Merrin J, Stopp J, Tasciyan S, Aguilera J, de Vries I, Hauschild R, Hons M, Piel M, et al. 2020. Cellular locomotion using environmental topography. *Nature* **582:** 582–585. doi:10.1038/s41586-020-2283-z

Ridley AJ, Schwartz MA, Burridge K, Firtel RA, Ginsberg MH, Borisy G, Parsons JT, Horwitz AR. 2003. Cell migration: integrating signals from front to back. *Science* **302:** 1704–1709. doi:10.1126/science.1092053

Roca-Cusachs P, Conte V, Trepat X. 2017. Quantifying forces in cell biology. *Nat Cell Biol* **19:** 742–751. doi:10.1038/ncb3564

Rossetti L, Grosser S, Abenza JF, Valon L, Roca-Cusachs P, Alert R, Trepat X. 2024. Optogenetic generation of leader cells reveals a force–velocity relation for collective cell migration. *Nat Phys* **20:** 1659–1669. doi:10.1038/s41567-024-02600-2

Ruprecht V, Wieser S, Callan-Jones A, Smutny M, Morita H, Sako K, Barone V, Ritsch-Marte M, Sixt M, Voituriez R, et al. 2015. Cortical contractility triggers a stochastic switch to fast amoeboid cell motility. *Cell* **160:** 673–685. doi:10.1016/j.cell.2015.01.008

Schwarz US, Soiné JR. 2015. Traction force microscopy on soft elastic substrates: a guide to recent computational advances. *Biochim Biophys Acta* **1853:** 3095–3104. doi:10.1016/j.bbamcr.2015.05.028

SenGupta S, Parent CA, Bear JE. 2021. The principles of directed cell migration. *Nat Rev Mol Cell Biol* **22:** 529–547. doi:10.1038/s41580-021-00366-6

Srivastava N, Kay RR, Kabla AJ. 2017. Method to study cell migration under uniaxial compression. *Mol Biol Cell* **28:** 809–816. doi:10.1091/mbc.e16-08-0575

Stroka KM, Jiang H, Chen SH, Tong Z, Wirtz D, Sun SX, Konstantopoulos K. 2014. Water permeation drives tumor cell migration in confined microenvironments. *Cell* **157:** 611–623. doi:10.1016/j.cell.2014.02.052

Sukumaran SK, Paul P, Guttal V, Holla B, Vemula A, Bhatt H, Bisht P, Mathew K, Nadella RK, Varghese AM, et al. 2022. Abnormalities in the migration of neural precursor cells in familial bipolar disorder. *Dis Model Mech* **15:** dmm049526. doi:10.1242/dmm.049526

Sun Y, Ferreira F, Reid B, Zhu K, Ma L, Young BM, Hagan CE, Tsolis RM, Mogilner A, Zhao M. 2024. Gut epithelial electrical cues drive differential localization of enterobacteria. *Nat Microbiol* **9:** 2653–2665. doi:10.1038/s41564-024-01778-8

Tan JL, Tien J, Pirone DM, Gray DS, Bhadriraju K, Chen CS. 2003. Cells lying on a bed of microneedles: an approach to isolate mechanical force. *Proc Natl Acad Sci* **100:** 1484–1489. doi:10.1073/pnas.0235407100

Théry M, Piel M. 2009. Adhesive micropatterns for cells: a microcontact printing protocol. *Cold Spring Harb Protoc* **2009:** pdb.prot5255. doi:10.1101/pdb.prot5255

Tischer D, Weiner OD. 2014. Illuminating cell signalling with optogenetic tools. *Nat Rev Mol Cell Biol* **15:** 551–558. doi:10.1038/nrm3837

Tong Z, Balzer EM, Dallas MR, Hung WC, Stebe KJ, Konstantopoulos K. 2012. Chemotaxis of cell populations through confined spaces at single-cell resolution. *PLoS One* **7:** e29211. doi:10.1371/journal.pone.0029211

Town JP, Weiner OD. 2023. Local negative feedback of Rac activity at the leading edge underlies a pilot pseudopod-like program for amoeboid cell guidance. *PLoS Biol* **21:** e3002307. doi:10.1371/journal.pbio.3002307

Valon L, Marín-Llauradó A, Wyatt T, Charras G, Trepat X. 2017. Optogenetic control of cellular forces and mechanotransduction. *Nat Commun* **8:** 14396. doi:10.1038/ncomms14396

Vignaud T, Copos C, Leterrier C, Toro-Nahuelpan M, Tseng Q, Mahamid J, Blanchoin L, Mogilner A, Théry M, Kurzawa L. 2021. Stress fibres are embedded in a contractile cortical network. *Nat Mater* **20:** 410–420. doi:10.1038/s41563-020-00825-z

Vorselen D, Wang Y, de Jesus MM, Shah PK, Footer MJ, Huse M, Cai W, Theriot JA. 2020. Microparticle traction force microscopy reveals subcellular force exertion patterns in immune cell-target interactions. *Nat Commun* **11:** 20. doi:10.1038/s41467-019-13804-z

Wilson CA, Tsuchida MA, Allen GM, Barnhart EL, Applegate KT, Yam PT, Ji L, Keren K, Danuser G, Theriot JA. 2010. Myosin II contributes to cell-scale actin network treadmilling through network disassembly. *Nature* **465:** 373–377. doi:10.1038/nature08994

Winkler J, Abisoye-Ogunniyan A, Metcalf KJ, Werb Z. 2020. Concepts of extracellular matrix remodelling in tumour progression and metastasis. *Nat Commun* **11:** 5120. doi:10.1038/s41467-020-18794-x

Wisdom KM, Adebowale K, Chang J, Lee JY, Nam S, Desai R, Rossen NS, Rafat M, West RB, Hodgson L, et al. 2018. Matrix mechanical plasticity regulates cancer cell migration through confining microenvironments. *Nat Commun* **9:** 4144. doi:10.1038/s41467-018-06641-z

Wisniewski EO, Mistriotis P, Bera K, Law RA, Zhang J, Nikolic M, Weiger M, Parlani M, Tuntithavornwat S, Afthinos A, et al. 2020. Dorsoventral polarity directs cell responses to migration track geometries. *Sci Adv* **6:** eaba6505. doi:10.1126/sciadv.aba6505

Wittmann T, Dema A, van Haren J. 2020. Lights, cytoskeleton, action: optogenetic control of cell dynamics. *Curr Opin Cell Biol* **66:** 1–10. doi:10.1016/j.ceb.2020.03.003

Wolf K, Te Lindert M, Krause M, Alexander S, Te Riet J, Willis AL, Hoffman RM, Figdor CG, Weiss SJ, Friedl P. 2013. Physical limits of cell migration: control by ECM space and nuclear deformation and tuning by proteolysis and traction force. *J Cell Biol* **201:** 1069–1084. doi:10.1083/jcb.201210152

Wong S, Guo WH, Wang YL. 2014. Fibroblasts probe substrate rigidity with filopodia extensions before occupying an area. *Proc Natl Acad Sci* **111:** 17176–17181. doi:10.1073/pnas.1412285111

Xu X, Xu S, Wan J, Wang D, Pang X, Gao Y, Ni N, Chen D, Sun X. 2023. Disturbing cytoskeleton by engineered nanomaterials for enhanced cancer therapeutics. *Bioact Mater* **29:** 50–71. doi:10.1016/j.bioactmat.2023.06.016

Yamada KM, Sixt M. 2019. Mechanisms of 3D cell migration. *Nat Rev Mol Cell Biol* **20:** 738–752. doi:10.1038/s41580-019-0172-9

Yankaskas CL, Bera K, Stoletov K, Serra SA, Carrillo-Garcia J, Tuntithavornwat S, Mistriotis P, Lewis JD, Valverde MA, Konstantopoulos K. 2021. The fluid shear stress sensor TRPM7 regulates tumor cell intravasation. *Sci Adv* **7:** eabh3457. doi:10.1126/sciadv.abh3457

Zhang Y, Li Y, Thompson KN, Stoletov K, Yuan Q, Bera K, Lee SJ, Zhao R, Kiepas A, Wang Y, et al. 2022. Polarized NHE1 and SWELL1 regulate migration direction, efficiency and metastasis. *Nat Commun* **13:** 6128. doi:10.1038/s41467-022-33683-1

Zhao M, Song B, Pu J, Wada T, Reid B, Tai G, Wang F, Guo A, Walczysko P, Gu Y, et al. 2006. Electrical signals control wound healing through phosphatidylinositol-3-OH kinase-gamma and PTEN. *Nature* **442:** 457–460. doi:10.1038/nature04925

Zhao R, Cui S, Ge Z, Zhang Y, Bera K, Zhu L, Sun SX, Konstantopoulos K. 2021. Hydraulic resistance induces cell phenotypic transition in confinement. *Sci Adv* **7:** eabg4934. doi:10.1126/sciadv.abg4934

Zhou X, Wang J, Chen J, Qi Y, Di N, Jin L, Qian X, Wang X, Chen Q, Liu X, et al. 2018. Optogenetic control of epithelial-mesenchymal transition in cancer cells. *Sci Rep* **8:** 14098. doi:10.1038/s41598-018-32539-3

Zhou Z, Van der Jeught K, Li Y, Sharma S, Yu T, Moulana I, Liu S, Wan J, Territo PR, Opyrchal M, et al. 2023. A T cell-engaging tumor organoid platform for pancreatic cancer immunotherapy. *Adv Sci (Weinh)* **10:** e2300548. doi:10.1002/advs.202300548

Guiding Cells with Light and Chemicals: A Toolbox for Dynamic Studies of Cell Migration

Abhijit Deb Roy,[1,2,3] Elmer Rho,[1] and Takanari Inoue[1]

[1]Department of Cell Biology and Center for Cell Dynamics, Johns Hopkins University School of Medicine, Baltimore, Maryland 21205, USA

[2]Center for Cell Analysis and Modeling, University of Connecticut School of Medicine, Farmington, Connecticut 06030, USA

[3]Department of Cell Biology, University of Connecticut School of Medicine, Farmington, Connecticut 06030, USA

Correspondence: abdebroy@uchc.edu; jctinoue@jhmi.edu

Cell migration is a fundamental biological process central to a number of physiological and pathophysiological events. Traditional genetic and pharmacological approaches have identified crucial molecular regulators of migration, yet they often lack the spatial and temporal resolution required to interrogate the highly dynamic signaling events that govern cell locomotion. Chemogenetic and optogenetic platforms—genetically encoded inducible systems activated by chemical or light stimuli, respectively—have emerged as powerful tools for achieving precise, on-demand control over protein function in living cells. These systems enable researchers to dissect molecular signaling pathways in real time and with subcellular precision, even as cells are actively migrating. Together with advances in de novo protein design, biosensors, and live-cell imaging, inducible molecular tools are transforming our ability to manipulate and elucidate the intricate mechanisms underlying cell motility. Looking forward, the application of these technologies in animal models will be crucial for gaining deeper physiological and pathophysiological insights.

Cell motility is one of the most dynamic physiological processes in the body. It allows cells to move from one location to another to carry out essential functions, including the elimination of pathogens and toxic agents, as well as the development of tissues and organs. Its highly dynamic nature has inspired decades of research. These studies have revealed that coordinated signal transduction and cytoskeletal action is critical for guiding cell migration. Effective migration—whether random or directional—depends on establishing and maintaining spatial asymmetry inside cells; this intracellular polarity involves the uneven distribution or activity of molecular components, including signaling molecules, cytoskeletal elements, and switch-like molecules.

Approaches such as ribonucleic acid (RNA) interference, genetic knockouts, and pharmacological perturbations have greatly advanced the identification of molecular players involved in cell migration. Visualization techniques applied

Cite this article as *Cold Spring Harb Perspect Biol* doi: 10.1101/cshperspect.a041754

to these players have further revealed how dynamically they behave inside migrating cells. However, these perturbation methods often lack the spatial and/or temporal precision needed to manipulate the rapid and localized molecular events that are proposed to drive motility.

To address these limitations, researchers have developed custom-designed molecular tools. These tools enable acute and specific manipulation of molecular activities or localizations, providing unprecedented control over cellular behavior. Inducible molecular actuators, in particular, allow rapid initiation or inhibition of signaling events with high temporal resolution. In doing so, they bypass the constraints of conventional genetic approaches. Although transcriptional and translational regulation has been widely and effectively used in migration studies, this review will focus on strategies that operate at the posttranslational level, directly targeting proteins. We will first outline the design principles of these tools and then highlight how they have been applied to uncover molecular mechanisms of cell motility that would otherwise remain intractable.

OVERVIEW OF INDUCIBLE MOLECULAR TOOLS

The ideal tools for dynamic molecular actions inside actively migrating cells require manipulation of target molecules with a temporal and spatial scale similar to what native molecules are regulated with. This means that, often, second timescale and submicron length scale are required. These needs were met by chemically and light-inducible tools, which can trigger intended molecular actions in response to external stimuli such as chemical ligands or light illumination. One of the first inducible chemogenetic systems was developed in the early 1990s, when Spencer et al. (1993) engineered the synthetic bivalent ligand FK1012 to induce oligomerization of chimeric T-cell receptors, thereby triggering their activation. Although not originally applied to cell motility, this system demonstrated the utility of such a synthetic tool in probing spatial-temporal dynamics of signaling events. This laid the groundwork for the field of chemically induced dimerization (CID), which leverages synthetic ligands to bring proteins into close proximity, induces protein–protein interactions, and thereby modulates cellular functions (Stanton et al. 2018). In addition, this early work inspired the development of a series of CID systems that use synthetic ligands, as well as light-inducible systems, to control protein functions and cell behaviors. Many of these inducible tools rely on the following broad strategies (Fig. 1) for their actions: (A) chemically induced heterodimerization, (B) light-induced heterodimerization, (C) homodimerization, (D) homo-oligomerization, (E) protein conformational change exposing a previously cryptic domain, (F) disassembly of a preformed protein complex, and (G) translocation from one intracellular compartment to another. We will discuss some of these inducible tools in the following segment.

CHEMICALLY INDUCIBLE DIMERIZATION (CID) AND MULTIMERIZATION SYSTEMS

One of the first CID systems, and arguably the most widely used, is the rapamycin-inducible dimerization of two protein domains: FK506- and rapamycin-binding protein (FKBP) and FKBP-rapamycin-binding domain (FRB). Upon rapamycin addition, FKBP and FRB rapidly dimerize with nanomolar affinity. Therefore, rapamycin triggers recruitment of proteins of interest (POIs) tagged with a dimerizing protein (FRB) to an intended subcellular location where the dimerization partner (FKBP) is anchored via a targeting sequence (DeRose et al. 2013). This approach has been employed to trigger localized signaling events with high temporal precision. The system's flexibility allows enhanced signaling through tandem repeats of FKBP or FRB domains, which can promote clustering and mimic nucleation-like processes (Inoue et al. 2005; DeRose et al. 2013; Nakamura et al. 2019). A modified FKBP (iFKBP) has been developed in rapamycin-based CID systems to allosterically control protein kinase activation (Karginov et al. 2010). This approach was further refined by development of the tool RapTRAP, which enabled kinase action on a specific substrate (Kar-

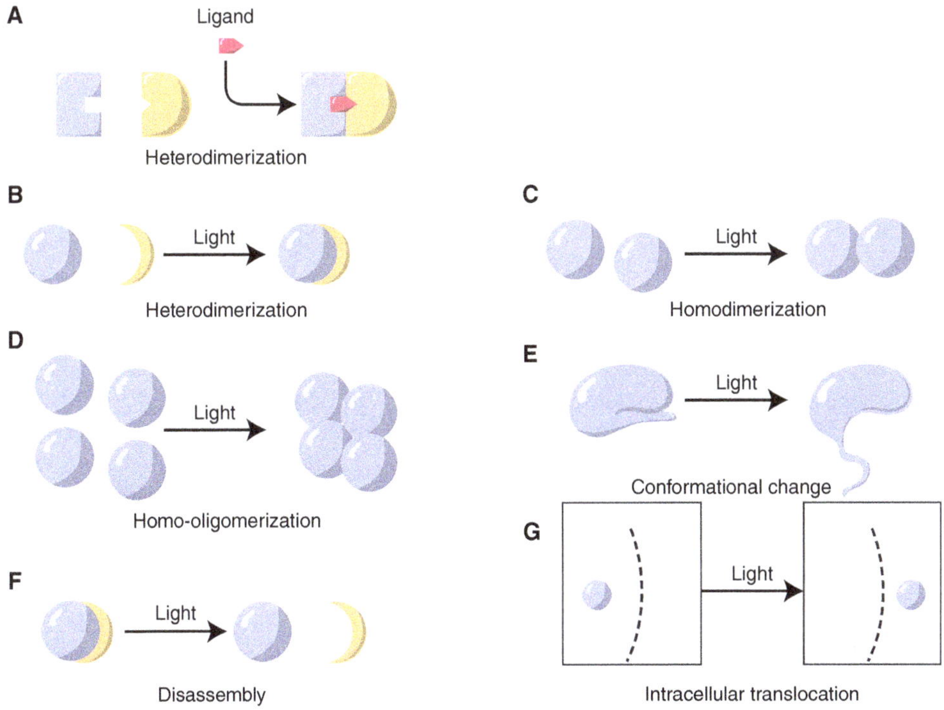

Figure 1. Strategies employed by inducible tools. (*A*) Chemically induced heterodimerization on addition of a synthetic ligand, (*B*) light-induced heterodimerization, (*C*) light-induced homodimerization, (*D*) light-induced homo-oligomerization, (*E*) light-induced conformational change, (*F*) light-induced disassembly, and (*G*) light-induced intracellular translocation.

ginov et al. 2014). Variants of the system, such as split FKBP/FRB constructs, have enabled the interrogation of three-component protein interactions with high temporal resolution (Wu et al. 2020). Additional modifications of the FKBP-FRB dimerization system have enabled inducible higher-order oligomerization of proteins.

In addition to the FKBP-FRB-rapamycin system, additional CID systems have been developed that allow multiplexed, orthogonal stimulations to control many signaling pathways and cell behaviors. Liang et al. (2011) used the plant hormone abscisic acid (ABA) to induce dimerization of the proteins PYR1-like protein 1 (PYL1, ABA receptor) and ABA-insensitive 1 (ABI1, protein phosphatase 2C 56). Miyamoto et al. (2012) used the plant hormone gibberellin to induce dimerization of gibberellin acid-insensitive (GAI) and gibberellin-insensitive dwarf 1 (GID1) proteins. Both systems function independently of rapamycin in mammalian cells, and the gibberellin-based CID was used to develop AND and OR logic gates focused on cytoskeletal dynamics in mammalian cells (Miyamoto et al. 2012).

Chemically inducible systems, broadly, do not easily provide a robust means to activate signaling pathways in a spatially restricted manner. To achieve spatially restricted control of signaling pathways, researchers have applied chemical gradients using microchannels and attained localized release of rapamycin using micropipettes (Lin et al. 2012, 2015; Razavi et al. 2024). A different approach to achieving spatially restricted CID was light-inducible uncaging of rapamycin in a spatially restricted manner. Karginov et al. (2011) modified rapamycin with a bulky acyl group, which sterically blocked its binding to iFKBP and FRB. This steric hindrance was relieved by ultraviolet (UV) light mediated uncaging of rapamycin, leading to dimerization of iFKBP and FRB. Umeda et al. (2011) applied a different strategy, modifying rapamycin with a bulky biotin–streptavidin conjugate, which blocked its entry into cells. On UV stimulation, rapamycin was released from the biotin–streptavidin conjugate and could freely diffuse into cells and induce FKBP-FRB dimerization.

LIGHT-INDUCIBLE SYSTEMS

More recently, several genetically encoded light-inducible optogenetic systems have been developed, which have enabled researchers to manipulate specific signaling molecules and examine their impact on signaling pathways and cell behavior. Although the first optogenetic tools were developed based on ion channels to control the membrane potential in excitable cells such as neurons, here we will mainly discuss tools that have been used to study cytoskeleton and cell motility. To our knowledge, the first use of optogenetics in cell migration studies used the light, oxygen, voltage-sensing domain 2 (LOV2) from *Avena sativa* phototropin1 protein, which underwent a conformational change on blue-light stimulation. This conformational change was harnessed to control the signaling pathways downstream from small GTPases (GTP hydrolyzing enzymes) in migrating cells (Wu et al. 2009).

The first light-inducible dimerization system was developed by Kennedy et al. using the blue-light-mediated dimerization of the *Arabidopsis thaliana* photoreceptor cryptochrome 2 (Cry2) PHR domain with the cryptochrome-interacting basic helix–loop–helix protein (CIB1) (Kennedy et al. 2010). Based on the observation that Cry2 also underwent homo-oligomerization on blue-light stimulation, mutational screening identified a Cry2-olig variant, which underwent robust homo-oligomerization on blue-light stimulation (Taslimi et al. 2014). The LOV2 from *A. sativa*, discussed earlier, was used to develop a light-inducible dimerization system named improved light-induced dimer (iLID) (Guntas et al. 2015). Briefly, the Jα helix of the LOV2 domain was fused to the bacterial protein SsrA (SsrA tag peptide from bacteria), such that in the dark, SsrA was sterically blocked from interacting with its binding partner: stringent starvation protein B (SspB). Blue-light stimulation relieved this steric hindrance, enabling heterodimerization of SsrA and SspB. Various mutations to modulate the strength of SsrA and SspB binding were further exploited to develop additional iLID variants with different dynamic ranges. The As-LOV2 domain was used to develop additional optogenetic tools, namely, light-inducible nuclear export system (LEXY) and light-inducible nuclear localization signal (LINuS), to respectively control nuclear export and import of proteins (Niopek et al. 2014, 2016). A different approach employing an engineered PDZ (PSD-95, Dlg1, and ZO-1) domain fused to the LOV2 Jα helix was used to develop a different dimerization system called tunable light-controlled interacting protein (TULIP) tags (Strickland et al. 2012). In contrast to light-inducible protein–protein interaction, the Zdk-LOV2 system was developed where the Zdk-LOV2 exists as a heterodimer in the dark and is unbound on blue-light stimulation. This light-inducible dissociation was further made use of to develop the LOVTRAP module to control protein activity (Wang et al. 2016).

In addition to these plant photoreceptor-based tools, fungal photoreceptors have also been used to develop optogenetic tools. The fungal photoreceptor vivid (VVD) from *Neurospora crassa* undergoes homodimerization on blue-light stimulation (Nihongaki et al. 2014). VVD was engineered to introduce positive or negative charges on the dimerization surface to create a heterodimerization system named Magnets (Kawano et al. 2015). LOV4 photoreceptor domain from *Botrytis cinerea* (BcLov4) was engineered to undergo cytoplasm to plasma membrane translocation on blue-light stimulation without a membrane tag as well as undergo oligomerization (Glantz et al. 2018). Unlike the previously described systems, BcLOV4 did not require an additional membrane target for translocation. The original BcLOV4 was further modified to abrogate the membrane translocation, while maintaining its light-inducible clustering in cytoplasm (Huang Dennis et al. 2024).

The optogenetic systems described thus far require blue-light stimulation. Levskaya et al. (2009) developed a red-light-inducible dimerization system using the phytochrome B (PhyB) and phytochrome interaction factor 3 (PIF3). Zhou et al. (2022) developed a red-light-inducible dimerization system called red/far-red-light-mediated and miniaturized Δphytochrome A–based photoswitch system (REDMAP) based on the plant photoreceptor phytochrome A (PhyA) and far-red elongated hypocotyl 1 (FHY1). Both the PhyB-PIF3 and the PhyA-FHY1 dimers could be dissociated by stimulating with far-red light. A far-red-light-stimulated dimerization system was also developed using the bacterial bathy phytochrome from *Rhodopseudomonas palustris* (RpBphP1) (Kaberniuk et al. 2016) and its natural binding partner RpPpsR2 (photosynthesis gene expression repressor from *R. palustris*), as well as an engineered binding partner quantum perception and signal 1 (QPAS1) (Redchuk et al. 2018). These red- and far-red light-inducible systems allow for optogenetic stimulation deeper into animal tissue, as well as multiplexing with the blue-light-stimulated optogenetic modules, thus enabling complex manipulations of cell behavior.

There are many more chemically inducible and light-inducible systems that we have not covered here. An exhaustive discussion on all inducible systems is beyond the scope of this paper. We have listed some of them in Tables 1 and 2.

APPLICATION OF INDUCIBLE SYSTEMS TO STUDY CELL MOTILITY

Directional cell migration requires precise spatial and temporal coordination of dynamic remodeling across multiple cytoskeletal systems—including actin, microtubules, and the plasma membrane—to sustain persistent front–back polarity. This polarity emerges through localized regulation of the biochemical composition, stability, and biophysical properties of these cytoskeletal components by a remarkably large network of signaling pathways and molecules, including ligand–receptor complexes, intracellular enzymes, and adaptor proteins (some of which are illustrated in Fig. 2).

Chemical and light-inducible systems have been implemented in cell migration and cytoskeleton studies for more than 30 years. These approaches enable manipulation of protein localization and activation with spatial and temporal control, which is particularly attractive in studying a highly dynamic and spatially regulated phenomenon such as cell motility. Many such tools have been developed and used to study the role of various proteins in cytoskeletal dynamics and cell motility. We will discuss some of these examples below.

INDUCIBLE SYSTEMS TARGETING THE ACTIN CYTOSKELETON AND SMALL GTPASES

The actin cytoskeleton is the fundamental regulator of eukaryotic cell motility, and the majority of cell migratory behavior is accomplished through a myriad of actin-based cellular processes, including lamellipodia, filopodia, and invadopodia. The actin cytoskeleton also engages with integrin-based adhesions to facilitate cell attachment to their extracellular matrix and with myosin motor proteins to control cell contractility. Small GTPases are critical regulators of cell motility through remodeling of the actin cytoskeleton. Small GTPases are often considered molecular switches which exist in an inactive guanosine diphosphate (GDP)-bound or an active guanosine triphosphate (GTP)-bound form. The activation of small GTPases is facilitated by a family of enzymes called guanine nucleotide exchange factors (GEFs), whereas the inactivation is mediated by another family of enzymes named GTPase-activating proteins (GAPs). In particular, Ras-related C3 botulinum toxin substrate 1 (Rac1), transforming protein RhoA, and cell division control protein 42 homolog (Cdc42) small GTPases have been extensively studied for their roles in actin dynamics, cell morphology, and motility. Holsinger et al. (1995) developed one of the very first inducible systems by using FKBP12 and FK1012 to recruit the Ras GEF Son of Sevenless (Sos) to the plasma membrane and examine the downstream signaling pathways. The first application of inducible systems to study actin dynamics employed the FKBP–FRB dimerization system to recruit a constitutively active Cdc42 mutant to the plasma membrane. The constitutively active Cdc42

Table 1. Summary of commonly used chemically inducible dimerization (CID) systems

System (ligand)	Protein domains	Key features	Common applications
Rapamycin (Inoue et al. 2005)	FKBP and FRB	High affinity (nM), rapid, modular, widely used	Signal transduction, cytoskeletal dynamics, chemotaxis
ABA (Liang et al. 2011)	PYL1 and ABI1	Orthogonal in mammalian cells, reversible	Synthetic signaling circuits, cytoskeletal control
GA_3 (Miyamoto et al. 2012)	GID1 and GAI	Orthogonal in mammalian cells, fast dimerization	Inducible subcellular targeting
Auxin (indole-3-acetic-acid) (Nishimura et al. 2009)	TIR1 and AID	Promotes degradation	Conditional and reversible protein depletion
Shield-1 (Banaszynski et al. 2006)	FKBP-derived destabilizing domain	Ligand stabilizes otherwise unstable proteins	Conditional protein expression and activation
TMP (Iwamoto et al. 2010)	DHFR	Small molecule–regulated stabilization	Transient modulation of protein levels
Methotrexate (Volpato et al. 2007, 2009; Fossati et al. 2008)	DHFR variants	Specific and tunable binding	Chemical control of protein–protein interactions
SNAP/HaloTag (synthetic ligands) (Zimmermann et al. 2014)	SNAP-tag, HaloTag, CLIP	Covalent, photocleavable to achieve reversibility	Protein labeling, subcellular localization, synthetic scaffolds

(ABA) Abscisic acid, (TMP) trimethoprim, (GA_3) gibberellin, (FKBP) FK506- and rapamycin-binding protein, (FRB) FKBP–rapamycin binding domain, (PYL1) PYR1-like protein 1, (ABI1) ABA-insensitive 1, (TIR1) transport inhibitor response 1, (AID) auxin-inducible degron, (DHFR) dihydrofolate reductase.

was targeted to the intracellular domain of the cluster of differentiation 25 (Cd25) receptor (interleukin-2 receptor α chain), which was clustered with antibody-coated beads to trigger localized actin polymerization and filopodia formation (Castellano et al. 1999). They further used this approach to recruit constitutively active Rac1 to demonstrate a role of localized actin polymerization in phagocytosis (Castellano et al. 2000).

Inoue et al. (2005) developed a significantly simplified and robust approach to control small GTPase function by recruiting constitutively active or dominant-negative mutants of Rac1, Rac2, RhoA, and Cdc42 to the plasma membrane using a Lyn kinase amino-terminal sequence as a plasma membrane targeting signal. This simplified approach of inducibly targeting proteins to specific cellular compartments would inspire many new molecular actuators (Komatsu et al. 2010). Inoue and Meyer (2008) refined this technique by recruiting the GEF domain of T-cell lymphoma invasion and metastasis 1 (Tiam1) to activate endogenous Rac1, leading to robust lamellipodia formation, and implemented this inducible activation of Tiam1 to examine the molecular cross talk between phosphoinositide 3-kinase (PI3K) and Rac1 signaling in neutrophil polarization in what is arguably the first study implementing inducible tools to specifically study cell migration. This was followed by localized photo-uncaging of rapamycin to recruit FKBP–Tiam1 GEF domain within subcellular regions, which was sufficient to induce lamellipodia in a localized manner (Umeda et al. 2011). Implementation of microfluidic channels to generate linear gradients of rapamycin and recruiting the FKBP–Tiam1 GEF domain induced cell polarization and chemotaxis-like directional migrating in the absence of any chemoattractants (Lin et al. 2012). Miao et al. (2019) used plasma membrane recruitment of FKBP–RacGEF to examine wave patterns of molecular signaling pathways that mediate protrusion formation in *Dictyostelium*. Inducible recruitment of GEF or GAP domains to the plasma membrane to control small GTPase function would eventually

Table 2. Summary of commonly used light-inducible (optogenetic) systems

System (protein domains)	Light color	Reversibility	Mechanism of action	Common applications
PA-Rac1, PA-Cdc42 (LOV2) (Wu et al. 2009)	Blue	Reversible	Conformational change	Directional cell motility
CRY2–CIB1 (Kennedy et al. 2010)	Blue	Reversible	Heterodimerization	Subcellular or organelle targeting, transcriptional regulation
Cry2-Olig (Taslimi et al. 2014)	Blue	Slower reversibility compared to CRY2–CIB1	Homo-oligomerization	Protein clustering, receptor tyrosine kinase activation
iLID–SspB (Guntas et al. 2015)	Blue	Fast reversibility	Heterodimerization	Subcellular or organelle targeting
LOVTRAP (LOV2, Zdk) (Wang et al. 2016)	Blue	Reversible	Light-induced dissociation	Sequestration and release of proteins
TULIP (LOV2, ePDZ, ligand) (Strickland et al. 2012)	Blue	Reversible	Heterodimerization	Subcellular or organelle targeting
Magnets (Kawano et al. 2015)	Blue	Reversible	Heterodimerization	Subcellular or organelle targeting
PixD–PixE (Yuan and Bauer 2008)	Blue	Reversible	Oligomerization	Oscillators, synthetic signaling circuits
PhyB–PIF (Levskaya et al. 2009)	Red/ far-red	Reversible with far-red stimulation	Heterodimerization	Subcellular or organelle targeting
BphP1–QPAS1 (Kaberniuk et al. 2016)	Near IR/red	Slow reversibility in dark, fast reversibility with red-light stimulation	Heterodimerization	Subcellular or organelle targeting
UVR8–COP1 (Crefcoeur et al. 2013)	UV-B	Reversible	Dissociation of UVR8 homodimer and heterodimerization with COP1	Subcellular or organelle targeting
LEXY (Niopek et al. 2016)	Blue	Reversible	Nuclear export	Nuclear export
LiNUS (Niopek et al. 2014)	Blue	Reversible	Nuclear import	Nuclear import
BcLOV4 (Glantz et al. 2018)	Blue	Reversible	Oligomerization, membrane association	Receptor tyrosine kinase activation

(CRY2) Cryptochrome 2, (LOV2) light, oxygen, voltage-sensing domain 2, (UV) ultraviolet, (TULIP) tunable light-controlled interacting protein, (PhyB) phytochrome B, (PIF3) phytochrome interaction factor 3, (LEXY) light-inducible nuclear export system, (LINuS) light-inducible nuclear localization signal (LINuS), (iLID) improved light-induced dimer, (SspB) stringent starvation protein B, (PA-Rac1) photoactivatable-Ras-related C3 botulinum toxin substrate 1, (PA-Cdc42) photoactivatable-cell division control protein 42 homolog.

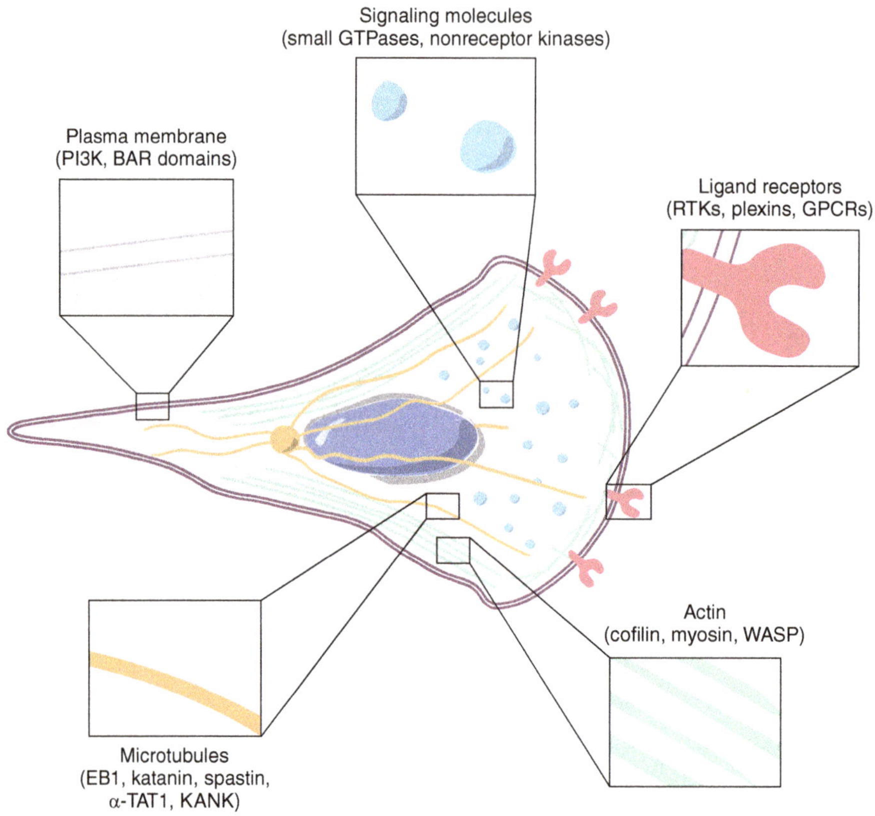

Figure 2. Examples of cellular components targeted by inducible tools to study cell motility.

become a widely used approach to manipulate actin dynamics in living cells.

The first use of optogenetics in cell migration studies was the development of photoactivatable-Rac1 (PA-Rac1) and photoactivatable-Cdc42 (PA-Cdc42). Wu et al. (2009) fused different Rac1 mutants to the LOV2 domain from *A. sativa* phototropin1 protein, which sterically blocked interaction of Rac1 or Cdc42 with its downstream effectors. On blue-light stimulation, the Jα helix of the LOV2 domain underwent a conformational change, relieving the steric hindrance and allowing interactions between Rac1 or Cdc42 and their effectors. These were also the first optogenetic tools engineered to control small GTPase activities. Subcellular stimulation of Rac1 signaling enabled localized lamellipodia formation and directional motility in HeLa cells and demonstrated negative regulation of RhoA activation by Rac1 signaling in a spatially restricted manner (Wu et al. 2009). Wang et al. (2010) applied PA-Rac1 to examine collective migration behavior in border cells in the *Drosophila* ovary. They observed that Rac1 activation or inactivation in one cell in the border cell cluster was sufficient to alter the direction of the entire group, thereby demonstrating a collective directional sensing based on Rac1 activity levels. Yoo et al. (2010) used the PA-Rac1 system to control neutrophil migration in zebrafish embryos.

Previously discussed CID-based approaches had demonstrated that plasma membrane recruitment of constitutively active or dominant-negative small GTPases, or GEF and GAP domains, was sufficient to manipulate specific small GTPase signaling pathways. Taking advantage of this experimental strategy, numerous optogenetic dimerization systems have enabled precise, light-controlled modulation of small

GTPase activity, advancing our understanding of cell motility, contractility, and morphogenesis across diverse cellular contexts. Plasma membrane recruitment of dominant-negative RhoA inhibited actomyosin contractility and when applied to examine mesoderm invagination during *Drosophila* embryogenesis demonstrated a critical role of actomyosin contractility in preventing premature tissue relaxation but not in the actual folding of the mesoderm (Guo et al. 2022). The AsLOV2-based iLID system was used for localized recruitment of Tiam1 or intersectin GEF domain to, respectively, activate Rac1 or Cdc42 and induce localized directional motility in cells to examine differential roles of extracellular matrices in stabilizing protrusions induced by Rac1 or Cdc42, demonstrating that Rac1 required exogenous fibronectin for stable protrusions, whereas Cdc42 did not (Guntas et al. 2015; Zimmerman et al. 2017). Patterned illumination enabled a graduated recruitment of these GEF domains to establish a gradient of Rac1 and Cdc42 activity and demonstrated that a sharp Rac1 activity gradient promotes cell speed, whereas a sharp Cdc42 gradient governs the direction of cell migration (de Beco et al. 2018). The robust and rapid cell morphological changes induced by recruitment of Rac1 or Cdc42 GEFs were used by different groups to characterize new optogenetic modules (Levskaya et al. 2009; Zhou et al. 2012). Plasma membrane recruitment of the GEF domain of PDZ-RhoGEF (Rho guanine nucleotide exchange factor 11) led to RhoA activation in epithelial cells, along with increased intercellular contractility, tissue compaction, and nuclear translocation of yes-associated protein (YAP). In contrast, recruitment of this GEF domain to mitochondria led to an overall decrease in cell contractility (Valon et al. 2017). Plasma membrane recruitment of various RhoA GEF domains demonstrated that the GEF concentration determines the changes in cell morphology: Low concentrations induce retractions, whereas high basal concentrations induce protrusions through sequestration of RhoA by the GEF-PH domain and activation of Cdc42 (Seze et al. 2025). Dagliyan et al. (2016) developed a series of photo-inactivated (PI) small GTPases (PI-Rac1, PI-Cdc42, and PI-RhoA) and PI GEFs (PI-Vav2, PI-Intersectin1, and PI-GEF-H1) using LOV2 insertion into these specific proteins (GEF-H1: Rho guanine nucleotide exchange factor 2; Vav2: guanine nucleotide exchange factor VAV2). Beyond the RhoA family of small GTPases, plasma membrane recruitment of the RasGEF RasGRF1 using the FKBP-FRB system induced membrane ruffling (Komatsu et al. 2010). More recently, Pal et al. (2023b,c) used Cry2-CIB-based optogenetic activation or inhibition of Ras by using the RasGEF, RAS guanyl-releasing protein 4 (RasGRP4), and RasGAP-activating-like protein 1 (RASAL), respectively, in *Dictyostelium* or neutrophils to demonstrate a critical role of localized activation of the Ras-mechanistic target of rapamycin complex 3-protein kinase B (Ras-mTORC3-Akt) signaling pathway in chemotaxis.

Aside from small GTPase-mediated actin remodeling, various groups have used other regulatory proteins with inducible systems to manipulate actin dynamics. Plasma membrane recruitment of Wiskott–Aldrich syndrome protein (WASP) triggered filopodia formation (Castellano et al. 1999), whereas clustering of the *Listeria* actin assembly (ActA)-inducing protein induced actin comet formation (Nakamura et al. 2023). An optogenetic myosin phosphatase based on myosin phosphatase target subunit 1 (MYPT1) inhibited nonmuscle myosin II activity, leading to protrusion formation and reduction in traction forces in migrating cells (Yamamoto et al. 2021). Plasma membrane recruitment of myosin heavy chain kinase C using FKBP-FRB CID inhibited bipolar thick filament assembly of myosin, leading to a rapid increase in Ras activity, revealing an inhibitory role of myosin on Ras signaling pathways (Kuhn et al. 2025). Multiple approaches of optogenetic activation of Cofilin demonstrated that cofilin activation was sufficient to induce actin remodeling and produce protrusions and invadopodia formation in cells (Hughes and Lawrence 2014; Stone et al. 2019).

INDUCIBLE SYSTEMS TARGETING THE PLASMA MEMBRANE

Both the phospholipid composition and the morphology of the plasma membrane have been implicated in initiating various signaling pathways relevant to cell motility. Manipulation of the phospholipid composition of the plasma membrane using inducible systems has been implemented by many groups to examine the role of phospholipids in cell motility. Plasma membrane recruitment of inositol polyphosphate 5-phosphatase (Inp54p) and PI3K using the FKBP-FRB dimerization system was used to interrogate the role of phospholipids in regulating KCNQ, potassium voltage-gated channel (Suh et al. 2006). The FKBP-FRB dimerization system was used to recruit a type IV 5-phosphatase domain to the plasma membrane to examine how rapid depletion of phosphatidylinositol 4,5-bisphosphate (PI(4,5)P_2) affected diverse cellular functions (Varnai et al. 2006). Inoue and Meyer (2008) used the FKBP-FRB system to recruit the iSH domain of p85 (phosphatidylinositol 3-kinase regulatory subunit) to the plasma membrane to activate PI3K and demonstrated that uniform activation of PI3K in HL-60 neutrophils is sufficient to induce polarization and motility. Recently, a noncatalytic role of PI3K in mesenchymal cell migration was revealed by plasma membrane recruitment of the p110-inter-proto-oncogene tyrosine-protein kinase Src domain homology 2 (iSH2) domain (p110: phosphatidylinositol 3-kinase catalytic subunit) (Matsubayashi et al. 2024). CID-based plasma membrane recruitment of Inp54b phosphatase was used to reveal how phosphatase and tensin homolog (PTEN) localization is regulated by the lipid composition of the plasma membrane (Rahdar et al. 2009), and cross talk between phospholipids and Ras/Rap signaling pathways in the diverse migratory modes in *Dictyostelium* (Miao et al. 2017). Ueno et al. (2011) used the FKBP-FRB system to elegantly alter the PI(4,5)P_2 levels through either creation of new PI(4,5)P_2 or liberation of previously bound PI(4,5)P_2 to demonstrate different actin remodeling phenotypes. Similar strategies using light-inducible optogenetic systems were implemented by many groups to further characterize the role of phospholipids in cell migration. Optogenetic recruitment of OCRL phosphatase (inositol polyphosphate 5-phosphatase) or PI3K to the plasma membrane using the Cry2-CIB1 system was sufficient to manipulate phospholipid content in subcellular regions, leading to localized retractions or protrusions (Idevall-Hagren et al. 2012). Optogenetic stimulation of PI3K using other optogenetic systems also induced localized protrusions and revealed a role of phosphatidylinositol 3,4,5-trisphosphate-dependent Rac exchanger 1 (P-Rex1) and ARHGAP15 in transient activation of Rac1 downstream from PI3K signaling (Toettcher et al. 2011; Kawano et al. 2015; Graziano et al. 2017). Optogenetic stimulation of the PI3K signaling could also induce epithelial-to-mesenchymal transformation in cancer cells (Zhou et al. 2018). Recently, optogenetic PI3K activation was implemented to demonstrate a critical cross talk between actin cortex and plasma membrane for long-range propagation of membrane tension (De Belly et al. 2023).

Beyond manipulating the phospholipid composition, researchers have also used inducible systems to alter the biophysical properties of the plasma membrane. FKBP-FRB-mediated recruitment of Bin-Amphiphysin-Rvs (BAR) domains to the plasma membrane, including N-BAR, F-BAR, and I-BAR domains rapidly induced curvature in plasma membrane and other organelles (Suarez et al. 2014). This was followed up by Takada et al. (2018) to demonstrate a synergistic interaction between BAR domain proteins and phospholipid flippases to remodel plasma membrane architecture. Optogenetic recruitment of charged peptides to the plasma membrane altered the electrostatic properties of the inner leaflet of the plasma membrane, leading to increased or decreased protrusions, demonstrating a role of electrostatic surface charges of the plasma membrane in mediating front–back polarity and directional migration (Banerjee et al. 2022).

INDUCIBLE SYSTEMS TARGETING MICROTUBULES

Microtubules, which are dynamic polymers of α-tubulin and β-tubulin, mediate directional cell motility through various avenues including molecular cross talk and structural interactions with other cytoskeletons. One of the first adoptions of inducible systems to control microtubule dynamics was using the LOV2-Zdk system to control microtubule stability. The LOVTRAP optogenetic module was used to develop a PI end-binding protein 1 (EB1), which dissociated from microtubule growing tips on blue-light stimulation (van Haren et al. 2018). This dissociation of EB1 from microtubule growing tips induced microtubule disassembly, leading to a change in direction of cell migration. iLID-SspB-mediated cross-linking of microtubule growing tip with F-actin stalled microtubule entry into the lamellipodial areas resulting in a microtubule-free zone in cell periphery (Adikes et al. 2018). Optogenetic recruitment of kinesin-13 led to microtubule disassembly, which was used to reveal kinesin-1 and myosin-V competition in *Drosophila* development (Lu et al. 2020). In addition to this, different chemogenetic and optogenetic systems were used by different groups to control the microtubule-severing proteins spastin and katanin to engineer disruption of microtubule network using the spastin protein (Liu et al. 2022; Meiring et al. 2022). In addition to these genetically encoded systems, there have also been a number of photoinducible chemical reagents that have been developed to control microtubule dynamics (Müller-Deku et al. 2020; Gao et al. 2021, 2022; Sailer et al. 2021). The role of posttranslational modifications of microtubules has intrigued researchers for decades. Recently, two optogenetic tools, Z-lock-α-tubulin acetyl transferase 1 (TAT1) and optoTAT were developed to control microtubule acetylation (Stone et al. 2019; Deb Roy et al. 2024). OptoTAT was used to examine the role of microtubule acetylation in directional motility and elucidate the underlying molecular cross talk with the actin cytoskeleton. Aureille et al. (2024) recently developed optoKANK using the iLID–SspB system to examine the cross talk between microtubules and focal adhesions in migrating cells (KN motif and ankyrin repeat domain-containing protein [KANK]).

INDUCIBLE SYSTEMS TARGETING OTHER SIGNALING PATHWAYS

Consistent with the wide array of molecules involved in cell motility, researchers have applied inducible systems to address questions pertinent to many other molecules of interest. CID-based recruitment of the nonreceptor tyrosine kinase ZAP70 (tyrosine-protein kinase ZAP-70) was used to demonstrate the necessity of plasma membrane recruitment and conformational flexibility in its activation (Graef et al. 1997). Insertion of iFKBP to inhibit the catalytic activity of focal adhesion kinase (FAK), Src, and p38 (mitogen-activated protein kinase 11) kinases, which could be rescued by addition of rapamycin (Karginov et al. 2010). Cry-olig optogenetic system was used to cluster and activate fibroblast growth factor (FGF) receptor tyrosine kinase, which demonstrated its role in cell polarization and directional migration (Kim et al. 2014). A similar approach was to develop an optogenetic ephrin type-B receptor 2 (EphB2) tyrosine kinase, which demonstrated its role in inducing cell retraction (Locke et al. 2017). The BcLOV4 photoreceptor was used to develop a series of optogenetic tools to activate all the epidermal growth factor (EGF) receptor tyrosine kinase family members (Pal et al. 2023a). The Cry2–CIB system was used to develop an optogenetic Plexin-B1 to demonstrate a role of Plexin-B1 signaling in contact inhibition of locomotion between osteoblasts and osteoclasts (Deb Roy et al. 2017). Optogenetic approaches to both activate and inhibit G-protein-coupled receptors were used to induce directional migration in immune cells (Karunarathne et al. 2013; O'Neill and Gautam 2014). Various chemogenetic and optogenetic dimerization approaches have been implemented to control integrin activation (Baaske et al. 2019), integrin-kindlin association (Liao et al. 2017), plasma membrane recruitment of talin to promote integrin association (Liao et al. 2021), and vimentin localization in cells (Pasolli et al. 2025). Recently, CID and optogenetic approaches were used to reveal a critical role of plasma membrane proximal glycolysis in cell migration (Zhan et al. 2025). The approaches discussed so far use exogenous expression of the molecules of interest, which may suffer from higher background signaling even in an unstimulated state. Various groups have reported use of nanobodies coupled with optogenetics to control the function of endogenous proteins, thereby inducing signaling pathways at scales closer to endogenous levels (Yu et al. 2019; Gil et al. 2020; Redchuk et al. 2020).

PERSPECTIVES

Inducible molecular tools have transformed biological research by enabling precise control over molecular events in cells. These technologies allow researchers to modulate protein activity, interactions, and localization with fine spatial and temporal resolution while maintaining molecular specificity. However, their use requires certain precautions. First, one or more genes must be introduced into the target sample. Second, expression of these tools may influence cellular functions of interest even before stimulation by light or chemical signals. Despite these limitations, the modular and orthogonal nature of many inducible systems makes them broadly applicable. This versatility is valuable not only for probing diverse molecules in migrating cells but also for investigating biological processes beyond cell migration.

Applying these tools to study cell migration in vivo is an attractive next step, as it enables examination of migratory behavior in the native environment. The main challenge lies in delivering stimulation: Light often fails to penetrate skin, and chemicals may not readily reach target tissues. To address this, red- and far-red-activated optogenetic systems hold promise for manipulating signaling pathways in deeper mammalian tissues, as well as in other commonly used model systems such as *Drosophila* and *Caenorhabditis elegans*. Red- and far-red-activated optogenetic systems could be especially helpful for studying cell migration in *C. elegans*, which exhibits blue-light avoidance (Sawin-McCormack et al. 1995; Kniazeva and Ruvkun 2025). However, *C. elegans* does not produce the cofactor biliverdin required for the red-light-stimulated PHYB-PIF3 and BphP1-QPAS1 optogenetic modules (Zhang et al. 2023), making these optogenetic modules incompatible with the model organism without additional modifications. Therefore, development of new inducible systems or modifying existing ones to avoid these limitations may greatly advance in vivo investigations of cell migration pathways.

Recent advances have expanded inducible systems beyond chemical or light stimulation. For example, Benman et al. (2025) engineered the BcLOV4 protein to develop "Melt"—a thermogenetic molecular tool that underwent oligomerization and membrane translocation in response to change in temperature. They used Melt to develop a series of thermogenetic tools to control EGF receptor (EGFR), Ras, and Cdc42 signaling. Chemicals and light do not always reach deep tissues, and controlling temperature in tissues is technically challenging and may have off-target effects. Therefore, targeted application of inducible tools using these stimuli is complicated by the properties of the tissues involved. Discovery of various proteins that are responsive to other external stimuli such as electric or magnetic field (Maeda et al. 2012; Xu et al. 2021; Kawasaki et al. 2023; Chen and Chen 2024; Belliveau et al. 2025) may pave the way for development of newer inducible systems orthogonal to chemical and light-inducible ones.

The development of predictive tools for de novo protein and peptide design, when integrated with inducible systems, has the potential to further empower studies of cell migration. Combined with advances in biosensors, live cell, and intravital microscopy approaches, inducible molecular actuators can significantly advance our understanding of the biochemical and biophysical pathways that guide more complete understanding of cell motility. Although inducible molecular actuators often painstakingly require careful design, validation, and calibration, they provide a powerful means to dissect such complex and dynamic biological behavior such as cell motility. Overall, inducible systems have become indispensable for uncovering mechanistic insights in cell migration, and we expect that their application will continue to expand the boundaries of experimental manipulation in cell migration studies.

ACKNOWLEDGMENTS

We thank Dr. Bernard L. Cook for assistance with the figures and Dr. Yuchuan Miao for thoughtful comments on the content.

REFERENCES

Adikes RC, Hallett RA, Saway BF, Kuhlman B, Slep KC. 2018. Control of microtubule dynamics using an optogenetic microtubule plus end-F-actin cross-linker. *J Cell Biol* **217:** 779–793. doi:10.1083/jcb.201705190

Aureille J, Prabhu SS, Barnett SF, Farrugia AJ, Arnal I, Lafanechère L, Low BC, Kanchanawong P, Mogilner A, Bershadsky AD. 2024. Focal adhesions are controlled by microtubules through local contractility regulation. *EMBO J* **43:** 2715–2732. doi:10.1038/s44318-024-00114-4

Baaske J, Mühlhäuser WWD, Yousefi OS, Zanner S, Radziwill G, Hörner M, Schamel WWA, Weber W. 2019. Optogenetic control of integrin-matrix interaction. *Commun Biol* **2:** 15. doi:10.1038/s42003-018-0264-7

Banaszynski LA, Chen L, Maynard-Smith LA, Ooi AGL, Wandless TJ. 2006. A rapid, reversible, and tunable method to regulate protein function in living cells using synthetic small molecules. *Cell* **126:** 995–1004. doi:10.1016/j.cell.2006.07.025

Banerjee T, Biswas D, Pal DS, Miao Y, Iglesias PA, Devreotes PN. 2022. Spatiotemporal dynamics of membrane surface charge regulates cell polarity and migration. *Nat Cell Biol* **24:** 1499–1515. doi:10.1038/s41556-022-00997-7

Belliveau NM, Footer MJ, Platenkamp A, Rodriguez C, Kim H, Prinz CK, van Loon AP, Lin Y, Eustis TE, et al. 2025. Galvanin (TMEM154) is an electric-field sensor for directed cell migration. bioRxiv doi:10.1101/2024.09.23.614580

Benman W, Huang Z, Iyengar P, Wilde D, Mumford TR, Bugaj LJ. 2025. A temperature-inducible protein module for control of mammalian cell fate. *Nat Methods* **22:** 539–549. doi:10.1038/s41592-024-02572-4

Castellano F, Montcourrier P, Guillemot JC, Gouin E, Machesky L, Cossart P, Chavrier P. 1999. Inducible recruitment of Cdc42 or WASP to a cell-surface receptor triggers actin polymerization and filopodium formation. *Curr Biol* **9:** 351–360. doi:10.1016/S0960-9822(99)80161-4

Castellano F, Montcourrier P, Chavrier P. 2000. Membrane recruitment of Rac1 triggers phagocytosis. *J Cell Sci* **113:** 2955–2961. doi:10.1242/jcs.113.17.2955

Chen J, Chen Y. 2024. Controlling cell behavior with engineered magnetosensitive proteins. *Biophys J* **123:** 20a. doi:10.1016/j.bpj.2023.11.229

Crefcoeur RP, Yin R, Ulm R, Halazonetis TD. 2013. Ultraviolet-B-mediated induction of protein–protein interactions in mammalian cells. *Nat Commun* **4:** 1779. doi:10.1038/ncomms2800

Dagliyan O, Tarnawski M, Chu P-H, Shirvanyants D, Schlichting I, Dokholyan NV, Hahn KM. 2016. Engineering extrinsic disorder to control protein activity in living cells. *Science* **354:** 1441–1444. doi:10.1126/science.aah3404

de Beco S, Vaidžiulytė K, Manzi J, Dalier F, di Federico F, Cornilleau G, Dahan M, Coppey M. 2018. Optogenetic dissection of Rac1 and Cdc42 gradient shaping. *Nat Commun* **9:** 4816. doi:10.1038/s41467-018-07286-8

De Belly H, Yan S, Borja da Rocha H, Ichbiah S, Town JP, Zager PJ, Estrada DC, Meyer K, Turlier H, Bustamante C, et al. 2023. Cell protrusions and contractions generate long-range membrane tension propagation. *Cell* **186:** 3049–3061.e15. doi:10.1016/j.cell.2023.05.014

Deb Roy A, Yin T, Choudhary S, Rodionov V, Pilbeam CC, Wu YI. 2017. Optogenetic activation of Plexin-B1 reveals contact repulsion between osteoclasts and osteoblasts. *Nat Commun* **8:** 15831. doi:10.1038/ncomms15831

Deb Roy A, Gonzalez CS, Shahid F, Yadav E, Inoue T. 2024. Optogenetically induced microtubule acetylation unveils the molecular dynamics of actin-microtubule crosstalk in directed cell migration. bioRxiv doi:10.1101/2024.12.01.626286

DeRose R, Miyamoto T, Inoue T. 2013. Manipulating signaling at will: chemically inducible dimerization (CID) techniques resolve problems in cell biology. *Pflüg Arch* **465:** 409–417. doi:10.1007/s00424-012-1208-6

Fossati E, Volpato JP, Poulin L, Guerrero V, Dugas DA, Pelletier JN. 2008. 2-Tier bacterial and in vitro selection of active and methotrexate-resistant variants of human dihydrofolate reductase. *J Biomol Screen* **13:** 504–514. doi:10.1177/1087057108318783

Gao L, Meiring JCM, Kraus Y, Wranik M, Weinert T, Pritzl SD, Bingham R, Ntouliou E, Jansen KI, Olieric N, et al. 2021. A robust, GFP-orthogonal photoswitchable inhibitor scaffold extends optical control over the microtubule cytoskeleton. *Cell Chem Biol* **28:** 228–241.e6. doi:10.1016/j.chembiol.2020.11.007

Gao L, Meiring JCM, Heise C, Rai A, Müller-Deku A, Akhmanova A, Thorn-Seshold J, Thorn-Seshold O. 2022. Photoswitchable epothilone-based microtubule stabilisers allow GFP-imaging-compatible, optical control over the microtubule cytoskeleton. *Angew Chem Int Ed* **61:** e202114614. doi:10.1002/anie.202114614

Gil AA, Carrasco-López C, Zhu L, Zhao EM, Ravindran PT, Wilson MZ, Goglia AG, Avalos JL, Toettcher JE. 2020. Optogenetic control of protein binding using light-switchable nanobodies. *Nat Commun* **11:** 4044. doi:10.1038/s41467-020-17836-8

Glantz ST, Berlew EE, Jaber Z, Schuster BS, Gardner KH, Chow BY. 2018. Directly light-regulated binding of RGS-LOV photoreceptors to anionic membrane phospholipids. *Proc Natl Acad Sci* **115:** E7720–E7727. doi:10.1073/pnas.1802832115

Graef IA, Holsinger LJ, Diver S, Schreiber SL, Crabtree GR. 1997. Proximity and orientation underlie signaling by the non-receptor tyrosine kinase ZAP70. *EMBO J* **16:** 5618–5628. doi:10.1093/emboj/16.18.5618

Graziano BR, Gong D, Anderson KE, Pipathsouk A, Goldberg AR, Weiner OD. 2017. A module for Rac temporal signal integration revealed with optogenetics. *J Cell Biol* **216:** 2515–2531. doi:10.1083/jcb.201604113

Guntas G, Hallett RA, Zimmerman SP, Williams T, Yumerefendi H, Bear JE, Kuhlman B. 2015. Engineering an improved light-induced dimer (iLID) for controlling the localization and activity of signaling proteins. *Proc Natl Acad Sci* **112:** 112–117. doi:10.1073/pnas.1417910112

Guo H, Swan M, He B. 2022. Optogenetic inhibition of actomyosin reveals mechanical bistability of the mesoderm epithelium during *Drosophila* mesoderm invagination. *eLife* **11:** e69082. doi:10.7554/eLife.69082

Holsinger LJ, Spencer DM, Austin DJ, Schreiber SL, Crabtree GR. 1995. Signal transduction in T lymphocytes using a conditional allele of Sos. *Proc Natl Acad Sci* **92:** 9810–9814. doi:10.1073/pnas.92.21.9810

Huang Dennis Z, Benman W, Dong L, Bugaj LJ. 2024. Rapid optogenetic clustering in the cytoplasm with BcLOVclust. *J Mol Biol* **436:** 168452. doi:10.1016/j.jmb.2024.168452

Hughes RM, Lawrence DS. 2014. Optogenetic engineering: light-directed cell motility. *Angew Chem Int Ed* **53:** 10904–10907. doi:10.1002/anie.201404198

Idevall-Hagren O, Dickson EJ, Hille B, Toomre DK, De Camilli P. 2012. Optogenetic control of phosphoinositide metabolism. *Proc Natl Acad Sci* **109:** E2316–E2323. doi:10.1073/pnas.1211305109

Inoue T, Meyer T. 2008. Synthetic activation of endogenous PI3K and Rac identifies an AND-gate switch for cell polarization and migration. *PLoS One* **3:** e3068. doi:10.1371/journal.pone.0003068

Inoue T, Heo WD, Grimley JS, Wandless TJ, Meyer T. 2005. An inducible translocation strategy to rapidly activate and inhibit small GTPase signaling pathways. *Nat Methods* **2:** 415–418. doi:10.1038/nmeth763

Iwamoto M, Björklund T, Lundberg C, Kirik D, Wandless TJ. 2010. A general chemical method to regulate protein stability in the mammalian central nervous system. *Chem Biol* **17:** 981–988. doi:10.1016/j.chembiol.2010.07.009

Kaberniuk AA, Shemetov AA, Verkhusha VV. 2016. A bacterial phytochrome-based optogenetic system controllable with near-infrared light. *Nat Methods* **13:** 591–597. doi:10.1038/nmeth.3864

Karginov AV, Ding F, Kota P, Dokholyan NV, Hahn KM. 2010. Engineered allosteric activation of kinases in living cells. *Nat Biotechnol* **28:** 743–747. doi:10.1038/nbt.1639

Karginov AV, Zou Y, Shirvanyants D, Kota P, Dokholyan NV, Young DD, Hahn KM, Deiters A. 2011. Light regulation of protein dimerization and kinase activity in living cells using photocaged rapamycin and engineered FKBP. *J Am Chem Soc* **133:** 420–423. doi:10.1021/ja109630v

Karginov AV, Tsygankov D, Berginski M, Chu PH, Trudeau ED, Yi JJ, Gomez S, Elston TC, Hahn KM. 2014. Dissecting motility signaling through activation of specific Src-effector complexes. *Nat Chem Biol* **10:** 286–290. doi:10.1038/nchembio.1477

Karunarathne WKA, Giri L, Patel AK, Venkatesh KV, Gautam N. 2013. Optical control demonstrates switch-like PIP_3 dynamics underlying the initiation of immune cell migration. *Proc Natl Acad Sci* **110:** E1575–1583. doi:10.1073/pnas.1220755110

Kawano F, Suzuki H, Furuya A, Sato M. 2015. Engineered pairs of distinct photoswitches for optogenetic control of cellular proteins. *Nat Commun* **6:** 6256. doi:10.1038/ncomms7256

Kawasaki H, Okano H, Ishiwatari H, Kishi T, Ishida N. 2023. A role of cryptochrome for magnetic field-dependent improvement of sleep quality, lifespan, and motor function in *Drosophila*. *Genes Cells* **28:** 496–502. doi:10.1111/gtc.13030

Kennedy MJ, Hughes RM, Peteya LA, Schwartz JW, Ehlers MD, Tucker CL. 2010. Rapid blue-light–mediated induction of protein interactions in living cells. *Nat Methods* **7:** 973–975. doi:10.1038/nmeth.1524

Kim N, Kim JM, Lee M, Kim CY, Chang KY, Heo WD. 2014. Spatiotemporal control of fibroblast growth factor receptor signals by blue light. *Chem Biol* **21:** 903–912. doi:10.1016/j.chembiol.2014.05.013

Kniazeva M, Ruvkun G. 2025. The effects of intense light on *Caenorhabditis elegans*. bioRxiv doi:10.1101/2025.06.05.658132

Komatsu T, Kukelyansky I, McCaffery JM, Ueno T, Varela LC, Inoue T. 2010. Organelle-specific, rapid induction of molecular activities and membrane tethering. *Nat Methods* **7:** 206–208. doi:10.1038/nmeth.1428

Kuhn J, Banerjee P, Haye A, Robinson DN, Iglesias PA, Devreotes PN. 2025. Complementary cytoskeletal feedback loops control signal transduction excitability and cell polarity. *Nat Commun* **16:** 7482. doi:10.1038/s41467-025-62799-3

Levskaya A, Weiner OD, Lim WA, Voigt CA. 2009. Spatiotemporal control of cell signalling using a light-switchable protein interaction. *Nature* **461:** 997–1001. doi:10.1038/nature08446

Liang FS, Ho WQ, Crabtree GR. 2011. Engineering the ABA plant stress pathway for regulation of induced proximity. *Sci Signal* **4:** rs2. doi:10.1126/scisignal.2001449

Liao Z, Kasirer-Friede A, Shattil SJ. 2017. Optogenetic interrogation of integrin αVβ3 function in endothelial cells. *J Cell Sci* **130:** 3532–3541. doi:10.1242/jcs.205203

Liao Z, Gingras AR, Lagarrigue F, Ginsberg MH, Shattil SJ. 2021. Optogenetics-based localization of talin to the plasma membrane promotes activation of β3 integrins. *J Biol Chem* **296:** 100675. doi:10.1016/j.jbc.2021.100675

Lin B, Holmes WR, Wang CJ, Ueno T, Harwell A, Edelstein-Keshet L, Inoue T, Levchenko A. 2012. Synthetic spatially graded Rac activation drives cell polarization and movement. *Proc Natl Acad Sci* **109:** E3668–E3677. doi:10.1073/pnas.1210295109

Lin B, Yin T, Wu YI, Inoue T, Levchenko A. 2015. Interplay between chemotaxis and contact inhibition of locomotion determines exploratory cell migration. *Nat Commun* **6:** 6619. doi:10.1038/ncomms7619

Liu GY, Chen S, Lee G, Shaiv K, Chen P, Cheng H, Hong S, Yang W, Huang S, Chang Y, et al. 2022. Precise control of microtubule disassembly in living cells. *EMBO J* **41:** e110472. doi:10.15252/embj.2021110472

Locke C, Machida K, Tucker CL, Wu Y, Yu J. 2017. Optogenetic activation of EphB2 receptor in dendrites induced actin polymerization by activating Arg kinase. *Biol Open* **6:** 1820–1830. doi:10.1242/bio.029900

Lu W, Lakonishok M, Liu R, Billington N, Rich A, Glotzer M, Sellers JR, Gelfand VI. 2020. Competition between kinesin-1 and myosin-V defines *Drosophila* posterior determination. *eLife* **9:** e54216. doi:10.7554/eLife.54216

Maeda K, Robinson AJ, Henbest KB, Hogben HJ, Biskup T, Ahmad M, Schleicher E, Weber S, Timmel CR, Hore PJ. 2012. Magnetically sensitive light-induced reactions in cryptochrome are consistent with its proposed role as a magnetoreceptor. *Proc Natl Acad Sci* **109:** 4774–4779. doi:10.1073/pnas.1118959109

Matsubayashi HT, Mountain J, Takahashi N, Deb Roy A, Yao T, Peterson AF, Saez Gonzalez C, Kawamata I, Inoue T. 2024. Non-catalytic role of phosphoinositide 3-kinase in mesenchymal cell migration through non-canonical induction of p85β/AP2-mediated endocytosis. *Nat Commun* **15:** 2612. doi:10.1038/s41467-024-46855-y

Meiring JCM, Grigoriev I, Nijenhuis W, Kapitein LC, Akhmanova A. 2022. Opto-katanin, an optogenetic tool for localized, microtubule disassembly. *Curr Biol* **32:** 4660–4674.e6. doi:10.1016/j.cub.2022.09.010

Miao Y, Bhattacharya S, Edwards M, Cai H, Inoue T, Iglesias PA, Devreotes PN. 2017. Altering the threshold of an excitable signal transduction network changes cell migratory modes. *Nat Cell Biol* **19:** 329–340. doi:10.1038/ncb3495

Miao Y, Bhattacharya S, Banerjee T, Abubaker-Sharif B, Long Y, Inoue T, Iglesias PA, Devreotes PN. 2019. Wave patterns organize cellular protrusions and control cortical dynamics. *Mol Syst Biol* **15:** e8585. doi:10.15252/msb.20188585

Miyamoto T, DeRose R, Suarez A, Ueno T, Chen M, Sun T, Wolfgang MJ, Mukherjee C, Meyers DJ, Inoue T. 2012. Rapid and orthogonal logic gating with a gibberellin-induced dimerization system. *Nat Chem Biol* **8:** 465–470. doi:10.1038/nchembio.922

Müller-Deku A, Meiring JCM, Loy K, Kraus Y, Heise C, Bingham R, Jansen KI, Qu X, Bartolini F, Kapitein LC, et al. 2020. Photoswitchable paclitaxel-based microtubule stabilisers allow optical control over the microtubule cytoskeleton. *Nat Commun* **11:** 4640. doi:10.1038/s41467-020-18389-6

Nakamura H, DeRose R, Inoue T. 2019. Harnessing biomolecular condensates in living cells. *J Biochem* **166:** 13–27. doi:10.1093/jb/mvz028

Nakamura H, Rho E, Lee CT, Itoh K, Deng D, Watanabe S, Razavi S, Matsubayashi HT, Zhu C, Jung E, et al. 2023. Actuator, a *Listeria*-inspired molecular tool for physical manipulation of intracellular organizations through de novo actin polymerization. *Cell Rep* **42:** 113089. doi:10.1016/j.celrep.2023.113089

Nihongaki Y, Suzuki H, Kawano F, Sato M. 2014. Genetically engineered photoinducible homodimerization system with improved dimer-forming efficiency. *ACS Chem Biol* **9:** 617–621. doi:10.1021/cb400836k

Niopek D, Benzinger D, Roensch J, Draebing T, Wehler P, Eils R, Di Ventura B. 2014. Engineering light-inducible nuclear localization signals for precise spatiotemporal control of protein dynamics in living cells. *Nat Commun* **5:** 4404. doi:10.1038/ncomms5404

Niopek D, Wehler P, Roensch J, Eils R, Di Ventura B. 2016. Optogenetic control of nuclear protein export. *Nat Commun* **7:** 10624. doi:10.1038/ncomms10624

Nishimura K, Fukagawa T, Takisawa H, Kakimoto T, Kanemaki M. 2009. An auxin-based degron system for the rapid depletion of proteins in nonplant cells. *Nat Methods* **6:** 917–922. doi:10.1038/nmeth.1401

O'Neill PR, Gautam N. 2014. Subcellular optogenetic inhibition of G proteins generates signaling gradients and cell migration. *Mol Biol Cell* **25:** 2305–2314. doi:10.1091/mbc.e14-04-0870

Pal AA, Benman W, Mumford TR, Huang Z, Chow BY, Bugaj LJ. 2023a. Optogenetic clustering and membrane translocation of the BcLOV4 photoreceptor. *Proc Natl Acad Sci* **120:** e2221615120. doi:10.1073/pnas.2221615120

Pal DS, Banerjee T, Lin Y, de Trogoff F, Borleis J, Iglesias PA, Devreotes PN. 2023b. Actuation of single downstream nodes in growth factor network steers immune cell migration. *Dev Cell* **58:** 1170–1188.e7. doi:10.1016/j.devcel.2023.04.019

Pal DS, Lin Y, Zhan H, Banerjee T, Kuhn J, Providence S, Devreotes PN. 2023c. Optogenetic modulation of guanine nucleotide exchange factors of Ras superfamily proteins directly controls cell shape and movement. *Front Cell Dev Biol* **11:** 1195806. doi:10.3389/fcell.2023.1195806

Pasolli M, Meiring JCM, Conboy JP, Koenderink GH, Akhmanova A. 2025. Optogenetic and chemical genetic tools for rapid repositioning of vimentin intermediate filaments. *J Cell Biol* **224:** e202504004. doi:10.1083/jcb.202504004

Rahdar M, Inoue T, Meyer T, Zhang J, Vazquez F, Devreotes PN. 2009. A phosphorylation-dependent intramolecular interaction regulates the membrane association and activity of the tumor suppressor PTEN. *Proc Natl Acad Sci* **106:** 480–485. doi:10.1073/pnas.0811212106

Razavi S, Wong F, Abubaker-Sharif B, Matsubayashi HT, Nakamura H, Nguyen NTH, Robinson DN, Chen B, Iglesias PA, Inoue T. 2024. Synthetic control of actin polymerization and symmetry breaking in active protocells. *Sci Adv* **10:** eadk9731. doi:10.1126/sciadv.adk9731

Redchuk TA, Kaberniuk AA, Verkhusha VV. 2018. Near-infrared light-controlled systems for gene transcription regulation, protein targeting and spectral multiplexing. *Nat Protoc* **13:** 1121–1136. doi:10.1038/nprot.2018.022

Redchuk TA, Karasev MM, Verkhusha PV, Donnelly SK, Hülsemann M, Virtanen J, Moore HM, Vartiainen MK, Hodgson L, Verkhusha VV. 2020. Optogenetic regulation of endogenous proteins. *Nat Commun* **11:** 605. doi:10.1038/s41467-020-14460-4

Sailer A, Meiring JCM, Heise C, Pettersson LN, Akhmanova A, Thorn-Seshold J, Thorn-Seshold O. 2021. Pyrrole hemithioindigo antimitotics with near-quantitative bidirectional photoswitching that photocontrol cellular microtubule dynamics with single-cell precision. *Angew Chem Int Ed* **60:** 23695–23704. doi:10.1002/anie.202104794

Sawin-McCormack EP, Sokolowski MB, Campos AR. 1995. Characterization and genetic analysis of *Drosophila melanogaster* photobehavior during larval development. *J Neurogenet* **10:** 119–135. doi:10.3109/01677069509083459

Seze JD, Bongaerts M, Boulevard B, Coppey M. 2025. Optogenetic control of a GEF of RhoA uncovers a signaling switch from retraction to protrusion. *eLife* **12:** RP93180. doi:10.7554/eLife.93180

Spencer DM, Wandless TJ, Schreiber SL, Crabtree GR. 1993. Controlling signal transduction with synthetic ligands. *Science* **262:** 1019–1024. doi:10.1126/science.7694365

Stanton BZ, Chory EJ, Crabtree GR. 2018. Chemically induced proximity in biology and medicine. *Science* **359:** eaao5902. doi:10.1126/science.aao5902

Stone OJ, Pankow N, Liu B, Sharma VP, Eddy RJ, Wang H, Putz AT, Teets FD, Kuhlman B, Condeelis JS, et al. 2019. Optogenetic control of Cofilin and αTAT in living cells using Z-lock. *Nat Chem Biol* **15:** 1183–1190. doi:10.1038/s41589-019-0405-4

Strickland D, Lin Y, Wagner E, Hope CM, Zayner J, Antoniou C, Sosnick TR, Weiss EL, Glotzer M. 2012. TULIPs: tunable, light-controlled interacting protein tags for cell biology. *Nat Methods* **9:** 379–384. doi:10.1038/nmeth.1904

Suarez A, Ueno T, Huebner R, McCaffery JM, Inoue T. 2014. Bin/Amphiphysin/Rvs (BAR) family members bend membranes in cells. *Sci Rep* **4:** 4693. doi:10.1038/srep04693

Suh BC, Inoue T, Meyer T, Hille B. 2006. Rapid chemically induced changes of PtdIns(4,5)P_2 gate KCNQ ion channels. *Science* **314:** 1454–1457. doi:10.1126/science.1131163

Takada N, Naito T, Inoue T, Nakayama K, Takatsu H, Shin H-W. 2018. Phospholipid-flipping activity of P4-ATPase drives membrane curvature. *EMBO J* **37:** e97705. doi:10.15252/embj.201797705

Taslimi A, Vrana JD, Chen D, Borinskaya S, Mayer BJ, Kennedy MJ, Tucker CL. 2014. An optimized optogenetic clustering tool for probing protein interaction and function. *Nat Commun* **5:** 4925. doi:10.1038/ncomms5925

Toettcher JE, Gong D, Lim WA, Weiner OD. 2011. Light control of plasma membrane recruitment using the Phy–PIF system. *Methods Enzymol* **497:** 409–423. doi:10.1016/B978-0-12-385075-1.00017-2

Ueno T, Falkenburger BH, Pohlmeyer C, Inoue T. 2011. Triggering actin comets versus membrane ruffles: distinctive effects of phosphoinositides on actin reorganization. *Sci Signal* **4:** ra87. doi:10.1126/scisignal.2002033

Umeda N, Ueno T, Pohlmeyer C, Nagano T, Inoue T. 2011. A photocleavable rapamycin conjugate for spatiotemporal control of small GTPase activity. *J Am Chem Soc* **133:** 12–14. doi:10.1021/ja108258d

Valon L, Marín-Llauradó A, Wyatt T, Charras G, Trepat X. 2017. Optogenetic control of cellular forces and mechanotransduction. *Nat Commun* **8:** 14396. doi:10.1038/ncomms14396

van Haren J, Charafeddine RA, Ettinger A, Wang H, Hahn KM, Wittmann T. 2018. Local control of intracellular microtubule dynamics by EB1 photodissociation. *Nat Cell Biol* **20:** 252–261. doi:10.1038/s41556-017-0028-5

Varnai P, Thyagarajan B, Rohacs T, Balla T. 2006. Rapidly inducible changes in phosphatidylinositol 4,5-bisphosphate levels influence multiple regulatory functions of the lipid in intact living cells. *J Cell Biol* **175:** 377–382. doi:10.1083/jcb.200607116

Volpato JP, Fossati E, Pelletier JN. 2007. Increasing methotrexate resistance by combination of active-site mutations in human dihydrofolate reductase. *J Mol Biol* **373:** 599–611. doi:10.1016/j.jmb.2007.07.076

Volpato JP, Yachnin BJ, Blanchet J, Guerrero V, Poulin L, Fossati E, Berghuis AM, Pelletier JN. 2009. Multiple conformers in active site of human dihydrofolate reductase F31R/Q35E double mutant suggest structural basis for methotrexate resistance. *J Biol Chem* **284:** 20079–20089. doi:10.1074/jbc.M109.018010

Wang X, He L, Wu YI, Hahn KM, Montell DJ. 2010. Light-mediated activation reveals a key role for Rac in collective guidance of cell movement in vivo. *Nat Cell Biol* **12:** 591–597. doi:10.1038/ncb2061

Wang H, Vilela M, Winkler A, Tarnawski M, Schlichting I, Yumerefendi H, Kuhlman B, Liu R, Danuser G, Hahn KM. 2016. LOVTRAP: an optogenetic system for photoinduced protein dissociation. *Nat Methods* **13:** 755–758. doi:10.1038/nmeth.3926

Wu YI, Frey D, Lungu OI, Jaehrig A, Schlichting I, Kuhlman B, Hahn KM. 2009. A genetically encoded photoactivatable Rac controls the motility of living cells. *Nature* **461:** 104–108. doi:10.1038/nature08241

Wu HD, Kikuchi M, Dagliyan O, Aragaki AK, Nakamura H, Dokholyan NV, Umehara T, Inoue T. 2020. Rational design and implementation of a chemically inducible heterotrimerization system. *Nat Methods* **17:** 928–936. doi:10.1038/s41592-020-0913-x

Xu J, Jarocha LE, Zollitsch T, Konowalczyk M, Henbest KB, Richert S, Golesworthy MJ, Schmidt J, Déjean V, Sowood DJC, et al. 2021. Magnetic sensitivity of cryptochrome 4 from a migratory songbird. *Nature* **594:** 535–540. doi:10.1038/s41586-021-03618-9

Yamamoto K, Miura H, Ishida M, Mii Y, Kinoshita N, Takada S, Ueno N, Sawai S, Kondo Y, Aoki K. 2021. Optogenetic relaxation of actomyosin contractility uncovers mechanistic roles of cortical tension during cytokinesis. *Nat Commun* **12:** 7145. doi:10.1038/s41467-021-27458-3

Yoo SK, Deng Q, Cavnar PJ, Wu YI, Hahn KM, Huttenlocher A. 2010. Differential regulation of protrusion and polarity by PI3K during neutrophil motility in live zebrafish. *Dev Cell* **18:** 226–236. doi:10.1016/j.devcel.2009.11.015

Yu D, Lee H, Hong J, Jung H, Jo Y, Oh BH, Park BO, Heo WD. 2019. Optogenetic activation of intracellular antibodies for direct modulation of endogenous proteins. *Nat Methods* **16:** 1095–1100. doi:10.1038/s41592-019-0592-7

Yuan H, Bauer CE. 2008. Pixe promotes dark oligomerization of the BLUF photoreceptor PixD. *Proc Natl Acad Sci* **105:** 11715–11719. doi:10.1073/pnas.0802149105

Zhan H, Pal DS, Borleis J, Deng Y, Long Y, Janetopoulos C, Huang CH, Devreotes PN. 2025. Self-organizing glycolytic waves tune cellular metabolic states and fuel cancer progression. *Nat Commun* **16:** 5563. doi:10.1038/s41467-025-60596-6

Zhang H, Papadaki S, Sun X, Wang X, Drobizhev M, Yao L, Rehbock M, Köster RW, Wu L, Namikawa K, et al. 2023. Quantitative assessment of near-infrared fluorescent proteins. *Nat Methods* **20:** 1605–1616. doi:10.1038/s41592-023-01975-z

Zhou XX, Chung HK, Lam AJ, Lin MZ. 2012. Optical control of protein activity by fluorescent protein domains. *Science* **338:** 810–814. doi:10.1126/science.1226854

Zhou X, Wang J, Chen J, Qi Y, Nan D, Jin L, Qian X, Wang X, Chen Q, Liu X, et al. 2018. Optogenetic control of epithelial-mesenchymal transition in cancer cells. *Sci Rep* **8:** 14098. doi:10.1038/s41598-018-32539-3

Zhou Y, Kong D, Wang X, Yu G, Wu X, Guan N, Weber W, Ye H. 2022. A small and highly sensitive red/far-red optogenetic switch for applications in mammals. *Nat Biotechnol* **40:** 262–272. doi:10.1038/s41587-021-01036-w

Zimmerman SP, Asokan SB, Kuhlman B, Bear JE. 2017. Cells lay their own tracks—optogenetic Cdc42 activation stimulates fibronectin deposition supporting directed migration. *J Cell Sci* **130:** 2971–2983. doi:10.1242/jcs.205948

Zimmermann M, Cal R, Janett E, Hoffmann V, Bochet CG, Constable E, Beaufils F, Wymann MP. 2014. Cell-permeant and photocleavable chemical inducer of dimerization. *Angew Chem Int Ed* **53:** 4717–4720. doi:10.1002/anie.201310969

A Mechanometabolism Toolbox for Studying Cell Migration

Katherine M. Young,[1] Santiago Lopez,[1] Keefer Boone,[1] Mehmet Acikel,[2] Xavi Gallart,[1] and Cynthia Reinhart-King[1]

[1]Department of Bioengineering, [2]Department of BioSciences, Rice University, Houston, Texas 77251, USA

Correspondence: cynthia.reinhart-king@rice.edu

Cell migration is greatly affected by both the physical properties of the motile cell itself and the environment through which the cell is moving. In addition to studying cellular and extracellular mechanical properties in the context of cell migration, there is a growing interest in understanding the intersection between migration, mechanics, and metabolism. In this work, we discuss the many techniques and approaches researchers are currently using to study cellular mechanics, extracellular mechanics, and metabolism in the context of cell migration. Our goal is to bring exposure to new approaches in the fields of mechanobiology and mechanometabolism and highlight the importance of studying cell migration through a mechanical lens.

The ability of cells to migrate throughout the body plays an important role in many healthy and diseased biological processes, including the immune response, tissue development, wound healing, angiogenesis, and cancer metastasis. Across these multiple processes, the mechanical properties, such as stiffness and viscoelasticity, of both cells and the environments they are moving through influence the ability of cells to migrate (Zaman et al. 2006; Luo et al. 2016). The stiffness of a cell is determined by the structure of the internal cytoskeleton made up of actin filaments, intermediate filaments, and microtubules, as well as by the stiffness of the nucleus and composition of the membrane (Wang 1998; Wang and Stamenović 2000; Caille et al. 2002; Fischer et al. 2020; Efremov et al. 2023). The stiffness of the extracellular matrix (ECM) environment is determined primarily by the presence and cross-linking of collagen and elastin and is affected by other ECM components such as tenascin, fibronectin, laminin, hyaluronic acid, and proteoglycans (Yuan et al. 2023; Mai et al. 2024). Cells generate force to move throughout the body, and they sense the mechanical properties of the surrounding environment through many established mechanotransduction pathways (Wolfenson et al. 2013; Robert-Paganin et al. 2020; Chai et al. 2023; Saraswathibhatla et al. 2023; Young and Reinhart-King 2023). In this work, we discuss the known role of both cellular and extracellular stiffness in migration and the new approaches that are currently being used to study these roles. We additionally discuss the emerging field of mechanometabolism in migration, or the study of the

Cite this article as *Cold Spring Harb Perspect Biol* doi: 10.1101/cshperspect.a041755

intersection of mechanical properties and metabolic energy usage during cell migration.

STUDYING THE ROLE OF CELLULAR MECHANICS IN MIGRATION

For many biological processes, the stiffness of cells has been measured and correlated with motility and migration. For example, during an immune response, more motile, activated T cells are softer than naive T cells (Thauland et al. 2017). During tissue development, when stem cells migrate to different regions for differentiation, pluripotent cells are softer than differentiated cells (Pillarisetti et al. 2011; Bongiorno et al. 2014, 2016, 2018). In the processes of wound healing and angiogenesis, migrating cells at the edge of tissue regrowth or vessel branching events are often softer than nonmotile cells (Mihai et al. 2012; Huveneers and Phng 2024). Finally, cancer cells of many different origins often share the quality of being softer than noncancerous counterparts, and cancer cells that are more migratory and metastatic are even softer than nonmetastatic cancer cells (Guck et al. 2005; Xu et al. 2012; Maxwell and Roskelley 2014; Luo et al. 2016; Rianna and Radmacher 2016; Alibert et al. 2017; Liu et al. 2020). While there are some examples to the contrary (Bastatas et al. 2012; Chaki et al. 2013; Liu et al. 2015; Luo et al. 2015), there is a general trend that more motile cells are less stiff than nonmigratory cells.

Tools for Measuring Cell Stiffness

To study the role that cellular mechanical properties play in migration, we first have to be able to measure the mechanical properties of cells. Some examples of approaches to measure cell stiffness include atomic force microscopy (AFM) and newer cell nanoindentation tools, micropipette aspiration, optical tweezers (OTs), and microfluidic techniques, such as deformability cytometry (Fig. 1). These many tools have enabled the study of whole cell mechanics as well as nuclear mechanics (see Box 1) and migration by allowing researchers to compare the stiffness of cell populations of different migratory ability.

While AFM was first developed for topographical imaging of structures on the subnanometer scale length, the AFM remains the most widely used tool for cell mechanical measurements due to the ease of sample preparation, ability to take measurements at near-physiological conditions, and precision of application and measurements of piconewton-scale forces (Binnig et al. 1986; Deng et al. 2018). Force spectroscopy with AFM works by bringing a microcantilever probe tip in contact with a cell or material of interest and measuring the deflection of the cantilever as force is applied to the sample (Fig. 1A). The apparent Young's modulus, or stiffness, of the sample is determined by fitting the collected force-indentation curve to a contact mechanics model, typically the Hertzian model developed in 1881, although others have continued to develop new models for the extraction of mechanical information from these force curves that account for different model assumptions (Abuhattum et al. 2022). Recent improvements to this technique center around the integration of AFM with other imaging techniques. For example, to study the fenestrations of liver sinusoidal endothelial cells, Szafranska et al. (2022) combined and compared AFM with scanning electron microscopy, structured illumination microscopy, stimulated emission depletion microscopy, and conventional microscopy to understand the mechanical, topological, and chemical landscape of this biological system. AFM has also been used to link single-cell mechanics, gene expression, and migratory ability. A new methodology that combines AFM with the use of a micropipette for single-cell isolation and multiplexed single-cell reverse transcription quantitative polymerase chain reaction (RT-qPCR) was used to show correlations between cell stiffness, cell migratory ability, and a prometastatic gene expression profile in ovarian cancer cells (Young et al. 2023). Cell stiffness was negatively correlated with the expression of ECM remodeling genes, such as *TGM2* and *SERPINE1*, and markers of the epithelial-to-mesenchymal transition, such as *VIM* and *ZEB1*, which were overexpressed in the most

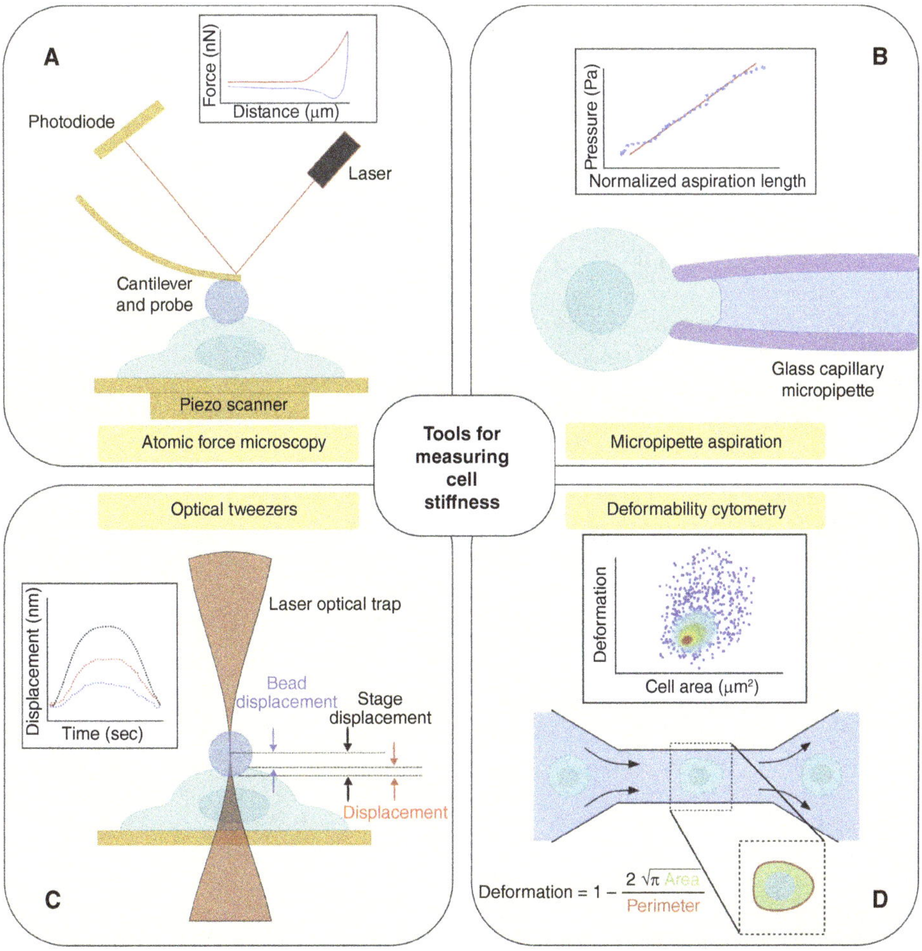

Figure 1. Tools for measuring cell stiffness. Classical approaches for studying cell stiffness include (*A*) atomic force microscopy, (*B*) micropipette aspiration, (*C*) optical tweezers, and (*D*) microfluidic deformability cytometry. Each of these tools has been updated over time to explore the role of cellular mechanics in the process of cell migration. In *A*, the red portion of the force–distance curve represents the approach portion of the collected data, and the blue line represents the retraction portion of the collected data. In *B*, the blue dots represent the measured data, and the red line represents the linear model fit of the data. In *C*, the blue data represent the displacement of the bead over time, the black data represent the motorized stage displacement, and the red data represent the calculated displacement of the cell. In *D*, the example data are represented by a density plot to visualize the distribution of cells with different areas and deformability. (Schematics of cell deformation and deformation calculation are based on data in Otto et al. 2015.)

migratory cell line. In addition to updates to conventional AFM for the measurement of cell and tissue mechanical properties, new nanoindentation tools that rely on optical interferometry have been developed to increase the throughput of sample probing and increase the area of measurement to allow for the characterization of both large samples and cells in 96 well plates (Colombo et al. 2023; Thomas-Chemin et al. 2025).

While AFM and nanoindentation rely on measurements of cells under compression, micropipette aspiration uses negative pressure to place cells under tension to measure their

BOX 1. FOCUS ON NUCLEAR MECHANICS AND MIGRATION

The nucleus plays an important role in cell mechanics and migration because of its interdependence with the cytoskeleton and extracellular matrix, its role as a highly limiting parameter in migration, and its mechanotransduction signaling via the nuclear envelope. As the largest and stiffest organelle, its mechanics directly affect surrounding organelles and the cellular microenvironment. During metastasis, a large and stiff nucleus presents a substantial obstacle during migration through dense interstitial spaces and narrow capillaries, causing many cancer cells to sense and deform their nucleus to fit through narrow spaces (Wolf et al. 2013; Davidson et al. 2014; Mukherjee et al. 2020). Some contributing factors that determine the migratory behavior of metastatic cells are their nuclear deformability and the coupling between the nucleus and cytoskeleton. Mechanical deformability of the nucleus has been shown to be heavily influenced by the nuclear envelope (Stefanello et al. 2021; Hansen and Holaska 2023). Still, nuclear deformability varies depending on cell type and chromatin configuration. Specifically, the well-studied nucleostructural proteins, lamin A, C, and B2, have an important effect on nuclear stiffness (Denais and Lammerding 2014; Vortmeyer-Krause et al. 2020). Research has shown that low lamin A/C levels enhance confined cell migration and metastasis of breast cancer cells (Bell et al. 2022). However, this observation is not the case for other cancer cell types like prostate and colorectal cancers (Bell et al. 2022). Lamins interact with various components of the cell (e.g., inner nuclear membrane proteins, nuclear pore components, DNA, chromatin, transcription factors, actin, and titin) and, in breast cancer cells, have been shown to influence processes such as nuclear deformability, extracellular remodeling, cell adhesion, metabolism, and the phosphatidylinositol 3-kinase (PI3K)/Akt signaling pathway (Denais and Lammerding 2014; Bell et al. 2022). Additionally, there is variation in terms of how chromatin remodeling and condensation can influence nuclear stiffness (Heo et al. 2020). After treating highly invasive MDA-MB-231 breast cancer cells with trichostatin A (TSA), a histone deacetylase inhibitor, chromatin is decondensed, and both cytoskeletal and nuclear stiffness increased (Fischer et al. 2020). In contrast, in less migratory MCF-7 breast cancer cells, TSA treatment induced a softer nucleus and cytoskeleton. Both cell types demonstrated increased invasion after TSA treatment. When MDA-MB-231 cells were treated with both TSA and latrunculin A, the nucleus did not appear to stiffen, implying perinuclear actin may impair nuclear softening in MDA-MB-231 cells after chromatin decondensation (Fischer et al. 2020). The results from this paper highlight the importance of exploring the relationships between the actomyosin cytoskeleton structures and nucleus, as they are a major contributor to whole cell stiffness and nuclear response to chromosomal remodeling during migration through confined spaces.

mechanical properties. A typical micropipette aspiration system works by applying a negative pressure to pull a suspended cell into a glass capillary micropipette while recording images of the cell entry into the micropipette (Fig. 1B). Assuming the cell can be modeled as a homogenous elastic membrane holding a homogenous Newtonian liquid, the Young–Laplace equation can be applied to calculate the Young's modulus from the pressure change, pipette geometry, and protrusion length of the cell into the pipette (Lee and Liu 2014). Using this technique, researchers have recently identified that aging results in a progressive increase of T-cell apparent stiffness and viscosity that is coupled with a reduced rate of spontaneous migration related to DNA hypomethylation, lamin B1 defects, decrease of cytoskeletal proteins, and accumulation of age-marker lipofuscin (González-Bermúdez et al. 2022). One of the major limiting factors of micropipette aspiration is the throughput of data collection (<1 cell per 10 min). Updated methodologies have overcome this limitation by incorporating the principles of micropipette aspiration into microfluidic devices, allowing for the increased throughput of measurements of both single cells and multicellular spheroids (Davidson et al. 2019; Wu et al. 2020; Boot et al. 2023). Additionally, advancements in computing have allowed researchers to improve the analysis of micropipette aspiration data through the training and deployment of neural networks (Abarca-Ortega et al. 2024).

Another technique, OTs, uses highly focused laser beams to trap and manipulate particles, atoms, and molecules (Fig. 1C). Specifically, its high-resolution position and force measurements allow for precise calculations of cell stiffness (Arbore et al. 2019). One area of focus in the field of mechanobiology has been the feedback between cell stiffness and substrate stiffness. When using OT to measure the stiffness of breast cancer cell lines of differing migratory ability, the less aggressive and less motile HBL-100 and MCF7 breast cells demonstrated cell softening on soft substrates compared to stiff substrates, while the more motile and aggressive MDA-MB-231 breast cancer cell line showed cell stiffening on soft substrates compared to stiff substrates (Yousafzai et al. 2017). Additionally, some researchers have explored OT in combination with microfluidics for cell stabilization to develop ways to overcome AFM and micropipette aspiration limitations, like controllability, physiological environment, bead incompatibility, numerical simulation complexity, and the inability to image protein recruitment directly (Honarmandi et al. 2011; Vasse et al. 2021; Hu et al. 2022). Other researchers have used OT to explore rheology and tunneling membrane nanotubes in relation to viscoelasticity and the cytoskeleton, respectively, with a particular focus on the formation of filopodia, which play an important role in cell migration (Ayala et al. 2016; Nussenzveig 2018).

Real-time deformability cytometry (RT-DC) is a powerful tool used to explore the elastic and mechanical properties of cells through hydrodynamic deformation of cells on timescales of roughly 1 msec (Mietke et al. 2015). As cells are deformed during flow through a microfluidic channel, the Young's modulus of the cell can be determined through a series of linear elastic theory approximations and hydrodynamical analysis of a high-speed video of the cell deformation (Fig. 1D). Microfluidic deformability cytometry has become a powerful and effective high-throughput tool with applications in various fields, as microfluidics reduce operational complexity and reagent consumption (Chen et al. 2023b). Traditional methods for measuring cell stiffness (AFM, micropipette aspiration, OTs, etc.) have a low throughput of <100 cells/h. RT-DC has a throughput that ranges from 100 cells/sec up to 65,000 cells/sec, depending on the type of deformability cytometry (Urbanska et al. 2020; Chen et al. 2023b). Deformability cytometry tools can be divided into categories based on their methods for generating stress. Constriction-based DC, extrusion or shear flow DC, hydrodynamic or extensional flow DC, electric stretching DC, optical stretching DC, and acoustic stretching DC all perturb the cells by applying between <1 and ~6 kPa of force (Urbanska et al. 2020; Chen et al. 2022). RT-DC has also been used to explore cell viscoelasticity, which has been linked to collective cell migration through E-cadherin-dependent phosphorylation of epidermal growth factor receptor (EGFR) (Mokbel et al. 2017; Gerum et al. 2022; Asghari et al. 2024; Fu et al. 2024). Additionally, RT-DC can distinguish between mature and immature bone marrow–derived dendritic cells, which, as a counterexample to many other biological processes, undergo a cell stiffening process during maturation that allows them to migrate through lymphatic vessels (Leblanc-Hotte et al. 2023). One limitation of this technique is that it cannot measure cells in contact with a substrate and has a limited field of view for measurement (Gensbittel et al. 2021; Chen et al. 2022). Potential areas for further exploration include using RT-DC as a cell sorting technique similar to flow-assisted cell sorting, combining types of cell deformability measurement methods, and improving data processing throughput of the detection technology (Nawaz et al. 2020; Herbig et al. 2022; Chen et al. 2023c).

Tools for Adjusting Cell Stiffness

In addition to measuring cell stiffness to study its role in cell migration, being able to adjust cell stiffness allows for the study of a causal link between cell mechanics and migration. Varying chemical compounds that affect the cytoskeleton can be used as treatment to modulate cell stiffness and migration (Table 1). Specifically, compounds that help regulate actomyosin activ-

Table 1. Effect of different chemical compounds on cell stiffness and migratory/invasive ability

Chemical compound	Mechanism of action	Cell type tested on	Effect on cell stiffness	Effect on migratory and invasive ability	References
Blebbistatin	Myosin II inhibitor	HepG2	Softened	Increased	Wang et al. 2024
Cytochalasin D	F-actin inhibitor	HepG2	Softened	Increased	Wang et al. 2024
Jasplakinolide	F-actin polymerization activator	HepG2, MHCC97-L	Stiffened	Decreased	Wang et al. 2024
Narciclasine	Rho/ROCK/LIMK/ cofilin activator	HepG2, MHCC97-L	Stiffened	Decreased	Wang et al. 2024
PD98059	MAPK1 inhibitor	VSMC, SW480	Stiffened	Decreased	Liu et al. 2024; Tian et al. 2024
SB203580	p38 MAPK inhibitor	T24	Unknown	Increased	Xie et al. 2024
SP600125	JNK inhibitor	VSMC, HepG2, Huh-7	Stiffened	Increased	Tian et al. 2024; Wang et al. 2024; Xie et al. 2024
Y-27632	ROCK inhibitor	Huh-7, HepG2	Softened	Increased	Wang et al. 2024

(MAPK1) Mitogen-activated protein kinase 1, (ROCK) rho-associated coiled-coil-containing protein kinase, (JNK) c-Jun amino-terminal kinase, (VSMC) vascular smooth muscle cell.

ity have been found to affect the migration and invasion of hepatocellular carcinoma (HCC) cells. Blebbistatin (Bleb), cytochalasin D, and Y-27632 are known chemical compounds that inhibit myosin, F-actin, and rho-associated coiled-coil-containing protein kinase (ROCK) activity, inhibiting the ability to form actomyosin complexes. Their treatment on HCC cells has been shown to significantly reduce stiffness and increase migratory and invasive ability (Wang et al. 2024). The treatment of jasplakinolide (Jasp) and narciclasine (Narci) on HCC cells to strengthen the cytoskeleton through F-actin polymerization and Rho activation showed an increase in cellular stiffness and a suppression of cell migration and invasion (Wang et al. 2024). This study also found that softening HCC cells reduced protein kinase c-Jun amino-terminal kinase (JNK) activity by downregulating upstream protein kinase kinases MKK4 and MKK7, while stiffening cells upregulated these genes and increased JNK phosphorylation. Treatment using the mitogen-activated protein kinase 1 (MAPK1) inhibitor PD98059 and the JNK inhibitor SP600125 on vascular smooth muscle cells (VSMCs) increased cell stiffness (Tian et al. 2024). SW480 colorectal cancer cells transfected with protein phosphatase protein phosphatase 2 scaffold subunit Aβ (PPP2R1B) small interfering RNA (siRNA) were also treated with PD98059, and it was found to suppress the cell invasion and migration regulated by the PPP2R1B silencing (Liu et al. 2024). A T24 bladder cancer cell line was treated with a natural alkaloid, securinine (SEC), and it significantly lowered the cells' migratory and invasive ability. To see whether SEC inhibits tumor growth by activating MAPK1 and JNK signaling pathways, the SEC-treated cells were further treated with SP600125 and the MAPK1 inhibitor SB203580 to confirm this hypothesis (Xie et al. 2024). This further treatment did in fact reduce the SEC's inhibitory effect on the T24 cells' motility, proving that SEC inhibits tumor growth by activating MAPK1 and JNK signaling pathways. Other common chemical compounds used to adjust cell mechanical properties, including stiffness and contractility, include myosin light chain kinase inhibitor (ML-7), latrunculin A (F-actin inhibitor), calphostin C (protein kinase C inhibitor), nocodazole (microtubule inhibitor), paclitaxel (microtubule stabilizer), and calyculin A (protein phosphatase inhibitor) that have also been shown to affect cell migration in a high-throughput screen (Nehwa et al. 2020).

While chemical alteration of cell mechanics can allow for a controlled study of the effect of cell mechanics on migration, these chemical perturbations can have other effects beyond a change to cell stiffness, which may also contribute to the changes observed in cell behavior. Another option for studying the causal link between cell mechanics and migration involves sorting cells by their mechanical properties. The primary method for sorting cells has involved microfluidic devices. These tools can be designed with unique geometries and flow profiles, featuring various outlets and inlets. The primary type of microfluidic cell sorting is size-dependent; however, strategies have been developed to explore other mechanical properties as well. New developments have included adhesion-based microfluidic devices (Chrit et al. 2024), usage of non-Newtonian/power-law fluids (Chrit et al. 2023), active cell separation approaches (dielectrophoresis, acoustophoresis, and magnetophoresis), and passive cell separation approaches (microfilters, bifurcation, pinch flow fraction, deterministic lateral displacement, and inertial microfluidics) (Farahinia et al. 2021). One microfluidic sorting device was used to process ovarian cancer patient ascites fluid to enrich for metastatic cells that had moved away from the ovaries into the peritoneal cavity (Stone et al. 2021). The small processing volumes inherent to microfluidic devices can serve as an obstacle when trying to process a large number of cells. Some researchers have overcome this challenge by increasing the concentration of cells processed or increasing the flow rate, but there are limitations to these strategies. The more promising approach involves designing microfluidic devices using strategies like parallelization, which enhance yield and reduce processing time (Young et al. 2022). In the future, improving throughput and mechanical sorting capabilities of microfluidic devices will be an important step in increasing efficiency and potential applications to study a causal link between cell stiffness and migratory ability.

STUDYING THE ROLE OF EXTRACELLULAR MECHANICS IN MIGRATION

Beyond studying the role of cellular mechanics in cell migration, much attention has been focused on the effect of mechanically different extracellular environments on cell movement. Within the ECM, some of the major components are fibronectin, a glycoprotein that modulates cell–ECM interactions through the formation of stable focal adhesions (FAs) (Lu et al. 2020), laminins, proteins located within the basal lamina of the ECM that help in the binding interaction between cellular receptors and integrins (Aumailley 2013), proteoglycans and glycosaminoglycans (GAGs), which trap growth factors and cytokines in the ECM (Park 2022), and lastly collagen, the most abundant protein in the ECM that helps to maintain the structure of the tissue through its fibrous networks and multiple cross-links (Kular et al. 2014). Accumulation of these components increases the stiffness of the ECM microenvironment and changes the nature of development and disease progression. In the tumor microenvironment, during the growth of the primary tumor, the abundance of ECM components increases through deposition by cancer-associated fibroblasts (Yang et al. 2023a) and ECM remodeling by cancer cells themselves (Winkler et al. 2020). Changes in the ECM help cancer cells to move through the microenvironment to metastasize to different organs in the body. ECM stiffness is also important during different biological processes such as development, wound healing, and immune cell migration.

While it has generally been observed that cells tend to spread more and generate larger traction forces on stiffer substrates but also tend to move more slowly than cells on soft substrates, there is also work that suggests that cells migrate toward areas of "optimal stiffness" where cells can generate the maximal traction, a quality that will differ by cell and tissue type (Lange and Fabry 2013; Janmey et al. 2020; Yu et al. 2021; Isomursu et al. 2022). Durotaxis is the tendency of cells to move toward stiff substrates, while negative durotaxis is the tendency of cells to move toward softer substrates. Whether a cell undergoes durotaxis, negative durotaxis, or adurotaxis, where cells are agnostic to their mechanical environment, is dependent on the cell type and the expression of mechanosensitive signals (Ji and Huang 2023). For example, in a study investigating the migratory differences between strongly and weakly adherent cancer cells in a durotaxis gradient, strongly adherent cells move along the durotaxis gradient driven by their actomyosin activity while weakly adherent cells were adurotactic (Yeoman et al. 2021). Durotaxis studies are typically conducted in a 2D environment; however, there have been attempts to also model this process in 3D to better model physiologically relevant contexts (Puleo et al. 2019; Mathieu et al. 2024). In physiological conditions, cells experience environmental stiffnesses that vary greatly throughout the body, ranging from hundreds of Pascals in soft tissues like the brain up to millions of Pascals in stiff tissues like the bone (Janmey and Miller 2011; Wells 2013; Abbas et al. 2019). Cells respond to these varying tissue-level stiffness differences; for example, cancer cells tend to metastasize to specific organs, a phenomenon defined as organotropism, which has been linked to tissue architecture and mechanical properties (Azubuike and Tanner 2023). Brain-specific metastatic cancer cells show improved metastatic colonization after priming in a soft environment that matches the brain tissue stiffness, demonstrating a mechanical memory that could play an important role in organ-specific migration patterns (Tang et al. 2025). Durotaxis and organotropism represent two examples of how extracellular mechanics affect and direct cell migration.

Tools for Measuring Extracellular Structure and Mechanics

To visualize and quantify physical properties of the ECM and its interactions with cells, microscopy platforms and advanced techniques become essential to ensure accurate analysis (Fig. 2). A challenge is measuring the properties noninvasively at the scale that is sensed by cells. Brillouin microscopy uses optical elastography to measure the viscoelastic mechanical properties of biological samples in 3D at a subcellular resolution without actively changing the chemical composition or coming in contact with the sample (Fig. 2A; Prevedel et al. 2019). Limitations include the need for the local refractive index and material density of samples, which can be difficult to measure in situ (although it was recently shown that the effect of refractive index can be neglected in excised zebrafish tissues) (Schlüßler et al. 2018), long data acquisition times, and harmful illumination to the sample. A recent development surpasses the limitation of phototoxicity of the pump–probe interaction of the system by lowering the laser power to allow imaging of mouse embryos, zebrafish larva, and *Caenorhabditis elegans* embryos (Yang et al. 2023b). While Brillouin microscopy has been used to elucidate cellular mechanics, such as showing that the nuclei of tumor cells soften during migration through an endothelial layer and fibroblasts' response to different substrate stiffness, it has also been used to measure the mechanical properties of ECM in live zebrafish (Bevilacqua et al. 2019; Roberts et al. 2021; Cheburkanov and Yakovlev 2025). The noninvasive nature of this technique opens up the possibility for the further exploration of extracellular and cellular mechanics in more physiologically relevant contexts.

Another popular modality to study ECM structure is the use of second harmonic generation (SHG) microscopy. Second harmonic microscopy uses a nonlinear model between the intensity of the illumination and the signal of the system to visualize several fibrillar structures in the ECM, such as collagen (Fig. 2B; Zipfel et al. 2003). The size, orientation, and spacing of fibrillar structures can be quantified based on SHG imaging (Chen et al. 2012). This technique was recently used to show that hyaluronan and proteoglycan link protein 1 (HAPLN1) produced by cancer-associated fibroblasts (CAFs) promotes gastric cancer cell invasion by decreasing the number, density, width, and length of collagen fibers while increasing fiber alignment (Zhang et al. 2022). Raman spectroscopy can also quantify the spatial distribution of specific ECM molecules. When light is in contact with a biological sample, photons are shifted in

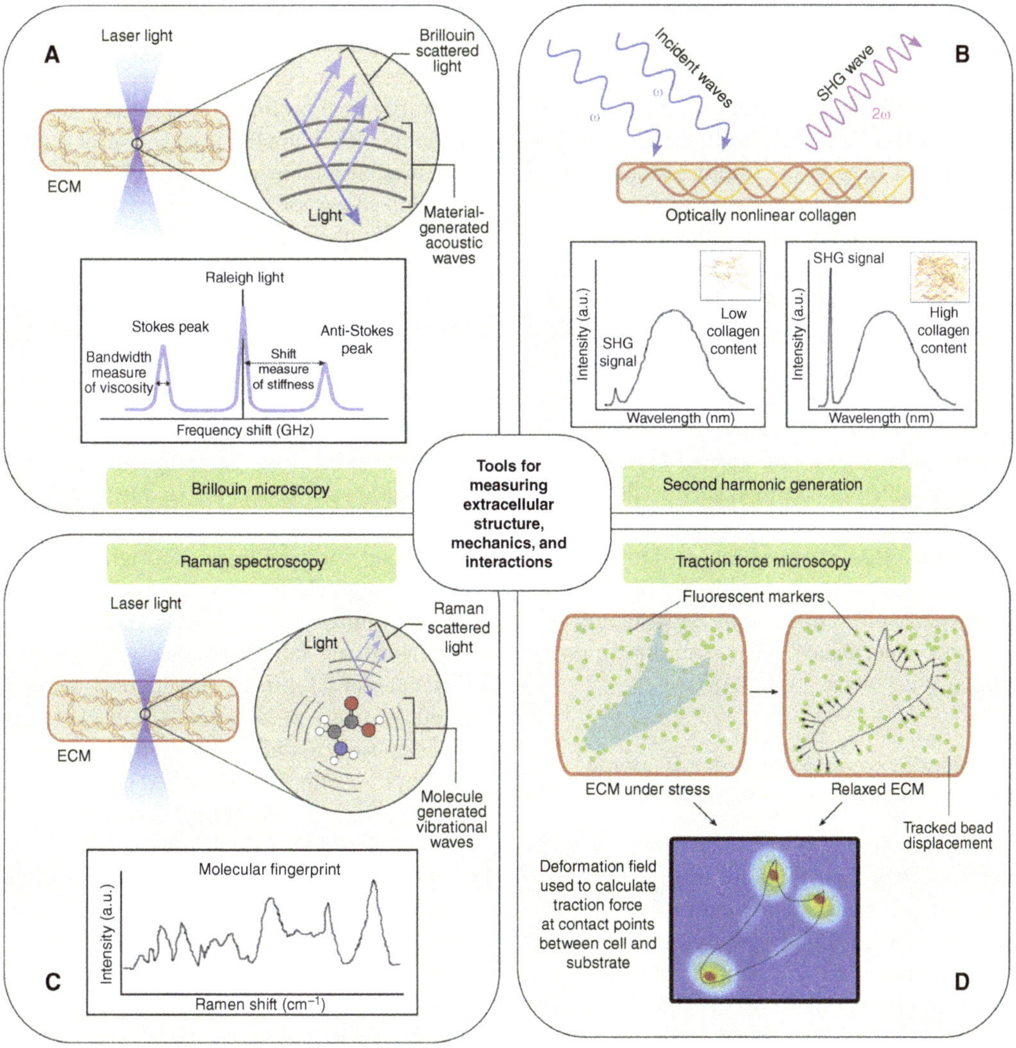

Figure 2. Tools for measuring extracellular structure, mechanics, and interactions. Techniques for studying extracellular structure and mechanics include (*A*) Brillouin microscopy, (*B*) second harmonic generation microscopy, (*C*) Raman spectroscopy, and (*D*) traction force microscopy. Each of these tools has been updated over time to explore the role of extracellular mechanics in the process of cell migration. (ECM) Extracellular matrix, (SHG) second harmonic generation.

frequency based on their contact with the vibrational modes of molecules in the sample (Fig. 2C). This shift is then attributed to specific proteins found in the ECM like GAGs, elastin, and collagen (Sherlock et al. 2021). A recent study combined SHG and Raman microspectroscopy to investigate how the molecular organization of collagen in tendons changes with age, which could affect tendon cell migration to the core in response to loading (Van Gulick et al. 2019; Stańczak et al. 2024).

In addition to visualization of the ECM, researchers have also used mechanical testing to probe the physical properties of the ECM. One of the most highly used systems to measure ECM mechanics is AFM. AFM force volume mapping has been used to derive a spatial and mechanical map of collagen deposition in human and mouse tissue (Calò et al. 2020). Another recent improvement to AFM includes the introduction of STIFMap, or the spatially transformed inferential force map, which uses computer vision to automate and extend AFM indentation to large tissue samples along with a convolutional neural network to predict the mechanical properties of tumor cells and surrounding tissue to link tissue mechanics with markers of mechanosignaling and the epithelial-to-mesenchymal transition (Stashko et al. 2023). Another group used AFM peak force imaging and mechanical mapping to analyze the topography of decellularized matrices to find differences between pathological and normal fibroblast invasion on different commercially available matrices used for surgical implantation (Viji Babu et al. 2019). AFM has also been used to measure in vivo mechanical cues in a developing *Xenopus laevis* embryo to show how tissue stiffening coordinates collective cell migration during morphogenesis (Barriga et al. 2018).

Modeling Microenvironments of Varying Stiffness and Cellular Interaction

To mimic the ECM microenvironment and measure forces that cells exert during migration, researchers have developed materials and metrics that allow for in vitro analysis. Many engineered biomaterials and naturally derived substrates have been used to model the ECM, including polyacrylamide (PA) gels, polyethylene glycol, polydimethylsiloxane, polypeptide hydrogels, polyacrylonitrile nanofiber scaffolds, gelatin, decellularized ECM, collagen, and fibrin (Emig et al. 2021; Liu et al. 2021; Marhuenda et al. 2021; Zhou et al. 2021; Chang et al. 2022; Guo et al. 2022; Jang et al. 2022; Lachowski et al. 2022; Wang et al. 2022; Chen et al. 2023a; Young and Reinhart-King 2023). One of the common techniques to study mechanical interactions between cells and different stiffness ECM environments is traction force microscopy (TFM). TFM uses fluorescent microscopy to visualize and quantify deflection of fluorescent markers on the substrate to approximate the forces imposed by the cells on the substrate (Fig. 2D). While the use of TFM is not a new technique, researchers have recently developed new approaches to expand its application. To improve quantifying 3D traction force, one group developed a technique to quantify traction force using refractive index tomography that allows visualization of the 3D mass distribution of the cell on a planar hydrogel, which allows for high-speed imaging able to capture and quantify the forces exerted during the rapid contraction of a cell during a rear-detachment mode of migration (Lee et al. 2024). Another group developed photonic crystals that can detect vertical cellular forces on the micro and nano level in situ (Li et al. 2023). Because of the stability of the photonic crystal substrate, long-term time-lapse imaging enabled the quantification of vertical forces during cell migration, which revealed a gradual rise of vertical force during cell spreading and a highly dynamic and strong cell–substrate interaction. In recent years, microfluidic devices and programmable hydrogels have been used to mimic the ECM microenvironment of different diseases without the use of animals. These platforms have been used to study cell–cell and cell–matrix interactions along with the effects of cytokine and chemokine gradients on 3D cancer migration and invasion (Mehta et al. 2022). For example, one group developed an integrated "microbuckets" and hydrogel platform that is programmed to have a controlled release of transforming growth factor β to help direct cancer cell spheroid invasion in vitro (Liu et al. 2022).

Tools for Modulating ECM Stiffness

Since ECM stiffness changes accompany multiple diseases, there has been intense interest in both academia and industry in studying the effect of drugs targeting ECM cross-links and the intracellular pathways leading to ECM deposition (Lampi and Reinhart-King 2018). Enzymatic cross-linking through lysyl oxidase (LOX) (Herchenhan et al. 2015) and nonenzymatic cross-linking through advanced glycation end products (AGEs), which are created by a reaction between sugars and amino acid groups found in proteins of the ECM, have both been targets. The use of β-aminopropionitrile (BAPN), a LOX breaker, has improved the outcome in studies involving atherosclerosis (Kothapalli et al. 2012), cancer (Bordeleau et al. 2017), and liver fibrosis (Liu et al. 2016). Aminoguanidine and algaebrium are both drugs that tackle AGEs in either the formation of AGEs or the breaking of these cross-links and have seen favorable results in diseases like atherosclerosis (Forbes et al. 2004; Watson et al. 2011). Although in vitro data have been promising, clinical trials of these drugs have been unfavorable, and thus, improvements are still needed on these drugs to both enhance efficacy and minimize off-target effects.

Aside from targeting cross-linking, both the targeting of the cells producing excess ECM stiffness and the cellular response to stiffness have been considered. As one example, targeting ECM production by CAFs and the mechanosensitive pathways within tumor cells has been suggested as a method to improve immunotherapy effectiveness in tumors (Mai et al. 2024). Understanding the concepts and components behind ECM stiffness, its impact on cellular activity, and the tools available to study this interaction can lead to new discoveries and solutions mediated by this phenomenon.

Mechanometabolism and Migration

Beyond studying the role of cellular and extracellular stiffness in migration, the field of mechanobiology has expanded its focus to include the study of mechanometabolism, the intersection of mechanics and metabolic energy usage. Cell migration is an energy-intensive behavior and, therefore, it is important to understand how cells generate energy for the task. While some trends have been identified of cells switching metabolic states during different migration modes (see the section Shifting Metabolic Profiles in Studying Mechanics and Migration), there is still much room for the expansion of our understanding of how cells change their metabolic profile to make migration more efficient. We predict that cellular metabolism acts as a linking factor between cellular and extracellular mechanics, allowing cells to adjust to their current environment to most efficiently generate energy, force, and motion. Numerous tools and probes tailored to specific metabolic pathways and activities have been developed to study these processes, from fluorescent probes to instrumental techniques (Table 2). The most significant recent advances in tools for the study of mechanometabolism in migration have enabled the study of single-cell behaviors.

Molecular Probes for Monitoring Metabolism

Adenosine triphosphate (ATP) is a universal energy currency in cells, and its measurement provides critical insights into a cell's metabolic state and energy production. ATPlite is a luciferase-based bioluminescent assay where ATP reacts with luciferin to produce light proportional to ATP concentration. It is widely used for high-throughput cell viability, proliferation, and cytotoxicity analysis (Kangas et al. 1984). ATPlite was used to show increasing cellular energy levels on substrates of increasing stiffness, which also resulted in increased traction force and FA formation (Schunk et al. 2024). While it offers high sensitivity and straightforward execution, it is limited to end-point measurements and requires lysis of cell populations, making it unsuitable for dynamic studies. These limitations have been partially overcome with the creation of ATeam, a genetically encoded fluorescent biosensor based on fluorescence resonance energy transfer (FRET) (Yoshida et al. 2017). ATeam senses ATP through conformational changes upon binding, enabling real-time monitoring of ATP in specific subcellular compartments such as mitochondria and the cytosol (Yoshida et al. 2017). Although it allows for live-cell imaging with subcellular resolution, it requires genetic manipulation to express the sensor. PercevalHR is another fluorescent biosensor that monitors the ATP/adenosine diphosphate (ADP) ratio, providing insights into cellular energy balance (Tantama et al. 2013; Zanotelli et al. 2018). Its ratiometric signal minimizes variability due to sensor expression levels, although it is sensitive to changes in pH, requiring additional probes for correction. The PercevalHR probe has been used to study the intersection of metabolism and tissue mechanics finding the ATP:ADP ratio increases in cells in denser matrices where migration is impaired and that

Table 2. Molecular probes for monitoring cellular metabolism

Metabolic pathway	Biosensor	Detected variable	Signal production	References
ATP production	ATPlite	ATP concentration	Bioluminescence	Kangas et al. 1984; Schunk et al. 2024
	ATeam	ATP concentration	FRET	Yoshida et al. 2017
	PercevalHR	ATP:ADP ratio	Fluorescence	Tantama et al. 2013; Zanotelli et al. 2018; Zhang and Reinhart-King 2023
	iATPSnFR2	ATP concentration	Fluorescence	Marvin et al. 2024
Mitochondrial activity	TMRM	Mitochondrial membrane potential	Fluorescence	Wu et al. 2021; Mosier et al. 2023
	SIR-PDI-OH	Mitochondrial membrane potential	Near-infrared fluorescence	Peng et al. 2024
Glycolysis	2-NBDG	Glucose uptake	Fluorescence	Zou et al. 2005; Wang et al. 2023
Lipid metabolism	BODIPY	Lipid distribution and oxidation	Fluorescence	Madonna et al. 2021; Nóbrega-Pereira et al. 2023
Reactive oxygen species	PeroxiSPY	Peroxisomal redox status	Fluorescence	Wanders et al. 2023; Korotkova et al. 2024

(ATP) Adenosine triphosphate, (ADP) adenosine diphosphate, (FRET) fluorescence resonance energy transfer, (BODIPY) boron dipyrromethene, (2-NBDG) 2-(*N*-(7-nitrobenz-2-oxa-1,3-diazol-4-yl)amino)-2-deoxyglucose glucose analog.

minimization of bioenergetic costs drives leader–follower behavior during collective migration (Zanotelli et al. 2018; Zhang and Reinhart-King 2023). iATPSnFR2 is a single-fluorophore ATP sensor based on circularly permuted superfolder green fluorescent protein (cpSFGFP) (Marvin et al. 2024). It undergoes a fluorescence shift upon binding ATP, enabling real-time measurement of ATP levels with reduced pH sensitivity compared to earlier sensors. While it offers a more straightforward design and reliable ratiometric measurements, it requires careful calibration for absolute ATP quantification. This new probe will allow for more precise measurement of energy utilization in subcellular compartments, such as at FAs and in mitochondria, during migration in different mechanical environments.

Mitochondrial membrane potential is a crucial indicator of mitochondrial health and function, and it can be assessed using specific fluorescent probes. Tetramethylrhodamine methyl ester (TMRM) is a cationic dye that accumulates in active mitochondria in a membrane potential-dependent manner, making it a widely used tool for detecting mitochondrial depolarization. It enables real-time, quantitative analysis of mitochondrial function through microscopy (Mosier et al. 2023). However, it is sensitive to dye concentration and requires careful optimization to avoid quenching effects. TMRM staining has been used to show increased oxidative phosphorylation (OXPHOS) with increasing cell elongation and matrix density as well as more trafficking of active mitochondria to the front of cells migrating through confined spaces, when compared to unconfined cells, resulting in increased migration speed (Wu et al. 2021; Mosier et al. 2023). SIR-PDI-OH is a recently developed near-infrared fluorogenic probe that explicitly targets mitochondria and provides enhanced photostability and sensitivity for long-term imaging (Peng et al. 2024). Unlike TMRM, it allows for deeper tissue penetration and a better signal-to-noise ratio, making it particularly useful for in vivo mitochondrial imaging. This probe, in combination with intravital imaging, will enable the exploration of mechanometabolism during cell migration in a true physiological context.

Glycolysis is a key metabolic pathway for glucose breakdown, and glycolytic activity can be monitored using specific probes. 2-(*N*-(7-nitrobenz-2-oxa-1,3-diazol-4-yl)amino)-2-deoxyglucose glucose analog (2-NBDG) is a fluorescent glucose analog that is transported into cells via glucose transporters but cannot be further metabolized. It measures glucose uptake in live cells using microscopy or flow cytometry, offering real-time, nonradioactive, and quantitative analysis (Zou et al. 2005). However, it does not provide insights into downstream glycolytic flux. This probe was used to show that glucose uptake was increased for cells collectively migrating through smaller confined channels, which required higher actomyosin contractility and tensile forces (Wang et al. 2023).

Lipid metabolism and oxidative stress can also be assessed using specific fluorescent probes. Boron dipyrromethene (BODIPY) is a versatile lipophilic dye that integrates into lipid structures, allowing visualization of lipid droplets, membrane dynamics, and lipid oxidation. It is widely used in live-cell imaging and flow cytometry due to its high photostability and tunable spectral properties (Madonna et al. 2021). This dye was used to show that low-density lipoprotein cholesterol induces triple-negative breast cancer migration through affecting mitochondrial metabolism (Nóbrega-Pereira et al. 2023). However, while BODIPY effectively tracks lipid distribution and turnover, it does not provide direct insights into lipid metabolism or oxidation rates. PeroxiSpy is a new specialized probe designed to detect peroxisomal activity and reactive oxygen species (ROS). It selectively localizes to peroxisomes, enabling the real-time assessment of peroxisomal redox status and oxidative stress in live cells (Korotkova et al. 2024). Unlike general ROS indicators, PeroxiSpy provides targeted peroxisomal measurements, which will allow for the direct study of peroxisomal diseases such as Zellweger syndrome, marked by cortical neuronal migration defects (Wanders et al. 2023).

Tools for Measuring Cell Metabolism

In addition to fluorescent probes, several instrumental techniques provide complementary methods for measuring metabolism. The Seahorse XF Analyzer measures oxygen consumption rate (OCR) and extracellular acidification rate (ECAR) to assess mitochondrial respiration and glycolysis, respectively (Dranka et al. 2011). This tool evaluates metabolic reprogramming in live cells with noninvasive, real-time measurements and high throughput, although it requires specialized equipment and data interpretation. Using the Seahorse XF Analyzer along with treatments with carbonyl cyanide 4-(trifluoromethoxy) phenylhydrazone (FCCP) to uncouple OXPHOS from the electron transport chain, it has been shown how mitochondrial respiration regulates human mesenchymal cell wound healing migration and micromotion (Lee et al. 2016).

Mass spectrometry (MS) quantifies metabolic intermediates, lipid profiles, and isotope-labeled tracers with high sensitivity, offering comprehensive metabolic flux analysis and pathway profiling (Rampler et al. 2021). Although highly specific and sensitive, it involves expensive instrumentation and complex sample preparation. This tool has been used to study the metabolic heterogeneity of fast and slow subpopulations of glioblastoma cells identifying significant differences in lipid profiles, suggesting fast-migrating cells consume more lipids to promote migration as compared to slow-migrating cells (Cheng et al. 2024).

Nuclear magnetic resonance (NMR) detects metabolic intermediates using magnetic properties of nuclei in a magnetic field. It identifies and quantifies metabolites and metabolic fluxes in live cells or tissues, providing noninvasive structural and kinetic information (Chadeau-Hyam et al. 2010). However, it has lower sensitivity compared to MS and requires specialized equipment. This technique has been used to identify hypoxanthine as an oncometabolite that could increase the migratory and invasive abilities of less metastatic breast cancer cells through induction of the epithelial-to-mesenchymal transition, proangiogenesis gene expression, and affecting lipid metabolism (Shakartalla et al. 2024).

The Role of Mitochondria in Extracellular Mechanosensing and Migration

Metabolism is not solely dictated by intrinsic cellular programming, but is profoundly influenced by extrinsic mechanical forces and environmental cues, particularly in the context of tissue stiffness and cell migration. Mitochondrial activity is profoundly influenced by cellular interactions with the tissue microenvironment, particularly the mechanical properties of the surrounding ECM. For example, metastatic breast cancer cells grown on softer substrates demonstrate a preference for mitochondrial fission, leading to a fragmented mitochondrial network through dynamin-related protein 1 (DRP1) and nuclear factor erythroid 2-related factor 2 (NRF2) signaling (Romani et al. 2022). The authors tied these stiffness-induced mitochondrial dynamics to cancer cell dormancy in soft metastatic organs by showing in a mouse model that inhibition of DRP1 and NRF2 could restore chemotherapeutic sensitivity and prevented disseminated cell activation. Similarly, mesenchymal stem cells seeded on stiffer PA gels exhibit increased mitochondrial fusion mediated through kindlin-2, PINCH-1, and DRP1 inhibition (Chen et al. 2021). Pancreatic adenocarcinoma cells cultured on stiff matrices also exhibit a notable increase in mitochondrial fusion, reinforcing the idea that ECM stiffness drives mitochondrial adaptations that support tumor progression (Papalazarou et al. 2020). In contrast, human mammary epithelial cells (MECs) seeded on stiffer PA gels exhibit increased mitochondrial fragmentation, related to a mitochondrial stress response, suggesting cellular mechanosensing is tied to mitochondrial dysfunction observed in aging and cancer (Tharp et al. 2021). These findings highlight the critical role of ECM mechanics in regulating mitochondrial dynamics, with softer environments generally favoring fragmentation and glycolytic metabolism, while stiffer matrices promote fusion and metabolic adaptation to stress.

Beyond studying mitochondrial fusion and fission dynamics in response to mechanical cues, there has also been a specific focus on connecting mitochondrial intracellular transportation with migratory ability. In highly motile and metastatic cells, mitochondria responsible for ATP generation are actively transported to the leading edge of membrane protrusions in migrating cells (Cunniff et al. 2016). This mitochondrial trafficking is facilitated by Miro1, which regulates mitochondrial positioning to support cellular movement (Desai et al. 2013; Schuler et al. 2017). The transport is facilitated by the upregulation of DRP1. In immune cells (Simula et al. 2018) and cancer cells (Desai et al. 2013), DRP1 enhances mitochondrial fission, allowing for greater intracellular mitochondrial motility. Further knowledge of mitochondrial dynamics and trafficking in response to their extracellular mechanical environment will reveal new targets for controlling cell migration.

Shifting Metabolic Profiles in Studying Mechanics and Migration

While mitochondrial activity is integral to the cellular response to mechanical stimuli and contributes significantly to cell motility, limiting the metabolic scope to the mitochondria disregards other key players in the metabolic profile that governs cellular behavior. Some evidence suggests that cells transitioning from stiff to soft environments preferentially engage in OXPHOS for energy production (Romani et al. 2022), whereas others propose a shift toward glycolysis under similar conditions (Papalazarou et al. 2020). It has additionally been shown that laminar flow promotes OXPHOS, while oscillatory flow induces a glycolytic shift in endothelial (Doddaballapur et al. 2015; Feng et al. 2017) and breast cancer cells (Park et al. 2022). The effects of mechanical stimuli on glutamine and fatty acid metabolism have been reviewed to a much lesser extent but have been studied in CAFs and adipocytes. YAP/TAZ mechanosignaling in CAFs has been implicated in the upregulation of glutamine metabolism in response to increased ECM stiffness (Yamaguchi and Taouk 2020). Additionally, fatty acid metabolism is upregulated in adipocytes cultured on stiffer matrices, indicating a mechanosensitive metabolic adaptation (Ky et al. 2023). These findings underscore the complexity of mechanotransduction-driven metabolic reprogramming and highlight the necessity for further investigation into the interplay between ECM stiffness, metabolic flexibility, and cellular function.

 Cite this article as *Cold Spring Harb Perspect Biol* doi: 10.1101/cshperspect.a041755

Cell migration requires a finely tuned balance between energy production and consumption to sustain movement. Cells frequently upregulate glycolysis to meet the dynamic energy demands of migration, which facilitates rapid ATP generation to support cytoskeletal remodeling and adhesion turnover (Danhier et al. 2017). Furthermore, glycolysis-derived lactate can be taken up by neighboring cells, thereby contributing to a metabolically supportive microenvironment that sustains collective cell migration (Marelli-Berg and Jangani 2018). Although glycolysis is a primary energy source for motile cells, OXPHOS also contributes to cell migration, particularly with respect to maintaining long-term energy homeostasis and sustaining ROS signaling, which plays a crucial role in FA turnover and directional migration (Yu et al. 2017). The metabolic preferences of migrating cells are tightly modulated by mechanotransduction and bioenergetic signaling networks, including pathways such as FA kinase, PI3K/Akt, and adenosine monophosphate-activated protein kinase (AMPK), which integrate mechanical stimuli and energetic demands to fine-tune cellular motility (Boscaro et al. 2020). These findings underscore the intricate interplay between metabolic flexibility and mechanotransduction in regulating cell migration, highlighting how dynamic shifts in metabolic profile enable cells to adapt their bioenergetic state.

CONCLUDING REMARKS

The physical properties of cells, their surrounding mechanical environment, and, as we are more clearly coming to realize, their metabolic response to different mechanical properties all play a crucial role in the ability of cells to move. Numerous approaches have emerged to study cellular stiffness, extracellular stiffness, and mechanometabolism in the context of cell migration, so that more researchers can continue to tackle the open questions in the field. Beyond understanding the correlation of cellular and extracellular stiffness with cell migratory ability, we need to identify how to harness this information to direct or prevent cell migration depending on the biological context. While we have covered some of the tools being used in the laboratory for adjusting cell and tissue stiffness, there is a need to further explore how to transform these tools for a clinical setting to adjust mechanical properties in the body to stop cancer cell migration leading to metastasis or redirect migration during development or wound healing to improve health outcomes. Additionally, there is a need to further elucidate the role of the cellular component in the extracellular environment in remodeling tissue stiffness. As just one example, the field's understanding of CAFs is continuing to evolve as we learn more about particular subpopulations of CAFs that play different roles in tissue remodeling, immune activation, and immune suppression in the tumor microenvironment. Further understanding of how these cells and other stromal components reshape the mechanical environment is necessary to fully comprehend how cells move through their surroundings. Other areas of interest in the field of mechanobiology and migration include the exploring the feedback systems between cell and tissue stiffness, expanding on the concept and mechanisms of mechanical memory, where cell behavior and movement is affected by "priming" in different mechanical environments, and targeting the intersection of mechanics, metabolism, and movement to adjust cell behavior in healthy and diseased states. The new tools and techniques discussed in this work will enable many others to continue researching cell migration through a mechanobiology lens to improve human health. (For a glossary of terms, abbreviations, and acronyms, see Box 2.)

ACKNOWLEDGMENTS

We acknowledge the many researchers in the mechanobiology field whose work was not discussed and cited in this work due to limitations of space and scope. This work is supported in part by the National Institutes of Health (NIH) through GM131178 and CA271715. K.M.Y. is supported by the NIH National Cancer Institute (NCI) F32CA281162 fellowship. S.L. is supported by the National Science Foundation Graduate Research Fellowship Program.

BOX 2. GLOSSARY OF TERMS, ABBREVIATIONS, AND ACRONYMS

2-NBDG—2-(*N*-(7-nitrobenz-2-oxa-1,3-diazol-4-yl)amino)-2-deoxyglucose glucose analog
ADP—adenosine diphosphate
AFM—atomic force microscopy
AGE—advanced glycation end-product
AKT—serine/threonine protein kinase
AMPK—adenosine monophosphate-activated protein kinase
ATeam—FRET-based ATP detection biosensor
ATP—adenosine triphosphate
ATPlite—luminescence-based ATP detection assay
Bleb—blebbistatin
BODIPY—boron dipyrromethene
CAF—cancer-associated fibroblast
Calphostin C—protein kinase C inhibitor
Calyculin A—protein phosphatase inhibitor
cpSFGFP—circularly permuted superfolder green fluorescent protein
DC—deformability cytometry
DRP1—dynamin-related protein 1 mitochondrial fission protein
ECAR—extracellular acidification rate
ECM—extracellular matrix
EGFR—epidermal growth factor receptor
FCCP—carbonyl cyanide 4-(trifluoromethoxy) phenylhydrazone
FRET—fluorescence resonance energy transfer
GAG—glycosaminoglycan
HAPLN1—hyaluronan and proteoglycan link protein 1
HBL-100—healthy human breast epithelial cell line
HCC—hepatocellular carcinoma
iATPSnFR2—intensity-based ATP sensing fluorescent reporter
Jasp—jasplakinolide
JNK—c-Jun amino-terminal kinase
Latrunculin A—F-actin inhibitor
LOX—lysyl oxidase
MAPK1—mitogen-activated protein kinase 1, also known as ERK or p38
MCF10a—human breast nontumorigenic cell line
MCF7—human breast adenocarcinoma cell line
MDA-MB-231—human breast adenocarcinoma cell line
MEC—mammary epithelial cell
MKK4—mitogen-activated protein kinase kinase 4
MKK7—mitogen-activated protein kinase kinase 7
ML-7—myosin light chain kinase inhibitor
MS—mass spectrometry
Narc—narciclasine
NMR—nuclear magnetic resonance
Nocodazole—microtubule inhibitor
NRF2—nuclear factor erythroid 2-related factor 2
OCR—oxygen consumption rate
OT—optical tweezers
OXPHOS—oxidative phosphorylation
p38—mitogen-activated protein kinase 1
PA—polyacrylamide
Paclitaxel—microtubule stabilizer

Continued

PD98059—MAPK1 inhibitor
PercevalHR—fluorescent cytoplasmic ATP:ADP ratio biosensor
PeroxiSPY—peroxisome specific fluorescent probe
PI3K—phosphoinositide-3-kinase
PINCH-1—LIM zinc finger domain-containing 1—adaptor protein
PPP2R1B—protein phosphatase 2 scaffold subunit Aβ
ROCK—rho-associated coiled-coil-containing protein kinase
ROS—reactive oxygen species
RT-DC—real-time deformability cytometry
RT-qPCR—real-time quantitative polymerase chain reaction
SEC—securinine
SERPINE1—serpin family E member 1
SHG—second harmonic generation
SiR-PDI-OH—zwitterionic fluorescent mitochondrial membrane potential reversible probe
STIFMap—spatially transformed inferential force map
SP600125—JNK inhibitor
SW480—human colorectal adenocarcinoma cell line
T24—human bladder transitional cell carcinoma cell line
TFM—traction force microscopy
TGM2—transglutaminase 2
TMRM—tetramethylrhodamine methyl ester
TSA—trichostatin A
VIM—vimentin
VSMC—vascular smooth muscle cell
YAP/TAZ—yes-associated transcription factor/WW domain-containing transcription regulator—mechanotransductive transcriptional coactivators
ZEB1—zinc finger E-box-binding homeobox 1

REFERENCES

Abarca-Ortega A, González-Bermúdez B, Plaza GR. 2024. Enhancing micropipette aspiration with artificial-intelligence analysis. *Biophys J* **123:** 2860–2868. doi:10.1016/j.bpj.2024.04.006

Abbas Y, Carnicer-Lombarte A, Gardner L, Thomas J, Brosens JJ, Moffett A, Sharkey AM, Franze K, Burton GJ, Oyen ML. 2019. Tissue stiffness at the human maternal–fetal interface. *Hum Reprod* **34:** 1999–2008. doi:10.1093/humrep/dez139

Abuhattum S, Mokbel D, Müller P, Soteriou D, Guck J, Aland S. 2022. An explicit model to extract viscoelastic properties of cells from AFM force-indentation curves. *iScience* **25:** 104016. doi:10.1016/j.isci.2022.104016

Alibert C, Goud B, Manneville JB. 2017. Are cancer cells really softer than normal cells? *Biol Cell* **109:** 167–189. doi:10.1111/boc.201600078

Arbore C, Perego L, Sergides M, Capitanio M. 2019. Probing force in living cells with optical tweezers: from single-molecule mechanics to cell mechanotransduction. *Biophys Rev* **11:** 765–782. doi:10.1007/s12551-019-00599-y

Asghari M, Ivetich SD, Aslan MK, Aramesh M, Melkonyan O, Meng Y, Xu R, Colombo M, Weiss T, Balabanov S, et al. 2024. Real-time viscoelastic deformability cytometry: high-throughput mechanical phenotyping of liquid and solid biopsies. *Sci Adv* **10:** eabj1133. doi:10.1126/sciadv.abj1133

Aumailley M. 2013. The laminin family. *Cell Adhes Migr* **7:** 48–55. doi:10.4161/cam.22826

Ayala YA, Pontes B, Ether DS, Pires LB, Araujo GR, Frases S, Romão LF, Farina M, Moura-Neto V, Viana NB, et al. 2016. Rheological properties of cells measured by optical tweezers. *BMC Biophys* **9:** 5. doi:10.1186/s13628-016-0031-4

Azubuike UF, Tanner K. 2023. Biophysical determinants of cancer organotropism. *Trends Cancer* **9:** 188–197. doi:10.1016/j.trecan.2022.11.002

Barriga EH, Franze K, Charras G, Mayor R. 2018. Tissue stiffening coordinates morphogenesis by triggering collective cell migration in vivo. *Nature* **554:** 523–527. doi:10.1038/nature25742

Bastatas L, Martinez-Marin D, Matthews J, Hashem J, Lee YJ, Sennoune S, Filleur S, Martinez-Zaguilan R, Park S. 2012. AFM nano-mechanics and calcium dynamics of prostate cancer cells with distinct metastatic potential. *Biochim Biophys Acta* **1820:** 1111–1120. doi:10.1016/j.bbagen.2012.02.006

Bell ES, Shah P, Zuela-Sopilniak N, Kim D, Varlet AA, Morival JLP, McGregor AL, Isermann P, Davidson PM, Elacqua JJ, et al. 2022. Low lamin A levels enhance confined cell migration and metastatic capacity in breast cancer. *Oncogene* **41:** 4211–4230. doi:10.1038/s41388-022-02420-9

Bevilacqua C, Sánchez-Iranzo H, Richter D, Diz-Muñoz A, Prevedel R. 2019. Imaging mechanical properties of sub-micron ECM in live zebrafish using Brillouin microscopy. *Biomed Opt Express* **10:** 1420–1431. doi:10.1364/BOE.10.001420

Binnig G, Quate CF, Gerber C. 1986. Atomic force microscope. *Phys Rev Lett* **56:** 930–933. doi:10.1103/PhysRevLett.56.930

Bongiorno T, Kazlow J, Mezencev R, Griffiths S, Olivares-Navarrete R, McDonald JF, Schwartz Z, Boyan BD, McDevitt TC, Sulchek T. 2014. Mechanical stiffness as an improved single-cell indicator of osteoblastic human mesenchymal stem cell differentiation. *J Biomech* **47:** 2197–2204. doi:10.1016/j.jbiomech.2013.11.017

Bongiorno T, Chojnowski JL, Lauderdale JD, Sulchek T. 2016. Cellular stiffness as a novel stemness marker in the corneal limbus. *Biophys J* **111:** 1761–1772. doi:10.1016/j.bpj.2016.09.005

Bongiorno T, Gura J, Talwar P, Chambers D, Young KM, Arafat D, Wang G, Jackson-Holmes EL, Qiu P, McDevitt TC, et al. 2018. Biophysical subsets of embryonic stem cells display distinct phenotypic and morphological signatures. *PLoS One* **13:** e0192631. doi:10.1371/journal.pone.0192631

Boot RC, Roscani A, van Buren L, Maity S, Koenderink GH, Boukany PE. 2023. High-throughput mechanophenotyping of multicellular spheroids using a microfluidic micropipette aspiration chip. *Lab Chip* **23:** 1768–1778. doi:10.1039/D2LC01060G

Bordeleau F, Mason BN, Lollis EM, Mazzola M, Zanotelli MR, Somasegar S, Califano JP, Montague C, LaValley DJ, Huynh J, et al. 2017. Matrix stiffening promotes a tumor vasculature phenotype. *Proc Natl Acad Sci* **114:** 492–497. doi:10.1073/pnas.1613855114

Boscaro C, Carotti M, Albiero M, Trenti A, Fadini GP, Trevisi L, Sandonà D, Cignarella A, Bolego C. 2020. Non-genomic mechanisms in the estrogen regulation of glycolytic protein levels in endothelial cells. *FASEB J* **34:** 12768–12784. doi:10.1096/fj.202001130R

Caille N, Thoumine O, Tardy Y, Meister JJ. 2002. Contribution of the nucleus to the mechanical properties of endothelial cells. *J Biomech* **35:** 177–187. doi:10.1016/S0021-9290(01)00201-9

Calò A, Romin Y, Srouji R, Zambirinis CP, Fan N, Santella A, Feng E, Fujisawa S, Turkekul M, Huang S, et al. 2020. Spatial mapping of the collagen distribution in human and mouse tissues by force volume atomic force microscopy. *Sci Rep* **10:** 15664. doi:10.1038/s41598-020-72564-9

Chadeau-Hyam M, Ebbels TMD, Brown IJ, Chan Q, Stamler J, Huang CC, Daviglus ML, Ueshima H, Zhao L, Holmes E, et al. 2010. Metabolic profiling and the metabolome-wide association study: significance level for biomarker identification. *J Proteome Res* **9:** 4620–4627. doi:10.1021/pr1003449

Chai XX, Liu J, Yu TY, Zhang G, Sun WJ, Zhou Y, Ren L, Cao HL, Yin DC, Zhang CY. 2023. Recent progress of mechanosensitive mechanism on breast cancer. *Prog Biophys Mol Biol* **185:** 1–16. doi:10.1016/j.pbiomolbio.2023.09.003

Chaki SP, Barhoumi R, Berginski ME, Sreenivasappa H, Trache A, Gomez SM, Rivera GM. 2013. Nck enables directional cell migration through the coordination of polarized membrane protrusion with adhesion dynamics. *J Cell Sci* **126:** 1637–1649. doi:10.1242/jcs.119610

Chang Y, Zhang J, Huo X, Qu X, Xia C, Huang K, Xie F, Wang N, Wei X, Jia Q. 2022. Substrate rigidity dictates colorectal tumorigenic cell stemness and metastasis via CRAD-dependent mechanotransduction. *Cell Rep* **38:** 110390. doi:10.1016/j.celrep.2022.110390

Cheburkanov V, Yakovlev VV. 2025. Brillouin microscopy analysis of the fibroblast mechanical response to substrate's stiffness. *Soft Matter* **21:** 4710–4718. doi:10.1039/D5SM00315F

Chen X, Nadiarynkh O, Plotnikov S, Campagnola PJ. 2012. Second harmonic generation microscopy for quantitative analysis of collagen fibrillar structure. *Nat Protoc* **7:** 654–669. doi:10.1038/nprot.2012.009

Chen K, Wang Y, Deng X, Guo L, Wu C. 2021. Extracellular matrix stiffness regulates mitochondrial dynamics through PINCH-1- and kindlin-2-mediated signalling. *Curr Res Cell Biol* **2:** 100008. doi:10.1016/j.crcbio.2021.100008

Chen H, Guo J, Bian F, Zhao Y. 2022. Microfluidic technologies for cell deformability cytometry. *Smart Med* **1:** e20220001. doi:10.1002/SMMD.20220001

Chen G, Deng Y, Xia B, Lv Y. 2023a. In situ regulation and mechanisms of 3D matrix stiffness on the activation and reversion of hepatic stellate cells. *Adv Healthc Mater* **12:** 2202560. doi:10.1002/adhm.202202560

Chen Y, Guo K, Jiang L, Zhu S, Ni Z, Xiang N. 2023b. Microfluidic deformability cytometry: a review. *Talanta* **251:** 123815. doi:10.1016/j.talanta.2022.123815

Chen Y, Jiang L, Zhang X, Ni Z, Xiang N. 2023c. Viscoelastic-sorting integrated deformability cytometer for high-throughput sorting and high-precision mechanical phenotyping of tumor cells. *Anal Chem* **95:** 18180–18187. doi:10.1021/acs.analchem.3c03792

Cheng KW, Su PR, Feller KJA, Chien MP, Hsu CC. 2024. Investigating the metabolic heterogeneity of cancer cells using functional single-cell selection and nLC combined with multinozzle emitter mass spectrometry. *Anal Chem* **96:** 624–629. doi:10.1021/acs.analchem.3c03688

Chrit FE, Barton J, Sulchek T, Alexeev A. 2023. Enhancing stiffness-based cell sorting using power-law fluids in ridged microchannels. *Phys Fluids* **35:** 043327. doi:10.1063/5.0145921

Chrit FE, Li P, Sulchek T, Alexeev A. 2024. Adhesion-based high-throughput label-free cell sorting using ridged microfluidic channels. *Soft Matter* **20:** 1913–1921. doi:10.1039/D3SM01117H

Colombo F, Villiou M, Taheri F, Fröhlich L, Taale M, Albert V, Jiang Q, Selhuber-Unkel C. 2023. Fluctuations of dry and total mass of cells exposed to different molecular weights of polyethylene glycol. *Adv NanoBioMed Res* **3:** 2200156. doi:10.1002/anbr.202200156

Cunniff B, McKenzie AJ, Heintz NH, Howe AK. 2016. AMPK activity regulates trafficking of mitochondria to the leading edge during cell migration and matrix invasion. *Mol Biol Cell* **27:** 2662–2674. doi:10.1091/mbc.e16-05-0286

Danhier P, Bański P, Payen VL, Grasso D, Ippolito L, Sonveaux P, Porporato PE. 2017. Cancer metabolism in space and time: beyond the Warburg effect. *Biochim Biophys Acta Bioenerg* **1858:** 556–572. doi:10.1016/j.bbabio.2017.02.001

Davidson PM, Denais C, Bakshi MC, Lammerding J. 2014. Nuclear deformability constitutes a rate-limiting step during cell migration in 3-D environments. *Cell Mol Bioeng* **7:** 293–306. doi:10.1007/s12195-014-0342-y

Davidson PM, Fedorchak GR, Mondésert-Deveraux S, Bell ES, Isermann P, Aubry D, Allena R, Lammerding J. 2019. High-throughput microfluidic micropipette aspiration device to probe time-scale dependent nuclear mechanics in intact cells. *Lab Chip* **19:** 3652–3663. doi:10.1039/C9LC00444K

Denais C, Lammerding J. 2014. Nuclear mechanics in cancer. *Adv Exp Med Biol* **773:** 435–470. doi:10.1007/978-1-4899-8032-8_20

Deng X, Xiong F, Li X, Xiang B, Li Z, Wu X, Guo C, Li X, Li Y, Li G, et al. 2018. Application of atomic force microscopy in cancer research. *J Nanobiotechnology* **16:** 102. doi:10.1186/s12951-018-0428-0

Desai SP, Bhatia SN, Toner M, Irimia D. 2013. Mitochondrial localization and the persistent migration of epithelial cancer cells. *Biophys J* **104:** 2077–2088. doi:10.1016/j.bpj.2013.03.025

Doddaballapur A, Michalik KM, Manavski Y, Lucas T, Houtkooper RH, You X, Chen W, Zeiher AM, Potente M, Dimmeler S, et al. 2015. Laminar shear stress inhibits endothelial cell metabolism via KLF2-mediated repression of PFKFB3. *Arterioscler Thromb Vasc Biol* **35:** 137–145. doi:10.1161/ATVBAHA.114.304277

Dranka BP, Benavides GA, Diers AR, Giordano S, Zelickson BR, Reily C, Zou L, Chatham JC, Hill BG, Zhang J, et al. 2011. Assessing bioenergetic function in response to oxidative stress by metabolic profiling. *Free Radic Biol Med* **51:** 1621–1635. doi:10.1016/j.freeradbiomed.2011.08.005

Efremov YM, Shimolina L, Gulin A, Ignatova N, Gubina M, Kuimova MK, Timashev PS, Shirmanova MV. 2023. Correlation of plasma membrane microviscosity and cell stiffness revealed via fluorescence-lifetime imaging and atomic force microscopy. *Cells* **12:** 2583. doi:10.3390/cells12212583

Emig R, Knodt W, Krussig MJ, Zgierski-Johnston CM, Gorka O, Groß O, Kohl P, Ravens U, Peyronnet R. 2021. Piezo1 channels contribute to the regulation of human atrial fibroblast mechanical properties and matrix stiffness sensing. *Cells* **10:** 663. doi:10.3390/cells10030663

Farahinia A, Zhang WJ, Badea I. 2021. Novel microfluidic approaches to circulating tumor cell separation and sorting of blood cells: a review. *J Sci Adv Mater Devices* **6:** 303–320. doi:10.1016/j.jsamd.2021.03.005

Feng S, Bowden N, Fragiadaki M, Souilhol C, Hsiao S, Mahmoud M, Allen S, Pirri D, Ayllon BT, Akhtar S, et al. 2017. Mechanical activation of hypoxia-inducible factor 1α drives endothelial dysfunction at atheroprone sites. *Arterioscler Thromb Vasc Biol* **37:** 2087–2101. doi:10.1161/ATVBAHA.117.309249

Fischer T, Hayn A, Mierke CT. 2020. Effect of nuclear stiffness on cell mechanics and migration of human breast cancer cells. *Front Cell Dev Biol* **8:** 393. doi:10.3389/fcell.2020.00393

Forbes JM, Yee LTL, Thallas V, Lassila M, Candido R, Jandeleit-Dahm KA, Thomas MC, Burns WC, Deemer EK, Thorpe SR, et al. 2004. Advanced glycation end product interventions reduce diabetes-accelerated atherosclerosis. *Diabetes* **53:** 1813–1823. doi:10.2337/diabetes.53.7.1813

Fu C, Dilasser F, Lin S-Z, Karnat M, Arora A, Rajendiran H, Ong HT, Mui Hoon Brenda N, Phow SW, Hirashima T, et al. 2024. Regulation of intercellular viscosity by E-cadherin-dependent phosphorylation of EGFR in collective cell migration. *Proc Natl Acad Sci* **121:** e2405560121. doi:10.1073/pnas.2405560121

Gensbittel V, Kräter M, Harlepp S, Busnelli I, Guck J, Goetz JG. 2021. Mechanical adaptability of tumor cells in metastasis. *Dev Cell* **56:** 164–179. doi:10.1016/j.devcel.2020.10.011

Gerum R, Mirzahossein E, Eroles M, Elsterer J, Mainka A, Bauer A, Sonntag S, Winterl A, Bartl J, Fischer L, et al. 2022. Viscoelastic properties of suspended cells measured with shear flow deformation cytometry. *eLife* **11:** e78823. doi:10.7554/eLife.78823

González-Bermúdez B, Kobayashi H, Abarca-Ortega A, Córcoles-Lucas M, González-Sánchez M, De la Fuente M, Guinea GV, Elices M, Plaza GR. 2022. Aging is accompanied by T-cell stiffening and reduced interstitial migration through dysfunctional nuclear organization. *Immunology* **167:** 622–639. doi:10.1111/imm.13559

Guck J, Schinkinger S, Lincoln B, Wottawah F, Ebert S, Romeyke M, Lenz D, Erickson HM, Ananthakrishnan R, Mitchell D, et al. 2005. Optical deformability as an inherent cell marker for testing malignant transformation and metastatic competence. *Biophys J* **88:** 3689–3698. doi:10.1529/biophysj.104.045476

Guo Y, Mei F, Huang Y, Ma S, Wei Y, Zhang X, Xu M, He Y, Heng BC, Chen L, et al. 2022. Matrix stiffness modulates tip cell formation through the p-PXN-Rac1-YAP signaling axis. *Bioact Mater* **7:** 364–376. doi:10.1016/j.bioactmat.2021.05.033

Hansen E, Holaska JM. 2023. The nuclear envelope and metastasis. *Oncotarget* **14:** 317–320. doi:10.18632/oncotarget.28375

Heo SJ, Song KH, Thakur S, Miller LM, Cao X, Peredo AP, Seiber BN, Qu F, Driscoll TP, Shenoy VB, et al. 2020. Nuclear softening expedites interstitial cell migration in fibrous networks and dense connective tissues. *Sci Adv* **6:** eaax5083. doi:10.1126/sciadv.aax5083

Herbig M, Tessmer K, Nötzel M, Nawaz AA, Santos-Ferreira T, Borsch O, Gasparini SJ, Guck J, Ader M. 2022. Label-free imaging flow cytometry for analysis and sorting of enzymatically dissociated tissues. *Sci Rep* **12:** 963. doi:10.1038/s41598-022-05007-2

Herchenhan A, Uhlenbrock F, Eliasson P, Weis M, Eyre D, Kadler KE, Magnusson SP, Kjaer M. 2015. Lysyl oxidase activity is required for ordered collagen fibrillogenesis by tendon cells. *J Biol Chem* **290:** 16440–16450. doi:10.1074/jbc.M115.641670

Honarmandi P, Lee H, Lang MJ, Kamm RD. 2011. A microfluidic system with optical laser tweezers to study mechanotransduction and focal adhesion recruitment. *Lab Chip* **11:** 684–694. doi:10.1039/C0LC00487A

Hu S, Ye J, Zhao Y, Zhu C. 2022. Advanced optical tweezers on cell manipulation and analysis. *Eur Phys J Plus* **137:** 1024. doi:10.1140/epjp/s13360-022-03190-9

Huveneers S, Phng L-K. 2024. Endothelial cell mechanics and dynamics in angiogenesis. *Curr Opin Cell Biol* **91:** 102441. doi:10.1016/j.ceb.2024.102441

Isomursu A, Park KY, Hou J, Cheng B, Mathieu M, Shamsan GA, Fuller B, Kasim J, Mahmoodi MM, Lu TJ, et al. 2022. Directed cell migration towards softer environments. *Nat Mater* **21:** 1081–1090. doi:10.1038/s41563-022-01294-2

Jang M, Oh SW, Lee Y, Kim JY, Ji ES, Kim P. 2022. Targeting extracellular matrix glycation to attenuate fibroblast activation. *Acta Biomater* **141:** 255–263. doi:10.1016/j.actbio.2022.01.040

Janmey PA, Miller RT. 2011. Mechanisms of mechanical signaling in development and disease. *J Cell Sci* **124:** 9–18. doi:10.1242/jcs.071001

Janmey PA, Fletcher DA, Reinhart-King CA. 2020. Stiffness sensing by cells. *Physiol Rev* **100:** 695–724. doi:10.1152/physrev.00013.2019

Ji C, Huang Y. 2023. Durotaxis and negative durotaxis: where should cells go? *Commun Biol* **6:** 1–7. doi:10.1038/s42003-023-05554-y

Kangas L, Grönroos M, Nieminen AL. 1984. Bioluminescence of cellular ATP: a new method for evaluating cytotoxic agents in vitro. *Med Biol* **62:** 338–343.

Korotkova D, Borisyuk A, Guihur A, Bardyn M, Kuttler F, Reymond L, Schuhmacher M, Amen T. 2024. Fluorescent fatty acid conjugates for live cell imaging of peroxisomes. *Nat Commun* **15:** 4314. doi:10.1038/s41467-024-48679-2

Kothapalli D, Liu SL, Bae YH, Monslow J, Xu T, Hawthorne EA, Byfield FJ, Castagnino P, Rao S, Rader DJ, et al. 2012. Cardiovascular protection by ApoE and ApoE-HDL linked to suppression of ECM gene expression and arterial stiffening. *Cell Rep* **2:** 1259–1271. doi:10.1016/j.celrep.2012.09.018

Kular JK, Basu S, Sharma RI. 2014. The extracellular matrix: structure, composition, age-related differences, tools for analysis and applications for tissue engineering. *J Tissue Eng* **5:** 2041731414557112. doi:10.1177/2041731414557112

Ky A, McCoy AJ, Flesher CG, Friend NE, Li J, Akinleye K, Patsalis C, Lumeng CN, Putnam AJ, O'Rourke RW. 2023. Matrix density regulates adipocyte phenotype. *Adipocyte* **12:** 2268261. doi:10.1080/21623945.2023.2268261

Lachowski D, Matellan C, Gopal S, Cortes E, Robinson BK, Saiani A, Miller AF, Stevens MM, del Río Hernández AE. 2022. Substrate stiffness-driven membrane tension modulates vesicular trafficking via Caveolin-1. *ACS Nano* **16:** 4322–4337. doi:10.1021/acsnano.1c10534

Lampi MC, Reinhart-King CA. 2018. Targeting extracellular matrix stiffness to attenuate disease: from molecular mechanisms to clinical trials. *Sci Transl Med* **10:** eaao0475. doi:10.1126/scitranslmed.aao0475

Lange JR, Fabry B. 2013. Cell and tissue mechanics in cell migration. *Exp Cell Res* **319:** 2418–2423. doi:10.1016/j.yexcr.2013.04.023

Leblanc-Hotte A, Audiger C, Chabot-Roy G, Lombard-Vadnais F, Delisle JS, Peter YA, Lesage S. 2023. Immature and mature bone marrow-derived dendritic cells exhibit distinct intracellular mechanical properties. *Sci Rep* **13:** 1967. doi:10.1038/s41598-023-28625-w

Lee LM, Liu AP. 2014. The application of micropipette aspiration in molecular mechanics of single cells. *J Nanotechnol Eng Med* **5:** 0408011–0408016. doi:10.1115/1.4029936

Lee YW, Tung TH, Gomez S, Lo CM. 2016. Mitochondrial respiration regulates proliferation, micromotion, and wound healing migration of human mesenchymal stem cells. *FASEB J* **30:** 1300.10. doi:10.1096/fasebj.30.1_supplement.1300.10

Lee M, Jeong H, Lee C, Lee MJ, Delmo BR, Heo WD, Shin JH, Park Y. 2024. High-resolution assessment of multidimensional cellular mechanics using label-free refractive-index traction force microscopy. *Commun Biol* **7:** 1–11. doi:10.1038/s42003-024-05788-4

Li Q, Chen Z, Zhang Y, Ding S, Ding H, Wang L, Xie Z, Fu Y, Wei M, Liu S, et al. 2023. Imaging cellular forces with photonic crystals. *Nat Commun* **14:** 7369. doi:10.1038/s41467-023-43090-9

Liu CY, Lin HH, Tang MJ, Wang YK. 2015. Vimentin contributes to epithelial–mesenchymal transition cancer cell mechanics by mediating cytoskeletal organization and focal adhesion maturation. *Oncotarget* **6:** 15966–15983. doi:10.18632/oncotarget.3862

Liu SB, Ikenaga N, Peng Z-W, Sverdlov DY, Greenstein A, Smith V, Schuppan D, Popov Y. 2016. Lysyl oxidase activity contributes to collagen stabilization during liver fibrosis progression and limits spontaneous fibrosis reversal in mice. *FASEB J* **30:** 1599–1609. doi:10.1096/fj.14-268425

Liu Z, Lee SJ, Park S, Konstantopoulos K, Glunde K, Chen Y, Barman I. 2020. Cancer cells display increased migration and deformability in pace with metastatic progression. *FASEB J* **34:** 9307–9315. doi:10.1096/fj.202000101RR

Liu Z, Mo H, Liu R, Niu Y, Chen T, Xu Q, Tu K, Yang N. 2021. Matrix stiffness modulates hepatic stellate cell activation into tumor-promoting myofibroblasts via E2F3-dependent signaling and regulates malignant progression. *Cell Death Dis* **12:** 1–9. doi:10.1038/s41419-020-03229-8

Liu Q, Muralidharan A, Saateh A, Ding Z, ten Dijke P, Boukany PE. 2022. A programmable multifunctional 3D cancer cell invasion micro platform. *Small* **18:** 2107757. doi:10.1002/smll.202107757

Liu W, Tang J, Gao W, Sun J, Liu G, Zhou J. 2024. PPP2R1B abolishes colorectal cancer liver metastasis and sensitizes Oxaliplatin by inhibiting MAPK/ERK signaling pathway. *Cancer Cell Int* **24:** 90. doi:10.1186/s12935-024-03273-w

Lu J, Doyle AD, Shinsato Y, Wang S, Bodendorfer MA, Zheng M, Yamada KM. 2020. Basement membrane regulates fibronectin organization using sliding focal adhesions driven by a contractile winch. *Dev Cell* **52:** 631–646.e4. doi:10.1016/j.devcel.2020.01.007

Luo Q, Wu K, Zhang B, Song G. 2015. Mechano growth factor E peptide promotes rat bone marrow-derived mesenchymal stem cell migration through CXCR4-ERK1/2. *Growth Factors* **33:** 210–219.

Luo Q, Kuang D, Zhang B, Song G. 2016. Cell stiffness determined by atomic force microscopy and its correlation with cell motility. *Biochim Biophys Acta* **1860:** 1953–1960. doi:10.1016/j.bbagen.2016.06.010

Madonna MC, Duer JE, Lee JV, Williams J, Avsaroglu B, Zhu C, Deutsch R, Wang R, Crouch BT, Hirschey MD, et al. 2021. In vivo optical metabolic imaging of long-chain fatty acid uptake in orthotopic models of triple-negative breast cancer. *Cancers (Basel)* **13:** 148. doi:10.3390/cancers13010148

Mai Z, Lin Y, Lin P, Zhao X, Cui L. 2024. Modulating extracellular matrix stiffness: a strategic approach to boost cancer immunotherapy. *Cell Death Dis* **15:** 1–16. doi:10.1038/s41419-024-06697-4

Marelli-Berg FM, Jangani M. 2018. Metabolic regulation of leukocyte motility and migration. *J Leukoc Biol* **104:** 285–293. doi:10.1002/JLB.1MR1117-472R

Marhuenda E, Fabre C, Zhang C, Martin-Fernandez M, Iskratsch T, Saleh A, Bauchet L, Cambedouzou J, Hugnot JP, Duffau H, et al. 2021. Glioma stem cells invasive phenotype at optimal stiffness is driven by MGAT5 dependent mechanosensing. *J Exp Clin Cancer Res* **40:** 139. doi:10.1186/s13046-021-01925-7

Marvin JS, Kokotos AC, Kumar M, Pulido C, Tkachuk AN, Yao JS, Brown TA, Ryan TA. 2024. iATPSnFR2: a high-dynamic-range fluorescent sensor for monitoring intracellular ATP. *Proc Natl Acad Sci* **121:** e2314604121. doi:10.1073/pnas.2314604121

Mathieu M, Isomursu A, Ivaska J. 2024. Positive and negative durotaxis—mechanisms and emerging concepts. *J Cell Sci* **137:** jcs261919. doi:10.1242/jcs.261919

Maxwell C, Roskelley C. 2014. *Genomic instability and cancer metastasis: mechanisms, emerging themes, and novel therapeutic strategies*. Springer, New York.

Mehta P, Rahman Z, ten Dijke P, Boukany PE. 2022. Microfluidics meets 3D cancer cell migration. *Trends Cancer* **8:** 683–697. doi:10.1016/j.trecan.2022.03.006

Mietke A, Otto O, Girardo S, Rosendahl P, Taubenberger A, Golfier S, Ulbricht E, Aland S, Guck J, Fischer-Friedrich E. 2015. Extracting cell stiffness from real-time deformability cytometry: theory and experiment. *Biophys J* **109:** 2023–2036. doi:10.1016/j.bpj.2015.09.006

Mihai C, Bao S, Lai JP, Ghadiali SN, Knoell DL. 2012. PTEN inhibition improves wound healing in lung epithelia through changes in cellular mechanics that enhance migration. *Am J Physiol Lung Cell Mol Physiol* **302:** L287–L299. doi:10.1152/ajplung.00037.2011

Mokbel M, Mokbel D, Mietke A, Träber N, Girardo S, Otto O, Guck J, Aland S. 2017. Numerical simulation of real-time deformability cytometry to extract cell mechanical properties. *ACS Biomater Sci Eng* **3:** 2962–2973. doi:10.1021/acsbiomaterials.6b00558

Mosier JA, Fabiano ED, Ludolph CM, White AE, Reinhart-King CA. 2023. Confinement primes cells for faster migration by polarizing active mitochondria. *Nanoscale Adv* **6:** 209–220. doi:10.1039/D3NA00478C

Mukherjee A, Barai A, Singh RK, Yan W, Sen S. 2020. Nuclear plasticity increases susceptibility to damage during confined migration. *PLoS Comput Biol* **16:** e1008300. doi:10.1371/journal.pcbi.1008300

Nawaz AA, Urbanska M, Herbig M, Nötzel M, Kräter M, Rosendahl P, Herold C, Toepfner N, Kubánková M, Goswami R, et al. 2020. Intelligent image-based deformation-assisted cell sorting with molecular specificity. *Nat Methods* **17:** 595–599. doi:10.1038/s41592-020-0831-y

Nehwa FJ, Matsui TS, Honghan L, Matsunaga D, Sakaguchi Y, Deguchi S. 2020. Multi-well plate cell contraction assay detects negatively correlated cellular responses to pharmacological inhibitors in contractility and migration. *Biochem Biophys Res Commun* **521:** 527–532. doi:10.1016/j.bbrc.2019.10.160

Nóbrega-Pereira S, Santos F, Oliveira Santos M, Serafim TL, Lopes AP, Coutinho D, Carvalho FS, Domingues RM, Domingues P, Bernardes de Jesus B, et al. 2023. Mitochondrial metabolism drives low-density lipoprotein-induced breast cancer cell migration. *Cancer Res Commun* **3:** 709–724. doi:10.1158/2767-9764.CRC-22-0394

Nussenzveig HM. 2018. Cell membrane biophysics with optical tweezers. *Eur Biophys J* **47:** 499–514. doi:10.1007/s00249-017-1268-9

Otto O, Rosendahl P, Mietke A, Golfier S, Herold C, Klaue D, Girardo S, Pagliara S, Ekpenyong A, Jacobi A, et al. 2015. Real-time deformability cytometry: on-the-fly cell mechanical phenotyping. *Nat Methods* **12:** 199–202. doi:10.1038/nmeth.3281

Papalazarou V, Zhang T, Paul NR, Juin A, Cantini M, Maddocks ODK, Salmeron-Sanchez M, Machesky LM. 2020. The creatine-phosphagen system is mechanoresponsive in pancreatic adenocarcinoma and fuels invasion and metastasis. *Nat Metab* **2:** 62–80. doi:10.1038/s42255-019-0159-z

Park PW. 2022. Extracellular matrix: surface proteoglycans. In *Encyclopedia of respiratory medicine*, 2nd ed. (ed. Janes SM), pp. 157–163. Academic, Oxford.

Park HA, Brown SR, Jansen J, Dunn T, Scott M, Mnatsakanyan N, Jonas EA, Kim Y. 2022. Fluid shear stress enhances proliferation of breast cancer cells via downregulation of the c-subunit of the F1FO ATP synthase. *Biochem Biophys Res Commun* **632:** 173–180. doi:10.1016/j.bbrc.2022.09.084

Peng F, Ai X, Bu X, Sun J, Zhao Z, Yang Z, Qin X, Gao B. 2024. Dual-modality imaging with a zwitterionic fluorescent probe for reversible monitoring of mitochondrial membrane potential dynamics. *ACS Appl Mater Interfaces* **16:** 47216–47225. doi:10.1021/acsami.4c09001

Pillarisetti A, Desai JP, Ladjal H, Schiffmacher A, Ferreira A, Keefer CL. 2011. Mechanical phenotyping of mouse embryonic stem cells: increase in stiffness with differentiation. *Cell Reprogram* **13:** 371–380. doi:10.1089/cell.2011.0028

Prevedel R, Diz-Muñoz A, Ruocco G, Antonacci G. 2019. Brillouin microscopy: an emerging tool for mechanobiology. *Nat Methods* **16:** 969–977. doi:10.1038/s41592-019-0543-3

Puleo JI, Parker SS, Roman MR, Watson AW, Eliato KR, Peng L, Saboda K, Roe DJ, Ros R, Gertler FB, et al. 2019. Mechanosensing during directed cell migration requires dynamic actin polymerization at focal adhesions. *J Cell Biol* **218:** 4215–4235. doi:10.1083/jcb.201902101

Rampler E, Abiead YE, Schoeny H, Rusz M, Hildebrand F, Fitz V, Koellensperger G. 2021. Recurrent topics in mass spectrometry-based metabolomics and lipidomics—standardization, coverage, and throughput. *Anal Chem* **93:** 519–545. doi:10.1021/acs.analchem.0c04698

Rianna C, Radmacher M. 2016. Cell mechanics as a marker for diseases: biomedical applications of AFM. *AIP Conf Proc* **1760:** 020057. doi:10.1063/1.4960276

Robert-Paganin J, Pylypenko O, Kikuti C, Sweeney HL, Houdusse A. 2020. Force generation by myosin motors: a structural perspective. *Chem Rev* **120:** 5–35. doi:10.1021/acs.chemrev.9b00264

Roberts AB, Zhang J, Raj Singh V, Nikolić M, Moeendarbary E, Kamm RD, So PTC, Scarcelli G. 2021. Tumor cell nuclei soften during transendothelial migration. *J Biomech* **121:** 110400. doi:10.1016/j.jbiomech.2021.110400

Romani P, Nirchio N, Arboit M, Barbieri V, Tosi A, Michielin F, Shibuya S, Benoist T, Wu D, Hindmarch CCT, et al. 2022. Mitochondrial fission links ECM mechanotransduction to metabolic redox homeostasis and metastatic chemotherapy resistance. *Nat Cell Biol* **24:** 168–180. doi:10.1038/s41556-022-00843-w

Saraswathibhatla A, Indana D, Chaudhuri O. 2023. Cell–extracellular matrix mechanotransduction in 3D. *Nat Rev Mol Cell Biol* **24:** 495–516. doi:10.1038/s41580-023-00583-1

Schlüßler R, Möllmert S, Abuhattum S, Cojoc G, Müller P, Kim K, Möckel C, Zimmermann C, Czarske J, Guck J. 2018. Mechanical mapping of spinal cord growth and repair in living zebrafish larvae by brillouin imaging. *Biophys J* **115:** 911–923. doi:10.1016/j.bpj.2018.07.027

Schuler MH, Lewandowska A, Caprio GD, Skillern W, Upadhyayula S, Kirchhausen T, Shaw JM, Cunniff B. 2017. Miro1-mediated mitochondrial positioning shapes intracellular energy gradients required for cell migration. *Mol Biol Cell* **28:** 2159–2169. doi:10.1091/mbc.e16-10-0741

Schunk CT, Wang W, Sabo LN, Taufalele PV, Reinhart-King CA. 2024. Matrix stiffness increases energy efficiency of endothelial cells. *Matrix Biol* **133:** 77–85. doi:10.1016/j.matbio.2024.08.004

Shakartalla SB, Ashmawy NS, Semreen MH, Fayed B, Al Shareef ZM, Jayakumar MN, Ibrahim S, Rahmani M, Hamdy R, Soliman SSM. 2024. 1H-NMR metabolomics analysis identifies hypoxanthine as a novel metastasis-associated metabolite in breast cancer. *Sci Rep* **14:** 253. doi:10.1038/s41598-023-50866-y

Sherlock BE, Chen J, Mansfield JC, Green E, Winlove CP. 2021. Biophotonic tools for probing extracellular matrix mechanics. *Matrix Biol Plus* **12:** 100093. doi:10.1016/j.mbplus.2021.100093

Simula L, Pacella I, Colamatteo A, Procaccini C, Cancila V, Bordi M, Tregnago C, Corrado M, Pigazzi M, Barnaba V, et al. 2018. Drp1 controls effective T cell immune-surveillance by regulating T cell migration, proliferation, and cMyc-dependent metabolic reprogramming. *Cell Rep* **25:** 3059–3073.e10. doi:10.1016/j.celrep.2018.11.018

Stańczak M, Kacprzak B, Gawda P. 2024. Tendon cell biology: effect of mechanical loading. *Cell Physiol Biochem* **58:** 677–701. doi:10.33594/000000743

Stashko C, Hayward M-K, Northey JJ, Pearson N, Ironside AJ, Lakins JN, Oria R, Goyette M-A, Mayo L, Russnes HG, et al. 2023. A convolutional neural network STIFMap reveals associations between stromal stiffness and EMT in breast cancer. *Nat Commun* **14:** 3561. doi:10.1038/s41467-023-39085-1

Stefanello ST, Luchtefeld I, Liashkovich I, Pethö Z, Azzam I, Bulk E, Rosso G, Döhlinger L, Hesse B, Oeckinghaus A, et al. 2021. Impact of the nuclear envelope on malignant transformation, motility, and survival of lung cancer cells. *Adv Sci (Weinh)* **8:** e2102757. doi:10.1002/advs.202102757

Stone NE, Raj A, Young KM, DeLuca AP, Chrit FE, Tucker BA, Alexeev A, McDonald J, Benigno BB, Sulchek T. 2021. Label-free microfluidic enrichment of cancer cells from non-cancer cells in ascites. *Sci Rep* **11:** 18032. doi:10.1038/s41598-021-96862-y

Szafranska K, Neuman T, Baster Z, Rajfur Z, Szelest O, Holte C, Kubisiak A, Kus E, Wolfson DL, Chlopicki S, et al. 2022. From fixed-dried to wet-fixed to live—comparative super-resolution microscopy of liver sinusoidal endothelial cell fenestrations. *Nanophotonics* **11:** 2253–2270. doi:10.1515/nanoph-2021-0818

Tang K, Zheng Y, Hu G, Xin Y, Li K, Zhang C, Chen X, Zhang B, Li X, Hu B, et al. 2025. Local soft niches in mechanically heterogeneous primary tumors promote brain metastasis via mechanotransduction-mediated HDAC3 activity. *Sci Adv* **11:** eadq2881. doi:10.1126/sciadv.adq2881

Tantama M, Martínez-François JR, Mongeon R, Yellen G. 2013. Imaging energy status in live cells with a fluorescent biosensor of the intracellular ATP-to-ADP ratio. *Nat Commun* **4:** 2550. doi:10.1038/ncomms3550

Tharp KM, Higuchi-Sanabria R, Timblin GA, Ford B, Garzon-Coral C, Schneider C, Muncie JM, Stashko C, Daniele JR, Moore AS, et al. 2021. Adhesion-mediated mechanosignaling forces mitohormesis. *Cell Metab* **33:** 1322–1341.e13. doi:10.1016/j.cmet.2021.04.017

Thauland TJ, Hu KH, Bruce MA, Butte MJ. 2017. Cytoskeletal adaptivity regulates T cell receptor signaling. *Sci Signal* **10:** eaah3737. doi:10.1126/scisignal.aah3737

Thomas-Chemin O, Janel S, Boumehdi Z, Séverac C, Trevisiol E, Dague E, Duprés V. 2025. Advancing high-throughput cellular atomic force microscopy with automation and artificial intelligence. *ACS Nano* **19:** 5045–5062. doi:10.1021/acsnano.4c07729

Tian L, Zhao S, Ding F, Zhang R. 2024. ITIH4 reversed the effects of thrombin on VSMCs stiffness via JNK and ERK signaling pathway. *Exp Cell Res* **441:** 114189. doi:10.1016/j.yexcr.2024.114189

Urbanska M, Muñoz HE, Bagnall JS, Otto O, Manalis SR, Di Carlo D, Guck J. 2020. A comparison of microfluidic methods for high-throughput cell deformability measurements. *Nat Methods* **17:** 587–593. doi:10.1038/s41592-020-0818-8

Van Gulick L, Saby C, Morjani H, Beljebbar A. 2019. Age-related changes in molecular organization of type I collagen in tendon as probed by polarized SHG and Raman microspectroscopy. *Sci Rep* **9:** 7280. doi:10.1038/s41598-019-43636-2

Vasse GF, Buzón P, Melgert BN, Roos WH, van Rijn P. 2021. Single cell reactomics: real-time single-cell activation kinetics of optically trapped macrophages. *Small Methods* **5:** 2000849. doi:10.1002/smtd.202000849

Viji Babu PK, Rianna C, Mirastschijski U, Radmacher M. 2019. Nano-mechanical mapping of interdependent cell and ECM mechanics by AFM force spectroscopy. *Sci Rep* **9:** 12317. doi:10.1038/s41598-019-48566-7

Vortmeyer-Krause M, te Lindert M, te Riet J, te Boekhorst V, Marke R, Perera R, Isermann P, van Oorschot T, Zwerger M, Yang F, et al. 2020. Lamin B2 follows lamin A/C-mediated nuclear mechanics and cancer cell invasion efficacy. bioRxiv doi:10.1101/2020.04.07.028969v1

Wanders RJA, Baes M, Ribeiro D, Ferdinandusse S, Waterham HR. 2023. The physiological functions of human peroxisomes. *Physiol Rev* **103:** 957–1024. doi:10.1152/physrev.00051.2021

Wang N. 1998. Mechanical interactions among cytoskeletal filaments. *Hypertens* **32:** 162–165. doi:10.1161/01.HYP.32.1.162

Wang N, Stamenović D. 2000. Contribution of intermediate filaments to cell stiffness, stiffening, and growth. *Am J Physiol Cell Physiol* **279:** C188–C194. doi:10.1152/ajpcell.2000.279.1.C188

Wang W, Hapach LA, Griggs L, Smart K, Wu Y, Taufalele PV, Rowe MM, Young KM, Bates ME, Johnson AC, et al. 2022. Diabetic hyperglycemia promotes primary tumor progression through glycation-induced tumor extracellular matrix stiffening. *Sci Adv* **8:** eabo1673. doi:10.1126/sciadv.abo1673

Wang XH, Liu Y, Kang B, Xu JJ, Chen HY. 2023. Cell mechanics and energetic costs of collective cell migration under confined microchannels. *Chin Chem Lett* **34:** 107789. doi:10.1016/j.cclet.2022.107789

Wang J, Zhang B, Chen X, Xin Y, Li K, Zhang C, Tang K, Tan Y. 2024. Cell mechanics regulate the migration and invasion of hepatocellular carcinoma cells via JNK signaling. *Acta Biomater* **176:** 321–333. doi:10.1016/j.actbio.2024.01.024

Watson AMD, Soro-Paavonen A, Sheehy K, Li J, Calkin AC, Koitka A, Rajan SN, Brasacchio D, Allen TJ, Cooper ME, et al. 2011. Delayed intervention with AGE inhibitors attenuates the progression of diabetes-accelerated atherosclerosis in diabetic apolipoprotein E knockout mice. *Diabetologia* **54:** 681–689. doi:10.1007/s00125-010-2000-9

Wells RG. 2013. Tissue mechanics and fibrosis. *Biochim Biophys Acta* **1832:** 884–890. doi:10.1016/j.bbadis.2013.02.007

Winkler J, Abisoye-Ogunniyan A, Metcalf KJ, Werb Z. 2020. Concepts of extracellular matrix remodelling in tumour progression and metastasis. *Nat Commun* **11:** 5120. doi:10.1038/s41467-020-18794-x

Wolf K, Te Lindert M, Krause M, Alexander S, Te Riet J, Willis AL, Hoffman RM, Figdor CG, Weiss SJ, Friedl P. 2013. Physical limits of cell migration: control by ECM space and nuclear deformation and tuning by proteolysis and traction force. *J Cell Biol* **201:** 1069–1084. doi:10.1083/jcb.201210152

Wolfenson H, Lavelin I, Geiger B. 2013. Dynamic regulation of the structure and functions of integrin adhesions. *Dev Cell* **24:** 447–458. doi:10.1016/j.devcel.2013.02.012

Wu Y, Cheng T, Chen Q, Gao B, Stewart AG, Lee PVS. 2020. On-chip surface acoustic wave and micropipette aspiration techniques to assess cell elastic properties. *Biomicrofluidics* **14:** 014114. doi:10.1063/1.5138662

Wu Y, Zanotelli MR, Zhang J, Reinhart-King CA. 2021. Matrix-driven changes in metabolism support cytoskeletal activity to promote cell migration. *Biophys J* **120:** 1705–1717. doi:10.1016/j.bpj.2021.02.044

Xie L, Liang S, Jiwa H, Zhang L, Lu Q, Wang X, Luo L, Xia H, Li Z, Wang J, et al. 2024. Securinine inhibits the tumor growth of human bladder cancer cells by suppressing Wnt/β-catenin signaling pathway and activating p38 and JNK signaling pathways. *Biochem Pharmacol* **223:** 116125. doi:10.1016/j.bcp.2024.116125

Xu W, Mezencev R, Kim B, Wang L, McDonald J, Sulchek T. 2012. Cell stiffness is a biomarker of the metastatic potential of ovarian cancer cells. *PLoS One* **7:** e46609. doi:10.1371/journal.pone.0046609

Yamaguchi H, Taouk GM. 2020. A potential role of YAP/TAZ in the interplay between metastasis and metabolic alterations. *Front Oncol* **10:** 928. doi:10.3389/fonc.2020.00928

Yang D, Liu J, Qian H, Zhuang Q. 2023a. Cancer-associated fibroblasts: from basic science to anticancer therapy. *Exp Mol Med* **55:** 1322–1332. doi:10.1038/s12276-023-01013-0

Yang F, Bevilacqua C, Hambura S, Neves A, Gopalan A, Watanabe K, Govendir M, Bernabeu M, Ellenberg J, Diz-Muñoz A, et al. 2023b. Pulsed stimulated Brillouin microscopy enables high-sensitivity mechanical imaging of live and fragile biological specimens. *Nat Methods* **20:** 1971–1979. doi:10.1038/s41592-023-02054-z

Yeoman B, Shatkin G, Beri P, Banisadr A, Katira P, Engler AJ. 2021. Adhesion strength and contractility enable metastatic cells to become adurotactic. *Cell Rep* **34:** 108816. doi:10.1016/j.celrep.2021.108816

Yoshida T, Alfaqaan S, Sasaoka N, Imamura H. 2017. Application of FRET-based biosensor "ATeam" for visualization of ATP levels in the mitochondrial matrix of living mammalian cells. *Methods Mol Biol* **1567:** 231–243. doi:10.1007/978-1-4939-6824-4_14

Young KM, Reinhart-King CA. 2023. Cellular mechanosignaling for sensing and transducing matrix rigidity. *Curr Opin Cell Biol* **83:** 102208. doi:10.1016/j.ceb.2023.102208

Young KM, Shankles PG, Chen T, Ahkee K, Bules S, Sulchek T. 2022. Scaling microfluidic throughput with flow-balanced manifolds to simply control devices with multiple inlets and outlets. *Biomicrofluidics* **16:** 034104. doi:10.1063/5.0080510

Young KM, Xu C, Ahkee K, Mezencev R, Swingle SP, Yu T, Paikeday A, Kim C, McDonald JF, Qiu P, et al. 2023. Correlating mechanical and gene expression data on the single cell level to investigate metastatic phenotypes. *iScience* **26:** 106393. doi:10.1016/j.isci.2023.106393

Yousafzai MS, Coceano G, Bonin S, Niemela J, Scoles G, Cojoc D. 2017. Investigating the effect of cell substrate on cancer cell stiffness by optical tweezers. *J Biomech* **60:** 266–269. doi:10.1016/j.jbiomech.2017.06.043

Yu L, Chen X, Sun X, Wang L, Chen S. 2017. The glycolytic switch in tumors: how many players are involved? *J Cancer* **8:** 3430–3440. doi:10.7150/jca.21125

Yu Y, Ren LJ, Liu XY, Gong XB, Yao W. 2021. Effects of substrate stiffness on mast cell migration. *Eur J Cell Biol* **100:** 151178. doi:10.1016/j.ejcb.2021.151178

Yuan Z, Li Y, Zhang S, Wang X, Dou H, Yu X, Zhang Z, Yang S, Xiao M. 2023. Extracellular matrix remodeling in tumor progression and immune escape: from mechanisms to treatments. *Mol Cancer* **22:** 48. doi:10.1186/s12943-023-01744-8

Zaman MH, Trapani LM, Sieminski AL, Mackellar D, Gong H, Kamm RD, Wells A, Lauffenburger DA, Matsudaira P. 2006. Migration of tumor cells in 3D matrices is governed by matrix stiffness along with cell-matrix adhesion and proteolysis. *Proc Natl Acad Sci* **103:** 10889–10894. doi:10.1073/pnas.0604460103

Zanotelli MR, Goldblatt ZE, Miller JP, Bordeleau F, Li J, VanderBurgh JA, Lampi MC, King MR, Reinhart-King

CA. 2018. Regulation of ATP utilization during metastatic cell migration by collagen architecture. *Mol Biol Cell* **29:** 1–9. doi:10.1091/mbc.E17-01-0041

Zhang J, Reinhart-King CA. 2023. Analysis of energy-driven leader-follower hierarchy during collective cancer cell invasion. *Methods Mol Biol* **2608:** 247–262. doi:10.1007/978-1-0716-2887-4_15

Zhang T, Li X, He Y, Wang Y, Shen J, Wang S, You Q, Zhai J, Shen L. 2022. Cancer-associated fibroblasts-derived HAPLN1 promotes tumour invasion through extracellular matrix remodeling in gastric cancer. *Gastric Cancer* **25:** 346–359. doi:10.1007/s10120-021-01259-5

Zhou DW, Fernández-Yagüe MA, Holland EN, García AF, Castro NS, O'Neill EB, Eyckmans J, Chen CS, Fu J, Schlaepfer DD, et al. 2021. Force-FAK signaling coupling at individual focal adhesions coordinates mechanosensing and microtissue repair. *Nat Commun* **12:** 2359. doi:10.1038/s41467-021-22602-5

Zipfel WR, Williams RM, Webb WW. 2003. Nonlinear magic: multiphoton microscopy in the biosciences. *Nat Biotechnol* **21:** 1369–1377. doi:10.1038/nbt899

Zou C, Wang Y, Shen Z. 2005. 2-NBDG as a fluorescent indicator for direct glucose uptake measurement. *J Biochem Biophys Methods* **64:** 207–215. doi:10.1016/j.jbbm.2005.08.001

Guiding Cell Migration with Electric Fields: Mechanisms and Applications of Galvanotaxis

Min Zhao,[1,2] Yan Zhang,[3] Kan Zhu,[1] and Brian Reid[1]

[1]Department of Ophthalmology & Vision Science, University of California, Davis, Davis, California 95616, USA

[2]Department of Dermatology, University of California, Davis, Davis, California 95616, USA

[3]Department of Occupational and Environmental Health, Hangzhou Normal University School of Public Health and Nursing, Hangzhou 311121, China

Correspondence: minzhao@ucdavis.edu

Electric field–guided cell migration, known as galvanotaxis or electrotaxis, has garnered great interest as an engineering manipulation but has not been widely considered physiologically relevant. Here we provide experimental evidence proving galvanotaxis is a fundamental biological process, like chemotaxis, and show that the application of electric fields provides a powerful engineering approach. We will review our understanding of (1) endogenous electric fields naturally found in biological systems; (2) galvanotaxis of different cell types; and (3) sensing and signaling mechanisms of galvanotaxis. We reason that the bioelectrical mechanism is likely to be part of the environmental cues that cells and tissues integrate to make motility decisions.

Galvanotaxis (or electrotaxis), the directed movement of organisms or cells in response to an electric field (EF), is an experimentally demonstrable phenomenon conserved in diverse cell types across phylogeny. This intriguing behavior has been observed in cells for over one and a half centuries first in protists (Verworn 1890; Nagel 1895), later in bacteria (Adler and Shi 1988), and in cultured cells from vertebrates (Cooper and Schliwa 1985; Robinson 1985). Observations span a broad range, from bacteria and fungi to animal and human cells both in vitro and, more recently, in vivo. Decades of research suggest that galvanotaxis is a fundamentally important biological process. The fact that galvanotaxis is observed in a wide range of cell types across kingdoms of life, species, and tissue origins suggests that it is likely to be an evolutionarily conserved mechanism, which may confer a selective advantage for a variety of cells in complex and challenging environments.

Experiments in various developmental systems demonstrated the presence of endogenous EFs and cells responding to applied EFs (Levin 2012a,b, 2021). Almost all cell types that are important in wound healing manifested galvanotaxis (Nishimura et al. 1996; Zhao et al. 1996a,b, 2006). It is thus accepted that galvanotaxis is a mechanism that mobilizes and guides

Cite this article as *Cold Spring Harb Perspect Biol* doi: 10.1101/cshperspect.a041744

epithelial cells toward the wound center to cover the wound bed (Nuccitelli 2003b; Hunckler and de Mel 2017; Zhao et al. 2022; Rabbani et al. 2024).

This review aims to demonstrate that galvanotaxis is a crucial and versatile mechanism driving cell migration in diverse biological contexts. We will briefly discuss how endogenous EFs are naturally generated in developmental systems and at wounds to emphasize physiological relevance. We will present an overview of galvanotaxis of different cell types, cell groups, and epithelial tissues and provide examples of galvanotaxis in vivo. We will also present some intriguing cellular responses, such as galvanotaxis of cell fragments. Comparing galvanotaxis with other well-established mechanisms of directional migration, we will present experimental evidence supporting polarized signaling in cells as the mechanism and include the latest notion of excitable networks in signaling integration. Finally, we will select some examples of experimental tools to study galvanotaxis. Please refer to some reviews relevant to other cellular bioelectric mechanisms (Mycielska and Djamgoz 2004; Bonazzi and Minc 2014; Barriga and Theveneau 2020; Levin 2021; SenGupta et al. 2021; Zhao et al. 2022; Nwogbaga et al. 2023).

PRINCIPLE AND MECHANISMS OF THE GENERATION OF ENDOGENOUS EFs

Electrical activities in living cells, tissues, and organisms are generated mainly through the movement and segregation of ions and charged molecules, which generates resting and action potentials. At the tissue level, the same principle applies. The movement of ions and segregation of ions across a membrane, which can be epithelium or endothelium, generates EFs at the tissue level. Across most if not all epithelia, there is an electrical potential difference, termed *trans*-epithelial potential (TEP) (Fig. 1A). Such potential differences have been measured in the skin, cornea, kidney, respiratory system, and gut (McCaig et al. 2005; Saw et al. 2022). In cystic fibrosis, TEP measurement is a clinically used diagnostic index (Gowen et al. 1991).

The TEP is established through the transport of ions through pumps (mainly the sodium–potassium pump—Na^+/K^+-ATPase) and channels. Na^+/K^+-ATPase in most epithelia is expressed at the basal side (Fig. 1B). When activated, the pump consumes energy and transports three Na^+ out of cells and two K^+ in. Tight junctions between epithelial cells are highly electrically resistant, thus making the epithelium an effective segregation membrane preventing the free diffusion of ions. Polarized expressions of these pumps increase the concentration of Na^+ at the basal side. The TEP measures in a range of 5–50 mV or more depending on the type of epithelia. This translates to an EF strength across the epithelium of ~500 mV/mm.

When the epithelial barrier is compromised (e.g., at a wound), the TEP drives the flow of positive charges into the wound, forming a wound EF (Fig. 1C). Return currents flow in the superficial layer and through ion pumps in the intact epithelium to the basal side. In principle, the intact epithelium is like an epithelial battery: transport of ions generates power, and epithelial tight junctions form a high resistance barrier like the off switch (Fig. 1D, left). Upon injury, or changes of resistivity in the epithelium, the transepithelial potential drives current along the newly formed low-resistance pathway. The switch is on, the circuit is complete, and new laterally orientated EFs are established (Fig. 1D, right, dashed arrow).

Endogenous wound EFs, measured in skin and cornea wounds across various experimental animal models, typically range from 150 to 200 mV/mm (Barker et al. 1982; Chiang et al. 1992; Nuccitelli 2003b). Human wound EFs in old subjects are significantly lower than those in young subjects (Nuccitelli et al. 2011). Corneal wounds in diabetic animals produce significantly weaker EFs than nondiabetic controls (Shen et al. 2016).

During development, the endogenous EF can be attributed to the spatiotemporally regulated expression and activity of voltage-gated calcium channels and gap junctions (Jaffe and Stern 1979; Shi and Borgens 1995; Li et al. 2018a). Endogenous EFs of 18–37 mV/mm could be established mechanically by conver-

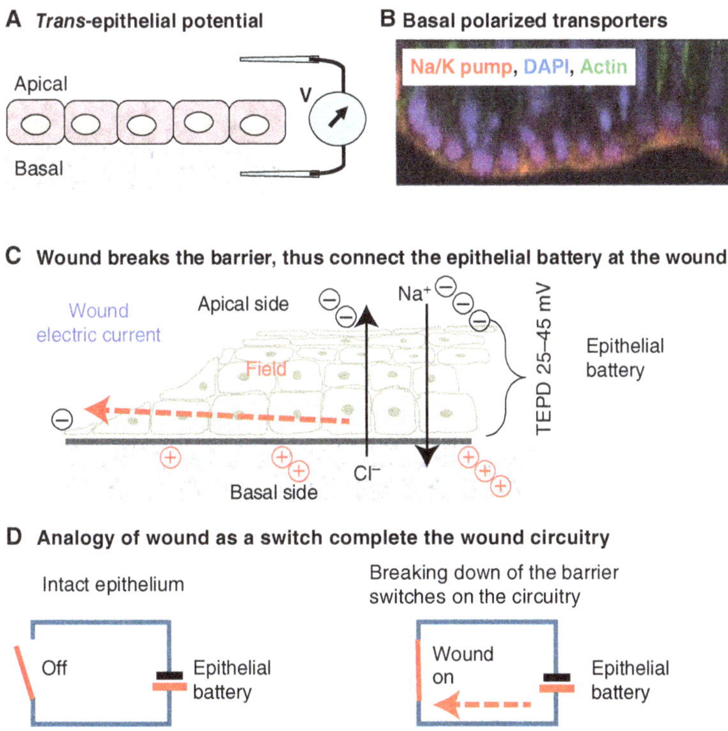

Figure 1. The biological basis of naturally occurring endogenous electric fields (EFs). (*A*,*B*) *Trans*-epithelial potentials (TEPs) are generated by the concerted activity of ion channels and pumps in polarized epithelial cells that also establish tight junctions (electrical resistance). An immunohistochemical image of a monolayer of Madin–Darby canine kidney (MDCK) cells shows polarized expression of the Na/K pump (red) at the basolateral side of the epithelium. (*C*) Generation of wound EFs. Intact corneal epithelium (*right*) transports Cl^- to the apical side and Na^+ to the basal side across the epithelium, maintaining a TEP of 25–45 mV, positive at the basal side relative to the apical side. Injury to the cornea (*left*) collapses the TEP locally at the wound to zero, thus resulting in EFs orientated from outside the wound to the wound center (red dashed arrow). (*D*) Schematic diagrams show an epithelium as a battery that maintains TEP. When the electrical resistance of epithelium is compromised (as at the wound), the wound circuit is completed resulting in electric currents and EFs from the wound edge toward the wound bed. (*B* is reprinted from Tran et al. 2013, under the terms of the Creative Commons Attribution License.)

gent extension movements of the ectoderm, which generate a membrane tension gradient that opens stretch-activated ion channels (Ferreira et al. 2025).

DIFFERENT TYPES OF INDIVIDUAL CELLS SHOW ROBUST GALVANOTAXIS

Microorganisms were the first group of cells in which galvanotaxis was experimentally demonstrated. The protists showed remarkable galvanotaxis (Verworn 1890; Nagel 1895). Soil amoeba, *Dictyostelium* cells, and many protists (e.g., ciliates such as paramecium) manifest robust galvanotaxis or galvanotropism (Eckert 1972; Nuccitelli and Jaffe 1974; Korohoda et al. 2000; Zhao et al. 2002b; Ogawa et al. 2006; Shanley et al. 2006). *Dictyostelium* cells and the amoeba proteus migrate remarkably well to the cathode in applied EFs (Fig. 2A,B; Korohoda et al. 2000; Gao et al. 2011, 2015; Li et al. 2018b; Guido et al. 2024). Cells extend pseudopodia and lamellipodia and align the cell body long axis toward the cathode, resembling the behavior of cells undergoing chemotaxis in chemoattractant gradients.

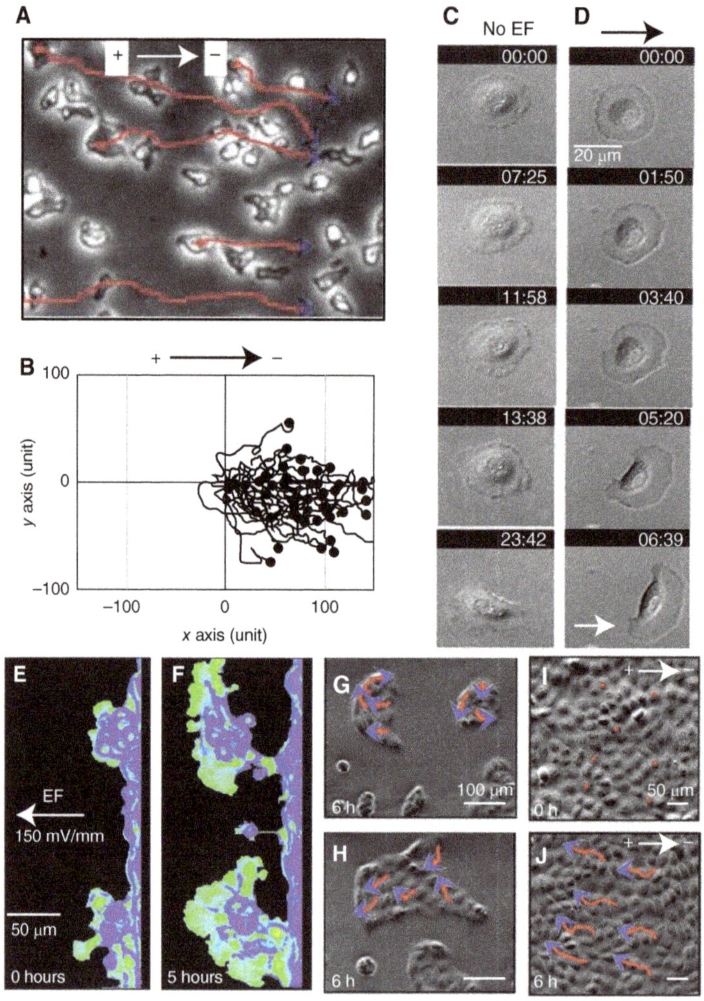

Figure 2. Electric field (EF)-guided cell migration. Galvanotaxis of individual cells, groups of cells, and epithelial sheets. (*A–D*) Galvanotaxis of individual cells. (*A*,*B*) Galvanotaxis of *Dictyostelium* cells showing the migration trajectories of selected cells. Red lines and blue arrows are cell trajectories and end points over 15 min. Composite trajectories of migrating cells with the starting points placed at the origin and end points as the round dots (*B*). (*C*,*D*) EFs initiate and accelerate polarization and direct migration of fish keratocytes. Time-lapse images show spontaneous polarization (*C*) and EF-initiated polarization (*D*, 4 V/cm, field direction positive to negative as indicated by black arrow). Subsequently, this cell underwent directional migration to the cathode (white arrow pointing to the *right*, i.e., parallel to field direction). Time is in mm:ss. Scale bar, 20 µm. (*E*,*F*) Galvanotaxis of cell groups. An EF-induced cathode-directed expansion and migration of groups of bovine corneal epithelial cells. The static reference marker is on the *right* edge. Considerable extension of cell membrane lamellipodia (green) toward the cathode is seen, while two cell groups move toward the cathode (the *left*). Dark blue/purple shows cell bodies. Scale bar, 50 µm. (*G*,*H*) Galvanotaxis of a group of cells becomes stronger when the size of cell groups increases. Red lines with blue arrowheads represent migration paths and direction at 200 mV/mm for 6 h. Scale bar, 100 µm. (*I*,*J*) Robust galvanotaxis in a monolayer epithelial sheet. Madin–Darby canine kidney (MDCK) II cells in a sheet show robust collective galvanotaxis in an EF. Red lines with blue arrowheads represent migration paths and directions in 200 mV/mm for 6 h. Scale bar, 50 µm. (*A*,*B* are reprinted from Song et al. 2021, under the terms of the Creative Commons Attribution 4.0 International License; *C*,*D* are reprinted, with permission, from Sun et al. 2018; *E*,*F* are adapted from Zhao et al. 1996a, © Association for Research in Vision and Ophthalmology; *G–J* are reprinted, with permission, from Li et al. 2012.)

Bacteria orient themselves and migrate toward either the positive or negative pole of an EF, a behavior considered crucial for their survival and adaptation (Adler and Shi 1988; Rajnicek 1993; Rajnicek et al. 1994). For example, experimental results suggest that an EF drives opposing directional migration of *Salmonella enterica* serotype Typhimurium (pathogen) to the cathode and *Escherichia coli* (commensal bacteria) to the anode (Adler and Shi 1988; Sun and Zhao 2025). Such selective galvanotaxis might facilitate *Salmonella* colonization and invasion at the portals of entry (scarce areas of the intestinal lining vulnerable to bacterial invasion, i.e., the follicle-associated epithelium over the Peyer's patches) (Sun et al. 2024). It is well established that M cells that reside at the Peyer's patch are the preferred entry cells for *Salmonella* and many other pathogens (Jepson and Clark 1998; Vallon-Eberhard et al. 2006; Turner 2009; Griffin and McSorley 2011). *Salmonella* uses specific virulence factors to interact with and invade M cells, facilitating their entry into the host, and allowing them to translocate into the underlying tissue and disseminate (Jones et al. 1994; Jepson and Clark 2001; Griffin and McSorley 2011; Tahoun et al. 2012; Li 2022). Fungi have galvanotropism behavior in their hyphal growth, enabling them to navigate through their environment and optimize nutrient acquisition (McGillivray and Gow 1986; Minc and Chang 2010).

A wide variety of vertebrate cell types, including human cells, exhibit robust galvanotaxis (Cooper and Schliwa 1985; Robinson 1985; McCaig et al. 2005). Epithelial cells, endothelial cells, fibroblasts, neutrophils, lymphocytes, macrophages, stem cells, and neuroblasts, for example, all respond to EFs with directional migration (Franke and Gruler 1990; Zhao et al. 2002a; Bai et al. 2004; Zhang et al. 2011; Feng et al. 2012; Frederich et al. 2017; Lin et al. 2017; Sarkar et al. 2019; Moarefian et al. 2021; SenGupta et al. 2021; Ende et al. 2024). Individual fish keratocytes in culture self-polarize and migrate actively (Fig. 2C,D). An EF significantly speeds up symmetry breaking and cell polarization (approximately three to four times faster than no EF control) and directional migration ensues (Fig. 2B–D; Sun et al. 2018).

Cell morphology changes significantly in EFs. Most cells orient and elongate/stretch themselves perpendicular to the EF lines (Fig. 2D). The morphological change becomes more obvious when the intensity of the EF increases. This is considered to be the action of cellular self-protection in high-intensity EFs—the cell eliminates the impact of overloaded EFs by adapting the shape of the cell to minimize the voltage across the cell (Robinson 1985; Nishimura et al. 1996; Zhao et al. 2002a). However, this is not always the case: *Dictyostelium* and neutrophils orientate with the cell body's long axis parallel to the field lines (Fig. 2A). These shape changes and alignment of the long axis of the cell body appear to be consistent with the preferred motile morphology of the cell type (Fig. 2A,D).

Galvanotaxis can be cathodal (to the negative pole; cathode) or anodal (to the positive pole; anode). Most cells studied show cathodal galvanotaxis, but there are cell types that migrate to the positive pole (Nishimura et al. 1996; Mycielska and Djamgoz 2004; Sun et al. 2013; Cortese et al. 2014; Leal et al. 2023; Ferreira et al. 2025). Cells isolated from different parts of the same tissue can also show dramatic differences in galvanotaxis, for example, ocular lens epithelial cells (LECs) from central and peripheral regions migrate in opposite directions (Wang et al. 2003).

EFs of different strengths can affect cells differently in the same medium (Zhao et al. 1997, 2014). Cells at different stages of differentiation may have varying sensitivities to EFs. Studies have revealed that ependymal neural precursor cells migrate to the cathode, while well-differentiated astrocytes lose the galvanotaxis response (Babona-Pilipos et al. 2011). Similarly, highly metabolizing cancer cells such as highly metastatic prostate cancer cells and lung adenocarcinoma cells are found to be more responsive to EFs than less malignant tumor cells (Pu et al. 2007; Tsai et al. 2013). The concentration of ions (Ca^{2+} and K^{+}) and serum, pH, and many other components in the culture medium or extracellular matrix are also found to affect galvanotaxis (Sheridan et al. 1996; Rajnicek et al. 1998; Zhao et al. 1999; Sun et al. 2004; Allen et al. 2013).

GALVANOTAXIS AS A NATURAL BIOLOGY PROCESS AND AN ENGINEERING TOOL

Powerful galvanotactic effects coupled with advanced bioelectronics have seen a recent surge in enthusiasm for using galvanotaxis (and the effects of EFs on other cell behaviors) in manipulating cells for tissue engineering and therapeutic uses (Cohen et al. 2014; Wolf et al. 2022; Barman et al. 2023; Jiang et al. 2023; Luo et al. 2023; Shirzaei Sani et al. 2023; Shim et al. 2024). This is an obvious and very productive avenue. At the same time, it is important to understand the biology per se of galvanotaxis and to delineate carefully where biology transitions to engineering to exploit galvanotactic manipulation of cell migration.

Much of the evidence for galvanotaxis comes from in vitro studies, where cells are exposed to artificial EFs. While these studies are valuable for understanding the underlying mechanisms, they may not fully replicate the complex conditions found in vivo or in the natural environment of cells. The strength and direction of EFs, as well as other parameters (duration, fluctuation, and ionic components) used in experiments, can vary widely, and it is not always clear how these parameters compare to the natural EFs present in the body. This makes it challenging to extrapolate findings from the laboratory to real-world biological contexts. More experiments in vivo will therefore be needed for the establishment of robust physiological relevance of galvanotaxis in intact organisms.

For some cell types, like protists, it is unclear whether galvanotaxis is a natural biological process. *Dictyostelium* cells for example normally need field strengths in the range of 1000–2000 mV/mm to achieve optimal galvanotaxis (~10–100 times larger than EFs that induce galvanotaxis of some mammalian cells), although this may be in part due to the lower conductivity of the medium, because the ionic strength of the medium for *Dictyostelium* is about 20× lower. EFs of such strength in the normal environments of these cells are yet to be identified. The powerful and highly manipulatable galvanotactic effects, however, are useful in investigating molecular and genetic elements of the directional migration of these cells (Zhao et al. 2002b; Sato et al. 2009; Gao et al. 2011; Hoeller et al. 2016).

There are several points worth noting when considering the physiological relevance of galvanotaxis. These include the following: How strong the endogenous naturally occurring EFs are likely to be, and can they be experimentally verified? Do relevant cells show galvanotaxis in EFs of similar strength? Can loss or gain of galvanotaxis be tested in vivo or in the natural environments of these cells? For some cells, galvanotaxis at lower voltages than are in the physiological range is likely to be biologically relevant. Examples with strong evidence to support the physiological relevance of galvanotaxis are its role in development and wound healing (please see the next two subsections).

Galvanotaxis in Development

Extracellular endogenous EFs are readily measurable in developing embryos (Jaffe and Nuccitelli 1977; Jaffe and Stern 1979; Shi and Borgens 1995; Ferreira et al. 2025). Disruption of the endogenous EFs disrupts development (Borgens and Shi 1995; Nuccitelli 2003a). A recent comprehensive review elegantly illustrates the role of EFs in vivo (Levin 2021). Researchers, however, do not have a consistent view of whether cells respond to endogenous EFs in vivo (Stern 1986), despite strong electrical currents leaving the primitive streak of chick embryos forming EFs during development (Jaffe and Stern 1979). An early study using cultured chick mesoderm cells, however, appeared to show that those cells migrated to both the cathode and the anode; therefore, it was concluded that the cells did not respond to EFs (Stern 1981). We have repeated some experiments with chick mesoderm cells. Our unpublished results suggested that there are likely mixed subpopulations of chick mesoderm cells in culture; some migrate to the cathode, some to the anode, the same as Stern observed. These subpopulations in mixed cultures would thus show directional migration to the positive or negative poles. When the field polarity was reversed during the experiments, the galvanotaxis of such subpopulations became evident, because cells migrating to the cathode or anode now reversed migration direction to the new cathode or new anode, respectively. These results thus favor the idea that galvanotaxis is a physiological phenomenon in chick embryo development, suggesting further experiments are needed to conclusively determine that.

The threshold of the intensity of EFs required for galvanotaxis for different cell types from embryos varies. For amphibian nerve cells, the threshold of galvanotaxis is ~7 mV/mm (Hinkle et al. 1981). Fields of this magnitude and up to two to three times stronger are present in at least some parts of amphibian embryos (Shi and Borgens 1995).

Galvanotaxis in Wound Healing

Another compelling example supporting galvanotaxis as a natural biological process is wound healing. When tissue is damaged, a natural EF is generated at the wound site. There is well-established theoretical and experimental evidence showing how endogenous wound EFs naturally come into being (Sta Iglesia and Vanable 1998; Reid et al. 2007; Nuccitelli et al. 2008, 2011). Keratinocytes and corneal epithelial cells that are critical to wound healing show galvanotaxis well within the field strength that can be measured in vivo at wounds. For example, keratinocytes respond to an EF as low as 8 mV/mm (Nishimura et al. 1996), whereas fields up to 150 mV/mm can be measured at skin wounds (McCaig et al. 2005; Mukerjee et al. 2006; Nuccitelli et al. 2008; Zhao et al. 2014). In the cornea, wounds readily generate endogenous EFs above 42 mV/mm up to several hundreds of mV/mm (Chiang et al. 1992; Reid et al. 2005). Corneal epithelial cells readily respond to a field ~20 mV/mm (Zhao et al. 2006, 2014). Applied EFs of similar strength direct migration of not only single cells (Fig. 2C,D) but also monolayer and stratified epithelial sheets in situ (Fig. 2E–H) (please see the section Collective Galvanotaxis of Groups of Cells, Epithelial Sheets, and Tissues below). Importantly, when applied against other coexisting guidance cues, such as free edge and injury, EFs of the strength measured at wounds direct cell migration as an overriding guidance cue (Zhao et al. 2006, 2022; Zhao 2009) (please see the section Galvanotaxis In Situ and In Vivo).

COLLECTIVE GALVANOTAXIS OF GROUPS OF CELLS, EPITHELIAL SHEETS, AND TISSUES

The migration of epithelial sheets is a fundamental process in embryonic development (Weijer 2009; Scarpa and Mayor 2016). For example, during gastrulation, epithelial sheets undergo invagination and migration to form germ layers. Similarly, neural tube closure, a key step in central nervous system development, involves the folding and fusion of the neural plate into the neural tube. Endogenous EFs have been measured at these development stages (Jaffe and Stern 1979; Shi and Borgens 1995; Nuccitelli 2003a; Levin 2012a, 2021).

Collective galvanotaxis of epithelial sheets, in which a monolayer or stratified multilayer epithelium migrates directionally into the wound to heal, is essential for wound healing (Rodrigues et al. 2019). Epithelial sheets migrate in wound healing, with the cells maintaining physical and functional connectivity with each other (Zhao et al. 2003, 2022; Zhang et al. 2025).

Collective galvanotaxis has been demonstrated in epithelial cells, neural crest cells, and cancer cells (Zhao et al. 1996a, 2006). Epithelia are the source of endogenous EFs in vivo, known as the epithelial battery. Epithelial sheets, perhaps counterintuitively and different from chemotaxis, respond to EFs better, with lower thresholds and better-coordinated directional migration, than single cells (Cooper and Schliwa 1986; Zhao et al. 1996a). For instance, bovine cornea epithelial groups behave like a unit, in which cells coordinate in extending lamellipodia together (green in Fig. 2E,F), while the cell bodies remain at the back staying close to each other (dark blue in Fig. 2E,F, compare the morphology of the two clusters of cells and in between a single cell undergoing galvanotaxis). One noticeable feature of the cell groups is that the groups appear in overall morphology like a single cell. The lamellipodia fan out together and nuclei cluster together, the overall morphology resembling that of a single epithelial cell (compare Fig. 2F to Fig. 2D at 5:20 and 6:39).

Cell groups migrate with a directional translocation speed (17 µm/h), which is nearly double that of isolated individual cells under the same conditions (9 µm/h) (Zhao et al. 1996a). Such significantly increased efficiency in galvanotaxis is also observed in MDCK cells. Cells with tight cell–cell junctions show more efficient galvanotaxis (Fig. 2G–J; Li et al. 2012); as the number of cells in the cluster grows, the efficiency of galvanotaxis increases. Galvanotaxis of patterned cell sheets further revealed that cell groups carry out galvanotaxis as a "well-coordinated unit" in motility (Cohen et al. 2014; Zhang et al. 2022) and at molecular (Sun et al. 2023) and mechanical levels (Curtze et al. 2004; Li et al. 2012; Cho et al. 2018; Wolf et al. 2022). It is worth noting that increased efficiency collectively in galvanotaxis is like collective chemotaxis of neural crest cells. Clusters of neural crest cells manifest a significantly increased chemotaxis response to chemoattractant Sdf1 when the cell numbers in that group increase (Theveneau et al. 2010).

GALVANOTAXIS OF CELL FRAGMENTS AND GROWTH CONES

Keratocytes, as well as leukocytes, may spontaneously detach part of their cytoplasm to form cell fragments that are devoid of nuclear and major organelles, similar to platelets (Keller and Bessis 1975; Sun et al. 2013; Zhu et al. 2016). The cell fragments are motile in culture (Verkhovsky et al. 1999; Ofer et al. 2011). When cultured in an EF, the cell fragments show strong galvanotaxis (Sun et al. 2013; Zhu et al. 2016). What is remarkable and surprising is that cell fragments isolated from primary fish keratocytes migrate to the anode, in the opposite direction of galvanotaxis of their parental cells, which migrate to the cathode (Fig. 3A,B). The opposite migration directions of the parental cells and their fragments in the same field suggest that they both sense the field, but the steering mechanism is different. A balance of "competing" mechanisms of the phosphoinositide 3-kinase (PI3K) signaling and myosin II contraction resulted in cathodal migration of cells, and anodal galvanotaxis of cell fragments (Sun et al. 2013). Thus, the effect of myosin II was predominant in the directional migration of fragments, while PI3 kinase signaling and actin polymerization led to the migration of whole cells.

Growth of neurites from neuroblasts of *Xenopus*, mice, and rats can be guided by an EF (Jaffe and Poo 1979; McCaig et al. 2002). When an EF induces galvanotaxis of rat hippocampal neurons to the cathode, the growth cones and neurites extend to the cathode (Fig. 3C; Yao et al. 2008, 2009). When retinal ganglion cells are largely immobilized in explants from early postnatal mouse retina, axons are guided to the cathode, and when the polarity of the field is reversed, growth direction of the axon is reversed to the new cathode direction (Fig. 3D; Gokoffski et al. 2019).

GALVANOTAXIS IN SITU AND IN VIVO

Stratified multilayer epithelia of cornea show collective galvanotaxis in situ. Cornea organ culture has been used to study epithelial migration in wound healing. In this model, when an EF is applied, epithelial tissues respond by robust galvanotaxis (Fig. 4A–C). An applied EF with the field direction (positive to negative) pointing to the wound center caused the epithelial migration distance to triple (Fig. 4B,C, compared to Fig. 4A). What is remarkable is that when the polarity is applied against the default direction (pointing away from the wound center), epithelial tissues migrate away from the wound (Fig. 4A,B). This effect in directing epithelial tissue to migrate away from the wound center against other coexisting cues is significant, suggesting that galvanotaxis overrides well-accepted guidance mechanisms like injury stimulation, wound void, contact inhibition release, and coexisting other cues (Zhao et al. 2006, 2022).

It has been challenging to apply well-controlled EFs in vivo, due to the complex electrical conductivity and resistance of biological tissues. There are quite a few recent publications exploiting electrical stimulation in wound healing suggesting enhanced cell migration (Jiang et al. 2023; Luo et al. 2023; Shirzaei Sani et al. 2023),

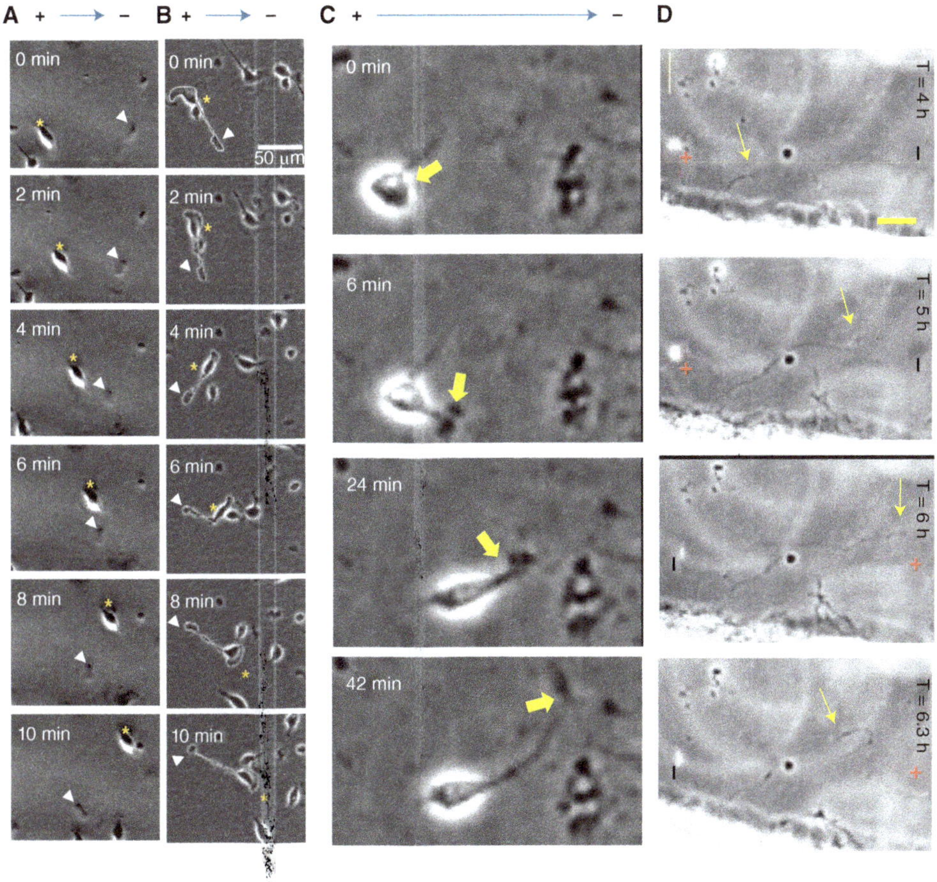

Figure 3. Galvanotaxis of cell fragments and galvanotropism of neurites. (*A*,*B*) EFs guide the migration of cell fragments in the opposite direction to parental cells. Cell fragments refer to a small piece of a cell's cytoplasm, essentially a part of the cell that has broken off, and the most common example is a platelet that lacks a nucleus. Two examples of cell fragment galvanotaxis free from parental cell (*A*) and tethered to the parent (*B*) are shown. The representative cells are marked with *, and the white arrows point to fragments. The parental cells migrate to the cathode, while the fragments move to the anode (adapted, with permission, from Sun et al. 2013). EFs = 4 V/cm. Scale bar, 50 µm. (*C*) EFs induced cathodal galvanotaxis of a rat hippocampal neuron, and a neurite extension, to the cathode. EF = 300 mV/mm. Yellow arrows point to neurites (*C* is reprinted from Yao et al. 2008). (*D*) Reversal of EFs induced neurites of rat retinal ganglion cells to change direction. EF = 100 mV/mm. The polarity of EF was reversed by 180° at 6 h. The yellow arrow points to a representative growth cone. Yellow scale bar, 50 µm. Note at 6.3 h, the neurite has changed orientation to now grow toward the new cathode. (*D* is reprinted from Gokoffski et al. 2019, © Association for Research in Vision and Ophthalmology.)

but this research did not directly study cell migration per se.

One obstacle to delivering direct current (DC) to relatively highly conductive tissues such as the brain is the detrimental effects of electrochemical products from the electrodes. In addition to temperature rise due to heat generation, local changes in pH and ionic concentrations are harmful to cells (Zhao et al. 2020). When a DC EF was applied by inserting electrodes into the thin tissue of mouse ears, green fluorescent protein (GFP)-tagged immunocytes in the skin preferentially migrated toward the cathode, suggesting a role for galvanotaxis in regulating the recruitment of immune cells (Lin et al. 2008; Sun et al. 2019). Feng et al.

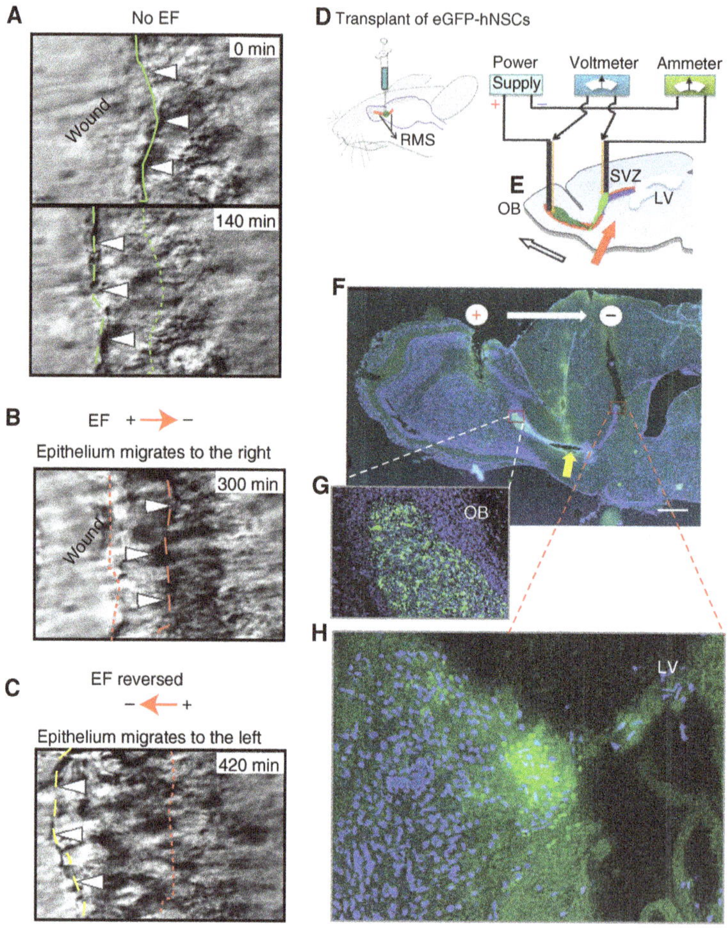

Figure 4. Galvanotaxis of epithelial tissue in situ and neural stem cells in vivo. (*A–C*) Electric Fields (EFs) provide an overriding direction signal in guiding epithelial migration during wound healing in an organ culture model. (*A*) Time-lapse images show the migration of stratified corneal epithelium in situ to heal a wound (toward the *left* indicated by the arrowheads). (*B*) EFs applied with polarity opposite to the default healing direction direct the wound edge to migrate away from the wound center (toward the *right* indicated by the arrowheads). (*C*) Migration of epithelium in wound healing is significantly enhanced by reversed EF direction into the wound center (toward *left*). EF = 200 mV/mm. Coarse-colored dashed lines indicate the wound edge at the indicated time. White arrows point toward the migration direction of wound edges. (*A–C* are reprinted from Zhao et al. 2006, Copyright © 2006, Springer Nature Limited.) (*D–H*) EFs guide the migration of human neural stem cells (hNSCs) in the rat brain. (*D*) An illustration of transplantation of hNSCs expressing enhanced green fluorescent protein (EGFP) (EGFP-hNSCs) (in green) in the middle of rostral migration stream (RMS, in red) in rat brain. (*E*) hNSCs transplanted in the RMS have a default migration direction toward the olfactory bulb (OB) (in dark green, migration direction indicated by white arrow). (SVZ) Subventricular zone (blue), (LV) lateral ventricle. Application of EFs in vivo induced directional migration of transplanted hNSCs to the SVZ (in bright green, migration direction indicated by red arrow). Delivery and monitoring of electrodes and circuits are schematically indicated. (*F,G*) Immunohistochemical imaging of brain sections reveals clear EGFP signals at RMS. Electrical stimulation (positive at OB and negative at LV) induced a significant number of hNSCs to migrate to the SVZ, against coexisting guidance cues that normally direct cells to the OB (*H*). A larger number of hNSCs maintained migration to the OB (*G*). Electrodes were implanted, and electrical stimulation was applied for 10 h after 24 h of EGFP-hNSC transplantation. Scale bars, 1 mm. The yellow arrowhead indicates the cell transplantation site; 4′,6-diamidino-2-phenylindole (DAPI) is blue, and EGFP is green. (*D–H* are reprinted, with permission, from Feng et al. 2017.)

(2012, 2017) used carbon electrodes to deliver intermittent DC pulse to the live rat brain. Together with continuous monitoring by electroencephalography of the EFs generated in brain tissue, a tolerant stimulation scheme was developed. Tracking of transplanted human neural stem cells demonstrated that the intermittent DC EFs mobilized and guided the migration of stem cells in the brain in vivo. Remarkably, the guidance effects of the EFs were powerful enough that the migration of a significant number of human neural stem cells could be guided to migrate upstream away from the olfactory bulb against intrinsic cues in the rostral migration stream (Fig. 4D–H; Feng et al. 2017).

Taking advantage of a model that is more genetic and tractable in vivo, a recent study provides strong experimental evidence of the importance of galvanotaxis of neural crest cells during tissue morphogenesis in developing embryos (Ferreira et al. 2025). These cells are among the most motile cells in development, and their migration is essential for craniofacial development, forming the peripheral nervous system, and the development of the heart and other critical organs in vertebrates. Measurements using a vibrating probe and glass microelectrodes revealed a unique map of EFs during development. Electric currents enter the flank ectoderm and exit the neural fold, establishing a subectodermal EF of 7–37 mV/mm (Ferreira et al. 2025). Neural crest cells in culture when subjected to an applied EF of similar strength showed anodal galvanotaxis. An EF applied against the endogenous EF significantly inhibited displacement of the neural crest cells, whereas an EF applied in the direction of the endogenous EF enhanced the migration in vivo. Loss and gain-of-function analysis of voltage-sensitive phosphatase 1 (Vsp1) in the cells demonstrated that it is a specific sensor for EFs.

Importantly, the emergence of EFs in this model appears to come from the establishment of membrane tension gradients during the development of the embryo. Inhibition of planar cell polarity significantly reduced the formation of dorsoventral membrane tension gradients and endogenous EFs. Graft experiments suggested that defective migration of neural crest cells can be attributed to the decreased EFs resulting from inhibition of the planar cell polarity, thus suggesting a complex link between mechanobiology and endogenous EFs (Ferreira et al. 2025). This is an in vivo system that has both the EF-generating component—the ectoderm—and the EF-responding component—neural crest cells. In embryo development, tension variation in the flanking ectoderm exerts mechanical forces to both cortical and cell membrane. Thus, endogenous EFs are mechanically established by opening stretch-activated ion channels, resulting from the formation of dorsoventral membrane tension gradients via the convergent extension movements of the ectoderm (Ferreira et al. 2025), whereas Vsp1 as a phosphoinositide phosphatase in the neural crest cells is a specific sensor of the EFs. Its activity has been shown to couple to its transmembrane domain as an intrinsic voltage sensor (Murata et al. 2005). Genetic manipulation of Vsp1 in neural crest cells in vitro and in vivo demonstrates its essential role in galvanotaxis of neural crest cells (Ferreira et al. 2025).

PRINCIPLES AND MOLECULAR MACHINERY OF GALVANOTAXIS

Four common principles, or pillars, underpinning directed migration have been proposed as a framework for comparing galvanotaxis to chemotaxis, haptotaxis, durotaxis, and topotaxis (SenGupta et al. 2021). These include the following: first, the generation of a directional signal in the milieu of cells (i.e., endogenous EFs); second, sensing of the signal; third, the transmission of the signal from the sensing modules to the machinery that executes the actual movement; and fourth, conversion of the signaling into the application of asymmetric force relative to the substrate.

Sensing Physiological EFs

Galvanotaxis Is Different from Electrophoresis

Very strong fields are likely to exert significant electrical forces on cells—thus causing electrophoresis (i.e., positively charged cells move to the cathode), whereas negatively charged cells move to the anode. Such effects have been used for cell electrophoresis, particularly when cells are freely floating in solution (Cook 2016). Electrophoresis of cells is largely a nonbiological "physics" phenomenon. Most intact animal cells are negatively charged due to abundant sialic acids, which have a ubiquitous presence on their surface. However, most animal cells migrate to the cathode in physiological EFs, thus arguing against a pure electrical effect for galvanotaxis at physiological field strengths, because if electrophoresis were the sole mechanism, most cells would migrate to the anode. *E. coli* and *Salmonella typhimurium* are both negatively charged, and when fixed with formaldehyde, they both move to the anode by electrophoresis (Shi et al. 1996; Wilson et al. 2001). The galvanotaxis of live bacteria propelled by flagella, however, is in the opposite direction. *E. coli* galvanotaxis to the anode, whereas *S. Typhimurium* is to the cathode (Adler and Shi 1988; Sun et al. 2024; Sun and Zhao 2025). Adler's model of galvanotaxis predicts cathodal movement and anodal movement of floating bacteria with flagella. The speed of anodal migration is roughly consistent with the effects of electrophoresis and flagella propelling "combined," whereas the speed of cathode migration is that of the effects of flagella propelling "minus" the effect of electrophoresis (Adler and Shi 1988; Shi et al. 1996; Sun et al. 2024).

Galvanotaxis Is Chemotaxis Independent

Most biomolecules are electrically charged in physiological pH. Growth factors (e.g., epidermal growth factor) and cytokines (e.g., interleukins) could therefore be redistributed to form gradients in EFs. Several experiments help to exclude the possibility that galvanotaxis is merely mediated by chemotaxis. When potential chemoattractant gradients are disrupted by continuous perfusion of the medium, galvanotaxis persists (Nishimura et al. 1996; Pu and Zhao 2005).

Chemotaxis null mutants of both bacteria and *Dictyostelium* maintained galvanotaxis (Adler and Shi 1988; Shi et al. 1996; Zhao et al. 2002b; Sun et al. 2024). Upon knockout of the Gβ subunit, an essential molecule to chemotaxis in *Dictyostelium* cells, galvanotaxis remains (Zhao et al. 2002b; Hoeller et al. 2016). Cell membrane potential, the concentration of potassium ions, and pH in the culture medium significantly influence galvanotaxis (Gao et al. 2000, 2011; Zhao et al. 2012), but not chemotaxis (Van Duijn et al. 1990). Thus, although chemical concentration gradients and EFs coexist, and chemotaxis and galvanotaxis share signaling pathways involved with cell polarization and migration (Gao et al. 2015), strong evidence argues against the dependence of galvanotaxis on chemotaxis (see sections Requirement of Membrane Potential for Galvanotaxis and Asymmetric Intracellular Signaling Orientates Cell Polarization below).

Polarized Distribution of Membrane Proteins Underlies EF Sensing

The cell membrane is extremely high in electrical resistance to direct current EFs, acting as a "Faraday cage," shielding the cell interior from external EFs (Mossop et al. 2004). Components of the cell membrane, therefore, are likely to be the initial step of cellular EF sensing.

Redistribution of membrane molecules in EFs as shown by experiments spanning several decades point to a role in galvanotaxis of polarized redistribution of molecules on the cell membrane through electrophoresis or electroosmosis (Jahn 1961; Poo and Robinson 1977; Poo 1981; Zhao et al. 2002a; Allen et al. 2013; Sarkar et al. 2019). This effect includes a great number of different membrane molecules, including heparan sulfate (HS) (Huang et al. 2017), Concanavalin A (Con A) receptor (Poo and Robinson 1977), acetylcholine receptor (Orida and Poo 1978; Stollberg and Fraser 1988), epidermal growth factor receptor (Fig. 5A; Fang et al. 1999; Zhao et al. 1999), Fcε receptor (immunoglobulin E [IgE] Fc receptors) (Giugni et al. 1987), glycoproteins and glycolipids (Poo and Robinson 1977), integrin (Pullar et al. 2006), and a more recently identified "galvanin" (Belliveau et al. 2024). Polarization of membrane proteins has thus been proposed as a general mechanism to explain galvanotaxis

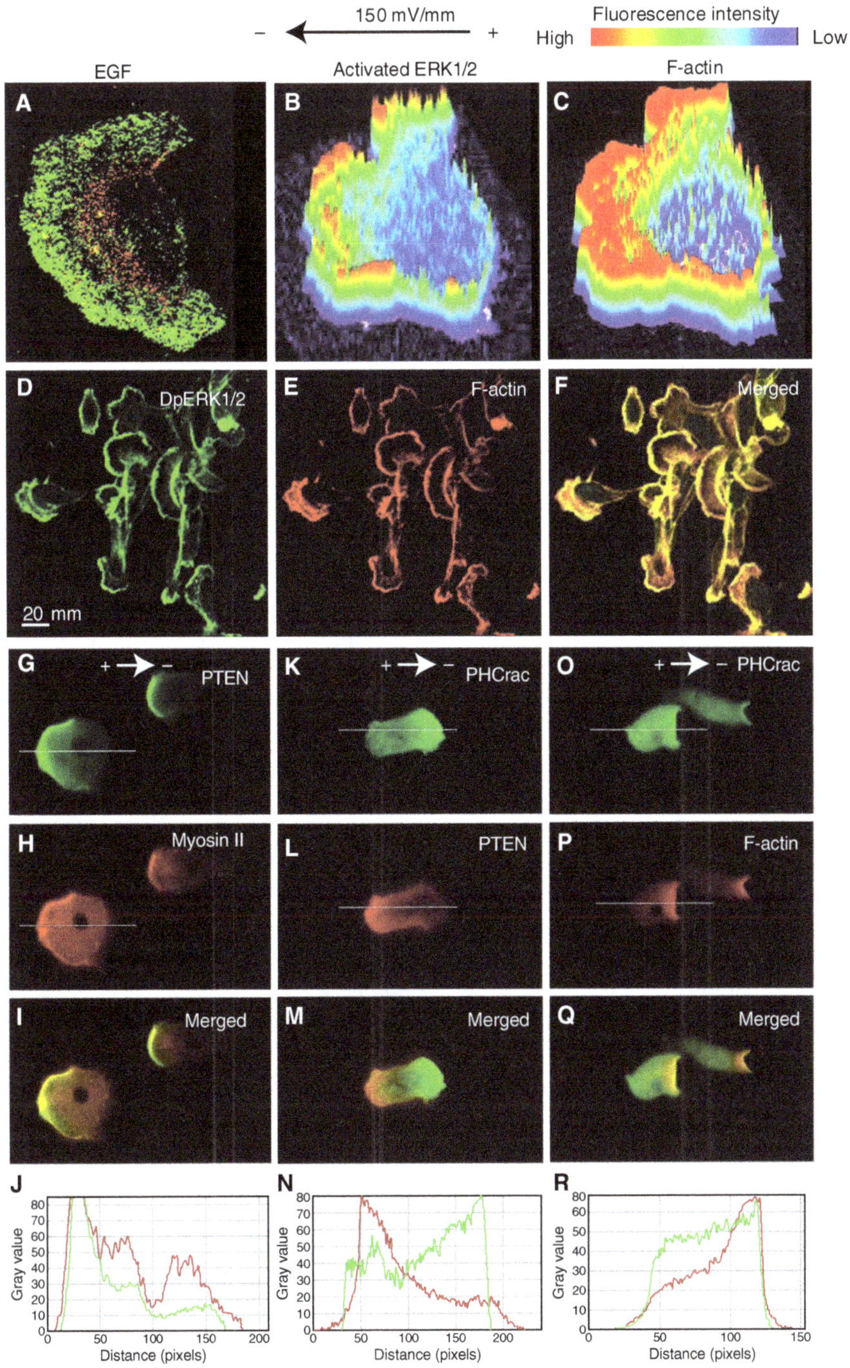

Figure 5. Polarized intracellular signaling underlies galvanotaxis. (*A–F*) Polarized epidermal growth factor (EGF) receptors, intracellular signaling, and actin polymerization in corneal epithelial cells in an electric field (EF). (*A*) Immunocytochemistry image shows EGF receptor (EGFR) accumulated to the cathodal side of cells (green at *left*; EF = 150 mV/mm, 1.5 h). (*B,D*) Extracellular signal-regulated kinase (ERK) (green) was activated preferentially at the cathodal side, together with an accumulation of F-actin (red) (*C,E*), which were closely colocalized (*F*). (*A–F* are reprinted, with permission, from Zhao et al. 2002a.) (*G–Q*) EFs induce asymmetrical redistribution of signaling molecules in *Dictyostelium* cells during galvanotaxis: PTEN and myosin II to the rear (*G–I,L,M*), PHCrac and F-actin to the leading edge (*K,M,O–Q*). EF = 30 V/cm. (*G–R* are reprinted from Song et al. 2021, under the terms of the Creative Commons Attribution 4.0 International License.)

across different cell types (Kulkarni et al. 2025). The fact that such diverse groups of membrane proteins are being suggested to be "sensing" the EFs indicates a very different sensing mechanism from that of chemotaxis.

Polarized Activation of Ion Channels Underlies EF Sensing

Asymmetric activation of ion channels may cause the membrane potential change. The anodal side of a cell may have Ca^{2+} influx, and the cathodal side a Ca^{2+} efflux, causing net movement of the cell toward the cathode (Mycielska and Djamgoz 2004; Guo et al. 2015). Voltage-gated Na^{+} channels might be involved in the metastasis of many types of cancer (Djamgoz et al. 2001, 2019). Epithelial sodium channel (ENaC), a non-voltage-gated channel, was proposed to mediate galvanotaxis of keratinocytes (Yang et al. 2013). Manipulating cell membrane potential or ion channel activity indeed affects galvanotaxis (Brown and Loew 1994; Trollinger et al. 2002; Huang et al. 2009b; Gao et al. 2011, 2015; Allen et al. 2013; Nakajima et al. 2015). The involvement of ion channels is likely another redundant mechanism in sensing, signaling, and executing galvanotaxis.

Fast activation of ion channels and slow relocation of membrane proteins in EFs may complement each other in galvanotaxis. Indeed, in 3T3 fibroblasts, a biphasic galvanotaxis mechanism (i.e., initial ion channel activation followed by membrane receptor redistribution) has been suggested (Lasota et al. 2024).

Requirement of Membrane Potential for Galvanotaxis

Other sensing possibilities are likely present in certain cell types. For example, depolarization of membrane potential on the cathode side and hyperpolarization on the anode side of a cell in an EF, albeit very small (Jahn 1961; Jaffe and Nuccitelli 1977; Gao et al. 2011). This change in membrane potential was demonstrated by imaging of the spatial distribution and transmembrane potential changes in human carcinoma cells in strong fields (Gross et al. 1986). The mechanisms of the effects of such small nonuniform membrane potential changes on cells, particularly on tiny cell fragments and growth cones (Fig. 3), are not fully understood.

Voltage-sensing phosphatase 1 (Vsp1), a phosphoinositide phosphatase that converts $PtdIns(3,4,5)P_3$ to $PtdIns(4,5)P_2$ in a manner regulated by a voltage sensor domain, can affect other channels in feedback loops and also control the cytoskeleton (Murata et al. 2005; Okamura et al. 2018; Yu et al. 2024). Vsp1 contains a transmembrane voltage-sensing domain, which allows it to detect changes in membrane potential. The transmembrane domain is homologous to the S1–S4 segments of voltage-gated channels and its cytoplasmic domain is similar to phosphatase and tensin homolog. The threshold of membrane potential to activate Vsp1 is much lower and earlier of charge movement than regular voltage-gate ion channels to induce a conformational change in the phosphatase domain (Murata et al. 2005). Vsp1 was shown to be a key component of the molecular mechanism enabling neural crest cells to specifically transduce EFs into a directional cue in vivo (Ferreira et al. 2025).

More strong evidence supporting the role for membrane potential in galvanotaxis comes from a more drastic experiment—electroporation—which punches holes in cells. Electroporation significantly reduces membrane potential by inducing holes in the cell membrane. Collapse of membrane potential in this way abolished galvanotaxis and then cell membrane gradually reseals, the membrane potential recovers, and cells regain galvanotaxis response (Gao et al. 2011). Manipulating membrane potential through changes of extracellular pH, on the contrary, did not affect chemotaxis in *Dictyostelium* cells (Van Duijn et al. 1990; Gao et al. 2011).

The Mechanisms for Extracellular EF Sensing Are Highly Likely Redundant

As a conserved cellular response across cells in different kingdoms of life, it seems unlikely that a single molecule, or even a small group of molecules, on the membrane functions as the dedicated sensor(s) for galvanotaxis, as is often the case in chemotaxis. Highly redundant molecules on the membrane that are not dependent on binding to specific ligands make it very difficult if not impossible to perform genetic manipulation to completely nullify galvanotaxis, but only significantly decrease it (Zhao et al. 2002b; Pu et al. 2007; Gao et al. 2015; Nakajima et al. 2015; Song et al. 2021; Belliveau et al. 2024). Such "nonspecific" or "less specific" sensing perhaps underlies why a large complex of membrane molecules (i.e., lipid rafts) have been suggested in mediating galvanotaxis (Zhao et al. 2002a; Lin et al. 2017). We suggest that the mechanisms for extracellular EF sensing are highly redundant and can be compensated through multiple molecules and molecular complexes through bioelectrical and biochemical responses, which could also be multiphasic (Nakajima et al. 2015; Sun et al. 2023; Belliveau et al. 2024; Ciesielska et al. 2024; Lasota et al. 2024; Copos et al. 2025; Gorzkowska et al. 2025).

ASYMMETRIC INTRACELLULAR SIGNALING ORIENTATES CELL POLARIZATION

Asymmetric intracellular signaling orientates cell polarization, a process that directs cells to organize their internal structures and functions in response to external cues (SenGupta et al. 2021). When cells detect external EFs, they follow the same or similar intracellular signaling paradigm that governs their directional migration, closely mirroring the principles observed in chemotaxis (Fig. 5; Devreotes et al. 2017; Kuhn et al. 2021; SenGupta et al. 2021). This similarity suggests that cells use a conserved strategy to interpret and respond to diverse environmental signals, whether they are chemical or electrical in nature.

Numerous intracellular signaling pathways are activated when cells are exposed to EFs, with key players including PI3 kinase/Akt, Src-family kinases, and mitogen-activated protein (MAP) kinases (Zhao et al. 2002a, 2006; Sato et al. 2009; Nakajima et al. 2015; Song et al. 2021). These pathways are not activated uniformly across the cell; rather, they exhibit polarized activation that defines distinct regions: the leading edge, which drives forward movement, and the trailing edge, which retracts or stabilizes the rear. This polarized signaling is a hallmark of directional migration and aligns with observations in chemotaxis (Van Haastert and Devreotes 2004; Sato et al. 2009). For instance, in corneal epithelial cells exposed to an EF, extracellular signal-regulated kinases (ERK1/2) are activated more at the leading edge (Fig. 5B,D,F). This localized activation highlights how intracellular signals spatially coordinate cellular responses to external stimuli.

Further evidence of this signaling paradigm comes from studies in *Dictyostelium* cells, where PI3 kinase (PI3K) activation is marked by the distribution of PHCrac, a downstream effector, concentrating at the leading edge (Fig. 5K–M; Zhao et al. 2002b; Song et al. 2021). Conversely, the phosphatase Pten localizes to the trailing edge, establishing a molecular gradient that reinforces directional movement (Fig. 5K–M). This asymmetry in signaling molecule distribution ensures that cellular resources and activities are allocated efficiently to support migration.

The motor machinery of the cell also exhibits polarization in response to EFs, closely resembling patterns seen in chemotaxis (Zhao et al. 2002a,b, 2006; Pu and Zhao 2005; Cao et al. 2011; Nakajima et al. 2015; Song et al. 2021). Filamentous actin (F-actin), a key component of the cytoskeleton, accumulates at the leading edge (Fig. 5C,E,F), where it facilitates protrusion and forward movement. In contrast, myosin II, a motor protein responsible for contraction, concentrates at the trailing edge (Fig. 5I–R), aiding in retraction and maintaining cell shape during migration. This coordinated distribution of F-actin and myosin II underscores the mechanical basis of polarized movement, driven by the underlying signaling asymmetries.

Asymmetric intracellular signaling thus serves as a fundamental mechanism for orienting cell polarization and directing migration in response to external EFs (Zhao et al. 2002a; Shanley et al. 2006; Nakajima et al. 2015; SenGupta et al. 2021). By activating specific pathways like PI3K/Akt, Src-family kinases, and MAP kinases in a spatially organized manner, cells establish a leading-edge rich in protrusive F-actin and a trailing edge stabilized by myosin

II. These processes, strikingly similar to those in chemotaxis, reveal a unified framework through which cells interpret and navigate their environments, whether guided by EFs or chemical gradients (SenGupta et al. 2021).

EXCITABLE NETWORKS INTEGRATE, AMPLIFY, AND COORDINATE GALVANOTAXIS

Recent research points to "excitable networks," dynamic systems that act like a cellular decision-maker, amplifying and coordinating environmental signals to guide cell migration (Iglesias and Devreotes 2012; Devreotes et al. 2017). These networks include two main parts: the signal transduction excitable network (STEN), which handles signal processing with molecules like Ras and PI3K, and the cytoskeletal excitable network (CEN), which manages the cell's structure, especially the actin cytoskeleton, to form protrusions for movement. These networks can generate wave-like signals, helping cells polarize and move directionally (Iglesias and Devreotes 2012; Devreotes et al. 2017; Miao et al. 2019; Li et al. 2020; Banerjee et al. 2022). These excitable networks provide a dynamic framework for understanding how cells perceive and respond to environmental cues, unifying the signaling mechanisms for chemotaxis and galvanotaxis. Indeed, EFs induce such excitable network activities (Guo et al. 2021; Zhang et al. 2022; Zhu et al. 2022), which may mediate galvanotaxis (Iglesias and Devreotes 2012; Hoeller et al. 2016; Devreotes et al. 2017; Kuhn et al. 2021; Bull et al. 2022; Yang et al. 2022).

TOOLS/TECHNOLOGIES FOR GALVANOTAXIS

Recent advances in bioelectronics provide many exciting new tools and protocols for studying galvanotaxis and application of galvanotaxis in vivo. Classical galvanotaxis chambers are designed to expose cells to a constant electric current within a channel formed by two parallel glass coverslips in a Petri dish. Cells or tissue are seeded in the resulting trough, and a closed EF is established by connecting the cell culture media to agar salt bridges that link to the electrodes. Many new engineered devices have been developed (Song et al. 2007; Huang et al. 2009a; Cohen et al. 2014; Banks et al. 2015; Feng et al. 2017; Zajdel et al. 2020). Microfabricated devices integrating microchannels and fluidic components are now used in electrotactic studies to establish precise microenvironments that control conditions like pH, nutrient levels, temperature, and temporospatial controlled electrical stimuli (Fig. 6A–C; Zhao et al. 2014). A transwell-based electrotactic assay has been specifically designed for nonadherent cells, such as lymphocytes (Fig. 6D; Lin et al. 2008). Furthermore, a device featuring multiple electrodes, controlled by a computer, allows for two-dimensional manipulation through a series of independent migration command pulses along the x- or y-axis (Fig. 6E–G; Zajdel et al. 2020).

FUTURE PROSPECTS

Elucidating the mechanisms for sensing physiological EFs remains a central and challenging task. While recent studies have implicated numerous cell membrane proteins as potential galvanotaxis sensors, a unified mechanistic understanding will likely involve the excitable networks, the STEN and the CEN (Devreotes et al. 2017; Miao et al. 2019; Song et al. 2021; Banerjee et al. 2022; Yang et al. 2022). Advanced genetic screening (Belliveau et al. 2024), dynamic signaling imaging (Banerjee et al. 2022; Bull et al. 2022; Yang et al. 2022), and network analysis (both experimental and modeling approaches) (Iglesias and Devreotes 2012; Zhan et al. 2020; Banerjee et al. 2022) are expected to bring about a unifying sensing paradigm that we believe is more likely an integrating network through excitability. The dynamic, wave-like activity observed in migrating cells enables them to integrate and respond to diverse sensing components to respond to the electric cues effectively (Devreotes et al. 2017).

The powerful galvanotaxis response with great manipulability coupled with the advancement of bioelectronics will offer a useful tool for studying directional cell migration, as well as for engineering tissues. The latter can be exploited

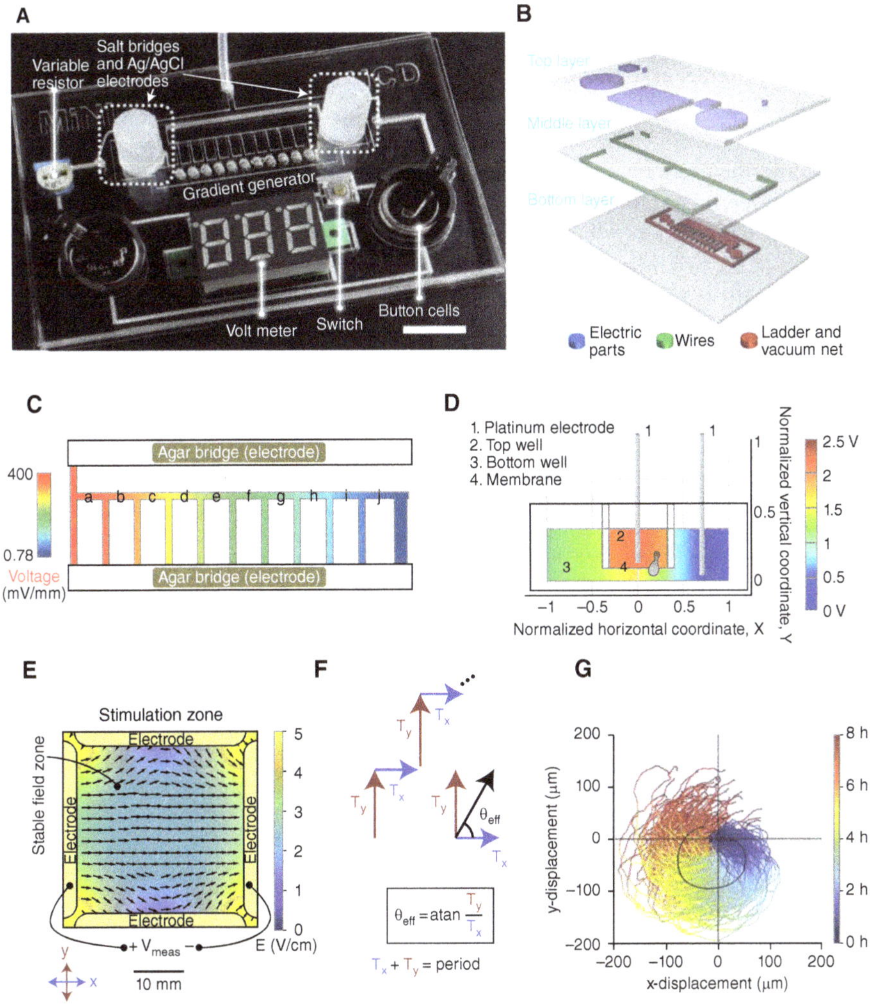

Figure 6. Example of new engineering tools for galvanotaxis of cells. (*A*) An integrated ElectroTaxis-on-a-Chip (ETC) device (scale bar, 1 cm). (*A*,*B*) The three-layer structure of the device. (*B*) This ETC device facilitates high-throughput experiments of galvanotaxis by incorporating all necessary components for inducing cell galvanotaxis—excluding microscopy—onto a platform the size of a credit card. (*C*) Drawing inspiration from the R-2R resistor ladder topology used in digital signal processing infinitely expandable microfluidic generator capable of producing electric field (EF) gradients of great numbers for galvanotaxis experiments was developed. The strongest field is shown on the *left* in red; the minimum field on the *right* in dark blue. (*A* is reprinted from Zhao et al. 2014; *C* is reprinted from Zhao et al. 2020.) (*D*) A transwell-based electrotactic assay with the EFs applied to the transwell (2.5 V across the transwell) by positioning two electrodes on the *top* and *bottom* of the transwell. Cells are allowed to migrate from the *top* well to the *bottom* well through pores in the membrane. This method is designed for nonadherent cells, such as lymphocytes. Simulated electrical potentials are color-coded indicating the voltage. (*D* is reprinted, with permission, from Lin et al. 2008.) (*E*–*G*) A device with four electrodes. (*E*) Computational simulations show EFs generated by the device when the horizontal electrode pair is activated. The EF intensity is ~200 mV/mm. (*F*) The stimulation procedure divides a 20 sec stimulation interval into two segments: horizontal stimulation (Tx) and vertical stimulation (Ty). The ratio of these segments determines the effective stimulation angle that the cells experience. (*G*) The trajectories of central segments from two keratinocyte monolayers during an 8 h circular stimulation period are shown, with black arrows indicating the direction of stimulation. (*E*–*G* are adapted from Zajdel et al. 2020, under the terms of the Creative Commons Attribution 4.0 International license.)

with other cellular effects of EFs, such as cell orientation, division, and even the assembly of cells into specific patterns, and thus has the potential for promoting or inhibiting tissue construction, as well as blood vessel and neuron growth, and could therefore be a powerful new tool for tissue engineering. At the monolayer cell level, a patterned EF could result in the patterned dynamics of the cell migration in monolayer (Cohen et al. 2014; Zhang et al. 2022).

Investigating galvanotaxis in vivo is challenging but will reveal perhaps more robust examples in wound healing, regeneration, development, and cancer metastasis (Djamgoz et al. 2019; Levin 2021). Tractable model systems such as the migration of neural crest cells in *Xenopus* embryos are expected to bring significant mechanistic insights (Ferreira et al. 2025).

COMPETING INTEREST STATEMENT

M.Z. is a member of the Scientific Advisory Board (SAB) of Morphoceuticals, Inc. The other authors declare that they have no known competing financial interests or personal relationships that could have appeared to influence the work reported in this paper. The content of this article was written by the authors listed.

ACKNOWLEDGMENTS

The work in the Zhao laboratory was supported by National Institutes of Health (NIH) grant 1R21AI156409-01, by Defense Advanced Research Projects Agency (DARPA) HR001119S0027 (program PI: M. Rolandi), an Air Force Office of Scientific Research Multidisciplinary University Research Initiative (AFOSR MURI) grant (FA9550-16-1-0052, Program Leader: W. Losert, University of Maryland), a Defense University Research Instrumentation Program (DURIP) (FA9550-22-1-0149), NIH grant 1R01EY019101, and a National Eye Institute (NEI) Core Grant (P-30 EY012576). The Zhao laboratory thanks the Burns family, Mr. and Mrs. Meyers, and H. Schroeter for their generous donations that bolster research efforts. We thank the original authors for permitting us to reuse figures in this paper.

REFERENCES

Adler J, Shi W. 1988. Galvanotaxis in bacteria. *Cold Spring Harb Symp Quant Biol* **53:** 23–25. doi:10.1101/SQB.1988.053.01.006

Allen GM, Mogilner A, Theriot JA. 2013. Electrophoresis of cellular membrane components creates the directional cue guiding keratocyte galvanotaxis. *Curr Biol* **23:** 560–568. doi:10.1016/j.cub.2013.02.047

Babona-Pilipos R, Droujinine IA, Popovic MR, Morshead CM. 2011. Adult subependymal neural precursors, but not differentiated cells, undergo rapid cathodal migration in the presence of direct current electric fields. *PLoS One* **6:** e23808. doi:10.1371/journal.pone.0023808

Bai H, McCaig CD, Forrester JV, Zhao M. 2004. DC electric fields induce distinct preangiogenic responses in microvascular and macrovascular cells. *Arterioscler Thromb Vasc Biol* **24:** 1234–1239. doi:10.1161/01.ATV.0000131265.76828.8a

Banerjee T, Biswas D, Pal DS, Miao Y, Iglesias PA, Devreotes PN. 2022. Spatiotemporal dynamics of membrane surface charge regulates cell polarity and migration. *Nat Cell Biol* **24:** 1499–1515. doi:10.1038/s41556-022-00997-7

Banks TA, Luckman PS, Frith JE, Cooper-White JJ. 2015. Effects of electric fields on human mesenchymal stem cell behaviour and morphology using a novel multichannel device. *Integr Biol (Camb)* **7:** 693–712. doi:10.1039/c4ib00297k

Barker AT, Jaffe LF, Vanable JW Jr. 1982. The glabrous epidermis of cavies contains a powerful battery. *Am J Physiol* **242:** R358–R366.

Barman SR, Chan SW, Kao FC, Ho HY, Khan I, Pal A, Huang CC, Lin ZH. 2023. A self-powered multifunctional dressing for active infection prevention and accelerated wound healing. *Sci Adv* **9:** eadc8758. doi:10.1126/sciadv.adc8758

Barriga EH, Theveneau E. 2020. In vivo neural crest cell migration is controlled by "mixotaxis." *Front Physiol* **11:** 586432. doi:10.3389/fphys.2020.586432

Belliveau NM, Footer MJ, Platenkamp A, Kim H, Eustis TE, Theriot JA. 2024. Galvanin is an electric-field sensor for directed cell migration. bioRxiv doi:10.1101/2024.09.23.614580

Bonazzi D, Minc N. 2014. Dissecting the molecular mechanisms of electrotactic effects. *Adv Wound Care* **3:** 139–148. doi:10.1089/wound.2013.0438

Borgens RB, Shi R. 1995. Uncoupling histogenesis from morphogenesis in the vertebrate embryo by collapse of the transneural tube potential. *Dev Dyn* **203:** 456–467. doi:10.1002/aja.1002030408

Brown MJ, Loew LM. 1994. Electric field-directed fibroblast locomotion involves cell surface molecular reorganization and is calcium independent. *J Cell Biol* **127:** 117–128. doi:10.1083/jcb.127.1.117

Bull AL, Campanello L, Hourwitz MJ, Yang Q, Zhao M, Fourkas JT, Losert W. 2022. Actin dynamics as a multiscale integrator of cellular guidance cues. *Front Cell Dev Biol* **10:** 873567. doi:10.3389/fcell.2022.873567

Cao L, Pu J, Zhao M. 2011. GSK-3beta is essential for physiological electric field-directed Golgi polarization and optimal electrotaxis. *Cell Mol Life Sci* **68:** 3081–3093. doi:10.1007/s00018-010-0608-z

Chiang M, Robinson KR, Vanable JW Jr. 1992. Electrical fields in the vicinity of epithelial wounds in the isolated bovine eye. *Exp Eye Res* **54:** 999–1003. doi:10.1016/0014-4835(92)90164-N

Cho Y, Son M, Jeong H, Shin JH. 2018. Electric field-induced migration and intercellular stress alignment in a collective epithelial monolayer. *Mol Biol Cell* **29:** 2292–2302. doi:10.1091/mbc.E18-01-0077

Ciesielska P, Lasota S, Bobis-Wozowicz S, Madeja Z. 2024. The role of TGF-β in the electrotactic reaction of mouse 3T3 fibroblasts in vitro. *Acta Biochim Pol* **71:** 12993. doi:10.3389/abp.2024.12993

Cohen DJ, Nelson WJ, Maharbiz MM. 2014. Galvanotactic control of collective cell migration in epithelial monolayers. *Nat Mater* **13:** 409–417. doi:10.1038/nmat3891

Cook GM. 2016. Glycobiology of the cell surface: its debt to cell electrophoresis 1940–65. *Electrophoresis* **37:** 1399–1406. doi:10.1002/elps.201500476

Cooper M, Schliwa M. 1985. Electrical and ionic controls of tissue cell locomotion in DC electric fields. *J Neurosci Res* **13:** 223–244. doi:10.1002/jnr.490130116

Cooper M, Schliwa M. 1986. Motility of cultured fish epidermal cells in the presence and absence of direct current electric fields. *J Cell Biol* **102:** 1384–1399. doi:10.1083/jcb.102.4.1384

Copos C, Sun YH, Zhu K, Zhang Y, Reid B, Draper B, Lin F, Yue H, Bernadskaya Y, Zhao M, et al. 2025. Galvanotactic directionality of cell groups depends on group size. *Proc Natl Acad Sci* **122:** e2416440122. doi:10.1073/pnas.2416440122

Cortese B, Palamà IE, D'Amone S, Gigli G. 2014. Influence of electrotaxis on cell behaviour. *Integr Biol* **6:** 817–830. doi:10.1039/C4IB00142G

Curtze S, Dembo M, Miron M, Jones DB. 2004. Dynamic changes in traction forces with DC electric field in osteoblast-like cells. *J Cell Sci* **117:** 2721–2729. doi:10.1242/jcs.01119

Devreotes PN, Bhattacharya S, Edwards M, Iglesias PA, Lampert T, Miao Y. 2017. Excitable signal transduction networks in directed cell migration. *Annu Rev Cell Dev Biol* **33:** 103–125. doi:10.1146/annurev-cellbio-100616-060739

Djamgoz MBA, Mycielska M, Madeja Z, Fraser SP, Korohoda W. 2001. Directional movement of rat prostate cancer cells in direct-current electric field: involvement of voltage-gated Na^+ channel activity. *J Cell Sci* **114:** 2697–2705. doi:10.1242/jcs.114.14.2697

Djamgoz MBA, Fraser SP, Brackenbury WJ. 2019. In vivo evidence for voltage-gated sodium channel expression in carcinomas and potentiation of metastasis. *Cancers (Basel)* **11:** 1675. doi:10.3390/cancers11111675

Eckert R. 1972. Bioelectric control of ciliary activity. *Science* **176:** 473–481. doi:10.1126/science.176.4034.473

Ende K, Santos F, Guasch J, Kemkemer R. 2024. Migration of human T cells can be differentially directed by electric fields depending on the extracellular microenvironment. *iScience* **27:** 109746. doi:10.1016/j.isci.2024.109746

Fang KS, Ionides E, Oster G, Nuccitelli R, Isseroff RR. 1999. Epidermal growth factor receptor relocalization and kinase activity are necessary for directional migration of keratinocytes in DC electric fields. *J Cell Sci* **112:** 1967–1978. doi:10.1242/jcs.112.12.1967

Feng JF, Liu J, Zhang XZ, Zhang L, Jiang JY, Nolta J, Zhao M. 2012. Guided migration of neural stem cells derived from human embryonic stem cells by an electric field. *Stem Cells* **30:** 349–355. doi:10.1002/stem.779

Feng JF, Liu J, Zhang L, Jiang JY, Russell M, Lyeth BG, Nolta JA, Zhao M. 2017. Electrical guidance of human stem cells in the rat brain. *Stem Cell Rep* **9:** 177–189. doi:10.1016/j.stemcr.2017.05.035

Ferreira F, Moreira S, Zhao M, Barriga EH. 2025. Stretch-induced endogenous electric fields drive directed collective cell migration in vivo. *Nat Mater* **24:** 462–470. doi:10.1038/s41563-024-02060-2

Franke K, Gruler H. 1990. Galvanotaxis of human granulocytes: electric field jump studies. *Eur Biophys J* **18:** 334–346. doi:10.1007/BF00196924

Frederich BJ, Timofeyev V, Thai PN, Haddad MJ, Poe AJ, Lau VC, Moshref M, Knowlton AA, Sirish P, Chiamvimonvat N. 2017. Electrotaxis of cardiac progenitor cells, cardiac fibroblasts, and induced pluripotent stem cell-derived cardiac progenitor cells requires serum and is directed via PI3′K pathways. *Heart Rhythm* **14:** 1685–1692. doi:10.1016/j.hrthm.2017.06.038

Gao J, Sun X, Yatsula V, Wymore R, Mathias R. 2000. Isoform-specific function and distribution of Na/K pumps in the frog lens epithelium. *J Membr Biol* **178:** 89–101. doi:10.1007/s002320010017

Gao RC, Zhang XD, Sun YH, Kamimura Y, Mogilner A, Devreotes PN, Zhao M. 2011. Different roles of membrane potentials in electrotaxis and chemotaxis of *Dictyostelium* cells. *Eukaryot cell* **10:** 1251–1256. doi:10.1128/EC.05066-11

Gao R, Zhao S, Jiang X, Sun Y, Zhao S, Gao J, Borleis J, Willard S, Tang M, Cai H, et al. 2015. A large-scale screen reveals genes that mediate electrotaxis in *Dictyostelium discoideum*. *Sci Signal* **8:** ra50. doi:10.1126/scisignal.aab056

Giugni TD, Braslau DL, Haigler HT. 1987. Electric field-induced redistribution and postfield relaxation of epidermal growth factor receptors on A431 cells. *J Cell Biol* **104:** 1291–1297. doi:10.1083/jcb.104.5.1291

Gokoffski KK, Jia X, Shvarts D, Xia G, Zhao M. 2019. Physiologic electrical fields direct retinal ganglion cell axon growth in vitro. *Invest Ophthalmol Vis Sci* **60:** 3659–3668. doi:10.1167/iovs.18-25118

Gorzkowska J, Kozak W, Bobis-Wozowicz S, Cherepashuk I, Madeja Z, Lasota S. 2025. The dynamics of chemoattractant receptors redistribution in the electrotaxis of 3T3 fibroblasts. *Cell Commun Signal* **23:** 173. doi:10.1186/s12964-025-02165-4

Gowen CW Jr, Gowen MA, Knowles MR. 1991. Colonic transepithelial potential difference in infants with cystic fibrosis. *J Pediatr* **118:** 412–415. doi:10.1016/S0022-3476(05)82158-4

Griffin AJ, McSorley SJ. 2011. Development of protective immunity to Salmonella, a mucosal pathogen with a systemic agenda. *Mucosal Immunol* **4:** 371–382. doi:10.1038/mi.2011.2

Gross D, Loew LM, Webb WW. 1986. Optical imaging of cell membrane potential changes induced by applied electric fields. *Biophys J* **50:** 339–348. doi:10.1016/S0006-3495(86)83467-1

Guido I, Olszok N, Diehl D, Bodenschatz E. 2024. Electrotaxis of *Dictyostelium discoideum*, migration in an electric field. In *Cell motility and chemotaxis: methods and protocols* (ed. Beta C, Martinez-Torres C), pp. 107–117. Springer, New York.

Guo L, Xu C, Li D, Zheng X, Tang J, Bu J, Sun H, Yang Z, Sun W, Yu X. 2015. Calcium ion flow permeates cells through SOCs to promote cathode-directed galvanotaxis. *PLoS One* **10:** e0139865. doi:10.1371/journal.pone.0139865

Guo L, Zhu K, Pargett M, Contreras A, Tsai P, Qing Q, Losert W, Albeck J, Zhao M. 2021. Electrically synchronizing and modulating the dynamics of ERK activation to regulate cell fate. *iScience* **24:** 103240. doi:10.1016/j.isci.2021.103240

Hinkle L, Mccaig CD, Robinson KR. 1981. The direction of growth of differentiating neurones and myoblasts from frog embryos in an applied electric field. *J Physiol* **314:** 121–135. doi:10.1113/jphysiol.1981.sp013695

Hoeller O, Toettcher JE, Cai H, Sun Y, Huang CH, Freyre M, Zhao M, Devreotes PN, Weiner OD. 2016. Gβ regulates coupling between actin oscillators for cell polarity and directional migration. *PLoS Biol* **14:** e1002381. doi:10.1371/journal.pbio.1002381

Huang CW, Cheng JY, Yen MH, Young TH. 2009a. Electrotaxis of lung cancer cells in a multiple-electric-field chip. *Biosens Bioelectron* **24:** 3510–3516. doi:10.1016/j.bios.2009.05.001

Huang L, Cormie P, Messerli MA, Robinson KR. 2009b. The involvement of Ca^{2+} and integrins in directional responses of zebrafish keratocytes to electric fields. *J Cell Physiol* **219:** 162–172. doi:10.1002/jcp.21660

Huang YJ, Schiapparelli P, Kozielski K, Green J, Lavell E, Guerrero-Cazares H, Quinones-Hinojosa A, Searson P. 2017. Electrophoresis of cell membrane heparan sulfate regulates galvanotaxis in glial cells. *J Cell Sci* **130:** 2459–2467. doi:10.1242/jcs.203752

Hunckler J, de Mel A. 2017. A current affair: electrotherapy in wound healing. *J Multidiscip Healthc* **10:** 179–194. doi:10.2147/JMDH.S127207

Iglesias PA, Devreotes PN. 2012. Biased excitable networks: how cells direct motion in response to gradients. *Curr Opin Cell Biol* **24:** 245–253. doi:10.1016/j.ceb.2011.11.009

Jaffe LF, Nuccitelli R. 1977. Electrical controls of development. *Annu Rev Biophys Bioeng* **6:** 445–476. doi:10.1146/annurev.bb.06.060177.002305

Jaffe LF, Poo MM. 1979. Neurites grow faster towards the cathode than the anode in a steady field. *J Exp Zool* **209:** 115–128. doi:10.1002/jez.1402090114

Jaffe LF, Stern CD. 1979. Strong electrical currents leave the primitive streak of chick embryos. *Science* **206:** 569–571. doi:10.1126/science.573921

Jahn TL. 1961. The mechanism of ciliary movement. I: Ciliary reversal and activation by electric current; the Ludloff phenomenon in terms of core and volume conductors. *J Protozool* **8:** 369–380. doi:10.1111/j.1550-7408.1961.tb01231.x

Jepson MA, Clark MA. 1998. Studying M cells and their role in infection. *Trends Microbiol* **6:** 359–365. doi:10.1016/S0966-842X(98)01337-7

Jepson MA, Clark MA. 2001. The role of M cells in *Salmonella* infection. *Microbes Infect* **3:** 1183–1190. doi:10.1016/S1286-4579(01)01478-2

Jiang Y, Trotsyuk AA, Niu S, Henn D, Chen K, Shih CC, Larson MR, Mermin-Bunnell AM, Mittal S, Lai JC, et al. 2023. Wireless, closed-loop, smart bandage with integrated sensors and stimulators for advanced wound care and accelerated healing. *Nat Biotechnol* **41:** 652–662. doi:10.1038/s41587-022-01528-3

Jones BD, Ghori N, Falkow S. 1994. *Salmonella typhimurium* initiates murine infection by penetrating and destroying the specialized epithelial M cells of the Peyer's patches. *J Exp Med* **180:** 15–23. doi:10.1084/jem.180.1.15

Keller HU, Bessis M. 1975. Migration and chemotaxis of anucleate cytoplasmic leukocyte fragments. *Nature* **258:** 723–724. doi:10.1038/258723a0

Korohoda W, Mycielska M, Janda E, Madeja Z. 2000. Immediate and long-term galvanotactic responses of *Amoeba proteus* to DC electric fields. *Cell Motil Cytoskeleton* **45:** 10–26. doi:10.1002/(SICI)1097-0169(200001)45:1<10::AID-CM2>3.0.CO;2-T

Kuhn J, Lin Y, Devreotes PN. 2021. Using live-cell imaging and synthetic biology to probe directed migration in *Dictyostelium*. *Front Cell Dev Biol* **9:** 740205. doi:10.3389/fcell.2021.740205

Kulkarni S, Tebar F, Rentero C, Zhao M, Sáez P. 2025. Competing signaling pathways controls electrotaxis. *iScience* **28:** 112329. doi:10.1016/j.isci.2025.112329

Lasota S, Zimolag E, Bobis-Wozowicz S, Pilipiuk J, Madeja Z. 2024. The dynamics of the electrotactic reaction of mouse 3T3 fibroblasts. *Biochim Biophys Acta Mol Cell Res* **1871:** 119647. doi:10.1016/j.bbamcr.2023.119647

Leal J, Shaner S, Jedrusik N, Savelyeva A, Asplund M. 2023. Electrotaxis evokes directional separation of co-cultured keratinocytes and fibroblasts. *Sci Rep* **13:** 11444. doi:10.1038/s41598-023-38664-y

Levin M. 2012a. Molecular bioelectricity in developmental biology: new tools and recent discoveries: control of cell behavior and pattern formation by transmembrane potential gradients. *Bioessays* **34:** 205–217. doi:10.1002/bies.201100136

Levin M. 2012b. Morphogenetic fields in embryogenesis, regeneration, and cancer: non-local control of complex patterning. *Biosystems* **109:** 243–261. doi:10.1016/j.biosystems.2012.04.005

Levin M. 2021. Bioelectric signaling: reprogrammable circuits underlying embryogenesis, regeneration, and cancer. *Cell* **184:** 1971–1989. doi:10.1016/j.cell.2021.02.034

Li Q. 2022. Mechanisms for the invasion and dissemination of *Salmonella*. *Can J Infect Dis Med Microbiol* **2022:** 2655801. doi:10.1155/2022/2655801

Li L, Hartley R, Reiss B, Sun Y, Pu J, Wu D, Lin F, Hoang T, Yamada S, Jiang J, et al. 2012. E-cadherin plays an essential role in collective directional migration of large epithelial sheets. *Cell Mol Life Sci* **69:** 2779–2789. doi:10.1007/s00018-012-0951-3

Li A, Cho JH, Reid B, Tseng CC, He L, Tan P, Yeh CY, Wu P, Li Y, Widelitz RB, et al. 2018a. Calcium oscillations coordinate feather mesenchymal cell movement by SHH dependent modulation of gap junction networks. *Nat Commun* **9:** 5377. doi:10.1038/s41467-018-07661-5

Li Y, Gu Y, Wang H, Liu Z, Song B, Yin T. 2018b. Electric pulses can influence galvanotaxis of *Dictyostelium discoideum*. *Biomed Res Int* **2018:** 2534625. doi:10.1155/2018/2534625

Li X, Miao Y, Pal DS, Devreotes PN. 2020. Excitable networks controlling cell migration during development and disease. *Semin Cell Dev Biol* **100:** 133–142. doi:10.1016/j.semcdb.2019.11.001

Lin F, Baldessari F, Gyenge CC, Sato T, Chambers RD, Santiago JG, Butcher EC. 2008. Lymphocyte electrotaxis in vitro and in vivo. *J Immunol* **181:** 2465–2471. doi:10.4049/jimmunol.181.4.2465

Lin BJ, Tsao SH, Chen A, Hu SK, Chao L, Chao PG. 2017. Lipid rafts sense and direct electric field-induced migration. *Proc Natl Acad Sci* **114:** 8568–8573. doi:10.1073/pnas.1702526114

Luo R, Liang Y, Yang J, Feng H, Chen Y, Jiang X, Zhang Z, Liu J, Bai Y, Xue J, et al. 2023. Reshaping the endogenous electric field to boost wound repair via electrogenerative dressing. *Adv Mater* **35:** e2208395. doi:10.1002/adma.202208395

McCaig CD, Rajnicek AM, Song B, Zhao M. 2002. Has electrical growth cone guidance found its potential? *Trends Neurosci* **25:** 354–359. doi:10.1016/S0166-2236(02)02174-4

McCaig CD, Rajnicek AM, Song B, Zhao M. 2005. Controlling cell behavior electrically: current views and future potential. *Physiol Rev* **85:** 943–978. doi:10.1152/physrev.00020.2004

McGillivray AM, Gow NAR. 1986. Applied electrical fields polarize the growth of mycelial fungi. *J Gen Microbiol* **132:** 2515–2525. doi:10.1099/00221287-132-9-2515

Miao Y, Bhattacharya S, Banerjee T, Abubaker-Sharif B, Long Y, Inoue T, Iglesias PA, Devreotes PN. 2019. Wave patterns organize cellular protrusions and control cortical dynamics. *Mol Syst Biol* **15:** e8585. doi:10.15252/msb.20188585

Minc N, Chang F. 2010. Electrical control of cell polarization in the fission yeast *Schizosaccharomyces pombe*. *Curr Biol* **20:** 710–716. doi:10.1016/j.cub.2010.02.047

Moarefian M, Davalos RV, Burton MD, Jones CN. 2021. Electrotaxis-on-Chip to quantify neutrophil migration towards electrochemical gradients. *Front Immunol* **12:** 674727. doi:10.3389/fimmu.2021.674727

Mossop BJ, Barr RC, Zaharoff DA, Yuan F. 2004. Electric fields within cells as a function of membrane resistivity—a model study. *IEEE Trans Nanobioscience* **3:** 225–231. doi:10.1109/TNB.2004.833703

Mukerjee EV, Isseroff RR, Nuccitelli R, Collins SD, Smith RL. 2006. Microneedle array for measuring wound generated electric fields. *Conf Proc IEEE Eng Med Biol Soc* **2006:** 4326–4328. doi:10.1109/IEMBS.2006.260205

Murata Y, Iwasaki H, Sasaki M, Inaba K, Okamura Y. 2005. Phosphoinositide phosphatase activity coupled to an intrinsic voltage sensor. *Nature* **435:** 1239–1243. doi:10.1038/nature03650

Mycielska ME, Djamgoz MB. 2004. Cellular mechanisms of direct-current electric field effects: galvanotaxis and metastatic disease. *J Cell Sci* **117:** 1631–1639. doi:10.1242/jcs.01125

Nagel WA. 1895. Ueber galvanotaxis. *Archiv Physiol Menschen Tiere* **59:** 603–642.

Nakajima K, Zhu K, Sun YH, Hegyi B, Zeng Q, Murphy CJ, Small JV, Chen-Izu Y, Izumiya Y, Penninger JM, et al. 2015. KCNJ15/Kir4.2 couples with polyamines to sense weak extracellular electric fields in galvanotaxis. *Nat Commun* **6:** 8532. doi:10.1038/ncomms9532

Nishimura KY, Isseroff RR, Nuccitelli R. 1996. Human keratinocytes migrate to the negative pole in direct current electric fields comparable to those measured in mammalian wounds. *J Cell Sci* **109:** 199–207. doi:10.1242/jcs.109.1.199

Nuccitelli R. 2003a. Endogenous electric fields in embryos during development, regeneration and wound healing. *Radiat Prot Dosimetry* **106:** 375–383. doi:10.1093/oxfordjournals.rpd.a006375

Nuccitelli R. 2003b. A role for endogenous electric fields in wound healing. *Curr Top Dev Biol* **58:** 1–26. doi:10.1016/S0070-2153(03)58001-2

Nuccitelli R, Jaffe LF. 1974. Spontaneous current pulses through developing fucoid eggs. *Proc Natl Acad Sci* **71:** 4855–4859. doi:10.1073/pnas.71.12.4855

Nuccitelli R, Nuccitelli P, Ramlatchan S, Sanger R, Smith PJ. 2008. Imaging the electric field associated with mouse and human skin wounds. *Wound Repair Regen* **16:** 432–441. doi:10.1111/j.1524-475X.2008.00389.x

Nuccitelli R, Nuccitelli P, Li C, Narsing S, Pariser DM, Lui K. 2011. The electric field near human skin wounds declines with age and provides a noninvasive indicator of wound healing. *Wound Repair Regen* **19:** 645–655. doi:10.1111/j.1524-475X.2011.00723.x

Nwogbaga I, Kim AH, Camley BA. 2023. Physical limits on galvanotaxis. *Phys Rev E* **108:** 064411. doi:10.1103/PhysRevE.108.064411

Ofer N, Mogilner A, Keren K. 2011. Actin disassembly clock determines shape and speed of lamellipodial fragments. *Proc Natl Acad Sci* **108:** 20394–20399. doi:10.1073/pnas.1105333108

Ogawa N, Oku H, Hashimoto K, Ishikawa M. 2006. A physical model for galvanotaxis of *Paramecium* cell. *J Theor Biol* **242:** 314–328. doi:10.1016/j.jtbi.2006.02.021

Okamura Y, Kawanabe A, Kawai T. 2018. Voltage-sensing phosphatases: biophysics, physiology, and molecular engineering. *Physiol Rev* **98:** 2097–2131. doi:10.1152/physrev.00056.2017

Orida N, Poo MM. 1978. Electrophoretic movement and localisation of acetylcholine receptors in the embryonic muscle cell membrane. *Nature* **275:** 31–35. doi:10.1038/275031a0

Poo M. 1981. In situ electrophoresis of membrane components. *Annu Rev Biophys Bioeng* **10:** 245–276. doi:10.1146/annurev.bb.10.060181.001333

Poo MM, Robinson KR. 1977. Electrophoresis of concanavalin A receptors along embryonic muscle cell membrane. *Nature* **265:** 602–605. doi:10.1038/265602a0

Pu J, Zhao M. 2005. Golgi polarization in a strong electric field. *J Cell Sci* **118:** 1117–1128. doi:10.1242/jcs.01646

Pu J, McCaig CD, Cao L, Zhao Z, Segall JE, Zhao M. 2007. EGF receptor signalling is essential for electric-field-directed migration of breast cancer cells. *J Cell Sci* **120:** 3395–3403. doi:10.1242/jcs.002774

Pullar CE, Baier BS, Kariya Y, Russell AJ, Horst BA, Marinkovich MP, Isseroff RR. 2006. Beta4 integrin and epidermal growth factor coordinately regulate electric field-mediated directional migration via Rac1. *Mol Biol Cell* **17:** 4925–4935. doi:10.1091/mbc.e06-05-0433

Rabbani M, Rahman E, Powner MB, Triantis IF. 2024. Making sense of electrical stimulation: a meta-analysis for wound healing. *Ann Biomed Eng* **52:** 153–177. doi:10.1007/s10439-023-03371-2

Rajnicek AM. 1993. Bacterial galvanotropism: mechanisms and applications. *Sci Prog* **77:** 139–151.

Rajnicek AM, McCaig CD, Gow NA. 1994. Electric fields induce curved growth of *Enterobacter cloacae*, *Escherichia coli*, and *Bacillus subtilis* cells: implications for mechanisms of galvanotropism and bacterial growth. *J Bacteriol* **176:** 702–713. doi:10.1128/jb.176.3.702-713.1994

Rajnicek AM, Robinson KR, McCaig CD. 1998. The direction of neurite growth in a weak DC electric field depends on the substratum: contributions of adhesivity and net surface charge. *Dev Biol* **203:** 412–423. doi:10.1006/dbio.1998.9039

Reid B, Song B, McCaig CD, Zhao M. 2005. Wound healing in rat cornea: the role of electric currents. *FASEB J* **19:** 379–386. doi:10.1096/fj.04-2325com

Reid B, Nuccitelli R, Zhao M. 2007. Non-invasive measurement of bioelectric currents with a vibrating probe. *Nat Protoc* **2:** 661–669. doi:10.1038/nprot.2007.91

Robinson KR. 1985. The responses of cells to electrical fields: a review. *J Cell Biol* **101:** 2023–2027. doi:10.1083/jcb.101.6.2023

Rodrigues M, Kosaric N, Bonham CA, Gurtner GC. 2019. Wound healing: a cellular perspective. *Physiol Rev* **99:** 665–706. doi:10.1152/physrev.00067.2017

Sarkar A, Kobylkevich BM, Graham DM, Messerli MA. 2019. Electromigration of cell surface macromolecules in DC electric fields during cell polarization and galvanotaxis. *J Theor Biol* **478:** 58–73. doi:10.1016/j.jtbi.2019.06.015

Sato MJ, Kuwayama H, van Egmond WN, Takayama AL, Takagi H, van Haastert PJ, Yanagida T, Ueda M. 2009. Switching direction in electric-signal-induced cell migration by cyclic guanosine monophosphate and phosphatidylinositol signaling. *Proc Natl Acad Sci* **106:** 6667–6672. doi:10.1073/pnas.0809974106

Saw TB, Gao XM, Li MC, He JA, Le AP, Marsh S, Lin KH, Ludwig A, Prost J, Lim CT. 2022. Transepithelial potential difference governs epithelial homeostasis by electromechanics. *Nat Phys* **18:** 1122–1128. doi:10.1038/s41567-022-01657-1

Scarpa E, Mayor R. 2016. Collective cell migration in development. *J Cell Biol* **212:** 143–155. doi:10.1083/jcb.201508047

SenGupta S, Parent CA, Bear JE. 2021. The principles of directed cell migration. *Nat Rev Mol Cell Biol* **22:** 529–547. doi:10.1038/s41580-021-00366-6

Shanley LJ, Walczysko P, Bain M, MacEwan DJ, Zhao M. 2006. Influx of extracellular Ca^{2+} is necessary for electrotaxis in *Dictyostelium*. *J Cell Sci* **119:** 4741–4748. doi:10.1242/jcs.03248

Shen YY, Pfluger T, Ferreira F, Liang JB, Navedo MF, Zeng QL, Reid B, Zhao M. 2016. Diabetic cornea wounds produce significantly weaker electric signals that may contribute to impaired healing. *Sci Rep* **6:** 26525. doi:10.1038/srep26525

Sheridan DM, Isseroff RR, Nuccitelli R. 1996. Imposition of a physiologic DC electric field alters the migratory response of human keratinocytes on extracellular matrix molecules. *J Invest Dermatol* **106:** 642–646. doi:10.1111/1523-1747.ep12345456

Shi R, Borgens RB. 1995. Three-dimensional gradients of voltage during development of the nervous system as invisible coordinates for the establishment of embryonic pattern. *Dev Dyn* **202:** 101–114. doi:10.1002/aja.1002020202

Shi W, Stocker BA, Adler J. 1996. Effect of the surface composition of motile *Escherichia coli* and motile *Salmonella* species on the direction of galvanotaxis. *J Bacteriol* **178:** 1113–1119. doi:10.1128/jb.178.4.1113-1119.1996

Shim G, Breinyn IB, Martínez-Calvo A, Rao S, Cohen DJ. 2024. Bioelectric stimulation controls tissue shape and size. *Nat Commun* **15:** 2938. doi:10.1038/s41467-024-47079-w

Shirzaei Sani E, Xu C, Wang C, Song Y, Min J, Tu J, Solomon SA, Li J, Banks JL, Armstrong DG, et al. 2023. A stretchable wireless wearable bioelectronic system for multiplexed monitoring and combination treatment of infected chronic wounds. *Sci Adv* **9:** eadf7388. doi:10.1126/sciadv.adf7388

Song B, Gu Y, Pu J, Reid B, Zhao Z, Zhao M. 2007. Application of direct current electric fields to cells and tissues in vitro and modulation of wound electric field in vivo. *Nat Protoc* **2:** 1479–1489. doi:10.1038/nprot.2007.205

Song B, Gu Y, Jiang W, Li Y, Ayre WN, Liu Z, Yin T, Janetopoulos C, Iijima M, Devreotes P, et al. 2021. Electric signals counterbalanced posterior vs anterior PTEN signaling in directed migration of *Dictyostelium*. *Cell Biosci* **11:** 111. doi:10.1186/s13578-021-00580-x

Sta Iglesia DD, Vanable JW Jr. 1998. Endogenous lateral electric fields around bovine corneal lesions are necessary for and can enhance normal rates of wound healing. *Wound Repair Regen* **6:** 531–542. doi:10.1046/j.1524-475X.1998.60606.x

Stern CD. 1981. Behaviour and motility of cultured chick mesoderm cells in steady electrical fields. *Exp Cell Res* **136:** 343–350. doi:10.1016/0014-4827(81)90013-6

Stern CD. 1986. Problems and paradigms: do lonic currents play a role in the control of development? *Bioessays* **4:** 180–184. doi:10.1002/bies.950040410

Stollberg J, Fraser SE. 1988. Acetylcholine receptors and concanavalin A-binding sites on cultured *Xenopus* muscle cells: electrophoresis, diffusion, and aggregation. *J Cell Biol* **107:** 1397–1408. doi:10.1083/jcb.107.4.1397

Sun Y, Zhao M. 2025. Bacterial galvanotaxis: a neglected biological process. *Trends Microbiol* **33:** 713–717. doi:10.1016/j.tim.2025.02.013

Sun S, Wise J, Cho M. 2004. Human fibroblast migration in three-dimensional collagen gel in response to noninvasive electrical stimulus. I: Characterization of induced three-dimensional cell movement. *Tissue Eng* **10:** 1548–1557. doi:10.1089/ten.2004.10.1548

Sun Y, Do H, Gao J, Zhao R, Zhao M, Mogilner A. 2013. Keratocyte fragments and cells utilize competing pathways to move in opposite directions in an electric field. *Curr Biol* **23:** 569–574. doi:10.1016/j.cub.2013.02.026

Sun YH, Sun Y, Zhu K, Reid B, Gao X, Draper BW, Zhao M, Mogilner A. 2018. Electric fields accelerate cell polarization and bypass myosin action in motility initiation. *J Cell Physiol* **233:** 2378–2385. doi:10.1002/jcp.26109

Sun Y, Reid B, Ferreira F, Luxardi G, Ma L, Lokken KL, Zhu K, Xu G, Sun Y, Ryzhuk V, et al. 2019. Infection-generated electric field in gut epithelium drives bidirectional migration of macrophages. *PLoS Biol* **17:** e3000044. doi:10.1371/journal.pbio.3000044

Sun Y, Reid B, Zhang Y, Zhu K, Ferreira F, Estrada A, Sun Y, Draper BW, Yue H, Copos C, et al. 2023. Electric field-guided collective motility initiation of large epidermal cell groups. *Mol Biol Cell* **34:** ar48. doi:10.1091/mbc.E22-09-0391

Sun Y, Ferreira F, Reid B, Zhu K, Ma L, Young BM, Hagan CE, Tsolis RM, Mogilner A, Zhao M. 2024. Gut epithelial electrical cues drive differential localization of enterobacteria. *Nat Microbiol* **9:** 2653–2665. doi:10.1038/s41564-024-01778-8

Tahoun A, Mahajan S, Paxton E, Malterer G, Donaldson DS, Wang D, Tan A, Gillespie TL, O'Shea M, Roe AJ, et al. 2012. Salmonella transforms follicle-associated epithelial cells into M cells to promote intestinal invasion. *Cell Host Microbe* **12:** 645–656. doi:10.1016/j.chom.2012.10.009

Theveneau E, Marchant L, Kuriyama S, Gull M, Moepps B, Parsons M, Mayor R. 2010. Collective chemotaxis requires contact-dependent cell polarity. *Dev Cell* **19:** 39–53. doi:10.1016/j.devcel.2010.06.012

Tran V, Zhang X, Cao L, Li H, Lee B, So M, Sun Y, Chen W, Zhao M. 2013. Synchronization modulation increases transepithelial potentials in MDCK monolayers through Na/K pumps. *PLoS One* **8:** e61509. doi:10.1371/journal.pone.0061509

Trollinger DR, Isseroff RR, Nuccitelli R. 2002. Calcium channel blockers inhibit galvanotaxis in human keratinocytes. *J Cell Physiol* **193:** 1–9. doi:10.1002/jcp.10144

Tsai HF, Huang CW, Chang HF, Chen JJ, Lee CH, Cheng JY. 2013. Evaluation of EGFR and RTK signaling in the electrotaxis of lung adenocarcinoma cells under direct-current electric field stimulation. *PLoS One* **8:** e73418. doi:10.1371/journal.pone.0073418

Turner JR. 2009. Intestinal mucosal barrier function in health and disease. *Nat Rev Immunol* **9:** 799–809. doi:10.1038/nri2653

Vallon-Eberhard A, Landsman L, Yogev N, Verrier B, Jung S. 2006. Transepithelial pathogen uptake into the small intestinal lamina propria. *J Immunol* **176:** 2465–2469. doi:10.4049/jimmunol.176.4.2465

Van Duijn B, Vogelzang SA, Ypey DL, Van Der Molen LG, Van Haastert P. 1990. Normal chemotaxis in *Dictyostelium discoideum* cells with a depolarized plasma membrane potential. *J Cell Sci* **95:** 177–183. doi:10.1242/jcs.95.1.177

Van Haastert PJ, Devreotes PN. 2004. Chemotaxis: signalling the way forward. *Nat Rev Mol Cell Biol* **5:** 626–634. doi:10.1038/nrm1435

Verkhovsky AB, Svitkina TM, Borisy GG. 1999. Self-polarization and directional motility of cytoplasm. *Curr Biol* **9:** 11–20. doi:10.1016/S0960-9822(99)80042-6

Verworn M. 1890. Die polare Erregung der Protisten durch den galvanischen Strom. *Archiv Physiol Menschen Tiere* **46:** 267–303.

Wang E, Zhao M, Forrester JV, McCaig CD. 2003. Bi-directional migration of lens epithelial cells in a physiological electrical field. *Exp Eye Res* **76:** 29–37. doi:10.1016/S0014-4835(02)00257-9

Weijer CJ. 2009. Collective cell migration in development. *J Cell Sci* **122:** 3215–3223. doi:10.1242/jcs.036517

Wilson WW, Wade MM, Holman SC, Champlin FR. 2001. Status of methods for assessing bacterial cell surface charge properties based on zeta potential measurements. *J Microbiol Methods* **43:** 153–164. doi:10.1016/S0167-7012(00)00224-4

Wolf AE, Heinrich MA, Breinyn IB, Zajdel TJ, Cohen DJ. 2022. Short-term bioelectric stimulation of collective cell migration in tissues reprograms long-term supracellular dynamics. *PNAS Nexus* **1:** pgac002. doi:10.1093/pnasnexus/pgac002

Yang HY, Charles RP, Hummler E, Baines DL, Isseroff RR. 2013. The epithelial sodium channel mediates the directionality of galvanotaxis in human keratinocytes. *J Cell Sci* **126:** 1942–1951. doi:10.1242/jcs.113225

Yang Q, Miao Y, Campanello LJ, Hourwitz MJ, Abubaker-Sharif B, Bull AL, Devreotes PN, Fourkas JT, Losert W. 2022. Cortical waves mediate the cellular response to electric fields. *eLife* **11:** e73198. doi:10.7554/eLife.73198

Yao L, Shanley L, McCaig C, Zhao M. 2008. Small applied electric fields guide migration of hippocampal neurons. *J Cell Physiol* **216:** 527–535. doi:10.1002/jcp.21431

Yao L, McCaig CD, Zhao M. 2009. Electrical signals polarize neuronal organelles, direct neuron migration, and orient cell division. *Hippocampus* **19:** 855–868. doi:10.1002/hipo.20569

Yu Y, Zhang L, Li B, Fu Z, Brohawn SG, Isacoff EY. 2024. Coupling sensor to enzyme in the voltage sensing phosphatase. *Nat Commun* **15:** 6409. doi:10.1038/s41467-024-50319-8

Zajdel TJ, Shim G, Wang L, Rossello-Martinez A, Cohen DJ. 2020. SCHEEPDOG: programming electric cues to dynamically herd large-scale cell migration. *Cell Syst* **10:** 506–514.e3. doi:10.1016/j.cels.2020.05.009

Zhan H, Bhattacharya S, Cai H, Iglesias PA, Huang CH, Devreotes PN. 2020. An excitable Ras/PI3K/ERK signaling network controls migration and oncogenic transformation in epithelial cells. *Dev Cell* **54:** 608–623.e5. doi:10.1016/j.devcel.2020.08.001

Zhang J, Calafiore M, Zeng Q, Zhang X, Huang Y, Li RA, Deng W, Zhao M. 2011. Electrically guiding migration of human induced pluripotent stem cells. *Stem Cell Rev* **7:** 987–996. doi:10.1007/s12015-011-9247-5

Zhang Y, Xu G, Wu J, Lee RM, Zhu Z, Sun Y, Zhu K, Losert W, Liao S, Zhang G, et al. 2022. Propagation dynamics of electrotactic motility in large epithelial cell sheets. *iScience* **25:** 105136. doi:10.1016/j.isci.2022.105136

Zhang Y, Huang S, Cao Y, Li L, Yang J, Zhao M. 2025. New opportunities for electric fields in promoting wound healing: collective electrotaxis. *Adv Wound Care (New Rochelle)* **14:** 418–428. doi:10.1089/wound.2024.0003

Zhao M. 2009. Electrical fields in wound healing-An overriding signal that directs cell migration. *Semin Cell Dev Biol* **20:** 674–682. doi:10.1016/j.semcdb.2008.12.009

Zhao M, Agius-Fernandez A, Forrester JV, McCaig CD. 1996a. Directed migration of corneal epithelial sheets in physiological electric fields. *Invest Ophthalmol Vis Sci* **37:** 2548–2558.

Zhao M, Agius-Fernandez A, Forrester JV, McCaig CD. 1996b. Orientation and directed migration of cultured corneal epithelial cells in small electric fields are serum dependent. *J Cell Sci* **109:** 1405–1414. doi:10.1242/jcs.109.6.1405

Zhao M, McCaig CD, Agius-Fernandez A, Forrester JV, Araki-Sasaki K. 1997. Human corneal epithelial cells reorient and migrate cathodally in a small applied electric field. *Curr Eye Res* **16:** 973–984. doi:10.1076/ceyr.16.10.973.9014

Zhao M, Dick A, Forrester JV, McCaig CD. 1999. Electric field-directed cell motility involves up-regulated expression and asymmetric redistribution of the epidermal growth factor receptors and is enhanced by fibronectin and laminin. *Mol Biol Cell* **10:** 1259–1276. doi:10.1091/mbc.10.4.1259

Zhao M, Pu J, Forrester JV, McCaig CD. 2002a. Membrane lipids, EGF receptors, and intracellular signals colocalize and are polarized in epithelial cells moving directionally in a physiological electric field. *FASEB J* **16:** 857–859. doi:10.1096/fj.01-0811fje

Zhao M, Jin T, McCaig CD, Forrester JV, Devreotes PN. 2002b. Genetic analysis of the role of G protein–coupled receptor signaling in electrotaxis. *J Cell Biol* **157:** 921–928. doi:10.1083/jcb.200112070

Zhao M, Song B, Pu J, Forrester JV, McCaig CD. 2003. Direct visualization of a stratified epithelium reveals that wounds heal by unified sliding of cell sheets. *FASEB J* **17:** 397–406. doi:10.1096/fj.02-0610com

Zhao M, Song B, Pu J, Wada T, Reid B, Tai G, Wang F, Guo A, Walczysko P, Gu Y, et al. 2006. Electrical signals control wound healing through phosphatidylinositol-3-OH kinase-gamma and PTEN. *Nature* **442:** 457–460. doi:10.1038/nature04925

Zhao M, Chalmers L, Cao L, Vieira AC, Mannis M, Reid B. 2012. Electrical signaling in control of ocular cell behaviors. *Prog Retin Eye Res* **31:** 65–88. doi:10.1016/j.preteyeres.2011.10.001

Zhao S, Zhu K, Zhang Y, Zhu Z, Xu Z, Zhao M, Pan T. 2014. ElectroTaxis-on-a-Chip (ETC): an integrated quantitative high-throughput screening platform for electrical field-directed cell migration. *Lab Chip* **14:** 4398–4405. doi:10.1039/C4LC00745J

Zhao Z, Zhu K, Li Y, Zhu Z, Pan L, Pan T, Borgens RB, Zhao M. 2020. Optimization of electrical stimulation for safe and effective guidance of human cells. *Bioelectricity* **2:** 372–381. doi:10.1089/bioe.2020.0019

Zhao M, Rolandi M, Isseroff RR. 2022. Bioelectric signaling: role of bioelectricity in directional cell migration in wound healing. *Cold Spring Harb Perspect Biol* **14:** a041236. doi:10.1101/cshperspect.a041236

Zhu K, Sun Y, Miu A, Yen M, Liu B, Zeng Q, Mogilner A, Zhao M. 2016. cAMP and cGMP play an essential role in galvanotaxis of cell fragments. *J Cell Physiol* **231:** 1291–1300. doi:10.1002/jcp.25229

Zhu K, Guo L, Patel P, Albeck J, Qing Q, Losert W, Zhao M. 2022. Protocol for electrical modulation of ERK signaling dynamics in cell lines. *STAR Protoc* **3:** 101752. doi:10.1016/j.xpro.2022.101752

From Signals to Protrusions: Modeling Excitable Systems in Cell Migration

Pablo A. Iglesias[1,2,3,4] **and Parijat Banerjee**[5]

[1]Department of Electrical and Computer Engineering, Johns Hopkins University, Baltimore, Maryland 21218, USA

[2]Department of Cell Biology, Johns Hopkins School of Medicine, Baltimore, Maryland 20205, USA

[3]Department of Biomedical Engineering, Johns Hopkins School of Medicine, Baltimore, Maryland 20205, USA

[4]Department of Applied Mathematics and Statistics, Johns Hopkins University, Baltimore, Maryland 21218, USA

[5]Department of Physics and Astronomy, Johns Hopkins University, Baltimore, Maryland 21218, USA

Correspondence: pi@jhu.edu

Cell migration is a fundamental biological process critical for development, immune response, and wound healing, but its dysregulation contributes to pathological conditions such as cancer metastasis. Recent research has demonstrated that migration is driven by excitable signal transduction and cytoskeletal networks, which function as separate but coupled systems. The signal transduction excitable network (STEN) propagates excitatory signals, while the cytoskeletal excitable network (CEN) generates cytoskeletal protrusions. Although distinct, these networks interact dynamically: STEN regulates CEN, while CEN provides feedback to STEN, influencing cell polarization and directionality. Computational models incorporating nonlinear dynamics and reaction–diffusion systems have successfully recapitulated these interactions, shedding light on their role in pseudopod formation, chemotaxis, and mechanosensation. This review discusses recent experimental and theoretical advances, highlighting how excitable systems underlie cell motility and how mathematical modeling helps to understand their role.

Cell migration is a tightly regulated process guided by a diverse set of signaling cues from the cell's environment. Different cell types display distinct migration behaviors tailored to their specific functions and contexts. For example, directed cell migration is essential for physiological processes such as embryonic development (Szabó and Mayor 2018), wound healing (de Oliveira et al. 2016), and immune responses (Kolaczkowska and Kubes 2013). However, directed migration can lead to pathological conditions. For instance, in metastasis, cancer cells detach from the primary tumor, migrate to distant organs, and form secondary tumors (Roussos et al. 2011). Similarly, abnormal migration of dendritic cells can result in autoimmune disorders, where the immune system mistakenly attacks healthy tissues (Tiberio et al. 2018).

Cite this article as *Cold Spring Harb Perspect Biol* doi: 10.1101/cshperspect.a041746

Thus, unraveling the mechanisms underlying cell migration is essential for developing therapeutic strategies to address diseases linked to aberrant cell movement, such as cancer metastasis.

Much of what is known about cell migration in eukaryotic cells can be traced to experiments from the social ameba *Dictyostelium* (Rappel and Loomis 2009; Nichols et al. 2015; Devreotes et al. 2017; Matsuoka et al. 2024). These cells exhibit ameboid motion, which is characterized by the dynamic extension of pseudopods: protrusions formed by localized actin polymerization. Ameboid motion can arise in two conditions. In the absence of an external cue guiding its path, pseudopods are extended in various directions, randomly leading to cell movement in an undirected and stochastic manner. This type of motion allows cells to explore their environment, for example, when seeking nutrients. Alternatively, in the presence of specific cues (chemical, mechanical, etc.), cells polarize by establishing distinct front and rear regions. In this case, pseudopods are created persistently at the front of the cell, driving motion toward the source of these cues. Interestingly, detailed statistical analysis suggests that the pseudopods used in both settings are the same (Andrew and Insall 2007; van Haastert 2024).

Ameboid movement requires actin polymerization. Twenty-five years ago, Vicker (2000, 2002) investigated the temporal and spatial pattern of actin polymerization in *Dictyostelium* cells and observed that actin filaments formed propagating waves. He posited that actin propagated as self-organized, reaction–diffusion waves of reversible F-actin assembly and that these waves were closely linked to pseudopod extensions. The discovery of similar waves in neutrophils sparked wider interest (Fig. 1A; Weiner et al. 2007). Moreover, these new results also pointed out similarities between the observed actin waves and those of action potentials in neuronal signaling. For example, when two waves collide, they annihilate each other—thus highlighting the connection between F-actin propagation and excitable systems (Fig. 1A–C). Actin waves have subsequently been observed in many cellular contexts (Case and Waterman 2011; Ryan et al. 2012; Wu et al. 2013, 2018; Bement et al. 2015; Winans et al. 2016; Coscia et al. 2024).

Although signaling molecules also demonstrate similar wave propagation (Arai et al. 2010; Xiong et al. 2010), it is now appreciated that the signaling components that regulate actin polymerization form an independent excitable network (Huang et al. 2013; Gerhardt et al. 2014; van Haastert et al. 2017; Miao et al. 2019). While the networks are coupled in cells, they can be decoupled, showing distinct behavior. We use the terms signal transduction excitable network (STEN) and cytoskeletal excitable network (CEN) to differentiate between the two systems. In what follows, however, we will mostly talk about generic excitable systems and return to the distinctions later.

Understanding cell migration requires integrating biochemical signaling, cytoskeletal dynamics, and mechanical forces. While experiments provide crucial insights, mathematical models have enabled researchers to dissect these complex interactions quantitatively, identify key regulatory parameters, and predict emergent behaviors that might be difficult to isolate experimentally (Iglesias and Devreotes 2008; Mogilner et al. 2012; Rappel and Edelstein-Keshet 2017; Othmer 2019; Beta et al. 2023). Because ameboid motion is driven by excitable systems, which are highly complex systems incorporating nonlinear dynamics with spatial and stochastic effects, mathematical models offer a powerful framework for understanding ameboid migration.

MATHEMATICAL MODELS OF EXCITABILITY

Excitable Systems

The study of excitable systems gained significant momentum in the 1950s with the groundbreaking work of Hodgkin and Huxley (1952), who conducted experiments to uncover the mechanisms underlying the propagation of electrical signals in neurons. They demonstrated that a neuron's excitability—its capacity to generate action potentials—arises from voltage-dependent changes in membrane conductance, specif-

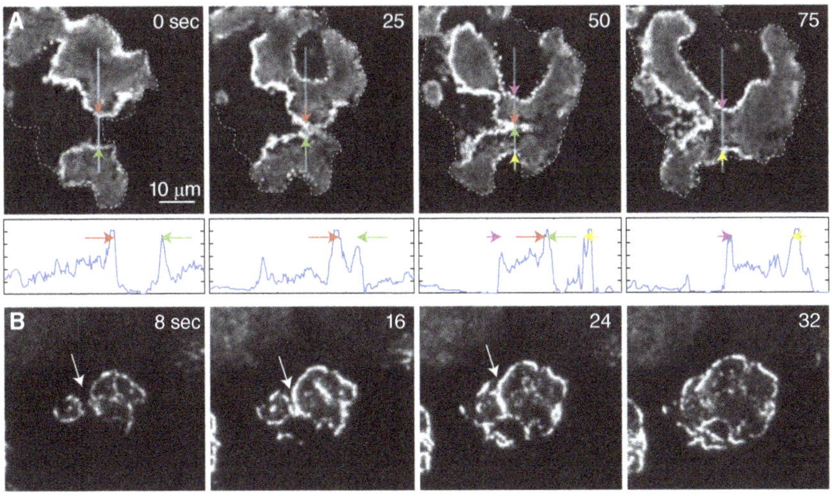

Figure 1. Wave annihilation and refractory periods are characteristic features of excitable systems. (*A*) Electrofused *Dictyostelium* cell expressing LIM domain protein E (LimE)-RFP, showing wave propagation and annihilation. The dotted line shows the outline of the cell. The *bottom* panels show the LimE-RFP intensity along the red line. Note that the two peaks seen approaching each other at 0 sec (marked by the red and green arrows) and almost touching at 25 sec come together at a single peak at 50 sec and are no longer there at 75 sec. The yellow and mauve arrows point to other intensity peaks. Note that the wave activity is spontaneously generated; there is no external stimulus. (*B*) HL-60 cell expressing Hem-1-YFP, continually exposed to chemoattractant (20 nM *N*-formylmethionyl-leucyl-phenylalanine [fMLP]) and imaged using total internal reflection fluorescence (TIRF) microscopy at one frame every 2 sec. Hem-1 waves within a single cell annihilate upon collision (see arrows). (Adapted from Weiner et al. 2007, Video S13, under the terms of the Creative Commons Attribution License.)

ically the selective permeability to ions. Their findings revealed a positive feedback loop in which an initial depolarization (a small increase in voltage) opens sodium (Na^+) channels, causing further depolarization and the characteristic rapid spike of an action potential. In addition to this positive feedback, they also found a separation of timescales, where Na^+ channels open rapidly and transiently, potassium (K^+) channels open more slowly. Moreover, by helping to repolarize the membrane and terminate the action potential, they act as a negative feedback loop to the initial depolarization.

To complement their experimental insights, Hodgkin and Huxley formulated a mathematical model that quantitatively described the ionic currents governing neuronal activity. Now known as the Hodgkin–Huxley (HH) model, it consists of four coupled nonlinear differential equations that describe the membrane potential and the dynamics of sodium and potassium ion channels, including their activation and inactivation states. The HH model not only transformed the understanding of neural excitability but also laid the foundation for modern computational neuroscience and the study of excitable systems across diverse biological and physical systems.

Owing to the model's complexity, several simplifications have been proposed. One of the most notable is the FitzHugh–Nagumo (FHN) model, introduced by FitzHugh (1961) as a simplified two-variable differential equation and complemented a year later when Nagumo et al. (1962) developed an equivalent electrical circuit using a tunnel diode that recreated the same dynamics. In the FHN model, the two variables represent generic activator and inhibitor species. The FHN model retains critical features of the HH model. The activator includes fast positive feedback whose amplification is mitigated by the presence of a slow, negative feedback loop. This slower timescale is important because it allows the positive feedback term to increase activity before the slow negative feedback, provided by the inhibitor, drives the system back to its basal value. Finally, the activator is also influenced by an external stimulus term that alters the set point—the steady-state balance between activation and inhibition—which determines the level of excitability of the system. As we show below, the regulation of this point is an integral part of understanding how excitable systems regulate cell motility.

Excitable Systems and Cell Signaling in Migrating Cells

Although the FHN model is useful for understanding the mathematical basis of excitability, it has limitations when applied to excitable signaling or cytoskeletal dynamics in migrating cells. One key drawback is that the model does not constrain variables to remain nonnegative, making it unsuitable for representing concentrations or abundances of biochemical species. Furthermore, the FHN model lacks a direct connection to biological signals. To address these issues, Bhattacharya and Iglesias (2018) introduced a modified activator–inhibitor system that retains many features of the FHN model while incorporating biologically relevant modifications (Fig. 2A):

$$\begin{aligned} \frac{dF}{dt} &= -(a_1 + a_2 R)F + \frac{a_3 F^2}{a_4^2 + F^2} + a_5 \\ \frac{dR}{dt} &= \varepsilon(c_1 F - c_2 R) \end{aligned}.$$

These equations describe the changes in time of two species. The symbols for the activator (F) and inhibitor (R) are mnemonics representing the "front" and "refractory" species of the system, respectively, highlighting their biological roles during cell migration. As we will see later, the activator species localizes to the front of migrating cells. The spatial distribution of the refractory species, however, is more nuanced. We avoid calling it the "inhibitor" species as there are several inhibitory species in the system.

Because the F–R model is a second-order system (consisting of two differential equations), it can be studied using phase-plane analysis, a graphical method for examining the qualitative behavior of two-dimensional systems of ordinary differential equations (Fig. 2B). This approach provides insights into system dynamics without requiring explicit solutions. In the phase-plane analysis, the system's trajectory is visualized by plotting arrows that represent the evolution of two variables (F and R) over time. Two critical features of this method are the nullclines—curves where the rate of change of one variable is zero (either constant F: $dF/dt = 0$ or constant R: $dR/dt = 0$). The intersections of these nullclines correspond to equilibrium points, where the system is at rest. In this model, the front nullcline has an inverted "N" shape, reflecting distinct segments associated with front-species inactivation and positive feedback dynamics (Keener and Sneyd 1998). As in the FHN model, the refractory nullcline is a straight line with a slope given by c_1/c_2. The refractory species operates on a slow timescale, owing to the parameter $\varepsilon \gg 1$ (Fig. 2B). The front equation consists of three terms. The first two represent negative regulation of the front species, including both basal levels of inactivation ($-a_1 F$) and negative feedback from the refractory element ($-a_2 RF$). The model uses a saturating function for positive feedback:

$$\frac{a_3 F^2}{a_4^2 + F^2}.$$

The choice of a Hill coefficient of two means that small changes in front-species concentration lead to sudden transitions in system behavior. This maintains the characteristic "N" shape of the front nullcline, a hallmark of excitable systems. Figure 2C shows the fast response of the front (F) and the slower response of the refractory elements (R). Note how they both show a peak, but that of F precedes that of R. In experiments, when only one species is observed, it may be difficult to ascertain whether it belongs to F or R as both start at low, basal levels and rise quickly to a peak before returning to their prestimulus levels (Fig. 2C).

 Cite this article as *Cold Spring Harb Perspect Biol* doi: 10.1101/cshperspect.a041746

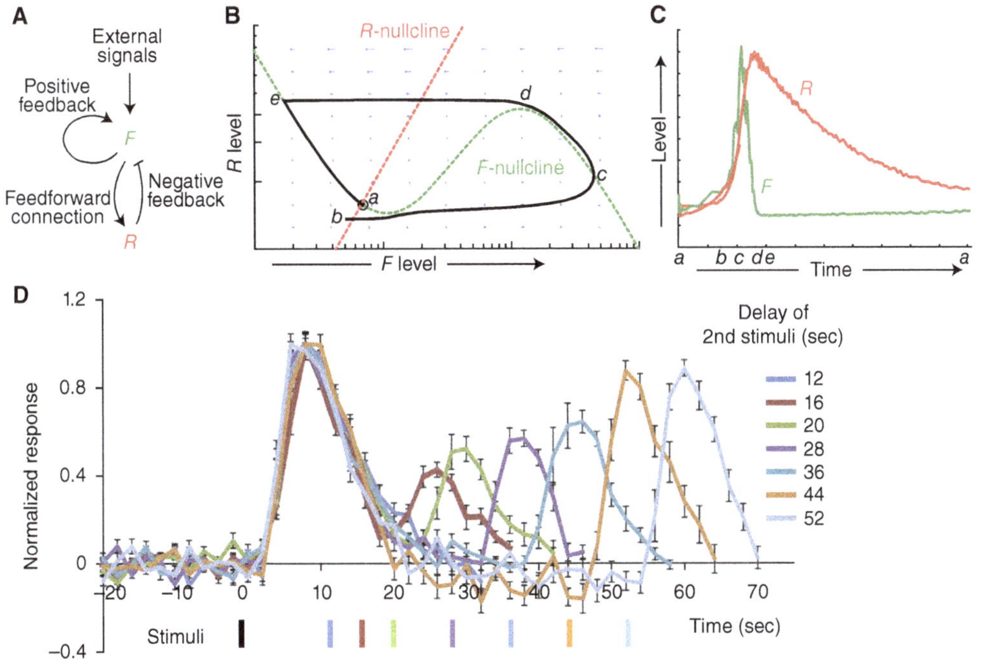

Figure 2. Front-refractory excitable network. (*A*) Activator–inhibitor system consisting of a front (*F*) component and a refractory (*R*) element. (*B*) Phase-plane representation of the front-refractory system shown in *A*. The green and red lines indicate the *F*- and *R*-nullclines, respectively. Blue arrows depict the short-term trajectory of the system at different values of *F* and *R*. The unique stable equilibrium (point *a*) is marked with a black circle. The black line traces the temporal evolution of the system from a given initial condition. A perturbation away from equilibrium (point *b*) causes a rapid increase in *F*, moving the system rightward. During this phase, *R* accumulates slowly due to timescale separation. As the trajectory crosses the *F*-nullcline (point *c*), *F* begins to decrease, while *R* continues increasing, guiding the system along the *F*-nullcline. Upon reaching the local peak of the nullcline (point *d*), the system rapidly transitions to lower *F* levels, while *R* reaches a maximum before gradually decreasing. This phase corresponds to a rapid shift in the state of *F*. Eventually, the system settles on the leftmost arm of the *F*-nullcline (point *e*), where *F* increases and *R* decreases, leading to a gradual return to equilibrium (*a*). (*C*) Time evolution of *F* and *R*, both normalized. (*D*) Responses of *Dictyostelium* cells to two 2 sec chemoattractant (cAMP) stimuli separated by increasing time intervals. The black bar represents the first stimulus, while the colored bars indicate the delay before the second stimulus. (Reprinted from Huang et al. 2013. Used with permission.)

Refractory Behavior

We have referred to *R* as the refractory element of the excitable system. It is worth outlining the properties of this element. The refractory period in excitable systems is the interval following a triggering during which the system is either completely unable to fire again or requires a significantly stronger stimulus to do so. In the HH model, this arises from the inactivation of voltage-gated sodium channels and the delayed opening of potassium channels, which hyperpolarizes the membrane. In the *F*–*R* model, the refractory period can be seen in the phase plane (Fig. 2B). Following a perturbation away from the equilibrium (from point *a* to point *b*, in Fig. 2B), the trajectory undergoes a large excursion away from the equilibrium (*b* to *c*) during which the front element quickly accumulates with little change in the refractory species. After crossing the nullcline, the refractory element begins slowly accumulating (*c* to *d*). Eventually, enough refractory element has accumulated that the system begins a rapid return toward the equilibrium (*d* to *e*). After crossing the F-nullcline, the system returns to its steady state (*e* to *a*). During this last phase, known as the refractory period, the system is less responsive to perturbations, as changes in the set point cannot induce further large deviations, mirroring the refractory period of the HH model.

Experimentally, the presence of a refractory period can be observed by applying two short stimuli separated by varying time intervals (Fig. 2D). When the system is at rest, the first stimulus triggers the excitable behavior. If the second stimulus is delivered too soon, during the refractory state, no response is observed. However, as the interval increases, the system regains its ability to respond. Using two brief pulses of cAMP as stimuli, Huang et al. (2013) demonstrated that RasGTP in *Dictyostelium* cells exhibits a relative refractory period with an apparent half-time of ~7.5 sec (Fig. 1D). Tanabe et al. (2018) demonstrated that different pathways in *Dictyostelium* have different refractory periods; similarly, Miao et al. (2019) showed that the signaling and CENs have different refractory periods and that the coupling of the fast cytoskeletal and slow signaling networks can lead to waves in which the signaling network shows a single peak flanked by two peaks of actin polymerization. The presence of the excitable system has also been instrumental in understanding some of the underlying network organization. For example, Tang et al. (2014) alternated the two impulses using different chemoattractants (cAMP and folic acid). When the same chemoattractant was used for both pulses, the cells showed evidence of a refractory period. More interestingly, cells did not respond to cAMP during the refractory period induced by folic acid, suggesting that both chemoattractants activate a common signaling excitable network. Similar experiments using chemoattractants and shear flow-induced responses showed that chemical and mechanical stimuli activate a common excitable system (Artemenko et al. 2016).

The properties of the refractory period also shed light on how cells adapt to external cues. When stimulated by persistent stimuli, cells demonstrate adaptive behavior—over time, they regain the ability to respond to new stimuli. Varying the length of the first stimulus showed that cells adapted to both short and long initial pulses, but the recovery time to the latter was delayed (Tang et al. 2014). Moreover, the response to the second stimulus was smaller as the length of the initial stimulus increased. These experiments were used to contrast predictions made by different models of how cells coordinate receptor occupancy with excitable behavior. The experimental results agreed with the local-excitation, biased-excitable network model (Xiong et al. 2010; Iglesias and Devreotes 2012), which posits that adaptation happens before the signal reaches the signaling excitable network. The results, however, rejected models in which the external stimulus directly biases the excitable behavior and adaptation happens through negative feedback (Meinhardt 1999; Neilson et al. 2011).

The *F*–*R* system of equations effectively captures many characteristics of excitable behavior in motile cells, but it remains a simplified approximation. Numerous cellular components exhibit front-associated patterns of activation (see Box 1). For instance, RasGTP and PI (3,4,5)P_3 consistently show transient increases in membrane localization upon stimulation—whether induced by chemoattractants, mechanical cues, or spontaneous activation. These increases typically peak within ~20 sec, and are predominantly confined to the front of a migrating cell, collectively represented by the mathematical term *F*. Other molecules, such as suppressor of cAMP receptor/WASP-family verprolin-homologous protein (SCAR/WAVE) and RacGTP, also exhibit similar front-localized dynamics, reinforcing their role in driving protrusive activity and motility. Responses like PKB (Akt) activation display a similar temporal pattern but lag front-associated markers (Miao et al. 2019). Moreover, synthetic recruitment of PKB to the membrane suppresses excitability (Miao et al. 2019). Together, these two behaviors mirror that of the refractory element in the model. Consequently, we associate membrane levels of PKB with *R*, the refractory component in the *F*–*R* system.

Conversely, certain components behave oppositely to front markers: They are abundant on

BOX 1. GLOSSARY

Chemoattractants

- cAMP (cyclic AMP). A chemoattractant in *Dictyostelium*, activating the cAMP receptor and downstream signaling pathways for directed migration.
- fMLP (*N*-formylmethionyl-leucyl-phenylalanine). A bacterial-derived chemoattractant that activates G-protein-coupled receptors, leading to PI3K activation and actin polymerization in neutrophils.
- EGF (epidermal growth factor). A growth factor that binds to the EGF receptor (EGFR), activating the Ras-PI3K pathway, promoting cell polarization and motility in mammalian cells.

Components of the signal transduction excitable system (STEN)

- Ras (Rat sarcoma). A small GTPase that regulates PI3K activity, leading to PI(3,4,5)P_3 production and cytoskeletal reorganization. The Ras-binding domain (RBD), which serves as a biosensor for activated Ras, is found at the front of migrating cells.
- PI5K (phosphatidylinositol-5-kinase). Enzyme that converts PI(4)P to PI(4,5)P_2. PI5K is phosphorylated by PKB (Fets et al. 2014). PI5K and Ras activities follow a mutually inhibitory localization (Deng et al. 2024).
- PI3K (phosphoinositide 3-kinase). Enzyme that catalyzes the conversion of PI(4,5)P_2 to PI(3,4,5)P_3, promoting leading-edge formation and actin polymerization—seen at the front of chemotaxing cells (Janetopoulos et al. 2004).
- PTEN (phosphatase and tensin analog). PI3K antagonist that mediates the conversion of PI(3,4,5)P_3 to PI(4,5)P_2—localizes to the rear of migrating *Dictyostelium* cells.
- PI(3,4)P_2 (phosphatidylinositol (3,4)-bisphosphate)—involved in actin remodeling and endocytosis. PI(3,4)P_2 and RasGTP are mutually antagonistic (Li et al. 2018).
- PI(4,5)P_2 (phosphatidylinositol (4,5)-bisphosphate). A membrane lipid that regulates the cytoskeleton and membrane trafficking. It serves as a precursor for PI(3,4,5)P_3 and is cleaved by PLC (phospholipase C).
- PI(3,4,5)P_3 (phosphatidylinositol (3,4,5)-trisphosphate). A phospholipid that accumulates at the leading edge of migrating cells. It recruits signaling proteins like PKB.
- PKB (protein kinase B, also known as Akt). A downstream effector of PI(3,4,5)P_3 signaling that promotes cell survival and regulates actin cytoskeleton dynamics—found at the front of migrating cells (Meili et al. 1999).
- TorC2 (target of rapamycin complex 2). Activates PKB, promoting actin polymerization and cell polarity.
- PLC (phospholipase C). Cleaves PI(4,5)P_2, leading to calcium release and activation of protein kinase C.
- Rap (Ras-related protein). A small GTPase that regulates the TorC2 and PI3K pathways (Scavello et al. 2017).
- RalGDS (Ral guanine nucleotide dissociation stimulator). A downstream effector of Ras, activating Ral GTPases, which regulate actin dynamics and vesicle trafficking. Fluorescently tagged RalGDS serves as a biosensor for Rap. Temporally, RalGDS is one of the earliest STEN events observed (Miao et al. 2017).

Components of the cytoskeletal excitable system (CEN)

- Rac (Ras-related C3 botulinum toxin substrate). A small GTPase that activates the SCAR/WAVE complex, promoting actin branching and lamellipodia formation. RacGEF1 activates Rac by

Continued

promoting the exchange of GDP for GTP. Activated Rac1 is found at the leading edge of migrating *Dictyostelium* cells (Filić et al. 2012).

- PAK1 (p21-activated kinase 1). A regulator of actin polymerization and cell adhesion. In *Dictyostelium*, PAK1 localizes to leading edges of migrating cells and to endocytotic cups (Filić et al. 2012).
- LimE (Lim domain protein E). An actin-binding protein that marks newly polymerized actin filaments in *Dictyostelium* cells.
- SCAR/WAVE (suppressor of cAMP receptor/WASP-family verprolin-homologous protein). A nucleation-promoting factor that activates the Arp2/3 complex, leading to branched actin filament formation.
- Coronin. An actin-associated protein that regulates actin polymerization and disassembly. It interacts with the Arp2/3 complex and promotes actin turnover in migrating cells. It is found at the leading edge of cells (Gerisch et al. 1995).
- Sca1 (Scar-like protein 1). Part of the SCAR/WAVE complex, promoting actin branching via the Arp2/3 complex.

the membrane at rest but transiently decrease upon stimulation. For example, phosphatase and tensin analog (PTEN), phosphatidylinositol (4,5)-bisphosphate ($PI(4,5)P_2$), and myosin II are enriched at the cell's rear and undergo redistribution during activation. These rear-associated components are critical for contractile activity and serve as stabilizers of the back. Experimental findings suggest that these back (e.g., $PI(3,4)P_2$) and front (e.g., RasGTP) elements are mutually inhibitory (see Fig. 3; Li et al. 2018). To incorporate these "back" elements into the model, we collectively represent them by defining a new variable, B, whose dynamics are given by

$$\frac{dB}{dt} = -(b_1 + b_2F)B + b_3.$$

At steady state, $B = b_3/(b_1 + b_2F) \approx (b_3/b_2)/F$, indicating that B is inversely proportional to F. This inverse relationship captures the suppression of back-associated elements by front-associated activity. To complete the mutual inhibition between F and B, we modify the positive feedback term in the activator equation. Specifically,

$$\frac{a_3F^2}{a_4^2 + F^2},$$

is replaced with

$$\frac{\tilde{a}_3}{\tilde{a}_4^2 + B^2}.$$

This modification ensures that an increase in B reduces the positive feedback driving F, reflecting the biological antagonism between front and back components. Simulations of the modified system reveal reciprocal behavior between F and B, with increases in one component leading to suppression of the other.

The integration of F, B, and R provides a cohesive framework for understanding the spatial and temporal dynamics of excitable signaling in motile cells. Note, however, that the introduction of B as a third variable renders the two-dimensional phase-plane analysis invalid. However, if the timescales of the front and back inhibition are much faster than that of the refractory element, we can approximate B using its steady-state value. This approximation allows us to retain the F–R phase-plane framework while still capturing the influence of the back element.

Wave Behavior

Although Figure 4 shows excitable wave activity, to this point, we have described excitability in a spatially homogeneous system (Fig. 4). Excitable

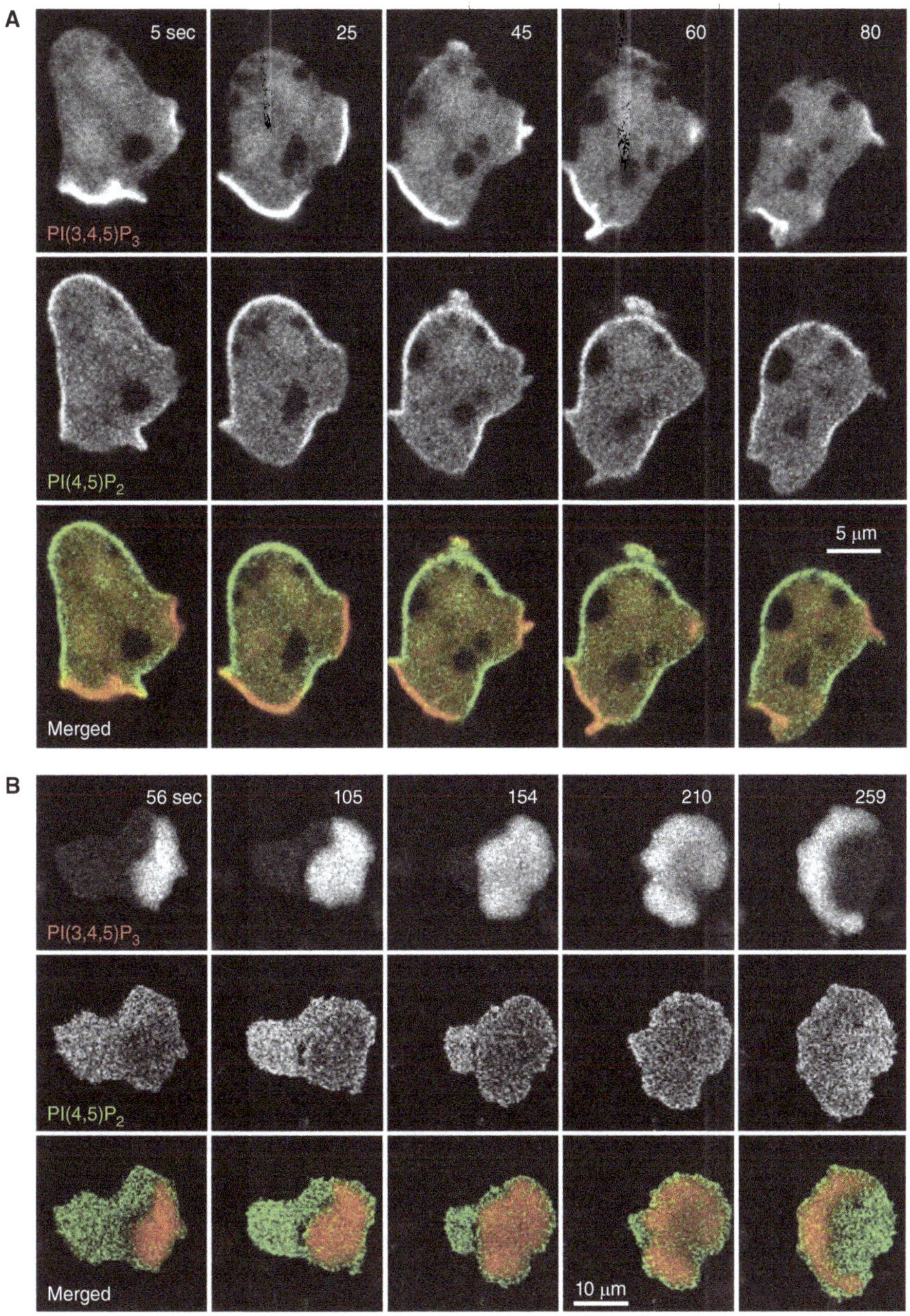

Figure 3. Front–back complementarity in *Dictyostelium* cells. Live-cell images of a *Dictyostelium* cell expressing the phosphatidylinositol (4,5)-bisphosphate ($PI(4,5)P_2$) and phosphatidylinositol (3,4,5)-trisphosphate ($PI(3,4,5)P_3$) sensors, showing the complementary localization of these two lipids. $PI(3,4,5)P_3$, a front-associated species, localizes at the leading edge of ventral waves and protrusions during migration. Panel *A* shows a confocal slice; panel *B* shows the substrate attached side. Note that the cell, initially moving to the right, changes direction. (Adapted from Supplementary Video S1, Banerjee et al. 2022. Used with permission.)

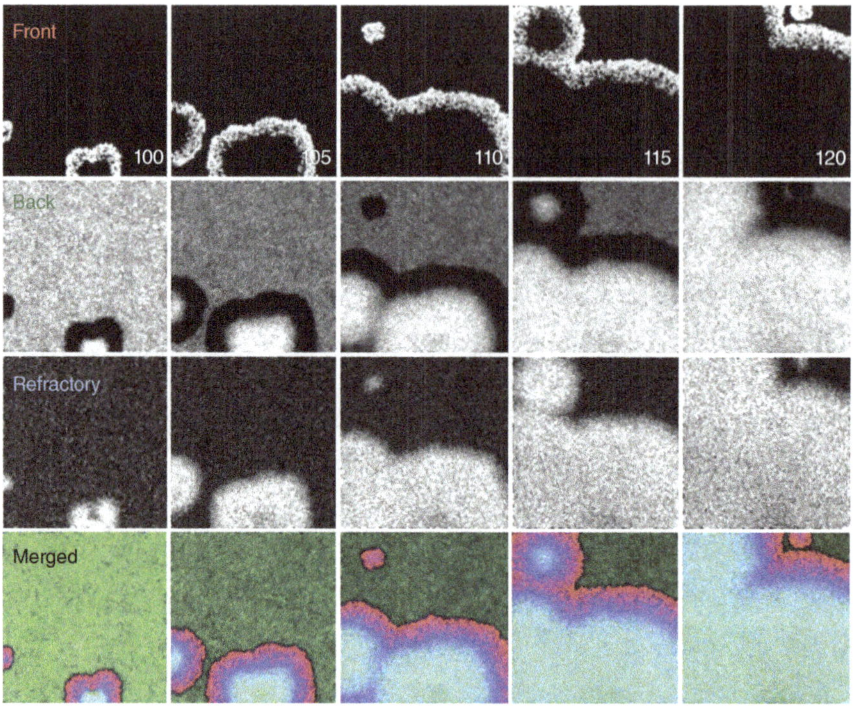

Figure 4. Wave behavior in the FBR model. Temporal dynamics of the front (red), back (green), and refractory (blue) species during wave propagation, shown individually and as a merged representation. The frame number is in arbitrary units. Experimentally measured wave speeds in *Dictyostelium* range from ~5 to 10 μm per min (Huang et al. 2013).

waves emerge when local activation spreads spatially. Mathematically, this can be achieved by introducing diffusion terms to the equations describing the various species, enabling them to spread through space. Thus, if the system fires at a given location, the front species increases quickly, diffusing to neighboring regions. This activity increase in the neighboring regions, coupled with the localized positive feedback, results in a sharp increase in the activity of the medium, creating a wave front that propagates in space. The velocity of this wave is largely dependent on the diffusion coefficient of the front species and the location of the equilibrium relative to the local maxima and minima of the nullcline (Tyson and Keener 1988; Bhattacharya et al. 2020). Owing to the slower timescale of the refractory species, the rate of change is much slower than that of the front species. This delayed negative feedback is crucial for wave propagation, as a simultaneous response would prevent wave formation.

For the wave to stop on its own, the refractory element must eventually overtake the front species in space, even while lagging in time. This can take place if the refractory species diffusion coefficient is higher than that of the front species. As the wave propagates, the high diffusion rate of the refractory element causes its concentration to increase across adjacent points on the perimeter more quickly. Incrementally, this raises the threshold for subsequent activations, slowing the wave front at each step before it eventually stops.

Another way of stopping excitable waves is through the interaction of two wave fronts. The presence of a refractory period ensures the unidirectional propagation of wave fronts. The wave front is caused by the triggering of adjacent regions in the medium that are at the basal state.

Following these new firings, those regions enter a refractory period that inhibits further firings. This effectively blocks the wave from moving backward and ensures that the wave moves away from the site of the initial trigger.

The refractory period also accounts for the wave annihilation that is a hallmark of excitable systems (Fig. 1A–C). When two waves moving in opposite directions approach each other, they leave regions of refractoriness behind them. When they collide, they cannot move through because those regions are in a refractory state from the passage of the other wave front. In two-dimensional media, particularly systems that have spatial heterogeneities, colliding waves may not annihilate. For example, when two waves collide but do so at an oblique angle, new wave patterns, including rotating and spiral waves, can result (Pertsov et al. 1993; Zykov et al. 2017; Bhattacharya and Iglesias 2019).

Figure 4 shows wave propagation based on the more complete *F–B–R* model. In this simulation, diffusion and stochastic perturbation terms (explained below) were added to the three differential equations for *F*, *B*, and *R* given above. Although the initial condition was set at the system equilibrium, the random perturbations make the system "fire" at random times and locations. For example, a new firing occurs near the top left portion of frame 110. This region of high *F* activity then propagates radially throughout the two-dimensional field; note how it forms a small empty circle in frame 115. Several aspects of the propagation are evident. The "front" species leads the wave front. It is also relatively short-lived, so it appears to have a narrow distribution—this causes the empty circle noted above. The "back" species is almost completely complementary, leading to so-called "shadow" waves—when a back species is observed in live-cell imaging, it is the absence of the species that is seen propagating. Thus, the *B* species fills the hole left behind by the *F* species. The pattern of the refractory species is more complicated. It appears as a propagating wave with its front trailing that of the front species. However, the timescale separation means that it is much longer lived. So, it propagates forward as the front species does (we also see a circle develop). However, it then leaves behind a long trail that tends to correlate spatially with the "back" species marker. This means that the *R*-species circle is not empty but fills up. We also see the annihilation of two wave fronts between frames 105 and 110. One other interesting aspect of excitable systems, seen in the back species channel, is that following the passing of a wave, the level of the back species is higher than that which precedes the shadow wave front. For example, compare the brightness of the back channel in the upper half of frame 110, before the shadow wave passes to that of the lower half.

Thresholds and Noise-Induced Firings

A hallmark of excitable systems is their threshold-like, all-or-nothing behavior (Izhikevich 2007). When the system resides at a stable equilibrium, small external perturbations cause only minor deviations from this state. However, a sufficiently large perturbation drives the system on a significant trajectory in the phase plane, arising from the interplay of the fast activator and slow inhibitor dynamics. This results in a large, characteristic excursion that constitutes the system's response (Fig. 2).

Experimentally, this has been demonstrated in the signaling network of *Dictyostelium* cells by stimulating cells with chemoattractant (Huang et al. 2013; Nishikawa et al. 2014). Sufficiently large stimuli exhibited spatially uniform responses around the cell, while weaker stimuli caused partial responses. In regions where the cell responded, the intensity of the response was equivalent to that of the strong stimulus; however, other parts of the cell showed no response. The differences in behavior can be interpreted as arising from local variations in threshold levels.

To understand how this threshold behavior emerges, it is useful to consider the effect of stochastic perturbations. Chemical systems inherently fluctuate owing to the random nature of molecular interactions and the probabilistic events of biochemical reactions. These stochastic effects are particularly pronounced in systems with small molecule numbers, where chance encounters and energy barriers for reactions introduce variability. It is usually assumed that proteins present in high copy numbers exhibit relatively small fluctuations; however, systems that operate near a threshold can experience significant effects from these stochastic variations, potentially altering the cells' behavior or phenotype.

Stochastic dynamics can be modeled in several ways. One exact method is the Gillespie algorithm, which simulates reactions by probabilistically computing reaction times and selecting reactions based on propensity functions (Gillespie 1976). This approach captures the discrete nature of molecular interactions but becomes computationally prohibitive for systems with large molecule counts. An alternative, approximate method is the Langevin approach, which models stochastic fluctuations through stochastic differential equations (Gillespie 2000). These equations extend deterministic models by incorporating random noise terms, enabling the analysis of systems under stochastic influences while remaining computationally efficient. For example, the front-species equation of the F–R model is modified as:

$$\frac{dR}{dt} = -(a_1 + a_2 R)F + \frac{a_3 F^2}{a_4^2 + F^2} + a_5 + \sigma(F,\ R)\eta(t).$$

Here, the first three terms mirror those of the deterministic equation, and the last term represents the addition of a state-dependent stochastic component. The function $\eta(t)$ represents a zero mean, Gaussian white-noise process (unit power spectrum), and $\sigma(F, R)$ specifies the magnitude of the perturbation.

The interaction between noise and the nonlinear dynamics of a system is complex (Horsthemke and Lefever 1984; Naoki et al. 2008; Iglesias 2013). If we assume that $\varepsilon \rightarrow 0$, then the system can be analyzed in a quasi-steady state where the refractory state is constant ($dR/dt = 0$) making $R \approx (c_1/c_2)F$. Under this assumption, the system can be described by a single stochastic differential equation:

$$\frac{dF}{dt} = -(a_1 + a_2(c_1/c_2)F)F + \frac{a_3 F^2}{a_4^2 + F^2} + a_5 + \sigma(F)\eta(t).$$

By integrating the deterministic terms on the right-hand side of this equation, we can rewrite this as

$$\frac{dF}{dt} = -\frac{dU(F)}{dF} + \sigma\eta(t).$$

Here, $U(F)$ is a potential function of the state F. If the system operates in an excitable regime, $U(F)$ takes the form of a double-well potential with two stable and one unstable equilibrium. In such cases, the rate at which the system escapes one equilibrium basin and transitions to the other can be described using Kramers' theory (Pollak and Talkner 2005), which states that the escape rate—a transition from one stable equilibrium to the other—depends on key properties of the potential function

$$\text{rate} \propto \sqrt{U''_s/U''_u|}\exp(-2\Delta U/\sigma),$$

where U''_s and U''_u describe the curvatures of the potential at the bottom (F_s) and top (F_u) of the well, and $\Delta U = U(F_u) - U(F_s)$ is the height of the barrier. This energy barrier, relative to the noise level, is the main determinant of the system's firing rate. Intuitively, ΔU provides a measure of the activation threshold: the higher the barrier, the lower the likelihood of a transition due to noise. Although this is an approximation, in terms of the phase plane, we can view the threshold as the distance from the equilibrium point to the bottom of the activator nullclines (Fig. 5A).

Triggering Excitable Behavior through External Stimuli

Because the size of the threshold determines the rate at which firings take place, it is a key determinant of the cell's excitable behavior. We now consider how this threshold can be altered, and in doing so, change the behavior of the cell.

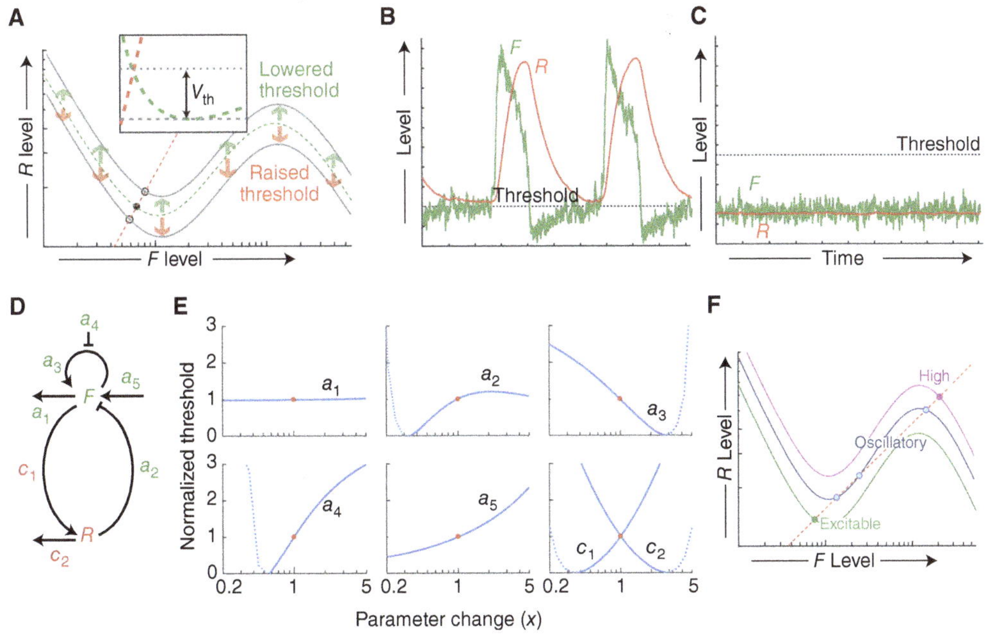

Figure 5. Altering the threshold modulates system excitability. (*A*) The activation threshold of the excitable system is determined by the distance between the equilibrium point (black circle) and the local minimum of the *F*-nullcline (*inset*). Raising (green arrows) or lowering (red arrows) the *F*-nullcline (dotted green line), for example, by modifying the rate of *F* production, shifts the equilibrium and alters the corresponding threshold. (*B*,*C*) Effect of threshold modulation on system activity. Lowering the threshold (*B*) increases the frequency of excitatory events, whereas raising the threshold (*C*) reduces their occurrence. (*D*) Sensitivity of the threshold to parameter changes in the *F*–*R* model. Each plot shows the relative threshold size as a parameter is varied (relative to its nominal value). The dotted line indicates a loss of equilibrium stability. (*E*) Effect of large equilibrium shifts. In the excitable regime, a single equilibrium exists with a defined activation threshold. (*F*) If this threshold is eliminated, the system can transition to an oscillatory state with multiple equilibria. Further parameter changes can drive the system into a state of sustained activity.

We first consider extrinsic perturbations of the system. This can be thought of, for example, as driven by changes to the parameters a_3 and/or a_5 in the F equation. Suppose that a cell is stimulated through the addition of an external stimulus, such as a chemoattractant, and that this has the effect of increasing a_5. From the form of the equation, we see that this has the effect of increasing the production rate of the inhibitor. In the phase-plane, this raises the activator nullcline, making the threshold smaller, and increasing the likelihood of firings (Fig. 5A,B). Alternatively, a decrease in the value of these parameters lowers the nullcline and raises the threshold, making firings highly unlikely (Fig. 5A,C). It should be emphasized that although altering external signal parameters can affect the threshold, it is not the only way. Figure 5D shows the interactions of the F–R model and how the different model parameter values affect the strength of the various feedback loops. In Figure 5E, we show how a change in each of the parameters influences the threshold. For example, increasing the strength of the positive feedback loop (i.e., higher a_3) lowers the threshold. Similarly, decreasing the strength of the negative feedback (lower c_2 or a_2; higher c_1) also lowers the threshold leading to more frequent firings. Note, however, that in some cases the relationship can be complex. For example, small increases in a_2 raise the threshold, but eventually, it becomes smaller (Fig. 5E).

This argument above shows that a chemoattractant stimulus can induce the firing of the excitable system. Experimentally, when a uniform dose is applied to a *Dictyostelium* cell, it undergoes distinct morphological changes. It first stops moving and briefly contracts—a response known as cringing—before spreading and resuming movement (Futrelle et al. 1982). Simulations of chemoattractant-induced movement exhibit the same behavior (Shi et al. 2013). They show that after stimulation, the excitable network fires globally, causing the cell to attempt extension in all directions, which results in rounding. The rise in activator levels then leads to increased inhibitor production. Once the inhibitor dominates, signaling is suppressed throughout the cell. This global firing is followed by a refractory period, reinforcing the rounding before the cell spreads again.

If the chemoattractant stimulus is spatially graded, then the effect is to change the corresponding parameter in a spatially heterogeneous way. We have previously proposed that the chemoattractant is processed through a spatial local excitation, global inhibition (LEGI) mechanism (Levchenko and Iglesias 2002). This has the effect that, at equilibrium, the front of the cell (where receptor occupancy is highest) has a lower threshold, whereas, at the rear of the cell, it is higher. This leads to a spatially biased excitable system with an increased production rate of protrusions on the side of the cell facing the chemoattractant and reduced activity at the rear. Simulations based on this principle reproduce directed migration with an efficiency that depends on the steepness of the gradient (Shi et al. 2013).

Altering the Threshold of the Excitable System through Synthetic Perturbations

We have shown how changing the nature of the nullclines through an external stimulus triggers a firing of the excitable system. If these changes are temporary, then a wave is triggered, but the effect is limited. Theoretically, it is known that permanent changes to the equilibrium point of the excitable system, for example, by varying the strength of the two feedback loops, can induce qualitative changes in excitable behavior, leading to phenomena such as oscillations, multistability, or the presence of permanently high steady states. In Figure 5E, these changes are seen in the dotted lines. For example, a smaller value for the parameter a_4 lowers the threshold. Eventually, for significantly high changes, the threshold is zero, which corresponds to a change in the stability of the equilibrium. In practice, we can move from excitable to oscillatory or multistable equilibria or even situations in which the single stable equilibrium is at a permanently "high" state (Fig. 5F).

In cells, acute changes like those suggested above could be done either through genetic mutations, or perturbations based on chemically or light-induced dimerization of key regulators. For example, Miao et al. (2017) used chemically induced dimerization (CID) to recruit the $PI(4,5)P_2$ phosphatase Inp54p, thereby lowering $PI(4,5)P_2$ levels synthetically in cells. Owing to the complementary inhibition between $PI(4,5)P_2$ and RasGTP (Li et al. 2018), this had the effect of increasing the positive feedback loop. Following this perturbation, the ameboid cells switched to fan-like or oscillatory behavior —consistent with the expectation that the excitable equations crossed through a Hopf bifurcation point (Kostova et al. 2004). To demonstrate that the perturbation did alter the threshold, the $PI(3,4,5)P_3$ response to the chemoattractant was measured before and after perturbation. Following Inp54p recruitment, cells that initially failed to respond to low cAMP doses became responsive to the same dose, while responses to saturating levels remained unchanged, indicating a reduced threshold for activation. These findings demonstrated that perturbations reducing $PI(4,5)P_2$ levels increase the excitability of the signaling network, facilitating expanded waves and migratory transitions.

Theoretical analysis also suggests the presence of a saddle-node bifurcation, leading to a new stable equilibrium with high activity (Kostova et al. 2004; Xiong et al. 2010). *Dictyostelium* cells lacking PTEN (hence with reduced $PI(4,5)P_2$) and expressing activated Ras GTPase-family proteins (increased RasGTP) transformed into a persistently activated state

(Edwards et al. 2018) that is consistent with a phase-plane portrait in which the intersection of the nullclines exists in a state of elevated F and R (Fig. 5F). Cells in this hyperactivated state showed intense, persistent spreading throughout the perimeter, expanding greatly but becoming quite thin. Eventually, these cells fragmented causing cell death.

CID has been used to modify other nodes of the network with corresponding changes in the shape and kinetics of the excitable waves (Miao et al. 2019). For example, the recruitment of PKB (the refractory element in the F–R model), decreased signaling wave speed and range, consistent with its role as the inhibitor, and a corresponding increase in the threshold. On the other hand, the elevated PKB levels raised the level of the cytoskeletal activator, thereby lowering the threshold for actin excitability and resulting in a greater number of actin waves with a small range.

These perturbations of the excitable networks have now been carried out in a wider variety of cell types including neutrophils and macrophages (Pal et al. 2023), epithelial cells, and cancer-derived cells (Zhan et al. 2020) as well as using optogenetic techniques to recruit effectors (Pal et al. 2023). The latter has several advantages: it is quite fast, can be targeted spatially to only portions of the cell, and it is reversible. Of particular interest were experiments in which synthetic activation of Ras was found to enhance wave activity in metastatic cancer cells, trigger new wave activity in breast epithelial cells, and increase protrusions in both cell types. Elevated wave activity is strongly correlated with the invasive behavior of increasingly metastatic breast cancer cell lines. Additionally, a slight reduction in $PI(4,5)P_2$ shifted the dose–response curve to epidermal growth factor (EGF) stimulation, lowering the threshold of the Ras-PI3K signaling network and increasing ERK sensitivity to EGF (Zhan et al. 2020).

Polarization

Polarization is a fundamental characteristic of cells, enabling them to establish spatial asymmetry in the distribution of molecular components and functions (Devreotes and Janetopoulos 2003). This phenomenon arises from the organization of signaling molecules and cytoskeletal elements into distinct domains, creating a functional axis. While polarization often occurs in response to chemoattractant gradients, it is not merely a reaction to external stimuli but an inherent cellular property. Polarization allows cells to maintain directionality, facilitate cell dispersal, and enhance environmental sensing (van Haastert 2021).

The extent of polarization in cells becomes evident when exposed to changes in the direction of chemoattractant gradients (Devreotes and Janetopoulos 2003). Unpolarized cells respond promptly by extending new pseudopods in the direction of the new gradient. In contrast, polarized cells reorient toward the gradient while maintaining their leading edge. This differential sensitivity supports the hypothesis that polarization is mediated by alterations in the activation threshold of the excitable network.

The establishment and maintenance of cellular polarity rely on complex feedback mechanisms that synchronize internal organization with external stimuli. One aspect of this polarization is the ability of cells to develop a memory. As argued above, the refractory period of excitable systems indicates that the cellular response depends on the previous stimulation history. Thus, cells "remember" their stimulation history. This can be in the form of chemoattractant gradients (Skoge et al. 2014) or periodic stimulation (Karmakar et al. 2024); the latter can aid cells navigating when exposed to chemoattractant waves (Uwamichi et al. 2023). However, cells also develop a memory independent of any external signaling. Evidence highlights the critical role of an active cytoskeleton in maintaining polarity (Janetopoulos et al. 2004; van Haastert et al. 2018; van Haastert 2020). In *Dictyostelium* cells, two distinct polarity mechanisms drive movement: short-term memory associated with pseudopod positioning and long-term memory tied to a stable polarity axis (van Haastert 2021). Short-term memory, lasting ~20 sec, depends on localized modifications of actin-regulating proteins like SCAR/WAVE, promoting pseudopod reformation at previous locations. In contrast, long-term memory, persisting for several minutes, involves the cyclic guanosine monophosphate (cGMP) pathway and myosin filament activity, suppressing pseudopod formation in the cell's rear and preserving a stable front-rear polarity.

Incorporating Polarity into Models

Polarization has been modeled as a product of mutually antagonistic signals that regulate front and back states (Iglesias and Devreotes 2008; Mori et al. 2008; Jilkine and Edelstein-Keshet 2011; Ku et al. 2012; Matsuoka and Ueda 2018). In these models, local responses to attractants or stochastic signals initiate strong positive feedback loops. This amplification of small, noise-induced heterogeneities leads to the spatial segregation of antagonistic components, resulting in distinct front and back states. Experimental evidence supports these theoretical frameworks showing the interplay between positive and negative feedback mechanisms in establishing and maintaining cellular polarity (van Haastert et al. 2018; Pal et al. 2023; Kuhn et al. 2024; Lin et al. 2024).

To integrate polarity models within the framework of excitable systems, a positive feedback loop from the cytoskeleton into the signaling network can be added (Charest et al. 2010; Khamviwath et al. 2013; Shi et al. 2013; van Haastert et al. 2018; Miao et al. 2019). Local activation of the cytoskeleton reduces the activation threshold of the signaling network in the corresponding region, increasing the likelihood of subsequent RasGTP activation. While this positive feedback loop ensures that cytoskeletal activity increases the likelihood of subsequent local activity, without further regulation, this activity will grow unabated leading to hyperactivity. Introducing a negative feedback loop stabilizes polarization by preventing the emergence of multiple fronts (Shi et al. 2013).

While models of this inhibitory loop have traditionally assumed a dispersed pattern—so-called global inhibition (Meinhardt 1999; Shi et al. 2013; van Haastert 2020)—recent experimental results have shed some light on the spatial distribution of this inhibition. For example, during neutrophil polarization, the WAVE complex exhibits spatially biased dynamics, driven by a combination of intrinsic and extrinsic cues, with positive feedback between actin polymerization and the SCAR/WAVE complex localization playing a crucial role in establishing cell polarity (Millius et al. 2009). Similar feedback has been observed in *Dictyostelium* cells (van Haastert 2021). On the other hand, the response of HeLa cells to synthetically generated Rac activation suggests that local inhibition at the cell's rear, mediated by cGMP signaling and myosin filament activity, is more likely than local activation (Holmes et al. 2012). Experiments in *Dictyostelium* also point to the inhibitory role of myosin filaments at the side and in the rear on the formation of new, branched F-actin (Bosgraaf and Van Haastert 2002; van Haastert 2021). This inhibition appears to act on the signaling network (Kuhn et al. 2024). While complementary local positive and negative feedback loops can polarize cells, there is also evidence of global inhibitory feedback, acting through the cytoskeleton (Kabacoff et al. 2007) or through membrane tension (Batchelder et al. 2011; Houk et al. 2012; De Belly et al. 2023), possibly involving membrane curvature-sensing proteins (Tsujita et al. 2015).

Interaction between Signaling and Cytoskeleton

As argued earlier although signaling and cytoskeletal components both exhibit wave propagation, they function within distinct excitable networks (Fig. 6A). The STEN consists of Ras, Rap, PIP_3, and PKBs and serves as a signaling module that initiates and propagates excitatory signals. It exhibits single-peak wave structures. In contrast, the CEN—composed of Rac, F-actin, RacGEF1, formins, and coronin—generates cytoskeletal protrusions and acts as a mechanical force generator (Huang et al. 2013; Taniguchi et al. 2013, Miao et al. 2019; Chua et al. 2024).

While STEN and CEN can operate independently, they are functionally coupled in cells. STEN regulates CEN through PKBs, which help trigger and organize cytoskeletal waves (Miao et al. 2019). Conversely, CEN provides

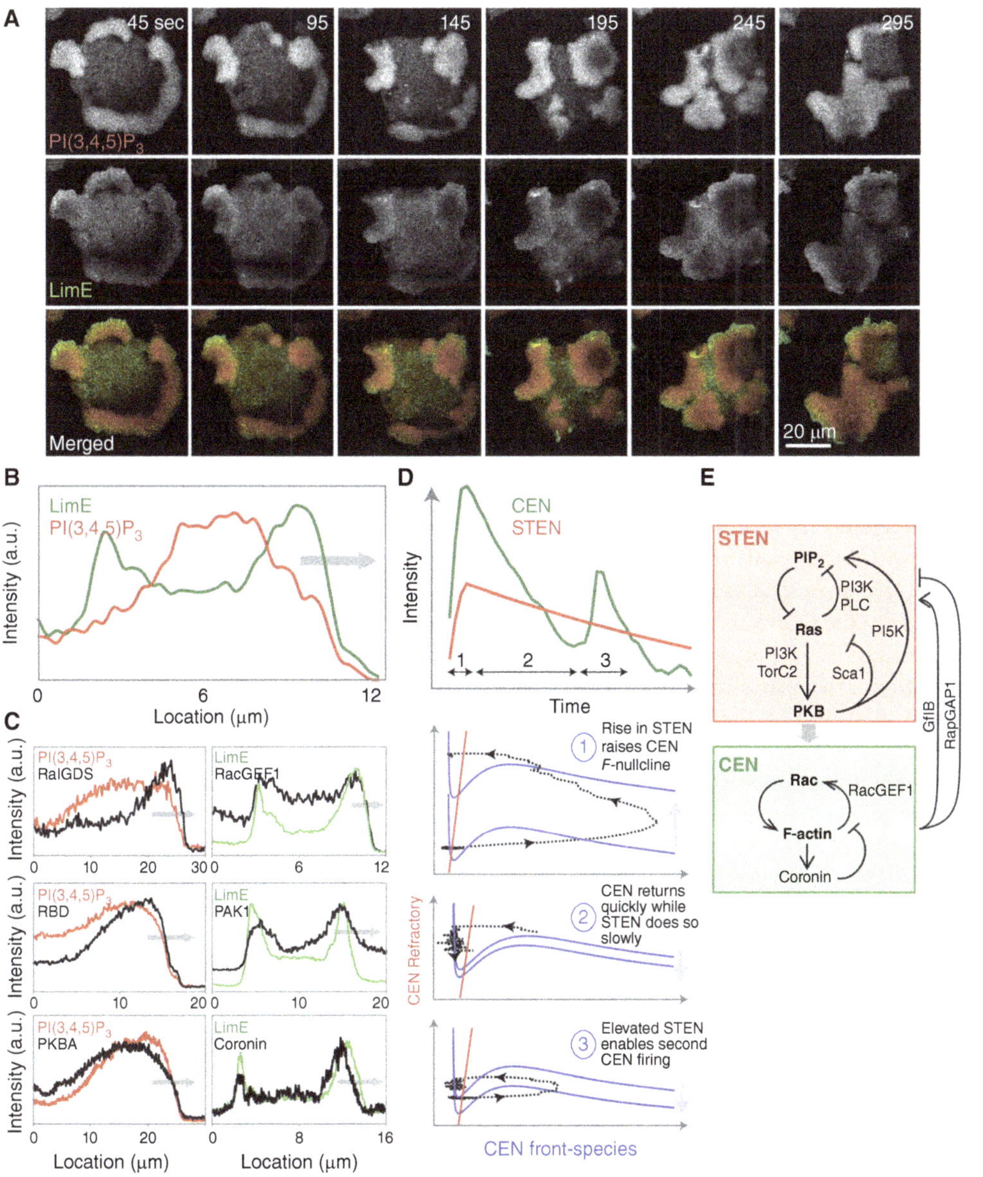

Figure 6. Cytoskeletal excitable network (CEN) and signal transduction excitable network (STEN) exhibit distinct profiles during wave propagation. (*A*) Wave propagation in a *Dictyostelium* cell expressing biosensors for the STEN (PI(3,4,5)P_3; biosensor is PH-domain) and the CEN (LimE). (*B*) Cross-sectional profile of the wave shows that the CEN marker (LimE, green) exhibits a double-peaked structure, with higher intensity at the cell front. In contrast, the STEN wave front (PI(3,4,5)P_3, red) is single-peaked and positioned inside the CEN wave. (*C*) Components of STEN and CEN follow similar peak patterns but differ in their timing relative to the fiducial markers PI(3,4,5)P_3 (PH biosensor) and LimE, respectively. (*D*) Simulation of the coupled STEN–CEN network (red and green, respectively), highlighting single- and double-peaked wave dynamics. Phase-plane portraits show how STEN activation alters the CEN front-element nullcline, leading to CEN excitation. Although STEN is triggered first, CEN reaches its initial peak earlier owing to its faster dynamics. As CEN returns to its new equilibrium, persistent STEN activation results in a secondary, patchier CEN firing. (*E*) Schematic representation of the STEN–CEN network and its putative interactions. (Images for the panels in this figure have been adapted from Miao et al. 2019; reprinted under the terms of the Creative Commons Attribution 4.0 License.)

feedback to STEN via RacGEF1 and RapGAP1, a mechanism that establishes cellular memory, enabling cell polarization (Charest et al. 2010). Experimental studies using micropatterned substrates have shown that CEN functions as the mechanosensory element. The feedback from CEN to STEN then directs cell migration (Driscoll et al. 2014; Honda et al. 2021; Yang et al. 2023).

The two networks operate on distinct timescales: STEN functions more slowly, while CEN responds rapidly, driving cytoskeletal rearrangements. This can be seen by observing intensity levels across a propagating wave front. Whereas CEN components show a double peak structure, with a large peak at the wave front and a smaller, sometimes punctate peak at the rear, STEN species show a single peak (Fig. 6A,B; Gerhardt et al. 2014; Miao et al. 2019). Note that although the timing of the individual peaks changes (Fig. 6C), the number of peaks is consistent across species within the respective system, suggesting that it is a feature of the particular excitable network.

Computational models demonstrate that these differences in timescales explain the phase shift between the two waves. In a propagating wave, the fast CEN wave peaks first, ahead of STEN. As CEN rapidly completes its phase-plane cycle, STEN remains elevated, allowing CEN to fire a second time—although with reduced intensity. By the time this second activation completes its trajectory, STEN returns to equilibrium, preventing a third peak (Fig. 6D–F). This timescale separation also explains a puzzling feature of the coupled STEN–CEN network. While chemoattractant signaling is channeled to STEN, which is then transmitted to CEN, it is the latter that peaks first. However, the simulations show that this is possible if the firing of CEN is faster than that of STEN.

Parameter Selection

All models involve simplification (Mogilner et al. 2012), and even in simple cases, the number of parameters can grow rapidly. In general, there is a trade-off between model accuracy and parameter identifiability (Lockley et al. 2015). This challenge is particularly pronounced in excitable systems of cell motility, where threshold-driven dynamics make the system highly sensitive to small parameter variations. Additionally, the interplay between stochastic signaling and diffusion further amplifies this sensitivity.

Various techniques have been proposed for parameter estimation in stochastic differential equations (Altmeyer and Reiß 2021; Altmeyer et al. 2022). Most of these methods rely on minimizing, in a least-squares sense, the difference between the deterministic component of the stochastic, partial differential equation (SPDE) and its derivative. For example, Pasemann et al. (2021) fit a stochastic reaction–diffusion FHN-like model to actin waves observed in *Dictyostelium*. Recently, Abubaker-Sharif et al. (2024) compared two data-driven methodologies—Kramers–Moyal regression and wave features optimization—as approaches for estimating the parameters of an SPDE description of excitable waves. While both techniques performed well under ideal conditions, wave feature optimization was more resilient in more realistic scenarios, such as limited data and model mismatch.

Although not specifically designed to estimate signaling systems regulating cell motility, Croft et al. (2015) applied a nonlinear least-squares approach using a Levenberg–Marquardt optimization method to estimate parameters in cell motility models of cellular morphology from imaging data.

OUTLOOK

The discovery that cell motility in eukaryotic cells is regulated by excitable systems has resolved several long-standing questions. For example, demonstrating how small changes in system parameters can drastically alter the activation threshold has clarified how cells amplify weak external gradients to drive chemotaxis. Similarly, the characteristic trajectories of excitable systems explain why the statistical properties of pseudopod dynamics remain nearly indistinguishable between randomly migrating cells and those undergoing directed movement.

The presence of two coupled excitable networks—STEN and CEN—highlights the system's complexity, even at a theoretical level. Signaling governs cytoskeletal organization, which in turn generates forces and shapes cellular morphology—processes further modulated by interactions with the extracellular environment. A purely experimental approach is insufficient to fully elucidate these dynamics, yet computational models also face significant challenges due to the system's inherent complexity, often requiring simplifications. For instance, this review has not addressed integrated models that couple signaling and cytoskeletal activity to cellular morphology. Holmes et al. (2012) and Di Napoli et al. (2020) review computational techniques for simulating cell deformations.

Moving forward, the integration of mathematical modeling and computational simulations with experimental studies will be essential for advancing our understanding of cell motility. Continued interdisciplinary efforts will be critical for bridging theoretical predictions with biological observations, ultimately unraveling the fundamental principles governing cell movement.

ACKNOWLEDGMENTS

The work of our laboratory has been a longstanding collaboration with the Devreotes laboratory (Johns Hopkins School of Medicine). We thank members of the Iglesias and Devreotes laboratory for many years of fruitful collaboration. We are grateful to Bedri Abubaker-Sharif (Johns Hopkins) for providing the unpublished video in Figure 1A and to Miguel Iglesias (Princeton) for comments on the manuscript. This work was supported in part by grant R01 GM149073.

REFERENCES

Abubaker-Sharif B, Devreotes PN, Iglesias PA. 2024. Machine learning sparse reaction-diffusion models from stochastic dynamics and spatiotemporal patterns. bioRxiv doi:10.1101/2024.10.02.616367

Altmeyer R, Reiß M. 2021. Nonparametric estimation for linear SPDEs from local measurements. *Ann Appl Probab* **31:** 1–38. doi:10.1214/20-AAP1581

Altmeyer R, Bretschneider T, Janák J, Reiß M. 2022. Parameter estimation in an SPDE model for cell repolarization. *SIAM/ASA J Uncertain Quantif* **10:** 179–199. doi:10.1137/20M1373347

Andrew N, Insall RH. 2007. Chemotaxis in shallow gradients is mediated independently of PtdIns 3-kinase by biased choices between random protrusions. *Nat Cell Biol* **9:** 193–200. doi:10.1038/ncb1536

Arai Y, Shibata T, Matsuoka S, Sato MJ, Yanagida T, Ueda M. 2010. Self-organization of the phosphatidylinositol lipids signaling system for random cell migration. *Proc Natl Acad Sci* **107:** 12399–12404. doi:10.1073/pnas.0908278107

Artemenko Y, Axiotakis L Jr, Borleis J, Iglesias PA, Devreotes PN. 2016. Chemical and mechanical stimuli act on common signal transduction and cytoskeletal networks. *Proc Natl Acad Sci* **113:** E7500–E7509. doi:10.1073/pnas.1608767113

Banerjee T, Biswas D, Pal DS, Miao Y, Iglesias PA, Devreotes PN. 2022. Spatiotemporal dynamics of membrane surface charge regulates cell polarity and migration. *Nat Cell Biol* **24:** 1499–1515. doi:10.1038/s41556-022-00997-7

Batchelder EL, Hollopeter G, Campillo C, Mezanges X, Jorgensen EM, Nassoy P, Sens P, Plastino J. 2011. Membrane tension regulates motility by controlling lamellipodium organization. *Proc Natl Acad Sci* **108:** 11429–11434. doi:10.1073/pnas.1010481108

Bement WM, Leda M, Moe AM, Kita AM, Larson ME, Golding AE, Pfeuti C, Su KC, Miller AL, Goryachev AB, et al. 2015. Activator-inhibitor coupling between Rho signalling and actin assembly makes the cell cortex an excitable medium. *Nat Cell Biol* **17:** 1471–1483. doi:10.1038/ncb3251

Beta C, Edelstein-Keshet L, Gov N, Yochelis A. 2023. From actin waves to mechanism and back: how theory aids biological understanding. *eLife* **12:** e87181. doi:10.7554/eLife.87181

Bhattacharya S, Iglesias PA. 2018. The threshold of an excitable system serves as a control mechanism for noise filtering during chemotaxis. *PLoS One* **13:** e0201283. doi:10.1371/journal.pone.0201283

Bhattacharya S, Iglesias PA. 2019. Controlling excitable wave behaviors through the tuning of three parameters. *Biol Cybern* **113:** 61–70. doi:10.1007/s00422-018-0771-0

Bhattacharya S, Banerjee T, Miao Y, Zhan H, Devreotes PN, Iglesias PA. 2020. Traveling and standing waves mediate pattern formation in cellular protrusions. *Sci Adv* **6:** eaay7682. doi:10.1126/sciadv.aay7682

Bosgraaf L, van Haastert PJM. 2002. A model for cGMP signal transduction in *Dictyostelium* in perspective of 25 years of cGMP research. *J Muscle Res Cell Motil* **23:** 781–791. doi:10.1023/A:1024431813040

Case LB, Waterman CM. 2011. Adhesive F-actin waves: a novel integrin-mediated adhesion complex coupled to ventral actin polymerization. *PLoS One* **6:** e26631. doi:10.1371/journal.pone.0026631

Charest PG, Shen Z, Lakoduk A, Sasaki AT, Briggs SP, Firtel RA. 2010. A Ras signaling complex controls the RasC-TORC2 pathway and directed cell migration. *Dev Cell* **18:** 737–749. doi:10.1016/j.devcel.2010.03.017

Chua XL, Tong CS, Su M, Xǔ XJ, Xiao S, Wu X, Wu M. 2024. Competition and synergy of Arp2/3 and formins in nucleating actin waves. *Cell Rep* **43:** 114423. doi:10.1016/j.celrep.2024.114423

Coscia SM, Moore AS, Wong YC, Holzbaur ELF. 2024. Mitochondrially-associated actin waves maintain organelle homeostasis and equitable inheritance. *Curr Opin Cell Biol* **88:** 102364. doi:10.1016/j.ceb.2024.102364

Croft W, Elliott CM, Ladds G, Stinner B, Venkataraman C, Weston C. 2015. Parameter identification problems in the modelling of cell motility. *J Math Biol* **71:** 399–436. doi:10.1007/s00285-014-0823-6

De Belly H, Yan S, Borja da Rocha H, Ichbiah S, Town JP, Zager PJ, Estrada DC, Meyer K, Turlier H, Bustamante C, et al. 2023. Cell protrusions and contractions generate long-range membrane tension propagation. *Cell* **186:** 3049–3061.e15. doi:10.1016/j.cell.2023.05.014

Deng Y, Banerjee T, Pal DS, Banerjee P, Zhan H, Borleis J, Iglesias PA, Devreotes PN. 2024. PIP5K-Ras bistability initiates plasma membrane symmetry breaking to regulate cell polarity and migration. bioRxiv doi:10.1101/2024.09.15.613115

de Oliveira S, Rosowski EE, Huttenlocher A. 2016. Neutrophil migration in infection and wound repair: going forward in reverse. *Nat Rev Immunol* **16:** 378–391. doi:10.1038/nri.2016.49

Devreotes PN, Janetopoulos C. 2003. Eukaryotic chemotaxis: distinctions between directional sensing and polarization. *J Biol Chem* **278:** 20445–20448. doi:10.1074/jbc.R300010200

Devreotes PN, Bhattacharya S, Edwards M, Iglesias PA, Lampert T, Miao Y. 2017. Excitable signal transduction networks in directed cell migration. *Annu Rev Cell Dev Biol* **33:** 103–125. doi:10.1146/annurev-cellbio-100616-060739

DiNapoli KT, Robinson DN, Iglesias PA. 2020. Tools for computational analysis of moving boundary problems in cellular mechanobiology. *Wiley Interdiscip Rev Syst Biol Med* **13:** e1514. doi:10.1002/wsbm.1514

Driscoll MK, Sun X, Guven C, Fourkas JT, Losert W. 2014. Cellular contact guidance through dynamic sensing of nanotopography. *ACS Nano* **8:** 3546–3555. doi:10.1021/nn406637c

Edwards M, Cai H, Abubaker-Sharif B, Long Y, Lampert TJ, Devreotes PN. 2018. Insight from the maximal activation of the signal transduction excitable network in *Dictyostelium discoideum*. *Proc Natl Acad Sci* **115:** E3722–E3730. doi:10.1073/pnas.1710480115

Fets L, Nichols JME, Kay RR. 2014. A PIP5 kinase essential for efficient chemotactic signaling. *Curr Biol* **24:** 415–421. doi:10.1016/j.cub.2013.12.052

Filić V, Marinović M, Faix J, Weber I. 2012. A dual role for Rac1 GTPases in the regulation of cell motility. *J Cell Sci* **125:** 387–398. doi:10.1242/jcs.089680

FitzHugh R. 1961. Impulses and physiological states in theoretical models of nerve membrane. *Biophys J* **1:** 445–466. doi:10.1016/S0006-3495(61)86902-6

Futrelle RP, Traut J, McKee WG. 1982. Cell behavior in *Dictyostelium discoideum*: preaggregation response to localized cyclic AMP pulses. *J Cell Biol* **92:** 807–821. doi:10.1083/jcb.92.3.807

Gerhardt M, Ecke M, Walz M, Stengl A, Beta C, Gerisch G. 2014. Actin and PIP3 waves in giant cells reveal the inherent length scale of an excited state. *J Cell Sci* **127:** 4507–4517. doi:10.1242/jcs.156000

Gerisch G, Albrecht R, Heizer C, Hodgkinson S, Maniak M. 1995. Chemoattractant-controlled accumulation of coronin at the leading edge of *Dictyostelium* cells monitored using a green fluorescent protein-coronin fusion protein. *Curr Biol* **5:** 1280–1285. doi:10.1016/S0960-9822(95)00254-5

Gillespie DT. 1976. A general method for numerically simulating the stochastic time evolution of coupled chemical reactions. *J Comp Phys* **22:** 403–434. doi:10.1016/0021-9991(76)90041-3

Gillespie DT. 2000. The chemical Langevin equation. *J Chem Phys* **113:** 297–306. doi:10.1063/1.481811

Hodgkin AL, Huxley AF. 1952. A quantitative description of membrane current and its application to conduction and excitation in nerve. *J Physiol* **117:** 500–544. doi:10.1113/jphysiol.1952.sp004764

Holmes WR, Lin B, Levchenko A, Edelstein-Keshet L. 2012. Modelling cell polarization driven by synthetic spatially graded Rac activation. *PLoS Comput Biol* **8:** e1002366. doi:10.1371/journal.pcbi.1002366

Honda G, Saito N, Fujimori T, Hashimura H, Nakamura MJ, Nakajima A, Sawai S. 2021. Microtopographical guidance of macropinocytic signaling patches. *Proc Natl Acad Sci* **118:** e2110281118. doi:10.1073/pnas.2110281118

Horsthemke W, Lefever R. 1984. *Noise-induced transitions: theory and applications in physics, chemistry, and biology*. Springer, Berlin.

Houk AR, Jilkine A, Mejean CO, Boltyanskiy R, Dufresne ER, Angenent SB, Altschuler SJ, Wu LF, Weiner OD. 2012. Membrane tension maintains cell polarity by confining signals to the leading edge during neutrophil migration. *Cell* **148:** 175–188. doi:10.1016/j.cell.2011.10.050

Huang CH, Tang M, Shi C, Iglesias PA, Devreotes PN. 2013. An excitable signal integrator couples to an idling cytoskeletal oscillator to drive cell migration. *Nat Cell Biol* **15:** 1307–1316. doi:10.1038/ncb2859

Iglesias PA. 2013. Excitable systems in cell motility. In *IEEE 52nd Annual Conference on Decision and Control*, Firenze, Italy, pp. 757–762. doi: 10.1109/CDC.2013.6759973

Iglesias PA, Devreotes PN. 2008. Navigating through models of chemotaxis. *Curr Opin Cell Biol* **20:** 35–40. doi:10.1016/j.ceb.2007.11.011

Iglesias PA, Devreotes PN. 2012. Biased excitable networks: how cells direct motion in response to gradients. *Curr Opin Cell Biol* **24:** 245–253. doi:10.1016/j.ceb.2011.11.009

Izhikevich EM. 2007. *Dynamical systems in neuroscience: the geometry of excitability and bursting*. MIT, Cambridge, MA.

Janetopoulos C, Ma L, Devreotes PN, Iglesias PA. 2004. Chemoattractant-induced phosphatidylinositol 3,4,5-trisphosphate accumulation is spatially amplified and adapts, independent of the actin cytoskeleton. *Proc Natl Acad Sci* **101:** 8951–8956. doi:10.1073/pnas.0402152101

Jilkine A, Edelstein-Keshet L. 2011. A comparison of mathematical models for polarization of single eukaryotic cells in response to guided cues. *PLoS Comput Biol* **7:** e1001121. doi:10.1371/journal.pcbi.1001121

Kabacoff C, Xiong Y, Musib R, Reichl EM, Kim J, Iglesias PA, Robinson DN. 2007. Dynacortin facilitates polarization of chemotaxing cells. *BMC Biol* **5:** 53. doi:10.1186/1741-7007-5-53

Karmakar R, Karanam A, Tang MH, Rappel WJ. 2024. Eukaryotic chemotaxis under periodic stimulation shows temporal gradient dependence. *Phys Rev Lett* **133:** 068401. doi:10.1103/PhysRevLett.133.068401

Keener JP, Sneyd J. 1998. *Mathematical physiology*. Springer, New York.

Khamviwath V, Hu J, Othmer HG. 2013. A continuum model of actin waves in *Dictyostelium discoideum*. *PLoS One* **8:** e64272. doi:10.1371/journal.pone.0064272

Kolaczkowska E, Kubes P. 2013. Neutrophil recruitment and function in health and inflammation. *Nat Rev Immunol* **13:** 159–175. doi:10.1038/nri3399

Kostova T, Ravindran R, Schonbek M. 2004. FitzHugh–Nagumo revisited: types of bifurcations, periodical forcing and stability regions by a Lyapunov functional. *Int J Bifurcation Chaos* **14:** 913–925. doi:10.1142/S0218127404009685

Ku C-J, Wang Y, Weiner OD, Altschuler SJ, Wu LF. 2012. Network crosstalk dynamically changes during neutrophil polarization. *Cell* **149:** 1073–1083. doi:10.1016/j.cell.2012.03.044

Kuhn J, Banerjee P, Haye A, Robinson DN, Iglesias PA, Devreotes PN. 2024. Complementary cytoskeletal feedback loops control signal transduction excitability and cell polarity. *Nat Commun* doi:10.1101/2024.02.13.580131

Levchenko A, Iglesias PA. 2002. Models of eukaryotic gradient sensing: application to chemotaxis of amoebae and neutrophils. *Biophys J* **82:** 50–63. doi:10.1016/S0006-3495(02)75373-3

Li X, Edwards M, Swaney KF, Singh N, Bhattacharya S, Borleis J, Long Y, Iglesias PA, Chen J, Devreotes PN. 2018. Mutually inhibitory Ras-PI(3,4)P_2 feedback loops mediate cell migration. *Proc Natl Acad Sci* **115:** E9125–E9134. doi:10.1073/pnas.1809039115

Lin Y, Pal DS, Banerjee P, Banerjee T, Qin G, Deng Y, Borleis J, Iglesias PA, Devreotes PN. 2024. Ras suppression potentiates rear actomyosin contractility-driven cell polarization and migration. *Nat Cell Biol* **26:** 1062–1076. doi:10.1038/s41556-024-01453-4

Lockley R, Ladds G, Bretschneider T. 2015. Image based validation of dynamical models for cell reorientation. *Cytometry A* **87:** 471–480. doi:10.1002/cyto.a.22600

Matsuoka S, Ueda M. 2018. Mutual inhibition between PTEN and PIP3 generates bistability for polarity in motile cells. *Nat Commun* **9:** 4481. doi:10.1038/s41467-018-06856-0

Matsuoka S, Iwamoto K, Shin DY, Ueda M. 2024. Spontaneous signal generation by an excitable system for cell migration. *Front Cell Dev Biol* **12:** 1373609. doi:10.3389/fcell.2024.1373609

Meili R, Ellsworth C, Lee S, Reddy TB, Ma H, Firtel RA. 1999. Chemoattractant-mediated transient activation and membrane localization of Akt/PKB is required for efficient chemotaxis to cAMP in *Dictyostelium*. *EMBO J* **18:** 2092–2105. doi:10.1093/emboj/18.8.2092

Meinhardt H. 1999. Orientation of chemotactic cells and growth cones: models and mechanisms. *J Cell Sci* **112:** 2867–2874. doi:10.1242/jcs.112.17.2867

Miao Y, Bhattacharya S, Edwards M, Cai H, Inoue T, Iglesias PA, Devreotes PN. 2017. Altering the threshold of an excitable signal transduction network changes cell migratory modes. *Nat Cell Biol* **19:** 329–340. doi:10.1038/ncb3495

Miao Y, Bhattacharya S, Banerjee T, Abubaker-Sharif B, Long Y, Inoue T, Iglesias PA, Devreotes PN. 2019. Wave patterns organize cellular protrusions and control cortical dynamics. *Mol Syst Biol* **15:** e8585. doi:10.15252/msb.20188585

Millius A, Dandekar SN, Houk AR, Weiner OD. 2009. Neutrophils establish rapid and robust wave complex polarity in an actin-dependent fashion. *Curr Biol* **19:** 253–259. doi:10.1016/j.cub.2008.12.044

Mogilner A, Allard J, Wollman R. 2012. Cell polarity: quantitative modeling as a tool in cell biology. *Science* **336:** 175–179. doi:10.1126/science.1216380

Mori Y, Jilkine A, Edelstein-Keshet L. 2008. Wave-pinning and cell polarity from a bistable reaction-diffusion system. *Biophys J* **94:** 3684–3697. doi:10.1529/biophysj.107.120824

Nagumo J, Arimoto S, Yoshizawa S. 1962. An active pulse transmission line simulating nerve axon. *Proc IRE* **50:** 2061–2070. doi:10.1109/JRPROC.1962.288235

Naoki H, Sakumura Y, Ishii S. 2008. Stochastic control of spontaneous signal generation for gradient sensing in chemotaxis. *J Theor Biol* **255:** 259–266. doi:10.1016/j.jtbi.2008.08.012

Neilson MP, Veltman DM, van Haastert PJM, Webb SD, Mackenzie JA, Insall RH. 2011. Chemotaxis: a feedback-based computational model robustly predicts multiple aspects of real cell behaviour. *PLoS Biol* **9:** e1000618. doi:10.1371/journal.pbio.1000618

Nichols JM, Veltman D, Kay RR. 2015. Chemotaxis of a model organism: progress with *Dictyostelium*. *Curr Opin Cell Biol* **36:** 7–12. doi:10.1016/j.ceb.2015.06.005

Nishikawa M, Hörning M, Ueda M, Shibata T. 2014. Excitable signal transduction induces both spontaneous and directional cell asymmetries in the phosphatidylinositol lipid signaling system for eukaryotic chemotaxis. *Biophys J* **106:** 723–734. doi:10.1016/j.bpj.2013.12.023

Othmer HG. 2019. Eukaryotic cell dynamics from crawlers to swimmers. *Wiley Interdiscip Rev Comput Mol Sci* **9:** e1376. doi:10.1002/wcms.1376

Pal DS, Banerjee T, Lin Y, de Trogoff F, Borleis J, Iglesias PA, Devreotes PN. 2023. Actuation of single downstream nodes in growth factor network steers immune cell migration. *Dev Cell* **58:** 1170–1188.e7. doi:10.1016/j.devcel.2023.04.019

Pasemann G, Flemming S, Alonso S, Beta C, Stannat W. 2021. Diffusivity estimation for activator-inhibitor models: theory and application to intracellular dynamics of the actin cytoskeleton. *J Nonlinear Sci* **31:** 59. doi:10.1007/s00332-021-09714-4

Pertsov AM, Davidenko JM, Salomonsz R, Baxter WT, Jalife J. 1993. Spiral waves of excitation underlie reentrant activity in isolated cardiac muscle. *Circ Res* **72:** 631–650. doi:10.1161/01.RES.72.3.631

Pollak E, Talkner P. 2005. Reaction rate theory: what it was, where is it today, and where is it going? *Chaos* **15:** 26116. doi:10.1063/1.1858782

Rappel WJ, Edelstein-Keshet L. 2017. Mechanisms of cell polarization. *Curr Opin Syst Biol* **3:** 43–53. doi:10.1016/j.coisb.2017.03.005

Rappel WJ, Loomis WF. 2009. Eukaryotic chemotaxis. *Wiley Interdiscip Rev Syst Biol Med* **1:** 141–149. doi:10.1002/wsbm.28

Roussos ET, Condeelis JS, Patsialou A. 2011. Chemotaxis in cancer. *Nat Rev Cancer* **11:** 573–587. doi:10.1038/nrc3078

Ryan GL, Petroccia HM, Watanabe N, Vavylonis D. 2012. Excitable actin dynamics in lamellipodial protrusion and retraction. *Biophys J* **102:** 1493–1502. doi:10.1016/j.bpj.2012.03.005

Scavello M, Petlick AR, Ramesh R, Thompson VF, Lotfi P, Charest PG. 2017. Protein kinase A regulates the Ras, Tap1 and TORC2 pathways in response to the chemoattractant camp in *Dictyostelium*. *J Cell Sci* **130:** 1545–1558. doi:10.1242/jcs.177170

Shi C, Huang CH, Devreotes PN, Iglesias PA. 2013. Interaction of motility, directional sensing, and polarity modules recreates the behaviors of chemotaxing cells. *PLoS Comput Biol* **9:** e1003122. doi:10.1371/journal.pcbi.1003122

Skoge M, Yue H, Erickstad M, Bae A, Levine H, Groisman A, Loomis WF, Rappel WJ. 2014. Cellular memory in eukaryotic chemotaxis. *Proc Natl Acad Sci* **111:** 14448–14453. doi:10.1073/pnas.1412197111

Szabó A, Mayor R. 2018. Mechanisms of neural crest migration. *Annu Rev Genet* **52:** 43–63. doi:10.1146/annurev-genet-120417-031559

Tanabe Y, Kamimura Y, Ueda M. 2018. Parallel signaling pathways regulate excitable dynamics differently to mediate pseudopod formation during eukaryotic chemotaxis. *J Cell Sci* **131:** jcs214775. doi:10.1242/jcs.214775

Tang M, Wang M, Shi C, Iglesias PA, Devreotes PN, Huang CH. 2014. Evolutionarily conserved coupling of adaptive and excitable networks mediates eukaryotic chemotaxis. *Nat Commun* **5:** 5175. doi:10.1038/ncomms6175

Taniguchi D, Ishihara S, Oonuki T, Honda-Kitahara M, Kaneko K, Sawai S. 2013. Phase geometries of two-dimensional excitable waves govern self-organized morphodynamics of amoeboid cells. *Proc Natl Acad Sci* **110:** 5016–5021. doi:10.1073/pnas.1218025110

Tiberio L, Del Prete A, Schioppa T, Sozio F, Bosisio D, Sozzani S. 2018. Chemokine and chemotactic signals in dendritic cell migration. *Cell Mol Immunol* **15:** 346–352. doi:10.1038/s41423-018-0005-3

Tsujita K, Takenawa T, Itoh T. 2015. Feedback regulation between plasma membrane tension and membrane-bending proteins organizes cell polarity during leading edge formation. *Nat Cell Biol* **17:** 749–758. doi:10.1038/ncb3162

Tyson JJ, Keener JP. 1988. Singular perturbation theory of traveling waves in excitable media (a review). *Physica D* **32:** 327–361. doi:10.1016/0167-2789(88)90062-0

Uwamichi M, Miura Y, Kamiya A, Imoto D, Sawai S. 2023. Random walk and cell morphology dynamics in *Naegleria gruberi*. *Front Cell Dev Biol* **11:** 1274127. doi:10.3389/fcell.2023.1274127

van Haastert PJM. 2020. Unified control of amoeboid pseudopod extension in multiple organisms by branched F-actin in the front and parallel F-actin/myosin in the cortex. *PLoS One* **15:** e0243442. doi:10.1371/journal.pone.0243442

van Haastert PJM. 2021. Short- and long-term memory of moving amoeboid cells. *PLoS One* **16:** e0246345. doi:10.1371/journal.pone.0246345

van Haastert PJM. 2024. Pseudopod tracking and statistics during cell movement in buffer and chemotaxis. *Methods Mol Biol* **2828:** 185–204. doi:10.1007/978-1-0716-4023-4_14

van Haastert PJM, Keizer-Gunnink I, Kortholt A. 2017. Coupled excitable Ras and F-actin activation mediates spontaneous pseudopod formation and directed cell movement. *Mol Biol Cell* **28:** 922–934. doi:10.1091/mbc.e16-10-0733

van Haastert PJM, Keizer-Gunnink I, Kortholt A. 2018. The cytoskeleton regulates symmetry transitions in moving amoeboid cells. *J Cell Sci* **131:** jcs208892. doi:10.1242/jcs.208892

Vicker MG. 2000. Reaction-diffusion waves of actin filament polymerization/depolymerization in *Dictyostelium* pseudopodium extension and cell locomotion. *Biophys Chem* **84:** 87–98. doi:10.1016/S0301-4622(99)00146-5

Vicker MG. 2002. F-actin assembly in *Dictyostelium* cell locomotion and shape oscillations propagates as a self-organized reaction-diffusion wave. *FEBS Lett* **510:** 5–9. doi:10.1016/S0014-5793(01)03207-0

Weiner OD, Marganski WA, Wu LF, Altschuler SJ, Kirschner MW. 2007. An actin-based wave generator organizes cell motility. *PLoS Biol* **5:** e221. doi:10.1371/journal.pbio.0050221

Winans AM, Collins SR, Meyer T. 2016. Waves of actin and microtubule polymerization drive microtubule-based transport and neurite growth before single axon formation. *eLife* **5:** e12387. doi:10.7554/eLife.12387

Wu M, Wu X, De Camilli P. 2013. Calcium oscillations-coupled conversion of actin travelling waves to standing oscillations. *Proc Natl Acad Sci* **110:** 1339–1344. doi:10.1073/pnas.1221538110

Wu Z, Su M, Tong C, Wu M, Liu J. 2018. Membrane shape-mediated wave propagation of cortical protein dynamics. *Nat Commun* **9:** 1–12. doi:10.1038/s41467-017-02469-1

Xiong Y, Huang CH, Iglesias PA, Devreotes PN. 2010. Cells navigate with a local-excitation, global-inhibition-biased excitable network. *Proc Natl Acad Sci* **107:** 17079–17086. doi:10.1073/pnas.1011271107

Yang Q, Miao Y, Banerjee P, Hourwitz MJ, Hu M, Qing Q, Iglesias PA, Fourkas JT, Losert W, Devreotes PN. 2023. Nanotopography modulates intracellular excitable systems through cytoskeleton actuation. *Proc Natl Acad Sci* **120:** e2218906120. doi:10.1073/pnas.2218906120

Zhan H, Bhattacharya S, Cai H, Iglesias PA, Huang CH, Devreotes PN. 2020. An excitable Ras/PI3K/ERK signaling network controls migration and oncogenic transformation in epithelial cells. *Dev Cell* **54:** 608–623.e5. doi:10.1016/j.devcel.2020.08.001

Zykov V, Krekhov A, Bodenschatz E. 2017. Fast propagation regions cause self-sustained reentry in excitable media. *Proc Natl Acad Sci* **114:** 1281–1286. doi:10.1073/pnas.1611475114

Organized Randomness: How Hierarchical Coupling of an Excitable System and a Bistable Switch Shape Spontaneous Cell Migration

Masahiro Ueda,[1] Hiroaki Takagi,[2] and Satomi Matsuoka[1]

[1]Laboratory of Single Molecule Biology, Graduate School of Frontier Biosciences, The University of Osaka, Suita, Osaka 565-0871, Japan

[2]Department of Physics, School of Medicine, Nara Medical University, Nara, 634-8521, Japan

Correspondence: ueda.masahiro.fbs@osaka-u.ac.jp

In various eukaryotic organisms, amoeboid cells undergoing spontaneous migration generate membrane protrusions and establish front–rear polarity in an autonomous manner through signal transduction networks that couple excitable Ras dynamics to downstream bistable modules. Recent genetic and live-cell imaging studies in *Dictyostelium* cells have identified Ras GTPase and the newly characterized guanine nucleotide exchange factor RasGEFX as core drivers of this excitability, amplifying stochastic basal Ras activity into discrete signaling patches and propagating waves. These excitatory events initiate pseudopod formation even in the absence of external guidance cues and engage the bistable PIP_3-PTEN switch to stabilize cell polarity. The same Ras excitable network also drives constitutive macropinocytosis, revealing a shared mechanism for membrane protrusion and fluid uptake. This hierarchical excitable–bistable architecture, coupled with motile machinery, converts intrinsic molecular noise into structured, non-Brownian exploratory behavior, representing an adaptive strategy that enables cells to exploit stochasticity for navigation in complex natural environments.

The organization principles by which spontaneous cellular behaviors emerge from internal molecular reactions remain elusive. Many cell types exhibit spontaneous random migration, reflecting internal reaction dynamics that autonomously generate signaling events to drive cell motility (Oosawa 2001, 2007). Amoeboid motility promotes efficient tissue surveillance in neutrophils and dendritic cells, whereas in cancer cells it facilitates invasion and dissemination (Friedl and Wolf 2003; van Haastert and Devreotes 2004; Friedl and Weigelin 2008; Petrie et al. 2009; Harris et al. 2012). More generally, fluctuation-driven behavioral variability enables cells to explore complex environments and navigate obstacles even without external guidance cues. The social amoeba *Dictyostelium discoideum* provides a useful model for dissecting the intrinsic mechanisms underlying spontaneous cellular behavior, as it exhibits random migration and constitutive macropinocytosis under uniform conditions lacking nutrients, attractants, or adhesive gradients (Maeda et al. 2008; Takagi et al. 2008; van Haastert and Bosgraaf 2009; Devreotes et al. 2017; King and Kay 2019; Matsuoka et al. 2024; Tu et al. 2025). A

Cite this article as *Cold Spring Harb Perspect Biol* doi: 10.1101/cshperspect.a041750

challenge is to understand how noisy biochemical networks generate highly ordered yet stochastic patterns of cellular activity.

Extensive studies of chemotactic signaling have revealed a conserved reaction network comprising hundreds of gene products, with a core architecture shared between *Dictyostelium* and mammalian leukocytes. Among these systems, *Dictyostelium* offers the most detailed spatial and temporal characterization, and its stimulus-responsive properties have been extensively reviewed (Devreotes et al. 2017; Van Haastert et al. 2021; Banerjee et al. 2025). Here, we instead focus on stimulus-independent dynamics underlying spontaneous random motility. Even in uniform conditions without chemoattractant gradients or other guidance cues, *Dictyostelium* cells establish front–rear polarity and persistent migration, characterized by an actin-rich protrusive front and a myosin II–enriched contractile rear (Van Haastert et al. 2021; Uyeda et al. 2025). Corresponding asymmetries are observed in upstream signaling components: active Rat sarcoma-guanosine triphosphate (Ras-GTP), phosphoinositide 3-kinase (PI3K), phosphatidylinositol 3,4,5-trisphosphate (PIP_3), protein kinase B (Akt/PKB), and the Scar/WAVE complex localize to the leading edge, whereas PTEN, $PI(4,5)P_2$, $PI(3,4)P_2$, and CynA accumulate at the rear (see Box 1; Li et al. 2020). These spatial patterns arise from reciprocal antagonism between "frontness" and "backness" modules, forming a self-reinforcing bistable switch that segregates signaling activities and stabilizes cell polarity (Li et al. 2018; Matsuoka and Ueda 2018; Deng et al. 2024; Matsuoka et al. 2024).

Bistability alone, however, is insufficient to explain key features of spontaneous cell migration. In particular, the initial symmetry-breaking event that converts a quiescent cell into a polarized one, and the directional reorientation that generates complex migratory trajectories require stochastic triggers to escape homogeneous states or reset established polarity. Excitable dynamics within the signaling network provide an intrinsic mechanism of such triggers. Microscopic fluctuations can be amplified by excitable system into macroscopic, all-or-none activation events, thereby initiating de novo polarity (Devreotes et al. 2017; Beta et al. 2023; Matsuoka et al. 2024; Banerjee et al. 2025). Indeed, *Dictyostelium* cells spontaneously generate Ras-GTP patches and traveling waves on the plasma membrane under uniform conditions (Arai et al. 2010; Xiong et al. 2010; Huang et al. 2013; Miao et al. 2017, 2019; Fukushima et al. 2019). These events recruit PI3K, generate PIP_3-enriched domains, and drive actin polymerization, resulting in spontaneous pseudopod formation. Similar excitable dynamics have been reported in fibroblasts, macrophages, neutrophils, and cancer cells (Li et al. 2020; Zhan et al. 2020; Banerjee et al. 2025), supporting the view that Ras excitability is a conserved mechanism for symmetry breaking and protrusion formation across eukaryotic cells.

Despite extensive characterization of Ras excitability during random migration, its molecular origin has long remained elusive. The *Dictyostelium* genome encodes at least 25 RasGEFs and 14 RasGAPs, several of which mediate chemoattractant-induced Ras activation (Sasaki et al. 2004, 2007; Wilkins et al. 2005; Charest and Firtel 2006; Zhang et al. 2008; Xu and Jin 2022; Xu et al. 2022). However, none of these regulators were uniquely essential for spontaneous Ras activation. A recent excitability-based genetic screen exploiting intrinsic Ras wave dynamics identified RasGEFX as a key regulator of spontaneous Ras excitation (Iwamoto et al. 2025). Loss of RasGEFX abolishes spontaneous Ras wave generation, PIP_3 patch formation, random migration, and macropinocytosis, whereas RasGEFX overexpression enhances Ras excitability, pseudopod formation, and macropinocytic ruffling. These findings position RasGEFX as a key driver of the excitable Ras module.

In this review, we propose a hierarchical, three-layered signaling architecture that underlies spontaneous random migration (Fig. 1). An upstream excitable system centered on Ras GTPase and RasGEFX amplifies molecular fluctuations into transient signaling domains, which engage the downstream bistable PIP_3-PTEN switch. This coupling unifies noise-driven symmetry breaking with robust stabilization of pseudopods for migration and macropinocytosis cups for fluid uptake. By integrating theoretical models, genetic

BOX 1. GLOSSARY AND ABBREVIATIONS

Small GTPases

A set of hydrolase enzymes, also known as small G proteins, which bind to and hydrolyze a guanosine triphosphate (GTP) into a guanosine diphosphate (GDP). Typically, they are active when bound to GTP and inactive when bound to GDP, switching between on and off states. The exchange of GDP for GTP is catalyzed by guanine nucleotide exchange factors (GEFs), and GTP hydrolysis is facilitated by GTPase-activating proteins (GAPs). There are five major families of small GTPases with similar sequences and functions, two of which are the Ras family, which includes Ras and Ras-related protein (Rap), and the Rho family, which includes Rho, Rac and cell division control protein 42 (Cdc42).

Phosphatidylinositol

A glycerol-based phospholipid consisting of two fatty acid chains and inositol. Of the five hydroxyl groups on the inositol ring, those at positions 3, 4, and 5 can be phosphorylated by kinases and dephosphorylated by phosphatases, resulting in seven different phosphorylated forms called phosphoinositides. Phosphatidylinositol 4-phosphate (PI(4)P), phosphatidylinositol 3,4-bisphosphate (PI (3,4)P_2), phosphatidylinositol 4,5-bisphosphate (PI(4,5)P_2) and phosphatidylinositol 3,4,5-trisphosphate (PI(3,4,5)P_3) are found as minor components on the cytoplasmic side of the cell membrane. They play important roles in signal transduction through dynamic phosphorylation and dephosphorylation by phosphatidylinositol-4-phosphate 5-kinases (PIP5K), phosphoinositide 3 kinase (PI3K), phosphatase and tensin homolog (PTEN) and other enzymes.

Cyclic adenosine monophosphate (cAMP), cyclic guanosine monophosphate (cGMP)

Cyclic nucleotides derived from adenosine triphosphate (ATP) and GTP, respectively, and both function as second messengers.

Pleckstrin-homology (PH) domain

A structurally well-characterized domain of ~120 aa found in a wide variety of signaling proteins involved in intracellular trafficking, cell signal transduction, and cytoskeletal remodeling. The PH domain from Akt/PKB preferentially binds PI(3,4,5)P_3 in *Dictyostelium discoideum*, and, when tagged with a fluorescent protein such as green fluorescent protein (GFP), serves as a genetically encoded fluorescent probe for PI(3,4,5)P_3.

Ras-binding domain (RBD)

A structurally well-characterized domain of ~85 amino acids in the amino-terminal region of human c-Raf. The RBD has a ubiquitin-like fold and selectively binds only GTP-bound Ras proteins.

Abbreviations for the other signaling molecules

(PKB) Protein kinase B
(Scar) suppressor of cAMP receptor
(Wave, WASP) (Wiskott–Aldrich syndrome protein) family Verprolin homologous protein
(CynA) Callipygian
(TORC2) target of rapamycin complex 2
(PAR) partitioning defective

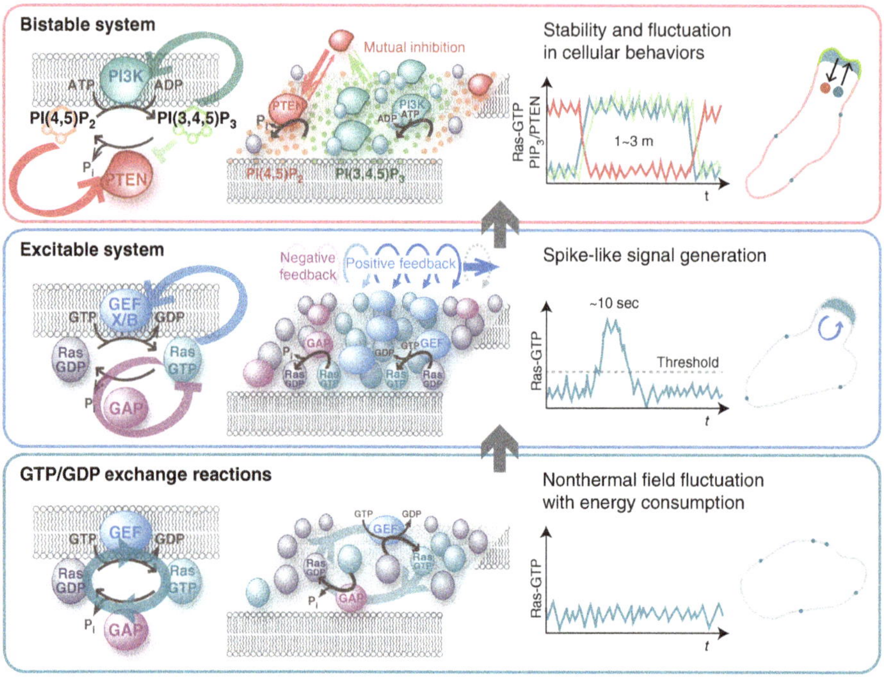

Figure 1. Hierarchical excitable–bistable architecture underlying spontaneous cell migration. (*Lower* panel) Guanosine triphosphate [GTP]/guanosine diphosphate [GDP] exchange reactions. Basal Ras activity generated by continuous RasGEF- and RasGAP-mediated GTPase cycling. These energy-consuming reactions produce active, nonthermal fluctuations in Rat sarcoma-guanosine triphosphate (Ras-GTP) levels, providing a stochastic molecular input. (*Middle* panel) Excitable system. An excitable Ras signaling module in which stochastic Ras activation is selectively amplified by RasGEFX/B-dependent positive feedback, generating discrete Ras-GTP patches or propagating waves. (*Upper* panel) Bistable system. A bistable polarity module exemplified by the PIP_3-PTEN switch, which stabilizes front–rear asymmetry by capturing transient Ras excitations and converting them into persistent polarized states. Hierarchical coupling of these three layers enables spontaneous cell migration by transforming intrinsic molecular noise into structured, non-Brownian motile behavior in the absence of external guidance cues.

analyses, and live-cell imaging, we discuss how these excitable–bistable networks generate highly ordered yet stochastic cellular behaviors and consider how excitability-driven, non-Brownian search strategies may promote cellular adaptation in complex natural environments.

EXCITABLE SYSTEMS IN BIOLOGY: NOISE AMPLIFICATION AND SYMMETRY BREAKING

Excitable systems are dynamical systems that reside in a stable, low-activity state but generate large, stereotyped responses when perturbed beyond a defined threshold. The classic example is the Hodgkin–Huxley model of neuronal action potentials (Hodgkin and Huxley 1952). In this system, a small suprathreshold depolarization activates voltage-gated Na^+ channels, producing a rapid all-or-none spike followed by a refractory period during which the membrane is transiently unresponsive. This behavior defines three hallmark properties of excitable systems: (1) a sharp threshold, (2) a large-amplitude, nearly invariant response, and (3) a refractory state that prevents immediate reexcitation and enforces unidirectional wave propagation.

The FitzHugh–Nagumo model provides conceptual insights into nonlinear excitable behavior, which reduces the Hodgkin–Huxley equations into a minimal activator–inhibitor framework (FitzHugh 1961; Nagumo et al. 1962). Fast positive feedback in the activator causes a rapid increase in activity once the system crosses the excitation threshold, whereas slower negative feedback suppresses the response and restores the resting state. This separation of timescales between rapid activation and delayed recovery underlies the characteristic features of excitable systems, including thresholded responses, spike-like excitations, and refractory periods (Keener and Sneyd 1998; Lindner et al. 2004; Iglesias and Devreotes 2012; Bhattacharya et al. 2020). When these activator–inhibitor dynamics are spatially extended and coupled by diffusion, they generate propagating waves and oscillatory patterns. Fast positive feedback enables the initiation and propagation of waves, while delayed inhibition enforces refractoriness behind the wavefront. These mechanisms underlie circular and spiral waves in two-dimensional systems, as exemplified by the Belousov–Zhabotinsky reaction, cardiac action potential propagation, and intracellular Ca^{2+} waves (Winfree 1972; Tyson and Fife 1980; Lechleiter et al. 1991; Keener and Sneyd 1998; Berridge et al. 2003; Beta and Kruse 2017).

A fundamental property of excitable systems, particularly in noisy biological systems, is their ability to amplify intrinsic stochastic fluctuations. When biochemical parameters place the system close to the excitation threshold, molecular noise alone can occasionally push the system across the boundary, triggering spontaneous, stimulus-independent activation events (Lindner et al. 2004; Beta and Kruse 2017; Beta et al. 2023). In this way, excitability functions as a noise amplifier that converts microscopic fluctuations into macroscopic stochastic excitations, providing a general mechanism for spontaneous symmetry breaking (Fig. 1, middle and lower panels). Such noise-triggered events can explain transient activation patches, traveling waves, or oscillations, as observed in Ras-GTP, PIP_3 and actin dynamics in *Dictyostelium* and many mammalian cell types (Killich et al. 1993; Vicker 2002; Weiner et al. 2007; Asano et al. 2008; Bretschneider et al. 2009; Arai et al. 2010; Xiong et al. 2010; Iglesias and Devreotes 2012; Huang et al. 2013; Taniguchi et al. 2013; Graziano and Weiner 2014; Devreotes et al. 2017; Fukushima et al. 2019). The initial stochastic excitation typically emerges at a random location on the plasma membrane, generating a localized activator domain that can be stabilized by downstream bistable modules, such as the PIP_3-PTEN switch, thereby establishing a new axis of front–rear polarity (Fig. 1, upper panel; Li et al. 2018; Matsuoka and Ueda 2018; Deng et al. 2024). Repeated noise-driven excitations at different locations on the membrane can redirect pseudopod formation, leading to persistent random migration.

Thus, excitability converts intrinsic molecular noise into discrete macroscopic signaling events, thereby providing motile cells with spontaneous, self-organized behaviors. Moreover, basal excitable dynamics preconfigure the signaling capabilities, such that external cues, including shallow chemoattractant gradients, electric fields, shear stress, or adhesion gradients, can bias the location and timing of excitation events (Janetopoulos et al. 2004; Andrew and Insall 2007; Iglesias and Devreotes 2012; Artemenko et al. 2016; Devreotes et al. 2017; Li et al. 2020). In this way, undirected random motility is transformed into directional cellular responses.

SPONTANEOUS CELLULAR DYNAMICS IN *DICTYOSTELIUM* WITHOUT EXTERNAL CUES

The social amoeba *D. discoideum* provides a useful model for studying cell motility across different physiological states due to its well-characterized developmental transition from solitary vegetative growth to multicellular aggregation (Kessin 2001; Eichinger and Rivero 2016). During vegetative growth, cells behave independently, feeding on bacteria by phagocytosis and acquiring fluid nutrients through macropinocytosis. These cells exhibit spontaneous motility characterized by a power-law distribution of travel distances, consistent with broad-area search strategies (Li et al. 2008, 2011; Maeda et al. 2008; Takagi et al. 2008; van Haastert and Bosgraaf 2009). Upon starvation, cells enter an early aggregation phase (0–4 h), during which spontaneous motility intensifies. Migration speed increases from ∼5–7 to 10–12 µm/min, directional persistence increases, and trajectories become more ballistic. After ∼5 h, cells acquire chemotactic responsiveness to cAMP. Even in shallow or steep gradients, they continue to generate pseudopodia that deviate from the gradient direction (Swanson and Taylor 1982; Andrew and Insall 2007; Bosgraaf and van Haastert 2009). This behavior indicates that the intrinsic spontaneous activities remain active and are biased by external cues, which is consistent with the operation of an internal excitable signaling network.

Quantitative analysis of cell trajectories across physiological states, from vegetative growth to 5.5 h of starvation, reveals the statistical characteristics of spontaneous motility (Fig. 2A; Takagi et al. 2008). Under all examined conditions, cell trajectories exhibit anomalous superdiffusion, with mean square displacement (MSD) scaling as $\langle r(t)^2 \rangle \sim t^{2H}$.

The Hurst exponents (H) increase from 0.62 in vegetative cells to 0.83 in starved cells, indicating more ballistic motion with developmental progression (Fig. 2B; Wu et al. 2026). In vegetative cells, velocity distributions are non-Gaussian with heavy-tailed power-law decay, indicative of intermittent high-speed bursts of movements characteristic of Lévy-walk-like foraging. As development progresses, directional persistence increases, and velocity autocorrelation functions exhibit biexponential decay with fast (5–11 sec) and slow (1.8–3.8 min) components that correspond to rapid pseudopod dynamics and longer-timescale polarity maintenance, respectively (Fig. 2C). Specifically, the decay constants are ∼11 sec and 1.8 min in vegetative cells, and ∼5 sec and 3.8 min in starved cells. The fast component is consistent with observations of alternating left–right pseudopod splitting at ∼15 sec intervals (Bosgraaf and van Haastert 2009; van Haastert 2020). These multiscale signatures suggest that spontaneous motility arises from coupling between a fast excitable system that initiates pseudopods and a slower bistable module that stabilizes polarity, enabling both local exploration and persistent migration.

These features can be reproduced by a generalized Langevin equation incorporating velocity-dependent friction, memory, and multiplicative noise (Takagi et al. 2008):

$$\frac{d\vec{v}(t)}{dt} = -\beta(\vec{v}(t))\vec{v}(t) + \alpha\vec{V}(t) + \sigma(\vec{v}(t))\vec{\eta}(t), \tag{1}$$

where $\vec{v}(t)$ is the cell velocity, $\beta(v)$ is a velocity-dependent decay rate, and $\sigma(v)\eta(t)$ represents multiplicative Gaussian noise. The memory term $\vec{V}(t)$ is given by

$$\vec{V}(t) = \alpha \int_{-\infty}^{t} e^{-\gamma(t-t')}\vec{v}(t')dt',$$

where α is the memory strength, and γ is the memory decay rate. The noise amplitude exhibits a cubic dependence, which is essential to generate power-law velocity distributions. Simple Ornstein–Uhlenbeck models fail to capture the observed statistics (Uhlenbeck and Ornstein 1930; Selmeczi et al. 2005; Bödeker et al. 2010). These quantitative characteristics, including superdiffusion, heavy-tailed velocity distributions, multiscale persistence, and noise amplification, collectively define random motility in *Dictyostelium* as a canonical example of excitability-driven behavior.

When an electric field is applied, *Dictyostelium* cells exhibit strong cathodal galvanotaxis. But their individual trajectories remain zigzagged, and migration speed remains comparable to that in spontaneous conditions (Fig. 2A; Zhao et al. 2006; Sato et al. 2007, 2009). Directedness (cos θ) typically increases to 0.7–0.8 but does not approach unity. This again indicates that the electric field biases pseudopod orientation without suppressing intrinsic stochastic motility. This behavior can be captured by adding a bias term $\vec{f}(E)$ to Equation (1) (Takagi et al. 2014):

$$\vec{V}(t) = \alpha \int_{-\infty}^{t} e^{-\gamma(t-t')}(\vec{v}(t') + \vec{f}(E))dt',$$

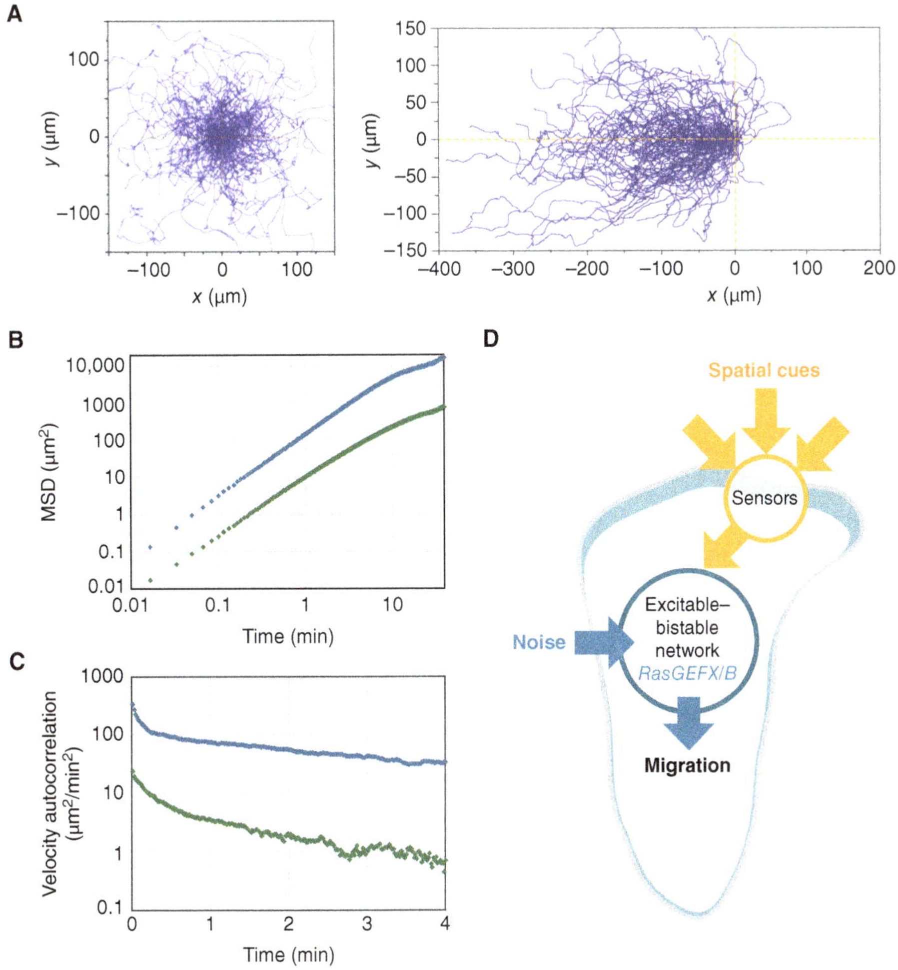

Figure 2. Statistical characterization of spontaneous cell migration. (*A*) Representative migration trajectories of wild-type *Dictyostelium* cells in the absence of external guidance cues (*left*) and under an electric field of 10 V/cm (*right*). (Panel *A* reprinted from Sato et al. 2009 with permission from the authors.) (*B*) Mean-squared displacement (MSD) of cell trajectories at 0 h (green) and 5.5 h (blue) after starvation. Both conditions deviate from linear scaling expected for Brownian motion, indicating non-Brownian migration with distinct persistence properties. (*C*) Velocity autocorrelation functions for cells at 0 h (green) and 5.5 h (blue) after starvation. The presence of multiple decay timescales reflects internally generated directional memory associated with spontaneous protrusive dynamics. (Panels *B* and *C* adapted from Takagi et al. 2008 courtesy of CC BY.) (*D*) Schematic model showing how spontaneous motility arises from internally generated signaling activity, and how external guidance cues bias the spontaneous signaling events to generate directed movement.

with

$$\vec{f}(E) \propto \frac{V_{\max}E^2}{E^2 + K_E^2} \cdot \frac{\vec{E}}{E},$$

where E is the electric field strength, K_E is the half-saturation field strength, and $V_{\max}$ is the maximum migration speed (with experimentally determined values $K_E = 2.6$ V/cm and $V_{\max} = 9.2$ μm/min; Sato et al. 2007). A similar phenomenon is observed in chemotaxis: in cAMP gradients, cells move with a strong directional bias while maintaining intrinsically noisy trajectories, yielding chemotactic indices of only

0.7–0.9 (Fisher et al. 1989; Song et al. 2006; Andrew and Insall 2007; Kamimura et al. 2016). Together, these observations support the view that external cues modulate the internal excitable–bistable machinery that drives spontaneous motility (Fig. 2D).

EXCITABLE Ras WAVES AND THE IDENTIFICATION OF RasGEFX AS A CORE DRIVER

The concept of excitable signaling in cell motility predates the direct visualization of Ras activity and initially emerged from live-cell imaging of the downstream lipid second messenger PIP_3 and the actin cytoskeleton. Early studies using the PH-domain probe PH-GFP in *Dictyostelium* revealed that uniform cAMP stimulation elicits a transient, all-or-none burst of PIP_3 on the plasma membrane, followed by rapid adaptation despite sustained stimulation (Parent et al. 1998; Meili et al. 1999; Parent and Devreotes 1999). Under continuous stimulation, cells further exhibit self-organizing PIP_3 patches that trigger actin-enriched protrusions even after the adaptation response has plateaued (Postma et al. 2004). These digital responses, together with the appearance of traveling PIP_3 waves under actin-disrupted and caffeine-treated conditions, provided early evidence that the chemotactic signaling network operates as an excitable system (Arai et al. 2010; Xiong et al. 2010). External perturbations, including impulsive and sustained stimuli, can induce PIP_3 domains that are indistinguishable from those that arise spontaneously, and a refractory period between successive PIP_3 activation events also confirmed the excitable nature of the system (Nishikawa et al. 2014). Mathematical models incorporating positive and negative feedback loops successfully reproduced spontaneous symmetry breaking and the emergence of traveling signaling domains (Arai et al. 2010; Xiong et al. 2010; Shibata et al. 2012, 2013; Huang et al. 2013). This excitable framework explains stochastic protrusive dynamics during random migration and provides a mechanistic basis for how shallow chemotactic gradients bias the location and timing of PIP_3 domain initiation.

Despite extensive characterization of PIP_3 dynamics, the high degree of parallelism and redundancy in chemotactic signaling pathways has long obscured the identification of the molecular components that constitute the minimal excitability-generating unit (Devreotes et al. 2017). A major advance came from shifting attention upstream of PI3K to Ras GTPase. Ras dynamics exhibit all the canonical features of excitability, including all-or-none activation, spontaneous firing, and continuous traveling waves (Fig. 3). Ras activation consistently precedes PIP_3 accumulation by several seconds, indicating that Ras lies upstream of PIP_3 in the excitable cascade (Miao et al. 2017, 2019; Fukushima et al. 2019; van Haastert 2020; Zhan et al. 2020). Critically, spontaneous Ras waves arise in the absence of upstream stimuli not only in wild-type cells but also in *pi3k1-5*-null and *pten*-null strains, demonstrating that neither PIP_3 production nor its turnover is required for Ras excitability (Fukushima et al. 2019; Matsuoka et al. 2024). Furthermore, Ras waves persist even when PI3K, guanylate cyclase, TORC2, and phospholipase A2 are inhibited simultaneously, indicating that Ras excitability does not depend on major chemotactic signaling pathways (Kortholt et al. 2011; Fukushima et al. 2019). These observations have established the Ras GTPase along with its regulators RasGEFs and RasGAPs as an upstream excitable module.

Genetic and live-cell imaging analyses identified RasGEFX as a key component of the excitable module in *Dictyostelium* (Fig. 4; Iwamoto et al. 2025). Loss of RasGEFX abolishes propagating Ras waves despite the presence of other RasGEFs, revealing its essential, nonredundant role in spontaneous wave generation. RasGEFX deletion impairs stochastic Ras initiation events, likely by reducing the probability that basal fluctuations cross the excitability threshold. Conversely, RasGEFX overexpression shortens wave periods and increases excitation frequency, whereas reduced expression prolongs oscillation periods. These effects propagate downstream, such that elevated RasGEFX enhances pseudopod formation, while its deletion severely impairs protrusive activity. Thus, RasGEFX tunes the temporal dynamics of the excitable Ras

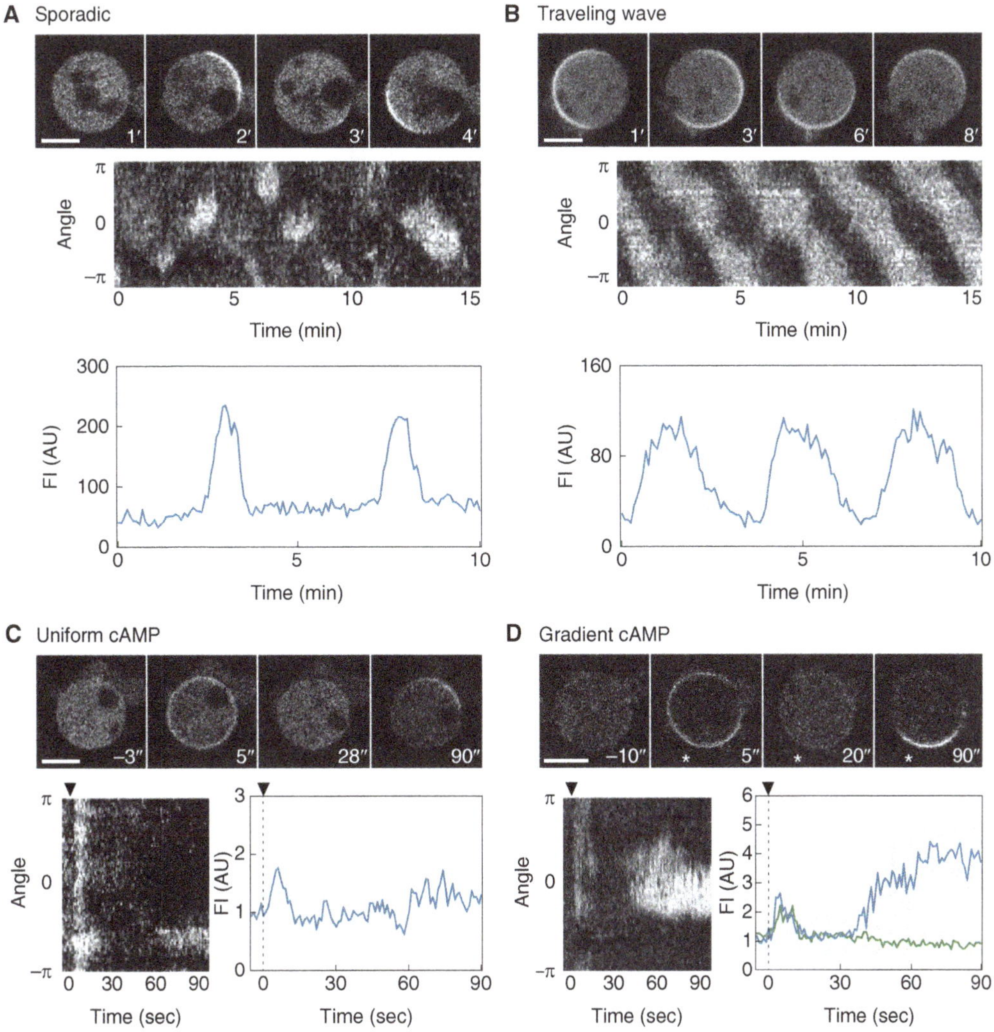

Figure 3. Excitable dynamics of Ras signaling. (*A*) Transient formation of localized Rat sarcoma-guanosine triphosphate (Ras-GTP)-enriched domains on the plasma membrane, reflecting spontaneous Ras activation events characteristic of an excitable system (*top*). Kymographs show the temporal changes in fluorescence intensity along the circular cell periphery in confocal images, which corresponds to the cell membranes (*middle*). A representative example of the time course of fluorescence intensity (FI) in arbitrary units (AUs) is shown (*bottom*). (*B*) Propagating Ras activation waves observed under caffeine treatment. Caffeine promotes the transition from isolated Ras-enriched domains to traveling waves, revealing the wave-supporting regime of the Ras excitable network. (Panel *B* adapted with permission from the author from Iwamoto et al. 2025 courtesy of CC BY-NC-ND 4.0.) (*C*) Uniform, transient activation of Ras across the entire plasma membrane following global cAMP stimulation, indicative of a threshold-dependent, all-or-none excitable response. (*D*) PIP_3 dynamics in response to a cAMP gradient applied with a glass micropipette whose tip is placed at the location indicated by an asterisk. An initial global, transient excitation of PIP_3 is followed by its stabilization and enrichment on the side facing the higher chemoattractant concentration, reflecting coupling between excitable Ras signaling and downstream bistable polarity circuits. Fluorescence intensities at the near (blue) or far (green) side of a cell from the micropipette containing cAMP. Scale bar, 5 µm. (Panel *D* is a still taken from a movie in Kamimura et al. 2016, with permission from the authors.)

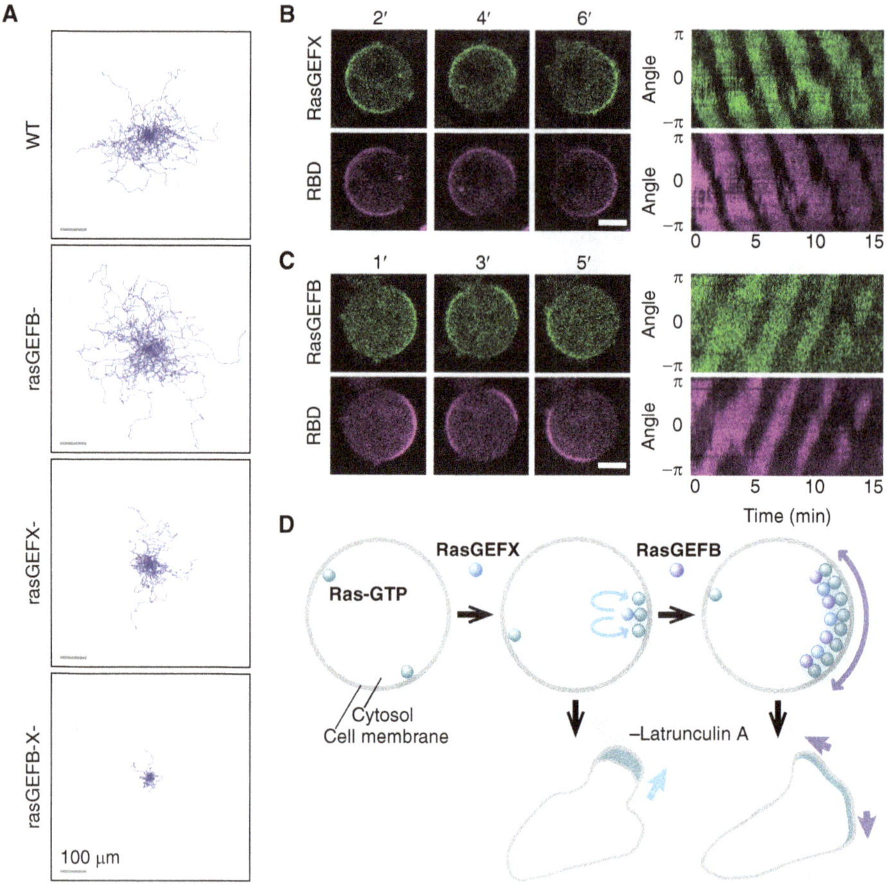

Figure 4. Distinct roles of RasGEFX and RasGEFB in regulating Ras excitability and spontaneous cell motility. (*A*) Representative migration trajectories of wild-type (WT), RasGEFB knockout (rasGEFB-), RasGEFX knockout (rasGEFX-), and RasGEFX/B double-knockout (rasGEFB-X-) cells in the absence of latrunculin A. Whereas WT and single-knockout cells retain varying degrees of spontaneous motility, RasGEFX/B DKO cells exhibit severely impaired movement with little or no spontaneous migration, indicating that RasGEFX is essential in combination with RasGEFB for Ras-driven spontaneous motility. (*B*) Colocalization of RasGEFX with active Ras-GTP (RBD) during propagating Ras activation waves in the presence of latrunculin A added for the suppression of cell motility. RasGEFX participates in the amplification and temporal regulation of excitable Ras signaling. Scale bar, 5 μm. (*C*) Colocalization of RasGEFB with Ras-GTP (RBD) during propagating Ras activation waves. RasGEFB has a role in regulating the spatial size of Ras activation domains within the excitable network. Scale bar, 5 μm. (*D*) Schematic summary showing functional specialization within the Ras excitable system. RasGEFX primarily regulates the temporal dynamics of Ras excitability, thereby controlling the frequency of pseudopod formation. RasGEFB regulates the spatial scale of Ras activation and the size of pseudopods. (*A–C* adapted with permission from the author from Iwamoto et al. 2025 courtesy of CC BY-NC-ND 4.0.)

module, thereby controlling the protrusion frequency.

Motility analyses further reveal that RasGEFX acts together with RasGEFB and RasGEFU to regulate Ras excitability during spontaneous cell migration (Fig. 4A; Iwamoto et al. 2025). Cells lacking RasGEFX exhibit heterogeneous phenotypes: many become nearly immotile, whereas others retain reduced motility. Double deletion of RasGEFX with either RasGEFB or RasGEFU abolishes spontaneous migration in almost all cells, revealing that these

RasGEFs have cooperative yet nonredundant roles. RasGEFB regulates the spatial organization of Ras activity, controlling the size of Ras-enriched domains and pseudopods. RasGEFU, originally identified as the cGMP-binding protein D (GbpD), which is involved in Rap1-mediated adhesion (Kortholt et al. 2006, 2010; Kortholt and van Haastert 2008), supports the conversion of Ras-driven protrusions into traction forces. Live-cell imaging further shows that RasGEFX is recruited to propagating Ras-GTP domains, where it likely reinforces local activation via positive feedback (Fig. 4B,C). RasGEFB copropagates with Ras-GTP and modulates domain size, whereas RasGEFU is broadly distributed along the membrane, consistent with its adhesion-related function (Iwamoto et al. 2025). Together, these membrane recruitment dynamics, wave propagation behaviors, and genetic dependencies support a model in which RasGEFX and RasGEFB constitute the core positive feedback machinery that converts stochastic Ras activation into excitable waves that drive cell motility (Fig. 4D).

Despite these advances, many uncertainties remain. Although the persistence of Ras waves in PI3K- or PTEN-null cells suggests that PIP_3 is not essential for wave initiation or propagation, Ras waves emerge more efficiently and propagate more robustly when PIP_3 signaling is intact (Fukushima et al. 2019). This observation implies the existence of a PIP_3-dependent positive-feedback loop that reinforces Ras excitability (Sasaki et al. 2007). The molecular basis of this reinforcement remains unclear. PIP_3 may act directly on Ras regulators, indirectly via actin-mediated feedback, or by stabilizing membrane microdomains and surface charge to support wave propagation (Charest and Firtel 2006; Banerjee et al. 2022, 2023; Shin et al. 2023). Further questions concern the identity and regulation of the negative feedback processes that terminate Ras excitation, which are likely to involve RasGAPs/RapGAPs (Li et al. 2018). Another question is which specific Ras/Rap GTPase drives spontaneous excitability during unstimulated migration. Although chemotactic signaling engages RasC, RasD, RasG, and Rap1 (Kortholt and van Haastert 2008; Devreotes et al. 2017; Liu et al. 2018), the GTPase species responsible for spontaneous Ras activation remains undefined. Notably, the commonly used reporter RBD-Raf1 binds to multiple Ras family GTPases (Kae et al. 2004) and can detect membrane-associated active species even in *rasC*$^-$/*rasG*$^-$/*rapA*$^-$ triple mutants (Iwamoto et al. 2025), suggesting the involvement of additional paralogs. Furthermore, all RasGEFX/B/U mutants retain chemotaxis to cAMP (Iwamoto et al. 2025), implying that distinct RasGEFs mediate Ras activation during directional sensing. Candidate regulators include RasGEFA, RasGEFT (GbpC), RasGEFR, and multiple RasGAPs (Insall et al. 1996; Kae et al. 2007; Kortholt and van Haastert 2008; Zhang et al. 2008; Kortholt et al. 2012; Xu and Jin 2022; Xu et al. 2022; Lin et al. 2024, 2025), but a comprehensive framework explaining how excitable Ras dynamics during spontaneous migration are integrated with chemotactic signaling remains an unsolved issue.

BISTABLE SIGNALING AND THE ESTABLISHMENT OF CELLULAR POLARITY

Cellular polarity often emerges from biochemical networks capable of adopting alternative stable states. This allows cells to segregate molecular activities spatially. Such bistable networks typically depend on local positive feedback coupled with mutual inhibitory interactions. Positive feedback amplifies small spatial asymmetries, enabling a molecule that is slightly enriched at one site to recruit or activate factors that further promote its own accumulation. Mutual inhibition, mediated either by diffusible global inhibitors or by spatially segregated antagonistic molecules, suppresses competing states elsewhere in the cell. These mechanisms generate discrete and self-maintaining polarity states that are robust to molecular noise (Wang et al. 2008). Incorporating bistability in reaction-diffusion systems provides spatial exclusivity and the long-term maintenance of polarity despite stochastic biochemical fluctuations (Beta et al. 2008; Mori et al. 2008). This mechanism is conserved across diverse systems, including yeast Cdc42 polarity, anterior–posterior patterning of PAR proteins in the *Caenorhabditis elegans* zygote, and front–rear organization of

Rac and Rho GTPases in migrating mammalian cells such as fibroblast and epithelial cells (Ridley et al. 2003; Goldstein and Macara 2007; Hoege and Hyman 2013; Thompson 2013; Chiou et al. 2017).

Cell polarity in *D. discoideum* represents one of the most extensively characterized examples of bistability embedded within a multilayered signaling network. Phosphatase and tensin homolog (PTEN) preferentially and more stably binds to $PI(4,5)P_2$-enriched regions, reinforcing rear identity (Iijima and Devreotes 2002; Yoshioka et al. 2020). Conversely, Ras-GTP accumulation is expanded and prolonged in the presence of PI3K activity, showing front state reinforcement (Funamoto et al. 2002; Sasaki et al. 2007; Fukushima et al. 2019). PIP_3 suppresses PTEN membrane association and promotes membrane dissociation, conferring reciprocal antagonism and a bistable switch between the front and rear states (Matsuoka and Ueda 2018). This mutual inhibition, combined with two local positive feedbacks, provides the mechanistic basis for a self-reinforcing polarity.

Bistability in *Dictyostelium* extends beyond phosphoinositide metabolism to Ras signaling networks. Active Ras-GTP inhibits $PI(4,5)P_2$ synthesis by suppressing PIP5K and simultaneously promotes PIP_3 production by activating PI3K, thereby reinforcing the front signaling domain (Li et al. 2018; Deng et al. 2024). Conversely, $PI(4,5)P_2$ and $PI(3,4)P_2$ suppress Ras activation at the rear, forming a mutual inhibition that sharpens the boundary between active and inactive Ras membrane regions. These interactions suggest that polarity is controlled by bistable circuits that span lipid signaling and small GTPase regulation. This implies that early symmetry-breaking events likely emerge based on Ras-centered antagonisms embedded within the broader hierarchical excitable–bistable signaling network.

A THREE-LAYERED EXCITABLE–BISTABLE ARCHITECTURE FOR CELL MOTILITY

Bistability alone cannot explain how motile cells transition spontaneously from a symmetric state to a polarized state or how they reorient their migration direction. These behaviors arise from the excitable nature of Ras signaling that operates upstream of the bistable polarity modules (Fig. 1). Spontaneous Ras excitation persists even when PIP_3 dynamics are eliminated (Kortholt et al. 2011; Fukushima et al. 2019; van Haastert 2020). This directional dependency supports a hierarchical organization in which Ras, together with RasGEFs and RasGAPs, constitutes the excitable module that initiates polarity (Matsuoka et al. 2024; Iwamoto et al. 2025). Without excitable triggers, cells remain apolar, lacking a Ras-GTP-enriched domain as seen in RasGEFX-null cells, and exhibit a uniform PTEN distribution (Janetopoulos et al. 2004; Matsuoka and Ueda 2018; Iwamoto et al. 2025). When a newly generated Ras excitation flips the downstream PIP_3-PTEN bistable switch, a stable PIP_3-enriched front is established, mechanistically explaining spontaneous polarization. Repeated Ras excitations at different locations on the membrane can destabilize or reorient an existing front, resulting in spontaneous reorientation and random cell migration. In this framework, Ras excitability provides sensitivity and stochasticity, while downstream bistable modules confer persistence and memory. This three-layered organization balances the competing requirements of motile cells: robust polarity maintenance and flexible reorganization.

Consistent with this view, *Dictyostelium* exhibits multi-timescale motility. Fast dynamics (~10 sec) correspond to short-lived protrusions, whereas slower dynamics (~1–3 min) reflect persistent movement with polarity (Takagi et al. 2008; Bosgraaf and van Haastert 2009; van Haastert 2020). Although the molecular origins of these timescales are not fully understood, they can be interpreted as emergent properties of a Ras-driven excitable system coupled to bistable PIP_3-PTEN and Ras-PIP_2/PIP5K modules. Ras excitation events generate stochastic protrusions, whereas bistable feedback loops maintain selected protrusions, leading to persistent polarization and migration. This interpretation is consistent with the early observations by Swanson and Taylor (1982), who showed that directed migration in *Dictyostelium* arises from the spatial coordination of locally generated protrusive activities.

Dictyostelium cells can respond to extremely shallow chemoattractant gradients, on the order of ~2% across the cell body (Mato et al. 1975; Fisher et al. 1989; Song et al. 2006; Ohtsuka et al. 2021). At such low gradient strengths, directional sensing operates near fundamental physical limits. Because cells sample only a finite number of chemoattractant molecules over biologically relevant timescales, ligand-receptor binding is inherently stochastic. Near the threshold, the input signal is dominated by noise rather than by deterministic concentration differences (Berg and Purcell 1977). Consistent with this view, single-molecule imaging has revealed spatial and temporal fluctuations in receptor–ligand interactions, indicating that gradient information is encoded as probabilistic biases across the cell surface (Ueda et al. 2001; Miyanaga et al. 2007). These physical constraints are formalized by signal-to-noise ratio (SNR)-based description of chemotactic signaling, demonstrating that stochastic receptor inputs impose a fundamental limit on the accuracy of directional sensing (Ueda and Shibata 2007; Shibata and Ueda 2008). Such theoretical limits can explain the dependence of chemotactic accuracy observed experimentally (Fisher et al. 1989; Ueda and Shibata 2007). This view is consistent with the biased random-walk model proposed by van Haastert and Postma (2007) and experimentally confirmed by Amselem et al. (2012).

Excitable–bistable signaling architectures can provide an effective computational mechanism for extracting directional information from noisy receptor inputs. The excitable module acts as a nonlinear filter, amplifying rare fluctuations that exceed a threshold and suppressing subthreshold noise (Bhattacharya and Iglesias 2018). In this way, stochastic differences in receptor occupancy can be digitized into all-or-none Ras activation events. The resulting Ras-GTP-enriched domain biases downstream signaling, in which bistable modules stabilize the domain along chemical gradients. Once established, this directional bias generates the front–rear asymmetry with memory and persistence (Nakajima et al. 2014; Skoge et al. 2014). This hierarchical processing enables cells to balance their high sensitivity to noisy spatial cues with the robust maintenance of polarized migration. The nonlinear amplification, such as excitable–bistable signaling, suggest that cells can operate near the fundamental limits of molecular sensing (Berg and Purcell 1977; Endres and Wingreen 2008). Notably, a local excitation and global inhibition (LEGI) mechanism has been proposed for gradient sensing (Levchenko and Iglesias 2002; Xiong et al. 2010; Takeda et al. 2012). This mechanism is assumed to operate between noisy receptor inputs and downstream excitable signaling, extracting gradient information across a wide range of chemoattractant concentration ranges. The LEGI-type schemes can transmit information about spatial differences downstream through the adaptive subtraction of average concentration information by global inhibitors. In this view, LEGI-mediated adaptation and excitable Ras dynamics act in concert, enabling cells to convert shallow receptor inputs into digitized directional cues across wide chemotaxis ranges. Detailed analyses of receptor-mediated G-protein dynamics underlying wide-range chemotaxis are described elsewhere and are beyond the scope of this review (Janetopoulos et al. 2001; Kamimura et al. 2016; Miyanaga et al. 2018; Kamimura and Ueda 2021; Xu and Jin 2022; Xu et al. 2022).

A SHARED EXCITABLE TRIGGER FOR CELL MIGRATION AND MACROPINOCYTOSIS

Excitable signaling networks provide a shared framework for understanding how cells generate pseudopods and macropinocytic cups. Pseudopods form when excitable Ras/PIP_3 patches extend with substrate adhesion, whereas macropinocytic cups develop when Ras/PIP_3 activation spreads without adhesion (Nichols et al. 2015; Devreotes et al. 2017; King and Kay 2019; Miao et al. 2019; Kay et al. 2024; Tu et al. 2025). Thus, pseudopod extension and macropinocytosis are distinct geometric outcomes of a shared excitable signaling mechanism. Ras/PIP_3 dynamics are also sensitive to physical microtopography. Substrate microtopography biases the geometry of excitable domains (Honda et al. 2021). Anisotropic environments favor elongated Ras/PIP_3 patches and pseudopod formation, whereas isotropic environments allow for unrestricted radial expansion and promote the formation and closure of macropinocytic cups. These observations show the mechanosensitivity of the Ras excitable system (also see Artemenko et al. 2016). Although the molecular mechanisms underlying this sensitivity remain unclear, it is known that PIP_3 activation increases in knockout strains of Talin involved in adhesion (Yamazaki et al. 2020).

RasGEFX acts as a component linking Ras excitability to macropinocytosis. RasGEFX localizes to macropinocytic caps, suggesting that it amplifies Ras activation locally during cup formation (Iwamoto et al. 2025). Loss of RasGEFX impairs macropinocytosis and pseudopod formation, whereas increased RasGEFX expression enhances both processes. In contrast, RasGEFB primarily shapes the spatial size of Ras-GTP-enriched domains, generating large patches and pseudopods with its overexpression, but RasGEFB suppresses macropinocytosis. This is consistent with the geometric constraints required for cup closure. RasGEFU (GbpD), a Rap1-specific GEF that enhances PI3K-mediated PIP_3 production, promotes adhesion-rich pseudopods but does not efficiently drive macropinocytosis (Kortholt and van Haastert 2008; Iwamoto et al. 2025). The partial persistence of macropinocytosis in RasGEFX-null cells and in mutants lacking various RasGAPs indicates that multiple RasGEFs and RasGAPs operate in parallel or partially redundant modules to ensure context-dependent control of Ras excitability (Li et al. 2018; Williams et al. 2019; Xu et al. 2022; Tu et al. 2025).

BIOLOGICAL ADVANTAGES OF EXCITABILITY-DRIVEN SPONTANEOUS MIGRATION

A series of theoretical studies has advanced a dynamical view in which spontaneous cell motility emerges from the interplay between local biochemical feedback and cell-scale behavior (Camley and Rappel 2017; Rappel and Edelstein-Keshet 2017; Belly and Weiner 2024). Mechanistic simulations have shown that feedback between actin-driven protrusion and its inhibitory "cortical factors" can spontaneously generate front–rear asymmetry and cyclic protrusive dynamics even in the absence of external cues (Nishimura et al. 2009). In this model, actin polymerization locally dilutes inhibitory cortical factors at protrusive sites, thereby enhancing further protrusion, while other regions rich in cortical factors maintain contractility. The coupling of local positive feedback and global negative feedback allows random turns and persistent motility even without guidance cues (Nishimura et al. 2009). Trajectories obtained from the simulations display the superdiffusive migration with non-Gaussian displacement distributions and prolonged directional persistence (Nishimura et al. 2012). Experimental analyses of spontaneous cell migration support these theoretical predictions. Live-cell tracking revealed long-tailed velocity distributions, biexponential decay of velocity autocorrelation functions, and deviations from classical random-walk behavior, as described above (Takagi et al. 2008, 2014; Bödeker et al. 2010). The multi-timescales measured experimentally match those produced by theoretical models (Nishimura et al. 2009, 2012). Incorporating excitable–bistable dynamics into the mechanical feedback model may provide a more comprehensive description of amoeboid migration by linking molecular-scale excitability and bistability to cell-scale polarity maintenance and long-range persistent movement.

The widespread occurrence of non-Brownian motility raises the possibility that such behavior reflects an evolutionarily selected mode of operation (Takagi et al. 2008; Harris et al. 2012; Wu et al. 2026). The molecular constituents of the signaling network, their excitable–bistable architecture, and the resulting stochastic spatiotemporal dynamics can be interpreted as adaptive solutions for functioning in intrinsically noisy and unpredictable environments. Due to the combined process of random turns and long-lived persistent motions, the simulated amoebae can navigate complex environments with obstacles in an efficient way (Fig. 5), sug-

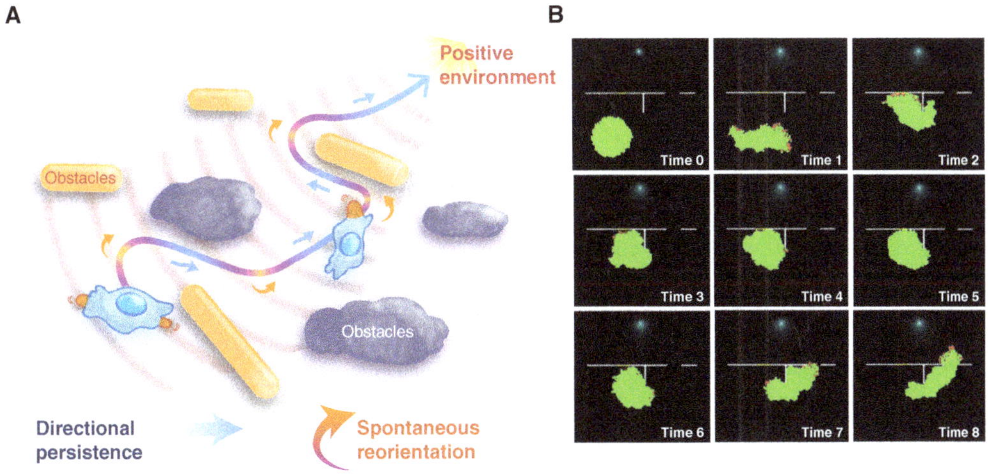

Figure 5. Non-Brownian migration enables obstacle avoidance. (*A*) Schema illustration outlining the mechanisms of non-Brownian spontaneous migration enabling navigation in complex environments. Intrinsic fluctuations in cell movement generate trajectories with intermittent directional persistence, allowing cells to approach a distant goal while circumventing physical obstacles. (*B*) Simulation of amoeboid movement based on the Nishimura et al. (2012) model. Time-lapse images show the cell exhibiting protrusive dynamics and non-Brownian motion. Blue patterns indicate the extracellular chemical signals. Extracellular signals can permeate a yellow-colored linear bar in the wall, but the cell does not pass. Persistent but fluctuating motility allows the cell to reorient upon obstacle encounter and ultimately advance beyond the barrier.

gesting the biological advantage for motile cells to adopt the non-Brownian dynamics. Internally generated directional memory enables amoeboid cells to pass through physical obstacles without continuous sensory input, although the obstacle sizes remain comparable to those of the cells. Explaining how such organized and functionally advantageous behavior arises requires a hierarchical framework that links molecular processes to cellular dynamics. As proposed by Oosawa (2001, 2007), spontaneous cellular activities are driven by active, energy-consuming molecular fluctuations. In excitable Ras signaling networks, basal Ras activity generated by RasGEF- and RasGAP-mediated GTPase cycles is likely to provide a source of such nonthermal and active fluctuations (Fig. 1). These energy-dependent stochastic events are amplified by excitable system and stabilized through downstream bistable modules. Through this hierarchical signaling system, cells can transform molecular-scale active fluctuations into structured, non-Brownian migratory behavior.

CONCLUSION

Recent studies have revealed that spontaneous cell migration and macropinocytosis are driven by a shared excitable signaling network. This excitable Ras module converts intrinsic stochastic fluctuations into discrete activation events, and when coupled to a downstream bistable module in phosphoinositide metabolism, it forms a hierarchical signaling system capable of generating both rapid exploratory protrusions and long-lived polarized states. Through this coupling of excitable and bistable modules, cells can achieve the stochastic initiation of protrusions and the robust maintenance of front–rear asymmetry simultaneously, giving rise to non-Brownian migratory behavior with multi-timescale spontaneous motility. Clarifying how these excitable and bistable modules are wired, tuned, and coordinated across different physiological contexts is essential for understanding how organized randomness is implemented in biological hierarchical systems, thereby enabling living cells to exploit stochasticity for efficient exploration and adaptation in complex natural environments.

COMPETING INTEREST STATEMENT

The authors declare that the research was conducted in the absence of any commercial or financial relationships that could be construed as a potential conflict of interest.

ACKNOWLEDGMENTS

We thank all members of the Single Molecule Biology Laboratory, The University of Osaka for helpful discussions. We appreciate NBRP Nenkin and DictyBase for plasmids and cell strains. This work was supported by Japan Science and Technology Agency grant no. JPMJCR21E1 to M.U., no. JPMJKP25V4 to M.U., no. JPMJPR1879 to S.M., and JSPS KAKENHI grant no. 19H05798 to S.M.

REFERENCES

Amselem G, Theves M, Bae A, Beta C, Bodenschatz E. 2012. Control parameter description of eukaryotic chemotaxis. *Phys Rev Lett* **109:** 108103. doi:10.1103/PhysRevLett.109.108103

Andrew N, Insall RH. 2007. Chemotaxis in shallow gradients is mediated independently of PtdIns 3-kinase by biased choices between random protrusions. *Nat Cell Biol* **9:** 193–200. doi:10.1038/ncb1536

Arai Y, Shibata T, Matsuoka S, Sato MJ, Yanagida T, Ueda M. 2010. Self-organization of the phosphatidylinositol lipids signaling system for random cell migration. *Proc Natl Acad Sci* **107:** 12399–12404. doi:10.1073/pnas.0908278107

Artemenko Y, Axiotakis L, Borleis J, Iglesias PA, Devreotes PN. 2016. Chemical and mechanical stimuli act on common signal transduction and cytoskeletal networks. *Proc Natl Acad Sci* **113:** E7500–E7509. doi:10.1073/pnas.1608767113

Asano Y, Nagasaki A, Uyeda TQ. 2008. Correlated waves of actin filaments and PIP_3 in *Dictyostelium* cells. *Cell Motil Cytoskeleton* **65:** 923–934. doi:10.1002/cm.20314

Banerjee T, Biswas D, Pal DS, Miao Y, Iglesias PA, Devreotes PN. 2022. Spatiotemporal dynamics of membrane surface charge regulates cell polarity and migration. *Nat Cell Biol* **24:** 1499–1515. doi:10.1038/s41556-022-00997-7

Banerjee T, Matsuoka S, Biswas D, Miao Y, Pal DS, Kamimura Y, Ueda M, Devreotes PN, Iglesias PA. 2023. A dynamic partitioning mechanism polarizes membrane protein distribution. *Nat Commun* **14:** 7909. doi:10.1038/s41467-023-43615-2

Banerjee T, Deng Y, Pal DS, Zhan H, Iglesias PA, Devreotes PN. 2025. Signaling and actin waves at a glance. *J Cell Sci* **138:** jcs263634. doi:10.1242/jcs.263634

Belly HD, Weiner OD. 2024. Follow the flow: actin and membrane act as an integrated system to globally coordinate cell shape and movement. *Curr Opin Cell Biol* **89:** 102392. doi:10.1016/j.ceb.2024.102392

Berg HC, Purcell EM. 1977. Physics of chemoreception. *Biophys J* **20:** 193–219. doi:10.1016/S0006-3495(77)85544-6

Berridge MJ, Bootman MD, Roderick HL. 2003. Calcium signalling: dynamics, homeostasis and remodelling. *Nat Rev Mol Cell Biol* **4:** 517–529. doi:10.1038/nrm1155

Beta C, Kruse K. 2017. Intracellular oscillations and waves. *Annu Rev Condens Matter Phys* **8:** 239–264. doi:10.1146/annurev-conmatphys-031016-025210

Beta C, Amselem G, Bodenschatz E. 2008. A bistable mechanism for directional sensing. *New J Phys* **10:** 083015. doi:10.1088/1367-2630/10/8/083015

Beta C, Edelstein-Keshet L, Gov N, Yochelis A. 2023. From actin waves to mechanism and back: how theory aids biological understanding. *eLife* **12:** e87181. doi:10.7554/eLife.87181

Bhattacharya S, Iglesias PA. 2018. The threshold of an excitable system serves as a control mechanism for noise filtering during chemotaxis. *PLoS One* **13:** e0201283. doi:10.1371/journal.pone.0201283

Bhattacharya S, Banerjee T, Miao Y, Zhan H, Devreotes PN, Iglesias PA. 2020. Traveling and standing waves mediate pattern formation in cellular protrusions. *Sci Adv* **6:** eaay7682. doi:10.1126/sciadv.aay7682

Bödeker HU, Beta C, Frank TD, Bodenschatz E. 2010. Quantitative analysis of random ameboid motion. *Europhys Lett* **90:** 28005. doi:10.1209/0295-5075/90/28005

Bosgraaf L, van Haastert PJM. 2009. The ordered extension of pseudopodia by amoeboid cells in the absence of external cues. *PLoS One* **4:** e5253. doi:10.1371/journal.pone.0005253

Bretschneider T, Anderson K, Ecke M, Müller-Taubenberger A, Schroth-Diez B, Ishikawa-Ankerhold HC, Gerisch G. 2009. The three-dimensional dynamics of actin waves, a model of cytoskeletal self-organization. *Biophys J* **96:** 2888–2900. doi:10.1016/j.bpj.2008.12.3942

Camley BA, Rappel WJ. 2017. Physical models of collective cell motility: from cell to tissue. *J Phys D Appl Phys* **50:** 113002. doi:10.1088/1361-6463/aa56fe

Charest PG, Firtel RA. 2006. Feedback signaling controls leading-edge formation during chemotaxis. *Curr Opin Genet Dev* **16:** 339–347. doi:10.1016/j.gde.2006.06.016

Chiou JG, Balasubramanian MK, Lew DJ. 2017. Cell polarity in yeast. *Annu Rev Cell Dev Biol* **33:** 77–101. doi:10.1146/annurev-cellbio-100616-060856

Deng Y, Banerjee T, Pal DS, Banerjee P, Zhan H, Borleis J, Igleias PA, Devreotes PN. 2024. PIP5K-Ras bistability initiates plasma membrane symmetry breaking to regulate cell polarity and migration. bioRxiv doi:10.1101/2024.09.15.613115

Devreotes PN, Bhattacharya S, Edwards M, Iglesias PA, Lampert T, Miao Y. 2017. Excitable signal transduction networks in directed cell migration. *Annu Rev Cell Dev Biol* **33:** 103–125. doi:10.1146/annurev-cellbio-100616-060739

Eichinger L, Rivero F, ed. 2016. *Dictyostelium discoideum protocols*, 2nd ed. In *Methods in molecular biology*, Vol. 983. Humana, New York.

Endres RG, Wingreen NS. 2008. Accuracy of direct gradient sensing by single cells. *Proc Natl Acad Sci* **105:** 15749–15754. doi:10.1073/pnas.0804688105

Fisher PR, Merkl R, Gerisch G. 1989. Quantitative analysis of cell motility and chemotaxis in *Dictyostelium discoideum* by using an image processing system and a novel chemotaxis chamber providing stationary chemical gradients. *J Cell Biol* **108:** 973–984. doi:10.1083/jcb.108.3.973

FitzHugh R. 1961. Impulses and physiological states in theoretical models of nerve membrane. *Biophys J* **1:** 445–466. doi:10.1016/S0006-3495(61)86902-6

Friedl P, Weigelin B. 2008. Interstitial leukocyte migration and immune function. *Nat Immunol* **9:** 960–969. doi:10.1038/ni.f.212

Friedl P, Wolf K. 2003. Tumour-cell invasion and migration: diversity and escape mechanisms. *Nat Rev Cancer* **3:** 362–374. doi:10.1038/nrc1075

Fukushima S, Matsuoka S, Ueda M. 2019. Excitable dynamics of Ras triggers spontaneous symmetry breaking of PIP_3 signaling in motile cells. *J Cell Sci* **132:** jcs224121. doi:10.1242/jcs.224121

Funamoto S, Meili R, Lee S, Parry L, Firtel RA. 2002. Spatial and temporal regulation of 3-phosphoinositides by PI 3-kinase and PTEN mediates chemotaxis. *Cell* **109:** 611–623. doi:10.1016/s0092-8674(02)00755-9

Goldstein B, Macara IG. 2007. The PAR proteins: fundamental players in animal cell polarization. *Dev Cell* **13:** 609–622. doi:10.1016/j.devcel.2007.10.007

Graziano BR, Weiner OD. 2014. Self-organization of protrusions and polarity during eukaryotic chemotaxis. *Curr Opin Cell Biol* **30:** 60–67. doi:10.1016/j.ceb.2014.06.007

Harris TH, Banigan EJ, Christian DA, Konradt C, Tait Wojno ED, Norose K, Wilson EH, John B, Weninger W, Luster AD, et al. 2012. Generalized Lévy walks and the role of chemokines in migration of effector $CD8^+$ T cells. *Nature* **486:** 545–548. doi:10.1038/nature11098

Hodgkin AL, Huxley AF. 1952. A quantitative description of membrane current and its application to conduction and excitation in nerve. *J Physiol* **117:** 500–544. doi:10.1113/jphysiol.1952.sp004764

Hoege C, Hyman AA. 2013. Principles of PAR polarity in *Caenorhabditis elegans* embryos. *Nat Rev Mol Cell Biol* **14:** 315–322. doi:10.1038/nrm3558

Honda G, Saito N, Fujimori T, Hashimura H, Nakamura MJ, Nakajima A, Sawai S. 2021. Microtopographical guidance of macropinocytic signaling patches. *Proc Natl Acad Sci* **118:** e2110281118. doi:10.1073/pnas.2110281118

Huang CH, Tang M, Shi C, Iglesias PA, Devreotes PN. 2013. An excitable signal integrator couples to an idling cytoskeletal oscillator to drive cell migration. *Nat Cell Biol* **15:** 1307–1316. doi:10.1038/ncb2859

Iglesias PA, Devreotes PN. 2012. Biased excitable networks: how cells direct motion in response to gradients. *Curr Opin Cell Biol* **24:** 245–253. doi:10.1016/j.ceb.2011.11.009

Iijima M, Devreotes P. 2002. Tumor suppressor PTEN mediates sensing of chemoattractant gradients. *Cell* **109:** 599–610. doi:10.1016/s0092-8674(02)00745-6

Insall RH, Borleis J, Devreotes PN. 1996. The aimless RasGEF is required for processing of chemotactic signals through G-protein-coupled receptors in *Dictyostelium*. *Curr Biol* **6:** 719–729. doi:10.1016/S0960-9822(09)00453-9

Iwamoto K, Matsuoka S, Ueda M. 2025. Excitable Ras dynamics-based screens reveal RasGEFX is required for macropinocytosis and random cell migration. *Nat Commun* **16:** 117. doi:10.1038/s41467-024-55389-2

Janetopoulos C, Jin T, Devreotes P. 2001. Receptor-mediated activation of heterotrimeric G-proteins in living cells. *Science* **291:** 2408–2411. doi:10.1126/science.1055835

Janetopoulos C, Ma L, Devreotes PN, Iglesias PA. 2004. Chemoattractant-induced phosphatidylinositol 3,4,5-trisphosphate accumulation is spatially amplified and adapts, independent of the actin cytoskeleton. *Proc Natl Acad Sci* **101:** 8951–8956. doi:10.1073/pnas.0402152101

Kae H, Lim CJ, Spiegelman GB, Weeks G. 2004. Chemoattractant-induced Ras activation during *Dictyostelium* aggregation. *EMBO Rep* **5:** 602–606. doi:10.1038/sj.embor.7400151

Kae H, Kortholt A, Rehmann H, Insall RH, van Haastert PJ, Spiegelman GB, Weeks G. 2007. Cyclic AMP signalling in *Dictyostelium*: G-proteins activate separate Ras pathways using specific RasGEFs. *EMBO Rep* **8:** 477–482. doi:10.1038/sj.embor.7400936

Kamimura Y, Ueda M. 2021. Different heterotrimeric G protein dynamics for wide-range chemotaxis in eukaryotic cells. *Front Cell Dev Biol* **9:** 724797. doi:10.3389/fcell.2021.724797

Kamimura Y, Miyanaga Y, Ueda M. 2016. Heterotrimeric G-protein shuttling via Gip1 extends the dynamic range of eukaryotic chemotaxis. *Proc Natl Acad Sci* **113:** 4356–4361. doi:10.1073/pnas.1516767113

Kay RR, Lutton JE, King JS, Bretschneider T. 2024. Making cups and rings: the "stalled-wave" model for macropinocytosis. *Biochem Soc Trans* **52:** 1785–1794. doi:10.1042/BST20231426

Keener J, Sneyd J. 1998. *Mathematical physiology*. Springer, New York.

Kessin RH. 2001. *Dictyostelium: evolution, cell biology, and the development of multicellularity*. Cambridge University Press, Cambridge.

Killich T, Plath PJ, Wei X, Bultmann H, Rensing L, Vicker MG. 1993. The locomotion, shape and pseudopodial dynamics of unstimulated *Dictyostelium* cells are not random. *J Cell Sci* **106:** 1005–1013. doi:10.1242/jcs.106.4.1005

King S, Kay RR. 2019. The origins and evolution of macropinocytosis. *Philos Trans R Soc Lond B Biol Sci* **374:** 20180158. doi:10.1098/rstb.2018.0158

Kortholt A, van Haastert PJM. 2008. Highlighting the role of Ras and Rap during *Dictyostelium* chemotaxis. *Cell Signal* **20:** 1415–1422. doi:10.1016/j.cellsig.2008.02.006

Kortholt A, Rehmann H, Kae H, Bosgraaf L, Keizer-Gunnink I, Weeks G, Wittinghofer A, van Haastert PJ. 2006. Characterization of the GbpD-activated Rap1 pathway regulating adhesion and cell polarity in *Dictyostelium discoideum*. *J Biol Chem* **281:** 23367–23376. doi:10.1074/jbc.M600804200

Kortholt A, Bolourani P, Rehmann H, Keizer-Gunnink I, Weeks G, Wittinghofer A, van Haastert PJ. 2010. A Rap/phosphatidylinositol 3-kinase pathway controls pseudopod formation. *Mol Biol Cell* **21:** 936–945. doi:10.1091/mbc.e09-03-0177

Kortholt A, Kataria R, Keizer-Gunnink I, Van Egmond WN, Khanna A, van Haastert PJM. 2011. *Dictyostelium* chemotaxis: essential Ras activation and accessory signalling pathways for amplification. *EMBO Rep* **12:** 1273–1279. doi:10.1038/embor.2011.210

Kortholt A, Van Egmond WN, Plak K, Bosgraaf L, Keizer-Gunnink I, van Haastert PJM. 2012. Multiple regulatory mechanisms for the *Dictyostelium* Roco protein GbpC. *J Biol Chem* **287:** 2749–2758. doi:10.1074/jbc.M111.315739

Lechleiter J, Girard S, Peralta E, Clapham D. 1991. Spiral calcium wave propagation and annihilation in *Xenopus laevis* oocytes. *Science* **252:** 123–126. doi:10.1126/science.2011747

Levchenko A, Iglesias PA. 2002. Models of eukaryotic gradient sensing: application to chemotaxis of amoebae and neutrophils. *Biophys J* **82:** 50–63. doi:10.1016/S0006-3495(02)75373-3

Li L, Nørrelykke SF, Cox EC. 2008. Persistent cell motion in the absence of external signals: a search strategy for eukaryotic cells. *PLoS One* **3:** e2093. doi:10.1371/journal.pone.0002093

Li L, Cox EC, Flyvbjerg H. 2011. "Dicty dynamics": *Dictyostelium* motility as persistent random motion. *Phys Biol* **8:** 046006. doi:10.1088/1478-3975/8/4/046006

Li X, Edwards M, Swaney KF, Singh N, Bhattacharya S, Borleis J, Long Y, Iglesias PA, Chen J, Devreotes PN. 2018. Mutually inhibitory Ras-PI(3,4)P_2 feedback loops mediate cell migration. *Proc Natl Acad Sci* **115:** E9125–E9134. doi:10.1073/pnas.1809039115

Li X, Miao Y, Pal DS, Devreotes PN. 2020. Excitable networks controlling cell migration during development and disease. *Semin Cell Dev Biol* **100:** 133–142. doi:10.1016/j.semcdb.2019.11.001

Lin Y, Pal DS, Banerjee P, Banerjee T, Qin G, Deng Y, Borleis J, Iglesias PA, Devreotes PN. 2024. Ras suppression potentiates rear actomyosin contractility-driven cell polarization and migration. *Nat Cell Biol* **26:** 1062–1076. doi:10.1038/s41556-024-01453-4

Lin Y, Parajón E, Yuan Q, Ye S, Qin G, Deng Y, Borleis J, Koyfman A, Iglesias PA, Konstantopoulos K, et al. 2025. Ras-mediated dynamic and biphasic regulation of cell migration. *Proc Natl Acad Sci* **122:** e2503847122. doi:10.1073/pnas.2503847122

Lindner B, García-Ojalvo J, Neiman A, Schimansky-Geier L. 2004. Effects of noise in excitable systems. *Phys Rep* **392:** 321–424. doi:10.1016/j.physrep.2003.10.015

Liu Y, Lacal J, Firtel RA, Kortholt A. 2018. Connecting G protein signaling to chemoattractant-mediated cell polarity and cytoskeletal reorganization. *Small GTPases* **9:** 360–364. doi:10.1080/21541248.2016.1235390

Maeda YT, Inose J, Matsuo MY, Iwaya S, Sano M. 2008. Ordered patterns of cell shape and orientational correlation during spontaneous cell migration. *PLoS One* **3:** e3734. doi:10.1371/journal.pone.0003734

Mato JM, Losada A, Nanjundiah V, Konijn TM. 1975. Signal input for a chemotactic response in the cellular slime mold *Dictyostelium discoideum*. *Proc Natl Acad Sci* **72:** 4991–4993. doi:10.1073/pnas.72.12.4991

Matsuoka S, Ueda M. 2018. Mutual inhibition between PTEN and PIP_3 generates bistability for polarity in motile cells. *Nat Commun* **9:** 4481. doi:10.1038/s41467-018-06856-0

Matsuoka S, Iwamoto K, Shin DY, Ueda M. 2024. Spontaneous signal generation by an excitable system for cell migration. *Front Cell Dev Biol* **12:** 1373609. doi:10.3389/fcell.2024.1373609

Meili R, Ellsworth C, Lee S, Reddy TB, Ma H, Firtel RA. 1999. Chemoattractant-mediated transient activation and membrane localization of Akt/PKB is required for efficient chemotaxis to cAMP in *Dictyostelium*. *EMBO J* **18:** 2092–2105. doi:10.1093/emboj/18.8.2092

Miao Y, Bhattacharya S, Edwards M, Cai H, Inoue T, Iglesias AP, Devreotes PN. 2017. Altering the threshold of an excitable signal transduction network changes cell migratory modes. *Nat Cell Biol* **19:** 329–340. doi:10.1038/ncb3495

Miao Y, Bhattacharya S, Banerjee T, Abubaker-Sharif B, Long Y, Inoue T, Iglesias AP, Devreotes PN. 2019. Wave patterns organize cellular protrusions and control cortical dynamics. *Mol Syst Biol* **15:** MSB188585. doi:10.15252/msb.20188585

Miyanaga Y, Matsuoka S, Yanagida T, Ueda M. 2007. Stochastic signal inputs for chemotactic response in *Dictyostelium* cells revealed by single molecule imaging techniques. *BioSystems* **88:** 251–260. doi:10.1016/j.biosystems.2006.07.011

Miyanaga Y, Kamimura Y, Kuwayama H, Devreotes PN, Ueda M. 2018. Chemoattractant receptors activate, recruit and capture G proteins for wide range chemotaxis. *Biochem Biophys Res Commun* **507:** 304–310. doi:10.1016/j.bbrc.2018.11.029

Mori Y, Jilkine A, Edelstein-Keshet L. 2008. Wave-pinning and cell polarity from a bistable reaction-diffusion system. *Biophys J* **94:** 3684–3697. 10.1529/biophysj.107.120824

Nagumo J, Arimoto S, Yoshizawa S. 1962. An active pulse transmission line simulating nerve axon. *Proc IRE* **50:** 2061–2070. doi:10.1109/JRPROC.1962.288235

Nakajima A, Ishihara S, Imoto D, Sawai S. 2014. Rectified directional sensing in long-range cell migration. *Nat Commun* **5:** 5367. doi:10.1038/ncomms6367

Nichols JM, Veltman D, Kay RR. 2015. Chemotaxis of a model organism: progress with *Dictyostelium*. *Curr Opin Cell Biol* **36:** 7–12. doi:10.1016/j.ceb.2015.06.005

Nishikawa M, Hörning M, Ueda M, Shibata T. 2014. Excitable signal transduction induces both spontaneous and directional cell asymmetries in the phosphatidylinositol lipid signaling system for eukaryotic chemotaxis. *Biophys J* **106:** 723–734. doi:10.1016/j.bpj.2013.12.023

Nishimura SI, Ueda M, Sasai M. 2009. Cortical factor feedback model for cellular locomotion and cytofission. *PLoS Comput Biol* **5:** e1000310. doi:10.1371/journal.pcbi.1000310

Nishimura SI, Ueda M, Sasai M. 2012. Non-Brownian dynamics and strategy of amoeboid cell locomotion. *Phys Rev E Stat Nonlin Soft Matter Phys* **85:** 041909. doi:10.1103/PhysRevE.85.041909

Ohtsuka D, Ota N, Amaya S, Matsuoka S, Tanaka Y, Ueda M. 2021. A sub-population of *Dictyostelium discoideum* cells shows extremely high sensitivity to cAMP for directional migration. *Biochem Biophys Res Commun* **554:** 131–137. doi:10.1016/j.bbrc.2021.03.095

Oosawa F. 2001. Spontaneous signal generation in living cells. *Bull Math Biol* **63:** 643–654. doi:10.1006/bulm.2001.0236

Oosawa F. 2007. Spontaneous activity of living cells. *BioSystems* **88:** 191–201. doi:10.1016/j.biosystems.2006.05.006

Parent CA, Devreotes PN. 1999. A cell's sense of direction. *Science* **284:** 765–770. doi:10.1126/science.284.5415.765

Parent CA, Blacklock BJ, Froehlich WM, Murphy DB, Devreotes PN. 1998. G protein signaling events are activated at the leading edge of chemotactic cells. *Cell* **95:** 81–91. doi:10.1016/s0092-8674(00)81784-5

Petrie R, Doyle A, Yamada K. 2009. Random versus directionally persistent cell migration. *Nat Rev Mol Cell Biol* **10:** 538–549. doi:10.1038/nrm2729

Postma M, Roelofs J, Goedhart J, Loovers HM, Visser AJWG, van Haastert PJM. 2004. Sensitization of *Dictyostelium* chemotaxis by phosphoinositide-3-kinase-mediated self-organizing signalling patches. *J Cell Sci* **117:** 2925–2935. doi:10.1242/jcs.01143

Rappel WJ, Edelstein-Keshet L. 2017. Mechanisms of cell polarization. *Curr Opin Syst Biol* **3:** 43–53. doi:10.1016/j.coisb.2017.03.005

Ridley AJ, Schwartz MA, Burridge K, Firtel RA, Ginsberg MH, Borisy G, Parsons JT, Horwitz AR. 2003. Cell migration: integrating signals from front to back. *Science* **302:** 1704–1709. doi:10.1126/science.1092053

Sasaki AT, Chun C, Takeda K, Firtel RA. 2004. Localized Ras signaling at the leading edge regulates PI3K, cell polarity, and directional cell movement. *J Cell Biol* **167:** 505–518. doi:10.1083/jcb.200406177

Sasaki AT, Janetopoulos C, Lee S, Charest PG, Takeda K, Sundheimer LW, Meili R, Devreotes PN, Firtel RA. 2007. G protein-independent Ras/PI3K/F-actin circuit regulates basic cell motility. *J Cell Biol* **178:** 185–191. doi:10.1083/jcb.200611138

Sato MJ, Ueda M, Takagi H, Watanabe TM, Yanagida T, Ueda M. 2007. Input–output relationship in galvanotactic response of *Dictyostelium* cells. *BioSystems* **88:** 261–272. doi:10.1016/j.biosystems.2006.06.008

Sato MJ, Kuwayama H, van Egmond WN, Takayama ALK, Takagi H, van Haastert PJM, Yanagida T, Ueda M. 2009. Switching direction in electric-signal-induced cell migration by cyclic guanosine monophosphate and phosphatidylinositol signalling. *Proc Natl Acad Sci* **106:** 6667–6672. doi:10.1073/pnas.0809974106

Selmeczi D, Mosler S, Hagedorn PH, Larsen NB, Flyvbjerg H. 2005. Cell motility as persistent random motion: theories from experiments. *Biophys J* **89:** 912–931. doi:10.1529/biophysj.105.061150

Shibata T, Ueda M. 2008. Noise generation, amplification and propagation in chemotactic signaling systems of living cells. *BioSystems* **93:** 126–132. doi:10.1016/j.biosystems.2008.04.003

Shibata T, Nishikawa M, Matsuoka S, Ueda M. 2012. Modeling the self-organized phosphatidylinositol lipid signaling system in chemotactic cells using quantitative image analysis. *J Cell Sci* **125:** 5138–5150. doi:10.1242/jcs.108373

Shibata T, Nishikawa M, Matsuoka S, Ueda M. 2013. Intracellular encoding of spatiotemporal guidance cues in a self-organizing signaling system for chemotaxis in *Dictyostelium* cells. *Biophys J* **105:** 2199–2209. doi:10.1016/j.bpj.2013.09.024

Shin DY, Takagi H, Hiroshima M, Matsuoka S, Ueda M. 2023. Sphingomyelin metabolism underlies Ras excitability for efficient cell migration and chemotaxis. *Cell Struct Funct* **48:** 145–160. 10.1247/csf.23045

Song LL, Nadkarni SM, Bödeker HU, Beta C, Bae A, Franck C, Rappel WJ, Loomis WF, Bodenschatz E. 2006. *Dicyostelium discoideum* chemotaxis: threshold for directed motion. *Eur J Cell Biol* **85:** 981–989. doi:10.1016/j.ejcb.2006.01.012

Skoge M, Yue H, Erickstad M, Bae A, Levine H, Groisman A, Loomis WF, Rappel WJ. 2014. Cellular memory in eukaryotic chemotaxis. *Proc Natl Acad Sci* **111:** 14448–14453. doi:10.1073/pnas.1412197111

Swanson JA, Taylor DL. 1982. Local and spatially coordinated movements in *Dictyostelium discoideum* amoebae during chemotaxis. *Cell* **28:** 225–232. doi:10.1016/0092-8674(82)90340-3

Takagi H, Sato MJ, Yanagida T, Ueda M. 2008. Functional analysis of spontaneous cell movement under different physiological conditions. *PLoS One* **3:** e2648. doi:10.1371/journal.pone.0002648

Takagi H, Sato MJ, Yanagida T, Ueda M. 2014. Saibouko-gaku. *J Cell Technol* **33:** 48–54 (in Japanese). doi:10.15105/j01677.2014121048

Takeda K, Shao D, Adler M, Charest PG, Loomis WF, Levine H, Groisman A, Rappel WJ, Firtel RA. 2012. Incoherent feedforward control governs adaptation of activated ras in a eukaryotic chemotaxis pathway. *Sci Signal* **5:** ra2. doi:10.1126/scisignal.2002413

Taniguchi D, Ishihara S, Oonuki T, Honda-Kitahara M, Kaneko K, Sawai S. 2013. Phase geometries of two-dimensional excitable waves govern self-organized morphodynamics of amoeboid cells. *Proc Natl Acad Sci* **110:** 5016–5021. doi:10.1073/pnas.1218025110

Thompson BJ. 2013. Cell polarity: models and mechanisms from yeast, worms and flies. *Development* **140:** 13–21. doi:10.1242/dev.083634

Tu H, Wang W, Cai H. 2025. Macropinocytosis: molecular mechanisms and regulation. *Curr Opin Cell Biol* **95:** 102563. doi:10.1016/j.ceb.2025.102563

Tyson JJ, Fife PC. 1980. Target patterns in a realistic model of the Belousov–Zhabotinsky reaction. *J Chem Phys* **73:** 2224–2237. doi:10.1063/1.440418

Ueda M, Shibata T. 2007. Stochastic signal processing and transduction in chemotactic response of eukaryotic cells. *Biophys J* **93:** 11–20. doi:10.1529/biophysj.106.100263

Ueda M, Sako Y, Tanaka T, Devreotes PN, Yanagida T. 2001. Single molecule analysis of chemotactic signaling in *Dictyostelium* cells. *Science* **294:** 864–867. doi:10.1126/science.1063951

Uhlenbeck GE, Ornstein LS. 1930. On the theory of Brownian motion. *Phys Rev* **36:** 823–841. doi:10.1103/PhysRev.36.823

Uyeda TQP, Yamazaki Y, Kijima ST, Noguchi TQP, Ngo KX. 2025. Multiple mechanisms to regulate actin functions: "fundamental" versus lineage-specific mechanisms and hierarchical relationships. *Biomolecules* **15:** 279. doi:10.3390/biom15020279

Van Haastert PJM. 2020. Symmetry breaking during cell movement in the context of excitability, kinetic fine-tuning and memory of pseudopod formation. *Cells* **9:** 1809. doi:10.3390/cells9081809

Van Haastert PJM, Bosgraaf L. 2009. Food searching strategy of amoeboid cells by starvation induced run length extension. *PLoS One* **4:** e6814. doi:10.1371/journal.pone.0006814

Van Haastert PJM, Devreotes PN. 2004. Chemotaxis: signalling the way forward. *Nat Rev Mol Cell Biol* **5:** 626–634. doi:10.1038/nrm1435

Van Haastert PJM, Postma M. 2007. Biased random walk by stochastic fluctuations of chemoattractant-receptor interactions at the lower limit of detection. *Biophys J* **93:** 1787–1796. doi:10.1529/biophysj.107.104356

Van Haastert PJM, Keizer-Gunnink I, Pots H, Ortiz-Mateos C, Veltman D, van Egmond W, Kortholt A. 2021. Forty-five years of cGMP research in *Dictyostelium*: understanding the regulation and function of the cGMP pathway for cell movement and chemotaxis. *Mol Biol Cell* **32:** ar8. doi:10.1091/mbc.E21-04-0171

Vicker MG. 2002. Eukaryotic cell locomotion depends on the propagation of self-organized reaction-diffusion waves and oscillations of actin filament assembly. *Exp Cell Res* **275:** 54–66. doi:10.1006/excr.2001.5466

Wang J, Xu L, Wang E. 2008. Potential landscape and flux framework of nonequilibrium networks: robustness, dissipation, and coherence of biochemical oscillations. *Proc Natl Acad Sci* **105:** 12271–12276. doi:10.1073/pnas.0800579105

Weiner OD, Marganski WA, Wu LF, Altschuler SJ, Kirschner MW. 2007. An actin-based wave generator organizes cell motility. *PLoS Biol* **5:** e221. doi:10.1371/journal.pbio.0050221

Wilkins A, Szafranski K, Fraser DJ, Bakthavatsalam D, Müller R, Fisher PR, Glöckner G, Eichinger L, Noegel AA, Insall RH. 2005. The Dictyostelium genome encodes numerous RasGEFs with multiple biological roles. *Genome Biol* **6:** R68. doi:10.1186/gb-2005-6-8-r68

Williams TD, Paschke PI, Kay RR. 2019. Function of small GTPases in *Dictyostelium* macropinocytosis. *Philos Trans R Soc Lond B Biol Sci* **374:** 20180150. doi:10.1098/rstb.2018.0150

Winfree AT. 1972. Spiral waves of chemical activity. *Science* **175:** 634–646. doi:10.1126/science.175.4022.634

Wu PH, Phillip JM, Du W, Forjaz A, Nair PR, Wirtz D. 2026. Methods to analyze cell migration data: fundamentals and practical guidelines. *Nat Methods* **23:** 43–55. doi:10.1038/s41592-025-02935-5

Xiong Y, Huang C, Iglesias PA, Devreotes PN. 2010. Cells navigate with a local-excitation, global-inhibition-biased excitable network. *Proc Natl Acad Sci* **107:** 17079–17086. doi:10.1073/pnas.1011271107

Xu X, Jin T. 2022. Ras inhibitors gate chemoattractant concentration range for chemotaxis through controlling GPCR-mediated adaptation and cell sensitivity. *Front Immunol* **13:** 1020117. doi:10.3389/fimmu.2022.1020117

Xu X, Pots H, Gilsbach BK, Parsons D, Veltman DM, Ramachandra SG, Li H, Kortholt A, Jin T. 2022. C2GAP2 is a common regulator of Ras signaling for chemotaxis, phagocytosis, and macropinocytosis. *Front Immunol* **13:** 1075386. doi:10.3389/fimmu.2022.1075386

Yamazaki SI, Hashimura H, Morimoto YV, Miyanaga Y, Matsuoka S, Kamimura Y, Ueda M. 2020. Talin B regulates collective cell migration via PI3K signaling in *Dictyostelium discoideum* mounds. *Biochem Biophys Res Commun* **525:** 372–377. doi:10.1016/j.bbrc.2020.02.060

Yoshioka D, Fukushima S, Koteishi H, Okuno D, Ide T, Matsuoka S, Ueda M. 2020. Single-molecule imaging of PI(4,5)P_2 and PTEN in vitro reveals a positive feedback mechanism for PTEN membrane binding. *Commun Biol* **3:** 92. doi:10.1038/s42003-020-0818-3

Zhan H, Bhattacharya S, Cai H, Iglesias PA, Huang CH, Devreotes PN. 2020. An excitable Ras/PI3K/ERK signaling network controls migration and oncogenic transformation in epithelial cells. *Dev Cell* **54:** 608–623.e5. doi:10.1016/j.devcel.2020.08.001

Zhang S, Charest PG, Firtel RA. 2008. Spatiotemporal regulation of Ras activity provides directional sensing. *Curr Biol* **18:** 1587–1593. doi:10.1016/j.cub.2008.08.069

Zhao M, Song B, Pu J, Wada T, Reid B, Tai G, Wang F, Guo A, Walczysko P, Gu Y, et al. 2006. Electrical signals control wound healing through phosphatidylinositol-3-OH kinase-γ and PTEN. *Nature* **442:** 457–460. doi:10.1038/nature04925

Modeling and Simulating Single and Collective Cell Motility

Jupiter Algorta, Ali Fele-Paranj, Jack M. Hughes, and Leah Edelstein-Keshet

Department of Mathematics, University of British Columbia, Vancouver, BC V6T 1Z2, Canada
Correspondence: keshet@math.ubc.ca

We survey a combination of classical and recent experimental and modeling developments in eukaryotic cell migration, from the subcellular processes and regulation, to single and collective cell dynamics. We showcase several examples that demonstrate simulations at several hierarchies: subcellular actin waves, corresponding migratory cell behavior, and collective behavior of several multicellular systems. We argue that the use of shared open-source software packages (such as Morpheus, in our case) to simulate multiscale models would be a boon to the community, allowing us to recreate, investigate, and build on existing work. A brief summary of currently available software is provided.

Technological progress in microscopy, in new methods of observing, measuring, manipulating, and controlling cells has resulted in large data sets rich in content that encode subcellular, cellular, and tissue dynamics. Given that ample data are available, a remaining hurdle is making sense of that data, and bringing our understanding to a level that synthesizes information at multiple scales, and in multiple types of cells and conditions. The methods of physics, mathematical modeling, and simulations can supplement biological tools to help with this goal and to link events at one scale with those at smaller and larger scales.

Many recent reviews have summarized the current state of experiments and models of single and collective cell migration. Some examples include Camley and Rappel (2017), Alert and Trepat (2020), Buttenschön and Edelstein-Keshet (2020), Stehbens et al. (2024), and others. Many of these reviews highlight recent experimental findings and modeling efforts, with discussion of how various approaches contribute to the field, and what challenges remain. In this work, we provide a few updates, some examples of case studies, and a perspective of a personal journey from the subcellular cytoskeleton dynamics and its regulation, to how these intracellular systems play out in single and group cell migration.

SUBCELLULAR DYNAMICS

Early research on molecular aspects of cell migration focused on the dynamics of the actin cytoskeleton and forces generated by filamentous actin (F-actin). The biochemistry of many proteins that affect and bind to actin was studied experimentally and dynamically (Pollard et al. 2001; Pollard 2016). Physicists contributed tools

Cite this article as *Cold Spring Harb Perspect Biol* doi: 10.1101/cshperspect.a041796

from soft matter physics to study the "active gels" that represent the actin cytoskeleton (Jülicher et al. 2007), and links between in vitro and in silico experiments helped to understand how actin can produce mechanical and turning responses (Manhart et al. 2019).

In the past two decades, some of the attention has shifted to the complex regulatory pathways that control actin assembly by formins, actin filament branching and nucleation (via WASP/WAVE and Arp2/3), and association of actin with myosin to produce contractile forces (Svitkina and Borisy 1999; Skruber and Mullins 2024). Signaling networks that regulate these processes play a vital role (for review, see Devreotes and Horwitz 2015). The small GTPases (Rac, Rho, and Cdc42), their mutual connectivity, and their activators (guanine nucleotide exchange factors [GEFs]) and inactivators (GTPase activating proteins [GAPs]) have been studied experimentally with new fluorescent constructs (Bishop and Hall 2000; Itoh et al. 2002), modeled mathematically using reaction–diffusion equations (Jilkine et al. 2007; Otsuji et al. 2007), and manipulated in various cell types using optogenetics (Hadjitheodorou et al. 2021; Town and Weiner 2023). The effect of mechanical stimuli (such as membrane tension, and pulling forces, including those exerted by cells) on the regulatory circuits has also been of great interest (Houk et al. 2012).

Mathematical models have contributed various insights to the mix, including the following: (1) Some of the inherent biology of GTPases, namely, cycling between active and inactive forms (see Fig. 1A,B), membrane residence of the active form (low diffusion rate) and the approximately constant pool of GTPase (mass conservation) can, under the appropriate assumptions, lead to self-organized polar distributions that define a clear front and back of a cell. This was denoted "wave-pinning" by Mori et al. (2008), since it amounts to a stalling wave of GTPase activity. This could then explain spontaneous cell polarization given an appropriate stimulus. (2) Cell geometry contributes to the polarization, with curved portions of the lamellipod stabilizing hotspots of GTPase activity (Marée et al. 2012; Camley et al. 2017). This is an outcome inherent to the reaction–diffusion physics, and independent of membrane-curvature protein affinity. (3) A single GTPase has such properties, but in a small and relatively sensitive parameter regime. Networks of GTPases, including the mutually antagonistic Rac and Rho make the system more robust by increasing the appropriate parameter regimes (Holmes and Edelstein-Keshet 2015). (4) Linking together GTPases via mutual antagonism, autocatalysis, or positive feedback greatly increases the variety of possible cell shapes and behaviors, even for a single parameter setting (Zmurchok and Holmes 2020). Thus, variable cell behavior could be an outcome of cell history, rather than genetic or physiological state. Restated, we can explain cell heterogeneity mathematically by invoking various "initial conditions" or distinct stimuli from the environment, aside from the genetic disparity that is associated with distinct parameter sets.

While GTPases such as Rac and Cdc42 promote the nucleation of F-actin, it has also been shown that F-actin can recruit GAPs that inactivate GTPases, creating a negative feedback loop as shown in Figure 1C. These interactions, explored both experimentally and theoretically, can lead to a variety of spatiotemporal patterns of "actin waves" (Weiner et al. 2007; Holmes et al. 2012; Allard and Mogilner 2013; Bement et al. 2015; Barnhart et al. 2017; Moldenhawer et al. 2022), including static polar states, traveling waves, and spiral waves in 2D (see Fig. 1D–F).

Models of the reaction–diffusion type (parabolic partial differential equations [PDEs]) have provided valuable insight into how these patterns are generated and how they interact with one another inside cells. For example, Holmes et al. (2012) showed that the strength of the negative feedback from F-actin to its nucleation-promoting factor (e.g., Rac) controls the types of solution observed: low versus high negative feedback leads to polar versus traveling wave solutions (Hughes et al. 2024). The significance is that such distinct dynamical patterns of F-actin in the cell can create distinct cellular motility behavior, including directed motion, turning, or other dynamics (see Fig. 2). Experi-

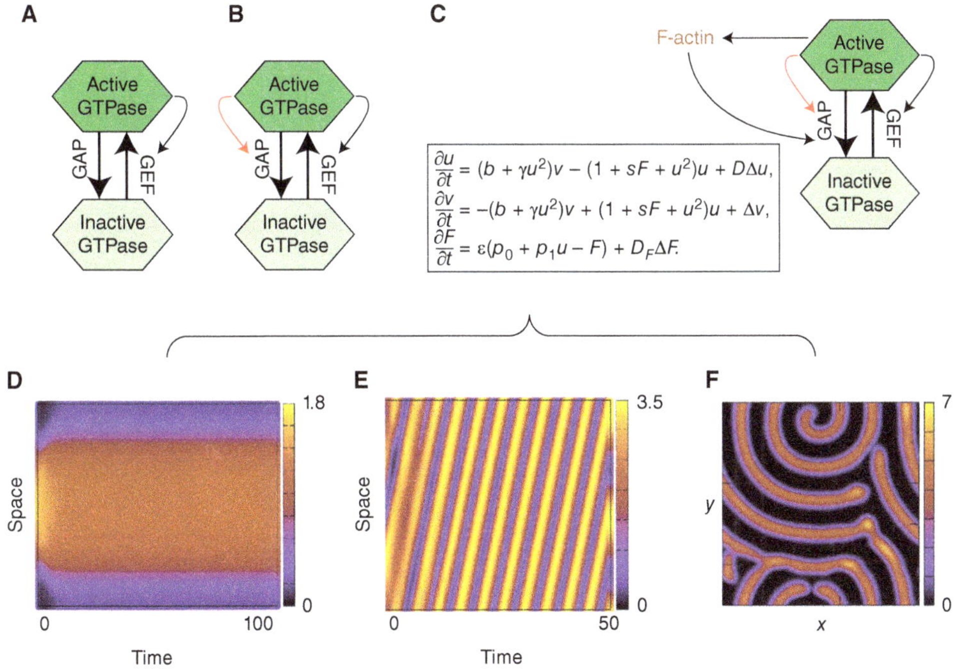

Figure 1. (*Top*) Schematic diagrams for a reaction–diffusion model of GTPase with and without interactions with filamentous actin (F-actin). (Equations: model reaction–diffusion equations for the active and inactive GTPase (u, v) and F-actin, F as in the partial differential equation [PDE] bifurcation analysis by Hughes et al. [2024]). (*Bottom*) Some of the polar and wave dynamics predicted by such models. (*A*,*B*) Two variants of the "wave-pinning" model (Mori et al. 2008) that produce polar patterns. Positive feedback (activation by GEFs) and negative feedback (inactivation by GAPS, red arrow in *B*) are shown. (*C*) A model with negative feedback from F-actin that produces more exotic patterns of the active GTPase, as shown in the kymographs (*D*,*E*), and the 2D heatmap in *F*. (*D*) A wave-pinning pattern in 1D that represents robust cell polarity—a static region of active GTPase creates the cell "front" (*E*) traveling waves of active GTPase in 1D (*F*) spiral waves in 2D (*E*). See Figure 2 for corresponding 2D cell shape dynamics based on the patterns (*C*) and (*D*) along the edge of a "motile cell" (see Supplemental Material for more information). (Reprinted from Hughes et al. 2024 under the CC BY 4.0 license (*top*); reprinted from https://identifiers.org/morpheus/M2071, Morpheus.Lab @ TUD Dresden University of Technology, under the CC BY 4.0 license (*bottom*).

ments have led to the discovery of details of the feedback (e.g., Bement et al. 2024), and this was followed by physical and mathematical modeling (Goryachev et al. 2016). A recent paper attempts to unify some of the proposed mechanisms (Beta et al. 2023). Bifurcation analysis to map out the universe of possible behaviors is ongoing (Liu et al. 2021; Buttenschön and Edelstein-Keshet 2022; Hughes et al. 2024). Some of this is largely of mathematical interest. This kind of treatment also sheds light on how cells can switch between one mode of behavior and another, for example, polarized directed migration, and random turning and searching.

Understanding such PDE models (beyond repetitive simulations and parameter sweeps) is technically challenging. Full PDE bifurcation analysis is needed to determine behavior for broad parameter regimes and values that mark critical transitions (bifurcation points), but this is much more complicated for PDEs than for time-dependent ordinary differential equations (ODEs). PDEs have additional symmetries that require more constraints on the problem (Beyn and Thümmler 2007). Numerically, PDE systems require careful spatial discretization and appropriate approximation of derivatives. These also result in much longer computational run-

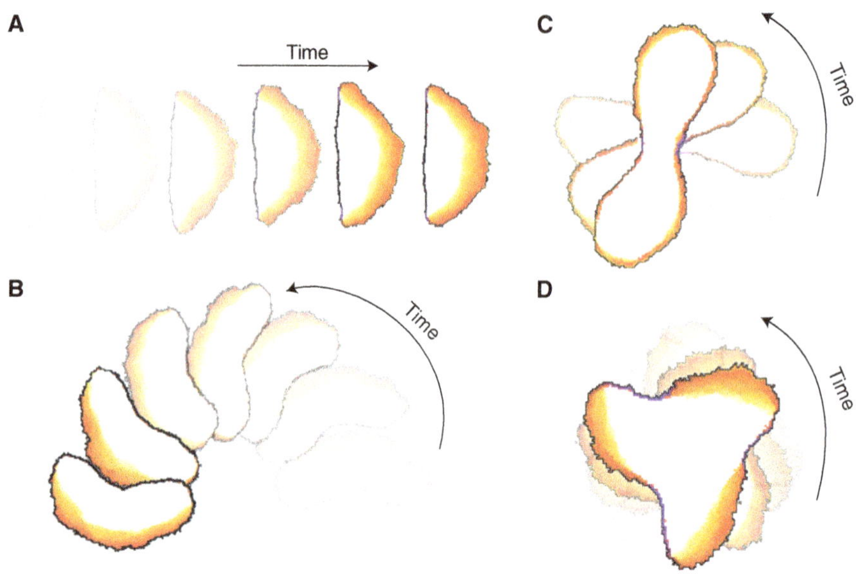

Figure 2. Shapes and motion of 2D "motile cell" corresponding to intracellular dynamics as described in Figure 1C. Implementing the partial differential equation (PDE) model of Figure 1C along the edge of a 2D cellular Potts model (CPM) cell (in Morpheus) together with protrusion (due to F-actin) results in distinct cell motility behaviors. (*A*) A polar cell results from a static zone of Rac activity, corresponding to the model and parameter values used for the kymograph in Figure 1D. (*B*) For the same model with parameters as in Figure 1E, a single zone of active Rac moves around the cell perimeter, causing the protrusion to shift clockwise. This makes the cell turn and move a long a circular trajectory. (*C,D*) If the traveling wave has two or more "lobes," multiple protrusions circulate around the perimeter. This cell does not move directionally: it appears to "ruffle" as lamellipodial waves circumnavigate its edge. (*B–D*) Traveling waves of various frequencies moving along the edge of the cell, corresponding to the same model, with parameter values that produced Figure 1D. Regions of high F-actin are assumed to create protrusion in the CPM cell (see Supplemental Material for more information). (Figure reprinted from https://identifiers.org/morpheus/M2072, Morpheus.Lab @ TUD Dresden University of Technology, under the CC BY 4.0 license.)

time compared to ODE systems. Local perturbation analysis (LPA) (Holmes et al. 2015) is a shortcut tool for detecting bifurcations of homogeneous steady states, but no information is gained on solution behavior away from the bifurcation point.

SINGLE-CELL MIGRATION

For migration, a eukaryotic cell needs a balance between protrusive (traction) forces, adhesion to its surrounding scaffold (substrate or extracellular matrix [ECM]), and contractile forces to break adhesion bonds and detach its rear (Cramer 2010). Inside the cell, the front assembly of actin and rear contraction of actomyosin have emerged as a dominant paradigm in the migration of keratocytes and neutrophils (see, e.g., Allen et al. 2020). But other modes of migration have been quantified, including blebbing (rupture of the cell cortex) (see Paluch and Raz 2013). Stimuli that lead to migration include chemotaxis, topotaxis, durotaxis, haptotaxis, and galvanotaxis (that are directed up gradients of a chemical, environmental topography, stiffness, adhesion, or electrical potential) are all well studied (Allen et al. 2013; Ricoult et al. 2015; Park et al. 2018; Sunyer and Trepat 2020). Most of these influences are sensed by cell-surface receptors and ultimately funneled to the actin-regulatory GTPase circuits.

The various components involved in cell motility have been studied experimentally. These include the turnover and flow of actin

(Pollard and Borisy 2003). Originally, cell observations in vitro were relegated to motion on flat surfaces. The importance of interactions between cells and the ECM (a fibrous 3D meshwork produced, degraded, and remodeled by cells) has been recognized. This resulted in experiments that aim to replicate aspects of 3D cell motility (e.g., Park et al. 2017). The investigations were extended to mechanical forces due to actomyosin (Keren et al. 2008), adhesion dynamics (Kaverina et al. 2022), and their interplay with biophysical effects such as substrate stiffness (Lo et al. 2000), 2D versus 3D motion (Pawluchin and Galic 2022), motion through the ECM (Friedl and Bröcker 2000), and curvature-dependent motion along nanofibers (Mukherjee et al. 2019), among many other topics. In a few cases, a link is made to the regulatory GTPases (e.g., Wittmann and Waterman-Storer 2001; Ridley 2015; Byrne et al. 2016).

We have seen a proliferation of inventive ways to deconstruct cell decisions. Experimental novelties in recent years included complex stimuli such as conflicting gradients (Shi et al. 2013; Lin et al. 2015), mazes (Um et al. 2019), adhesive micropatterns (Brock et al. 2003), and many more. Of recent developments, optogenetics has proven to be a remarkable tool for presenting accurate, carefully guided cues from which cell responses can be quantified (e.g., Valon et al. 2017; Hadjitheodorou et al. 2021).

The mathematical modeling approach to cell motility is reviewed in Mogilner (2009, 2019). Models and simulations of single-cell motility have been carried out in 1D, 2D, and 3D spatial coordinates (for review, see Buttenschön and Edelstein-Keshet 2022). The simplicity of 1D spatial models means that they can often be formulated in terms of PDEs for the contents of a cell, with boundary conditions for its opposed edges, or a periodic domain along the cell perimeter, with or without stochasticity (e.g., Xiong et al. 2010). Applied mathematics and scientific computing has developed many tools for understanding problems formulated in this way. Computational platforms such as Morpheus (Starruß et al. 2014), reviewed below, are starting to make PDE-based cell motility simulations available. For example, the PDE model of Figure 1C, simulated along the periodic edge of a "cell" results in directed motility, ruffling, and/or protrusions, as shown in Figure 2. While this is a great oversimplification of cell geometry, simple models can provide insights into how various effects (chemical, mechanical, size, noise) combine to generate cell protrusion, contraction, polarity, gradient sensing, directed migration, and reversal. Simulations in 2D and 3D can then be more directly targeted to exploring interesting details.

As far as a link from intracellular states to the cell's migration, one challenge has been to allow for cell shape deformation while solving equations for the reacting and diffusing components, for actin, and for mechanical forces. The methods for capturing cell shapes in 2D and 3D include phase field (Camley et al. 2013), cellular Potts model (CPM) (Marée et al. 2006; Rens and Edelstein-Keshet 2021), finite-element method (FEM) (Rubinstein et al. 2005; Cusseddu et al. 2019), and others. Currently, most simulations are custom built by a given investigator or laboratory and designed to answer specific questions studied by that laboratory. The development of software code requires expertise, troubleshooting, and fine-tuning. Typically, the code migrates with expert postdocs that developed it, having to be recreated from scratch when they depart. Unfortunately, there is no multipurpose, easily shareable suite of consensus software for cell motility modeling that is widely available to the community and adopted by many laboratories. Hence, aside from having to reinvent the wheel, the reproducibility of published results is still relatively poor, even when the code in a paper is published; documentation of decisions about how the models are built is rarely sufficient to reproduce the published results.

Several challenges are still to be addressed in modeling single-cell behavior. A partial list includes (1) understanding the essential core components and their interconnectivity, and fine-tuning due to a host of effectors and feedbacks; (2) refining the appropriate description of cell material properties, suitable for formulating the classical active-matter and solid mechanics modeling frameworks; (3) refining the model description of diffusion inside the crowded cell interior and on its membrane, as well as the association–dissociation of molecules with the cell membrane ("bulk-surface" models); (4) representing the dynamically deforming cell shape in 3D and its effect on internal reaction-diffusion dynamics; and (5) creating a curated consensus platform for cell shape dynamics that is versatile enough to be used for multiple distinct investigations, while preserving the core cell characteristics.

SINGLE-CELL DECISIONS

To bridge to the next step of multicellular systems, it is helpful to explore how cells make decisions and change their migratory pattern. For example, how does a polarized cell react to a reversal or side stimulus that impinges on the polarity mechanism (Hadjitheodorou et al. 2021)? Until recently, this topic has had relatively little attention. Experimentally, cells migrating in one direction have been restimulated to reverse their polarity by chemical (Lin et al. 2015) or optogenetic (Hadjitheodorou et al. 2021; Town and Weiner 2023) methods, but a full systematic exploration with various cell types and conditions is yet to be done. Mathematically, the problem of cell repolarization was investigated by PDE bifurcation analysis (Buttenschön and Edelstein-Keshet 2022) to show how the intensity and width of an "optogenetic stimulus" could repolarize a minimal GTPase circuit. Theoretically, the dynamics of cell turning was shown to depend on the interaction of diffusible regulators with the curvature of a deforming cell's edge (Marée et al. 2012; Camley et al. 2017).

DYNAMICS OF CELL PAIRS

To understand the collective behavior of cells in tissues, we need to understand how cells influence one another, and how their internal dynamics affect their interactions with neighboring cells. When a cell encounters another, will it synchronize and adapt its polarity to match? Will it retract due to contact inhibition of locomotion (CIL)? Or will it lose polarity? Such questions were investigated in experiments (Desai et al. 2013; Jain et al. 2020) for cell–cell collisions and cell trains. More broadly, the idea of exploring cell pairs, cell trains, and small cell groups as intermediate steps to understanding collective behavior was eloquently proposed in De Pascalis and Etienne-Manneville (2017).

It is clear that addressing the above questions is best accomplished by targeted and controlled experiments on cell pairs that amass sufficient data to allow for statistically significant measurement of effective forces. A recent paper by physicists, Brückner et al. (2021), does precisely this, with an elegant setup for observing many ($N = 90$–150) collisions of cell pairs confined to micropatterns. Machine learning (ML) was used to "learn" and quantify attraction, repulsion, and frictional interactions, demonstrating differences between cancerous and normal cells. Such experiments hold great promise for future multicellular understanding and should be extended to a variety of cell types and conditions. While cells probably interact with multiple neighbors in nontrivial ways, understanding the simplified pairwise interactions can be illuminating.

The in vivo migration of cell pairs was studied experimentally in the tunicate Ciona and modeled using CPM simulations in 2D and 3D (Bernadskaya et al. 2021). The authors could thereby dissect the influences of proposed cell–cell versus cell–substrate adhesion, internal pressure, protrusion/retraction forces, and surface tension. They showed that the shape of the cell pair is consistent with the leading cell being mainly protrusive, and the follower cell being mainly contractile.

Simulations of cell pairs were carried out by Kulawiak et al. (2016) using the phase-field method for cell shape, a minimal cell polarity module (modified wave-pinning), and a mechanism for CIL based on the production of a local inhibitor. This minimal set of ingredients could already account for static and motile cell chains, CIL, and cells walking past one another. The advantage of such simple models is that they can be queried to investigate how cell properties (such as strength of propulsion, adhesion) and environment (confinement region size), in combination, affect the outcome.

COLLECTIVE BEHAVIOR OF MULTIPLE CELLS

Collective cell migration became a hot topic when it was observed that tumors can spread through invasion by large cell groups, clusters, or even tissue protrusion (Friedl and Gilmour 2009). A central question at the collective cell migration level is what new features (emergent properties) arise at this scale, and how do they depend on the single-cell repertoires and on the cell–cell interactions (Bi et al. 2015). At this level, many new events enter the picture, including cell heterogeneity, differential cell–cell adhesion, cell division and death, directed cell migration, and mechanical pulling by myriad cells. The subject is rich in possibilities, only a few of which have been explored so far.

On the experimental side, observing collective behavior in vivo is challenged by tissue opaqueness, difficulties of observing the tissue at both the cell scale (to resolve and track individual cells) and the macroscopic scale (to see the emergent behavior). For this reason, in vitro experiments with cell cultures, primary stem cells, tumor spheroids, and other groups of cells have been an important focus so far. Here, one can control the geometry, the initial state, the environment, and specific perturbations to probe the response of the system. In some cases, this is carried out with cell lines (normal vs. cancerous), with transformed cells (synthetic biology), or with primary cells (human pluripotent stem cells [hPSCs]). Some of the central questions driving the research are (1) what combinations of cell motility, division, and death account for observed growth and/or spread of a tissue (e.g., in wound healing or tumor growth), (2) what are minimal requirements that could account for tissue self-organization into specific layers or structures of specialized cells, and (3) how do cell clusters stay together, and how do they manage to collectively navigate to a target site? A combination of experiments and models have addressed such questions.

Models that link internal cell states to the interactions of cells in a tissue and forces between those cells have included minimal models (e.g., one or two regulators and cell size). An impressive example of the effect of extracellular signal-regulated kinase (ERK) signaling on cell-surface tension and cell size is the mechanochemical work in Boocock et al. (2021). The authors quantified ERK activity in a Madin–Darby Canine Kidney (MDCK) cell monolayer in 2D and showed that waves of activity transmitted in the epithelium could be explained by a minimal (three variable) continuum model. The model depicts feedback between ERK and local tissue stretching. This beautiful experiment-modeling work has a similar flavor, but goes well beyond the purely theoretical model by Zmurchok et al. (2018).

The problem of macroscopic spread has been addressed by continuum PDE models for cell population growth under a variety of assumptions. At the most basic level, cell proliferation is described by density-dependent (logistic) growth, coupled to random cell motility (essentially, a diffusion operator) in the classic Fisher equation (Simpson et al. 2007; Nardini and Bortz 2018). The attractive aspect of continuum approaches is the availability of PDE tools for characterizing solutions (traveling waves, steady-state distributions, fingering, etc.) and regimes of behavior (using stability and bifurcation analyses) (see Stepien and Swigon 2014), for example. PDEs can be derived from first principles based on cell-level assumptions (e.g., Othmer and Hillen 2000). Furthermore, tools for parameter reduction based on dimensional analysis and efficient methods for parameter estimation from experimental data are available. A disadvantage is that the level of detail that can be captured is limited, there is a tendency to discount cell heterogeneity, and the interpretation of parameters is not typically directly related to underlying molecular mechanisms inside the cells.

At the opposite extreme are computational models for simulating very large and possibly heterogenous cell populations with internal cell states and external nutrients, stimuli, or drugs that affect cellular growth, division, apoptosis, and so forth. These are predominately studied by simulations. Examples of this type of software include PhysiCell (Ghaffarizadeh et al. 2018) and its extensions, described briefly below. At the same time, there remain challenging aspects of collective cell migration that have not been fully addressed. One of these is the long-range cell guidance and cell–cell interactions mediated by the ECM. Cells migrate along and transmit long-ranged pulling forces via the ECM (Crossley et al. 2024).

Collective cell simulations have been designed recently to encompass multiscale models, where internal cell states are implemented (generally by bulk ODEs or Boolean regulatory networks, rather than the more costly spatially explicit PDEs for the internal cell states) (see, for example, Scianna and Preziosi 2012). The internal states then affect cell shape, division, death, cell–cell adhesion, migration speed, and other aspects. Out of many tools that facilitate simulations of such behaviors, a personal favorite of this group of authors is Morpheus, described in greater detail in the following section. Here, we highlight case studies of such multiscale collective cell behavior.

Example 1. Neural Crest Cell Swarms

Merchant et al. (2018) modeled neural crest cell (NCC) migration in a confined population of between 2 and 50 nonadherent cells. This paper employed a cell-edge-based chemical system (a Rac-Rho GTPase system diffusing on the periodic cell perimeter), with assumptions for cell–cell coattraction and CIL to show a link between internal regulatory dynamics and collective cell behavior. The model, simulated in custom-built code (Merchant et al. 2018), demonstrated that an additional assumption was needed, the persistence of cell polarity, to obtain robust collective migration. NCC migration has also been modeled by agent-based (ABM), PDEs, and nonlocal PDE models (for review, see Giniūnaitė et al. 2020).

Example 2. Self-Organizing Synthetic Cell Aggregates

Synthetic biology has provided useful experimental examples of self-organization in vitro. Toda et al. (2018) engineered cell lines in which Delta-ligand—Notch-receptor association of neighboring cells would activate genes for E-cadherin, an adhesion molecule that makes cells stick together. Suitable fine-tuning of the in vitro synthetic gene circuitry resulted in symmetry breaking, with cells differentiating into two states, "sender" and "receiver." The former has Delta dominating, and the latter has Notch dominating. Hence, by virtue of the gene circuit, the system is bistable (i.e., has two distinct eventual fates). The cells are driven to one of two states as a result of cell–cell interactions.

Delta–Notch signaling is well known and a number of models for this system are available in the form of ODEs with ABM patterning in 1D and 2D "tissue" (Collier et al. 1996), or as sets of ODEs (Boareto et al. 2015) with bifurcation analysis. By combining the known cell gene dynamics within multiscale simulations of interacting cells, Mulberry and Edelstein-Keshet (2020) recapitulate in silico the in vitro observations of Toda et al. (2018) using CPM cells in Morpheus. Starting from either two cell types (sender and receiver) or a single-cell type, the aggregation and contact of cells induce Notch activity. Neighboring cells exert lateral inhibition and there is a spontaneous symmetry breaking that results in two dynamic cell fates. The cell's genetic state dictates its adhesion to other cells, and relative adhesion results in local cell sorting that shuffles neighbors, breaking and making new contacts. This, in turn, engages Delta–Notch binding on new partners, and causes changes in cell states. As shown in Figure 3, the aggregate develops distinct cell fates—with the more adhesive (E-cad cells, shown in green) engulfed by the less adhesive (red) cells.

It is interesting to note that the Delta–Notch genetic circuit has the inherent property of bistability. Nevertheless, the capacity of this circuit to form a layered structure emerges not only from its bistable circuit, but also from the spatially distributed cell–cell interactions. Cells that converge to a position inside the core have to be more adhesive to one another than to those enveloping them, and even less prone to contact with the surrounding medium. At the same time, cell sorting acting on its own is notoriously slow, and tends to get trapped in various undesirable configurations. Hence, the combination of cell state influencing affinity to

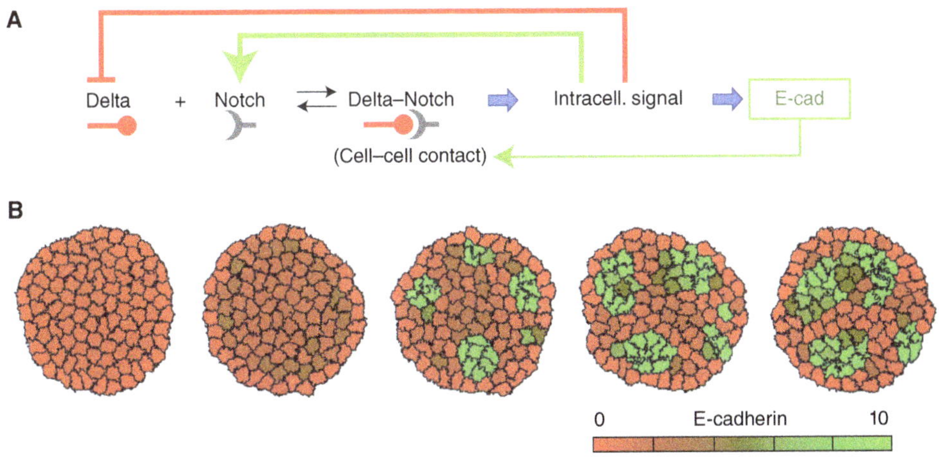

Figure 3. A multiscale model of synthetic Delta–Notch experiments by Toda et al. (2018), created in the Morpheus cellular Potts model (CPM) platform. (*A*) Schematic diagram of the signaling system (based on a model in Mulberry and Edelstein-Keshet 2020). (*B*) Starting with a single-cell type, a cluster of 100 cells develops internal self-organization, showing a time sequence (*left* to *right*) at $t = 10, 30, 60, 120, 180$ Monte Carlo time steps (MCS). (Panel reprinted, with permission, from Mulberry and Edelstein-Keshet 2020.) Delta–Notch-receptor binding on neighboring cells activates internal signaling that amplifies Notch and suppresses Delta receptor expression, as well as increasing E-cadherin. The latter adhesive protein enhances cell–cell contact (see Supplemental Material for more detail).

neighbors (adhesion) and cell contact influencing genetic state (Delta–Notch signaling) greatly speeds up a process of self-organization. Cell-based simulations can reveal emergent properties that are not apparent from the cell wiring on its own.

Example 3. Wound-Healing and YAP-Regulated Metastasis

Cells are sensitive to many features of their environment, including topography and mechanical cues. Park et al. (2019) observed the moving front in epithelial sheets of MDCK cells. The epithelia were tracked on both flat and nanoridge arrays (NRAs) substrates. On NRA topography, there was the enhanced activity of the mechanosensitive transcription factor YAP (Yes-associated protein), which resulted in uneven fronts, fingering (i.e., initiation and growth of multicellular protrusions from the edge of the tissue), and dissemination of ameboid cell clusters that broke away. The authors demonstrated that YAP is in a mutually inhibitory pathway with effectors of E-cadherin, and a mutually reinforcing pathway with the GTPase Rac (Fig. 4). Hence, up-regulation of YAP near the sheet front reduces cell adhesion and accelerates migration.

Mukhtar et al. (2022) adapted and implemented an ODE model of Park et al. (2019) in a CPM Morpheus simulation. The simple YAP-Rac-E cadherin ODE model that governs the state (and hence behavior) of each cell is bistable, and stimulation by the edge-sensing topography is able to change the outcome from low to high YAP activity. The emergence of fingering and dissemination of clusters then follows once adhesion and motility are in suitable state-dependent ranges. The model predicts an abrupt switch in YAP activity at some distance away from the front edge. Figure 4 demonstrates the time sequence and breakage of cell clusters, as well as the speed of migrating cells.

Example 4. Early Mammalian Embryogenesis

A multiscale model for mammalian embryo development (from zygote to 128 cells) is showcased in Figure 5. A crucial aspect of this devel-

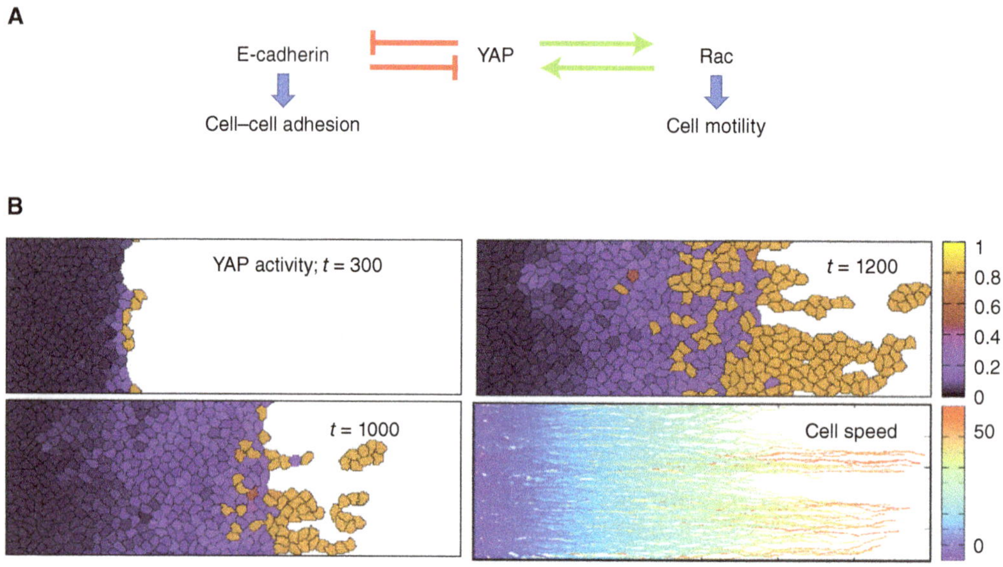

Figure 4. A multiscale model for Yes-associated protein (YAP) signaling in an epithelial sheet, demonstrating fingering behavior. (*A*) A schematic diagram of the simplified signaling circuit proposed in Park et al. (2018) and implemented in Mukhtar et al. (2022). (*B*) A time sequence of the sheet edge (at $t = 300$, 1000, 1200 Monte Carlo steps) with YAP activity indicated on a heatmap. YAP is in a mutual negative (positive) circuit with E-cadherin (Rac). Hence, active YAP loosens cell adhesion and accelerates cell motility. The tissue edge forms multicellular protrusions that resemble "fingers," some of which break away (see Supplemental Material for more information). (*B* reprinted from https://identifiers.org/morpheus/M7121, Morpheus.Lab @ TUD Dresden University of Technology, under the CC BY 4.0 license.)

opment is the self-organization of differentiated cells into a proper configuration. This model represents two stages: (1) First, from the zygote state, cell–cell contact signaling drives differentiation into two cell types, trophectoderm (TE) and inner cell mass (ICM). (2) Second, diffusible FGF4 ligand secreted by ICM cells into the medium regulates the transition from ICM to epiblast (Epi) and primitive endoderm (PrE) cells.

Preexisting models (De Mot et al. 2016; Cang et al. 2021) are highly detailed, making them excellent starting points for new users to implement and explore the system. These are medium-large models. For stage 1, cell states satisfy two ODEs with eight parameters. For stage 2, there are five ODEs with 39 parameters. Due to the careful detail presented in the above papers, assembling a multiscale CPM Morpheus simulation (where each cell has dynamic internal states) proved to be feasible for an undergraduate summer research project. Even so, translating previous models into a spatial multiscale model in a new platform always poses a challenge, requiring adjustment of the spatial assumptions, framework, and parameter sets. The user-friendly GUI (graphical user interface) of Morpheus made these challenges less daunting. Results shown in Figure 5 are consistent with those of the two cited papers and with known human embryo development.

EXISTING CHALLENGES

Assembling multiscale tissue simulations can help to further our understanding of collective cell behavior. But some challenges still make this a nontrivial undertaking, even when quantitative information or models of the single cell are available. One recurring question is which aspects of cell behavior should be included in scaling up from one cell to many. It is generally required to reduce the level of detail in such

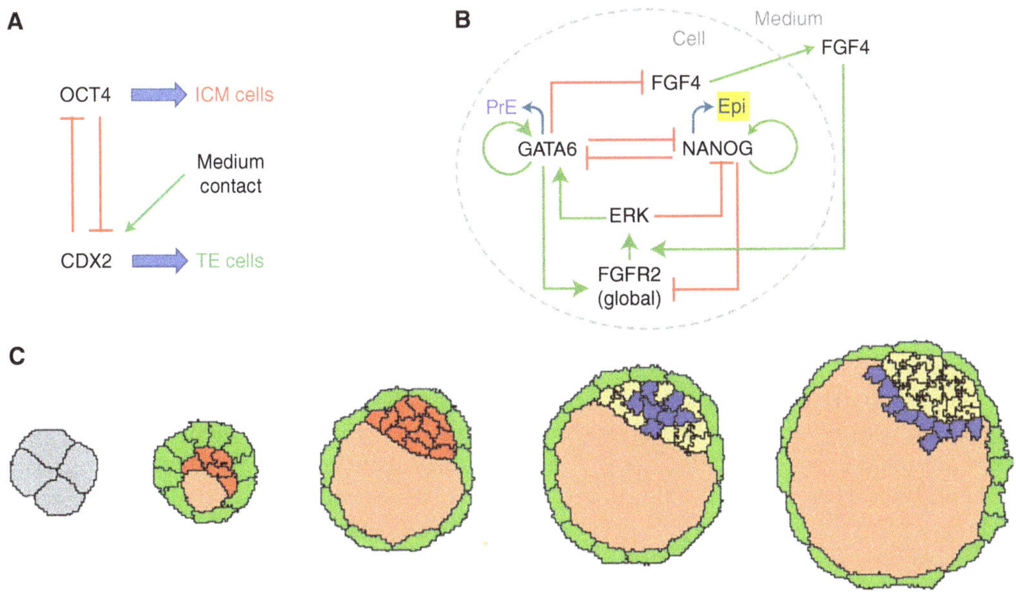

Figure 5. Stages of cell differentiation in mammalian blastocyst development. (*A*) A zygote (not shown) divides into four undifferentiated cells (gray) that divide further and differentiate. An intracellular Oct4-CDX4 transcription factor circuit regulates differentiation into trophectoderm (TE, green, exterior) and an inner cell mass (ICM, red, interior). A cell-free lumen (orange) also forms and grows. (*B*) GATA6, NANOG, and ERK signaling in ICM cells further lead to specialization into epiblast (Epi, yellow) and primitive endoderm (PrE, blue) cells. FGF4 is a diffusible ligand secreted by ICM cells that influences the differentiation of nearby cells. (*C*) Morpheus cellular Potts model (CPM) simulation time sequence (*left* to *right*, $t = 40, 210, 570, 830, 1400$ Monte Carlo time steps [MCS]) shows the dynamics of the self-organized structure. Difference in cell–cell adhesion between cell types affects the evolving embryo (see Supplemental Material for more information). (Figure based on models in Cang et al. 2021 and De Mot et al. 2016; see these publications for details about the original models.)

scale-ups to keep a coarse-grained model tractable, and to allow the cell–cell interactions to be explored efficiently. A second issue is how to build on current knowledge (rather than reinventing the wheel). Too often, existing models in scientific papers suffer from missing information and undocumented assumptions buried in proprietary code. Free software platforms reviewed below are a great boon to the community and considerably increase the efficiency of research. But what is still needed is a curated, updated archive of biomodules where model equations, values of parameters, initial conditions, boundary conditions, and other specific details are (1) documented and explained, and (2) available not only in print (or buried in code) but also in text files that are easily imported and used. Overall challenges for modeling multicellular tissues are described in the recent work of Fletcher and Osborne (2022).

SIMULATION PLATFORMS

Depending on the modeling of interest, a variety of software solutions exist. To simulate ODE models and perform necessary analyses (linear stability analysis, bifurcation analysis, sensitivity analysis, model reduction, etc.), a wide variety of tools have been developed. A longstanding free package, XPPAUTO (sites.pitt.edu/~phase/bard/bardware/xpp/xpp.html) (Ermentrout and Mahajan 2003) has been popular in the math-bio community since the late 1980s (see Klipp et al. 2016 for a comprehensive list of about 57 open-source/commercial software solutions and their capabilities). Also, the blog Stochastic

Lifestyle (www.stochasticlifestyle.com/comparison-differential-equation-solver-suites-matlab-r-julia-python-c-fortran) provides a useful comparison of available software suites categorized by programming language (Rackauckas 2017).

Due to the complexities of PDE models, there are fewer integrated software solutions. The web-based VisualPDE (visualpde.com) (Walker et al. 2023) is a great recent tool to simulate and visualize fairly complex PDE systems. It offers an interesting and growing set of examples, along with a complete explanation of each model, serving as both a great learning tool and a useful starting point for adapting existing models. Limitations include nonscalability for more complex simulations, lack of scripting, and restricted available domain shapes. Other options to simulate and visualize PDE models include FeNiCsX (jsdokken.com/dolfinx-tutorial), Matlab "PDE modeler" (www.mathworks.com/help/pde/ug/pdemodeler-app.html), freeFEM (Hecht 2012), FlexPDE (www.pdesolutions.com), and COMSOL Multiphysics (www.comsol.com). Easy-to-use libraries based on finite difference methods, such as PyPDE (Zwicker 2020) and Scikit-diff (Cellier and Ruyer-Quil 2019) often come in handy for simulations of intermediate complexity. Unlike software tools for simulating PDEs, bifurcation analysis of PDE models is still not "out of the box." Often, more dedicated libraries are needed, such as BifurcationKit.jl (Rankin et al. 2014), MATCONT (Dhooge et al. 2003), and pde2path (pde2path.uol.de) (Uecker et al. 2014; see Blyth et al. 2020 for a more detailed discussion of such software tools, along with a tutorial).

For discrete ABM, Abar et al. (2017) provide an extensive list of available software. Hybrid modeling that integrates discrete and continuous models harnesses the best of both worlds. This has become a popular approach recently and has been very successful in mathematical biology research. Various tools such as Morpheus (Box 1) (morpheus.gitlab.io) (Starruß et al. 2014)—demonstrated in our examples above—as well as CHASTE (chaste.github.io) (Mirams et al. 2013), CompuCell3D (compucell3d.org) (Swat et al. 2012), Artistoo (Wortel and Textor 2021), and CellularPotts.jl (Gregg and Benos 2024) exist that integrate the cellular Potts models with continuous ODE/PDE models. Each comes with its limitations and strengths. Comparing these software solutions can be done by considering many different aspects, such as scalability, performance, user-friendliness, GUI support for rapid model development, and usability with scripting to harness its full power (for example, see Fig. 2; Fletcher and Osborne 2022 for such a comparison).

Although the working principle is the same for various CPM software, differences in implementation and structure can make it challenging to reproduce results obtained with one software using another software package, even with the "same set of parameters." Nonetheless, slight modification generally leads to the same qualitative results (see Fig. 6). To put these software tools on the spectrum of ease of use for pedagogical purposes, Artistoo has demonstrated significant power. Developed purely with JavaScript, it runs natively in browsers, making it accessible for students. Additionally, the detailed repository of predesigned models serves

BOX 1. MORPHEUS, A USER-FRIENDLY MULTISCALE SIMULATION PLATFORM

Morpheus (Starruß et al. 2014) is an open-source freely available software platform (downloadable from morpheus.gitlab.io) for simulating models based on ordinary and partial differential equations (of the reaction–diffusion type), single and multiple cell shapes (via the cellular Potts formalism), directed cell motility, chemotaxis, division, death, and differentiation, cell–cell interactions, adhesion, and dynamic internal cell states. The software comes with many built-in examples that can be run and adapted by the user directly on a convenient graphical user interface (GUI). Each simulation run is automatically documented (xml file) for reproducibility. Morpheus is relatively easy to learn and serves as a great research and teaching tool.

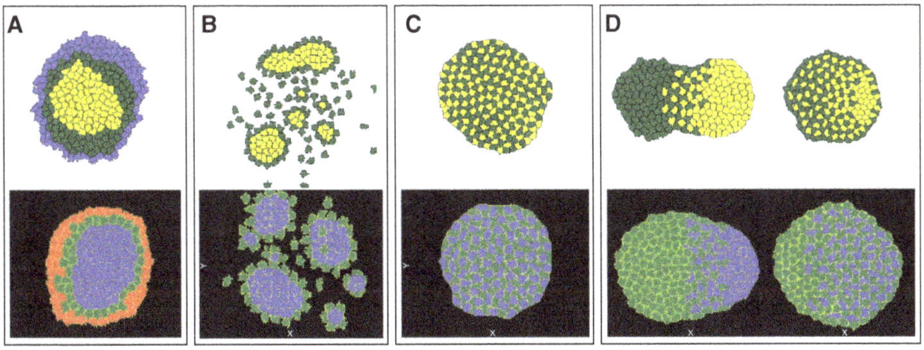

Figure 6. A comparison of cellular Potts model (CPM) results in Morpheus (*top*) and CompuCell3D (*bottom*) using four "benchmark" cell-sorting simulations in 2D. (*A*) Three-layer structure with core, intermediate, and peripheral cell types. Adhesion energy "cost" (J_{ij}) for cell–cell contact between cells of types (i,j) were in ranges $J_{cm} > J_{ci}$, $J_{im} > J_{ip}$, J_{pm}, and $J_{pi} < J_{pc}$, where c, i, and p represent core, inner, and peripheral cells and m is medium. (*B*) Fragmentation: Cell type 1 forms clusters due to low cell–cell contact energy J_{11}, while type 2 forms a single layer around the clusters due to high cell-medium contact energy, $J_{1\mathrm{m}}$. (*C*) Checkerboard pattern: Two cell types have $J_{1\mathrm{m}} \sim J_{2\mathrm{m}}$ and $J_{12} < J_{11}$, J_{22}. (*D*) Cluster fusion: Two clusters of cells merge and create a checkerboard pattern (see Supplemental Material for more information).

as an excellent starting point. However, ease of use and learning often come at the cost of nonscalability for more complex tasks and scripts.

Compared to the software suites that support ODE/PDE models natively, CPM/hybrid software is often limited in its capability to handle the continuous part of the model (i.e., ODEs, PDEs). Essential tools like sensitivity analysis, steady-state solution analysis, bifurcation analyses, numerical schemes to support stiff models, and support for more complicated PDE models and deforming domains are either not implemented or are very basic and need expansion.

The trend in modern software ecosystems is integrated software suites that are extendable through various plugins. Some software solutions like CHASTE use this paradigm. For instance, the microvessel CHASTE plugin (Grogan et al. 2017) is designed to be used for angiogenesis models. This framework can be promising for more scalable software solutions where various modules can be developed as plugins by a dedicated user.

PhysiCell, an open-source software (Ghaffarizadeh et al. 2018) simulates cell growth dynamics, migration, and interaction with secreted chemicals in 3D. This software has become a popular tool for tumor modeling. An extension (PhysiBoss; see, e.g., Ponce-de-Leon et al. 2023) implements internal cell regulatory networks and preserves the 3D visualization capability of thousands of spherical cell agents. A benefit of open-source platforms like PhysiCell is that distinct groups can help to extend the capabilities of the system. For example, PhysiCOOL is an extension that facilitates parameter space exploration (Gonçalves et al. 2023).

DISCUSSION

Huge progress has been made over the last decade in our ability to experimentally observe, measure, model, and simulate cell biology systems, from the intracellular to the multicellular levels. The interconnections between micro-, meso-, and macroscopic levels make this into a rich and diverse field of study.

Some key challenges remain to be addressed. How should we each document and keep track of many decisions made while modeling to make our models shareable and reproducible? The description of models in most scientific papers (see Supplemental Information) is still usually deficient, as any student attempting to learn from previous work finds out rapidly. This means that while the cell biology field is reasonably collegial in sharing cell types, media, or probes, the modeling of cell biology lags behind this generous spirit. This deficiency means that it is hard to build on work by others. It would be especially useful to agree on and/or standardize some basic modeling guidelines and icons or nomenclature to account for those, and to document the stepwise process that went into constructing a complex model, not just the final version. We often learn more from the failure of early variants that could not account for some observed behavior than from the full complexity that could.

We are yet to converge on the best ways to describe the material properties of cells, links between stimuli and response, and how cells make decisions in the context of competing stimuli. A unified model that allows us to manipulate cell properties and include or exclude aspects of interest could be very influential. Another challenge is to decipher what aspects of the implementation (CPM, vertex-based, ABM, etc.) affect the types of results obtained in a given simulation. Benchmark comparisons similar to those of Osborne et al. (2017) would be particularly useful, not only within a given platform (CHASTE in the latter case), but across computational platforms. This would help to point to software that is particularly useful for a given context—for example, vertex-based collective cell simulations are great for epithelia, but perhaps less ideal for describing breakage of clusters and metastasis than CPM or cell-centered models.

Like the "human atlas" for tissue and for biomolecular interactions (Börner et al. 2022; Jain et al. 2023), tissue banks, and repositories of seeds, it would be highly useful to have a community-wide resource like a searchable archive of models that run across software programs. Ideally, entries would account for cell-type-dependent and environment-dependent parameters of a given model. One would be able to "upload" a model for the migration of neutrophils and compare it against *Dictyostelium* or other cell types merely with a few mouse clicks. We have much existing knowledge to draw upon, but not yet an organized way to easily share it.

More generally, the field of cell motility modeling would benefit from a more closely collaborative research community, akin to communities that arose around model organisms (Dietrich et al. 2014; Ankeny and Leonelli 2016). A process of organic self-selection would then lead to overall standardization of the types of information, data, computational methods, and metrics commonly (if not universally) adopted. The success of such a community depends on willingness to organize joint conferences, train young scientists to use and contribute to common resources, and develop databases of models and software. It also depends on stakeholder buy-in that emphasizes and rewards these collaborative needs at funding agencies (Dietrich et al. 2014; Ankeny and Leonelli 2016).

The interrelationships of distinct model types, and formal ways of deriving the PDEs and integro-PDEs from underlying cell behaviors are an important way of extending mathematical analysis to a given setting. But we still need to work on linking cell-based simulations to stochastic differential equations (SDEs) and/or PDEs in a rational way. Simulations in and of themselves can fail to address important questions: Is the simulated behavior "generic" or a coincidence in some specific corner of parameter space? What other behaviors are possible in other corners of that space? How does one regime (say "normal") behavior transition into pathological dynamics (say, "metastatic"), or how do cells migrating in one direction decide to turn or reverse? Such questions are not easy to address by parameter sweeps or trial and error in simulations that have numerous free parameters (for additional information, see Supplemental Material).

ACKNOWLEDGMENTS

L.E.-K. is supported by a Natural Sciences and Engineering Research Council (NSERC, Canada) Discovery Grant that also pays part or all of the RA support for J.M.H., J.A., and A.F.-P. J.M.H. has been supported by an NSERC CGS-D Scholarship. A.F.-P. was also partly supported by Networking Health Ltd. research support to Prof. Arman Rahmim (UBC). We thank Lutz Brusch and Jörn Starruß, the developers of Morpheus, for making this valuable resource freely available and for much help in learning how to use it. We thank the Pacific Institute for Mathematical Sciences for hosting our meetings to develop and discuss this review.

REFERENCES

Abar S, Theodoropoulos GK, Lemarinier P, O'Hare GM. 2017. Agent based modelling and simulation tools: a review of the state-of-art software. *Comput Sci Rev* **24:** 13–33. doi:10.1016/j.cosrev.2017.03.001

Alert R, Trepat X. 2020. Physical models of collective cell migration. *Annu Rev Condens Matter Phys* **11:** 77–101. doi:10.1146/annurev-conmatphys-031218-013516

Allard J, Mogilner A. 2013. Traveling waves in actin dynamics and cell motility. *Curr Opin Cell Biol* **25:** 107–115. doi:10.1016/j.ceb.2012.08.012

Allen GM, Mogilner A, Theriot JA. 2013. Electrophoresis of cellular membrane components creates the directional cue guiding keratocyte galvanotaxis. *Curr Biol* **23:** 560–568. doi:10.1016/j.cub.2013.02.047

Allen GM, Lee KC, Barnhart EL, Tsuchida MA, Wilson CA, Gutierrez E, Groisman A, Theriot JA, Mogilner A. 2020. Cell mechanics at the rear act to steer the direction of cell migration. *Cell Syst* **11:** 286–299. doi:10.1016/j.cels.2020.08.008

Ankeny RA, Leonelli S. 2016. Repertoires: a post-Kuhnian perspective on scientific change and collaborative research. *Stud Hist Philos Sci* **60:** 18–28. doi:10.1016/j.shpsa.2016.08.003

Barnhart EL, Allard J, Lou SS, Theriot JA, Mogilner A. 2017. Adhesion-dependent wave generation in crawling cells. *Curr Biol* **27:** 27–38. doi:10.1016/j.cub.2016.11.011

Bement WM, Leda M, Moe AM, Kita AM, Larson ME, Golding AE, Pfeuti C, Su KC, Miller AL, Goryachev AB, et al. 2015. Activator-inhibitor coupling between Rho signaling and actin assembly make the cell cortex an excitable medium. *Nat Cell Biol* **17:** 1471–1483. doi:10.1038/ncb3251

Bement WM, Goryachev AB, Miller AL, von Dassow G. 2024. Patterning of the cell cortex by Rho GTPases. *Nat Rev Mol Cell Biol* **25:** 290–308. doi:10.1038/s41580-023-00682-z

Bernadskaya YY, Yue H, Copos C, Christiaen L, Mogilner A. 2021. Supracellular organization confers directionality and mechanical potency to migrating pairs of cardiopharyngeal progenitor cells. *eLife* **10:** e70977. doi:10.7554/eLife.70977

Beta C, Edelstein-Keshet L, Gov N, Yochelis A. 2023. From actin waves to mechanism and back: how theory aids biological understanding. *eLife* **12:** e87181. doi:10.7554/eLife.87181

Beyn WJ, Thümmler V. 2007. Phase conditions, symmetries and PDE continuation. In *Numerical continuation methods for dynamical systems* (ed. Krauskopf B, Osinga HM, Galán-Vioque J). Springer, Berlin.

Bi D, Lopez JH, Schwarz JM, Manning ML. 2015. A density-independent rigidity transition in biological tissues. *Nat Phys* **11:** 1074–1079. doi:10.1038/nphys3471

Bishop AL, Hall A. 2000. Rho GTPases and their effector proteins. *Biochem J* **348:** 241–255. doi:10.1042/bj3480241

Blyth M, Renson L, Marucci L. 2020. Tutorial of numerical continuation and bifurcation theory for systems and synthetic biology. arXiv doi:10.48550/arXiv.2008.05226

Boareto M, Jolly MK, Lu M, Onuchic JN, Clementi C, Ben-Jacob E. 2015. Jagged–Delta asymmetry in Notch signaling can give rise to a sender/receiver hybrid phenotype. *Proc Natl Acad Sci* **112:** E402–E409. doi:10.1073/pnas.1416287112

Boocock D, Hino N, Ruzickova N, Hirashima T, Hannezo E. 2021. Theory of mechanochemical patterning and optimal migration in cell monolayers. *Nat Phys* **17:** 267–274. doi:10.1038/s41567-020-01037-7

Börner K, Bueckle A, Herr BW, Cross LE, Quardokus EM, Record EG, Ju Y, Silverstein JC, Browne KM, Jain S, et al. 2022. Tissue registration and exploration user interfaces in support of a human reference atlas. *Commun Biol* **5:** 1369. doi:10.1038/s42003-022-03644-x

Brock A, Chang E, Ho CC, LeDuc P, Jiang X, Whitesides GM, Ingber DE. 2003. Geometric determinants of directional cell motility revealed using microcontact printing. *Langmuir* **19:** 1611–1617. doi:10.1021/la026394k

Brückner DB, Arlt N, Fink A, Ronceray P, Rädler JO, Broedersz CP. 2021. Learning the dynamics of cell–cell interactions in confined cell migration. *Proc Natl Acad Sci* **118:** e2016602118. doi:10.1073/pnas.2016602118

Buttenschön A, Edelstein-Keshet L. 2020. Bridging from single to collective cell migration: a review of models and links to experiments. *PLoS Comput Biol* **16:** e1008411. doi:10.1371/journal.pcbi.1008411

Buttenschön A, Edelstein-Keshet L. 2022. Cell repolarization: a bifurcation study of spatio-temporal perturbations of polar cells. *Bull Math Biol* **84:** 114. doi:10.1007/s11538-022-01053-z

Byrne KM, Monsefi N, Dawson JC, Degasperi A, Bukowski-Wills JC, Volinsky N, Dobrzyński M, Birtwistle MR, Tsyganov MA, Kiyatkin A, et al. 2016. Bistability in the Rac1, PAK, and RhoA signaling network drives actin cytoskeleton dynamics and cell motility switches. *Cell Sys* **2:** 38–48. doi:10.1016/j.cels.2016.01.003

Camley BA, Rappel WJ. 2017. Physical models of collective cell motility: from cell to tissue. *J Phys D Appl Phys* **50:** 113002. doi:10.1088/1361-6463/aa56fe

Camley BA, Zhao Y, Li B, Levine H, Rappel WJ. 2013. Periodic migration in a physical model of cells on micropatterns. *Phys Rev Lett* **111:** 158102. doi:10.1103/PhysRevLett.111.158102

Camley BA, Zhao Y, Li B, Levine H, Rappel WJ. 2017. Crawling and turning in a minimal reaction-diffusion cell motility model: coupling cell shape and biochemistry. *Phys Rev E* **95:** 012401. doi:10.1103/PhysRevE.95.012401

Cang Z, Wang Y, Wang Q, Cho KW, Holmes W, Nie Q. 2021. A multiscale model via single-cell transcriptomics reveals robust patterning mechanisms during early mammalian embryo development. *PLoS Comput Biol* **17:** e1008571. doi:10.1371/journal.pcbi.1008571

Cellier N, Ruyer-Quil C. 2019. scikit-finite-diff, a new tool for PDE solving. *J Open Source Software* **4:** 1356. doi:10.21105/joss.01356

Collier JR, Monk NA, Maini PK, Lewis JH. 1996. Pattern formation by lateral inhibition with feedback: a mathematical model of Delta-Notch intercellular signalling. *J Theor Biol* **183:** 429–446. doi:10.1006/jtbi.1996.0233

Cramer LP. 2010. Forming the cell rear first: breaking cell symmetry to trigger directed cell migration. *Nat Cell Biol* **12:** 628–632. doi:10.1038/ncb0710-628

Crossley RM, Johnson S, Tsingos E, Bell Z, Berardi M, Botticelli M, Braat QJ, Metzcar J, Ruscone M, Yin Y, et al. 2024. Modeling the extracellular matrix in cell migration and morphogenesis: a guide for the curious biologist. *Front Cell Dev Biol* **12:** 1354132. doi:10.3389/fcell.2024.1354132

Cusseddu D, Edelstein-Keshet L, Mackenzie JA, Portet S, Madzvamuse A. 2019. A coupled bulk-surface model for cell polarisation. *J Theor Biol* **481:** 119–135. doi:10.1016/j.jtbi.2018.09.008

De Mot L, Gonze D, Bessonnard S, Chazaud C, Goldbeter A, Dupont G. 2016. Cell fate specification based on tristability in the inner cell mass of mouse blastocysts. *Biophys J* **110:** 710–722. doi:10.1016/j.bpj.2015.12.020

De Pascalis C, Etienne-Manneville S. 2017. Single and collective cell migration: the mechanics of adhesions. *Mol Biol Cell* **28:** 1833–1846. doi:10.1091/mbc.e17-03-0134

Desai RA, Gopal SB, Chen S, Chen CS. 2013. Contact inhibition of locomotion probabilities drive solitary versus collective cell migration. *J R Soc Interface* **10:** 20130717. doi:10.1098/rsif.2013.0717

Devreotes P, Horwitz AR. 2015. Signaling networks that regulate cell migration. *Cold Spring Harb Perspect Biol* **7:** a005959. doi:10.1101/cshperspect.a005959

Dhooge A, Govaerts W, Kuznetsov YA. 2003. MATCONT: a MATLAB package for numerical bifurcation analysis of ODEs. *ACM Trans Math Softw* **29:** 141–164. doi:10.1145/779359.779362

Dietrich MR, Ankeny RA, Chen PM. 2014. Publication trends in model organism research. *Genetics* **198:** 787–794. doi:10.1534/genetics.114.169714

Ermentrout B, Mahajan A. 2003. Simulating, analyzing, and animating dynamical systems: a guide to XPPAUT for researchers and students. *Appl Mech Rev* **56:** B53. doi:10.1115/1.1579454

Fletcher AG, Osborne JM. 2022. Seven challenges in the multiscale modeling of multicellular tissues. *WIREs Mech Dis* **14:** e1527. doi:10.1002/wsbm.1527

Friedl P, Bröcker EB. 2000. The biology of cell locomotion within three-dimensional extracellular matrix. *Cell Mol Life Sci* **57:** 41–64. doi:10.1007/s000180050498

Friedl P, Gilmour D. 2009. Collective cell migration in morphogenesis, regeneration and cancer. *Nat Rev Mol Cell Biol* **10:** 445–457. doi:10.1038/nrm2720

Ghaffarizadeh A, Heiland R, Friedman SH, Mumenthaler SM, Macklin P. 2018. PhysiCell: an open source physics-based cell simulator for 3-D multicellular systems. *PLoS Comput Biol* **14:** e1005991. doi:10.1371/journal.pcbi.1005991

Giniūnaitė R, Baker RE, Kulesa PM, Maini PK. 2020. Modelling collective cell migration: neural crest as a model paradigm. *J Math Biol* **80:** 481–504. doi:10.1007/s00285-019-01436-2

Gonçalves IG, Hormuth DA, Prabhakaran S, Phillips CM, García-Aznar JM. 2023. PhysiCOOL: a generalized framework for model Calibration and Optimization Of modeLing projects. *GigaByte* **2023:** gigabyte77. doi:10.46471/gigabyte.77

Goryachev AB, Leda M, Miller AL, von Dassow G, Bement WM. 2016. How to make a static cytokinetic furrow out of traveling excitable waves. *Small GTPases* **7:** 65–70. doi:10.1080/21541248.2016.1168505

Gregg RW, Benos PV. 2024. Cellularpotts. jl: simulating multiscale cellular models in Julia. *Bioinformatics* **40:** btad773. doi:10.1093/bioinformatics/btad773

Grogan JA, Connor AJ, Markelc B, Muschel RJ, Maini PK, Byrne HM, Pitt-Francis JM. 2017. Microvessel chaste: an open library for spatial modeling of vascularized tissues. *Biophys J* **112:** 1767–1772. doi:10.1016/j.bpj.2017.03.036

Hadjitheodorou A, Bell GR, Ellett F, Shastry S, Irimia D, Collins SR, Theriot JA. 2021. Directional reorientation of migrating neutrophils is limited by suppression of receptor input signaling at the cell rear through myosin II activity. *Nat Commun* **12:** 6619. doi:10.1038/s41467-021-26622-z

Hecht F. 2012. New development in FreeFem++. *J Numer Math* **20:** 251–266. doi:10.1515/jnum-2012-0013

Holmes WR, Carlsson AE, Edelstein-Keshet L. 2012. Regimes of wave type patterning driven by refractory actin feedback: transition from static polarization to dynamic wave behaviour. *Phys Biol* **9:** 046005. doi:10.1088/1478-3975/9/4/046005

Holmes WR, Mata MA, Edelstein-Keshet L. 2015. Local perturbation analysis: a computational tool for biophysical reaction-diffusion models. *Biophys J* **108:** 230–236. doi:10.1016/j.bpj.2014.11.3457

Houk AR, Jilkine A, Mejean CO, Boltyanskiy R, Dufresne ER, Angenent SB, Altschuler SJ, Wu LF, Weiner OD. 2012. Membrane tension maintains cell polarity by confining signals to the leading edge during neutrophil migration. *Cell* **148:** 175–188. doi:10.1016/j.cell.2011.10.050

Hughes JM, Modai S, Edelstein-Keshet L, Yochelis A. 2024. Travelling waves and wave pinning (polarity): switching between random and directional cell motility. arXiv doi:10.48550/arXiv.2410.12213

Itoh RE, Kurokawa K, Ohba Y, Yoshizaki H, Mochizuki N, Matsuda M. 2002. Activation of rac and cdc42 video imaged by fluorescent resonance energy transfer-based single-molecule probes in the membrane of living cells. *Mol Cell Biol* **22:** 6582–6591. doi:10.1128/MCB.22.18.6582-6591.2002

Jain S, Cachoux VM, Narayana GH, de Beco S, D'alessandro J, Cellerin V, Chen T, Heuzé ML, Marcq P, Mège RM, et al. 2020. The role of single-cell mechanical behaviour and polarity in driving collective cell migration. *Nat Phys* **16:** 802–809. doi:10.1038/s41567-020-0875-z

Jain S, Pei L, Spraggins JM, Angelo M, Carson JP, Gehlenborg N, Ginty F, Gonçalves JP, Hagood JS, Hickey JW, et al. 2023. Advances and prospects for the Human BioMolecular Atlas Program (HuBMAP). *Nat Cell Biol* **25:** 1089–1100. doi:10.1038/s41556-023-01194-w

Jilkine A, Marée AF, Edelstein-Keshet L. 2007. Mathematical model for spatial segregation of the Rho-family GTPases based on inhibitory crosstalk. *Bull Math Biol* **69:** 1943–1978. doi:10.1007/s11538-007-9200-6

Jülicher F, Kruse K, Prost J, Joanny JF. 2007. Active behavior of the cytoskeleton. *Phys Rep* **449:** 3–28. doi:10.1016/j.physrep.2007.02.018

Kaverina I, Krylyshkina O, Small JV. 2022. Regulation of substrate adhesion dynamics during cell motility. *Int J Biochem Cell Biol* **34:** 746–761. doi:10.1016/s1357-2725(01)00171-6

Keren K, Pincus Z, Allen GM, Barnhart EL, Marriott G, Mogilner A, Theriot JA. 2008. Mechanism of shape determination in motile cells. *Nature* **453:** 475–480. doi:10.1038/nature06952

Klipp E, Liebermeister W, Wierling C, Kowald A. 2016. *Systems biology: a textbook.* Wiley, Hoboken, NJ.

Kulawiak DA, Camley BA, Rappel WJ. 2016. Modeling contact inhibition of locomotion of colliding cells migrating on micropatterned substrates. *PLoS Comput Biol* **12:** e1005239. doi:10.1371/journal.pcbi.1005239

Lin B, Yin T, Wu YI, Inoue T, Levchenko A. 2015. Interplay between chemotaxis and contact inhibition of locomotion determines exploratory cell migration. *Nat Commun* **6:** 6619. doi:10.1038/ncomms7619

Liu Y, Rens EG, Edelstein-Keshet L. 2021. Spots, stripes, and spiral waves in models for static and motile cells: GTPase patterns in cells. *J Math Biol* **82:** 1–38. doi:10.1007/s00285-021-01560-y

Lo CM, Wang HB, Dembo M, Wang YL. 2000. Cell movement is guided by the rigidity of the substrate. *Biophys J* **79:** 144–152. doi:10.1016/S0006-3495(00)76279-5

Manhart A, Icheva TA, Guerin C, Klar T, Boujemaa-Paterski R, Thery M, Blanchoin L, Mogilner A. 2019. Quantitative regulation of the dynamic steady state of actin networks. *eLife* **8:** e42413. doi:10.7554/eLife.42413

Marée AF, Jilkine A, Dawes A, Grieneisen VA, Edelstein-Keshet L. 2006. Polarization and movement of keratocytes: a multiscale modelling approach. *Bull Math Biol* **68:** 1169–1211. doi:10.1007/s11538-006-9131-7

Marée AF, Grieneisen VA, Edelstein-Keshet L. 2012. How cells integrate complex stimuli: the effect of feedback from phosphoinositides and cell shape on cell polarization and motility. *PLoS Comput Biol* **8:** e1002402. doi:10.1371/journal.pcbi.1002402

Merchant B, Edelstein-Keshet L, Feng JJ. 2018. A Rho-GTPase based model explains spontaneous collective migration of neural crest cell clusters. *Dev Biol* **444:** S262–S273. doi:10.1016/j.ydbio.2018.01.013

Mirams GR, Arthurs CJ, Bernabeu MO, Bordas R, Cooper J, Corrias A, Davit Y, Dunn SJ, Fletcher AG, Harvey DG, et al. 2013. Chaste: an open source C++ library for computational physiology and biology. *PLoS Comput Biol* **9:** e1002970. doi:10.1371/journal.pcbi.1002970

Mogilner A. 2009. Mathematics of cell motility: have we got its number? *J Math Biol* **58:** 105–134. doi:10.1007/s00285-008-0182-2

Mogilner A. 2019. Alex Mogilner (Q&A). *Curr Biol* **29:** R915–R917. doi:10.1016/j.cub.2019.07.077

Moldenhawer T, Moreno E, Schindler D, Flemming S, Holschneider M, Huisinga W, Alonso S, Beta C. 2022. Spontaneous transitions between amoeboid and keratocyte-like modes of migration. *Front Cell Dev Biol* **10:** 898351. doi:10.3389/fcell.2022.898351

Mori Y, Jilkine A, Edelstein-Keshet L. 2008. Wave-pinning and cell polarity from a bistable reaction-diffusion system. *Biophys J* **94:** 3684–3697. doi:10.1529/biophysj.107.120824

Mukherjee A, Behkam B, Nain AS. 2019. Cancer cells sense fibers by coiling on them in a curvature-dependent manner. *IScience* **19:** 905–915. doi:10.1016/j.isci.2019.08.023

Mukhtar N, Cytrynbaum EN, Edelstein-Keshet L. 2022. A multiscale computational model of YAP signaling in epithelial fingering behavior. *Biophys J* **121:** 1940–1948. doi:10.1016/j.bpj.2022.04.010

Mulberry N, Edelstein-Keshet L. 2020. Self-organized multicellular structures from simple cell signaling: a computational model. *Phys Biol* **17:** 066003. doi:10.1088/1478-3975/abb2dc

Nardini JT, Bortz DM. 2018. Investigation of a structured Fisher's equation with applications in biochemistry. *SIAM J Appl Math* **78:** 1712–1736. doi:10.1137/16M1108546

Osborne JM, Fletcher AG, Pitt-Francis JM, Maini PK, Gavaghan DJ. 2017. Comparing individual-based approaches to modelling the self-organization of multicellular tissues. *PLoS Comput Biol* **13:** e1005387. doi:10.1371/journal.pcbi.1005387

Othmer HG, Hillen T. 2000. The diffusion limit of transport equations derived from velocity-jump processes. *SIAM J Appl Math* **61:** 751–775. doi:10.1137/S0036139999358167

Otsuji M, Ishihara S, Co C, Kaibuchi K, Mochizuki A, Kuroda S. 2007. A mass conserved reaction–diffusion system captures properties of cell polarity. *PLoS Comput Biol* **3:** e108. doi:10.1371/journal.pcbi.0030108

Paluch EK, Raz E. 2013. The role and regulation of blebs in cell migration. *Curr Opin Cell Biol* **25:** 582–590. doi:10.1016/j.ceb.2013.05.005

Park J, Holmes WR, Lee SH, Kim HN, Kim DH, Kwak MK, Wang CJ, Edelstein-Keshet L, Levchenko A. 2017. Mechanochemical feedback underlies coexistence of qualitatively distinct cell polarity patterns within diverse cell populations. *Proc Natl Acad Sci* **114:** E5750–E5759. doi:10.1073/pnas.1700054114

Park J, Kim DH, Levchenko A. 2018. Topotaxis: a new mechanism of directed cell migration in topographic ECM gradients. *Biophys J* **114:** 1257–1263. doi:10.1016/j.bpj.2017.11.3813

Park J, Kim DH, Shah SR, Kim HN, Kshitiz KP, Quiñones-Hinojosa A, Levchenko A. 2019. Switch-like enhancement of epithelial-mesenchymal transition by YAP through feedback regulation of WT1 and Rho-family GTPases. *Nat Commun* **10:** 2797. doi:10.1038/s41467-019-10729-5

Pawluchin A, Galic M. 2022. Moving through a changing world: single cell migration in 2D vs. 3D. *Front Cell Dev Biol* **10:** 1080995. doi:10.3389/fcell.2022.1080995

Pollard TD. 2016. Actin and actin-binding proteins. *Cold Spring Harb Perspect Biol* **8:** a018226. doi:10.1101/cshperspect.a018226

Pollard TD, Borisy GG. 2003. Cellular motility driven by assembly and disassembly of actin filaments. *Cell* **112:** 453–465. doi:10.1016/S0092-8674(03)00120-X

Pollard TD, Blanchoin L, Mullins RD. 2001. Actin dynamics. *J Cell Sci* **114:** 3–4. doi:10.1242/jcs.114.1.3

Ponce-de-Leon M, Montagud A, Noël V, Meert A, Pradas G, Barillot E, Calzone L, Valencia A. 2023. PhysiBoSS 2.0: a sustainable integration of stochastic Boolean and agent-based modelling frameworks. *NPJ Syst Biol Appl* **9:** 54. doi:10.1038/s41540-023-00314-4

Rackauckas C. 2017. A comparison between differential equation solver suites in MATLAB, R, Julia, Python, C, Mathematica, Maple, and Fortran. www.stochasticlifestyle.com/comparison-differential-equation-solver-suites-matlab-rjulia-python-c-fortran

Rankin J, Avitabile D, Baladron J, Faye G, Lloyd DJ. 2014. Continuation of localized coherent structures in nonlocal neural field equations. *SIAM J Sci Comput* **36:** B70–B93. doi:10.1137/130918721

Rens EG, Edelstein-Keshet L. 2021. Cellular tango: how extracellular matrix adhesion choreographs Rac-Rho signaling and cell movement. *Phys Biol* **18:** 066005. doi:10.1088/1478-3975/ac2888

Ricoult SG, Kennedy TE, Juncker D. 2015. Substrate-bound protein gradients to study haptotaxis. *Front Bioeng Biotechnol* **3:** 40. doi:10.3389/fbioe.2015.00040

Ridley AJ. 2015. Rho GTPase signalling in cell migration. *Curr Opin Cell Biol* **36:** 103–112. doi:10.1016/j.ceb.2015.08.005

Rubinstein B, Jacobson K, Mogilner A. 2005. Multiscale two-dimensional modeling of a motile simple-shaped cell. *Multiscale Model Simul* **3:** 413–439. doi:10.1137/04060370X

Scianna M, Preziosi L. 2012. Multiscale developments of the cellular Potts model. *Multiscale Model Simul* **10:** 342–382. doi:10.1137/100812951

Shi C, Huang CH, Devreotes PN, Iglesias PA. 2013. Interaction of motility, directional sensing, and polarity modules recreates the behaviors of chemotaxing cells. *PLoS Comput Biol* **9:** e1003122. doi:10.1371/journal.pcbi.1003122

Simpson MJ, Merrifield A, Landman KA, Hughes BD. 2007. Simulating invasion with cellular automata: connecting cell-scale and population-scale properties. *Phys Rev E* **76:** 021918. doi:10.1103/PhysRevE.76.021918

Skruber K, Mullins D. 2024. The majority of monomers enter branched actin networks from a membrane surface. *Biophys J* **123:** 465a–466a. doi:10.1016/j.bpj.2023.11.2818

Starruß J, De Back W, Brusch L, Deutsch A. 2014. Morpheus: a user-friendly modeling environment for multiscale and multicellular systems biology. *Bioinformatics* **30:** 1331–1332. doi:10.1093/bioinformatics/btt772

Stehbens SJ, Scarpa E, White MD. 2024. Perspectives in collective cell migration—moving forward. *J Cell Sci* **137:** jcs261549. doi:10.1242/jcs.261549

Stepien TL, Swigon D. 2014. Traveling waves in a one-dimensional elastic continuum model of cell layer migration with stretch-dependent proliferation. *SIADS J Appl Dyn Syst* **13:** 1489–1516. doi:10.1137/130941407

Sunyer R, Trepat X. 2020. Durotaxis. *Curr Biol* **30:** R383–R387. doi:10.1016/j.cub.2020.03.051

Svitkina TM, Borisy GG. 1999. Arp2/3 complex and actin depolymerizing factor/cofilin in dendritic organization and treadmilling of actin filament array in lamellipodia. *J Cell Biol* **145:** 1009–1026. doi:10.1083/jcb.145.5.1009

Swat MH, Thomas GL, Belmonte JM, Shirinifard A, Hmeljak D, Glazier JA. 2012. Multi-scale modeling of tissues using CompuCell3D. *Methods Cell Biol* **110:** 325–366. doi:10.1016/B978-0-12-388403-9.00013-8

Toda S, Blauch LR, Tang SK, Morsut L, Lim WA. 2018. Programming self-organizing multicellular structures with synthetic cell-cell signaling. *Science* **361:** 156–162. doi:10.1126/science.aat0271

Town JP, Weiner OD. 2023. Local negative feedback of Rac activity at the leading edge underlies a pilot pseudopod-like program for amoeboid cell guidance. *PLoS Biol* **21:** e3002307. doi:10.1371/journal.pbio.3002307

Uecker H, Wetzel D, Rademacher JD. 2014. pde2path-a Matlab package for continuation and bifurcation in 2D elliptic systems. *Numer Math Theory Methods Appl* **7:** 58–106. doi:10.4208/nmtma.2014.1231nm

Um E, Oh JM, Park J, Song T, Kim TE, Choi Y, Shin C, Kolygina D, Jeon JH, Grzybowski BA, et al. 2019. Immature dendritic cells navigate microscopic mazes to find tumor cells. *Lab Chip* **19:** 1665–1675. doi:10.1039/C9LC00150F

Valon L, Marín-Llauradó A, Wyatt T, Charras G, Trepat X. 2017. Optogenetic control of cellular forces and mechanotransduction. *Nat Commun* **8:** 14396. doi:10.1038/ncomms14396

Vanderlei B, Feng JJ, Edelstein-Keshet L. 2011. A computational model of cell polarization and motility coupling mechanics and biochemistry. *Multiscale Model Simul* **9:** 1420–1443. doi:10.1137/100815335

Walker BJ, Townsend AK, Chudasama AK, Krause AL. 2023. VisualPDE: rapid interactive simulations of partial differential equations. *Bull Math Biol* **85:** 113. doi:10.1007/s11538-023-01218-4

Weiner OD, Marganski WA, Wu LF, Altschuler SJ, Kirschner MW. 2007. An actin-based wave generator organizes cell motility. *PLoS Biol* **5:** e221. doi:10.1371/journal.pbio.0050221

Wittmann T, Waterman-Storer CM. 2001. Cell motility: can Rho GTPases and microtubules point the way? *J Cell Sci* **114:** 3795–3803. doi:10.1242/jcs.114.21.3795

Wortel IM, Textor J. 2021. Artistoo, a library to build, share, and explore simulations of cells and tissues in the web browser. *eLife* **10:** e61288. doi:10.7554/eLife.61288

Xiong Y, Huang CH, Iglesias PA, Devreotes PN. 2010. Cells navigate with a local-excitation, global-inhibition-biased excitable network. *Proc Natl Acad Sci* **107:** 17079–17086. doi:10.1073/pnas.1011271107

Zmurchok C, Holmes WR. 2020. Simple Rho GTPase dynamics generate a complex regulatory landscape associated with cell shape. *Biophys J* **118:** 1438–1454. doi:10.1016/j.bpj.2020.01.035

Zmurchok C, Bhaskar D, Edelstein-Keshet L. 2018. Coupling mechanical tension and GTPase signaling to generate cell and tissue dynamics. *Phys Biol* **15:** 046004. doi:10.1088/1478-3975/aab1c0

Zwicker D. 2020. py-pde: a python package for solving partial differential equations. *J Open Source Software* **5:** 2158. doi:10.21105/joss.02158

Exploring Self-Organization of Development in the Zebrafish Posterior Lateral Line Primordium with Computer Models

Ajay Chitnis

Division of Developmental Biology, Eunice Kennedy Shriver National Institute of Child Health and Human Development, Bethesda, Maryland 20892, USA

Correspondence: chitnisa@nih.gov

The zebrafish posterior lateral line (pLL) primordium migrates under the skin, from the ear to the tip of the tail, periodically forming and depositing neuromasts to spearhead formation of the pLL system. Cellular, molecular, genetic, and physical manipulation of the pLL primordium coupled with high-resolution imaging of the migrating primordium has helped us understand how interactions between cells coordinate pattern formation, morphogenesis, and collective migration of the pLL primordium. These studies reveal, more broadly, how myriad self-organizing processes operating at multiple scales through physical, chemical, and genetic interaction determine robust emergence of fate, form, and function in the growing animal. This review will summarize how we have used agent-based models and cellular Potts models to integrate what has been learned from experiments and investigate how signaling and mechanical interactions determine emergent patterns, fate, morphogenesis, and collective migration of the primordium.

The zebrafish posterior lateral line (pLL) primordium migrates under the skin, from the ear to the tip of the tail, periodically forming and depositing neuromasts to spearhead formation of the pLL system (Fig. 1A). Neuromasts are the sensory organs of the lateral line, a sensory system that has evolved to allow fish and aquatic amphibians to detect the pattern of water flow over their body surface and give a sense of "touch at a distance" (Dijkgraaf 1963). The primordium has emerged as an excellent model system to understand how physical, chemical, and genetic interactions determine pattern, cell fate, morphogenesis, and collective migration during animal development (Ghysen and Dambly-Chaudiere 2007).

This review summarizes how we have used agent-based models (ABMs) (Wilensky 1999) and cellular Potts models (CPMs), also known as Glazier–Graner–Hogeweg models (Graner and Glazier 1992; Savill and Hogeweg 1997), to integrate what has been learned from cellular, molecular, genetic, and physical manipulation and high-resolution imaging of the migrating primordium to understand how interactions between cells coordinate pattern formation, morphogenesis, and collective migration of the pLL primordium. While the focus of much of this

Cite this article as *Cold Spring Harb Perspect Biol* doi: 10.1101/cshperspect.a041797

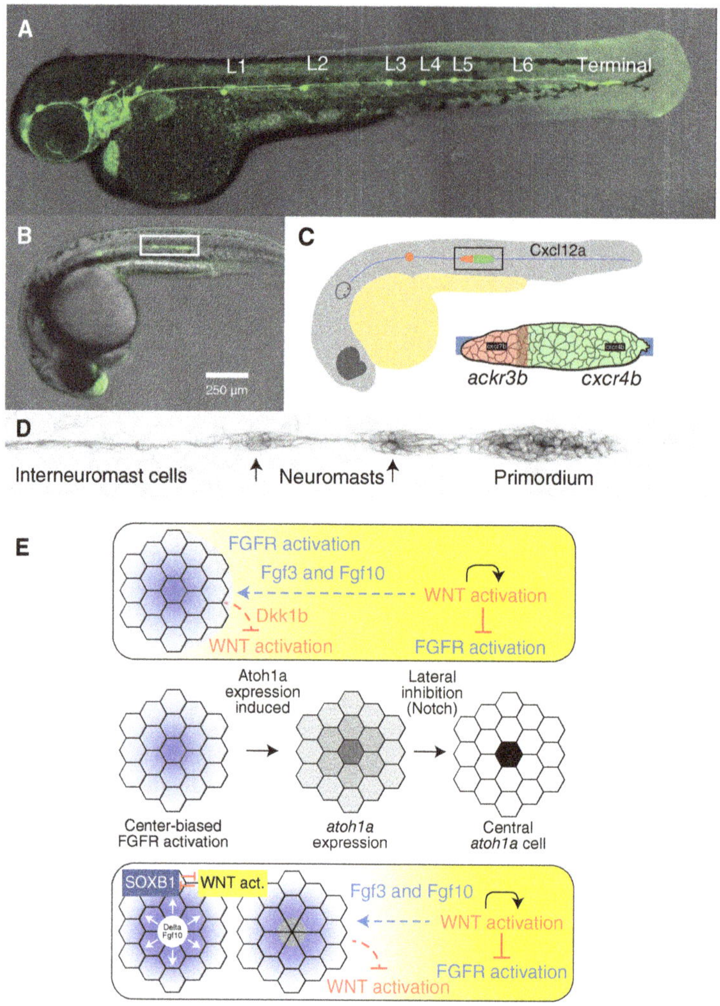

Figure 1. The posterior lateral line primordium (pLLp) migrates from the ear to the tip of the tail, along a path defined by expression of Cxcl12a, periodically forming and depositing neuromasts. (*A*) Tg(cldnb:lyn-gfp) embryo at 54 h showing the pattern of neuromast deposition (L1–L6 and terminal neuromasts). (*B*) Green fluorescence of Tg(cldnb:lyn-gfp) superimposed on a Nomarski image of an embryo at ~24 h postfertilization (hpf) shows the pLLp and a depositing neuromast in the rectangular *inset*. (*C*) Schematic showing pLLp migrating along a stripe of Cxcl12a expression. Expanded rectangular *inset* shows leading green *cxcr4b*-expressing and trailing red *ackr3b*-expressing cells. (*D*) Protoneuromasts are deposited as neuromasts by the migrating primordium, while cells not incorporated into neuromasts are deposited in between as interneuromast cells. (*E*) Sequential steps in periodic formation of protoneuromasts. Local activation of Wnt coupled with its long-range inhibition by Fgf-dependent Dkk1b expression creates a center-biased domain of FGF activity, which seeds protoneuromast formation (*upper* panel). FGF activity determines center-biased atoh1a expression, where lateral inhibition mediated by Notch restricts *atoh1a* to a central cell (*middle* panel)—the *atoh1a*-expresing cell becomes a new source of Delta and Fgf that activates neighboring cells to consolidate epithelial rosette formation. In addition, SoxB1 transcription factors, Sox2 and Sox3, inhibit Wnt activity to facilitate restriction of Wnt activity to a smaller leading zone creating the conditions for emergence of a new protoneuromast (*lower* panel). (Panels *A* and *B* are adapted from Dalle Nogare and Chitnis 2020 and reprinted under the Creative Commons CC0 public domain dedication.)

Cite this article as *Cold Spring Harb Perspect Biol* doi: 10.1101/cshperspect.a041797

volume is cell migration, mechanisms that determine the specification of cell fate and morphogenesis are inextricably tied to the mechanisms that determine collective migration of cells in the primordium, and the models discussed here examine the integration of these events. The models described have been developed to visualize how interactions defined by experiments have the potential to determine emergent fate, pattern, and behavior of the primordium.

The primary pLL primordium is a cluster of ~140 cells that migrates as a cohesive column along the horizontal myoseptum from near the ear to the tip of the tail (Fig. 1B,C; Ghysen and Dambly-Chaudiere 2007; Dalle Nogare et al. 2017). Cells underlying the horizontal myoseptum are the source of the chemokine Cxcl12a, which serves as a migratory cue for directed migration of the primordium (Fig. 1C; Li et al. 2004; Sapède et al. 2005). A striking feature of this system is that the chemokine is expressed relatively uniformly along the horizontal myoseptum, and its polarized expression along the path is not responsible for directed migration of the primordium. Instead, caudal migration is determined by the polarized expression of two chemokine receptors, Cxcr4b and Ackr3b (previously known as Cxcr7b) (Haas and Gilmour 2006; Dambly-Chaudière et al. 2007; Valentin et al. 2007) within the migrating primordium. Trailing cells expressing Ackr3b bind, internalize, and degrade Cxcl12a, creating a local gradient of Cxcl12a (Donà et al. 2013; Venkiteswaran et al. 2013). Cxcr4b, whose expression is initiated in the leading domain, allows primordium cells to respond to this local gradient and determine directed caudal migration of the primordium. This mechanism allows the primordium to migrate indefinitely along a path defined by relatively uniform Cxcl12a expression. We first describe a model that shows how a self-generated Cxcl12a gradient directs migration of the polarized primordium (see Fig. 2; Dalle Nogare et al. 2014). We discuss both how the model recapitulates specific experimental observations and how its failure to recapitulate specific observations contributed to our recognition of the complementary role of fibroblast growth factors (Fgfs), produced by leading cells, in directing migration of trailing cells.

The polarized expression of *cxcr4b* and *ackr3b* in the leading and trailing parts of the primordium is determined by Wnt and Fgf signaling in the primordium (Aman and Piotrowski 2008). Wnt activity dominates in the leading zone and is lower in the trailing domain (Fig. 1E, yellow). Wnt activity determines expression of Fgf ligands, Fgf10 and Fgf3; however, these ligands cannot activate Fgf signaling in the leading zone as Wnt signaling also determines expression of factors that prevent the Fgf receptors and their downstream effectors from responding effectively to the Fgfs they produce. Instead, these Fgfs activate Fgf receptors in the trailing zone, where there is less inhibition from Wnt signaling. While Wnt signaling inhibits *ackr3b* expression in the leading zone, it is thought that Fgfs induce their expression in the trailing zone via noncanonical FGFR signaling (Dries et al. 2020). Conversely, an FGF signaling-dependent factor inhibits *cxcr4b* expression in the trailing zone, while absence of Fgf signaling permits *cxcr4b* expression in the leading zone (Aman and Piotrowski 2008). In this manner, polarized Wnt and Fgf signaling determines polarized expression of Cxcr4b, Ackr3b, and directed migration of the primordium along a path defined by the relatively uniform expression of *cxcl12a*.

In addition to determining directed migration, polarized Wnt and Fgf signaling coordinates periodic formation of nascent neuromasts (protoneuromasts), their maturation, and eventual deposition by the migrating primordium (Fig. 1E). Each protoneuromast is characterized by a central *atoh1a*-expressing sensory hair cell progenitor (SHCP) (Fig. 1E; Itoh and Chitnis 2001; Lecaudey et al. 2008), which is surrounded by cells that form an epithelial rosette. Fgf signaling initiates formation of protoneuromasts (Lecaudey et al. 2008; Nechiporuk and Raible 2008) by promoting expression of *atoh1a* and factors that keep *atoh1a* expression restricted to a central cell. It also promotes epithelialization of cells and expression of factors like *shroom3* that promote apical constriction and reorganization of cells to form epithelial rosettes (Ernst et al. 2012). By inhibiting Fgf signaling, Wnt signaling inhibits protoneuromast formation in the leading zone. However, Fgfs delivered to the trailing zone activate Fgf signaling to initiate protoneuromast formation, starting from the trailing end.

As the protoneuromast begins to mature, its central *atoh1a*-expressing cell becomes a source of Fgf10 and DeltaD (Fig. 1E; Matsuda and Chitnis 2010). Together, they activate Fgf and Notch signaling in their neighbors to stabilize epithelial rosette formation in the maturing protoneuromast, independent of the Wnt-dependent Fgf signals that initiated rosette formation. In addition, *atoh1a*, whose expression was initiated by Wnt-dependent Fgf signals, becomes independent of these signals by promoting its own expression (Matsuda and Chitnis 2010). These changes in the maturing protoneuromast prepare it for eventual deposition as a neuromast, independent of the migrating primordium, where its formation was initiated. In addition, maturation of the trailing protoneuromast is associated with expression of SoxB1 factors, Sox2 and Sox3, that inhibit Wnt signaling in the trailing part of the primordium (Palardy et al. 2025). Consequently, the initially broad domain of Wnt activity becomes restricted to a smaller leading domain, and formation of a new protoneuromast is initiated in its wake (Fig. 1E).

As the primordium migrates, its cells proliferate, resulting in it becoming longer. However, as the Wnt system shrinks, trailing cells lose their capacity to migrate, and they are deposited as neuromasts, if the cells had been effectively incorporated into the epithelial rosette of a maturing protoneuromast, or as interneuromast cells, if not. The interneuromast cells are deposited continuously like a string of beads between the periodically deposited neuromasts (Fig. 1D). In toto time-lapse imaging of the migrating primordium (Dalle Nogare et al. 2017) showed that, despite proliferation, the primordium loses more cells than it makes, and it progressively shrinks during its migration. These studies, coupled with studies tracking the progressive shrinking of the Wnt system with a transgenic Wnt reporter line, showed that the relative length of the Wnt system, on average, is always ~60% the length of the migrating column of primordium cells (Dalle Nogare et al. 2017). Eventually, as the Wnt system shrinks below some threshold and Fgf signaling comes to dominate the system, the remaining primordium cells resolve to make two to three terminal neuromasts. These observations were used to develop a model that uses initial size, proliferation rate, the rate at which the Wnt system shrinks, and migratory speed to effectively predict the pattern of neuromast deposition by the migrating primordium (see Fig. 3A; Dalle Nogare et al. 2017) .

Although there is not yet a definitive explanation for why the length of the Wnt system scales with the length of the primordium, it could, in part, be related to the observation that leading Wnt active cells are the source of diffusible Fgf signals that serve as migratory cues for trailing cells. We hypothesized that the reach of such diffusible factors would depend on the size of the source of diffusible signals. A third model shows how this could account for primordium length scaling with the length of a leading domain (see Fig. 3B).

A striking feature of a protoneuromast is that an *atoh1a*-expressing SHCP is specified at its center, and this cell is surrounded by cells that reorganize to form an epithelial rosette. Fgf signaling initiates *atoh1a* expression in nascent protoneuromasts, which gives cells the potential to become SHCPs. However, their expression of Delta ligands allows each Atoh1a-expressing cell to activate Notch in their neighbors, inhibiting them from expressing *atoh1a*. This lateral inhibition, mediated by Notch signaling, results in *atoh1a* becoming restricted to a central cell. However, if all cells in a nascent protoneuromast inhibited each other equally, a central cell would receive inhibition from all its neighbors and be very unlikely to be the one to retain *atoh1a* expression following lateral inhibition. On the other hand, if *atoh1a* were initiated in a center-biased pattern, lateral inhibition could indeed result in *atoh1a* becoming restricted to the central cell. Consistent with this prediction, it was shown that *atoh1a* and Fgf signaling, which initiates *atoh1a* expression, are initiated in a center-biased pattern in nascent protoneuromasts (Chitnis et al. 2012). This discovery, however, raised a new question about how previously described interactions between Wnt and Fgf signaling might determine center-biased Fgf signaling. A potential answer to this question is described in a fourth model that shows how local activation of Wnt activity coupled with its Fgf-dependent long-range inhibition has the potential to constitute a "local activation and long-range inhibition"-based patterning system (Meinhardt and Gierer 2000) that could determine self-organization of center-biased Fgf signaling centers in the wake of a shrinking leading domain where Wnt activity dominates (see Fig. 4).

Although a prepattern of Fgf signaling centers helps seed the periodic formation of protoneuromasts with their central hair cell progenitors and surrounding epithelial rosettes, the number and size of the epithelial rosettes is initially unstable; it can be influenced by manipulations that alter mechanical tension along the length of the primordium, and rosette formation can be uncoupled from the periodic prepattern of Fgf signaling that determines *atoh1a* expression. When migration of leading cells is selectively slowed and the migrating primordium becomes compressed, previously formed rosettes can fuse to form fewer larger rosettes. Conversely, selective slowing of trailing cells results in stretching of the primordium and splitting of previously formed rosettes to form smaller ones (Mukherjee et al. 2025). These observations can be accounted for by a mechanics-based model, where local interactions associated with apical constriction and cell adhesion promote clustering and formation of rosettes, while tension along the length of the primordium, influenced by the relative efficacy of leading and trailing cell migration, opposes such aggregation. ABM (see Fig. 5) and CPM (see Fig. 6) were built to show how these interactions represent a mechanical form of local activation and long-range inhibition that can determine spontaneous clustering of primordium cells to form epithelial rosettes, in parallel with a signaling prepattern that specifies exactly where they should form. These studies show how signaling and mechanics are likely to work together to influence morphogenesis in the migrating primordium.

Although many models have now been developed to examine specific behaviors in the migrating primordium (Meier-Schellersheim et al. 2009; Di Costanzo et al. 2015; Knutsdottir et al. 2017; Zinn-Björkman and Adler 2018; Colombi et al. 2021), this review is limited to a subset of models developed by the author. All these models, except one, are ABMs, developed using the NetLogo programming environment (Box 1; Wilensky 1999). The last is a CPM using the CompuCell3D programming environment (Swat et al. 2012). In the next few sections, a brief description of these models is given, explaining what they were developed to explore, how they were built, what was learned, what some of the limitations of the models are, and how they might be extended in the future.

MODELS

Model 1: Directed Migration in Response to a Self-Generated Chemokine Gradient

What It Explores

This model (Dalle Nogare et al. 2014) examines how polarized expression of Cxcr4b and Cxcr7b (Ackr3b) in a compartment of leading and trailing cells, respectively, determines directed migration of the primordium along a path defined by a stripe of Cxcl12a expression (Fig. 2). It was originally built to see whether simulations would recapitulate experimentally observed behavior of a leading fragment of the primordium following its severance from the remaining part of the primordium by laser. Once released, the severed fragment moves forward and backward randomly; this behavior typically persists until the trailing primordium contacts the severed leading fragment; they rejoin and directed caudal migration of the entire primordium resumes.

How It Was Built

The primordium is represented by a column of two *breeds* of *turtles*, *cxcr4-cells* and *cxcr7-cells*, in a leading and trailing compartment, bound to each other by spring-like *links* (Fig. 2A). The *turtles* move in response to a Cxcl12a, secreted by a stripe of *patches* that run along the length of the path. Parameters regulating spring-like behavior of *links* between adjacent cells can be

BOX 1. NetLogo AND ABMs

Agent-based models (ABMs) are simulation platforms that help to visualize how interactions between "agents" defined by specific rules can determine the dynamic self-organization of pattern, organization, and behavior of the agents. Traditionally, they have been used to show how, for example, a simple set of rules can determine flocking of birds and schooling of fish, the spontaneous development of traffic jams, or how a critical density of trees can define a transition point that dramatically alters the likelihood of a forest fire spreading. The examples above are phenomena that emerge because of relatively simple interactions between moving or stationary agents within a defined grid. *Turtles, Termites, and Traffic Jams: Explorations in Massively Parallel Microworlds* by Mitchell Resnick (1994) provides an excellent introduction to ABMs and their ability to account for emergence of these phenomena using a precursor to NetLogo, StarLogo. NetLogo, the modeling environment used, was developed by Uri Wilensky, currently at Northwestern University, Chicago.

NetLogo allows users to define rules governing behavior of three types of agents: (1) *turtles*, agents that can move, divide, "die," and form spring-like *links* with other *turtles* and can be used to represent cells or different types of cells (*breeds*); (2) *patches*; square spaces in the grid over which *turtles* move, which can represent extracellular environment and can be used to simulate diffusion of signaling molecules produced by *turtles* and deposited on the *patches*, and (3), *links*, spring-like agents that connect *turtles*. They can be used to represent mechanical coupling between cells resulting from the combination of intercellular adhesion and apical constriction. These *links* have *turtle* breed-specific parameters including a *spring-constant*, *spring-length*, and *repulsion-constant*.

Together, these agents provide the opportunity to model myriad phenomena in developmental biology at multiple scales. While *turtles* are typically used to represent cells and *patches*, the environment within which they migrate, *patches* can also be used to represent a sheet of stationary cells. In models where either *turtles* or *patches* represent cells, these agents can be assigned variables that define genes or signaling components that are part of a regulatory network operating within a cell or between cells. Simple algebraic statements in the form of Hill equations help define how genes regulate each other (see Box 2). Both *turtles* and *patches* can be the source of diffusible factors that can influence gene regulation or the behaviors of migratory cells. Rules can be written to define how dynamic changes in the level of specific factors, *turtles* or *patches* in the environment, or connections defined by *links*, influence the state, movement, aggregation, division, or "death" of cells represented by *turtles*. Different "breeds" of *turtles* can be defined to represent distinct cell types, which respond, follow rules, and behave in a cell-type-specific manner. *links* were introduced in NetLogo as agents that represent connections between nodes in a network represented by *turtles*. However, we have leveraged their spring-like properties to model cell adhesion and mechanical interactions within or between cells. Together, these features make NetLogo a powerful environment for modeling a wide range of developmental processes.

This modeling environment is relatively easy to use; the software is free and runs on most computers. It is easy to develop simulations, and sliders allow one to explore how changes in various parameters affect emergent behaviors. They are for the most part "toy" models where rules that define cell behavior or gene regulatory networks assumed to operate in each cell have been defined based on what has been learned from experiments. As there is limited quantitative information available, the primary strength of these models is their ability to help visualize how potential patterns and behaviors emerge based on known interactions and how they might be altered when specific parameters are altered.

BOX 2. MODELING GENE REGULATORY NETWORKS—A SIMPLE PRIMER

Simple algebraic statements are used to update the level of gene expression in a *turtle* or a *patch* at each iteration or "tick" of a program. For example:

Update of GeneA = current GeneA + GeneA produced − GeneA degraded

Production of *GeneA* could be promoted by *GeneB* and inhibited by *GeneC*. These interactions can be represented in the form of Hill equations.

Although these can have different forms, the following helps to understand the basic structure of these equations.

$$GeneA\ produced = k_1 \times GeneB^m/1 + k_2 \times GeneC^n$$

where k_1 and k_2 are constants that define the strength of the influence of the respective genes, m and n represent the nonlinear nature of their influence, and genes that promote production are in the numerator and those that inhibit are in the denominator.

$$GeneA\ degraded = GeneA \times d$$

where d is the fraction by which *GeneA* is degraded at each iteration.

Diffusion in NetLogo

In NetLogo, *patches* are assigned variables that represent diffusible factors. To model their "diffusion" at each iteration, a *patch* adds a defined fraction of its "diffusible factor" to its neighboring *patches*. This operation will then be repeated, and each of the *patches* will once again add the same defined fraction of their current level of the factor to its neighbors and the factor diffuses. To model faster diffusion, this operation is repeated more times at each "tick" of the program and fewer times for slower diffusion.

varied to see how the degree of mechanical coupling between individual cells influences collective migration. Trailing *cxcr7-cells* degrade Cxcl12a, which allows them to create a local gradient of Cxcl12a. At each iteration, cells move ahead a defined distance (speed) if the Cxcl12a in the patch ahead either exceeds a defined level (threshold Cxcl12a) or the difference in Cxcl12a at the current patch and patch ahead exceeds a defined threshold (threshold gradient). Degradation of *Cxcl12a* by trailing Cxcr7b-expressing cells self-regulates to maintain a linear Cxcl12a gradient (Lau et al. 2020); this suggests that while some threshold level of Cxcl12a may still be required for the migratory response, the gradient is likely to determine how effectively cells respond.

When the model was originally built, it was thought that only edge cells were required to respond to Cxcl12a to determine migration (Haas and Gilmour 2006). More recent work suggests, however, that all Cxcr4b-expressing cells are required for effective collective migration, not just those at the primordium edge (Colak-Champollion et al. 2019). To compare, simulations can be run in two modes, with either all cells having the potential to respond or only edge cells.

links between cells along the primordium can be cut to release a fragment at a defined fractional distance from the trailing end. The cleaved fragment has the potential to reform *links* with cells, leaving the remaining primordium behind. The interface of the model (Fig. 2A) shows the simplicity with which parameters and simulations can be set up to see effects of changing the width of the Cxcl12a-expressing stripe, its production and diffusion rate, and the rate at which the leading *cxcr4-cell* and trailing *cxcr7-cell* domains degrade Cxcl12a.

What Did We Learn

The model shows how a self-generated gradient can account for the directed migration of the primordium, and it helped define constraints on the nature of the gradient (Supplemental

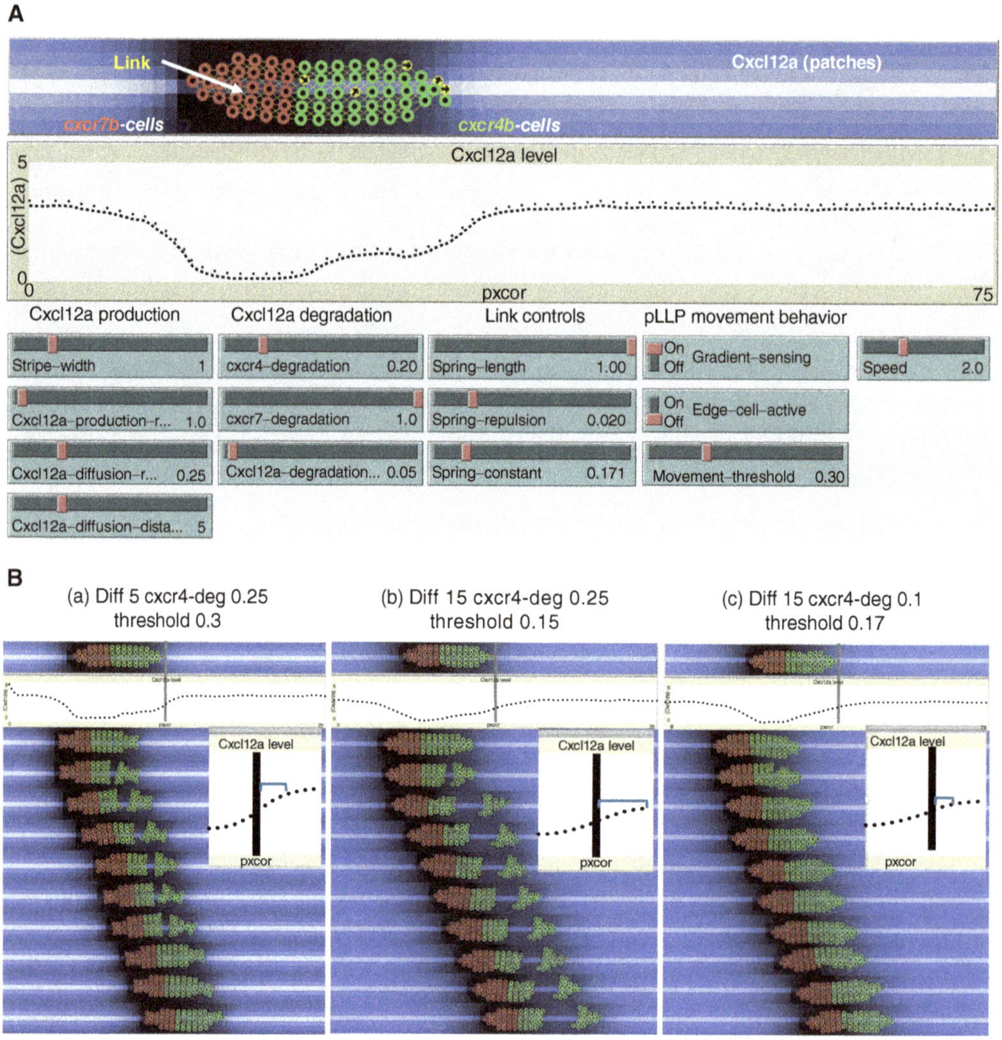

Figure 2. A NetLogo agent-based model (ABM) used to visualize emergent migratory behavior of the primordium following release of a leading fragment of the primordium. (*A*) Screenshot of the model interface showing primordium with leading *cxcr4b-cells* (green) and trailing *cxcr7b-cells* (red) migrating along a self-generated gradient of Cxcl12a expression (blue). At each iteration, turtles face a random direction, and, if the level of Cxcl12 ahead of them meets defined conditions to trigger movement, they turn yellow and move ahead at the defined "speed" with the arrow indicating the direction. However, as all turtles are mechanically coupled with spring-like *links*, there is collective migration (see Supplemental Movie S1). (*B*) Three model runs with different (diff.) self-generated gradients that result in the released fragment: (*a*) briefly moving away and eventually rejoining the primordium, (*b*) continuing to move forward without rejoining, and (*c*) rejoining without moving away. Each behavior is associated with a different length (bracket) over which the gradient can effectively determine migratory behavior. The dotted line is the Cxcl12a level, and the vertical bar marks the leading edge of the primordium. Each run is associated with a different range of Cxcl12a diffusion (diff), a different level of degradation of Cxcl12a by leading *cxcr4-cells* (cxcr4-deg), or threshold of difference in level of Cxcl12a required to trigger migratory behavior (threshold) indicated for *a*, *b*, and *c* (see Supplemental Movies 2 and 3).

 Cite this article as *Cold Spring Harb Perspect Biol* doi: 10.1101/cshperspect.a041797

Movies 1 and 2). The back-and-forth movement of a released leading fragment was recapitulated in the model using either the assumption that only edge cells respond to the absolute level of Cxcl12a (Dalle Nogare et al. 2014) or, when revisiting the model in preparing this review, with the assumption that all *Cxcr4-cells* respond to a threshold gradient of Cxcl12a. Cxcl12a production, diffusion, and degradation by trailing cells can be adjusted to determine a linear gradient. Separation of a leading fragment following laser ablation was typically followed by its back-and-forth movement before it rejoined the trailing fragment. The primary constraint for recapitulating this experimentally observed behavior is the effective length of the self-generated gradient. It must be such that its influence does not extend much beyond the leading end of the primordium (Fig. 2B; Supplemental Movie 3). If it did, the released leading fragment would keep moving away from the approaching trailing fragment, and it would be unlikely to reconnect. On the other hand, if the length over which the gradient influences migration was too short, the released fragment would not move away before rejoining, and unlike what was observed in vivo, no back-and-forth movement would be observed.

While simulations using the model effectively recapitulate behavior of a released leading fragment, there was a particular behavior of the remaining trailing primordium in vivo that could not be recapitulated by the model; the trailing primordium would sometimes appear to leap toward the leading cells (Dalle Nogare et al. 2014). This alerted us to the fact that there was something missing in the model that was determining this behavior of the trailing fragment. Further investigation revealed that while a self-generated Cxcl12a gradient steers the migration of the leading domain, cells in the leading zone are the source of diffusible Fgf signals that serve as directional cues for trailing cells (Dalle Nogare et al. 2014). These two mechanisms work in tandem to determine collective migration of the primordium.

What Are Its Limitations

The model does not address how interactions between primordium cells determine clustering of cells to form epithelial rosettes, nor how primordium cells lose their capacity for migration and deposit cells as neuromasts or interneuromast cells. Furthermore, the original model was not subsequently modified to incorporate Fgf signaling-dependent migration of trailing cells. Such an integrated model, which also incorporates adhesive mechanisms that determine periodic cell cluster formation within the primordium and deposition, has been developed by Colombi et al. (2021). It effectively recapitulates collective migration of the primordium, formation of cell clusters within the migrating primordium, and their deposition. It also effectively simulates many of the behaviors described for leading and trailing fragments observed with time-lapse video following the release of the leading fragment by laser ablation. Another model uses experimental data from our laser ablation experiments to develop a quantitative analytical model that makes predictions about diffusion constants of Cxcl12a and FGF in the lateral line system (Zinn-Björkman and Adler 2018).

Model 2: Predicting the Pattern of Neuromast Deposition

What It Explores

As the ratio between the length of the Wnt system and the length of the primordium remains on average 60% and the Wnt system shrinks at a defined rate, length of the primordium at each point along its migration can be predicted based on the expected length of the Wnt system (Dalle Nogare et al. 2017). This relationship allowed development of a model that could predict the pattern of neuromast formation and deposition by the migrating primordium. In toto imaging allowed determination of where in the wake of the Wnt system new protoneuromasts form, how many cells wide they are on average size, and what the distance between sequentially formed protoneuromasts is. In toto imaging also showed that cells proliferate at a defined rate in a random pattern along the length of the primordium (Dalle Nogare et al. 2017). Data from in toto imaging experiments were used to define initial cell number and primordium length, the rate at which the Wnt domain shrinks, mean cell cycle length, average radius of protoneuromasts, minimum distance between neuromasts, and migration speed. This information was used as an input for a model that could predict the pattern of neuromast deposition (Dalle Nogare and Chitnis 2017).

How It Was Built

The measured primordium size was used to initiate a column of turtles (Fig. 3A). The Wnt domain was initially active along the entire length but was made to shrink based on the experimentally measured rate. As cells emerged out of the Wnt active domain, cells within a defined radius of a central cell, specified as a hair cell progenitor, were defined as protoneuromast cells. As trailing cells reached a distance from the leading end beyond the primordium length predicted by the current length of the Wnt domain, these cells lost their capacity to migrate; trailing cells in protoneuromasts were deposited as neuromasts, and surrounding cells were deposited as interneuromast cells. Cells within the migrating primordium divided at a rate determined by a stochastic cell cycle model with random cell length assigned using in vivo mean cell cycle length and standard deviation data.

What Did We Learn

The model shows how the rate at which protoneuromasts are made and deposited is dependent on both the cell proliferation rate and the rate at which the Wnt system shrinks (Aman et al. 2011; Matsuda et al. 2013); both determine when primordium cells will emerge out of the Wnt active domain to form protoneuromasts and when they will reach a trailing distance that exceeds the predicted or allowed length of the primordium (Fig. 3A; Supplemental Movie 4; note the pLL primordium is the frame of reference in Supplemental Movie 4). The model also effectively predicts the average pattern of neuromast deposition in wild-type and in *lef1* mutant embryos, where a lower proliferation rate and progressive slowing of the primordium result in deposition of fewer closer-spaced neuromasts and premature termination of migration (Dalle Nogare and Chitnis 2017).

Model 3: Why Length of the Primordium Scales with the Length of the Wnt System

What It Explores

The previous model was based on the empirical observation that the Wnt system is always ~60% the length of the primordium; however, it was not clear why this relationship exists. This model examines one hypothesis for why this relationship exists. Wnt active cells in a leading domain are the source of diffusible Fgf signals that serve as directional cues for trailing cells (Dalle Nogare et al. 2014). It was hypothesized that trailing cells that are sufficiently far from leading cells would stop migrating and be deposited when the level of diffusible Fgf accessible to them drops below some threshold required to direct migration. Furthermore, it was expected that the size of the Wnt system determines the net level of Fgf produced by leading cells. As the Wnt system shrinks, the size of the source of such diffusible factors would become smaller, as would the effective reach of such factors. A model was built to test the hypothesis that such a relationship could account, at least in part, for why the length of the migrating primordium scales with the size of the Wnt system (Fig. 3B).

How It Was Built

In this model, a column of cells is the source of a diffusible factor (arrow-shaped cells). The fraction of cells that produce the signal progressively shrinks toward the leading end. Trailing cells in the column (round cells) remain if they receive a defined threshold of the diffusible factor; if not, they are removed (deposited).

What Did We Learn

A plot that tracks the level of the diffusible factor along the length of the column of cells shows how the maximum level of the diffusible factor, as well as its reach, reduces as the source shrinks (Fig. 3B; Supplemental Movie 5). Another plot shows how, as both the length of the diffusible

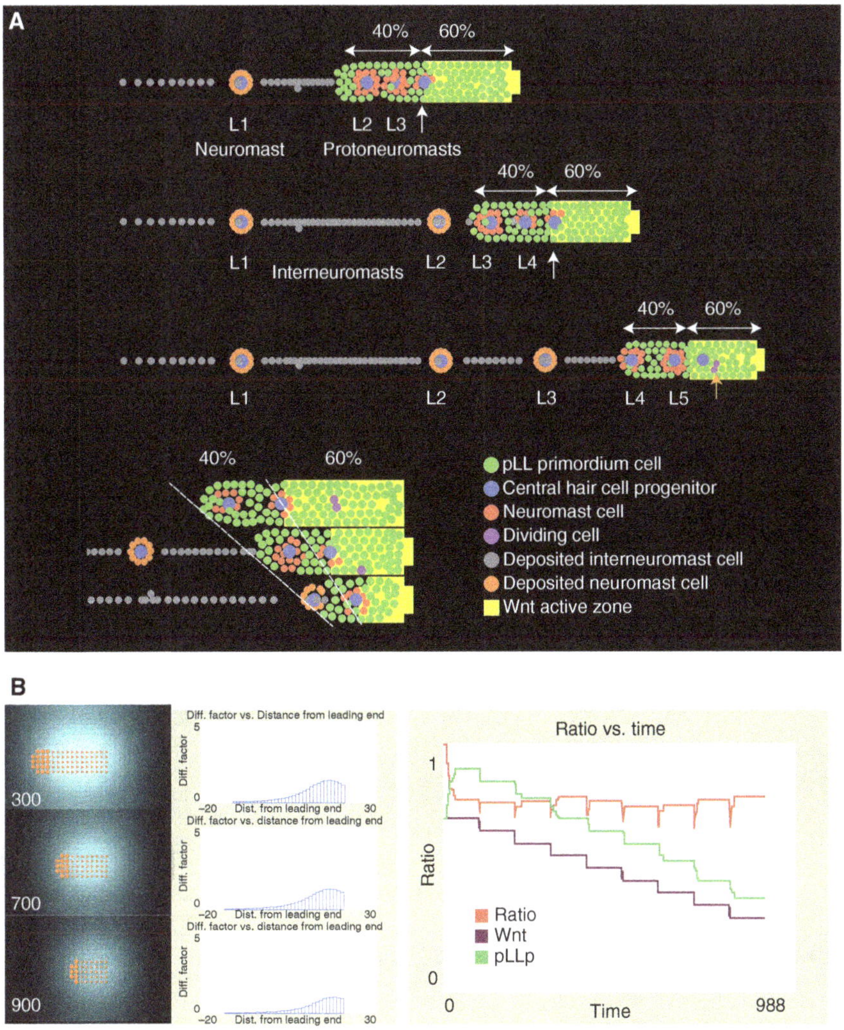

Figure 3. Primordium length scales with the length of the leading Wnt active domain. (*A*) Three frames from a model run showing the progressive shrinking of the Wnt-active zone together with the posterior lateral line primordium (pLLp) coupled with periodic neuromast deposition. The model predicts the pattern of protoneuromast formation and deposition based on the empirical observation that the Wnt system is 60% the size of the migrating primordium and that new protoneuromast forms in the wake of a shrinking Wnt system. When primordium cells (green) emerge from the shrinking Wnt domain (yellow), a central sensory hair cell progenitor is specified (blue), and *links* are formed with surrounding cells with a defined radius to form a protoneuromast (red). When primordium cells fall outside the permitted length of the primordium, they are deposited as neuromasts (orange), if previously in protoneuromasts, and as interneuromast cells (gray), if not. White arrow shows the trailing edge of the Wnt domain, and the orange arrow points to dividing cells. Dashed white lines and stacked images of the primordium at sequential stages of migration show how the relative proportion of leading Wnt active zone (60%) and trailing nonactive zones (40%) remains constant (see Supplemental Movie 4). (*B*) Scaling of primordium length with the progressively reduced size of the Wnt system could be related to the progressively reduced reach of a diffusible (diff.) factor required by primordium cells for collective migration. *Left* panel shows three frames at 300, 700, and 900 time steps from a model run. Leading arrowhead-shaped turtles represent cells in the Wnt active zone that are the source of a diffusible factor (cyan). Filled orange circles are cells in the trailing zone, which continue to migrate together with leading cells as long as the diffusible factor at their location is above a defined threshold. The histograms in the *middle* panel show the changing level of the diffusible factor along the length of the model primordium at corresponding time steps. The graph in the *right* panel shows how the ratio of Wnt domain size:primordium length remains relatively constant with progressive shrinking of the Wnt system (see Supplemental Movie 5). (Dist.) Distance.

factor source and the entire length of the column of cells shrink, the ratio between them remains relatively stable. The precise ratio is determined by the rate of diffusible factor production, degradation, and diffusion, and the threshold level is required to retain trailing cells.

Model 4: How Wnt–Fgf Interactions Create Center-Biased Fgf Signaling Centers

What It Explores

This model shows how the previously described regulatory interactions between Wnt and FGF signaling systems (Aman and Piotrowski 2008) could operate in the context of a gradient of Wnt activity to determine periodic formation of center-biased Fgf signaling centers in the wake of a progressively shrinking domain of Wnt activity (Dalle Nogare and Chitnis 2020).

How It Was Built

The gene regulatory network (GRN) (Fig. 4) operates in a column of *turtles* representing the primordium cells. The system is initiated with a Wnt activity gradient, whose level is maximum at the leading end (Fig. 4B). Wnt activity promotes expression of Wnt and FGF, which is placed on *patches* below and differentially diffused at each iteration. In the example shown, Wnt diffuses one cell distance away, while FGF diffuses five cells away at each iteration. As a result, Wnt locally promotes Wnt activity, while

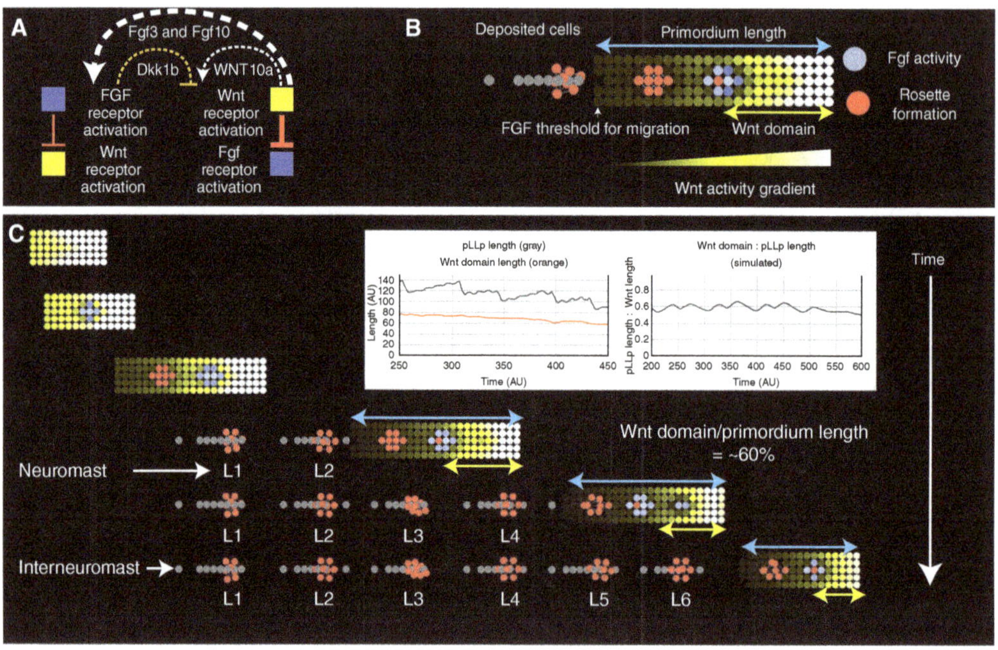

Figure 4. Interactions between the Wnt and Fgf signaling systems constitute a patterning system that can determine periodic formation of center-biased Fgf signaling centers in the wake of a progressively shrinking Wnt system to coordinate periodic formation and deposition of neuromasts. (*A*) Schematic of the gene regulatory network incorporated in the model. (*B*) Frame from model showing an annotated primordium and deposited neuromast and interneuromast cells. (*C*) Seven frames from a model run; the first shows initial size of the primordium and graded Wnt activity; second, primordium growth and formation of the first center-biased domain of Fgf activity (blue); the third shows that the first Fgf activity domain has formed a protoneuromast (red) while the next Fgf domain forms. Subsequent frames show deposition of L2, L4, and L6 neuromasts as the primordium progressively shrinks. During this time, deposition of cells is determined by drop of Fgf below a threshold required for migration. As shown with the previous model, the graphs show how the ratio of Wnt domain size to primordium size remains relatively constant (see Supplemental Movie 6).

simultaneously inhibiting FGF-activation. Fgf, on the other hand, diffuses a longer distance to promote FGF activity. This, in turn, drives expression of a diffusible Wnt inhibitor, Dkk1b, and inhibits Wnt activation. Cells proliferate within this column by addition of new cells at the leading end, and all cells move together at a defined speed. As the primordium moves, the maximum level of Wnt-activity permitted progressively reduces, and the Wnt domain progressively shrinks toward the leading end. As Fgf signaling centers form, persistence of Fgf activity beyond a defined threshold in primordium cells is a trigger for cells to form *links* with each other, simulating the formation of epithelial rosettes. When Fgf in *patches* drops below the threshold defined to sustain migration, trailing cells stop migrating and are deposited as neuromasts if they had been incorporated into rosettes or as interneuromast cells, if not.

What Did We Learn

The model shows how the GRN has the potential to account for spontaneous formation of center-biased Fgf signaling centers that seed periodic formation and deposition of neuromasts in the wake of a shrinking Wnt system (Fig. 4; Supplemental Movie 6). The model also shows how deposition of trailing cells when Fgf levels fall below a defined threshold can account for primordium length scaling with the length of the leading Wnt system.

What Are Its Limitations

The GRN represented in the model only represents interactions that initiate the formation of center-biased Fgf signaling centers in nascent protoneuromasts in the wake of a progressively shrinking Wnt system. It does not incorporate subsequent interactions that use this prepattern of Fgf signaling to determine specification of a central *atoh1a*-expressing cell, nor how *atoh1a* cell plays a critical role in determining effective maturation of the protoneuromasts prior to their deposition.

While interactions between Wnt and Fgf signaling systems can account for self-organization of center-biased Fgf signaling centers in the wake of a progressively shrinking gradient of Wnt activity, the GRN does not account for the generation of the Wnt gradient nor for the biological mechanism that determines its progressive shrinking. These features had to be hardwired into the program for it to recapitulate periodic neuromast formation and deposition, and in this manner, the model identifies features whose biological basis needs to be better understood.

The progressive deposition of trailing cells when they no longer have access to some threshold level of Fgf secreted by Wnt active cells helps account for scaling of the primordium length with the shrinking Wnt system. However, mechanisms that determine deposition of trailing cells remain poorly understood. Loss of migratory potential is also likely to be related to progressive loss of additional Wnt-dependent factors that help sustain collective migration. It is also likely to be related to maturation of protoneuromasts (Durdu et al. 2014; Palardy et al. 2025) and involve additional mechanisms that regulate and balance adhesive interactions within the migrating primordium; some that operate at a tissue scale to determine cohesive migration as a column of cells, while others promote integrity of protoneuromast epithelial rosettes and their separation from the migrating primordium.

Model 5: How Mechanical Interactions Determine Periodic Aggregation without a Prepattern

What It Explores

This ABM, built using NetLogo (Mukherjee et al. 2025), shows how the combination of adhesion and apical constriction can determine self-organization of periodic cell clusters. In the model, spring-like interactions between primordium cells promote aggregation of cells, while tension along the length of the primordium, associated with relatively slower migration of trailing cells, pulls them apart, inhibiting aggregation.

How It Was Built

The primordium is represented as a column of *turtles* that form spring-like *links* with their neighbors within a defined radius (Fig. 5). The column of *turtles* is divided into a leading compartment with *WNTers* (green), representing *turtles* in a Wnt domain, while *FGFers* (yellow) are in a trailing Fgf domain. *WNTers* move with a defined speed, while *FGFers* move as a result of mechanical coupling by *links* or are assigned an independent lower speed, creating a tension that stretches the primordium. At each iteration, the *turtles* form *links* with neighbors that are within a defined radius. This pulls them closer promoting their aggregation. The initial aggregation can bring additional *turtles* within the radius that allows *link* formation. As *links* form with these additional *turtles* in the next iteration of the program, the aggregate grows. Left unimpeded, the aggregates would continue to grow; however, tension associated with migration pulls cells apart stretching the *links*, and if they extend beyond a defined length, the *links* break and aggregation is inhibited.

The length of the domain with *WNTers* shrinks at a defined rate leaving trailing cells in a Fgf active zone with *FGFers*. As in the previous models (Figs. 3 and 4), the length of the primordium is set based on the current length of the Wnt active zone, so that the *WNTer* domain is always 0.6 the length of the primordium. Cells in a trailing zone beyond the permitted length of the primordium are designated as *Depositers* (red), and their *links* are set to break at a lower threshold, facilitating their deposition by the migrating primordium. To simulate regional differences, spring parameters are set for relatively tighter *links* in the *FGFers* compared to *WNTers*.

What Did We Learn

The model shows how cells spontaneously aggregate to form periodic clusters within the migrating primordium (Fig. 5; Supplemental Movie 7). Increased adhesion, represented by stronger spring constants or shorter *links*, determines formation of fewer larger clusters. On the other hand, when the leading domain migrates faster, stretching of the primordium results in formation of fewer smaller aggregates. This model of the primordium is like the first model described in Figure 2, in which *turtles* were connected by *links*. However, in that model, cell clusters did not spontaneously form. A key difference is that in the first model, *links* were more static; once established they remained in place.

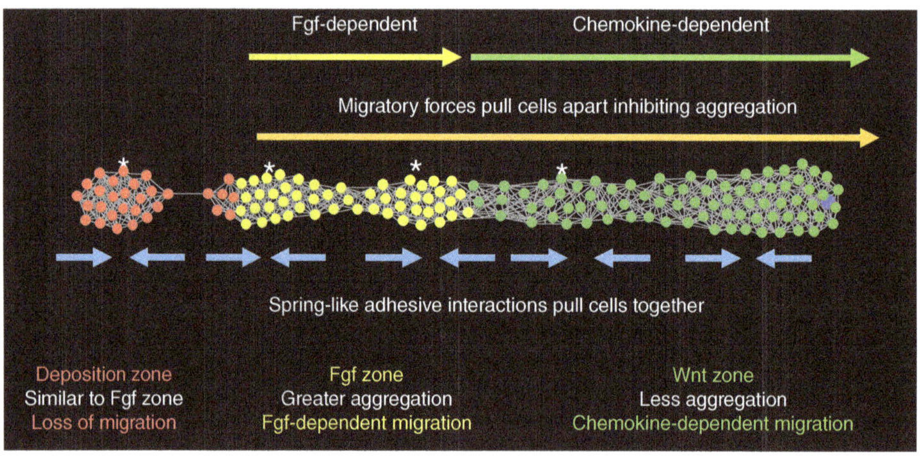

Figure 5. Agent-based model shows how mechanical interactions can determine periodic cell aggregation. Spring-like *links* connect turtles in a leading Wnt zone (green), Fgf zone (yellow), and a trailing depositing zone (red). The model shows how the combination of forces associated with iteratively formed *links* that pull cells together to promote aggregation, coupled with tension associated with migration that pulls cells apart breaking *links*, can result in self-organization of periodic clustering in the primordium. Asterisks indicate self-organized cell clusters (see Supplemental Movie 7).

In this model, *links* with neighbors are dynamically regulated; at every iteration, *turtles* can make additional *links* with new neighbors that are pulled within the defined radius required for *link* formation. At the same time, stretch of *links* beyond a defined length results in their loss. This dynamic ability to make and break *links*, which results in selective growth of some clusters and dissipation of others, is critical to self-organization of periodic clustering.

What Are Its Limitations

In this ABM, cells are represented as dimensionless nodes connected by *links*, which represent the influence of both cell adhesion and apical constriction in vivo. CPMs, described below, were developed to overcome this limitation (Box 3). In this environment, cells have features with which cell–cell adhesion and apical constriction can be independently represented. An additional limitation of this model is that it is exclusively based on mechanical interactions, and it ignores the role of the Fgf activity prepattern in influencing where aggregates form. Future models will integrate both mechanisms to visualize how cell signaling and mechanics operate together to influence the number and size of cell clusters.

Model 6: A Cellular Potts Model that Shows How Cluster Number Is Affected by Migration

What It Explores

In CPMs, distinct types of cells can be specified, each with their own target area, perimeters, adhesive properties, and migratory behaviors. In addition, spring-like elements *link* cells or compartments within cells to simulate the function of cytoskeletal elements that determine polarization of cells and apical constriction. As before, a model of the migrating primordium was built to visualize how mechanical interactions mediated by adhesion and apical constriction could determine self-organization of cell clusters as in the previous model (Mukherjee et al. 2025). However, this model was different from previous models in two important ways. While previous two-dimensional (2D) models represented the primordium from a "top" view, this model represented the primordium in a side view (Fig. 6A,B). With this perspective, epithelialized cells in the trailing compartment with apicobasal polarity could be represented as structures with distinct apical, lateral, and basal compartments, each with distinct target parameters and behaviors. Furthermore, as CPMs are especially effective at simulating emergent cellular organization and behavior determined by differential adhesion between different types of cells, this model simulated emergent behavior and cellular organization of the primordium, not only because of interactions within the primordium but also because of interactions with surrounding tissue as it migrates in confinement with the skin above and ECM of the muscle layer below.

How It Was Built

The model defined six types of cells (Fig. 6C). *Wnt* domain cells (blue) in the leading domain were represented as single-compartment cells without inherent polarization, and they moved forward in response to a force representing Cxcl12a-dependent migration. Apicobasally polarized *FGF* domain cells in a trailing domain were created with separate *apical*, *lateral*, and *basal* compartments, whose target parameters could be independently defined. Spring-like *links* between apical compartments were used to simulate apical constriction, and migratory behavior of the tripartite FGF cells was dependent on basal compartments, which moved forward in response to a force representing Fgf-dependent migration. Deeper *FGF* and *Wnt* domain cells were covered by an additional population of *superficial* flat cells (Dalle Nogare et al. 2020; Olson and Nechiporuk 2021) (yellow) that were themselves apposed to overlying *skin* cells (green). The primordium cells migrated over a population of ECM cells, representing the ECM of deeper *muscle* cells. As primordium cells migrate in a cohesive column, all primordium cells were assigned adhesivity that was higher between primordium cells than with overlying skin or ECM cells. In addition, as there is no obvious mixing of differ-

BOX 3. CELLULAR POTTS MODELS (CPMs)

While NetLogo agent-based models (ABMs) are very effective for modeling dynamic organization of cells, examining the dynamics of individual cell shape, especially when there are a very large number of cells, is much more difficult in this modeling environment. Cellular Potts models (CPMs) represent a special class of ABM that is effective at modeling the dynamics of cell shape as each individual cell is represented by a collection of pixels, whose dynamic rearrangement in position or number represents changes in shape, size, or movement of individual cells. Different types of cells can be specified in the model, each with their target cell size, boundary size, and potential to move in response to external influences like chemotactic factors. At each iteration of the program (Monte Carlo step [MCS]), a random pixel associated with the cell perimeter can protrude or "retract" resulting in a change in shape, cell size, boundary size, or position of the cell. This change is accepted or rejected based on the computed phenomenological "energy" (the Hamiltonian) of each cell, which is a measure of the difference between cell's parameters and their defined target value. Changes that move a cell closer to its target values reduce its "energy," and the change is accepted, while changes that move it further away increase its "energy," and the change is rejected or accepted at some defined low frequency. The concept of "contact energy" of a cell is used to model cell sorting resulting from differential adhesion between cells, where juxtaposition of pixels of adjacent cells with strong adhesion to each other is associated with low contact energy, while the juxtaposition of pixels of cells associated with low adhesion is associated with high contact energy. The repeated iteration of this process at each MCS where a move of a random pixel changes the perimeter of a cell, and the change is accepted or rejected based on where it reduces or increases the energy of the cell, allows one to observe potential dynamic changes in cell shape and aggregate behavior of large groups of cells following interactions within cells, between cells, and with the environment within which they move.

CPMs are based on minimizing a Hamiltonian (a phenomenological "energy") rather than explicit Newtonian forces. Furthermore, MCSs used to evolve the model do not naturally scale to actual physical time. These and additional issues have led to criticism of CPMs. Nevertheless, the models are quite effective at capturing a fundamental dynamic property of cells, where gene regulatory and signaling networks operate at different scales to keep cells within target parameters that define the unique features of a particular cell type. Although the feedback mechanisms that determine such homeostatic mechanisms are often poorly understood, CPMs show how their operation within cells together with the response of cells to other cells or features in their environmental influence can determine the dynamic evolution of form and cellular organization.

ent cell types in the primordium, WNT, FGF, and superficial cells had most adhesivity to other cells of the same type. Furthermore, as superficial cells and the basal compartments of FGF cells migrate in apposition to the skin above and ECM below, there was increased adhesivity between these cells but not so high as to interfere with movement of these cells. Finally, skin cells were lightly adherent to the underlying ECM cells, keeping these two layers apposed in the absence of the primordium.

What Did We Learn

Selective adhesion based on the principles outlined above resulted in emergent morphology and behavior that recapitulated specific aspects of primordium morphogenesis remarkably well (Fig. 6D–F; Supplemental Movie 8; Mukherjee et al. 2025). The primordium migrated as a cohesive structure wedged between the skin and ECM prying them apart as it moved forward, while the skin and ECM moved back in apposition following passage of the primordium. Within the primordium, FGF cells self-organized to form a stable number of clusters representing epithelial rosettes. Initiation of cell clusters in this context is stochastic and not determined by any prepattern. Initially, their number and size are unstable as each cell pulls on it neighbors to pull it closer. As some clusters grow, others fall apart, resulting eventually in the formation of a stable pattern. Finally, when

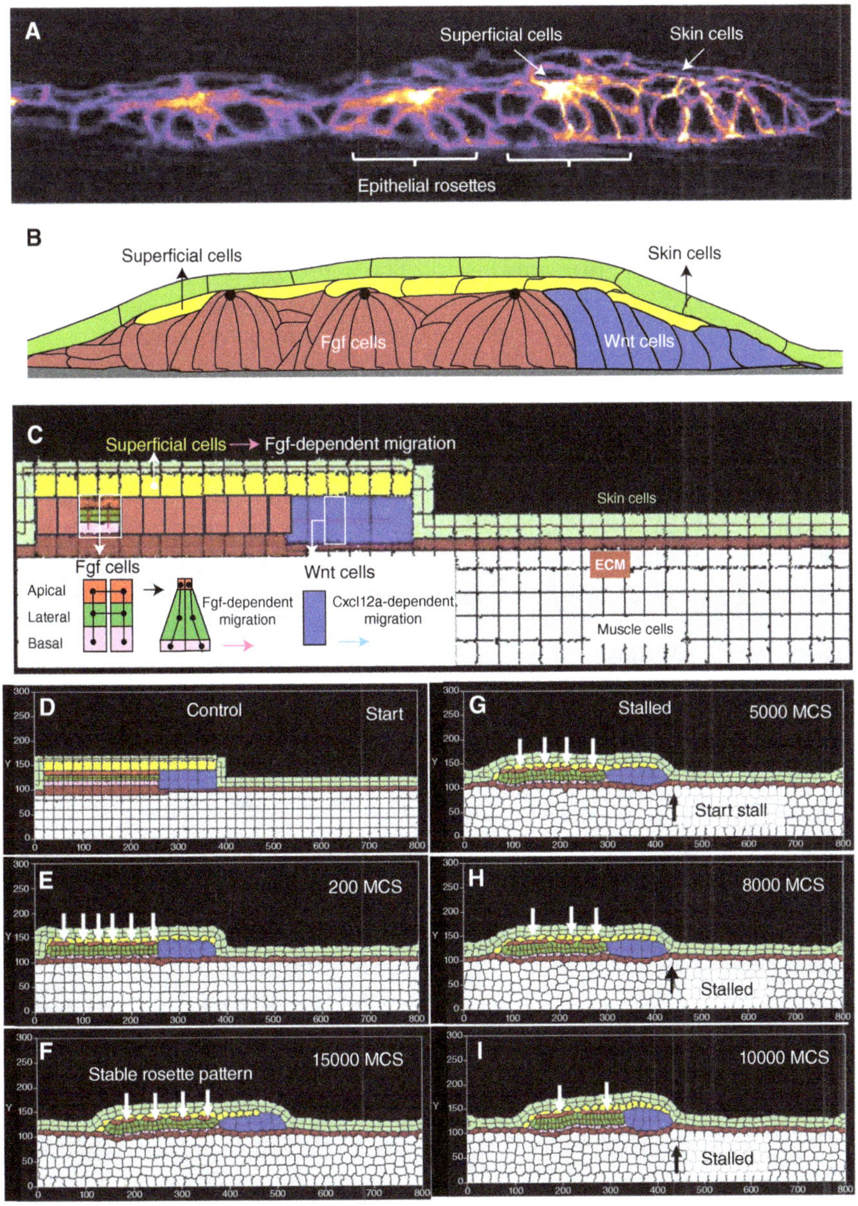

Figure 6. A cellular Potts model showing how mechanical interactions can determine periodic epithelial rosette formation. (*A*) Sagittal confocal slice of the migrating primordium in a Tg(cldnB:lyn-GFP) embryo showing overlying skin cells, superficial cells, and deeper trailing cells that have reorganized to form epithelial rosettes. GFP fluorescence intensity is visualized with the fire lookup table. (*B*) Schematic of cell types in the primordium. (*C*) Schematic of cell types in the primordium represented in the model. Fgf cells are represented as a composite of apical (red), lateral (green), and basal (pink) compartments. Black lines represent spring-like elements that *link* adjacent compartments; Fgf cell compartments represent cytoskeletal elements that influence polarized cell shape between adjacent apical compartments to simulate apical constriction. Superficial cells and basal compartments of Fgf cells move in response to Fgf, while Wnt cells move in response to chemokine signals. (*D–F*) Self-organization of stable "epithelial rosettes." White arrows show dynamic reorganization of cell clustering. (*G–I*) Stalling of Wnt cells (blue) after a stable pattern of four clusters has formed results in coalescing of rosettes to form three (*H*) and then two larger rosettes (*I*) (see Supplemental Movies 8 and 9). (MCS) Monte Carlo step.

leading Wnt domain cells transiently stalled and trailing cells kept moving, previously formed rosettes in the trailing zone fused to form fewer larger "rosettes" and subsequently split again to form more smaller rosettes when leading cell migration resumed (Fig. 6G–I; Supplemental Movie 9), recapitulating the dynamic reorganization of epithelial rosettes seen in vivo following similar experimental manipulations (Mukherjee et al. 2025).

What Are Its Limitations

The distinctive properties of apical, lateral, and basal compartments representing polarized FGF domain cells in this CPM helped simulate the self-organization of epithelial rosettes. However, many features incorporated in previous ABM are missing, including cell proliferation and deposition of cells. Furthermore, like the previous mechanics-based ABM, this CPM was developed to show how periodic clustering can emerge independent of Fgf signaling centers that normally seed the periodic formation of rosettes in the primordium. In vivo, signaling and mechanics interact to determine emergent behavior. Future models will be developed to better understand how integration of these processes contributes to robust, reproducible developmental trajectories that determine the emergence of pattern, form, and cell behavior.

CONCLUSION

The models described show how relatively simple ABM and CPM "toy" models can be used to understand how signaling and mechanical interactions between cells, defined by experimental observation, could determine emergent signaling patterns, cellular organization, and behavior in the migrating primordium. Together, the models show how polarized Wnt and Fgf signaling coordinates directed migration in response to a self-generated gradient, how scaling of primordium length with Wnt domain length helps predict the pattern of neuromast formation, and why primordium length may scale with the size of the Wnt domain. The models also show how interactions between Wnt and Fgf signaling have the potential to constitute a patterning system that coordinates periodic formation and deposition of neuromasts. Furthermore, they show how signaling and mechanical interactions between cells may work together to coordinate morphogenesis in the primordium. The limitations described in each section define directions for further development of models that were originally developed to address specific questions related to emergence of pattern, form, and migratory behavior of the primordium. Furthermore, the development of the models and comparison with in vivo behavior has helped define lacunae in our understanding and helped define biological questions that will be pursued in future studies. Finally, the models show how the primordium serves as a platform for understanding more broadly the principles that determine self-organization in development (for additional movies, see Supplemental Material).

ACKNOWLEDGMENTS

A previously unpublished NetLogo Model developed in the laboratory by Swetha Rao was used in this review (Fig. 3B). Chitnis laboratory members provided helpful feedback. This work was funded by the Eunice Kennedy Shriver National Institute of Child Health and Human Development, Division of Intramural Research.

REFERENCES

Aman A, Piotrowski T. 2008. Wnt/β-catenin and Fgf signaling control collective cell migration by restricting chemokine receptor expression. *Dev Cell* **15:** 749–761. doi:10.1016/j.devcel.2008.10.002

Aman A, Nguyen M, Piotrowski T. 2011. Wnt/β-catenin dependent cell proliferation underlies segmented lateral line morphogenesis. *Dev Biol* **349:** 470–482. doi:10.1016/j.ydbio.2010.10.022

Chitnis AB, Dalle Nogare D, Matsuda M. 2012. Building the posterior lateral line system in zebrafish. *Dev Neurobiol* **72:** 234–255. doi:10.1002/dneu.20962

Colak-Champollion T, Lan L, Jadhav AR, Yamaguchi N, Venkiteswaran G, Patel H, Cammer M, Meier-Schellersheim M, Knaut H. 2019. Cadherin-mediated cell coupling coordinates chemokine sensing across collectively migrating cells. *Curr Biol* **29:** 2570–2579.e7. doi:10.1016/j.cub.2019.06.061

Colombi A, Scianna M, Preziosi L. 2021. A hybrid integro-differential model for the early development of the zebrafish posterior lateral line. *J Theor Biol* **514:** 110578. doi:10.1016/j.jtbi.2020.110578

Dalle Nogare D, Chitnis AB. 2017. A framework for understanding morphogenesis and migration of the zebrafish posterior lateral line primordium. *Mech Dev* **148:** 69–78. doi:10.1016/j.mod.2017.04.005

Dalle Nogare D, Chitnis AB. 2020. NetLogo agent-based models as tools for understanding the self-organization of cell fate, morphogenesis and collective migration of the zebrafish posterior lateral line primordium. *Semin Cell Dev Biol* **100:** 186–198. doi:10.1016/j.semcdb.2019.12.015

Dalle Nogare D, Somers K, Rao S, Matsuda M, Reichman-Fried M, Raz E, Chitnis AB. 2014. Leading and trailing cells cooperate in collective migration of the zebrafish posterior lateral line primordium. *Development* **141:** 3188–3196. doi:10.1242/dev.106690

Dalle Nogare D, Nikaido M, Somers K, Head J, Piotrowski T, Chitnis AB. 2017. In toto imaging of the migrating zebrafish lateral line primordium at single cell resolution. *Dev Biol* **422:** 14–23. doi:10.1016/j.ydbio.2016.12.015

Dalle Nogare DE, Natesh N, Vishwasrao HD, Shroff H, Chitnis AB. 2020. Zebrafish posterior lateral line primordium migration requires interactions between a superficial sheath of motile cells and the skin. *eLife* **9:** e58251. doi:10.7554/eLife.58251

Dambly-Chaudière C, Cubedo N, Ghysen A. 2007. Control of cell migration in the development of the posterior lateral line: antagonistic interactions between the chemokine receptors CXCR4 and CXCR7/RDC1. *BMC Dev Biol* **7:** 23. doi:10.1186/1471-213X-7-23

Di Costanzo E, Natalini R, Preziosi L. 2015. A hybrid mathematical model for self-organizing cell migration in the zebrafish lateral line. *J Math Biol* **71:** 171–214. doi:10.1007/s00285-014-0812-9

Dijkgraaf S. 1963. The functioning and significance of the lateral-line organs. *Biol Rev Camb Philos Soc* **38:** 51–105. doi:10.1111/j.1469-185X.1963.tb00654.x

Donà E, Barry JD, Valentin G, Quirin C, Khmelinskii A, Kunze A, Durdu S, Newton LR, Fernandez-Minan A, Huber W, et al. 2013. Directional tissue migration through a self-generated chemokine gradient. *Nature* **503:** 285–289. doi:10.1038/nature12635

Dries R, Lange A, Heiny S, Berghaus KI, Bastmeyer M, Bentrop J. 2020. Cell proliferation and collective cell migration during zebrafish lateral line system development are regulated by Ncam/Fgf-receptor interactions. *Front Cell Dev Biol* **8:** 591011. doi:10.3389/fcell.2020.591011

Durdu S, Iskar M, Revenu C, Schieber N, Kunze A, Bork P, Schwab Y, Gilmour D. 2014. Luminal signalling links cell communication to tissue architecture during organogenesis. *Nature* **515:** 120–124. doi:10.1038/nature13852

Ernst S, Liu K, Agarwala S, Moratscheck N, Avci ME, Dalle Nogare D, Chitnis AB, Ronneberger O, Lecaudey V. 2012. Shroom3 is required downstream of FGF signalling to mediate proneuromast assembly in zebrafish. *Development* **139:** 4571–4581. doi:10.1242/dev.083253

Ghysen A, Dambly-Chaudiere C. 2007. The lateral line microcosmos. *Genes Dev* **21:** 2118–2130. doi:10.1101/gad.1568407

Graner F, Glazier JA. 1992. Simulation of biological cell sorting using a two-dimensional extended Potts model. *Phys Rev Lett* **69:** 2013–2016. doi:10.1103/PhysRevLett.69.2013

Haas P, Gilmour D. 2006. Chemokine signaling mediates self-organizing tissue migration in the zebrafish lateral line. *Dev Cell* **10:** 673–680. doi:10.1016/j.devcel.2006.02.019

Itoh M, Chitnis AB. 2001. Expression of proneural and neurogenic genes in the zebrafish lateral line primordium correlates with selection of hair cell fate in neuromasts. *Mech Dev* **102:** 263–266. doi:10.1016/S0925-4773(01)00308-2

Knutsdottir H, Zmurchok C, Bhaskar D, Palsson E, Dalle Nogare D, Chitnis AB, Edelstein-Keshet L. 2017. Polarization and migration in the zebrafish posterior lateral line system. *PLoS Comput Biol* **13:** e1005451. doi:10.1371/journal.pcbi.1005451

Lau S, Feitzinger A, Venkiteswaran G, Wang J, Lewellis SW, Koplinski CA, Peterson FC, Volkman BF, Meier-Schellersheim M, Knaut H. 2020. A negative-feedback loop maintains optimal chemokine concentrations for directional cell migration. *Nat Cell Biol* **22:** 266–273. doi:10.1038/s41556-020-0465-4

Lecaudey V, Cakan-Akdogan G, Norton WH, Gilmour D. 2008. Dynamic Fgf signaling couples morphogenesis and migration in the zebrafish lateral line primordium. *Development* **135:** 2695–2705. doi:10.1242/dev.025981

Li Q, Shirabe K, Kuwada JY. 2004. Chemokine signaling regulates sensory cell migration in zebrafish. *Dev Biol* **269:** 123–136. doi:10.1016/j.ydbio.2004.01.020

Matsuda M, Chitnis AB. 2010. Atoh1a expression must be restricted by Notch signaling for effective morphogenesis of the posterior lateral line primordium in zebrafish. *Development* **137:** 3477–3487. doi:10.1242/dev.052761

Matsuda M, Dalle Nogare D, Somers K, Martin K, Wang C, Chitnis AB. 2013. Lef1 regulates Dusp6 to influence neuromast formation and spacing in the zebrafish posterior lateral line primordium. *Development* **140:** 2387–2397. doi:10.1242/dev.091348

Meier-Schellersheim M, Klauschen F, Angermann B. 2009. Computational modeling of signaling networks for eukaryotic chemosensing. *Methods Mol Biol* **571:** 507–526. doi:10.1007/978-1-60761-198-1_33

Meinhardt H, Gierer A. 2000. Pattern formation by local self-activation and lateral inhibition. *Bioessays* **22:** 753–760. doi:10.1002/1521-1878(200008)22:8<753::AID-BIES9>3.0.CO;2-Z

Mukherjee A, Hilzendeger M, Rinvelt A, Fatma S, Schupp M, Dalle Nogare D, Chitnis AB. 2025. Signaling and mechanics influence the number and size of epithelial rosettes in the migrating zebrafish posterior lateral line primordium. bioRxiv doi: 10.1101/2025.05.17.650326

Nechiporuk A, Raible DW. 2008. FGF-dependent mechanosensory organ patterning in zebrafish. *Science* **320:** 1774–1777. doi:10.1126/science.1156547

Olson HM, Nechiporuk AV. 2021. Lamellipodia-like protrusions and focal adhesions contribute to collective cell migration in zebrafish. *Dev Biol* **469:** 125–134. doi:10.1016/j.ydbio.2020.10.007

Palardy G, Yoo KW, Fatma S, Mukherjee A, Wang C, Ravi P, Chitnis AB. 2025. Soxb1 family members inhibit Wnt signaling to promote maturation and deposition of stable neuromasts by the zebrafish posterior lateral line primordium. bioRxiv doi: 10.1101/2025.04.23.650055

Resnick M. 1994. *Turtles, termites, and traffic jams: explorations in massively parallel microworlds.* MIT Press, Cambridge, MA.

Sapède D, Rossel M, Dambly-Chaudière C, Ghysen A. 2005. Role of SDF1 chemokine in the development of lateral line efferent and facial motor neurons. *Proc Natl Acad Sci* **102:** 1714–1718. doi:10.1073/pnas.0406382102

Savill NJ, Hogeweg P. 1997. Modelling morphogenesis: from single cells to crawling slugs. *J Theor Biol* **184:** 229–235. doi:10.1006/jtbi.1996.0237

Swat MH, Thomas GL, Belmonte JM, Shirinifard A, Hmeljak D, Glazier JA. 2012. Multi-scale modeling of tissues using CompuCell3D. *Methods Cell Biol* **110:** 325–366. doi:10.1016/B978-0-12-388403-9.00013-8

Valentin G, Haas P, Gilmour D. 2007. The chemokine SDF1a coordinates tissue migration through the spatially restricted activation of Cxcr7 and Cxcr4b. *Curr Biol* **17:** 1026–1031. doi:10.1016/j.cub.2007.05.020

Venkiteswaran G, Lewellis SW, Wang J, Reynolds E, Nicholson C, Knaut H. 2013. Generation and dynamics of an endogenous, self-generated signaling gradient across a migrating tissue. *Cell* **155:** 674–687. doi:10.1016/j.cell.2013.09.046

Wilensky U. 1999. *NetLogo. Center for connected learning and computer-based modeling.* Northwestern University, Evanston, IL.

Zinn-Björkman L, Adler FR. 2018. Modeling factors that regulate cell cooperativity in the zebrafish posterior lateral line primordium. *J Theor Biol* **444:** 93–99. doi:10.1016/j.jtbi.2018.02.009

Modern Modeling of Single-Cell Migration: From Membrane Tension and Galvanotaxis to Machine Learning

Wenzheng Shi[1] and Alex Mogilner[1,2]

[1]Courant Institute; [2]Department of Biology, New York University, New York, New York 10012, USA

Correspondence: wenzhengshi@nyu.edu; mogilner@cims.nyu.edu

Cell migration phenomenon has inspired and benefited from computational modeling for decades. Here, we review recent applications of traditional bottom-up modeling to three aspects of cell migration: the role of membrane tension (MT) in organizing directional cell motility, the role of the electric field (EF) as the directional cue for migration, and the mechanics of three-dimensional migration. We then discuss nascent applications of machine learning (ML) to cell migration and galvanotaxis. We focus on the migratory mechanisms of the single cell and highlight the feedback between theory and experiment.

Cells migrate individually and in cohesive groups in many physiological processes, including wound healing, metastasis, immune response, and embryogenesis (SenGupta et al. 2021). A useful way to break down the phenomenon of cell migration is to think of it as the interaction of three dynamic processes—cell motility, polarization, and directional sensing (Shi et al. 2013). Namely, a cell must generate motility (protrusions and retractions around its periphery), and then it must polarize the protrusions to the front and retractions to the rear, and finally, it has to orient the rear–front axis in response to an external cue.

Several decades of using microscopy on single cells migrating in vitro on flat hard surfaces brought a detailed mechanistic understanding of each of these three processes (Blanchoin et al. 2014). In recent years, the focus has started to shift to multicellular migration (Shi et al. 2025) but studying individual migrating cells is not over. Even understanding of 2D cell migration is full of holes, and quantitative research on cells crawling through 3D tissues and the extracellular matrix (ECM) is still in its infancy (Yamada and Sixt 2019).

There are many excellent and comprehensive reviews of the cell biology of the migration phenomenon, for example (Blanchoin et al. 2014; Rappel and Edelstein-Keshet 2017; Van Helvert et al. 2018; SenGupta et al. 2021), which explain the interlinked action of actomyosin network mechanics and signaling pathways needed for motility, polarity, and directional sensing to self-organize and synergize. Even the number of reviews on the modeling of cell migration is great, we refer the reader to a few of them (Danuser et al. 2013; Buttenschön and

Cite this article as *Cold Spring Harb Perspect Biol* doi: 10.1101/cshperspect.a041745

Edelstein-Keshet 2020; DiNapoli et al. 2020). Due to the amount of attention paid to cell migration, choosing what else needs to be said is not an easy task. In this review, we decided to focus on three modern modeling problems, each related to the respective fundamental process of migration—mechanics of motility, migratory polarity, and direction sensing. We review several modeling frameworks applied to simulate single cells crawling through the 3D ECM, modeling the role of membrane tension (MT) as a global inhibitor in cell polarity maintenance, and nascent modeling of cells using electric fields (EFs) to navigate wounds. Lastly, we address initial efforts to change the modeling paradigm and embrace machine learning (ML) to predict traction forces (TFs) and learn migratory trajectories.

Two notes to mention are, in general, there are two modes of cell migration. One is called mesenchymal migration, in which a growing actin network generates protrusion, firm adhesion of the protrusion to the substrate, and actomyosin contraction to pull up the rear. Another is amoeboid motility (Callan-Jones 2022), in which cytoplasmic pressure is harnessed to generate a protrusion, and then actomyosin contraction causes a retrograde flow of an actin cortex, the friction of which against the ECM propels the cell. Here, we mostly discuss the mesenchymal migration. Second, even after narrowing the scope, we cannot discuss and cite a great number of relevant papers, for which we apologize.

MEMBRANE TENSION IN CELL MOTILITY

The well-understood mode of cell migration is 2D mesenchymal motility. However, one of its important aspects, the role of MT in maintaining motile cell polarity, has only recently become quantitatively elucidated. The number of publications on this topic is not too great, and several reviews of cell biological (Blanchoin et al. 2014; Van Helvert et al. 2018) and biophysical (Keren 2011; Diz-Muñoz et al. 2013; Sens and Plastino 2015; Sitarska and Diz-Muñoz 2020) aspects of this topic provide a very helpful guide to the MT role. Here, we focus on the plasma membrane and refer to it as "the membrane."

About 15 years ago, early models explained how motile cells could use the peculiar physics of the lipid membrane to serve as a global mechanical inhibitor: the membrane is a non-stretchable 2D fluid (Keren 2011), so if the membrane is tensed locally, the elevated MT spreads globally throughout the cell surface on a subsecond scale (Kozlov and Mogilner 2007). The main source of this local MT is pushed by the growing branched actin network on the inside of the membrane (Lieber et al. 2013). Nowhere does this simplistic concept work better than in one of the fastest cell types—fish epidermal keratocyte—in which the large motile appendage, lamellipodium, is a wide and flat actin network polymerizing against the unfolded and unconnected to the actin membrane. According to the simplest model of the lamellipodium (Ofer et al. 2011), actin, pushing from the inside, generates the globally spread MT (Fig. 1). This MT crushes the actin network weakened by gradual disassembly at the cell rear (Pontes et al. 2017) and stops actin at the sides from protruding, so that the actin network treadmills, polymerizing at the front, retracting at the rear, and disassembling throughout the lamellipodium. This model explains how this steady motility can be maintained even without the contractile action of myosin (Ofer et al. 2011). In the laboratory coordinate system, the actomyosin network is essentially immobile due to firm adhesions throughout the lamellipodium, away from the cell edges. Thus, in the moving cell, myosin is essentially swept to the rear, where its contractile action helps to crush the network allowing MT to retract the rear (Kozlov and Mogilner 2007; Gauthier et al. 2011) in sync with the protruding front. According to both experiments and models, high MT also helps compress and realign the actin network, enhancing the contractile action at the rear (Gauthier et al. 2011; Lomakin et al. 2015). Another model, based on experiments with motile nematode sperm cells, which use a network of fibers made of major sperm protein instead of actin, posited another function of MT—to compress the sides of the motile cell thereby aligning

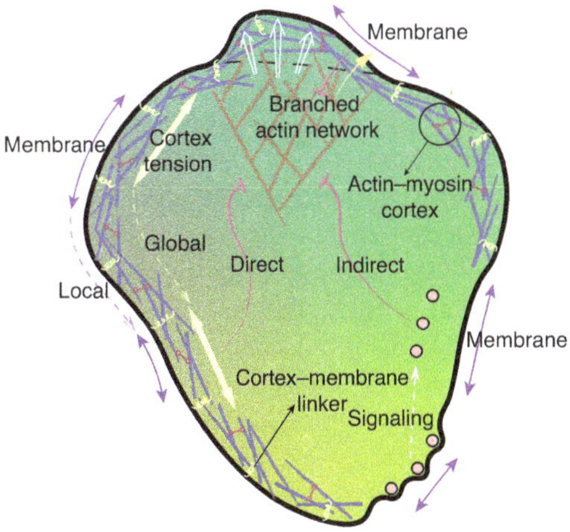

Figure 1. Membrane tension assists cell polarization at the onset of motility. Local growth of branched actin network pushes on the inside of the cell leading edge stretching membrane and generates tension. In-plane tension spreads locally in the plasma membrane, while tension propagates globally in the actin–myosin cortex. The membrane and cortex are interlinked, which, together with their complex mechanical properties and the nature of local mechanical action of the pushing force at the leading edge, fine-tunes the speed and range of the tension spread from the leading edge. The tension directly stalls actin protrusions at the cell sides and rear, thereby enabling actin–myosin contraction there. In addition, the tension could stretch out the membrane folds, releasing signaling molecules and triggering indirect regulatory pathways that selectively inhibit actin dynamics.

the growing fibers along the rear–front axis focusing the protrusion at the cell front (Batchelder et al. 2011).

As the cell crawls, the membrane envelope around the cell must be propelled forward. One respective mechanism is exocytosis at the leading edge (Gauthier et al. 2011) accompanied by endocytosis throughout the cell, together with forward motor-driven translocation of membrane vesicles in the cell (Bretscher 1984). A recent mathematical model of the cell moving in a 3D ECM integrated experimental data (Hetmanski et al. 2019) points out that MT is lower at the cell rear triggering endocytosis there. These endocytic vesicles recruit molecules activating myosin, so myosin retracts the cell rear lowering MT and creating the positive feedback loop. However, in the absence of endocytosis/exocytosis, the membrane can simply flow forward in its own plane in the motile cell. This flow is driven by the slight tension gradient: MT is elevated by the actin pushing at the front and is lower at the rear (Fogelson and Mogilner 2014; Schweitzer et al. 2014).

The first systematic investigation of the MT's polarizing effect was conducted in a seminal study on neutrophils (Houk et al. 2012). The authors (Houk et al. 2012), after establishing the global inhibitory effect of MT on protrusions away from the leading edge of these cells of the immune system, asked whether this effect is not the direct and immediate mechanical effect, as in the keratocyte models, but rather an indirect action of signaling molecules, activated by the leading edge protrusions, diffusing to the cell rear and triggering a reaction that inhibits the protrusions there. Elegant experiments with bilobed cell shapes with a long and narrow corridor between one protruding and one passive lobe and simulations of two plausible reaction–diffusion models on these shapes demonstrated that such indirect mechanisms would be much slower than observed, leaving the rapidly and globally spreading MT (Fig. 1) as the only factor explaining the measurements (Houk et al. 2012).

It turned out that not only direct mechanical tension but also additional mechanosensory feedback, indirectly linking MT with actin assembly, stalls the side and rear protrusions (Diz-Muñoz et al. 2016). Specifically, elevated MT acts globally through signaling molecules to limit actin nucleation (Fig. 1). How exactly these signaling molecules are affected by MT is not yet known, but a very recent study (Quiroga et al. 2023) hints at the following possibility: at low MT, the membrane bends and folds creating local evaginations, to which curvature-sensing proteins can bind and potentially deactivate the actin inhibitors. High MT stretches out the membrane causing unbinding of the curvature-sensing proteins reactivating the actin inhibitors (Fig. 1). Mathematical modeling suggests roles for both direct (mechanical) and indirect (mechanochemical) feedbacks in organizing cell polarity and motility (Diz-Muñoz et al. 2016): the indirect feedback leads to competition between multiple protrusions, essential for nimble path finding, while the direct feedback spatially restricts the protrusion to the single wide leading edge for persistent movement.

The early successes of the MT modeling were thrown into doubt by the more recent study (Shi et al. 2018), which demonstrated a relatively slow (minutes rather than seconds) and not global MT propagation around the source of MT tension. However, the latest study (De Belly et al. 2023), employing both modeling and experiment, offers an elegant and convincing explanation for the controversy. This study draws attention to the fact that in most cell types (keratocyte is but a rare exception), the membrane is underlined by the thin shell of the actomyosin cortex, linked to the membrane by multiple proteins (Fig. 1). De Belly et al. (2023) found that a global MT propagation occurs either after an optogenetic activation of local cortex contraction (not directly affecting the membrane) or during micropipette aspiration experiments that exert a tugging force on the cell envelope locally and directly affecting both membrane and cortex. In contrast, pulling a membrane tether (not directly affecting the cortex) did not lead to global MT propagation (Fig. 1). These results are explained by a mathematical model (De Belly et al. 2023), which treats the membrane as an elastic contour, the cortex as a viscous, and contractile contour and links between them as elastic links. We find it difficult to explain the results of this model's simulations intuitively, but they show that in certain parameter ranges, the model accounts for all three experimental results.

An alternative recent model (Barnoy et al. 2023; Dharan et al. 2025) looks at the same problem from a different angle. This model posits that the membrane is subdivided into submicron compartments by a network of transmembrane protein "picket-fences." These fences are hydrodynamic barriers for the in-plane membrane flow. According to the model, the propagation speed of MT perturbations on the cell surface is set by the intracellular pressure and hydrodynamic permeability of the fences and can vary by several orders of magnitude, so that MT spreads globally and fast at high cytoplasmic pressures and only locally and slowly at low cytoplasmic pressures.

Three recent modeling studies make additional predictions for the MT role. All these models simulate motile cells as free boundary bodies, so the boundary contours (cell edges) in 2D protrude and retract locally according to some dynamic rules. The first of these models (Chen et al. 2023) considers adhesion linkers that turn over and diffuse along the cell periphery, detaching at a force-dependent rate, while the local force is set by MT. The local protrusion/retraction rates are then set by the local linkers' density and MT. This model predicts an instability of a randomly perturbed, nonpolar cell shape, and that different MT levels can result in diverse motility modes, including gliding, zigzagging, rotating, and chaotic movements.

The second, most detailed model couples MT, actin, and myosin dynamics with the action of Rac/Rho molecules regulating actin and myosin (Tao et al. 2020). Simulations of this model predict that there is an optimal tension for rapid migration: the low tension allows multiple protrusions in random directions without persistent motile progress and the high-tension stalls protrusion everywhere. The latter conclusion is also reached by the third model (Winkler et al. 2016) in which the "phase field," which determines dynamic cell shape and area, is coupled to the evolving "polarization field" variable, describing the average local orientation of actin filaments inside the cell and to MT that effectively attenuates actin dynamics.

QUANTITATIVE UNDERSTANDING OF GALVANOTAXIS

Motile and polarized cells orient their rear–front axis in response to multiple external signals. Many such "taxis" behaviors were studied; the most researched of these is chemotaxis (Shi et al. 2013), a response to spatial chemical gradients, but other biased migration (i.e., topotaxis [response to ECM density gradients] and durotaxis [response to gradients in ECM stiffness]) also received attention (SenGupta et al. 2021). Here, we focus on galvanotaxis—the directional response of migrating cells to an EF (Chang and Minc 2014). Physiological EFs, generated by inhomogeneous transepithelial ion flows, are ubiquitous in wounds and developing embryos (Chang and Minc 2014). Normally, the cathode (negative EF terminal) is in the wound, and several types of migrating cells move to the cathode closing and repairing the wound in the healing process (Zhao 2009). Importantly, several experiments established that EF overrides other directional signals, such as chemotaxis (Zhao 2009; Sun et al. 2019). Mathematical modeling of cells navigating chemotactic gradient and EF simultaneously was reported in Vanegas-Acosta et al. (2012). However, very little modeling of galvanotaxis was done, and here we review these modeling efforts.

Galvanotaxis uses the same intracellular signaling pathways that guide motile cytoskeletal machinery as chemotaxis (Zhao et al. 2006), and the latter was extensively studied, both experimentally and theoretically (Shi et al. 2013). Thus, the main question about galvanotaxis is what the EF sensors are upstream of the intracellular signaling pathways. As the cell cytoplasm is an electric conductor, the low-frequency physiological EF does not penetrate the cell (Taghian et al. 2015), and the EF sensors must be on the cell surface. Four biologically relevant EF-sensing mechanisms are (Fig. 2; Allen et al. 2013) (1) EF-sensitive ion channels can be differently affected on the cathodal and anodal (negative and positive EF terminals, respectively) cell sides. This mechanism is known to orient the growth of bacteria and yeast cells (Chang and Minc 2014). There is also an interesting mathematical model showing that if, for example, ion channels on the cell front/rear are opened/closed, respectively, then ion influx creates a high osmotic pressure near the leading edge, which can generate a local water influx inflating the cell front and effectively driving the cell movement (Fig. 2; Li et al. 2015). However, for epithelial migrating cells, this mechanism was ruled out experimentally (Allen et al. 2013). (2) The cell surface has a net charge (McLaughlin and Poo 1981), so a small Coulomb force acts on the whole cell in EF, tugging the cell to either cathode or anode. A mathematical model demonstrated that such force can significantly bias the swimming of some cell types (Ogawa et al. 2006), yet again, for epithelial migrating cells this mechanism is highly unlikely (Allen et al. 2013). (3) Extracellular fluid is an electrolyte, and EF generates the so-called electro-osmotic flow: mobile counterions near the cell surface are pulled by EF to either the cathode or anode and the ions drag the extracellular fluid with them. The resulting force from this flow is too small to affect the cell directly (Allen et al. 2013), but this flow can drag mobile membrane proteins with charged residues outside the cell to

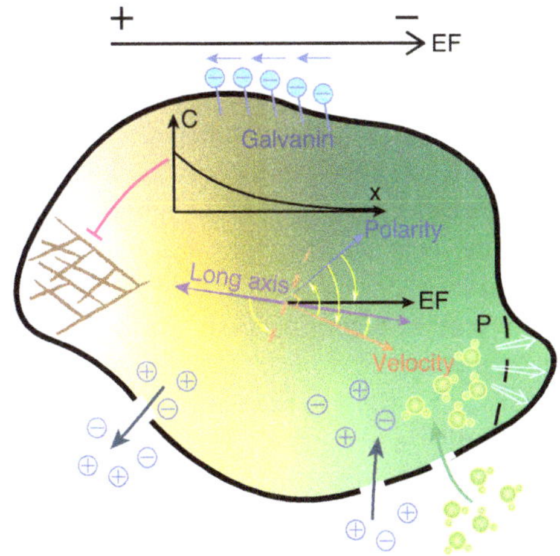

Figure 2. Electric field (EF) orients migrating cells. An electric field drives negatively charged mobile membrane proteins (Galvanin) to the anodal (positive terminal) side of the cell, generating a concentration (C) gradient of this protein and inhibiting protrusion at that side thereby enabling the cell to polarize toward the cathode (negative terminal). Another potential orientation mechanism is based on the differential opening/closing of voltage-sensitive membrane ion channels at the anodal/cathodal cell sides, respectively, leading to the respective outflow/inflow of ions. The resulting high ion concentration at the cathodal cell side creates local osmotic pressure and influx of water through aquaporin channels. Combined local pressure (P) generates protrusion toward the cathode. Hypothesized feedback in a model of cell orientation in an electric field are shown in the center: cell polarity vector aligns with the cathodal direction and turns to the velocity vector; in turn, velocity rotates to the polarity vector and pivots to the cell long axis. Lastly, the cell's long axis adjusts perpendicularly to the velocity.

one of the cell sides (McLaughlin and Poo 1981). (4) EF applies Coulomb force to the mobile membrane proteins with charged residues outside the cell causing the so-called electrophoresis of these proteins and their accumulation to one of the cell sides (Fig. 2; McLaughlin and Poo 1981; Allen et al. 2013). Biophysical experiments reported in Allen et al. (2013) demonstrated indirectly that the accumulation of charged mobile membrane proteins to one of the cell sides is the most likely EF-sensing mechanism.

Recently, a seminal study (Belliveau et al. 2024) appeared reporting experimental proof that a highly negatively charged mobile membrane protein Galvanin is biased to the anodal side of the migrating cell. As soon as a threshold accumulation of Galvanin at the anodal side is reached, the net protrusion at that side is stalled (Fig. 2), and the cell starts biased migration to the opposite, cathodal side. Galvanin is likely either a direct part of the signaling relay affecting actomyosin dynamics downstream or is a binding partner to a key intracellular signaling molecule.

Simple galvanotaxis-related mathematical models have addressed related redistribution of the charged mobile membrane proteins in EF. The earliest of such models showed that electroosmotic flow can counteract the Coulomb force and, depending on cell surface charges, can move the charged mobile proteins to either the cathodal or anodal side of the cell (McLaughlin and Poo 1981). A detailed model of electro-osmotic flow was considered in Sarkar et al. (2019). Allen et al. (2013) analyzed the drift-diffusion model for the charged protein moving along the membrane and derived formulas for steady protein concentration gradient in EF and the time scale for gradient development (Fig. 2; Allen et al. 2013).

Another mathematical model, supported by microscopy data (Lin et al. 2017), demonstrated that in addition to individual charged proteins, highly charged glycolipid redistribution in EF can be an alternative candidate as the primary EF sensor. This study showed that rather than individual glycolipids, charged lipid rafts polarized to the cathode in EF, while in the randomly migrating cells without EF, no preferential raft distribution was found. In EF, the rafts merged and increased in size while approaching the cathode, which decreased the raft mobility and stabilized their cathodal bias. Several signaling molecules were shown to colocalize with the rafts to the cathode, while their distributions were not spatially biased in the absence of EF.

Three quantitative models were applied to learn the principles of galvanotaxis from analyses of cell trajectories in EF. The earliest of these theories (Gruler and Nuccitelli 2000) modeled the orientation of the velocity vector of the motile cell undergoing a biased random walk in an angular space, so deterministic turning of this vector to the cathodal direction is juxtaposed with random velocity turns while the cell speed is independent of EF signal (Gruler and Nuccitelli 2000). Another recent model (Prescott et al. 2021) considered three additional (to the velocity bias [Gruler and Nuccitelli 2000]) possible ways by which the EF could bias the random cell turns: polarity bias of the cell, an increase of cell speed in the cathodal direction, and speed alignment with EF. The authors (Prescott et al. 2021) used experimental training data sets—cell trajectories with and without EF and a Bayesian approach to parameter inference—to compare all four possible ways of migration. They concluded that only the polarity bias effect of the EF explains the data. Finally, an ML model was applied, for the first time (Sargent et al. 2022), to forecast the direction of cell migration in EF. The ML network was trained to predict cell direction one time-step ahead, given measured cell directions at previous time points and using several hypothetical models, such as (1) velocity vector does not change in time, and (2) velocity vector is a linear function of velocity vectors of previous two time steps, etc. The model trained by the data of certain EF strengths makes successful predictions for migration in other EF strengths and in time-varying EFs.

Lastly, a very recent model of galvanotaxis, which considered not only overall cell movements but also the cell dynamic shape, was proposed (Fig. 2; Nwogbaga and Camley 2023). In this model, the following processes are postulated: the cell is characterized by separate velocity and polarity vectors that turn randomly but also toward each other; velocity tends to turn to the cell's long axis and to align to the polarity vector; the cell elongates perpendicularly to the velocity; and polarity tends to align with EF vector. Simulations of this model, fitted to experimental data on keratocyte galvanotaxis, predicted that stiff, slow cells react to EF slowly but follow EF reliably. When cells are exposed to EF that switch direction rapidly, cells follow the average EF directions, while if the field is switched more slowly, cells follow a "staircase" trajectory.

MECHANICS OF THREE-DIMENSIONAL MIGRATION

There are important physiological examples of 2D cell migration (Labuz et al. 2023), but frequently cells migrate through the ECM, a complex 3D porous structure dominated by collagen fibers (Caswell and Zech 2018). Understanding 3D motility mechanics is complicated by the remarkable plasticity of cells switching between amoeboid and mesenchymal motility in the ECM. For example, 3D cells generate a variety of protrusions; their nucleus, which is an afterthought in 2D, becomes a limiting mechanical factor; often, cells deploy metalloproteases to cleave ECM and enlarge pore diameters on their path (Wolf et al. 2013). Modeling of the 3D motility is also hindered by much scarcer 3D microscopy data: we have only a vague idea about the ultrastructure of the 3D actomyosin networks. One of the rare exceptions is an elegant experimental study (Wilson et al. 2013) illustrating how complex 3D motile designs can become. This study discovered that the leading edge of a cell confined in a channel generates two distinct F-actin networks: an adherent network that polymerizes perpendicular to the channel walls and a "free" network that grows from the free membrane at the cell front. The authors propose that inward growth of the adherent network prevents retrograde movement of the free network enabling polymerization of the latter to be converted to forward protrusion (Wilson et al. 2013).

Despite this complexity, computational modeling of 3D cell migration through the ECM is a vigorous subfield. Early simple models of 3D random walks (Parkhurst and Saltzman 1992) and of an unstructured cell and ECM with anisotropic TFs (Zaman et al. 2005) started this effort; reviews of such modeling can be found in Rangarajan and Zaman (2008). Here, we give a limited review of the recent evolution of detailed mechanistic models of 3D migrating cells that started to deal with specific cell and ECM geometries. These models can be grouped into five categories (Fig. 3). We briefly mention biological insights from these models (so far very modest) focusing instead on the modeling frameworks. Many other notable computational models of 3D cell migration (Schlüter et al. 2012; van Oers et al. 2014; Cao et al. 2015; Heck et al. 2020; Godeau et al. 2022; Merino-Casallo et al. 2022), to mention but a few, can be loosely classified into one of the five categories that we discuss below.

First, there is a conceptually simple motile structure of the motor-clutch mechanism (Campbell et al. 2021): a dynamic adhesion clutch with a force-dependent detachment rate connects ECM fibers with the protruding end of a treadmilling segment of an actin network. Inward, the disassembling end of the actin segment is pulled to the cell body by myosin motors that obey a force–velocity relation. Effectively 1D, the motor-clutch mechanism becomes an "arm" of a cell, pulling the cell body along an ECM fiber (Fig. 3). In the efficient "cell migration simulator" (Bangasser et al. 2017; Prahl et al. 2020), the model cell resembles a spider with one cell body pulled by several such motor-clutch arms on the spiderweb of the ECM (Fig. 3). Simulations of this model make expected predictions, for example, about an ECM stiffness optimum for rapid migration that can be shifted by altering the number of active molecular motors and clutches. Despite the simplicity of this model, it has recently brought insight into glioblastoma cell migration (Hou et al. 2024). The authors of this study first reduced the 11-dimensional parameter space of the cell migration simulator to three principal mechanical parameters: motor number (responsible for myosin activity), clutch number (describing adhesion strength), and actin polymerization rate. A comparison of simulations and experiments showed

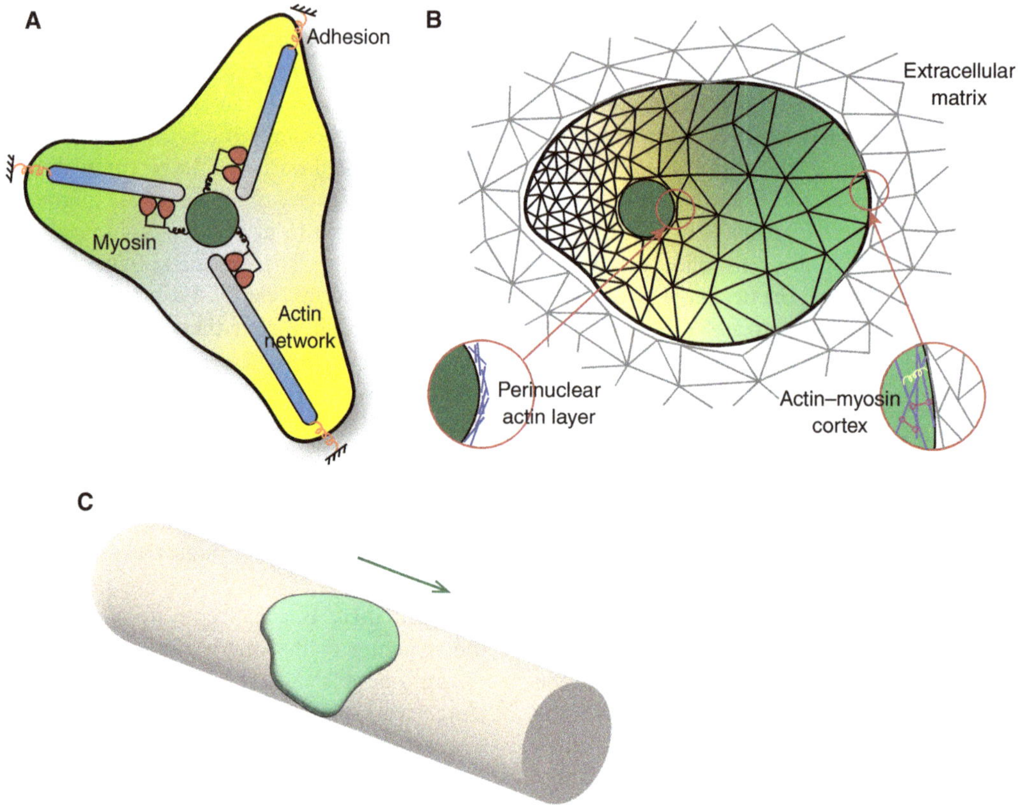

Figure 3. Mechanics of 3D cell motility. (*A*) Motor-clutch model of the cell motile mechanics: the cell has several motile appendages (arms) "grabbing and pulling" extracellular matrix (ECM) fibers. Each arm consists of an actin network that connects the cell body/nucleus, by myosin motors, to the adhesion complex (clutch). (*B*) Several models simulate the motile cell interior as an active link-node mesh, with links expanding on one side of the cell and contracting on the other. ECM is modeled as a passive link-node network. The latest models become complex and layered, differentiating between several types and dynamics of actin and myosin networks in the cell. (*C*) Phase-field models simulate the cell as a motile "active blob" migrating on a curved surface of an ECM fiber.

that glioblastoma cells balance motor/clutch ratios to enable effective migration. The authors identified 18 genes that correlated with the mechanical model parameters, suggesting that transcriptomic data alone could potentially predict the mechanics and speed of migration (Hou et al. 2024).

Second, there are 3D models of mesenchymal motile cells on top of 2D surfaces. The surfaces are not necessarily flat; for example, they can be cylindrical ECM fibers (if a cell crawls along a single fiber) or inner cylindrical walls (if a cell crawls inside a channel) (Winkler et al. 2019). In a way, one can call such models

$$2\frac{1}{2}\text{D}.$$

Three modeling strategies were used to simulate the mechanics of cell shape. (1) The most direct approach, faithful to the physical reality, is to consider a cell as a physical, elastic, and/or viscous body with a free boundary, the bulk of which deforms under the action of both passive and active (myosin-powered) forces, and the boundary of which protrudes and retracts as prescribed (Stolarska et al. 2009; Herant and Dembo 2010). Two main disadvantages of such models are the great computational challenge of simulating realistic mechanics on complex meshes and too many assumptions and parameters that are hard to constrain because of the lack of data on cell rheology. (2) Cellular Potts framework—representing the cell as a 3D set of pixels and minimizing hypothetic mechanical energy of the cell (the sum of the volume, surface, and adhesion energies) by stochastically adding and deleting surface pixels—was applied by adding protrusion and retraction forces to mimic a 3D motile cell (Fortuna et al. 2020). The advantage of this model type is in the relative ease of 3D simulations because the model is essentially a 2D model of a deformable cell surface embedded into the 3D space and clear intuitive qualitative connections to biological reality. The disadvantage is that it is not easy (but possible) (Rens and Edelstein-Keshet 2019) to connect the hypothetical energy to actual physical forces. (3) Phase-field models avoid simulating complex 3D meshes by representing dynamic cell shape as a deformable "blob," the boundaries of which smoothly protrude and retract according to evolving densities of reacting and diffusing abstract densities inside the blob (Fig. 3; Cao et al. 2019; Winkler et al. 2019). These densities can, for example, represent an activator/inhibitor pair (Cao et al. 2019) or a vector field responsible for local actin polarity (Winkler et al. 2019). The pluses and minuses of these models are somewhat like those of the cellular Potts models—ease of simulation tempered by vague connections to physical and biological realities. Impressively looking simulations of one of such models (Fig. 3) showed how ECM fiber curvature and topography modulate the cell's speed, shape, and actin organization (Winkler et al. 2019). Another model closely mimicked three observed morphodynamic behaviors of motile *Dictyostelium* cells by simply varying a couple of parameters (Cao et al. 2019).

Third, cellular Potts model cells were embedded into and made to transiently adhere to either a deformable discrete mesh of ECM fibers (Scianna et al. 2013; Tsingos et al. 2023) or deformable cells of the surrounding tissues (Bernadskaya et al. 2021). For example, Tsingos et al. (2023) made the contractile cellular Potts model cell pull with discrete adhesion sites on bead-spring ECM fiber networks. The model reproduced experimentally observed spatiotemporal fiber densification and displacement on characteristic viscoelastic ECMs.

Fourth, there are mechanical models of 2D cross-sections of 3D cells (Tozluoğlu et al. 2013; Zhu and Mogilner 2016; Maxian et al. 2020), in which the cell mechanics are modeled crudely with a simple elastic or viscoelastic cytoskeleton-like network. A pioneering model of this kind (Tozluoğlu et al. 2013), which was accompanied by experiments, represented the pressurized cell surrounded by a cell cortex membrane contour with the nodes connected by viscoelastic springs, which could either stretch, forming blebs that pried apart the ECM pores, or fold into lamellipodia-like protrusions adhering to the ECM elements. Another model of this kind (Zhu and Mogilner 2016) had a nucleus surrounded by an elastic node-link network, such that the links at one side of the nucleus contracted while expanding on the other side (Fig. 3). The cell membrane transiently adhered to the ECM node-link network. By varying parameters of this model, six different modes of locomotion were reproduced, for example, contracting the cytoskeletal mesh around the nucleus to squeeze the nucleus forward using it as a ram to push through the ECM. The disadvantages of these kinds of models are (1) reduced 2D representation of the 3D phenomenon, and (2) numerous ad hoc assumptions.

Fifth, there is a state-of-the-art, truly 3D, impressively detailed model (Fig. 3) of invadopodia (Kim et al. 2022), elongated ventral membrane protrusions that allow cancer cells to physically widen the ECM pores creating channels for migration (early, simpler models of invadopodia, i.e., Enderling et al. [2008], paved the way for Kim et al. [2022]). The authors of Kim et al. (2022) filled the 3D cell volume with the fine and layered mesh of nodes connected by lines. The lines in these structures are simulated with "spring-dashpot in parallel" elements. This viscoelastic mesh consists of triple surface layers of (1) cell membrane, (2) force transduction layer, and (3) actin cortex layer. Underneath this triple layer is the intracellular branched actin network and the double layer of the nuclear envelope (perinuclear actin layer and nuclear membrane surface). The virtual energy method was used to compute elastic forces between the actin networks, cross-linking molecules, and myosin motors at each time step; balancing them with viscous forces generated deformations. There is a protrusive phase driven by F-actin polymerization, a retractile phase due to the contractile action of perinuclear myosin motors, and an actin disassembly phase in the model. The ECM is also a fine viscoelastic mesh. Compared to stunning images of realistically looking model cells in Kim et al. (2022), the findings from the simulations are modest, qualitative, and expected (i.e., that invadopodia speeds become faster as the protrusive phase becomes longer, and that it is impossible to generate contractile force without cross-linkers).

MACHINE LEARNING APPROACHES TO CELL MIGRATION

Traditional bottom-up models reviewed above have a relatively low-dimensional parameter space and are appropriate for understanding simple features of modest amounts of data. However, modern high-throughput cell biological methods generate data sets that contain enormous volumes of quantitative data characterized by a very high-dimensional parameter space (Rafelski and Theriot 2024). One of the tasks of modern theoretical research is to project raw data onto low-dimensional space while extracting features from the data that are relevant to studied phenomena. ML is ideally suited for this task. For example, ML was recently used to identify lamellipodia and other types of protrusion from raw images (Moshkov et al. 2024), to predict migratory trajectories from dynamic cell shapes (Schoenauer Sebag et al. 2015), to find metastatic melanoma cells (Zaritsky et al. 2021), and to predict 3D migratory behavior from 2D cell motility features (Baskaran et al. 2020).

Here, we focus on two rapidly developing applications of ML to the problem of cell migration: prediction of TFs and analysis of cell trajectories. Cells move on elastic substrates by generating contractile and protruding forces and pulling on the substrates to propel themselves; maps of these forces reveal what motility mechanisms cells use (Mogilner and Savinov 2023). The ingenious traditional method to obtain TFs is to place a cell on an elastic film with embedded fluorescent beads, measure displacements of the beads, and solve an inverse problem of elasticity theory (reverse-engineer the forces from the displacements they cause) (Dembo and Wang 1999). This method, used for decades with spectacular results, suffers from the mathematical ill-posedness of the inverse mathematical problems, especially from the difficulty of separating mechanical and measurement noise from meaningful signals.

ML offers a conceptually straightforward and promising alternative way that harnesses supervised learning (using labeled training data to map a feature to a label) to measure TFs (Fig. 4). Simply speaking, the method could work as follows: (1) make a model to generate thousands of active force distributions, (2) in each case solve the well-posed direct mathematical problem of predicting the distributions of (a) reactive forces of the elastic substrate (these are TFs), and (b) deformations of the substrate, (3) train an ML network by thousands of (a–b) pairs, and (4) test the ML network by presenting it with measured substrate deformation distribution and asking it to predict respective TF distribution. This program was successfully implemented (Wang and Lin 2021) where the authors used a simple mathematical model of a dynamic motile cell (Satulovsky et al. 2008) to generate active forces along the cell periphery, apply these forces to the elastic substrate, compute the substrate's elastic deformations, and train a neural network by the substrate's deformation–force pairs' data. The trained neural network was tested by generating additional data, as above, presenting the substrate's deformations map, asking the network to predict the TFs, and comparing the predicted and computed TFs. The authors found, however, that the traditional TF calculation method and ML method outperform each other in different cases

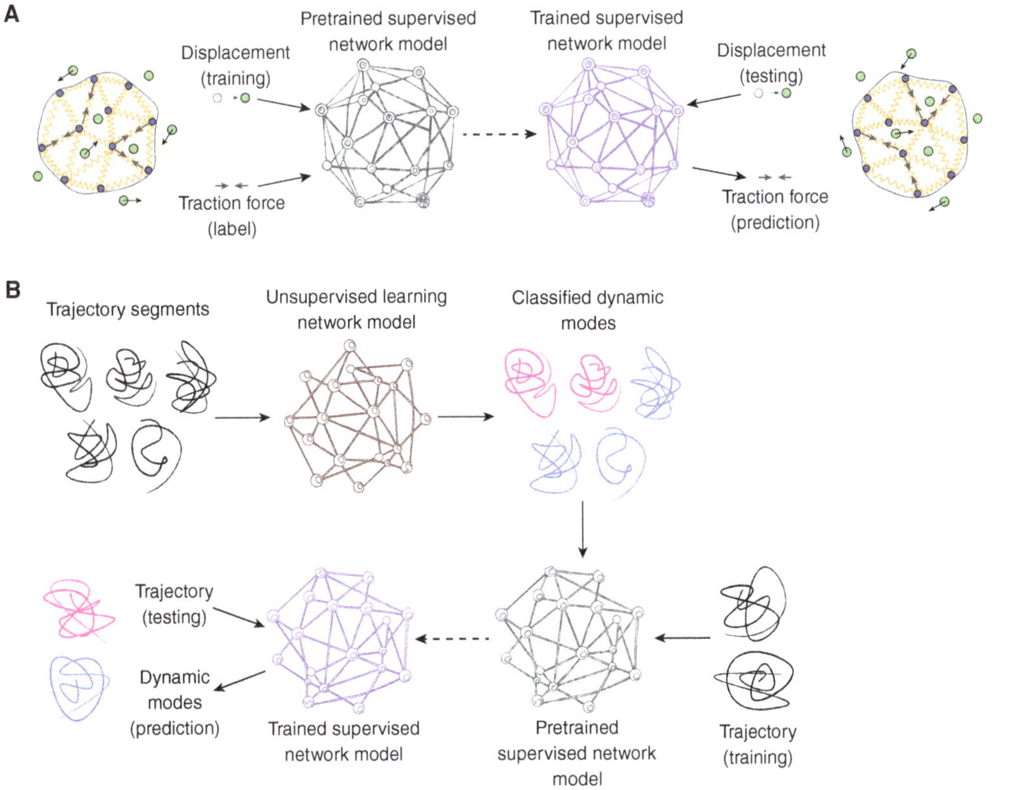

Figure 4. Applications of machine learning to study cell migration. (*A*) A neural network is trained to reverse-engineer traction forces from measured substrate displacements as follows: a traditional computational model is used to generate active intracellular forces that are applied to the deformable substrate generating both displacements and forces in the substrate. Each of a great number of intracellular force distributions generates a unique pair of displacement and force patterns in the substrate. The process of supervised learning presents the network with the training patterns of displacements and pairs each of these patterns with the corresponding pattern of forces (label). The trained network can then be tested by being presented with a displacement pattern not used in the training process and comparing the predicted traction force pattern with the traction force distribution corresponding to the displacement pattern, both of which come from one simulation of the computational model. (*B*) Training of a neural network to quantify cell migratory trajectories starts with an unsupervised learning process, in which the network uses geometric metrics to classify trajectory segments. This results in clustering the trajectories into several dynamic modes, which are then applied, together with previously unused trajectory data, in the second, supervised learning stage of the training process. The fully trained network is then able to predict switches between dynamic modes of motility along the experimentally obtained migratory trajectories.

and their accuracies are generally the same. Hence, ML offers no obvious advantage yet. Conceptually similar approaches are reported in Pielawski et al. (2020) and Duan and Huang (2022).

Another relevant study (Tao et al. 2024) proposed a different approach to circumvent the problems of destabilizing noise and of not knowing the exact substrate's rheology. Tao et al. (2024) found that ML can very accurately predict the cell shape and substrate displacements from the shapes and displacements of several previous time frames. Importantly, this prediction process averages the noise out. Then, TF can be computed by the traditional approach of solving the inverse elasticity theory problem starting from the predicted smooth displacement map.

Schmitt et al. (2024) went one step further. The authors of that state-of-the-art study asked the question: Can spatial distribution of a key protein predict TF directly? They adapted the so-called U-net neural network (Ronneberger et al. 2015; Falk et al. 2019) to predict TF using the spatial distribution of zyxin, a focal adhesion protein, as input. The procedure introduced in Schmitt et al. (2024) is to (1) measure spatial maps of zyxin distribution and substrate displacements, (2) compute TFs by using the traditional approach of solving the inverse elasticity theory problem with the measured displacements as the initial condition, and (3) train U-net on the resulting zyxin-TF data. It turned out that the trained network could then predict TFs from zyxin distributions without the displacement data. Interestingly, when other key proteins (i.e., actin, myosin, and paxillin) were used to train U-net, the prediction quality became worse; combinations of zyxin with a non-zyxin protein at most performed as well as zyxin alone.

Schmitt et al. (2024) went even further: besides using the trained U-net as a "black box," they developed a "physical bottleneck neural network," which incorporated a minimal elastic model combined with the neural network. This network allowed computing the "adhesion strength functional," Y, which is a function of both spatial coordinate and zyxin distribution, such that TF, **F**, can be computed from the displacement field, **u**, from a linear equation: $\mathbf{F} = Y\mathbf{u}$. Computing TF from this equation is not as accurate as using the "black box" neural network but is easier and offers mechanistic insight. Lastly, they used a neural network based on physical rules but agnostic to a particular physical hypothesis, called "Green's function neural network." Such a network derived a nonlinear functional relation between zyxin distribution as input and TF as output. Using this relation, TF can be predicted with almost 80% accuracy, while using just 10 model parameters compared with about 10^5 times greater number of parameters of the unconstrained "black box" U-net.

Analysis of cell migratory trajectories is also very amenable to the use of ML (Fig. 4). One of the relevant questions is whether and how the trajectories can be quantified and classified. In other words, while the whole set of trajectories constitutes multidimensional data, we would like to project these data onto a low-dimensional space, preserving the most essential and relevant features of the data and clustering the trajectories into several informative and intuitive classes. An example of implementation of this general task is a recent study (Song et al. 2023) where, first, unsupervised learning was used to recognize the trajectory type of dendritic cells moving in 2D. The authors computed several features (i.e., radius of gyration, end-to-end distance, and statistics of direction turning) quantifying the trajectories segmented into equal-length tracks and used the quantifiers as the input of an unsupervised ML model to cluster the trajectory segments into different modes. They found that the data are best categorized into three dynamic migratory modes: slow diffusion, and slow and fast persistent migration. Then, they trained the supervised ML model with the trajectory segments set as the input and the dynamic modes as the output, which are learned from the unsupervised part of the study and applied the trained network to analyze the entire trajectories. The analysis showed that the cells switched with certain frequencies between the three motility modes, depending on the cells' developmental stage.

Another study (Korabel et al. 2022) started from simulating multiple trajectories of cells engaged in the so-called fractional Brownian motion (in which the average displacement of a cell increases proportionally to time in power α, where α is the so-called anomalous diffusion exponent), and then used these data to train the neural network to identify the anomalous diffusion exponents. The authors applied the trained network to analyze hemocyte cell trajectories in 3D in live *Drosophila* embryos and found that the cells were engaged in super-diffusive motility on average ($1 < \alpha < 2$). Also, the motility was heterogeneous, in the sense that the anomalous diffusion exponent fluctuated, both in time and between different trajectories.

OUTLOOK

Integrative approaches to modeling cell migration are still mostly the future; for now, this research mostly follows a tested and reliable reductionist approach, according to which the migration process can be reduced to the problems of polarity, directionality, and motility. In conclusion of our review, we discuss the strengths and limitations of the different modeling endeavors and principal open questions, as well as the challenges of integrating multiprocess models and data.

Cell Polarity

The strength of the models of the MT-related cell polarization models that we discussed is excellent integration of theory and experiment: many of these models are introduced in experimental studies, where modeling was judiciously used to discriminate between alternative hypotheses and to interpret biophysical measurements.

One limitation of these models is that constitutive mechanical relations, for example between relevant stresses and strains, are rarely based on measurements, and so there is a lingering doubt whether assumed mechanical relations were chosen as such for better fits.

An interesting open question about polarity models relates to the curse of specificity in cell biology—so far, unifying theories of cell migration, of the sort that physics is accustomed to, remain elusive, and having an adequate polarization model of one motile cell type helps little in elucidating the directional motility onset of another cell type. The great future modeling challenge is to start revealing general design principles of integration of mechanochemical cell parts into diverse polarization events.

Directional Sensing

The models of galvanotaxis are in their infancy, compared to rich traditions of modeling cell mechanics and polarity. Considering that chemotaxis is very closely related to galvanotaxis and that computational modeling of chemotaxis is thriving for decades, the most promising direction would be focusing on one cell type and integrating microscopy, genetic, biophysical, and computer tools to elucidate three interacting modules of galvanotaxis: EF-sensing machinery on the cell surface, respective signal transduction pathways, and cytoskeletal response to respective signals.

Mechanics of Cell Motility

The main strength of the models of 3D cell motility that we reviewed is in the diversity of approaches to the cell mechanics, from considering the cell as a set of contractile adhesive "arms," to minimizing a sum of volume-, surface-, and adhesion-related energies of a "trembling blob," to simulating a mesh mimicking dynamic cytoskeletal scaffolds. This diversity allows us to look at the pliable modes of cell migration from different angles, each bringing an additional sliver of understanding.

The challenge we are facing is that increasingly detailed models become hopelessly cumbersome and unwieldy to be used to nimbly test multiple hypotheses.

Among many open questions about the future of mechanical modeling, the one we would like to highlight is how do we overcome the tremendous computational expense of solving coupled equations of nonlinear mechanics and stochastic cytoskeletal dynamics, exacerbated by the need to scan multidimensional parameter spaces?

Embracing Machine Learning

The strength of traditional bottom-up models, such as those we reviewed in the sections on polarity, motility, and directional sensing, is in bringing mechanistic insights into cell migration. These models, simply speaking, explain gears of subcellular machines in an intuitive way and generate clear hypotheses for future experiments.

However, the main limitation of such models is that even the most detailed of them are still too simplistic to embrace the complexity of the multiscale cell migration process. On the other hand, the data-driven approaches, which are often not constrained by laws of physics and chemistry, are excellent for embracing complexity, finding key features of cell migration, and predicting migration mechanics from "big data." Yet, these approaches rarely bring mechanistic understanding.

Likely, the way forward is to augment traditional models with data science and ML methods, as discussed in Choi et al. (2021) and Brückner and Broedersz (2024). Lastly, the big open question is: Is it possible to use nascent artificial intelligence (AI) to build multiple detailed traditional models from data, simulate them faster, and use data to discriminate between alternative models automatically, freeing human researchers from routine technical tasks.

COMPETING INTEREST STATEMENT

The authors declare no competing interests.

ACKNOWLEDGMENTS

This work was supported by the National Science Foundation (grant numbers DMS 20525 15, DMS 1953430).

REFERENCES

Allen GM, Mogilner A, Theriot JA. 2013. Electrophoresis of cellular membrane components creates the directional cue guiding keratocyte galvanotaxis. *Curr Biol* **23:** 560–568. doi:10.1016/j.cub.2013.02.047

Bangasser BL, Shamsan GA, Chan CE, Opoku KN, Tüzel E, Schlichtmann BW, Kasim JA, Fuller BJ, McCullough BR, Rosenfeld SS, et al. 2017. Shifting the optimal stiffness for cell migration. *Nat Commun* **8:** 15313. doi:10.1038/ncomms15313

Baskaran JP, Weldy A, Guarin J, Munoz G, Shpilker PH, Kotlik M, Subbiah N, Wishart A, Peng Y, Miller MA, et al. 2020. Cell shape, and not 2D migration, predicts extracellular matrix-driven 3D cell invasion in breast cancer. *APL Bioeng* **4:** 026105. doi:10.1063/1.5143779

Batchelder EL, Hollopeter G, Campillo C, Mezanges X, Jorgensen EM, Nassoy P, Sens P, Plastino J. 2011. Membrane tension regulates motility by controlling lamellipodium organization. *Proc Natl Acad Sci* **108:** 11429–11434. doi:10.1073/pnas.1010481108

Belliveau NM, Footer MJ, Platenkamp A, Kim H, Eustis TE, Theriot JA. 2024. Galvanin is an electric-field sensor for directed cell migration. bioRxiv doi:10.1101/2024.09.23.614580

Bernadskaya YY, Yue H, Copos C, Christiaen L, Mogilner A. 2021. Supracellular organization confers directionality and mechanical potency to migrating pairs of cardiopharyngeal progenitor cells. *eLife* **10:** e70977. doi:10.7554/eLife.70977

Blanchoin L, Boujemaa-Paterski R, Sykes C, Plastino J. 2014. Actin dynamics, architecture, and mechanics in cell motility. *Physiol Rev* **94:** 235–263. doi:10.1152/physrev.00018.2013

Bretscher MS. 1984. Endocytosis: relation to capping and cell locomotion. *Science* **224:** 681–686. doi:10.1126/science.6719108

Brückner DB, Broedersz CP. 2024. Learning dynamical models of single and collective cell migration: a review. *Rep Prog Phys* **87:** 056601. doi:10.1088/1361-6633/ad36d2

Buttenschön A, Edelstein-Keshet L. 2020. Bridging from single to collective cell migration: a review of models and links to experiments. *PLoS Comput Biol* **16:** e1008411. doi:10.1371/journal.pcbi.1008411

Callan-Jones A. 2022. Self-organization in amoeboid motility. *Front Cell Dev Biol* **10:** 1000071. doi:10.3389/fcell.2022.1000071

Campbell S, Zitnay R, Mendoza M, Bidone TC. 2021. Computational model of 3D cell migration based on the molecular clutch mechanism. bioRxiv doi:10.1101/2021.09.29.462287

Cao X, Lin Y, Driscoll TP, Franco-Barraza J, Cukierman E, Mauck RL, Shenoy VB. 2015. A chemomechanical model of matrix and nuclear rigidity regulation of focal adhesion size. *Biophys J* **109:** 1807–1817. doi:10.1016/j.bpj.2015.08.048

Cao YS, Ghabache E, Miao YC, Niman C, Hakozaki H, Reck-Peterson SL, Devreotes PN, Rappel WJ. 2019. A minimal computational model for three-dimensional cell migration. *J R Soc Interface* **16:** 20190619. doi:10.1098/rsif.2019.0619

Caswell PT, Zech T. 2018. Actin-based cell protrusion in a 3D matrix. *Trends Cell Biol* **28:** 823–834. doi:10.1016/j.tcb.2018.06.003

Chang F, Minc N. 2014. Electrochemical control of cell and tissue polarity. *Annu Rev Cell Dev Biol* **30:** 317–336. doi:10.1146/annurev-cellbio-100913-013357

Chen Y, Saintillan D, Rangamani P. 2023. Interplay between mechanosensitive adhesions and membrane tension regulates cell motility. *PRX Life* **1:** 023007. doi:10.1103/PRXLife.1.023007

Choi HJ, Wang C, Pan X, Jang J, Cao M, Brazzo JA III, Bae Y, Lee K. 2021. Emerging machine learning approaches to phenotyping cellular motility and morphodynamics. *Phys Biol* **18:** 10.1088/1478-3975/abffbe. doi:10.1088/1478-3975/abffbe

Danuser G, Allard J, Mogilner A. 2013. Mathematical modeling of eukaryotic cell migration: insights beyond experiments. *Annu Rev Cell Dev Biol* **29:** 501–528. doi:10.1146/annurev-cellbio-101512-122308

De Belly H, Yan S, da Rocha HB, Ichbiah S, Town JP, Zager PJ, Estrada DC, Meyer K, Turlier H, Bustamante C, et al. 2023. Cell protrusions and contractions generate long-range membrane tension propagation. *Cell* **186:** 3049–3061.e15. doi:10.1016/j.cell.2023.05.014

Dembo M, Wang YL. 1999. Stresses at the cell-to-substrate interface during locomotion of fibroblasts. *Biophys J* **76:** 2307–2316. doi:10.1016/S0006-3495(99)77386-8

Dharan R, Barnoy A, Tsaturyan AK, Grossman A, Goren S, Yosibash I, Nachmias D, Elia N, Sorkin R, Kozlov MM. 2025. Intracellular pressure controls the propagation of tension in crumpled cell membranes. *Nat Commun* **16:** 91. doi:10.1038/s41467-024-55398-1

DiNapoli KT, Robinson DN, Iglesias PA. 2020. Tools for computational analysis of moving boundary problems in cellular mechanobiology. *Wiley Interdiscip Rev Syst Biol Med* **10:** e1514. doi:10.1002/wsbm.1514

Diz-Muñoz A, Fletcher DA, Weiner OD. 2013. Use the force: membrane tension as an organizer of cell shape and motility. *Trends Cell Biol* **23:** 47–53. doi:10.1016/j.tcb.2012.09.006

Diz-Muñoz A, Thurley K, Chintamen S, Altschuler SJ, Wu LF, Fletcher DA, Weiner OD. 2016. Membrane tension acts through PLD2 and mTORC2 to limit actin network assembly during neutrophil migration. *PLoS Biol* **14:** e1002474. doi:10.1371/journal.pbio.1002474

Duan X, Huang J. 2022. Deep-learning-based 3D cellular force reconstruction directly from volumetric images. *Biophys J* **121:** 2180–2192. doi:10.1016/j.bpj.2022.04.028

Enderling H, Alexander NR, Clark ES, Branch KM, Estrada L, Crooke C, Jourquin J, Lobdell N, Zaman MH, Guelcher SA, et al. 2008. Dependence of invadopodia function on collagen fiber spacing and cross-linking: computational modeling and experimental evidence. *Biophys J* **95:** 2203–2218. doi:10.1529/biophysj.108.133199

Falk T, Mai D, Bensch R, Çiçek Ö, Abdulkadir A, Marrakchi Y, Böhm A, Deubner J, Jäckel Z, Seiwald K, et al. 2019. U-Net: deep learning for cell counting, detection, and morphometry. *Nat Methods* **16:** 67. doi:10.1038/s41592-018-0261-2

Fogelson B, Mogilner A. 2014. Computational estimates of membrane flow and tension gradient in motile cells. *PLoS One* **9:** e84524. doi:10.1371/journal.pone.0084524

Fortuna I, Perrone GC, Krug MS, Susin E, Belmonte JM, Thomas GL, Glazier JA, de Almeida RMC. 2020. CompuCell3d simulations reproduce mesenchymal cell migration on flat substrates. *Biophys J* **118:** 2801–2815. doi:10.1016/j.bpj.2020.04.024

Gauthier NC, Fardin MA, Roca-Cusachs P, Sheetz MP. 2011. Temporary increase in plasma membrane tension coordinates the activation of exocytosis and contraction during cell spreading. *Proc Natl Acad Sci* **108:** 14467–14472. doi:10.1073/pnas.1105845108

Godeau AL, Leoni M, Comelles J, Guyomar T, Lieb M, Delanoë-Ayari H, Ott A, Harlepp S, Sens P, Riveline D. 2022. 3D single cell migration driven by temporal correlation between oscillating force dipoles. *eLife* **11:** e71032. doi:10.7554/eLife.71032

Gruler H, Nuccitelli R. 2000. The galvanotaxis response mechanism of keratinocytes can be modeled as a proportional controller. *Cell Biochem Biophys* **33:** 33–51. doi:10.1385/CBB:33:1:33

Heck T, Vargas DA, Smeets B, Ramon H, Van Liedekerke P, Van Oosterwyck H. 2020. The role of actin protrusion dynamics in cell migration through a degradable viscoelastic extracellular matrix: insights from a computational model. *PLoS Comput Biol* **16:** e1007250. doi:10.1371/journal.pcbi.1007250

Herant M, Dembo M. 2010. Form and function in cell motility: from fibroblasts to keratocytes. *Biophys J* **98:** 1408–1417. doi:10.1016/j.bpj.2009.12.4303

Hetmanski JHR, de Belly H, Busnelli I, Waring T, Nair RV, Sokleva V, Dobre O, Cameron A, Gauthier N, Lamaze C, et al. 2019. Membrane tension orchestrates rear retraction in matrix-directed cell migration. *Dev Cell* **51:** 460–475.e10. doi:10.1016/j.devcel.2019.09.006

Hou J, McMahon M, Jubenville T, Sarkaria JN, Chen CC, Odde DJ. 2024. Cell migration simulator-based biomarkers for glioblastoma. *Neurooncol Adv* **6:** vdae184. doi:10.1093/noajnl/vdae184

Houk AR, Jilkine A, Mejean CO, Boltyanskiy R, Dufresne ER, Angenent SB, Altschuler SJ, Wu LF, Weiner OD. 2012. Membrane tension maintains cell polarity by confining signals to the leading edge during neutrophil migration. *Cell* **148:** 175–188. doi:10.1016/j.cell.2011.10.050

Keren K. 2011. Cell motility: the integrating role of the plasma membrane. *Eur Biophys J* **40:** 1013–1027. doi:10.1007/s00249-011-0741-0

Kim MC, Li R, Abeyaratne R, Kamm RD, Asada HH. 2022. A computational modeling of invadopodia protrusion into an extracellular matrix fiber network. *Sci Rep* **12:** 1231. doi:10.1038/s41598-022-05224-9

Korabel N, Clemente GD, Han D, Feldman F, Millard TH, Waigh TA. 2022. Hemocytes in embryos move via heterogeneous anomalous diffusion. *Commun Phys* **5:** 269.

Kozlov MM, Mogilner A. 2007. Model of polarization and bistability of cell fragments. *Biophys J* **93:** 3811–3819. doi:10.1529/biophysj.107.110411

Labuz EC, Footer MJ, Theriot JA. 2023. Confined keratocytes mimic in vivo migration and reveal volume-speed relationship. *Cytoskeleton (Hoboken)* **80:** 34–51. doi:10.1002/cm.21741

Li Y, Mori Y, Sun SX. 2015. Flow-driven cell migration under external electric fields. *Phys Rev Lett* **115:** 268101. doi:10.1103/PhysRevLett.115.268101

Lieber AD, Yehudai-Resheff S, Barnhart EL, Theriot JA, Keren K. 2013. Membrane tension in rapidly moving cells is determined by cytoskeletal forces. *Curr Biol* **23:** 1409–1417. doi:10.1016/j.cub.2013.05.063

Lin BJ, Tsao SH, Chen A, Hu SK, Chao L, Chao PG. 2017. Lipid rafts sense and direct electric field-induced migration. *Proc Natl Acad Sci* **114:** 8568–8573. doi:10.1073/pnas.1702526114

Lomakin AJ, Lee KC, Han SJ, Bui DA, Davidson M, Mogilner A, Danuser G. 2015. Competition for actin between two distinct F-actin networks defines a bistable switch for cell polarization. *Nat Cell Biol* **17:** 1435–1445. doi:10.1038/ncb3246

Maxian O, Mogilner A, Strychalski W. 2020. Computational estimates of mechanical constraints on cell migration through the extracellular matrix. *PLoS Comput Biol* **16:** e1008160. doi:10.1371/journal.pcbi.1008160

McLaughlin S, Poo MM. 1981. The role of electro-osmosis in the electric-field-induced movement of charged macromolecules on the surfaces of cells. *Biophys J* **34:** 85–93. doi:10.1016/S0006-3495(81)84838-2

Merino-Casallo F, Gomez-Benito MJ, Martinez-Cantin R, Garcia-Aznar JM. 2022. A mechanistic protrusive-based model for 3D cell migration. *Eur J Cell Biol* **101:** 151255. doi:10.1016/j.ejcb.2022.151255

Mogilner A, Savinov M. 2023. Crawling, waving, inch worming, dilating, and pivoting mechanics of migrating cells: lessons from Ken Jacobson. *Biophys J* **122:** 3551–3559. doi:10.1016/j.bpj.2023.03.023

Moshkov N, Bornholdt M, Benoit S, Smith M, McQuin C, Goodman A, Senft RA, Han Y, Babadi M, Horvath P, et al. 2024. Learning representations for image-based profiling of perturbations. *Nat Commun* **15:** 1594. doi:10.1038/s41467-024-45999-1

Nwogbaga I, Camley BA. 2023. Coupling cell shape and velocity leads to oscillation and circling in keratocyte galvanotaxis. *Biophys J* **122:** 130–142. doi:10.1016/j.bpj.2022.11.021

Ofer N, Mogilner A, Keren K. 2011. Actin disassembly clock determines shape and speed of lamellipodial fragments. *Proc Natl Acad Sci* **108:** 20394–20399. doi:10.1073/pnas.1105333108

Ogawa N, Oku H, Hashimoto K, Ishikawa M. 2006. A physical model for galvanotaxis of *Paramecium* cell. *J Theor Biol* **242:** 314–328. doi:10.1016/j.jtbi.2006.02.021

Parkhurst MR, Saltzman WM. 1992. Quantification of human neutrophil motility in three-dimensional collagen gels. Effect of collagen concentration. *Biophys J* **61:** 306–315. doi:10.1016/S0006-3495(92)81838-6

Pielawski N, Hu JJ, Strömblad S, Wählby C. 2020. In silico prediction of cell traction forces. In *2020 IEEE 17th International Symposium on Biomedical Imaging (ISBI)*, pp. 877–881. IEEE, New York.

Pontes B, Monzo P, Gole L, Le Roux AL, Kosmalska AJ, Tam ZY, Luo W, Kan S, Viasnoff V, Roca-Cusachs P, et al. 2017. Membrane tension controls adhesion positioning at the leading edge of cells. *J Cell Biol* **216:** 2959–2977. doi:10.1083/jcb.201611117

Prahl LS, Stanslaski MR, Vargas P, Piel M, Odde DJ. 2020. Predicting confined 1D cell migration from parameters calibrated to a 2D motor-clutch model. *Biophys J* **118:** 1709–1720. doi:10.1016/j.bpj.2020.01.048

Prescott TP, Zhu K, Zhao M, Baker RE. 2021. Quantifying the impact of electric fields on single-cell motility. *Biophys J* **120:** 3363–3373. doi:10.1016/j.bpj.2021.06.034

Quiroga X, Walani N, Disanza A, Chavero A, Mittens A, Tebar F, Trepat X, Parton RG, Geli MI, Scita G, et al. 2023. A mechanosensing mechanism controls plasma membrane shape homeostasis at the nanoscale. *eLife* **12:** e72316. doi:10.7554/eLife.72316

Rafelski SM, Theriot JA. 2024. Establishing a conceptual framework for holistic cell states and state transitions. *Cell* **187:** 2633–2651. doi:10.1016/j.cell.2024.04.035

Rangarajan R, Zaman MH. 2008. Modeling cell migration in 3D: status and challenges. *Cell Adh Migr* **2:** 106–109. doi:10.4161/cam.2.2.6211

Rappel WJ, Edelstein-Keshet L. 2017. Mechanisms of cell polarization. *Curr Opin Syst Biol* **3:** 43–53. doi:10.1016/j.coisb.2017.03.005

Rens EG, Edelstein-Keshet L. 2019. From energy to cellular forces in the cellular Potts model: an algorithmic approach. *PLoS Comput Biol* **15:** e1007459. doi:10.1371/journal.pcbi.1007459

Ronneberger O, Fischer P, Brox T. 2015. U-Net: convolutional networks for biomedical image segmentation. *Lect Notes Comput Sc* **9351:** 234–241. doi:10.1007/978-3-319-24574-4_28

Sargent B, Jafari M, Marquez G, Mehta AS, Sun YH, Yang HY, Zhu K, Isseroff RR, Zhao M, Gomez M. 2022. A machine learning based model accurately predicts cellular response to electric fields in multiple cell types. *Sci Rep* **12:** 9912. doi:10.1038/s41598-022-13925-4

Sarkar A, Kobylkevich BM, Graham DM, Messerli MA. 2019. Electromigration of cell surface macromolecules in DC electric fields during cell polarization and galvanotaxis. *J Theor Biol* **478:** 58–73. doi:10.1016/j.jtbi.2019.06.015

Satulovsky J, Lui R, Wang YL. 2008. Exploring the control circuit of cell migration by mathematical modeling. *Biophys J* **94:** 3671–3683. doi:10.1529/biophysj.107.117002

Schlüter DK, Ramis-Conde I, Chaplain MA. 2012. Computational modeling of single-cell migration: the leading role of extracellular matrix fibers. *Biophys J* **103:** 1141–1151. doi:10.1016/j.bpj.2012.07.048

Schmitt MS, Colen J, Sala S, Devany J, Seetharaman S, Caillier A, Gardel ML, Oakes PW, Vitelli V. 2024. Machine learning interpretable models of cell mechanics from protein images. *Cell* **187:** 481–494.e24. doi:10.1016/j.cell.2023.11.041

Schoenauer Sebag A, Plancade S, Raulet-Tomkiewicz C, Barouki R, Vert JP, Walter T. 2015. A generic methodological framework for studying single cell motility in high-throughput time-lapse data. *Bioinformatics* **31:** i320–i328. doi:10.1093/bioinformatics/btv225

Schweitzer Y, Lieber AD, Keren K, Kozlov MM. 2014. Theoretical analysis of membrane tension in moving cells. *Biophys J* **106:** 84–92. doi:10.1016/j.bpj.2013.11.009

Scianna M, Preziosi L, Wolf K. 2013. A cellular Potts model simulating cell migration on and in matrix environments. *Math Biosci Eng* **10:** 235–261. doi:10.3934/mbe.2013.10.235

SenGupta S, Parent CA, Bear JE. 2021. The principles of directed cell migration. *Nat Rev Mol Cell Biol* **22:** 529–547. doi:10.1038/s41580-021-00366-6

Sens P, Plastino J. 2015. Membrane tension and cytoskeleton organization in cell motility. *J Phys Condens Matter* **27:** 273103. doi:10.1088/0953-8984/27/27/273103

Shi C, Huang CH, Devreotes PN, Iglesias PA. 2013. Interaction of motility, directional sensing, and polarity modules recreates the behaviors of chemotaxing cells. *PLoS Comput Biol* **9:** e1003122. doi:10.1371/journal.pcbi.1003122

Shi Z, Graber ZT, Baumgart T, Stone HA, Cohen AE. 2018. Cell membranes resist flow. *Cell* **175:** 1769–1779.e13. doi:10.1016/j.cell.2018.09.054

Shi W, Gupta S, Copos C, Mogilner A. 2025. Collective mechanics of small migrating cell groups. *Semin Cell Dev Biol* **166:** 1–12. doi:10.1016/j.semcdb.2024.12.001

Sitarska E, Diz-Muñoz A. 2020. Pay attention to membrane tension: mechanobiology of the cell surface. *Curr Opin Cell Biol* **66:** 11–18. doi:10.1016/j.ceb.2020.04.001

Song T, Choi Y, Jeon JH, Cho YK. 2023. A machine learning approach to discover migration modes and transition dynamics of heterogeneous dendritic cells. *Front Immunol* **14:** 1129600. doi:10.3389/fimmu.2023.1129600

Stolarska MA, Kim Y, Othmer HG. 2009. Multi-scale models of cell and tissue dynamics. *Philos Trans R Soc A* **367:** 3525–3553. doi:10.1098/rsta.2009.0095

Sun Y, Reid B, Ferreira F, Luxardi G, Ma L, Lokken KL, Zhu K, Xu G, Sun Y, Ryzhuk V, et al. 2019. Infection-generated electric field in gut epithelium drives bidirectional migration of macrophages. *PLoS Biol* **17:** e3000044. doi:10.1371/journal.pbio.3000044

Taghian T, Narmoneva DA, Kogan AB. 2015. Modulation of cell function by electric field: a high-resolution analysis. *J R Soc Interface* **12:** 20150153. doi:10.1098/rsif.2015.0153

Tao K, Wang J, Kuang X, Wang W, Liu F, Zhang L. 2020. Tuning cell motility via cell tension with a mechanochemical cell migration model. *Biophys J* **118:** 2894–2904. doi:10.1016/j.bpj.2020.04.030

Tao Y, Ghagre A, Molter CW, Clouvel A, Al Rahbani J, Brown CM, Nowrouzezahrai D, Ehrlicher AJ. 2024. Inferring cellular contractile forces and work using deep morphology traction microscopy. *Biophys J* **123:** 3217–3230. doi:10.1016/j.bpj.2024.07.020

Tozluoğlu M, Tournier AL, Jenkins RP, Hooper S, Bates PA, Sahai E. 2013. Matrix geometry determines optimal cancer cell migration strategy and modulates response to interventions. *Nat Cell Biol* **15:** 751–762. doi:10.1038/ncb2775

Tsingos E, Bakker BH, Keijzer KAE, Hupkes HJ, Merks RMH. 2023. Hybrid cellular Potts and bead-spring modeling of cells in fibrous extracellular matrix. *Biophys J* **122:** 2609–2622. doi:10.1016/j.bpj.2023.05.013

Vanegas-Acosta JC, Garzón-Alvarado DA, Zwamborn AP. 2012. Mathematical model of electrotaxis in osteoblastic cells. *Bioelectrochemistry* **88:** 134–143. doi:10.1016/j.bioelechem.2012.08.002

Van Helvert S, Storm C, Friedl P. 2018. Mechanoreciprocity in cell migration. *Nat Cell Biol* **20:** 8–20. doi:10.1038/s41556-017-0012-0

van Oers RF, Rens EG, LaValley DJ, Reinhart-King CA, Merks RM. 2014. Mechanical cell-matrix feedback explains pairwise and collective endothelial cell behavior in vitro. *PLoS Comput Biol* **10:** e1003774. doi:10.1371/journal.pcbi.1003774

Wang YL, Lin YC. 2021. Traction force microscopy by deep learning. *Biophys J* **120:** 3079–3090. doi:10.1016/j.bpj.2021.06.011

Wilson K, Lewalle A, Fritzsche M, Thorogate R, Duke T, Charras G. 2013. Mechanisms of leading edge protrusion in interstitial migration. *Nat Commun* **4:** 2896. doi:10.1038/ncomms3896

Winkler B, Aranson IS, Ziebert F. 2016. Membrane tension feedback on shape and motility of eukaryotic cells. *Physica D* **318:** 26–33. doi:10.1016/j.physd.2015.09.011

Winkler B, Aranson IS, Ziebert F. 2019. Confinement and substrate topography control cell migration in a 3D computational model. *Commun Phys* **2:** 82. doi:10.1038/s42005-019-0185-x

Wolf K, Te Lindert M, Krause M, Alexander S, Te Riet J, Willis AL, Hoffman RM, Figdor CG, Weiss SJ, Friedl P. 2013. Physical limits of cell migration: control by ECM space and nuclear deformation and tuning by proteolysis and traction force. *J Cell Biol* **201:** 1069–1084. doi:10.1083/jcb.201210152

Yamada KM, Sixt M. 2019. Mechanisms of 3D cell migration. *Nat Rev Mol Cell Biol* **20:** 738–752. doi:10.1038/s41580-019-0172-9

Zaman MH, Kamm RD, Matsudaira P, Lauffenburger DA. 2005. Computational model for cell migration in three-dimensional matrices. *Biophys J* **89:** 1389–1397. doi:10.1529/biophysj.105.060723

Zaritsky A, Jamieson AR, Welf ES, Nevarez A, Cillay J, Eskiocak U, Cantarel BL, Danuser G. 2021. Interpretable deep learning uncovers cellular properties in label-free live cell images that are predictive of highly metastatic melanoma. *Cell Syst* **12:** 733–747.e6. doi:10.1016/j.cels.2021.05.003

Zhao M. 2009. Electrical fields in wound healing—an overriding signal that directs cell migration. *Semin Cell Dev Biol* **20:** 674–682. doi:10.1016/j.semcdb.2008.12.009

Zhao M, Song B, Pu J, Wada T, Reid B, Tai G, Wang F, Guo A, Walczysko P, Gu Y, et al. 2006. Electrical signals control wound healing through phosphatidylinositol-3-OH kinase-γ and PTEN. *Nature* **442:** 457–460. doi:10.1038/nature04925

Zhu J, Mogilner A. 2016. Comparison of cell migration mechanical strategies in three-dimensional matrices: a computational study. *Interface Focus* **6:** 20160040. doi:10.1098/rsfs.2016.0040

Modeling Collective Cell Migration in a Data-Rich Age: Challenges and Opportunities for Data-Driven Modeling

Ruth E. Baker,[1,3] Rebecca M. Crossley,[1,3] Carles Falcó,[1,3] and Simon F. Martina-Perez[2,3]

[1]Mathematical Institute, University of Oxford, Oxford OX2 6GG, Oxfordshire, United Kingdom

[2]School of Medicine and Biomedical Sciences, University of Oxford, John Radcliffe Hospital, Oxford OX3 9DU, Oxfordshire, United Kingdom

Correspondence: ruth.baker@maths.ox.ac.uk

Mathematical modeling has a long history in the context of collective cell migration, with applications throughout development, disease, and regenerative medicine. The aim of modeling in this context is to provide a framework in which to mathematically encode experimentally derived mechanistic hypotheses and then to test and validate them to provide new insights and understanding. Traditionally, mathematical models have consisted of systems of partial differential equations that model the evolution of cell density over time, together with the dynamics of any associated biochemical signals or the underlying substrate. The various terms in the model are usually chosen to provide simplified, phenomenological descriptions of the underlying biology and follow long-standing conventions in the field. However, with the recent development of a plethora of new experimental technologies that provide quantitative data on collective cell migration processes, we now have the opportunity to leverage statistical and machine learning tools to determine mathematical models directly from the data. This article aims to provide an overview of recently developed data-driven modeling approaches, outlining the main methodologies and the challenges involved in using them to interrogate real-world data relating to collective cell migration.

Collective cell migration involves the coordinated movement of large groups of cells as a cohesive unit, and it plays a crucial role in development, disease, and repair (Friedl and Gilmour 2009; Rorth 2011; Haeger et al. 2015; Stehbens et al. 2024). Such collective behavior is driven by complex cell–cell and cell–environment interactions, which can make it difficult to tease apart the dominant driving mechanisms. Mathematical modeling provides an ideal platform to encode different hypothesized mechanisms for collective cell migration and to tease apart the contributions of different processes therein. Traditionally, mathematical modeling of collective cell migration has involved the researcher writing down systems of partial differential equations (PDEs) that describe how cell density, as well as any biochemical signals (e.g., chemoattractants) or substrates (e.g., extracellular matrix), evolve in time. The

[3]All authors contributed equally to this work, and authorship is alphabetical by surname.

Cite this article as *Cold Spring Harb Perspect Biol* doi: 10.1101/cshperspect.a041757

choice of functional forms in the equations is usually based on heuristic attempts to describe the mechanisms at play, and they generally follow standard conventions in the field. More recently, modelers have also used individual-based descriptions of collective cell migration, with each cell modeled as a discrete entity whose location evolves according to either a differential equation or a set of rules that account for cell–cell and cell–environment interactions. These rules are, once again, chosen by the user and based on their own experiences and the practice of others in the field. There are several review articles that provide an excellent overview of previous modeling efforts in these contexts (Schumacher et al. 2016; Buttenschön and Edelstein-Keshet 2020; Giniūnaitė et al. 2020; Crossley et al. 2024). In the simplest case, which models the evolution of cell density in one spatial dimension, x, over time, t, a PDE model can be written as

$$\frac{\partial u}{\partial t} = \mathcal{F}(u; \boldsymbol{p}), \tag{1}$$

where F is a differential operator that acts on $u(x, t)$ (i.e., it is a function of x, t, u and its partial derivatives), and $\boldsymbol{p}$ explicitly denotes the model parameters. The equation is specified on some domain Ω and is closed by specifying relevant initial conditions, $u(x, 0) = u_0(x)$ for $x \in \Omega$, as well as boundary conditions. A canonical model used for studying collective cell migration is that put forward by Fisher, as well as Kolmogorov, Petrovskii, and Piskunov, and known as the Fisher–KPP equation (Murray 2003),

$$\frac{\partial u}{\partial t} = D\frac{\partial^2 u}{\partial x^2} + ru\left(1 - \frac{u}{K}\right), \quad x \in (-\infty, \infty), \; t > 0, \tag{2}$$

which describes the evolution of cell density due to random motility and cell proliferation, where the cell proliferation rate is a decreasing function of cell density (representing crowding effects). The parameter vector $\boldsymbol{p} = (D, r, K)$ consists of the diffusion coefficient, D, proliferation rate, r, and carrying capacity, K. Given suitable initial and boundary conditions, for example, $u(x, 0) = 1$ for $x < 0$ and $u(x, 0) = 0$ for $x \geq 0$ together with $u(x, t) \to 1$ as $x \to -\infty$ and $u(x, t) \to 0$ as $x \to \infty$, Equation (2) displays traveling wave solutions, which are representative of collective cell invasion (see Fig. 1A).

Many interdisciplinary studies on collective cell migration seek to test and validate mathematical models, and the mechanistic hypotheses encoded therein, via comparison between the model predictions and experimental observations. Historically, researchers were restricted to qualitative comparison between models and data, often using perturbation experiments to challenge model-based hypotheses. Over the past five years, however, there has been a huge transformation in the extent to which it is possible to routinely collect high-resolution spatiotemporal data on collective cell migration. As a result, many researchers now seek to calibrate their mathematical models directly to quantitative data—this allows them to estimate the model parameters and quantify the relative contributions of different mechanisms to collective migration, as well as provide a quantitative measure of the ability of their model to predict the data (Simpson et al. 2013a,b; Jin et al. 2016; Nardini et al. 2016). Moreover, the use of Bayesian statistics approaches, for example, enables researchers to quantify the uncertainty in any parameter estimates and, with this, the uncertainty in model-based predictions (Johnston et al. 2014; Vo et al. 2015; Browning et al. 2018; Falcó et al. 2023, 2025; Stepien 2024). Calibration of models to quantitative data also enables the process of model selection, in which mathematical models encoding different biological hypotheses can be quantitatively compared and selected between based on, for example, their ability to replicate the data, the extent to which model parameters can be confidently estimated from data, and model complexity (Warne et al. 2019; Liu et al. 2024).

Figure 1 illustrates use of the Fisher–KPP equation in modeling collective cell migration in the context of an in vitro wound healing assay. Here, the cells are grown to form a confluent monolayer in a dish before a portion of the cells are removed using a sharp-tipped instrument (Fig. 1B,C). Collective cell invasion is observed

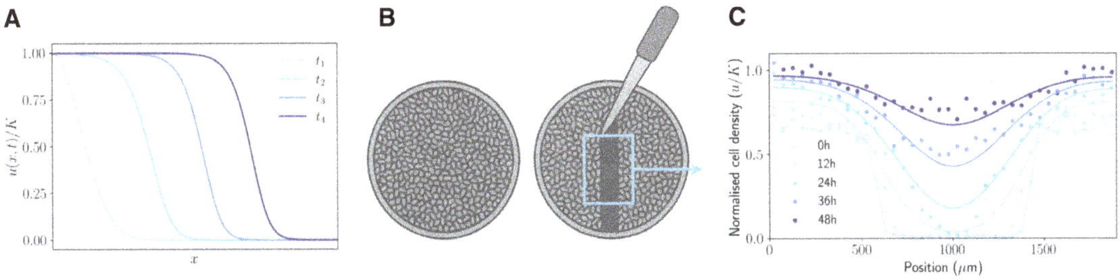

Figure 1. Traveling wave solutions of the Fisher–Kolmogorov, Petrovskii, and Piskunov (Fisher–KPP) equation and application to cell migration in scratch assays. (*A*) Numerical solutions of the Fisher–KPP equation exhibit traveling wave behavior, with wave fronts propagating at constant speed. (*B*) Schematic of a scratch assay, where a cell monolayer is wounded and imaged over time to track migration into the wound area. (*C*) Experimental data from Jin et al. (2016) are used to parametrize the Fisher–KPP model—$D = 1030\ \mu m^2/h$, $r = 0.064\ h^{-1}$, $K = 1.7 \times 10^{-3}$ cells/μm^2—showing good agreement between the model and observed cell density profiles over time.

as the population recolonizes the vacant scratched region. The Fisher–KPP equation can be fit to these data to provide estimates of the diffusion coefficient, D, proliferation rate, r, and carrying capacity, K. A comparison of the solution of the Fisher–KPP equation, given these parameter estimates, and the experimental data are shown in Fig. 1C—the model is able to provide accurate predictions of the cell density as the wound area is closed.

An alternative approach to mathematical modeling of collective cell migration that has been enabled by the recent data revolution is that of learning the models directly from the data, herein referred to as data-driven modeling. The aim of these methods is to infer the underlying mechanistic model directly from spatiotemporal data, by leveraging statistical and machine learning tools (Brunton and Kutz 2024; Yu and Wang 2024). By doing so, the hope is to eliminate the inherent bias of traditional mathematical modeling approaches, which often rely on heuristics and convention to choose the model structure and functional forms describing different biological mechanisms. Although still in their infancy, there are now a number of different methods that can be used for data-driven modeling of collective cell migration. This *Perspectives in Biology* article aims to provide the reader with a broad overview of the field of data-driven modeling, discussing the advantages and disadvantages of the different methods in the context of learning models of collective cell motility, as well as avenues for future research. We focus on learning PDE-based models, a process that we will often refer to as equation learning, with an emphasis on understanding what constitutes a good model, and how to handle noise, quantify uncertainty, and develop models in a computationally efficient manner. Recent examples of methods for learning PDE-based models in the context of cell migration can be found in Lagergren et al. (2020a,b), Nardini et al. (2020, 2021), Martina-Perez et al. (2021), Simpson et al. (2022), Nardini (2024), and van den Heuvel et al. (2024).

From our perspective, a good data-driven model is one that not only fits observed data with high quantitative accuracy but also one that captures the fundamental governing dynamics (physics) of the system, rather than merely interpolating the available data, so that it can generalize beyond the regime in which it was trained to make useful quantitative predictions. As such, one major challenge in equation learning is achieving a balance between accurately fitting the data and maintaining model parsimony to avoid overfitting. A good model should also be interpretable from a mechanistic perspective, meaning that its terms correspond to meaningful physical, chemical, or biological interactions. Such interpretability allows researchers to gain deeper insights into underlying mechanisms, refine hypotheses, and potentially make novel predictions about system behavior in untested scenarios. Moreover, a good model should have sensible mathematical properties, for example, it should preserve positivity of concentrations and densities and satisfy relevant physical laws such as conservation of mass or momentum.

Learning mathematical models directly from data inherently involves dealing with noise, measurement errors, and incomplete observations. A key question in selecting a method for equation learning is how much data a particular method requires to learn a model, as well as how robust the resulting model is to variations in the data and its quality (Fajardo-Fontiveros et al. 2023). Here, robustness needs to be considered both in terms of the inferred model structure and in the variability of parameter estimates. For example, when it comes to noisy data, an important question to ask is whether a given method consistently gives rise to the same terms in a differential equation model, and/or the extent to which parameter estimates vary across data replicates (Martina-Perez et al. 2021).

Finally, the computational complexity of data-driven methods is a critical consideration in selecting an approach, particularly when dealing with large-scale or high-dimensional data or complicated mathematical models. One significant factor influencing the computational complexity of a method is the need to compute numerical solutions of the model at each step of the learning algorithm—this can significantly increase computational costs. Such repeated numerical integration may even render a method computationally prohibitive from a practical perspective if many thousands (or even millions) of rounds of numerical integration are required. As such, it is important to analyze how computational costs scale with the amount of data and whether certain approximations or surrogate models can be used to accelerate learning while maintaining accuracy. The development of scalable algorithms that can handle large data sets efficiently will be key to making equation learning-based approaches accessible in real-world applications.

In this work, we discuss two major categories of data-driven modeling techniques in use today: symbolic regression–based approaches and neural differential equation–based approaches. In each case, we provide a brief outline of the approach, how it could potentially be used in the context of data-driven modeling of collective cell invasion (with reaction–diffusion models such as the Fisher–KPP model and its relatives assumed as the underlying mathematical framework) and its advantages and disadvantages. We aim to provide only a simple description of the fundamentals of each approach here, acknowledging that many researchers have made significant contributions in improving and extending them since their original conception. We conclude in the section "Discussion and Future Perspectives" with a brief discussion of the challenges involved in data-driven modeling and suggest avenues for possible future research.

SYMBOLIC REGRESSION–BASED APPROACHES

One of the most influential ideas in data-driven modeling is that of symbolic regression (Bongard and Lipson 2007; Schmidt and Lipson 2009), and one of the earliest works (Crutchfield and McNamara 1987) introduced a method for reconstructing the deterministic portion of equations of motion using a function basis. This approach gained particular attention through the development of the sparse identification of nonlinear dynamics (SINDy) algorithm, which learns systems of ordinary differential equations directly from data (Brunton et al. 2016a), and its extension to PDEs with PDE-FIND (Rudy et al. 2017). Symbolic regression generally assumes data measured on a grid of spatial and temporal points. The approach constructs a library of candidate terms that may appear in the governing equation, evaluates them at the grid points, and performs sparse regression to identify the most relevant terms for the final model.

More concretely, a simple application of PDE-FIND to collective cell motility would be to learn a PDE model for the evolution of cell density from noisy measurements of the cell density of the form $U^{\text{data}}(x_i, t_j)$, for $i = 1,\ldots,n$ and $j = 1,\ldots,m$. The data are assumed to be a noisy discretization of a function $u(x, t)$, which is the solution of an unknown PDE of the form given in Equation (1). PDE-FIND then attempts to reconstruct this equation by writing the right-hand side of the PDE as a linear combination of distinct library terms,

$$\frac{\partial u}{\partial t} = \mathcal{F}^{\text{SR}}(u; \boldsymbol{p}) = \sum_{i=1}^{N_\ell} \xi_i \mathcal{F}_i^{\text{SR}}(u), \qquad (3)$$

where the $\mathcal{F}_i^{\text{SR}}$ are differential operators, each with corresponding coefficient ξ_i. A key assumption in PDE-FIND is the right-hand side of the PDE contains only a few terms, and so from the large library of terms,

$$\{\mathcal{F}_1^{\text{SR}}, \ldots, \mathcal{F}_{N_\ell}^{\text{SR}}\},$$

that make up the right-hand side of Equation (3), only a few should have nonzero coefficients ξ_i. This makes it possible to use sparse regression approaches to find these nonzero coefficients and estimate them. In this formulation, the unknown parameters $\boldsymbol{p}$ of the model are captured in the coefficients ξ_i—the terms $\mathcal{F}_i^{\text{SR}}$ may contain parameters but these cannot be estimated as part of the equation learning process.

The first step of the PDE-FIND pipeline is to numerically approximate both the temporal derivative on the left-hand side of Equation (3) and the required spatial derivatives of the library terms on the right-hand side of Equation (3) directly from the data—this can be done, for example, using finite differences. The data and its derivatives are then combined in a matrix $F^{\text{SR}}(\boldsymbol{U}^{\text{data}})$, where each column of F^{SR} contains all of the values of a particular candidate function across the entire $n \times m$ grid and $\boldsymbol{U}^{\text{data}}$ is a matrix consisting of the complete data set. The result of approximating each of the terms in Equation (3) is a linear matrix equation representing the PDE evaluated at the data points,

$$\boldsymbol{U}_t^{\text{data}} = F^{\text{SR}}(\boldsymbol{U}^{\text{data}})\boldsymbol{\xi}, \qquad (4)$$

where $\boldsymbol{\xi} = [\xi_1, \ldots, \xi_{N_\ell}]^T$. Each row in this matrix equation represents the governing dynamics behind the data at one point in time and space.[4] PDE-FIND then proceeds to learn a *sparse* vector $\boldsymbol{\xi}^{\text{sparse}}$ as a solution of Equation (3) by optimizing the fit of the data to the PDE while penalizing for the large value of the ξ.

Mathematically, this process can be expressed by finding $\boldsymbol{\xi}^{\text{sparse}}$ that minimizes the loss function

$$\mathcal{L}^{\text{SR}}(\boldsymbol{\xi}) = \left\| \boldsymbol{U}_t^{\text{data}} - F^{\text{SR}}(\boldsymbol{U}^{\text{data}})\boldsymbol{\xi} \right\|_2^2 + \lambda \|\boldsymbol{\xi}\|_2^2, \qquad (5)$$

alternatively,

$$\boldsymbol{\xi}^{\text{sparse}} = \text{argmin}_{\boldsymbol{\xi}} \mathcal{L}^{\text{SR}}(\boldsymbol{\xi}). \qquad (6)$$

In Equation (5), $\lambda > 0$ is a free parameter that penalizes large coefficients—this is the method of *ridge regression* (Rudy et al. 2017; Kaiser et al. 2018; Nardini et al. 2021). PDE-FIND combines ridge regression optimization within an iterative *sequential thresholding* procedure, in which a solution to Equation (6) is found, and all library terms that have coefficients smaller than some tolerance ($|\xi_i| < \text{tol}$) are removed. The process is repeated until all coefficients are larger than the threshold or some maximum number of iterations is reached. The regularizing term $\|\boldsymbol{\xi}\|_2^2$ can be replaced with others, for example, $\|\boldsymbol{\xi}\|_1^2$, which corresponds to performing LASSO (Lagergren et al. 2020a; Nardini et al. 2021), or $\|\boldsymbol{\xi}\|_0^2$, which corresponds to hard thresholding. The optimal choice is largely problem-dependent, and no method has been proven to be definitively preferred over another (Kaiser et al. 2018; De Silva et al. 2020).

An example application of sparse regression to learn a PDE for collective cell invasion is shown in Fig. 2. A library of candidate terms is

[4]Generally, F^{SR} is overspecified, which means that the dynamics can be represented as linear combination of the columns of F^{SR}.

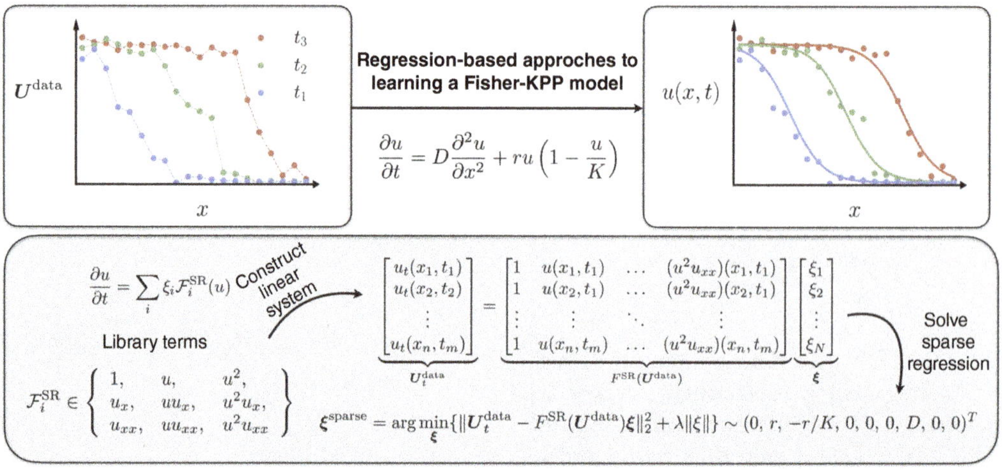

Figure 2. Sparse regression–based approach to learning a Fisher–Kolmogorov, Petrovskii, and Piskunov (Fisher–KPP) model. A library of candidate terms is constructed from the data, and the governing partial differential equation (PDE) is identified by solving a sparse regression problem, where only the most relevant terms (e.g., diffusion and logistic growth) are retained.

constructed, for example (Martina-Perez et al. 2021):

$$\left\{1, u, u^2, \frac{\partial u}{\partial x}, u\frac{\partial u}{\partial x}, u^2\frac{\partial u}{\partial x}, \frac{\partial^2 u}{\partial x^2}, u\frac{\partial^2 u}{\partial x^2}, u^2\frac{\partial^2 u}{\partial x^2}\right\}. \tag{7}$$

The parameters of the model, $\boldsymbol{p}$, appear in the library coefficients $\boldsymbol{\xi} = [\xi_1, \ldots, \xi_{N_\ell}]^T$, where each ξ_i can represent, for instance, a diffusion coefficient, a growth rate, a carrying capacity, etc., depending on the form of the library term. A linear matrix equation is then formed from data for the cell density, $u(x, t)$, where the time derivative u_t is expressed as a linear combination of these terms. By performing sparse regression, only the most relevant terms are retained—typically recovering the diffusive and logistic growth components of the Fisher–KPP model. This approach enables interpretable model discovery directly from data.

Extensions of this symbolic regression approach have evolved rapidly, incorporating diverse methodologies to enhance performance and applicability (Brunton and Kutz 2024). Some approaches focus on reduced coordinate representations, learning coordinate systems where nonlinear dynamics are rendered linear (Kutz et al. 2016; Brunton et al. 2022), balancing accuracy with simplicity. On the other hand, SINDy with control (SINDYc) extends SINDy to include external inputs and feedback control (Brunton et al. 2016b), while SINDy-BVP generalizes the framework to boundary value problems (Shea et al. 2021). AI Feynman (Udrescu and Tegmark 2020) combines neural network fitting with physics-inspired techniques such as dimensional analysis, symmetry, separability, and compositionality to simplify the regression process.

The main advantage of these methods is their interpretability and ability to produce tractable, sparse models. In addition, they are computationally efficient because they do not require solutions of the differential equation itself—this is a *soft constraint.* However, they are highly sensitive to noise, which can lead to incorrect or physically implausible models, for example, violating conservation laws. As such, one of the most significant challenges involved in using sparse regression approaches is dealing with noisy data and assessing the ensuing uncertainty in the learned equations. One way to mitigate the effects of noise is through the use of

Bayesian inference approaches. For example, Martina-Perez et al. (2021) developed Bayes-PDE-FIND to quantify the uncertainty in estimates provided by PDE-FIND, and the approach of Li et al. (2020) seeks to automatically distinguish noise from true dynamics using a model-free method. The challenges relating to noisy data are exacerbated when computing higher order derivatives as it is incredibly difficult to estimate them from noisy data. This issue can, in part, be addressed using *weak formulations* (Schaeffer and McCalla 2017; Gurevich et al. 2019; Reinbold et al. 2020, 2021; Messenger and Bortz 2021a,b). For example, Weak-SINDy (Messenger and Bortz 2021a,b), which has been used by Srivastava et al. (2025), for example, to infer interpretable PDE models of migration and proliferation from experimental wound healing data, avoids pointwise derivative approximations to improve model recovery from noise-free data and support more robust identification of PDEs, even in high-noise settings.

Addressing computational cost is another crucial aspect of advancing symbolic regression techniques. Using *fast Fourier transform*–based weak formulations (Schaeffer and McCalla 2017; Messenger and Bortz 2021a,b) significantly lowers computational demand, although symbolic regression methods inherently require large libraries to achieve good predictive power. Parallelization techniques, such as those in SINDy-PI (Kaheman et al. 2019), further enhance efficiency by systematically testing library terms for sparse model selection. This approach is orders of magnitude more noise-robust than previous methods and enables the identification of complex dynamical systems.

Ultimately, sparse regression–based approaches, although easily interpretable in terms of biological mechanism, do not lend themselves well to the kinds of data routinely collected in biology, which is generally both noisy and sparse. As such, they have rapidly made way for deep learning–based methods that focus on using neural networks to learn models directly from the data as they suffer far less from such issues. We review a number of such approaches in the following section.

NEURAL DIFFERENTIAL EQUATION–BASED APPROACHES

A neural differential equation is a type of differential equation where a neural network is used to define some, or all, of the governing dynamics of the system. Neural differential equations offer a powerful framework for combining deep learning with traditional differential equations, allowing for data-driven discovery of complex system dynamics (Kidger 2022). In this section, we outline methods that have arisen from the basic concept of a *neural PDE* (NPDE), which replaces the traditional right-hand side of the PDE given in Equation (1) with a neural network, to write

$$\frac{\partial u}{\partial t} = \mathcal{N}(u; \boldsymbol{\theta}), \tag{8}$$

where $\boldsymbol{\theta}$ is a vector containing the parameters of the neural network $\mathcal{N}$.

In the context of learning an NPDE for the evolution of cell density, $u(x, t)$, from noisy measurements of the cell density of the form $U^{\text{data}}(x_i, t_j)$, for $i = 1, \ldots, n$ and $j = 1, \ldots, m$, the parameters $\boldsymbol{\theta}$ are established through training the neural network parameters to minimize a loss function of the form

$$\mathcal{L}^{\text{NN}}(\boldsymbol{\theta}) = \left\| \boldsymbol{U}^{\text{data}} - \boldsymbol{u}^{\text{pred}}(\boldsymbol{\theta}) \right\|_2^2, \tag{9}$$

where, again, $\boldsymbol{U}^{\text{data}}$ is a matrix representing the data, and $\boldsymbol{u}^{\text{pred}}(\boldsymbol{\theta})$ is a matrix formed from the solution $u(x, t; \boldsymbol{\theta})$ of the NDE, Equation (8), evaluated at points (x_i, t_j). This means that, at each stage of training of the neural network, Equation (8) must be solved using current values of the parameters $\boldsymbol{\theta}$ to evaluate the loss function. The downside of this approach is therefore the computational cost of solving Equation (8), but the advantage is that the NPDE is satisfied exactly at each stage of training, that is, it is a hard constraint. In addition, although the method is very flexible, in the sense that it does not require any prior knowledge of the underlying biological mechanisms, it does not give rise to an interpretable model, and it contains large numbers of parameters, which means that it can be prone to overfitting of the data.

We now discuss two main data-driven modeling approaches that have arisen from this concept: Firstly, universal differential equations, which constitute a hybrid approach that integrates mechanistic models with neural networks to learn missing components of a model, and secondly, physics-informed neural networks (PINNs), which use neural networks to approximate the solution of differential equations while enforcing physical constraints.

Universal Partial Differential Equations

The idea underlying universal (partial) differential equations (UPDEs) is to leverage both prior scientific knowledge and data-driven methods to improve model accuracy and interpretability (Rackauckas et al. 2020; Philipps et al. 2025): it combines traditional PDE modeling approaches and NPDEs to write

$$\frac{\partial u}{\partial t} = \mathcal{F}(u; \boldsymbol{p}) + \mathcal{N}(u; \boldsymbol{\theta}). \tag{10}$$

The term $\mathcal{F}(u; \boldsymbol{p})$ is a differential operator that incorporates *known* parts of the model, with the parameters $\boldsymbol{p}$ to be estimated as part of the learning process, while the neural network, $\mathcal{N}(u; \boldsymbol{\theta})$, is used to infer *unknown* components of the dynamics from data through learning of the parameters $\boldsymbol{\theta}$.

Given the data of the form $\boldsymbol{U}^{\text{data}}$, the unknown dynamics together with the parameters $\boldsymbol{p}$ are estimated by minimizing a loss function of the form

$$\mathcal{L}^{\text{UPDE}}(\boldsymbol{p}, \boldsymbol{\theta}) = \left\| \boldsymbol{U}^{\text{data}} - \boldsymbol{u}^{\text{pred}}(\boldsymbol{p}, \boldsymbol{\theta}) \right\|_2^2, \tag{11}$$

where $\boldsymbol{u}^{\text{pred}}(\boldsymbol{p}, \boldsymbol{\theta})$ is a matrix representing the solution $u(x, t; \boldsymbol{p}, \boldsymbol{\theta})$ of the UPDE, Equation (10), at points (x_i, t_j). As for NPDEs, the UPDE is a *hard constraint*, and so Equation (10) must be solved numerically at each stage of the training process. The UPDEs approach can incorporate cases where the *known* model parameters, $\boldsymbol{p}$, are assumed pre-estimated (known), so that the loss function depends only on the neural network parameters $\boldsymbol{\theta}$, that is, $\mathcal{L}^{\text{UPDE}}(\boldsymbol{\theta})$, or mixed cases where some parameters are assumed already known and others are not.

In Fig. 3A, we show the UPDEs approach applied to a Fisher–KPP model. The structure of the PDE is preserved, with a neural network used to represent the unknown nonlinear reaction term. Given spatiotemporal data of the cell density, the network is trained together with any unknown parameters (e.g., the diffusion coefficient) by minimizing the mismatch between the predicted and observed data. This allows incorporation of mechanistic structure while leveraging the flexibility of neural networks to learn complex functional forms directly from data.

One of the key advantages of universal differential equations is their potential for reliable extrapolation, particularly when compared to purely data-driven neural network models. By embedding known physical or biological dynamics within part of the model, universal differential equations retain some grounding in established physical principles, making them more robust when predicting outside the training domain. They also offer notable flexibility, particularly for modeling complex nonlinear systems such as those encountered in cellular behavior. For example, the growing availability of high-dimensional data sets—like transcriptomics, encompassing tens of thousands of genes—presents an insurmountable challenge for traditional mechanistic modeling approaches. Universal differential equations provide a way forward by combining mechanistic modeling with neural networks that can digest and represent these large, intricate data sets. This hybrid architecture also allows for integration of multimodal data, such as genomics, proteomics, and metabolomics, within a unified framework that can learn from each data stream while remaining consistent with known dynamics.

However, these benefits come with tradeoffs. Universal differential equations can suffer from poor interpretability, especially when used to integrate multiple data sources. While they may outperform existing models in predictive accuracy, the contributions of the neural net-

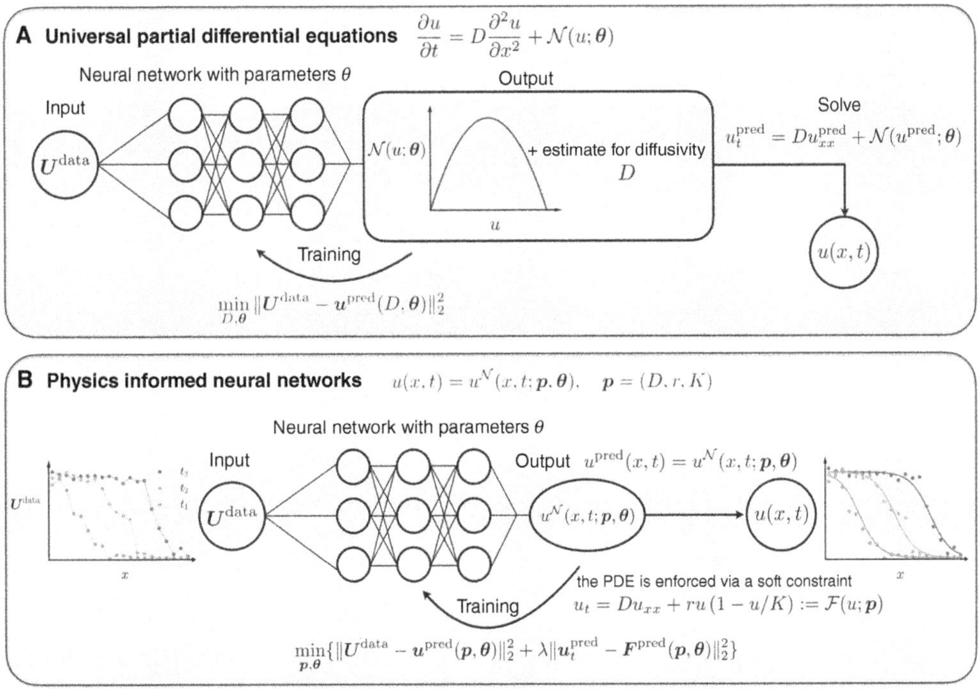

Figure 3. Neural network–based approaches to learning a Fisher–KPP model. (*A*) Universal differential equations approach: the neural network learns the nonlinear reaction term, while the PDE structure is preserved. (*B*) Physics-informed neural networks: the full solution is approximated by a neural network, and the PDE is enforced as a soft constraint during training.

work component of the model can be opaque and they have limited analytical tractability. As such, understanding how specific features drive model behavior or mapping them back to biological mechanisms can be challenging. Moreover, universal differential equations are subject to the same issues as other neural network models, including the risk of becoming ill-posed when data are sparse. This can lead to overfitting or produce predictions that are biologically implausible (de Rooij et al. 2025). To mitigate these risks, researchers often introduce constraints—such as enforcing nonnegativity (Philipps et al. 2024) or penalizing deviations from known behavior—but these strategies are still developing and may not fully resolve the underlying limitations.

Physics-Informed Neural Networks

PINNs take a very different approach to using neural networks to learn differential equation models (Raissi et al. 2019; Pang and Karniadakis 2020). They assume a PDE model of the form given in Equation (1), that is,

$$\frac{\partial u}{\partial t} = \mathcal{F}(u; \boldsymbol{p}), \tag{12}$$

where the parameters, $\boldsymbol{p}$, of the differential operator $\mathcal{F}(u; \boldsymbol{p})$ are unknown. The PINNs approach approximates the solution of Equation (12), $u(x, t; \boldsymbol{p})$, using a neural network $u^{\mathcal{N}}(x, t; \boldsymbol{p}, \boldsymbol{\theta})$, which has parameters $\boldsymbol{\theta}$. The *underlying physics* of the problem is captured through the loss function, which incorporates terms that require the neural network prediction to fit both the observed data and the PDE:

$$\mathcal{L}^{\text{PINN}}(\boldsymbol{p}, \boldsymbol{\theta}) = \left\|\boldsymbol{U}^{\text{data}} - \boldsymbol{u}^{\text{pred}}(\boldsymbol{p}, \boldsymbol{\theta})\right\|_2^2 + \lambda\left\|\boldsymbol{u}_t^{\text{pred}}(\boldsymbol{p}, \boldsymbol{\theta}) - \boldsymbol{F}^{\text{pred}}(\boldsymbol{p}, \boldsymbol{\theta})\right\|_2^2. \tag{13}$$

This time, $\boldsymbol{u}^{\text{pred}}(\boldsymbol{p}, \boldsymbol{\theta})$ is the matrix of neural network predictions, $u^{\mathcal{N}}(x, t; \boldsymbol{p}, \boldsymbol{\theta})$, $\boldsymbol{u}_t^{\text{pred}}(\boldsymbol{p}, \boldsymbol{\theta})$ is a matrix comprised of the neural network partial derivative $\partial u^{\mathcal{N}}/\partial t$, and $\boldsymbol{F}^{\text{pred}}(\boldsymbol{p}, \boldsymbol{\theta})$ is the matrix of the neural network PDE right-hand side $\mathcal{F}(u^{\mathcal{N}}; \boldsymbol{p})$, each at points (x_i, t_j).

One of the main advantages of PINNs is that the numerical solution of the PDE is not required at each step of training, and automatic differentiation can be used to efficiently evaluate the second term in the loss function,

$$\left\| \boldsymbol{u}_t^{\text{pred}}(\boldsymbol{p}, \boldsymbol{\theta}) - \boldsymbol{F}^{\text{pred}}(\boldsymbol{p}, \boldsymbol{\theta}) \right\|_2^2 .$$

This means that in the PINNs approach, the PDE is a soft constraint, and the hyperparameter $\lambda > 0$ balances the emphasis placed on the fit of the neural network to the data (first term in the loss function) compared to the extent to which the PDE is satisfied (second term in the loss function).

In Fig. 3B, we depict the PINNs approach for the Fisher–KPP equation, where the solution $u(x, t; \boldsymbol{p})$ is represented by a neural network. Given spatiotemporal data of the cell density, the PDE is imposed as a soft constraint during training of the neural network for the cell density through the inclusion of its residual in the loss function. This formulation enables simultaneous learning of the solution and the model parameters (the diffusion coefficient, growth rate, and carrying capacity).

A major strength of PINNs is their natural capacity to handle inverse problems: unknown parameters within a differential equation are treated as trainable variables, allowing the model to learn both dynamics and model parameters in single trainable neural network architecture. This flexibility is particularly useful in biological contexts, where direct estimation of model parameters is generally experimentally inaccessible. Moreover, PINNs (like universal differential equations) are well-suited to integrate multimodal data—such as those arising from genomics, proteomics, and metabolomics—by learning a latent function depending on many inputs, an increasingly important feature in the era of multiomics systems biology.

Despite its strengths, several limitations have tempered the enthusiasm surrounding PINNs. One key challenge lies in balancing the dual objectives of minimizing prediction error and enforcing the differential equation: This typically involves tuning the weighting parameter (λ) that scales the relative importance of data fidelity versus physics consistency. As noted by Lagergren et al. (2020a), poor tuning can result in models that either fit the data but violate physical constraints or satisfy the physics but poorly capture the observed dynamics. Johnston et al. (2024) provide a potential solution to this approach, in the context of the method of (Ramsay et al. 2007; Ramsay and Hooker 2017), which involves updating the weighting parameter iteratively. Ramsay et al. (2007) established methods for fitting models to data that avoid the use of numerical solutions of the model by using surrogate models (e.g., in the form of splines) and approximately enforcing the governing differential equation. In addition, recent advances such as PINNverse (Almanstötter et al. 2025) have enhanced the reliability of PINNs for inverse problems by dynamically balancing data and equation losses during training and improving convergence and robustness even when data are noisy or the forward problem is computationally expensive.

Combining UPDEs and PINNs

Recent developments have sought to combine the advantages of both the UPDEs and PINNS approaches: They relax the *mechanistic knowledge* requirement of the PINNs approach that the entire model structure is known in advance, letting some aspects of the right-hand side of the equation be learned using neural networks, and relax the *hard constraint* requirement of the UPDEs approach that the UPDE is satisfied at each stage of the training process.

The biologically informed neural networks (BINNs) approach enables the user to learn the functional forms of different terms on the right-hand side of Equation (12) (Lagergren et al. 2020a). For example, suppose that we anticipate that cell density evolves according to the PDE

$$\frac{\partial u}{\partial t} = \frac{\partial}{\partial x}\left(D(u)\frac{\partial u}{\partial x}\right) + u\, G(u),$$

where $D(u)$ describes the density-dependent diffusion coefficient and $G(u)$ describes the density-dependent net birth rate, but the functional forms of these functions are unknown. The BINNs approach learns these functional forms using neural networks by assuming that

$$\frac{\partial u}{\partial t} = \frac{\partial}{\partial x}\left(D^{\mathcal{N}}(u;\, \boldsymbol{\theta}_{\mathcal{D}})\frac{\partial u}{\partial x}\right) + u\, G^{\mathcal{N}}(u;\, \boldsymbol{\theta}_{\mathcal{G}}), \tag{14}$$

and approximating the solution of Equation (14) using a neural network $u^{\mathcal{N}}(x, t;\, \boldsymbol{\theta}_{\mathcal{D}}, \boldsymbol{\theta}_{\mathcal{G}}, \boldsymbol{\theta}_{\text{solution}})$. The loss function is of the form

$$\mathcal{L}^{\text{BINN}}(\boldsymbol{\theta}_{\mathcal{D}}, \boldsymbol{\theta}_{\mathcal{G}}, \boldsymbol{\theta}_{\text{solution}}) = \left\|\boldsymbol{U}^{\text{data}} - \boldsymbol{u}^{\text{pred}}(\boldsymbol{\theta}_{\mathcal{D}}, \boldsymbol{\theta}_{\mathcal{G}}, \boldsymbol{\theta}_{\text{solution}})\right\|_2^2 + \lambda\left\|\boldsymbol{u}_t^{\text{pred}}(\boldsymbol{\theta}_{\mathcal{D}}, \boldsymbol{\theta}_{\mathcal{G}}, \boldsymbol{\theta}_{\text{solution}}) - \boldsymbol{F}^{\text{pred}}(\boldsymbol{\theta}_{\mathcal{D}}, \boldsymbol{\theta}_{\mathcal{G}})\right\|_2^2, \tag{15}$$

where, as for UPINNs, $\boldsymbol{u}^{\text{pred}}$ is the matrix of neural network predictions $u^{\mathcal{N}}$ of the density, u, $\boldsymbol{u}_t^{\text{pred}}$ is a matrix comprised of the neural network partial derivative $\partial u^{\mathcal{N}}/\partial t$, and $\boldsymbol{F}^{\text{pred}}$ is the matrix arising from the right-hand side of the PDE, $\mathcal{F}$, each at point (x_i, t_j).

The main advantage of the BINNs approach is that not all functional forms of the terms in the governing PDE need to be known a priori. However, when part of the differential operator is modeled nonparametrically, some mechanistic interpretability can be lost, and, as for other neural differential equation–based approaches, analytical tractability (e.g., to linear stability analysis, traveling wave analysis) is limited.

In the universal PINNs (UPINNs) approach (Podina et al. 2023), on the other hand, the solution $u(x, t;\, \boldsymbol{p}, \boldsymbol{\theta}_{\text{model}})$ of the UPDE

$$\frac{\partial u}{\partial t} = \mathcal{F}(u;\, \boldsymbol{p}) + \mathcal{N}(u;\, \boldsymbol{\theta}_{\text{model}}), \tag{16}$$

is approximated by a neural network $u^{\mathcal{N}}(x, t;\, \boldsymbol{p}, \boldsymbol{\theta}_{\text{model}}, \boldsymbol{\theta}_{\text{solution}})$, so that the loss function becomes

$$\mathcal{L}^{\text{UPINN}}(\boldsymbol{p}, \boldsymbol{\theta}_{\text{model}}, \boldsymbol{\theta}_{\text{solution}}) = \left\|\boldsymbol{U}^{\text{data}} - \boldsymbol{u}^{\text{pred}}(\boldsymbol{p}, \boldsymbol{\theta}_{\text{model}}, \boldsymbol{\theta}_{\text{solution}})\right\|_2^2 + \lambda\left\|\boldsymbol{u}_t^{\text{pred}}(\boldsymbol{p}, \boldsymbol{\theta}_{\text{model}}, \boldsymbol{\theta}_{\text{solution}}) - \boldsymbol{FN}^{\text{pred}}(\boldsymbol{p}, \boldsymbol{\theta}_{\text{model}})\right\|_2^2. \tag{17}$$

Here, similarly to the PINNs approach, $\boldsymbol{u}^{\text{pred}}(\boldsymbol{p}, \boldsymbol{\theta}_{\text{model}}, \boldsymbol{\theta}_{\text{solution}})$ is the matrix of neural network predictions $u^{\mathcal{N}}$ of the density, u, $\boldsymbol{u}_t^{\text{pred}}(\boldsymbol{p}, \boldsymbol{\theta}_{\text{model}}, \boldsymbol{\theta}_{\text{solution}})$ is a matrix comprised of the neural network partial derivative $\partial u^{\mathcal{N}}/\partial t$, and $\boldsymbol{FN}^{\text{pred}}$ is the matrix arising from the right-hand side of the UPDE, $\mathcal{F}(u^{\mathcal{N}};\, \boldsymbol{p}) + \mathcal{N}(u^{\mathcal{N}};\, \boldsymbol{\theta}_{\text{model}})$, each at points (x_i, t_j).

The UPINNs approach combines the advantages of the UPDEs approach, in being able to learn missing physics, with that of the PINNs approach in avoiding solution of the PDE at each step. However, it also suffers from some of its issues, in lacking some interpretability and analytical tractability. This approach has not yet seen application in collective cell migration, although it has been applied in other areas of the biomedical sciences (Podina et al. 2025).

DISCUSSION AND FUTURE PERSPECTIVES

In this article, we have discussed a range of approaches for learning PDE-based models of collective cell migration from experimental data, focusing on symbolic regression–based and neural differential equation–based approaches. We have described the mathematical concepts underlying each approach and provided discussion of the advantages and disadvantages of each, as well as recent developments and applications to collective cell migration. We have grounded our examples in the Fisher–KPP equation, one of the simplest, and possibly the most widely used, model of collective cell migration.

In addition to PDE models, significant progress has also been made in learning interaction rules in individual-based models (Haupt 2004; Thede 2004; Bottou 2010; Lu et al. 2019, 2022a,b; Brückner et al. 2021; Li et al. 2021; Maggioni et al. 2025). Data-driven approaches to infer interaction kernels in this case have allowed researchers to identify behaviors such as repulsion, adhesion, and alignment in collective migration systems. These kernels have been learned using a variety of techniques including regularized least squares (Lu et al. 2019), nonparametric regression (Lu et al. 2022a), approximation theory (Miller et al. 2023), reproducing kernel Hilbert spaces (RKHS)-based methods (Lang and Lu 2022), and machine learning frameworks such as neural networks and integral operator learning (Lu et al. 2024). Recent work has also explored identifiability and inference in stochastic interacting particle systems using Bayesian frameworks (Lang and Lu 2023; Tang et al. 2024). There are a number of comprehensive reviews written on this subject, such as those by Lu et al. (2021) and Lang and Lu (2022).

Despite the successes in applying data-driven methods to learning mathematical models of biological systems, a number of technical and conceptual challenges remain. First, identifiability and uncertainty quantification remain critical bottlenecks in data-driven modeling. Questions about the uniqueness and robustness of the inferred model structure and parameters must be addressed, particularly in ill-posed inverse problems or when observational data are sparse or noisy (Li et al. 2021; Tang et al. 2024). Similarly, learning model components that are often overlooked—such as boundary conditions, nonlocal terms, or spatially dependent forcing functions—poses challenges that are only beginning to be explored (Long et al. 2019; Sun et al. 2020; Tartakovsky et al. 2020; Yuan et al. 2022).

Another area that requires active development concerns the implementation choices that influence the learning process. Sparse regression methods, for instance, rely on careful selection of basis functions and regularization parameters (Messenger et al. 2022), while neural network approaches for learning PDEs demand choices about the architecture, loss formulation, and inductive biases (Raissi et al. 2019). Recent hybrid approaches, such as those combining neural and symbolic methods—for example, through physics-informed regression splines (Karniadakis et al. 2021)—offer promising avenues for balancing expressivity of the model with interpretability.

The increasing availability of high-resolution biological data, particularly from technologies such as omics, microscopy, and barcoding, is transforming both the questions asked by experimental biologists, as well as the models developed by mathematical biologists. This data deluge offers unprecedented opportunities to formulate and validate mechanistic hypotheses. However, it also demands a shift in how mathematical modelers collaborate, design models, and interpret results. The design of experiments—ranging from the spatial and temporal resolution of data collection to the initial setup—can dramatically influence model identifiability and interpretability. Modelers can provide valuable input into experimental design, ensuring that the resulting data are optimally structured for robust inference and validation of mathematical models.

One of the most potentially impactful opportunities from the wealth of data is the ability to learn models of increased complexity, describing, for example, the evolution of internal state variables of individual cells, and to interrogate and model phenotypic heterogeneity at scale. Previously, distinctions between cell subpopulations or state transitions were difficult to quantify. Now, omics data enable not only identification of distinct cellular subtypes but also inference of their roles in dynamic processes, for instance, through differential gene expression or pathway activity. This growing data richness demands models that can incorporate cell heterogeneity explicitly and evolve alongside the increasing complexity of biological data (Brückner et al. 2021; Li et al. 2021).

In developing these models of increased complexity, a guiding principle should remain that models must be biologically grounded and interpretable, so that individual terms in the model can be attributed to particular biological processes, feedbacks, or interactions. In addition, the learned models should be physically grounded, preserving mathematical properties such as positivity of densities, conservation of mass, and relevant boundary or symmetry conditions. Integrating these constraints explicitly —either through the model architecture or via regularization—can improve generalizability and foster biological trust in the models produced.

Looking ahead, the continuing convergence of high-resolution data and computational inference techniques opens up exciting opportunities for learning nonlocal models, individual cell-level behaviors, and multiscale dynamics that reflect the true complexity of biological systems. However, progress will require not only methodological advances but also cultural shifts: closer and earlier collaboration with experimentalists, shared design of data collection strategies, and a commitment to developing models that are both data-driven and mechanistically interpretable. Ultimately, the future of data-driven mathematical biology lies in its ability to synthesize detailed experimental insight with sound mathematical structure. In adopting data-driven approaches, we will be better positioned to not only explain observed phenomena but also to predict and control complex biological systems.

CODE AVAILABILITY

Code to produce Fig. 1 is available on GitHub (https://github.com/carlesfalco/Fisher-KPP-Scratch-Assay). Repositories for the various data-driven modeling approaches described in this article include the following:

- SINDy (Brunton et al. 2016a)—PySINDy (https://github.com/dynamicslab/pysindy).
- PDE-FIND (Rudy et al. 2017)—PDEFIND (https://github.com/snagcliffs/PDE-FIND).
- Universal differential equations (Rackauckas et al. 2020)—UPDEs (https://github.com/ChrisRackauckas/universal_differential_equations).
- BINNs (Lagergren et al. 2020a)—BINNS (https://github.com/jlager/BINNs).

ACKNOWLEDGMENTS

R.E.B. is supported by a grant from the Simons Foundation (MP-SIP-00001828). R.M.C. is supported by the Engineering and Physical Sciences Research Council (EP/T517811/1). C.F. acknowledges support via a fellowship from “la Caixa” Foundation (ID 100010434) with code LCF/BQ/EU21/11890128. S.F.M.-P. thanks the Foulkes Foundation for funding.

REFERENCES

Almanstötter M, Vetter R, Iber D. 2025. PINNverse: accurate parameter estimation in differential equations from noisy data with constrained physics-informed neural networks. arXiv doi:10.48550/arXiv.2504.05248

Bongard J, Lipson H. 2007. Automated reverse engineering of nonlinear dynamical systems. *Proc Natl Acad Sci* **104:** 9943–9948. doi:10.1073/pnas.0609476104

Bottou L. 2010. Large-scale machine learning with stochastic gradient descent. In *Proceedings of COMP-STAT'2010: 19th International Conference on Computational Statistics*, Paris, France, August 22–27, 2010 Keynote, Invited and Contributed Papers, pp. 177–186. Springer, New York.

Browning AP, Haridas P, Simpson MJ. 2018. A Bayesian sequential learning framework to parameterize continuum models of melanoma invasion into human skin. *Bull Math Biol* **18:** 676–698. doi:10.1007/s11538-018-0532-1

Brückner DB, Arlt N, Fink A, Ronceray P, Rädler JO, Broedersz CP. 2021. Learning the dynamics of cell–cell interactions in confined cell migration. *Proc Natl Acad Sci* **118:** e2016602118. doi:10.1073/pnas.2016602118

Brunton SL, Kutz JN. 2024. Promising directions of machine learning for partial differential equations. *Nat Com Sci* **4:** 483–494. doi:10.1038/s43588-024-00643-2

Brunton SL, Proctor JL, Kutz JN. 2016a. Discovering governing equations from data by sparse identification of nonlinear dynamical systems. *Proc Natl Acad Sci* **113:** 3932–3937. doi:10.1073/pnas.1517384113

Brunton SL, Proctor JL, Kutz JN. 2016b. Sparse identification of nonlinear dynamics with control (SINDYc). *IFAC-Pap* **49:** 710–715. doi:10.1016/j.ifacol.2016.10.249

Brunton SL, Budišić M, Kaiser E, Kutz JN. 2022. Modern Koopman theory for dynamical systems. *SIAM Rev* **64:** 229–340. doi:10.1137/21M1401243

Buttenschön A, Edelstein-Keshet L. 2020. Bridging from single to collective cell migration: a review of models and links to experiments. *PLoS Comp Biol* **16:** e1008411. doi:10.1371/journal.pcbi.1008411

Crossley RM, Johnson S, Tsingos E, Bell Z, Berardi M, Botticelli M, Braat QJS, Metzcar J, Ruscone M, Yin Y, et al. 2024. Modeling the extracellular matrix in cell migration and morphogenesis: a guide for the curious biologist. *Front Cell Dev Biol* **12:** 1354132. doi:10.3389/fcell.2024.1354132

Crutchfield JP, McNamara BS. 1987. Equations of motion from a data series. *Complex Syst* **1:** 417–452.

de Rooij M, Erdős B, van Riel NAW, O'Donovan SD. 2025. Physiology-informed regularisation enables training of universal differential equation systems for biological applications. *PLoS Comp Biol* **21:** e1012198. doi:10.1371/journal.pcbi.1012198

De Silva BM, Higdon DM, Brunton SL, Kutz JN. 2020. Discovery of physics from data: universal laws and discrepancies. *Front Artif Intell* **3:** 25. doi:10.3389/frai.2020.00025

Fajardo-Fontiveros O, Reichardt I, De Los Ríos HR, Duch J, Sales-Pardo M, Guimerà R. 2023. Fundamental limits to learning closed-form mathematical models from data. *Nat Commun* **14:** 1043. doi:10.1038/s41467-023-36657-z

Falcó C, Cohen DJ, Carrillo JA, Baker RE. 2023. Quantifying tissue growth, shape and collision via continuum models and Bayesian inference. *J Roy Soc Interface* **20:** 20230184. doi:10.1098/rsif.2023.0184

Falcó C, Cohen DJ, Carrillo JA, Baker RE. 2025. Quantifying cell cycle regulation by tissue crowding. *Biophys J* **124:** 923–932. doi:10.1016/j.bpj.2024.05.003

Friedl P, Gilmour G. 2009. Collective cell migration in morphogenesis, regeneration and cancer. *Nat Rev Mol Cell Biol* **10:** 445–457. doi:10.1038/nrm2720

Giniūnaitė R, Baker RE, Kulesa PM, Maini PK. 2020. Modelling collective cell migration: neural crest as a model paradigm. *J Math Biol* **80:** 481–504. doi:10.1007/s00285-019-01436-2

Gurevich DR, Reinbold PA, Grigoriev RO. 2019. Robust and optimal sparse regression for nonlinear PDE models. *Chaos* **29:** 103113. doi:10.1063/1.5120861

Haeger A, Wolf K, Zegers MM, Friedl P. 2015. Collective cell migration: guidance principles and hierarchies. *Trends Cell Biol* **25:** 556–566. doi:10.1016/j.tcb.2015.06.003

Haupt RL. 2004. *Practical genetic algorithms*. Wiley, Hoboken, NJ.

Jin W, Shah ET, Penington CJ, McCue SW, Chopin LK, Simpson MJ. 2016. Reproducibility of scratch assays is affected by the initial degree of confluence: experiments, modelling and model selection. *J Theor Biol* **390:** 136–145. doi:10.1016/j.jtbi.2015.10.040

Johnston ST, Simpson MJ, McElwain DLS, Binder B, Ross JVJ. 2014. Interpreting scratch assays using pair density dynamics and approximate Bayesian computation. *Open Biol* **4:** 140097. doi:10.1098/rsob.140097

Johnston A, Baker RE, Simpson MJ. 2024. Efficient inference for differential equation models without numerical solvers. arXiv doi:10.48550/arXiv.2411.10494

Kaheman K, Kaiser E, Nair A, Kutz N, Brunton S. 2019. Identifying coarse-grained PDE models from data with the sparse identification of nonlinear dynamics. *APS Division of Fluid Dynamics Meeting Abstracts*. https://ui.adsabs.harvard.edu/abs/2019APS.DFDP10007K/abstract

Kaiser E, Kutz JN, Brunton SL. 2018. Sparse identification of nonlinear dynamics for model predictive control in the low-data limit. *Proc Math Phys Eng Sci* **474:** 20180335. doi:10.1098/rspa.2018.0335

Karniadakis GE, Kevrekidis IG, Lu L, Perdikaris P, Wang S, Yang L. 2021. Physics-informed machine learning. *Nat Rev Phys* **3:** 422–440. doi:10.1038/s42254-021-00314-5

Kidger P. 2022. On neural differential equations. arXiv doi:10.48550/arXiv.2202.02435

Kutz JN, Brunton SL, Brunton BW, Proctor JL. 2016. *Dynamic mode decomposition: data-driven modeling of complex systems*. SIAM, Philadelphia.

Lagergren JH, Nardini JT, Baker RE, Simpson MJ, Flores KB. 2020a. Biologically informed neural networks guide mechanistic modeling from sparse experimental data. *PLoS Comp Biol* **16:** e1008462. doi:10.1371/journal.pcbi.1008462

Lagergren JH, Nardini JT, Michael LG, Rutter EM, Flores KB. 2020b. Learning partial differential equations for biological transport models from noisy spatio-temporal data. *Proc Math Phys Eng Sci* **476:** 20190800. doi:10.1098/rspa.2019.0800

Lang Q, Lu F. 2022. Learning interaction kernels in mean-field equations of first-order systems of interacting particles. *SIAM J Sci Comput* **44:** A260–A285. doi:10.1137/20M1377072

Lang Q, Lu F. 2023. Identifiability of interaction kernels in mean-field equations of interacting particles. *Found Data Sci* **5:** 480–502. doi:10.3934/fods.2023007

Li J, Sun G, Zhao G, Lehman LWH. 2020. Robust low-rank discovery of data-driven partial differential equations. *Proc AAAI Conf Artif Intell* **34:** 767–774. doi:10.1609/aaai.v34i01.5420

Li Z, Lu F, Maggioni M, Tang S, Zhang C. 2021. On the identifiability of interaction functions in systems of interacting particles. *Stoch Process Appl* **132:** 135–163. doi:10.1016/j.spa.2020.10.005

Liu Y, Suh K, Maini PK, Cohen DJ, Baker RE. 2024. Parameter identifiability and model selection for partial differential equation models of cell invasion. *J R Soc Interface* **21:** 20230607. doi:10.1098/rsif.2023.0607

Long Z, Lu Y, Ma X, Dong B. 2019. PDE-Net 2.0: learning PDEs from data with a numeric-symbolic hybrid deep network. *J Comput Phys* **399:** 108925. doi:10.1016/j.jcp.2019.108925

Lu F, Zhong M, Tang S, Maggioni M. 2019. Nonparametric inference of interaction laws in systems of agents from trajectory data. *Proc Natl Acad Sci* **116:** 14424–14433. doi:10.1073/pnas.1822012116

Lu F, Maggioni M, Tang S. 2021. Learning interaction kernels in heterogeneous systems of agents from multiple trajectories. *J Mach Learn Res* **22:** 1–67.

Lu F, Maggioni M, Tang S. 2022a. Learning interaction kernels in stochastic systems of interacting particles from multiple trajectories. *Found Comput Math* **22:** 1013–1067. doi:10.1007/s10208-021-09521-z

Lu F, Lang Q, An Q. 2022b. Data adaptive RKHS Tikhonov regularization for learning kernels in operators. *PMLR* **190:** 158–172.

Lu F, An Q, Yu Y. 2024. Nonparametric learning of kernels in nonlocal operators. *J Peridynamics Nonlocal Model* **6:** 347–370. doi:10.1007/s42102-023-00105-9

Maggioni M, Lang Q, Wang X, Lu F. 2025. Interacting particle systems on networks: joint inference of the network and the interaction kernel. *SSRN* doi:10.2139/ssrn.523 4037

Martina-Perez S, Simpson MJ, Baker RE. 2021. Bayesian uncertainty quantification for data-driven equation learning. *Proc Math Phys Eng Sci* **477:** 20210426. doi:10.1098/rspa.2021.0426

Messenger DA, Bortz DM. 2021a. Weak SINDy for partial differential equations. *J Comp Phys* **443:** 110525. doi:10.1016/j.jcp.2021.110525

Messenger DA, Bortz DM. 2021b. Weak SINDy: Galerkin-based data-driven model selection. *Multiscale Model Simul* **19:** 1474–1497. doi:10.1137/20M1343166

Messenger DA, Wheeler GE, Liu X, Bortz DM. 2022. Learning anisotropic interaction rules from individual trajectories in a heterogeneous cellular population. *J R Soc Interface* **19:** 20220412. doi:10.1098/rsif.2022.0412

Miller J, Tang S, Zhong M, Maggioni M. 2023. Learning theory for inferring interaction kernels in second-order interacting agent systems. *Sampl Theory Signal Process Data Anal* **21:** 21. doi:10.1007/s43670-023-00055-9

Murray JD. 2003. *Mathematical biology. I: An introduction*, Vol. I, 3rd ed. Springer, Berlin.

Nardini JT. 2024. Forecasting and predicting stochastic agent-based model data with biologically informed neural networks. *Bull Math Biol* **86:** 130. doi:10.1007/s11538-024-01357-2

Nardini JT, Chapnick DA, Liu X, Bortz DM. 2016. Modeling keratinocyte wound healing dynamics: cell–cell adhesion promotes sustained collective migration. *J Theor Biol* **400:** 103–117. doi:10.1016/j.jtbi.2016.04.015

Nardini JT, Lagergren JH, Hawkins-Daarud A, Curtin L, Morris B, Rutter EM, Swanson KR, Flores KB. 2020. Learning equations from biological data with limited time samples. *Bull Math Biol* **82:** 119. doi:10.1007/s11538-020-00794-z

Nardini JT, Baker RE, Simpson MJ, Flores KB. 2021. Learning differential equation models from stochastic agent-based model simulations. *J R Soc Interface* **18:** 20200987. doi:10.1098/rsif.2020.0987

Pang G, Karniadakis GE. 2020. Physics-informed learning machines for partial differential equations: Gaussian processes versus neural networks. In *Emerging frontiers in nonlinear science*, pp. 323–343. Springer, Berlin.

Philipps M, Körner A, Vanhoefer J, Pathirana D, Hasenauer J. 2024. Non-negative universal differential equations with applications in systems biology. *IFAC-PapersOnLine* **58:** 25–30. doi:10.1016/j.ifacol.2024.10.005

Philipps M, Schmid N, Hasenauer J. 2025. Current state and open problems in universal differential equations for systems biology. *npj Syst Biol Appl* **11:** 101. doi:10.1038/s41540-025-00550-w

Podina L, Eastman B, Kohandel M. 2023. Universal physics-informed neural networks: symbolic differential operator discovery with sparse data. In *Proceedings of the 40th International Conference on Machine Learning, volume 202 of Proceedings of Machine Learning Research*, pp. 27948–27956. https://proceedings.mlr.press/v202/podina23a.html

Podina L, Ghodsi A, Kohandel M. 2025. Learning chemotherapy drug action via universal physics-informed neural networks. *Pharm Res* **42:** 593–612. doi:10.1007/s11095-025-03858-8

Rackauckas C, Ma Y, Martensen J, Warner C, Zubov K, Supekar R, Skinner D, Ramadhan A, Edelman A. 2020. Universal differential equations for scientific machine learning. arXiv doi:10.48550/arXiv.2001.04385

Raissi M, Perdikaris P, Karniadakis GE. 2019. Physics-informed neural networks: a deep learning frame-work for solving forward and inverse problems involving nonlinear partial differential equations. *J Comp Phys* **378:** 686–707. doi:10.1016/j.jcp.2018.10.045

Ramsay J, Hooker G. 2017. *Dynamic data analysis: modeling data with differential equations. Springer Series in Statistics.* Springer, New York.

Ramsay JO, Hooker G, Campbell D, Cao J. 2007. Parameter estimation for differential equations: a generalized smoothing approach. *J R Stat Soc B: Stat Methodol* **69:** 741–796. doi:10.1111/j.1467-9868.2007.00610.x

Reinbold PA, Gurevich DR, Grigoriev RO. 2020. Using noisy or incomplete data to discover models of spatiotemporal dynamics. *Phys Rev E* **101:** 010203. doi:10.1103/PhysRevE.101.010203

Reinbold PA, Kageorge LM, Schatz MF, Grigoriev RO. 2021. Robust learning from noisy, incomplete, high-dimensional experimental data via physically constrained symbolic regression. *Nat Comm* **12:** 23479. doi:10.1038/s41467-021-23479-0

Rorth P. 2011. Whence directionality: guidance mechanisms in solitary and collective cell migration. *Dev Cell* **20:** 9–18. doi:10.1016/j.devcel.2010.12.014

Rudy SH, Brunton SL, Proctor JL, Kutz JN. 2017. Data-driven discovery of partial differential equations. *Sci Adv* **3:** e1602614. doi:10.1126/sciadv.1602614

Schaeffer H, McCalla SG. 2017. Sparse model selection via integral terms. *Phys Rev E* **96:** 023302. doi:10.1103/PhysRevE.96.023302

Schmidt M, Lipson H. 2009. Distilling free-form natural laws from experimental data. *Science* **324:** 81–85. doi:10.1126/science.1165893

Schumacher LJ, Kulesa PM, McLennan R, Baker RE, Maini PK. 2016. Multidisciplinary approaches to understanding collective cell migration in developmental biology. *Open Biol* **6:** 160056. doi:10.1098/rsob.160056

Shea DE, Brunton SL, Kutz JN. 2021. SINDy-BVP: sparse identification of nonlinear dynamics for boundary value problems. *Phys Rev Res* **3:** 023255. doi:10.1103/PhysRevResearch.3.023255

Simpson MJ, Binder BJ, Haridas P, Wood BK, Treloar KK, McElwain DLS, Baker RE. 2013a. Experimental and modelling investigation of monolayer development with clustering. *Bull Math Biol* **75:** 871–889. doi:10.1007/s11538-013-9839-0

Simpson MJ, Treloar KK, Binder BJ, Haridas P, Manton KJ, Leavesley DI, McElwain DLS, Baker RE. 2013b. Quantifying the roles of cell motility and cell proliferation in a circular barrier assay. *J Roy Soc Interface* **10:** 20130007. doi:10.1098/rsif.2013.0007

Simpson MJ, Baker RE, Buenzli PR, Nicholson R, Maclaren OJ. 2022. Reliable and efficient parameter estimation using approximate continuum limit descriptions of stochastic models. *J Theor Biol* **549:** 111201. doi:10.1016/j.jtbi.2022.111201

Srivastava S, Kinnunen PC, Wang Z, Ho KKY, Humphries BA, Chen S, Linderman JJ, Luker GD, Luker KE, Garikipati K. 2025. Inference of weak-form partial differential equations describing migration and proliferation mechanisms in wound healing experiments on cancer cells. *PLOS: Comput Biol* doi:10.1371/journal.pcbi.1013607

Stehbens SJ, Scarpa E, White MD. 2024. Perspectives in collective cell migration—moving forward. *J Cell Sci* **137:** jcs261549. doi:10.1242/jcs.261549

Stepien TL. 2024. An approximate Bayesian computation approach for embryonic astrocyte migration model reduction. *Bull Math Biol* **86:** 126. doi:10.1007/s11538-024-01354-5

Sun L, Gao H, Pan S, Wang JX. 2020. Surrogate modeling for PDEs with random inputs using physics-informed neural networks. *J Comput Phys* **408:** 109770.

Tang S, Tuerkoen M, Zhou H. 2024. On the identifiability of nonlocal interaction kernels in first-order systems of interacting particles on Riemannian manifolds. *SIAM J Appl Math* **84:** 2067–2086. doi:10.1137/23M1599835

Tartakovsky AM, Marrero CO, Perdikaris P, Tartakovsky GD, Barajas-Solano D. 2020. Physics-informed deep neural networks for learning parameters and constitutive relationships in subsurface flow problems. *Adv Water Resour* **56:** 103548. doi:10.1029/2019WR026731

Thede SM. 2004. An introduction to genetic algorithms. *J Comput Sci Coll* **20:** 115–123. doi:10.1023/A:1025850513781

Udrescu SM, Tegmark M. 2020. AI feynman: a physics-inspired method for symbolic regression. *Sci Adv* **6:** eaay2631. doi:10.1126/sciadv.aay2631

van den Heuvel DJ, Buenzli PR, Simpson MJ. 2024. Pushing coarse-grained models beyond the continuum limit using equation learning. *Proc R Soc Lond A* **480:** 20230619. doi:10.1098/rspa.2023.0619

Vo BN, Drovandi CC, Pettitt AN, Simpson MJ. 2015. Quantifying uncertainty in parameter estimates for stochastic models of collective cell spreading using approximate Bayesian computation. *Math Biosci* **263:** 133–142. doi:10.1016/j.mbs.2015.02.010

Warne DJ, Baker RE, Simpson MJ. 2019. Using experimental data and information criteria to guide model selection for reaction–diffusion problems in mathematical biology. *Bull Math Biol* **81:** 1760–1804. doi:10.1007/s11538-019-00589-x

Yu R, Wang R. 2024. Learning dynamical systems from data: an introduction to physics-guided deep learning. *Proc Natl Acad Sci* **121:** 27948–27956. doi:10.1073/pnas.2311808121

Yuan L, Ni YQ, Deng XY, Hao S. 2022. A-PINN: auxiliary physics informed neural networks for forward and inverse problems of nonlinear integro-differential equations. *J Comp Phys* **462:** 111260. doi:10.1016/j.jcp.2022.111260

Index

D

E

F

O

P

Q

R

S

Y

Z

www.ingramcontent.com/pod-product-compliance
Lightning Source LLC
Chambersburg PA
CBHW082222140826
49882CB00005B/21
* 9 7 8 1 6 2 1 8 2 5 3 1 9 *